XIV International Conference on Electrical Bioimpedance and 11th Conference on Biomedical Applications of Electrical Impedance Tomography 2010

(ICEBI & EIT 2010)

Journal of Physics: Conference Series Volume 224

Gainesville, Florida, USA
4-8 April 2010

ISBN: 978-1-61738-482-0
ISSN: 1742-6588

Printed from e-media with permission by:

Curran Associates, Inc.
57 Morehouse Lane
Red Hook, NY 12571

Some format issues inherent in the e-media version may also appear in this print version.

Printed by Curran Associates, Inc. (2010)

For permission requests, please contact the Institute of Physics
at the address below.

Institute of Physics
Dirac House, Temple Back
Bristol BS1 6BE UK

Phone: 44 1 17 929 7481
Fax: 44 1 17 920 0979

techtracking@iop.org

Additional copies of this publication are available from:

Curran Associates, Inc.
57 Morehouse Lane
Red Hook, NY 12571 USA
Phone: 845-758-0400
Fax: 845-758-2634
Email: curran@proceedings.com
Web: www.proceedings.com

TABLE OF CONTENTS

KEYNOTE

1-A ELECTRODES AND INSTRUMENTATION 1

1-B ELECTRODES AND INSTRUMENTATION 2

1-C EIT HARDWARE

1-D MREIT

1-E EIT ALGORITHMS 1

1-F IMAGING NEURAL ACTIVITY

1-G LUNG EIT

1-H EIT MAMMOGRAPHY

1-I HYBRID AND EMERGING IMAGING METHODS 2

1-J MIT 1

1-K EIT APPLICATIONS 1

1-L MIT 2

1-R EIT APPLICATIONS 3

2-A THEORY AND MODELING

2-B HYBRID AND EMERGING IMAGING METHODS 1

2-C CELL SCALE IMPEDANCE

2-E TISSUE AND ORGAN IMPEDANCE

2-F SKIN IMPEDANCE

2-G IMPEDANCE CARDIOGRAPHY 1

2-I TISSUE AND SYSTEM PATHOLOGY

2-J BIOIMPEDANCE ANALYSIS 1

2-K BIOIMPEDANCE ANALYSIS 2

2-L POPULATION STUDIES

2-M IMPEDANCE CARDIOGRAPHY 2

2-N FLUID ANALYSIS

2-O BIOIMPEDANCE SPECTROSCOPY

POSTER SESSION 1: BIOIMPEDANCE ANALYSIS

POSTER SESSION 2: ELECTRICAL IMAGING TECHNIQUES

POSTER SESSION 3: CLINICAL APPLICATIONS/ELECTRODES & INSTRUMENTS

PREFACE

The XIVth International Conference on Electrical Bioimpedance, held in conjunction with the 11th Conference on Biomedical Applications of EIT (ICEBI & EIT 2010), took place from 4–8 April 2010 in the Reitz Union of the University of Florida, in Gainesville, USA. This was the first time since its inception in 1969 that the ICEBI was held in the United States. As in the last three conferences (Graz 2007, Gdansk 2004 and Oslo 2001) the ICEBI was combined with the Conference on Biomedical Applications of EIT – a mutually beneficial approach for those interested in the biophysics of tissue electrical properties and those developing imaging methods and measurement systems based thereon. This year's conference was particularly notable for the many papers presented on hybrid and emerging imaging techniques such as Electric Property Tomography (EPT), Magneto Acoustic Tomography using Magnetic Induction (MAT-MI) and Magnetic Resonance Electrical Impedance Tomography (MREIT); sessions on Cell Scale Impedance, Cardiac Impedance and Imaging Neural Activity. About 180 scientists from all over the world attended, including keynote speakers on topics of fundamental electromagnetic principles (Jaakko Malmivuo), Electrical Source and Impedance Imaging (Bin He), Bioimpedance applications in Nephrology, (Nathan Levin), and Lung EIT (Gerhard Wolf). The papers in this volume are peer-reviewed four-page works selected from over 150 presented in oral and poster sessions at the conference. The complete program is available from the conference website.

All papers published in this volume of *Journal of Physics: Conference Series* have been peer reviewed through processes administered by the proceedings Editors. Reviews were conducted by expert referees to the professional and scientific standards expected of a proceedings journal published by IOP Publishing.

Journal of Physics: Conference Series **224** (2010) 012001 doi:10.1088/1742-6596/224/1/012001

Principle of Reciprocity Solves the Most Important Problems in Bioimpedance and in General in Bioelectromagnetism

Jaakko Malmivuo

Tampere University of Technology, Tampere, Finland

jaakko.malmivuo@tut.fi

Abstract. Though the principle of reciprocity was invented by Hermann von Helmholtz already over 150 years ago, and though it is a very powerful tool in solving various important problems in bioelectromagnetism, it is not generally used. In impedance tomography the measurement sensitivity distribution has generally been misunderstood. This can be easily demonstrated with the principle of reciprocity. Some other applications of the principle of reciprocity are also discussed.

1. Principle of reciprocity

The principle of reciprocity is a general principle, which holds for all linear systems. It was first described by Hermann von Helmholtz over 150 years ago [1]. In brief, it may be described by noting, that the source and the measurement sites may be swapped without a change in the detected signal. For us, who work with bioelectromagnetic problems it was a great honor, that Helmholtz demonstrated this principle in connection to measuring a bioelectric source.

Helmholtz described the principle of reciprocity in the following way:

A galvanometer is connected to the surface of the body. Now every single element of a biological electromotive surface produces such a current in the galvanometer circuit as would flow through that element itself if its electromotive force were impressed on the galvanometer wire. If one adds the effects of all the electromotive surface elements, the effect of each of which are found in the manner described, he will have the value of the total current through the galvanometer.

This statement also includes the principle of superposition for the first time.

2. Sensitivity distribution in impedance tomography

2.1. Classical illustrations of measurement

In the beginning of impedance tomography analysis it was erroneously understood, that in the impedance measurement the voltage measurement with an electrode pair reflects the impedance of the area (or volume in three dimensions) bounded by the equipotential lines (or surfaces) generated by the current injection and passing through the voltage measurement electrodes. This concept is referred in several standard references even so late as 2005 [2, 3, 4, 5]

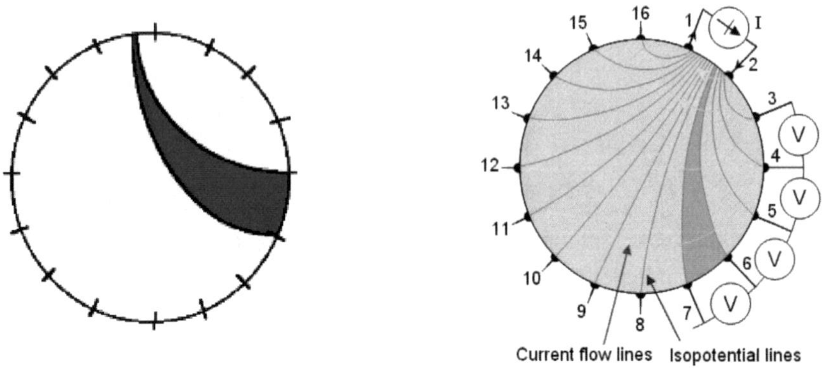

Figure 1. The logo of the Electrical Impedance Tomography (EIT) community (left). Classical but erroneous illustration of the distribution of measurement sensitivity in electrical impedance tomography (right).

2.2. Reciprocal approach

The principle of reciprocity clearly indicates that the classical concept of the region, from where the impedance is measured, can not be correct. According to reciprocity, the impedance measurement result should not change if the current feeding and voltage measurement are swapped. Therefore, the measurement region should be similar close to both electrode pairs. Furthermore, the measurement sensitivity may not be limited to an accurately bounded region, but the signal must be affected with some amount by all regions of the volume conductor.

We noted this and recalculated the measurement sensitivity distribution for neighboring, opposite, cross and adaptive methods [6]. The data is also presented as animations. This data clearly demonstrates both the symmetry and wide distribution of the impedance measurement sensitivity.

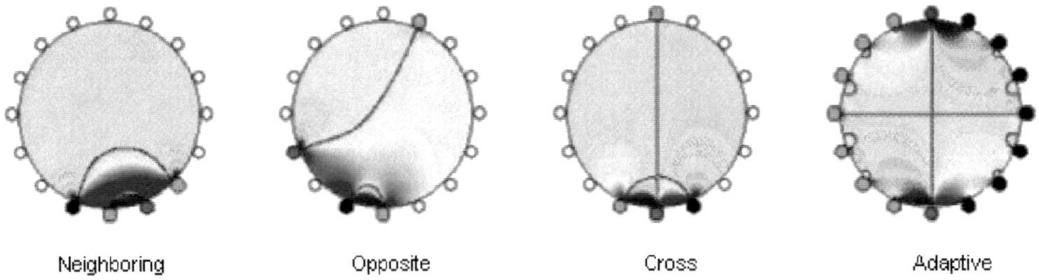

Figure 2. Distribution of the measurement sensitivity in one electrode configuration in the neighboring, opposite, cross and adaptive methods.

2.3. Positive and negative sensitivity

Figure 2 indicates the measurement sensitivity with a color code, where red is positive and blue is negative sensitivity. The concept of negative sensitivity complicates the analysis of the impedance signal. In positive sensitivity the impedance signal increases when the impedance of that region increases. In negative sensitivity the signal decreases with increasing impedance. The background for this phenomenon is, that when the lead fields (or their components) of the current and voltage leads are in the same direction the sensitivity is positive, when they are in opposite direction, the sensitivity is negative [4]. This phenomenon of positive and negative sensitivity is seen for instance in the measurement sensitivity analysis of different impedance cardiography leads, Figure 3 [7].

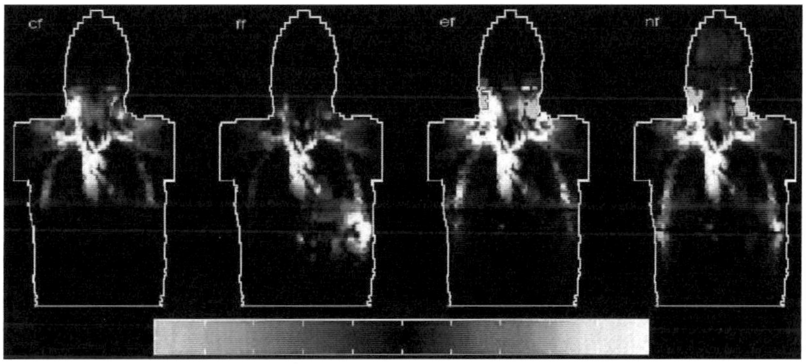

Figure 3. Distribution of measurement sensitivity in different impedance cardiography lead configurations. Red color indicates positive and blue color negative sensitivity. [7]

Note that the positive and negative sensitivity phenomenon is not unique in bioelectromagnetic problems. Similarly, if in measuring a bioelectric source the source is oriented in the direction of the detector's lead field, the sensitivity is positive. If the source is oriented in the opposite direction, the sensitivity is negative.

2.4. Resolution in impedance tomography

That the resolution of impedance tomography is so poor originates from the aforementioned distribution of the measurement sensitivity. In computer tomography with X-rays the resolution is excellent, because the detected amplitude of the signal depends on the density of the material on the narrow straight line of X-ray beam between the X-ray source and the detector. The accuracy of the back-projection is determined by this factor. If the classical understanding of the measurement sensitivity distribution had been correct, the accuracy of impedance tomography could had been improved close to the level of X-ray tomography by increasing the number of electrodes so that their distance would be extremely small.

3. Examples of other applications of the principle of reciprocity in bioelectromagnetism

The principle of reciprocity helps in solving several problems in bioelectromagnetism [8]. It also gives a very illustrative tool to understand the behavior of measurement sensitivity or stimulus intensity distribution problems. In the following some examples are given.

3.1. Distribution of measurement sensitivity and stimulus intensity

Feeding a unit current to the measurement electrodes in electric measurement or coil in magnetic measurement raises a current distribution in the volume conductor. This current, called the lead field, is exactly the same as the distribution of measurement sensitivity.

Because the lead field is a physical current field it is actually the distribution of stimulus current. Therefore the distributions of the measurement sensitivity and the stimulus current for a pair of electrodes or a coil are identical.

3.2. Reciprocally reciprocal method

The principle of reciprocity holds also for the reciprocal situation. As an application in finding the lead field for a magnetic measurement, the reciprocal current may be fed to the source dipole and signal detected with a superconducting SQUID magnetometer. Despite of this the measurement results may be considered as the reciprocal current had been fed to the coil and the signal detected with the electrodes in the volume conductor. The former way of thinking is made to understand the lead fields of the magnetic detector and the latter physical procedure is made because the SQUID magnetometer is more sensitive detector than dipolar electrodes in the volume conductor model [9].

3.3. Independence/interdependence of electric and magnetic measurements

The fundamental problem in biomagnetism is: do the magnetic measurements include information which is independent from the electric measurements?

On the basis of the Helmholtz theorem it was first believed, that the electric and magnetic measurements are independent [10]. If this were the case, with the magnetic measurement it would be possible to obtain as much new independent information as is obtained with electric ones. The practical measurements have, however, shown that this is not the case.

With the lead field theory it has been possible to demonstrate, that the electric and magnetic lead fields are independent, but the measured signals are only partially independent [4].

3.4. Reducing calculation in multichannel problems

For example in cortical potential imaging the cortical surface may consist up to 100,000 nodes and the number of electrodes may be 250. Applying the principle of reciprocity the number of calculations in cortical potential imaging decreases the number of calculations from 100,000 to 250. This means that the calculation time is decreased to 1/400.

4. Conclusion

The principle of reciprocity is not generally applied in bioelectromagnetic problems. Though the direct method is straightforward and may be applied in several problems, in many important problems the reciprocal method gives faster and more illustrative solution. In some problems it is the only possible method for obtaining the solution.

References

[1] Helmholtz HLF (1853): Ueber einige Gesetze der Vertheilung elektrischer Ströme in körperlichen Leitern mit Anwendung auf die thierisch-elektrischen Versuche. Ann. Physik und Chemie 89: 211-33, 354-77.

[2] Sakamoto K and Kanai H: Electrical impedance tomography, in: Michio Miyakawa, J. Ch. Bolomey (eds.): Non-Invasive Thermometry of the Human Body, CRC Press, Boca Raton, 1995, 272 p.

[3] Electrical Impedance Tomography (EIT) community. http://www.eit.org.uk/ referred on February 2010.

[4] Malmivuo J and Plonsey R: Bioelectromagnetism, Principles and Applications of Bioelectric and Biomagnetic Fields, Oxford University Press, 1995, 482 p. www.bem.fi/book/

[5] Holder DS (ed): Electrical impedance tomography, Institute of Physics Publishing, 2005. p. 108, 359 and 436.

[6] Kauppinen P, Hyttinen J and Malmivuo J: Sensitivity Distribution Visualizations of Impedance Tomography Measurement Strategies, International Journal of Bioelectromagnetism 2006, Vol. 8, No. 1 pp. VII/1 - VII/9, http://ijbem.k.hosei.ac.jp/2006-/volume8/number1/htm/kauppine.htm

[7] Kauppinen, P., Hyttinen, J. and Malmivuo, J. 1998. Sensitivity distributions of impedance cardiography using band and spot electrodes analyzed by a three-dimensional computer model. Annals of Biomedical Engineering 26 pp. 694-702.

[8] Malmivuo J: Main contributions to Bioelectromagnetism, http://www.bem.fi/malmivuo/

[9] Malmivuo J: On the Detection of the Magnetic Heart Vector - An Application of the Reciprocity Theorem. Helsinki University of Technology, Doctor of Technology (PhD) Thesis, Acta Polytechnica Scandinavica, Electrical Engineering Series. 39, 1976, 112 p.

[10] Plonsey R: Capability and limitations of electrocardiography and magnetocardiography. IEEE Trans. Biomed. Eng. BME-19:(3) 239-44, 1972.

Journal of Physics: Conference Series **224** (2010) 012002 doi:10.1088/1742-6596/224/1/012002

Impedance measurement set-up based on off-the-shelf PXI modules

F J Pettersen[1,2], Ø G Martinsen[2,1], S Grimnes[2,1], and J O Høgetveit[1,2]

[1]Dept. of Clinical and Biomedical Engineering, Oslo University Hospital, Norway
[2]Department of Physics, University of Oslo, Norway

E-mail: fred.johan.pettersen@rikshospitalet.no

Abstract. As bioimpedance measurements are being done in an increasing variety of measurands and environments, measurement equipment must be flexible in use and easy to handle for the researcher. Measurement equipment used today is excellent for a range of uses, but it is often bulky, heavy, and not very flexible. A new compact and flexible measurement set-up based on off-the-shelf modules is described. The system is based on commercially available PXI-modules, one module that may be custom made, and custom software. The basis of the system is a PXI chassis with a power supply unit. The chassis is equipped with a controller module running software controlling the system, a signal generator, and a multi-channel digitizer. All of this is commercially available. Both a custom made and a commercially available impedance interfaces were used. A LabVIEW program is controlling the measurement system, and provides a user interface. The LabVIEW program handles necessary signal conditioning and calibration. The measurement system is capable of doing frequency response measurements at frequencies up to 15 MHz. In addition to the two normal voltage pick-up electrodes used in four-electrode measurements, this system has the possibility to use 5 extra voltage pick-up electrodes. This simplifies exploration of segmental impedance, anisotropy, reciprocity, etc.

1. Introduction

Impedance measurements are done by measuring voltage across, and current through a sample at several frequencies, and then calculate the complex impedance using Ohm's law. This method needs some refinements to be practical for measuring impedance in biological materials. Dedicated frequency response analyzer systems typically use 2, 3, or 4 electrodes to reduce effect of the electrode interfaces, and to provide a method for volume selectivity [1]. To be able to pick out even very small signals at the frequencies of interest a method called lock-in amplification is used. Instruments with these features tend to be physically large, heavy, and they tend to have limited flexibility when it comes to signal processing [2]. To increase areas where bioimpedance measurements are possible, and to enhance bio-impedance measurement signal processing there is a need for a compact and flexible instrumentation platform.

2. Materials and method

2.1. *System overview.*

The measurement system is made as a PXI system. The system consists of a compact chassis that provides the physical housing of the measurement system, and power supply [3]. A controller module is used to control the system (CTRL). The controller is a PXI-module containing a PC that is running MS Windows and LabVIEW. These parts form the instrument infrastructure.

The instrument is made of one signal source providing an arbitrary voltage waveform (DAC), a multi-channel voltage digitizer (ADC), and an impedance interface. The signal source and the voltage digitizer are commercially available as PXI-modules. The impedance interface is not commercially available as a PXI-module, so two alternatives were used: One custom made, which can be used in the PXI chassis, and one commercially available non-PXI.

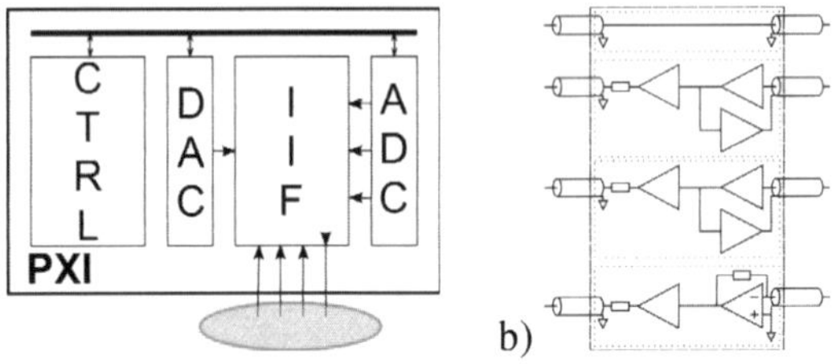

a) b)

Figure 1: a) Complete impedance measurement system in a PXI chassis, b) impedance interface. The impedance interface may be placed either inside or outside the PXI chassis.

2.2. *About PXI.*
PXI is a standard for modular PC-based instrumentation systems. A typical system consists of a PXI chassis equipped with power-supply and cooling plus one or more modules [4].

2.3. *Of-the-shelf PXI modules.*
The off-the-shelf modules used in this particular case are:

- NI PXI-1031: A general-purpose 4-slot chassis for PXI.
- NI PXI-8183: A PXI-controller, which is a PC-module that fits in the PXI-chassis.
- NI PXI-5402: A 20 MHz arbitrary waveform signal source.
- NI PXI-5105: A 60 MS/s multichannel digitizer with simultaneously sampled inputs.

All PXI parts used are from National Instruments.

2.4. *Comercially available external impedance interface.*
The Solartron 1294 impedance interface was used. One basic configuration of this device is shown in figure 1 b). The IIF-part in figure 1 a) is on the outside of the PXI box in figure 1 a) when the external impedance interface is used.

2.5. *Custom-made impedance interface.*
The custom-made impedance interface in figure 1 b) has several functions necessary to use the measurement system as a four-electrode measurement system.

The top signal path is not doing any signal processing, but passes the stimulus voltage through.

The bottom signal path is a current-to-voltage converter and a buffer. The return current from the signal source is converted to a voltage in this path, and the resulting voltage is buffered by a buffer with 50 Ω output impedance. The buffered signal is fed to the voltage digitizer.

The two signal paths in the middle are voltage pick-up signal paths. The input signal is connected to a unity-gain input buffer with very high input impedance. The output of this buffer is fed to a buffer with 50 Ω output impedance. The signal from the input buffer is connected to another buffer that drives the shield of the input coaxial cable. This gives us a driven-shield solution that presents a reduced parasitic capacitance load to the input signal. The two buffers used in the driven-shield path must have very low accumulated phase shift for the intended measurement bandwidth (15 MHz).

Several RC-filters are placed in the signal paths to ensure stability and to optimize signal quality.

2.6. *Software.*
The software part of the system is based on the graphical programming language LabVIEW from National Instruments.

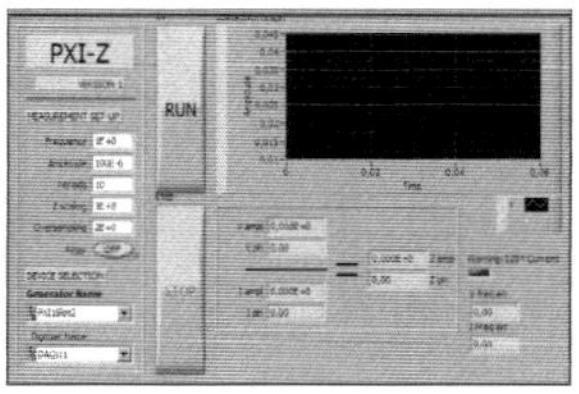

Figure 2: LabVIEW instrument that controls the PXI-based instrument and provides a user-interface.

The software generates stimuli, process the measured signals, and produce desired presentation of the results. The front panel of a measurement program is shown in figure 2.

3. Results

3.1. *Simulation results.*

The custom-made impedance front-end has been simulated to verify performance. The most important signal paths are the voltage pick-ups and the current to voltage converter. Since the driven-shield signal path forms a loop that can be unstable, it was simulated.

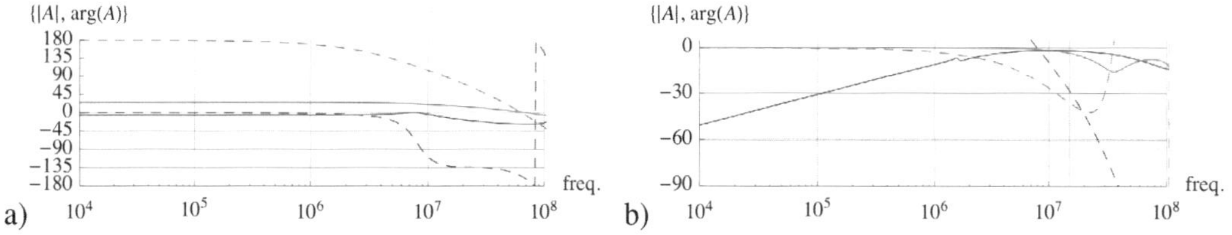

Figure 3: Solid lines: modulus, dashed lines: phase. a) Transfer function for voltage pick-up (blue lines) and current-to-voltage converter (red lines), b) transfer function for potentially unstable feedback-loop (blue lines) and transfer function for driven shield circuit (red lines).

The Bode plots for the signal paths for the measurement signals are shown in figure 3 a).

Figure 3 b) shows that the loop gain is less than 0 dB at all frequencies, and that the driven-shield circuit has a phase shift less than 36° and less than 3.6 dB attenuation for all frequencies up to 15 MHz.

3.2. *Measurement results.*

A simplified measurement system was made using PXI modules and a Solartron 1294 impedance interface. The LabVIEW instrument in figure 2 controlled the PXI system. A comparison of the new PXI-based system and the Solartron 1260+1294 combination was done on a known test-circuit. Figure 4 shows results at 13 frequencies.

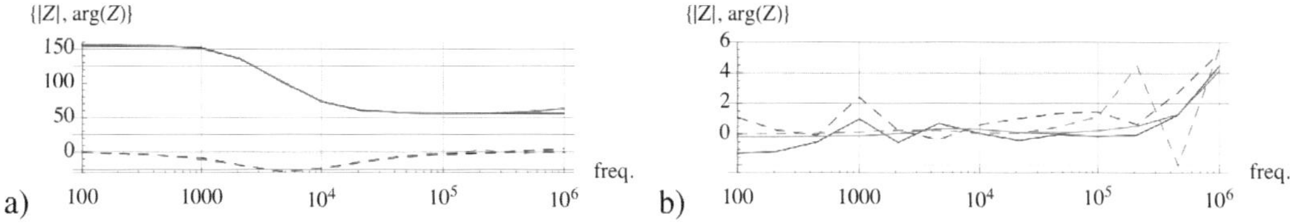

Figure 4: Solid lines: modulus, dashed lines: phase. a) Modulus and phase for calculated and measured impedances, b) Measurement error for PXI (blue) and Solartron (red).

4. Discussion

4.1. *Flexibility.*

The system is very flexible with regards to several properties: a) hardware choices, b) excitation set-up, and c) signal processing.

It is possible to expand the measurement by adding one or more voltage pick-ups. This makes it possible to explore several properties like anisotropy and segmental impedance simultaneously using more than two voltage pick-up electrodes. The impedance front-end can be replaced by similar front-ends offering different properties such as galvanic isolation, current-limiting, controlled-current stimuli, etc.

Since custom-made software defines the excitation, it is possible to define arbitrary waveforms for excitation. It is also possible to use any frequency-hopping or waveform-hopping scheme. It is also possible to use specialized waveforms such as those used by pacemakers. Such flexibility is not common in a frequency response analyzer like the Solartron 1260 [2].

It is possible to tailor the interface to most needs. Post-processing of the measurements can be done. Examples of post-processing are harmonic distortion analysis, noise analysis, linearity analysis, and comparison with existing data.

4.2. *Simulation results.*

The simulations shows that a) the feedback loop formed by the driven shield network and coaxial cable is stable, b) the phase shift in the circuits driving the shield has low enough phase shift to reduce capacitance seen by the voltage electrode, and c) there are no abrupt changes in gain or phase. The last point is important since it means that it is possible to calibrate the system by measuring the impedance of a reference over the relevant frequency range.

It is crucial for the driven-shield concept that the phase shift in the buffers driving the shield is as close to 0° as possible. Simulation shows that a phase shift of less than 36° is obtained for frequencies up to 15 MHz.

4.3. *Measurement results.*

The measurements show that the new PXI-based system is slightly poorer than the Solartron 1260+1294-combination in performance on the test-circuit. It is expected that this error can be reduced by software optimization (filters, lock-in, etc.).

4.4. *Performance.*

Measurement time can be reduced compared to available instruments due to the flexibility of the new measurement system. The cost of reduced measurement time may be increased programming time. Reduced measurement time may be especially useful when doing measurements *in vivo*.

Since the current and voltages are digitized and available, it is possible to utilize techniques such as lock-in, digital filtering, noise analysis, and harmonic distortion analysis.

5. Conclusions

A PXI-based impedance measurement system is useful since it is compact and flexible when looking at both hardware and software, and the first measurements demonstrates performance comparable to existing equipment. The presented system gives the scientist an opportunity to build an instrument that can be tailored to several specific tasks.

References

[1] Grimned S, Martinsen Ø G 2008 *Bioimpedance and Bioelecticity Basics, Second Edition* (Oxford: Elsevier)
[2] Solartron Instruments 1995 *1260 Impedance Ganin-Phase Analyzer Operating Manual* (Farnborough: Solartron Instruments).
[3] National Instruments 2007 *NI PXI-1031 Data Sheet* (Austin: National Instruments)
[4] PXI Systems Alliance web-site. http://www.pxisa.org/

Focused Impedance Method (FIM) and Pigeon Hole Imaging (PHI) for localized measurements – a review

K Siddique-e Rabbani
Department of Biomedical Physics & Technology, University of Dhaka, Dhaka 1000, Bangladesh

email: rabbani@univdhaka.edu

Abstract. This paper summarises up to date development in Focused Impedance Method (FIM) initiated by us. It basically involves taking the sum of two orthogonal tetra-polar impedance measurements around a common central region, giving a localized enhanced sensitivity. Although the basic idea requires 8 electrodes, versions with 6- and 4-electrodes were subsequently conceived and developed. The focusing effect has been verified in 2D and 3D phantoms and through numerical analysis. Dynamic stomach emptying, and ventilation of localized lung regions have been studied successfully suggesting further applications in monitoring of gastric acid secretion, artificial respiration, bladder emptying, etc. Multi-frequency FIM may help identify some diseases and disorders including certain cancers. FIM, being much simpler and having less number of electrodes, appears to have the potential to replace EIT for applications involving large and shallow organs. An enhancement of 6-electrode FIM led to Pigeon Hole Imaging (PHI) in a square matrix through backprojection in two orthogonal directions, good for localising of one or two well separated objects.

1. Introduction

Although Electrical Impedance Imaging (EIT) attempts to give images with pixel level resolution [1], such images are not always necessary. In many applications like the study of stomach emptying, lungs functional study, etc., a region of interest is usually marked out in the EIT image and the time progression of the sum of all the pixels within the region is followed for a dynamic study [2]. In some applications a single target object may be studied at different frequencies if its impedance has frequency dependence, at a significant difference from that of the background [3]. For the study of such single target applications a family of localised impedance measurement techniques having the capability of isolating a target object from its neighbours was conceived and developed by the author's group at Dhaka University, Bangladesh [4][5][6]. This work has recently been taken up by University of Warwick, UK [7] and Kyung Hee University, Korea [8], and the state of development so far is reviewed in this paper.

2. 6-electrode Focused Impedance Method (FIM)

The idea originated from placing two traditional Tetrapolar Impedance Measurement (TPIM) systems orthogonally around a region of interest as shown in figure 1 where A-A' and B-B' are the two orthogonal current drive electrodes. In a further step of innovation, two potential electrodes, u and v, placed diagonally in the central region, replace the four potential electrodes [4] that would have been required for the two TPIM configurations. Respective horizontal and vertical equipotential lines through u and v are shown as straight lines for simplicity. Figure 1 also shows a simple admittance matrix model where the shown admittances (Y_{11} to Y_{33}) are assumed to be isotropic, and contributions of zones outside are ignored. Measurement through current drive pair AA' will give the admittance ($kY_{21}+Y_{22}+kY_{23}$) where k is a constant factor, usually less than 1, which takes care of the sensitivity differences between the central zone and the outer ones. Similarly for measurement through the current drive pair BB', the admittance is ($kY_{12}+Y_{22}+kY_{32}$). An algebraic addition of the two measurements gives the total measured admittance as $Y_T = 2Y_{22} + k\,(Y_{21} + Y_{23} + Y_{12} + Y_{32})$. Since k<1, the central zone has more than twice the sensitivity of the outer ones. Thus the central zone will dominate in the combined measurement and it may be said to be 'focused'. In a practical situation if only Y_{22} changes because of a physiological activity, then $\Delta Y_T = 2\,\Delta Y_{22}$ and the contributions of the other regions would be eliminated. If one of the admittances Y_{21}, Y_{23}, Y_{12} or Y_{32} changes simultaneously, its contribution will be much less than $2\Delta Y_{22}$. studies on a 2D saline phantom showed a success in the focusing effect as expected, albeit with some asymmetrical smearing.

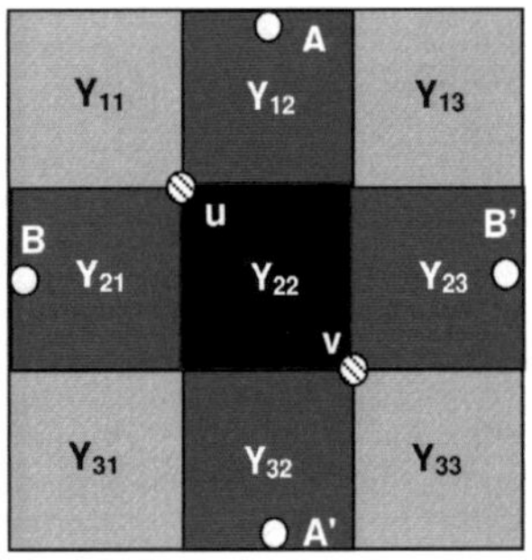

Figure 1. Basics of FIM and simplified admittance model

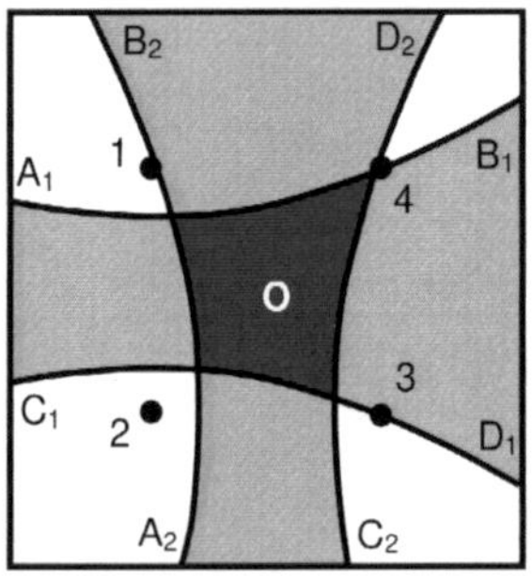

Figure 2. Basic idea of 4 electrode FIM and the focused zone at the centre.

3. 4-electrode FIM

Another simpler version of FIM was developed subsequently [5] using 4 electrodes (1 to 4) at four corners of a square region as shown in figure 2. Firstly current is driven through electrodes 1-2 and potential is measured across electrodes 3-4. The region between equipotentials A_1-B_1 and C_1-D_1 would be the sensitive zone in this measurement. Next the current is driven through electrodes 2-3 and potential is measured across electrodes 4-1 when the sensitive region would be that between equipotentials A_2-B_2 and C_2-D_2. When the two impedance measurements are added, the dark shaded central region would have a dominant contribution similar to that for the 6-electrode FIM discussed before. The measurement may be improved further by rotating the measurement sequence and summing all the values. A 2D phantom study also demonstrated a focusing effect as expected.

4. Numerical Analysis

Numerical analysis at Warwick University,UK, using Geselowitz's lead field method has confirmed the focussing effect of all the three FIM versions – having 8, 6, and 4 electrodes [7]. A maximum in the average sensitivity of a plane was observed at $^1/_3$ of the drive-receive electrode spacing for all the three configurations. Negative sensitivity regions down to $^1/_3$ of the drive-receive electrode spacing was observed for the 4 electrode system while these were down to ½ of the drive-receive electrode spacing for the others. The depth at which the positive peak beneath the electrodes became dominant occurred at 0.7, 1.25 and 0.15 of the receive electrode spacing for the 8, 6 and 4 electrode configurations respectively.

5. Physiological applications of FIM

FIM was applied to study gastric emtpying after a drink of saline, placing electrodes on the frontal surface[4] and the result is shown in figure 3. Region of interest marked around the stomach in EIT studies [4][9] have shown an exactly similar behaviour. In lung ventilation studies normal healthy subjects took a deep breath and then expired a little air at a time in steps and held the breath for a short while each time. During the breath holding period, FIM data were recorded for the lower right lobe of the lungs from the front and simulataneously corresponding volume of air expired was measured using a bellows type spirometer. From the above data, percentage change in focused impedance and change in air volume, with reference to the values at total expiration, were calculated. Figure 6 shows a very good correlation between these parameters [10]. Similar localised lung FIM measurement from different subjects having different Vital Capacity (VC - total air volume expired

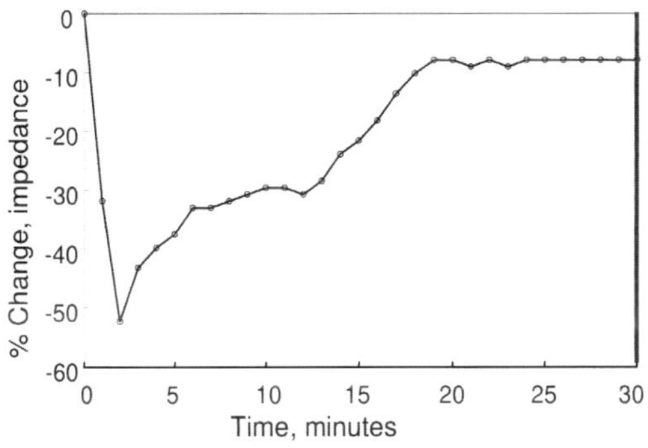

Figure 3. Gastric emptying study using FIM after a drink of saline

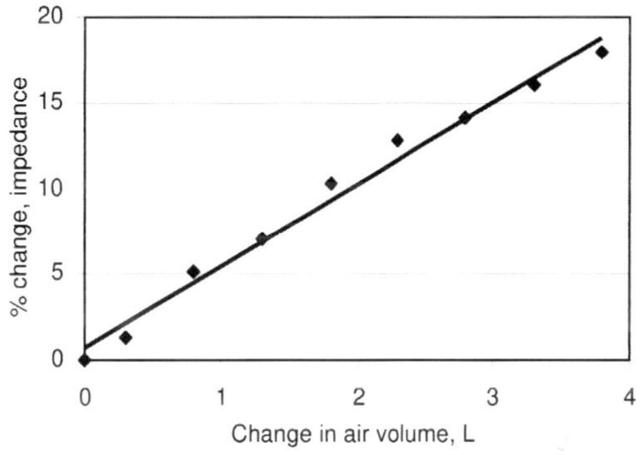

Figure 4. Lung ventilation study using FIM showing linear relationship to change in air volume.

on forced expiration following deep inspiration) showed an almost linear relationship with VC. This shows the utility of FIM in measuring lung ventilation, particularly in identifying localised ventilation disorders.

FIM may replace many investigation of large organs that have been proposed with 2D EIT through region of interest analysis. These include gastric emptying, gastric acid secretion, lungs ventilation and perfusion, monitoring of artificial repiration, bladder emptying, etc. Multifrequency FIM may have application in the diagnosis and detection of pneumonia and of cancerous tissues at shallow depths, including cervical and oral cancers. Success in detecting cervical cancer using tetrapolar arangement [11] suggests that FIM may also be successful in this application. Besides, FIM may be used in the measurement of tissue temperature in radiotherapy, and in monitoring of cancerous tissue ablation in non-reversible electroporation [12].

7. Pigeon Hole Impedance Imaging (PHI)

By extending the concept of the 6 electrode FIM the PHI technique was developed for providing a crude image, in the configuration of a pigeon hole [6]. Figure 5 shows an example with four diagonal potential electrodes which have replaced the two in 6-electrode FIM. In a simplified admittance model as shown in figure 6, combinations of current drives and potential measuring pairs give the total row

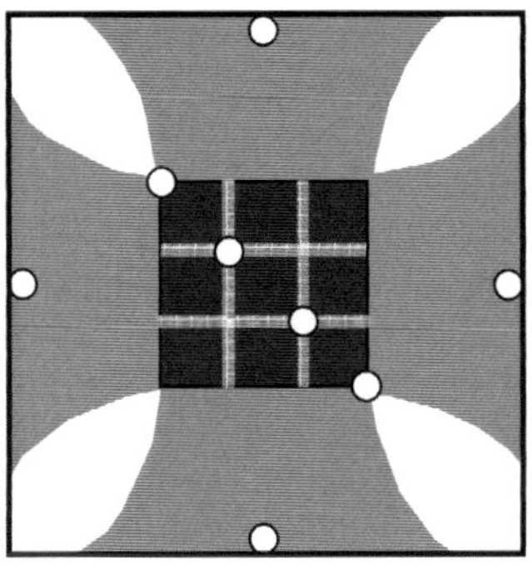

Figure 5. Basics of PHI (4x4 matrix shown)

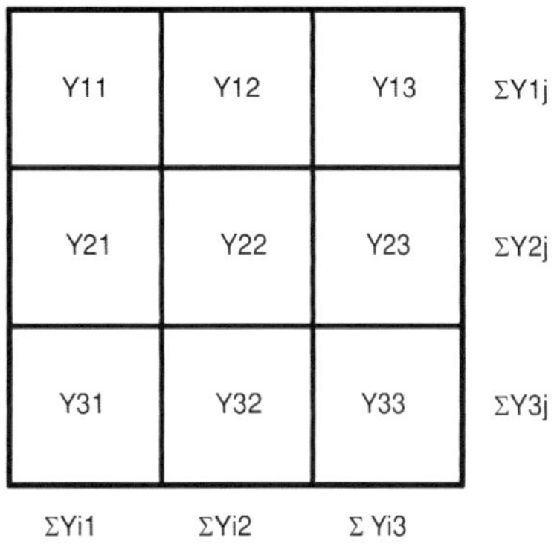

Figure 6. Simple admittance matrix model

and column admittances (ΣY_{1j} to ΣY_{3j}, ΣY_{i1} to ΣY_{i3}). Backprojections of these measured admittances along the two orthogonal directions will give values in which each of the matrix admittances will dominate, such as $Y_{11} \cong (\Sigma Y_{1j} + \Sigma Y_{i1})/3$, $Y_{12} \cong (\Sigma Y_{1j} + \Sigma Y_{i2})/3$, etc., thus giving the individual pixels of the 'image'. Experimental study using saline filled 2D and 3D phantoms gave the locations of a single object successfully. PHI can only be expected to differentiate objects that are widely separated, and mostly it would be useful in locating a single organ of interest in an individual.

8. Discussion

As mentioned in the beginning, in many applications a single organ is often the object of interest for which a sophisticated imaging system like EIT is not needed, and the focused impedance systems including PHI fill an important gap. Being simple in methodology and instrumentation the costs will be significantly lower compared to EIT systems, besides, having much less number of electrodes to manage, these systems offer additional advantages in terms of robustness and ease of application. The family of FIM's presented in this review may find potential clinical application sooner than EIT because of their simplicity. These would particularly be suitable for the Third World as the instruments can be designed and made locally thus making them affordable and sustainable for widespread use, benefitting a large number of people globally.

9. Acknowledgments

The author acknowledges many research students of his at Dhaka University who have contributed to the development of the FIM and PHI techniques. He also acknowledges Professor B H Brown of Sheffield for introducing this group to his pioneering EIT.

10. References

[1] Barber DC, Brown BH and Freeston IL 1983 *Electron. Lett.* **19** 933-5
[2] Harris ND, Suggett AJ, Barber DC and Brown BH 1987 *Clin. Phys. Physiol. Meas.* **8** (Suppl. A) 155-65
[3] Brown BH, Barber DC, Wang W, Lu L, Leathard AD, Smallwood RH, Hampshire AR, Mackay R and Hatzigalanis K 1994 *Physiol. Meas.* **15** (Suppl. 2A) 1-11
[4] Rabbani KS, Sarker M, Akond MHR and Akter T 1999 *"Electrical Bioimpedance methods", Annals of the New York Academy of Sciences,* **873** 408 - 20
[5] Rabbani KS, and Karal MAS 2008 *Annals of Biomedical Engg.* **36** 1072-77
[6] Afroj K, Alam N, Rahman M and Rabbani KS 2004 *J. of Bangladesh Medical Physics Association*, **3** 7-13
[7] Islam N, Rabbani KS and Wilson A 2009 *Proceedings, 10th Int Con, Biomed Appl of El Imp Tomography (EIT2009) and Workshop on Electromagnetic Inverse Problems,* http://www.maths.manchester.ac.uk/eit2009/abstracts/islam.pdf
[8] Wi H and Woo EJ 2009 10th International Conference on Biomedical Applications of Electrical Impedance Tomography (*EIT2009*), http://www.maths.manchester.ac.uk/eit2009/abstracts/wi1.pdf
[9] Avill R, Mangnall YF, Bird NC, Brown BH, Barber DC, Seagar AD, Johnson AG and Read NW 1987 *Gastroenterology* **92** 1019-26
[10] Kadir MA, Baig TN and Rabbani KS, *Proceedings, 10th International Conference on Biomedical Applications of Electrical Impedance Tomography (EIT2009),* http://www.maths.manchester.ac.uk/eit2009/abstracts/kadir.pdf
[11] Brown BH, Tidy JA, Boston K, Blackett AD, Smallwood RH and Sharp F 2000 *The Lancet* **355** 892-895
[12] Rubinsky P, Onik G and Mikus P 2007 *Technol Cancer Res Treat*, **6** 37-48.

Multifrequency simultaneous bioimpedance measurements using multitone burst signals for dynamic tissue characterization.

B.Sanchez and R.Bragos

Center for Research in Biomedical Eng. (CREB), Dept. of Electronic Eng., UPC, Barcelona, Spain

E-mail: bsanchez@eel.upc.edu

Abstract. In this paper we present the keypoints to perform multifrequency simultaneous bioimpedance measurements using multitone signals. Concerning the frequency distribution, tones are spread over 1kHz to 1MHz range using a custom frequency distribution which we called *Bilateral Quasi Logarithmic* (BQL). BQL concentrates a higher number of tones around the impedance relaxation and contains a frequency plan algorithm. It minimizes the intermodulation effects due to non-linearities behaviours of the DUT and electrodes by slightly shifting the original tones in order to guarantee a guard bandwith. Regarding the multitone phase distribution, a *Genetic Algorithm* (GA) has been developed to minimize multitone *Crest Factor* (CF). This allow us to maximize the resultant *Signal to Noise Ratio* (SNR) of the acquisition system. This paper also presents the relation between parameters such as sampling frequency and ADC bits with the SNR and the effect in the overall amplitude and phase error when using multitone signals as excitation waveforms. Finally, we present characterization results from a measurement system based on a modular PXI architecture.

1. Introduction

Multitone signals are useful for fast impedance spectroscopy characterization of biological systems whose electrical properties change along within time. Using this technique, is possible to obtain the impedance spectrum free of the errors induced by asynchronous undersampling of the mechanical modulation of the biological system under test. Essentially, a multitone is composed by the sum of M tones, each one with its own amplitude and phase (eq.1).

$$x\left[n\right] = \sum_{m=0}^{M-1} a_m \cos\left(2\pi\frac{f_m}{f_s}n + \varphi_m\right), \quad n = 0, ..., N-1 \tag{1}$$

1.1. Measures of Signal Quality: Crest Factor

When comparing signals it is necessary to have a quality measure in the form of a performance index. A classical index is the Crest Factor (CF), which gives an idea of signal's compactness but there are other options like *Performance Index for Perturbation Signals* (PIPS) [1] or *Time Factor* (TF) [2]. Thus, CF of a function $x\left(t\right)$ is defined as the ratio of its peak value (also known as Chebyshev norm) and its Root Mean Square value (RMS) [3] is given by:

$$CF(x) = \frac{l_\infty(x)}{l_2(x)} = \frac{\max\limits_{n\in[0,N-1]} |x[n]|}{\sqrt{\frac{1}{N}\sum\limits_{n=0}^{N-1} |x[n]|}} \quad (2)$$

The classical ways to minimize CF use heuristic approaches [4], [5]. Other approaches consisted on analytical methods. They use different iterative methods [6], [7] or employ natural selection and evolutionary-inspired operators like Genetic Algorithms (GA) [8], [9].

1.2. Frequency distribution

Typically, multitone frequencies are either distributed logarithmically or equally spaced [10]. However, both have important disadvantages when measuring impedance. In one hand, equidistant frequency distribution needs a high number of tones to cover a wide frequency range ands avoid losing spectral resolution. Moreover, intermodulation products will appear at the same measurement frequencies if the global system presents non-linearities, provoking errors in the amplitude and phase determination. In the case of logarithmic distribution there is a big number of tones at low and high frequencies content where the relaxation do not include much information.

One of the main issues is how to arrange multitone frequencies to minimize intormodulation products effects. One option is generate each tone as a result of a fundamental frequency multiplied by a sequence values. This solution is based on special sets of multipliers in order to accomplish as many as possible disjunct distortion signal frequencies and to test signal frequencies is presented in [11]. In our case, BQL algorithm algorithm combines exponential and logarithmic functions to concentrate more frequency components close to the relaxation slope band. Moreover, it spreads the remaining tones at low and high frequencies (fig.1) [9] and also redistributes original tones taking into account where the intermodulation frequencies would appear.

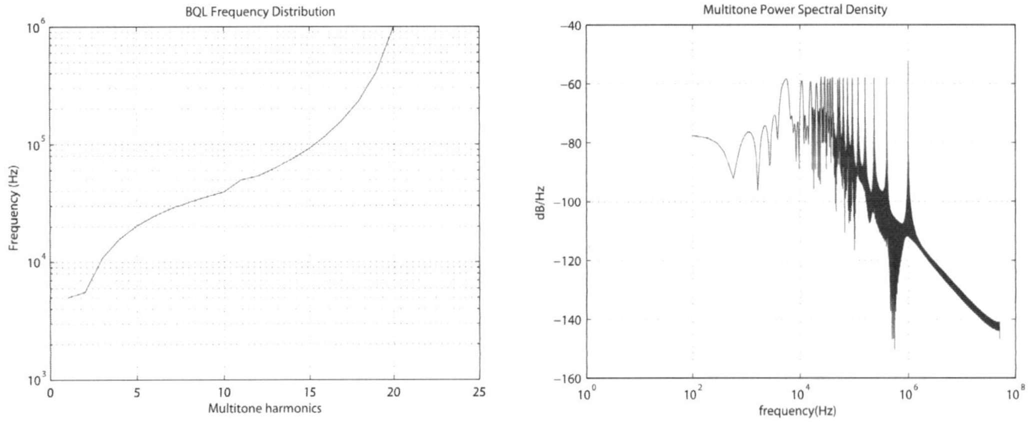

Figure 1. Example of BQL fundamental harmonics to characterize impedance relaxation around 50kHz.

2. Precision of Transfer Function Measurement using Multitone Signals

In single frequency impedance measurements is well known the relationship between the SNR with parameters such as ADC bits or sampling frequency. In the case of multitone measurements there are more hidden parameters that have to be taken into account. In fact, the SNR of an ideal b-bit AD converter measured over the Nyquist bandwidth ($DC - F_s/2$) when using a multitone input signal as excitation is decreased due to the Crest Factor and the number of tones M according to eq.3:

$$SNR_{multitone} = 6.02b + 1.76 - 20log_{10}\left(\frac{CF}{\sqrt{2}}\right) - 10log_{10}(M) \quad [dB] \tag{3}$$

Finally, has been shown in [12] the relationship between SNR when using multitone and the magnitude and phase measurement accuracy:

$$\epsilon_l = 20log_{10}\left(1 + 10^{\frac{-SNR_{multitone}}{20}}\right)[dB] \qquad \epsilon_\theta = \frac{180}{\pi}10^{\frac{-SNR_{multitone}}{20}} \; [°] \tag{4}$$

where ϵ_l and ϵ_θ are magnitude and phase accuracy. Eq.4 allows to find the minimum system that satisfies quality measurement requirements in terms of SNR or accuracy using multitone signals only with the a priori information of Crest Factor, number of bits and number of tones.

3. System characterization

The measurement system has been build around a PXI architecture from National Instruments. The system includes an embedded controller PXIe-8130, 2 channel digitizer card PXIe-5122 (100Ms/s, 64MB/channel, 14bits) and an arbitrary waveform card PXI-5422 (200Ms/s, 32MB, 16 bits). The system has been characterized measuring resistors and RC networks with a custom 4-wire wideband front-end. The set up has been carried out performing 100 measurements (2ms burst length) of a set of resistors (1% tolerance). Multitone signal was designed with 21 tones spread from 5kHz to 1.313MHz, 1mA of amplitude current and crest factor 2.7. Theoretical SNR obtained with this system parameters is 67.2dB which corresponds to a measurement error of 0.0038dB and 0.023° in magnitude and phase respectively. Magnitude and phase accuracy has been validated with the standard desviation obtained from the measurements. Also a preliminary dynamic scenario have been characterized based on yeast cell suspension settlement (fig.2) over interdigitated 4-electrode located at the bottom of a small bioreactor [13]. Multitone parameters used were the same as described before. System performed 100 measurements (10ms) every 5 minutes up to 12 hours, with a a total amount of 14400 measurements.

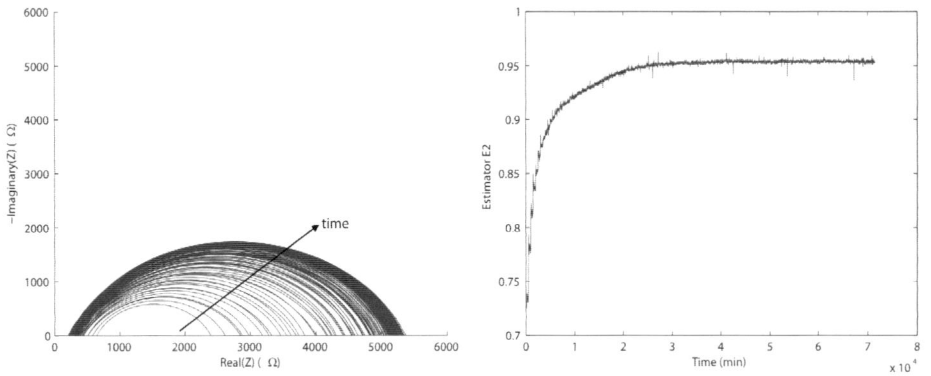

Figure 2. Time evolution of Cole-Cole arcs and cell growth estimator's described in [14].

At this moment, our work is focused on characterizing human stem cells for tissue engineered applications. The goal is to characterize stem cells differentiation into cardyomiogenic cells when applying electrical and mechanical stimuli in regenerative medicine.

4. Conclusions

One of the multitone signal strengths is that it is very rapid for those systems that have a short settling time and are not plagued by significant noise. In addition, SNR can be increased by aplying more energy in the desired spectral components in applications where the total amount of energy applied is not restricted. In applications where the energy is limited, a single multitone shot characterization obtains greater SNR than using Pseudo-Random Binary Sequences (PRBS). This is due the fact that the energy is focused on the measurement frequencies and it is not spread in the whole frequency band. One of the weaknesses of multitone signals is the crest factor. However, it is possible to minimize the impact of the crest factor over the measurement quality using common algorithms proposed in the literature.

Acknowledgments

This work has been suported by grant from Spanish Ministry MICINN SAF2008-05144-C02-02.

5. References

[1] K. R. Godfrey, H. A. Barker, A. J. Tucker 1999 *Comparison of perturbation signals for linear system identification in the frequency domain* vol. 146 (*IEEE Proc.-Control Theory Appl.*) pp. 535–548.

[2] J. Schouckens, P. Guillaume, R.Pintelon 1991 *Design of multisine excitations* vol. 40 (*International Conference on Control*) pp. 638–643.

[3] P. Guillaume, J. Schoukens, R.Pintelon and I. Kollar 1991 *Crest-factor minimization using nonlinear Chebyshev approximation methods* vol. 40 (*IEEE Transactions on Instruments and Measurement*) pp. 982–989.

[4] E. Beller and D. J. Newman 1965 *An l1 Extremal Problem for Polynomials* vol. 16 no. 3 (*Proceedings of the American Mathematical Society*) pp. 1287–1290.

[5] M. Schroeder 1970 *Synthesis of low-peak-factor signals and binary sequences with low autocorrelation* vol. 16 no. 1 (*IEEE Transactions of Information Theory*) pp. 85–89.

[6] A. Van den Bos 1987 *A new method for synthesis of low-peak-factor signals* vol. 35 no.1 (*IEEE Transactionss on Acoustics, Speech and Signal Processing*) pp. 120–122.

[7] E. Van der Ouderaa, J. Schoukens and J. Renneboog 1988 *Peak factor minimization using a time-frequency domain swapping algorithm* vol. 37 no.1 (*IEEE Transactions on Instruments and Measurement*) pp. 145–147.

[8] A. Horner and J. Beauchamp 1996 *A genetic algorithm-based method for synthesis of low peak amplitude signals* vol. 99 (*Journal of the Acoustical Society of America*) pp. 433–443.

[9] B. Sanchez and R. Bragos *Fast Electrical Impedance Spectroscopy for Moving Tissue Characterization Using Bilateral QuasiLogarithmic Multisine Bursts Signals* vol. 22 no. 9 (*4th European Conference of the International Federation for Medical and Biological Engineering*) pp. 1084–1087.

[10] A. Romsauerova, A. McEwan and D.S. Holder 2006 *Identification of a suitable current waveform for acute stroke imaging* vol.27 (*Physiol. Meas.*) pp. 211–219

[11] C. Evans and D. Rees 2000 *Nonlinear distortions and multisine signals. I. Measuring the. best linear approximation* vol. 49 no.3 (*IEEE Transactions on Instrumentation and Measurement*) pp. 602–609

[12] H. Kitayoshi, S. Sumida, K. Shirakawa and S. Takeshita 1985 *DSP synthesized signal source for analog testing stimulus and new test method* vol. 49 no.3 (*IEEE Transactions on Instrumentation and Measurement*) pp. 602–609.

[13] R. Bragos, E. Sarro, A. Fontova, A. Soley, J. Cairo, A. Bayes-Genis, and J. Rosell 2006 *Four versus two-electrode measurement strategies for cell growing and diferentiation monitoring using electrical impedance spectroscopy* (*Proc. 28th Annual International Conference of the IEEE Engineering in Medicine and Biology Society EMBS*) pp. 2106-2109.

[14] R. Bragos, X. Gamez, J. Cair, P. Riu, F. Godia 1999 *Biomass monitoring using impedance spectroscopy* vol.873 (*Annals of the NewYork Academy of Sciences*), pp 299–305.

Portable Bioimpedance Spectroscopy device and textile electrodes for mobile monitoring applications

L. Beckmann[1], M. Jacob[1], C. Hoog Antink[1], A. Cordes[1], R. Pikkemaat[1], N Jungbecker[2], T. Gries[2] and S. Leonhardt[1]
[1]Philips Chair of Medical Information Technology, RWTH Aachen University, Pauwelsstr. 20, 52074 Aachen, Germany
[2]Institut für Textiltechnik, RWTH Aachen University, Otto-Blumenthal-Str. 1, 52074 Aachen Germany

E-mail: beckmann@hia.rwth-aachen.de

Abstract. A balanced body composition is necessary for a person's health and performance. Therefore, it is important to control the body composition continuously since complications and diseases due to dehydration often appear gradually. Based on these facts a miniaturized mobile Bioimpedance Spectroscopy device was developed that can be integrated into clothing and allows the continuous monitoring of a person's body water. The implemented system has been tested using different body models. The first measurements showed very precise and stable results. Besides the portable measurement system, textile electrodes are needed for continuous long term monitoring. Therefore, special textile electrodes were developed, tested and evaluated. The electrodes are structured in a specific way leading to a rougher surface. Such a surface improves the interface impedance and therefore optimizes the connection between electronic hardware and body. For comparison, five different structured electrodes were manufactured and tested on a special test setup that allows reproducible interface-impedance measurements using a dummy made of agar-agar to simulate the skin. It could be shown that the surface structure significantly influences the interface impedance in a positive way as compared to standard plane textile electrodes. In the future, a combination of the miniaturized BIS electronic and the structured textile electrodes could allow reproducible long term monitoring of a person's body composition.

1. Introduction

Textile integrated measurement systems are very important applications in the future. Such systems allow long-term monitoring in order to improve the recovery of patients as well as to simplify early diagnosis. One scenario for long-term measurements is monitoring the body composition of elderly people. Such people gradually loose the physiological attraction of drinking and eating, as a consequence of the natural ageing process, which can lead to age anorexia and dehydration with severe consequences. Bioimpedance Spectroscopy (BIS) measurements may allow the continuous monitoring of body composition. By now, the accuracy of BIS measurements has been demonstrated in several clinical trials [1]. Even if BIS is a helpful application as early warning system, most measurement devices are still quite big and are not portable or integrated into clothes, which is necessary for mobile monitoring scenarios. In the past a few miniaturised hardware systems were developed and presented e.g. by Vuorela [2] and Beckmann [3]. But they are still quite big or restricted in terms of measurement resolution and frequency range. Besides the hardware for a complete integration, textile electrodes are important. Several researchers are currently developing and testing such electrodes for different applications. First test for BIS measurements were presented by Vuorela [2] and Medrano [4]. The major problem of using textile electrodes is the high interface impedance between electrode and skin due to the lack of hydrogel. Therefore, in this article textile electrodes are presented that improves the electrode-skin-contact due to special structured surfaces.

2. Hardware Realization

2.1. Hardware Design

A portable BIS measurement device was developed, that allows continuous body composition monitoring. The system consists of two small boards as shown in Fig. 1. Embedded into a small box the system can be worn on the body. The device is battery-driven and measures the impedance between 5 kHz and 1 MHz. The measurement current is generated using a signal generator and an additional current source. During a measurement the injected current and occurring voltage is measured and separated into real and imaginary part using a quadrature demodulator. The measurement data are converted and saved on a microcontroller. Finally, the data for each frequency are sent to a PC or PDA for further analysis. Additionally, the measurement device features a special safety circuit to guarantee adherence to the IEC 60601-1 electrical safety standards.

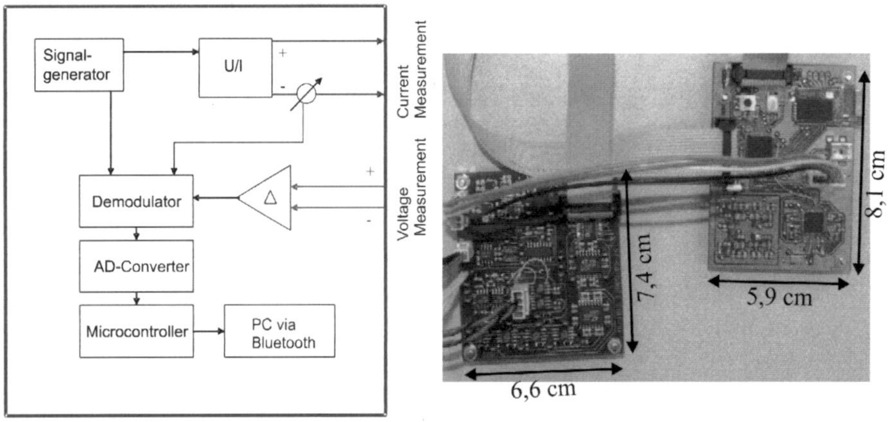

Figure 1. Block diagram and picture of the new developed hardware.

2.2. Results

The new developed measurement device was tested and evaluated using different test circuits. The test circuits consist of a Cole-Cole circuit built up with passive components simulating the whole body of a person or different body segments (e.g. thigh). Additionally, the circuits are equipped with further passive components to simulate the electrode-skin-contact. After sending the measured impedance values to the PC the Cole-Cole parameter were calculated and compared to the original data of the test circuit components. Table 1 shows the original values, the measured values as well as the calculated relative error.

For the whole body measurement the relative error for all Cole-Cole parameters is around 3,4%, especially R_e was measured very precise with a deviation of only 0.66%. For thigh measurements the error increases slightly to a maximum of 5,56%. Compared to whole body measurements the measurement signals during a thigh measurement are much smaller (R_e = 681 Ω compared to R_e = 54 Ω). Smaller signals are more difficult to detect and are stronger influenced by the existing noise. However, by adjusting the measurement amplifier to the specific thigh measurement range, even better results could be achieved in the future.

Table 1: Original and measured Cole-Cole parameters in two dummy measurements

Body Segment	**Parameter**	**Original Data**	**Measured Data**	**Relative Error**
Whole Body	R_e	681 Ω	685,6 Ω	0,66 %
	R_i	909 Ω	877,9 Ω	3,41 %
	C_m	3,3 nF	3,35 nF	1,52 %
Thigh	R_e	54 Ω	51,2 Ω	5,56 %
	R_i	56 Ω	57,7 Ω	2,68 %
	C_m	47 nF	49,2 nF	4,68 %

3. Textile Electrodes

3.1. Materials

Besides the small measurement hardware, textile electrodes are important ingredients for integrated portable measurement scenarios. Therefore, 5 different textile electrodes were developed and tested. Four of the newly developed textile electrodes are structured which leads to a rougher surface and improve the interface-impedance and optimizes the connection between hardware and body. All textile electrodes are made of silvered polyamide fibres mixed with other polyamides and spandex. The spandex fibres make the final textile very elastic and therefore very comfortable to wear. Fig. 2 shows the structured surfaces of the new developed textiles.

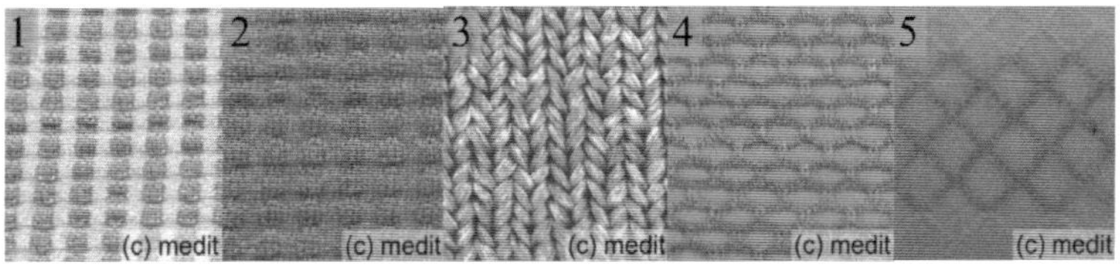

Figure 2. Surface of the new developed textile electrodes.

3.2. Results

For testing the suitability of the new electrodes, tests with five elderly volunteers were made. Figure 3B shows the interface impedance measured with the five different textile electrodes as well as with standard glued electrodes as a reference exemplarily for one volunteer. The standard glued electrodes are electrodes that are used for measurements with the Bioimpedance Analyser Xitron Hydra 4200, from Xitron Technologies, USA.

The measurements made with volunteers have one severe disadvantage. They are quite difficult to analyse in order to investigate the suitability of the electrodes for BIS measurements. Even if the interface impedance is different for every electrode, the measurements are very difficult to reproduce and to compare between the volunteers since the dielectric properties of human skin change from person to person and within hours. To avoid such measurement influences the electrodes were also tested using a special electrode test setup. The test setup contains a skin dummy made of agar-agar and allows measuring the interface impedance within a frequency range of 5 kHz – 1 MHz and for different contact pressures.

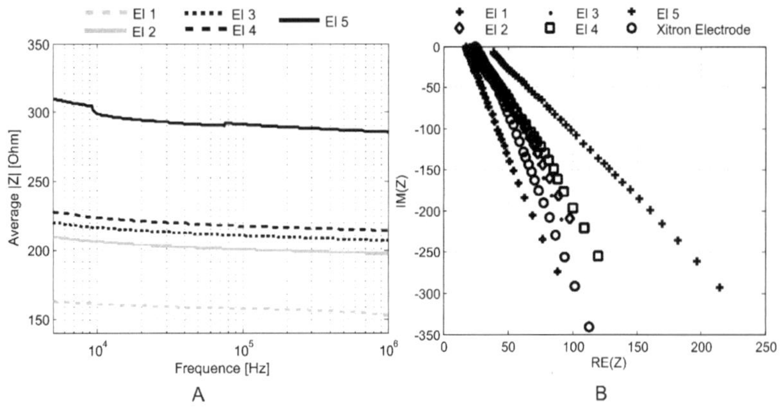

Figure 3. A) Interface impedance measured on a special test setup
B) Interface impedance measured on an elderly volunteer

More details about the setup can be found in [5]. Fig 3A shows the interface impedance measured with the test setup at a stable contact pressure of 9.73N. Obviously, the interface impedance decreases for higher frequencies but differs strongly between the 5 textile electrodes. Textile electrode 5 has the highest interface impedance around 300 Ω, textile electrode 1 the lowest impedance around 160 Ω. Three of the four structured textile electrodes have smaller interface impedances than the non structured reference electrode 3. This result shows that a structured surface improves the electrode-skin-contact and lead to smaller interface impedances, whereas the structure of textile electrode 1 seems to be best suitable for measurements.

4. Discussion

The textile integration of BIS measurement systems is becoming more and more important for various applications. Therefore, a miniaturized hardware as well as textile electrodes were presented in this paper. The measurement error of the hardware was beyond 3,1% and 5,6% respectively, which is quite good for such small hardware designs. Additionally, structured electrodes were developed and tested. It could be shown that the structured surface influences the electrode-skin-contact and improves the interface impedance. In the future, the newly developed hardware and the structured electrodes should be combined for specific measurement application to allow very precise long-term Bioimpedance monitoring.

5. Acknowledgement

This work is supported by the German Federal Ministry of Education and Research (BMBF) in the Microsystems program 2004-2009.

[1] Moissl U, Wabel P, Chamney P, et al. 2006 Body fluid volume determination via body composition spectroscopy in health and disease *Physiol. Meas 27; 921-933.*

[2] Vuorela T, et al. 2005 Textile electrode usage in a bioimpedance measurement *Intelligent Ambience and Well-Being, Finland*

[3] Beckmann L, Medrano G, Leonhardt S 2006 Tragbares Bioimpedanz-Spektroskopie Gerät. Dreiländertagung BMT 2006, ETH Zürich

[4] Medrano G, Beckmann L, Zimmermann N, et al. 2007 *4th International Workshop on Wearable and Implantable Body Sensor Networks, BSN, RWTH Aachen, Germany*

[5] Beckmann L, Neuhaus C, Medrano G, Jungbecker N, Walter M, Gries T, Leonhardt S 2010, Characterization of textile electrodes and conductors using standardized measurement setups, *Physiol. Meas. (accepted, in press)*

Human Interface Design using Button-type PEDOT Electrode Array in EIT

Hun Wi[1], Tong In Oh[1], Sun Yoon[2], Kap Jin Kim[2] and Eung Je Woo[1]

[1]Department of Biomedical Engineering, Kyung Hee University, Korea
[2]Department of Advanced Polymer and Fiber Materials, Kyung Hee University, Korea

E-mail: ejwoo@khu.ac.kr

Abstract. Animal and human experiments using a multi-channel EIT system requires a cumbersome procedure to attach multiple electrodes. We have to ensure good contact of all electrodes and manage many lead wires during experiments. The problem becomes more severe as we increase the number of electrodes. These may limit the applicability of the imaging method in practice. Noting this technical difficulty, there have been a few trials to design human interface means such as electrode belts, helmets or rings. In this study, we developed an electrode belt for long-term monitoring of human lung ventilation. The belt includes 16 embossed electrodes which make good contact with the skin. The electrode is made by conductive polymer and metallic thread. Soft cushion and wide contact area minimize uncomfortable sensation and reduce contact impedances. The electrodes are attached to an elastic fabric belt at equal spacing. We describe details of its design and fabrication. Using the electrode belt and recently developed multi-frequency EIT system KHU Mark2, we show time-difference chest images of three human subjects during normal breathing cycles.

Keyword : EIT, electrode belt, fabric electrode

1. Introduction

In electrical impedance tomography (EIT), we inject current into an imaging object. Measuring induced voltages on the boundary, we may reconstruct cross-sectional admittivity images of the object. Clinical applications may include lung imaging for ventilation monitoring, stomach emptying, head and cardiac imaging [1].

In most applications of EIT, it is important to attach electrodes quickly and stably. During imaging, we must keep good contact between electrodes and the skin regardless of posture. There have been several trials to develop human interface methods such as electrode belts, helmets or rings depending on intended applications [2]. In this paper, we propose a new design of human interface for EIT imaging.

We are especially targeting the monitoring of human lung ventilation using EIT technique. Though the shape of the human chest is approximately an ellipse, it is quite irregular and varies among subjects. We describe a developed flexible electrode belt of a conductive polymer which is equipped with multiple electrodes. For a long-term monitoring application, we adopted dry electrodes.

2. Method

2.1 Electrode

For wider applications, we avoided using a wet electrode with gel. As a wearable sensor, we used a conductive polymer called PEDOT to fabricate a dry electrode. PEDOT or PEDT is a conducting polymer based on 3,4-ethylenedioxylthiophene. Its fiber diameter is less than 700 nm as shown in figure 1.

(a) Fabric coated PEDOT (b) PEDOT fiber (×50,000 magnification)

Figure 1. Fabric material for a button-type dry electrode.

Among different conducting polymers, PEDOT has been widely used in recent years. PEDOT coatings possess high stability over different charge and discharge cycles and can be electro-generated directly on a conductive support such as Pt, Au, glassy carbon, indium tin oxide and so on in organic solvents or in aqueous solution [3,4]. Its advantages are high stability, moderate band gap and low redox potential. Poor solubility will be another advantage when we use it as electrode material. Compared with commercial ECG electrodes, it has a relatively high contact resistance of 100 to 200 Ω.

2.2 Electrode belt

Figure 2 shows a design of electrode belt with its length of 48 cm. When it is stretched, it may reach the maximal length of 96 cm. Sixteen electrodes are arranged along the center line of the elastic band. The gap between adjacent electrodes is 3 cm. When the belt is stretched, the gap can be increased up to 6 cm. We use a pair of hooks to connect both ends so that it encloses the chest. Figure 3(a) is the manufactured electrode belt and (b) shows the button-type PEDOT electrode. In order to easily replace the electrode, we adopted a button structure, which is connected to the band by using an eyelet. Figure 4 shows the dimension of the electrode and belt.

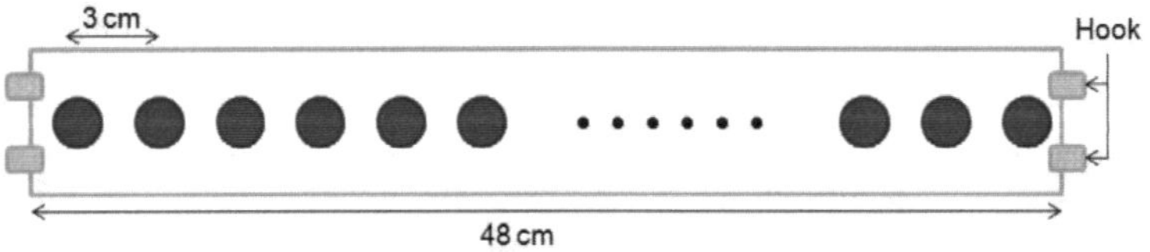

Figure 2. Electrode belt design.

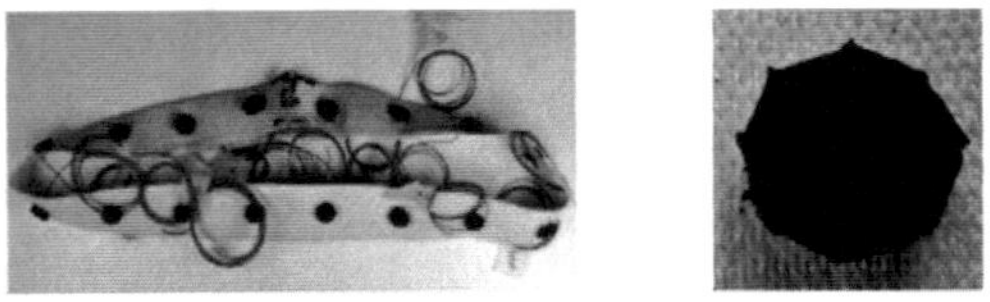

(**a**) Electrode belt. (**b**) Button electrode.

Figure 3. Belt and electrode.

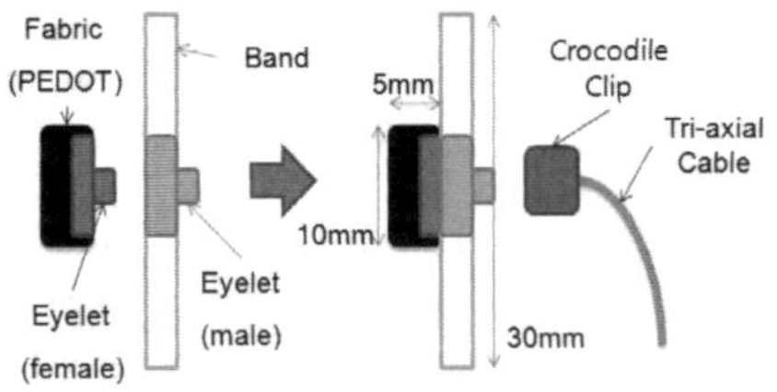

Figure 4. Button electrode design.

3. Results

3.1 Noise test

We measured the noise level of the electrode belt on a TX151 gel phantom. We used our electrode belt as well as commercial Ag-AgCl ECG electrodes (2237, 3M, USA). We measured voltage on each electrode on the gel phantom using a biopotential amplifier (MP35, Biopac Systems, USA). Figure 5 shows plots of measured voltages as a measure of noise levels. The Ag-AgCl ECG electrode has a noise level of 6 μV_{PP}. The electrode belt has a higher noise level of about 16 μV_{PP}.

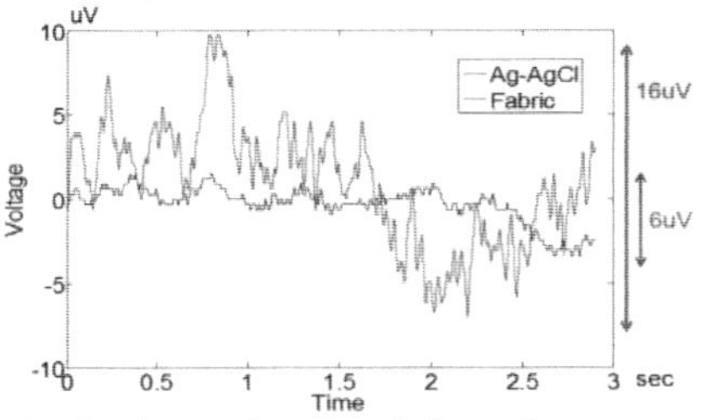

Figure 5. Noise measurements using fabric electrode belt and commercial ECG electrode on agar phantom.

Using both electrodes, we injected current and measured induced voltages as usually done in EIT data acquisitions. Figure 6 shows evaluated signal-to-noise ratio (SNR). We found that there is no significant difference between two electrodes in both short-term(1 sec.) and long-term(10 min.) SNRs.

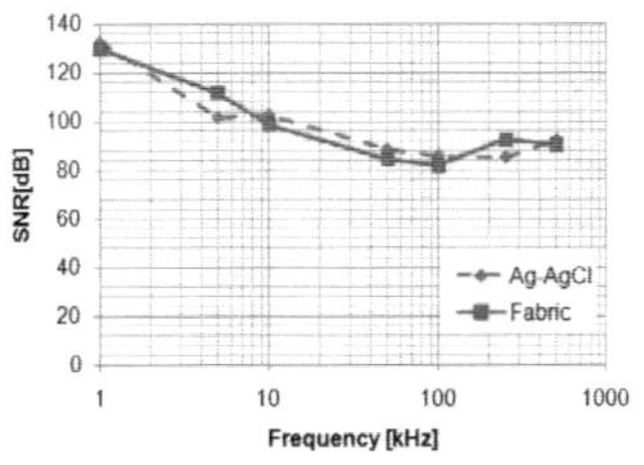

(a) Short-term SNR vs. frequency.

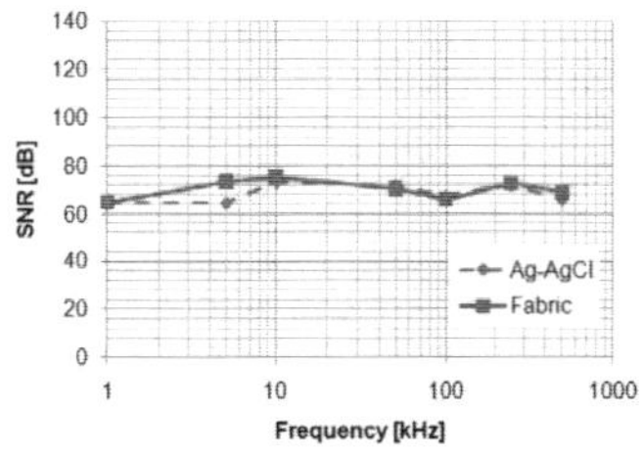

(b) Long-term SNR vs. frequency.

Figure 6. Comparison of SNRs using the KHU Mark2 EIT system.

3.2 Phantom experiment

Figure 7 show the TX151 gel phantom. We made 3 different TX151 phantoms. The background gel had 0.1% NaCl while two anomalies in blue and red had 0.02 and 0.5% NaCl, respectively. Figure 8 shows reconstructed time-difference images of the TX151 gel phantom using the electrode belt with sixteen PEDOT button-type electrodes.

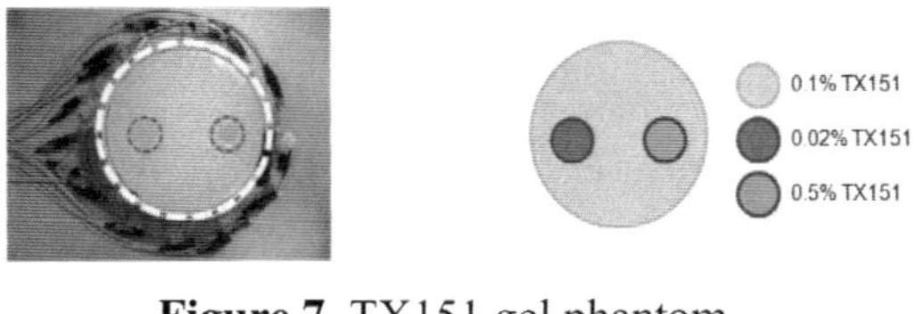

Figure 7. TX151 gel phantom.

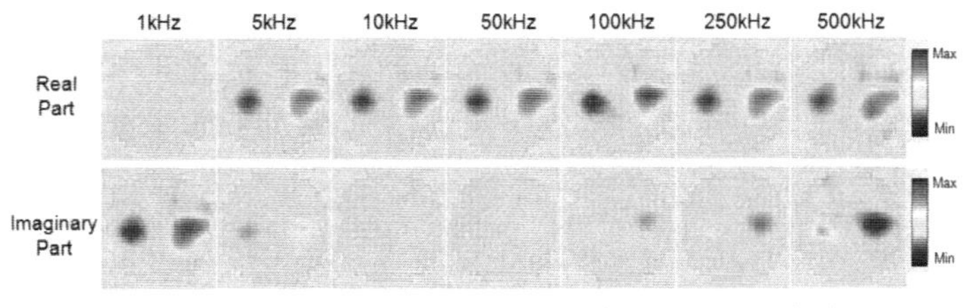

Figure 8. Time-difference images of the TX151 gel phantom

3.3 Human experiment for pulmonary monitor

Figure 9 is the picture of the flexible electrode belt attached on the human chest. During normal ventilation cycles, we collected boundary current-voltage data sets at 50 kHz operating frequency. Figure 10 shows time-difference images of the chest illustrating effects of lung ventilation.

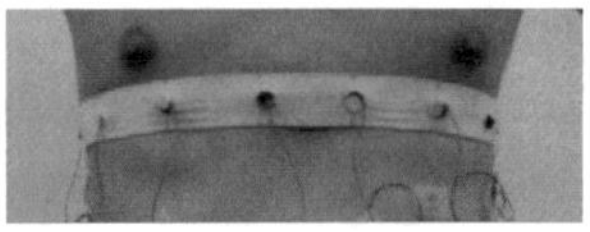

Figure 9. Electrode belt around the human chest.

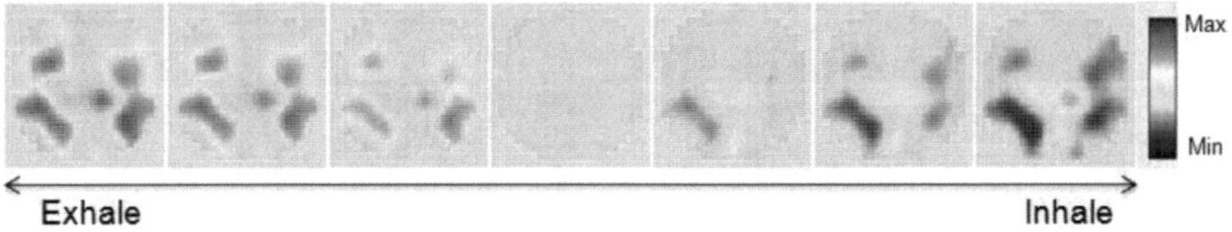

Figure 10. Time-difference images of the human chest using the electrode belt.

4. Conclusion and discussion

We designed the electrode belt with button-type PEDOT dry electrodes. Though it showed an increased resting noise level compared with commercial Ag-AgCl ECG electrodes, its SNR in EIT data was comparable to that of the commercial electrodes. When we used the electrode belt with the new EIT system KHU Mark2, the overall short-term SNR was acceptable for time-difference imaging. From both phantom and human imaging experiments, we could produce time-difference images without degrading the image quality. We plan to apply the flexible electrode belt to clinical EIT imaging of lung ventilation.

Acknowledgement

This work was supported by the SRC/ERC program (R11-2002-103) of MOST/NRF.

References

[1] D. S. Holder (2005) Electrical Impedance Tomography: Methods, History and Applications (Bristol, UK: IOP Publishing)

[2] A. T. Tidswell, A. P. Bagshaw, D. S. Holder, R. J. Yerworth, L. Eadie, S. Murray, L. Morgan and R. H. Bayford (2003) A comparison of headnet electrode arrays for electrical impedance tomography of the human head *Physiol. Meas.* **24**, 527-544.

[3] A. Bello, M. Giannetto, G. Mori, R. Seeber, F. Terzi and C. Zanardi (2007) Optimization of the DPV potential waveform for determination of ascorbic acid on PEDOT-modified electrodes *Sensors Actuat.* B **121**, 430-435.

[4] S. S. Kumar, J. Mathiyarasu, K. L. N. Phani and V. Yegnaraman (2006) Simultaneous determination of dopamine and ascorbic acid on poly (3,4-ethylenedioxythiophene) modified glassy carbon electrode *J. Solid State Electrochem.* **10**, 905-913.

[5] V. Cherepenin, A. Karpov, A. Korjenevsky, V. Kornienko, Y. Kultiasov, A. Mazaletskaya and D. Mazourov (2002) Preliminary static EIT images of the thorax in health and disease, *Physiol. Meas.* **23**, 33-41.

Journal of Physics: Conference Series **224** (2010) 012007 doi:10.1088/1742-6596/224/1/012007

Instrumentation for low frequency EIT studies of the human head and its validation in phantom experiments

Brian Esler[1], Thomas Lyons[2], Sergei Turovets[1,3] and Don Tucker[3]

[1]Neuroniformatics Center, University of Oregon, Eugene, OR 97403, USA
[2]Marker Gene Technologies, Inc, Eugene, OR 97403, USA
[3]Electrical Geodesics, Inc, Eugene, OR 97403, USA

E-mail: sergei@cs.uoregon.edu

Abstract. We describe instrumentation for low frequency (< 500 Hz) EIT studies of the human head and its calibration, testing and validation in the phantom experiments. Our EIT system prototype is based on a 256 channel commercial EEG system complimented by the current injection module and lock-in detection software. We have designed and built two types of head phantoms: i) a resistor network and ii) a cylinder tank filled with saline and gel insertions with chemically targeted conductivity values. We have developed a technology for fabricating, handling and storage of agar TX151 gel insertions. Independent and direct conductivity measurements of gel samples have been performed using a HP LCR meter in a four electrode conductivity cell specifically designed and built for this purpose. Measurements of saline conductivity were done with commercially available salinity / conductivity meters. Our inverse conductivity estimates in the phantom experiments with EIDORS and in-house software cross-validate the viability of the EIT-EEG system.

1. Introduction

Electrical Impedance Tomography (EIT) is a portable, noninvasive and inexpensive technique with potential clinical applications in breast cancer prescreening and detecting respiratory and gastric diseases [1]. Although, imaging of brain functions with EIT might be a challenging goal due to the shielding properties of the low-conductive adult skull, the combination of EEG and bounded EIT modalities is promising to make the human head models significantly more accurate for source localization by providing specific information on the average head tissue conductivities. In this paper we describe the instrumentation for such a combined modality based on the commercial EEG system and its calibration, testing and validation in the physical phantom experiments. The results of two types of the EIT validation experiments with a cylinder tank are reported: piecewise constant domain conductivity estimations, or bounded EIT, and difference imaging. The forward and inverse modeling has been performed within an open source software suite, EIDORS [3] and also with our in-house finite difference solver [4]. Our preliminary experiments demonstrate viability and great promise of the EIT-EEG system for improving source localization in an analysis of brain activity.

2. Methods

2.1 *Phantom design, instrumentation and data acquisition*

The first head phantom is a simple 1D ladder of 127 serially connected 10 Ohm resistors. It was made in the form of an encased test plug and attached back-to-back to the EEG amplification block.
The second one is an acrylic cylinder tank approximately the size of the typical adult human head (20 cm height and 18 cm diameter). 128 standard EEG Ag/AgCl electrodes with rubber o-rings were hermetically placed into the holes uniformly arranged on the lateral surface in 8 rows and 16 columns. The electrodes were wired to the EEG system through the standard EEG leads (Electric Geodesics, Inc.). The cylinder is filled by saline with conductivity close to CSF. A few agar gel insertions of the smaller size and lower conductivity (cylinder and rectangular rods) or Plexiglass rods can be placed inside the cylinder in arbitrary positions. For our EIT studies, we employed the existing NetAmps 300 EEG amplifier (Electrical Geodesics, Inc.) and an isolated current generator. The NetAmps 300 platform synchronously digitizes 256 analog channels at 20 kHz and 24 bits, and uses a field-programmable gate array (FPGA) to collate and transfer data in IEEE 1394 (Firewire) format to a computer. The NetAmps 300 platform uses "sigma-delta" type A-to-D converters, achieving high linearity, accuracy and resolution at low cost, so that each detector can have a dedicated A-to-D converter, without the need for high-speed multiplexing of different detector signals onto a single higher-speed A-to-D converter. The current source is based on a Howland-type design. It is battery-powered and isolated from the amplifier circuitry. Advantages of this topology over other designs include its small number of components, a single active device and the ability to adjust output resistance. The waveform of the injected current is monitored across a 1 kOhm resistor in series with the impedance load containing the head phantom and injector/sink electrodes. A scanning protocol, including specific sequence of source/sink electrodes, frequency, amperage (<100 μA), and duration of epochs is programmable via a graphic user interface and allows to swipe any of these parameters. The impressed potentials are recorded as regular EEG signals and their amplitude and phase are extracted through a software implementation of locked-in detection. The bandwidth of the current injector is set from 0.5 Hz to 500 Hz by the low pas hardware and digital filters in the NetAmp 300.

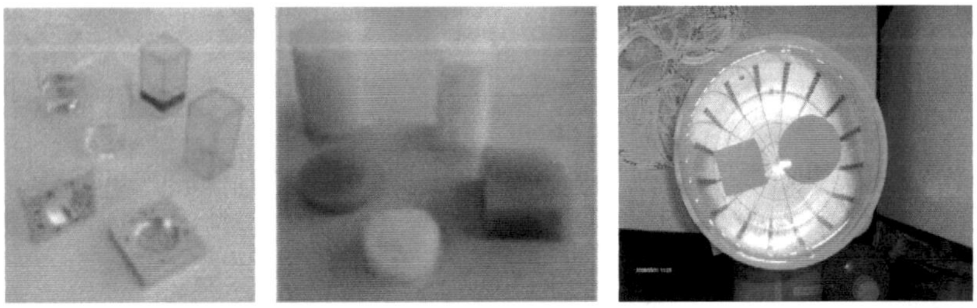

Fig.1 Cylinder tank (right) with TX151 gel phantoms (middle) and their molds (left).

2.2 *Fabrication of TX-151 Agar Gels*

TX-151 agar gels of varying shapes and sizes were fabricated from the same batch using a modified version of the procedure of Oh et al. [2] (Fig. 1). Gel samples of a standard size fitting to our 4-electrode conductivity cell were used to calibrate the relationship between salinity and conductivity by direct measurements (Fig. 2 and 3) in the frequency range 20 Hz -100 Hz with the HP LCR meter

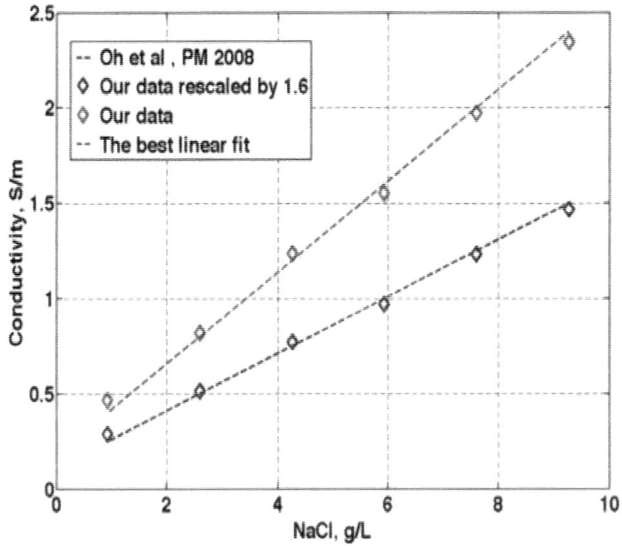

Fig. 2 Direct conductivity measurements of gel phantoms. Concentration of NaCl is the conductivity control parameter.

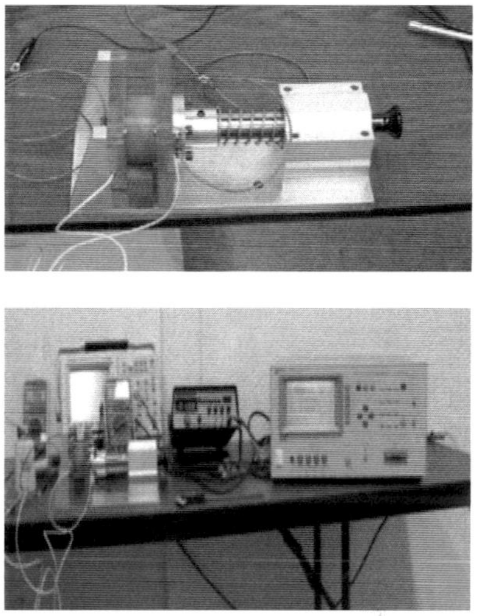

Fig. 3 Experimental setup (bottom) and the 4-electrode conductivity cell

(Model 4284A). For the standard samples used in direct measurements a machined aluminum cylindrical mold (6.41 cm in diameter and 2.54 cm tall) was used, while for other shapes plastic and glass molds were utilized. After solidification for 1 hour at 25°C, gels were removed from the molds, cut to their final dimensions and stored in hermetically sealed glass dishes at 4-8°C. The gels were found to be chemically, mechanically and electrically stable at least for 2 weeks, unless the samples were exposed to room temperature, air and /or saline for a prolonged period of time. A solution of NaCl was made up in de-ionized water based on the recipe given in [2]. In difference with [2], we found empirically at room temperature (22 C) that $\sigma \cong 0.15\,x$ S/m for saline and $\sigma \cong 0.23\,x$ S/m for TX151 gels where x (g/L) is NaCl concentration (Fig.2). The saline conductivity used in tank experiments was ~ 1.31 S/m as measured by the conductivity meter (Amber Science, Model 3082).

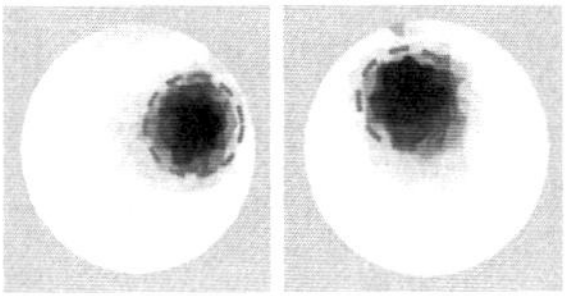

Fig. 4 EIDORS difference EIT reconstruction with a cylindrical gel insertion located at positions given by dashed circles.

3. Results

The 1D resistor network load has been used primarily for identifying bad channels in the system, calibration and device bandwidth measurements. The forward model for this resistor ladder head phantom is quite simple: $V_{ij}=|i\text{-}j|\;RI_{inj}$, where i,j are source-sink numbers, I_{inj} is injected current and R is "unknown" resistance of the ladder element. Applying a linear regression as the inverse procedure to the measured dependencies $V_{ij}(|i\text{-}j)|$ we found R=10.1 Ohm, which is within 1 % of the resistor component specification. FEM models for the tank were created using the EIDORS software [3]. The

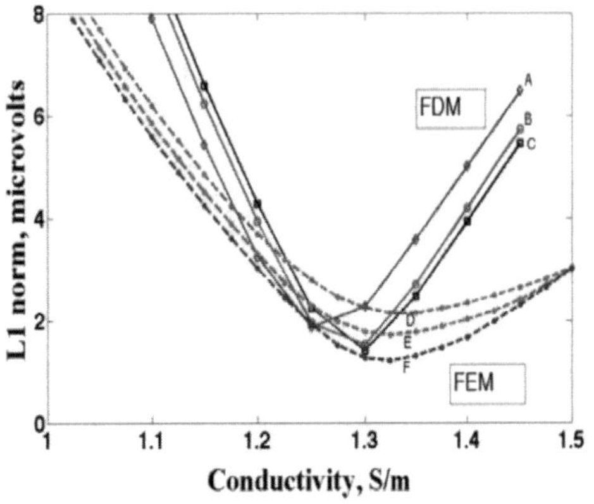

Fig. 5 Error function for three increased resolutions (A- F) in FDM and FEM.

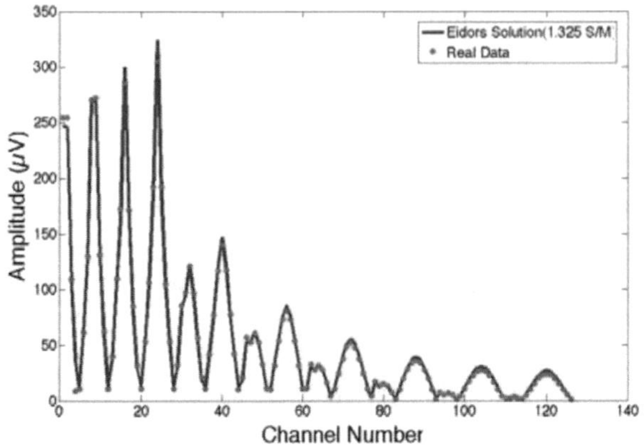

Fig. 6 EIDORS forward solution (1.32 S/m) displayed with the real data for one source–sink pair.

typical model used for our reconstructions contained 2755 nodes, 13824 element mesh with electrodes approximated as points along the perimeter. Our data sets included measurements for each of the 64 opposite source-sink pairs. The voltage measurements were taken relative to a common reference electrode. The images in Fig. 4 depict difference reconstructions for two different configurations and demonstrate the qualitative precision of the technique. As a precursor to creating difference reconstructions, we utilized the EIDORS forward solver to model the homogeneous cylinder (a simple bEIT), which we could then compare to our experimental data and simulation using a BiCG FDM solver described in our concurrent paper at this conference [4]. As can be seen in Fig. 5 and 6 the EIDORS model is quite adequate to the experimental setup and the simulation results fit rather well to the measured data in a simple case of homogeneous conductivity. The error minimum was about 1 μV at the optimal conductivity 1.32 S/m. Inverse search with our FDM has converged to the error function minimum at 1.3 S/m at the relatively sparse grid size 32x32x72 nodes. With TX151 gels of 0.48 S/m inserted the preliminary 2D inverse search produced estimates 1.2 S/m for saline and 0.6 S/m for insertions. In conclusion, we have demonstrated and tested one of the first multi-modal EEG-EIT systems. In difference with the previously proposed methods for regional conductivity estimation [5,6] and multi-modality EEG/EIT systems [1], we have implemented the EIT functionality as an adds-on product to the existing commercial dense array EEG system. In our future work we will explore utility of this system for bEIT in human trials with anatomically accurate head models based on FDM.

Acknowledgments

This study was supported in part by National Institute of Health grant number 5R44NS056758-04.

References

[1] Holder DS 2005 *Eletrical Impedance Tomography* (Bristol: Institute of Physics Publishing)
[2] Oh T I, Koo H, Lee K H et al 2008 *Physiol. Meas.* **29** 295-307
[3] Adler A and Lionheart R B 2006 Psysiol Meas **27** S25-S42
[4] Volkov V M, Zherdetskij A A, Turovets S I and Malony A D 2010 ICEB Gainesville, FL
[5] Ferree T C, Eriksen K J and Tucker D M 2000 *IEEE Trans Biomed Eng* **47** 1584–1592
[6] Gonçalves S, de Munck J C et al 2003 *IEEE Trans Biomed Eng* **50** 754–76

Performance of the Load-in-the-Loop Single Op-Amp Voltage Controlled Current Source from the Op-Amp Parameters

R Macías[1,2], F Seoane[2,3] and R Bragós[1]

[1] Department of Electronic Engineering, Universitat Politècnica de Catalunya. Campus Nord, C-4. Jordi Girona 1-3. 08034, Barcelona. SPAIN

[2] School of Engineering, University of Borås, SE-501 90 Borås, SWEDEN

[3] Department of Signal and Systems, Chalmers University of Technology, SE-41296, Gothenburg, SWEDEN

E-mail: rbb@eel.upc.edu

Abstract. In recent years, Electrical Bioimpedance (EBI) methods have gained importance. These methods are often based on obtaining impedance spectrum in the range of β-dispersion, *i.e. from a few kHz up to some MHz*. To measure EBI a constant current is often injected and the voltage across the tissue under study is recorded. Due to the performance of the current source influences the performance of the entire system, in terms of frequency range, several designs have been implemented and studied. In this paper the basic structure of a Voltage-Controlled Current Source based on a single Op-Amp in inverter configuration with a floating load, known as load-in-the-loop current source, is revisited and studied deeply. We focus on the dependence of the output impedance with the circuit parameters, *i.e. the feedback resistor and the inverter-input resistor*, and the Op-Amp main parameters, *i.e. open loop gain, CMRR and input impedance*. After obtaining the experimental results, using modern Op-Amps, and comparing to the theoretical and simulated ones, they confirm the design under study can be a good solution for multi-frequency wideband EBI applications because of higher values of the output impedance than 100kΩ at 1MHz are obtained. Furthermore, an enhancement of the basic design, using a current conveyor as a first stage, is proposed, studied and implemented.

1. Introduction

Given the number of applications of wideband EBI spectroscopy arisen recently in several medical fields such as skin cancer detection [1] or organ transplantation [2], researching and developing new EBI measurement systems with the upper limit frequency as high as possible is required. Hence and due to the current source is an essential block of these systems, several more complex approaches have been proposed aiming to improve their performances in terms of frequency range. Unfortunately, the performance of all these complex approaches degrades markedly near or below 1MHz.

On the other hand, advances in the development of integrated circuits have provided new wide-bandwidth Op-Amps. Therefore, in this paper a simple structure based on a simple, single Op-Amp based circuit topology to operate as a Voltage-Controlled Current Source (VCCS) in wide-bandwidth EBI applications is revisited. Furthermore, an enhancement of this structure is also proposed, studied and tested.

2. Theoretical Analysis

As previously mentioned, current sources are an essential block of EBI systems. Ideally, a current source generates an electrical signal with constant current amplitude independently of the load at any frequency. That means the value of the output impedance, Z_{out}, must be infinite at all frequencies [3]. In practice, the value of the Z_{out} is not infinite and it is frequency-dependent. Therefore, the main aim is to achieve current sources with a large output impedance value for a frequency range as wide as possible.

Thus, in this section the output current, I_{out}, and the output impedance of both structures are obtained mathematically and their dependences in terms of several parameters are studied. The first structure is a basic structure based on a single Op-Amp in inverting configuration with floating load, also known as load-in-the-loop current source [4]; see figure 1. The other is based on an enhancement of the first one. It consists of adding a first stage based on a VCCS implemented by a current conveyor (CCII) [4]; see figure 2.

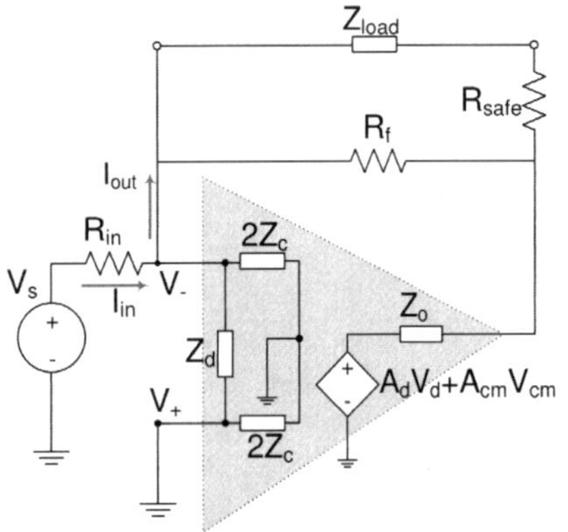

Figure 1. Load-in-the-Loop Current Source.

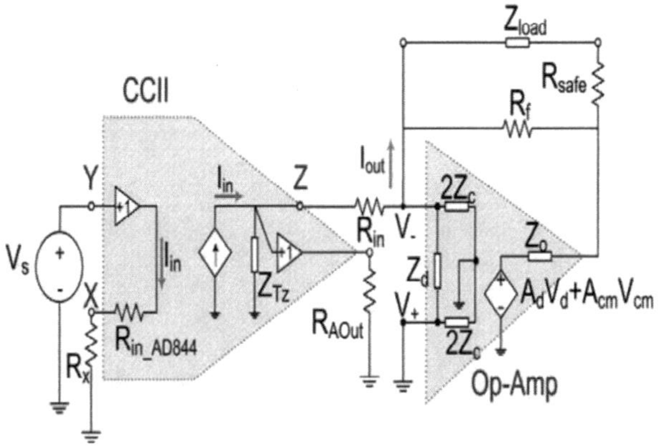

Figure 2. Current-Driven VCCS. It consists of the Load-in-the-Loop Current source driven by a CCII.

2.1. A Single Op-Amp VCCS: Load-in-the-Loop Current Source

To carry out a detailed and realistic analysis of the circuit, a non-ideal operational model of Op-Amp has been selected. It takes into account a finite value of the differential and common-mode input impedance, Z_d and Z_c respectively, a non-zero output impedance, Z_o, and a non-zero common-mode gain, A_{cm}. Furthermore, it also considers the differential-mode gain and the common-mode rejection ratio are frequency-dependent, $A_d(s)$ and *CMRR(s)* respectively.

Therefore, given the aforementioned considerations, the circuit under study can be analyzed.

2.1.1. The output current, I_{out}. It is given by the ratio between V_s and R_{in}; see (1). R_{in} is then the transconductance of the Load-in-the-loop current source.

$$I_{out} \approx I_{in} = \frac{V_s}{R_{in}} \tag{1}$$

2.1.2. The ouput impedance, $Z_{out}(s)$. To find the output impedance of the circuit, the voltage source, V_s, and the load, Z_{load}, are replaced for a short circuit and an auxiliary voltage source, V_x, respectively. Furthermore, the overall input impedance of the Op-Amp, Z_{in}, can be written as (2). Therefore, given the simplified circuit shown in figure 3, the expression for the output impedance can be written as (3),

$$Z_{in} = \left(Z_d \| 2Z_c\right) \tag{2}$$

$$Z_{out}(s) = R_{safe} + R_f \| \left(Z_o + \left(R_{in} \| Z_{in} \right) \cdot opz(s) \right) \tag{3}$$

where *opz(s)* is the Op-Amp Impedance factor and is defined as (4).

$$opz(s) = 1 + A_d(s) \cdot \left(1 - \frac{1}{2CMRR(s)} \right) \tag{4}$$

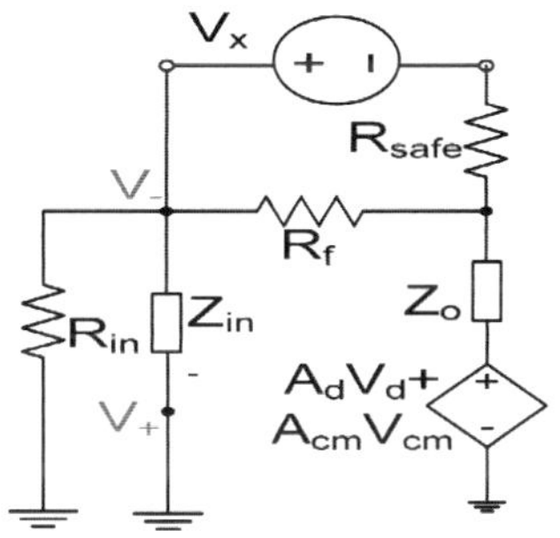

Figure 3. Equivalent circuit used to calculate the output impedance of the Load-in-the-Loop Current Source.

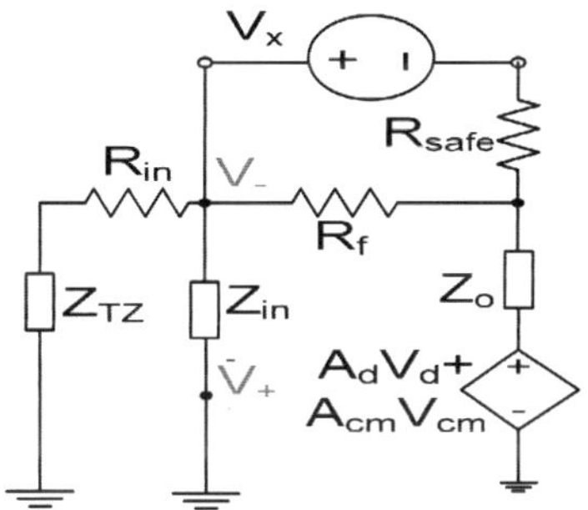

Figure 4. Equivalent circuit used to calculate the output impedance of the Current-Driven VCCS.

2.2. *Current-Driven VCCS*

The second proposed circuit consists basically on replacing the Thevenin source, V_s, by a Norton source, *i.e. the CCII shown in figure 2*. Thus, the output current, I_{out}, is generated independently of any element related to $Z_{out}(s)$. Furthermore, the output impedance of this circuit is essentially proportional to the features of the active devices instead of the input resistor, R_{in}. This allows to overcome the intrinsic limitation posed by the fact that R_{in} cannot be arbitrarily high given that it defines the transconductance of the VCCS.

2.2.1. The output current, I_{out}. It is approximately equal to the output current in node Z of the CCII; see figure 2. Therefore, it is the same current flowing through node X [5] and it is independent of the input resistor, R_{in}; see (5).

$$I_{out} \approx I_Z = I_{in} = \frac{V_s}{R_X + R_{in\_AD844}} \tag{5}$$

2.2.2. The output impedance, $Z_{out}(s)$. After simplifying the proposed circuit into the circuit shown in figure 4, the expression for the output impedance, $Z_{out}(s)$, can be written as (6)

$$Z_{out}(s) = R_{safe} + R_f \| \left(Z_o + \left(\left(R_{in} + Z_{TZ} \right) \| Z_{in} \right) \cdot opz(s) \right) \tag{6}$$

where Z_{TZ} is the output impedance of the CCII and its value is much higher than the input resistor, R_{in}.

3. Measurements and Results

To measure the output impedance of both configurations the Impedance Analyzer LCR HP4192A is used in Gain/Phase measurement mode and the technique used by Bertemes-Filho [6] is applied.

On the other hand, the integrated circuits used as the Op-Amp and the CCII are, respectively, LMH6654 by National Semiconductor and AD844 by Analog Devices. Furthermore, the values used for calculating the output impedance magnitude are shown in Table 1.

Table 1: Values & Expressions used for calculating the Z_{out} magnitude.

Symbol	Expression	Values		Symbol	Values
Z_d	R_{id} // C_{id}	R_{id} = 20kΩ	C_{id} = 0.55pF	Z_o	80Ω
Z_c	R_{cm} // C_{cm}	R_{cm} = 4MΩ	C_{cm} = 0.9pF	R_{in}	6.2kΩ
$A_d(s)$	$A_{d0}/(1+(s/\omega_d))$	A_{d0} = 67dB	$\omega_d = 2\pi\cdot 125$kHz	R_{safe}	390Ω
CMRR(s)	$CMRR_0/(1+(s/\omega_{cm}))$	$CMRR_0$ = 90dB	$\omega_{cm} = 2\pi\cdot 9$kHz	R_f	390kΩ
Z_{TZ}	R_{TZ} // C_{TZ}	R_{TZ} = 3MΩ	C_{TZ} = 4.5pF		

Therefore, the measured output impedance values from the experimental tests are shown in figure 5 and figure 6.

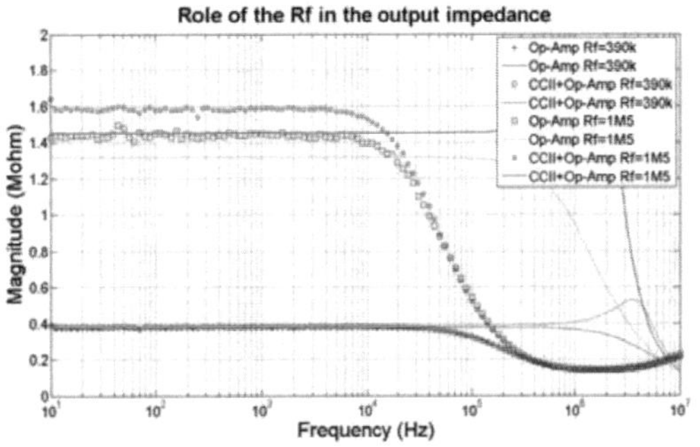

Figure 5. Plotted output impedance for different values of R_f. R_f = 390kΩ and 1.5MΩ. **N.B.** R_{in} = 6.2kΩ.

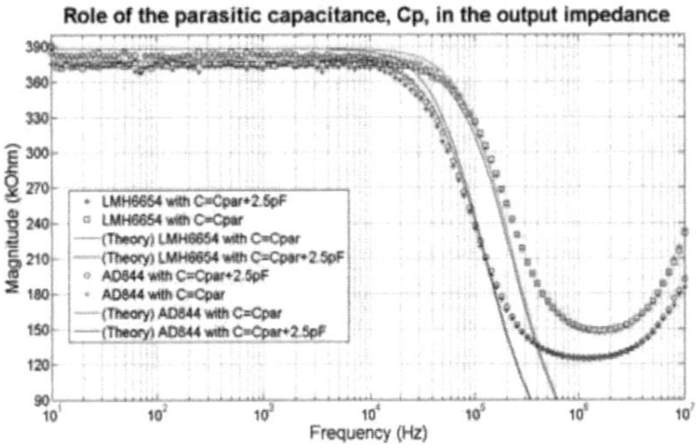

Figure 6. Adding a 2.5pF parasitic capacitance in the calculated Z_{out}, the measured and the simulated Z_{out} match up. Furthermore, if a 2.5pF capacitor is added to the implemented board, the cut-off frequency becomes smaller.

4. Discussion

4.1. Regarding the feedback resistor, R_f

Due to R_f is in parallel with the output of the current source, it acts as current divider of the output current. Furthermore, it limits the maximum value of the output impedance; see figure 5.

Thus, the higher R_f is, the higher the output impedance becomes. But, because of providing a path for the bias current of the Op-Amp is usually required in EBI systems, R_f cannot be removed.

4.2. Regarding the parasitic capacitance, C_p

As figure 5 shows, there is a good agreement between the measured and the calculated output impedance in the frequency range below 100 kHz. Above this range, the value of the cut-off frequency varies depending on the value of R_f. The higher R_f is, the smaller the cut-off frequency becomes. The main reason is because of the influence of parasitic capacitances, C_p, is not negligible at high frequencies. A 2.5pF value of C_p is found as the best value which fits with the measured output impedance.

Hence, the working frequency range for a VCCS is not given only by the Op-Amp features but also by the parasitic capacitances present at the output of the VCCS.

5. Conclusions

As this results show, implementing wideband current sources using a single Op-Amp circuit for spectroscopy measurements of EBI is a realistic approach. Using the Load-in-the-loop Current Source, wideband multifrequency applications can be implemented properly in terms of large output impedance at a very low cost. Furthermore, to achieve an output current of the VCCS more robust and constant and also independent of any element related to the output impedance, the improved circuit called Current-Driven VCCS is proposed.

Therefore, the overall output impedance for both configurations can be plotted as in figure 7. As observed, the overall output impedance is modeled as three impedances building up a parallel bridge: the Z_{in}, Z_i and R_f. The parameter Z_{in} was defined in (2) as the Op-Amp input impedance. Z_i is the equivalent input impedance from the inverting input of the Op-Amp to ground, *i.e.* R_{in} *in the Load-in-the-loop Current Source and* $R_{in}+Z_{TZ}$ *in the Current-Driven VCCS*. Furthermore, the *opz(s)* was also defined in (4) as the Op-Amp Impedance factor. The higher these three impedances and the *opz(s)* are, the higher the overall output impedance becomes. This can help to choose the right Op-Amp from the system requirements.

On the other hand, as previously mentioned, the effect of the parasitic capacitances is the critical factor limiting the performance of the current source at high frequencies. Hence, to improve the performance of the current source, parasitic capacitances should be avoided or minimized as far as possible from the earliest stage of the circuit design. Furthermore, in a real measurement system, a higher capacitance exists due to the coaxial cables connected between the current source and the load. This problem affects all current source structures and can be only overcome if the VCCS is placed near the electrodes or driven guards are used for the coaxial cables. Also dominant pole strategies can be employed to avoid oscillations in that last case.

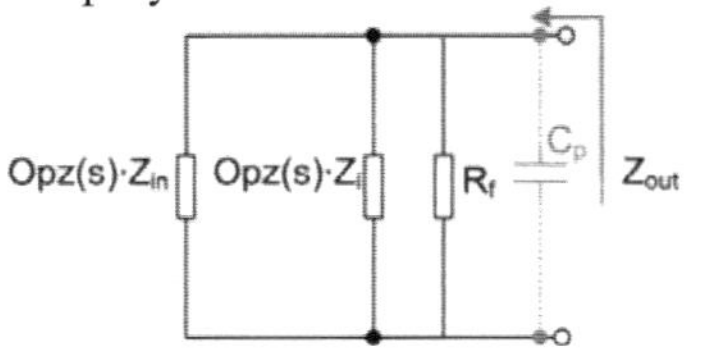

Figure 7. Equivalent circuit for the output impedance of both VCCS circuits. The parasitic capacitances' effect is added in discontinuous trace. **N.B.** the values of R_{safe} and Z_o are considered negligible.

References

[1] Aberg P, Nicander I, Hanson J, Geladi P, Holmgren U and Ollmar S 2004 Skin cancer identification using multifrequency electrical impedance – A potential screening tool. *IEEE Trans. Bio. Med. Eng.*, 51:(12), 2097-2102

[2] Ivorra A and Genesca M 2005 Bioimpedance dispersion width as a parameter to monitoring living tissues Electrical bioimpedance measurement during hypothermic rat kidney preservation for assessing ischemic injury, *In: Physiolo. Meas.*, 26:(2), 165-173

[3] Seoane F, Bragós R, Lindecrantz K and Riu P 2007 Current source design for electrical bioimpedance spectroscopy, *In: Encyclopedia of healthcare information systems*

[4] Seoane F, Bragós R and Lindecrantz K 2006 Current source for multifrequency broadband electrical bioimpedance spectroscopy systems. A novel approach. *Engineering in Medicine and Biology Society Conference, 2006. EMBS '06.* pp *5121 - 5125*

[5] Sedra S A and Roberts W G 1990 Current conveyor: theory and practice, *In: Analog IC Design: The Current-Mode Approach* pp 93-126

[6] Bertemes-Filho P, Brown B H and Wilson A J 2000 A comparison of modified Howland circuits as current generators with current mirror type circuits, *In: Physiological Measurement* 21:(1), 1-6

Journal of Physics: Conference Series **224** (2010) 012009 doi:10.1088/1742-6596/224/1/012009

Offset-free bidirectional current source for impedance measurement

Uwe Pliquett, Markus Schönfeldt, Andreas Barthel, Dieter Frense, Thomas Nacke

Institut für Bioprozess- und Analysenmesstechnik e.V., Rosenhof
D-37308 Heilbad Heiligenstadt, Germany

uwe.pliquett@iba-heiligenstadt.de

Abstract. In order to minimize the influence of electrode polarization and noise in impedance measurements, especially in the low frequency region, current injection shows several advantages compared to voltage application. Although the existence of a great variety of current sources, they meet our requirements only partially. We developed a voltage controlled current source for active front ends for a wide range of impedances, from Ω up to TΩ. A broad bandwidth from dc to 100 MHz was ensured by minimizing the parasitic capacitances. The current source outputs a symmetric signal without dc-offset. This is accomplished by a differential driver where the non-inverting branch was used for creating a current conveyor while the inverting branch host the feedback for offset control. We focused especially on the stability of the current source for usage with small metal electrodes in aqueous solutions.

1. Introduction

Bioimpedance measurement is a simple and innocuous method for electrical characterization of biological matter [1;2]. Especially in the low frequency region from 100 Hz up to 10 MHz the behavior of the lipid membranes in cell structures dominate the impedance spectrum [3]. The impedance measurement in this frequency range employs either a voltage excitation of the material under test (MUT) and monitors the current through the material or *vice versa*. Other methods, like bridge calibration, are not considered here.

The theoretical background for electrodes with geometrical dimensions on the order of centimeters is teaching book material. Both, potentiostatic or galvanostatic excitation is well established and a great variety of devices for impedance measurement exist. However, if the same theory and the same devices are used in conjunction with microelectrodes, a departure from the expected results is often observed. The reason is the non-ideal behavior of the front end together with the usually neglected electrochemistry at the electrodes, even below the decomposition voltage. Often, any change in current density at the electrodes is not taken into account and parasitic elements at the input of the voltage monitor are neglected as well. Here we show how to circumvent these problems at least partially.

Besides using a voltage monitor with extremely high input impedance, a symmetrical current source for dc-offset free excitation can compensate asymmetric influences of the electrochemical reactions at the electrodes.

2. Electrode polarization

In numerous applications the MUT is connected to the electronics via galvanically coupled electrodes. The most employed electrode materials are metals like iron (stainless steel), titanium, gold or platinum.

A double layer develops at the electrode-electrolyte interface [2]. At low frequencies, the impedance of this double layer has a similar magnitude as the impedance of the biological object. Any measurement done with two electrodes will result in the real impedance of the MUT plus the impedance of the electrodes. In order to overcome this, the four electrode interface, where an outer pair of electrodes is used for the application of the excitation signal (applicator electrodes) and the inner electrodes are used for monitoring the voltage dropping across the MUT (monitor electrodes) was introduced. In such an arrangement, any voltage dropping at the impedance of the applicator electrodes influences the measurement result only negligibly [4]. If the potential difference between both single monitor electrodes and the bulk electrolyte is equal, the measured voltage between the inner electrodes will truly be the voltage drop between the inner electrodes.

In this sense, it is quite unimportant whether a potentiostatic or galvanostatic excitation is used. As a matter of fact, even if neither, current nor voltage is controlled, impedance measurement would theoretically work just by measuring voltage and current.

When microelectrodes are used, the outer electrodes can influence the inner electrode by very close location. Moreover, due to the high curvature of such electrodes, local high field strength can overcome the limits for linearity resulting in harmonic distortion, even at low voltage. This means, especially for miniaturized electrodes with sharp geometry any change in the current density at the outer electrodes will influence the measured signal at the inner electrodes. Using a galvanostatic approach controls the current density and therefore prevents unpredictable interaction between applicator and monitor electrodes.

3. Current source for excitation

It is generally accepted that excitation signals for bio-impedance measurements should be dc-offset free in order to minimize electrochemical effects at the material [5]. Offset free means that the time average of the signal is zero. Although the current through the current carrying electrodes is always complementary, even if one electrode is connected to ground, the voltage at the monitor electrodes may be very different (Fig.1a). This implies of course no shunting pathways to ground or any other potential, i.e. the MUT is electrically fully insulated from the environment. Although this sounds simple and clear, researchers often forget about hidden grounding or electrical connections when doing impedance measurements outside the laboratory.

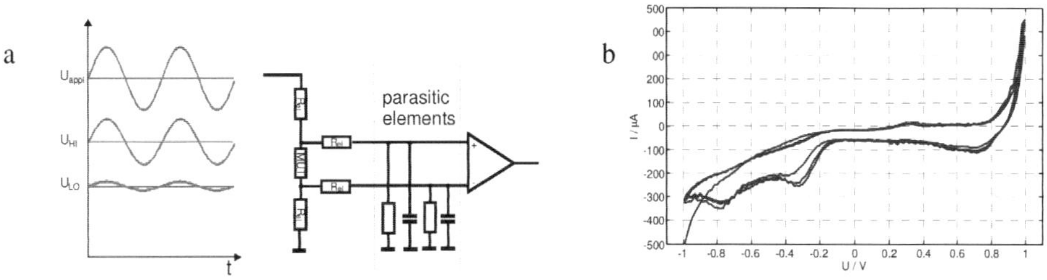

Fig.1a. Input signal at the monitor electrodes due to the excitation by an offset free but not symmetrical signal. Although the input impedance of the difference amplifier is ideally infinite, parasitic elements yield an asymmetric current at the monitor electrodes, which disturbs dramatically the compensation of the electrode polarization for a tetra-polar interface. (b) Voltagram of a gold electrode of 1 cm² in contract with 100 mM KCl

The potential drop between an electrode and the bulk electrolyte depends on the current density. Using cyclic voltammetry, reveals a small range, usually below 100 mV, where the electrode behaves linearly (Fig.1b). Exceeding this range, yields nonlinear behaviour of the electrode which generates harmonic distortion [6].

The influence of the double layer at the monitor electrodes can be partially compensated using a high symmetry of all signals with respect to ground. This yields the same electrode polarization at both monitor electrodes which is compensated by using the difference of both signals for further processing. This approach implies the usage of a symmetric excitation source, where the potential at the applicator electrodes is complementary with respect to ground. Moreover, a controlled current rather than the voltage yields lower distortion from the applicator electrodes.

4. Symmetric high bandwidth current source

The symmetry of the applicator electrodes with respect to ground ensures the same time averaged speed and extent of electrochemical reactions at the voltage monitoring electrodes in tetrapolar applications.

The newly developed current source is a successor of the well known Howland-Circuit which can be regarded as current conveyor. Other than the general Howland-Circuit which is a high side driver with respect to ground, the new approach uses the both outputs of a symmetric line driver (OPA1) in order to ensure a symmetric output (Fig.2). Only the non-inverting output is used for the feedback from the sampling resistor R_s. After buffering (OPA2), which ensures essentially no current is injected into the feedback loop, the signal is summarized with the input signal and feed into the non-inverting input. In order to compensate the voltage drop at the sampling resistor, a second resistor of the same size is added to the inverting output. The current though the output loop is: $I = U_{in} / R_s$. The high impedance instrumentation amplifier (OPA3) is used as current monitor.

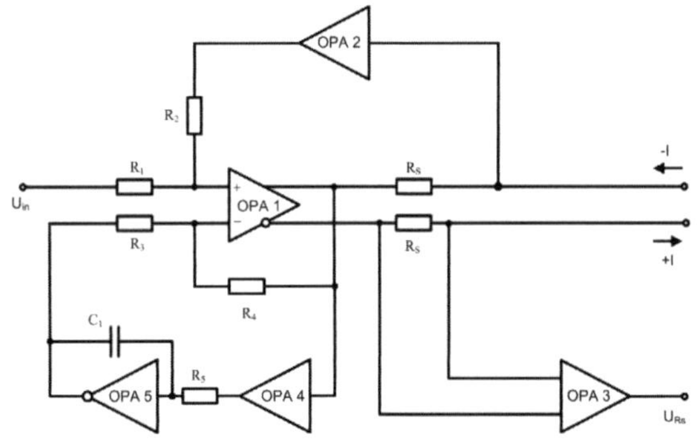

Fig. 2. Simplified schematic of the symmetric and offset free current source

A second feedback is used for offset compensation. Therefore the inverted output is buffered (OPA4) and integrated (OPA5/C1/R5) in order to monitor the actual offset. This signal is fed into the inverting input of OPA1. The loop gain is adjusted using R3/R4.

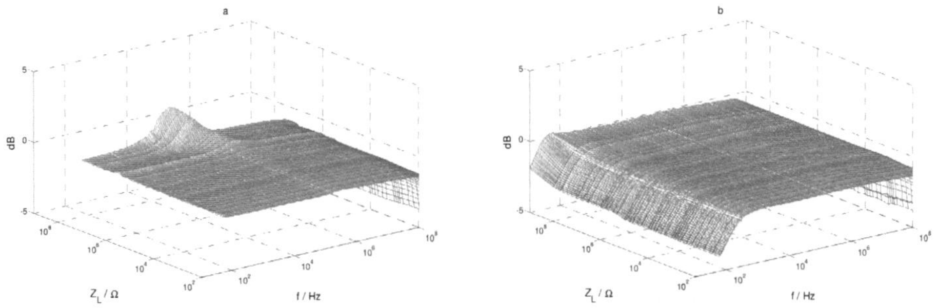

Fig.3. Voltage across the load resistor in dB with respect to the input voltage U_{in}. The sample resistor was (a) 1 MΩ and (b) 1 GΩ.

Measurements

Initial tests for symmetry and stability were done on SMD-resistors using an oscilloscope together with a square wave generator (1 kHz, 0.2 V). The stability of the source was proven over a wide range of sample and load resistors (10 Ω – 10 GΩ).

The frequency response of the current source was further tested in a range from 100 Hz up to 100 MHz. A HP4194A – network analyzer applied an U_{in} of 0.5 V and measured the voltage across the load resistor. In Fig.3 is the logarithmic ratio of the voltage across the load resistor and the input voltage in dB shown.

For test with real electrode systems we used an electrode as shown in Fig.4a. The current was injected at the outer electrodes and the voltage at the inner electrodes was monitored using an instrumentation amplifier like for monitoring the current. Different ionic solutions, from distilled water up to high ionic strength were used for testing. Fig.4b shows an example for distilled water. The voltage trace shows a time constant due to the high permittivity of the water.

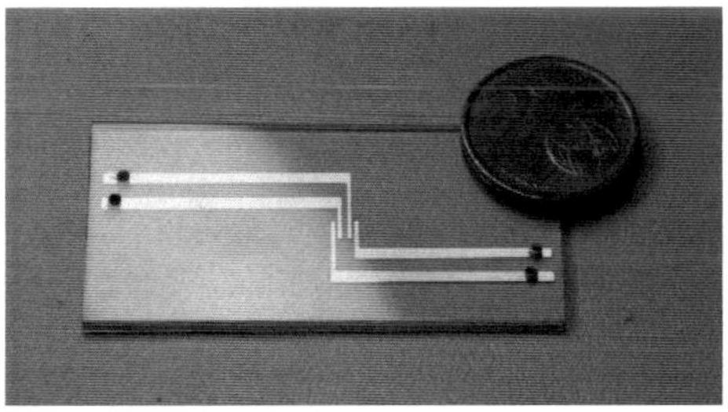

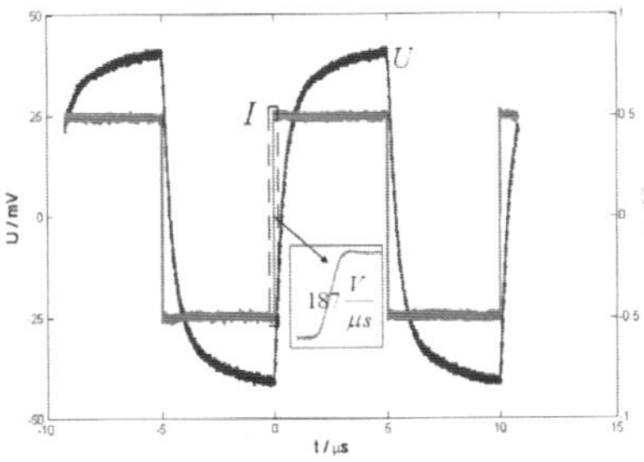

Fig.4a. Example of an electrode system as it is used in flow chambers. (b). Current and voltage trace for a square wave applied to U_{in} of the current source. The chamber filled with distilled water. A 1 kΩ sample resistor was used.

5. Conclusion

Impedance measurement using galvanically coupled electrodes relies on the transduction of the excitation signal from the electrode to the material under test. Artifacts due to the electrochemistry at the electrodes can be partially compensated using current excitation rather than voltage controlled signals. Monitoring the voltage across an additional pair of electrodes can greatly enhance the performance at low frequencies, but is influenced by parasitic elements in the setup, especially if one applicator electrode is grounded. The current source presented here will circumvent this problem due to its symmetric output. Moreover, the dc-offset free signal prevents electrochemical reactions at the applicator electrodes.

References

[1] K.R.Foster and H.P.Schwan, Dielectric properties of tissues and biological materials : A critical review. CRC Crit.Rev.Biomed.Eng. 17 (1989) 25-104.

[2] S.Grimnes and O.G.Martinsen, (2000) Bioimpedance and Bioelectricity Basics, Academic Press.

[3] H.P.Schwan, (1957) Electrical Properties of Tissue and Cell Suspensions, in: J.H.Lawrence and C.A.Tobias (Eds.) Academic Press, New York, 1957 p. 147

[4] H.P.Schwan, (1963) Determination of biological impedances, in: W.L.Nastuk (Ed.) Physical Techniques in Biological Research, Academic Press, New York, 1963 pp. 323-407

[5] R.Bragos, J.Rosell, and P.Riu, A wide-band AC-coupled current source for electrical impedance tomography. Physiol.Meas.1994) 91-100.

[6] M.Moussavi, H.P.Schwan, and H.H.Sun, Harmonic distortion caused by electrode polarisation. Med.Biol.Eng.Comput. 32 (1994) 121-125.

Model of an inductive sensor of cardiac activity attached to patient

Jüri Vedru[1] and Rauno Gordon[2]

[1] Institute of Physics, University of Tartu, Tähe 4, 51010 Tartu, Estonia
[2] Department of Electronics, Tallinn University of Technology, Ehitajate tee 5, 19086 Tallinn, Estonia

E-mail: juri.vedru@ut.ee; rauno@elin.ttu.ee

Abstract. An electric circuit model describing an inductive sensor of cardiac mechanical activity in its working condition has been developed. The sensor comprises a single-turn coil which is fed by 7.7 MHz constant current and induces probing eddy currents in the body. The inductor is considered to be attached to the thoracic surface of a normal human male, in front of the heart. A simple axial-symmetric model of the thorax, formed of tightly packed circular current tubes, has been used to calculate the resistances, self-inductances and mutual inductances within the human body. Then the inductor and the eddy currents in the body were reduced to a system of two inductively coupled coils; estimates of the parameters and frequency response of the system have been found; the active and reactive contributions of the human body to the resulting impedance of the inductor were calculated.

1. Introduction

In induction methods for electrical impedance measurement and tracking [1], [2], the probing currents are attached in the form of Foucault or eddy currents, using inductors. The same inductors are applied as signal receivers, thus forming sensors for inductive measurement. To improve the induction methods, identified models of such sensors are necessary. The present study is based on a concrete example of the Foucault cardiographic sensor which is applied to track heart mechanical activity [2], [3]. The objective of the present work was to develop a realistic model of the sensor loaded with a patient and, using the model, to study the factors which determine the origination of signal in the induction method for tracking the heart.

2. The inductive sensor

The example sensor comprises a shielded single-turn coil. The inductor coil is 13.5 cm in diameter and made of copper wire with a rectangular 10 mm × 2 mm cross-section. Together with a capacitor it forms a resonance circuit with a resonance frequency of 7.7 MHz. A single-frequency Foucault cardiograph incorporating the sensor operates at the resonance frequency of this resonance circuit. The inductor is electrically shielded by a grounded box-like surrounding. The shielding box (see figure 1) is made of 1 mm-thick foil-clad laminate that has been extra engraved to disable eddy currents in it. The sensor is assumed to be attached to the thoracic surface of a normal human male just in front of the heart.

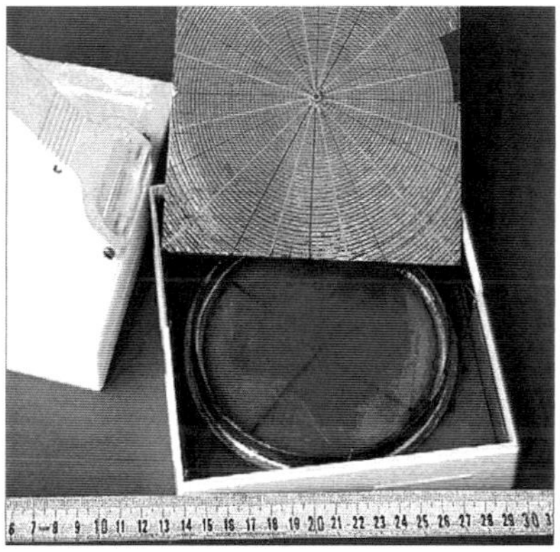

Figure 1. The example sensor: coil with the uncovered shielding in enclosure.

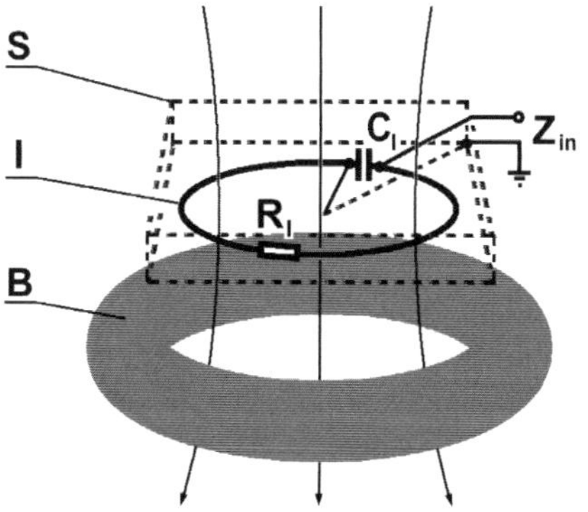

Figure 2. The scheme of the situation studied. B – the eddy currents region (symbolic); I – the inductor coil; S – electrical shielding; R_{I} – the coil resistance (symbolic), C_{I} – primary circuit capacitance, Z_{in} – points where input impedance is measured.

3. Model of the induction features in the system

A static model with cylindrically symmetric geometry (similar to the one used in [2], [3], see figure 3) was applied to calculate the inductor-eddy currents interaction. The eddy currents region of the body had to be modeled as a distributed system. To calculate the interaction, the cylinder representing the body region of eddy currents was approximated by a dense-packed set of *n* co-axial current tubes or coils that had equal small square cross-sections. The situation in figure 2 was further interpreted in terms of two inductive-coupled coils or circuits, in which there was a single-turn primary one (the inductor) and a secondary equivalent one. The impedance of the latter was considered to consist of resistive and inductive components.

At a fixed topography, a distribution of individual currents is established in the eddy currents region. It has an influence on the equivalent characteristics of the secondary circuits. Starting from equal power loss requirement, the total loss resistance of the secondary circuit R_{II} has to be calculated from the individual resistances of the secondary coils R_i:

$$R_{\mathrm{II}} = \sum_{i=1}^{n} K_i^2 \cdot R_i\,, \quad \text{where} \quad K_i = I_i \cdot \left(\sum_{k=1}^{n} I_k \right)^{-1}. \tag{1}$$

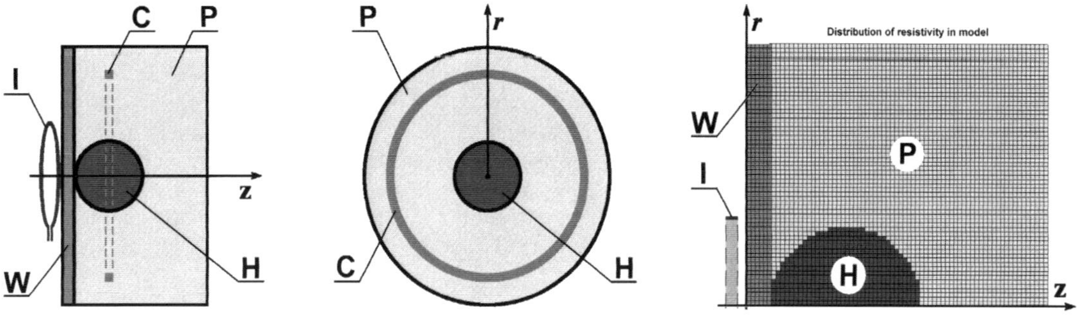

(a) (b) (c)

Figure 3. The model of the eddy currents region of the body. (a) and (b) – lateral and frontal cross-sections of the model, (c) – the region of calculation: a half of the section of the model (case of 60×72 secondary coils).
I – the inductor, W – thoracic wall, C – a secondary coil, P – pulmonary tissue, H – the heart.

Similarly, the equivalent self-inductance of the secondary circuit and the equivalent mutual inductance between the inductor and the secondary circuit were calculated. Before composing a single equivalent for the multiple secondary coils, the actual values of self-inductances L_i and mutual inductances between the secondary coils M_{ij} had to be calculated. Under the condition of diversity of the secondary (eddy) currents, the equal magnetic flux requirement has led to formulae, which express L_i and M_{ij} through the corresponding values $\tilde{L}_i$ and $\tilde{M}_{ij}$ for stand-alone coils:

$$L_i = nK_i \cdot \tilde{L}_i , \quad M_{ij} = n\sqrt{K_i K_j} \cdot \tilde{M}_{ij} . \tag{2}$$

The stand-alone values $\tilde{L}_i$ and $\tilde{M}_{ij}$ were calculated using available formulae from [4]. A similar approach was used to calculate the mutual inductances between the secondary coils and the inductor coil.

Then the equivalent total self-inductance [4] of the secondary circuit L_{II} was expressed

$$L_{\mathrm{II}} = \frac{1}{n^2} \cdot \left(\sum_{i=1}^{n} L_k + \sum_{i=1}^{n} \sum_{j=1}^{n} M_{ij} \right), \; i \neq j . \tag{3}$$

Skin effect causes concentration of the inductor currents into the opposite narrow sides of the inductor coil's surface. Therefore, the inductor is modeled as two equal threads of current. The equivalent mutual inductance between the primary and the secondary circuit $M_{\mathrm{I,II}}$ was expressed as

$$M_{\mathrm{I,II}} = \frac{1}{n \cdot 2} \cdot \sum_{i=1}^{n} \left(M_{i,\mathrm{prox}} + M_{i,\mathrm{dist}} \right), \tag{4}$$

where $M_{i,\mathrm{prox}}$ and $M_{i,\mathrm{dist}}$ are the mutual inductances between the secondary coils and either proximal or distal current threads in the inductor coil, calculated according to the formula (2).

Distribution of the currents in the secondary coils, which is necessary in formula (1) to find the coefficients K_i , was estimated ignoring the backward influence of the magnetic field of the induced currents. Calculation of the corresponding electromotive forces is presented in [5]. The currents were obtained using the Ohm's Law.

An inductive sensor transforms the secondary circuit impedance into additional series impedance $\Delta Z_{\mathrm{I}} = \Delta R_{\mathrm{I}} + \mathrm{i} \cdot \Delta X_{\mathrm{I}}$ in the primary circuit coil. This feature is described by the following formulae [6]:

$$\Delta R_{\mathrm{I}}(\omega) = \frac{\omega^2 M_{\mathrm{I,II}}^2}{R_{\mathrm{II}}^2 + \omega^2 L_{\mathrm{II}}^2} \cdot R_{\mathrm{II}} , \quad \Delta X_{\mathrm{I}}(\omega) = -\frac{\omega^2 M_{\mathrm{I,II}}^2}{R_{\mathrm{II}}^2 + \omega^2 L_{\mathrm{II}}^2} \cdot \omega L_{\mathrm{II}} \tag{5}$$

For the actualization of the model, the size dimensions of the inductor, as well as its electric properties (R_{I} = 33 mΩ, L_{I} = 228 nH, C_{I} = 1.80 nF) were predetermined by (or measured in) the example sensor. The capacitances from the shielding to the inductor and to the body, $C_{\mathrm{S}} \approx$ 40 pF, $C_{\mathrm{B}} \approx$ 140 pF, respectively, were measured in the example sensor. Due to its relatively small value, C_{S} in the present sensor only weakly influences the primary circuit's parallel capacitance. The other one, C_{B} , in a geometrically symmetric model should not act at all. Though in real asymmetric cases it would cause a certain amount of noise if changed, in the present model this is not the case.

The dimensions of the body disk and its contents (Table 1) were roughly derived from a normal adult person. Determination of the specific impedances for the regions of the body disk is an intricate task. Combining the values from various literature data, a fairly realistic approximation with pure active resistivities has been taken.

Table 1. Values of some parameters of the model.

Parameters				Values		
Diameters of the	body disk,	inductor coil,	the heart	40 cm	13.4 cm	12 cm
Thicknesses of the	body disk,	thoracic wall		24 cm	2 cm	
Distance between the	inductor and thoracic surface			0.7 cm		
Resistivities of the	heart,	thoracic wall,	pulmonary region	136 Ω·cm	150 Ω·cm	325 Ω·cm

4. Study of the model. Discussion

Calculations of the model with MATLAB gave the following properties of the system for the working frequency f_w = 7.686 MHz: R_{II} = 65 Ω, L_{II} = 45 nH ($\omega_w L_{II}$ = 2.2 Ω), $M_{I,II}$ = 39 nH ($\omega_w M_{I,II}$ = 1.9 Ω) that lead to ΔR_I = 0.054 Ω, ΔX_I = – 0.0018 Ω. Calculation of the input impedance Z_{in} of the loaded inductor, using these results, gave the frequency response shown in figure 4(a), the middle curve. It is a typical resonance curve with a quality factor Q = 125, max $|Z_{in}|$ = 1.40 kΩ. The latter agrees with the values measured using the example sensor on normal male volunteers in experiment [3], which ranged in region of 1…2 kΩ.

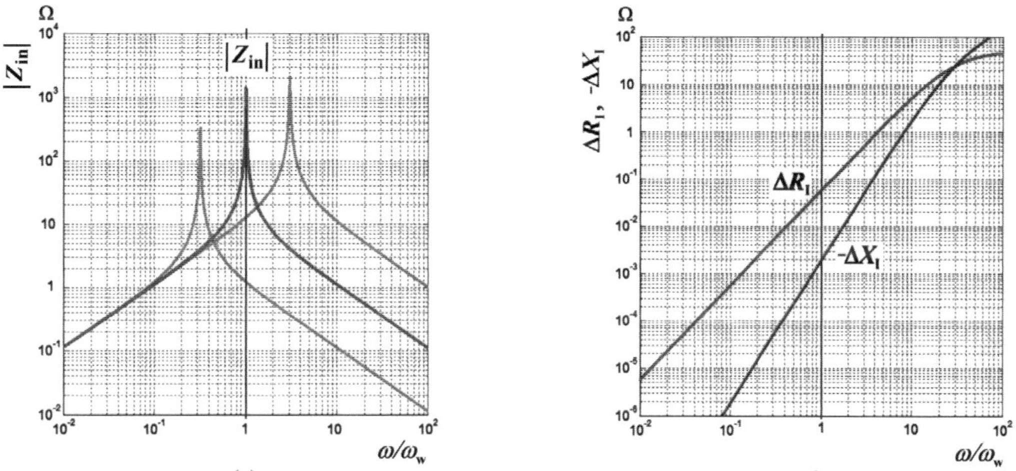

(a) (b)

Figure 4. The frequency responses of the loaded sensor.
(a) – for the modulus of the input impedance Z_{in} : middle curve – the actual case of the sensor, with C_I = 1.80 nF; the shadow side curves – with C_I increased or decreased 10 times.
(b) – for the additional primary circuit resistance ΔR_I and reactance ΔX_I.

Both the ΔR_I and ΔX_I are strongly frequency-dependent (figure 4(b); at the working frequency the slopes are 40 dB/decade and 60 dB/decade, respectively). Thus to improve the signal magnitude, higher resonance frequency would be instrumental. This way is limited by the parasitic capacitance C_S.

In the sensor under review, which is attached to the patient, of the object-related contributions to the inductor coil impedance, the resistive one prevails over the reactive one up to the frequencies which are 30 times above the present working frequency. Thus, from the signal taken with the sensor, detection of either the in-phase component or the modulus of the signal is reasonable. Though at the higher frequencies the prevalence would change, working at these frequencies is rendered impossible by the C_S.

We are going to use the developed model as an instrument for analyzing further the capabilities of the induction-based cardiac electrical bioimpedance recording.

5. References

[1] Tarjan P P and McFee R 1968 *IEEE Trans. Biomed. Eng.* **15** 266–78
[2] Humal L-H and Vedru J 1996 *Med. Biol. Eng. Comput.* **34** Suppl. 1 Part 2 183–4
[3] Trolla J and Vedru J 2001 *Proc. XI ICEBI (Oslo, Norway, 17-21 June 2001)* pp 649–52
[4] Kalantarov P L and Zeitlin L A 1986 *Calculation of Inductances: A Handbook* (Leningrad, Energoatomizdat; in Russian)
[5] Vedru J 2004 *Proc. XII ICEBI and V EIT (Gdansk, Poland, 20-24 June 2004)* pp 305–8
[6] Neumann L R and Kalantarov P L 1967 *Theoretical Foundations of Electrical Engineering, Part 3* (Tallinn, Valgus; in Estonian) pp. 88–90

Journal of Physics: Conference Series **224** (2010) 012011 doi:10.1088/1742-6596/224/1/012011

AD5933-based Spectrometer for Electrical Bioimpedance Applications

J. Ferreira[1], F. Seoane[1,2], A. Ansede[1,3] and R. Bragos[3]

[1] School of Engineering, University of Borås, SE-501 90 Borås, SWEDEN

[2] Department of Signal & Systems, Chalmers University of Technology, SE-41296, Gothenburg, SWEDEN

[3] Department of Electronic Engineering. Universitat Politècnica de Catalunya. Campus Nord, C-4. Jordi Girona 1-3. 08034, Barcelona. SPAIN

E-mail: fernando.seoane@hb.se

Abstract. To build an Electrical Bioimpedance (EBI) spectrometer using the Impedance Measurement System-On-Chip AD5933 together with a 4-Electrode Analog Front End (4E-AFE) has been proven practicable. Such small measurement devices can make possible several new applications of EBI technology, especially when combined with functional textiles, which can enable wearable applications for personal health and home monitoring. After the implementation and functional validation of the 4E-AFE-enabled spectrometer, the next natural step is to validate for which EBI applications the 4E-AFE-enabled system is suitable. To test the applicability of this novel spectrometer on several EBI applications, 2R1C equivalent models have been experimentally obtained and impedance spectroscopy measurements have been performed with the system under study and with the SFB7 EBI spectrometer manufactured by ImpediMed. The 2R1C circuit parameters have been estimated with the BioImp software from the spectra obtained with both EBI spectrometers and the estimated values have been compared with the original values used in each circuit model implementation. The obtained results indicated that the 4E-AFE-enabled system cannot beat the performance of the SFB7 in accuracy but it performs better in preciseness. In any case the overall performance indicates that the 4E-AFE-enabled system can perform spectroscopy measurements in the frequency range from 5 to 100 kHz.

1. Introduction

Over recent years, advances in different technological fields like electronics, physics and materials have made possible to embed complex functional systems into a single integrated circuit producing the System-on-Chip technology. As a result of such technological advances a whole impedance spectrometer is available on a single integrated circuit: the AD5933 [1] from Analog Devices. Such a level of integration together with advances made in textile technology, especially in the development of textile electrodes suitable for applications of Electrical Bioimpedance (EBI) [2], might enable wearable applications for personal health and home monitoring based in EBI measurements.

In this work the performance of a custom-made EBI spectrometer based on the AD5933 in combination with a 4-electrode Analog Front End [3] is evaluated when measuring 3 different impedance loads modeling three typical EBI applications: Total Body Composition (TBC), Respiration Rate (RR) and Lungs Composition (LC), also known as Pulmonary Edema [4] .

2. Materials and Methods

2.1. Spectrometers and 2R1C models

Two different spectrometers have been used in this work: the ImpediMed SFB7 and a custom-made device based in the AD5933 in combination with an analog front-end to enable 4-electrode impedance measurements [3].

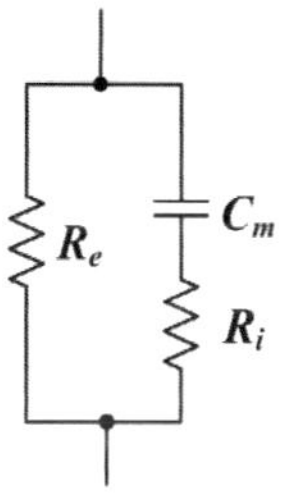

Figure 1. 2R1C parallel bridge model

The measurements have been performed on a 2R1C parallel electrical model representing biological tissue, see Figure 1. The values of the model parameter to build the 2R1C models, R_e, R_i and C_m, were extracted from experimental EBI measurements with the software BioImp [5]. The 2R1C have been build to produce an impedance complex spectrum similar to experimental impedance spectra obtained in the aforementioned applications. The resistors used in the model belong to the E-24 series and a potentiometer was used to trim the values of the circuit parameters, see Table 1, to obtain the intended impedance spectrum.

Table 1. 2R1C Modeling for TBC, RR and LC EBI Applications.

EBI Application	Electrical Equivalent 2R1C Values		
	R_e (Ω)	R_i (Ω)	C_m (nF)
TBC	917.5	665.4	3.42
RR	58.5	25.58	75.7
LC	81.5	22.15	47.7

2.2. Methods

Using both impedance spectrometers, 100 impedance measurements were performed in the frequency range 5 to 100 kHz on each experimentally-based 2R1C circuit model. For each EBI application the average complex spectrum from the measurements was obtained for each of the impedance spectrometers. A comparison between the resistance and the reactance spectra obtained from each spectrometer has been done.

Using the software BioImp applied on the measured impedance spectra obtained with each of the spectrometers, the value of the 2R1C circuit parameters was estimated and the obtained estimation error was studied to perform a comparison between the SFB7 and the AD5933-based spectrometer.

3. Results

3.1. Spectral Measurements

The reactance and resistance spectra from the impedance measurements for the different EBI applications are plotted in Figure 2. As it is denoted in all the spectral plots, the spectra obtained with both devices are perfectly overlapped for both the resistance and the reactance. Only a small deviation at high frequency can be observed on the reactance of for TBC and RR on Figure 2.b) and 2.d) respectively.

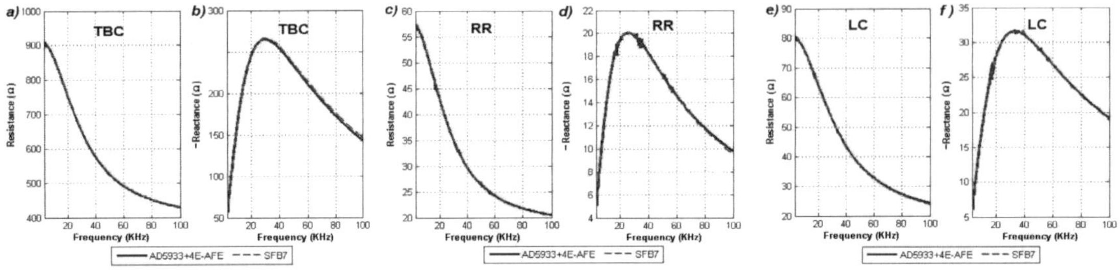

Figure 2. Resistance and reactance spectrum plots from the measurements obtained with the AD5933-based spectrometer in continuous and SFB7 in dashed trace, respectively.

3.2. 2R1C Circuit Parameters Estimation

The average values estimated for the 2R1C parameters from the impedance measurements obtained with both spectrometers are indicated in Table 2. For each of the applications, the relative differences respect the real values of the circuit components, as indicated in Table 1, are indicated by the relative error. The Standard Deviation obtained with both spectrometers for the estimation of each of the parameters is also indicated. For all the applications and all the parameters with the exception of the estimation of C_m for TBC, the SFB7 exhibit a smaller error than the AD5933-based spectrometer. Regarding to the Standard Deviation, it is the AD5933-based spectrometer the one that presents lower values for each of the parameters and all three applications.

Table 2. Error and Accuracy on 2R1C Parameters Estimation, SFB7 vs. AD5933+4E-AFE

	EBI Application	TBC		RR		LC	
Parameter		SFB7	AD+AFE	SFB7	AD+AFE	SFB7	AD+AFE
	Average Value (Ω)	917,88	914,15	58,33	57,89	81,37	81,07
R_e	Standard Deviation	0,1594	0,0423	0,0177	0,0077	0,0193	0,0131
	Relative Error	0,04%	0,36%	0,29%	1,03%	0,16%	0,53%
	Average Value (Ω)	664,49	676,3	25,60	26,00	22,25	22,95
R_i	Standard Deviation	0,5540	0,1280	0,0493	0,0238	0,0498	0,0193
	Relative Error	0,14%	1,64%	0,08%	1,63%	0,44%	3,63%
	Average Value (nF)	3,34	3,37	74,56	74,08	46,67	46,28
C_m	Standard Deviation	0,0009	0,0003	0,0444	0,0181	0,0149	0,0117
	Relative Error	2,22%	1,47%	1,50%	2,14%	2,16%	2,97%

4. Discussion

The spectral plots obtained with both spectrometers show that the AD5933-based spectrometer with the 4-electrode enabling front-end produced complex impedance spectra that are almost exact to the spectra obtained with the SFB7.

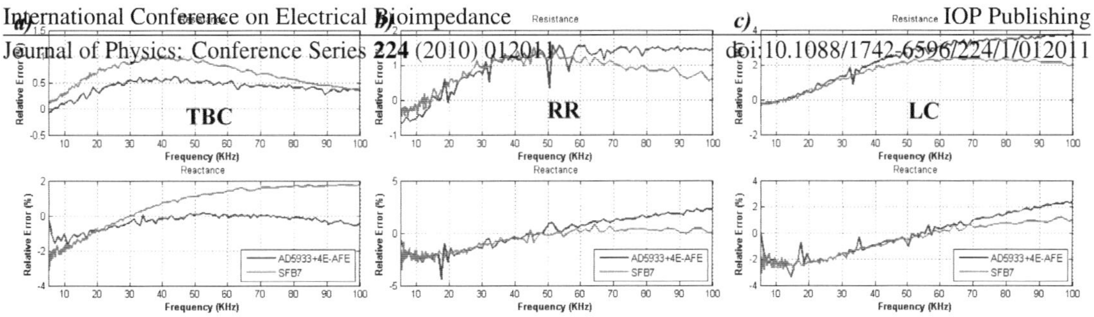

Figure 3. Relative measurement error in the resistance and the reactance obtained for the measurements with both EBI spectrometers for the three applications under study: TBC, RR and LC in a),b) and c) respectively

The analysis of the error data provided by Table 2 and Figure 3, suggest that the SFB7 presents a better accuracy than the AD5933-based spectrometer. In any case the estimation errors produced with both devices are generally low.

In the other hand according to Table 2, the preciseness shown by the custom-made impedance spectrometer based on the AD5933 is higher than the preciseness obtained with the SFB7. Both dispersions and errors obtained with both equipments are small compared with the high variability of the clinical measurements. E.g. in the measuring of Lungs Circulate Air Volume [5] the error measured with a neumotacometer and with Electrical Impedance Tomography (EIT) is around 20%, compared with the worst case studied in this work for RR the error drops below 4%.

5. Conclusion

The good complex spectroscopy performance exhibited by the AD5933-based spectrometer as well as the high accuracy and precision exhibited for the 2R1C model parameter estimation, which is critical in applications of total body composition assessment [6], strongly supports that EBI spectrometers based in SOC solutions like the AD5933 can be use in typical EBI applications like body composition analysis, respiration rate and pulmonary edema.

The reduced size of the AD5933, its small power consumption [7] together with the availability of off-the-shelf Bluetooth solutions indicate that the implementation of portable applications of EBI is close, even wearable if combined with textile garments [8] and electrodes [2].

6. References

[1] Analog Devices *AD5933 Product web site*. Accessed on 2010-01-15; Available from: http://www.analog.com/en/AD5933/productsearch.html.

[2] Márquez JC *et al.* 2009 Textile Electrodes in Electrical Bioimpedance Measurements - A Comparison with Conventional Ag/AgCl Electrodes *in Conf Proc IEEE Eng Med Biol Soc*, **1**: p. 4816-9

[3] Seoane F *et al.* 2008 An analog front-end enables electrical impedance spectroscopy system on-chip for biomedical applications. *Physiol. Meas.* **29**(6): p. S267-78.

[4] Mayer M *et al.* 2005 Monitoring of lung edema using focused impedance spectroscopy: a feasibility study. *Physiol. Meas.* **26**: p. 185-92.

[5] Ansede A 2009 A Feasibility Study of the Suitability of an AD5933-based Spectrometer for EBI Applications *Master Thesis Series* University of Borås.

[6] De Lorenzo A *et al.* 1997 Predicting body cell mass with bioimpedance by using theoretical methods: a technological review *J Appl. Physiol.* **82**(5): p. 1542-1558.

[7] Macias R 2009 Towards Wearable Spectroscopy Bioimpedance Applications Power Management for a Battery Driven Impedance Meter *Master Thesis Series* University of Borås.

[8] MyHeart IST 2002-507816. *Project web site* Accessed on 2010-01-15; Available from: http://www.hitech-projects.com/euprojects/myheart.

System front-end design for concurrent acquisition of electroencephalograms and EIT data

R Guardo, J Jehanne-Lacasse. A P Moumbe and H Gagnon

Institut de génie biomédical, École Polytechnique de Montréal, Montréal, Canada

E-mail: guardo@igb.polymtl.ca

Abstract.

There is recently considerable interest in medical imaging to combine recording of bioelectrical signals with imaging procedures. For example, electroencephalograms (EEGs) recorded during functional magnetic resonance imaging are increasingly being used for neurological and behavioural research. Concurrent acquisition of EEGs and electrical impedance tomography (EIT) data have been suggested as a non invasive technique that could help localize the area of the brain responsible for seizures in epileptic patients awaiting resective surgery. Despite reasonably distinct spectra, EEGs and EIT signals are difficult to record simultaneously because of their very different amplitudes. In this paper, we describe the front-end of a 24-channel system designed to acquire both signals from the same set of scalp electrodes using time-division multiplexing. We have developped a 10-layer 20x15 cm printed circuit board of the front-end and are currently performing circuit characterization tests. System performance parameters and *in vivo* images will be presented at the conference.

1. Introduction

Epilepsy is a chronic neurological disorder prevalent in approximately 1% of the general population. Long-term drug therapy is the main form of treatment, but more than 30% of patients are refractory to medication. Patients with uncontrolled seizures require considerable health care resources and experience a high number of disability days leading to unemployment or low income. Surgery is recommended when medication fails and the seizures are confined to an area of the brain where tissue can be safely removed. To localize the epileptogenic zone, patients undergo a pre-surgical evaluation that includes clinical tests, various imaging procedures and video-electroencephalography monitoring. When these tests fail, an invasive study is sometimes performed where intracranial electrodes are implanted under general anaesthesia. Chronic implantation of intracranial electrodes carries a risk of infection and haemorrhage.

Concurrent acquisition of electrical impedance tomography (EIT) data and electroencephalograms (EEGs) have been suggested to assist in the localization of the epileptogenic zone. The rationale for including EIT is the possibility it offers of imaging changes in blood volume and flow that occur during epileptic activity [1]. Identification by both EIT and EEG analyses of an area of the brain with altered hemodynamic and electrophysiological activity, combined to other clinical tests, could reduce the number of cases requiring an intracranial study prior to surgery. Even if an invasive study is done to confirm the results, the EIT-EEG analyses may reduce the number of electrodes implanted and optimize their positioning. The patient would benefit either from elimination of the intracranial study or a reduction of its duration.

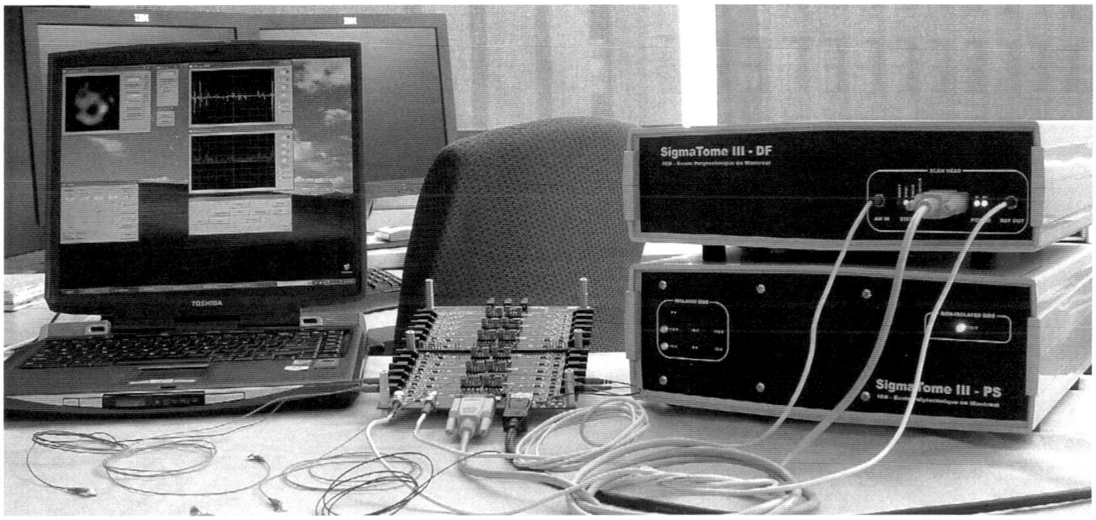

Figure 1. View of the combined EIT-EEG data acquisition system under development.

2. Methods

The EIT-EEG data acquisition system developped consists of a scan-head, a base-station and a PC. The scan-head (Figure 1, center) contains the front-end circuits for recording EIT signals and EEGs from 24 scalp electrodes. The base-station contains the isolation barrier, data converters, carrier synthesizers and demodulators, acquisition sequencing logic, and a USB-2.0 link. The PC hosts the user interface, displays EEGs and EIT images in real time, and archives the data.

To acquire EEGs and EIT signals with the same electrodes the following time-division approach was adopted. Acquisition intervals of 1 ms are divided in two phases, T_{EIT} and T_{EEG}, respectively reserved for EIT and EEG measurements. During T_{EIT}, the EIT circuits are enabled and the system performs one transimpedance measurement. Saturation of the EEG amplifiers is prevented by digitally controlled gain stages with very short settling times. When the T_{EEG} phase starts, the EIT current drivers are disabled and the EEG amplifiers returned to their normal (high) gain state. Outputs of the 24 EEG amplifiers are sampled simultaneously and the samples are sequentially digitized by a single A/D converter. Typically, each frame will consist of 250 transimpedance measurements. This leads to an EIT frame rate of 4 Hz and an EEG sampling rate of 1 kHz per channel. Since all timing is controlled by firmware, other operating modes are possible, including acquisition of EIT data only or EEGs only, with corresponding increases in frame or sampling rate.

Figure 2 shows a top-level block diagram of the scan head where the EIT and EEG signal conditioning circuits are abstracted in 24 Active Electrode (AE) modules. The double-ended arrows on the left of each AE module represent connectors for scalp electrodes. Auxiliary Circuits use 4 other electrodes to generate the Wilson Central Terminal (WCT) and the Right Leg Drive (RLD) signals. The System Control bus has 9 lines derived from 4 control signals generated by a sequencer in the base station. The Analogue busses comprise 5 signals. The R signal is the EIT reference carrier which controls the applied current. This carrier is generated by a Direct Digital Synthesizer in the base station. The A+ and A- signals are amplitude modulated carriers originating from the two AE modules selected as the EIT voltage sensing pair. The difference carrier (A+ - A-) is digitized by a 14-bit A/D converter and demodulated by a Digital Down Converter in the base station. The E bus carries the time-multiplexed output samples of the 24 EEG amplifiers. These are digitized by a 16-bit A/D converter in the scan head.

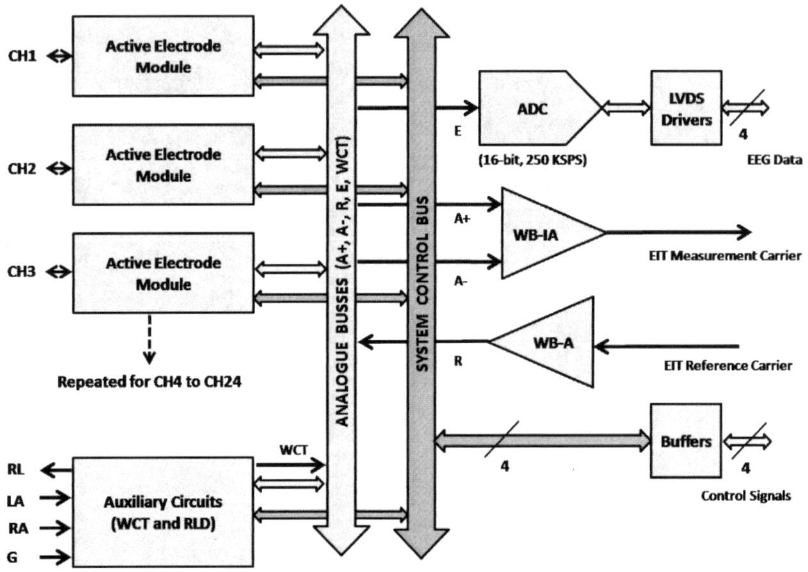

Figure 2. Top-level block diagram of the scan-head module.

Figure 3 shows the block diagram of one AE module. An instrumentation amplifier (Preamp) amplifies the difference between the electrode potential and WCT. Two microcontroller I/O lines set its gain to 10 V/V during T_{EEG} and 1 V/V during T_{EIT}. The preamp is followed by an EEG high-pass filter ($f_c = 0.5$ Hz) which rejects the electrode polarization voltage. This filter can be reset by the microcontroller to rapidly restore the signal after an overvoltage condition. The second stage of the EEG amplifier has a digitally controlled gain of 10 V/V during T_{EEG} and 0.1 V/V during T_{EIT}. The settling time of this stage is 500 ns. The third stage combines an instrumentation amplifier and analogue gates to form a sample-and-hold amplifier with a gain of 100 V/V. Data acquisition is initiated by a Control Bus signal which causes simultaneous sampling of all 24 EEG channels. The final stage of the EEG chain is a T-type analogue switch that connects the sample-and-hold output to the E bus.

The EIT signal conditioning chain comprises a current driver (CD) and a wideband preamplifier (WP). The CD consists of a voltage-controlled current source (VCCS) and a 3-way multiplexer. Each CD takes one of three states during the scan sequence: current source, current sink, and open circuit. During T_{EEG}, all current drivers are set to the open circuit state. The first stage of the WP uses the same instrumentation amplifier as the EEG chain. It is followed by a high-pass filter ($f_c = 1$ kHz) and a multiplexer. WPs take one of three states during the EIT scan sequence: follower, inverter, and open circuit. For the follower state, the multiplexer routes the filter output to the A+ line, whereas it routes it to the A- line for the inverter state. For the open state, the multiplexer disconnects the filter from both lines of the A-bus. These are connected to a wideband differential amplifier and a fast A/D converter in the base station. Since potential differences can be measured between any pair of electrodes and the CDs can apply balanced currents to any pair of electrodes, about 25,000 transimpedance measurements are possible. With N=24 electrodes, $N*(N-3)/2 = 252$ measurements are linearly independent.

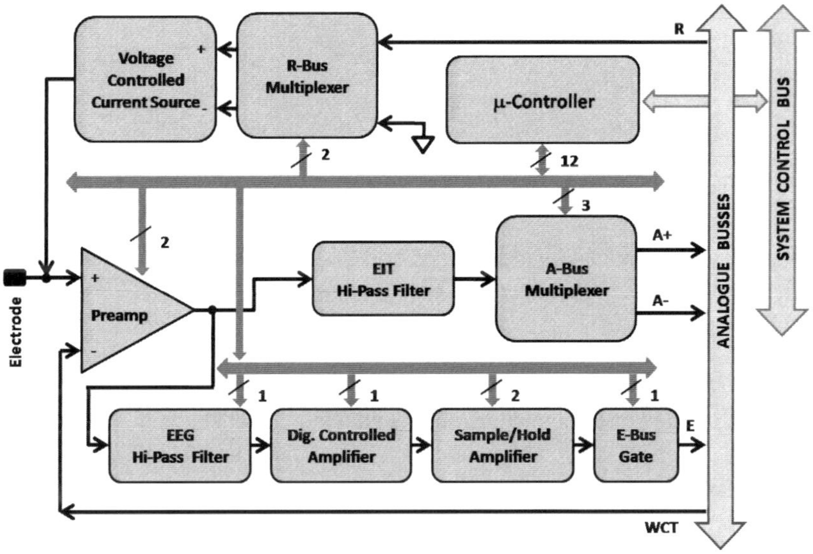

Figure 3. Block diagram of one active electrode (AE) module.

3. Results

The system is under development at the time of writing. Performance parameters of a partially populated printed circuit board of the scan head have been measured with a Network/Spectrum Analyzer. For the EIT signal processing chain, Bode plots of the following parameters were obtained for the 1 kHz to 10 MHz frequency range: a) transconductance, leakage conductance and output impedance of the current drivers and b) voltage gain, crosstalk, common-mode rejection ratio, and intrinsic noise density of the preamplifiers. The signal-to-noise ratio (SNR) and accuracy of the system have been measured with a precision 340-resistor mesh phantom using the test procedure described in [2]. For a 16 channel EIT setup, with 120 cm electrode leads for connecting the scan head to the phantom, we measured a mean SNR of 66.9 dB and a mean accuracy of 98.88 % at a frame rate of 4.7 Hz. Similar tests are under way for the EEG signal processing chain.

4. Discussion

The results of this project will lead to a new method of localizing an epileptogenic zone based on a combination of EIT and video-EEG monitoring. An EIT pilot study on patients with refractory partial epilepsy admitted to the epilepsy unit for a comprehensive pre-surgical evaluation should be able to prove its superiority over ictal SPECT (ascribable to the good temporal resolution of EIT) and show congruency with the reference standard.

5. References

[1] Holder D 2005 Electrical Impedance Tomography Methods History and Applications (Bristol and Philadelphia: Institute of Physics Publishing) chapter 4 pp 127166
[2] Robitaille N, Guardo R, Maurice I, Hartinger A E and Gagnon H 2009 *Physiol. Meas.* **30** S57-S71

Acknowledgments

This work was supported in part by the Natural Sciences and Engineering Research Council of Canada (NSERC) and the Canadian Institutes of Health Research (CIHR).

Performance Evaluation of KHU Mark2 Parallel Multi-frequency EIT System

D Y Kim, H Wi, P J Yoo, T I Oh and E J Woo

Department of Biomedical Engineering, Kyung Hee University, Yongin, Korea

E-mail: ejwoo@khu.ac.kr

Abstract. We describe a new parallel multi-frequency EIT system, KHU Mark2. It is based on the impedance measurement module (IMM), which comprises a single-ended constant current source and a voltmeter. Each IMM has an FPGA for its independent operations including current injection at multiple frequencies, voltage amplification, ADC, digital phase-sensitive demodulation and intra-networking with a main controller of the system. The main controller is based on a DSP and an isolated USB for its connection to a PC. There is an FPGA-based intranet controller, which arbitrates data exchanges between the DSP and multiple IMMs. Unlike its precursor, KHU Mark1, it is a true parallel system with no switching for both current injection and voltage sensing. The small size of the IMM results in a much reduced dimension of a multi-channel system. The KHU Mark2 can be assembled in any channels between 1 and 64. Depending on a chosen application, we custom design an analog backplane that interfaces multiple IMMs with electrodes. Special care was given to the system calibration to maximize its performance in frequency-difference EIT imaging as well as time-difference. Flexibility is the key improvement factor compared with the KHU Mark1. The new system can accommodate any current injection and voltage sensing protocol including the optimal injection current pattern. Reduced size and new internal architecture significantly improved mechanical as well as electrical stability of the system.

Keywords: EIT, IMM, KHU Mark2

1. Introduction

Electrical impedance tomography (EIT) aims to reconstruct image of an admittivity distribution inside the human body [1]. EIT systems measure boundary voltages due to multiple injection currents to reconstruct cross-sectional images of the admittivity distribution. EIT systems have been developed for imaging gastric function, pulmonary ventilation, cardiac function, brain functions and others. EIT is usually used for obtaining time difference images using a time-reference data. When it is impossible to obtain a time-reference data, we may adopt a frequency-difference technique to image acute stroke or cancer for example. EIT systems need a wider bandwidth for better choice of multiple frequencies and multiple parallel channels for faster operations. Depending upon a specific application, it should be able to inject in such a way that we can extract as much internal information as possible. This also requires a flexible electrode configuration.

Our previous EIT system KHU Mark1 has one current source and a switching circuit to inject current between a chosen pair of electrodes sequentially [2-4]. Though it has multiple independent voltmeters, two inputs of each voltmeter are connected to an adjacent pair of electrodes. In order to overcome these limitations, we developed the KHU Mark2. In this paper, we describe its development and basic performance.

2. Methods

2.1. System structure

The KHU Mark2 EIT system consists of the following parts: (i) a PC with an USB port and EIT software, (ii) a main controller, DSP(TMS320LF2812A, Texas Instruments, USA) with an USB interface, (iii) an intra-network controller on a digital backplane, (iv) impedance measurement modules (IMM) and (v) switching circuits on an analog backplane. Figure 1 shows its structure.

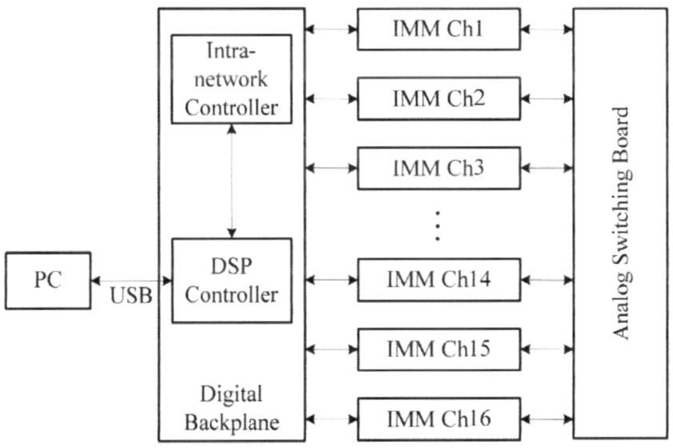

Figure 1. Structure of the KHU Mark2.

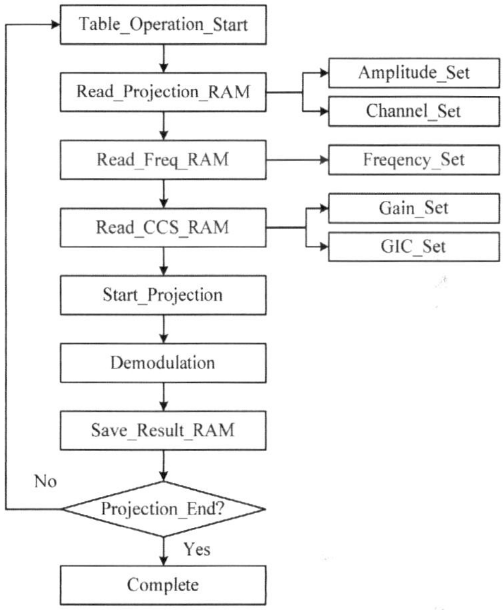

Figure 2. Table operation flow chart.

2.2. Impedance measurement module (IMM)

Our previous KHU Mark1 EIT system had one balanced current source and 8 or more voltmeters. The new KHU Mark2 is based on the impedance measurement module (IMM), which includes both a single-ended constant current source and a voltmeter. Using *N* IMMs, we construct an *N*-channel KHU Mark2 EIT system. Since IMMs are independent each other, the KHU Mark2 can inject current and measure voltage in any possible combinations. Using a high-capacity FPGA (EP3C10F256C8N, Altera, USA), the IMM performs all necessary functions including digital waveform generation, phase-sensitive demodulation, data exchange and controls.

In the KHU Mark1 system, we implemented the balanced current source by two single-ended current sources. The constant current source in the IMM of the KHU Mark2 system is basically identical to the single-ended current source in the KHU Mark2 system. The current source is based on the Howland circuit with 4 generalized impedance converters (GIC) [2-5]. We used these GICs at frequencies above 10 kHz to cancel out stray capacitances. Using digital potentiometers (DS1267, Dallas semiconductor, USA), we calibrate each current source for its minimum output impedance of 1 MΩ throughout the chosen bandwidth.

The voltmeter in the IMM consists of the following parts: (i) differential amplifier, (ii) band-pass filter, (iii) voltage amplifier, (iv) ADC and (v) digital phase-sensitive demodulator. Using an ADC (AD9235BRUZ-20, Analog Devices, USA), we fed the voltage data to the FPGA for digital phase-

Journal of Physics: Conference Series **224** (2010) 012013 doi:10.1088/1742-6596/224/1/012013

sensitive demodulations to produce both in-phase and quadrature components.

2.3 Main controller and data communication
The DSP-based main controller controls all functions of multiple IMMs to collect boundary current-voltage data. The sequence of events is as follows: (i) PC sends a command to the main controller, (ii) main controller decodes the command and sends instructions to the intra-network controller, (iii) intra-network controller transmits instructions to IMMs and then (iv) IMM receives the instructions and controls current injection and voltage measurement. Each IMM sends measured data back to PC in the reverse order.

2.4 Table operation
Unlike the KHU Mark1, we implemented table operations in the KHU Mark2, which automate data acquisition and control in a faster way. Instead of receiving one instruction per command packet, we download an instruction table to each IMM including a sequence of multiple instructions such as current amplitude, frequency, voltage gain, GIS settings and so on. First, PC sends a table of instructions to the main controller, which is transferred to the IMM. Second, PC sends '*Table_Operation_Start*' command. When each IMM receives this command via the main controller, it decodes the table stored in its FPGA. Third, the IMM sets up all hardware settings necessary for the table operation. Finally, when the IMM gets '*Scan_Start*' command, it sequentially performs all data collection instructions stored in the table.

2.5 Lead switching
The KHU Mark2 is flexible in terms of electrode configuration and data collection protocol. Figure 3 and 4 show two lead switching boards for two- and three-dimensional electrode configurations. Using the analog backplane in figure 3, we can conveniently implement the two-dimensional neighboring method. The analog backplane in figure 4 imposes no predetermined connection for versatile three-dimensional electrode configuration. We may inject current through multiple electrodes at the same time with the same or different frequencies and amplitudes. Voltage measurements between any electrode pairs are possible with this analog backplane.

Figure 3. Lead switching for two-dimensional electrode configuration .

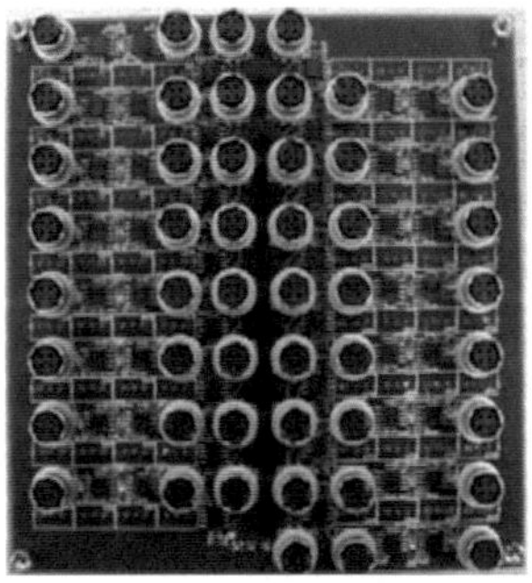

Figure 4. Lead switching for three-dimensional electrode configuration.

3. Results

We found that the KHU Mark2 is about 10 times faster than the KHU Mark1 by using the table operation. Figure 5 and 6 show time-difference and frequency-difference images of a saline phantom including an insulator and a conductor anomaly, respectively.

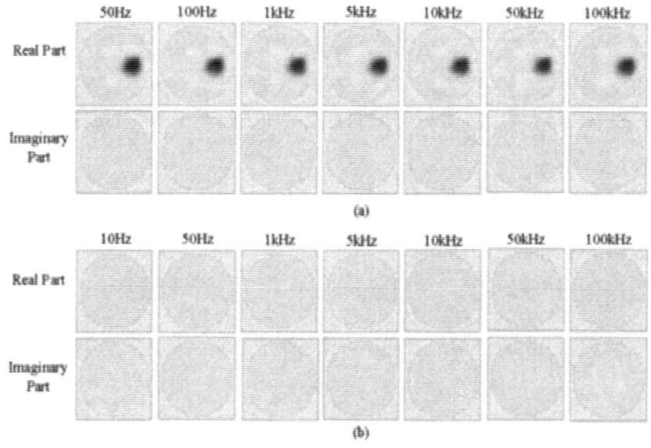

Figure 5. Insulator in saline. (a) Time-difference and (b) frequency-difference images.

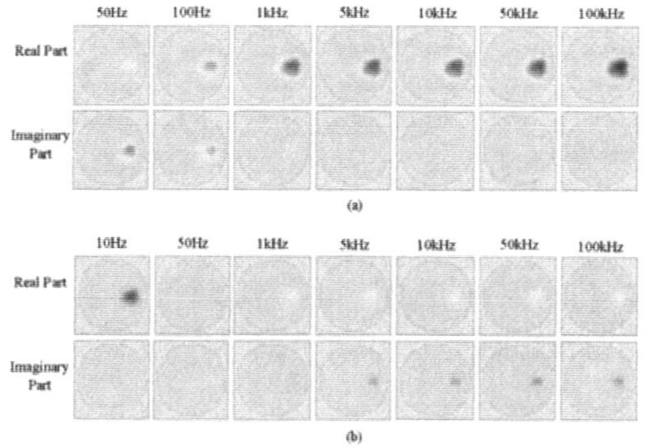

Figure 6. Conductor in saline. (a) Time-difference and (b) frequency-difference images.

4. Conclusion and Discussion

We developed the new parallel multi-frequency EIT system, KHU Mark2. Using multiple independent IMMs, current injection and voltage measurement can be done in any configuration. We significantly reduced the data acquisition time for a faster frame rate. To reduce the increased power consumption in the KHU Mark2, we plan to develop a SOC for the IMM. Our future studies include thorough performance evaluations and applications to chest and head imaging.

Acknowledgments

This work was supported by the SRC/ERC program (R11-2002-103) of MOST/NRF.

References

[1] D. S. Holder 2005 *Electrical Impedance Tomography: Methods, History and Applications* (IOP: Bristol, UK)

[2] T. I. Oh, E. J. Woo and D. Holder 2007 Multi-frequency EIT system with radially symmetric architecture: *KHU Mark1 Physiol. Meas.* **28**, S183–S196

[3] T. I. Oh, K. H. Lee, S. M. Kim, H. Koo, E. J. Woo and D. Holder 2007 Calibration methods for a multi-channel multi-frequency EIT system *Physiol. Meas.* **28** 1175-1188

[4] T. I. Oh, H. Koo, K. H. Lee, S. M. Kim, J. Lee, S. W. Kim, J. K. Seo and E. J. Woo 2008 Validation of a multi-frequency electrical impedance tomography (mfEIT) system KHU Mark1: impedance spectroscopy and time-difference imaging *Physiol. Meas.* **29**, 295-307

[5] A. S. Ross, G. J. Saulnier, J. C. Newell and D. Isaacson 2003 Current source design for electrical impedance tomography *Physiol. Meas.* **24**, 509-516

Journal of Physics: Conference Series **224** (2010) 012014 doi:10.1088/1742-6596/224/1/012014

Suitability of the INPHAZE impedance analyzer for Bio-impedance and EIT

Sugashine Jeganathan Andre van Schaik, Craig Jin and Alistair McEwan,
CARLAB, School of Electrical and Information Engineering,
The University of Sydney, NSW, Australia, 2006.
alistair.mcewan@sydney.edu.au

Abstract: The suitability of a new impedance analyzer 'INPHAZE' for bio-impedance and Electrical Impedance Tomography (EIT) is investigated using measurements on simple resistive and capacitive models, and readily available biological objects. The INPHAZE is an impedance spectrometer developed to measure the impedance of thin films and layers but has been used in various applications. We found that it is superior to the UCL mk2.5 EIT system on measurements of resistive models and biological objects; however it is significantly over 10x slower as with each measurement checked for stability and SNR. We suggest that the device may provide significant improvements for bio-impedance. However for EIT we suggest exploration of expansion to more channels and the effect of reducing measurement times.

1. Introduction

The purpose of this paper is to investigate the suitability of the INPHAZE impedance analyzer for bio impedance and Electrical Impedance Tomography EIT. The INPHAZE Impedance Spectrometer is a unique electrical impedance spectroscopy device developed at The University of Sydney. The INPHAZE device operates at level of accuracy that is orders of magnitude better than other impedance spectroscopy devices with impedance precision of 0.002% and phase resolution of 0.001degree over a large frequency range from 1mHz to 1MHz and impedance range of 0.1 to 10^{10} Ω [1-3]. The device performance was assessed in laboratory conditions using saline and simple resistive and capacitive models. The INPHAZE system consists of control panel software, Impedance Spectrometer Unit which yields high resolution impedance measurements, the amplifier unit which connects the signals between the Spectrometer Unit and the sample and the impedance analyzer software which is used to view the measurement data in a variety of plot types.

2. Materials and methods

The INPHAZE EIS system can easily be calibrated using the supplied Control Panel software. Each calibration result is stored in the user's personalized protocol file, so the system can be personally calibrated under different loads and conditions depending on each user's application. The INPHAZE EIS system is supplied with a range of high-precision resistors and capacitors for calibration and measurement purposes. The system has been calibrated with the supplied 100Ω resistor as the calibration load for all of the following experiments. Preceding these measurements the current flowing through the electrodes was measured as 80μA RMS by measuring the voltage across a 1kΩ resistor connected between the electrodes. This current level is within the IEC60-601 safety regulations for frequencies 1kHz and above.

2. 1 *Simple Resistor*

The system was initially calibrated using two 100 Ω resistors, one as the sample and another as the standard. The resulting protocol file was used to measure the frequency response of resistors of values 10.003Ω, 20.3Ω, 31.3Ω, 50.1Ω, 59.4Ω, 73.5Ω. The experiment was performed over a frequency range of 10 Hz to 1MHz in steps of 1, over three spectra.

2.2 *Contact Impedance Influence*

The four terminal setup in Fig 1 was used to measure the contact impedance of a resistive sample. With the contact resistors R1, R2, R3 and R4 kept constant at 1k Ω, resistors of values 100 Ω, 73.5Ω, 59.4Ω, 50.1Ω, and 38.9Ω were used as the sample resistor. The experiment was performed over a frequency range of 1k Hz to 1MHz.

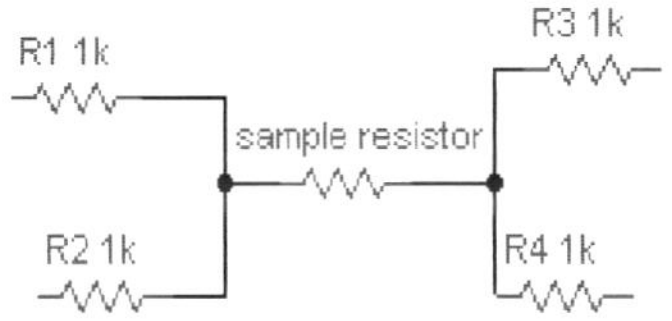

Fig 1: Arrangement used to measure the influence of contact impedance

2.3 *Saline Solution*

The impedance of various concentrations of salt solution was measured using the INPHAZE device by placing the salt solution in a 16 electrode cylindrical tank of 10cm diameter and connecting the I+ and V+ electrodes to two adjacent channels, and I- and V- electrodes to the diagonally opposite channels, a four terminal connection. Concentration of the salt solution was varied from 0.5g/L to 1g/L to 1.5g/L and for each concentration, the impedance was measured over a range of 1kHz to 1MHz.

2.4 *Biological Object*

The impedance of a simple biological material, a banana was measured by initially filling a plastic tube of 10 cm length and 2cm. At each end was a chloride silver disc electrode. All air bubbles were carefully eliminated. One end was permanently sealed, and the tube was filled with saline solution of 1g/L and the INPHAZE electrodes were connected to the tube, using a two terminal connection. This data was used to calibrate the measurements taken with banana. The measurement was then repeated by replacing a small quantity of the saline solution with a cylinder of banana of 1cm in diameter and 1.5 cm in length, followed by two pieces of banana of the same dimension, and then three. The measurement was taken over a frequency range of 20Hz to 1MHz.

3. Results

3.1 *Simple resistor*

The impedance vs. frequency response of all the simple resistors measured is shown in Fig 2

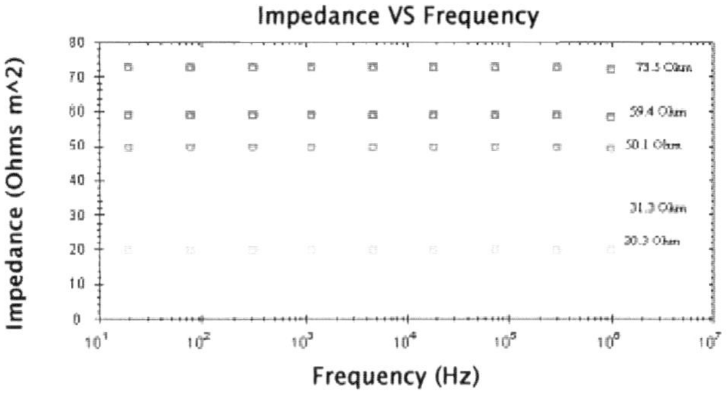

Fig 2: Plot of Impedance (Ω m^2) vs. Frequency (Hz) for simple resistors. The resistors used were 73.5, 59.4, 50.1, 31.3, 20.3 Ω from top to bottom.

3.2 Contact Impedance influence

The frequency response of the contact impedance influence measured using various resistor values for the sample resistor is shown in the plot in Fig 3.

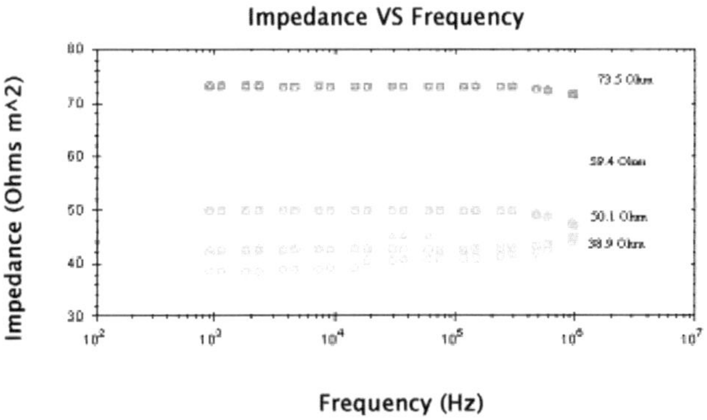

Fig 3: Plot of Impedance (Ω m^2) vs. Frequency (Hz) for contact impedance measurement. The resistors used were 73.5, 59.4, 50.1, 38.9, 31.3Ω from top to bottom.

3.3. Saline Solution

The frequency response of the impedance of the saline solutions of various concentrations is shown in the plot in Fig 4.

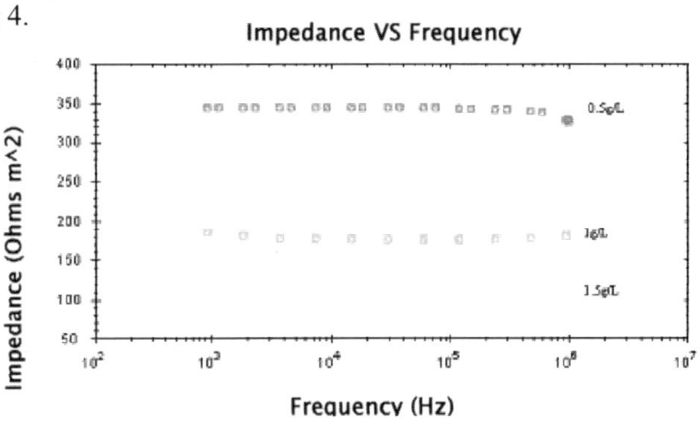

Fig 4: Plot oi impedance (Ω m^2) vs. Frequency (Hz) of saline solutions of various concentrations.

3.4. Biological Object

The frequency response of the impedance of banana calibrated with the saline solution is shown in the plot in Fig 5.

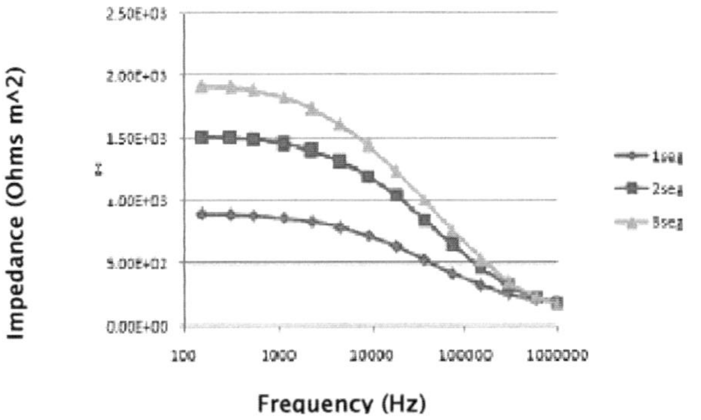

Fig 5: Plot of Impedance (Ω m^2) vs. Frequency (Hz) of saline solution with banana

4. Discussion

The suitability of the INPHAZE impedance analyzer for bio impedance and EIT measurements has been demonstrated. It is robust to the presence of contact impedance and able to measure the impedance spectra of biological objects. For the purpose this feasibility study, the amplifier unit included in the system was taken out of the INPHAZE Faraday Cage to resemble remote location applications. However, at the time of this study, the manufacturer had not released the INPHAZE External Amplifier Enclosure to the market yet. The scheduled release date of these external enclosures is mid-March 2010. With this enclosure, the amplifier unit would be properly shielded for remote setups, as they would be in the original Faraday Cage. As seen from the results, there is an effect, perhaps due to stray capacitance at frequencies above 100kHz, e.g. , a decrease by 6 % in the presence of a 1kΩ contact impedance with the 50.1Ω resistor. However, further studies are needed to confirm this as the sample load is only 50.1Ω and it should have overcome most of the impedance caused by the stray capacitance that is parallel to the sample load. In future studies, it would be advantageous to use the INPHAZE External Amplifier Enclosure to improve the measurement quality. And in cases where the effects of stray capacitances are observed, users can use the supplied Control Panel software to calibrate for the stray capacitances. It was suggested by the manufacturer to try the stray capacitance calibration after the amplifier unit has been put inside the dedicated enclosure, for a more stabilized result. As for the measurement taken with biological objects, varying concentrations of banana showed noticeable change in the impedance measured over a range of frequency. The centre frequency remains at 10kHz which compares well with previous measurements taken with the HP impedance analyzer (gold standard) [4]. The INPHAZE spectrometer is significantly over 10x slower as with each measurement checked for stability and SNR. We suggest that the device may provide significant improvements for bio-impedance. However for EIT we suggest exploration of expansion to more channels and an investigation into reduction of measurement times.

5. References

[1] L Gaedt, TC Chilcott, M Chan, AG Fane, HGL Coster (2002), "Electrical impedance spectroscopy characterization of conducting membranes", Journal of Membrane Science 195:169-180

[2] H Coster., Analysis of nanostructural layers using low frequency impedance spectroscopy, Part 2: http://www.Inphaze.com.au/downloads/tuts/Tut2_StructureRefinement.pps accessed 9/12/2009 pg 24.

[3] ELS Wong, TC Chilcott, M James, HGL Coster (2006), "Electrical Impedance Spectroscopy Characterizations of Alkyl-Functionalized Silicon", Biophysical Reviews and Letters, 1 (3): 1-7

[4] A. McEwan et al 2006 Design and calibration of a compact multi-frequency EIT system for acute stroke imaging J. Physiol. Meas. 27 S199-S210

[5] A. McEwan, A., G. Cusick, and D.S. Holder, 2007 A Review of Errors in Multi-frequency EIT Instrumentation J. Physiol. Meas. 28 S197-215

Performance assessment of EIT measurement systems

P J Riu and D Anton

Department of Electronic Engineering. Centre for Research on Biomedical Engineering. Universitat Politecnica de Catalunya, c./Jordi Girona, 1-3, Edifici C4, 08034 Barcelona, Spain

Email: priu@eel.upc.edu

Abstract. EIT acquisition systems, when regarded as measuring instruments, can be characterized by their inherent characteristics: reproducibility, repeatability and discrimination. The ultimate limit to all three is noise, whether long-term or short-term noise. In addition, for some applications accuracy can also play a role, although accuracy is not an inherent characteristic and requires a known test object to be defined. Because EIT systems are multichannel instruments and the information of all channels is being combined to produce an image, a procedure to standardize the performance assessment must be set up in order to allow for meaningful comparison of different systems or different settings in the same system. We propose the use of three parameters defined long ago but scarcely used among EIT systems designers: SER (Systematic Error Ratio), NER (Noise Error Ratio) and RER (Reciprocity Error Ratio). We applied these definitions to our latest EIT instruments, TIE5-sys, in a wide frequency range and with different settings in configurable acquisition parameters and provide the interpretation of the results obtained in terms of reproducibility, repeatability and discrimination. We also comment on the tradeoffs that show up when using the usual definition for these figures.

1. Introduction

Measurement systems can be characterized by its inherent characteristics: discrimination, repeatability and reproducibility [1] which are related to the internal structure of the system and to noise. In addition, accuracy may also be interesting in some cases. However, in order to define the accuracy for most measurement system, an external test object of declared value (standard) must be used [1].

Electrical Impedance Tomography (EIT) acquisition systems are complex instruments that need to take several individual measurements in order to produce a frame or image. These individual measurements will differ from each other in several ways: the internal configuration of switches may be different, or the profile of applied currents or voltages, the points of application of voltage or current in the object may be different, the collection points may be different, the gains or integration time in the measurement chain may also be different; to cite the most common expected changes.

Because of the large number of possible combinations, and because different systems using different approaches will have different combinations, it is not practical to give all the comprehensive information of noise or accuracy for each possible individual measurement, especially if this information will be used to make comparisons between systems.

The problem of system characterization and comparison arose from the very beginning in the development of EIT. During the European Concerted Action on Impedance Tomography (CAIT) a set of figures were proposed to solve this issue [2]. The definitions were never published in widely available format, and the available documents are plagued with typographic errors. We intend here to revise the figures, comment on their applicability and present the results of characterization of our recently developed EIT instrument.

2. Definition of figures of merit

We will start by explaining the CAIT approach [2] and comment on the implications of these choices. The CAIT approach originally intended to assess the noise performance and the accuracy.

2.1. Systematic errors

Systematic errors may have, at least, two different origins.

1. The individual measurement obtained with a particular combination of factors differs from what is expected. To assess these errors the Systematic Error Ratio (SER) is defined.
2. Two individual measurements, obtained with two different combination of factors, which must yield the same result (e.g. because of the reciprocity theorem) differ, irrespective of the fact that the actual value is close to the expected one. To assess these errors, the Reciprocity Error Ratio (RER) is defined.

2.1.1. *Systematic Error Ratio (SER).* SER is formally defined in equation 1. It is clear that in order to calculate this error, a known test object must be used. The test object can be built using components with an uncertainty smaller than the expected for the system, or might have been characterized using an instrument with a suitable uncertainty. In both cases, one must make sure that drifts are under control. For a frame of K individual measurements, with index k:

$$SER = \sqrt{\frac{1}{K}\sum_{k=1}^{K}\frac{(V_{mk} - V_{ck})^2}{V_{ck}^2}} \tag{1}$$

where: V_{mk} is the measured value for the measurement of index k
V_{ck} is the theoretical value for the measurement of index k

If random errors are known to have values similar to that of expected SER, then it is advisable to use an average value instead of a single V_{mk}:

$$V_{mk} = \frac{1}{N}\sum_{i=1}^{N}(V_{mk})_i \tag{2}$$

where N has an arbitrary value, large enough to make sure that the remaining random effect will not interfere with the calculation of SER.

2.1.2. *Reciprocity Error Ratio (RER).* RER is formally defined in equation 3. It is clear that in order to calculate this error, no known test object is needed. The only restriction is temporal stability during the time required to acquire a frame. For a frame of K individual independent measurements, with index k:

$$RER = \sqrt{\frac{1}{K}\sum_{k=1}^{K}\left(\frac{V_{k,ij}^{ab} - V_{ab}^{ij}}{V_{k,ij}^{ab}}\right)^2} \tag{3}$$

where: V_{ij}^{ab} is the measurement obtained when applying current trough electrodes i,j and detecting voltage through electrodes a,b.

Calculation of RER makes only sense for those systems that perform (or are able to) measurements that can be related using the reciprocity theorem. One particular case are systems using adjacent injection and adjacent voltage detection. For that case, if a number of E electrodes are being used, there are $E(E-1)/2$ independent measurements, of which E measurements are three-wire measurements that tend to be saturated, so are not used in practice. This leaves $E(E-3)/2$ useful measurements. In order to calculate RER we need to make twice the number of measurements and acquire the reciprocal of the independent ones. In that particular case, and assuming that measurements are identified by two indexes i,j indicating the injection and detection pair, RER can be expressed as:

$$RER = \sqrt{\frac{2}{E(E-3)}\left[\left(\sum_{i=1}^{E-2}\sum_{j=i+2}^{E}\left(\frac{V_{ij}-V_{ji}}{V_{ij}}\right)^2\right)-\left(\frac{V_{1E}-V_{E1}}{V_{1E}}\right)^2\right]} \quad (4)$$

2.2. Random errors

Short term random errors are associated with noise, either from the current source or the voltage detectors. Depending on the dominant source of error and the architecture of the system, the combined effects of the different contributions will have different correlation coefficients. It is not always easy to estimate these correlation coefficients. No known test object is required to make the calculation of noise power. For an individual measurement of index k, the RMS value of noise, relative to the measured value, is called the Noise Error Ratio (NER) (which is the inverse of the Signal to Noise Ratio (SNR)). NER for an individual measurement V_k can be obtained as:

$$NER_k = \frac{1}{\bar{V}_k}\sqrt{\frac{\sum_{i=1}^{N}\left(V_{ki}-\bar{V}_k\right)^2}{N}} \quad (5)$$

where N has an arbitrary value, large enough to make sure that the calculation is relevant. In order to give a single figure for the system, NER_k can be averaged, either linearly or RMS:

$$NER_L = \frac{1}{K}\sum_{k=1}^{K} NER_k \qquad NER_{RMS} = \sqrt{\frac{1}{K}\sum_{k=1}^{K} NER_k^2} \quad (6)$$

NER_L was used within the CAIT, but NER_{RMS} is more consistent with the given definitions for SER and RER.

3. Characterization of TIE5_sys

The definitions on section 2 were used to derive figures of merit for our most recent system, TIE5_sys [3]. Because of TIE5_sys is a broadband system, all the above parameters were obtained for several frequencies within the frequency range of operation (10 kHz – 1 MHz). All the measurements were performed in a resistor-mesh phantom, known as CARDIFF EIT PHANTOM (v1.0 1991) [3] made of 100 Ω ±1% SMD resistors. The shape of the phantom is depicted in figure 1. Table 1 summarizes some of the results, obtained using a fixed (maximum) gain for the Programmable Gain Amplifier (PGA), common mode feedback (CMF) connected and no electrode impedances. Actual theoretical values for the test object were measured using an HP4192A Impedance Analyzer.

Table 1. Sumary of results for SER, RER and NER_L as a function of frequency for TIE5_sys

	10 kHz	30 kHz	50 kHz	80 kHz	100 kHz	300 kHz	500 kHz	700 kHz	1 MHz
SER (%)	0.49	0.44	0.46	0.51	0.59	1.87	5.28	9.81	17.53
RER (%)	0.74	0.73	0.66	0.63	0.63	0.81	2.38	5.29	4.91
NER_L (%)	0.048	0.033	0.035	0.029	0.032	0.031	0.031	0.031	0.029

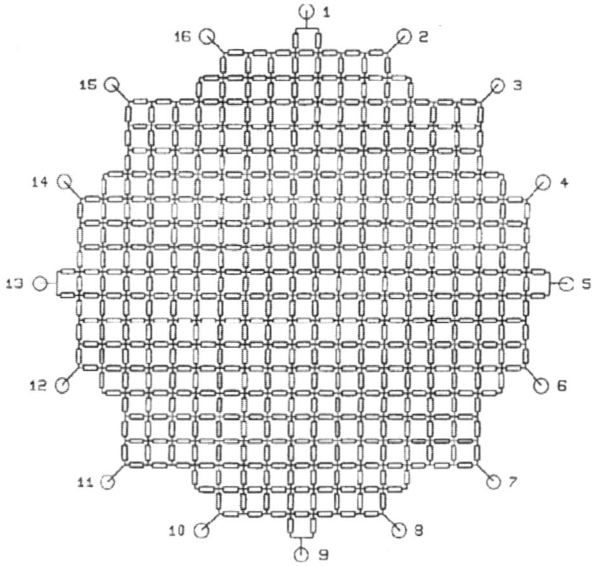

Figure 1. Shape of the CARDIFF EIT PHANTOM (v1.0 1991) used to characterize our system [3].

4. Discussion

The figures defined (SER, NER, RER) are suitable to characterize and compare EIT systems of (almost) any topology. However, to make valuable comparisons some agreement must be done on the measurement conditions, especially on the test object.

The shape and composition of the test object does have an impact not only on the calculation of SER, where the availability of a well known test object is required and its influence is obvious, but also on the computation of RER and NER. The shape and values of the test object will set the dynamic range of the measurements, thus setting a limit for the maximum achievable NER and RER. Using a test object consisting on a single resistor connected to all channels will produce the best results. This can be acceptable for comparison purposes but values obtained can be far from what is expectable in a real application. On the other hand, using a test object very tailored to an application can benefit some systems in front of others.

Using fixed or adaptive gain profiles is a common practice in EIT. However, the combination of gain profiles and a particular test object may bias the estimation of NER. For that reason using fixed gains and a "generic" test object produces values better suited for comparison.

References

[1] Sydenham P H, Hancock N H and Thorn R 1989 *Introduction to measurement science and engineering* (Chichester: John Wiley & Sons)

[2] Griffiths H, Leung H T L and Williams R J 1990 *Proc. EC Workshop on phantoms for Electrical Impedance Tomography*, Cardiff 23-24-11-1990

[3] Griffiths H 1988 *Clin. Phys. Physiol. Meas.* **9 A** 15

[4] Anton D, Balleza M, Fornos J, Kos B, Casan P, Riu P J 2007 *IFMBE Proceedings* **17** 564

Journal of Physics: Conference Series **224** (2010) 012016 doi:10.1088/1742-6596/224/1/012016

Common Mode Feedback Analysis for EIT Systems with Distributed Current Sources

D Anton and P J Riu

Department of Electronic Engineering. Centre for Research on Biomedical Engineering. Universitat Politecnica de Catalunya, c./Jordi Girona, 1-3, Edifici C4, 08034 Barcelona, Spain

Email: priu@eel.upc.edu

Abstract. The use of differential voltage measurements is widely used in EIT instruments. Instrumentation amplifiers are always affected by common mode voltages at their input. These voltages may have different origins, being the current sources and multiplexers the ones which contribute the most. However, even considering ideal current sources and multiplexers, some amount of common mode voltage will always be present due to the measurement principle. Errors produced by common mode voltages are very difficult to calibrate because they depend on the object being measured. At low frequencies it is easy to design instrumentation amplifiers with high CMRR values, but common mode voltages are also high, mostly because of unbalances in electrode impedances. At higher frequencies common mode voltages are smaller, but the CMRR of the amplifiers also decreases. Common mode feedback (CMF) to reduce common mode voltages has been largely used in EIT systems. There is a good theoretical understanding of CMF for systems using a single current source and a demultiplexer. However there is a new class of instruments which use distributed current sources. In these systems CMF is actually used to change the inherent unbalance among current sources instead of being applied to the output of the current source. We present a theoretical analysis of this new CMF technique and the results obtained when applying it to our latest EIT system

1. Introduction

The use of differential voltage measurements in EIT reduces the requirements of dynamic range for the acquisition chain. The presence of common mode voltages degrades the performance of the instrumentation amplifiers used to sense the differential voltages. Common mode voltages have different origins, some of which are instrumental: unbalances in current sources, multiplexers; capacitances to ground. However, even with an ideal instrument, some common mode voltage will appear because injection and detection electrode pairs are not symmetrical with respect to the object being measured for most of the used injection/detection combinations in EIT.

A careful design of the instrumentation amplifier will produce a high Common Mode Rejection Ratio (CMRR), but the unbalances in electrode impedances cannot be fully controlled and pose a limit to the achievable CMRR [1]. In addition, errors produced by common mode voltages are difficult to calibrate because they depend on the value of the impedance being measured, and of course, on the frequency [1].

The use of Common Mode Feedback (CMF) has been widely used to reduce these errors. Historically, two main approaches have been used: feedback is applied to the body using an extra electrode [2], or feedback is applied to a point in the current injection network [3]. Like any other feedback application, the network used to apply it must be optimized in order to reduce as much common mode voltage as possible while preserving stability. A comprehensive analysis is performed in [3] for different application points within the current injection subsystem.

A new current injection topology has been recently introduced for broadband systems working at high frequency [4], which avoids the use of current multiplexers. In this new topology, CMF cannot be applied to the current source, because there is a current source for each electrode. Feedback must be applied to some point in the signal generation network to produce an imbalance in the current sources that compensate the common mode voltage. A description of the new topology and a theoretical analysis of the CMF network are presented, along with practical results obtained in our most recent EIT instrument TIE-5sys.

2. Distributed current source topology

A simplified view of the distributed current source is shown in figure 1. A voltage/current converter, based on an AD8130, is connected to each electrode. The V/I converters are driven by a differential/differential amplifier based on an AD8138 through 2 multiplexers. The injection pair is formed by applying the positive output of the amplifier to one V/I converter (source) and the negative output to the other V/I converter (sink). A comprehensive description of the system can be found in [4].

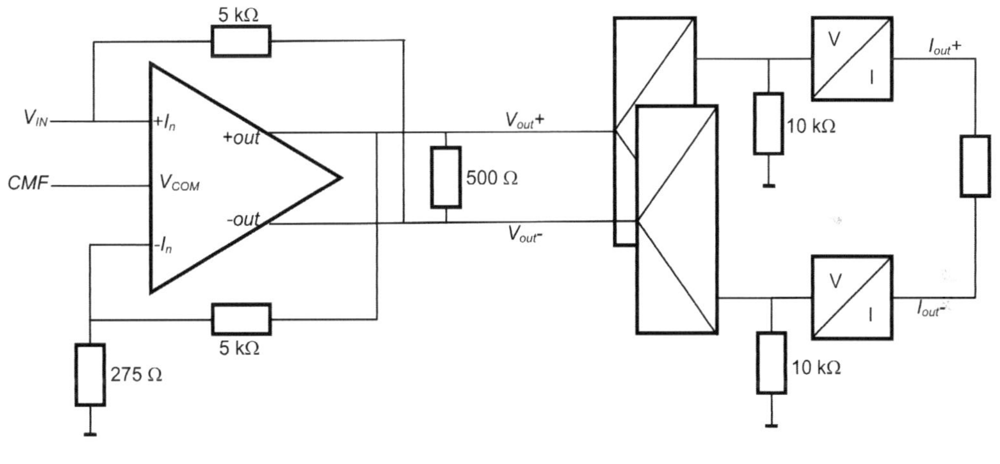

Figure 1. Schematic representation of the distributed current injection system.

The AD8138 provides with an additional input (V_{COM}) whose mission is to produce an unbalance at the output, thus unbalancing the current sources. CMF will be applied to that point. Because this point is not directly connected to the body, the analysis performed in [3] is not valid for this case.

3. Analysis of the feedback network

In order to analyze the stability of the common mode feedback we built a model of the complete injection subsystem, electrodes, body and detection buffers, involving one injection an one detection pair. Because the model is symmetrical, we only analyzed half of it. This model is presented in figure 2.

The different parts of the model have been grouped into logical entities to create the block diagram shown in figure 3. The transfer function of each block was derived, using the actual values for the different discrete components and values from manufacturer data sheets for the IC's. The Common

Mode Feedback amplifier consists on a single op-amp configured as a band-pass inverting amplifier. The high-pass filter is required in order to block any DC values that could revert on saturating the V/I converters. The low-pass filter is required in order to introduce a dominant pole to guarantee stability.

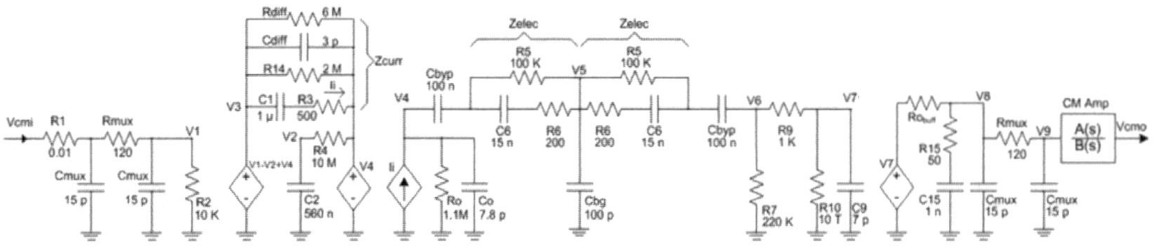

Figure 2. Equivalent schematic representation of one half of the injection/detection channel, showing the voltage demultiplexers, V/I converters, electrodes, buffers, active shielding and the Common Mode Feedback Amplifier

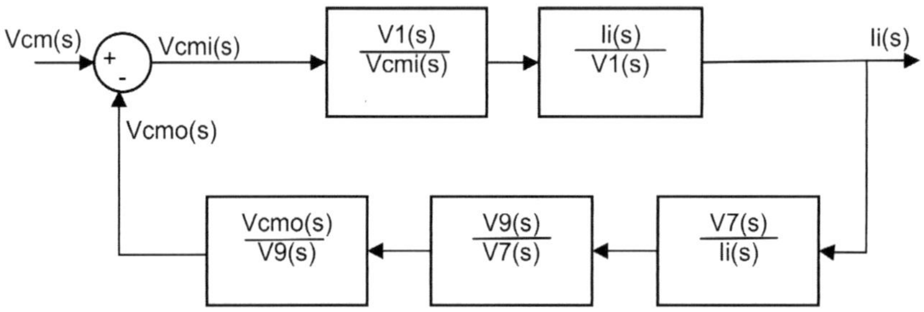

Figure 3. Simplified block diagram of the circuit in figure 2. Names of voltages and currents correspond to those in figure 2.

It is possible to derive an analytical expression for the open loop gain, but it is too cumbersome. Instead, it was numerically computed using actual values for components. The result can be seen in figure 4. This result was obtained using a gain of 0.1 (-20 dB) in the pass-band of the Common Mode Feedback amplifier. The obtained gain and phase margin make the system strongly stable. In fact, gain of the CMF amplifier can be increased by about 13 dB and the system will still preserve the minimum phase and gain margin to be stable. With this latter value, the total gain in the CMF loop will be of about 45 dB. In practice, the gain will have to be adjusted to account for stray capacitances and cable inductances which are difficult to model in an accurate way. A practical problem with the structure used for the CMF amplifier is that changing the gain will also change the frequency response, making the adjustment difficult, so a different implementation may be necessary in the future.

4. Results

To assess the effectiveness of the CMF several measurements were performed in the whole frequency range of operation of the system, with and without the CMF connected. The Systematic Error Ratio (SER) and the Reciprocity Error Ratio (RER)[5] were used to quantify the improvement in accuracy when the CMF was connected. Noise Error Ratio (NER) [5] was also obtained in order to make sure that the noise performance was not degraded when CMF is in operation. All measurements were performed in a resistor mesh test object. The gain of the CMF amplifier was set to 0.03, which was experimentally found to be the maximum allowed to prevent instabilities. Table 1 summarizes these results.

The use of CMF increases the accuracy significantly at low frequency. Above 200 kHz the improvement is less evident and becomes unnoticeable at 500 kHz. This may be expected because at that frequency the open loop gain is about 40 dB below the maximum. Also, as expected, the effect of CMF on noise performance is unnoticeable.

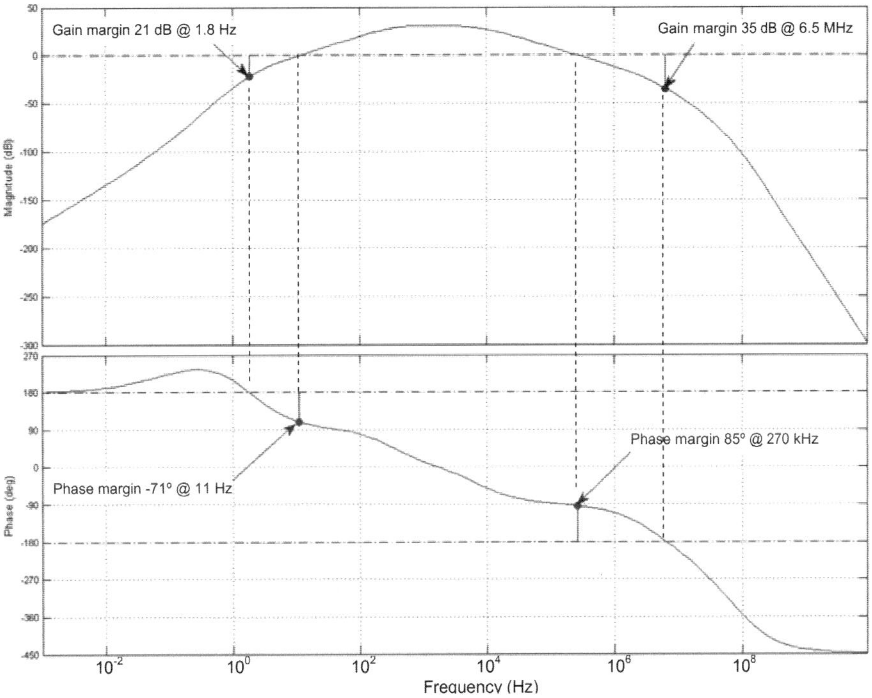

Figure 4. Gain and phase-shift for the open-loop gain of the complete model in figure 2.

Table 1. Summary of performance measurements to assess effectiveness of CMF

	SER (%)		RER (%)		NER (%)	
Freq.	CMF Off	CMF On	CMF Off	CMF On	CMF Off	CMF On
10 kHz	1.76	0.49	3.47	0.74	0.048	0.049
50 kHz	2.12	0.46	3.70	0.66	0.035	0.034
100 kHz	2.29	0.59	3.69	0.63	0.032	0.035
500 kHz	5.75	5.28	3.27	2.38	0.031	0.031
1 MHz	17.88	17.53	2.58	4.91	0.029	0.034

5. Conclusions

Using CMF produced an increase in accuracy in EIT systems. The distributed current source approach has the same tradeoffs between bandwidth, gain and stability in the CMF amplifier that was obtained in more classical approaches. However, the gain of the CMF amplifier on these systems needs to be low, even less than one, in order to preserve stability.

References

[1] Riu P J, Rosell J, Lozano A and Pallas-Areny R 1995 *Med. Biol. Eng. Comp.* **33** 784
[2] Rosell J. Murphy D, Pallas R and Rolfe P 1988 *Clin. Phys. Physiol. Meas.* **9** **A** 93
[3] Rosell J and Riu P J 1992 *Clin. Phys. Physiol. Meas* **13** **A** 11
[4] Anton D, Balleza M, Fornos J, Kos B, Casan P, Riu P J 2007 *IFMBE Proceedings* **17** 564
[5] Riu P J and Anton D 2010 Performance assessment of EIT Systems *J Phys: Conference Series (see this issue)*

Journal of Physics: Conference Series **224** (2010) 012017 doi:10.1088/1742-6596/224/1/012017

Performance of a FPGA-based Direct Digitising Signal Measurement module for MIT

R Patz[1], S Watson[1], C Ktistis[2], M Hamsch[3] and A J Peyton[2]

[1] Faculty of Advanced Technology, University of Glamorgan, Pontypridd, CF37 1DL, UK
[2] School of Electrical and Electronic Engineering, University of Manchester, UK
[3] Philips Research, Weisshausstrasse 2, D-52066 Aachen, Germany

Email: rpatz@glam.ac.uk

Abstract. A novel MIT Direct Digitising Signal Measurement (DDSM) module has been developed aiming to replace the centralised NI PXI system and PC processing of the Cardiff Mk2 MIT system, thus offering potentially faster measurement cycles. The proposed module replaces the signal acquisition and offers local processing. The core of the system is a Xilinx Spartan-3 FPGA, paired with a dual 14-bit ADC capable of 120 MS/s. The FPGA provides a flexible and fast platform for data acquisition and processing. The phase is measured using a two channel phase sensitive detection via I/Q demodulation. Built-in averaging reduces the data to a single signal period of 12 samples before multiplying and accumulating the data with the I/Q signals. The system provides the I/Q values for both channels directly, eliminating the long download and processing times required by the centralised NI PXI system currently used in the Cardiff Mk2 MIT system. The module acquires both the measurement and the reference signal and has a phase noise as low as 7.5m° using a measurement time constant of 6ms. The phase drift over 6 hours is considerable at 119m°. Details of the module, circuits and algorithms employed are provided as are the results of the performance measurements.

1. Introduction

Magnetic Induction Tomography (MIT) is a contactless and non-invasive method to obtain cross-sectional images of the conductivity, permittivity and/or permeability of materials. A number of applications for MIT have been identified in biomedical and industrial areas [1]. A biomedical application currently being investigated is the detection and classification of cerebral stroke [2].

A MIT system induces eddy currents via a primary field into an object and the perturbed magnetic field is detected using an array of coils. The detected magnetic field will include both, the primary and the secondary field. For low conductivity materials, such as biological tissues (<5S/m), the secondary signal will typically be in-quadrature to the primary field and much smaller [3]. The measurement of the small in-quadrature signal, via the phase shift, provides a major challenge in biomedical applications of MIT.

A typical MIT measurement system uses an array of excitation coils that are excited sequentially. The received signal is processed with the reference signal to obtain the phase shift for each detection channel. For the classification of cerebral stroke a phase sensitive detection with a high degree of phase stability is required to obtain usable results (phase noise in the order of 1m°) [2]. Direct digitization of the received signals has been used in recent MIT systems using National Instrument (NI) PXI systems [4, 5]. However, the sequential data transfer between the PXI system and the PC, where the data is processed, and the processing time on the PC limits the speed of the overall signal acquisition. A 14-channel version of the MIT system described in [5] requires 400ms for the data transfer plus a further 800ms for the data processing using an FFT (14-channels, Intel Core Duo) for a data length of 8.5ms (at 60 MS/s).

A novel MIT direct digitization signal measurement module (DDSM) has been developed with the intention to replace the centralized NI PXI system and PC processing [6]. The new module replaces the signal acquisition and offers local processing, thus offering potentially faster measurement cycles.

This paper describes the new module in terms of signal acquisition and signal processing. Phase noise and phase drift measurements are also described and their results are presented.

2. Methods

2.1. System Hardware

At the core of the DDSM is a Xilinx Spartan-3 FPGA with 1Mgates (XC3S1000-4) which is connected to a dual 120 MS/s 14-bit ADC (LTC2285). The external interface of the module is provided by a Microchip PIC microcontroller (PIC18F2480). The data transfer to the PC is done using the serial interface of the microcontroller and a PC application developed in LabWindows/CVI. Mathworks Matlab has been used to analyze the measured data.

Two types of measurement setups have been used for the DDSM. A single channel signal generator, TTi TDA 12101, has been used for the phase drift measurements, while a dual channel signal generator, TTi TGA 1242, has been employed for the phase noise vs. signal voltage level measurements. In both cases a 10MHz sine wave output has been fed to a resistive divider, followed by a set of differential amplifiers (LMH6626) before the ADC. The gain between the input of the amplifier and the ADC was 3.6. The input to the amplifier was varied between $56\mu V_{rms}$ and $65mV_{rms}$ ($0.3mV_{rms}$ – $350mV_{rms}$ at the signal generator). An ADC input range of $1V_{pp}$ was selected. Figure 1 shows the principle of this setup.

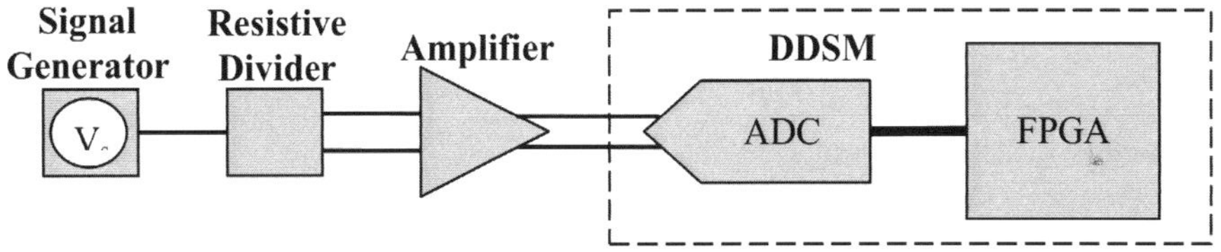

Figure 1. Signal chain for DDSM measurements.

2.2. Phase Measurement Algorithm

The FPGA provides a flexible and fast platform for data acquisition and processing. The algorithm used here is a real-time I/Q demodulation with built-in time-averaging. The ADC samples the two 10 MHz signals with 120 MSps simultaneously. Each of the 12 samples per signal period is accumulated during the acquisition cycle (time constant), leading to 12 large samples at the end of the acquisition cycle. These 12 samples are multiplied with the stored I/Q table (14-bit bit length), before being accumulated to give the I/Q value for the channel. The processing overhead is only 167ns per acquisition cycle. However, the pipelined design enables the DDSM to have only a gap of 25ns between cycles.

The algorithm was set up for a time constant of 1ms and 6ms. A total of 25 acquisition cycles were acquired without disabling the DDSM. This was repeated ten times for the phase noise measurement after a short delay (approx. 100ms). The measurement was only done once for the phase drift measurement.

3. Measurements

3.1. Phase linearity

The DDSM was connected to two output channels of the signal generator using identical cables. A signal level of $65mV_{rms}$ at the input of the amplifier ($350mV_{rms}$ at the signal generator) was set for both channels. The frequency was set to 10MHz and the phase was varied between 0° – 20°. The average of 16 phase measurements (1ms time constant) was used to determine the phase.

3.2. Phase noise vs. signal voltage level

The phase noise was measured with one of the channels amplitude adjusted between $56\mu V_{rms}$ and $65mV_{rms}$ ($0.3mV_{rms}$ – $350mV_{rms}$ at the signal generator). The second channel was kept constant at $65mV_{rms}$. The phase noise was defined as the standard deviation of 10 measurements. Two sets of measurements with different time constants (1ms and 6ms) were conducted.

3.3. Phase drift

The output of a single channel was split and connected to the two channels of the DDSM. The amplitude for both was kept at $65mV_{rms}$, and the phase was measured every minute for 5 hours. The phase drift was defined as the difference between the maximum and the minimum of the measured phase.

4. Results

4.1. Phase linearity

A linear regression fit between the set and measured phase was applied. The R^2 value was found to be 0.9998.

4.2. Phase noise vs. signal voltage level

The phase noise was measured against the input voltage level at the amplifiers. Figure 2 shows the results from the measurement. The minimum phase noise is 7.5m°, while the maximum phase noise was 119.4m°.

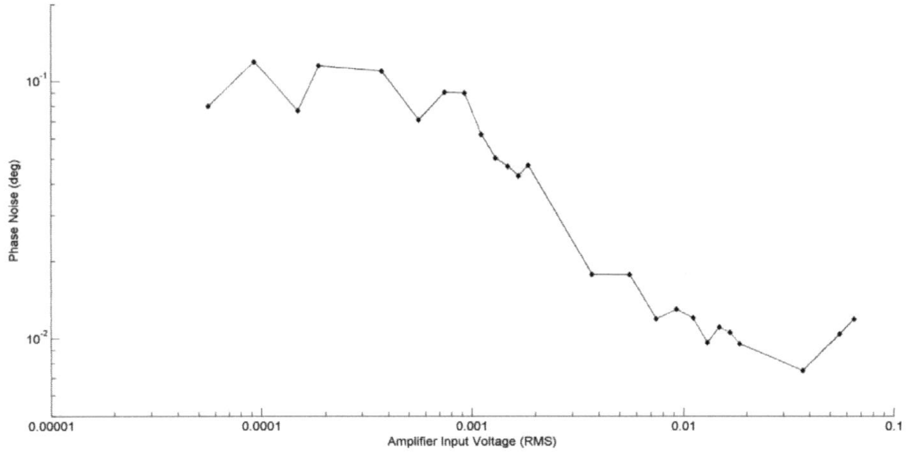

Figure 2. Measured phase noise vs. RMS input voltage level at amplifier (10MHz, Tc = 6ms, 120MS/s, 10 samples averaged).

4.3. Phase drift

The phase drift was measured over 6 hours and is shown in figure 3 together with the phase noise. The measured phase fluctuated considerably over the period and the maximum difference was observed to be 119m°.

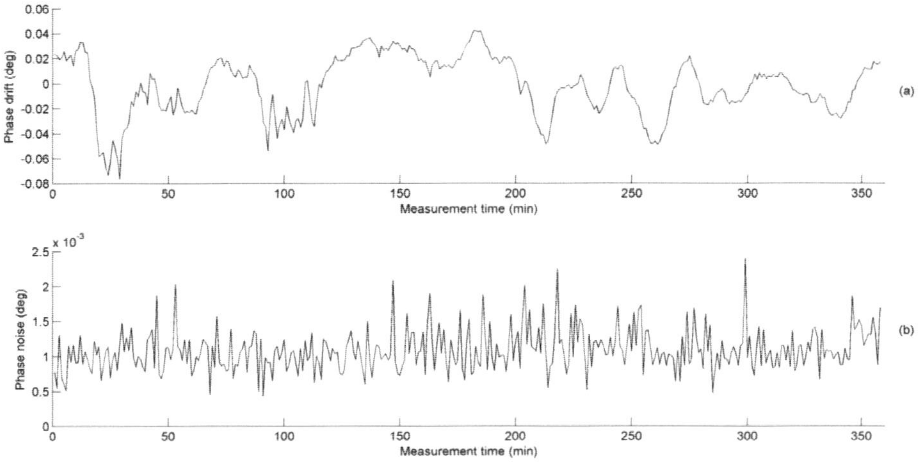

Figure 3. Phase drift over 6hrs (a) and associated phase noise (b) (10MHz, Tc = 6ms, 10 measurements averaged).

5. Discussion

The direct digitizing signal measurement module described achieves a phase noise as low as 7.5m° with a time constant of 6ms at 10MHz. The drift was considerable and fluctuated largely. The phase noise during the drift measurement was however consistently below 2.5m°. The NI PXI system reported by Wee et al. [5] achieved in comparison sub-milidegree phase noise and 3m° phase drift over 6 hours using a longer time constant (17ms instead of 6ms) and a variable gain amplifier (part of the NI PXI system). The low phase noise of the presented system during the phase drift measurements suggests that improvements can be expected through the optimization of the amplifier design.

Acknowledgement

This work is supported by grants from the Engineering and Physical Sciences Research Council, references EP/E009832/1, EP/E009158/1 and EP/E009697/1.

References

[1] Griffiths H 2001 Magnetic induction tomography *Measurement Science & Technology* **12** 1126 – 31

[2] Zolgharni M, Ledger P, Armitage D W, Griffiths H and Holder D S 2008 Detection of haemorrhagic cerebral stroke by magnetic induction tomography: FE and TLM numerical modelling. In: *2008 Electrical Impedance Tomography Conference,* (Dartmouth, USA)

[3] Watson S, Williams R J, Gough W and Griffiths H 2008 A magnetic induction tomography system for samples with conductivities below 10 S m-1 *Measurement Science and Technology*

[4] Scharfetter H, Kostinger A and Issa S 2007 Spectroscopic 16 channel magnetic induction tomography: The new Graz MIT system. In: *ICEBI 2007,* (Graz, Austria: Springer Berlin Heidelberg) pp 452 - 5

[5] Wee H C, Watson S, Patz R, Griffiths H and Williams R J 2008 A Magnetic Induction Tomography system with sub-millidegree phase noise and high long-term stability. In: *4th European Conference of the International Federation for Medical and Biological Engineering,* (Antwerp, Belgium: Springer Berlin Heidelberg) pp 744 - 7

[6] Patz R, Watson S, Ktistis C, Hamsch M, Williams R and Peyton A 2009 An FPGA-based signal measurement module for MIT. In: *10th International Conference on Biomedical EIT,* (Manchester, UK)

Journal of Physics: Conference Series **224** (2010) 012018 doi:10.1088/1742-6596/224/1/012018

Chemical Shift Artifact Correction in MREIT using Iterative Least Square Estimation Method

Z J Meng, A S Minhas, Y Q Han, W C Jeong, Y T Kim, H J Kim and E J Woo

Department of Biomedical Engineering, Kyung Hee University, Korea

E-mail: `ejwoo@khu.ac.kr`

Abstract. Human and animal imaging in magnetic resonance electrical impedance tomography (MREIT) demands high signal-to-noise ratio (SNR) data. We therefore perform MREIT experiments with a higher bandwidth per pixel. This leads to bigger chemical shift artifacts in MR images from fat regions. We may correct such artifacts in MREIT using a recently proposed method based on the three-point Dixon technique. This method is however not suitable for fast imaging pulse sequences. It has a poor SNR and also sometimes leads to swapping of water and fat signals in certain pixels when the field inhomogeneity phase unwrapping algorithm fails. This work demonstrates a new chemical shift artifact correction method in MREIT using a least square estimation method. Iterative separations of water and fat complex images obviate the phase unwrapping step. We present the separated water and fat images using the conventional and also the least square method. These two algorithms are compared in terms of the SNR and their water-fat separation capability. We propose the new method for future studies of fast MREIT imaging experiments.

1. Introduction

Magnetic resonance electrical impedance tomography (MREIT) enables us to perform high-resolution conductivity imaging of body tissues. MREIT data has to be collected with a high signal-to-noise ratio (SNR) for better accuracy of reconstructed conductivity images. This necessitates MR data acquisition with a high bandwidth per pixel. However, such setting may produce bigger chemical shift artifacts in MR images when we image animal or human subjects containing fatty tissues. The artifact creates misalignment of certain pixels in a fatty region. Reconstructed conductivity images in such regions are noisy and inaccurate.

Recently, few researchers proposed chemical shift artifact correction method in MREIT based on the three-point Dixon technique [1, 2]. The basic idea was to separate an MRI image into water and fat images, correct pixel misalignments in the fat image and then add the water and corrected fat images to produce an artifact-free MR image data. Reconstructed conductivity images from such corrected data are free from artifacts. However, using this method to separate water-fat images may lead to water-fat swapping in certain pixels if the unwrapping of the field inhomogeneity phase fails at those pixels. Besides, it is not suitable for fast imaging pulse sequences due to its longer time shift requirements to acquire Dixon's data [3].

In this work, we demonstrate the performance of a recently proposed iterative least square estimation method for water-fat separation and subsequently chemical shift artifact correction in MREIT. The algorithm obviates the phase unwrapping step by using an iterative algorithm with a parametric estimation of the phase due to field inhomogeneity. The separated water-fat images have a higher SNR. This method is suitable for fast pulse sequences like balanced steady state free precession (b-SSFP), which Minhas *et al* proposed to be useful in MREIT [4]. In this paper, we compare corrected images using the least square method and the Dixon method in terms of SNR and swapping of water-fat pixels. Future MREIT experiments using a fast imaging pulse sequence would utilize the proposed technique for chemical shift artifact corrections.

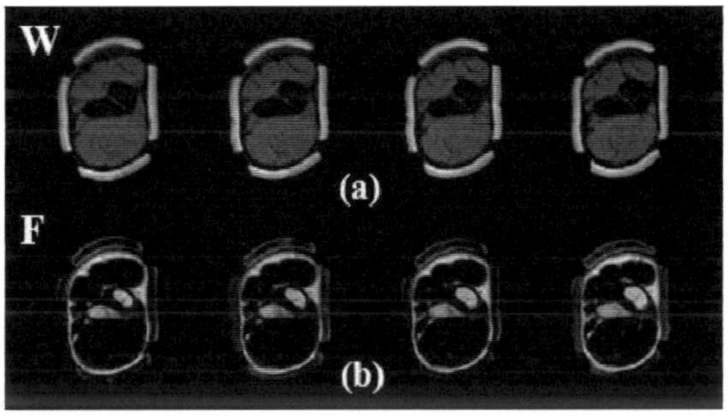

Figure 1. Multi-slice separated magnitude images of (a) water W and (b) fat F using the three-point Dixon method.

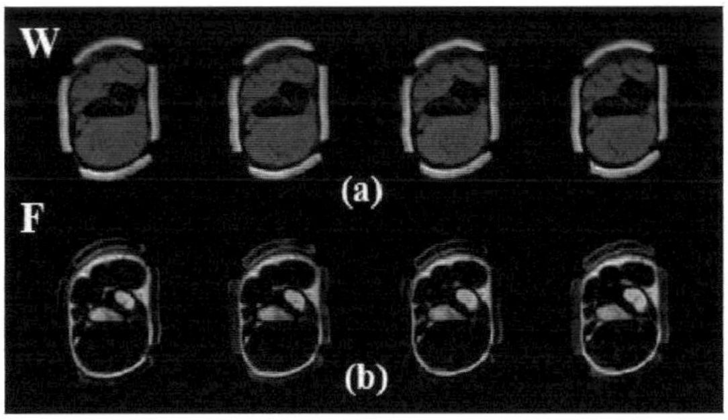

Figure 2. Multi-slice separated magnitude images of (a) water W and (b) fat F using the iterative least square method.

2. Methods

2.1. Water-fat separation using three Dixon data sets

We assume an imaging object containing a fatty region. The MR signal at every pixel is assumed to be a linear combination of water and fat signals. Three k-space Dixon data sets with the read gradient shift of $\tau = (0, +1.14, -1.14)\ ms$ were collected. This makes a phase shift of $\phi = (0, +\pi, -\pi)$ between water and fat signals. Using conventional post-processing method proposed by Glover *et al* [3], we estimate the field inhomogeneity phase from these data sets and cancels its effect from each data set. A simple addition and substraction of these new data sets provide us with separated water and fat images.

Another approach to post-process the Dixon data sets is the iterative least squares method proposed by Reeder *et al* [5]. Here, the water W and fat F signals are assumed to be complex and have a chemical shift of $\Delta f_{wf}\ (Hz)$ between them. The signal acquired at an echo time t_n is then given by,

$$S_n = (W + Fe^{j2\pi\Delta f_{wf}t_n})e^{j2\pi\psi t_n} \qquad (1)$$

where ψ is the field inhomogeneity induced phase. An iterative least square algorithm is used to estimate three parameters W, F and ψ at each pixel from n data sets collected at different echo times t_n. The data sets need not necessarily have the phase shift of $\pm m\pi, (m = 0, 1...)$ between them.

2.2. Imaging experiment

We used a swine leg from a local grocery store as an imaging object with fatty tissues. Four carbon-hydrogel electrodes with adhesive ($60 \times 60 \times 5$ mm^3) were attached around the leg for current injections. Current of 10 mA was sequentially injected in two mutually orthogonal directions. Using our 3 T MRI scanner (Medinus, Korea), we scanned the leg with a spin echo based multi-echo data acquisition method. We collected three Dixon data sets with read gradient shifts of $\tau = (0, +1.14, -1.14)$ ms with respect to the echo time TE. Eight slices in the central region were scanned with field-of-view of 150×150 mm^2, slice thickness of 4 mm, image size of 128×128 and an averaging of 10 cycles.

2.3. Chemical shift artifact correction

We set the bandwidth per pixel as 62.5 kHz, which produced one pixel shift in fatty tissues. We separated the water-fat images using the two techniques described in section 2.2. The fat images were shifted by one pixel from the bottom to top (frequency-encoding direction). These shifted fat images and water images were added to produce corrected data [1]. The z-component of current injection induced magnetic flux density B_z was then obtained from this corrected data.

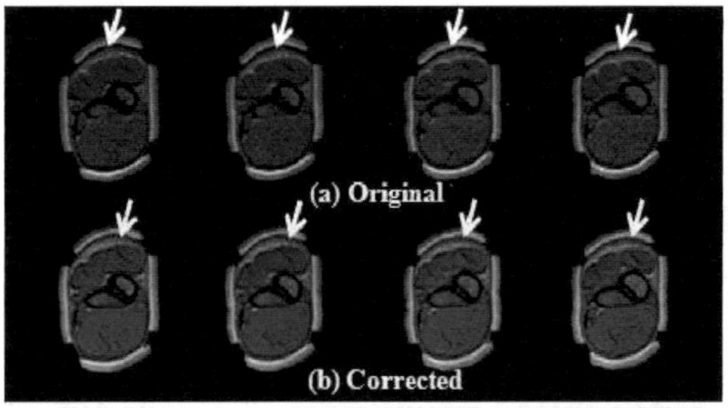

Figure 3. Multi-slice magnitude images (a) before and (b) after corrections using the three-point Dixon method.

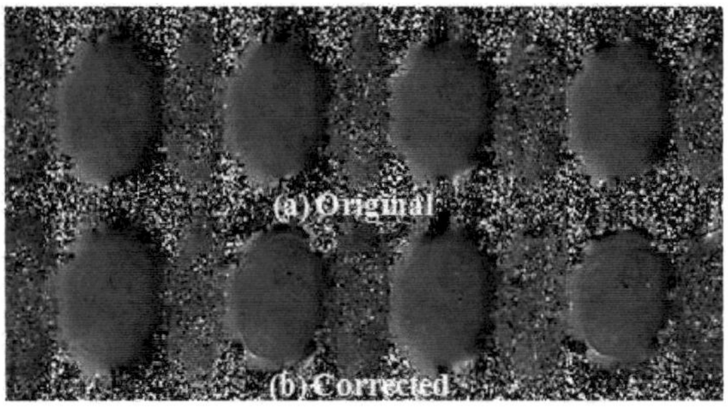

Figure 4. Multi-slice magnetic flux density B_z images for vertical current injection (a) before and (b) after corrections using the three-point Dixon method.

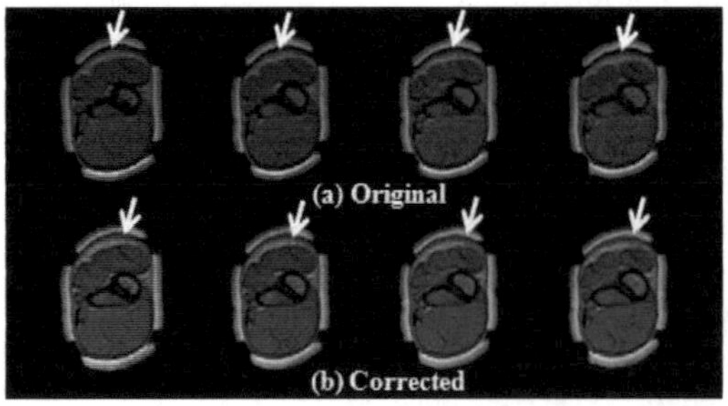

Figure 5. Multi-slice magnitude images (a) before and (b) after corrections using the iterative least square method.

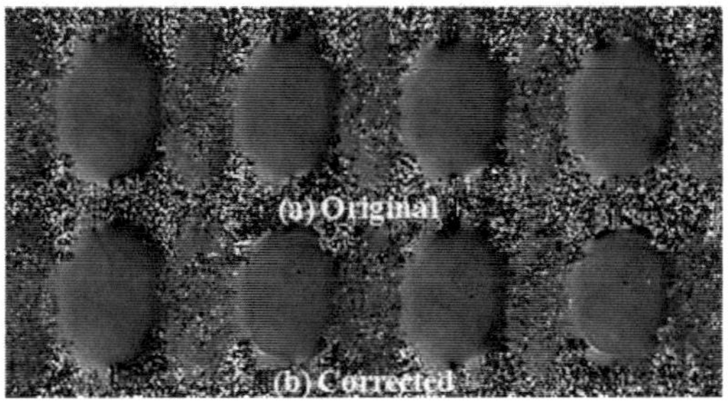

Figure 6. Multi-slice magnetic flux density B_z images for vertical current injection (a) before and (b) after corrections using the iterative least square method.

3. Results

Figure 1 and 2 show the results of separated water W and fat F images using the three-point Dixon method and the least square method, respectively. Both methods could separate water and fat images in a similar way. Figure 3 shows the original and corrected magnitude images in upper and lower rows, respectively using the three-point Dixon method. Figure 4 shows the corresponding phase images. Arrows in the upper row of figure 3 indicate the position of chemical shift near electrodes. Arrows in the lower row indicate corrections. Chemical shift is also observed in internal fatty regions of the swine leg. Corrected magnitude and phase images show better image contrast. Similar results for the least square method are shown in figures 5 and 6. With a quantitative measurement over a small region, a 10-15 % SNR improvement over the three-point Dixon method was observed in magnitude images using the least square method.

4. Discussion and Conclusion

We performed a comparative study of the three-point Dixon method and the least square method in MREIT for chemical shift artifact corrections. The results show that for a water-fat phase shift of $\pm\pi$, two methods give similar results in terms of the separation quality of water-fat images. However, the least square method provides a better SNR. Since we could not separate water and fat images properly using the least square method when the data was noisy, we speculate that it is sensitive to noise. This may be attributed to the failure in estimation of a proper field inhomogeneity phase. An improved field inhomogeneity phase estimation method suggested by Yu *et al* [6] may provide better results. We suggest using the least square method to correct chemical shift artifact in future MREIT studies using a fast pulse sequence.

Acknowledgments

This work was supported by the SRC/ERC program (R11-2002-103) of MEST/NRF.

References

[1] Minhas A S, Kim Y T, Jeong W C, Kim H J, Lee S Y and Woo E J 2009 Chemical shift artifact correction in MREIT *J. Biomed. Eng. Res.* **30** 461-8

[2] Hamamura M J, Nalcioglu O and Muftuler L T 2009 Correction of chemical shift artifacts in magnetic resonance electrical impedance tomography *Proc. Intl. Soc. Mag. Reson. Med. Conf., Honolulu, Hawai, USA* April 18-24 **17** 4658

[3] Glover G H and Schneider E 1991 Three-point dixon technique for true water/fat decomposition with B_0 inhomogeneity correction *Mag. Res. Med.* **18** 371-83

[4] Minhas A S, Woo E J and Lee S Y 2009 Magnetic flux density measurement with balanced steady state free precession pulse sequence for MREIT: a simulation study *Proc. IEEE EMBC09 (Minneapolis, MN, USA)* 2276-8

[5] Reeder S B, Wen Z, Yu H, Pineda A R, Gold G E, Markl M and Pelc N J 2004 Multicoil Dixon chemical species separation with an iterative Least-Squares estimation method *Mag. Res. Med.* **51** 35-45

[6] Yu H, Reeder S B, Shimakawa A, Brittain J H and Pelc N J 2005 Field map estimation with a region growing scheme for iterative 3-point water-fat decomposition *Mag. Res. Med.* **54** 1032-9

Simulation of MREIT using Balanced Steady State Free Precession (b-SSFP) Pulse Sequence

A S Minhas[1], E J Woo[1] and R Sadleir[2]

[1] Department of Biomedical Engineering, Kyung Hee University, Korea
[2] Department of Biomedical Engineering, University of Florida, Gainesville, FL, USA

E-mail: `ejwoo@khu.ac.kr`

Abstract. Magnetic resonance electrical impedance tomography (MREIT) utilizes the relation between conductivity and magnetic flux density induced by externally injected current to perform conductivity imaging of body tissues. A spin echo pulse sequence has been predominantly used in MREIT to acquire the z-component B_z of the induced magnetic flux density data from MR phase images. Spin echo based MREIT pulse sequences are most stable and successful in producing high-resolution conductivity images in postmortem and *in vivo* animal and human experiments. In some applications, localization of a physiological event is desirable. Examples may include detection of neural activities through conductivity changes. In such a case, it would be necessary to maximize the sensitivity. In this paper, we suggest using a balanced steady state free precession (b-SSFP) pulse sequence to localize a small conductivity change. The induced magnetic flux density B_z subject to an injection current makes an off-resonance phase in b-SSFP signals. We expect the high sensitivity of b-SSFP signals to any off-resonance phase change will be advantageous for detecting a small conductivity change. Using computer simulations, we show the feasibility of functional or time-difference MREIT using the b-SSFP pulse sequence.

1. Introduction

We perform conductivity imaging of an electrically conducting object in magnetic resonance electrical impedance tomography (MREIT). Injecting current into the object, it produces a distribution of an internal magnetic flux density $\mathbf{B} = (B_x, B_y, B_z)$. Using an MRI scanner with its main field in the z direction, we can obtain an image of B_z from MR phase data. The current injection must be done in synchronous with an adopted pulse sequence to properly accumulate B_z data in MR phase images. We can reconstruct a conductivity image of the object from multiple B_z data subject to multiple injection currents at different directions based on the relation between conductivity and B_z.

Studies in neurophysiology have detected about 2 to 5 % changes in conductivity values of neural tissues during neuronal activities [1]. The primary goal of this paper is to test the feasibility of detecting such a small conductivity change using MREIT. It would require fast data acquisitions and a high signal sensitivity to current injection. Balanced steady state free precession (b-SSFP) pulse sequence has been known for its high signal-to-noise ratio (SNR) per unit time and high sensitivity to any off-resonance phase. We have recently proposed its use in MREIT as an alternative to conventional spin echo based pulse sequences [2]. The phase change due to the induced B_z signal subject to a current injection during a time interval of T_c

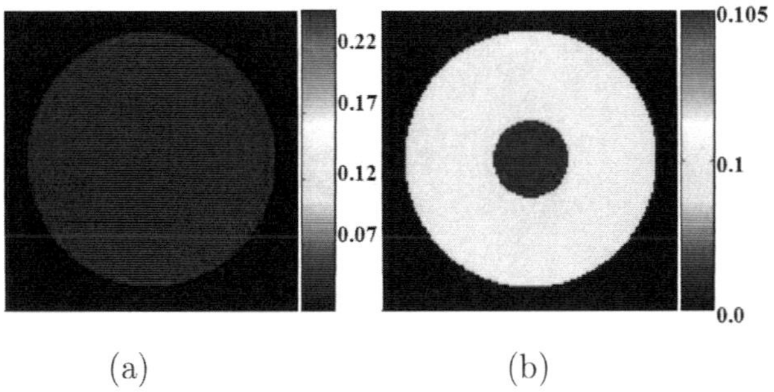

Figure 1. Cross-sections of a cylindrical domain with an anomaly at the center. (a) True MR magnitude image assuming the same proton density for both the anomaly and background. (b) True conductivity image with 5% conductivity contrast between the anomaly and background.

contributes to an extra off-resonance phase in b-SSFP signals. Since this will allow higher signal changes, we propose the b-SSFP in time-difference or functional MREIT.

In this study, we performed numerical simulations of the b-SSFP to test its feasibility in MREIT to detect 5% conductivity contrast. In previous MREIT studies, we have used only MR phase images to extract B_z data. Adopting the b-SSFP pulse sequence, we will investigate both MR magnitude and phase images since induced B_z data affect both of them in b-SSFP. We will propose utilizing the b-SSFP pulse sequence to further explore the feasibility of time-difference or functional MREIT as a new method in neuroimaging.

2. Methods

2.1. b-SSFP Signals

Right after a 90° RF pulse in the b-SSFP pulse sequence, x and y components of a transverse magnetization are given by

$$M_y(\theta) = M_0(1 - E_1)\sin\alpha(1 - E_2\cos\theta)/d \tag{1}$$

$$M_x(\theta) = M_0(1 - E_1)\sin\alpha E_2 \sin\theta/d \tag{2}$$

$$d = (1 - E_1\cos\alpha)(1 - E_2\cos\theta) - E_2(E_1 - \cos\alpha)(E_2 - \cos\theta) \tag{3}$$

where $E_1 = e^{-T_R/T_1}$ and $E_2 = e^{-T_R/T_2}$ [3], T_R is the repetition time, θ the off-resonance phase, M_0 the proton density and α is the flip angle. If we position the data acquisition window at the center of the pulsing period, the phase of the b-SSFP signal is given by

$$\phi(\theta) = \tan^{-1}(M_x/M_y) = \tan^{-1}\left(\frac{E_2\sin\theta}{1 - E_2\cos\theta}\right) + \frac{\theta}{2} \tag{4}$$

where $\theta = \gamma\Delta BT_R$. Injecting current for a duration of T_c in a pulsing period generates a z-component of a magnetic flux density B_z, which contributes to the off-resonance phase θ as

$$\theta = \gamma\Delta BT_R + \int_0^{T_c}\gamma B_z dt = \gamma\Delta BT_R + \gamma B_z T_c \tag{5}$$

where ΔB is the main field inhomogeneity and γ is the gyromagnetic ratio of hydrogen.

2.2. Numerical Simulation

We constructed a three-dimensional cylinder model with 120 mm diameter and 100 mm height and placed a cylindrical anomaly of 36 mm diameter and 30 mm height at its center. The anomaly and background conductivities were 0.105 S/m and 0.1 S/m respectively to simulate 5% conductivity contrast. Figure 1 shows the MR magnitude (for $\theta = 0$) and conductivity image of the cylindrical model with the anomaly. We attached four equally spaced electrodes

Journal of Physics: Conference Series **224** (2010) 012019 doi:10.1088/1742-6596/224/1/012019

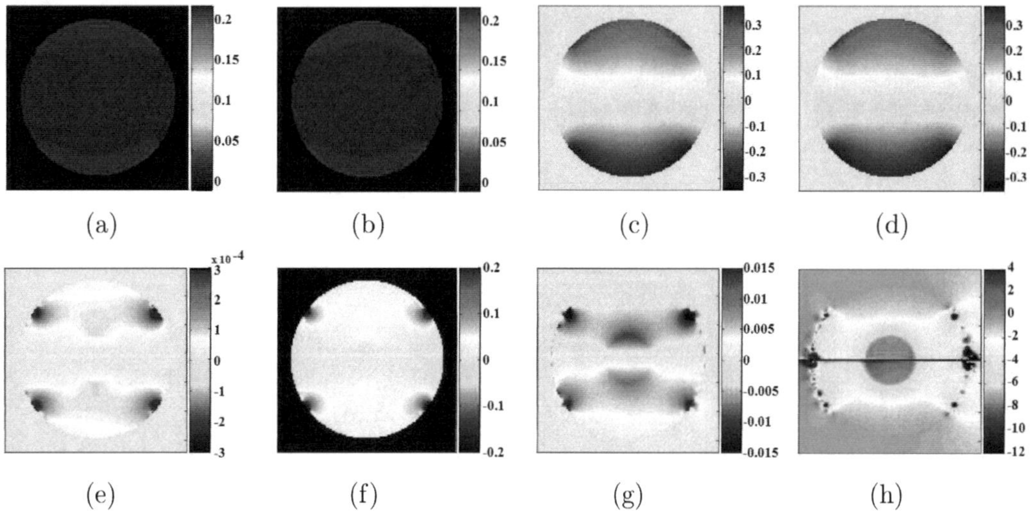

Figure 2. Simulated images using the b-SSFP for the horizontal current injection. (a) and (b) are MR magnitude images without and with the anomaly, respectively. (c) and (d) are corresponding MR phase images. (e) is the difference between (b) and (a). (f) is a scaled version of (e) to express percentile changes. (g) is the difference between (c) and (d). (h) is a scaled version of (g) to express percentile changes.

of size $80 \times 80 \times 5$ mm^3 around the cylindrical model. Imaging currents of 5 mA amplitude were injected horizontally and vertically between two opposing pairs of electrodes. The average current density underneath each electrode was 0.78 A/m^2. We first computed B_z images over a grid of 128×128 for both imaging currents using the three-dimensional MREIT forward solver implemented in COMSOL (Comsol AB, Burlington MA, USA). MR phase changes determined by these B_z data were plugged into the b-SSFP equations as the off-resonance phase term. Using the b-SSPF equations for MREIT described in section 2.1, we computed both MR magnitude and phase images. We assumed that $T_R = 70$ ms and $T_c = 63$ ms. We validated the numerical simulation method by checking various cases of known results.

3. Results

Figure 2 shows the results for the horizontal current injection in the middle slice of the cylindrical domain. MR magnitude images in figure 2(a) and (b) are from the model without and with the anomaly, respectively. These b-SSFP images are different from the true MR magnitude image in figure 1(a) since they include the effects of the off-resonance phase due to the injection current. Figure 2(e) shows the difference between two MR magnitude images in (b) and (a). Figure 2(f) is the same difference image expressed in percentile changes. We can see that the maximal change in the MR magnitude image is about 0.2% near edges of current injection electrodes where we have the largest phase change. Around the anomaly, the magnitude image change is about 0.05%. We note that the horizontal imaging current produces negligible changes in the MR magnitude image in a horizontal band at the center. Figure 2 (c) and (d) show two phase images in the middle imaging slice without and with the anomaly, respectively. Figure 2 (g) is the difference image between (d) and (c). Figure 2 (h) is the same difference image expressed in percentile changes. We found that the change in MR phase image around the anomaly region is about 8% or 0.46°. We note that the horizontal imaging current visualizes distinct phase changes along the horizontal (tangential) direction of the anomaly. Figure 3 shows the numerical simulation results for the case of vertical imaging current with similar interpretation.

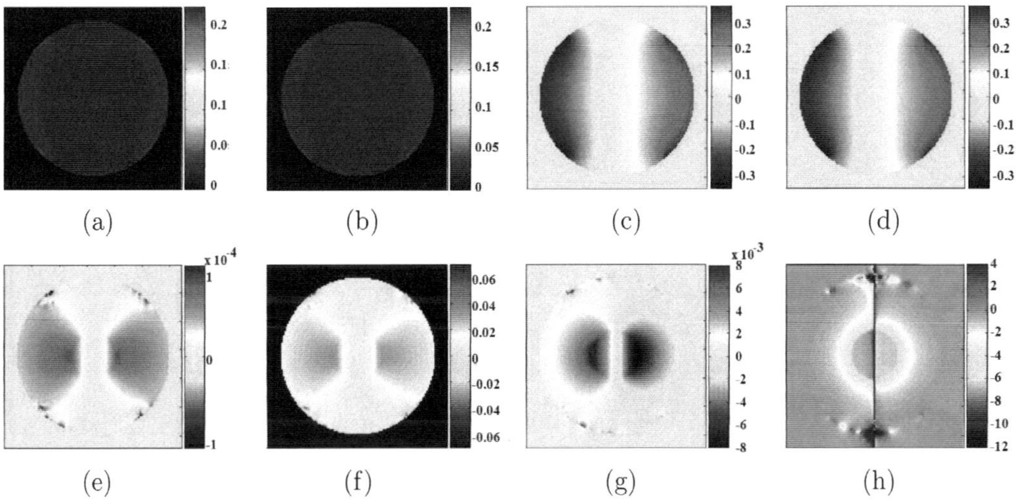

Figure 3. Simulated images using the b-SSFP for the vertical current injection. (a) and (b) are MR magnitude images without and with the anomaly, respectively. (c) and (d) are corresponding MR phase images. (e) is the difference between (b) and (a). (f) is a scaled version of (e) to express percentile changes. (g) is the difference between (c) and (d). (h) is a scaled version of (g) to express percentile changes.

4. Discussion and Conclusion

The edge effects appear prominent in figures 2 and 3. We may significantly reduce the edge effects since it is possible to design an electrode with almost uniform current density underneath it [4]. We may also reduce numerical errors around surface electrodes by using a finer mesh. By properly choosing the electrode size, the average current density of 0.78 A/m^2 underneath each electrode was kept lower than the estimated threshold value of 1.2 A/m^2 to stimulate a neuron. The anomaly is visible in a better way when we scaled the changes from absolute to percentile values. This scaling removes the sign in absolute phase changes and compensate effects of spatial variation in phase.

The fundamental requirement in functional MREIT is whether we can detect small conductivity changes with imaging current small enough not to stimulate a neuron. Numerical simulations described in this paper are encouraging for a successful detection of endogenous neural activities by using functional MREIT (fMREIT) or time-difference MREIT (tdMREIT). We propose b-SSFP based pulse sequences to perform functional imaging in MREIT.

Acknowledgments

This work was supported by the SRC/ERC program (R11-2002-103) of MEST/NRF.

References

[1] Meister M, Pine J and Baylor D A 1994 Multi-neuronal signals from the retina: acquisition and analysis *J. Neurosci. Meth.* **51** 95-106

[2] Minhas A S, Woo E J and Lee S Y 2009 Magnetic flux density measurement with balanced steady state free precession pulse sequence for MREIT: computer simulation study *Proc. IEEE EMBC09 (Minneapolis, MN, USA)* 2276-8

[3] Zur Y, Stokar S and Bendel B 1988 An analysis of fast imaging sequences with steady-state transverse magnetization refocusing *Mag. Reson. Med.* **6** 175-193

[4] Ksienski D A 1992 A minimum profile uniform current density electrode *IEEE Trans. Biomed. Eng.* **39** 682-92

In vivo conductivity imaging of canine male pelvis using a 3T MREIT system

H J Kim[1], W C Jeong[1], Y T Kim[1], A S Minhas[1], T H Lee[1], C Y Lim[2], H M Park[2], J K Seo[3] and E J Woo[1]

[1] Department of Biomedical Engineering, Kyung Hee University, Yongin, Korea
[2] Department of Veterinary Internal Medicine, College of Veterinary Medicine, Konkuk Univeristy, Seoul, Korea
[3] Department of Mathematics, Yonsei University, Seoul, Korea

E-mail: ejwoo@khu.ac.kr

Abstract. The prostate is an imaging area of growing concern related with aging. Prostate cancer and benign prostatic hyperplasia are the most common diseases and significant cause of death for elderly men. Hence, the conductivity imaging of the male pelvis is a challenging task with a clinical significance. In this study, we performed *in vivo* MREIT imaging experiments of the canine male pelvis using a 3T MRI scanner. Adopting carbon-hydrogel electrodes and a multi-echo pulse sequence, we could inject as much as 10 mA current in a form of 51 ms pulse into the pelvis. Collecting magnetic flux density data inside the pelvis subject to multiple injection currents, we reconstructed cross-sectional conductivity images using a MREIT software package CoReHA. Scaled conductivity images of the prostate show a clear contrast between the central and peripheral zones which are related with prostate diseases including cancer and benign prostatic hyperplasia. In our future work, we will focus on prostate cancer model animal experiments.

Keywords: MREIT, conductivity image, canine pelvis, prostate

1. Introduction

Magnetic resonance electrical impedance tomography (MREIT) is a new bio-imaging modality providing cross-sectional conductivity images from measurements of internal magnetic flux densities produced by externally injected currents [1-6]. Recent experimental studies of MREIT have shown its feasibility as a new bio-imaging method with a spatial resolution of a few millimetres [7-9].

Following the recent postmortem and *in vivo* imaging experiments of the canine brain, we have performed imaging experiments of the *in vivo* canine pelvis. The pelvic region with a relatively large field of view (FOV) and various organs imposes a new challenge when it is chosen as an imaging domain. Especially, the prostate is a growing concern imaging area related with aging. Prostate cancer and benign prostatic hyperplasia are the most common diseases and significant cause of death for elderly men [10]. This paper reports our first conductivity images of the *in vivo* canine pelvis. Showing MREIT images of the canine pelvis, we will discuss future research directions.

2. Methods

2.1. Animal preparation

To prevent dribbling during experiments, we injected 0.1 mg/kg of atrophine sulfate. Ten minutes later, we anesthetized the dogs with intramuscular injection of 0.2 ml/kg Tiletamine and Zolazepam (Zoletil 50, Virbac, France). Twenty minutes later, we clipped hair at four locations on the pelvis where we attached four carbon-hydrogel electrodes. With four electrodes attached, we positioned the dog inside the bore of our 3 T MRI scanner (Magnum 3, Medinus, Korea) as in figure 1. The experimental protocol was approved by the Institutional Animal Care and Use Committee (IACUC) of Konkuk University, Seoul, Korea.

Inside the shield room, we intubated the dog using an endotracheal tube of 8.5 mm diameter and began the general anesthesia using a veterinary anesthesia machine system (VME, MATRX, USA). We used 2% isoflurane mixed with oxygen at 800 ml/min flow rate. Ventilation was machine-controlled by using a ventilator (M-2002, Hallowell EMC, USA) with the respiration rate of 15 bpm and tidal volume of 200 ml. Using our custom-designed MREIT current source, we injected currents through different pairs of recessed electrodes to confirm the electrode contacts.

2.2. Imaging experiment

We injected the first current I_1 between one opposing pair of electrodes. The multi-echo pulse sequence was used with *TR*/*TE*=900/20 ms. The number of averaging was 24 and the injection current amplitude was 10 mA with the total pulse width of 51 ms. The slice thickness was 4 mm with no slice gap and the number of slices was 8 producing an imaging region with 32 mm thickness centered at the electrode plane. The image matrix size was 128×128 with the field-of-view of 180 ×180 mm^2 and the pixel size was 1.406×1.406 mm^2. After acquiring the first magnetic flux density (B_z) data set for I_1 in 8 axial slices, the second injection current I_2 with the same amplitude and pulse width was injected through the other pair of opposing electrodes to obtain the second data set.

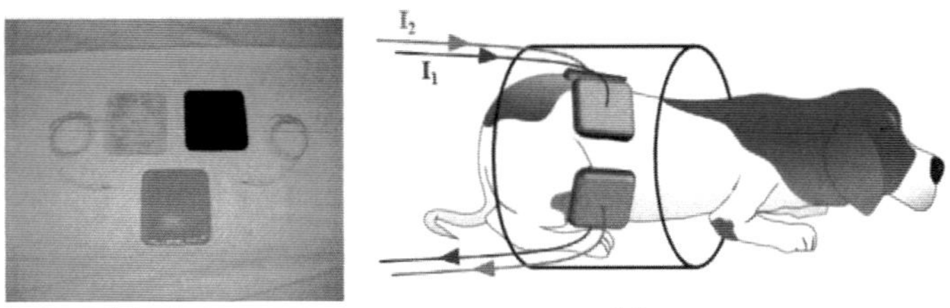

(a) (b)

Figure 1. (a) Carbon-hydrogel electrode and (b) imaging setup inside the bore.

2.3. Conductivity image reconstruction

Jeon *et al.* lately developed an MREIT software package with an efficient graphical user interface [11]. They called it CoReHA which stands for conductivity reconstructor using harmonic algorithms. Since CoReHA automated most parts of data processing tasks in MREIT, we utilized it to process *in vivo* canine pelvis experiment data and produce conductivity images. We used the harmonic B_z and local harmonic B_z algorithm for conductivity image reconstructions [12-14].

3. Results

Figure 2(a) is a typical MR magnitude image of a canine pelvis. Figure 2(b) and (c) are measured B_z

images in the same slice subject to current injections in two different directions. Figure 3 shows enlarged images of the canine prostate. Figure 3(a) is an anatomical structure of the prostate and (c) shows a reconstructed conductivity image in comparison with the corresponding MR magnitude image in (b). Conductivity images of the prostate showed clear contrast between central and peripheral zones which are related with prostate diseases including cancer and benign prostatic hyperplasia.

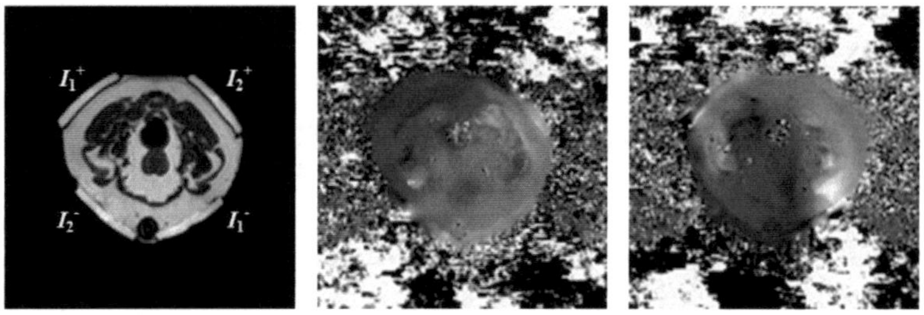

(a) (b) (c)

Figure 2. (a) Typical MR magnitude image of the canine pelvis. (b) and (c) are magnetic flux density images subject to the horizontal and vertical injection currents, respectively.

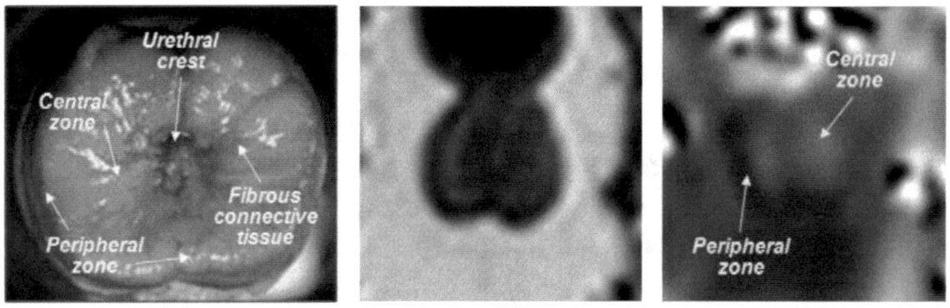

(a) (b) (c)

Figure 3. Magnified images of the prostate: (a) anatomical structure, (b) MR magnitude image and (c) reconstructed conductivity image.

4. Discussion

The MREIT technique has to be advanced beyond the stage of showing its feasibility. To support its clinical significance, we should demonstrate that the conductivity image provides meaningful diagnostic information that is not available from other imaging modalities. This requires accumulated experience and knowledge on how to interpret a conductivity image in relation with anatomy and pathology of a specific organ. Choosing the pelvis as an imaging region, we tried to provide such an example in this paper.

The prostate is a growing concern imaging area related with aging. Prostate cancer and benign prostatic hyperplasia are most common diseases and significant causes of death for elderly men. Conductivity imaging of the pelvis region is a challenging task. Its size is large to result in a reduced current density in some local regions. It contains various organs with different proton densities and there are local regions with low signal-to-noise ratios (SNR). Since the amount of noise in measured B_z data is inversely proportional to the MR magnitude image SNR, reconstructed conductivity images in such local regions are noisy. To handle this kind of technical problem, we may apply a denoising algorithm, inpainting method and/or the local harmonic B_z algorithm [6]. Based on our results, we can see conductivity contrast between central and peripheral zones, which are related with prostate diseases including cancer and benign prostatic hyperplasia. Considering clinical applications, future work should be focused on disease model animal experiments.

Acknowledgments

This work was supported by the SRC/ERC program (R11-2002-103) of MEST/NRF.

References

[1] Scott G C, Joy M L G, Armstrong R L and Henkelman R M 1991 Measurement of nonuniform current density by magnetic resonance *IEEE Trans. Med. Imaging* **10** 362–74
[2] Zhang N 1992 Electrical impedance tomography based on current density imaging *MS Thesis* University of Toronto
[3] Woo E J, Lee S Y and Mun C W 1994 Impedance tomography using internal current density distribution measured by nuclear magnetic resonance *Proc. SPIE* **2299** 377–85
[4] Ider Y Z and Birgul O 1998 Use of the magnetic field generated by the internal distribution of injected currents for electrical impedance tomography (MR-EIT) *Elektrik* **6** 215–25
[5] Kwon O, Woo E J, Yoon J R and Seo J K 2002 Magnetic resonance electrical impedance tomography (MREIT): simulation study of J-substitution algorithm *IEEE Trans. Biomed. Eng.* **49** 160–7
[6] Woo E J and Seo J K 2008 Magnetic resonance electrical impedance tomography (MREIT) for high-resolution conductivity imaging *Physiol. Meas.* **29** R1–R26
[7] Kim H J, Lee B I, Cho Y, Kim Y T, Kang B T, Park H M, Lee S Y, Seo J K andWoo E J 2007 Conductivity imaging of canine brain using a 3 T MREIT system: postmortem experiments *Physiol. Meas.* **28** 1341–53
[8] Kim H J, Oh T I, Kim Y T, Lee B I, Woo E J, Seo J K, Lee S Y, Kwon O, Park C, Kang B T and Park H M 2008 In vivo electrical conductivity imaging of a canine brain using a 3 T MREIT system *Physiol. Meas.* **29** 1145–55
[9] Jeon K, Minhas A S, Kim Y T, Jeong W C, Kim H J, Kang B T, Park H M, Lee C O, Seo J K and Woo E J 2009 MREIT conductivity imaging of postmortem canine abdomen using CoReHA *Physiol. Meas.* **30** 957–66
[10] Cohen R J, Shannon B A, Phillips M, Moorin R E, Wheeler T M and Garrett K L 2008 Central zone carcinoma of the prostate gland: a distinct tumor type with poor prognostic features *J. Urology* **179** 1762–7
[11] Jeon K, Kim H J, Lee C O, Woo E J and Seo J K 2009 CoReHA: conductivity reconstructor using harmonic algorithms for magnetic resonance electrical impedance tomography (MREIT) *J. Biomed. Eng. Res.* **30** 279–87
[12] Oh S H, Lee B I, Woo E J , Lee S Y, Cho M H, Kwon O and Seo J K 2003 Conductivity and current density image reconstruction using harmonic B_z algorithm in magnetic resonance electrical impedance tomography *Phys.Med. Biol.* **48** 3101–16
[13] Seo J K, Yoon J R,Woo E J and Kwon O 2003 Reconstruction of conductivity and current density images using only one component of magnetic field measurements *IEEE Trans. Biomed. Eng.* **50** 1121–4
[14] Seo J K, Kim S W, Kim S, Liu J J, Woo E J, Jeon K and Lee C O 2008 Local harmonic B_z algorithm with domain decomposition in MREIT: computer simulation study *IEEE Trans. Med. Imaging* **27** 1754–61

Journal of Physics: Conference Series **224** (2010) 012021 doi:10.1088/1742-6596/224/1/012021

Correcting for variability in mesh geometry in Finite Element Models

Andy Adler[1], and William R.B. Lionheart[2]
[1]Systems and Computer Engineering, Carleton University, Ottawa, Canada
[2]School of Mathematics, University of Manchester, UK

Abstract: We show that variability in the finite element model (FEM) geometry in the EIT forward model can result in image artefacts in the reconstructed images. We explain these artefacts as the result of changes in the projection of the anisotropic conductivity tensor onto the FEM system matrix. This introduces anisotropic components into the simulated voltages, which cannot be reconstructed onto an isotropic image, and appear as artefacts. In order to address this problem, we show that it is possible to incorporate a FEM vertex movement component into the formulation of the inverse problem. These results suggest that it may be important to consider artefacts due to FE mesh geometry in the formulation of EIT image reconstruction.

1 Introduction

In this paper, we describe the image reconstruction artefacts which occur in electrical impedance tomography (EIT) images due to limitations in the finite element models, and show an algorithmic approach to limit such effects. The earliest approaches to EIT image reconstruction were based on 2D circular approximations of the thorax[6]. However, since such analytical models cannot describe electrical propagation in realistic body shapes, finite element models were used. Other numerical models, such as those based on finite differences [9] have been used, but are less popular, primarily because FE mesh elements can be easily refined in regions of high electric field, typically near to electrodes. While the FEM literature is rich in terms of variety of model structure, most EIT research has used the simplest FEM implementations. Simplex elements are chosen (triangles in 2D and tetrahedrons in 3D) and conductivity is modelled as piecewise constant (so that changes in conductivity occur only at element boundaries). Such choices are reasonable: all FE meshing packages provide good support for simplex elements, and anatomically realistic conductivity changes do occur abruptly at organ boundaries. The most common models use first order elements. Physically, such elements may be modelled by a resistor network[8], which has been used as the basis for a physical resistor network model of the medium [4]. On the other hand, FEM model errors decrease linearly with element size for first order FEMs, while the rate of error decrease is larger for higher order models.

The most common approach to image reconstruction in EIT has been to parametrize the conductivity distribution, $\mathbf{x}$, based on the FEM element (or vertex) geometry. The conductivity (or conductivity change) distribution, $\hat{\mathbf{x}}$ is calculated from data (or difference data), $\mathbf{y}$ as

$$\hat{\mathbf{x}} = \arg\min_{\mathbf{x}} \|\mathbf{y} - F(\mathbf{x})\|^2_{\boldsymbol{\Sigma}_n^{-1}} + \|\mathbf{x} - \mathbf{x}_0\|^2_{\boldsymbol{\Sigma}_x^{-1}} \quad (1)$$

where $F(\mathbf{x})$ represents the FEM, $\boldsymbol{\Sigma}_n$ the noise covariance, and $\mathbf{x}_0$, and $\boldsymbol{\Sigma}_x$ the prior expected mean and covariance. In this way, reconstructed image is the one which, by varying the FEM conductivity parameters, best fits the data.

In this paper, we report that this approach to image reconstruction is very sensitive to any geometrical variability in the FEM. For example, small changes in the positions of internal nodes, such that the same conductivity distribution is represented, can result in large image reconstruction artefacts. This effect was unexpected, since the voltage distribution simulated by the FEM is accurate even for moderately dense models. This effect may explain a few puzzling results. For example, studies of electrode movement have often used FEM simulations to show that unacceptable artefacts occur for very small electrode displacements (under 1%) (eg. [2]); while, experimentally, electrode movements an order of magnitude greater still permit usable images.

2 Geometry variability example

In this section, we show how re-meshing around a circular object can introduce large artefacts into the reconstructed difference images. The easiest (and most common) way to simulate a target in a medium is to use a single FEM to select and then interpolate which elements are part of the target. There is no change to the underlying FEM, and thus no model noise in the images. A more accurate representation of a target is to create a target region within the FEM and to remesh around it. This means that the mesh changes between each target position, not only near the target, but throu[illegible]

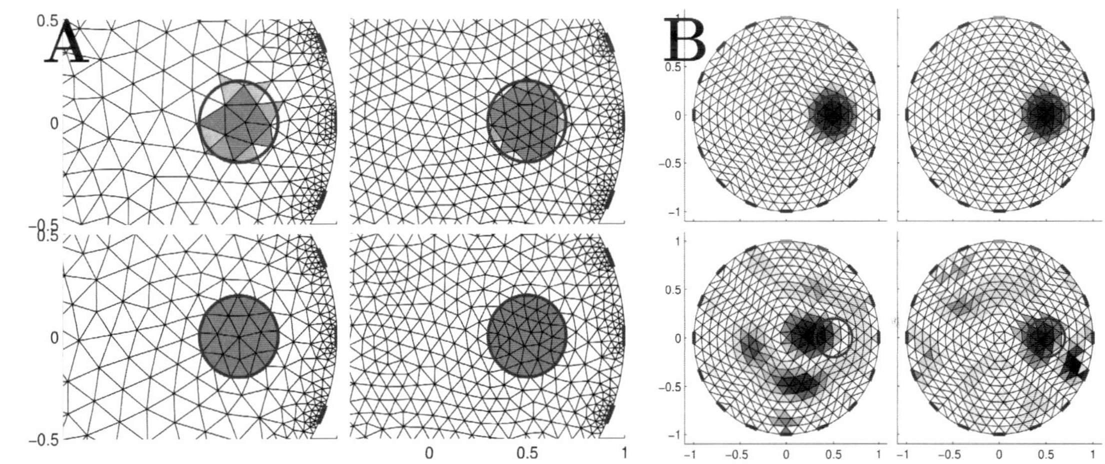

Figure 1: **A:** Simulation FEMs and simulated conductivity target positions in blue circle. Electrode nodes are indicated in green. An inner region surrounding a target is shown. *Left:* Coarse meshes (maximum mesh size of 0.16, 1441 triangles) *Right:* Fine meshes (maximum mesh size of 0.07, 1941 triangles) *Top:* Meshes with no adaptation for target. Element conductivity is defined by region membership. *Bottom:* Meshes adapted to target region. **B:** Reconstructed difference images calculated from the simulation FEMs (A) on a 576 element 2D mesh. In each case, the homogeneous data were simulated on the unadapted mesh. The target position is indicated by the blue circle.

To illustrate this process, Fig. 1 shows 2D circular FEMs with 16 electrodes with local refinement of the FEM near each electrode. Coarse and fine meshes are shown by controlling the maximum permissible element size. Two different strategies to specify the region of a simulated conductivity target region are shown. On top, the mesh is not adapted to the target. The conductivity of each element is selected based on the membership in the target region (an element with 50% of its area in the region will have a target conductivity of the average of the background and target region). On the bottom, the FEM is adapted to the target region, resulting in mesh geometry changes which propagate throughout the FEM. EIT data were simulated using a Sheffield-type adjacent stimulation and measurement protocol, and images are reconstructed on a simple regular mesh geometry. A one-step Gauss

Journal of Physics: Conference Series **224** (2010) 012021 doi:10.1088/1742-6596/224/1/012021

Newton reconstruction is used with a scaled diagonal image prior[3], and the regularization parameter is chosen such that the noise figure is 1.0[1].

Images in which the simulation mesh geometry matched exactly (top) show the expected location and shape. However, if the mesh geometry changes between difference data simulations (bottom) artefacts occur throughout the images. As shown in Fig. 1, reconstruction artefacts reduce as the FEM density increases. This is largely due to the increase in accuracy of the FEM model with decreasing mesh size. Similar artefacts occur for 3D simulations; however, it the effect is considerably larger because many more FEM elements are required to achieve the same level of refinement in 3D compared to 2D.

3 Compensation for image reconstruction errors

We hypothesize that these image reconstruction errors result largely from an effect equivalent to anisotropic conductivity in the simulated EIT measurements. A selection of a particular FEM geometry is equivalent to a particular representation of the (potentially anisotropic) conductivity tensors on each tetrahedron onto the FEM system matrix. Any change in the FEM geometry projects this conductivity tensor differently, resulting in slightly different anisotropic "content" in the simulated voltages. During image reconstruction, such anisotropic content cannot be explained by an inverse model parametrized only for isotropic conductivity. These differences then appear as noise, which is projected as artefacts into the reconstructed image.

Since a first-order FE model is equivalent to a resistor on each FE mesh edge, the maximum number of degrees of freedom in the FE model is the number of edges. Since there are more edges than elements (by approx $\frac{3}{2}$ in 2D), not all FEM matrices correspond to isotropic conductivities, and any distortion of the mesh will (typically) be consistent with an anisotropic conductivity on the original mesh.

This hypothesis suggests that the origin of these image errors is the geometrical constraints of the inverse model. The reconstruction process is not allowed to vary the FEM geometry to "explain" the measurements, but must only use isotropic conductivities on each element. Therefore, we can address this issue by parametrizing both the element conductivities and the positions of each vertex. This allows the image reconstruction to "jiggle" the node locations to avoid pushing artefacts into the conductivity images.

4 Image Reconstruction Formulation

We consider EIT difference measurements, $\mathbf{y} = F(\Delta\boldsymbol{\sigma})$, originating from isotropic conductivity changes $\Delta\boldsymbol{\sigma}$ and a FEM model, $F(\Delta\boldsymbol{\sigma}, \Delta\mathbf{p})$, with vertex positions $\Delta\mathbf{p}$. Based on this model, we calculate a conductivity Jacobian, $\mathbf{J}_c$ – the sensitivity of measurements to conductivity changes; and a movement Jacobian, $\mathbf{J}_m$ – the sensitivity of measurements to vertex position movements.

$$[\mathbf{J}_c]_{ij} = \frac{\partial F_i(\Delta\boldsymbol{\sigma}, \Delta\mathbf{p})}{\partial[\Delta\boldsymbol{\sigma}]_j}, \text{ and } [\mathbf{J}_m]_{ij} = \frac{\partial F_i(\Delta\boldsymbol{\sigma}, \Delta\mathbf{p})}{\partial[\Delta\mathbf{p}]_j}, \tag{2}$$

The standard linear solution, $\hat{\mathbf{x}}$, to (1) is

$$\hat{\mathbf{x}} = \left(\mathbf{J}^t\boldsymbol{\Sigma}_n^{-1}\mathbf{J} + \boldsymbol{\Sigma}_x^{-1}\right)^{-1}\mathbf{J}\boldsymbol{\Sigma}_n^{-1}\mathbf{y} = \boldsymbol{\Sigma}_x\mathbf{J}^t\left(\mathbf{J}\boldsymbol{\Sigma}_x\mathbf{J}^t + \boldsymbol{\Sigma}_n\right)^{-1}\mathbf{y}, \tag{3}$$

assuming $\mathbf{x}_0 = 0$, as normally done for difference EIT.

Normally, the parameters, $\hat{\mathbf{x}}$, represent only the conductivity change $\Delta\boldsymbol{\sigma}$. Instead, we represent our scheme using the conductivity change and vertex movement as parameters and: $\hat{\mathbf{x}} = [\Delta\boldsymbol{\sigma}^t|\Delta\mathbf{p}^t]^t$ and $\mathbf{J} = [\mathbf{J}_c|\mathbf{J}_m]$. In this case, the prior $\boldsymbol{\Sigma}_x$ may be decomposed into a conductivity changes $\boldsymbol{\Sigma}_c$ and movement $\boldsymbol{\Sigma}_m$ parts with no correlation assumed between these parameters (following [7]), yielding:

$$\hat{\mathbf{x}} = [\boldsymbol{\Sigma}_c\mathbf{J}_c|\boldsymbol{\Sigma}_m\mathbf{J}_m]^t \left(\mathbf{J}_c\boldsymbol{\Sigma}_c\mathbf{J}_c^t + \mathbf{J}_m\boldsymbol{\Sigma}_m\mathbf{J}_m^t + \boldsymbol{\Sigma}_n\right)^{-1}\mathbf{y} \tag{4}$$

Since we are only interested in displaying the reconstructed conductivity change, $\Delta\hat{\boldsymbol{\sigma}}$, the standard and proposed reconstruction may be shown:

$$\Delta\hat{\boldsymbol{\sigma}} = \begin{cases} \boldsymbol{\Sigma}_c\mathbf{J}_c^t\left(\mathbf{J}_c\boldsymbol{\Sigma}_c\mathbf{J}_c^t + \boldsymbol{\Sigma}_n\right)^{-1}\mathbf{y} & \text{standard} \\ \boldsymbol{\Sigma}_c\mathbf{J}_c^t\left(\mathbf{J}_c\boldsymbol{\Sigma}_c\mathbf{J}_c^t + \mathbf{J}_m\boldsymbol{\Sigma}_m\mathbf{J}_m^t + \boldsymbol{\Sigma}_n\right)^{-1}\mathbf{y} & \text{proposed} \end{cases} \tag{5}$$

5 Results and Discussion

The standard and proposed image reconstruc … d for the simulated data of Fig. 1, and are shown in … om row) shows a significant reduction in image artef … n of the conductivity target. $\boldsymbol{\Sigma}_m$ was chosen as the s … ude such that th …

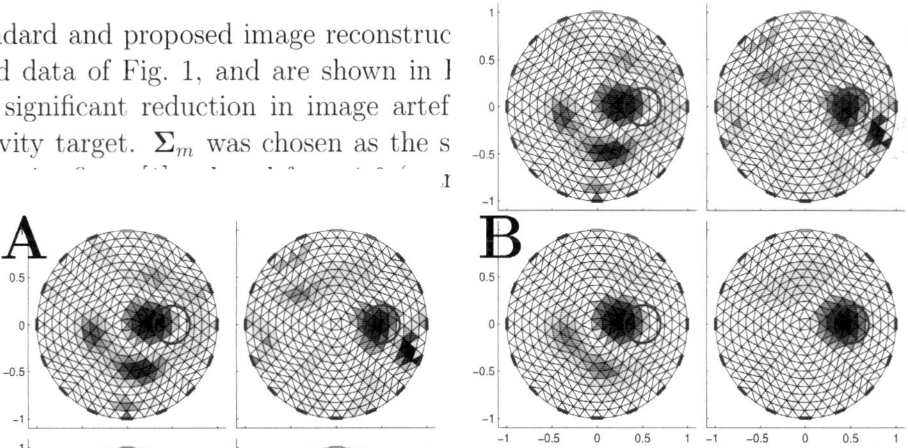

Figure 2: Reconstructed difference images calculated as in Fig. 1 for changes in FEM geometry. The target position is indicated by the blue circle. **A:** Standard approach using *Left:* Coarse Mesh, *Right:* Fine Mesh **B:** Proposed approach using *Left:* Coarse Mesh, *Right:* Fine Mesh.

Based on these results, we note the following observations. First, these results explain the strange effect that EIT simulations from adapted target meshes can show large artefacts. It also explains why simulations of electrode movement show much larger effects than are observed. Second, these results suggest that EIT image reconstruction should allow for anisotropic behaviour in the measured signals. Thus, this work calls for consideration of FE models and the assumption of isotropic conductivity in EIT.

References

[1] Adler A and Guardo R 1996 *IEEE T Med. Imaging* **15** 170-179
[2] Adler A, Guardo R and Berthiaume Y 1996 *IEEE T Biomed. Eng.* **43**(4) 414-20
[3] Cheney M, Isaacson D, Newell J C, *etal* 1990 *Int J Imaging Syst Technol* 2 66–75
[4] Gagnon H, Hartinger A E, Adler A, Guardo R 2008 *Proc. Conf. EIT,*
[5] Murai T, Kagawa Y 1985 *IEEE T Biomed Eng* 32:177-184
[6] Seagar A D Bates R H T 1985 *IEE Proceedings A* 132: 455–466
[7] Soleimani M, Gómez-Laberge C and Adler A 2006 *Physiol Meas* **27** S103–S13
[8] Silvester PP and Ferrari RL, 1990, Cambridge University Press
[9] Yang F Patterson R 2007 *Physiol. Meas.* 28 S153–S163

Finite Elements and Anisotropic EIT reconstruction

William R. B. Lionheart[1],Kyriakos Paridis[1]
[1]School of Mathematics, The University of Manchester, Manchester, UK.
E-mail: bill.lionheart@manchester.ac.uk

Abstract. We study the indeterminacy of the inverse problem for the finite element approximation to the anisotropic inverse problem of EIT

1. Introduction

It is well known that many body tissues such as muscle have an anisotropic conductivity. It is also known that the inverse problem for anisotropic EIT does not have a unique solution even with complete data (arbitrarily many, arbitrarily small electrodes) of arbitrary precision [1]. The usual numerical treatment of EIT uses a finite element (FE) model to solve the forward problem and the conductivity in this model is adjusted to fit the measured data. Extending the usual approach to the isotropic problem we approximate the potential as piecewise linear on tetrahedra elements and it seems natural to represent the anisotropic conductivity as a constant symmetric matrix on each tetrahedron.

There are two pit-falls with this approach. The finite element system matrix can have non zero elements only for pairs of vertex indices corresponding to an edge in the FE mesh. The diagonal elements are simply the negative of the sum of the off diagonal elements for that row or column so the maximum number of degrees of freedom in our FE model is the number of edges. Representing the anisotropic conductivity as a symmetric matrix on each elements gives $6n_t$ degrees of freedom (n_t is number of tetrahedra) – more than n_e the number of edges. This means that even if we knew the FE system matrix (let alone just the boundary data) we still could not uniquely determine the conductivity matrix on each tetrahedron uniquely.

However this is not the non-uniqueness that arises in the continuum problem. Let $\sigma(x) = (\sigma_{ij}(x))$ be an anisotropic conductivity on a domain $\Omega \subset \mathbb{R}^3$, and let $F : \Omega \to \Omega$ be a smooth invertible map (diffeomorphism) with $F(x) = x$ for $x \in \partial\Omega$ then the known non-uniqueness in EIT arises as the Neumann-Dirichlet map (transfer impedance) defined by

$$R_\sigma : \sigma\nabla u \cdot \mathbf{n} \mapsto u|_{\partial\Omega} \tag{1}$$

(where $\nabla \cdot \sigma\nabla u = 0$) satisfies

$$R_\sigma = R_{\tilde{\sigma}} \tag{2}$$

for another conductivity

$$\tilde{\sigma}(F(x)) = \frac{DF(x)\sigma(x)DF(x)}{\det DF(x)} \tag{3}$$

where $DF(x)$ is the Jacobian matrix of F. Using a fixed finite element mesh does not reveal this non-uniqueness in the inverse problem as a smooth map F does not preserve the mesh. So in this obvious approach we have chosen one of the infinite family of anisotropic conductivities consistent with the data simply by our choice of mesh.

In this paper we look at the singular value decomposition of the linearised forward problem (Jacobian) for anisotropic EIT, and we show for a variety of meshes the problem of determining edge conductances exhibits the same ill-conditioning as the isotropic problem, with no further drop in rank of the Jacobian. This indicates that we are as likely to be as successful in recovering edge conductances from EIT data as we are isotropic conductivities on elements. This shifts the problem of anisotropic EIT to finding a mesh, and an assignment of anisotropic conductivity matrices on elements, consistent with *a priori* data.

2. Finite element formulation

The finite element mesh consists of vertices $x_i \in \Omega$ $i = 1, ...n_v$ and tetrahedra T_k, $k = 1, ..., n_t$. the tetrahedra are convex hull of the sets of four distinct vertices, and they intersect at most in faces (that is the convex hull of the three vertices they share). The union $\bigcup_k T_k$ is a polyhedron approximating Ω. The nodal basis functions $\phi_i(x)$ are the piecewise linear functions such that $\phi(x_i) = 1$ and $\phi(x_j) = 0, j \neq i$. We approximate the potential as $u(x) = \sum_i u_i\phi_i(x)$, and assign a positive definite matrix σ^k to each tetrahedron. In this context the finite element system matrix $K \in \mathbb{R}^{n_v \times n_v}$ is given by

$$K_{ij} = \sum_{k:\{x_i,x_j\}\subset T_k} \nabla\phi_i \cdot \sigma^h \nabla\phi_j |T_k| \tag{4}$$

where $|T_k|$ is the volume of the tetrahedron and we note that on each tetrahedron $\nabla\phi_i$ is constant. For a boundary current density $J = \sigma\nabla u \cdot \mathbf{n}$ we define the current vector $I \in \mathbb{R}^{n_v}$ by

$$I_i = \int_{\partial\Omega} J\phi_i \, dx \tag{5}$$

and the FE system is

$$K\underline{u} = I \tag{6}$$

where $\underline{u}$ is the vector of u_i. One additional condition is required for a unique solution as the voltage is only determined up to an additive constant, one way to do this is to choose one ("grounded") vertex i_g and enforce $u_{i_g} = 0$ by deleting the i_g row and column from the system (6). It is clear from (4) that for a pair of vertices indexed by i, j that are not both in any tetrahedron, $K_{i,j} = 0$. For an isotropic conductivity, that is $\sigma^k = \gamma^k\mathbf{I}$ for scalars γ^k and $\mathbf{I}$ the identity matrix, the system (6) is equivalent to Ohm's and Kirchoff's law for a resistor network with distinct vertices labeled by i and j connected by a resistor with conductance

$$K_{ij} = 6 \sum_{k:\{x_i,x_j\}\subset T_k} \gamma_k \cot\theta_{ij}^k L_{ij} \tag{7}$$

(see Appendix) where θ_{ij}^k is the angle between the faces of T_k where they meet the edge *not* containing the vertices indexed by i or j and L_{ij} the length of that edge. Obviously some restrictions on the angles are necessary to ensure the conductances are nonnegative, if they are all acute that is certainly sufficient. If we allow negative conductances we can interpret K_{ij} as a an "edge conductance" even for anisotropic conductivity and possible non-acute angles. The total Ohmic power for the network will still be nonnegative for any solution. The mapping $(\sigma^k) \mapsto K$ is linear, with a domain of dimension $6n_t$ and range of dimension at most n_e. In a typical mesh $6n_t > n_e$.

3. Rank analysis of Jacobian

In practical EIT a system of electrodes is used that does not typically cover the whole of $\partial\Omega$ which means some conductivity information at the boundary could be inaccessible in between electrodes. We therefore consider the idealisation that any electrodes can be used. In the finite element context it means that any current vector I can be specified with zero sum and supported on boundary vertices, and all measurements of voltage made at boundary vertices. Without loss of generality we will set a current of -1 at i_g and then apply currents of 1 at each other boundary i vertex in turn. The voltage $V_{ij} = u(x_j)$ is measured at x_j. The Jacobian matrix gives the rate of change of each of these voltages when σ^k_{lm} is changed in tetrahedron T_k. This is

$$\frac{\partial V_{ij}}{\partial \sigma^k_{lm}} = -\int_{T_k} \frac{\partial u^i}{\partial x_l}\frac{\partial u^j}{\partial x_m}\, dx \tag{8}$$

where $u^i = \sum_q u^i_q \phi_q(x)$ is the finite element approximation to the potential for the current of 1 at vertex x_i.

Using Netgen[2] to generate meshes of geometric objects and a modification of code derived by Abascal[3] from Polydorides' EIDORS-3D[4] to calculate the Jacobian we studied the singular values of the anisotropic Jacobian, see table 1.

n_t	n_e	rank(J)
6	19	19
28	66	66
48	98	98
168	289	289
224	382	382
384	604	604
1344	1922	1922

(a) Cube

n_t	n_e	rank(J)
40	82	81
70	142	141
122	246	245
320	504	504
336	590	589
560	879	879
976	1529	1529

(b) Sphere

n_t	n_e	rank(J)
89	162	162
251	418	418
361	656	656
712	1049	1049

(c) Cylinder

Table 1: Meshes with n_t tetrahedra and n_e edges against the numerical rank of the Jacobian.

We note in figure 1 that sudden fall in the singular values after n_e, and the linear fall on a log scale before that, confirms that the edge conductances can be determined by the transfer impedance data, with a similar degree of illconditioning as the isotropic inverse problem. We note that this rank of the Jacobian is less than the $6n_t$ degrees of freedom in the anisotropic conductivity and as expected we cannot hope to recover these uniquely.

4. Discussion and Conclusion

The relation between finite element meshes and resistor network only applies to linear tetrahedral elements. It is also not necessarily appropriate to treat the conductivities on elements as each individually variable. In isotropic EIT it is quite usual to use a courser mesh for the conductivity than for the potential. In this case with fewer degrees of freedom for the conductivity, and perhaps basis functions reflecting *a priori* information, the loss of information we observed going from conductivity matrices to edge conductances is not so important. More important perhaps is the necessity in our numerical representation to reintroduce the non-uniqueness resulting from the undetermined diffeomorphism. One way to do this is to reconstruct edge conductances on a mesh, then use a priori information, for example about the orientation of muscle fibres, to constrain the mapping of the abstract resistor mesh, that has no geometric information, to the

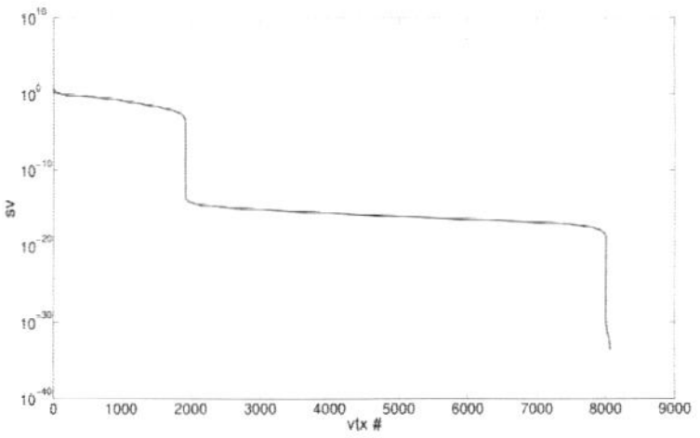

(a) Cube, 1344 tetrahedra

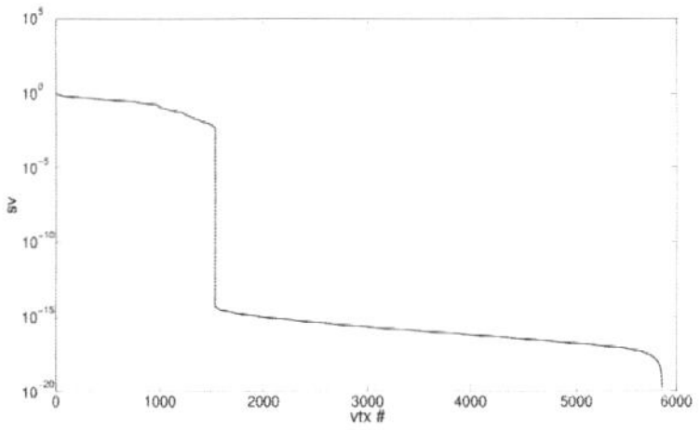

(b) Sphere, 976 tetrahedra

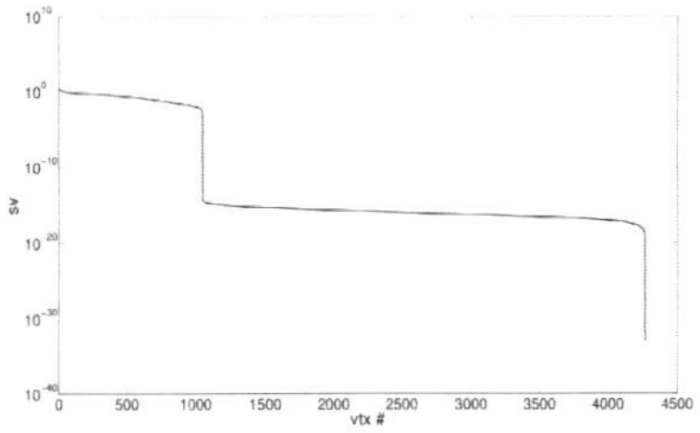

(c) Cylinder, 712 tetrahedra

Figure 1: Singular values of the Jacobian for selected meshes

vertex positions and hence conductivities. That is our intention for future work.

This work is supported by EPSRC grant EP/F033974/1.

References

[1] Lee J M and Uhlmann G 1989 *Comm. Pure Appl. Math.* **42** 1097–1112 ISSN 0010-3640 URL `http://dx.doi.org/10.1002/cpa.3160420804`

[2] Schöberl J 1997 *Comput. Visual. Sci* **1** 41–52

[3] Abascal J F P, Arridge S R, Atkinson D, Horesh R, Fabrizi L, Lucia M D, Horesh L, Bayford R H and SHolder D 2008 *NeuroImage* **43** 258–268

[4] Polydorides N and Lionheart W R B 2002 *Measurement Science and Technology* **13** 1871–1883 URL `http://stacks.iop.org/0957-0233/13/1871`

[5] Lee J R 1997 *J. Korea Soc. Math. Ed. Ser. B: Pure Appl. Math.* **4** 1–6 ISSN 1226-0657 URL `http://mathnet.kaist.ac.kr/mathnet/kms_tex/41198.pdf`

Appendix

The derivation of (7) is fairly elementary apart from one step. Consider one tetrahedron T with vertices $x_1, ..., x_4$. The unit normal to plane of vertices 2, 3, 4 is

$$\mathbf{N}_1 = \frac{(x_2 - x_3) \times (x_2 - x_4)}{|(x_2 - x_3) \times (x_2 - x_4)|} \tag{9}$$

Hence the nodal basis function on this tetrahedron is

$$\phi_1(x) = \frac{(x - x_2) \cdot (x_2 - x_3) \times (x_2 - x_4)}{(x_1 - x_2) \cdot (x_2 - x_3) \times (x_2 - x_4))} \tag{10}$$

and

$$\nabla\phi_1 \cdot \nabla\phi_2 |T| = \mathbf{N}_1 \cdot \mathbf{N}_2 \frac{4\text{Area}(234)\text{Area}(134)}{|T|} \tag{11}$$

Clearly $\mathbf{N}_1 \cdot \mathbf{N}_2 = \cos\theta_{12}$ where θ_{12} is the dihedral angle at the edge 34 and from [5] we see that

$$|T| = \frac{2}{3L_{12}} \text{Area}(234)\text{Area}(134)\sin\theta_{12}$$

where L_{12} is the length of the *opposite* edge 34. The result follows.

International Conference on Electrical Bioimpedance IOP Publishing
Journal of Physics: Conference Series **224** (2010) 012023 doi:10.1088/1742-6596/224/1/012023

Wavelet preconditioning for EIT[3]

P. Kantartzis[1], A. Kunoth[2], R. Pabel[2], P. Liatsis[1]

[1]Information Engineering and Medical Imaging Group, City University London, Northampton square, EC1V 0HB, London, UK

[2] Institut für Mathematik, Universität Paderborn, Warburger Str. 100, 33098 Paderborn, Germany

E-mail: p.kantartzis@city.ac.uk

Abstract. In this paper the forward problem of Electrical Impedance Tomography (EIT) is considered. To achieve a flexible treatment of boundaries, the original two-dimensional domain is extended to a simple square one and essential boundary conditions are imposed by means of Lagrange multipliers, resulting in a saddle point system. For discretisation, we employ a biorthogonal B-Spline wavelet basis for which it can be shown that in the suggested forward EIT formulation the condition number of the system matrix is asymptotically uniformly bounded, and therefore iterative solvers converge with a speed independent of the discretisation level.

1. Introduction

The forward problem in EIT is typically solved by iterative methods due to the large number of unknowns coupled with noise contaminated measurements. The speed of iterative (approximate) solvers, in turn, is determined by the spectral condition number, defined as $\kappa(A) := \|A\| \left\|A^{-1}\right\|$, of the system matrix A and the operator norm with respect to the Euclidean norm. Unfortunately, in modern numerical computations due to an increasing demand in the discretisation precision, large scale systems are encountered when standard finite elements (conventional piecewise polynomials of local support) are employed. Notably, the obtained condition numbers are inverse-proportional to the effectively shrinking pitch (grid-spacing). Thus, it is essential to counteract for the increasing condition number without, of course, spoiling the overall computational complexity. This procedure is called preconditioning and, ideally, yields an equivalent system with a uniformly bounded condition number of the system matrix. Here, we propose a conceptually different to the well-established optimal preconditioner approach (Multigrid). We follow the Domain Embedding Method (DEM) paradigm and embed the original domain Ω into a 'fictitious', simpler one, e.g., a unit square (denoted as $\square$) to allow for simpler yet efficient (uniform grid) discretisation. Essential boundary conditions are appended by Lagrange multipliers, leading to a saddle point problem. For its discretisation, we employ biorthogonal B-spline wavelets which provide an optimal preconditioning: norm equivalences between the underlying Sobolev spaces and the vector spaces of expansion coefficients show that a simple diagonal scaling achieves a uniform bound of the condition number with respect to the discretisation level. In addition, the multiscale structure of wavelets allows for a-posteriori steered adaptive solution schemes (if desired) for nonsmooth solutions.

[3] We gratefully acknowledge financial support by: the SFB/TR 32 "Pattern in Soil-Vegetation-Atmosphere Systems: Monitoring, Modelling, Data Assimilation" www.tr32.de (funded by the Deutsche Forschungsgemeinschaft (DFG) and by the DFG Research Training Group "Scientific Computing", PaSCo), the EPSRC (grant EP/G061580/1) and the State Scholarships Foundation of Greece (IKY).

2. The forward EIT problem

The forward problem in EIT is essentially a generalised Laplacian of the form $\nabla \cdot (\sigma \nabla u) = 0$ subject to boundary conditions to ensure existence and uniqueness of the solution. Following [1] in the formulation of the Complete Electrode Model (CEM), let a bounded domain $\Omega \subset \Box \subset \mathbb{R}^2$ be equipped with L electrodes attached on its Lipschitz boundary surface $\partial\Omega$. Denote by $\Gamma \subset \partial\Omega$ the union of areas under each electrode, assumed to be open connected subsets $\bigcup_{\ell=1}^{L} \Gamma_\ell = \Gamma$, whose closures are disjoint, $\bigcap_{\ell=1}^{L} \overline{\Gamma}_\ell = \emptyset$, and $\urcorner := \partial\Omega \setminus \Gamma$ is the union of the remaining areas. Boundary conditions on Γ are defined as

$$\sigma \nabla u \cdot \nu = \imath \qquad \text{on } \Gamma, \quad \text{and} \quad u + z_\ell \sigma \nabla u \cdot \nu = U_\ell \qquad \text{on } \Gamma_\ell$$

where $\sigma, u, U_\ell, \nu, \imath, z_\ell$ are the admittivity, the interior potential distribution, the surface potential on the $\ell-$th electrode, the outward unit normal vector, the current density and the surface impedance, respectively. Additional boundary conditions for the interelectrode gaps require that

$$\sigma \nabla u \cdot \nu = 0 \quad \text{on } \urcorner. \tag{1}$$

We denote the associated Sobolev space for u as $H_\Omega^1 := H^1(\Omega) \oplus \mathbb{C}^L$ with dual space $(H_\Omega^1)'$. Defining the sesquilinear form

$$a_\Omega((v, V), (w, W)) := \int_\Omega \sigma \nabla v \cdot \nabla \overline{w} \, d\Omega + \sum_{\ell=1}^{L} \int_{\Gamma_\ell} \frac{1}{z_\ell}(v - V_\ell)(\overline{w} - \overline{W}_\ell) \, ds_{\Gamma_\ell}, \tag{2}$$

the weak formulation of the EIT problem on the original domain Ω can be stated as follows: Given the stimulation patterns (right hand sides) $I = (I_1, \ldots, I_L)^T \in \mathbb{R}^L$, with I_ℓ denoting the current applied to the ℓ-th electrode, find $(u, U) \in H_\Omega^1$ such that

$$a((u, U), (v, V)) = \sum_{\ell=1}^{L} I_\ell \overline{V}_\ell \qquad \text{for all } (v, V) \in H_\Omega^1. \tag{3}$$

2.1. Saddle point formulation in the fictitious domain $\Box$

In order to allow for a flexible treatment of boundaries for the EIT problem, we extend the original domain Ω to a fictitious, simpler one, denoted as $\Box$, see, e.g., [2]. Assuming suitable extensions for all functions from Ω to $\Box$, the extension of $a_\Omega(\cdot, \cdot)$ to $\Box$ is defined as

$$a_\Box((v, V), (w, W)) := \int_\Box \sigma \nabla v \cdot \nabla \overline{w} \, d\Box + \sum_{\ell=1}^{L} \int_{\Gamma_\ell} \frac{1}{z_\ell}(v - V_\ell)(\overline{w} - \overline{W}_\ell) \, ds_{\Gamma_\ell}. \tag{4}$$

It is crucial here that the resulting weak problem formulations, using either $a_\Omega(\cdot, \cdot)$ or $a_\Box(\cdot, \cdot)$, are not equivalent, as the 'natural' boundary condition (1) translates into an 'essential' boundary condition for $\Box$ and, therefore, needs to be explicitly satisfied. Using Lagrange multipliers as in [3], an equivalent EIT formulation for (3) in $\Box$ can be recasted as a functional minimisation problem

$$\inf_{(v,V) \in H_\Box^1} \sup_{q \in (H^{1/2}(\urcorner))'} \frac{1}{2} a_\Box((v, V), (v, V)) - \sum_{\ell=1}^{L} I_\ell \overline{V}_\ell + b(v, q) \tag{5}$$

with $b(v, q) := \int_\urcorner q \sigma \nabla v \cdot \nu \, ds$. The standard first order optimality conditions are then

$$\begin{aligned} a_\Box((u, U), (v, V)) + b(v, p) &= \textstyle\sum_{\ell=1}^{L} I_\ell \overline{V}_\ell \\ b(u, q) &= 0. \end{aligned}$$

In operator form, this reads

$$\begin{bmatrix} A & B' \\ B & 0 \end{bmatrix} \begin{bmatrix} [u, U]^T \\ p \end{bmatrix} = \begin{bmatrix} \sum_{\ell=1}^{L} I_\ell \overline{V}_\ell \\ 0 \end{bmatrix} \tag{6}$$

where $A : H^1_\square \to (H^1_\square)'$ is self-adjoint, $B : H^1(\square) \to (H^{-1/2}(\urcorner))'$ is the trace operator and B' its adjoint. We have established existence and uniqueness of the solution of (6) on $\square$ along with the equivalence between the associated solutions in Ω and $\square$. We will publish these results together with remarks about the space $(H^{-1/2}(\urcorner))'$ elsewhere.

2.2. *Discretisation of the continuous problem using wavelets*

The wavelets we consider here are biorthogonal Riesz basis functions of the involved function spaces. This means that they allow a representation of the continuous saddle point problem (6) without having to specify a certain finest resolution level. Thus, the system is discretised but not yet made finite-dimensional. This makes it particularly apparent to explain our wavelet preconditioning. Our choice of wavelets are tensor products of the biorthogonal spline-wavelets from [4], adapted to the interval as in [5]. They are defined in terms of scaling functions $\phi_{j,k}$ and corresponding wavelet $\psi_{j,k}$ basis functions (for various resolution levels j and translations k). For notational convenience, we rewrite the fundamental two-scale relation $\phi(2^0 x) = \sum_m c_m \phi(2^1 x - m)$ in a compact matrix-vector form for resolution pitch 2^{-j} as $\mathbf{\Phi}_j = \mathbf{M}_{j,0}^T \mathbf{\Phi}_{j+1}$ and $\mathbf{\Psi}_j = \mathbf{M}_{j,1}^T \mathbf{\Phi}_{j+1}$ where the matrices $\mathbf{M}_{j,0}$,$\mathbf{M}_{j,1}$ contain appropriately scaled coefficients c_m, and d_n for the wavelets. Let $\mathbf{\Psi} = \{\mathbf{\Psi}_j\}_{j=j_0}^\infty$, $\tilde{\mathbf{\Psi}} = \{\tilde{\mathbf{\Psi}}_j\}_{j=j_0}^\infty$ be the collection of all functions on all levels, often called a multiresolution basis. Now every function $v \in H^1_\square$ can be expanded as

$$v = \mathbf{d}^T \mathbf{\Psi} =: \sum_{j=j_0-1}^{\infty} \mathbf{d}_j^T \mathbf{\Psi}_j := \mathbf{c}_{j_0}^T \mathbf{\Phi}_{j_0} + \sum_{j=j_0}^{\infty} \mathbf{d}_j^T \mathbf{\Psi}_j \tag{7}$$

whose coefficients $\mathbf{c}_{j_0}, \mathbf{d}_j$ correspond to inner products defined with respect to the duals $\tilde{\mathbf{\Phi}}_{j_0}$, $\tilde{\mathbf{\Psi}}_j$, and j_0 is the coarsest (minimum) level. From the Riesz basis property, one can infer that $\|\mathbf{d}^T \mathbf{\Psi}\|_{H^1(\square)} \sim \|\mathbf{D}\mathbf{d}\|_{\ell_2}$, where $\mathbf{D} := 2^j \delta_{j,j'} \delta_{n,n'}$ is a diagonal matrix and $a \sim b$ means that a, b are equivalent up to constants.

2.3. *Proposed multilevel preconditioning for EIT and numerical results*

For the EIT setting in saddle point form (6), A is an isomorphism on the kernel of B, and the corresponding relation $\|Au\|_{H^{-1}_\square} \sim \|u\|_{H^1_\square}$ can be expanded using $\tilde{\mathbf{\Psi}}$ as in [3],

$$\|Au\|_{H^{-1}_\square} = \|\langle Au, \mathbf{\Psi}\rangle \tilde{\mathbf{\Psi}}\|_{H^{-1}_\square} \sim \|\mathbf{D}^{-1}\langle Au, \mathbf{\Psi}\rangle\|_{\ell_2}, \tag{8}$$

where $\langle \cdot, \cdot \rangle$ denotes the dual form between $H^1_\square$ and its dual. Expanding u in (8) as in (7) yields

$$\|\mathbf{D}^{-1}\langle Au, \mathbf{\Psi}\rangle\|_{\ell_2} \sim \|\mathbf{D}^{-1}\langle A\mathbf{\Psi}, \mathbf{\Psi}\rangle \mathbf{d}\|_{\ell_2} = \|\mathbf{D}^{-1}\langle A\mathbf{\Psi}, \mathbf{\Psi}\rangle \mathbf{D}^{-1}\mathbf{D}\mathbf{d}\|_{\ell_2} \tag{9}$$

where $A_{\mathbf{\Psi}} := \langle A\mathbf{\Psi}, \mathbf{\Psi}\rangle$ is the representation of the operator A in terms of the infinite wavelet basis $\mathbf{\Psi}$. The main norm equivalence result is now

$$\|Au\|_{H^{-1}_\square} \sim \|u\|_{H^1_\square} = \|\mathbf{d}^T \tilde{\mathbf{\Psi}}\|_{H^1_\square} \sim \|\mathbf{D}^{-1} A_\Psi \mathbf{D}^{-1}\mathbf{D}\mathbf{d}\|_{\ell_2} \sim \|\mathbf{D}\mathbf{d}\|_{\ell_2}. \tag{10}$$

A similar relation for A^{-1} can be shown. As a result of both, one obtains a uniformly bounded spectral condition number for the resulting stiffness matrix [3],

$$\kappa(A^D_{\mathbf{\Psi}}) = O(1), \quad \text{where } A^D_{\mathbf{\Psi}} := \mathbf{D}^{-1} A_\Psi \mathbf{D}^{-1}. \tag{11}$$

Table 1. Spectral condition numbers for the boundary operator B_{Ψ} with 8 interelectrode gaps.

$l \setminus$	$j=07,\ \zeta=03$	$j=07,\ \zeta=04$	$j=07,\ \zeta=05$
01	2.69e+000	8.22e+000	4.90e+016
02	2.69e+000	8.22e+000	3.80e+016
03	2.12e+000	4.27e+000	2.48e+016
04	2.12e+000	4.26e+000	1.38e+016
05	2.72e+000	8.07e+000	1.42e+016
06	2.71e+000	8.08e+000	7.22e+015
07	2.12e+000	4.26e+000	3.38e+016
08	2.12e+000	4.27e+000	3.14e+016

Of course, for numerical simulations the operator A_{Ψ}^{D} has to be made finite-dimensional. From (11) one can immediately conclude that also any finite-dimensional portion of A_{Ψ}^{D} has a uniformly bounded condition number.

As for the boundary operator B, the well-posedness of the saddle point formulation (6) requires that the condition

$$\inf_{q \in (H^{1/2}(\urcorner))'} \sup_{v \in H^1_{\square}} \frac{|\langle Bv, q\rangle|}{\|v\|_{H^1_{\square}} \|q\|_{(H^{1/2}(\urcorner))'}} \geq \beta \tag{12}$$

holds for some constant $\beta > 0$. Assuming that B is discretised with wavelet basis functions for $\square$ and for $\urcorner$, for convenience abbreviated as B_{Ψ}, a similar yet slightly more complicated analysis for preconditioning holds where the scaling operator $\mathbf{D}$ is adjusted according to the underlying domain and trace spaces. However, condition (12) does not hold any more automatically for any finite-dimensional subspaces of $H^1_{\square}$ and $(H^{1/2}(\urcorner))'$. The corresponding, so-called 'LBB-condition' [3, 6] is a stability condition which requires, in the setting considered here, that the discretisation level ζ on the boundary is not too high when compared to the discretisation level j on $\square$. We have tested the behaviour of the operator B in wavelet coordinates for the case of Ω a disk of radius $R = 1/4$, centered at (.5,.5) with 8 electrodes attached ($z_{\ell} = 10$ Ohm), embedded into a fictitious unit square. We present the resulting spectral condition numbers of B_{Ψ} in Table 1 for different values of j and ζ, and l denotes the interelectrode gaps (arcs). It is evident from Table 1 that the LBB condition holds numerically as long as $\zeta + 2 < j$ (first and second column in Table 1). Finally, it can be shown that optimal preconditioning together with nested iteration yields an optimal $O(N)$ algorithm, N being the total amount of unknowns [2].

3. Conclusions

In this paper, a saddle point formulation using CEM and optimal preconditioning by wavelets was proposed for the solution of the forward EIT problem. It was shown that a diagonal scaling applied to operators in wavelet coordinates, A_{Ψ} and B_{Ψ}, yield uniformly bounded condition numbers, allowing for efficient iterative solvers. Moreover, for the boundary operator B and for $\zeta + 2 < j$, the computed condition number of B_{Ψ} attained very low absolute values, i.e., stability.

References

[1] Somersalo E, Cheney M and Isaacson D 1992 *SIAM J. App. Math.* **52** 1023–1040
[2] Kunoth A 2001 *Wavelet Methods-Elliptic boundary value problems and control problems* (B.G. Teubner)
[3] Kunoth A 1994 *Multilevel preconditioning* (Aachen: Shaker)
[4] Cohen A, Daubechies I and Feauveau J C 1992 *Comm. Appl. Math.* **45** 485–560
[5] Dahmen W, Kunoth A and Urban K 1999 *Appl. Comp. Harm. Anal.* **6** 132–196
[6] Brezzi F and Fortin M 1991 *Mixed and Hybrid Finite Element Methods* vol 15 (New York: Springer)

Tracking resistivity changes using suboptimal fading extended Kalman filter in electrical resistance tomography

B S Kim[1], A Rashid[2], A K Khambampati[2], D Liu[2], S Kim[1,3] and K Y Kim[2]

[1] Applied Radiological Science Research Institute, Jeju National University, Jeju 690-756, South Korea
[2] Department of Electronic Engineering, Jeju National University, Jeju 690-756, South Korea
[3] Department of Nuclear and Energy Engineering, Jeju National University, Jeju 690-756, South Korea

E-mail: kyungyk@jejunu.ac.kr, bongsk3@gmail.com

Abstract. This paper presents a suboptimal fading extended Kalman filter to track fast resistivity changes inside the heart as well as the lungs during the cardiac cycle. The performance of the proposed algorithm is evaluated through the numerical simulation of a cardiac cycle. The results demonstrate that the proposed method is more robust as compared to the extended Kalman filter in tracking fast changes in the resistivity distribution.

1. Introduction

Electrical resistance tomography (ERT) is a fast and non-invasive imaging modality in which a cross sectional image of resistivity distribution inside the human body can be obtained based on the applied currents and the resulting voltages through the electrodes attached to the skin of the body [1]. In order to enhance temporal resolution, dynamic techniques have been proposed for situations where the resistivity changes rapidly. In dynamic imaging, the inverse problem is formulated as a nonlinear state estimation problem in which the time-varying state is estimated with the aid of the linearized Kalman filter [2] or extended Kalman filter [3-4].

Usually the exact knowledge of the state evolution model in ERT is not available, so the so-called *random-walk* model is used instead. The modeling uncertainty of the *random-walk* model is usually not so small that there may be significant negative effects on the quality of the reconstructed image. Therefore, when the process noise covariance matrix is chosen improperly, these dynamic algorithms give poor estimation performance.

In this paper, an suboptimal fading extended Kalman filtering (SFEKF) algorithm [5] is proposed to track fast time-varying resistivity inside the heart as well as the lungs during the cardiac cycle. The performance can be improved with the use of the SFEKF in which a fading factor is introduced to reduce the modeling uncertainty and leads to a better estimation of the state-error covariance matrix. In particular, pre-grouping technique is employed to stabilize the inverse solver. The performance of the proposed algorithm is evaluated through the numerical simulation of the cardiac cycle. The results demonstrate that the proposed method is more robust as compared to the EKF in tracking fast changes in the resistivity distribution.

2. Method

The physical relationship between the resistivity and the voltage is governed by a partial differential equation with appropriate boundary conditions based on the complete electrode model [6]. For more details of the formulation of the forward problem, see Vauhkonen [1].

2.1. *Integration of the state*

The internal structure of the human body can be obtained *a priori* from anatomical imaging [2,4]. The major regions considered are the heart, the left and right lungs, backbone and background (other tissues), called regions of interest (ROIs). The elements in each ROI can be grouped together before inverse procedure. The pre-grouping can be written in the form $\tilde{\rho} \equiv \mathbf{G}\rho$ where $\rho \in \mathbb{R}^{N_f \times 1}$ and $\tilde{\rho} \in \mathbb{R}^{N_g \times 1}$ are the resistivities in the forward and inverse meshes, respectively, and $\mathbf{G} \in \mathbb{R}^{N_g \times N_f}$ is a grouping matrix.

2.2. *Formulation of the dynamic model in the inverse problem*

In the state estimation problem, the dynamic model is necessary, which consists of the state and observation equations. Here, the state equation is assumed to be of the following linear form

$$x_{k+1} = \mathbf{F}_k\, x_k + w_k \tag{1}$$

where x_k is the state vector at time k, $x_k \equiv \left[\tilde{\rho}_k^1, \tilde{\rho}_k^2, \ldots, \tilde{\rho}_k^{N_g}\right]^T$, $\mathbf{F}_k = \mathbf{I}_{N_g}$ (the identity matrix) is the state transition matrix for the *random-walk* model and N_g is the number of states. It is assumed that w_k is white Gaussian noise with known covariance matrix $\mathbf{Q}_k = E\left[w_k\, w_k^T\right]$.

Next, let $V_k \equiv \left[V_k^1, V_k^2, \ldots, V_k^L\right]^T \in \mathbb{R}^{L\times 1}$ be the actual boundary voltage measured by the kth current pattern. Then the observation equation can be described as the nonlinear mapping $V_k = U_k\left(\rho_k\right) + v_k$. Linearizing it about the predicted state $x_{k|k-1}$, we have

$$V_k = U_k + \mathbf{J}_k\left(x_k - x_{k|k-1}\right) + \bar{v}_k \tag{2}$$

where U_k is the calculated voltage and $\mathbf{J}_k \in \mathbb{R}^{L\times N_g}$ is the Jacobian matrix [3]. Therefore, we obtain the following linearized observation equation as

$$y_k = \mathbf{J}_k\, x_k + \bar{v}_k, \quad y_k \equiv V_k - U_k + \mathbf{J}_k\, x_{k|k-1} \tag{3}$$

where $\bar{v}_k$ is assumed to be white Gaussian noise with known covariance $\mathbf{R}_k = E\left[\bar{v}_k\, \bar{v}_k^T\right]$.

2.3. *Suboptimal fading extended Kalman filter*

The key of the suboptimal fading extended Kalman filter (SFEKF) is to introduce a fading factor into the EKF, to adjust the state-error covariance matrix and the Kalman gain matrix. The derivation of the SFEKF is presented in Zhou *et al* [5] in detail. With the Gaussian assumptions the recursive SFEKF algorithm in one single sampling cycle can be expressed as follows:

$$\hat{x}_{k|k-1} = \mathbf{F}_{k-1}\, \hat{x}_{k-1|k-1}, \quad r_k = y_k - \mathbf{J}_k\, \hat{x}_{k|k-1} \tag{4}$$

$$\lambda_k = \begin{cases} \lambda_0, & \lambda_0 \geq 1 \\ 1, & \lambda_0 < 1 \end{cases}, \quad \lambda_0 = \frac{\mathrm{tr}\left[\mathbf{N}_k\right]}{\mathrm{tr}\left[\mathbf{M}_k\right]}, \quad \mathbf{W}_k = \begin{cases} r_1\, r_1^T, & k = 1 \\ \frac{\eta\, \mathbf{W}_{k-1} + r_k\, r_k^T}{1+\eta}, & k \geq 2 \end{cases} \tag{5}$$

$$\mathbf{M}_k = \mathbf{J}_k\, \mathbf{F}_{k-1}\, \mathbf{P}_{k-1|k-1}\, \mathbf{F}_{k-1}^T\, \mathbf{J}_k^T, \quad \mathbf{N}_k = \mathbf{W}_k - \mathbf{J}_k\, \mathbf{Q}_{k-1}\, \mathbf{J}_k^T - \mathbf{R}_k \tag{6}$$

$$\mathbf{P}_{k|k-1} = \lambda_k\, \mathbf{F}_{k-1}\, \mathbf{P}_{k-1|k-1}\, \mathbf{F}_{k-1}^T + \mathbf{Q}_{k-1} \tag{7}$$

$$\mathbf{K}_k = \mathbf{P}_{k|k-1}\mathbf{J}_k^T \left[\mathbf{J}_k \mathbf{P}_{k|k-1}\mathbf{J}_k^T + \mathbf{R}_k\right]^{-1} \tag{8}$$

$$\hat{x}_{k|k} = \hat{x}_{k|k-1} + \mathbf{K}_k r_k, \quad \mathbf{P}_{k|k} = \left(\mathbf{I}_{N_g} - \mathbf{K}_k \mathbf{J}_k\right) \mathbf{P}_{k|k-1} \tag{9}$$

where λ_k is the suboptimal fading factor, $0 \leq \eta \leq 1$ is a forgetting factor, $\mathbf{P}_{k|k-1}$ is the state-error covariance matrix and $\mathbf{K}_k$ is the gain matrix. Here, the striking feature is that this algorithm makes the state-error covariance to be minimum and then adaptively adjusts the gain matrix forcing the residual r_k to be orthogonal. This makes to alleviate the modeling uncertainty.

3. Numerical simulation

In order to illustrate the estimation performance of the proposed algorithm, we simulated the cardiac cycle of the human heart. It is assumed that the volumes of the heart and the lungs are fixed during the simulation, which corresponds to breath holding. However, the blood volume causes the average resistivities of the heart and the lungs to change slightly. For the generation of the cardiac cycle, it is assumed that the resistivities of each ROI in figure 1(a) are increasing or decreasing in time due to the changes in the blood volume during the cardiac cycle as shown in figure 1(b). The average resistivity values are 1.78 Ωm for the heart, 10.35 Ωm for the left lung, 10.44 Ωm for the right lung, 10 Ωm for the backbone and 4 Ωm for the background [2],[4].

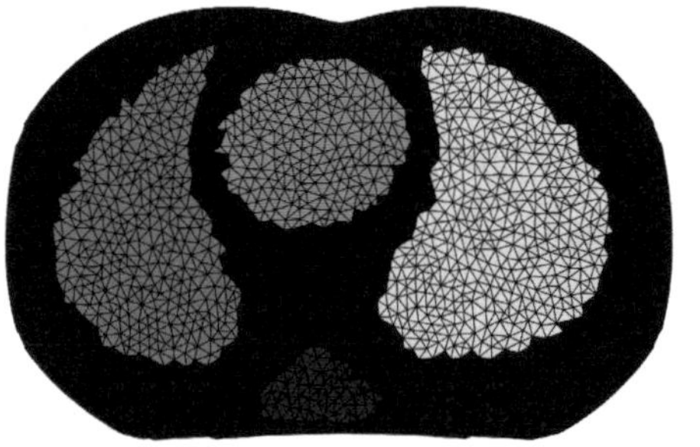

(a)

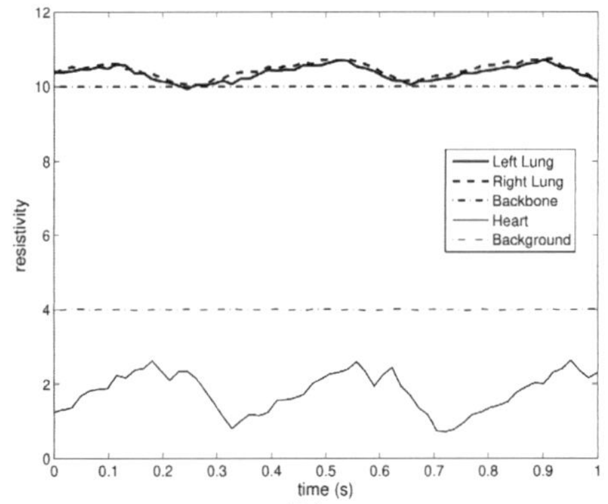

(b)

Figure 1. (a) The meshes used for forward solver with 4204 elements and inverse solver with 5 ROIs which correspond to the heart, lungs, backbone and background. (b) Simulated resistivity changes in the heart and in the lungs. The resistivity of the background is set to change vary slightly. The resistivity of the backbone is set to constant. The range of the resistivity values is given in Ωm.

The meshes used in the forward and inverse solvers were shown in figure 1(a), which shapes are intended to simulate the normal human thorax. In the forward computations we used the forward mesh with 4204 elements (N_f) and the elements were pre-grouped into 5 regions (N_g) in the inverse computations. Trigonometric current patterns were injected through 32 electrodes (L). It was also assumed that the measurement noise covariance matrix ($\mathbf{R}_k$) was diagonal and $\mathbf{R}_k = 10^{-2}\mathbf{I}_L$, and the initial value for the state-error covariance matrix ($\mathbf{P}_{0|0}$) was set to $\mathbf{I}_{N_g}$.

To estimate fast time-varying resistivities, we used the *random-walk* model for the state evolution model. Three different process noise covariance matrices ($\mathbf{Q}_k$) were used in the conventional EKF and SFEKF to evaluate their effects on the estimation performance, that is, EKF1 and SFEKF1, EKF2 and SFEKF2, and EKF3 and SFEKF3 correspond to $\mathbf{Q}_k = 2\mathbf{I}_{N_g}$, $\mathbf{Q}_k = 0.1\mathbf{I}_{N_g}$ and $\mathbf{Q}_k = 0.01\mathbf{I}_{N_g}$, respectively.

The estimated trajectories with three EKFs and SFEKF3 are shown in figure 2. The bold line is true resistivity, and other lines (–×–, –○–, –△– and –+–) correspond to the estimated resistivities by EKF1, EKF2, EKF3 and SFEKF3, respectively. From figure 2, it is noted that the tracking accuracy of the EKF heavily depends on the magnitude of the process noise covariance in the *random-walk* model due to its modeling uncertainty. Also, it should be pointed out that the estimation performances of three SFEKFs are very similar to each other and are not shown, and the result of the EKF1 is close to that of the SFEKF. As can be seen clearly, the performance of the SFEKF is enhanced than that of the EKF. The results for the other ROIs are similar and are not shown.

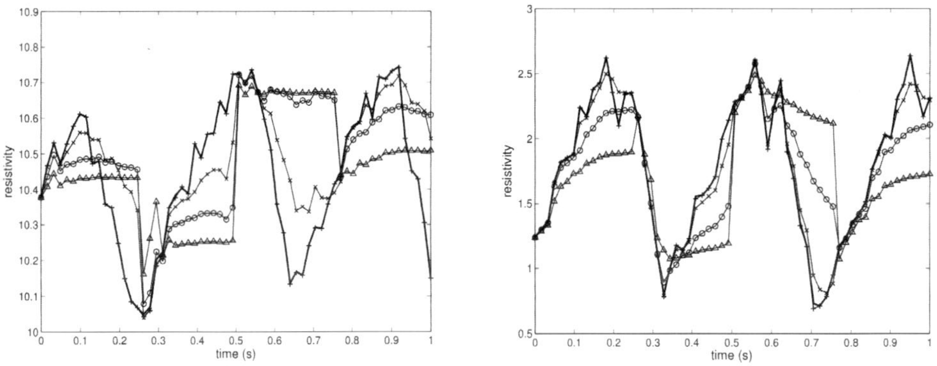

(a) (b)

Figure 2. The results from (a) the right lung and (b) the heart. True resistivity (—), EKF1 (-×-), EKF2 (-○-), EKF3 (-△-) and SFEKF3 (-+-).

4. Conclusions

In this paper, we formulated the dynamic inverse problem based on ERT. The inverse problem was treated as a nonlinear state estimation problem and the unknown time-varying resistivity was estimated on-line with the aid of the SFEKF which reduced the modeling uncertainty. In particular, pre-grouping technique was also employed to stabilize the inverse solver. The numerical simulation was provided to illustrate the reconstruction performance in terms of the spatial and temporal resolutions. The results showed that the proposed algorithm was more robust as compared to the EKF in tracking fast changes in the resistivity distribution.

Acknowledgments

This work was supported by the National Research Foundation of Korea Grant funded by the Korean Government [NRF-2009-353-D00060]. The part of researchers participating in this study were supported by the grant from "the 2nd phase BK21 project".

References

[1] Vauhkonen M 1997 Electrical impedance tomography and prior information *PhD Thesis* University of Kuopio, Finland online at http://venda.uku.fi/~mvauhkon/
[2] Vauhkonen M, Karjalainen P A and Kaipio J P 1998 A Kalman filter approach to track fast impedance changes in electrical impedance tomography *IEEE Trans. Biomed. Eng.* **45** 486–93
[3] Kim K Y, Kim B S, Kim M C, Lee Y J and Vauhkonen M 2001 Image reconstruction in time-varying electrical impedance tomography based on the extended Kalman filter *Meas. Sci. Technol.* **12** 1032–9
[4] Trigo F C, Gonzalez-Lima R and Amato M B P 2004 Electrical impedance tomography using the extended Kalman filter *IEEE Trans. Biomed. Eng.* **51** 72–81
[5] Zhou D H, Xi Y G and Zhang Z J 1991 A suboptimal multiple fading extended Kalman filter *Acta Autom Sinica* **17** 689–95
[6] Somersalo E, Cheney M and Isaacson D 1992 Existence and uniqueness for electrode models for electric current computed tomography *SIAM J. Appl. Math.* **52** 1023–40

fEITER – a new EIT instrument for functional brain imaging

J L Davidson[1], P Wright[1], S T Ahsan[1], R L Robinson[1], C J D Pomfrett[2] and H McCann[1]

[1] School of Electrical and Electronic Engineering, University of Manchester, UK
[2] School of Medicine, University of Manchester, UK

E-mail: j.davidson-2@manchester.ac.uk

Abstract. We report on human tests of the new EIT-based system fEITER (*functional Electrical Impedance Tomography of Evoked Responses*), targeted principally at functional brain imaging. It is designed and built to medical standard BS EN 60601-1:2006 and clinical trials have been approved by the MHRA in the UK. fEITER integrates an EIT sub-system with an evoked response sub-system capable of providing visual, auditory or other stimuli, and the timing of each stimulus is recorded within the EIT data to a resolution of 500 microseconds. The EIT sub-system operates at 100 frames per second using 20 polar/near-polar current patterns distributed among 32 scalp electrodes that are arranged in a 3-dimensional array on the subject. Presently, current injection is fixed in firmware at 1 mA pk-pk and 10 kHz. Performance testing on inanimate subjects has shown voltage measurement SNR better than 75 dB, at 100 frames per second. We describe the fEITER system and give example topographic results for a human subject under no-stimulus (i.e. reference) conditions and on application of auditory stimuli. The system's excellent noise properties and temporal resolution show clearly the influence of basic physiological phenomena on the EIT voltages. In response to stimulus presentation, the voltage data contain fast components (~100 ms) and components that persist for many seconds.

1. Introduction

Electrical Impedance Tomography (EIT) is a candidate technique for functional brain imaging through its sensitivity to conductivity changes within the subject. Most notably, this has been demonstrated over relatively long time-scales (tens of seconds) by Holder and co-workers at University College London (UCL), who attribute their measurements to the haemodynamic response to neural activity [e.g. 1]. Such early studies demonstrated the imaging capability of EIT using relatively modest frame rates. More recently, the same group recorded impedance changes with scalp electrodes during visual Evoked Response (ER) experiments using slow square wave current injection [2], claiming to detect fast neural activity through voltages measured on a small number of electrode sites and with averaging of repeated stimuli. The detection of sub-second neural activity associated with ER experiments using EIT was also reported in 2003-2006 during a pilot study [e.g. 3] at the University of Manchester (UoM) using an adapted industrial process tomography instrument. The pilot study motivated the development of an EIT system specifically designed for functional brain imaging which delivers high SNR at fast frame rates and is suitable for use in a clinical environment.

In this paper, we report on the development of the instrument called fEITER (functional Electrical Impedance Tomography of Evoked Responses). fEITER captures EIT data at 100 frames per second (fps) and is presently being used in a clinical trial (ISRCTN93596854) at Manchester Royal Infirmary.

2. The fEITER system

The fEITER system implements four key specification targets. Firstly, image reconstruction of 1% impedance changes in the brain due to neural function requires measurement sensitivity in the order of 80 dB [4]. Secondly, to probe the temporal domain of neural function, a frame rate of 100 fps is required. Thirdly, the complete system must comply with the principles of medical standard BS EN 60601-1:2006 and be suitable for use in both the Operating Room (OR) and Intensive Care Unit (ICU). Finally, the system must withstand the processes of electrosurgery and defibrillation which may be present within the OR. The system (Figure 1) is based on 32 voltage measurement channels and 20 near-polar sinusoidal current patterns per frame of data, and comprises of separate headbox and base unit assemblages joined by power and data cables. A CED μ1401 integrated into the base unit provides randomized stimulus trains to evoke a neurological responses using, e.g. audible clicks.

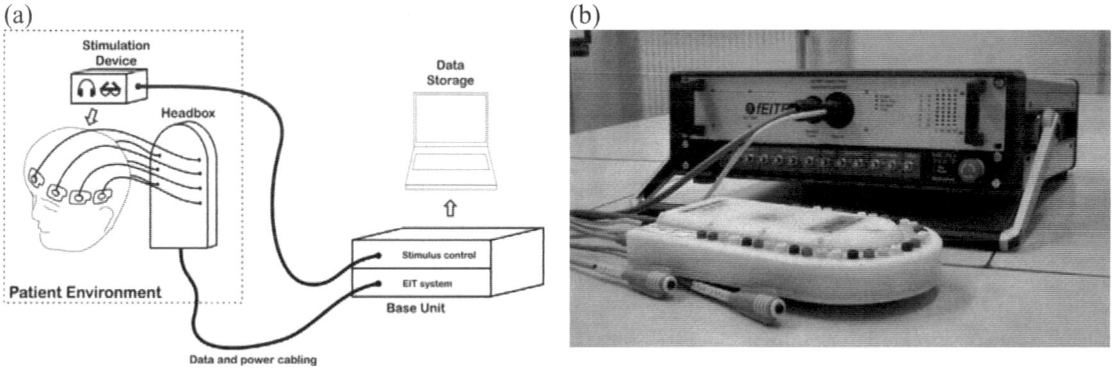

Figure 1. The fEITER system; (a) high-level schematic and (b) photograph showing the base and headbox units in the background and foreground respectively.

The EIT sub-system is a class II medical device and provides at least 5 kV galvanic isolation thereby allowing connection to both the CED μ1401 and a battery powered data collection laptop without the need of external separation devices. The patient-applied part (headbox) of the sub-system is controlled by a local FPGA (Field Programmable Gate Array, Xilinx Virtex-4 SX35) that provides Direct Digital Synthesis (DDS) to drive high-precision current sources and carries out Phase Sensitive Demodulation (PSD) of the EIT measured voltages. The headbox uses two constant-current sources [5] which can be switched via CMOS switches between any combination of drive and receive electrodes. Presently, the injected current is fixed by hardware to 1 mA pk-pk, and firmware sets the frequency of the current to 10 kHz. The current is approximately one third of the maximum permitted auxiliary current of 1 mA rms at 10 kHz as defined by the medical standard BS EN 60601-1:2006. Both the driven and received currents, and their waveform symmetry properties, are sensed by resistors in the headbox and monitored in real-time by firmware in the FPGA, to ensure that compliance with the standard is actively enforced. Parallel voltage measurements are captured between all pairs of adjacent electrodes by 16-bit ADCs running at 500 ksamples/s. The base unit of the EIT sub-system performs a serial-to-parallel conversion of incoming data from the headbox using a local Xilinx CPLD (Xilinx XC95144 family). The resultant parallel 8 bits of data are read at 500 kHz by a National Instruments USB DAQCard-6221 controlled using LabVIEW. Numerous tests with saline-filled tank tests have been carried out in order to assess the performance of the EIT sub-system, showing a precision around 80 dB in the acquired voltage measurements at 100 fps [6].

3. First human tests

(a)

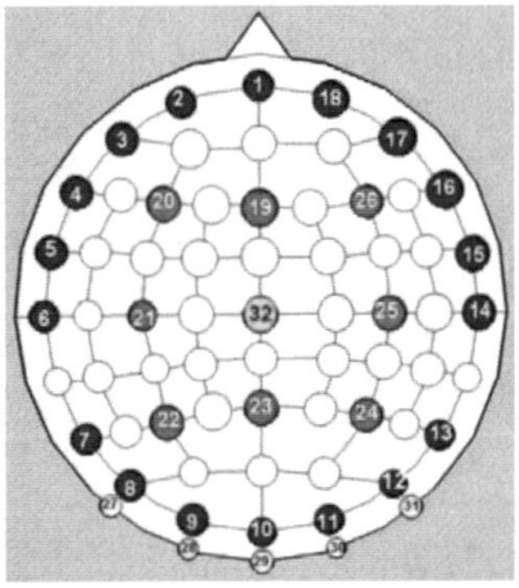

(b)

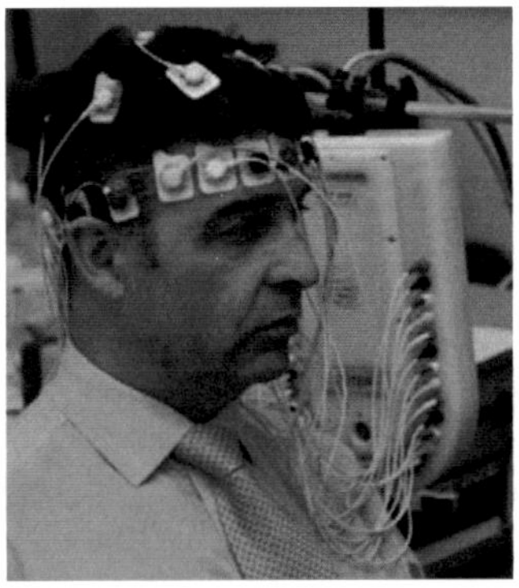

(c)

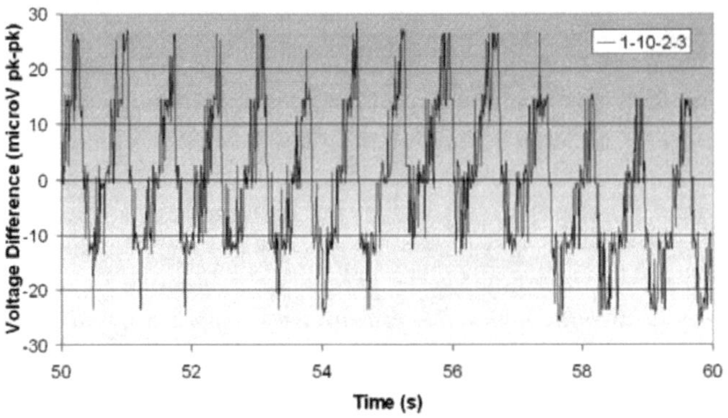

(d)

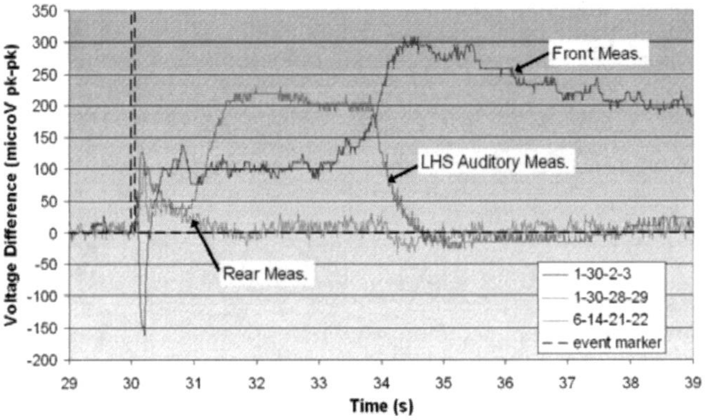

(e)

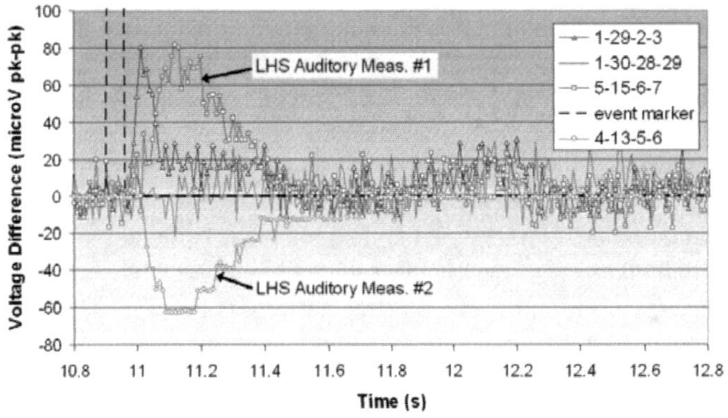

Figure 2. First human tests using fEITER; (a) electrode placement arrangement, (b) view of wired-up subject, (c) frontal voltage pair measurement during a reference condition showing 'saw-tooth' waveform indicative of trans-cranial blood flow, (d) example topographic measurements for a 'startle-type' auditory test over a total time-scale of 10 seconds and (e) example topographic measurements for a single event auditory tone test over a total time-scale of 2 seconds.

3.1. Methodology

Prior to a comprehensive clinical trial, numerous simple stimuli tests have been carried out with 2 subjects from the experimenter team at UoM with no objection from the local ethics committee. In all tests, the subject was wired up using 32 Ag/AgCl-type Zipprep electrodes (Aspect Medical Systems Inc., USA) for tomographic measurements based on a subset of the International 10-20 EEG System as shown in Figure 2(a). Additionally, a reference electrode was attached below the left ear-lobe. EIT data were captured continuously in 1-minute blocks at 100 fps (546 independent voltage pair measurements per frame) for reference conditions, and before, during and after stimulus presentations.

3.2. Example results

A typical result for a frontal measurement during a reference condition over a period of 10s is shown in Figure 2(c). With reference to Figure 2(a), the data correspond to current injection between electrode pair 1 and 10, with differential voltage measurement between electrode pair 2 and 3. A saw-tooth like waveform can be clearly seen, approaching 50 μV pk-pk. The frequency of the waveform is consistent with the heart-rate of the subject during the test, which has been subsequently verified. When the effect of this waveform is taken into account, the rms noise voltages are around 6 μV.

Figure 2(d) shows example topographic data evoked by a sharp bang (release of a party-popper) giving rise to a gross auditory stimulus. The data are for a single trial for 3 measurements corresponding to 1-30-2-3 (forehead), 1-30-28-29 (posterior head) and 6-14-21-22 which is close to the Left-Hand-Side (LHS) auditory cortex. The onset of sound was detected at 30.01s via the CED μ1401 and is shown as the event marker. Clear differences in the latencies and amplitudes in the measurement data can be seen. The data shows both fast components (~100 ms) and relatively slower components persisting for several seconds. The signal measured close to the auditory cortex has a very short-term (~100 ms) spike, followed by a plateau rising at about 1s and lasting approx. 3s. Haemodyamic (total blood volume) processes in visual cortex have been measured to start at about 1s by Sirotin *et al.* [7] using Intrinsic Signal Optical Imaging (ISOI), but their technique has a temporal resolution of about 130 ms and can't test the fast component of our signal. The frontal measurements are more complex, showing a 3-stage process with the last following after the slow auditory response, but measurements at the rear of the head show a simple decay to baseline after the 100 ms spike.

Figure 2(e) shows topographic data for a single presentation of an auditory tone delivered binaurally to the subject using inner-ear headphones, as part of a 22-event sequence comprising of pseudo-randomly spaced stimuli at 2.5 kHz, each of 50 ms in duration with varying intensities corresponding to a loudness range estimated between 60-90 dB SPL. On the presented time-scale, the temporal spacing (10 ms) of each measured data point can be seen, demonstrating the excellent temporal resolution of fEITER. The latencies of the peak measured signals closest to the auditory cortex are approximately 200 ms from the onset of the presented tone. This has been observed several times using fEITER and is consistent with the neural encoding associated with the late cortical auditory response typically observed using EEG methods in Auditory Evoked Potential (AEP) studies.

4. Conclusions and future work

Initial human tests using fEITER with simple auditory stimuli show the measurement of both fast and slow components in the EIT voltage signals on single event presentations. Image reconstruction of these data is in progress. We aim to improve our understanding and implications of the origin and mechanisms of the observed EIT data changes from more comprehensive tests, including an MHRA approved clinical trial at Manchester Royal Infirmary which is presently underway.

References

[1] Tidswell A T, Gibson R H, Bayford and D S Holder 2001 Three-dimensional electrical impedance tomography of human brain activity *Neuroimage* **13** 283-294

[2] Gilad O and Holder D S 2009 Impedance changes recorded with scalp electrodes during visual evoked responses: implications for EIT of fast neural activity *Neuroimage* **47** 514-522

[3] McCann H *et al* 2006 Sub-second functional imaging by Electrical Impedance Tomography in *Proc. of 28th Annual Int. Conf. of the IEEE EMBS* (New York City, NY, USA) 4269-4272

[4] Polydorides N, Lionheart W R B and McCann H 2002 Krylov subspace iterative techniques: on detection of brain activity with electrical impedance tomography *IEEE Trans. Med. Imag.*, **21** 596-603

[5] Rafiei-Naeini M and McCann H 2008 Low-noise current excitation sub-system for medical EIT *Physiol. Meas.* **29**(6) S173-S184

[6] Davidson J L *et al* 2009 fEITER – a new instrument for OR and ICU applications in *Proc. of 10th International Conference on Biomedical Applications of Electrical Impedance Tomography* (Manchester,UK) Available: http://www.maths.manchester.ac.uk/eit2009

[7] Sirotin Y B *et al* 2009 Spatiotemporal precision and hemodynamic mechanism of optical point spreads in alert primates *Proc. Natl. Acad. Sci.* **106**(43) 18390-18395

Journal of Physics: Conference Series **224** (2010) 012026 doi:10.1088/1742-6596/224/1/012026

A dual modality of cone beam CT and electrical impedance tomography for lung imaging

T Pengpan, C N Mitchell and M Soleimani

Invert and Imaging Centre, Department of Electronic and Electrical Engineering, University of Bath, Bath, BA2 7AY, UK

E-mail: m.soleimani@bath.ac.uk

Abstract. Electrical impedance tomography (EIT) is a new and emerging tomographic imaging system, which is able to provide high temporal resolution with low spatial resolution. In past few years, the EIT has been extensively studied for lung monitoring. Cone beam CT (CBCT) is an imaging system used in radiation therapy unit. The CBCT suffers from motion artefacts, making treatment planning inaccurate. This problem is very severe in lung cancer therapy. In this study a dual modality of EIT-CBCT has been proposed. Feasibility of such a dual system has been investigated using preliminary simulated phantom data. The motion information is extracted from the EIT images and has been used to compensate for the motion in CBCT through our recently developed iterative image reconstruction for CBCT. The results show that blurring artefacts around a tumour can be reduced for the motion compensated images using this dual modality combination. The dual EIT-CBCT imaging benefits from complementary high spatial resolution of CBCT and high temporal EIT resolution.

1. Introduction

Imaging systems have been integrated into radiation therapy systems for improving the accuracy of tumour localization during the treatment. CBCT is one of the main imaging systems that have been used in conjunction to linear accelerators[1]. One of the most challenging problems in CBCT is the patient and organ movement during imaging leading to motion artefacts and target tissue misalignments [2-5]. Radiotherapy of the lung tumour is heavily affected by respiratory movement [4, 6], making radiation therapy of lung cancer very ineffective. High precision radiotherapy is required for accurate tumour localization just prior and during the treatment [7]. Many methods have been investigated for correcting the respiratory artefacts. One of them is motion compensated CBCT, which uses estimated motions from patients for compensating in image reconstruction process. Motion artefacts are reduced and the quality of motion compensated image can be improved. A further problem is that CBCT cannot monitor respiratory lung and tumour movement during the radiation treatment session so that a movement monitoring system can be helpful then.

In this paper, a new approach using EIT imaging as motion monitoring system has been investigated. EIT is an imaging technique that can provide images of interior of the thorax by external measurement of electrical impedance at the surface electrodes. EIT has already been used in critical care units for lung monitoring [8]. The proposed approach is to combine EIT with CBCT taking advantage of high temporal resolution of the EIT and high spatial resolution of CBCT. Our preliminary investigation results and method follows here.

2. Motion compensation in CBCT

A three dimensional simulated phantom ($256 \times 256 \times 256$ voxels) has been created and used in this study. The phantom consists of two big rectangular objects with a small rectangular object. Both objects have different x-ray absorption properties. Projection data were obtained over 360 degrees and, then, were reconstructed with algebraic reconstruction technique (ART) of 10 iterations for 0.015 relaxation parameter [11]. A moving phantom has been created by applying a sinusoidal signal of 20 mm (figure 1(a)) in anterior – posterior (AP) direction during projection data acquisition. This motion information was compensated throughout the image reconstruction process by sorting the reconstructed image to original position during iterations. We assumed that we know the motion with some degree of certainties.

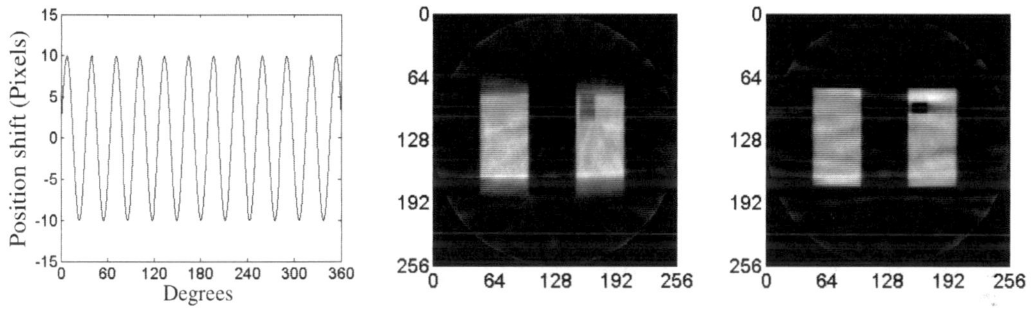

(a) (b) (c)

Figure 1. (a) Motion signal, (b) Reconstructed image of simulated phantom at z = 135 (a) with 20 mm of AP motion and (c) Motion – compensated image.

Figure 1(b) shows the reconstructed image ignoring a priori knowledge of the motion. It can be seen in figure 1(c) that motion blur is reduced in the reconstructed image with motion compensation from the fifth iteration of ART. The boundaries of a small rectangular object in motion compensated images can be seen more clearly than the boundaries of it on the image. We have tested this in several conditions and the results shows that even motion information with an error can be very helpful in reducing the artifacts in CT images. This suggests that a motion extracted from an EIT image could be good enough to improve this blurring effect due to the organ motions, which will be shown in the next section.

3. Combined CBCT/EIT study

In proposed dual modality approach, the EIT is used to dynamically image the boundaries of large organ (lungs). This provides information on the motion that can be fed into CBCT iterative image reconstruction. A simulated phantom consisting of two spherical objects with 30 mm of radius was moved by 40 mm of motion in AP direction during imaging by EIT system. Figure 2 shows the EIT images in different times, the position of spheres were also in different positions. Several methods are used for motion estimation such as optical flow methods and block matching techniques [9]. In this study centres of spheres were defined from the first EIT image as original positions and, then, were followed in further images for extracting movement data. The average of the motion data detected from both spheres was calculated and used for compensation. Projection data were collected over 360 degrees. The same reconstruction algorithm parameters as motion compensation study were used.

Figure 3 (a) and (b) are the CBCT image of two spheres without motion and with motion, respectively. Figure 3(c) is a motion compensated image by using motion signal extracted from EIT images. It can be seen that image blur is clearly improved on the image with motion compensation comparing with the motion image. The boundaries of spherical objects on the compensated image (c) are sharper than the boundaries of them on the motion image (b). This can be seen in 1D plot of column 68 of all these images in figure 4.

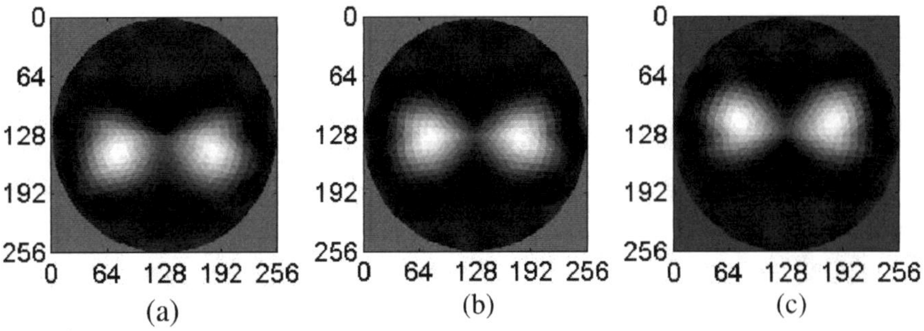

Figure 2. EIT images of spherical object at (a) the lowest position, (b) the original position and (c) the highest position.

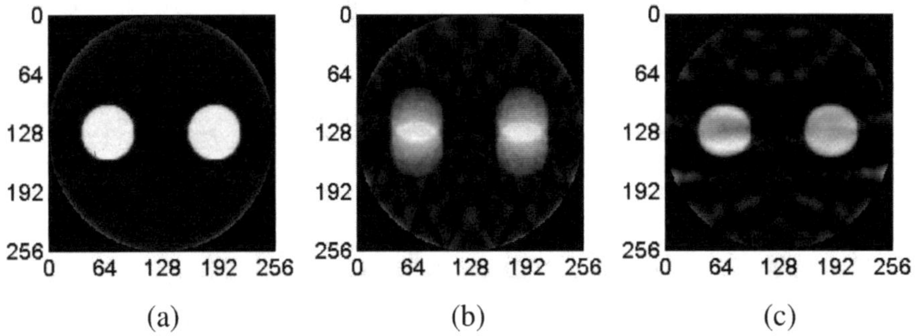

Figure 3. CBCT images of two spheres (a) without motion, (b) with 40 mm of motion in AP direction and (c) motion compensation by using EIT motion signal.

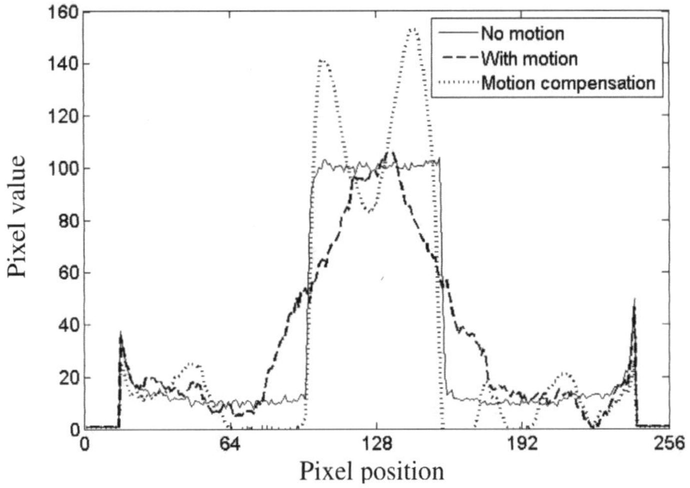

Figure 4. Image row plots for figure 3 at y=68 (centre of left sphere).

4. Conclusion

Radiation therapy is a primary tool for the treatment of lung cancer which is heavily affected by respiratory movement. The respiration monitoring system is required for tracking the respiration movement. Motion compensation by using EIT motion data through iterative algorithm of CBCT can reduce motion artefact for CBCT images. Boundaries of objects could be clearly seen on motion compensated images. Image quality is improved. Estimated motion information taken from EIT images can be used to compensate the motion for CBCT system. The fact that the EIT provides an internal image of movement makes this multimodality superior to other alternative techniques that are based on estimating motions from ECG and respiratory signals. This paper shows preliminary results in EIT-CT combination. The information about the motion of interior structures are extracted from low resolution EIT images and entered in CBCT image reconstruction through an iterative algorithm developed for volumetric CBCT. We have tested the motion detection algorithm using experimental data from magnetic induction tomography. Further validation is needed using in-vivo lung EIT images. In our continued effort we will establish a patient specific EIT model using anatomical information obtained from diagnostic CT. It is important to notice that the EIT only provides information about the (dynamical) movement of the larger organs (lungs) and do not have to capture more detailed structures. A shape based algorithm will be used to efficiently recover this boundary movement from EIT data [10]. Although the method discussed here is proposed for CBCT, the motion information generated by EIT can be used for conventional CT imaging.

References:

[1] Dietrich L and et al. 2006 Linac-integrated 4D cone beam CT: first experimental results *Physics in Medicine and Biology* **51** 2939

[2] Goitein M and Goitein 2004 Organ and Tumor Motion: An Overview, *Seminars in Radiation Oncology* **14** 2

[3] Bortfeld T, Jokivarsi K, Goitein M, Kung J and Jiang S B 2002 Effects of intra-fraction motion on IMRT dose delivery: statistical analysis and simulation. *Phys Med Biol.* **47** 2203–20

[4] Balter J M, Ten Haken R K, Lawrence T S, Lam K L and Robertson J M 1996 Uncertainties in CT-based radiation therapy treatment planning associated with patient breathing *International Journal of Radiation Oncology Biology Physics* **36** 167-74

[5] Chen G T Y, Kung J H and Beaudette K P 2004 Artifacts in computed tomography scanning of moving objects *Seminars in Radiation Oncology* **14** 19-26

[6] Balter J M, Lam K L, McGinn C J, Lawrence T S and Ten Haken R K 1998 Improvement of CT-based treatment-planning models of abdominal targets using static exhale imaging *International Journal of Radiation Oncology Biology Physics* **41** 939-43

[7] Isola A A and et al. 2008 Motion-compensated iterative cone-beam CT image reconstruction with adapted blobs as basis functions *Physics in Medicine and Biology* **53** 6777

[8] Richard J C, Pouzot C, Gros A, Tourevieille C, Lebars D, Lavenne F, Frerichs I and Guérin C 2009 Electrical impedance tomography compared to positron emission tomography for the measurement of regional lung ventilation: an experimental study. BioMed Central Ltd.)

[9] Coatrieux J L 2005 shape and function from motion in medical imaging: part I *IEEE ENGINEERING IN MEDICINE AND BIOLOGY MAGAZINE*

[10] Soleimani M, Lionheart W R B and Dorn O 2006 Level set reconstruction of conductivity and permittivity from boundary electrical measurements using experimental data *Inverse Problems in Science and Engineering* **14** 193 - 210

[11] Qiu W, Tong JR, Marchant T M, Mitchell C J, C N Spencer PJ and Soleimani M. New iterative cone beam CT reconstruction software: parameter optimisation and convergence study. Submitted.

Effects of body position on lung density estimated from EIT data

Makoto Noshiro[1], Kei Ebihara[1], Ena Sato[1], Satoru Nebuya[1], Brian H. Brown[2]

[1]Graduate School of Medical Science, Kitasato University, Sagamihara, Japan
[2]Department of Medical Physics and Clinical Engineering, University of Sheffield, Sheffield, UK

makotono@kitasato-u.ac.jp

Abstract. Normal subjects took the sitting, supine, prone, right lateral and left lateral positions during the measurement procedure. One minute epochs of EIT data were collected at the levels of the 3rd, 4th, 5th and 6th intercostal spaces in each position during normal tidal breathing. Lung density was then determined from the EIT data using the method proposed by Brown[5]. Lung density at the electrode level of the 6th intercostal space was different from that at almost any other levels in both male and female subjects, and lung density at the electrode levels of the 4th and 5th intercostal spaces in male subjects did not depend upon position.

1. INTRODUCTION

In principle, Electrical Impedance Tomography (EIT) can produce tomographic images of tissue resistivity. However, the technique has many problems which arise mainly from the 3-D spread of current, the need for many electrodes and the fact that body size and shape affect the measured potentials at least as much as tissue resistivity distribution. Several research workers have partially avoided the problem of body shape by imaging only changes in tissue resistivity as opposed to the absolute value of resistivity. However, whilst this difference technique can provide interesting data on physiological changes, such as those which arise during the respiratory and cardiac cycles, it is of little use in characterizing the absolute resistivity of tissue. For example, differences in absolute resistivity are associated with malignancy, ischemia and lung water [1, 2, 3, 4].

Relatively recently a method was described to determine the electrical resistivity of the lungs in neonates [5, 6], and was extended to adults [7]. This method was based upon a minimization of the differences between potentials calculated using a computer model of the thorax and potentials measured from 8 electrodes placed around the thorax.

Because the soft tissue of the lungs contains material with a relatively high relative permittivity, lung resistivity is a function of frequency. The main source of the high permittivity is cell membranes. Lung tissue at high frequency will consist of only two components, air with almost infinite resistivity, and condensed matter with an almost homogeneous resistivity determined by that of the intracellular and extracellular fluids. If these resistivities are known, then it is possible to calculate both lung density and air volume.

Lung disease that affects lung water will change lung density so that lung density is expected to be used for the evaluation of the recovery from lung disease in patients treated in intensive care units. In clinical practice, EIT data are collected from patients in the supine, prone or lateral positions. Accordingly, control EIT data should be collected from normal subjects in these positions. However, normal lung density only in the sitting position has been reported to date. In the present study, normal lung density was estimated at different levels of electrode placement in different body positions.

2. METHOD

2.2 Determination of absolute lung resistivity and lung density

The method of determination of lung tissue resistivity was as described by Brown et al [5] for the neonatal thorax. In summary, this method returns the best estimate of the absolute value of lung tissue resistivity by comparing the measured data to computed data sets. The computed data sets were generated via a 3-D finite difference analysis of the thorax. The elements within the 3-D model were ascribed resistivity values in accordance with their anatomical tissue type, as defined by CT imaging. The finite difference analysis was performed for a range of lung resistivities from 1 to 80 Ωm in order to generate the comparison data sets. Elements which did not form part of the lung tissue were given fixed values of resistivity. This fixation minimized the effects of the mediastinum on determined lung resistivity. Because of the long computational times involved the model data sets were pre-computed and then made available in a look-up table.

The 3D finite difference model of the thorax (Dmodel) was developed, based upon CT cross sections of a normal subject. (Segmented model provided by George Zubal, Yale). The features within the model to which values of resistivity can be assigned include fat, muscle, bone, blood and the lungs. The model is used to determine a data set of transfer impedance measurements scaled to take into account the circumference, chest depth and chest width of a particular subject.

Having calculated the data sets Dmodel, scaled for the shape of the particular adult, a set of EIT difference images was then produced by filtered back projection [6] for the measured data set, with reference to the Dmodel sets. One minute epochs (1500 frames) of EIT data were collected during tidal breathing in each subject. The 1500 frames within each epoch were averaged to provide a single data set Dmean. This averaging was performed over time. The changes with frequency between 4kHz and 813kHz (24 frequencies) were used to provide absolute lung resistivity spectra. The mean data for the three highest frequencies of the 24, i.e. 512kHz, 645kHz and 813kHz, were used to calculate the lung resistivities.

A region of interest was then placed over the lung regions in the set of images and the mean difference over the region of interest between the modelled and the measured data sets determined. Absolute lung resistivity was derived by determining which value of lung resistivity in the model minimized the difference between Dmean and Dmodel.

In order to relate absolute lung resistivity to lung tissue density (*LD*), the structure of the lung tissue and the electrical properties of the components must be known. Nopp et al [7] developed a detailed model for the human lung and used this to determine lung tissue resistivity as a function of frequency.

Lung filling factor (*FF*) is the ratio by volume of air to condensed matter within the lung tissue. The value can be used to set the volume of air contained within the lungs. The value was changed from 0.2 to 20 in steps of 0.2 and in every case the magnitude of the lung tissue resistivity at a frequency of 1MHz was calculated to provide absolute lung resistivity. Now, if the density (ρ_t) of the condensed matter of the lungs is known, then *FF* can be related to the overall density of lung tissue as follows:

$$LD = \frac{\text{lung weight}}{\text{air volume} + \text{tissue volume}} \quad (1)$$

$$FF = \frac{\text{air volume}}{\text{tissue volume}} \quad (2)$$

Putting equation (2) into equation (1) gives

$$LD = \frac{\rho_t}{FF + 1} \quad (3)$$

Absolute lung resistivity was determined as a function of *FF* from the Nopp model [7] and hence lung density related to absolute lung resistivity.

2.2. Experimental procedure

The subjects who participated in experiment were 21 normal male and 21 female volunteers who gave informed consent to the experimental procedures. The equipment used was a Sheffield Mk 3.5 system. Eight disposable Ag/AgCl hydrogel electrodes were spaced equally around the thorax at the levels of the 3rd, 4th, 5th and 6th intercostal spaces. The subjects took the sitting, supine, prone, right lateral and left lateral positions during the measurement procedure. One minute epochs of EIT data were collected in each position during normal tidal breathing.

We considered the difference between the means of lung density at the 4 electrode levels in the same position to be statistically significant at a p value of less than 0.05 using Steel-Dwass test. Comparisons of the means of lung density at the same electrode level in the 5 positions were made in the same way.

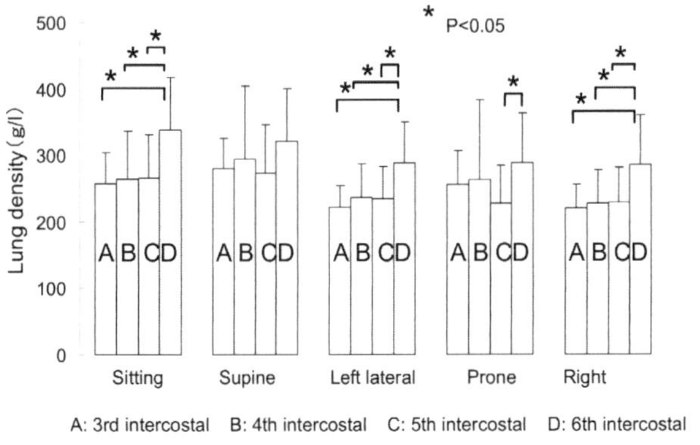

Fig. 1 Lung density at different electrode levels in male subjects

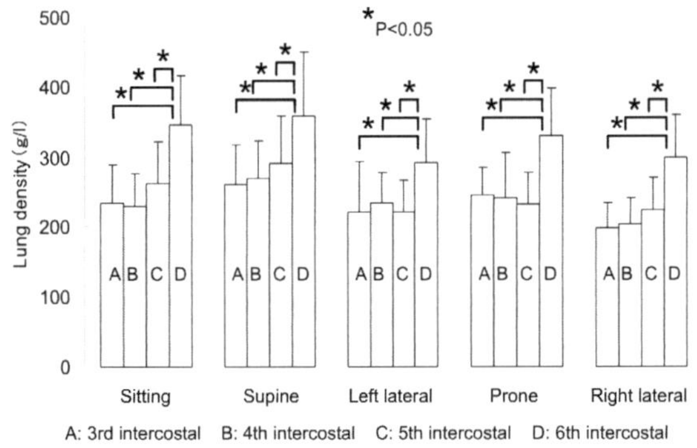

Fig. 2 Lung density at different electrode levels in female subjects

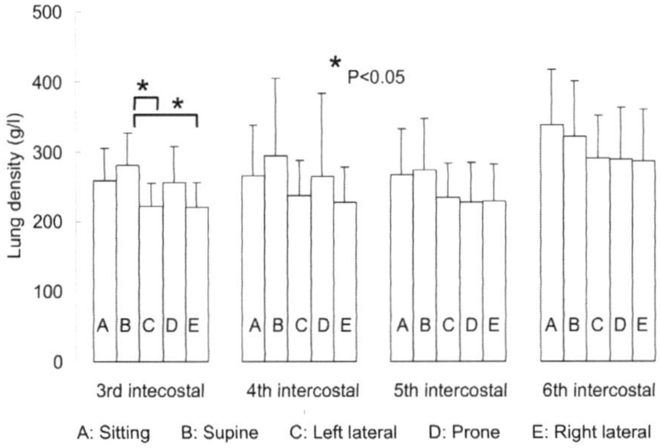

Fig. 3 Lung density in different positions in male subjects

3. Results

Figure 1 shows lung density determined in the 21 normal male subjects, divided according to their positions. In the sitting, right lateral and left lateral positions, significant differences were found in lung density between the electrode level of the 6th intercostal space and any other levels. In the prone position, a significant difference was found in lung density between the electrode levels of the 5th and 6th intercostal spaces. In the supine position, no significant differences were found between any levels.

Figure 2 shows lung density determined in the 21 normal female subjects, divided according to their positions. In all the positions, significant differences were found in lung density between the electrode level of the 6th intercostal space and any other levels.

Figure 3 shows male lung density identical to that in Fig. 1 but divided according to the electrode levels. At the electrode level of the 3rd intercostal space, significant differences were found in lung density between the supine and right lateral positions and between the supine and left lateral positions.

Figure 4 shows female lung density identical to that in Fig. 2 but divided according to the electrode levels. At the electrode level of the 6th intercostal space, no significant differences were found in lung density between any positions. At the electrode levels of the 3rd, 4th, and 5th intercostal spaces, significant differences were found in lung density between several positions.

4. Discussion

These results indicate that lung density at the electrode level of the 6th intercostal space is different from that at almost any other levels in both male and female subjects, and that lung density at the electrode levels of the 4th and 5th intercostal spaces in male subjects can be regarded as independent of position.

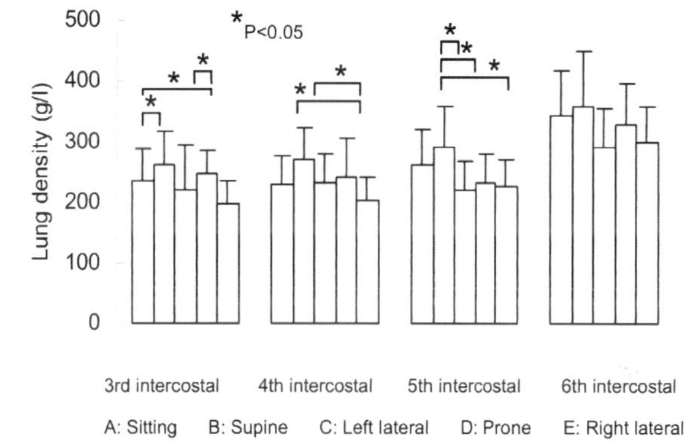

Fig. 4 Lung density in different positions in female subjects

One of the reasons why body position and electrode level affect lung density might be that the body position changes the lungs in shape and that both the body position and the electrode level were associated with the distance between the lungs and the electrodes. Chest CT imaging is necessary to confirm this hypothesis. There are also likely to be gravitational affects on lung density. There might be a further reason for the effects of the electrode level on lung density. The 3D finite difference model of the thorax used for the determination of lung density was based upon the anatomical structure at a specific electrode level. The model might not fully hold for other electrode levels. More models should be developed to enable accurate determination of lung density at other electrode levels.

References

[1] Brown B H, Tidy J, Boston K, Blackett A D, Smallwood R H and Sharp F, 2000, *The relationship between tissue structure and imposed electrical current flow in cervical neoplasia*, The Lancet, 355, 892-895.

[2] Gheorghiu M, Gersing E and Gheorgiu E, 1999, *Quantitative analysis of impedance spectra of organs during ischemia*, Ann. N Y Acad. Sc. 20, 873, 65-71.

[3] Noble T J, Harris N D, Morice A H, Milnes P and Brown B H, 2000, *Diuretic induced changes in lung water assessed by electrical impedance tomography*, Physiol. Meas. 21, 155-163.

[4] Frerichs I, Hahn G, Schroder T and Hellige G, 1998, *Electrical Impedance Tomography in Monitoring Lung Injury*, Intensive Care Medicine, 24(8), 829-836.

[5] Brown B H, Primhak R A, Smallwood R H, Milnes P, Narracott A J and Jackson M J, 2002, *Neonatal lungs –can absolute lung resistivity be determined non-invasively?*, Med. Biol. Eng. & Comp. 40, 388-394.

[6] Barber D C, Seagar A D, 1987, *Fast reconstruction of resistance images*, Clinical Physics & Physiol. Measurement, 8 (A), 47-54

[7] Nopp P, Harris N D, Zhao T-x and Brown B H, 1997, *Model for the dielectric properties of human lung tissue against frequency and air content*, Med. Biol. Eng. & Comp. 35, 695-702.

Separation of ventilation and perfusion related signals within EIT-data streams

R Pikkemaat[1] **and S Leonhardt**[1]

[1] Chair for Medical Information Technology, RWTH Aachen University, Aachen, Germany

E-mail: pikkemaat@hia.rwth-aachen.de

Abstract. Electrical impedance tomography is a widely established technique for ventilation monitoring in the case of clinical research. To assess the capability of a patient's gas exchange, it is necessary to determine the regional supply of ventilation V as well as the local perfusion Q. Therefore, many approaches have been investigated to extract both signals from EIT-data, which would allow a V/Q-mapping. The challenge of this issue depends on the very weak perfusion related signal compared to the signal of ventilation. First approaches to visualize the perfusion were performed inducing apnea, ECG-gating or were based on Fourier series. Rather new techniques use principle component analysis (PCA) defining eigenvectors as a signal in time and thus try to determine "typical" ventilatory as well perfusion related signal streams. In this work, the algorithms of separation will be analyzed and compared (using apnea needs no separation and will not be discussed in this work).

1. Introduction

Several approaches have been developed to separate the perfusion and ventilation related signals in EIT-data streams in order to provide an estimation of the V/Q-distribution within the lungs. This ratio can be used to assess the potential gas exchange and thus to evaluate the quality of the applied ventilation modality. Studies have shown that a V/Q ratio of 0.8 to 1.0 is optimal for the gas exchange between both media. Unfortunately, it is not possible to estimate the local V/Q ratio dynamically using classical imaging techniques. Hence, this specific issue is a very interesting aspect in EIT research, due to the great potential of this technique as a bedside monitoring device (small, cheap, no adverse effects) and the very fast data acquisition. To provide dynamic V/Q-mapping, the problem of separating both signals from each other has to be solved. In this paper different methods of extracting the perfusion related signals (PCA, Fourier-spectrum) or amplify the perfusion related part of the mixed EIT-signal (ECG-gated EITs) are compared to each other and discussed.

2. Materials and Methods

For data acquisition a pig has been ventilated using pressure controlled ventilation with a rather high PEEP of $20cmH_2O$. During the whole procedure EIT has been recorded with a frame rate of 40 frames per second (fps). We used a "Dräger EIT Evaluation Kit 2" (EEK2), which has been kindly provided by Dräger Medical AG and Co. KG. Besides the EIT, an ECG has been recorded with a sampling rate of 200Hz. In both data streams we used markers to define beginning and end of the measurement, which enabled us to synchronize both streams offline. Calculations and visualizations have been performed using Matlab.

3. ECG-gated EIT

The mixed EIT data stream is defined as $I(m,n,t)$, in which (m,n) denotes the spatial position (pixel) and t defines the point in time. The method of ECG-gating amplifies the cardiac/perfusion related signals (PRS). This technique uses the QRS-complex of the synchronously measured ECG-signal as trigger (see [2]). Each value of PRS built by ECG-gating $P_{ECG}(m,n,t)$ is given by a mean of N_{QRS} values measured at a time t after the corresponding trigger T_{QRS}. If the length of an ECG-gated data stream is defined as T and includes $F = T/fps + 1$ frames, the ECG-gated EIT data stream can be defined as a $mxnxF$-Matrix $P_{ECG}(m,n,t)$ beginning at a trigger T_{QRS_1}:

$$P_{ECG}(m,n,t) = \frac{\sum_{i=1}^{N_{QRS}} I(m,n,T_{QRS_i}+t)}{F} \tag{1}$$

Assuming that the perfusion related signal (PRS) has a deterministic characteristic and the ventilation related signals (VRS) are completely uncorrelated with PRS, the signal strength of the PRS gets amplified by a factor of $\sqrt{N_{QRS}}$ in comparison to the strength of VRS. In this work, we used a period $T = max(\Delta T_{QRS_i})$ at which ΔT_{QRS_i} is defined as $\Delta T_{QRS} = T_{QRS_{i+1}} - T_{QRS_i}$ and $N_{QRS} = 200$. To get a data stream containing several heart beats the single ECG-gated sequences were connected, while skipping the last frames if the gated sequence is longer than the current ΔT_{QRS_i}. So the real heart rate isn't affected by this algorithm.

4. Fourier-based approach

The EIT-measurements include a mix of PRS and VRS. In this approach the fact that the heart rate lies in another range of frequency as the ventilation rate. Based on this assumption, the decomposition of the EIT-data stream into both signals could be performed by defining the PRS and VRS related domain (magnitude and phase) in the spectrum of the transformed EIT-signal (see [4], [3]), extract these parts and get both signals by using an inverse Fourier-Transformation.

Unfortunately, the spectrum exhibits not only a strong peak in magnitude at the ventilator frequency but also the harmonics. Due to the reason that the PRS is much smaller than the VRS, it may happen that a harmonic of the VRS lies on the same frequency-band as the PRS. So it is rather difficult to separate them in the spectrum without having any shifts of energy between the VRS-harmonic and the PRS. Especially if one tries to assess the cardiac output quantitatively from this data, the shifts will lead to erroneous values. In this work we considered tomograms depicting the distribution of energy (magnitude) and phase angle at one certain frequency (similar considerations have been made by Kerrouche et al. [3]).

Considering Fig. 1 to Fig. 4, it is observable that the VRS has nearly no difference in phase angle within the lung-region. The PRS exhibits only phase synchronicity in the heart region. By contrast, the signal related to pulmonal perfusion exhibits the whole range of phase angle and points the blood ejecting into the pulmonary tissue out.

The strong magnitude in the heart region at both frequencies (0.42Hz and 1.47Hz) could be explained by a compression of the heart due to the increase air pressure. Another possibility is that the heart gets pushed out of the slice, which is mainly responsible for the measured EIT-image and thus gets smaller in the tomogram.

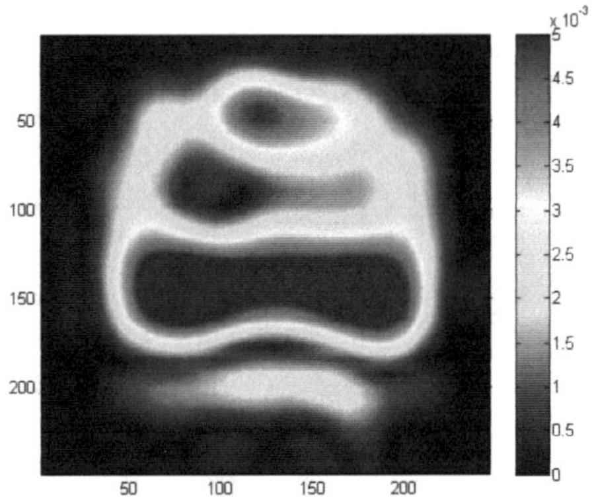

Figure 1. Magnitude in AU related to the ventilation at f=0.42 Hz.

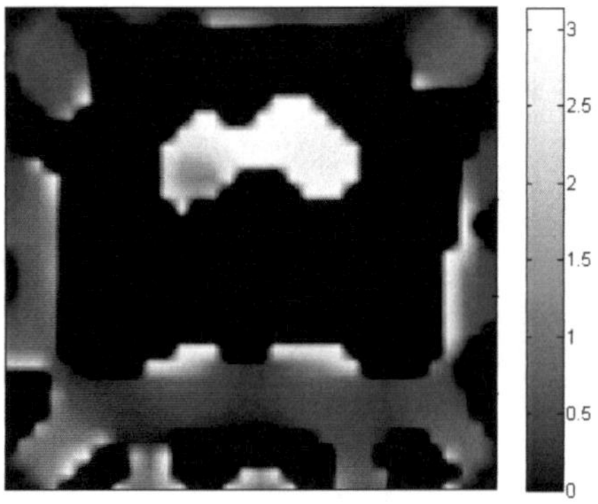

Figure 2. Phase angle in *rad* related to the ventilation at f=0.42 Hz.

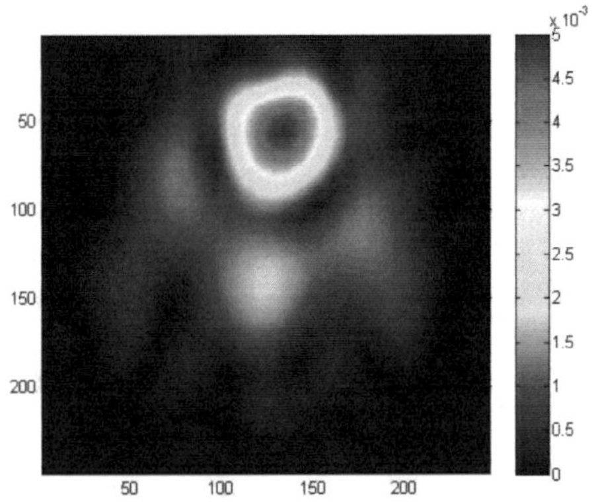

Figure 3. Magnitude in AU related to the ventilation at f=1.47 Hz.

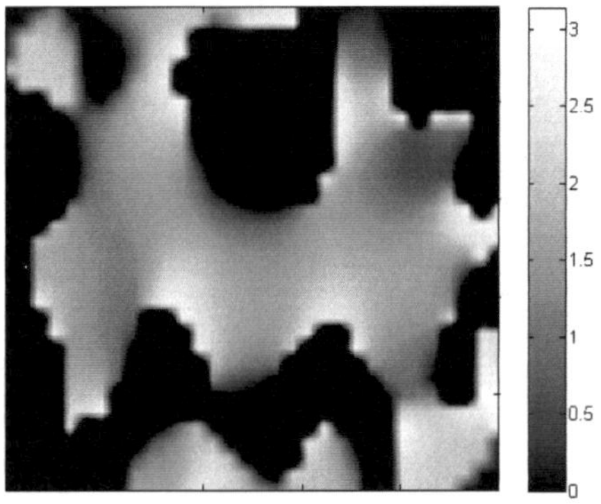

Figure 4. Phase angle in *rad* related to the ventilation at f=1.47 Hz.

5. Separation using PCA

Both strategies mentioned so far are not suitable to generate the PRS out of the whole EIT-data stream in real-time. Since the ECG-gating lasts N_{QRS} heart beats to get one period it reacts quite slowly to hemodynamic changes. Using a reduced N_{QRS} the calculated signal will be able to show faster dynamics at the cost of signal strength of the PRS. A similar problem occurs while using the Fourier-based algorithm. Before applying a fast-Fourier-Transformation (FFT) one has to define a certain span of time and thus this technique is only suitable for offline analyses.

A possibly more adequate strategy is to use principle component analyses. In this technique the EIT-data is analyzed for a certain period of time (comparable to the Fourier-based algorithm). In this case the signal is decomposed into its temporal principal components given by template functions in time denoting the biggest variance of EIT-signal in an F-dimensional space, where F is the number of applied frames (s. Kerrouche et al. [3], Deibele et al.[1]).

In this work we used a similar algorithm to the work of Deibele et al. This algorithm tries to combine multiple principal components to one PRS based on data, from which the VRS is already subtracted and band-pass filter was already applied. Since the ventilated parts are nearly phase-synchronous at the ventilator frequency, as could be seen in the Fourier-based algorithm (s. Fig. 2), the error of modeling the VRS by only one principal component is rather small.

The benefit of this technique in comparison to the Fourier-based one is that once the template functions have been calculated they can be used to determine the strength of the VRS and PRS within the mixed EIT signal very fast and thus this algorithm is more suitable for real-time calculations or possibly is able to show faster dynamics (e.g. changes in stroke volume). Of course, this will only give accurate results, if the characteristic shape of the PRS and VRS will not change too much, otherwise it is necessary to calculate the template function regularly. For more details see [1].

6. Comparison of PCA-, Fourier- and ECG based PRS-separation

Let $p_{ECG}(m,n,t)$ be the signal vector of the ECG-gated PRS $P_{ECG}(m,n,t)$ of length of 1 given by:

$$p_{ECG}(m,n,t) = \frac{P_{ECG}(m,n,t)}{\sqrt{\sum_{t=0}^{T} P_{ECG}(m,n,t)^2}} \tag{2}$$

In analogy to $p_{ECG}(m,n,t)$ the normed signal vector in temporal dimension of the PCA-based and the Fourier-based PRS are defined as $p_{PCA}(m,n,t)$ and $p_{Fourier}(m,n,t)$, respectively the temporal correlation r is the given by:

$$r_{ECG-PCA}(m,n) = \sum_{t=0}^{T} p_{ECG}(m,n,t) \cdot p_{PCA}(m,n,t) \tag{3}$$

The results of the correlation per pixel are shown in Fig. 5 to 7.

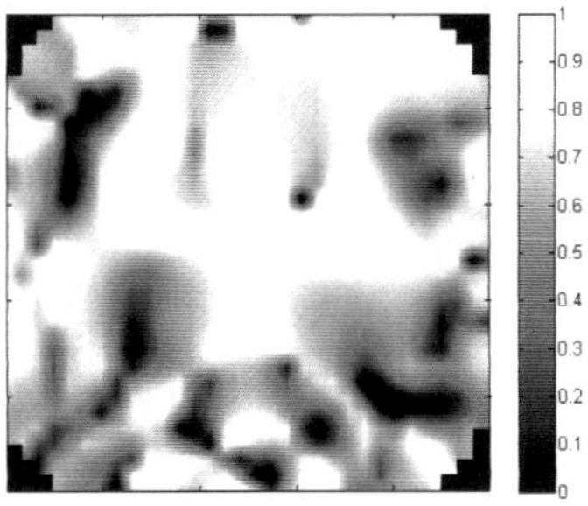

Figure 5. Temporal correlation between ECGg and Fourier based PRS; Mean r=0.78

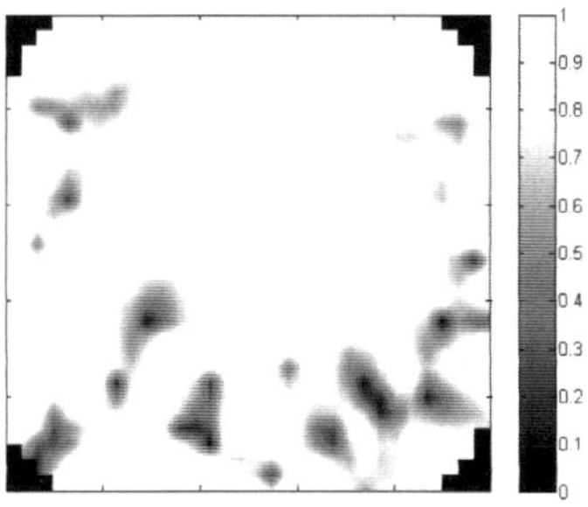

Figure 6. Temporal correlation between Fourier and PCA based PRS. Mean r=0.93

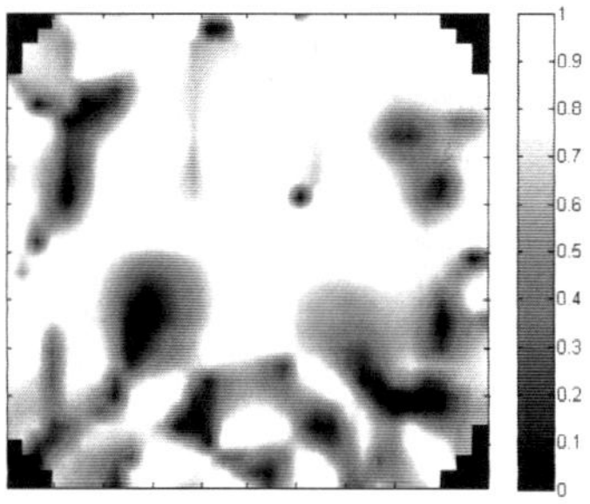

Figure 7. Temporal correlation between ECGg and PCA based PRS; Mean r=0.82

There are significant differences between the mean correlation r as well as the homogeneity of the correlation-image. Especially the ECG-gated PRS does not correlate as good to the PRS of the other techniques as $P_{PCA}(m,n,t)$ and $P_{Fourier}(m,n,t)$ to each other. The correlation-image of $P_{PCA}(m,n,t)$ and $P_{Fourier}(m,n,t)$ is rather homogenous and the mean correlation is quite high compared to both other correlations. The correlation in the heart region is generally homogeneous and quite high among all three PRS.

7. Results

All considered PRS show overall quite similar characteristics. In detail the Fourier based separation shows nearly equal results as the PCA-based algorithm. Especially in periods of fast dynamics, there are partly considerable differences in time. The temporal changes were mainly detected in areas distant to the heart region, where the amplitude of the perfusion related signal is very small.

The ECG-gated algorithms as implemented here exhibits weak points in connecting single gated periods to one continuous data stream. Due to the stability of this algorithm (it works on almost every data set), this technique is still a meaningful fallback solution. Both other techniques gave nearly equal PRS, but for being suitable to follow faster dynamics the Fourier-based separation has to include higher harmonics of the PRS and thus a more complex Fourier-based approach has to be developed. Without knowing the heart rate or other information the PCA-based algorithm showed reasonable results and is able to follow faster temporal dynamics, because the template functions have generally a non limited dynamic. As long as the signal's characteristic does not change too much, the template functions can be used to calculate the PRS fast and thus this technique might be promising for real-time separation.

Acknowledgments

I'd like to thank the Dräger Medical AG and Co. KG for supplying an EIT evaluation kit (EEK2) for our animal trial.

References

[1] J. Deibele, H. Luepschen, and Leonhardt S. Dynamic separation of pulmonary and cardiac changes in electrical impedance tomography. *Phys. Meas.*, 29:1–14, 2008.

[2] B.M. Eyboglu and B.H. Brown. Methods of cardiac gating applied potential tomography. *Clin. Phys. Meas.*, 9:43–48, 1988.

[3] N. Kerrouche, C.N. McLeod, and W.R. Lionheart. Time series of eit chest images using singular value decomposition and fourier transfrom. *Phys. Meas.*, 22:147–157, 2001.

[4] M. Zadehkoochak, B.H. Blott, T.K. Hames, and George R.F. Pulmonary perfusion and ventricular ejection imaging by frequency domain filtering of eit images. *Clin. Phys. Meas.*, 13:191–196, 1992.

Measurement of lung function using Electrical Impedance Tomography (EIT) during mechanical ventilation

Satoru Nebuya [1], Tomotaka Koike [2], Hiroshi Imai [3], Makoto Noshiro [1], Brian H. Brown [4], Kazui Soma [3]

[1]Graduate School of Medical Science, Kitasato University, Sagamihara, Japan
[2]Rehabilitation Centre, Kitasato University Hospital, Sagamihara, Japan
[3]Dept. of Emergency and Critical Care Medicine, Kitasato University, Sagamihara, Japan
[4]Dept. of Medical Physics and Clinical Engineering, University of Sheffield, Sheffield, UK

nebuya@kitasato-u.ac.jp

Abstract. The consistency of regional lung density measurements as estimated by Electrical Impedance Tomography (EIT), in eleven patients supported by a mechanical ventilator, was validated to verify the feasibility of its use in intensive care medicine. There were significant differences in regional lung densities between the normal lung and diseased lungs associated with pneumonia, atelectasis and pleural effusion (Steel-Dwass test, $p < 0.05$). Temporal changes in regional lung density of patients with atelectasis were observed to be in good agreement with the results of clinical diagnosis. These results indicate that it is feasible to obtain a quantitative value for regional lung density using EIT.

1. Introduction

Mechanical ventilation is necessary in intensive care medicine. Though management of respiration with mechanical ventilation can improve the prognosis for acute phase patients, it has recently been known to induce ventilator-induced lung injury (VILI) or a ventilator-associated lung injury (VALI). Also, acute respiratory distress syndrome (ARDS) is associated with inhomogeneous lesions in the lungs. Therefore, there is a need for local regional monitoring of lung lesions at the bedside. Electrical impedance tomography (EIT) has been tried for the monitoring of lung lesions of patients [1-4]. Brown et al have suggested a measurement method of calculating absolute lung resistivity [5] and of lung density [6] from EIT data sets. However, the clinical feasibility of these methods has not been studied on patients connected to a mechanical ventilator.

In this paper, regional lung density and lung disease in patients undergoing mechanical ventilation are compared, in order to verify the clinical feasibility of determining lung density from EIT data sets.

2. Method

2.1 Determination of lung density using absolute lung resistivity

Absolute lung resistivity (*AbR*) was determined using the method devised by Brown et al [5]. In summary, this method returns the best estimate of the absolute value of lung tissue resistivity by comparing the measured EIT data to computed data sets. Lung Filling Factor (*FF*) is the ratio by volume of air to condensed matter within the lung tissue. A function between *AbR* and *FF* was provided using a numerical lung model developed by Nopp et al [7]. If we know the density of the condensed matter of the lungs then we can relate *FF* to the overall density of lung tissue (Lung density: *LD*) as follows:

$$LD = \frac{Lung\_Weight}{Air\_Volume + Tissue\_Volume} \quad (1)$$

$$FF = \frac{Air\_Volume}{Tissue\_Volume} \quad (2)$$

Substituting from equation (1) into equation (2)

$$LD = \frac{Lung\_Weight}{FF.Tissue\_Volume + Tissue\_Volume} = \frac{\rho_t}{FF + 1} \quad (3)$$

and

$$FF = \frac{\rho_t}{LD} - 1 \quad (4)$$

We will use $\rho_t = 1050 kg \bullet m^{-3}$ [8]

AbR was determined as a function of FF from the Nopp model [6] and hence related to AbR.

2.2 Measurement of regional lung density using absolute lung resistivity

EIT data was measured 207 times in total from 11 male patients connected to a mechanical ventilator in the intensive care unit at Kitasato University Hospital. Table 1 shows the patients' profiles.

Lung densities were estimated for four lung regions defined as right anterior, left anterior, right posterior and left posterior as shown in Fig.1. The 828 lung regions were grouped into four categories, namely normal lung (332) and three pulmonary diseases that were pneumonia (104), atelectasis (274) and pleural effusion (87), on the basis of clinical diagnosis based on X-ray and CT images. The lung densities for these four categories were statistically compared using the Steel-Dwass test ($p < 0.05$). In addition, continuous changes in regional lung densities were measured for eight days from one patient and were compared with the clinical diagnosis.

The study was approved by the university ethics committee and informed consent was obtained from the patient's relative.

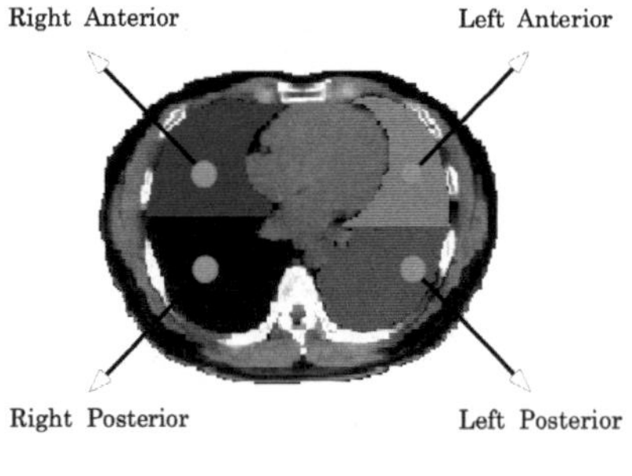

Fig. 1 Definition of lung regions at the level of the EIT electrode plane.

Table 1 Patients' profiles (N=11)

Height (cm)	165±0.04
BMI	23.0±3.3
Gender (M : F)	11:00
RASS index	-4.27±1.0

Diagnosis	
Post CPA	2
Severe pancreatitis	2
Septic shock	4
Subarachnoid hemorrhage	1
Brain contusion	1
Severe pneumonia	1
Total number	11

BMI; Body Mass Index
RASS; Richmond Agitation and Sedation Scale
CPA; Cardio Pulmonary Arrest

Table 2 Numbers of measurements, position, and pulmonary disease divided four areas (Eleven cases, 207 measurement, and 828 lung areas).

Posture	Total
Supine	101
Right lateral decubitus	48
Left lateral decubitus	51
Others	7

Lung area	RA	LA	RP	LP	Total
Normal	129	100	33	70	332
Pneumonia	15	71	12	6	104
Atelectasis	46	12	122	94	274
Pleural effusion	4	21	27	35	87
Total	194	204	194	205	797

RA: Right anterior, LA: Left anterior,
RP: Right posterior and LP: Left posterior.

3. Results

Table 2 gives statistics on the EIT measurements that were made in the clinically diagnosed groups. Approximately half of the EIT data were measured in the supine position and 40% of regional lung areas were diagnosed as normal.

Figure 2 shows the relationship between pulmonary disease and the measurements of lung density in the supine position. There were significant differences in regional lung density between the four groups($p < 0.05$), except for pneumonia and atelectasis, and the mean values of the regional lung densities in the three pulmonary diseases were higher than for those in normal lung. In addition, when the posture of the patients was changed from supine to the left lateral decubitus position, there was a significant difference in the regional lung densities between pneumonia and atelectasis but no significant difference between atelectasis and pleural effusion (Fig.3). Similar results were also obtained for the right lateral decubitus position.

Furthermore, regional lung density measurements in one patient diagnosed as having atelectasis decreased day by day (Fig.4). At Day 1, no obvious activity of the alveoli in the right lung had been observed by the clinician. The results indicated that only the left lung was working and the chest X-ray, taken on the same day, was in good agreement with this result. On the other hand, activity of the alveoli in the right anterior of the lung had clearly increased and the chest X-ray was in agreement with this on Day 8. Lung densities in the right lung decreased from higher values on Day 1 to the values measured from normal lung regions as shown Fig.3 (221 ± 49.4 kg・m^{-3}).

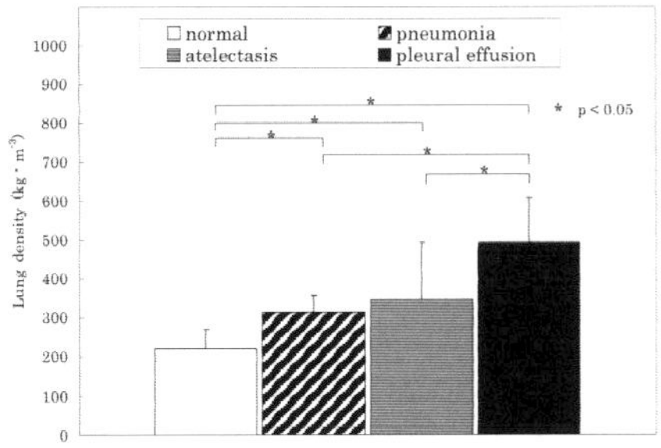

Fig. 2 Relationship between pulmonary disease and lung density in the supine position

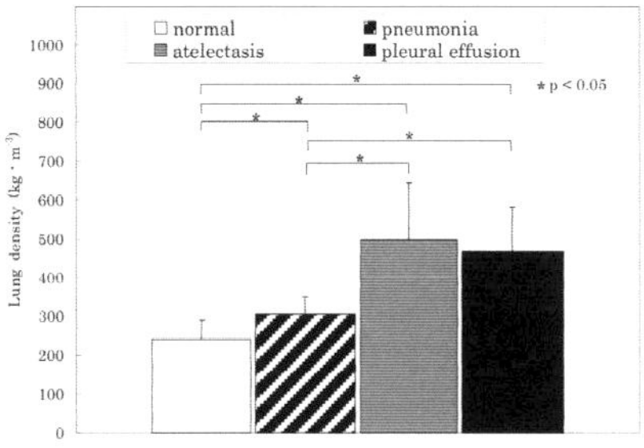

Fig. 3 Relationship between pulmonary disease and lung density in the left lateral decubitus position

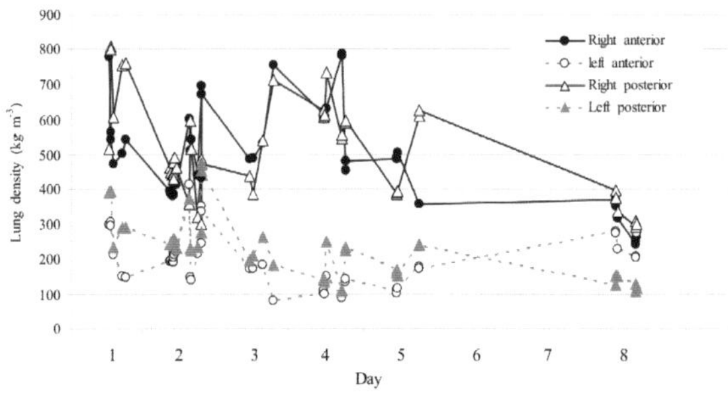

Fig. 4 Changes in regional lung densities over eight days for a patient with pneumonia and atelectasis of right lung who was attached to a ventilator.

4. Discussion

Statistical differences were found between the regional lung densities of normal and pulmonary diseases. However, there were no significant differences between pneumonia and pleural effusion in the supine position. To discriminate both diseases, measurement in the left lateral decubitus position was also needed. The reason for the observed large changes in regional lung density associated with pleural effusion was considered to be the movement of lung water with changes in posture of the patient.Furthermore, regional lung density measurements in one patient diagnosed as having atelectasis decreased day by day (Fig.4). This trend was observed to be in good agreement with the results of clinical diagnosis based on X-ray and CT images.

These results indicate that it is feasible to obtain a quantitative value for regional lung density using the EIT technique with no risk of exposure to harmful radiation as in X-ray and CT. EIT may be a useful technique for daily monitoring of the lung function of patients in an intensive care unit, although the CT has the advantage of providing much better anatomical diagnosis.

References

[1] Costa E L, Borges J B, Melo A, Suarez-Sipmann F, Toufen C Jr, Bohm S H and Amato M B, 2009, *Bedside estimation of recruitable alveolar collapse and hyperdistension by electrical impedance tomography*, Intensive Care Med., 35(6), 1132-1137.

[2] Bikker I G, Leonhardt S, Bakker J and Gommers D, 2009, *Lung volume calculated from electrical impedance tomography in ICU patients at different PEEP levels*, Intensive Care Med., 35(8), 1362-1367.

[3] Arad M, Zlochiver S, Davidson T, Shoenfeld Y, Adunsky A and Abboud S, 2009, *The detection of pleural effusion using a parametric EIT technique*, Physiol Meas., 30(4), 421-428.

[4] Zhao Z, Möller K, Steinmann D, Frerichs I and Guttmann J, 2009, *Evaluation of an electrical impedance tomography-based Global Inhomogeneity Index for pulmonary ventilation distribution*, Intensive Care Med., 35(11), 1900-1906.

[5] Brown B H, Primhak R A, Smallwood R H, Milnes P, Narracott A J and Jackson M J, 2002, *Neonatal lungs –can absolute lung resistivity be determined non-invasively?*, Med. Biol. Eng. & Comp. 40, 388-394.

[6] Brown B H, Milnes P and Mills G H, *Indirect measurement of lung density and air volume from electrical impedance tomography (EIT) data*, 2006, Proc. 7th Conf. Biomed. Appli. Electrical Impedance Tomography, Seoul, 85-89.

[7] Nopp P, Harris N D, Zhao T X and Brown B H, 1997, *Model for the dielectric properties of human lung tissue against frequency and air content*, Med & Biol Eng & Comp 35, 695-702.

[8] Boyd E, 1962, *in Altman and Dittmer (Eds) Growth, Including Reproduction and Morphological Development*, Biological Handbooks, Federation of American Society for Experimental Biology, Washington, 346-348.

Tidal Volume (TV) post-process obtained with Electrical Impedance Tomography on a group of chronic obstructive pulmonary disease (COPD) patients. Use of adjust equations

Marco Balleza[1], Daniel Anton[1], Pere Casan[2], Pere Riu[1]

1. Universitat Politécnica de Catalunya, Centre de Recerca en Enginyeria Biomèdica, Dept. Eng. Electrònica, Barcelona, España
2. Hospital Central de Asturias, Oviedo, Asturias, España

Email. jballeza@eel.upc.edu

Abstract. Equations used to estimate ventilation out of EIT images, validated on healthy volunteers show a significant bias and a larger variance when were applied on a COPD patients group. The differences in estimation values were found to be highly correlated with anthropometric parameters. Two groups of 13 and 4 COPD male patients (FEV1/FVC<70% and FVC $\geq$ 80%) were used in this study. We have measured different anthropometric parameters like age, weight, height and skinfolds. The EIT system (TIE4sys) and a pneumotach were simultaneously connected to monitor tidal volume. The main anthropometric parameters values of 13 COPD patients were: age: 67±9 years, height: 1.65±0.05 m, weight: 72±11 kg, BMI: 26.4±3.3; and the subscapular skinfold thickness was 23±9mm. The mean tidal volume estimated with TIE4sys and the pneumotach were: 0.580±0.212 L and 0.774±0.173 L r = 0.861 (p<0.01). The mean difference was 0.196±0.096 L (p<0.01). On this group we have found out an adjust equation and we have validated it on an independent group of 4 COPD patients. The equation was Diff=-1478+15.6(weight). The mean tidal volume values obtained with pneumotach and TIE4sys on the second group of COPD patients (M:4) were: 0.798±0.395 L and 0.732±0.327 L. The mean of the differences was 0.066±0.114L. The differences of determinations estimated with pneumotach and TIE4sys can be attributed to changes of anthropometric characteristics like subscapular skinfold.

1. Introduction

The monitoring of respiratory pattern in a non-invasive way is not a completely standardized procedure in pneumology. There are different monitoring devices like inductive plethysmography bands (e.g. Respitrace™) [1] which has fallen into disuse because they presented calibration problems. On the other hand, there is another monitoring device based on the transmission of respiratory vibrations which is not yet tested enough for this application [2]. Suitability of Electrical Impedance Tomography (EIT) applied to ventilation measurement has been demonstrated for the Unilateral Pulmonary function (UPF) [3][4] and more recently for ventilatory pattern monitoring in healthy people [5]. The EIT procedure was validated for the estimation of tidal volume (TV) magnitudes in healthy people. Calibration equations depend on different anthropometric parameters like height, weight, BMI, etc. These parameters affect the impedance determinations obtaining with EIT device thus the calibration equations valid for healthy volunteers may not be adequate for COPD patients.

The aim of this work is to further obtain a new correction equation by using a new group of COPD patients with a stricter inclusion criterion.

2. Materials and methods

2.1. Pneumotachometer

The pneumotachometer used to monitor the tidal volume was a Med Graphics preVent™ Pneumotach from Medical Graphics Corporation (St. Paul, MN, USA). This monitoring device records signals (volume – time) in real time,

2.2. TIE4sys device

The TIE4sys device is a prototype designed in the Department of Electronics Engineering in the Universitat Politècnica de Catalunya, Barcelona, Spian. It uses 16 electrodes and can obtain up to 17 frames/second. The electrodes (Red Dot 2560 – 3M, London; Ontario, Canada) were placed around the thoracic box on the sixth intercostals space following the protocol described in [3]. This device was programmed to apply an electrical current of 1 mA_{RMS} at 48 kHz. This electrical current was applied through all pairs of adjacent electrodes. The differential voltage at all pairs of adjacent electrodes was recorded for each injection pair, so redundant measurements were obtained allowing us calculate the reciprocity error, which is an estimation of the systematic errors and malfunctions (such as electrodes with a poor contact) present. Images are obtained by using a weighted back-projection algorithm and a median filter.

2.3. Volunteers

Three groups were used in this experiment. The first group consisted on 11 healthy volunteers males. This group of volunteers was used to establish the main relevant anthropometrical parameters and obtain a calibration equation relating ventilation in order to adjust the volume signal obtained out with EIT system. All volunteers were selected to be non smokers and they had spirometry values within the normal range. The second and third group consisted on 13 males and 4 males diagnosed with COPD and (FEV1/FVC<70% and FVC≥80%). The second group (M_{COPD}:13) was used to validate the calibration equation and obtain an adjust equation in order to adjust the determinations of volume obtained with TIE4sys. And the third one (M_{COPD}:4) was used to validate the adjust equation.

2.4. Procedures

The two monitoring devices (TIE4sys and pneumotach) were connected to take simultaneous measurements. The performance of each device is independent and there is no interference between them. The calibration procedures for each one was: 1) Pneumotach: a 3 litre needle, according to the protocol of the laboratory and manufacturer instructions. 2) TIE4sys: Acquisition speed and reciprocity error are checked before starting to record measurements. Several respiratory cycles (5 to 10) are then acquired at tidal volume to be averaged and constitute the reference for the backprojection. The tidal volume was recorded for several 30 s slots (between 5 and 8 respiratory cycles), with 3 min of resting between slots, until 25 and 30 cycles were obtained for each patient. Prior to start the measurements protocol, anthropometric data for each volunteer was collected, including sex, age, height, weight, BMI, skinfold thickness, thoracic parameters and diameters. In the case of COPD patients, the pulmonary function test data was collected.

2.5. Statistical Comparison

All data are expressed as average and standard deviation values. The comparison between the determinations obtained by pneumotach and TIE4sys device were done using a non-parametric t-test, the Spearman rank correlation coefficient and a Bland & Altman plot. Values of $p<0.05$ were considered to be significant.

3. Results

The anthropometric main values of the 11 healthy volunteer males were: age: 27±9 years, height: 1.80±0.1 m, weight: 78±13 kg, BMI: 24.2±3.1 kg/m^2; and subscapular skinfold thickness 23±9mm. The most relevant parameters for males were found to be the subscapular skinfold (SS_Sk) and the weight (W). The resulting equation is then:

$$\Delta V_{MALE} = \frac{\Delta IC}{2889 - 25.7(SS\_Sk) - 16(W)} \tag{1}$$

The determination coefficient of this equation was $R^2 = 959$ ($p<0.01$)
The anthropometric main values of the 13 COPD patients were: age: 67±9 years, height: 1.65±0.05 m, weight: 72±11 kg, BMI: 26.4±3.3 kg/m^2; and the subscapular skinfold thickness was 23±9mm. The spirometry main values of 13 COPD patients were: FVC: 83±13%; FEV1: 47±15%; FEV1/FVC: 40±9% of reference value.
The mean value for the tidal volume obtained by pneumotach and TIE4sys were: 0.774±0.173 L and 0.580±0.212 L, respectively. The mean of the differences between the determinations obtained with TIE4sys and pneumotach was 0.194±0.096 l ($p<0.01$). The acceptability of the differences between determinations was 0.003 L and 0.418L, as shown in figure 1a.
In order to reduce the error between TV estimations obtained with the pneumotach and TIE4sys (EIT device) we have obtained a linear equation which involve an anthropometric parameter: the weigth:

$$\Delta V_{ADJUST} = \frac{IC}{-1478 + 15.6(W)} \tag{2}$$

The coefficiente of determination $R^2 = 0.548$.

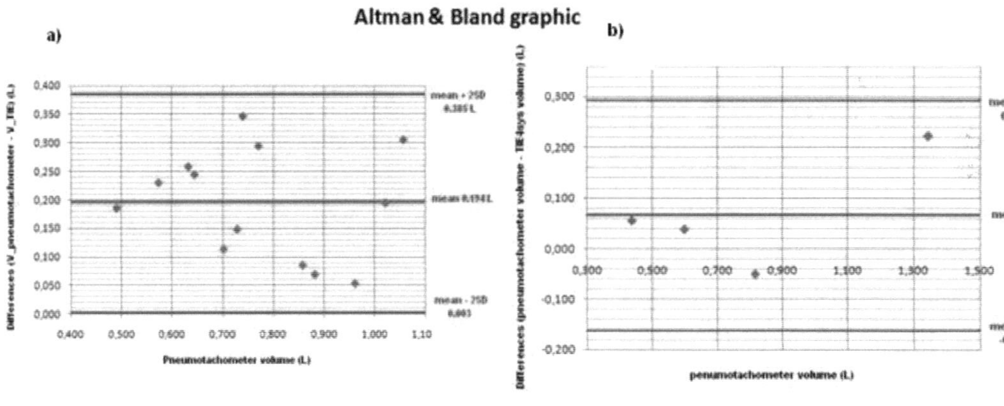

Figure 1. a) Bland & Altman plot for the differences between pnaumotachometer and EIT estimations for TV on 13 COPD patients; b) Bland & Altman for the differences between pneumotachometer and EIT estimations for TV on 4 COPD patients

In this way, the TV estimation for COPD patients is corrected as:

$$\Delta V_{TOTAL} = \frac{\Delta IC}{\left(2889 - 25.7(SS\_Sk) - 16(W)\right) + \left(-1478 + 15.6(W)\right)} \tag{3}$$

The anthropometric main values of the 4 COPD patients were: 66±14 years, height: 1.66±0.05 m, weight: 72±11 kg, BMI: 26.0±2.9 kg/m^2; and the subscapular skinfold thickness was 19±7mm. The spirometry main values of 4 COPD patients were: FVC: 74±5%; FEV1: 33±12%; FEV1/FVC: 31±9%.

On this group we have applied the calibration equation adding the adjust equation. The mean value for tidal volume obtained by pneumotach and TIE4sys were: 0.798±0.395 L and 0.732±0.327 L, respectively. The mean of the differences between the determinations obtained with TIE4sys and pneumotach was 0.066±0.114 L ($p<0.01$). The acceptability of the differences between determinations was 0.293 L and -0.161L, as shown in figure 1b.

4. Discussion

The data obtained in this paper suggest that TV estimation out of EIT in-vivo is affected by body shape and composition. This affirmation is based on the fact that differences between healthy volunteers and COPD patients show a strong relationship with composition related parameters (e.g. skinfold measurements). The visual reduction in deviation is noticeable as depicted on the figure 1b. All the patients are in the acceptability range. One of them presented a difference over 200 ml due to patient's respiratory difficulties and presented an estimation of volume with the pneumotachomer of 1.325 L. If this patient were removed from the sample, then the root mean standard error (RMSE) would be about 5%. In order to continue validating our adjust equation we need to increase our sample of COPD patients. This will be the next step in our research.

5. Conclusions

The impedance changes recorded with EIT during ventilation depend on both the external changes in thorax shape and size. It was expectable that the calibration equation derived for healthy volunteers could not be applied to COPD patients without further changes. This was already shown in a previous research with a group of less controlled COPD patients, including both obstructive and restrictive patterns [6]. Hopefully it seems feasible to correct the equations based on some anthropometrical parameters and reduce the estimation error down to a level acceptable for a monitoring device.

References

[1] Stick SM, Ellis E, LeSouëf PN and Sly PD. (1992) Validation of respiratory inductance plethysmography ("Respitrace") for the measurement of tidal volume breathing parameters in newborn. Pediatr Pulmonol 13;187-91

[2] Phillip Dellinger R, Smith J, Cinel I, Tay C, Rajanala S,, Glickman YA and Parrillo JE (2007) Regional distribution of acoustic-based lung vibration as a function of mechanical ventilation mode. Critical Care 11:R26

[3] Serrano RE, de Lema B, Casas O, Feixas T, Calaf N, Camacho V, Carrión I, Casan P, Sanchis J and Riu PJ (2002). Use of electrical impedance tomography (EIT) for the assessment of unilateral pulmonary function. Physiol Meas 23:211-220.

[4] Serrano RE, Riu PJ and Casan P (2004). Assessment of the unilateral pulmonary function by means of electrical impedance tomogephy using a reduce electrode set. Physiol Meas 25:803-813.

[5] Balleza M, Fornos J, Calaf N, Feixas T, González M, Antón D, Riu PJ, Casan P. (2007) Monitoring of Breathing Pattern at Rest by Electrical Impedance Tomography. Arch Bronconeumol 43:300-303.

[6] Balleza M, Fornos J, Calaf N, Feixas T, Gonzalez M, Anton D, Riu PJ, Casan P. (2007) Ventilatory Pattern monitoring with Electrical Impedance Tomography (EIT). Validation for Healthy Subjects. IFMBE Proceedings 17: 572-575.

Ventilation mapping of chest using Focused Impedance Method (FIM)

M Abdul Kadir[1], Humayra Ferdous[2], Tanvir Noor Baig[2] and K Siddique-e-Rabbani[1]
[1]Department of Bio medical Physics & Technology, University of Dhaka, Dhaka-1000, Bangladesh
[2]Department of Physics, University of Dhaka, Dhaka-1000, Bangladesh

email: rabbani@univdhaka.edu

Abstract: Focused Impedance Method (FIM) provides an opportunity for localized impedance measurement down to reasonable depths within the body using surface electrodes, and has a potential application in localized lung ventilation study. This however needs assessment of normal values for healthy individuals. In this study, localized ventilation maps in terms of electrical impedance in a matrix formation around the thorax, both from the front and the back, were obtained from two normal male subjects using a modified configuration of FIM. For this the focused impedance values at full inspiration and full expiration were measured and the percentage difference with respect to the latter was used. Some of the measured values would have artefacts due to movements of the heart and the diaphragm in the relevant anatomical positions which needs to be considered with due care in any interpretation.

1. Introduction

Monitoring of physiological events by impedance has become a subject matter of great interest for the last few decades. Dielectric properties of lungs tissue change as a function of air content [1], therefore lungs ventilation makes a potential area of application of electrical impedance techniques. Electrical Impedance Tomography (EIT) has been applied in the study of lungs ventilation successfully to explore lungs physiology as well as to monitor and diagnose lungs disorders and injury [2][3]. A relatively new development in the field of bio-impedance measurement is the Focused Impedance Method (FIM) [4][5], developed at the authors' laboratory at Dhaka University, in which two pairs of current drive electrodes perpendicular to each other over a common zone at the center are used to apply alternating currents of constant amplitude separately. Two potential measuring electrodes with a smaller separation, placed at the centre at an angle of 45^0 to either of the current drive directions, are used to measure the potentials resulting from each of the current drives. The sum of the two measured potentials has a dominant contribution from the central region, approximately a square with the potential electrodes at the diagonal points. Therefore this gives localized sensitivity at the central region which has been termed as the focused region. Because of a 3D sensitivity, FIM electrodes, placed on the skin surface, may be used to monitor changes at a depth, and has been successful in measuring gastric emptying [5]. FIM applied to a localized region of lung showed a linear relationship to change in expired volume of air in a subject [6]. It needs to be seen if FIM could be used to detect and quantify changes in localised lung ventilation, for potential use in detection and diagnosis of localized lung ventilation disorders. In this preliminary study, the FIM impedance change between maximum inspiration and maximum expiration at different localized segments of the thorax of some healthy human subjects was measured using the 6 electrode FIM system to get a ventilation mapping matrix, with the above aim.

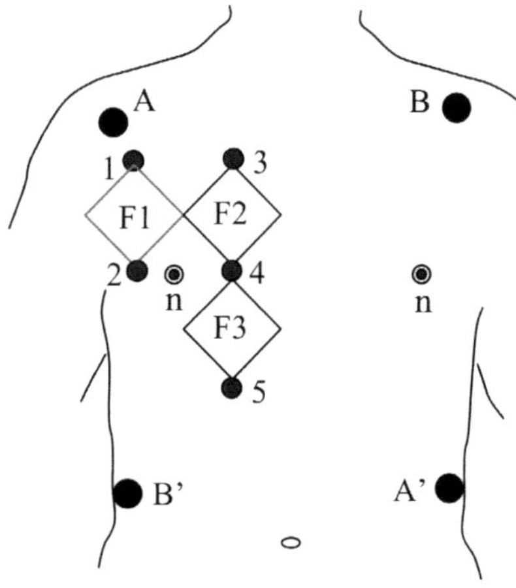

Figure 1. Electrode configuration and focused zones for ventilation mapping of human thorax. Current electrode pairs: A-A', B-B'. Potential electrodes pairs: 1-2, 3-4, 4-5, etc. giving focused zones F1, F2, F3, etc. respectively.

2. Methods and Measurements

Here the FIM has been applied with some modifications in order to facilitate measurement and standardization as shown in figure 1. A 10kHz alternating current of constant amplitude (approx. 0.5mA) was passed, in sequence, through two orthogonal electrode drive pairs A-A' and B-B' as shown, by switching the pairs manually. These electrodes were placed such that they make the largest possible approximate square on the chest. Using a hand-held probe, two potential measuring electrodes, 10cm apart, were then placed vertically at different places, such as at 1&2, 3&4, etc. on the chest. For each placement, the two potentials developed due to the two orthogonal current drives were recorded and summed, which essentially gave a value proportional to the focused impedance within the respective zones F1, F2, etc., since the current amplitudes were constant. Measurements were taken with the subject taking a deep breath (maximum inspiration) and then with the subject breathing out as much as possible (maximum expiration). The difference was expressed as a percentage of the value at maximum expiration, which is described as the %change of focused impedance in this work. Each measurement was repeated once and averaged to minimise the effect of changing blood volume in the heart. This %change in impedance is essentially a ventilation parameter of the lungs in the respective focused region, down to a certain depth. The procedure was then repeated with the potential electrode pair placed at different points on the chest in a predefined matrix. Similar measurements were carried out on the back as well. The breast nipple level (indicated by n-n) was taken as an anatomical reference for placement of potential electrodes. Measurements were made with the subject standing upright. The total volume of air exhaled by the subject in each of these procedures is a function of the lungs volume and is known as the Vital Capacity (VC), which was measured using a bellows type spirometer and recorded. Measurements were carried out on two normal healthy young male subjects having no complaints regarding respiration.

3. Results and observations

The %change of focused impedance between maximum inspiration and maximum expiration at localized segments of the thorax were calculated for each subject to get the ventilation mapping matrix, from the front and from the back, and are presented in figures 2 and 3 respectively. These are superimposed on the outline of a human thorax giving the outline of the lungs and the heart as well. The positions of the current electrodes are shown as grey circles. The matrix positions obtained from the back were flipped horizontally to give the same right and left orientation. Some relevant subject information including their vital capacity (VC) are given in the respective figure captions. Two extreme left and right regions on ventilation mapping matrix represent the impedance changes on lateral curved parts of the thorax. The back side offered measurements on a few extra matrices at the top, and are also shown. The positions of the nipples of the breasts are shown as black dots.

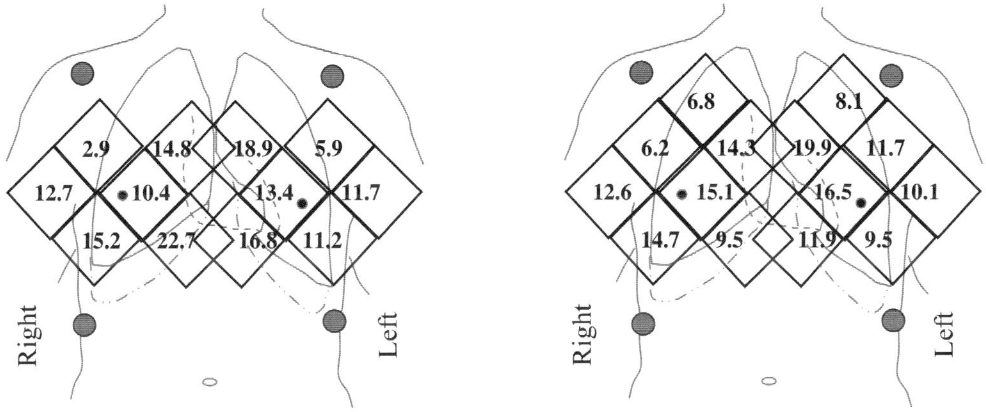

a) measured from front b) measured from back

Figure 2. Ventilation mapping of a subject from front (a) and back (b). Matrices from back has been arranged to match the view from the front. Subject age: 26 yrs, height: 1.71m, weight: 52 kg, Vital Capacity: 4.8 litres, non-smoker.

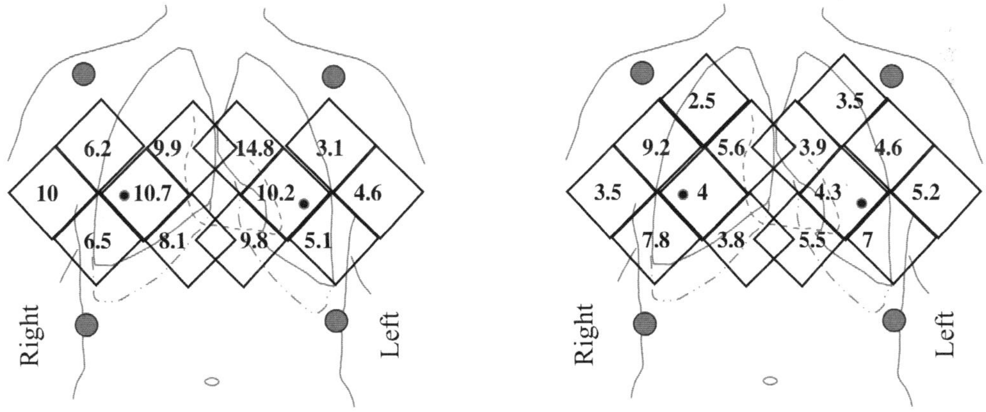

a) measured from front b) measured from back

Figure 3. Ventilation mapping of another subject from front (a) and back (b). Subject age: 20yrs, height: 1.68m, weight: 56 kg, Vital Capacity: 4.1 litres, non-smoker.

4. Discussions

The present work was taken up to have an overall idea of localized impedance change throughout the thorax due to lung ventilation using the newly developed 6-electrode FIM system. It is expected that the change in air content will be reduced if part of the lung is filled with water, blood of injury, or any other substance. In such cases, the impedance change measured in that region would be significantly reduced and this can possibly be detected by FIM. However to predict such distinction between localized zones of a healthy lung and a diseased lung a standard set of values are needed, and this should be based on measurements taken on a large number of healthy normal subjects.

The changes in air contents during ventilation in different regions of the lungs are different depending on the anatomical structure of the lungs. The expected change of impedance due to change in air content is visible from the maps in figures 2 and 3. The values in the corresponding matrices along the level of the nipple in the frontal measurement are similar for both the subjects; however for matrices above and below, there appears to be some difference. For subject 1 values of corresponding zones from the front and from the back are similar for nipple level, but are different at other levels. On the other hand, values from the front and back for subject 2 are different even at nipple level. Further work is required to resolve, or understand the cause for such discrepancies. The differences in VC between the two subjects need to be noted. Besides, there could be experimental error due to variation in the effort a person exerts in full inspiration and full expiration and posture of the subject. The inter-electrode separation plays a big role in the measurement [7] and has to be considered to set up a standard. The contribution of perfused blood should also be taken into account, however, by averaging the readings over a few seconds with breath-hold, this effect may be eliminated.

FIM looks at a limited region of the lungs, however, there is some spread in sensitivity including negative values beyond the focused zone [7] which may become significant in certain situations, and should be considered to get an improved result and standardization. Again, if the sensitive zone includes a boundary of the lungs, particularly on the lower side, it may cause a large error since the changes will be due to several factors including the movement of the diaphragm, besides lung ventilation. Therefore these areas need to be avoided. Similarly the cardiac region is best avoided because of the movement of the heart between the breathing cycles. Measurements from the sides and back may be useful for measurements around the heart. Again for female subjects, frontal measurements of lungs may be difficult because of fatty tissues, and measurement from the back may be desired.

Although EIT appears to offer pixel level resolution, 2D EIT images are cluttered by objects in 3D, and have limited accuracy. Therefore, for applications such as studies of lungs, the much simpler FIM may offer an acceptable solution for clinical applications, and therefore it seems justified to take up further study in this direction.

5. Acknowledgements

Acknowledgements are due to the two volunteers who took part in this study.

6. References:

[1] Nopp P, Rapp E, Pfutzner H, Nakesch H and Ruhsam C 1993 *Phys Med Biol* **38**:699-716
[2] Harris ND, Suggett AJ, Barber DC and Brown BH 1987 *Clin. Phys. Physiol. Meas.* ***8 Sup1.A*** 155-165
[3] Frerichs I., Hahn G., Schroder T., Hellige G 1998 *Intensive Care Med* **24**: 829-836
[4] Rabbani KS, Sarker M, Akond MHR. and Akter T 1998 *Proceedings of the X International conference on Electrical Bioimpedance,* Barcelona, Spain. 31-34
[5] Rabbani KS, Sarker M, Akond MHR. and Akter T 1999 *Ann. N Y Acad. Sci.* **873**: 408-420
[6] Kadir MA, Baig TN and Rabbani KS 2009 *Proceedings (on-line), 10th International Conference on Biomedical Applications of Electrical Impedance Tomography (EIT 2009), Manchester, UK,* http://www.maths.manchester.ac.uk/eit2009/abstracts/kadir.pdf
[7] Howlader S, Baig TN and Rabbani KS 2010 *Proceedings, International Conference on Electrical Bioimpedance,* Florida, USA

Electrical Impedance Potential Mammography for Visualization of Objects (Electrochemical Tests)

A Karpov[1,2], M Korotkova[1,2], Yu Tsofin[2] and V Tsyplyonkov[2] and M Machin[2]
[1]Clinical Hospital 9, Yaroslavl, Russia
[2]"SIM Technika" PKF, Yaroslavl, Russia

E-mail: kb9@mail.ru
sim-tech@net76.ru

Abstract. The article presents the result of the electrochemical testing of the potential electric impedance mammograph "MEIK" (version 5.6) which reconstructs the image using the reciprocal projection method. The testing was carried out on several categories: biological objects visualization, definition size, location, depth and shape of the biological objects, concentration and electrical conductivity, resolution capability. The research has been conducted in the original tank filled with homogeneous and heterogeneous medium. The research of the visualization of abiological and bialogical objects, objects dimensions, objects location, objects structure. According to the data of the electrochemical testing, the resolution capability of the objects with high electrical conductivity ranges from 1 to 3 mm and of the objects with low electrical conductivity – from 3 to 5 mm. Due to the given electrochemical testing results potencial electric impedance mammograph "MEIK" (version 5.6) can be rated as the medical visualization device.

1. Introduction
Nowadays the method of electric impedance potential mammography quite rapidly penetrates into medical practice. Thanks to the completion of clinical trials high informativity of this method was shown both for oncological diagnostic and for diagnostic of nonneoplastic diceases of the memmary gland. Quite wonderfully clinical trials were conducted prior to laboratory tests. The article demonstrates some results of the electrochemical tests of electrical impedance potential mammograph MEIK v. 5.6, in which backprojection method is used to reconstruct the image.

2. Methodology
Impedance tomograms were obtained by means of electrical impedance potential mammograph MEIK v. 5.6 (50 kHz, 0.5 mA) with single-plane placement of electrodes. An original tank with a base electrode was used for the placement of solutions and objects of interest. Computer scanning was performed each 8 mm to the maximum depth 46 mm (seven scan planes). The results were processed by means of descriptive statistics methods.

3. Backprojection algorithm
The backprojection algorithm is based on the simple fact described below. Changing the conductivity of elemental volume of a medium results in a change of surface potential difference located in a bounded area. This area is a projection of the original elementary volume to the measurement surface

along the equipotential surfaces. Our modification of the backprojection algorithm is based on the work. Under this approach, the conductivity $\sigma(r)$ in point r is estimated as:

$$\sigma(\bar{r}) = 1 + w_z(r_z) \cdot S(\bar{r}),$$

$$S(\bar{r}) = \sum_j P(\bar{r}, j) \cdot \frac{R(\bar{r}, j, Q)}{R(\bar{r}, j, Q')}, \quad R(\bar{r}, j, Q) = \int_{c(\bar{r}, j)} w(|\bar{r} - \bar{q}|) \cdot Q[D(j, \bar{q})] dq, \quad D(j, \bar{q}) = \frac{E_0^j(\bar{q}) - E^j(\bar{q})}{E_0^j(\bar{q})},$$

where $w_z(s)$- weight function that takes into account the decrease of sensitivity with depth; j - number of inducing electrode; $P(r, j)$ - regularization operator, which takes into account the change in sensitivity depending on the distance from the inducing electrode; $Q(s)$, $Q'(s)$ - regularization operators, taking into account the different reliability of the different values of relative change of electric field; $c(r, j)$ - intersection of the measurement surface with equipotential surface built for the j-th induction electrode and passing through the point r; $w(s)$ - weighting function that takes into account the decrease of sensitivity with distance from the point r; E^j - electric field to the measured surface potential; E_0^j- reference electric field for surface potential, measured for homogeneous medium. Electric field is calculated as surface potential gradient projection to the line of integration.

4.Result

The article demonstrates the results of the tests which were conducted in a homogeneous medium, which is moderately hard water (less than 10 mmol/l) or amylum aqueous solution (50 g/l). The objects were placed at 1-3 cm depth. Electrochemical tests were conducted for such categories: visualization of abiological and bialogical objects, objects dimensions, objects location, objects structure.

4.1 Visualization of Abiological Objects of Different Electrical Conductivity

The objects of dif ferent electrical conductivity (a coin and a rubber ball) are represented in figure 1. The object which has high electrical conductivity (the coin, 25 mm in diameter) can be visualized as a hypoechoic mass. The object which possesses low electrical conductivity (the rubber ball, 2 cm in

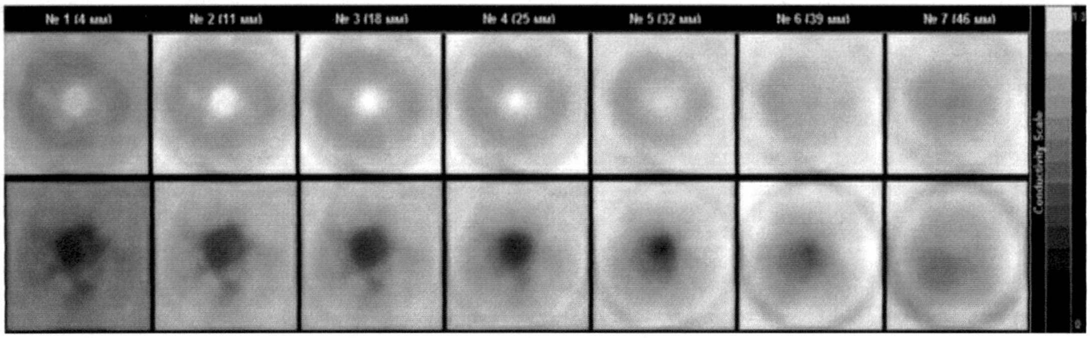

Figure 1. Visualization of abiological objects of different electrical conductivity. In the upper row there is the image of the coin (the object of high electrical conductivity). In the upper row there is the image of the rubber ball (the object of low electrical conductivity).

diameter) can be visualized as a hyperechoic mass. Certain blurring of the objects' contours, which impedes its sizing and locating. Additional image processing (inversion, color gamma) allows acquiring a more sharp contour of the object.

4.2 Visualization of Abiological Objects of Different Size

The figure 2 represents the images of coins of different size and thickness, made from alloy of steel, nickel and copper. The depth of the image plane is 11 mm. More sharp contours of the objects are

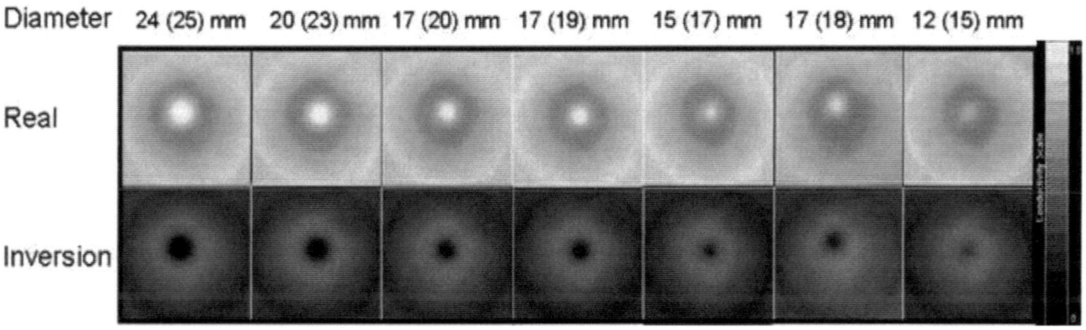

Figure 2. Electrical impedance images of coins of different size and thickness, made from alloy of steel, nickel and copper (the real dimensions of the coins are indicated in brackets in the first row).

acquired by means of inversion. The size of the images is somewhat smaller than the real size of the coins (the real sizes of the coins are indicated in brackets in the figure 2).

4.3 Visualization of Biological Objects of Different Electrical Conductivity

The figure 3 represents the biological objects of different electrical conductivity. The upper row represents the image of an onion cooked sized 27x31 mm. The object can be visualized in the center of

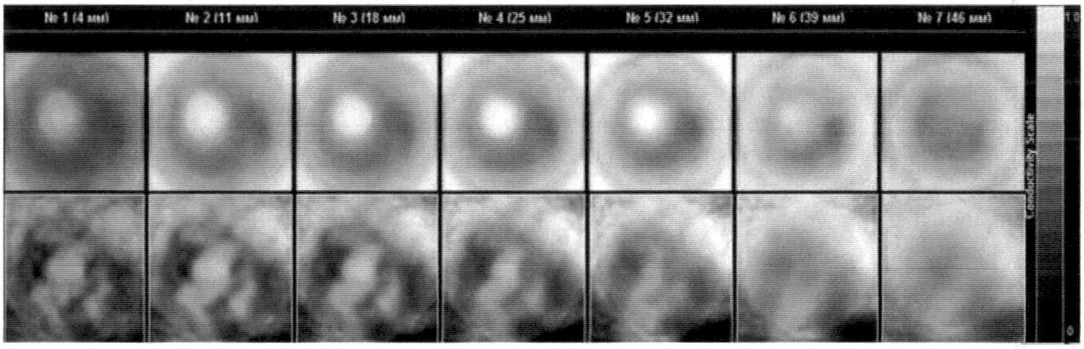

Figure 3. Visualization of biological objects of different electrical conductivity: The upper row represents the image of an onion cooked sized 27x31 mm, the lower row depicts an apple pit sized 7x3 mm.

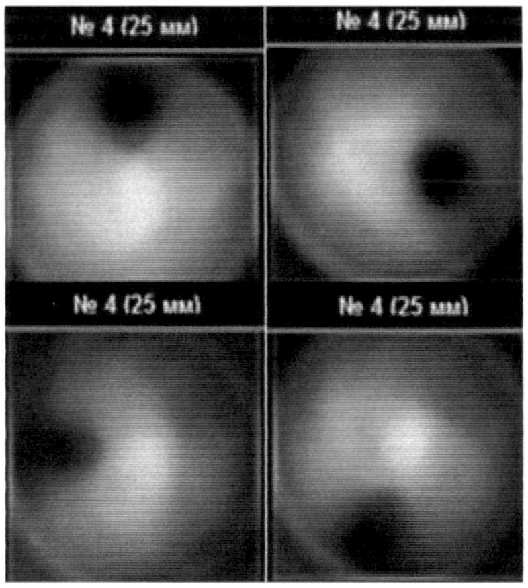

Figure 4. Electrical impedance image of the cooked onion which is moved clockwise (12, 3, 6 and 9 on the clock dial).

the image on 5 scan planes as a light round-shaped area (31x34 mm) with high electrical conductivity (hypoimpedance effect). The lower row depicts an apple pit sized 7x3 mm. The object can be visualized in the image on 3 scan planes at 9 on the clock dial as a dark round-shaped area (10x7 mm) with low electrical conductivity (hyperimpedance effect).

4.4 Objects Location

Changing of the location of the objects causes the changes of electrical conductivity. The figure 4 represents the image of the biological object (an onion cooked) which is moved clockwise. The object sized 34x31 mm was placed at 20 mm depth. The electrical impedance images are represented by the 4th inverse scan plane (25 mm). The object can be visualized as a dark round-shaped area (31x27 mm). Changing of the object location causes corresponding changes of the location of the high electrical conductivity area.

4.5 Visualization of Biological Objects of Different Size

Biological objects of different size, as well as abiological objects, can be visualazed by means of electrical impedance potential mammography. But there is a significant difference: real dimensions of biological objects are smaller than the size of their electrical impedance images, while in the case of abiological objects the situation is quite the contrary. There are actual and inverse electrical impedance images of two onions (raw and cooked) in the figure 5. A light round-shaped area sized 27x30 mm can be visualized on the image at 3 on the clock dial which corresponds with the location of the cooked onion sized 20x25 mm. A light round-shaped area sized 24x17 mm can be visualized on the image at 9 on the clock dial which corresponds with the location of the raw onion sized 20x13 mm.

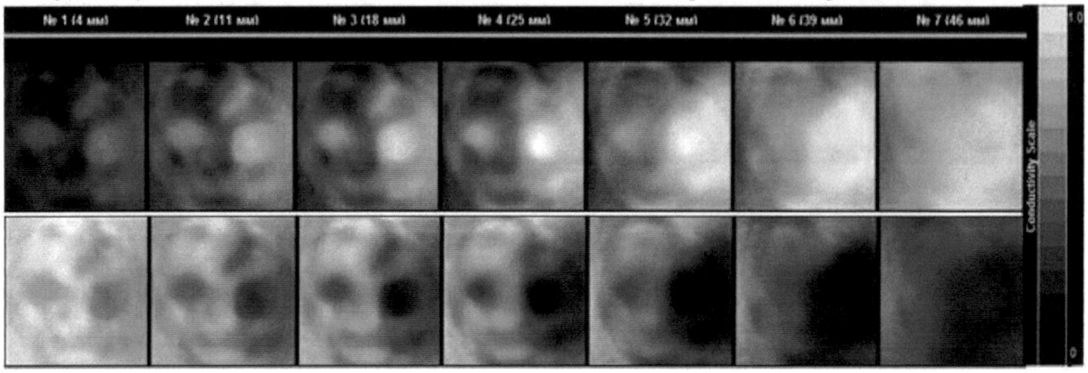

Figure 5. Actual (upper row) and inverse (lower row) electrical impedance images of two onions (raw and cooked) of different size.

4.6 Biological Object Structure

If a biological object is close to the array of electrodes (10-20 mm), then it's possible to acquire the electrical impedance image of its internal structure. The figure 6 in the upper row represents the electrical impedance image of a cucumber segment (11 mm depth) and its real image. The hyperimpedance cortical layer can be easily defined. In the center of the image there is the hypoimpedance pulp layer with hyperimpedance septa. The figure in the lower row represents the electrical impedance image (11 mm depth) and the ultrasound image of the mammary tumour. The hypoimpedance liquid component on the periphery of the mass and the heterogeneous hyperimpedance internal structure can be easily observed.

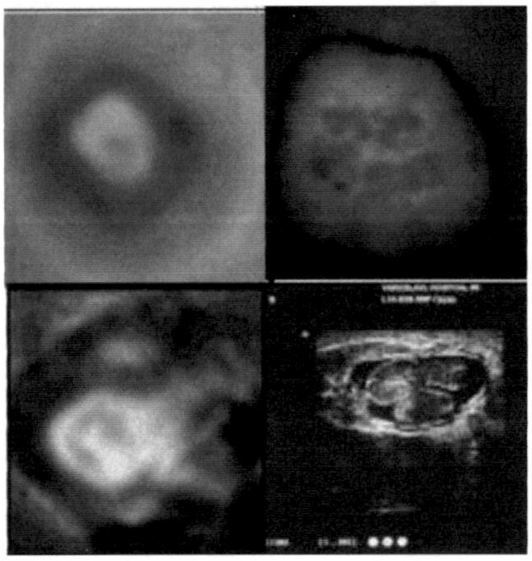

Figure 6. Electrical impedance image of the internal structure of a biological object: the ucumber in the upper row and the breast cancer in the lower row.

5. Conclusion

The obtained during the tests results allow us to consider electrical impedance potential mammography as a means of medical visualization. Whether it is good or bad is a topic in its own right. But the possibility to estimate electrical properties, dimensions and location of biological objects, as well as the capability to identify their internal structure is evidence of its usability. In the meantime the electrochemical tests should be continued both in homogeneous and in heterogeneous mediums.Further development of mathematical algorithms, instruments and software will allow improving image quality, minimizing "long-tailed trains" influence, ensuring adequate accuracy etc.

Journal of Physics: Conference Series **224** (2010) 012033 doi:10.1088/1742-6596/224/1/012033

Analysis of forward solvers for electrical impedance tomography in a mammography geometry

Ethan K. Murphy, David Isaacson, Gary J. Saulnier, Jonathan Newell

Rensselaer Polytechnic Institute, Troy, NY 12180, USA.

E-mail: murphe3.rpi.edu

Abstract. We explain how the problem of improving image reconstruction in electrical impedance mammography can be studied by comparing measured data with predictions from the continuum model, ave-gap model, and complete electrode model. The data is measured by the RPI ACT4 system using 60 electrodes arranged in a mammography geometry. Each model's accuracy can be determined by comparing the computed eigenvalues of its Neumann-to-Dirichlet map with those found from experimentally measured N-to-D maps. It will be explained how the different models can be used to reconstruct images of the electrical conductivity and permittivity inside a region bounded by a pair of electrode arrays in a mammography geometry. Reconstructions from experimental data are presented.

1. Introduction

In electrical impedance tomography the goal is to reconstruct electrical properties within a region from voltage and current measurements on the boundary. In reconstruction algorithms it is necessary to have a forward solution that produces accurate voltages from a known set of current patterns (CP) and known admittivity throughout the domain. One can calculate error terms between measured and simulated voltages to get an idea of the quality of the forward solver. Following the work in [4], we look at the eigenvalues of the Neumann-to-Dirichlet map, R_γ. Since $\mathbf{V}^n = R_\gamma \mathbf{T}^n$, where $\mathbf{V}^n$ and $\mathbf{T}^n$ are voltage and current patterns on the electrodes, one can easily understand why the eigenvalues of R_γ are known as the characteristic resistances, ρ_n. By comparing characteristic resistances between measured and simulated data we have a simple and intuitive way to evaluate the quality of our simulated data.

In our study we focus on a mammography geometry, shown in Figure 1, which can be modeled as a box with electrode arrays on the top and bottom. We calculate the analytic solution for this geometry. Assuming continuous current densities at the top and bottom we can find the exact eigenvalues of R_γ. In addition, we use the Average-Gap and complete electrode models to calculate the voltages. We describe how one can then find the eigenvalues from these voltages. This analysis is then used to tune a couple of parameters of the forward models, and a reconstruction is shown.

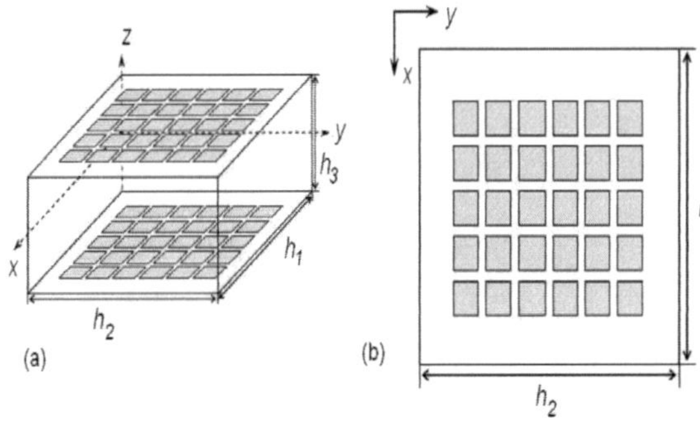

Figure 1. Illustration of the electrode array in the mammography geometry. There are 60 electrodes, 30 on each side.

2. Eigenvalue analysis of the data

The problem is modeled by the conductivity equation

$$\begin{aligned} \nabla \cdot \gamma \nabla u(x,y,z) &= 0, && 0 \le x \le h_1, 0 \le y \le h_2, -\tfrac{h_3}{2} \le z \le \tfrac{h_3}{2}, \\ \gamma \tfrac{\partial u}{\partial \nu} &= j^{\mathbb{T}}(x,y), && z = \tfrac{h_3}{2}, \\ \gamma \tfrac{\partial u}{\partial \nu} &= j^{\mathbb{B}}(x,y), && z = -\tfrac{h_3}{2}, \\ \gamma \tfrac{\partial u}{\partial \nu} &= 0, && x = 0, x = h_1, y = 0, y = h_2, \end{aligned} \tag{1}$$

where $j^{\mathbb{T}}$ and $j^{\mathbb{B}}$ represent the current densities at the top and bottom of the electrode arrays, respectively. The solution, for $\gamma = 1$, can be found to be

$$u(x,y,z) = b_0 z + \sum_{\substack{n,m=0 \\ n+m>0}}^{\infty} cos(\lambda_n x) cos(\lambda_m y) \left[a_{n,m} cosh(\lambda_{n,m} z) + b_{n,m} sinh(\lambda_{n,m} z)\right], \tag{2}$$

where $\lambda_n = \frac{n\pi}{h_1}$, $\lambda_m = \frac{m\pi}{h_2}$, and $\lambda_{n,m} = \sqrt{\lambda_n^2 + \lambda_m^2}$. This represents the solution using the continuum model, i.e. there is no electrode model and we assume that there is a current density applied to the entire top and bottom of the domain.

More accurate voltages can be calculated by properly incorporating electrode effects. We use the Average-Gap model given by

$$\mathbf{T}_l^n = \int_{e_l} \gamma \frac{\partial u}{\partial \nu} dS, \quad \mathbf{T}_l^n = 0, \quad (x,y,z) \notin \cup_{l=1}^{L} e_l, \quad \mathbf{V}_l^n = \frac{1}{|e_l|} \int_{e_l} u dS, \tag{3}$$

which is discussed in [1], and we use the complete electrode model (CEM), which uses the same boundary conditions (BC) as (3) except the last BC is replaced by $\mathbf{V}_l^n = u + z_l \gamma \frac{\partial u}{\partial \nu}$, [2]. The term z_l represents the contact impedance between the electrode and the surface.

Let us say that the matrix $\mathbf{T}$ is defined as the following

$$\mathbf{T}_l^n = \begin{cases} j^{\mathbb{T}}_{p_n,q_n}(x_l, y_l) &= M cos(\lambda_{p_n} x_l) cos(\lambda_{q_n} y_l), \\ j^{\mathbb{B}}_{p_n,q_n}(x_l, y_l) &= (-1)^{s_n} M cos(\lambda_{p_n} x_l) cos(\lambda_{q_n} y_l), \end{cases} \tag{4}$$

where p_n, q_n, and s_n (= 0 or 1) are coefficients of a certain prescribed ordering. Using the continuum model, i.e. combining (2) and (4), we get that $\mathbf{V}^n = R_\gamma \mathbf{T}^n = \rho_n \mathbf{T}^n$, where $\rho_n := \frac{1}{\gamma k_n}$, where

$$k_n := \begin{cases} \lambda_{p_n,q_n} tanh(\lambda_{p_n,q_n} \frac{h_3}{2}), & \text{if } s_n = 0, \\ \lambda_{p_n,q_n} coth(\lambda_{p_n,q_n} \frac{h_3}{2}). & \text{if } s_n = 1. \end{cases} \tag{5}$$

In the cases of using the Ave-Gap model, CEM, or measured data to calculate the characteristic resistances, the first step is to approximate R_γ. It should be noted that the vectors $\mathbf{T}^n$ are not necessarily eigenvectors of R_γ. We therefore assume that we have an unknown set of eigenvectors denoted by $\mathbf{t}_j$ for $j = 1...L-1$. Since $\mathbf{T}$, if scaled properly, is an orthonormal matrix we can express $\mathbf{t}_j$ in terms of $\mathbf{T}$ in the following way $\mathbf{t}_j = \sum_{k=1}^{L-1} \tau_{j,k}\mathbf{T}^k$, where we define $\tau_{j,k} := \langle \mathbf{T}^k, \mathbf{t}_j \rangle$. Starting with $R_\gamma \mathbf{t}_j = \rho_j \mathbf{t}_j$ one can find that $[R_\gamma]\,\tau_j = \rho_j \tau_j$ where we define $[R_\gamma]_{\alpha,k} := \langle \mathbf{T}^\alpha, \mathbf{V}^k \rangle$. Thus to find ρ_j we simply need to construct $[R_\gamma]$ and calculate its eigenvalues.

3. Reconstruction method: NOSER

In order to make reconstructions we use a Newton one-step error reconstructor (NOSER) method, which was originally described in [3] for a 2D circular problem. The algorithm works the same for u^n from Ave-Gap or CEM. We start by assuming that we have a solution, v^m, that satisfies the homogeneous problem, $\gamma = \gamma_0$, for the mth CP. Then by integrating over the domain, comparing solutions, and a step of integration by parts one can find that

$$\begin{aligned} \int_\Omega (\gamma - \gamma_0)\nabla u^n \cdot \nabla v^m dP &= \int_{\partial\Omega} v^m \gamma \frac{\partial u^n}{\partial \nu} - u^n \gamma_0 \frac{\partial v^m}{\partial \nu} ds, \\ \int_\Omega \delta\gamma \nabla u_0^n \cdot \nabla v^m dP &\approx D_{n,m}, \end{aligned}$$

where in the second line we assume u and γ are close to constants and higher order terms are ignored. Then we discretize the domain into a coarse set of N_v voxels. This gives us a linear equation to solve $\hat{J}\delta\gamma = \hat{D}$, where $\hat{J}$ is constructed from the Jacobian which is defined to be

$$J_{n,m,k} := \int_{B_k} \nabla u_0^n \cdot \nabla v^m dP,$$

where B_k represents the domain of the kth voxel. Our solution is given by $\gamma_1 = \gamma_0 + \delta\gamma$.

4. Results

The eigenvalue analysis allows us to easily tune parameters for particular geometries or electrode arrays. Since the eigenvalues drop off quickly in our illustrations we multiply them by a term k_n defined in (5). This makes the continuum model, $k_n\rho_n$, be described by a horizontal line. In our illustration we show the tuned results from our mammography tank filled with saline solution of 200 mS/m. We allow for there to be an extra spacing outside of the electrode array to compensate for currents that may travel outside the array, see Figure 1. The best parameters found for this spacing was 1 cm. When using the CEM the best contact impedance was found to be $z = 0.00015/\gamma_0\ \Omega m^2$. The characteristic resistances are shown in Figure 2. The best maximum relative errors were 12.1% and 4.7% using the Ave-Gap model and CEM, respectively. The error was defined by $E := max_n\left[|\rho_n^{meas} - \rho_n^{sim}|/|\rho_n^{meas}|\right]$. In Figure 3, we show a reconstruction using the tuned CEM voltages of a copper cube of 10 mm sides placed near the lower electrode array at 33 kHz.

5. Conclusion

In this work it is seen that by investigating the eigenvalues of the Neumann-to-Dirichlet map one can tune forward solvers in order to produce good reconstructions. This method provides a simple way to analyze the quality of simulated voltages in comparison to measured voltages, and will be a useful tool in investigating more complex forward models with inevitably more parameters that need to be chosen.

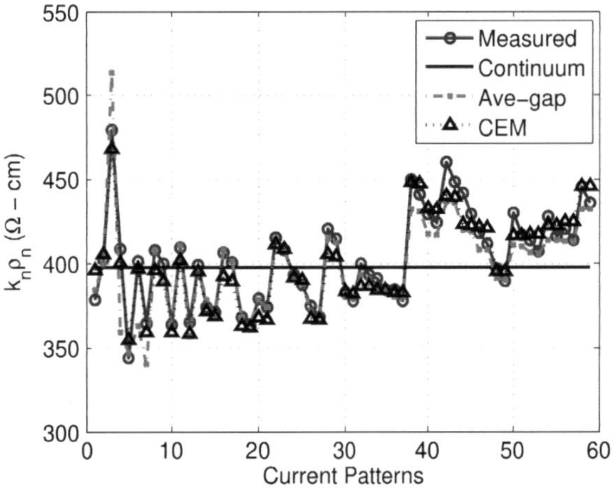

Figure 2. Illustration showing the characteristic resistances using voltages from measurements, the continuum model, the Ave-Gap model, and the CEM. The Ave-Gap and CEM use a spacing of 1 cm around the electrode array, and CEM uses a contact impedance of $z = 0.00015/\gamma_0\ \Omega m^2$.

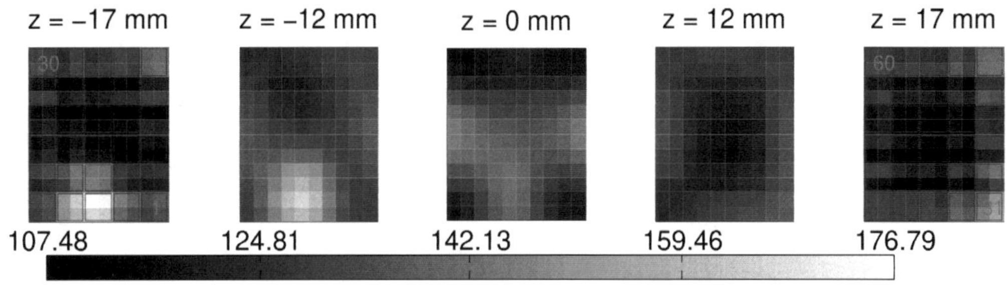

Figure 3. Cross-sections of a reconstruction shown at various heights of a copper cube with 10 mm sides placed within the mammography tank near the lower electrode array using CEM with $z = 0.00015/\gamma_0\ \Omega m^2$ and the NOSER algorithm.

References

[1] G. Boverman, T.-J. Kay, R. Kulkarni, B.S. Kim, D. Isaacson, G.J. Saulnier, J.C. Newell. Robust linearized image reconstruction multifrequency EIT of the breast. *IEEE Trans. on Med. Imag.*, 27(10):1439–1448, 2008.

[2] B.S. Kim, G. Boverman, J.C. Newell, G.J. Saulnier, D. Isaacson. The complete electrode model for EIT in a mammography geometry. *Physiol. Meas.*, 28(7):S57–S69, 2007.

[3] Cheney M, Isaacson D, Newell J C, Simske S and Goble J NOSER: an algorithm for solving the inverse conductivity problem. *Int. J. Imaging Syst. Technol.* 2:66-75, 1990.

[4] E. Somersalo, M. Cheney, and D. Isaacson. Existence and uniqueness for electrode models for electric current computed tomography. *SIAM J. Appl. Math.* 52(4):1023–1040, 1992.

Generation and performance of patient-specific forward models for breast imaging with EIT

A Tizzard[1], A Borsic[2], R Halter[2] and R Bayford[1]

[1]Department of Natural Sciences, Middlesex University, London, U.K.

[2]Thayer School of Engineering, Dartmouth College, 8000 Cummings Hall Hanover, NH 03755, US

E-mail: a.tizzard@mdx.ac.uk

Abstract. It has now been well established that accurate geometric conformity of the forward model for EIT reconstruction has significant benefits for artefact reduction and localisation of conductivity changes within the domain. The problems of generation of patient specific forward models need to be addressed as segmentation of volumetric data from CT or MRI is inadequate for time-critical clinical use. This group has pioneered methods of generating patient-specific surface models from known landmarks and electrode positions and have used this data to warp finite element models for EIT reconstruction. This paper presents a further application of these methods to use known electrode positions for breast imaging to generate an accurate B-Spline surface model of a subject and to warp an existing finite element model to the surface using elastic deformation. Results will show that a forward model can be generated, conforming more realistically to actual subject geometry, that will further enhance the performance of the reconstruction algorithm offering significant benefits to clinical EIT breast imaging.

1. Introduction

The geometric conformance between the subject and forward model is an important consideration for improving reconstruction of images in Electrical Impedance Tomography (EIT). It has been demonstrated that perturbation shape is improved and image artifacts are reduced when a geometrically accurate finite element (FE) mesh is generated for the forward model. This has been reported for both brain function [1] and lung function [2]. The most common approach for generating anatomically correct geometry is from segmented CT and MRI datasets. This is both labour-intensive and time-consuming and can only be considered as a means of providing generic models [3]. The alternative, which can produce more patient-specific models, is to warp (morph) simple geometry or generic models using surfaces built from known points on the patient such as electrode positions [4, 5].

The research group at the Dartmouth-Hitchcock Medical Center (Lebanon, NH) has developed a broadband high frequency EIT system for breast imaging [6] which has been used to image over 100 women in a clinical trial [7]. The forward models used in this work are FE meshes based on actual electrode positions but have idealized surface geometry based on a linear interpolation comprising simple geometric primitives such cylinders, cones, and ellipses to form a 3D shape that roughly resembles a breast shape matching the position of up to 4 rings of 16 electrodes. This paper reports on a work in progress that applies mesh warping methods described above to generate more accurate forward models based on ellipsoidal B-Spline surfaces.

2. Methods

2.1. Data Acquisition

EIT breast data is acquired using a dedicated data acquisition system and physical patient interface as shown in figure 1. The data acquisition system presents 64 channels for applying potentials and measuring resulting currents on a maximum of 64 electrodes, arranged to form four concentric rings of 16 electrodes each, stacked vertically with a separation of 3.25 cm. The physical patient interface consists of an examination bed where the patient lies prone. A hole in the surface allows the breast to be pendant under gravity below the level of the bed, where the data acquisition system is housed. Electrodes are mounted on the tip of motorized rods, which can be actuated to bring the electrodes in contact with the breast. All electrodes in a ring are actuated together to form a circular ring of a given diameter which is recorded by the system and later used to form the initial patient specific mesh. Depending on the size of the breast one or more electrode rings can be brought in contact and used; typically, 2 or 3 rings are used, that is 32 or 48 electrodes. The system adapts the number of applied voltage patterns based on the number of electrodes, and normally a protocol that uses the maximum number of linearly independent sinusoidal patterns is used, resulting in n-1 patterns for n electrodes.

Figure 1: Physical patient interface - examination table with a circular breast opening showing the top ring of 16 electrodes (white capsules mounted on rods) in contact with a gelatine breast phantom. Some electrodes of the second ring are visible in the lower part of the opening.

2.2. Mesh Warping

One example of a FE mesh taken from the study described in [6] comprising 55037 elements, 11414 nodes and 3 electrode rings was used as the source model. A number of assumptions are made regarding the original geometry of the subject's breast: 1) the original breast shape can be closely approximated as a semi-ellipsoid; 2) the major axis of the ellipsoid lies on the z-axis; 3) the minor axis lies at the breast/body interface and 4) that the ellipsoid passes through the largest z values of the electrodes in the 2nd and 3rd rings. The first ring is at z = 0.5 cm and the electrodes are 0.8 cm diameter. The ellipsoid is assumed therefore to pass through z = 7.4 and z =10.65 cm. The first three assumptions are in line with those of other investigators in the field of breast mammography, for example [8]. The fourth is the best approximation is the absence of undeformed geometry data and serves to demonstrate the principle.

The vectors defining the orthogonal projection of the surface nodes to the ellipsoid are evaluated using methods previously described in [4] and these are used to define the displacement boundary conditions for the warping algorithm [5]. The resulting ellipsoid model is then warped back using the original electrode positions to define the displacement boundary conditions. To minimize element quality degradation, the latter warping process was executed four times; one for each ring of electrodes and a fourth to achieve the final model. Poisson's ratio was taken to be 0.45. The whole procedure, coded in MATLAB, took under two minutes on a 3.16 GHz Intel Xeon processor under Windows XP.

2.3. Reconstruction

Image reconstruction is formulated as a traditional non-linear least squares inverse problem with

Tikhonov regularization as in equation (1). The forward model is based on a finite element implementation and uses the Complete Electrode Model; both conductivity and permittivity are reconstructed at the same time. While the meshes shown in figure 2 are used for computing the forward potentials, an interpolation scheme is used to represent conductivity on a coarser grid, having 1,200 elements. This arrangement reduces the number of parameters to be estimated, and it is based on lumping together the conductivity values of several elements in the fine mesh, to form a desired number of "coarse pixels".

$$\sigma_{rec} = \underset{\sigma}{argmin} \frac{1}{2}\|V(\sigma) - V_{meas}\|^2 + \alpha\frac{1}{2}\|L\sigma\|^2 \qquad (1)$$

3. Results

3.1. Mesh warping

The mesh warping procedure produced a final model that was visually more representative of the patient's breast, *in-situe*, with all electrodes applied. Element quality was reduced by the process: minimum stretch of the original model was 0.215 and for the final warped model it was 0.140. Stretch, in this case, is defined as the ratio of the inscribed radius in an element to its maximum edge length, normalised to lie between 0 and 1. The results are illustrated in figure 3.

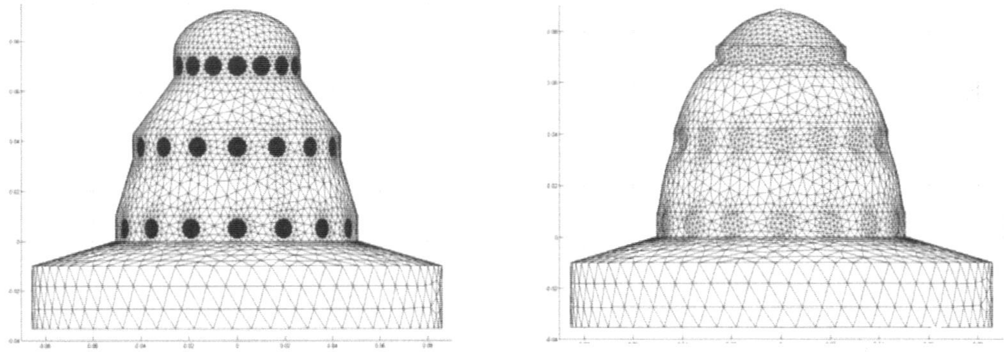

Figure 2: Warping process a) the original patient-specific model and b) final improved forward model

3.2. Reconstruction

EIT data from a volunteer subject with a known cancerous mass of 1.3 cm diameter at 5 cm from the nipple was reconstructed using two different FE meshes. Figure 2(a) shows the first mesh built using the standard procedure used at Dartmouth described in §1. The same data was reconstructed over the warped mesh shown in figure 2(b). Contrast enhanced MRI images of the subject are shown in Figure 3.

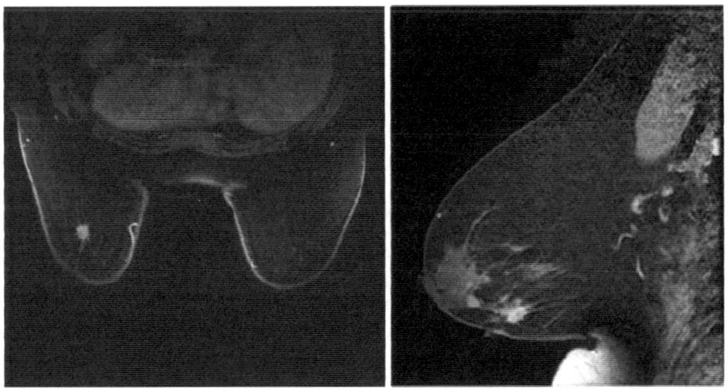

Figure 3: Contrast enhanced breast MRI images of a subject with a known cancerous mass. The 1.3cm mass, shown in brighter colour, is at a distance of 5cm from the nipple a) axial plane b) sagittal plane.

The two different EIT reconstructions are shown in figure 4; an increased conductivity, shown in red, corresponds to the location of the tumor.

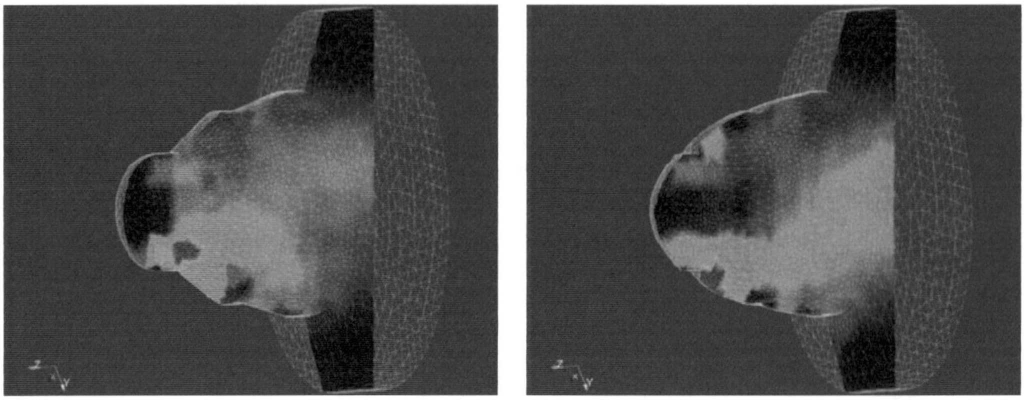

Figure 4: Cross sections in the saggital plane of 3D EIT reconstructions from the subject a) reconstruction with simplified breast mesh b) reconstruction with warped breast mesh.

4. Discussion and Conclusions

While the estimation of the original patient breast form was based on a relatively weak assumption, the selection of an ellipsoidal form is justified based on previous investigator's work. The outcome of this case study clearly demonstrates, in principle, that a more precise and valid forward model can be generated within good time scales for clinical use. While it is difficult to assert that image improvements are significant from this case study, the results show promise that benefits could be obtained.

This leads the way to future investigations. The ideal scenario would be to obtain more accurate geometric data of the subject, *in-situe*, and prior to electrode application. There are many ways that this can be simply achieved within the existing system, for example: photographically or by acquiring electrode positions on the breast before applying any significant force.

References

[1] Tizzard, A and Bayford, R H 2007 *Proc.IFMBE.13 ICEBI* ed H. Schlarfetter and R. Merva. (Graz, Austria: Springer-Verlag) pp 368-71

[2] Bayford, R, Kantartzis, P, Tizzard, A, Yerworth, R, Liatsis, P and Demosthenous, A 2008 Development of a neonate lung reconstruction algorithm using a wavelet AMG and estimated boundary form *Physiol. Meas.* **29** S125-38

[3] Tizzard, A, Horesh, L, Yerworth, R J, Holder, D S and Bayford, R H 2005 Generating accurate finite element meshes for the forward model of the human head in EIT *Physiol. Meas.* **26** S251-61

[4] Tizzard, A, Holder, D S and Bayford, R H 2005 *Proc.IFMBE* (Prague, Cz: IFMBE)

[5] Tizzard, A and Bayford, R H 2007 Improving the Finite Element Forward Model of the Human Head by Warping using Elastic Deformation *Physiol. Meas.* **28** S163-82

[6] Halter, R J, Hartov, A and Paulsen, K D 2008 A broadband high-frequency electrical impedance tomography system for breast imaging *IEEE Trans. Biomed. Eng.* **55** 650-9

[7] Poplack, S P, Tosteson, T D, Wells, W A, Pogue, B W, Meaney, P M, Hartov, A, Kogel, C A, Soho, S K, Gibson, J J and Paulsen, K D 2007 Electromagnetic breast imaging: results of a pilot study in women with abnormal mammograms *Radiology* **243** 350-9

[8] Pu, J, Zheng, B, Leader, J K and Gur, D 2008 An ellipse-fitting based method for efficient registration of breast masses on two mammographic views *Med. Phys.* **35** 487

Multi-Excitation Magnetoacoustic Tomography with Magnetic Induction (MAT-MI)

X Li and B He

Department of Biomedical Engineering, University of Minnesota, USA

Email: binhe@umn.edu

Abstract. Magnetoacoustic tomography with magnetic induction (MAT-MI) is an approach proposed to do non-invasive electrical conductivity imaging of biological tissue using magnetic induction and ultrasound measurements. In the present study, based on the analysis of the relationship between the conductivity distribution and the generated MAT-MI acoustic source, we propose a new multi-excitation algorithm to achieve better reconstruction of the conductivity distribution. In this algorithm, multiple magnetic excitations using different coil configurations are employed and the ultrasound measurements corresponding to each excitation are collected to derive the conductivity distribution inside the sample. The corresponding modified algorithm is also proposed for MAT-MI imaging when only limited bandwidth acoustic measurements data are available. Computer simulations have been done to demonstrate the performance of the proposed algorithm. It is shown that using this algorithm and unlimited bandwidth data, we can accurately reconstruct the internal conductivity contrast of the object. With limited bandwidth data and using the modified algorithm we can also reconstruct the relative conductivity contrast of the object instead of only the boundaries at the conductivity heterogeneity. Benefits come with this new algorithm include better differentiation of different tissue types with conductivity contrast using MAT-MI approach, specifically for the potential breast cancer screening application in the future.

1. Introduction

Noninvasive electrical impedance imaging of biological tissues has drawn lots of research interest in recent years. Related imaging techniques include electrical impedance tomography (EIT) [1], magnetic induction tomography (MIT) [2] and magnetic resonance electrical impedance tomography (MREIT) [3]. An alternative approach is through the coupling between electromagnetic field and acoustic field as reported in magnetoacoustic tomography (MAT) [4], [5] and Hall Effect imaging (HEI) [6]. In order to avoid the "shielding effect" associated with surface electrode current injection, magnetoacoustic tomography with magnetic induction (MAT-MI) [7]-[10] was proposed. MAT-MI utilizes magnetic induction and the same Lorentz force coupling mechanism as in MAT/HEI.

2. Theory

We consider a sample with isotropic conductivity $\sigma(\mathbf{r})$ in a uniform static magnetic field $\mathbf{B}_0 = B_{0z}\hat{z}$. With N different excitation coil setups, using the magnetic vector potential $\mathbf{A}^j(\mathbf{r},t)$ and electrical scalar potential $\varphi^j(\mathbf{r})$, magnetic induction in MAT-MI can be formulated as in equation (1) [8]

$$\nabla\cdot(\sigma(-\frac{\partial \mathbf{A}^j}{\partial t}-\nabla\varphi^j))=0 \qquad j=1,\cdots,N \tag{1}$$

With known σ and $\mathbf{A}^j$, we can solve for φ^j in the whole domain using finite element method [11].

In MAT-MI, the wave equation governing the generated pressure is given in equation (2) [5]

$$\nabla^2 p_j-\frac{1}{c_s^2}\frac{\partial^2 p_j}{\partial t^2}=\nabla\cdot(\mathbf{J}^j\times\mathbf{B}_0) \qquad j=1,\cdots,N \tag{2}$$

where p_j and $\mathbf{J}^j$ are the pressure and induced current density corresponding to the jth excitation and c_s is the acoustic speed in the media.

With the obtained acoustic measurements of p_j, we can reconstruct the acoustic source map $AS^j=\nabla\cdot(\mathbf{J}^j\times\mathbf{B}_0)$ using back projection method [7]. Expanding the acoustic source term, we have

$$AS^j=(\sigma(-B_{1z}^j)+(\frac{\partial\sigma}{\partial x},\frac{\partial\sigma}{\partial y})\cdot(E_y^j,-E_x^j))B_{0z} \qquad j=1,\ldots,N \tag{3}$$

where E_x^j and E_y^j are the x and y components of the induced electrical field vector $\mathbf{E}^j$, respectively. Using matrix form equation (3) becomes

$$\mathbf{Ux}=\mathbf{b} \tag{4}$$

where

$$\mathbf{U}=\begin{bmatrix}E_y^1 & -E_x^1\\ \vdots & \vdots\\ E_y^N & -E_x^N\end{bmatrix} \qquad \mathbf{x}=\begin{bmatrix}\frac{\partial\sigma}{\partial x} & \frac{\partial\sigma}{\partial y}\end{bmatrix}^T \text{ and } \mathbf{b}=\begin{bmatrix}\frac{AS^1(\mathbf{r})}{B_{0z}}+\sigma B_{1z}^1 & \cdots & \frac{AS^N(\mathbf{r})}{B_{0z}}+\sigma B_{1z}^N\end{bmatrix}^T$$

We can then solve $\mathbf{x}$ using the least square method. To computer σ from $\mathbf{x}$ in each imaging slice, an integration technique can be used [12]. The detailed algorithm description will be reported elsewhere.

One of the major technical limitations of the MAT-MI approach comes from the limited bandwidth acoustic measurements. Denoting the partly reconstructed acoustic source map as $A\tilde{S}^j$, we have

$$A\tilde{S}^j\approx((\frac{\partial\sigma}{\partial x},\frac{\partial\sigma}{\partial y})\cdot(E_y^j,-E_x^j))B_{0z} \qquad j=1,\ldots,N \tag{5}$$

A similar matrix equation can be obtained with a modified right hand side $\tilde{\mathbf{b}}$ and similar reconstruction procedure can be applied to estimate the conductivity distribution of the object.

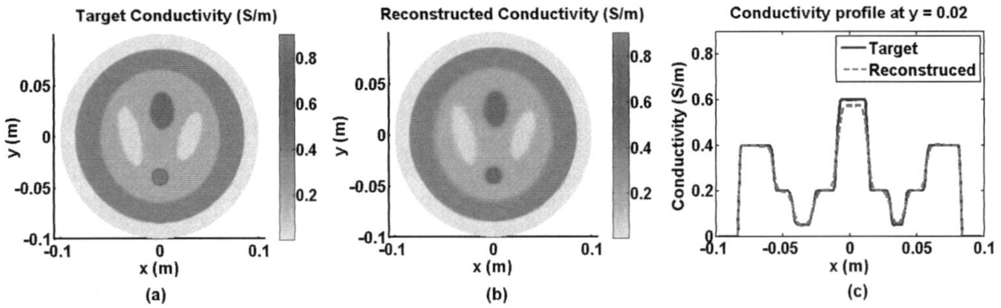

Figure 1. (a) Target conductivity image. (b) Reconstructed conductivity image using unlimited bandwidth data and under noise free condition. (c) Conductivity profiles along y = 0.02 m.

3. Results

To validate the proposed multi-excitation MAT-MI method, we first did a computer simulation using unlimited bandwidth data and under noise free condition. Three groups of coils are used to sequentially send magnetic excitations, i.e. $N = 3$.The reconstruction result is shown in figure 1. The correlation coefficient (CC) between the target image and the reconstructed image is 99.1% and the relative error (RE) is 8.5%. This result shows that a much better performance can be obtained using the current algorithm as compared to other previous algorithms [8], [10].

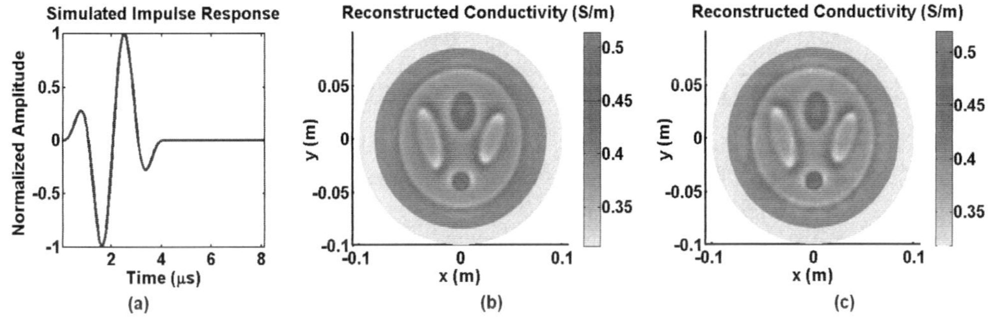

Figure 2. (a) Simulated impulse response function. (b) and (c) are reconstructed conductivity images using limited bandwidth data with SNR to be 1000 and 100, respectively.

We have also conducted simulation study to test the performance of the modified reconstruction algorithm when only limited bandwidth measurement data are available. Figure 2 (a) shows the simulated impulse response function. We use it to simulate the bandwidth limited acoustic measurements. The reconstructed conductivity images with two different SNR levels of 1000 and 100 are shown in figure 2 (b) and (c), respectively. Note the different color scales used in these images as compared with figure 1 (a). Only the relative conductivity contrast was reconstructed in these images.

Acknowledgements
This work was supported in part by NIH R21EB006070, NSF BES-0602957, NIH RO1EB007920, RO1HL080093, and RO1EB006433.

References
[1] Cheney M, Isaacson M and Newell J C 1999 *SIAM Review* **41** 85-101
[2] Griffiths H 2001 *Meas. Sci. Technol.* **12** 1126-31
[3] Woo E J and Seo J K 2008 *Physiol. Meas.* **29** R1-R26
[4] Towe B C and Islam M R 1988 *IEEE Trans. Biomed. Eng.* **35** 892-894
[5] Roth B J, Basser P J and Wikswo J P 1994 *IEEE Trans. Biomed. Eng.* **41** 723–28
[6] Wen H 2000 *Ultrason. Imaging* **22** 123-36
[7] Xu Y and He B 2005 *Phys. Med. Biol.* **50** 5175-87
[8] Li X, Xu Y and He B 2007 *IEEE Trans. Biomed. Eng.* **54** 323-30
[9] Xia R, Li X and He B 2007 *Appl. Phys. Lett.* **91**, 083903
[10] Xia R, Li X and He B 2009 *IEEE Trans. Med. Imaging* **28** 669-75
[11] Wang W and Eisenberg S R 1994 *IEEE Trans. Magn.* **30** 5105-23
[12] Oh S H, Lee B I, Woo E J, Lee S Y, Cho M H, Kwon O and Seo J K 2003 *Phys. Med. Biol.* **48** 3101-16

International Conference on Electrical Bioimpedance IOP Publishing
Journal of Physics: Conference Series **224** (2010) 012036 doi:10.1088/1742-6596/224/1/012036

The possibility of detecting magnetic moment generated by ultrasound in polarized tissues

Yuan Xu, Elena Renzhiglova

Department of Physics, Ryerson University, ON M5B2K3, Canada

E-mail: yxu@ryerson.ca

Abstract. We propose a method to combine focused ultrasound and electromagnetic field to image the permittivity of biological tissues. It will be shown that magnetic moment can be generated in a polarized tissue by propagating ultrasound. The value of this magnetic moment is proportional to the relative permittivity of the tissue. If the magnetic moment is detectable by a pick-up coil, the described method has the potential for imaging the electrical permittivity of biological tissues. In this paper, we derive the expression for the ultrasound-generated magnetic moment and the voltage in the pick-up coil. We also provide an estimation of the voltage signal.

1. introduction

1.1. Permittivity of biological tissues

Biological tissues possess unique electric properties, which have been successfully used in several imaging modalities. The current modalities, both clinically approved and under study, aim to image current density or conductivity of biological tissues [1, 2, 3]. These modalities can as well provide information on relative permittivity of biological tissues, which can be a reliable contrast mechanism for clinical imaging, due to the high variances of its value between different tissues [4].

Biological tissues have very large values of relative permittivity (on the order of 10^6) at low frequency range. Their dielectric spectra have several dispersion regions noted as α, β and γ. The α-dispersion observed at frequencies below 10 kHz may arise from counterion polarization around charged surfaces, such as cellular membranes, and from interfacial polarization on intracellular membrane systems and intercellular gap junctions [5, 6]. The region of α-dispersion and below is characterized by the largest values of relative permittivity for a particular tissue.

β dispersion occurs at intermediate frequencies and is due to the capacitive charging of cellular membranes, as well as those of membrane-bound intracellular bodies. Lastly, γ dispersion is due to the polarization of tissue water.

The electric permittivity can be imaged in Electrical Impedance Tomography (EIT) [2], Magnetic Induction Tomography (MIT) [3] and Magnetic Resonance Electrical Impedance Tomography (MREIT) [2, 7]. At present, the resolution of the existing impedance imaging modalities varies from 10% of the electrode/coil array diameter in EIT/MIT to a few millimeters in MREIT.

In this paper, we consider the possibility of imaging the electric permittivity of the tissues by combining ultrasound and electric fields. We will show in the following section that a magnetic moment can be generated in a polarized dielectric sample (e.g., tissue) with ultrasound.

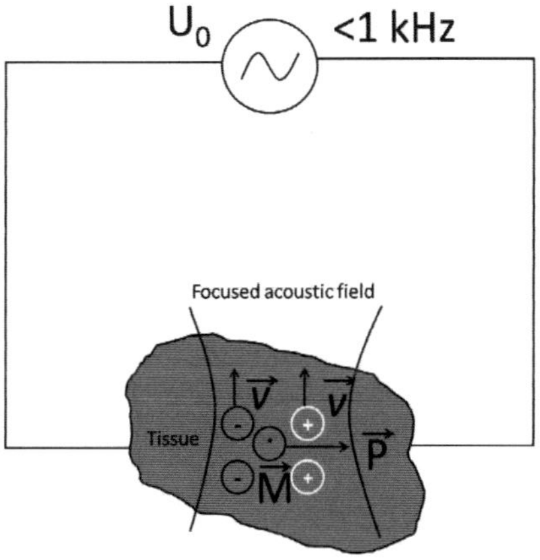

Figure 1. Generating a magnetic moment in a polarized tissue with ultrasound

When the frequency of the electric field applied to the tissue is in the α-dispersion range, biological tissues have huge permittivity and the magnetic moments induced by a diagnostic level ultrasound is estimated to be in the same order as the magnetic moment in a normal MRI imaging condition. The proposed method can be a potential mechanism for developing a low-cost high-spatial-resolution imaging modality of permittivity of biological tissues.

1.2. Theoretical formulation of the ultrasound-induced magnetic moment in a polarized tissue
Consider that a DC or low-frequency voltage with amplitude of U_0 and frequency ω_V is applied through a pair of electrodes to piece of tissue (Fig. 1). The electric field will polarize the tissue and create an electric dipole volume density **P** pointing to the right. Assume that a focused ultrasound field with frequency ω_U significantly larger than ω_V is propagating upward in the sample. The acoustic field causes the electric dipoles in the polarized tissue to vibrate vertically around their equilibrium positions with a velocity **v** at the ultrasound frequency. As $\omega_U \gg \omega_V$, the electric dipole density **P** can be assumed constant over the duration of the ultrasonic pulse. The positive charges will produce a current pointing upward, while the negative charges will produce a current with the same magnitude but pointing downward. Two currents pointing to opposite directions with the same amplitude will generate a magnetic moment. Since the dipoles vibrate around their equilibrium positions, the magnetic moment is oscillatory, and, therefore, it can be detected with a pick-up coil.

Now we will express the generated magnetic moment **M** in terms of electric dipole density **P** and vibration velocity **v**. For an arbitrary current distribution, the associated magnetic moment can be found as:

$$\mathbf{M} = \frac{1}{2}\int_V (\mathbf{r} \times \mathbf{J})\, d\mathbf{r}, \tag{1}$$

where **J** is the current density at position **r** [2]. For each dipole, those are two current densities $\mathbf{J}_+ = nq_+\mathbf{v}$ and $\mathbf{J}_- = nq_-\mathbf{v}$ associated with the displacement of the pair of charges q_+ and q_- in a dipole, whose positions are $\mathbf{r}_+$ and $\mathbf{r}_-$, respectively. Now, assuming that $\mathbf{J}_- = -\mathbf{J}_+$ because of $q_+ = -q_-$, we have:

$$\begin{aligned}
\mathbf{M_{tot}} &= \frac{1}{2}\int\limits_V (\mathbf{r}_+ \times \mathbf{J}_+ + \mathbf{r}_- \times \mathbf{J}_-)\, d\mathbf{r} = \frac{1}{2}\int\limits_V (\mathbf{r}_+ \times \mathbf{J}_+ - \mathbf{r}_- \times \mathbf{J}_+)\, d\mathbf{r} \\
&= \frac{1}{2}\int\limits_V \mathbf{J}_+ \times \mathbf{l_p}\, d\mathbf{r} = \frac{1}{2}\int\limits_V \mathbf{v} \times nq_+\mathbf{l_p}\, d\mathbf{r} \\
&= \frac{1}{2}\int\limits_V \mathbf{v} \times \mathbf{P}\, d\mathbf{r} = \int\limits_V \mathbf{M}\, d\mathbf{r},
\end{aligned} \tag{2}$$

where $\mathbf{P} = nq_+\mathbf{l_p}$ is the electric dipole density, n is the volume number density of positive and negative charges, $\mathbf{l_p} = \mathbf{r}_+ - \mathbf{r}_-$ is the separation between the positive and negative charges in a dipole and $\mathbf{M} = \frac{1}{2}(\mathbf{v} \times \mathbf{P})$ is the magnetic moment density. The magnetic moment $\mathbf{M}$ can be detected by a pick-up coil, which is similar to the signal detection in MRI. Therefore, we can write the following expression for the voltage signal in our experiment [8]:

$$U(t) = -\int_{object} \frac{\partial}{\partial t}\mathbf{M}(\mathbf{r},t) \cdot \mathbf{B_r}(\mathbf{r})\, d\mathbf{r} \tag{3}$$

where $\mathbf{M}$ ($\mathbf{r}$,t) is the time-varying magnetic field that exists throughout the focused ultrasound beam inside the sample and $\mathbf{B_r}$ is the magnetic field produced by a unit direct current in the coil.

2. Results

2.1. Simplified expression for the voltage signal

To estimate $\mathbf{M}$, we assumed that the applied low-frequency voltage creates a homogeneous electric field whose amplitude is simply $\frac{U_0}{d}$. Here U_0 is the amplitude of the voltage applied to the sample and d is the distance between the electrodes. This allows us writing the following expression for the electric dipole density:

$$P \approx \frac{\epsilon_0 \epsilon_r U_0}{d}, \tag{4}$$

since $\epsilon_r \gg 1$ for biological tissue. Here, ϵ_0 is the permittivity of vacuum, and ϵ_r is the relative permittivity.

2.2. Comparison of magnetization moment density induced by ultrasound with that in MRI

For the estimations,we used the following values: ω = 6.28 MHz, U=36 V, ϵ_r=10^7 (10 Hz, muscle tissue, longitudinal direction [4]), d=20 cm.

We can calculate magnetization in the sample induced by ultrasound. By substituting Eq. 4 in Eq. 2, for the parameters given above, the magnetization is 132 mA·m^2. This value can be compared to the value of the magnetization induced in MRI. For a spin system composed of hydrogen nuclei ^{1}H under magnetic field of strength B_0, the magnetization can be calculated as follows:

$$M_0 = \frac{B_0 \gamma^2 \hbar^2}{4kT} P_D, \tag{5}$$

where γ is the gyromagnetic ratio in rad·s^{-1}·T^{-1}, $\hbar$ is the Planck's constant, k is the Boltzmann's constant, T is the temperature in °K, P_D is the proton density, i.e. the number of hydrogen atoms in a unit volume [8]. Assuming $\frac{\gamma}{2\pi}$= 42.58 MHz·T^{-1}, B_0 = 1.5 T, T=300 °K and P_D of 1.1·10^5 M·m^{-3}, we get a magnetic moment with a magnitude of 30 mA·m^2.

3. Discussion

In the described method, same RF coils as in MRI can be used to detect the ultrasound-induced magnetic moment. However, the signal detection described here has an important difference from signal detection in MRI due to the difference in its bandwidth. In MRI, the signals have very narrow bandwidths [8]. Therefore, the coils are working at resonance frequency to increase the sensitivity. In our experiments, to obtain a good spatial resolution, ultrasound signals have broad bandwidth. Therefore, we also need the resonance frequency of the coil to be above the signal spectrum to detect the broad band signal. If the coil's resonance frequency is below the frequency of the signal, the coil behaves like a capacitor partially shunting the voltage dropping on the inductance. In comparison to EIT, MIT and MREIT, the advantage of our method is that it combines high resolution and low cost. The high resolution is obtained through the use of focused ultrasound to generate magnetic moment in a very localized volume inside the tissue. In addition, our method requires minimal electrical contact with the tissue in the case of signal detection with coil.

4. Conclusion and future work

We showed that an oscillating magnetic moment can be generated in a biological tissue by propagating ultrasound. Its value is proportional to the relative permittivity. We derived the expression for signal in the detection configuration of using a pick-up coil placed above the sample.

References

[1] Hasanov K, Ma A, Nachman A and Joy M 2008 *IEEE T. Biomed. Eng.* **27** 1301
[2] Holder D 2005 *Electrical impedance tomography: methods, history and applications* (London: IOP) pp xxi, 239
[3] Zou Y and Guo Z 2003 *Med. Eng. Phys.* **25** 79
[4] Miklavcic D, Pavselj N and Hart F 2006 *Electric properties of tissues* (New York: Wiley Encyclopedia of Biomedic al Engineering, John Wiley and Sons, Inc)
[5] Gabriel C 2006 *Dielectric properties of biological materials* (Taylor Francis Group LLC) p 77
[6] Asami K 2007 *J. Phys. D: Appl. Phys.* **40** 3718
[7] Sekino M Tatara S and Ohsaki H 2008 *IEEE T. Magn.* **44** 4460
[8] Prince J L and Links J M 2006 *Medical imaging signals and systems* (Upper Saddle River: Pearson Prentice Hall) pp 381-388
[9] Hayder S, Hrbek A and Xu Y 2008 *Physiol. Meas.* **29** S41

Development of the system for visualization of electric conductivity distribution in human brain and its activity by the magnetic induction tomography (MIT) method

S Sapetsky[1], V Cherepenin[1], A Korjenevsky[1], V Kornienko[1] and A Vartanov[2]

[1] Kotel'nikov Institute of Radio-engineering and Electronics of the RAS, Moscow, Russia

[2] Psychology Department of the Moscow State University, Moscow, Russia

E-mail: mds64@yandex.ru

Abstract. Currently rapid development of functional activity researches of human brain sets the problem of reliable and non-invasive detection of mental processes and states. At present we know some traditional methods of rapid and contactless acquisition of brain activity characteristics, such as functional tomography (fMRI) and magnetoencephalography. But these methods have low temporal resolution, complicated and ambiguous association of measured values with information processes in brain. So possibility of MIT application is investigated. Estimation of possibility of such changes registration is performed. Investigations of magnetic field configuration, schematics of transmit-receive modules and numerical algorithms are in progress. It may allow us to register high speed conductivity changes in brain tissues.

1. Introduction

Currently rapid development of functional activity researches of human brain sets the problem of reliable and non-invasive detection of mental processes and states. At present we know some traditional methods of rapid and contactless acquisition of brain activity characteristics, such as functional tomography (fMRI) and magnetoencephalography. But these methods impose some limitations on cognitive investigations due to low temporal resolution, complicated and ambiguous association of measured indirect values (i.e. blood flow changes) to real information processes in brain. Although electro encephalographic methods of analysis give us high temporal resolution, they do not allow localization of brain activity reliably and with acceptable accuracy. We are investigating possibility of MIT for brain activity visualization. This makes demands for such systems, related with spatial resolution, sufficient operation speed and receivers sensitivity to achieve maximum contrast of different biological structures. While there is no reliable data currently about relation between electrical impedance and fast activity in brain, we hope that the developed system will make it possible to detect fast variations of brain tissues conductivity related with information processes in brain.

2. Methods

2.1. MIT method

The developed method of contactless registration of brain activity is based on magnetic induction tomography technique (MIT) [1, 2], using interaction of external quasistatic high-frequency magnetic field and conductive media (brain tissues). Amplitude of used magnetic field is low, much smaller than natural Earth magnetic field. It amounts about 0.1–10 nT and depends on distance from inductor coil. These values are fully safe for human health.

Inductor coil generates magnetic field, which induces eddy currents in material. Receiver coils register these magnetic field variations. Signal produced by eddy currents are usually much smaller than primary one. Earlier we have demonstrated [3] that influence of conductive object leads to phase variation of detected signal (small quadrature component appears). It enables us to reconstruct conductivity. Therefore phase shift is proportional to the field frequency and conductivity:

$$\Delta\varphi \approx \omega \int_L W\sigma dl.$$

We may employ filtered back projection method [4] or artificial neural network method [5] for solving inverse problem and image reconstruction.

Spatial resolution of MIT system depends on number of independent inductor-detector combinations and coils size. Conductivity resolution depends on total working volume and on noise level of measuring system. At higher frequencies greater noise is acceptable. The measuring time is defined practically by electronics speed of operation and data transfer rate.

2.2. Current state of the MIT system

At present time multichannel MIT measuring system has designed and created. It is composed of 16 transmitter and receiver coils arranged at the 40 cm diameter circle electromagnetic screen. Acquisition time is about 0.5 sec now. It is easy to reduce it significantly by improvement of equipment and measurement and data transfer algorithm. Spatial resolution is about 5 cm. We have reconstructed conductivity distribution image inside biological objects: human thorax and brain for the first time by means of this system [6].

2.3. Lines of further development

In brief it's necessary to:

- increase coils quantity for better spatial resolution,
- further improve measuring modules for increasing of operation speed and accuracy of phase measurements,
- update software and develop of image reconstruction methods.

Obtaining of maximum contrast MIT tomograms which reflect local hi-frequency electrical activity, particularly at frequency 40-60 Hz (brain gamma-rhythm), corresponding to post synaptic potentials and spike activity of neurons, may be reached by means of optimal combination of such parameters as magnitude and configuration of excitation magnetic field, operating frequency and sensitivity of receivers. These parameters may vary depending on tissue under investigation and changes owing to biochemical and physiological processes. Thereby we plan to work out new experimental prototype of multifrequency MIT tomograph with better spatial resolution (about 1 cm) due to 64 coils with smaller size. Temporal resolution will be not worse than 5 msec. Estimation of optimal parameters for brain activity visualization will be carried out on experimental models and real objects during psychophysiological experiments.

2.4. Design of the new MIT system

Coils, as before, will be situated at circle electromagnetic screen. Such configuration provides protection from external electromagnetic fields and the best system sensitivity for brain investigation. We plan to increase coil quantity 4 times as many (from 16 to 64). And coils diameter will 4 times less. We need to increase power in inductor coil for phase measurement accuracy enhancement (see figure 1).

We will solve this task by means of power high-frequency amplifiers, which are capable to give higher power to output coil (inductor). Optimal quantity of turns will be chosen from operating frequency value and characteristics of specific amplifier. Microcontroller produces input signal for the amplifier and controls its working regime. It is necessary to use more sensitive receivers and thorough filtering in input and following cascades of receivers.

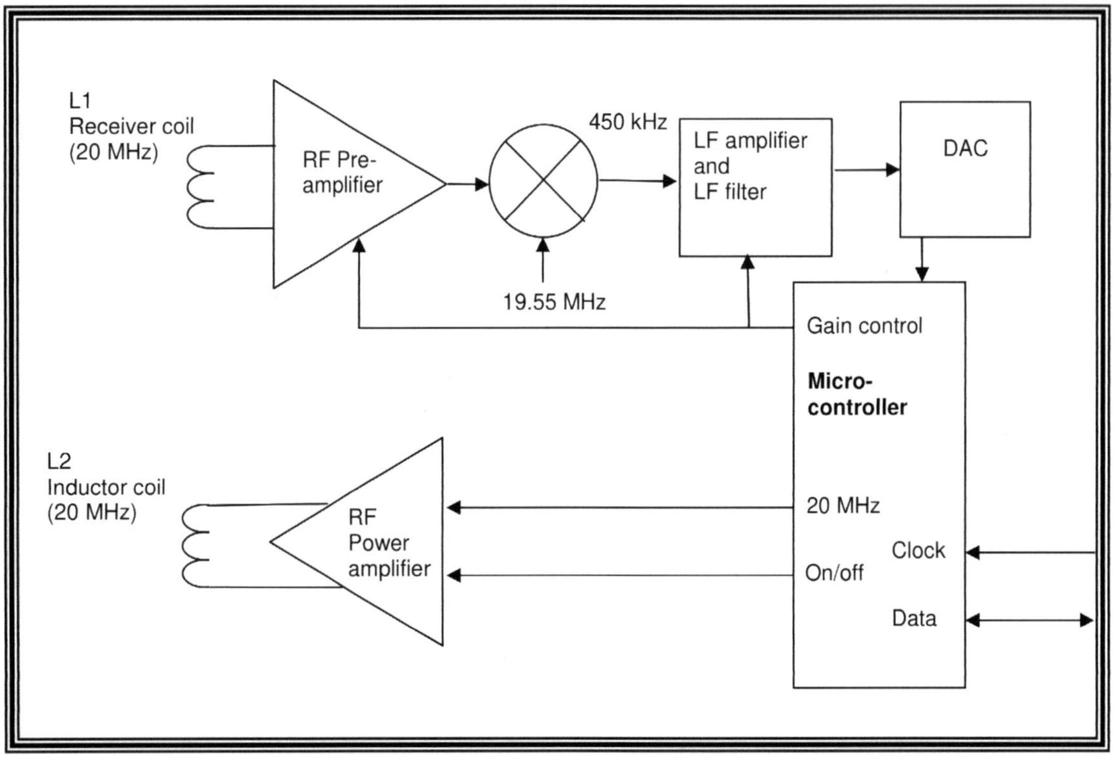

Figure 1. Block diagram of the MIT channel module.

It is necessary to obtain high temporal resolution (or operation speed) during measuring and data saving only. Operating frequency should be in 10-20 MHz range. For better results of amplifying, filtering and digitizing of the receiver coil signal we need to convert it to lower frequency (e.g. 450 kHz). At this frequency minimal data acquisition time is about 1.3 ms (for 10 periods of averaging). This is suitable value for investigations of 40-60 Hz processes. For significant reducing of acquisition time digitizing of receiver's signals should be carried out in parallel at all 64 modules. Using of superheterodyne receivers allows us to depart from using expensive hi-speed DACs and processors. But we can measure sufficiently quickly even at lower frequency.

Measurements results in each channel are accumulated in memory of corresponding microcontroller. It will be preferable to calculate two quadratures in the microcontroller rather than transfer ADC counts in real-time. It allows us to reduce data volume significantly. When storing data as 32-bit digits, 1 data set (frame) occupied 60 Kbytes only. After finishing of all measuring session, data from all modules are passed to PC through USB and saved here. Data processing and reconstruction will run later, not in real time.

Acknowledgment
The work is carried out with the financial support of the Russian Foundation for Basic Research, Grant No 09-02-12222.

References

[1] Korjenevsky A, Cherepenin V and Sapetsky S 2000 Magnetic induction tomography: experimental realization *Physiol. Meas.* **21(1)** 89-94
[2] H Griffiths 2001 Magnetic induction tomography *Meas. Sci. Technol.* **12** 1126-31
[3] Korjenevsky A and Cherepenin V 1997 Induction tomography: theory, computer simulation and elements of measuring system *Med. Biol. Eng. Comp., World Congr. Medical Physics and Biomedical Engineering* **35 suppl. part 1** (Nice, France) 330
[4] Korjenevsky A and Cherepenin V 1999 Progress in realization of magnetic induction tomography *Annals of the New York Academy of Sciences* **873** 346-52
[5] Korjenevsky A V 2002 Neural network algorithms for solving inverse problems in radio frequency tomography *Neyrokomp'utery: razrabotka i primemenie* N 9-10 26-33
[6] Sapetsky S A and Korjenevsky A V 2004 Magnetic induction tomography: visualization of extensive objects *Proc. XII Int. Conf. Electrical Bio-Impedance & V Electrical Impedance Tomography* (Gdansk, Poland) 695-98

Journal of Physics: Conference Series **224** (2010) 012039 doi:10.1088/1742-6596/224/1/012039

A non-contact method for imaging the posterior chest using magnetic induction principles that allows to monitor pulmonary oedema

D Gürsoy and H Scharfetter

Institute of Medical Engineering, Graz University of Technology, Kronesgasse 5, A-8010 Graz, Austria

E-mail: guersoy@tugraz.at

Abstract.
Real time monitoring of lung function is of particular importance for the patients who are in the intensive care unit, and thus spend long durations of time in a supine position. This kind of recumbent positioning of the patients gives rise to a markedly increased fluid accumulation in the posterior lung regions associated with the gravity dependency. In order to monitor the temporal behavior of the accumulation, we proposed a non-contact semi-tomography method which uses magnetic induction principles. In the proposed method, an eddy current density is induced within the dorsal tissues including the posterior lungs via the transmitter coils which are embedded into the patient bed, and the magnetic field strength is measured similarly using an array of sensor coils in a non-contact manner. For the assessment of the method, we used a patient specific, MRI-guided realistic chest model and presented the reconstructed time-differential images.

1. Introduction

Real time monitoring of lung function is of particular importance for the patients who are in the intensive care unit, and thus spend long durations of time in a supine position. This kind of recumbent positioning of the patients gives rise to a markedly increased fluid accumulation in the posterior lung regions associated with the gravity dependency. In earlier studies, electrical impedance tomography was reported as a diagnostic tool in order to monitor the temporal behavior of the accumulation [1], [2]. However, using electrodes has some formidable drawbacks due to the poor skin-electrode interface. Therefore, we proposed an alternative method which uses magnetic induction principles. The non-contact measurement methodology has been known for years and at first proposed over 40 years ago to monitor the cardiac and respiratory cycles of the patients [3] and is gaining interest again [4] to be used for imaging purposes. In the proposed method, an eddy current density is induced within the dorsal tissues including the posterior lungs via the transmitter coils which are embedded into the patient bed, and the magnetic field strength is measured similarly using an array of sensor coils in a non-contact manner to be used in image reconstruction. For the assessment of the method, we used a patient specific, MRI-guided realistic chest model and presented the reconstructed time-differential images using simulated data.

2. Methods

2.1. *Image reconstruction*

The relationship between the conductivity parameters and the measurement data is nonlinear and requires the solution of a set of electromagnetic equations. The corresponding formulation is based on the quasi-stationary approximation of Maxwell's equations and has been reported in earlier studies [5], [6]. Let us define this relation as, $v = \psi(\sigma)$, where σ is a piecewise-constant function that defines the conductivity of the medium and v represents the voltage measurements. The image reconstruction is known as estimating the conductivity σ from the measurements v.

2.1.1. Linearization Due to the nonlinear nature of the problem, absolute imaging fails to perform satisfactorily, and thus differential imaging is preferred. The corresponding reconstruction problem has commonly been formulated using the reciprocity theorem. According to the theorem, the sensitivities of the measurements to the conductivity perturbations at each voxel can be approximated as follows,

$$\left.\frac{\partial\psi}{\partial\sigma}\right|_{\sigma_0} \approx \Delta\sigma^{-1}\int_{\Omega_c}\Delta\sigma\mathbf{E}_1(\sigma_0)\cdot\mathbf{E}_2(\sigma_0)dV, \tag{1}$$

where $\mathbf{E}_1$ and $\mathbf{E}_2$ respectively denote the direct and reciprocal electric fields generated in the conducting domain Ω_c for a particular conductivity distribution σ_0 and $\Delta\sigma$ represents the conductivity deviation from σ_0. Assuming admissible conductivity deviations, by using (1), the following system of linear equations can be formed,

$$\mathbf{S}\Delta\sigma = \Delta v, \tag{2}$$

where $\mathbf{S}$ is referred to as the sensitivity matrix. Each row corresponds to a excitation-measurement coil pair and each column to a voxel.

2.1.2. Regularized Gauss-Newton Method The regularized least-squares solution for $\Delta\sigma$ can be obtained by minimizing the L_2-norm of the residuals between the estimated and measured data as defined by means of a minimization problem and a single iteration of the Gauss-Newton algorithm gives a fairly convenient solution for $\Delta\sigma$ [7]. The corresponding solution can be obtained from,

$$\Delta\sigma = \left(\mathbf{S}^T\mathbf{S} + \lambda\mathbf{R}^T\mathbf{R}\right)^{-1}\mathbf{S}^T\Delta v, \tag{3}$$

where $\mathbf{S}^T\mathbf{S}$ is an approximation of the Hessian, $\mathbf{R}$ and λ are the regularization matrix and regularization parameter, respectively [8]. A proper choice of λ can be obtained, e.g. according to the Morozov's discrepancy principle criterion [9].

2.2. *Simulation Setup*

2.2.1. Chest Model The model for the simulation was obtained by segmenting the MRI images to construct a three dimensional volume (see, figure 1). Using diffused imaging modalities, reconstruction of small details is infeasible; therefore, the model was constructed roughly by considering only the extended tissue groups such as the lungs, heart and the surrounding muscle tissue. The conductivity values for the tissues were assigned based on a parameterized Cole equation. The conductivities of the muscle, lung and heart tissues were taken as 0.35 S m^{-1}, 0.18 S m^{-1} and 0.21 S m^{-1}, respectively, at 200 kHz operating frequency.

2.2.2. Fluid accumulation The fluid accumulation that causes the so-called gravitational dependent collapse was assumed to take place in the posterior lungs. The collapsed region is a mixture of oedematous fluid together with the deflated lung tissue causing the conductivity of that region to be higher than the healthy tissue. We assumed 0.6 S m^{-1} for that region [1].

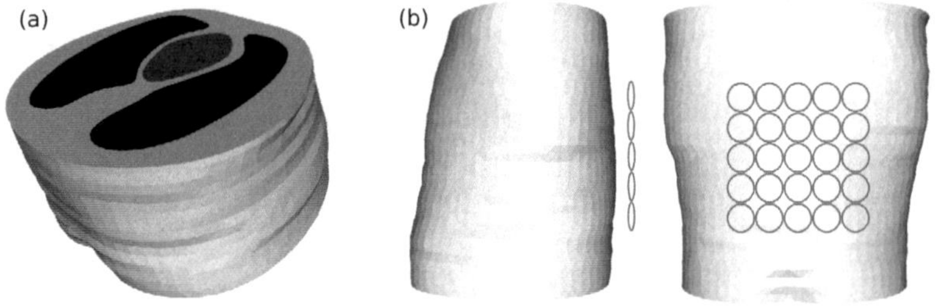

Figure 1. (a) The thorax model constructed using the tissue boundaries segmented from MRI images. (b) The arrangement of the 5×5 coil matrix according to the body.

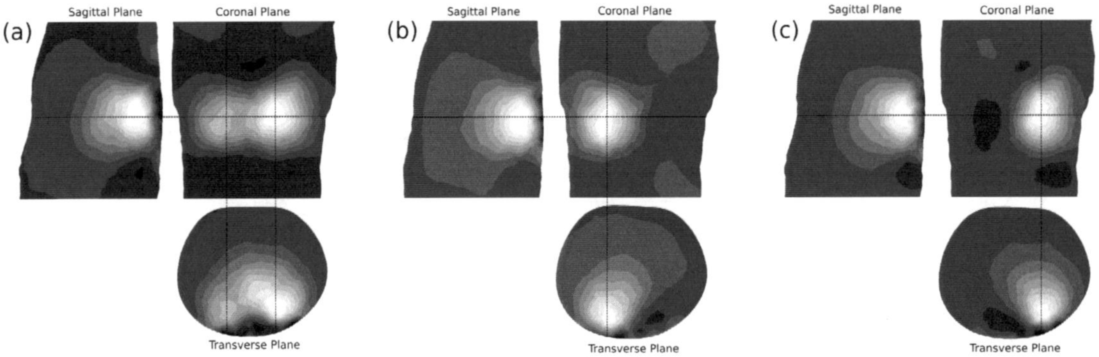

Figure 2. Reconstructed images of fluid accumulation in (a) both lungs, (b) left lung only and (c) right lung only.

2.2.3. Coil Configuration The 5×5 coil matrix was assumed to be embedded in the patient's bed and the corresponding arrangement is depicted in figure 1. The coils are assumed to act both as a transmitter and a receiver and have a radius of 20 mm. The approximate distance from the coil center to the body surface is 1 cm.

3. Results

Assume that the patient is in supine position and the measurements are taken at different times, t_1 and t_2. Using time-differential voltage measurements, i.e., $\Delta v = v(t_2) - v(t_1)$, one can monitor the change of fluid accumulation in time, i.e., $\Delta\sigma = \sigma(t_2) - \sigma(t_1)$. The initial measurements were taken from the healthy lungs at t_1 and the measurements at t_2 are taken assuming that the fluid accumulation has reached a state in which 2 cm of the lower lungs have collapsed. 1% uncorrelated Gaussian noise was added to the voltage data in order to simulate the noise of the receiver channels.

The reconstructed images of the fluid accumulation which has happened in both lungs simultaneously is shown in figure 2-(a). From the images, both collapsed regions can be identified in the regions close to the surface and in central regions they overlap and cannot be distinguished clearly. The remaining images in figure 2 represent the cases that the accumulation develops faster in one lung than the other, i.e., (b) and (c) corresponding to the collapse in the left lung and right lung only. The images show similar blurry characteristics and tend to shift towards the central regions which explains the overlap of reconstructions in figure 2-(a).

4. Discussions and conclusisons

In this work, the measurements were assumed to be insensitive to the cardiac and respiratory cycles. The heart is placed in the ventral portion of the chest and, knowing that the sensitivity decays rapidly with depth, it is reasonable to neglect the corresponding effects. Breathing, on the other hand, affects the measurements much more than the cardiac cycles. However, the conductivity change between the deflated and inflated lungs are way lower than the conductivity changes caused by the lung collapse and thus the effects are neglected.

Compensating the effects of the patient's movement during data acquisition is critical for this system [10]. For the patient in a supine position, the chest motion due to respiration is assumed to be only on the ventral surface and the movement artefacts are expected to be minimal for this positioning. However, there may still be a need to determine the patient's placement in the bed accurately.

This preliminary work points out the potential of the method for imaging of the fluidic changes in the dorsal parts of the lungs. Since the sensitivity of the measurements drops rapidly with depth below the skin, only the inhomogeneities that are very close to the skin can be monitored. The future work may include finding optimal coil matrices so that the spatial resolution of the images are increased for particular locations [11] which may provide better diagnostic information.

Acknowledgments

This work was supported by the SFB project F32-N18 granted by the Austrian Science Fund.

References

[1] Luepschen H, Riesen D, Beckmann L, Hameyer K and Leonhardt S 2008 *Magnetics, IEEE Transactions on* **44** 1450–1453
[2] Preis C, Luepschen H, Leonhardt S and Gommers D 2009 *Clinical Physiology and Functional Imaging* **29** 159–162
[3] Tarjan P P and McFee R 1968 *Biomedical Engineering, IEEE Transactions on* **BME-15** 266–278
[4] Steffen M, Heimann K, Bernstein N and Leonhardt S 2008 *Physiological Measurement* **29** 291–306
[5] Gençer N G and Tek M N 1999 *Medical Imaging, IEEE Transactions on* **18** 617–627
[6] Morris A, Griffiths H and Gough W 2001 *Physiological Measurement* **22** 113–119
[7] Merwa R, Hollaus K, Brunner P and Scharfetter H 2005 *Physiological Measurement* **26** 241–250
[8] Ziolkowski M, Gratkowski S and Palka R 2009 *International Journal of Applied Electromagnetics and Mechanics* **30** 245–253
[9] Hansen P C 1998 *Rank-deficient and discrete ill-posed problems* (Philadelphia: SIAM)
[10] Gürsoy D and Scharfetter H 2009 *Physiological Measurement* **30** 165–174
[11] Gürsoy D and Scharfetter H 2009 *Biomedical Engineering, IEEE Transactions on* **56** 1435–1441

Tracking of object movements for artefact suppression in Magnetic Induction Tomography (MIT)

H Scharfetter, S Issa, D Gürsoy

Institute of Medical Engineering, Graz University of Technology, Kronesgasse 5, A-8010 Graz, Austria
E-mail: hermann.scharfetter@tugraz.at

Abstract: One main difficulty in MIT is that movements of the object to be imaged can cause signal changes which are in the same order of magnitude or even higher than the wanted signals from the interior of the body. As patient fixation is unwanted and not always possible the best way to correct for movement artefacts is tracking of changes of the surface boundary. As additional sensors are, in general, undesirable we suggest retrieving the tracking information directly from the MIT signals. The basic idea is to place on the surface of the body a set of strategically placed active markers which consist of small loops of a very thin wire which can be opened and shorted via a tiny switch. When the loop is open it does not allow eddy currents to flow and therefore it is invisible in the reconstructed image. When the switch is closed, strong eddy currents flow and the signal essentially yields information on the marker positions. Our switches are remotely controlled MOSFETs mounted in a zone of low sensitivity of the coils so that they do not cause additional eddy currents. Image reconstruction then once provides the body information and the marker positions separately.

1. Introduction

Magnetic induction tomography (MIT) is an imaging modality which aims at the contact-less mapping of the complex electrical conductivity inside an object, e. g. a human body [1]. This is achieved by placing the object inside an array of transmit coils (TXC) which, by applying an AC magnetic field, cause eddy currents to flow. The resulting secondary magnetic field depends on the conductivity distribution inside of the coil array. The conductivity distribution is reconstructed from the voltages induced by the total magnetic field in an array of receiver coils (RXC). In medical MIT changes of the induced voltages by conductivity changes inside the object are by many orders of magnitude lower than the voltage induced by the primary field and hence there are many difficulties to be overcome both in hardware design as well as in image reconstruction.

One main difficulty is that the space inside of the coil array consists of a conducting and a non-conducting subdomain corresponding to the test object and the surrounding air, respectively. A particular problem is the boundary between the object and the surrounding air, because the eddy currents are confined to flow within the object and the voltages induced in the RXC depend strongly on the currents close to this boundary. In practice one is usually interested only in a small perturbation inside a background object (human head or thorax), while actually the total signal is essentially produced by the eddy currents close to the surface of the background object. A conducting perturbation with a diameter of 10% of that of a thorax and a conductivity of twice that of the background typically yields a signal which is 3 orders of magnitude less than that of the thorax itself. This in turn means that a small displacement of the background object or a small change of its boundary shape can induce a much larger voltage change than a diagnostically significant perturbation and image reconstruction becomes extremely difficult. Thus absolute MIT is very difficult but also time-differential ('dynamic') imaging may fail if the object movement is so fast that there is a significant shift between two states [2]. The artefact can, however, be filtered out if the movements of the outer object boundary are known [2]. Consequently the boundary should be tracked as exactly as possible, which could be achieved e. g. with a camera system or with an array of distance sensors. However, additional sensors are, in general, undesirable. Thus it would be a great advantage to get the tracking information from the MIT sensors (the coils). We suggest to place on the surface of the body a set of markers which produce a signal strong enough to reconstruct their positions and hence approximate changes of the object surface. The signal, however, must not perturb the object signal which is being used for the image reconstruction.

2. Methods

2.1. Experimental setup

In order to clearly separate the signals from markers and object we implemented as marker a small loop (radius 1 cm) of a very thin wire (diameter 0.05mm) which can be opened and shorted via a tiny CMOS-switch as schematically drawn in fig. 1. When the loop is open it does not allow eddy currents to flow and is invisible in the reconstructed image. When the switch is closed, strong eddy currents flow and the marker position can be reconstructed. Fig. 2 shows a possible arrangement in form of an elastic belt for tracking a thorax surface. In our experimental setup we mounted 16 markers on the surface of a plastic cylinder with a diameter of 200 mm. The positions were chosen such that the markers were in front of the gradiometers employed in the Graz MIT Mk2-system [3]. The distance from the gradiometers was 1 cm and the exact position was in the zone of maximum sensitivity of the coils, respectively. The on/off operation of the 16 markers was realized by means of 16 N-channel MOSFETs (BSN20, Philips). Due to their small plastic SMD-package and placement 10 cm apart from the marker coils outside the sensitive region of the transceivers they do not contribute to the formation of eddy currents. The measurement frequency was 500 kHz.

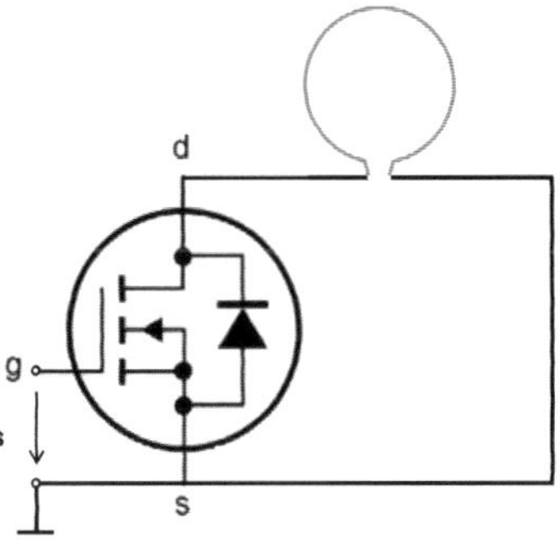

Figure 1. Active marker consisting of a wire loop and a CMOS switch for shorting the loop.

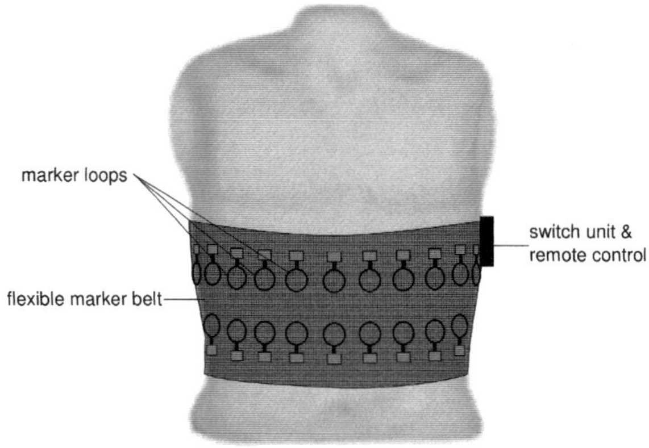

Figure 2. Schematic of an active marker system in form of an elastic belt with several loop/switch units which are controlled remotely from the data acquisition control unit.

Every frame was acquired twice, once with the switch closed and once with the switch opened. Then we carried out an experiment in order to investigate the visibility of the markers in their on and off-state, respectively, as well as the SNR of the marker signal. In total 45 frames were acquired whereas the first 15 were not used as measurement signals but served for phase calibration according to [4].

2.2. Simulated reconstruction of marker positions
The electromotive force in a coil in the presence of a magnetic field **B** can be expressed as,

$$e = -j\omega \int_S \vec{B} \cdot d\vec{S} \,,$$

where dS represents the surface element bounded by the coil contour pointing outward to that surface and ω is the angular frequency. According to the reciprocity theorem, the electromotive force in the receiver coil due to an active marker can be expressed as follows,

$$v = \eta e_T e_R \,,$$

where e_T and e_R represents the electro motive force in the marker coil created by the transmitter and receiver coils, respectively and v is the measured voltage in the receiver coil due to the marker signal. η is the admittance of the marker coil. Let N_T and N_R denote the number of transmitter and receiver coils respectively and N_M represents the finite number of possible marker configurations. Thus, by computing v for N_M different marker placements, it is possible to express a (N_T x N_R) by N_M sensitivity matrix **S**. Referring to ζ as the unknown marker placement, the corresponding marker reconstruction can be established by using the iterations as follows,

$$\zeta_{n+1} = \zeta_n + (S_n^T S_n + \lambda R^T R)^{-1} S_n^T (v_n - v_{meas})$$

where $S^T S$ is an approximation of the Hessian and R and λ are the regularization matrix and regularization parameter, respectively. v_n is the forward solution in step n and v_{meas} is the measured voltage.

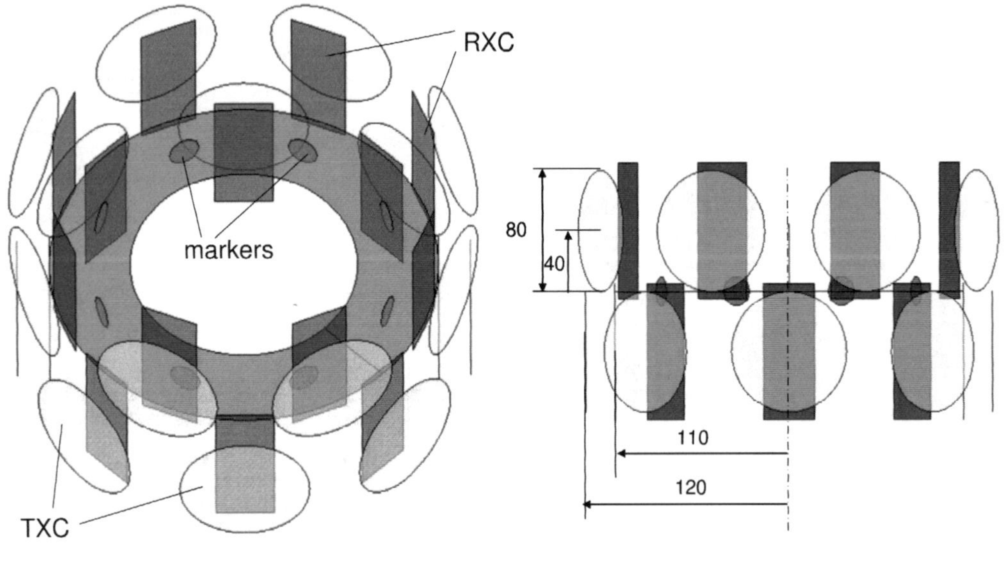

Figure 2. Geometry of the simulation model. All measures in mm.

For the simulations the geometry of the Graz MIT Mk2-system was approximated as shown in fig. 2. In contrast to the experiment, however, only 8 markers were positioned 1 cm in front of the gradiometer ring in the median plane. In order to get realistic results noise was added to the forward simulated voltages corresponding to an SNR of 100 dB of the marker signal. As shown in the section 'results' this is a realisti cally attainable value. As signal we defined the voltage difference between marker switched on an d marker switched off while noise was the standard deviation of t he voltage when the marker was switched off.

3. Results

3.1. Experimental results

The marker signal was clearly visible in both the real and imaginary part of the signals when switched on. The imaginary part was higher than the real one because the on-resistance of the MOSFET (max 15 Ω) dominated the loop impedance and made it more resistive than reactive. The SNR was about 91 dB when the switch was closed and below 0 dB when the switch was open.

3.2. Reconstruction results

The markers could be clearly reconstructed as shown in fig. 3. With the assumed SNR of 100 dB the localization error (deviation between true position and maximum of the reconstructed blob) was less than 1 mm.

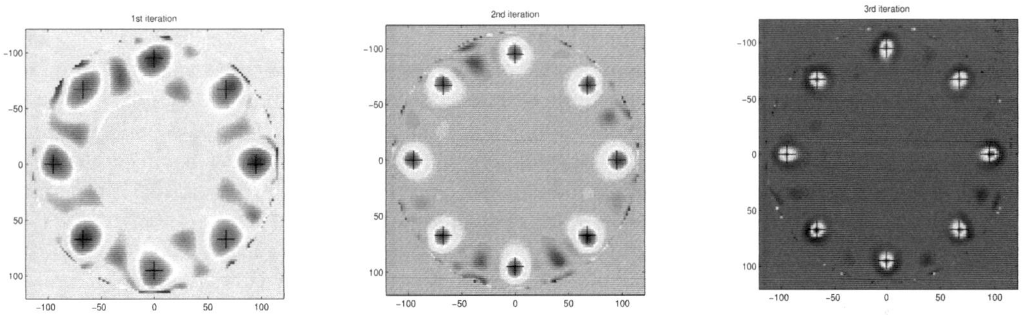

Figure 3. Reconstructed position of the markers after one, two and three iterations. The cros ses mark the true positions

4. Discussion

The achieved simulated images show that a re asonably accurate re construction of the marker s can be achieved when assuming an SNR which is close to th at which was determined experimentally. Thus it appears f easible to track object boundaries by onl y using t he MIT sig nal. However, further investigations must reveal the most appropriate marker designs and measurement frequencies so as to achieve optimum results. Furtherm ore alternative switches, e. g. PIN-diodes should be co nsidered in order to further lower the on-resistance and the ease of marker control.

Acknowledgments
This work was supported by the SFB project F32-N18 (Austrian Science Fund).

5. References

[1] Griffiths H 2005 *In: Electrical Impedance Tomography: Methods, History and Applications,* **ed Holder DS (UK, IOP Publishing)** p 213-38
[2] Gürsoy D, Scharfetter H 2009 *Physiol Meas* **30** S165-S174
[3] Scharfetter H, Köstinger A, Issa S 2008 *Physiol. Meas.* **29** S431-S443

Single-shot dual frequency excitation for magnetic induction tomography (MIT) at frequencies above 1 MHz

H Scharfetter

Institute of Medical Engineering, Graz University of Technology, Kronesgasse 5, A-8010 Graz, Austria
E-mail: hermann.scharfetter@tugraz.at

Abstract. In order to make MIT fast it is recommendable to excite several transmit coils in parallel and to acquire all receive channels simultaneously. The separation between the transmit channels can be achieved e.g. by slightly separating the excitation frequencies, so that it is possible to separate their contributions in the receivers by synchronous demodulation. One major problem is the low output impedance of the driver amplifiers so that each transmit coil acts as short circuited when seen from the other transceivers and hence perturbs the primary field. This causes a number of complications both in the reconstruction software as well as from the viewpoint of SNR optimization. Therefore we propose a special passive coupling network which transforms the low driver impedance into a high termination impedance for the coils at two frequencies (1 MHz and 10 MHz). We designed a dual-frequency transformation networks which operates at 1 and 10 MHz and which provides a termination impedance >5kΩ. The feeding current is typically 10mA per Volt driving voltage.

1. Introduction

Magnetic induction tomography (MIT) is an imaging modality which aims at the contact-less mapping of the complex electrical conductivity inside an object, e. g. a human body [1]. This is achieved by placing the object inside an array of transmit coils (TXC) which, by applying an AC magnetic field, cause eddy currents to flow. The resulting secondary magnetic field depends on the conductivity distribution inside of the coil array. The conductivity distribution is reconstructed from the voltages induced by the total magnetic field in an array of receiver coils (RXC). In medical MIT changes of the induced voltages by conductivity changes inside the object are by many orders of magnitude lower than the voltage induced by the primary field and hence there are many difficulties to be overcome both in hardware design as well as in image reconstruction.

In principle single-shot data acquisition can be achieved fairly easily, e. g. by driving all TXCs at the same time. Different coils operate at slightly different frequencies, so that it is possible to separate their contributions in the receivers by synchronous demodulation. One major problem is introduced by the driving power amplifier (PA) of the TXCs: If its output impedance is low, each transmit coil acts as if it were short circuited when seen from the receivers and perturbs the primary field considerably. Thus the field perturbation has to be considered in the reconstruction algorithm, which requires the explicit modelling of the coils and sophisticates the numerical solvers. Moreover magnetic coupling between the TXCs can propagate noise generated by the power amplifiers to the RXCs to a wider extend than if the TXCs were magnetically decoupled. This problem has been tackled in [2] by driving the coils by current sources below 1 MHz. However, the performance of current sources built with amplifiers and feedback networks gets poor when approaching high frequencies due to stability problems. In this paper an alternative approach is suggested which is based on the transformation of the low output impedance of the PA into a high impedance at the TXC side.

2. Methods

Two goals must be met in order to achieve a proper operation:

(1) provide a high termination impedance to the TX coil
(2) drive the same coil at two or more different frequencies

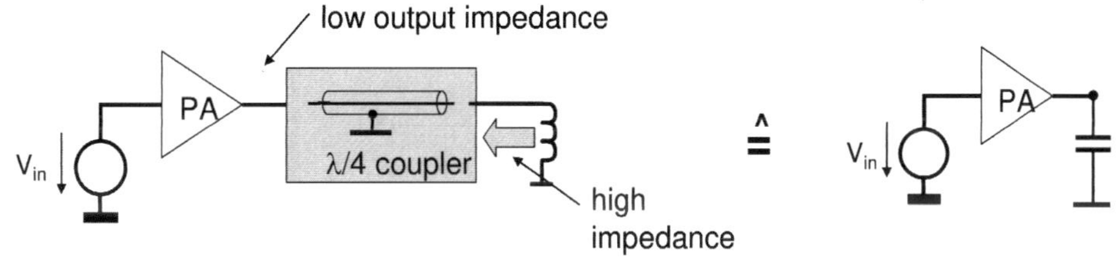

Figure 1. left: transformation of the low output impedance of the PA into a high termination impedance for the coil. right: transformation of the coil inductance into a capacitive load by the coupler.

If the spacing between the frequencies is large enough (e. g. 1 MHz – 10 MHz) goal 1 can be achieved by 2 λ/4 line transformers or the corresponding passive π-networks made with lumped reactive elements. Such networks transform a low impedance at one port into a high impedance at the second port at their resonance frequencies according to $Z_1 = Z_0^2/Z_2$ where Z_1, Z_2 are the impedances at ports 1 and 2 and Z_0 is the characteristic impedance of the coupler. Therefore using a simple push-pull amplifier with low output impedance can drive a current through the transformer while the coil 'sees' a high impedance and thus does not provide a significant field perturbation. The idea is illustrated in fig. 1 (left). The network transforms the inductance of the TXC into a load capacitance for the PA which may be unfavorable for safe operation (fig. 1, right). It is therefore advisable to drive the coil in resonance by introducing a series capacitor.

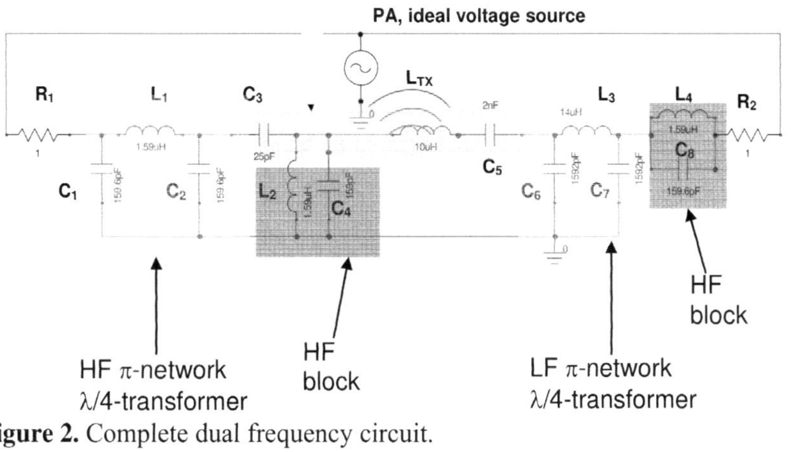

Figure 2. Complete dual frequency circuit.

Goal 2 can be achieved by providing two driving paths for the same TXC, one for low frequency (LF) and another one for high frequency (HF). Fig. 2 shows such a network which operates at 1 MHz (LF) and 10 MHz (HF).

The LF signal travels through the LF-π-network, the LF-series resonator (formed by the TXC and C_5) and L_2 in the HF blocking filter L_2/C_4. The path through the HF feeding network is blocked by the small capacitance C_3 which provides a high impedance at the low frequency. The TXC sees a high impedance, because the LF π-network shows a high impedance and the HF resonance capacitor C_3 has high impedance at LF so that the HF-feeding path is cut off.

The analysis of the HF path can be done in an analogous manner. Here the TX series resonator is made up by L_{TX} and the three capacitors C_3, C_5 and C_6. However, as C_3 is comparatively small, the resonance frequency is essentially determined by C_3. C_6 provides a convenient path to ground while the HF π-network establishes the connection to the PA. The TX coil again sees a high impedance, because the HF π-network shows high impedance and the network formed by L_2 and C_4 represents a parallel resonator at HF, thus cutting off the signal path.

Needless to say that the impedance transformers can also be made from transmission lines, but then the length may become impractically large and the characteristic impedance cannot be chosen arbitrarily when relying on commercially available cables (50Ω and 75Ω). For 10 MHz e. g. the cable length of a commercial 50Ω line must be approx. 5 m. This is still feasible, especially if the cable simultaneously acts as the feeder line. At lower frequencies a circuit with lumped elements is certainly preferable. Transmission line couplers, however, have the nice property that they operate at all odd multiples of the quarter wavelength and hence are intrinsically suitable for certain multifrequency applications where such multiples can be used as carriers.

The network in fig. 2 was simulated assuming a driver output impedance of 1 Ω, mimicked by the resistances R_1 and R_2. Simulations were done with Ansoft Designer SV, ver 2.2.0 (evaluation version, Ansoft Corp.). Then the network was built up and characterized with a network analyzer (ZVL3, Rohde and Schwarz).

3. Results

Fig. 3 shows the simulated termination impedance as seen from the TXC when simulating the network presented in fig 5 assuming a driver output impedance of 1Ω. The impedance at 1 MHz is 5.5kΩ while at 10 MHz a value of >8kΩ is reached. The values depend on both the driver output impedance and the characteristic impedance Z_0 of the π-network. Higher Z_0 yields higher termination impedance but at the same time reduces the coil current at constant driver voltage. Thus higher impedances can only be reached when simultaneously increasing the driving voltage.

Fig. 4 shows real data measured with the network analyzer (center frequency 6 MHz, span 12 MHz, measurement bandwidth 100 Hz, feeding power 20dBm) when the input port was shorted. The achieved peak values are somewhat higher than in the simulation due to ideal shorting and a somewhat higher coil quality than assumed in the simulation. The current in the TX coil is given as $I=V_0/Z_0$ whereby for the realized couplers Z_0 is given by $2\pi Lf_0$, L being L_1 an L_3 and f_0 being the resonance frequencies, i. e. 1 MHz and 10 MHz, respectively. V_0 is the output voltage of the feeding amplifier. In our case Z_0 was 100 Ω at both frequencies, so that I is 100mA when the when the PA drives 10 V.

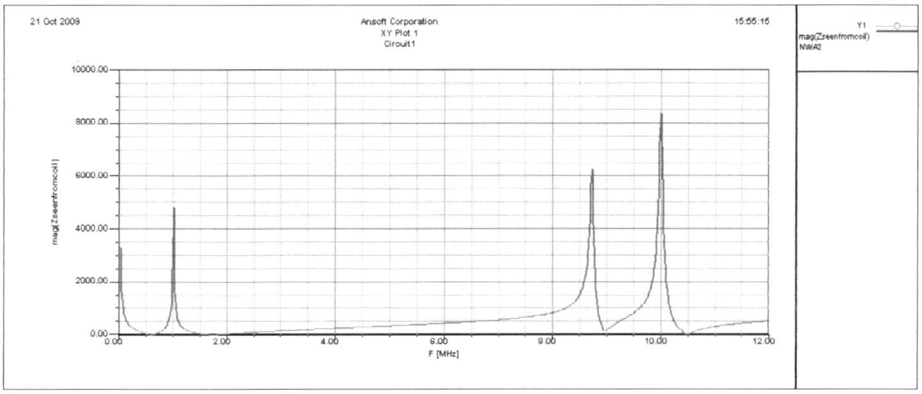

Figure 3: Termination impedance (magnitude) seen by the TXC as a function of frequency.

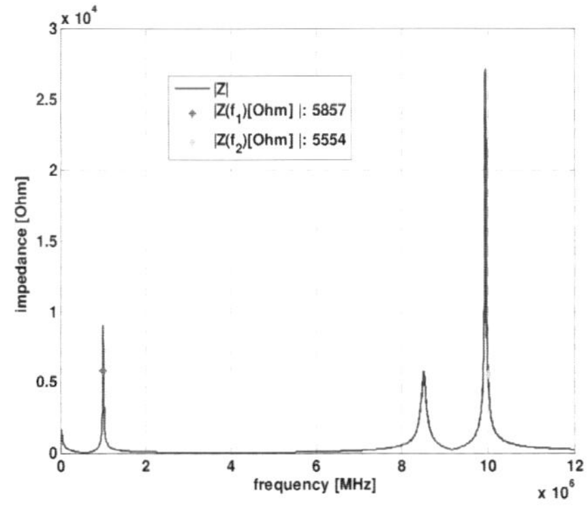

Figure 4: Termination impedance (magnitude) measured at the real network with a netwok analyzer when the input port was shorted. The two stars mark the actual readings at exactly 1 and 10 MHz.

4. Discussion

Our results show the feasibility of achieving high-impedance termination of the TX coils at two frequencies with a simple passive network. Thus, driving currents up to several hundreds of mA with low-impedance PAs is possible and the field perturbation can be kept low. However, there arise also some drawbacks of this design: (1) thermal drifts of the electrical parameters may change the resonance frequencies and hence the performance of the circuit at the nominal operating frequencies. (2) Unless ideally tuned the network introduces a phase shift so that there is no well-defined phase relationship between driving voltage and received signal. Thermal drifts make it necessary to frequently re-calibrate the reference phase for demodulation. In principle this goal can be achieved by using the method suggested in [3].

Acknowledgments
This work was supported by the SFB project F32-N18 (Austrian Science Fund).

5. References

[1] Griffiths H 2005 *In: Electrical Impedance Tomography: Methods, History and Applications,* **ed Holder DS (UK, IOP Publishing)** p 213-38
[2] Scharfetter H, Köstinger A, Issa S 2008 *Physiol. Meas.* **29** S431-S443
[3] Scharfetter H, Issa S 2008 Proc. 8th EMBEC 752-55

Journal of Physics: Conference Series **224** (2010) 012042 doi:10.1088/1742-6596/224/1/012042

Dynamic boundary estimation of human heart within a complete cardiac cycle using electrical impedance tomography

A Rashid[1], B S Kim[2], A K Khambampati[3], Dong Liu[1], S Kim[4] and K Y Kim[1]

[1] Department of Electronic Engineering, Jeju National University, Jeju, South Korea
[2] Applied Radiological Science Research Institute, Jeju National University, Jeju, South Korea
[3] BK21 Clean Energy Training & Education Center, Jeju National University, Jeju, South Korea
[4] Department of Nuclear and Energy Engineering, Jeju National University, Jeju, South Korea

E-mail: kyungyk@jejunu.ac.kr

Abstract. This paper presents an EKF based boundary estimation algorithm to estimate the shape and size of human heart ventricle during a complete cardiac cycle. First-order kinematic model is used as a state evolution model. The boundary of the heart is expressed as coefficients of truncated Fourier series and the conductivity distribution inside the thorax region is assumed to be known *a priori*. The proposed method is tested with the use of a realistic chest shape FEM mesh.

1. Introduction

Efficient and robust representations of cardiac morphology and morphodynamics are instrumental to conduct any large scale statistical analysis of regional heart shape and wall motion characteristics [1]. Electrical impedance tomography (EIT) is a non-invasive imaging modality which offers high temporal resolution making it a formidable technique to track cardiac related changes inside human body [2],[3],[4]. In EIT, a cross sectional image of the conductivity distribution inside human body can be obtained through the measurement of boundary potentials across the electrodes attached to the skin of the body. The poor spatial resolution of EIT makes it difficult to estimate the organ boundaries, thus undermining its anatomical significance. Key haemodynamic parameters related to heart physiology such as the cardiac output and ejection fraction can be better estimated if the boundary of the heart (ventricle) is estimated instead of its conductivity [4]. Dynamic algorithms such as linearized Kalman filter (LKF) or extended Kalman filter (EKF) can be used for this purpose [4],[5],[6],[7]. However, the estimation performance of these dynamic algorithms is still non-optimal due to the modeling uncertainty of the state evolution model, such as the *random-walk* model, used by them [3],[4],[5]. Since the size of target in cardiac motion varies smoothly, the estimation performance can be further improved with the use of a more suitable state evolution model such as the kinematic model [6],[7]. This paper presents an EKF based boundary estimation algorithm to estimate the cross sectional shape and size of the left ventricle of human heart during a complete cardiac cycle. First-order kinematic model is used as a state evolution model. The organ boundaries are expressed as coefficients of truncated Fourier series and the conductivity distribution inside the thorax region is assumed to be known *a priori* [8]. The proposed method is tested with the use of a realistic chest shape FEM mesh.

2. The boundary representation

The outer boundary $\partial\Omega$ of the body is assumed to be known. The organ boundary $C_\ell(s)$ is assumed to be sufficiently smooth to be expressed as coefficients of truncated Fourier series of order N_θ [4,],[8]

$$C_\ell(s) = \begin{pmatrix} x_\ell(s) \\ y_\ell(s) \end{pmatrix} = \sum_{n=1}^{N_\theta} \begin{pmatrix} \gamma_n^{x_\ell} \theta_n^x(s) \\ \gamma_n^{y_\ell} \theta_n^y(s) \end{pmatrix}, \ell = 1,2,...,S \quad \textbf{(1)}$$

where $\theta_n(s)$ is the periodic and smooth basis function of the form

$$\begin{aligned} \theta_1^\alpha &= 1 \\ \theta_{2n-1}^\alpha &= \sin(2n\pi s) \quad n = 1,2,... \\ \theta_{2n}^\alpha &= \cos(2n\pi s) \quad n = 1,2,... \end{aligned} \quad \textbf{(2)}$$

Here, $s \in [0,1]$, α denotes x or y and S is the number of boundaries. Expanding (1) and (2), the boundaries C_ℓ can be represented as γ of the shape coefficients, i.e.,

$$\gamma = (\gamma_1^{x_1},...,\gamma_{N_\theta}^{x_1},\gamma_1^{y_1},...,\gamma_{N_\theta}^{y_1},......\gamma_1^{x_S},...,\gamma_{N_\theta}^{x_S},\gamma_1^{y_S},...,\gamma_{N_\theta}^{y_S})^{\mathrm{T}} \quad \textbf{(3)}$$

where $\gamma \in \Re^{2SN_\theta}$. It is assumed that the boundaries of the lungs, the backbone as well as the outer heart are known *a priori*, while an approximate knowledge of the right ventricle boundary is assumed. The inverse problem, therefore, is a simplified case of only estimating the left ventricle elliptic boundary, i.e., C_ℓ with $S = 1$, and $N_\theta = 3$ has to be estimated

$$\gamma = (\gamma_1^x,...,\gamma_3^x,\gamma_1^y,...,\gamma_3^y) \quad \textbf{(4)}$$

3. Electrical impedance tomography: A Kalman filter approach

EIT is governed by a partial differential equation, with appropriate boundary condition such as the complete electrode model, establishing a relationship between voltage measurements and the conductivity distribution [3],[6]. The dynamic state estimation problem in EIT is based on a state space-space representation of the system, comprising of state evolution and observation models [5]

$$\gamma_{k+1} = f_k \gamma_k + w_k \quad \textbf{(5)}$$

$$U_k = g_k(\gamma_k) + v_k \quad \textbf{(6)}$$

where the subscript k is the discrete time index, f_k is the state transition model and g_k is the non-linear observation model. w_k and v_k denote the process and measurement noise respectively and are assumed to be white Gaussian noise. First-order kinematic model has been used as a state evolution model to estimate the radii (γ_2^x, γ_3^y) of the elliptic ventricle boundary, while the *random-walk* model is used for the rest of the parameters. The evolution model for the radiuis γ^α, $\alpha = x, y$, can be described as

$$\gamma_{k+1}^\alpha = \gamma_k^\alpha + \dot{\gamma}_k^\alpha \Delta T + w_k \quad (7)$$

$$\dot{\gamma}_{k+1}^\alpha = \dot{\gamma}_k^\alpha + w_k \quad \textbf{(8)}$$

where $\dot{\gamma}_k^\alpha$ is the velocity with which the boundary is expanding/contacting and ΔT denotes the time difference between two consecutive measurements k and $k+1$.

3.1. Extended Kalman Filter

The EIT observation equation (6) is nonlinear in nature. Suboptimal estimates can be obtained by using linear approximations. The most common approach for this purpose is the extended Kalman filtering in which linearization is obtained with respect to the previous state estimate [5-6]. The complete set of equations for the extended Kalman filter looks like

$$\gamma_{k+1|k} = f_k \gamma_{k|k} \quad \textbf{(9)}$$

$$C_{k+1|k} = f_k C_{k|k} f_k^T + Q \quad \textbf{(10)}$$

$$K_{k+1} = C_{k+1|k} J_{k+1}^T [J_{k+1} C_{k+1|k} J_{k+1}^T + R]^{-1} \quad \textbf{(11)}$$

$$C_{k+1|k+1} = (I_N - K_{k+1}J_{k+1})C_{k+1|k} \quad \textbf{(12)}$$

$$\gamma_{k+1|k+1} = \gamma_{k+1|k} + K_{k+1}[\overline{U}_{k+1} - J_{k+1}\cdot\gamma_{k+1|k}] \quad \textbf{(13)}$$

$$\overline{U}_k = U_k(\gamma_{k|k-1}) + \mathrm{J}_k(\gamma_k - \gamma_{k|k-1}) + v_k \quad \textbf{(14)}$$

where $J_k = \dfrac{\partial g_k}{\partial \gamma} \in \Re^{L\times N}$ is Jacobian obtained after linearizing the observation equation about the prior state estimate i.e., $\gamma = \gamma_{k|k-1}$. $\overline{U}_k \in \Re^{L\times P}$ is the linearized approximation of U_k in (6), $C_{k/k} \in \Re^{N\times N}$ is the error covariance matrix, $Q \in \Re^{N\times N}$ and $R \in \Re^{L\times L}$ are process noise and measurement noise covariance respectively. $K_k \in \Re^{N\times L}$ is Kalman gain and $I_N \in \Re^{N\times N}$ is an identity matrix. N is the number FEM mesh nodes, L is the number of electrodes and P is the number of current patterns. For details refer to [6].

4. Results and discussions

A realistic cross sectional chest shape phantom is considered to carry out the numerical simulations to verify the performance of the proposed method. Inhomogeneous FEM mesh structures, concentrated in the region of the heart, with 5768 and 5626 triangular elements have been used to solve the forward and inverse problems respectively. The dynamic boundary of the left ventricle of the heart has been recovered during complete cardiac cycle, while the boundaries of the lungs and the backbone are fixed and assumed to be known. The simulated scenario depicts the situation in which a patient is told to hold his breadth during EIT measurements. More accurate imaging techniques such as MRI can be used to recover the static boundaries. Clinically correct conductivities are assigned to the region, i.e., the conductivities of the heart muscle, blood, lungs, backbone and the background are 0.45, 0.75, 0.08, 0.1 and 0.2 S/m respectively [9]. A normal heart rate of 70 beats/min is assumed.

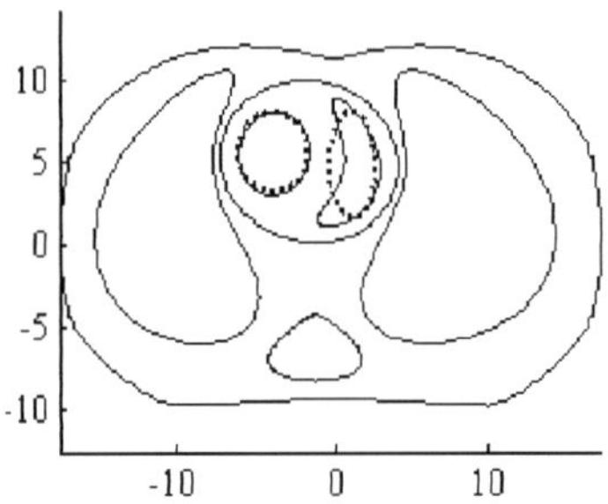

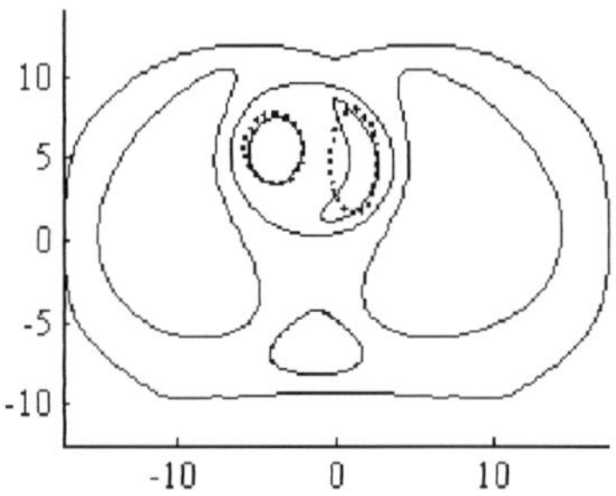

Figure 1. End diastolic (Left) and end systolic (Right) boundary. The broken lines represent the estimated boundary and the approximately known boundary for the left and right ventricles respectfully. The boundaries of the outside heart, lungs and the backbone are known *a priori*.

The first four modes of (cosine and sine) trigonometric current patterns are injected into 16 electrodes attached to the chest boundary, considering that they are more sensitive compared to the other modes. The complete electrode model with effective electrode contact impedance of 0.005 Ωcm^2 is used. In EIT, the voltage measurements are often noisy in nature. Therefore, zero-mean Gaussian noise with 0.5% standard deviation relative to the corresponding measured voltages has been added to the simulated voltage data. Using this methodology, 25 measurement frames (16 frames for diastole and nine frames for systole) are obtained for each cardiac cycle. The simulations have been carried out for three consecutive cardiac cycles to verify the robustness of the method. The state matrix for the kinematic models consists of six variables corresponding to the Fourier coefficients and two variables corresponding to the evolution velocities of the radii of the elliptic boundary. The state noise covariance is chosen as $Q = \mathrm{diag}[10^{-4}, 4\times10^{-3}, 10^{-4}, 10^{-4}, 10^{-4}, 4\times10^{-3}, 4\times10^{-3}, 4\times10^{-3}]$ while the coefficients of the error covariance and observation noise covariance are initialized to 10^{-4} and 8×10^{-2} respectively.

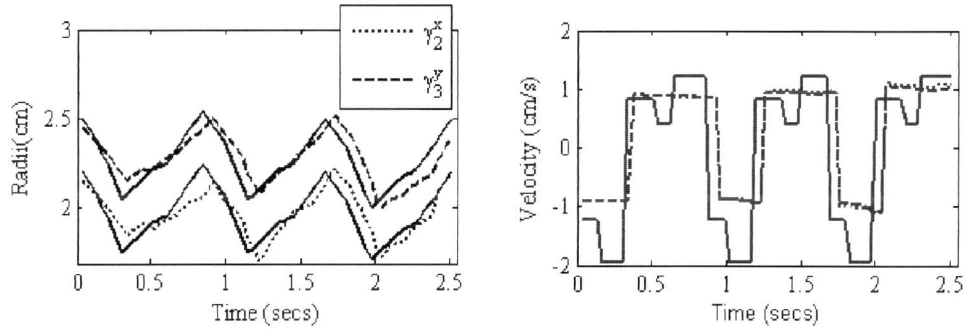

Figure 2. Evolution of the Fourier coefficients γ_2^x and γ_3^y which account for the radii of the elliptic ventricle in the x and y directions respectively (left), with their respective evolution velocities at each stage of the cardiac cycle (right). The solid lines are for the true values, whereas the dotted and broken lines represent the evolution of estimated parameters.

End diastolic and end systolic boundaries, as estimated in the third cycle, are shown in figure 1, while the evolution of the Fourier coefficients γ_2^x and γ_3^y, which account for the radii of the elliptic ventricle in the x and y directions respectively, has been shown in the figure 2. The results demonstrate reasonable estimation performance of the proposed methodology. A key assumption used while applying the kinematic model is the approximate knowledge of the diastolic and systolic cycle durations. This information can be easily obtained by using any other heart beat detection technique.

5. Conclusion

An EKF based 2D EIT measurement technique has been used to recover the boundary of the left heart ventricle during complete cardiac cycle, expressed as truncated Fourier series coefficients. Assuming a smooth expansion and contraction of the heart, first-order kinematic model is used as a state evolution model. Zero-mean Gaussian-distributed measurement noise has been added to simulate voltage measurement errors. Results obtained from a three-cycle, normal heart-rate simulation demonstrate the robustness of the method. Future work includes the extension of the method to 3D EIT for *in vivo* measurement and the simultaneous reconstruction of organ boundaries and their conductivities.

Acknowledgments

This work is supported by the grant from Korea Science and Engineering Foundation (KOSEF) funded by the Korea government (MEST) (No. R01-2007-000-20155-0). Part of the researchers participating in this study have been supported by the grant from the 2nd phase BK21 project.

References

[1] Youngl A A and Frangi A F 2008 *Exp. Physio.* **194** 578
[2] McArdle F J 1992 Investigation of cardiosynchronous images of the heart and head using Applied Potential Tomography *PhD Thesis* Univ. of Sheffield
[3] Vauhkonen M 1997 Electrical impedance tomography and prior information *PhD Thesis* University of Kuopio, Finland online at http://venda.uku.fi/~mvauhkon
[4] Vauhkonen M, Karjalainen P A and Kaipio J P 1998 A Kalman filter approach applied to the tracking of fast organ boundaries *Proc. 20th Annual Int. Conf. of the IEEE Engineering in Medicine and Biology Society* **20** 1048
[5] Kim K Y, Kim B S, Kim M C, Lee Y J and Vauhkonen M 2001 *Meas. Sci. Technol.* **12** 1032
[6] Tossavainen O P, Vauhkonen M, Kolehmainen V and Kim K Y 2006 *Chem. Eng. Sci.* **6** 7717
[7] Li X R 1997 *IEEE Ttrans. Aerospace and Electronic Systems* **33** 1212
[8] V Kolehmainen, Voutilainen A and Kaipio J P 2001 *Inverse Problems* **17** 1937
[9] Faes T J C, van der Meij H A, Munck J C and Heethaar R M 1999 *Physiol. Meas.* **20** R1–R10

A hand-held probe for combined ultrasound and electrical impedance tomography

R. Kulkarni[1], T-J. Kao[2], G. Boverman[3], G. J. Saulnier[1], D. Isaacson[4], T. L. Szabo[5], and J. C. Newell[2]

[1] Department of Electrical Engineering, Rensselaer Polytechnic Institute
[2] Department of Biomedical Engineering, Rensselaer Polytechnic Institute
[3] USC, Information Sciences Institute
[4] Department of Mathematics , Rensselaer Polytechnic Institute
[5] Department of Biomedical Engineering, Boston University

E-mail: saulng@rpi.edu

Abstract. The admittivity differences observed between normal and malignant tissue make electrical impedance tomography (EIT) a candidate technique for breast cancer detection. One way to improve tumor detection and classification is to combine EIT with other imaging modalities like ultrasound. We present in this paper a combined ultrasound/EIT hand-held probe which can simultaneously collect data from each modality. We develop a 6×6 ultrasound-like hand-held probe geometry using the Fourier decomposition method to obtain the forward solution in the linearized EIT reconstruction algorithm. We use a saline-filled tank to collect experimental target data for increasing target depths away from the electrode array. We demonstrate the satisfactory performance of the array with electrode dimensions 7 mm square, with a gap of 2 mm between electrodes and an overall array size of 54×54 mm in detecting a 1 cm^2 target up to depths of 25 mm. We also show successful detection of an agar target having lesser contrast, at lesser depths.

1. Introduction

The forward problem in electrical impedance tomography is to determine the voltages/currents on the surface of the body, given the admittivity distribution inside the body and currents/voltages applied to the surface. The inverse problem is to reconstruct the admittivity distribution inside the body given the voltage/current applied and measured on the surface of the body [1]. At Rensselaer, EIT data is collected using the fourth generation Adaptive Current Tomograph (ACT 4) [2]. Reconstructed images are made using the Newton's One Step Error Reconstructor (NOSER) algorithm for image reconstruction as described by Isaacson [3]. This algorithm uses one step of the Newton's method with constant admittivity as an initial guess [4].

EIT can be used in combination with other imaging modalities like ultrasound to improve tumor detection. This paper presents a dual-mode probe that can be used to collect ultrasound and EIT data simultaneously and in register. This probe places a 6×6 planar array of electrodes in front of an ultrasound probe, with the central electrodes being sonolucent. We solve the forward problem for a 6×6 hand-held geometry using a Fourier decomposition approach and show its satisfactory performance in detecting a copper target immersed up to depths of 25 mm in a saline tank. We also show results for an insulating target in a saline tank at lesser depths of up to 7 mm.

2. Dual ultrasound/EIT hand-held probe array

Figure 1 is a photograph of the dual-mode EIT/ultrasound hand-held probe. The probe uses a 6×6 array of EIT electrodes. A standard ultrasound probe is inserted inside the housing, with the active end of the ultrasound probe coupled to the back of the sonolucent central two columns of the array using ultrasound gel. These sonolucent electrodes are constructed by depositing a thin layer of gold on Kapton. The outer columns of electrodes are constructed as printed circuit boards (PCB). The overall probe is placed in contact with the patient using ultrasound gel which provides the needed electrical and acoustic coupling. The electrode array is a square grid with 6 electrodes in the x and y direction. The dimensions of the electrodes are 7 mm square, with a gap of 2 mm between electrodes. The array size is 54×54 mm.

Figure 1. EIT/ultrasound dual hand-held probe.

3. Fourier Forward Solution

The hand-held probe geometry can be modeled as an infinite half space, with the x and y co-ordinates going from $-\infty$ to $+\infty$ whereas the z co-ordinates go from 0 to $+\infty$. We express the forward solution as a decomposition of its Fourier components as done previously in the mammography geometry [4], [5], [6]. We solve the problem using the method of separation of variables. The forward voltage obtained by decomposition using the eigenfunctions of the ave-gap model [7], can be expressed as:

$$U(x,y,z) = \sum_{n=0}^{F} \sum_{m=0}^{F} u_{n,m} \cos \frac{n\pi x}{h_x} \cos \frac{m\pi y}{h_y} e^{-\lambda_{n,m} z} \tag{1}$$

Here $U(x,y,z)$ is the electric potential, F is the number of Fourier coefficients and $\lambda_{n,m} = \sqrt{\frac{n\pi}{h_x}^2 + \frac{m\pi}{h_y}^2}$.We solve for the unknown coefficient $u_{n,m} = \frac{j_{n,m}}{\gamma_0 \lambda_{n,m}}$where, j is the current density and γ_0 is constant admittivity from which the varying admittivity γ perturbs only slightly. This gives the forward solution in a homogeneous medium, with assumed constant admittivity γ_0.

The reconstruction algorithm is a linearized algorithm that assumes an admittivity distribution $\gamma(x,y,z)$ that differs only slightly from a constant value γ_0 for the homogeneous model [3].

4. 6×6 hand-held probe geometry

We built a PCB electrode array with the same dimensions as the 6×6 hand-held probe array for the purpose of experimentation. Figure 2 shows the geometry of the PCB electrode array. We test the implementation of our algorithm with an experimental tank with the 6×6 electrode array in one of the side walls. The tank dimensions are 41 cm, 58 cm, and 83.5 cm, in the x, y

Figure 2. The 6x6 array geometry.

Figure 3. Experimental tank set-up for the 6×6 hand-held probe array.

and z directions, respectively. The tank was filled with saline solution with conductivity of 190 mS/m. Figure 3 shows the experimental tank set-up. We obtained the voltage data from this tank for the set of optimal current patterns for the 6×6 array geometry with a 1 cm^3 copper target placed at different depths away from the electrode array. The reconstruction mesh used was a 576 voxel mesh, with 9 slices of 8×8 voxel grid each. Each slice was 5 mm thick except the first two slices which were 2 mm and 3 mm respectively. Reconstructions were obtained up to a depth of 40 mm. We do not display the first 2 mm slice adjacent to the electrode array in the reconstructions shown in Figure 4 due to its proximity to the electrode array which causes the reconstructed conductivity in this region to be considerably high in all voxels in the slice. We observe satisfactory depth resolution of the target for the reconstructions at all depths.

Next we explore the performance for a rod shaped insulating cylindrical object with diameter of 7 mm. The target is 7 mm from the array. Figure 5 shows the position in which the insulating target is held and reconstructions in the appropriate location. and as having lower conductivity than the medium.

5. Future work

In the future we would like to collect data with the hand-held probe in register with ultrasound data and analyze the results obtained from the two modalities. Of particular interest is the potential usefulness of the dual-mode data in the detection of breast cancer.

References

[1] Isaacson D and Edic P 1992 *IEEE EMBC* **14** 1693

[2] Saulnier G J, Ning L, Tamma C, Xia H, Kao T, Newell J and Isaacson D 2007 *29 th IEEE EMBS Annual International Conference* (Lyon, France)

[3] Isaacson D, Cheney M, Newell J, Simske S and Goble J 1990 *Imaging Systems and Technology* **2** 66–75

[4] Choi M, Kao T J, Isaacson D, Saulnier G and Newell J 2004 *Engineering in Medicine and Biology Society* **2** 1310–1313

[5] Kulkarni R, Boverman G, Isaacson D, Saulnier G J, Kao T J and Newell J C 2009 *Physiological Measurement* **30** S19–S34

[6] Kulkarni R 2009 *Analytical forward models for breast cancer detection using electrical impedance tomography* Ph.D. thesis Troy, NY

[7] Mueller J, Isaacson D and Newell J 1999 *IEEE trans. on Biomedical Eng* **46** 1379–1386

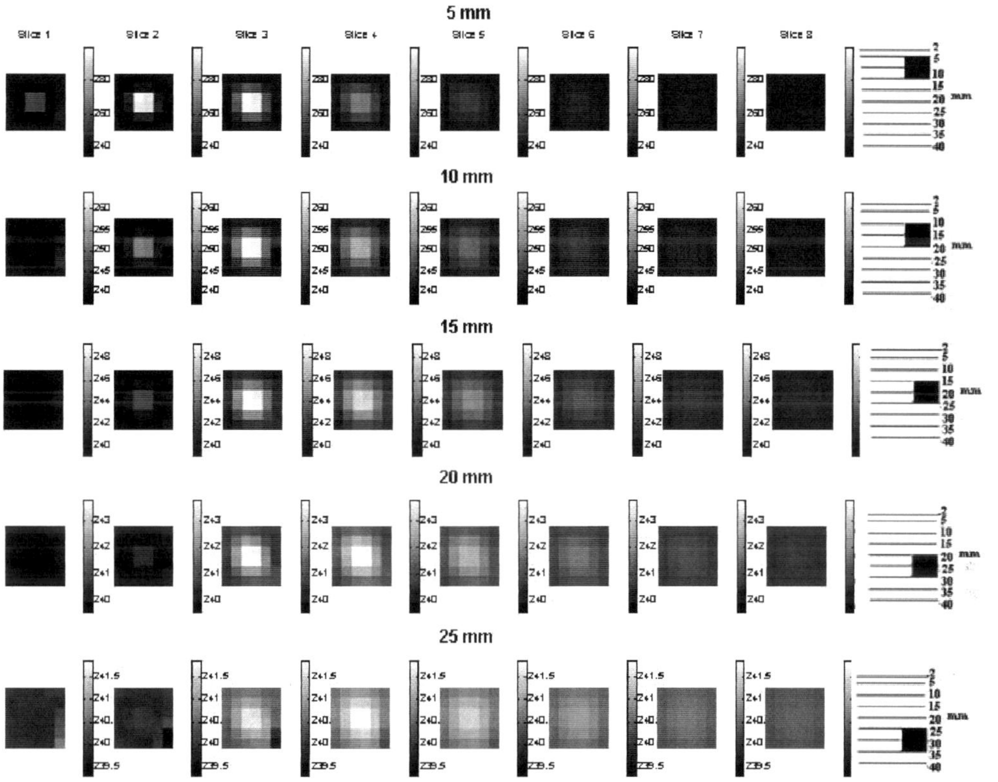

Figure 4. Reconstructions using 6 × 6 array with a target at 5 different depths as shown in the right-most column. The palette is the same across each row, but has less value as the target depth increases.

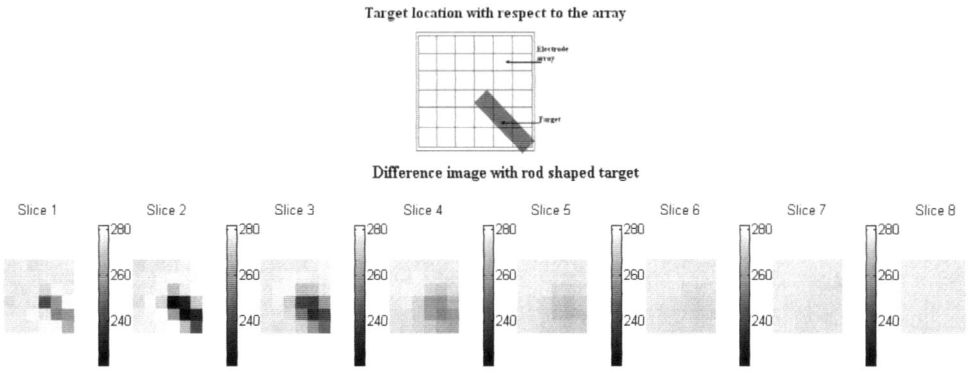

Figure 5. Difference reconstruction of an insulating target using the 6 × 6 array.

EM algorithm applied for estimating non-stationary region boundaries using electrical impedance tomography

A K Khambampati[1], A Rashid[2], B S Kim[3], Dong Liu[2], S Kim[4], K Y Kim[2]
[1] BK21 Clean Energy Training and Education Center, Jeju National University, South Korea, 690-756
[2]Department of Electrical and Electronic Engineering, Jeju National University, South Korea, 690-756
[3] Applied Radiological Science Research Institute, Jeju National University, South Korea, 690-756
[4] Department of Nuclear and Energy Engineering, Jeju National University, South Korea, 690-756

E-mail: kyungyk@jejunu.ac.kr

Abstract. EIT has been used for the dynamic estimation of organ boundaries. One specific application in this context is the estimation of lung boundaries during pulmonary circulation. This would help track the size and shape of lungs of the patients suffering from diseases like pulmonary edema and acute respiratory failure (ARF). The dynamic boundary estimation of the lungs can also be utilized to set and control the air volume and pressure delivered to the patients during artificial ventilation. In this paper, the expectation-maximization (EM) algorithm is used as an inverse algorithm to estimate the non-stationary lung boundary. The uncertainties caused in Kalman-type filters due to inaccurate selection of model parameters are overcome using EM algorithm. Numerical experiments using chest shaped geometry are carried out with proposed method and the performance is compared with extended Kalman filter (EKF). Results show superior performance of EM in estimation of the lung boundary.

1. Introduction

In many clinical and research studies long term ventilation has been applied and investigated. Accurate non-invasive method is useful for apnea detection in infants and adults, ventilation in intensive care units and lung monitoring for respiratory complexes [1]. EIT has been applied as an imaging technique to measure the resitivity changes for monitoring apnea and edema [2]. During the respiratory cycle, the air flows in and out of the lungs resulting in resistivity change. If the resisitivity distribution inside the lung, thorax and the other tissues are assumed to be known *a priori* then the unknown information can be size and shape of lung. The volume changes in the lung can be monitored and thus can provide valuable information when applying artificial ventilation. The boundaries are assumed to be smooth, therefore, can be represented using Fourier series [3]. The Fourier coefficients determine the size and shape of the boundary to be estimated. The fast moving boundary can be estimated using dynamic estimation methods.

In dynamic estimation, the inverse problem is treated as a state estimation problem and the time varying boundary coefficients are the state variables to be estimated [4, 5]. For the application of

Kalman-type estimators like EKF, UKF, the model parameters (dynamics of the evolution, the initial states, and the noise covariance of process and measurement models) have to be predefined [5]. In real situations, the dynamics of the evolution and noise matrices are complex and it is difficult to model them in a prior form. In situations, when there is uncertainty in determining the model parameters, the estimation performance of the Kalman-type filters is affected. Therefore, in this study, we apply EM as an inverse algorithm to reduce model uncertainties in estimating the closed boundaries [6].

2. Forward problem : boundary representation

Let us assume that the outer boundary of the body, that is, $\partial\Omega$ is known. If phase boundaries $C_k \in \Omega (k = 1, \cdots, \widetilde{P})$ of the objects are sufficiently smooth, they can be approximated as linear combination of known functions ([3, 5]),

$$C_k(s) = \begin{pmatrix} x_k(s) \\ y_k(s) \end{pmatrix} = \sum_{\omega=1}^{N_\theta} \begin{pmatrix} \gamma_\omega^{x_k} \theta_\omega^x(s) \\ \gamma_\omega^{y_k} \theta_\omega^y(s) \end{pmatrix}, \tag{1}$$

where $\theta_\omega^x(s)$ and $\theta_\omega^y(s)$ are periodic differentiable basis functions, $\widetilde{P}$ is the number of distinct regions and N_θ is the number of basis functions. In this paper, we express the phase boundaries as Fourier series in two-dimensional coordinates with respect to parameter s, i.e., basis functions are of the form,

$$\theta_1^\alpha(s) = 1 \tag{2}$$

$$\theta_\omega^\alpha(s) = \sin(2\pi \frac{\omega}{2} s), \quad \omega = 2, 4, 6, \cdots, N_\theta - 1 \tag{3}$$

$$\theta_\omega^\alpha(s) = \cos(2\pi \frac{(\omega - 1)}{2} s), \quad \omega = 1, 3, 5, \cdots, N_\theta, \tag{4}$$

where $s \in [0,1]$ and α denotes either x or y. Furthermore, using (1), the boundaries coefficients are identified with the vector $\gamma \in \Re^{M = 2\widetilde{P}N_\theta}$, that is,

$$\gamma = (\gamma_1^{x_1}, \ldots, \gamma_{N_\theta}^{x_1}, \gamma_1^{y_1}, \ldots, \gamma_{N_\theta}^{y_1}, \ldots, \ldots, \gamma_1^{x_{\widetilde{P}}}, \ldots, \gamma_{N_\theta}^{x_{\widetilde{P}}}, \gamma_1^{y_{\widetilde{P}}}, \ldots, \gamma_{N_\theta}^{y_{\widetilde{P}}})^T \tag{5}$$

3. Estimation of Fourier coefficients using EM algorithm

EM algorithm is a kind of maximum likelihood estimator which estimates the state variables and model parameters by minimizing the likelihood function using E and M step. In E step, the hidden variables (γ_k) are estimated with the assumed model parameters ($\theta = [\mu, \Pi, F, Q, R]$). In M step, using estimated γ_k, model parameters θ are estimated such that they increase the log-likelihood function. The EM algorithm is formulated for boundary estimation based on Kalman smoother. In E step, the expectation of the hidden variables is obtained using RTS Kalman smoother using n observations. The computed hidden variables are used to estimate the model parameters in the M step as follows [6],

$$F = BA^{-1} \tag{6}$$

$$Q = n^{-1}(C - BA^{-1}B^T) \tag{7}$$

$$R = n^{-1} \sum_{k=1}^{n} J_k^T p_{k|n} J_k + \{(V_k - J_k \gamma_{k|n})(V_k - J_k \gamma_{k|n})^T\} \tag{8}$$

$$\mu(r+1) = \gamma_{1|n}, \quad \Pi(r+1) = p_{1|n}, \tag{9}$$

where

$$A = \sum_{k=1}^{n} (P_{k-1|n} + \gamma_{k-1|n} \gamma_{k-1|n}^T) \tag{10}$$

$$B = \sum_{k=1}^{n} \left(P_{k,k-1|n} + \gamma_{k|n} \gamma_{k-1|n}^T\right) \tag{11}$$

$$C = \sum_{k=1}^{n} \left(P_{k|n} + \gamma_{k|n} \gamma_{k|n}^T\right). \tag{12}$$

4. Results and discussion

In order to test the EM algorithm, a 16-electrode chest shaped mesh with size and shape as that of an average human being is chosen. The conductivities of the thorax and the other organs inside chest are assumed to be known *a priori* figure 1(a) [1]. Opposite method is used as a current injection method and to simulate the dynamic scenario it is assumed that lung boundary changes after every single current injection. Different meshes with 5370 elements and 5220 elements are used in forward and inverse solver to avoid inverse crime. Four frames of data are considered in which the first two frames is for breathing or inhaling where the lung expands and two frames is for breathing out where the lung contracts. To represent the true dynamics of expansion and contraction, first-order kinematic model is used and the resulting scenario is shown in figure 1(b). In inverse computation, it is assumed that we do not have the prior knowledge of the evolution therefore random-walk model is used. To simulate with that of real conditions, random noise of zero-mean Gaussian noise having STD 1% of the value of the corresponding voltage is added to the computed voltage.

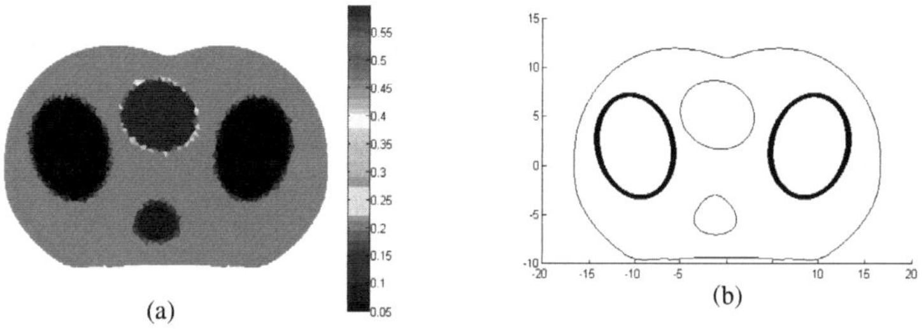

Figure 1. Chest shaped mesh used in numerical results (a) conductivity profile (b) generated scenario.

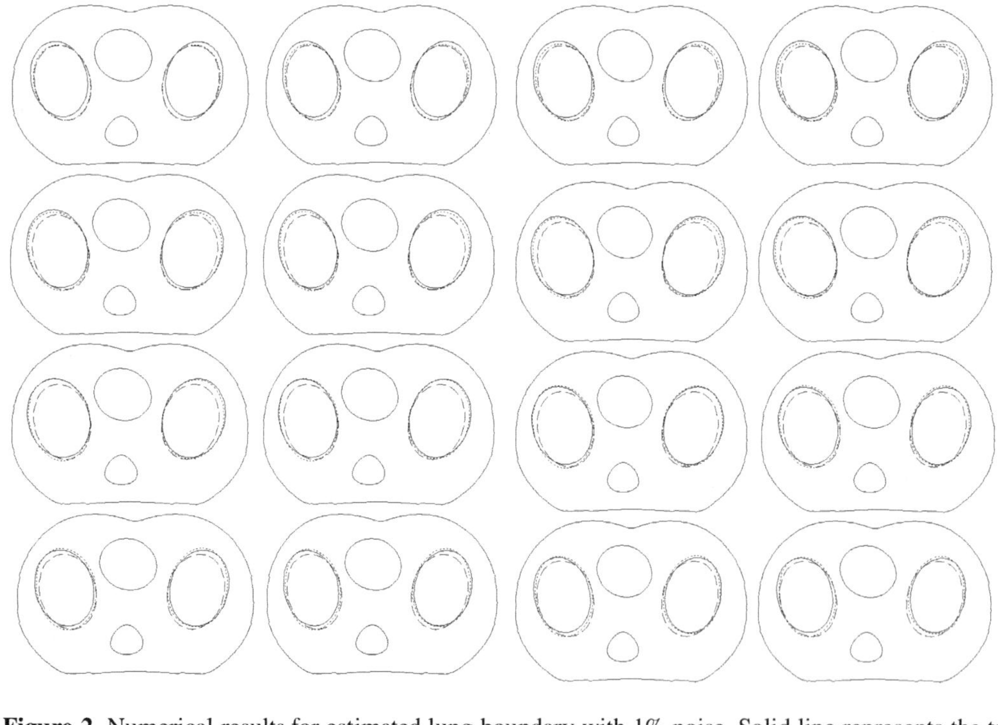

Figure 2. Numerical results for estimated lung boundary with 1% noise. Solid line represents the true boundary, dotted line is with EM algorithm and dashed line is using EKF.

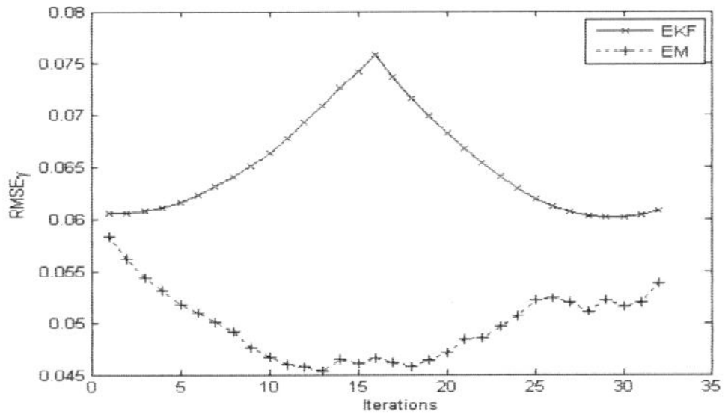

Figure 3. RMSE of estimated Fourier coefficients

Reconstructed lung boundary for 1 % relative noise is shown in figure 2. Alternative images are displayed in the reconstructed boundary. From figure 2, it can be noticed that EM has better estimation of the lung boundary as compared to EKF. The parameters used in the simulation with EKF and EM are given in Table 1. EM was able to track the fast changes of expansion and contraction of lung in one respiration cycle. The superior performance of EM is due to the estimation of model parameters (initial states, state evolution matrix and noise covariance matrix). The transition period is less as compared to EKF. For quantitative comparison, RMSE for boundary coefficients is computed and is shown in figure 3. It can be seen that EM has lower RMSE values as compared to EKF in all the iterations.

Table 1. Initial settings of model parameters used in EM and EKF.

F	Q	R	$P_{0\|0}$
$I_M \in \Re^{M\times M}$	$0.5I_M$	$500I_M$	$0.5I_M$

5. Conclusions

This paper presents an application of 2D-EIT using EM algorithm to estimate non-stationary lung boundary during respiratory cycle. The boundary of lungs and other organs are assumed to be smooth and therefore are represented using truncated Fourier series. It is assumed that the conductivities of thorax and other organs inside chest are constant and are known *a priori.* In respiratory cycle, it is assumed that the boundaries of heart and backbone are known and the unknown boundary is that of lung. The lung expands and contracts during the respiratory cycle and these changes are estimated using EM algorithm. The EM algorithm is formulated for boundary estimation in EIT and the model parameters along with the boundary coefficients are estimated. Numerical experiments are carried out to test the performance of the proposed method. Through the results, it is found that EM has better estimation of lung boundaries and thus can be used as an imaging tool for artificial ventilation.

Acknowledgement
This work was supported in part by the Korea Science and Engineering Foundation (KOSEF) grant funded by the Korea government (MEST) (No. R01-2007-000-20155-0) and part of the researchers in this study are supported by the grant from the 2nd phase BK21 project.

References
[1] Woo E J, Hua P, Webster J G and Tompkins W J 1992 *IEEE Trans. Biomed. Eng.* **39** 756-760
[2] Barber D C and Brown B H 1984 *J phys. E: Sci Instrum* **17** 723-733
[3] Kolehmainen V, Arridge S R, Lionheart W R B, Vauhkonen M and Kaipio J P 1999 *Inverse Probems* **15** 1375-1391
[4] Gelb A 1981 *Applied Optimal Estimation* MIT press
[5] Ijaz U Z, Khambampati A K, Lee J S, Kim S, and Kim K Y 2008 *J. Comput. Phys.* **227** 7089-7112
[6] Shumway R H and Stoffer D S 1981 *Technical report No. 27*, Division of statistics, University of California, Davis.

Development of a Trans-admittance Mammography (TAM) using 60×60 Electrode Array

Mingkang Zhao[1], Qin Liu[1], Tong In Oh[1], Eung Je Woo[1] and Jin Keun Seo[2]

[1] Department of Biomedical Engineering, Kyung Hee University, Korea
[2] Department of Mathematics, Yonsei University, Korea

E-mail: ejwoo@khu.ac.kr

Abstract. We have developed a trans-admittance mammography (TAM) system as a supplementary or alternative method of the X-ray mammography to diagnose the breast cancer. Mechanical structure of the system is similar to the X-ray mammography with the breast placed between two plates. The pair of plates is movable to accommodate breasts with different sizes and rotatable to provide multiple images with different projection angles. Without using ionizing radiation, it acquires a projection image of tissue admittivity values. One plate is a flat solid electrode where we apply a constant sinusoidal voltage with a variable frequency. The other is equipped with 60×60 array of current-sensing electrodes, of which potentials are kept at the signal reference level. The electrode array is connected to six switching modules and each module routes current signals from 600 electrodes to two ammeter modules. Each ammeter module includes six channels of ammeters and each one of them comprises an independent current-to-voltage converter, voltage amplifier, ADC and digital phase-sensitive demodulator. Each ammeter sequentially measures exit currents from 50 electrodes chosen by the corresponding switching module. An FPGA controls six ammeters to collect real- and imaginary-parts of trans-admittance data from 300 electrodes. A separate FPGA arbitrates data and command exchanges between a DSP-based main controller and ammeter modules. It also generates a sinusoidal voltage signal to be applied to the breast. All the 3600 complex current data from 12 ammeter modules are transferred to the main controller, which is interfaced to a PC through an isolated USB. The system is provided with a program to display real- and imaginary-parts of measured trans-admittance maps. The measured maps at multiple frequencies are incorporated into a frequency-difference anomaly detection algorithm. In this paper, we describe the design and construction of the system.

Keywords: Trans-admittance, mammography, conductivity image

1. Introduction

X-ray mammography is the primary method to diagnose breast cancer. In order to detect the cancer at an early stage, it is recommended for women to take a scan regularly. X-ray mammography uses ionizing radiation with a dose of 1 to 2.5 mGy per view. This ionizing radiation could be hazardous and induce malignancy. The risk of malignancy increases with the number of scans. In using X-ray mammography, it is therefore necessary to find a balance between its benefit and the risk. For this reason, it would be beneficial to develop other method of detecting breast cancer without ionizing radiation.

There have been several efforts to develop new methods to detect breast and impedance imaging is one of them [3]. However, impedance imaging methods still need to be improved to be used as a supplementary or alternative method to X-ray mammography. In this paper, we describe a new impedance imaging system for breast imaging, which we call the trans-admittance mammography (TAM).

2. System structure

The structure of the TAM system is similar to that of the X-ray mammography system as shown in figure 1(a). On the solid plate electrode, we apply a constant sinusoidal voltage with a variable frequency in the range of 1 to 500 kHz. We measure exit currents through the breast using 60×60 array of current-sensing electrodes. The TAM system comprises one voltage electrode, 3600 current-sensing electrodes, 6 switching modules, 12 ammeter modules and a main controller as shown in figure 1(b).

To measure an exit current on each current-sensing electrode, we need to connect the electrode to an ammeter. In order to measure exit currents from 3600 current-sensing electrodes, we implemented 72 ammeters so that each ammeter should be switched to one of 50 electrodes. Routing among 3600 electrodes, switching circuits and 72 ammeters is a technically challenging task and must be optimized for better performance. We came up with constructing 12 ammeter modules with each ammeter module containing 6 ammeters. Using 72 ammeters, this TAM system can measure 72 exit currents simultaneously. To handle switching among 72 ammeters and 3600 electrodes, we used 6 switching modules. A PC controls the TAM system through an isolated USB port. The DSP-based main controller inside the TAM system controls the constant voltage source and all ammeter modules.

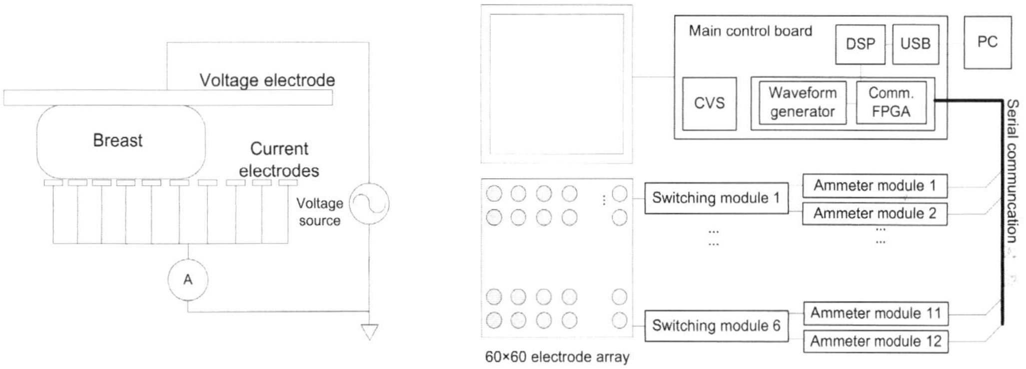

(a) **(b)**

Figure 1. (a) Basic concept and (b) structure of TAM system

3. Current-sensing electrode and switching module

The most distinct feature of the TAM system is the electrode plate with 60×60 current-sensing electrodes. Each current-sensing electrode is a gold-coated circle with 2 mm radius. The gap between adjacent electrodes in both x- and y-direction is 3 mm. On the back side of the electrode plate, we placed 18 BGA-type SMD connectors. Each connecter has 200 pins and each pin is connected to one current-sensing electrode.

Figure 2 shows how switching modules work. Each switching module handles 600 current-sensing electrodes and 12 ammeters. One ammeter sequentially measures an exit current from 50 current-sensing electrodes. During switching and also current sensing, each current-sensing electrode must be at the ground voltage all the time. Since one switching module needs 600 switches, we chose ADG734BRU (Analog Devices, USA) because of its small size (4 switches in one chip) and low on resistance. An FPGA on each ammeter module includes a switch controller to manage 600 switches.

The sequence of operation is stored inside the FPGA and it is initiated by the start signal from the main controller.

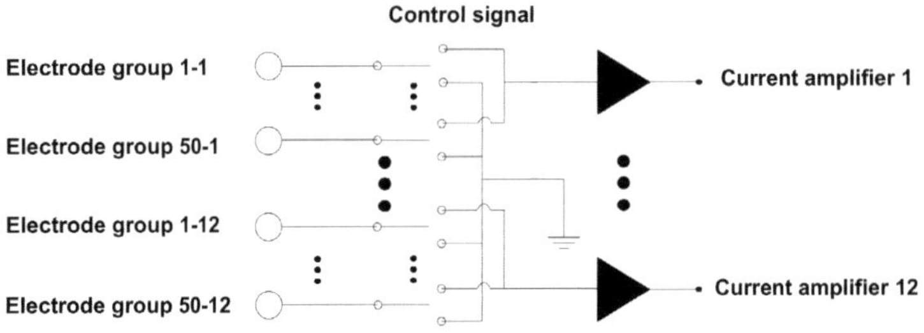

Figure 2. Switching module.

4. Ammeter module

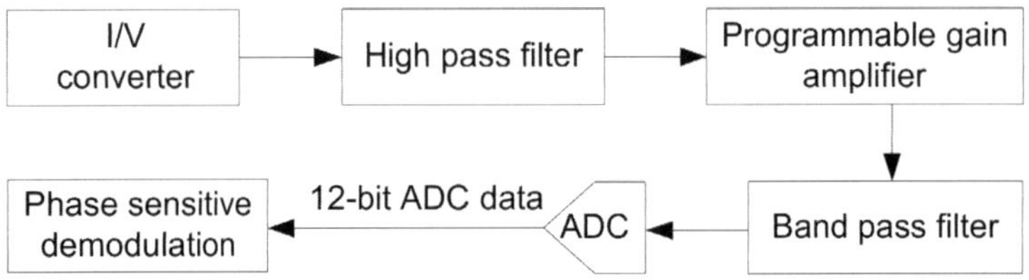

Figure 3. Block diagram of ammeter channel

The TAM system includes 12 ammeter modules. In one ammeter module, there are 6 independent ammeters. Figure 3 shows the block diagram of the ammeter. The exit current passes through the current-to-voltage (I/V) converter, which produces a voltage signal proportional to the exit current. The rest of the ammeter is the same as the voltmeter use in the multi-frequency EIT system KHU Mark2 [5]. The voltage amplifier adjusts its gain so that its output signal spans 90% of the input range of a high speed ADC. Two phase-sensitive demodulators inside the FPGA calculate the real- and imaginary-part of the complex exit current, which is equivalent to the corresponding trans-admittance.

5. Main controller

The main controller in figure 4 handles data communication among the PC and all ammeter modules. It also controls the constant voltage source for its amplitude and frequency. It comprises an FPGA (EP3C10F256C8N, Altera, USA), a DSP, a USB controller, a 16-bit DAC (AD9783, Analog Devices, USA) and several analog components for the voltage source. The output from the DAC is passed through a filter and amplifier to become a sinusoidal voltage signal applied on the breast. The DSP controls the entire system and the FPGA arbitrates data exchanges with 12 ammeter modules. The main controller is interfaced to a PC through the isolated USB controller.

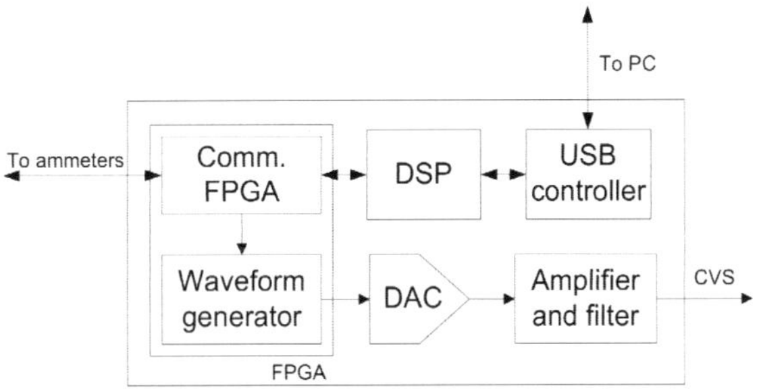

Figure 4. Main controller.

(a)

(b)

Figure 5. (a) Developed TAM system and its ammeter module.

6. Conclusion

Figure 5 shows the developed TAM system using 60×60 current-sensing electrode array. Its structure is similar to the X-ray mammography system. Using electrical current instead of ionizing radiation, the TAM system produces projection images of an admittivity distribution inside the breast. Since we can rotate the pair of electrode plates with the breast between them, it is possible to generate multiple projection images along multiple angles. In our future study, we plan to incorporate a frequency-difference anomaly detection algorithm into the TAM system and evaluate its performance.

Acknowledgments

This work was supported by the SRC/ERC program (R11-2002-103) of MOST/NRF.

References

[1] A. Malich, T. Boehm, M. Facius, M. G. Freesmeyer, M. Fleck, R. Anderson and W. A. Kaiser (2001) Differentiation of mammographically suspicious lesions: evaluation of breast ultrasound, MRI mammography and electrical impedance scanning as adjunctive technologies in breast cancer detection *Clinical Radiology*. **56**, 278-283.
[2] M. Reddy and R. Given-Wilson (2006) Screening for breast cancer *Women's Health Medicine* **3**, 22-27.
[3] Y. Zou and Z. Guo (2003) A review of electrical impedance techniques for breast cancer detection *Medical Engineering & Physics* **25**, 79–90.
[4] T. I. Oh, J. H. Lee, J. K. Seo, S. W. Kim and E J Woo (2007) Feasibility of breast cancer lesion detection using a multi-frequency trans-admittance scanner (TAS) with 10 Hz to 500 kHz bandwidth *Physiol. Meas*. **28,** S71–S84.
[5] H. Wi and E. J. Woo (2009) Design of KHU mark2 multi-frequency EIT System *Proc. Int. Conf. on Biomedical Applications of Electrical Impedance Tomography (Mancheter)*

The Cardiff Mk2b MIT head array: optimising the coil configuration

H Griffiths[1], M Zolgharni[1], P D Ledger[2] and S Watson[3]
Schools of Medicine[1] and Engineering[2], Swansea University, Swansea, SA2 8PP, UK
[3.]School of Electronics, University of Glamorgan, Pontypridd, CF37 1DL, UK

E-mail: huw.griffiths@abm-tr.wales.nhs.uk

Abstract A hemispherical MIT helmet coil array for imaging cerebral haemorrhage has been designed using a realistic 12-tissue finite-difference model of the head including a large peripheral haemorrhage (volume 49 ml). The coil array was first optimised by reaching a compromise between the quality of the reconstructed images and the financial cost of the digital detection system. The practical implementation of the helmet is partially complete.

1. Introduction

Recently, several groups have proposed the use of a hemispherical MIT array of coils for imaging cerebral stroke. Such an array will conform more closely to the shape of the head than a conventional cylindrical array and should increase the inductive coupling to the brain and hence the sensitivity to conductivity changes within it. Zolgharni *et al* [1] modelled a hemispherical coil array consisting of 56 coils using a finite-difference method combined with an anatomically-realistic head model comprising 12 tissue types. Frequency-difference images (1 – 10 MHz) were reconstructed from the modelled data with different levels of the added phase noise and two types of *a-priori* boundary errors that are likely to arise in a practical imaging system. It was concluded that a noise level of 3 m° (standard deviation) was adequate for obtaining good visualisation of a large peripheral haemorrhage (volume ≈ 50 ml). The simulations further showed that quite small discrepancies between the true skin surface and that used for computing the sensitivity matrix could give rise to significant artefacts on the images. Eichardt *et al* [2] have recently confirmed the better sensitivity distribution of the hemispherical arrangement than the cylindrical, using either 16 or 31 pairs of exciter and sensor coils. Zheng *et al* [3] have reported the construction of a hemispherical helmet MIT device using 15 axial gradiometers. Measurements (40-120 kHz) were made of a bowl filled with saline solution representing the brain and an immersed agar block of higher conductivity representing a haemorrhage. The inverse problem was not solved but 2D pseudo-images of the conductivity distribution at different frequencies were generated by interpolation.

In this paper we report the first stage of the practical realisation of the array of Zolgharni *et al* [1].

2. Concept of the helmet array

Following our earlier simulations and the work of others, the helmet was planned as an array of coils (separate exciters and sensors) positioned on a hemisphere, inside an outer, concentric, hemispherical, electromagnetic screen: this type of screen is known to be effective in rejecting extraneous interference and minimising inter-coil capacitive coupling [4]. An inner, plastic, hemispherical shell was to be fixed inside the coil array so as to completely enclose and protect the coils.

3. Determining the diameter of the plastic shell

The inner plastic shell needs to be small enough to allow the coils to be close to the scalp, in order to maximise the MIT signals, and large enough to accommodate most adult heads. The 1988 U.S. Army Anthropometry Survey [5] provides head measurements from 1774 males and 2208 females. The most relevant dimension is the anterior–posterior distance from between the eyebrows to the back of the cranium. The measurement fell within 209.5 mm for 96.5% of the males and 99.95% of the females. A further margin of 10 mm was allowed for the later addition of cushioning, leading to an inside diameter of 230 mm. The coils were positioned further out by another 10 mm, on a hemisphere 250 mm in diameter.

4. Determining the number of coils

The simulations of Zolgharni *et al* [1] configured the 56 coils as 28 exciters and 28 sensors. The new MIT electronics developed by our group can achieve a sufficiently low level of phase noise (1 m°) but is not compatible with the use of 56 coils. The number of exciters must be a multiple of 16 as the power multiplexers have 16 channels. The number of sensors must be a multiple of 7 because each PXI measurement board has 8 inputs, one of which is used for a phase reference [6]. The number of coils in the array determines the maximum coil diameter that will fit onto the hemisphere. Three arrays were considered: 30 coils (16 exciters, 14 sensors, coil diameter 45 mm); 46 coils (32 exciters, 14 sensors, coil diameter 35 mm); 60 coils (32 exciters, 28 sensors, coil diameter 30 mm). The realistic 12-tissue head model was discretized into a finite-difference mesh of 2.5 mm voxels for computing the simulated MIT signals. For speed of computation, the electromagnetic screen was initially not taken into account. The sensitivity matrix was computed by setting the conductivity to a uniform value of 1 S m^{-1} throughout the head.

For the 30-coil array, the number of possible ways of arranging the sensors and exciters is ${}_{30}C_{14}$ = 30!/16!14! = 145,422,675. Admitting only those arrangements symmetrical about the sagittal and coronal plans passing through the centre of the helmet reduces the number to 35. These were tested for quality according to the method of Gursoy and Scharfetter [7]. The quality factor, Q_r, (equation 12 in their paper) was computed for each coil arrangement: Q_r is a measure of the orthogonality of the rows of the sensitivity matrix and reflects the independence of the different sensor/exciter combinations. For the helmet, the values of Q_r spanned only a small range, 0.775-0.799, indicating that there was no clear best arrangement. A distributed arrangement was therefore chosen for this and for the 46- and 60-coil arrays (fig 1).

MIT measurements were then simulated at frequencies of 1 and 10 MHz and 1 m° (SD) Gaussian phase noise was added. Frequency-difference images were reconstructed by a single-step Tikhonov-regularized method. The value of the regularisation parameter, λ, was chosen so as to minimise the error

$$E = \frac{\left\| \Delta\sigma_{\text{true}} - \Delta\sigma_{\text{rec}} \right\|}{\left\| \Delta\sigma_{\text{true}} \right\|}$$

where $\Delta\sigma_{\text{true}}$ is the true volume distribution of conductivity changes between 1 and 10 MHz and and $\Delta\sigma_{\text{rec}}$ is the reconstructed image. Full details of the forward and inverse solutions and the tissue conductivities are provided elsewhere [8]. From a qualitative assessment of the images, the visualisation of the simulated haemorrhage improves when the number of coils is increased from 30 to 46, but there is less improvement by increasing the number of coils further to 60 (fig 1). The 46-coil array was chosen for practical implementation. Adding a further 14 sensors to form a 60-coil array would have increased the cost of the PXI digital detection system substantially.

5. Setting the diameter of the electromagnetic screen

Our previous work has shown that a standoff distance of 60-80% of the coil diameter gave reasonable suppression of inter-coil capacitive coupling without excessively damping the inductive signals [4]. A standoff distance of 32 mm was chosen (91% of the diameter) as this allowed the use of a commercially available size of spun aluminium hemisphere (outside diameter 320 mm, thickness 3 mm - fig 2a). Fig 2b shows the first stage of construction of the helmet containing the 46 coil formers fixed inside the aluminium screen and the vacuum-formed inner plastic shell.

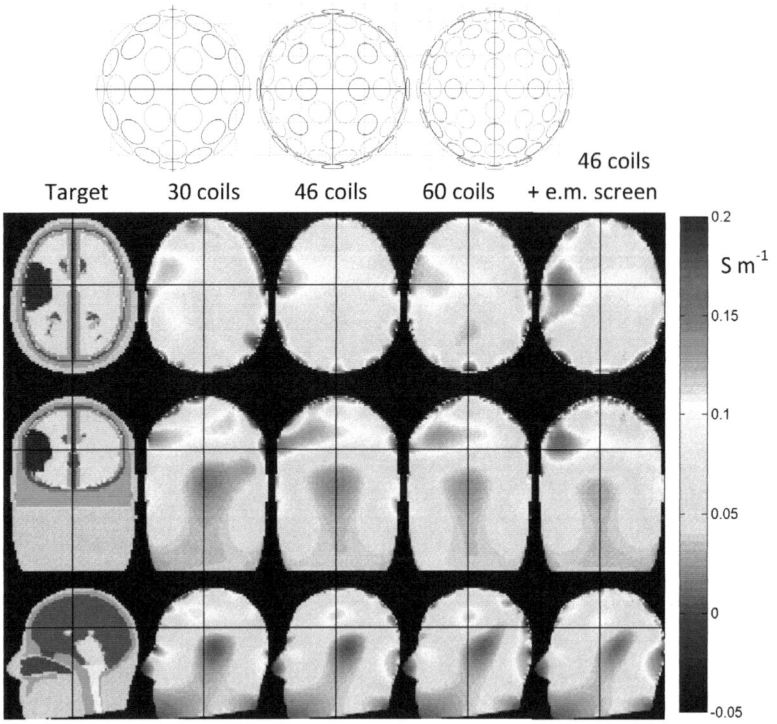

Fig 1. Columns left to right: 1) true distribution of conductivity changes modelled (1-10 MHz), including peripheral haemorrhage in right hemisphere of brain: top = transverse, middle = coronal, bottom = sagittal cross sections (sagittal not to scale). 2), 3) and 4) difference images reconstructed from data simulated for 30-, 46- and 60-coil arrays with no e.m. screen. Above each column is plan-view diagram of coil positions: red = exciter, black = sensor. 5) images for 46-coil array including modelling the e.m. screen. $\lambda = 7.5\times10^{-11}$ for column 2 and 2.5×10^{-10} for columns 3-5.

6. Simulating the electromagnetic screen

The electromagnetic screen cannot be simulated explicitly by the finite difference model which assumes a skin depth large compared with the dimensions of the conductive region. Whilst this is true for the head, it is not for the screen (skin depth ≈ 85 μm at 1 MHz). The effect of the screen can however be included by making use of the expression derived by Ktistis *et al* [9] for the voltage change, ΔV, in a sensor coil, a, when an exciter coil, b, is activated with a current, I, at a frequency, ω.

$$\Delta V = \frac{j\omega}{I}\int_v \mathbf{A}_a \cdot \mathbf{J}_b dv$$

where $\mathbf{J}_b$ is the current density induced in the target (the head) and the integral is performed over the volume of the target, v. $\mathbf{A}_a$ is the vector potential induced within the empty coil array, with no conductive target present, when now coil a is used as the exciter. The vector potential field within the empty coil array was pre-computed by Comsol Multiphysics for each of the 46 coils activated in turn as the exciter. $\mathbf{A}_a$ is then known directly and $\mathbf{J}_b$ is computed with the finite difference model using the vector potential, $\mathbf{A}_b$, pre-computed for coil b as the exciter. Consistent with the assumption of a large skin depth in the tissue is that in computing the eddy-current distribution, the finite difference model makes the approximation that the vector potential is always equal to the applied value that exists with no tissue present: the model neglects the contribution to the vector potential produced by the eddy currents in the tissue themselves. The reason for using the finite-difference model at all, rather than the potentially-more-accurate Comsol Multiphysics, was its much shorter computation time, by a factor of approximately 60.

Qualitatively, the frequency-difference image reconstructed from the data simulated including the e.m. screen allows at least as good a visualisation of the stroke as before (fig 1, column 5).

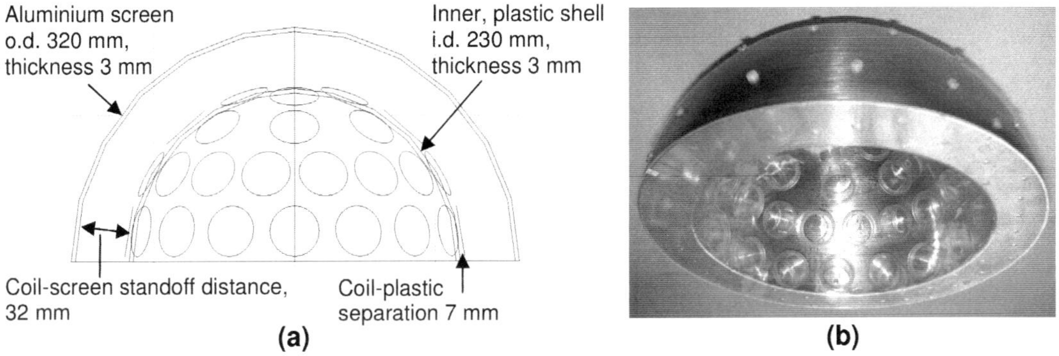

Fig 2. (a) Diagram of final dimensions of helmet, (b) photograph of helmet prior to adding the electronics

7. Conclusions

Simulations have shown that for a realistic level of random phase noise, it should be possible to visualise a large peripheral cerebral haemorrhage by frequency difference MIT imaging (1-10 MHz) with the 46-coil helmet. The construction of the helmet continues. The simulations assumed that the boundary of the conducting region (skin surface) was known precisely. Development of a boundary-measuring system meeting the requirements set by Zolgharni *et al* [8] is an essential next step towards clinical application of the MIT helmet.

Acknowledgements

We thank Mr C. Craven and Mr D. Etheridge for the mechanical construction of the helmet.

References

[1] Zolgharni M, Ledger P D and Griffiths H 2009c Imaging cerebral haemorrhage with MIT: frequency-difference imaging with a customized coil array *Proc. 10th Conf. on Biomedical Applications of Electrical Impedance Tomography (Manchester, UK)*

[2] Eichardt R, Igney C H, Kahlert J, Hamsch M, Vauhkonen M and Haueisen J 2009 Sensitivity Comparisons of Cylindrical and Hemi-spherical Coil Setups for Magnetic Induction Tomography *IFMBE Proceedings* **25** 269-72

[3] Zheng X, Haijun L, Wei H, Chuanhong H, Xiaodong S and Zhanglong Z 2009 A multi-channel magnetic induction tomography measurement system for human brain model imaging *Physiol. Meas.* **30** S175-86

[4] Watson S, Williams RJ, Gough W, Griffiths H, 2008, A magnetic induction tomography system for samples with conductivities less than 10 S m^{-1}. *Mea. Sci Technol.* **19** 045501 11pp

[5] U.S. Army Anthropometry Survey (ANSUR) 1988 http://www.dtic.mil/dticasd/docs-a/anthro_military.html

[6] Wee H C, Watson S, Patz R, Griffiths H and Williams R J 2008 A magnetic induction tomography system with sub-millidegree phase noise and high long-term phase stability *IFMBE Proceedings* **22** 744-7 ISBN 978-3-540-89207-6

[7] Gursoy D and Scharfetter H 2009 Optimum receiver array design for magnetic induction tomography *IEEE Trans BME* **56** no 5 1435-41

[8] Zolgharni M, Griffiths H and Ledger P D, 2010 Frequency-difference MIT imaging of cerebral haemorrhage with a hemispherical coil array: numerical modelling *Physiol. Meas* In press

[9] Ktistis C, Armitage D W and Peyton A J 2008 Calculation of the forward problem for absolute image reconstruction in MIT *Physiol. Meas* **28** S455-S464

Assessing the feasibility of detecting a Hemorrhagic type stroke using a 16 channel Magnetic Induction System

B Dekdouk[1], C Ktistis, D.W Armitage, A.J Peyton

[1]School of Electrical and Electronic Engineering, University of Manchester, M60 1QD

E-mail: bachir.dekdouk@postgrad.manchester.ac.uk

Abstract. Magnetic induction tomography (MIT) has been proposed as a possible method for imaging stroke in the human brain. Hemorrhagic stroke is characterized by local blood accumulation in the brain and exhibits a greater change in conductivity with frequency compared to other tissues which is observed in the frequency range of interest [1-10] MHz. In this study, we investigate the feasibility of detecting hemorrhagic stroke using absolute and frequency difference imaging. For this purpose, a model of the head originally obtained from MRI and X-ray data was used, to which a large stroke (50 ml) was added. In addition, a model of a 16 channel circular array MIT system was employed. The received coil induced voltages were computed using a custom eddy current solver, based on the finite difference method. For absolute imaging, the induced voltages at the receiver coils were calculated from various coil combinations at 10 MHz frequency together with anticipated systematic errors and biases (orientation and displacement of the coils, movement of the head). The induced voltage noise due to these systematic inaccuracies was compared with the voltage change due to the stroke. In order to decrease the impact of this noise, frequency difference was also considered, whereby measurements were performed at another frequency (1MHz) and subtracted. Comparison results are presented and a realistic picture is delivered with to regard the required mechanical stability and electronics accuracy for this particular medical application

1. Introduction

Magnetic Induction Tomography (MIT) is a non-intrusive modality that has been proposed as a possible method for imaging changes in electrical properties of biological tissues. The stroke is a life threatening pathological condition involving an accumulation of fluid in the brain. When the condition is due to blood leakage, the stroke is characterized as a hemorrhage. This blood accumulation causes the local conductivity in the region to change; hence MIT can be suitable for detecting or monitoring the condition. Recently, considerable research was devoted to investigate this MIT application [1]. In this paper we investigate the possibility of detecting a hemorrhage type stroke by examining the signal due to the stroke against that of systematic errors that can happen in a real MIT system.

2. Results and discussion

Figure 1a) and b) shows the simulation arrangement which includes a head model and an MIT system. This latter consists of a cylindrical shield with radius of 175 mm, height of 250 mm, and a circular coil array which consists of 16 excitation coils and 16 receiver coils (diameter 50 mm) modeled as filamentary and arranged in two concentric circles at radii of 142.5 mm and 132.5 mm surrounding the target. Unit current is passed through the excitation coil and induced voltages are measured through the inner receiver coils. A model of the head (Figure 1a) consisting of 7 biological tissues (Scalp, Skull, CSF, gray matter, white matter, spinal/optic nerves, and Eye balls) is placed in the scanning region so as the stroke line up with the sensor plane. The electrical conductivities for the corresponding tissues were obtained from [2].The voltages induced on the receiver coils are computed using a custom eddy current software developed by our group at the University of Manchester. The package employs a pre-computed primary field $\boldsymbol{A}_0$ by an FE package (e.g. COMSOL) and solves the MIT forward problem using the impedance method [3].

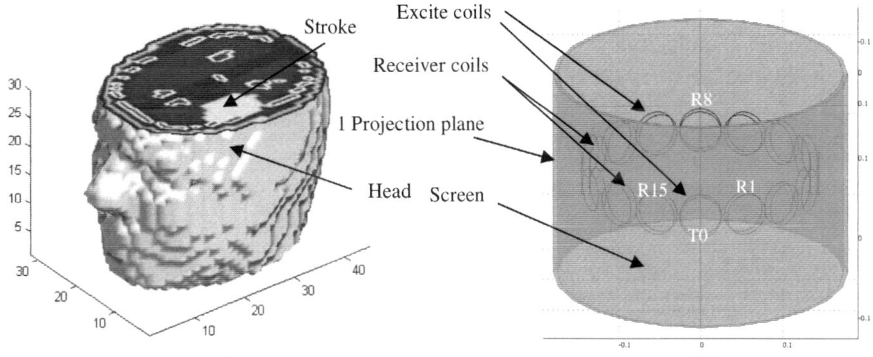

a) b)

Figure 1. MIT system with a) Simulated Target (Head and stroke), and b) MIT coil array

In a first analysis, one transmitter is excited and the differences in the voltage induced because of the presence of the head between the transmitter being in the correct orientation (solution 1) and the transmitter being rotated by one and two degrees (solution 2) are calculated for every receiver channel (equation (1)). The values are normalized to the rms of the induced voltages produced by the noise free case and plotted against channel number (Figure 2a). Receiver coil numbers are in order; number R1 and R15 are adjacent to transmitter T0, and R8 is the opposite receiver.

$$dV = \frac{\Delta \boldsymbol{V}_{\text{solution1}} - \Delta \boldsymbol{V}_{\text{solution2}}}{RMS(\Delta \boldsymbol{V}_{\text{solution2}})} \tag{1}$$

a) b)

Figure 2. Relative voltage errors caused by the coil rotation/displacement

From the graph (Figure 2a), we can see that the worst effected coils are the coils near the transmitter where the opposite coil has nearly zero error. On the same graph, the voltages induced by the background magnetic field are also plotted. The relative changes in the background measurements are smaller than the error in the secondary induced voltages except for the adjacent coils. The lines do not follow the same trend; hence it is difficult to use the background to compensate for the error on the secondary field.

On the other hand, figure 2b presents the results of the simulations for displacing the transmitter coil by 1 and 2mm along its axis towards the centre of the object space. As can be seen, similar observations made for the mis-ortientation of the coil earlier, remain valid for when the transmitter coil is displaced. In a second simulation, the effect of the displacement of the head to measurements differences is quantified. The head is moved by 2 mm toward R_4 and the relative voltage differences are plotted on the graph below (Figure 3). In this case, we can see that MIT is not as sensitive to the movement of the head as to the movement of the coils. The changes of the signal caused by the 50 ml stroke (inclusion) are also mapped in the figure for comparison. The relative signal from the stroke is in the same order of magnitude compared to the relative voltage errors caused by the movements of the coil and the head. This means that for a system that is built with these tolerances it will be very difficult to reconstruct the stroke.

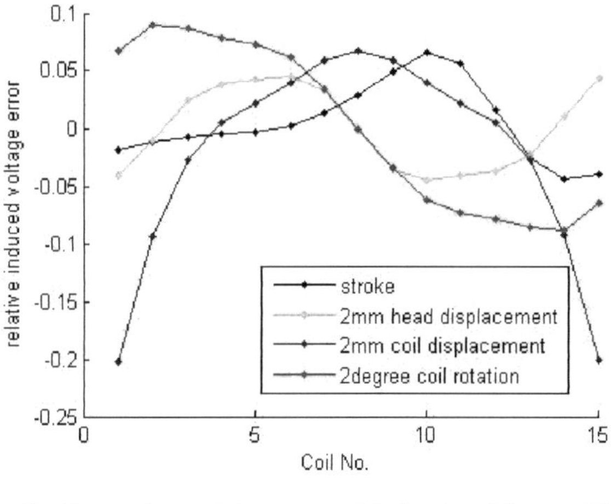

Figure 3. Comparison of the errors with the signal from a 50ml stroke

Recently frequency difference imaging has been suggested as attractive for imaging the stroke since biological tissues exhibit conductivity changes with frequency. Workers in the field of MIT reported this procedure whereby dual frequency measurements are used can reduce systematic errors investigated above [4]. In order to test if frequency difference can reduce the error caused by coil positioning, the following computational experiment was conducted. The induced voltages due to the presence of the head were computed at frequencies of 1 and 10 MHz for the case when the transmitter was in the correct position (solution 1) and displaced (i.e. rotated/moved - solution 2). The voltage differences were calculated in a similar fashion to the absolute case and have been normalized to the rms value of the frequency difference measurements for the correct position according to the equation given below:

$$dV = \frac{(\Delta V_{f_1} - (f_1/f_2)^2 \Delta V_{f_2})_{solution1} - (\Delta V_{f_1} - (f_1/f_2)^2 \Delta V_{f_2})_{solution2}}{RMS((\Delta V_{f_1} - (f_1/f_2)^2 \Delta V_{f_2})_{solution1})} \quad (2)$$

In figure 4a), we can observe that there is no improvement on the relative error caused by the movement of the coil.

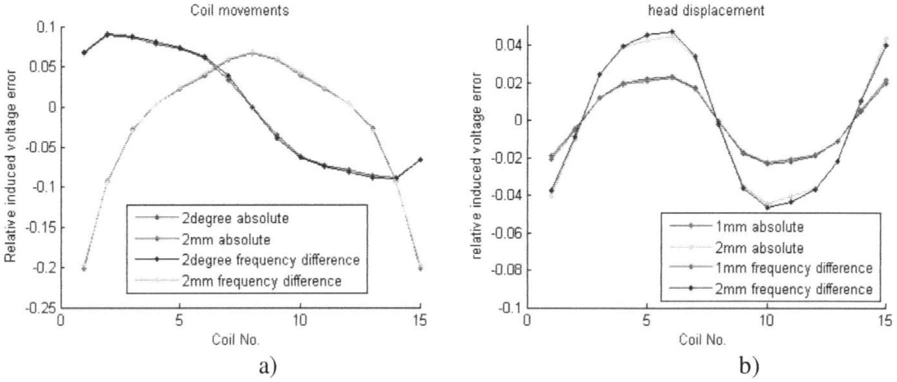

Figure 4. Comparison of errors between absolute and frequency difference measurements

The error cancelation when frequency difference measurements are used is also tested for head displacement (Figure 4b). Interestingly, the absolute and frequency difference relative error show similar pattern, which infers that there is no advantage of using frequency difference for head displacement error cancelation.

3. Conclusion

Nowadays the acquisition electronics used for MIT systems have been developed to a degree that different types of errors are more obvious in real MIT arrays. The positioning of the coils in a MIT system is very critical as well as the accurate knowledge of the target location and the shape. The tolerances of the coil array should be less than one tenth of a millimeter in order for the accuracy of the measurements to be the same order to the noise generated by the electronics.

4. References

[1] Merwa R. Hollaus K. Oszkar B and Scharfetter H 2004 Detection of brain oedema using MIT: a feasibility study of the likely sensitivity and detectability. Physiol. Meas. 25 347-54

[2] Gabriel S Lau R W Gabriel C 1996 The dielectric properties of biological tissues. II. Measurements in the frequency range 10 Hz to 20GHz Phys. Biol. Med., 41, 2251-2269.

[3] Dekdouk B Yin W Ktistis C Armitage D W and Peyton A J 2009 A Method to solve the Forward Problem in Magnetic Induction Tomography based on the Weakly Coupled Field Approximation

[4] Gursoy D and Scharfetter H 2009 Reconstruction artefacts in magnetic induction tomography due to patient's movement during data acquisition Physiol. Meas. 30 (2009) S165–S174 IEEE Trans on Biomedical Eng

Acknowledgments

The authors would like to thank the UK Engineering and Physical Sciences Research Council (EPSRC) for their financial support under contract EP/E009158/1. The authors gratefully acknowledge Dr David S. Holder for providing the FEM head model

The application of a priori structural information based regularization in image reconstruction in Magnetic Induction Tomography

B Dekdouk[1], C Ktistis, W Yin, D.W Armitage, A.J Peyton

[1]School of Electrical and Electronic Engineering, University of Manchester, M60 1QD

Abstract. Magnetic induction tomography (MIT) is a non-invasive contactless modality that could be capable of imaging the conductivity distribution of biological tissues. In this paper we consider the possibility of using absolute MIT voltage measurements for monitoring the progress of a peripheral hemorrhagic stroke in a human brain. The pathology is modelled as a local blood accumulation in the white matter. The solution of the MIT inverse problem is nonlinear and ill-posed and hence requires the use of a regularisation method. In this paper, we describe the construction and present the performance of a regularisation matrix based on a priori structural information of the head tissues obtained from a very recent MRI scan. The method takes the MRI scan as an initial state of the stroke and constructs a learning set containing the possible conductivity distributions of the current state of the stroke. This data is used to calculate an approximation of the covariance matrix and then a subspace is constructed using principal component analysis (PCA). It is shown by simulations the method is capable of producing a representative reconstruction of a stroke compared to smoothing Tikhonov regularization in a simplified model of the head.

1. Introduction

Magnetic induction tomography (MIT) is a low frequency electromagnetic modality, which aims to reconstruct the conductivity changes from coupled field measurements taken by inductive sensors [1]-[2]. MIT is a subject of research for medical clinical applications. One particular application of interest is imaging local blood accumulation in the brain also termed stroke. In this paper we show that absolute imaging for detecting the stroke using standard smoothing operators is a difficult task since the implemented a-priori information is not representative of the complex structure of the head. However, using images taken by MRI or CT, a regularization matrix can built which carries useful structural information about the head. It is shown this information makes the MIT inverse problem less under-determined and can make MIT useful for post diagnosis monitoring.

2. MIT system and forward problem

In order to generate the simulated measurement data and calculate the induced voltages, a custom eddy current software developed at the University of Manchester was employed [3]. With a pre-computed primary magnetic vector potential $\boldsymbol{A}_0$ in empty space, the package solves the eddy current problem in the target and calculates the induced voltages in the receiver coils. The incident field $\boldsymbol{A}_0$ is

[1] Email: bachir.dekdouk@postgrad.manchester.ac.uk.

obtained via a simulation of the MIT system in a commercial FE package (COMSOL) (Figure 1b). As can be seen, the MIT system model comprises of 16 pairs of exciter and receiver coils arranged in two circles of radii 141.5mm and 131.5mm respectively. Given a conductivity distribution (Figure 1a) excitation coils are sequentially driven with a time varying sinusoidal unit current at a frequency of 10 MHz, and 256 computed induced voltages are collected for 16 excitations. In an attempt to increase the number of independent measurements, additional planes of data are computed by displacing the coil array vertically along the z direction in 20 mm steps.

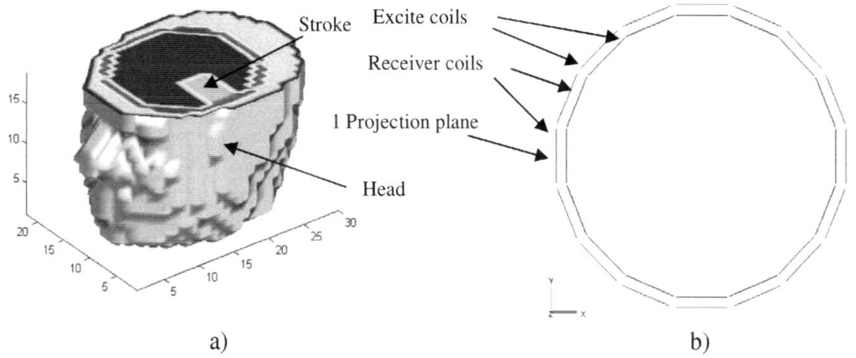

Figure 1. MIT system with a) Simulated Target (Head and stroke), and b) coil array

3. Theory

For conductivity image reconstruction the non-linear MIT inverse problem is solved in an iterative scheme. In this process, the sensitivity matrix $\boldsymbol{J} \in \mathbb{R}^{m \times n}$ is updated in every iteration. The formulation used for computing the sensitivity is an extension of the Gezelowitz theorem to the general electromagnetic case. For optimisation, the damping Gauss Newton method is employed, where the corresponding iteration is described by the following equations:

$$(\boldsymbol{J}_k^{\mathrm{T}}\boldsymbol{J}_k + \lambda_k \boldsymbol{R})\boldsymbol{d}_{dGN} = -\boldsymbol{J}_k^{\mathrm{T}}(\boldsymbol{F}(\boldsymbol{\sigma}_k)\text{-}\boldsymbol{D}) - \lambda_k \boldsymbol{R}\boldsymbol{\sigma}_k \quad (1)$$

$$\boldsymbol{\sigma}_{k+1} = \boldsymbol{\sigma}_k + \boldsymbol{d}_{dGN} \quad (2)$$

where the regularization parameter λ_k is updated in a damping process identical to the Levenberg Marquardt method, and the a-priori information about the conductivity distribution is implemented via the regularisation matrix $\boldsymbol{R}$. Here, $\boldsymbol{D}$ is the simulated measurement data, $\boldsymbol{F}(\boldsymbol{\sigma}_k)$ is the computed forward solution vector and $\boldsymbol{\sigma}_k$ is the absolute conductivity vector at the current iterate (k).

3.1. Regularisation

The inverse conductivity solution was calculated with two different regularization methods presented below:

a) 2nd Order Laplacian smoothing operator $\boldsymbol{R}_n = \boldsymbol{L}^{\mathrm{T}}\boldsymbol{L}$, where $\boldsymbol{L}$ (Neighboring matrix) is a discrete difference operator between neighboring conductivity voxels. Typically, $\boldsymbol{L}_{ij} = -1$ for $i \neq j$ when two elements are neighbors, and $\boldsymbol{L}_{ii} = -\sum_{j:j\neq i} \boldsymbol{L}_{ij}$ b) Structural Subspace regularization matrix (SSRM) $\boldsymbol{R}_{SSRM}$: A learning set of expectable conductivity distributions is constructed (i.e. $\sigma_i, i = 1,...,Q,$) near which the true conductivity distribution is assumed to be. By assuming the relative probabilities for these estimated conductivity distributions are uniform an approximation for the covariance matrix $\boldsymbol{C}_\sigma$ can be calculated as: $\boldsymbol{C}_\sigma = Q^{-1}\boldsymbol{\sigma}_i\boldsymbol{\sigma}_i^T$. Then, the eigenvalues λ_i and the corresponding eigenvectors ω_i are computed from $\boldsymbol{C}_\sigma$. A subspace S_ω is constructed by taking the M first eigenvectors corresponding to the largest singular values (i.e. $\boldsymbol{S}_\omega = span\{\boldsymbol{\omega}_m \mid 1 \le m \le M\}, M \ll Q$). The orthogonal projector onto the subspace S_ω can be written as $\boldsymbol{P} = \boldsymbol{W}\boldsymbol{W}^{\mathrm{T}}$, where $\boldsymbol{W} = (\boldsymbol{\omega}_1, \ldots, \boldsymbol{\omega}_M) \in \mathbb{R}^{N \times M}$. The regularization matrix is constructed as $\boldsymbol{R}_{SSRM} = \boldsymbol{L}^{\mathrm{T}}\boldsymbol{L}$, where $\boldsymbol{L} = \boldsymbol{I} - \boldsymbol{W}\boldsymbol{W}^{\mathrm{T}}$. For more details on the method please see [4]

4. Computational experiments

The non-linear image reconstruction was tested with a model approximating the shape of the head. A spherical layer crudely approximating the cerebral spinal fluid (CSF) was added together with a rectangular perturbation simulating the abnormality Figure 1a). The head was meshed into 9012 cubic voxels of 7.5mm^3 resolution which were distributed as (white matter: 7900, CSF: 1008, stroke region: 104) elements. The internal tissues were assigned conductivities as (white matter: 0.16, CSF: 1.5, stroke region: 1.1) Sm^{-1}. Note the CSF conductivity has been reduced to 1.5 Sm^{-1} from the value reported in the literature [5] of 2 Sm^{-1} in order to compensate for the increased thickness caused by mesh resolution. 3D conductivity maps were reconstructed from 11 planes of measurements, which amount to 2816 independent data. The inverse problem was calculated with a-priori information implemented using the aforementioned techniques namely: 2nd order Laplacian operator and SSRM. When using this latter, we assumed the initial state of the perturbation is given by Figure 2 a). The actual state of the perturbation to be reconstructed at the current time is shown in Figure 2 c). A learning set of 21 possible conductivity distributions was constructed to simulate different states of the perturbation randomly chosen between the minimal and maximal states Figure 2 a) and b). Vertical slices of reconstructed images using these regularization methods are shown in Figure 2 e) and d).

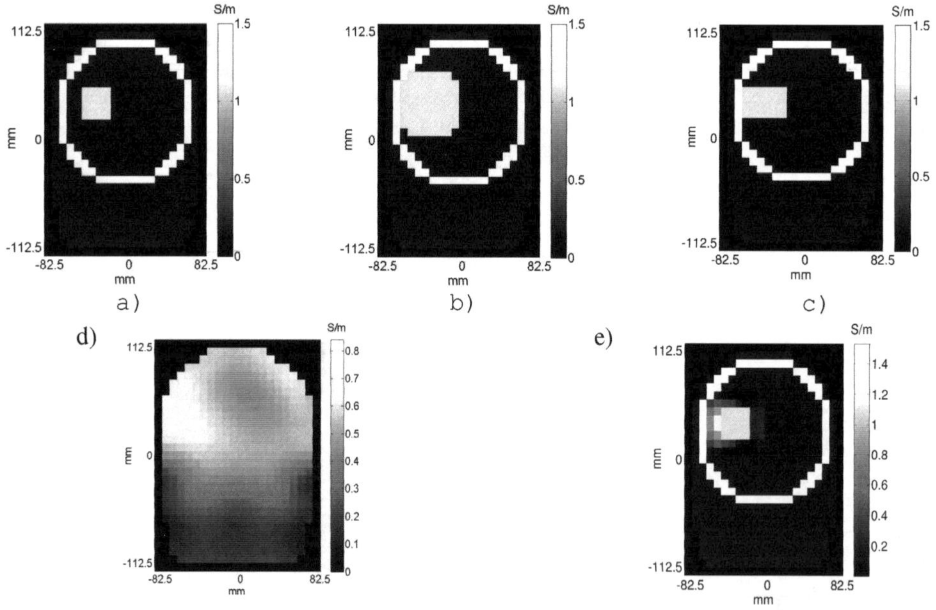

Figure 2. Case of head with two tissues (white matter, CSF) and stroke region (blood). a) Minimal and b) maximal state of the stroke. c) True image. Reconstructed conductivity using d) 2nd Order Laplacian Smoothing matrix, e) SSRM matrix, from noise free data

To further test the SSRM method's robustness, an extra tissue simulating the SCALP (skin and muscle) was modeled with conductivity of 0.6 Sm^{-1}. We attempted to reconstruct two different states of the stroke region as shown by Figure 3 a) and b). In case a) this time, a learning set of 84 conductivity estimates was constructed which simulates 21 different states and 4 different conductivities of the stroke region. In case b), Gaussian noise with 1% of the mean voltage data was added to the simulated measurement data. Reconstructed images are shown below:

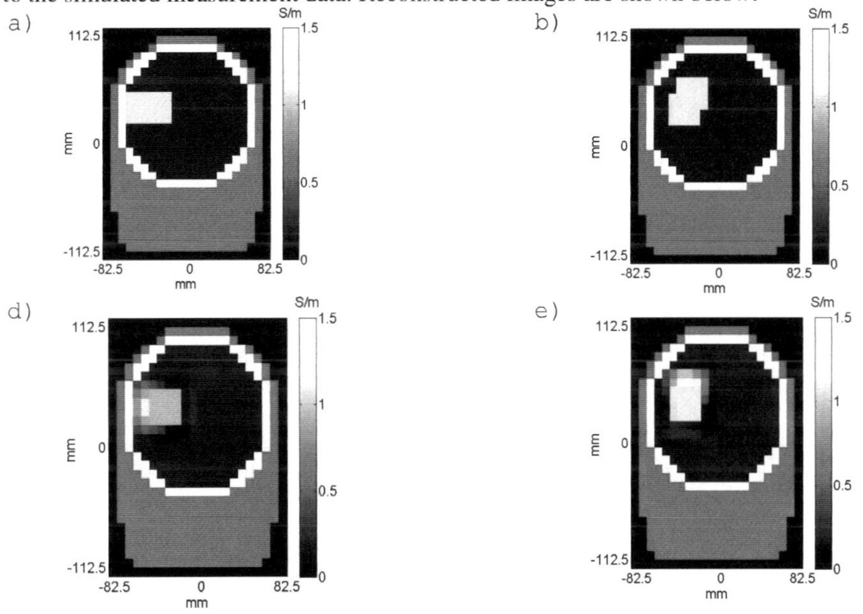

Figure 3. Case of head with three tissues (SCALP, CSF:, white matter:) and stroke (blood). a)-b) True images. c) reconstructed image from 84 conductivity and state estimates-d) reconstructed image from noise contaminated data (SNR=40dB)

5. Conclusion

The paper presents the synthesis of regularisation operator based on structural prior information of the head tissues. It was demonstrated with simulations on a head model with three tissues the stroke region can recovered using absolute imaging provided that the prior information containing the distribution of the head tissues is accurate.

6. Reference:

[1] Griffiths H 2001 Magnetic induction tomography. Meas. Sci. Technol. 26: 1126–1131

[2] Tapp H S and Peyton A J 2003 A state of the art review of electromagnetic tomography, *Proc. 3rd World Congress on Industrial Process Tomography*: 340–346

[3] Dekdouk B Yin W Ktistis C Armitage D W and Peyton A J 2009 A Method to solve the 2Forward Problem in Magnetic Induction Tomography based on the Weakly Coupled Field Approximation IEEE Trans on Biomedical Eng

[4] Vauhkonen M Vadasz D Karjalainen P A Somersalo E and Kaipio J P 1998 Tikhonov regularization and prior information in electrical impedance tomography, *IEEE Trans Med Imaging* 17

[5] Gabriel S Lau R W Gabriel C 1996 The dielectric properties of biological tissues. II. Measurements in the frequency range 10 Hz to 20GHz Phys. Biol. Med., 41, 2251-2269.

Acknowledgments

The authors would like to thank the UK Engineering and Physical Sciences Research Council (EPSRC) for their financial support under contract EP/E009158/1. The authors gratefully acknowledge Dr David S. Holder for providing the FEM head model

International Conference on Electrical Bioimpedance IOP Publishing
Journal of Physics: Conference Series **224** (2010) 012049 doi:10.1088/1742-6596/224/1/012049

Shape corrections for 3D EIT

Kyriakos Paridis[1], **William R. B. Lionheart**[1]

[1]School of Mathematics, The University of Manchester, Oxford Road, Manchester M13 9PL, UK.

E-mail: K.Parides@maths.man.ac.uk

Abstract. Movement of the boundary in biomedical Electrical Impedance Tomography (EIT) has been always a source of error in image reconstruction. In the case of pulmonary EIT, where the patient's chest shape changes during respiration, this is inevitable, so it is essential to be able to correct for shape changes and consequently avoid artifacts. Assuming that the conductivity is isotropic, an assumption that is reasonable for lung tissue but admittedly violated for muscle, the boundary shape up to a Möbius transformation (conformal mapping) as well as the conductivity can theoretically be determined by 3D EIT data. While in two dimensions the space of conformal mappings are infinite dimensional, in the three dimensional case the Möbius transformations are given by a finite number of parameters. In this paper, we concentrate on the three dimensional case and take a linear approximation. We will give results of numerical studies analogous to the two dimensional work of Boyle et al on the effect of electrode movement and shape error in 3D EIT.

1. Introduction

Movement of the boundary in EIT causes artifacts in the images. This issue arises more often in pulmonary EIT, where the patient's chest moves continuously due to breathing and posture changes [1]. During the last years some improvement has been made towards a better reconstruction when deformation of the boundary occurs. Lionheart [2] has shown that non-conformal changes in boundary shapes and electrode locations can be uniquely determined. Moreover, Soleimani et. al. [3] showed that in some cases a combination of image reconstruction model of both conductivity and shape changes can be used to recover conductivity and shape changes.

Generally, distortions of a domain spoil the assumption of an isotropic conductivity distribution [2], but this is not always the case. The distortions that are conformal maps, that is functions which preserve angles, do not lead to an anisotropic conductivity [4]. In two dimensions there are infinite conformal maps, while in three dimensions these maps are a finite dimensional set, known widely as the Möbius transformations, and their linearizations, the infinitesimal Möbius transformations.

In this paper, we present numerical results supporting the idea of recovering the non-conformal part of the linearized distortion from the EIT data. Our results show that reconstruction with shape correction gives a distorted conductivity of the true isotropic conductivity.

Journal of Physics: Conference Series **224** (2010) 012049 doi:10.1088/1742-6596/224/1/012049

2. Conformal Vector Fields

Let $\mathbf{V}$ be a sufficiently smooth vector field. A distortion can be linearized by adding a vector field $\mathbf{V}$ to each point. In the case of distortions that preserve the angles (that is, conformal mappings) then $\mathbf{V}$ is called conformal vector field. A vector field $\mathbf{V}$ is conformal if and only if

$$\frac{\partial V_i}{\partial x_j} + \frac{\partial V_j}{\partial x_i} = (\nabla \cdot \mathbf{V})\, \delta_{i,j}. \tag{1}$$

is satisfied. In two dimensions it is straight forward to show that (1) satisfies the Cauchy-Riemann equations. As we have mentioned, in the three dimensional space there is a finite dimensional set of conformal mappings, the Möbius transformations.

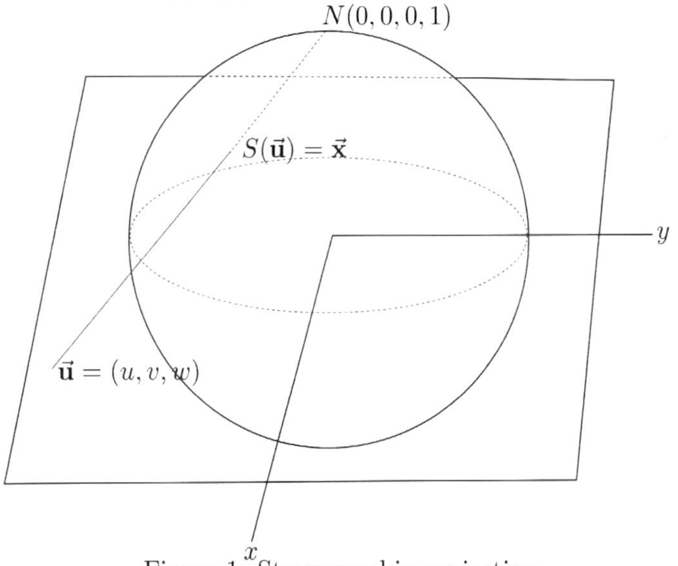

Figure 1: Stereographic projection

Consider the 4-dimensional sphere S^3. Using stereographic projection from the hypersphere on $\mathbb{R}^3$,then applying a rotation and finally backprojecting we construct a conformal vector field, since both stereographic projections and rotations preserve angles.

Let $S^3 \subset \mathbb{R}^4$ be the 4-dimensional sphere and $N(0,0,0,1)$ be the north pole. Suppose $P(x,y,z,t) \in S^3$ is an arbitrary point on the hypersphere. It is easy to calculate the stereographic projection $S : \mathbb{R}^3 \to S^3$, $(u,v,w) \mapsto (x,y,z,t)$

$$\left\{ \begin{array}{l} x = \frac{2u}{u^2+v^2+w^2+1} \\ y = \frac{2v}{u^2+v^2+w^2+1} \\ z = \frac{2w}{u^2+v^2+w^2+1} \\ t = \frac{u^2+v^2+w^2-1}{u^2+v^2+w^2+1} \end{array} \right. \tag{2}$$

Let $T : S^3 \to \mathbb{R}^3$ denote the inverse mapping. Let $R_\theta \in SO(4)$ be the rotation matrix about the $x - t$ hyperplane and let F_θ represent a curve, where θ is the angle of rotation. Then,

$$F_\theta = T R_\theta S(\vec{x}), \tag{3}$$

where $\vec{x} = (x,y,z,t) \in S^3$. A Möbius transformation vector is given by

$$\left.\frac{dF_\theta}{d\theta}\right|_{\theta=0} = DT \left.\frac{dR_\theta}{d\theta}\right|_{\theta=0} S(\vec{\mathbf{x}}) = \begin{pmatrix} \frac{-u^2+v^2+w^2-1}{2} \\ -uv \\ -uw \end{pmatrix}. \tag{4}$$

where DT is the Jacobian of the inverse mapping.

Permuting the axes generates other independent Möbius vector fields, and we can thus construct a basis of the conformal vector fields. The theory predicts that any vector field orthogonal to the space of Möbius vector fields can be determined from complete boundary data while a distortion by a Möbius vector field will produce data for which there is a consistent isotropic conductivity, which is distorted from the true one by the Möbius vector field.

As a first test of this in the context of finitely many electrodes and a finite element mesh we will simply verify that data subject to a conformal distortion results in a recognisable but distorted reconstruction of the true conductivity but a non-conformal distortion of the same size produces a reconstruction with more artifacts.

3. Numerical Experiments

In order to verify the theoretical results we used the familiar demonstration model demo_real from EIDORS [5]. Specifically, we used a finite element mesh with electrode positions (green) and simulated inhomogeneities (blue and red) (Figure 2). For our tests we modified the existing code to be able to apply conformal and non-conformal distortions on the inhomogeneities.

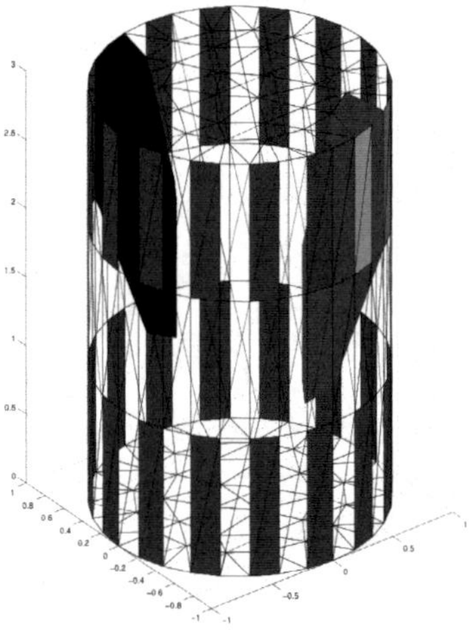

Figure 2: Finite element mesh with electrodes and simulated inhomogeneities.

In detail, we started by applying the following non-conformal perturbation

$$(x, y, z) \to \left(x + \epsilon \left(x \frac{z+3}{3} \right), y + \epsilon \left(y \frac{-z+6}{3} \right), z \right), \tag{5}$$

where ϵ is small enough and it was chosen so that the L^2-norms of the perturbations are equal. The non-conformal transformation was chosen so that it transforms circles to ellipses, that is a transformation which is "very" non-conformal. For the conformal distortion we used the Möbius vector (4) we have constructed before. The applied perturbation is given by

$$(x, y, z) \to \left(x + \epsilon \frac{-x^2 + y^2 + z^2 - 1}{2}, y - \epsilon xy, z - \epsilon xz \right). \tag{6}$$

The reconstruction was a standard regularized linear reconstruction using the undistorted mesh.

The simulated results shown in Figure 3 confirm the theory. It is clear that non-conformal movements cause significant artifacts in the conductivity reconstruction. On the other hand, conformal distortions do not affect the conductivity reconstruction, at least not as importantly as with non-conformal distortions.

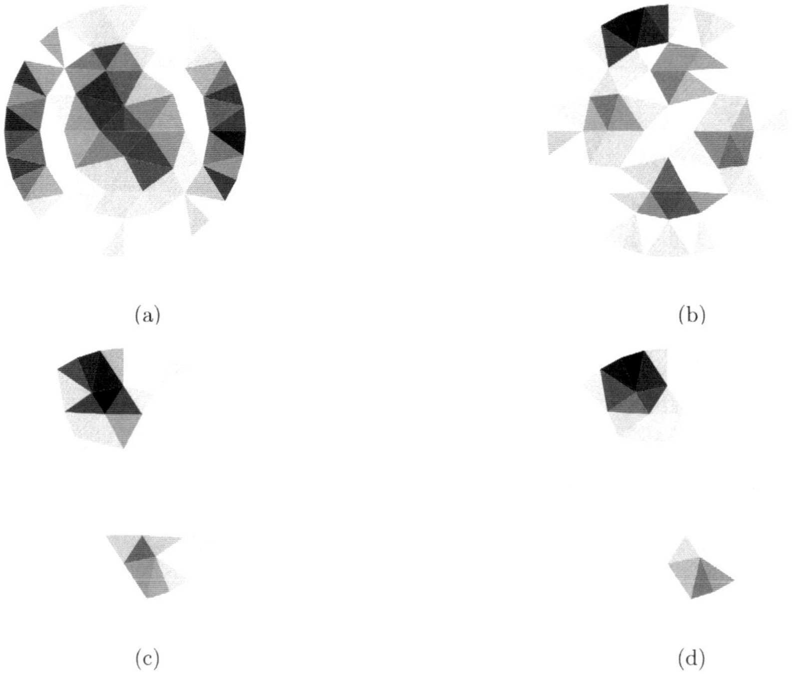

(a) (b) (c) (d)

Figure 3: Reconstructed conductivity distributions at z=1 (*left*) and z=2 *right*. *Top* (3a), (3b): Non-conformal distortion applied. *Bottom* (3c), (3d) Conformal distortion applied.

4. Discussion and Conclusion

This paper deals with the effect of conformal and non-conformal distortions on the conductivity reconstruction in Electrical Impedance Tomography. The simulated results show an important difference in the conductivity reconstruction for the two distortions and suggest that conformal vector fields (Möbius transformations) tend to give a better reconstruction, avoiding artifacts.

It is hoped that the idea in this paper can be used to reduce artifacts in chest EIT images caused by variable chest shape. The idea is to start with an initial realistic chest shape, but to compensate for the breathing component by calculating a non-conformal shape perturbation which is adjusted along with the conductivity to fit the data at each time frame. The error will be a conformal map, which can either be determined by a small number of mechanical measurements, or if undetermined will result in a distortion of the conductivity image that will still be clinically useful.

Acknowledgments

This work was supported by the EPSRC, UK research grant EP/F033974/1.

References

[1] Adler A, Guardo R and Berthiaume Y 1996 *Biomedical Engineering, IEEE Transactions on* **43** 414–420 ISSN 0018-9294

[2] Lionheart W R B 1998 *Inverse Problems* **14** 139–147 URL `http://stacks.iop.org/0266-5611/14/139`

[3] Soleimani M, Gomez-Laberge C and Adler A 2006 *Physiological Measurement* **27** S103–S113 URL `http://stacks.iop.org/0967-3334/27/S103`

[4] Boyle A, Lionheart W R, Gomez-Laberge C and Adler A 2008 *Proceedings of the Electrical Impedance Tomography Conference 2008 (EIT 2008), Dartmouth College, NH, USA, 16-18 June 2008* (NH, USA) pp 175–178

[5] Adler A and Lionheart W 2006 *Physiological Measurement* **27** S25–S42 ISSN 0967-3334 6th Conference on Biomedical Applications of Electrical Impedance Tomography, London, England, Jun 22-24, 2005

Compensation of errors due to incorrect model geometry in electrical impedance tomography

A Nissinen[1], V Kolehmainen[1] and J P Kaipio[1,2]

[1] Department of Physics and Mathematics, University of Eastern Finland, P.O.Box 1627, FIN-70211 Kuopio, Finland.
[2] Department of Mathematics, University of Auckland, Private Bag 92019, Auckland Mail Centre, Auckland 1142, New Zealand.

E-mail: antti.nissinen@uef.fi

Abstract. Electrical impedance tomography (EIT) is a highly unstable problem with respect to measurement and modeling errors. This instability is especially severe when absolute imaging is considered. With clinical measurements, accurate knowledge about the body shape is usually not available, and therefore an approximate model domain has to be used in the computational model. It has earlier been shown that large reconstruction artefacts result if the geometry of the model domain is incorrect. In this paper, we adapt the so-called approximation error approach to compensate for the modeling errors caused by inaccurately known body shape. This approach has previously been shown to be applicable to a variety of modeling errors, such as coarse discretization in the numerical approximation of the forward model and domain truncation. We evaluate the approach with experimental data from a thorax phantom, with absolute imaging considered. We show that the related modeling errors can be efficiently compensated for by the approximation error approach. We also show that recovery from simultaneous discretization related errors is feasible, allowing the use of computationally efficient reduced order models.

1. Introduction

The reconstruction of the conductivity in electrical impedance tomography (EIT) problem is a non-linear inverse boundary value problem which is highly unstable with respect to measurement and modeling errors. The effect of the measurement errors can be reduced by using an accurate measurement system and by careful modeling of the statistics of the measurement error.

Modeling errors, on the other hand, are often related, for example, to discretization of the forward model, truncation of the computational domain and unknown boundary data. Furthermore, a common modeling error in medical EIT is related to inaccurate knowledge on the shape of the target body. Most of the available reconstruction methods assume that the boundary of the target body is known. In principle, the shape of the patient's thorax could be obtained from other imaging modalities. However, such information is often not available and therefore the reconstruction has to be computed using an approximate model domain.

The traditional way to circumvent the problem of inaccurately known body shape has been to use difference imaging. In spite of the difference imaging modality being able to suppress some of the effects of model uncertainties, it has been shown that the breathing artefacts are still present in the reconstructions.

In this paper, we propose the reduction of the reconstruction errors caused by inaccurately known body shape in EIT by using the Bayesian approximation error approach, which was originally proposed in [1]. The key idea in the approximation error approach is, loosely speaking, to represent not only the measurement error, but also the effects of the computational model errors and uncertainties as an auxiliary additive noise process in the observation model. The realization of the modeling error is obviously unknown since its value depends on the actual unknown conductivity and body shape. However, the statistics of the related approximation error can be estimated using the prior distribution models of the conductivity and parameterization of the body shape. The approximation error statistics are then used in the reconstruction process for compensating the inaccurately known body shape. In this paper, the proposed approach is evaluated with a 3D example with experimental data from a thorax shaped measurement tank.

2. Methods

2.1. Forward model and notation

We model the measurements with the complete electrode model. The numerical approximation of the forward model is usually based on the finite element (FEM) approximation. In the following, we use notation $U_h(\sigma,\gamma) \in \mathbb{R}^m$ for the FEM based forward solution. The dependence of the forward model on the domain Ω is expressed by the parameterization γ of the boundary $\partial\Omega$ and the subindex h denotes the discretization level parameter.

The measurement noise in EIT experiments is commonly modeled as Gaussian additive noise which is mutually independent with the unknown conductivity. This leads to measurement model

$$V = U_h(\sigma,\gamma) + e, \qquad e \sim \mathcal{N}(e_*, \Gamma_e) \tag{1}$$

where $V \in \mathbb{R}^m$ is the vector of the measured voltages, $\sigma \in \mathbb{R}^N$ is the conductivity vector, and $e \in \mathbb{R}^m$ is a Gaussian distributed measurement noise with mean $e_* \in \mathbb{R}^m$ and covariance matrix Γ_e.

Note that if the boundary $\partial\Omega$ of the target body is not known accurately (γ is incorrect) or the FEM discretization is too coarse (h is too large), the error in the FEM approximation $U_h(\sigma,\gamma)$ may become significant compared to the measurement error e.

2.2. Approximation error approach

In this section, we explain shortly how the modelling error caused by using the *model domain* $\tilde{\Omega}$ instead of the actual (unknown) domain Ω in the computational (forward) model can be embedded in the likelihood model. For more detailed discussion, see [2].

Let $U_\delta(\bar{\sigma},\gamma)$ denote a (sufficiently) accurate forward model. Here the parameters γ of the boundary $\partial\Omega$ and discretization level parameter δ are such that the error in the FEM approximation is smaller than the measurement error. Let $U_h(\sigma,\tilde{\gamma})$ denote a approximate forward model where the discretization level parameter $h > \delta$ and $\tilde{\gamma}$ are the parameters of the boundary $\partial\tilde{\Omega}$ of the model domain.

The relation of the representation of the conductivities in models $U_\delta(\bar{\sigma},\gamma)$ and $U_h(\sigma,\tilde{\gamma})$ is of the form

$$P : \Omega \mapsto \tilde{\Omega}, \quad P\bar{\sigma} = \sigma, \tag{2}$$

where P is a mapping that describes the deformation of the conductivity from the domain Ω into $\tilde{\Omega}$. In the numerical examples considered in this study, P is a linear deformation mapping that preserves angle and relative distance from the center of the domain.

In the approximation error approach, the *accurate measurement model* is written in the form

$$\begin{aligned} V &= U_h(\sigma,\tilde{\gamma}) + (U_\delta(\bar{\sigma},\gamma) - U_h(\sigma,\tilde{\gamma})) + e \\ &= U_h(\sigma,\tilde{\gamma}) + \varepsilon(\bar{\sigma},\gamma) + e = U_h(\sigma,\tilde{\gamma}) + \eta, \end{aligned} \tag{3}$$

where $\varepsilon(\bar{\sigma}, \gamma)$ represents the modelling error due to the discretization and incorrect boundary, and we denote $\eta = \varepsilon + e$. The objective in the modelling error approach is to estimate the statistics of the error η and construct the likelihood model using this statistics instead of using only the statistics of the measurement error.

To obtain a computationally feasible and efficient approximation, we make the Gaussian approximation for the joint distribution $\pi(\sigma, \eta)$. This is the core of the most common implementation of the approximation error approach, in particular when computational efficiency is sought. To simplify the analysis further, we make an approximation where the mutual dependence of parameters σ and η is ignored. This approximation is referred to as an *enhanced error model.* In problems such as EIT that involve nonlinear forward models, the statistics of the approximation error have to be estimated by stochastic simulation.

For the Monte Carlo simulation, we generate a set of $N_s = 150$ draws from the prior models $\pi(\gamma)$ and $\pi(\bar{\sigma})$. For the draws for the domains and the boundary shapes, N_s CT images of different individuals were segmented. This resulted in the ensemble of domains $\{\Omega^{(\ell)}, \ell = 1, 2, \ldots, N_s\}$ and the corresponding parametric representations for the boundaries $\{\gamma^{(\ell)}\}$. The samples of the conductivity $\bar{\sigma}^{(\ell)}$ were drawn from a proper Gaussian smoothness prior [1]: $\bar{\sigma} \sim \mathcal{N}(\bar{\sigma}_*, \Gamma_{\bar{\sigma}})$. Given the accurate and target forward solutions, the samples $\varepsilon^{(\ell)}$ of the approximation error were obtained as $\varepsilon^{(\ell)} = U_\delta(\bar{\sigma}^{(\ell)}, \gamma^{(\ell)}) - U_h(\sigma^{(\ell)}, \tilde{\gamma})$. The mean ε_* and covariance Γ_ε of the approximation error were estimated as a sample average.

When the likelihood model is based on the estimated statistics of the error η and Gaussian prior is used, the computation of the MAP estimate amounts to solving the minimization problem

$$\sigma_{\mathrm{MAP}} = \arg\min_{\sigma \geq 0} \left\{ \| L_\eta (V - U_h(\sigma, \tilde{\gamma}) - \varepsilon_* - e_*) \|^2 + \| L_\sigma (\sigma - \sigma_*) \|^2 \right\}, \tag{4}$$

where Cholesky factors are $L_\eta^{\mathrm{T}} L_\eta = (\Gamma_\varepsilon + \Gamma_e)^{-1}$ and $L_\sigma^{\mathrm{T}} L_\sigma = \Gamma_\sigma^{-1}$. We refer to the MAP estimate (4) as MAP with the *approximation error model* (MAP-AEM).

In the conventional approach, approximation errors are ignored and the mean ε_* and covariance Γ_ε in (4) are assumed to be zero. This approach is refered to *MAP with conventional error model* (MAP-CEM).

3. Results

The experimental data was measured from a vertically symmetric measurement tank Ω, see Fig. 1. The height of the tank Ω and the model domain $\tilde{\Omega}$ was 5cm. Thus, no model error due to truncation of the computational domain in the vertical direction was present. To construct the phantom, heart and lung shaped inclusions were made of agar and placed in the measurement tank filled with saline of conductivity 3.0 $\mathrm{mS\,cm^{-1}}$. The conductivity of the lung and heart targets were 0.73 $\mathrm{mS\,cm^{-1}}$ and 5.8 $\mathrm{mS\,cm^{-1}}$, respectively.

The reconstructions are shown in Fig. 1, together with a photograph of the measurement phantom. The images that are shown, are the central horizontal cross sections of the reconstructed 3D conductivity, that is, 2.5 cm from both the top and bottom layers. The MAP-CEM estimate with the accurate forward model $U_\delta(\sigma, \gamma)$ can be considered as a reference estimate since no discretization errors or geometrical modeling errors are present. As can be seen, the MAP-CEM estimate with the reduced forward model $U_h(\sigma, \tilde{\gamma})$ show intolerable errors. In contrast, the MAP-AEM estimate with reduced model $U_h(\sigma, \tilde{\gamma})$ is able to capture the geometry of the organs well. Also, the actual values of the conductivity distribution match well.

Although it is clear that the MAP-CEM estimate with the actual geometry is a better estimate than the MAP-AEM with $U_h(\sigma, \tilde{\gamma})$, the computational time with the latter is only about 21% of the former. Of course, the central issue is that the actual domain does not need to be known.

Estimate	Forward model	Reconstruction	N_n	CPU time
EIT phantom		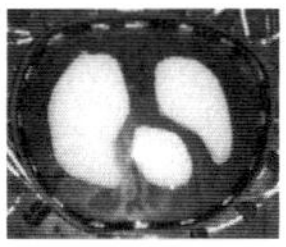		
MAP-CEM	$U_\delta(\sigma, \gamma)$	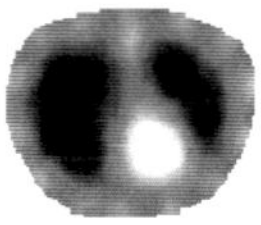	23949	307 s
MAP-CEM	$U_h(\sigma, \tilde{\gamma})$	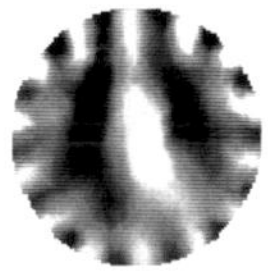	1283	65 s
MAP-AEM	$U_h(\sigma, \tilde{\gamma})$	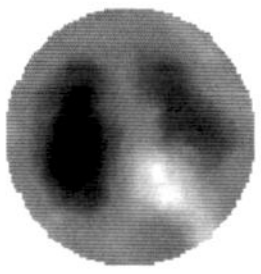	1283	65 s

Figure 1. The measurement phantom and MAP estimates. The images show the central horizontal cross sections from the 3D reconstructions. N_n is the number of nodes in the forward mesh.

4. Conclusions

In this paper, we considered electrical impedance tomography and the recovery from incorrectly modelled boundary shape. An incorrectly model boundary has been known to induce significant errors to the estimates, typically making absolute imaging impossible.

In this paper, we applied the recently proposed approximation error approach for the modelling of the errors induced by the unknown boundary shape. We considered a real tank measurement case, corresponding to a (absolute) thorax imaging problem. For the construction of the prior distribution for the thorax shape, CT images were used. The results show that the approximation error approach is well suited for the recovery from simultaneous uncertainty in boundary shape and the errors caused by using highly approximative forward solvers. The results of this paper, together with earlier results concerning the handling of the truncation of the computational domain, suggest that absolute EIT imaging might be a clinically relevant possibility.

References

[1] J. P. Kaipio and E. Somersalo *Statistical and computational inverse problems.* Applied Mathematical Sciences 160, Springer-Verlag, New York, 2005.

[2] A. Nissinen, V. Kolehmainen and J. Kaipio Compensation of modelling errors due to unknown domain boundary in electrical impedance tomography. *IEEE Trans. Med. Im.*, submitted, 2010.

Journal of Physics: Conference Series **224** (2010) 012051 doi:10.1088/1742-6596/224/1/012051

Electrode Models under Shape Deformation in Electrical Impedance Tomography

Alistair Boyle and Andy Adler

Systems and Computer Engineering, Carleton University, Ottawa, Canada

E-mail: alistair.js.boyle@gmail.com, adler@sce.carleton.ca

Abstract. Electrical Impedance Tomography (EIT) applies current and measures the resulting voltage on the surface of a target. In biomedical applications, this current is applied, and voltage is measured through electrodes attached to the surface. Electrode models represent these connections in the reconstruction, but changes in the contact impedance or boundary relative to the electrode area can introduce artifacts. Using difference imaging, the effects of boundary deformation and contact impedance variation were investigated.

The Complete Electrode Model (CEM) was found to be affected by conformal deformations. Contact impedance variability was found to be a significant source of artifacts in some cases.

In the context of Electrical Impedance Tomography (EIT), the effect of shape deformation on electrode models is considered and it is shown that under certain conditions significant artifacts can occur. The initial proofs of solution existence and uniqueness used a Continuum Model for the electrodes, implying complete knowledge of all boundary data.[1, 2] More recently, models allowing for regular gaps in the boundary data (Gap Model), or more physically realistic models such as the Shunt Electrode Model (SEM) and Complete Electrode Model (CEM) have also been utilized. The CEM adds a complex impedance for each electrode which models the metal electrode, conductive gel and chemical interaction at the skin-electrode interface.[3, 4] (Figure 1)

The Finite Element Method (FEM), used in the numerical solution of EIT images, requires boundary conditions based on these mathematical models. The simplest FEM boundary condition to implement is the Point Electrode Model (PEM), applying current and measuring voltage at single nodes on the boundary. An alternative electrode model implements the mathematical SEM, forcing all nodes associated with an electrode to the same voltage. The SEM is appropriate when contact impedances are so small that the matrices become ill-conditioned.[5] To reconstruct accurate images from in-vivo data, an implementation of the CEM is generally preferred.[3] In the EIT inverse problem, under homogeneous conductivity conditions, solving for the CEM's contact impedances has been successful.[6, 7, 8]

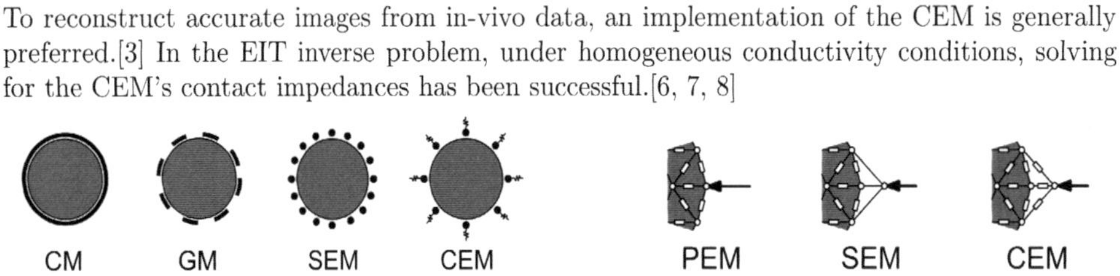

(a) Mathematical models (b) FEM models

Figure 1: Electrode models: CM - Continuous Model, GM - Gap Model, PEM - Point Electrode Model, SEM - Shunt Electrode Model, CEM - Complete Electrode Model

Journal of Physics: Conference Series **224** (2010) 012051 doi:10.1088/1742-6596/224/1/012051

Typically, electrode related image-reconstruction issues do not arise in simulation because a common electrode model is used in the forward problem and its inverse solution. The electrode contact impedance and area is assumed constant throughout and, therefore, cancels in difference imaging. If in-vivo deformations affect electrode contact impedance and area, simulations that do not apply appropriate variation are likely to get optimistic results.

1. Contact Impedance and Electrode Area

Electrode contact impedance is commonly defined in units of impedance and length or area ($\Omega \cdot$ m in 2D, $\Omega \cdot \text{m}^2$ in 3D). For an electrode model spanning multiple edges on the FEM mesh's boundary, there must be a distribution of the contact impedance amongst those edges. One method is to use a linear shape function that assigns the contact impedance based on the length between nodes in two-dimensions, or area of a boundary element in three-dimensions, as implemented in EIDORS.[9] The location of the FEM nodes for the electrode must accurately reflect the total area of the electrode to achieve the correct overall impedance.

With chest EIT, in-vivo deformations occur when a patient breaths.[10, 11] These deformation can be decomposed into independent conformal and not-conformal components. The electrodes themselves, though, are generally fabric or plastic backed and can not stretch to match the deformation exactly. As a result, an error in the area of the electrodes is likely if a conformal deformation such as a dilation occurs. The result will be artifacts in the image due to a change in the boundary's definition.

To investigate this, the behavior of the Point and Complete Electrode Models were explored under two types of conformal deformation: a 10% dilation, and a more complex deformation defined in Figure 2. Simulations were performed on a two-dimensional circular tank with homogeneous conductivity (1m initial tank radius, 33439 elements, 16 0.2m dia. electrodes, background conductivity of 1 S/m). A circular and rectangular target were simulated (conductivity 2 S/m), appearing in the second frame, and the measurements were reconstructed using a course mesh (7207 elements, 16 0.2m dia. electrodes, Gauss-Newton inverse solver μ =1e-5, Tikhonov image prior). Image artifacts were quantified using the Artifact Amplitude Measure (AAM), the sum of the squared normal error of an image's node voltages. Comparison of the deformation simulations is possible because reconstructions occur on identical undeformed meshes. (CEM and PEM models require different meshes.)

$$\text{AAM}_n = \sum \left[\frac{\text{image} - \text{no-noise image}}{\text{no-noise image}}\right]^2 \tag{1}$$

It was found that the PEM was not affected by conformal changes, whether the electrode model fixed the area or it changed to match the boundary. For the CEM, changes that were symmetric, i.e. dilation, did not result in significant artifacts if the electrode deformed with the boundary change. When the CEM was deformed in a complex conformal manner, deformation of the reconstructed image was observed as a form of artifact. When the area of the CEM was fixed and a dilation occurred, "ringing" artifacts were observed.(Table 1)

Table 1: Electrode Model Behavior under Deformation

	Deformation			
Model	Domain	Electrode	AAM_n	Comment
PEM	dilation	matching	0	
	complex		0.0807	
	dilation	fixed	0	
CEM	dilation	matching	0.0010	
	complex		2.013	artifacts (deformed)
	dilation	fixed	5.5	artifacts (ringing)

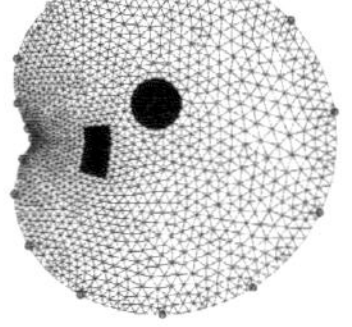

Figure 2: Complex conformal deformation: $z = x + \mathrm{i}y$; $z \to z + 0.4z^2$

2. Contact Impedance Variation

Depending on current injection patterns, there can be a significant requirement for correct specification of electrode impedance when an absolute image reconstruction is to be attempted.[12] Difference imaging reconstructs an image from two "frames" of data which removes the need to resolve the component of both contact and internal conductivity common to the two "frames." It is commonly assumed in the reconstruction process that the contact impedance remains constant, and thus, all measurement changes are due to internal conductivity changes. As electrodes age, their contact impedance changes, but changes that are not correlated will be minimized by difference imaging with frames taken at short intervals. (Typical EIT systems, such as the Goë-MF II EIT system, acquire at 13 frames-per-second.) Changes that are correlated do not affect the reconstruction with PEMs, while for CEMs, ringing was observed when impedances decreased, analogous to the fixed area electrodes under a dilation deformation. (Impedance increases have a minimal effect.) Independent of the electrdoe model, a large change in contact impedance for a single electrode will appear in the reconstructed image as an artifact near the electrode. The effects are less clear where contact impedance changes in a manner that is uncorrelated amongst the electrodes. For the purposes of this work, this will be termed "electrode noise."

Simulations of the effect of uncorrelated electrode contact impedance changes, as might be found with in-vivo electrodes, were performed. For comparison, a reconstruction with no electrode noise or measurement noise and a similar reconstruction with only measurement noise (-50dB SNR) were simulated. (Figure 3a)

These initial reconstructions were compared to reconstructions with only electrode noise at various levels (Figure 3b–d). Electrode contact impedance z_c $[\Omega \cdot \mathrm{m}]$ was assigned using an exponential Gaussian distribution

$$z_c = 10^{\mathcal{N}(\mu, \sigma^2)} \quad (2)$$

where $\mathcal{N}$ is a Gaussian distribution with a given mean μ and variance σ^2.

With electrode noise $\mu = 0$ and $\sigma^2 = 0.25$ ($\simeq 1\ \Omega \cdot \mathrm{m}$), no noticeable reconstruction artifacts were observed, similar to the no noise reconstruction. With an increase in electrode contact impedance variability, $\mu = -0.75$ and $\sigma^2 = 1.5$, image artifacts resembling those seen at -50dB SNR measurement noise were observed. Electrode impedances varying such that $\mu = -1.5$ and $\sigma^2 = 3$ resulted in significant artifacts throughout the image.

Both the mean of the electrode noise and the variance were varied to observe their affect on the reconstructed image. Reconstructions with an increase in the mean of the electrode noise, irrespective of variance, did not exhibit an increase in artifacts. (Figure 3e)

3. Discussion and Summary

A method was developed for understanding and quantifying the effect of errors in electrode area and contact impedance that occur in two-dimensional reconstructions when the boundary is deformed. Results show that the CEM produces artifacts when conformal deformations are applied. The results obtained for contact impedance variation simulations generally agree with previously published results [13] which indicated that variation of as little as 20% can result in an image that has artifacts significant enough to render the image "almost meaningless." The results also show that, in most cases, applying some level of measurement noise may have the same effect as electrode contact impedance and area variation and should be an appropriate approximation.

It would be reasonable to expect that, as electrode contact impedance variability increases, the magnitude and quantity of image artifacts increases whether the electrode's impedance has increased or decreased. Somewhat unexpectedly, increases in contact impedances result in no observable artifacts in the reconstructed image. This behavior might be explained by

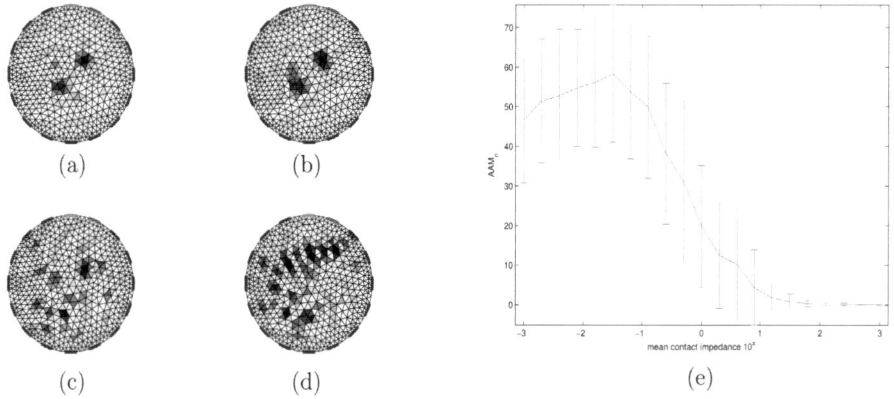

Figure 3: Electrode noise: electrodes form 1/3 of the boundary where the domain is 1 meter in radius (a) reconstruction, -50dB SNR, $\text{AAM}_n = 7.0$; (b) $\mu = 0$, $\sigma^2 = 0.25$, $\text{AAM}_n = 5.9$; (c) $\mu = -0.75$, $\sigma^2 = 1.5$, $\text{AAM}_n = 24.5$; (d) $\mu = -1.5$, $\sigma^2 = 3$, $\text{AAM}_n = 74.1$; (e) contact impedance ($-3 \leq \mu \leq 3$, $\sigma^2 = 1$) versus artifact amplitude (AAM_n) with results over 100 different electrode contact impedance configurations per-contact impedance mean

considering the electrode model as a resistor network attached to the FEM, itself a low impedance resistor network. Large contact impedances mean that the voltage measured at the electrode is approximately the average of the boundary voltage connected to the electrode; however, a small contact impedance will result in an electrode voltage that is highly dependent on the surrounding conductivity. In this environment, electrode voltages measured across small contact impedances will be heavily affected by reconstructed conductivity artifacts near the boundary and are therefore more likely to introduce these artifacts in the inverse problem.

In general biomedical and industrial applications, achieving a minimal contact impedance is desirable to maximize measurement sensitivity. These EIT simulations show that, with contact impedance variability, reconstruction artifacts can be a significant factor in image quality as contact impedance is reduced. To diminish this effect, one potential avenue would be to increase the electrode contact impedance. Another alternative would be to simultaneously reconstruct the electrode contact impedance and interior conductivity distribution, employing some form of regularization. Further in-vivo studies of electrode contact impedance under boundary movement are required to determine if electrode impedances do in fact vary by a sufficient amount to warrant further attempts at mitigating the effect of impedance variability.

References

[1] Calderón A P 2006 *Computational & Applied Mathematics* **25** 133–138 (reprint, orig. 1980)
[2] Nachman A I 1996 *The Annals of Mathematics* **143** 71–96
[3] Cheng K S, Isaacson D, Newell J C and Gisser D G 1989 *IEEE Trans. Biomed. Eng.* **36** 918–924
[4] Somersalo E, Cheney M and Isaacson D 1992 *SIAM J. Appl. Math.* **52** 1023–1040
[5] Holder D (ed) 2005 *Electrical impedance tomography: Methods, history and applications* (IOP Publishing)
[6] Vilhunen T, Kaipio J, Vauhkonen P, Savolainen T and Vauhkonen M 2002 *Meas. Sci. and Tech.* **13** 1848–1854
[7] Heikkinen L M, Vilhunen T, West R M and Vauhkonen M 2002 *Meas. Sci. and Tech.* **13** 1855–1861
[8] Hua P, Woo E, Webster J, Tompkins W, Inc S G and Estates H 1993 *IEEE Trans. Biomed. Eng.* **40** 335–343
[9] Adler A and Lionheart W R B 2006 *Physiol. Meas.* **27** S25–S42 ISSN 0967-3334
[10] Adler A, Guardo R and Berthiaume Y 1996 *IEEE Trans. Biomed. Eng.* **43** 414–420 ISSN 0018-9294
[11] Coulombe N, Gagnon H, Marquis F, Skrobik Y and Guardo R 2005 *Physiol. Meas.* **26** 401–411
[12] Kolehmainen V, Vauhkonen M, Karjalainen P and Kaipio J 1997 *Physiol. Meas.* **18** 289–303
[13] Boone K G and Holder D S 1996 *Medical and Biological Engineering and Computing* **34** 351–354

In Vivo Measurements of Structure/Electrode Position Changes during Respiration for Electrical Impedance Tomography

Jie Zhang[1], Lihong Qin[2], Tadashi Allen[1], Robert Patterson[3]

1. Department of Radiology, University of Minnesota, Minneapolis, MN 55455
2. Department of Therapeutic Radiology-Radiation Oncology, University of Minnesota, Minneapolis, MN 55455, USA
3. Department of Physical Medicine and Rehabilitation, University of Minnesota, Minneapolis, MN 55455

E-mail: zhang132@umn.edu

Abstract. For pulmonary applications of EIT systems, the electrodes are placed around the chest in a 2D ring, and the images are reconstructed based on the assumptions that the object is rigid and the measured resistivity change in EIT images is only caused by the actual resistivity change of tissue. Structural changes are rarely considered. Previous studies have shown that structural changes which result in tissue/organ and electrode position change tend to introduce artifacts to EIT images of the thorax. Since EIT reconstruction is an ill-posed inverse problem, any inaccurate assumptions of object may cause large artifacts in reconstructed images. Accurate information on structure/electrode position changes is necessary to understand factors contributing to the measured resistivity changes and to improve EIT reconstruction algorithm. In this study, in vivo structure/electrode position changes from a healthy male volunteer are investigated during respiration cycle at two levels, the nipple line level and the level approximately 5 cm below. For each level, sixteen fiduciary markers are equally spaced around the surface, the same as the electrode placement for EIT measurements. A MR scanner with respiration-gated ability is used to acquire images of the thorax. MR thoracic images are prospectively acquired corresponding temporally to specific time periods within respiration cycle (FRC, mid tidal volume, tidal volume). The chest expansions in anterior-posterior and lateral directions and inside tissue/organ position changes are then analyzed. The electrode position changes corresponding to different phases of respiration cycle are also measured.

1. Introduction

Electrical impedance tomography (EIT) has shown promise in monitoring pulmonary ventilation and fluid volume changes[1,2]. As the largest organ in the thorax, the lung resistivity change may be more accurately measured by EIT images compared to other thoracic organs or tissues[3]. The advantages of EIT are non-invasive, relatively inexpensive and non-cumbersome compared with other imaging techniques such as Computed Tomography (CT), Magnetic Resonance Imaging (MRI), etc. However, the poor spatial resolution of EIT significantly limits its applications. Different from other imaging techniques, where tissues/organs and the changes of anatomic structures can be perceived, EIT image shows resistivity distribution or resistivity changes relative to a reference frame within a body from which no specific organ/tissues can be identified directly.

For pulmonary application of EIT systems, the electrodes are placed around the chest in a 2D ring, and the images are reconstructed based on the assumption that the object is rigid. It has often been assumed that the measured resistivity change in EIT images is only caused by the actual resistivity change of tissue or in some cases the volume change of the vessel. Structural changes and the resulted electrode position changes are rarely considered. During inspiration, the chest expands as much as 10% in the anterior–posterior (AP) direction[4], and the tissues/organs inside the thorax move down with the diaphragm. Since image reconstruction of EIT is an ill-posed inverse problem, EIT is noise sensitive. Any changes in the structures and electrode positions will cause the errors in the measurements therefore introduce larger artefacts in the image.

Simulation study using a thoracic 3D model as virtual phantom has been played a major role in understanding factors that contribute to the measured resistivity changes of EIT image. Adler *et al* [5] studied the influence of chest expansion during respiration on EIT images using a 2D thorax finite element model. The results showed that chest expansion contributed to the measured change in conductivity and accounted for 2% to 20% of the reconstructed image amplitude in a broad zone at the center of the image. Another study using a 3D thorax finite different model showed that both chest expansion and tissue movements during respiration significantly contributed to the lung resistivity changes[6,7]. The averaged resistivity changes in the lung region caused by chest expansion were between 0.65% and 18.31%. The artifacts in the center ranged from –2% to 31% of the image magnitude.

The investigation using thoracic models is potentially limited due to lack of in vivo measurements of structural changes as well as corresponding electrode position changes. Any small error in the assumed electrode position changes may result in false estimations of artefacts in EIT image. Therefore, accurate information on structure/electrode position changes is necessary to understand factors contributing to the measured resistivity changes and to improve EIT reconstruction algorithm. In this study, we investigate in vivo structure/electrode position changes during respiration cycle.

2. Material and Methods

A Philips 1.5 T clinical MR system (Philips Medical Systems, Philips) located at the University of Minnesota-Fairview hospital was used to acquire images of the thorax. Sixteen fiduciary markers (IZI Medical Products Inc., Baltimore, MD, USA) were equally spaced around the surface of a healthy male volunteer (41 years old, 68 kg, 173cm), the same as the electrode placements for EIT measurements[8]. No current was injected and no voltage was measured. The markers were used only for position identification purpose. The electrodes were placed at two levels, the nipple line level and the level approximately 5 cm below, as shown in Fig. 1a. MR thoracic images were prospectively acquired corresponding temporally to specific time periods within respiration cycle (Functional Residual Capacity (FRC), mid tidal volume, tidal volume (TV)). Respiration signal was detected by a sensor put on the top of the thorax and under RF coil. Parallel acquisition technique was employed with the use of Sense™ XL torso RF coil.

Thirty-two cross-sectional MR images covering from the neck to the lower abdomen were obtained corresponding to each phase (FRC, mid TV, TV) for analysis. The image resolution is 1.3 mm x 1.3 mm x 8 mm. The chest expansions in AP and lateral directions at the two levels were measured and compared to those at the phase of FRC. Inside tissue/organ position changes were also demonstrated. By tracing marker positions, the potential electrode position changes during actual EIT measurements corresponding to different phases of respiration cycle were presented.

3. Results

Fig. 1 shows the positions of fiduciary markers (a) and their corresponding changes during respiration cycle. Fig. 1(b) shows the respiration cycle and the phases (FRC, mid TV, TV) where MR images of the thorax were obtained. The comparisons between TV and FRC for the nipple line level and the level approximately 5 cm below are shown in Fig. 1 (c) and (d), respectively.

For supine position, the electrode position changes are mainly observed in the electrodes placed at anterior regions of the chest. The chest expansion during respiration in AP and lateral directions are 20.8 cm and 32.9 cm at TV, 20.7 cm and 33.1 cm at mid TV, 20.6 cm and 33.2 cm at FRC, respectively for the upper level. For the lower level, the chest expansions are 20.4 cm and 31.0 cm at TV, 20.2 cm and 31.0 cm at mid TV, 20.1 cm and 31.1 cm at FRC, respectively. The changes are smaller than the chest expansion used for previous modeling study[5] and those measurements with breath holding at different air volumes[6,7]. Some factors may influence in vivo measurements of the structure changes. The scan time of MRI is long for image acquisition, which reduces the time gap between two adjacent phases. The MRI uses prospective respiration gating technique, a few respiration cycles are needed to acquire the images of the thorax. Variability in the rate of respiration cycle may cause variation in the gating points. In addition, the use of XL torso RF coil that was placed on the top of the thorax might also limit chest expansion.

Fig. 2 shows inside tissue/organ movements at the FRC and TV during respiration cycle. Large movement is observed which appears to be a potential issue for EIT measurements, especially for lower electrode level.

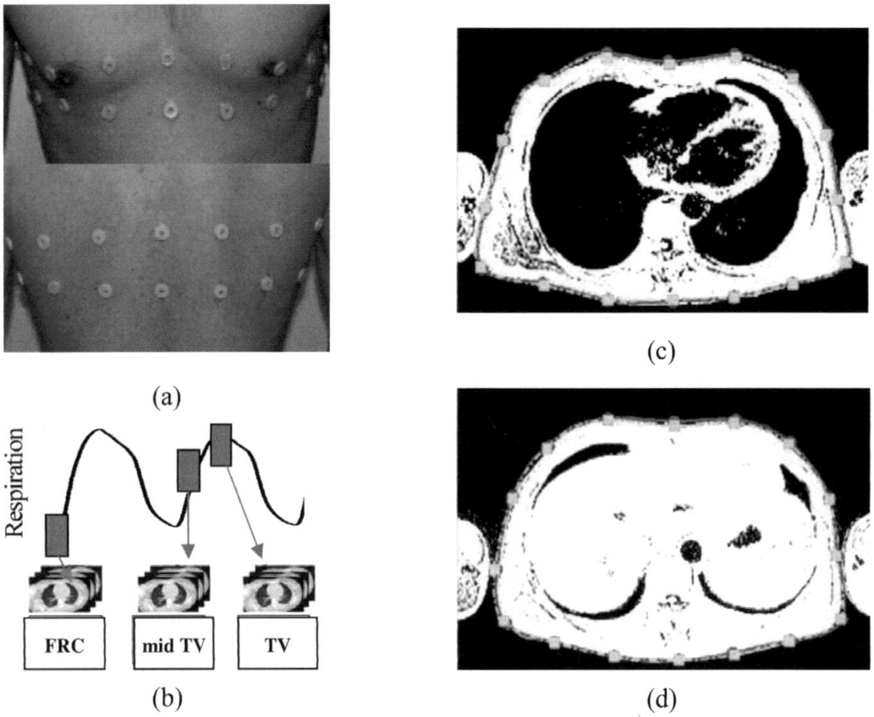

Fig. 1 (a) Fiduciary markers placed at the two levels, the nipple line level and the level approximately 5 cm below. For each level, 16 markers are equally spaced around the thorax, the same as the electrode placement for EIT measurements; (b) respiration gating positions, FRC (left bar), TV (right bar) and mid TV (middle bar) for image acquisition; and Fiduciary marker position changes during respiration between TV (red line) and FRC (green line) for the nipple line level (c) and for the level approximately 5 cm below (d).

4. Conclusion

Chest expansion/electrode position changes are obviously one of the potential sources that may cause errors in EIT images. We investigated in vivo structure and electrode position changes during respiration in a healthy male volunteer. The measurement implies the error caused by chest expansion/electrode position changes may not be significant as estimated by the use of model.

However, tissue/organ position may introduce large artefact to EIT images of the thorax, depending on the electrode levels.

To secure this investigation, we also proposed to acquire CT images from patients. CT images may provide more accurate electrode position changes during respiration with its fast acquisition time and the use of retrospective respiration gating technique. The contribution of chest expansion/ electrode position changes and tissue/organ movements to EIT images may be more accurately determined using 3D thoracic models developed from volume CT images with actual electrode position changes at different phases during respiration cycle.

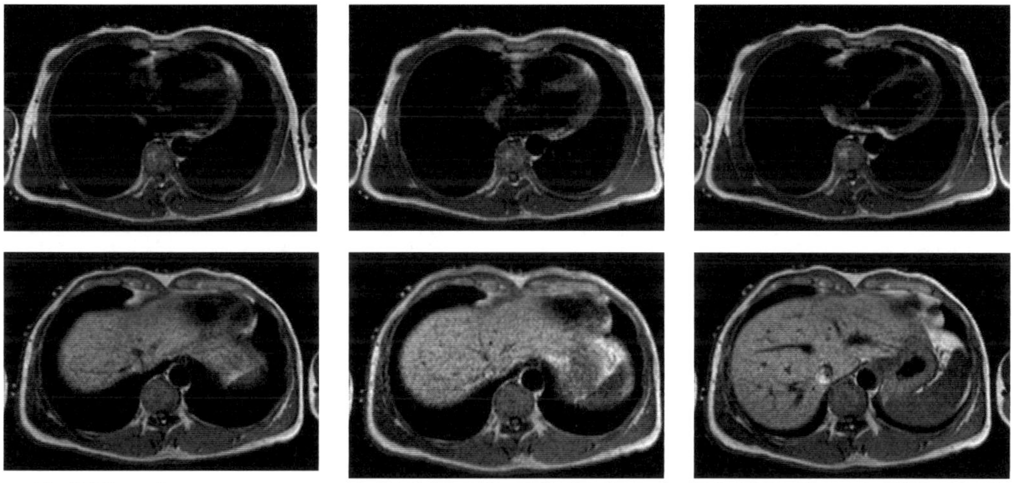

Fig.2 Tissue/organ movements at the tidal volume (left column), mid tidal volume (middle column), and FRC (right column) during respiration cycle for the nipple line level (upper images) and the level approximately 5 cm below (lower images).

Acknowledgement
We would like to acknowledge Dr. Hsiang Hsiung, Dan Busian and Lee Sternberg for assistance in acquiring MR images.

References

[1] Frerichs I, Hinz J, Herrmann P, Weisser G, Hahn G, Dudykevych T, Quintel M, Hellige G. 2002 Detection of local lung air content by electrical impedance tomography compared with electron beam CT. J ApplPhysiol;93: p660–666.
[2] Frerichs I, 2000 Electrical impedance tomography (EIT) in applications related to lung and ventilation: a review of experimental and clinical activities. Physiol. Meas.,21; R1–R21.
[3] Patterson P R and Zhang J 2003 Evaluation of an EIT reconstruction algorithm using finite difference human thorax models as phantoms Physiol. Meas.;24; p467–75.
[4] Detroyer A and Loring S H 1985 Respiratory muscle action Handbook of Physiology (Bethesda, MD: American Physiological Society) p3.
[5] Adler A, Guardo R and Berthiaume Y 1996 Impedance imaging of lung ventilation: do we need to account for chest expansion? IEEE Trans. Biomed. Eng.,43;p414–20
[6] Zhang J and Patterson RP 2005 EIT images of lung ventilation: what contributes to the resistivity changes? Physiol. Meas., 26; s81-s92.
[7] Zhang J Determination of absolute lung resistivity using electrical impedance tomography PhD Dissertation University of Minnesota, 2004.
[8] Barber D C and Seagar A D 1987 Fast reconstruction of resistance images Clin. Phys. Physiol. Meas. 8 47–54.

Forward Solving in Electrical Impedance Tomography with Algebraic Multigrid Wavelet Based Preconditioners

A. Borsic[1], **R.Bayford**[2]

[1]Thayer School Of Engineering, Dartmouth College, 8000 Cummings Hall Hanover, NH 03755, US

[2]Department of Natural Sciences, Middlesex University, London

E-mail: Andrea.Borsic@Dartmouth.edu

1. Abstract
Electrical Impedance Tomography is a soft–field tomography modality, where image reconstruction is formulated as a non-linear least-squares model fitting problem. The Newton-Rahson scheme is used for actually reconstructing the image, and this involves three main steps: forward solving, computation of the Jacobian, and the computation of the conductivity update. Forward solving relies typically on the finite element method, resulting in the solution of a sparse linear system. In typical three dimensional biomedical applications of EIT, like breast, prostate, or brain imaging, it is desirable to work with sufficiently fine meshes in order to properly capture the shape of the domain, of the electrodes, and to describe the resulting electric filed with accuracy. These requirements result in meshes with 100,000 nodes or more. The solution the resulting forward problems is computationally intensive. We address this aspect by speeding up the solution of the FEM linear system by the use of efficient numeric methods and of new hardware architectures. In particular, in terms of numeric methods, we solve the forward problem using the Conjugate Gradient method, with a wavelet–based algebraic multigrid (AMG) preconditioner. This preconditioner is faster to set up than other AMG preconditoiners which are not based on wavelets, it does use less memory, and provides for a faster convergence. We report results for a MATLAB based prototype algorithm an we discuss details of a work in progress for a GPU implementation.

2. Background
Electrical Impedance Tomography is a form of soft field tomography where image reconstruction is typically formulated as a non-linear least squares Tikhonov regularized inverse problem:

$$\sigma_{\mathbf{rec}} = \arg\min_{\sigma} \frac{\mathbf{1}}{\mathbf{2}} \left\|\mathbf{V}(\sigma) - \mathbf{V}_{\text{meas}}\right\|^{\mathbf{2}} + \alpha\frac{\mathbf{1}}{\mathbf{2}} \left\|\mathbf{L}\left(\sigma - \sigma^{*}\right)\right\|^{\mathbf{2}} \tag{1}$$

where $\sigma_{\mathbf{rec}}$ is the vector of conductivities to be estimated σ is the vector of model conductivities, $\mathbf{V}(\sigma)$ is the vector of electrode voltages resulting from the forward solver, $\mathbf{V}_{\text{meas}}$ is the vector of measured electrode voltages, α is the Tikhonov factor, L is a regularization matrix, and σ^* is a reference conductivity distribution. Equation (1) is usually solved in a iterative fashion, by application of the Gauss Newton method:

$$\delta\sigma_{\mathbf{n}} = -(\mathbf{J}_{\mathbf{n}}^{\mathbf{T}}\mathbf{J}_{\mathbf{n}} + \alpha\ \mathbf{L}^{\mathbf{T}}\mathbf{L})^{-\mathbf{1}}[\mathbf{J}_{\mathbf{n}}^{\mathbf{T}}(\mathbf{V}(\sigma_{\mathbf{n}}) - \mathbf{V}_{\text{meas}}) - \alpha\mathbf{L}^{\mathbf{T}}\mathbf{L}(\sigma_{\mathbf{n}} - \sigma^{*})] \tag{2}$$

where n is the iteration number, $\delta\sigma_{\mathbf{n}}$ is the conductivity update for iteration n and J_n is the Jacobian of the forward operator $\mathbf{V}(\sigma)$ calculated for $\sigma = \sigma_{\mathbf{n}}$. EIT reconstruction consists therefore in the repeated application of (2) to update an initial conductivity guess, typically a uniform conductivity distribution, forming therefore a reconstructed image. Three main computational steps are associated with (2): the computation of the forward solution $\mathbf{V}(\sigma_{\mathbf{n}})$, the computation of the Jacobian J_n, and the solution of the dense linear system represented by (2) itself. Typically the solution of the forward problem and the computation of the Jacobian matrix are the most computationally intensive steps of the reconstruction and different papers in literature address the problem of speeding up forward solutions. Techniques such Finite Elements and Finite Differences are normally used for discretizing the imaging domain and for solving the associated PDE, resulting in a sparse linear system. The system of sparse equations can be solved either by direct or by iterative methods. Iterative solvers, based on the Conjugate Gradients Method with preconditioning, have been studied by Soleimani [1] and by Horesh [2]. Soleimani considered an Algebraic Multigrid Method based on the traditional Algebraic Multigrid Method (AMG) of Ruge and Stüben [3]. The AMG (see next Section for more detail) preconditioning showed in this case considerable gains in terms of speed of convergence with respect to preconditioning the Conjugate Gradients (CG) method with incomplete Cholesky factorizations. Horesh considered an inverse-based multi-level preconditioning from the ILUPACK library, showing again significant gains compared to traditional preconditioning as the incomplete Cholesky factorization. Borsic [4] considered direct solvers. In the context of tomographic applications, where the same forward problem is solved for several excitations, direct solvers can prove to be faster than iterative solvers. Direct solvers are based on matrix factorizations, and once this is accomplished, multiple solutions for different right hand sides can be computed with very little cost. For three-dimensional problems, with meshes with approximately 100,000 nodes, Borsic showed that PARDISO [5] can be significantly faster than the iterative methods proposed in literature. Direct solvers have the inconvenience that they require more memory than iterative solvers, as they need to store the factored matrix, and they lend themselves less well to parallelization. Bayford [6] proposed the use of algebraic multigrid methods for preconditioning iterative solvers, where the interpolation between different levels of the grid uses wavelet-based operators (WAMG) [7]. Such an algorithm lends itself very well to parallelization, as iterative algorithms are trivial to parallelize and as wavelet-based multigrid preconditioners have also properties that are desirable. Traditional AMG methods [3] require the inspection of the whole system matrix to determine certain weight factors. If the system matrix is stored across different computing nodes, this requires a certain amount of inter communication. WAMG methods do not require this inspection, and the resulting communication between different parallel execution units. In our work we consider therefore WAMG methods as ideal candidates for the implementation on GPUs, thanks to their good parallelization properties.

3. Algebraic Multigrid Preconditioning for Conjugate Gradients

Algebraic Multigrid Methods can be used as preconditioners for iterative methods like the Conjugate Gradients algorithm. In Figure 1 we report the pseudo code for the Preconditioned Conjugate Gradients algorithm (PCG) [8]. In explicit preconditioning, the conjugate search direction $d_{(k)}$ is computed by solving $M^{-1}r_{(i)}$ (see Figure 1), where M is the preconditioner and $r_{(i)}$ is the residual at step i. AMG methods can used to precondition PCG by computing $M^{-1}r_{(i)}$ with an AMG v-cycle [3]. The v-cyle, whose code is reported in Figure 2, is a multigrid method for approximating the solution of a linear system. The v-cycle is based on the setup of an interpolation scheme based on restriction operators P^T and prolongation operators P that allow to respectively interpolate variables onto a coarser grid or to project them onto a finer grid. The operators allow also to establish a hierarchy of matrices $M_{k+1} = P_k^T M_k P_k$, which are a coarser and coarser (smaller in size) representations of M_k. In general the multigrid method

```
Preconditioned Conjugate Gradients Algorithm:    Solve Ax = b
```

$$r_{(0)} = b - Ax_{(0)}$$
$$d_{(0)} = M^{-1}r_{(0)}$$
$$\alpha_{(i)} = \frac{r_{(i)}^T M^{-1} r_{(i)}}{d_{(i)}^T A d_{(i)}}$$
$$x_{(i+1)} = x_{(i)} + \alpha_{(i)} d_{(i)}$$
$$r_{(i+1)} = r_{(i)} - \alpha_{(i)} A d_{(i)}$$
$$\beta_{(i)} = \frac{r_{(i+1)}^T M^{-1} r_{(i+1)}}{r_{(i)}^T M^{-1} r_{(i)}}$$
$$d_{(i+1)} = M^{-1} r_{(i+1)} + \beta_{(i+1)} d_{(i)}$$

Figure 1. Pseudo code for the Preconditioned Conjugate Gradient algorithm.

presents very rapid convergence properties compared to other iterative methods, and can be used on its own to solve systems of equations or as a preconditioner, as in our case. Figure 3 reports convergence results from a MATLAB based implementation of the wavelet AMG (WAMG) preconditioned PCG solver. The graphs represent the plot of the residual during the iterations of the PCG algorithm. As a comparison we report convergence graphs for PCG with incomplete Cholesky preconditioning with drop tolerances of $10e^{-1}$, $10e^{-2}$, and $10e^{-3}$ and for the unpreconditioned CG algorithm. The left part of Figure 3 is relative to the forward solution of a EIT problem on a mesh with approximately 59,000 nodes, and the right part to a mesh with approximately 95,000 nodes. WAMG shows a much faster convergence than the other preconditioners, solving the problems in less than 20 iterations. The setup of the PCG with incomplete Cholesky factorization requires lengthier times and much more memory than WAMG. For example, for the mesh with 95,000 nodes, WAMG requires 2.50MB of additional memory for the structures of the v-cycle, while the incomplete Cholesky factorization, with a drop tolerance of $10e^{-3}$, requires 25.27MB of additional memory to store M. The system matrix of this problem stores in 10.72MB. WAMG requires thus 33% additional memory, while the incomplete Cholesky factorization, with a small threshold, requires 235% of additional memory to store the preconditioner, which is an undesirable effect of the "fill-in" phenomenon, where the factored matrix has a less sparser structure then the original matrix. The WAMG method requires thus shorter setup times and less memory than incomplete Cholesky factorizations, and results in better convergence rates.

4. Concluions

Given the positive results form the MATLAB based prototype of the WAMG PCG, and thanks to its good parallelization properties, we are currently implementing the algorithm on a GPU architecture. We are using an NVIDIA Tesla S1070, a mutli-GPU computing solution with 4 GPUs, each with 240 cores, which is capable of running 3840 threads at the same time. GPUs have excellent performances for single precision computation, and they are less powerful in double precision, but they can still offer significant gains compared to CPUs also in double precision. We are in the process of implementing the main linear algebra functions that are needed to implement the full WAMG PCG algorithm. The main operations involved are element-wise products of vectors, sparse-matrix times vector, and sparse-matrix times sparse-matrix operations. For the linear algebra blocks we have implemented at this stage we have obtained speed gains in excess of 10 times compared to running the same code on a octal Xeon workstation. These results are obtained so far on a single GPU of the 4 GPUs on board of the Tesla S1070. We expect that

AMG V-Cycle: Solve $d_k = M_k^{-1} r_k$

procedure v-cycle(M_k, r_k, d_k, k)
d_k = smoother(M_k, r_k, d_k)
if (k = max-grid-level)
    $z_k = r_k - A_k d_k$
    $M_{k+1} = P_k^T M_k P_k$
    $d_{k+1} = 0$
    v-cycle($M_{k+1}, P_k^T zk, d_{k+1}, k+1$)
    $d_k = d_k + P_k d_{k+1}$
    $d_k =$ smoother(M_k, r_k, d_k)

Figure 2. Pseudo code for the AMG V-Cycle.

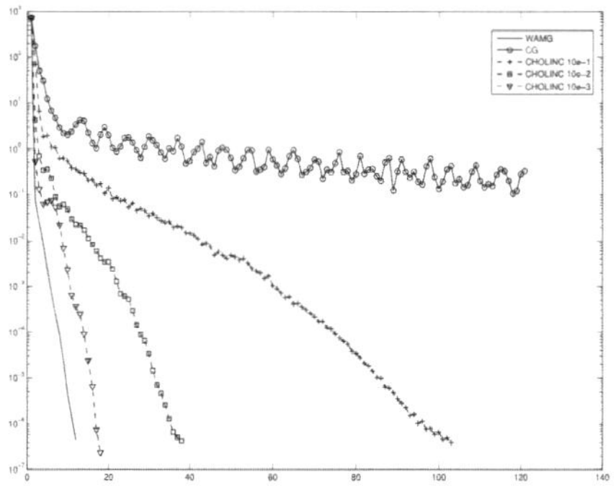

(a) Convergence plot, 59K unknowns problem.

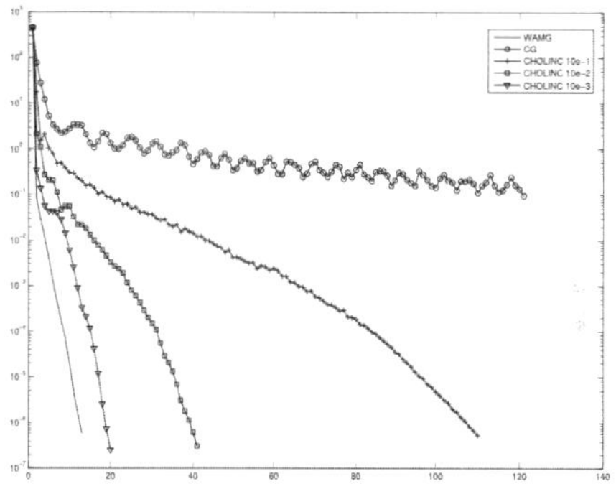

(b) Convergence plot, 95K unknowns problem.

Figure 3. Convergence plots for two forward EIT solutions, on the left for a mesh with 59,000 nodes, and on the right for a mesh with 95,000 nodes. The graphs show the reduction for the residual for the Conjugate Gradients method with no preconditioning (CG), with incomplete Cholesky preconditioning with drop tolerances of $10e^{-1}$, $10e^{-2}$, and $10e^{-3}$ (CHOLINC), and with wavelet AMG preconditioning (WAMG). WAMG shows better performance compared to CG and CHLINC, while requiring smaller setup times and less memory.

the final multi-GPU implementation will result in an FEM solver with superior computational performances thanks to the numerical and parallelization properties that WAMG PCG shows.

References

[1] Soleimani M, Powell C E and Polydorides N 2005 *IEEE Trans Med Imaging* **24** 577–582
[2] Horesh L, Schweiger M, Bollhofer M, Douiri A, Holder D S and Arrige S R 2005 *Int. J. Inform. Sys. Scie.* **2** 532–556
[3] Ruge J and Stuben K 1986 *Multigrid methods* **5** 73–130
[4] Borsic A, Hartov A and Paulsen K D 2008 *9th International Conference on Biomedical Applications of Electrical Impedance Tomography (EIT 2008), Hanover NH*
[5] Schenk O, Gärtner K, Fichtner W and Stricker A 2001 *Future Generation Computer Systems* **18** 69–78
[6] Bayford R, Kantartzis P, Tizzard A, Yerworth R, Liatsis P and Demosthenous A 2008 *Phys. Meas.* **29** S125–S138
[7] Pereira F, Verardi S and Nabeta S 2006 *Applied Mathematics and Computation* **182** 1098–1107
[8] Shewchuk J 1994

3D EFT imaging with planar electrode array: numerical simulation

T.Tuykin and A.Korjenevsky

Kotel'nikov Institute of Radio-engineering and Electronics of the RAS, Moscow, Russia

E-mails: tt@cplire.ru and korjenevsky@cplire.ru

Abstract. Electric field tomography (EFT) is the new modality of the quasistatic electromagnetic sounding of conductive media recently investigated theoretically and realized experimentally. The demonstrated results pertain to 2D imaging with circular or linear arrays of electrodes (and the linear array provides quite poor quality of imaging). In many applications 3D imaging is essential or can increase value of the investigation significantly. In this report we present the first results of numerical simulation of the EFT imaging system with planar array of electrodes which allows 3D visualization of the subsurface conductivity distribution. The geometry of the system is similar to the geometry of our EIT breast imaging system providing 3D conductivity imaging in form of cross-sections set with different depth from the surface. The EFT principle of operation and reconstruction approach differs from the EIT system significantly. So the results of numerical simulation are important to estimate if comparable quality of imaging is possible with the new contactless method. The EFT forward problem is solved using finite difference time domain (FDTD) method for the 8x8 square electrodes array. The calculated results of measurements are used then to reconstruct conductivity distributions by the filtered backprojections along electric field lines. The reconstructed images of the simple test objects are presented.

1. Introduction

Electric Field Tomography (EFT) is new imaging technique which enables contact-less visualization of spatial distribution of electrical properties inside conductive object. The object is investigated by alternating quasistatic electric field, with wavelength much longer then object size. Due to finite conductivity free charge carriers of object media can not redistribute immediately to compensate external field (Maxwell-Wagner relaxation [1, 2]) and produce weak secondary electric field, which depends on conductivity, permittivity, geometry of the object and frequency of sounding field. This secondary field is in quadrature with the primary field, so measuring of phase shift between perturbed and primary field using spatially distributed receiving and transmitting electrodes in conjunction with appropriate reconstruction algorithm enables to reconstruct spatial distribution of media electrical properties. In common case visualization combines both electrical characteristics (permittivity and conductivity), shows spatial distribution of object phase shift ability. But when frequency of investigation field is equals to relaxation frequency of media ($\omega_r = 1/\varepsilon_0\varepsilon\rho$, where ε is permittivity of the medium, ρ is specific resistance of the medium), the phase shift is maximal and its value do not depend on conductivity of media, and depend only on permittivity, geometry of media [1, 2]. And there is theoretical possibility to detach spatial distribution of conductivity and permittivity.

In contrast to electrical capacitance tomography [3] the EFT is intended for imaging of conductive medium, such as biological tissues and visualizes not only permittivity. Just as for magnetic induction tomography (MIT) [4], the EFT potential application areas include medicine and industrial process control, and due to contact-less operation it can be used for brain imaging and in security systems. In previous paper [5] experimental visualization of the test objects by electric field tomography method was shown at first time. The results of numerical simulation and comparison of 2D EFT system with different electrodes organization (round and linear) was presented in the paper [6]. It was shown that system with round electrodes organization provides image suitable for further visual analysis, but system with planar electrodes organization reconstructs image with worse quality, sufficient only for detection purposes. The main cause is a lack of field lines registered by planar electrodes system after interaction with object. Planar electrodes array geometry for measuring system more convenient in use for certain application than system with round electrodes array organization: there is no need to surround investigated objects by system for measurements and planar system can be easily mounted or hidden in wall, floor or cell. Clinical application experience of electrical impedance tomography system [7] with 2D flat electrode system for 3D visualization shown practical possibility to collect sufficient data set for visualization with further quantities analysis by planar electrodes array. This paper describes numerical experiment for 2D planar electrodes system for 3D EFT visualization.

2. Methods and materials

2.1. FEM and FDTD simulation

For the description of an electric field in an inhomogeneous conducting medium in the quasi-static approximation (the wavelength of sounding field is greater than system size), it is possible to use combination of the Poisson equation and the continuity equation (conservation law of charge) [1]:

$$\begin{cases} \nabla\cdot(\varepsilon\nabla\varphi) = -4\pi\rho \\ \nabla\cdot(\sigma\nabla\varphi) = \partial\rho/\partial t \end{cases}, \tag{1}$$

where φ is the potential of an electric field, ρ is the density of free charges and ε and σ are the permittivity and conductivity of a medium, respectively. Considering processes at angular frequency ω and eliminating the density of free charges ρ from system (1), we obtain a homogeneous equation with complex permittivity:

$$\nabla\cdot((\varepsilon - i4\pi\sigma/\omega)\nabla\phi) = 0. \tag{2}$$

This equation (2), with the corresponding boundary conditions, including the grounded shield and the active electrode with a given potential, describes the behavior of the electric field inside and outside the object and can be used for simulation a system [1]. The finite element method implementation to (2) is a common way to simulate a quasi-static electromagnetic tomography system. Description of simulated system is given in the chapter 2.4.

Other way to simulate system is to solve Maxwell equations:

$$\begin{cases} \mathrm{rot}(\mathbf{H}) = \partial\mathbf{D}/\partial t + \mathbf{J}, \\ \mathrm{rot}(\mathbf{E}) = -\partial\mathbf{B}/\partial t, \end{cases} \tag{3}$$

where **E** is electric field intensity, **H** is magnetic field intensity, **D** is electric flux density, **B** is magnetic flux density. If the medium is linear and dispersionless we have $\mathbf{D} = \varepsilon\varepsilon_0\mathbf{E}$, $\mathbf{B} = \mu\mu_0\mathbf{H}$, $\mathbf{J} = \sigma\mathbf{E}$, where ε, μ, and σ are permittivity, permeability and conductivity of the medium. It is rather difficult task to FEM method. Usually it can be solved by finite-difference time-domain method. This method need more time to calculate, but suitable not only for quasi-static. And the results obtained trough the method taking into account wave propagation effects. Also the method enables to simulate wideband frequency result in one simulation run for further spectral analysis. The FDTD method has been used in the previous paper [6].

2.2. FEM model description
The model consists of 2D flat electrode array (8x8 electrodes) and cylinder of fat-like tissue with conductivity 0.03 Sm/m and permittivity 10, see figure 1. Data simulation has been done for frequency 10 MHz.

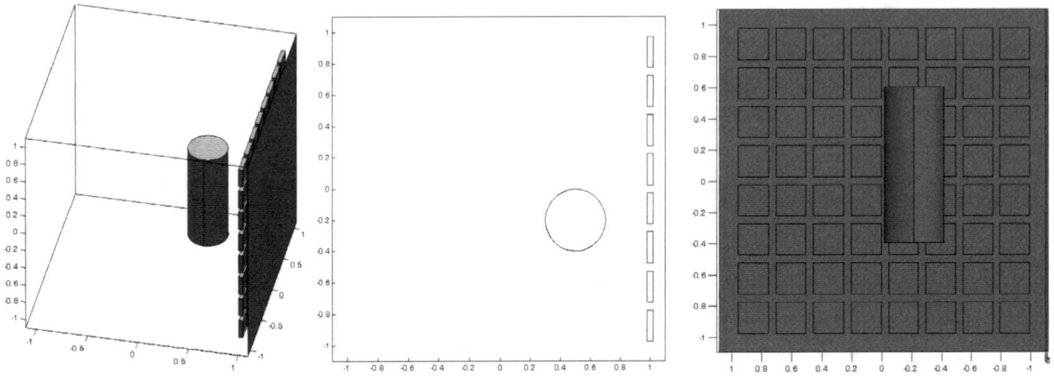

Figure 1. Model for quasi-static FEM simulation: electrodes array 8x8 and fat-like object (conductivity 0.03 S/m and permittivity 10).

2.3. Image reconstruction by weighted back projection along electric field line
The vector of values of medium phase shift ability $\vec{R}$ in nodes of cubic grid is calculated through symbolic expression: $\vec{R} = -\hat{B}\hat{F}\vec{\Delta}$, where $\hat{B}$ is operator of weighted back projection along non-perturbed electric field lines, $\hat{F}$ is filter operator (standard Ramachandran-Lakshminarayanan filter is used), $\vec{\Delta}$ is vector of measured (simulated) phase shifts for all combination of transmitting and receiving electrodes [1]. Operator $\hat{B}$, in addition to proper backprojection, provides correction of sensitivity non-uniformity inside the measuring system through multiplying projected data by the corresponding geometrical weighting factor.

3. Results
Images have been successfully reconstructed from simulated data set. The 3D distribution of conductivity is presented as the set of four 2D cross-sectional images with increasing depth. The cross-section planes are parallel to the electrode array. The resulting images are presented on figure 2. The cylinder object shown on the figure 1 is clearly identifiable on the images (darker shadows correspond to larger phase shift). The axis of the cylinder object is on the depth of 2 in units of the figure 2, so the cross-section planes 2 and 3 (depth 1.5 and 2.5 units) contain the object (dark area). The white area in the plane 1 and grey spots in the plane 4 are reconstruction artefacts.

4. Conclusions
EFT system with planar array for 3D subsurface imaging has been simulated. Images representing 3D spatial distribution of electrical properties have been reconstructed with proper object size and position.

In further research more electrodes in the system should be used to obtain images with higher quality. We consider using 16x16 electrodes array for experimental measuring system. This will enable spatial resolution similar to the resolution enabled by existing breast imaging EIT system. Contactless operation provided by the EFT method will extend fields of application significantly.

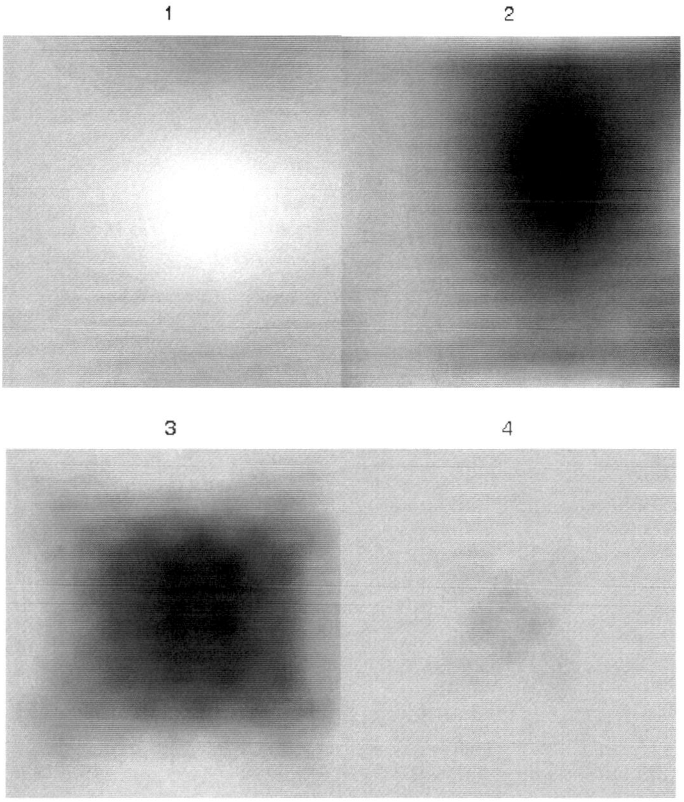

Figure 2. Reconstructed images from FEM simulated data set corresponding to geometry shown on fig. 1: 1 is cross-section parallel to the electrode array at depth equal to 0.5 units (the unit is step between electrodes), 2 – 1.5 units, 3 – 2.5 units and 4 – 3.5 units depth.

Acknowledgment
The work was partially supported by the Russian Foundation for Basic Research, Grant No 09-02-12222.

References

[1] Korjenevsky A V 2004 Electric field tomography for contactless imaging of resistivity in biomedical applications *Physiol. Meas.* **25** 391
[2] Korjenevsky A V 2005 Maxwell-Wagner relaxation in electrical imaging *Physiol. Meas.* **26** S101
[3] Huang S M, Plaskowski A, Xie C G and Beck M S 1988 Capacitance-based tomographic flow imaging system *Electron. Lett.* **24** 418
[4] Korjenevsky A, Cherepenin V and Sapetsky S 2000 Magnetic induction tomography: experimental realization *Physiol. Meas.* **21** 89
[5] Korjenevsky A and Tuykin T 2009 Experimental demonstration of the electric field tomography *Proc. 10th Int. Conf. Biomedical Applications of Electrical Impedance Tomography (EIT2009)* (Manchester)
[6] Korjenevsky A V and Tuykin T S 2008 Phase measurement for electric field tomography *Physiol. Meas.* **29** S151
[7] Cherepenin V, Karpov A, Korjenevsky A, Kornienko V, Kultiasov Y, Ochapkin M, Trochanova O and Meister D 2002 Three-dimensional EIT imaging of breast tissues: system design and clinical testing *IEEE Trans. Medical Imaging* **21** 662

An iterative FEM reconstructor for the mammography geometry

D. B. Ardrey[1], E. K. Murphy[2], D. Isaacson[3], G. J. Saulnier[1] and J. C. Newell[2]

Rensselaer Polytechnic Institute, [1]Department of Electrical, Computer and Systems Engineering, [2] Department of Biomedical Engineering, [3] Department of Mathematical Sciences

E-mail: ardred@rpi.edu

Abstract. Electrical Impedance Tomography applied in the mammography geometry has shown promise in the early detection of breast cancer. However, improvements are continually sought to improve the accuracy of static reconstructions. For this reason we present an initial framework outlining the implementation of a 3D FEM reconstructor for the mammography geometry using the complete electrode model.

1. Introduction

Electrical Impedance Tomography (EIT) is a method which allows the admittivity distribution on the interior of a body to be calculated based solely on voltage and current measurements on the surface of the body. Due to Jossinet's findings that cancerous breast tissue displays significant admittivity differences from non-cancerous breast tissue [1], EIT shows promise as a modality for the early detection of breast cancer. At Rensselaer, we have focused on the implementation of the mammography geometry, which allows for the simultaneous application of EIT and X-Ray mammography. There have been a number of algorithms created for this geometry ([2] [3] [4]) and initial patient testing has shown some promise in correctly classifying cancerous tissue [5].

One of the difficulties in applying EIT to breast cancer detection is that difference images are not available and we must strive for accurate static reconstructions. For this reason we are continuously attempting to improve the accuracy of our algorithms. This papers represents the first attempt at implementing a three-dimensional finite element model (FEM) reconstruction algorithm, which we can apply iteratively to improve the accuracy or our reconstructions.

2. Forward model

In X-ray mammography, the breast is compressed between two plates and an x-ray image of the breast is taken. We attach radiolucent arrays of electrodes [6] to both the top and bottom plates. Thus, the breast can be modeled as a rectangular parallelepiped with zero current density on all the boundaries except for electrodes located on the top and bottom faces, on which a desired current is applied. Please see Figure 1.

We model the electrodes using the complete electrode model (CEM). The CEM takes into account both the shunting effect of the electrodes and the contact impedances between the electrodes and tissue. Our goal is to compute the potential distribution (u) for a known admittivity distribution (γ) and applied current density. The problem is represented by the following set of equations:

$$\begin{aligned}
\nabla\cdot(\gamma\nabla u) &= 0, x\in\Omega, && \text{(a)}\\
u + z_l\gamma\frac{\partial u}{\partial\nu} &= U_l, \quad x\in e_l,\ l=1,2,...,L, && \text{(b)}\\
\int_{e_l}\gamma\frac{\partial u}{\partial\nu}dS &= T_l, \quad x\in e_l,\ l=1,2,...,L, && \text{(c)}\\
\gamma\frac{\partial u}{\partial\nu} &= 0, \quad x\notin\cup_{l=1}^{L}e_l, && \text{(d)}
\end{aligned} \tag{1}$$

where z_l is effective contact impedance between the lth electrode and tissue, T_l and U_l are, respectively, the applied current and measured voltage for the lth electrode. In addition, the following two conditions for the injected current and measured voltages are needed to ensure existence and uniqueness of the result:

$$\sum_{l=1}^{L} T_l^n = 0, \quad \sum_{l=1}^{L} U_l^n = 0. \tag{2}$$

We follow the implementation of a 3D FEM for the conductivity equation with CEM given in [7]. It states that for any (v,V), $v\in H^1(\Omega)$, $V\in\mathbb{R}^L$,

$$B_s((u,U),(v,V)) = \sum_{l=1}^{L} T_l V_l, \tag{3}$$

where B_s is the bilinear form associated with the complete model, given by

$$B_s((u,U),(v,V)) = \int_\Omega \gamma\nabla u\cdot\nabla v dx + \sum_{l=1}^{L}\frac{1}{z_l}\int_{e_l}(u-U_l)(v-V_l)dS. \tag{4}$$

We wish to compute a solution to the FEM problem, (u^h, U^h), which approximates the true potential distributions, (u,U). The approximate solution is defined as:

$$u^h = \sum_{i=1}^{N_n}\alpha_i\varphi_i \tag{5}$$

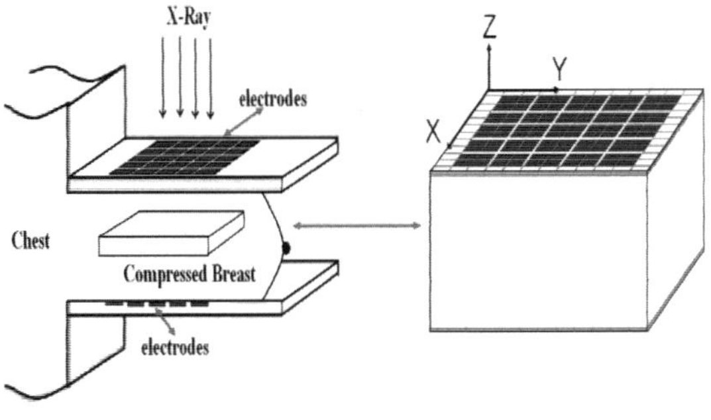

Figure 1. The mammography geometry

and

$$U^h = \sum_{j=1}^{L-1} \beta_j n_j, \tag{6}$$

where φ_i is assumed to be a linear basis element and defined

$$\varphi_j(N_i) = \begin{cases} 1 & \text{if } i = j \\ 0 & \text{if } i \neq j \end{cases}, \tag{7}$$

N_i is assumed to be the ith node of the FEM mesh, and $n_1 = [1, -1, 0, ..., 0]^T$, $n_2 = [1, 0, -1, 0, ..., 0]^T \in \mathbb{R}^L$, etc. This guarantees that (2) is not violated.

After a few steps one arrives at a linear system given by

$$Ab = \begin{pmatrix} B & C \\ C^T & D \end{pmatrix} b = \begin{pmatrix} \mathbf{0} \\ \sum_{l=1}^{L} T_l (n_j)_l \end{pmatrix} = f, \tag{8}$$

where $b = (\alpha,\ \beta)^T \in \mathbb{R}^{N_n+L-1}$ and $A \in \mathbb{R}^{(N_n+L-1)\times(N_n+L-1)}$ where $\mathbf{0} = (0, 0, ..., 0)^T \in \mathbb{R}^{N_n}$. For details on how the contents of A are calculated please see [7].

3. Reconstruction algorithm

To develop a reconstruction method we want to compute the unknown admittivity distribution, γ_1 from the applied current (T), the measured voltage (U) and the potential distribution and voltages (v, V) that result from T being applied to known admittivity distribution, γ_0. We note that $\nabla \cdot (\gamma_0 \nabla v) = 0$ and thus the following also equates to zero:

$$\begin{aligned} \textstyle\int_\Omega \nabla \cdot (\gamma_1 \nabla u) v - \nabla \cdot (\gamma_0 \nabla v) u dP &= 0, \\ \textstyle\int_\Omega \nabla \cdot (v(\gamma_1 \nabla u) - u(\gamma_0 \nabla v)) &= \textstyle\int_\Omega \gamma_1 \nabla v \cdot \nabla u - \gamma_0 \nabla u \cdot \nabla v dP, \\ \textstyle\int_S v\gamma_1 \frac{\partial u}{\partial \nu} - u\gamma_0 \frac{\partial v}{\partial \nu} &= \textstyle\int_\Omega \delta\gamma \nabla u \cdot \nabla v dP. \end{aligned} \tag{9}$$

where $\delta\gamma = \gamma_1 - \gamma_0$. Note that v is assumed to have similar boundary conditions to that of u. Next, assume that γ_1 is close to γ_0 and that $u(\gamma) \approx u^0(\gamma_0) + \mathcal{O}(\delta\gamma)$. Then from (9) we get

$$\int_S v\gamma_1 \frac{\partial u}{\partial \nu} - u\gamma_0 \frac{\partial v}{\partial \nu} = \int_\Omega \delta\gamma \nabla u \cdot \nabla v dP + \mathcal{O}(\delta\gamma^2) \tag{10}$$

We ignore the term of order $\delta\gamma^2$ and denote our data $D_{i,j}$ as

$$D_{i,j} = \int_S v_j \gamma_1 \frac{\partial u_i}{\partial \nu} - u_i \gamma_0 \frac{\partial v_j}{\partial \nu} \tag{11}$$

We can now use (1) to plug in known values as follows:

$$D_{i,j} = \sum_{\ell=1}^{L} \int_{e_\ell} \left(V_\ell^j - z_\ell \gamma_0 \frac{\partial v_j}{\partial \nu} \right) \gamma_1 \frac{\partial u_i}{\partial \nu} - \left(U_\ell^j - z_\ell \gamma_1 \frac{\partial u_i}{\partial \nu} \right) \gamma_0 \frac{\partial v_j}{\partial \nu} dS \tag{12}$$

The terms containing the surface impedance, z_ℓ, cancel out and we are left with:

$$\begin{aligned} D_{i,j} &= \sum_{\ell=1}^{L} V_\ell^j \int_{e_\ell} \gamma \frac{\partial u_i}{\partial \nu} dS - U_\ell^j \int_{e_\ell} \gamma_0 \frac{\partial v_j}{\partial \nu} dS \\ &= \sum_{\ell=1}^{L} V_\ell^j T_\ell^i - U_\ell^i T_\ell^j \end{aligned} \tag{13}$$

Our Jacobian, $J_{i,j,k}$ is given by

$$\delta\gamma_k J_{i,j,k} = \delta\gamma_k \int_{\Omega_k} \delta\gamma_k \nabla u_i^0 \cdot \nabla v_j dP, \tag{14}$$

where B_k represents the integral over the kth element. We assume here that we are working with a FEM implementation, therefore we assume that

$$u_i^0 = \sum_{r=1}^N \alpha_{i,r}\phi_r(P),$$

$$v_j = \sum_{s=1}^N \alpha_{j,s}\phi_s(P),$$

Then we can simplify J to be

$$\begin{aligned} J_{i,j,k} &= \int_{\Omega_k} \delta\gamma_k \nabla u_i^0 \cdot \nabla v_j dP, \\ &= \sum_{r,s=1}^N \delta\gamma_k \alpha_{i,r}\alpha_{j,s} \int_{\Omega_k} \nabla\phi_r \cdot \nabla\phi_s dP, \\ &= \delta\gamma_k V^T F^{(r)} V, \end{aligned} \tag{15}$$

where we have defined matrices $\mathbf{V}$ of size $N \times K$, where K are the number of voltage/current patterns and $F^{(k)}$ of size $N \times N$ to be the following

$$V_{i,r} = \alpha_{i,r},$$

and

$$F_{r,s}^{(k)} = \int_{\Omega_k} \nabla\phi_r \cdot \nabla\phi_s dP.$$

We can then regularize to compute $\delta\gamma$ and thus γ_1. We can then update D and J using γ_1 and iterate the entire process.

4. Conclusions

This paper has presented a first formulation of an iterative reconstruction algorithm for the mammography geometry using FEM and the CEM. Implementation and results are in progress and will be available by the time of the meeting.

References

[1] Jossinet J and Schmidtt M 1999 *Annals of the New York Academy of Sciences* **873** 30–41
[2] Choi M, Kao T, Isaacson D, Saulnier G and Newell J 2007 *IEEE Trans. Biomed. Engr.* **54** 700–710
[3] Kim B, Boverman G, Newell J, Saulnier G and Isaacson D 2007 *Physiological Measurement* **28** 57–69
[4] Boverman G, Kao T, Kulkarni R, Kim B, Isaacson D, Saulnier G and Newell J 2008 *IEEE Trans. Medical Imaging* **1** 1–9
[5] Kao T, Boverman G, Kim B, Isaacson D, Saulnier G, Newell J, Moore R and Kopans D 2007 *IEEE Trans. Medical Imaging (accepted)*
[6] Kao T, Saulnier G, Xia H, Tamma C, Newell J and Isaacson D 2007 *Physiological Measurement* **28** 291–299
[7] Vauhkonen P J, Vauhkonen M, Savolainen T and Kaipio J P 1999 *IEEE Trans. Biomed. Engr.* **46** 1150–1161

International Conference on Electrical Bioimpedance IOP Publishing
Journal of Physics: Conference Series **224** (2010) 012056 doi:10.1088/1742-6596/224/1/012056

Distinguishability in EIT using a hypothesis-testing model

Andy Adler[1], **Pascal Gaggero**[2] **and Yasheng Maimaitijiang**[1]
[1]Carleton University, Ottawa, Canada
[2]Centre Suisse d'Electronique et de Microtechnique, Landquart, Switzerland

Abstract: In this paper we propose a novel formulation for the distinguishability of conductivity targets in electrical impedance tomography (EIT). It is formulated in terms of a classic hypothesis test to make it directly applicable to experimental configurations. We test to *distinguish* conductivity distributions $\boldsymbol{\sigma}_2$ from $\boldsymbol{\sigma}_1$, from which EIT measurements are obtained with added white Gaussian noise with covariance $\boldsymbol{\Sigma}_n$. In order to *distinguish* the distributions, we must reject the null hypothesis H_0: $\hat{\text{x}} = 0$, which has a probability based on the z-score: $z = \frac{\bar{x}}{\sigma_x}$. This result shows that distinguishability is a product of the impedance change amplitude, the measurement strategy and the inverse of the noise amplitude. This approach is used to explore different current stimulation strategies.

1 Introduction

Electrical impedance tomography (EIT) attempts to reconstruct the impedance distribution within a body from electrical stimulation and measurement at a series of electrodes attached to body surface. One key figure of merit for a given EIT system is its distinguishability, which measures an index of the feasibility with which two different impedance distributions ($\boldsymbol{\sigma}_1$ and $\boldsymbol{\sigma}_2$) may be distinguished. This measure incorporates several factors: the size, impedance difference between $\boldsymbol{\sigma}_1$ and $\boldsymbol{\sigma}_2$, and the characteristics of the EIT systems such as signal to noise ratio (SNR), the position, number of electrodes, and stimulation and measurement patterns on the electrodes.

In this paper we propose a novel formulation for the distinguishability of conductivity targets in EIT. This work compliments the classic formulation of the problem by Isaacson[1], which calculated the measurement precision needed to distinguish between two different conductivity distributions. A similar formulation was proposed by Lionheart *et al* to find optimal current patterns considering electrical safety [2]. In this work, we propose the formulation of distinguishability in terms of a classic hypothesis test, which makes it easier to apply to experimental configurations. We develop the formulation for the distinguishability and demonstrate its applicability to a simulated EIT system.

2 Image Reconstruction

An EIT system, characterized by $F(\cdot)$, is used to make a set of measurements represented by an $M \times 1$ vector

$$\mathbf{v} = F(\boldsymbol{\sigma}) + \mathbf{n} \tag{1}$$

from an impedance distribution characterized by an $N \times 1$ parameter vector $\boldsymbol{\sigma}$, where $F(\cdot)$ describes the EIT measurement process. In all measurement systems there is a contribution of random noise $\mathbf{n}$, which we characterize as zero-mean, independent and Gaussian. This characterization is justified as follows: the noise has zero-mean because any average bias can theoretically be measured and incorporated into the EIT model $F(\cdot)$; the independent noise means that there is no correlation between the noise and the two samples we wish to distinguish. This is a reasonable assumption in the most EIT hardware designs if measurements are separated in time by several frames. Correlations between individual measurements with the measurement vector $\mathbf{v}$ may exist and be modelled with this approach. The assumption of Gaussian noise is designed to make the statistical computations easier; however, there is some evidence to indicate that EIT noise is not Gaussian[3]. Specifically, large errors occur far more often than would be predicted by a Gaussian with a time-invariant covariance $\mathbf{\Sigma}_n$.

Problem Formulation: we wish to *distinguish* conductivity distributions $\boldsymbol{\sigma}_2$ from $\boldsymbol{\sigma}_1$ through which EIT measurements $\mathbf{v}_1 = F(\boldsymbol{\sigma}_1) + \mathbf{n}_1$, and $\mathbf{v}_2 = F(\boldsymbol{\sigma}_2) + \mathbf{n}_2$ are performed, where $\mathbf{n}_1$ and $\mathbf{n}_2$ represent the instantiations of zero-mean white Gaussian noise with covariance $\mathbf{\Sigma}_n$. Since we generally want to distinguish small changes in conductivity, we linearize around $\boldsymbol{\sigma}_1$ to obtain a conductivity change $\mathbf{x} = \boldsymbol{\sigma} - \boldsymbol{\sigma}_2$ which is a linear function of measurements

$$\mathbf{y} = \mathbf{v} - \mathbf{v}_1 = \mathbf{J}\mathbf{x} + \mathbf{n}, \tag{2}$$

where $\mathbf{J}$ is the Jacobian (sensitivity) defined as

$$[\mathbf{J}]_{i,j} = \left.\frac{\partial F_i(\boldsymbol{\sigma})}{\partial \boldsymbol{\sigma}_j}\right|_{\boldsymbol{\sigma}=\boldsymbol{\sigma}_1}. \tag{3}$$

This linearization assumes that reasonably small impedance changes $\mathbf{x}$ occur, which is generally a reasonable assumption if we are investigating the distinguishability limits for most EIT systems.

From measurements $\mathbf{y}$, an impedance change image estimate $\hat{\mathbf{x}}$ is reconstructed, from a linearized difference EIT reconstruction algorithm as $\hat{\mathbf{x}} = \mathbf{R}\mathbf{y}$, defined from the norm

$$\hat{\mathbf{x}} = arg\,\underset{\mathbf{x}}{min}\,\|\mathbf{y} - \mathbf{J}\mathbf{x}\|_{\mathbf{\Sigma}_n^{-1}} + P(\mathbf{x}), \tag{4}$$

where $P(\cdot)$ represents a penalty or regularization term. Such a linear reconstruction matrix can describe the majority of difference EIT reconstruction algorithms, such as the Sheffield Backprojection and many regularization based schemes.

In general, we are interested in investigating the image output within a specific region of interest (ROI), defined by $\mathbf{I}_R$, where $\mathbf{I}_R$ is a vector of the size of the image parameter space $N \times 1$ in which each element $[\mathbf{I}_R]_i$ is zero, if i represents a region outside of the ROI, and the area of the parameter region i, if i is in the ROI. If a fraction of region i is in the ROI, $[\mathbf{I}_R]_i$ is weighted by the fraction of membership. Based on the ROI, we define a scalar impedance change x_R in the ROI as the weighted average of the impedance change $\mathbf{x}$ in the ROI, or $x_R = \mathbf{I}_R\mathbf{x}$. The best estimate of parameter x based on measurements $\mathbf{y}$ is defined as $\hat{\mathrm{x}}$.

$$\hat{\mathrm{x}} = \mathbf{I}_R\hat{\mathbf{x}} = \mathbf{I}_R\mathbf{R}\mathbf{y} = \mathbf{R}_R\mathbf{y}, \tag{5}$$

where $\mathbf{R}_R = \mathbf{I}_R\mathbf{R}$ is defined to be the reconstruction matrix of the ROI of size $1 \times M$. Since $\hat{\mathrm{x}}$ is a single parameter, no regularization term $P(\cdot)$ is required, and the maximum likelihood solution matrix, $\mathbf{R}_R$, is given by:

$$\mathbf{R}_R = \left(\mathbf{J}_R^t\mathbf{\Sigma}_n^{-1}\mathbf{J}_R\right)^{-1}\mathbf{J}_R^t\mathbf{\Sigma}_n^{-1}, \tag{6}$$

where $\mathbf{J}_R = \mathbf{J}\mathbf{I}_R^t$.

3 Formulation of Distinguishability

In order to *distinguish* $\boldsymbol{\sigma}_1$ from $\boldsymbol{\sigma}_2$ we must reject the null hypothesis H_0: $\hat{\mathrm{x}} = 0$. The probability of H_0 is based on the z-score:

$$z = \frac{\bar{x}}{\sigma_x}, \tag{7}$$

where

$$\bar{x} = E[\hat{\mathrm{x}}] = E[\mathbf{R}_R(\mathbf{y}+\mathbf{n})] = \mathbf{R}_R\mathbf{y} = \mathbf{R}_R\mathbf{J}\mathbf{x} = \mathbf{I}_R\mathbf{x},$$

and

$$\begin{aligned}\mathrm{Var}\ (x) &= \sigma_x^2 = E[\|\hat{\mathrm{x}} - \bar{x}\|^2] = E[\|\mathbf{R}_R\mathbf{n}\|^2] \\ &= E[\mathbf{R}_R\mathbf{n}\mathbf{n}^t\mathbf{R}_R^t] = \mathbf{R}_R E[\mathbf{n}\mathbf{n}^t]\mathbf{R}_R^t = \mathbf{R}_R\boldsymbol{\Sigma}_n\mathbf{R}_R^t.\end{aligned}$$

Given the maximum likelihood solution for $\hat{\mathrm{x}}$, we further simplify σ_x^2 as

$$\begin{aligned}\sigma_x^2 &= \mathbf{R}_R\boldsymbol{\Sigma}_n\mathbf{R}_R^t \\ &= \left(\mathbf{J}_R^t\boldsymbol{\Sigma}_n^{-1}\mathbf{J}_R\right)^{-1}\mathbf{J}_R^t\boldsymbol{\Sigma}_n^{-1}\boldsymbol{\Sigma}_n\boldsymbol{\Sigma}_n^{-1}\mathbf{J}_R\left(\mathbf{J}_R^t\boldsymbol{\Sigma}_n^{-1}\mathbf{J}_R\right)^{-1} \\ &= \left(\mathbf{J}_R^t\boldsymbol{\Sigma}_n^{-1}\mathbf{J}_R\right)^{-1}\left(\mathbf{J}_R^t\boldsymbol{\Sigma}_n^{-1}\mathbf{J}_R\right)\left(\mathbf{J}_R^t\boldsymbol{\Sigma}_n^{-1}\mathbf{J}_R\right)^{-1} = (\mathbf{J}^t\boldsymbol{\Sigma}_n^{-1}\mathbf{J})^{-1}.\end{aligned} \tag{8}$$

The z score may be calculated as

$$z = \frac{\bar{x}}{\sigma_x} = \frac{\hat{\mathrm{x}}}{\left(\mathbf{J}_R^t\boldsymbol{\Sigma}_n^{-1}\mathbf{J}_R\right)^{-\frac{1}{2}}} = \hat{\mathrm{x}}\sqrt{\mathbf{J}_R^t\boldsymbol{\Sigma}_n^{-1}\mathbf{J}_R}. \tag{9}$$

This result shows that the distinguishability is a product of the target ($\mathbf{x}$), the measurement strategy ($\mathbf{J}$) and the inverse of the noise amplitude ($\boldsymbol{\Sigma}_n$).

For a practical algorithm, we need to define the ROI of the reconstruction algorithm. This area should be large enough to include the most of the amplitude of the reconstructed response, but to avoid areas which show large reconstruction artefacts.

4 Distinguishability of Current Patterns

In order to illustrate use of this approach, we perform a test to determine the stimulation protocol with the largest distinguishability. Given a 16 electrode Sheffield-type pair drive EIT system, it is traditional to stimulate across adjacent pairs of electrodes (which we label [0-1]), opposite pairs (labelled [0-8]) or any other offset. Using a 2D circular Finite Element Model (FEM), we simulate each scenario for target positions from the medium centre to boundary. Results (Fig. 1) suggest that distinguishability varies with the target position, but a stimulation offset between 3 and 5 is a good compromise.

This formulation may be extended to separate the choice of simulation patterns, the medium geometry and current propagation. Most EIT systems apply a set of current patterns represented as the columns of $\mathbf{C}$ to make a set of p voltage measurements $\mathbf{V} = \mathbf{T}(\boldsymbol{\sigma})\mathbf{C}$, where $\mathbf{T}(\boldsymbol{\sigma})$ is the *transfer impedance* matrix of the medium with impedance distribution $\boldsymbol{\sigma}$ (in units of Ω). For the small change $\mathbf{x} = x_R\mathbf{I}_R$, we can define a transfer impedance change

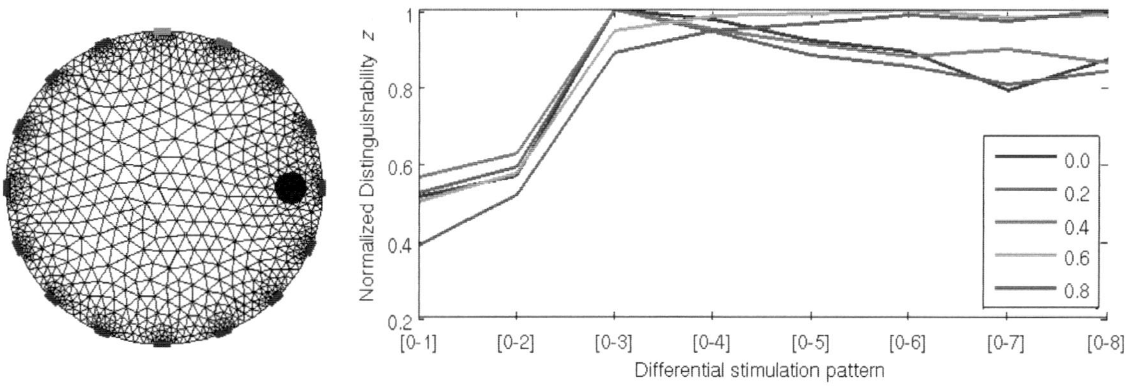

Figure 1: *Left:* FEM of a small conductive target in a 2D circular medium with 16 electrodes (green). The target moves from centre to edge (shown). *Right:* Normalized distinguishability z as a function of target position for pair drive stimulations with electrode offsets (X *axis*). Thus, $[0-1]$ indicates adjacent stimulation, while $[0-8]$ indicates opposite stimulation.

$\mathbf{T}_\Delta = \mathbf{T}(\boldsymbol{\sigma}+\mathbf{x}) - \mathbf{T}(\boldsymbol{\sigma})$. From $\mathbf{T}_\Delta$, the elements of the Jacobian $\mathbf{J}_R$ may be calculated from $\mathbf{MT}_\Delta\mathbf{C}$, where $\mathbf{M}$ represents the measurement scheme (which is typically the difference between adjacent electrodes). In this case, $\mathbf{J}_R$ is a vector $M \times 1$ while $\mathbf{MT}_\Delta\mathbf{C}$ is $\frac{N}{S} \times S$. If the noise covariance can be assumed to be the same for each current pattern (which we represent as $\mathbf{S}_n$), then

$$\boldsymbol{\Sigma}_n = \mathbf{S}_n \otimes \mathbf{I}_S, \tag{10}$$

where S is the number of stimulation patterns, $\mathbf{I}_S$ is the identity matrix. In this case

$$\mathbf{J}^t\boldsymbol{\Sigma}_n^{-1}\mathbf{J} = tr\ (\mathbf{C}^t{\mathbf{T}_\Delta}^t\mathbf{M}^t\mathbf{S}_n^{-1}\mathbf{MT}_\Delta\mathbf{C}), \tag{11}$$

where $\otimes$ represents the Kronecker product. Note that ${\mathbf{T}_\Delta}^t = \mathbf{T}_\Delta$ due to reciprocity. Thus, each diagonal element $[\mathbf{W}]_{ii} = [\mathbf{C}^t\mathbf{T}_\Delta\mathbf{M}^t\mathbf{S}_n^{-1}\mathbf{MT}_\Delta\mathbf{C}]_{ii}$ represents the contribution from the ith current stimulation, and

$$z = \hat{\mathrm{x}}\sqrt{\mathbf{W}_{11}^2 + \mathbf{W}_{22}^2 + \cdots \mathbf{W}_{pp}^2}, \tag{12}$$

This approach may then be extended to calculate the optimal current patterns $\mathbf{C}$ for a particular target position.

In summary, we propose a formulation for the distinguishability in EIT in terms of a hypothesis test, and show that distinguishability is a product of the impedance change amplitude, the measurement strategy and the inverse of the noise amplitude.

5 References

[1] Isaacson D 1986, Distinguishability of conductivities by electric current computed tomography *IEEE T Med Imaging* 5:91-95

[2] Lionheart WR, Kaipio J, McLeod CN 2001, Generalized optimal current patterns and electrical safety in EIT *Physiol. Meas.* 22 85-90

[3] Hahn G, Just A, Dittmar J and Hellige G 2008, Systematic errors of EIT systems determined by easily-scalable resistive phantoms *Physiol. Meas.* 29 S163-72

International Conference on Electrical Bioimpedance IOP Publishing
Journal of Physics: Conference Series **224** (2010) 012057 doi:10.1088/1742-6596/224/1/012057

Fisher information analysis and preconditioning in electrical impedance tomography

Sven Nordebo[1], Richard Bayford[2], Bengt Bengtsson[3], Andreas Fhager[4], Mats Gustafsson[5], Parham Hashemzadeh[2], Börje Nilsson[1], Thomas Rylander[4] and Therese Sjöden[1]

[1]School of Computer Science, Physics and Mathematics, Linnaeus University, 351 95 Växjö, Sweden. [2]Department of Natural Sciences, Middlesex university, Hendon London NW4 4BT, UK. [3]Romele Elektronik AB, Bigarrågatan 1, 24761 Veberöd, Sweden. [4]Department of Signals and Systems, Chalmers University of Technology, 412 96 Göteborg, Sweden. [5]Department of Electrical and Information Technology, Lund University, Box 118, 221 00 Lund, Sweden.

E-mail: sven.nordebo@lnu.se, R.Bayford@mdx.ac.uk, bb@maas.se, andreas.fhager@chalmers.se, mats.gustafsson@eit.lth.se, P.Hashemzadeh@mdx.ac.uk, borje.nilsson@lnu.se, rylander@chalmers.se, therese.sjoden@lnu.se.

Abstract. In this contribution, it is described how the Fisher information can be computed by using adjoint field techniques, and integrated with the gradient calculations used for optimization in electrical impedance tomography. In particular, the Fisher information can be used as a preconditioner to obtain improved convergence properties and a regularization for quasi-Newton optimization algorithms in electrical impedance tomography. Experimental data have been used to study the possibility of combining a good reconstruction quality with a low system complexity, which is achieved by using a four-electrode measurement technique to avoid the need of high-precision electrode modeling and allowing a very coarse FEM grid. Here, the Fisher information based preconditioning implies a regularization and an algorithm that works well on a coarse FEM grid with very small electrodes modeled as point sources.

1. Introduction

Gradient based least squares optimization and adjoint field techniques have great advantages of generality and simplicity, and can be used to solve a variety of inverse problems with applications in electromagnetics, acoustics, geoscience, etc., as long as there are fast direct solvers available, see *e.g.*, [4, 5, 11]. In Electrical Impedance Tomography (EIT), the gradient expressions and related computations of the Jacobian are oftenly based directly on the degrees of freedom that are present in the actual FEM formulation, *cf.*, *e.g.*, the so called *standard method* [2, 12, 13]. In this contribution, it is described how the Fisher information can be computed by using adjoint field techniques, and integrated with the gradient calculations used for optimization in electrical impedance tomography. The formulation is general and independent of the actual numerical method. As a numerical example, the Fisher information analysis is employed here in a preconditioning/regularization scheme for quasi-Newton optimization algorithms [8, 9], which has been formulated here in the context of a four-electrode measurement technique for EIT [6].

Accurate modeling of the electrodes is an important issue in EIT, particularly in the medical applications, see *e.g.*, [1, 12]. Hence, the complete electrode model was developed due to the need of an improved precision in the forward modeling, see *e.g.*, [12] for a survey on this topic. In this contribution, general gradient expressions are given that are based on the adjoint field approach, and which can be shown to be consistent with the commonly used shunt model and the complete electrode model [10, 12].

It should be noted that even though accurate electrode modeling is an important issue in EIT, it is also well known that the application of four-electrode measurement techniques [1, 6] is an efficient way of reducing the errors due to electrode impedances (both modeling and instrumentation errors), and hence to reduce the complexity of the numerical solver as well as of the measurement equipment. In this contribution, the Fisher information based preconditioner [8] is used to devise an algorithm that works well on a coarse FEM grid with very small electrodes.

2. Fisher information analysis

Consider the following frequency dependent parameter model for the complex conductivity [10]

$$\sigma = \sigma_{\mathrm{R}} + \mathrm{i}\omega\epsilon_0\epsilon_{\mathrm{r}} = \gamma_0(\beta + \mathrm{i}\nu\epsilon_{\mathrm{r}}) \tag{2.1}$$

where ω_0 is the center frequency, $\nu = \omega/\omega_0$, $\gamma_0 = \omega_0\epsilon_0$, ϵ_0 the permittivity of free space and where $\beta = \sigma_{\mathrm{R}}/\omega_0\epsilon_0$ and ϵ_{r} are the dimensionless real conductivity and permittivity parameters, respectively. Let θ denote the parameter vector $\theta = [\beta \ \epsilon_{\mathrm{r}}]^{\mathrm{T}}$. By employing the definition of the Fisher information integral operator according to [9], together with a Gaussian statistical measurement model as in [8, 9] the following expression for the Fisher information integral kernel is obtained

$$\mathcal{I}(\boldsymbol{r}', \boldsymbol{r}'') = 2\,\mathrm{Re}\sum_n\sum_m\sum_q \frac{|I_0^{(mnq)}|^2}{R_{mnq}} \delta_{\theta(\boldsymbol{r}')} Z^*_{mnq} \delta_{\theta^{\mathrm{T}}(\boldsymbol{r}'')} Z_{mnq} \tag{2.2}$$

where Z_{mnq} denotes the impedance parameter corresponding to a particular four-electrode current excitation configuration (n), voltage measurement configuration (m) and frequency (q). Here $\boldsymbol{r}', \boldsymbol{r}''$ denotes the spatial variables, $\delta_{\theta(\boldsymbol{r})} = [\delta_{\beta(\boldsymbol{r})} \ \delta_{\epsilon_{\mathrm{r}}(\boldsymbol{r})}]^{\mathrm{T}}$ a vector gradient operator [9], $I_0^{(mnq)}$ the excitation current and R_{mnq} the variance of the measurement noise.

Using the adjoint-field techniques [4, 5, 11], it can be shown that the gradients of Z_{mnq} with respect to β and ϵ_{r} can be calculated as

$$\begin{cases} \delta_{\beta(\boldsymbol{r})} Z_{mnq} = -\gamma_0 \boldsymbol{E}^{(n)} \cdot \boldsymbol{E}^{(m)} \\ \delta_{\epsilon_{\mathrm{r}}(\boldsymbol{r})} Z_{mnq} = -\mathrm{i}\nu\gamma_0 \boldsymbol{E}^{(n)} \cdot \boldsymbol{E}^{(m)}, \end{cases} \tag{2.3}$$

where $\boldsymbol{E}^{(n)}$ and $\boldsymbol{E}^{(m)}$ are the calculated electric fields with (unit) current excitation at the sensor electrode pairs corresponding to the current excitation (n) and the voltage measurement (m), respectively.

Let Z^{M}_{mnq} denote the measured impedance parameters. The gradient corresponding to the misfit functional

$$F(\sigma) = \sum_n\sum_m\sum_q \frac{|I_0^{(mnq)}|^2}{R_{mnq}} |Z_{mnq} - Z^{\mathrm{M}}_{mnq}|^2 \tag{2.4}$$

is then given by

$$\delta_{\theta(\boldsymbol{r})} F = 2\,\mathrm{Re}\sum_n\sum_m\sum_q \frac{|I_0^{(mnq)}|^2}{R_{mnq}} (Z_{mnq} - Z^{\mathrm{M}}_{mnq})^* \delta_{\theta(\boldsymbol{r})} Z_{mnq} \tag{2.5}$$

where $\delta_{\theta(\boldsymbol{r})} Z_{mnq}$ is given by (2.3). Note that it is the same impedance gradients $\delta_{\theta(\boldsymbol{r})} Z_{mnq}$ given by (2.3), that are used in the calculation of the Fisher information $\mathcal{I}(\boldsymbol{r}', \boldsymbol{r}'')$ in (2.2), as well as in the calculation of the total gradient $\delta_{\theta(\boldsymbol{r})} F$ in (2.5).

Let θ_i denote either of the real parameter functions β and $\epsilon_{\rm r}$ where $i = 1, 2$, respectively, and let $\mathcal{I}_{\theta_i}(\boldsymbol{r}', \boldsymbol{r}'')$ denote the corresponding (parameter-diagonal) Fisher information defined by (2.2). To obtain a preconditioner suitable for quasi-Newton optimization algorithms as described in [8], a linear parameter scaling is incorporated where

$$\theta_i^{\rm sc}(\boldsymbol{r}) = \theta_i(\boldsymbol{r})\sqrt{\mathcal{I}_{\theta_i^{\rm b}}(\boldsymbol{r}, \boldsymbol{r})}, \tag{2.6}$$

and where the parameter function $\theta_i^{\rm b}$ corresponds to some known background. Hence, the Fisher information for the scaled parameter function is given by $\mathcal{I}_{\theta_i^{\rm sc}}(\boldsymbol{r}', \boldsymbol{r}'') = \frac{1}{\sqrt{\mathcal{I}_{\theta_i^{\rm b}}(\boldsymbol{r}', \boldsymbol{r}')\mathcal{I}_{\theta_i^{\rm b}}(\boldsymbol{r}'', \boldsymbol{r}'')}}\mathcal{I}_{\theta_i}(\boldsymbol{r}', \boldsymbol{r}'')$. Clearly, if $\theta_i = \theta_i^{\rm b}$ then the diagonal Fisher information is $\mathcal{I}_{\theta_i^{\rm sc}}(\boldsymbol{r}, \boldsymbol{r}) = 1$ for $i = 1, 2$ and for all $\boldsymbol{r}$ in the imaging domain. Finally, the gradient corresponding to the misfit functional (2.4) is given by

$$\delta_{\theta_i^{\rm sc}(\boldsymbol{r})}F = \frac{1}{\sqrt{\mathcal{I}_{\theta_i^{\rm b}}(\boldsymbol{r}, \boldsymbol{r})}}\delta_{\theta_i(\boldsymbol{r})}F. \tag{2.7}$$

3. Experiments with very small electrodes

The purpose of the experiments described here is to study EIT imaging performance when using very small electrodes. This problem is motivated by the EIT application for non-destructive imaging of living trees [7] where contact electrodes in the form of very thin needles are used to penetrate through the bark layer. Subject to the required constraints of low hardware complexity as well as low computational complexity (fast solvers), a four-electrode measurement system was considered together with a very coarse FEM grid. A FEM numerical solver was employed using 1412 triangles and 753 nodes. The very small electrodes were modeled as single nodes in the FEM mesh.

A two-dimensional imaging domain was considered consisting of a cylindrical container of radius $a = 0.132$ m, filled with tap-water to a height of 50 mm. As electrodes were used 16 thin wires (diameter 0.5 mm) placed vertically and evenly distributed on the inner surface of the container. The objects consisted of one or two small cylindrical bottles of glass with diameter $d = 46$ mm, placed at different positions inside the imaging domain, see figure 1b. The measurement system consisted of a frequency synthesis generator, a 16-channel analog multiplexing unit, A/D converters, a DSP-based frequency analysis and a PC for control and data acquisition. A four-electrode measurement strategy was chosen based on 15 current excitation configurations and 12 corresponding voltage measurements. All measurements were made at the four frequencies 390, 1562, 7812, and 31250 Hz. Suitable background parameters $\epsilon_{\rm r} = 80$ and $\sigma_{\rm R} = 29.9$ mS/m were found by calibration without objects. A Fisher information analysis revealed that the parameter sensitivity close to the electrodes was about 20 dB in excess of the sensitivity in the general interior of the cylinder, *cf.*, figure 1a.

The following three algorithms were tested. The first algorithm is a quasi-Newton algorithm based on gradient calculations and the BFGS formula to update the inverse of the Hessian [3]. The second algorithm is an improvement of the quasi-Newton algorithm using the Fisher information based preconditioning (2.6). The third algorithm is a linearized one-step Newton algorithm (pseudo-inverse), *cf.*, [10]. In figure 1b is illustrated the four measurement cases together with the reconstructed images of the conductivity $\sigma_{\rm R}$. As can be seen in these images, both the first and the third algorithm have a tendency to over-emphasize the sensor positions, something that can be interpreted as a modeling artifact due to the very high sensitivity close to the point sources. However, the Fisher information based preconditioning is able to alleviate this artifact by proper parameter scaling (regularization), and hence to define an algorithm that works well on a coarse FEM grid with very small electrodes.

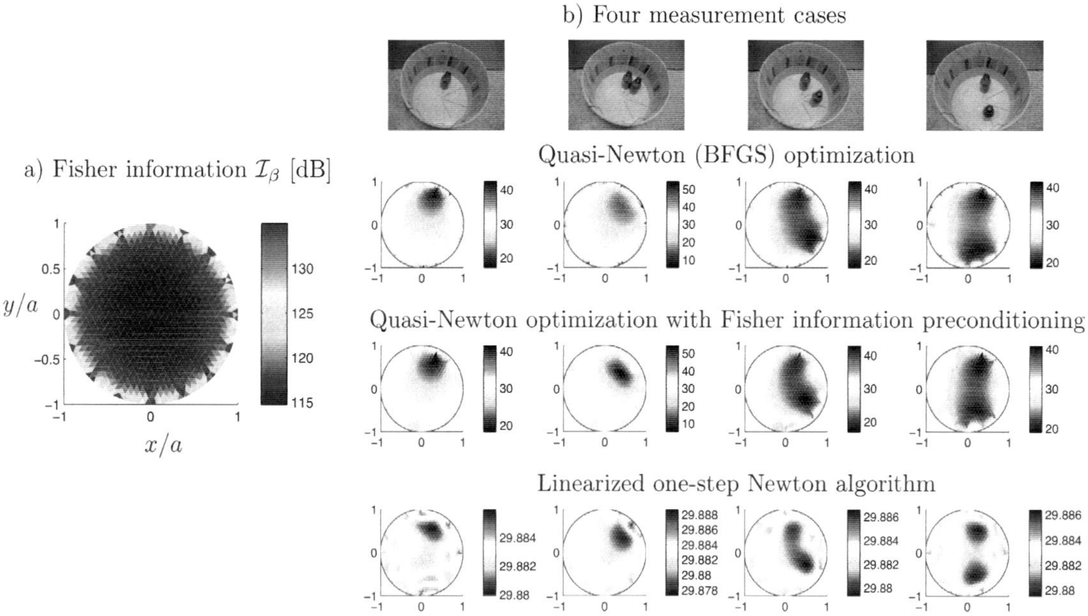

Figure 1. a) Fisher information $\mathcal{I}_\beta(\boldsymbol{r},\boldsymbol{r})$ for the background parameters. Note the excessively high sensitivity at the positions of the small electrodes. b) Four measurement cases and a comparison of using three different reconstruction algorithms for the conductivity parameter $\sigma_{\rm R}$ (in mS/m). Here, the calibrated background corresponds to $\epsilon_r = 80$ and $\sigma_{\rm R} = 29.9$ mS/m.

References

[1] R. Bayford. Bioimpedance tomography (electrical impedance tomography). *Annu. Rev. Biomed. Eng.*, **8**, 63–91, 2006.

[2] M. Cheney, D. Isaacson, and J. C. Newell. Electrical impedance tomography. *SIAM Review*, **41**(1), 85–101, 1999.

[3] R. Fletcher. *Practical Methods of Optimization.* John Wiley & Sons, Ltd., Chichester, 1987.

[4] M. Gustafsson. *Wave Splitting in Direct and Inverse Scattering Problems.* PhD thesis, Lund University, Department of Electromagnetic Theory, P.O. Box 118, S-221 00 Lund, Sweden, 2000. `http://www.eit.lth.se`.

[5] G. T. Herman, H. K. Tuy, K. J. Langenberg, and P. C. Sabatier. *Basic Methods of Tomography and Inverse Problems.* Adam Hilger Ltd., 1988.

[6] D. S. Holder. *Electrical Impedance Tomography.* IOP Publishing Ltd, 2005.

[7] G. Nicolotti, L. Socco, R. Martinis, A. Godio, and L. Sambuelli. Application and comparison of three tomographic techniques for detection of decay in trees. *Journal of Arboriculture*, **29**(2), 66–78, 2003.

[8] S. Nordebo, A. Fhager, M. Gustafsson, and M. Persson. A systematic approach to robust preconditioning for gradient based inverse scattering algorithms. *Inverse Problems*, **24**(2), 025027, 2008.

[9] S. Nordebo, A. Fhager, M. Gustafsson, and B. Nilsson. Fisher information integral operator and spectral decomposition for inverse scattering problems. Technical Report LUTEDX/(TEAT-7180)/1–21/(2009), Lund Institute of Technology, Department of Electrical and Information Technology, P.O. Box 118, S-211 00 Lund, Sweden, 2009. Submitted to Inverse Problems in Science and Engineering.

[10] E. Somersalo, M. Cheney, and D. Isaacson. Existence and uniqueness for electrode models for electric current computed tomography. *SIAM J. Appl. Math.*, **52**(4), 1023–1040, 1992.

[11] M. V. T. Rylander, P. Hashemzadeh. Reconstruction of metal protrusion on flat ground plane, 2009. Submitted to IET Radar, Sonar & Navigation.

[12] P. Vauhkonen, M. Vauhkonen, T. Savolainen, and J. P. Kaipio. Three-dimensional electrical impedance tomography based on the complete electrode model. *IEEE Trans. Biomed. Eng.*, **46**(9), 1150–1160, sep 1999.

[13] T. J. Yorkey, J. G. Webster, and W. J. Tompkins. Comparing reconstruction algorithms for electrical impedance tomography. *IEEE Trans. Biomed. Eng.*, **34**(11), 843–852, nov 1987.

An Investigation of the Impedance Properties of Gold Nanoparticles

Martina F. Callaghan, Torben Lund, Parham Hashemzadeh, Ivan M. Roitt, Richard H. Bayford

Department of Natural Sciences, Middlesex University, London, U.K.

M.Callaghan@mdx.ac.uk

Abstract. Over recent years there has been rapid growth in the research being carried out on nanoparticles. In the field of medical imaging, this interest has focussed primarily on the potential for drug delivery and using nanoparticles as a contrast agent, e.g. super-paramagnetic iron-oxide (SPIO) particles in MRI. More recently gold nanoparticles have been used in radiotherapy treatment of tumours to provide dose enhancement by increasing the efficacy of the radiation absorption. Nanoparticles coated with molecules such as glucose or cancer-specific antibodies can be directed towards specific cancer cells *in vivo*. Such targeting combined with the properties of nanoparticles shows great promise for localised therapy of tumours while leaving neighbouring healthy tissue unaffected. However, on the nanoparticle scale of sub-100nm the weighting of various factors and inter-atomic interactions which determine the bulk properties of a material changes. Many properties of the bulk material no longer hold. As such, each aspect of nanoparticle behaviour must be investigated afresh to explore the full extent of their potential. The property of nanoparticles we wish to explore and characterise is impedance. Bulk gold is well known to be highly conductive. If this were to remain the case on the nanoscale, it could be highly effective as a contrast agent for electrical impedance tomography, particularly when combined with tumour targeting.

1. Introduction

To explore the possibility of detecting gold nanoparticles via their impedance response we have carried out initial experiments to determine their conductivity using solutions of gold nanoparticles with diameter under 5nm and colloidal gold, which is comprised of much larger particles.

For use in cancer research, gold nanoparticles should preferably be less than 5nm in diameter. This is because nanoparticle diffusion across biological membranes, and consequently their distribution in the body, is size dependent. Nanoparticles less than 10nm will be able to migrate from the blood stream into tumours. Furthermore, nanoparticles less than 5nm can pass the blood-brain barrier, permeate blood vessels and be dialyzed in the kidney allowing unwanted particles to be flushed out of the body [1], [2]. This means that colloidal gold is too large to access tumours after intravenous injection or to be passed through the kidneys, potentially limiting its applicability in the treatment of cancer. It is included here to maximise sensitivity and explore any size-dependence of gold nanoparticle conductivity.

2. Methods

2.1. Gold Particle Solutions

Two gold nanoparticle solutions were used (Midatech Ltd, UK). The first was comprised of a 1.6nm diameter core of approximately 102 gold atoms. These gold nanoparticles (GNPs) are coated with approximately 43 glutathione (gsh) ligands such that the total gsh-GNP diameter approaches 5nm [3]. The second solution was comprised of larger citrate encapsulated colloidal gold particles.

2.2. Impedance Measurements

Impedance measurements were made using a Solartron 1260 impedance analyser (Advanced Measurement Technology, UK) across a frequency range of 1Hz to 1MHz with an applied voltage of 1V. Previous experiments (data not shown here) in which an aliquot of GNP solution was added to saline solutions of known impedance had proved insufficiently sensitive to detect any impedance change from the GNPs. Consequently, a test rig was designed and built in-house to hold a sufficiently low volume to examine the gold particle solutions directly without the need for any surrounding solution so as to maximize sensitivity to the impedance response of the nanoparticles.

2.2.1. *Test Rig Construction and Calibration*. A well to hold 200μl aliquots of solution was created by overlapping glass microscope slides. 0.5mm thick silver electrodes, across which the voltage was applied, were inserted at either end of this well. Contact was made at two points along the length of the well to facilitate a four-electrode impedance measurement [4]. Prior to investigating the unknown impedance properties of the various gold particle solutions, the rig was calibrated to ensure linearity between the measured impedance and the conductivity of commercially available known-conductivity solutions (Hanna Instruments, UK). Five solutions with conductivities ranging from 8.4mS/m to 8S/m were used in the calibration process. The solution temperature was recorded and the corresponding conductivity was determined from the solution's technical sheet. A linear fit was performed to determine the geometric factor for the rig. This was in turn used to estimate the absolute conductivity of the gold particle solutions.

2.3. Inductively Coupled Plasma Atomic Emission Spectroscopy (ICP-AES)

Upon completion of the impedance measurements, ICP-AES was carried out on the solutions to determine their gold atom concentrations and convert these to molar concentrations. The molar conductivity of the solution was determined by taking the ratio of the solution's estimated conductivity to its molar concentration.

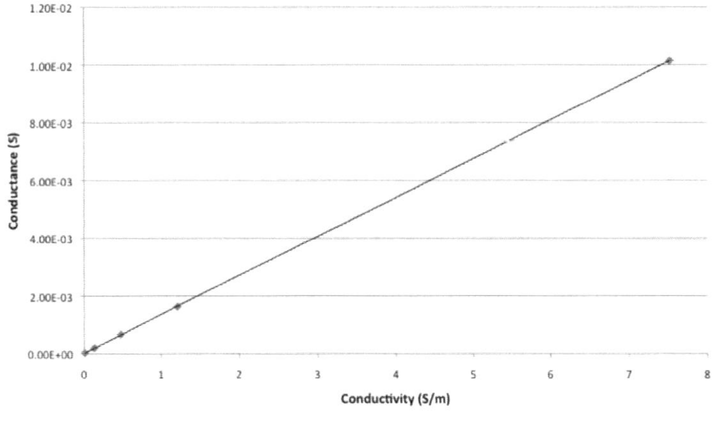

Figure 1. Calibration curve for the test rig used to measure the conductivity of the gold particle solutions. The measured conductance (blue points) was fitted to the conductivity, with a goodness of fit of 0.99999 (black line).

3. Results

The acquired impedance data was converted to conductance. Data above 100kHz was excluded to minimize any errors that could be introduced by the effects of stray capacitance.

Figure 1 shows the average conductance for each of the five calibration solutions versus the temperature-corrected conductivity (0.0079S/m, 0.1332S/m, 0.4711S/m, 1.215S/m and 7.52S/m at the measured temperature of 22°C) along with the linear trend line derived from fitting to these data.

Figure 2 shows the measured conductance of the gold nanoparticle solutions compared to that of highly pure deionised (DI) water and moderately pure DI water. The highly purified DI water is an aliquot of that used to manufacture the gold particle solutions and was therefore taken as a baseline with which to compare the increase in conductance caused by the gold particles. The glutathione coated GNPs show a 24-fold increase in conductance while the colloidal gold particles result in a 150-fold increase. The conductance measured for two of the known conductivity solutions are also presented to indicate the equivalent absolute conductivity of the gold nanoparticle solutions.

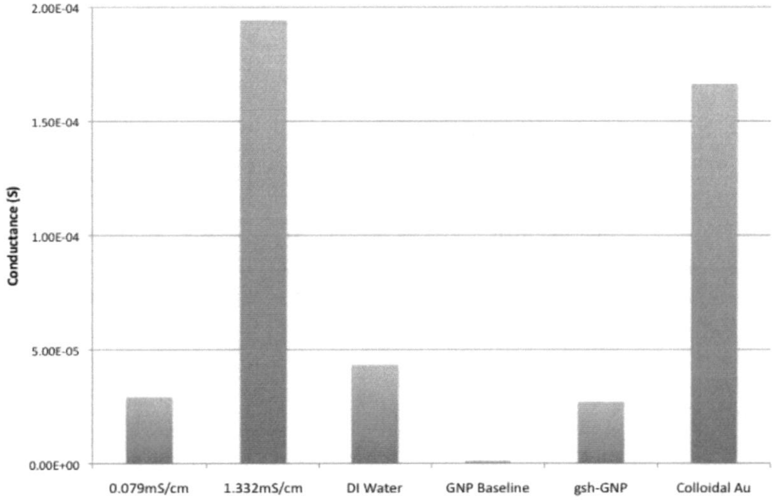

Figure 2. Comparison of the conductance between the gold particle solutions, DI water and two known conductivity solutions.

The molar concentration of gold atoms found using ICP-AES was 0.49mM for the colloidal gold solution and 11.78mM for the glutathione coated gold nanoparticle solution. The estimated molar conductivities, with respect to gold atom concentration, are 219mSm^2/mol for the colloidal gold and 1.49mSm^2/mol for the glutathione coated gold nanoparticles. These values are compared to those of published values for common electrolytic ions [5] in figure 3.

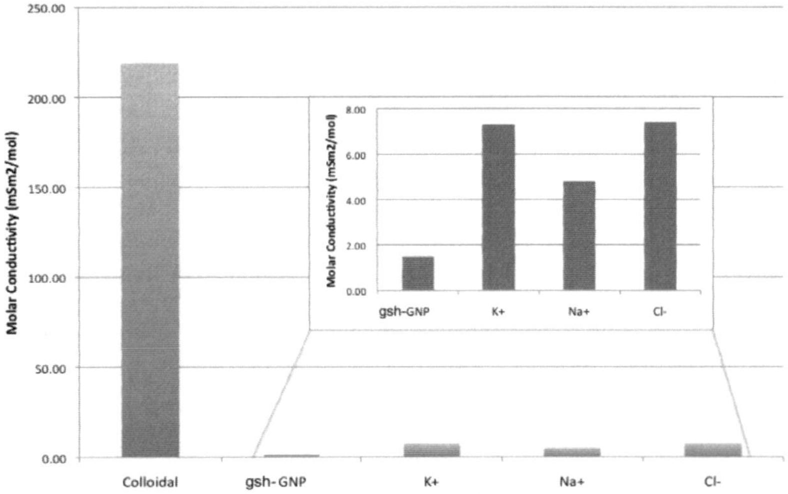

Figure 3. Comparison of the estimated molar conductivities of the gold particle solutions with the molar conductivities of common electrolytic ions found in the body.

4. Discussion

The typical concentration of GNP solutions is so small that measuring their effect on bulk impedance is at the detection limit for typical systems. The presence of other charge carriers, e.g. the ions in saline solutions or any residual impurities in deionised water samples, can dominate the impedance response and is therefore critical in determining whether or not there is sufficient sensitivity to detect the effect of the gold particles on bulk impedance. By using a rig designed specifically to look at small solution volumes (200μl) sufficient measurement sensitivity was achieved to investigate the conductivity of the gold nanoparticles. Although in absolute terms the conductance of these solutions is very low, the presence of the gold nanoparticles did result in significant amplification of the conductance with respect to baseline. The low absolute conductance is a reflection of the low concentration of gold nanoparticles in the solutions. However, the molar conductivities of the gold atoms in the solutions compare favourably with those of typical electrolytic ions found *in vivo* and is in fact many times greater in the case of the colloidal gold particles.

5. Future

In future we wish to explore the mechanisms behind the conductivity of the gold nanoparticles as well as their effect on bio-impedance within a cellular environment. Key factors in determining the impact of gold nanoparticles on bio-impedance *in vitro* and *in vivo*, where the number of mobile ions will be large, will be the amount of cellular uptake and the resultant concentration of gold nanoparticles that can be achieved. Their effect may well also depend on what ligand is used to coat the GNP, e.g. charged or uncharged, and on which cellular compartment takes up the gold nanoparticles, e.g. whether they are internalized or remain on the cell surface. Understanding these mechanisms may allow us to exploit properties of the gold nanoparticles in cancer detection and/or treatment.

References

[1] H.S. Choi, W. Liu, P. Misra, E. Tanaka, J.P. Zimmer, B. Itty Ipe, M.G. Bawendi, and J.V. Frangioni, "Renal clearance of quantum dots.," *Nature biotechnology*, vol. 25, 2007, pp. 1165-70.

[2] J.F. Hainfeld, D.N. Slatkin, and H.M. Smilowitz, "The use of gold nanoparticles to enhance radiotherapy in mice," *Physics in Medicine and Biology*, vol. 49, 2004, pp. N309-N315.

[3] P.D. Jadzinsky, G. Calero, C.J. Ackerson, D.A. Bushnell, and R.D. Kornberg, "Structure of a thiol monolayer-protected gold nanoparticle at 1.1 A resolution.," *Science (New York, N.Y.)*, vol. 318, 2007, pp. 430-3.

[4] K. Lee, Y. Kim, T. Oh, and E. Woo, "Complex Conductivity Spectra of Seven Materials and Phantom Design for EIT," *IFMBE PROCEEDINGS*, Springer, 2007, p. 344.

[5] R. Pethig, "Dielectric properties of biological materials: biophysical and medical applications," *IEEE Transactions on Electrical Insulation*, 1984, p. 453–474.

Weighted frequency-difference EIT measurement of hemisphere phantom

Sujin Ahn[1], Tong In Oh[3], Sung Chan Jun[1], Jeehyun Lee[2], Jin Keun Seo[2], and Eung Je Woo[3]
[1]Department of Information & Communications, Gwangju Institute of Science and Technology, Gwangju, Korea
[2]Department of Mathematics, Yonsei University, Seoul, Korea
[3]Department of Biomedical Engineering, Kyung Hee University, Yongin, Korea

E-mail: ejwoo@khu.ac.kr

Abstract. We have proposed a new frequency difference method using a weighted voltage difference (WFD-EIT) between two frequencies [1, 2]. Previous studies demonstrated its feasibility through numerical experiments and two-dimensional phantom experiments. In this study, we validate the WFD-EIT algorithm on a three-dimensional hemisphere phantom using a multi-frequency EIT system KHU Mark1. We built the hemisphere phantom with 17 stainless-steel electrodes on its inner surface. We filled the phantom with a biological material having a frequency-dependent admittivity such as carrot pieces mixed in saline. Using boundary voltage data from the deformed phantom, we reconstructed weighted frequency difference images on the computational model domain with a hemisphere shape. We discuss comparative reconstruction performance results including time difference (TD), simple frequency difference (FD), and weighted frequency difference (WFD). Animal and human head imaging experiments with the weighted frequency-difference EIT method are under investigation.

1. Introduction

The weighted frequency-difference (WFD) EIT algorithm is promising since it may extract a useful information related to an anomaly in a background with a frequency-dependent admittivity. We expect that it is advantageous to overcome common errors and artefacts in frequency-difference EIT imaging [1, 2]. There always occurs some mismatch between experimental and computational domains. In performing human experiments, boundary movements are caused by motion and breathing. Seo *et al* showed that the WFD method is robust against boundary geometry error using numerical simulations [1]. In this work, we investigate how the WFD algorithm deals with a deformation of the boundary shape. We present results of numerical simulations and imaging experiments using a deformed hemisphere as an imaging object.

2. Experimental design and methods

We prepared a hemispherical phantom filled with either saline or carrot pieces suspended in saline. As an anomaly, we used a cylindrical piece of banana and potato. The diameter of each anomaly was 20% of the diameter of the phantom. Images were reconstructed using time-difference (TD), simple frequency-difference (FD) and weighted frequency difference (WFD) methods from simulation data at 1, 5, 10, 50 and 100 kHz. We used a single-step algorithm based on a sensitivity matrix and the truncated singular value decomposition (TSVD). For the weighted frequency-difference method, the reference frequency of $\omega_0/2\pi$ = 1 kHz was used to estimate an equivalent homogeneous admittivity. We acquired experimental data from the hemisphere phantom using the KHU Mark1 multi-frequency EIT system operating from 1 kHz to 500 kHz [3].

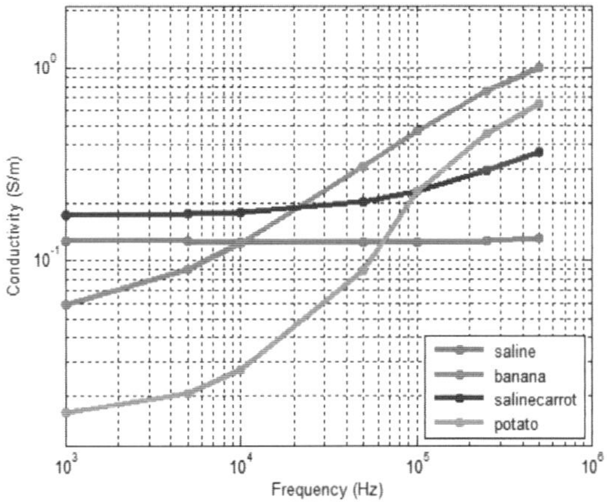

Figure 1. Conductivity spectra of saline, mixture of carrot pieces with saline, potato and banana

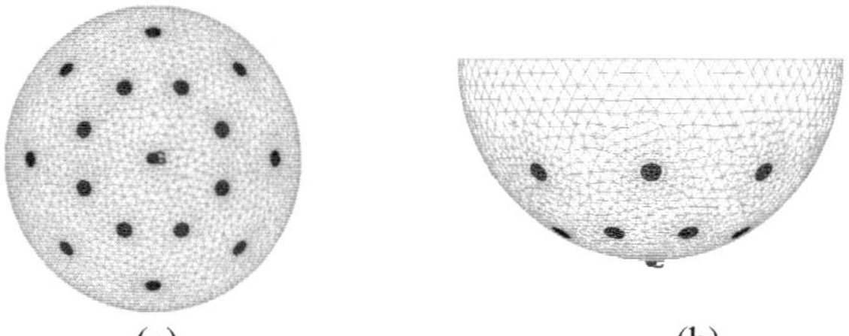

(a) (b)

Figure 2. Computational hemisphere model : (a) *xy*-view and (b) *yz*-view.

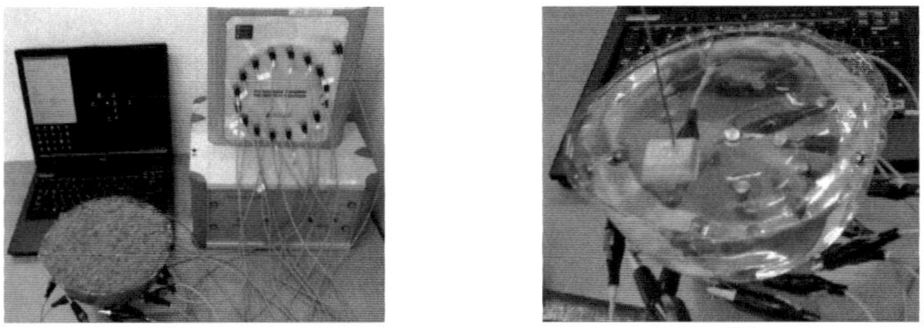

(a) (b)

Figure 3. (a) Hemisphere phantom filled with carrot pieces in saline and the KHU Mark1 EIT system. (b) Same phantom filled with saline and banana anomaly

2.1 Direct impedance measurement

We measured conductivity spectra of 5% saline, mixture of carrot pieces with 0.1% saline, potato, and banana using a BIS system from 1 kHz to 500 kHz. Figure 1 shows their conductivity spectra. Conductivity values of banana increased with frequency and crossed that of saline at around 10 kHz, and conductivity of potato crossed that of carrot pieces in saline at around 100 kHz.

2.2 Tank preparation

There were 17 electrodes located on the surface of a hemisphere model $\Omega=\{(x, y, z)|x^2+y^2+z^2\leq 9^2, z\leq 0\}$. The eight electrodes per layer were equally spaced around its circumference as shown in figure 2(a). One ground electrode was located on the apex. In numerical simulations, we generated simulated data on an elongated hemisphere in the x-direction as $\Omega_1=\{(x, y, z)|(x/1.1)^2+(y/0.9)^2+z^2\leq 9^2, z\leq 0\}$ with the same electrode configuration as Ω. We reconstructed images in the original hemispherical model Ω to see effects of modelling error. For the phantom experiments, we also used a deformed phantom Ω_1 to measure the boundary voltage data, then reconstructed images using the hemisphere model Ω. We filled the phantom with either 5% saline or a mixture of carrot pieces (60%) and 0.1% saline (40%). We placed an anomaly of cylindrically-shaped banana or potato at the position of (-5.0, 0) cm in Ω_1.

3. Results

3.1. Numerical simulations

We could observe that geometrical modelling error affects the appearance of the anomaly in reconstructed images. Compared with the FD method, the WFD method showed better performance with reduced artefact in reconstructed images (Figure 4).

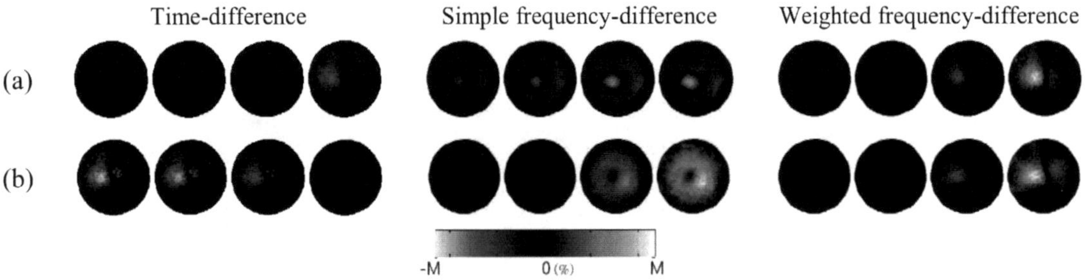

Figure 4. Reconstructed images by numerical simulations. We reconstructed four images for each case at 5, 10, 50, and 100 kHz. (a) The banana anomaly in the saline background (b) The potato anomaly in the mixture of carrot pieces and saline as the background. The pixel values mean percentile changes of conductivity.

3.2. Phantom experiments

Figure 5 shows results of phantom experiments. In the saline background, both FD and WFD methods produced comparable images. However, the location of the anomaly using the FD method was slightly shifted toward the center. The anomaly position in images by using the WFD method was closer to the true position. When the background was the mixture of carrot pieces with saline, the WFD method yielded a visible anomaly conductivity contrast while the FD method failed to locate the anomaly. Images of the TD method distinguished the anomaly well for all cases.

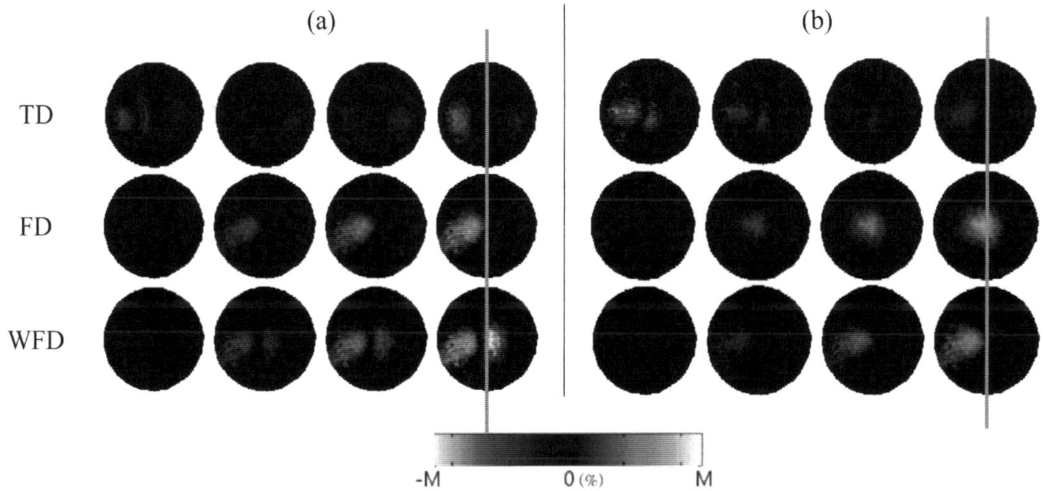

Figure 5. Reconstructed images from phantom experiments. We reconstructed four images at 5, 10, 50, and 100 kHz for each case. (a) The banana anomaly in the saline background. (b) The potato anomaly in the mixture of carrot pieces and saline background. The green vertical line denotes the y-axis at (0, 0).

4. Conclusion

We demonstrated how boundary geometry errors between the imaging object and a model domain affect image reconstructions for three algorithms including the time difference, simple frequency difference and weighted frequency difference. Compared with the simple frequency-difference method, the weighted frequency-difference method yields better images showing its robustness against modelling errors.

5. Acknowledgement

This work was supported by the SRC/ERC program of MOST/KOSEF (R11-2002-103), J K Seo and J Lee were supported by the WCU program (R31-2008-000-10049-0), and S C Jun was supported by NRF grant (2009-0071225) and the BioImaging Research Center at GIST.

6. References

[1] J K Seo et al., Frequency-difference electrical impedance tomography (fdEIT): algorithm development and feasibility study, Physiol. Meas. 29 (2008) 929–944.
[2] S C Jun et al., Frequency-difference EIT (fdEIT) using weighted difference and equivalent homogeneous admittivity: validation by simulation and tank experiment. Physiol. Meas. 30 (2009) 1087–1099.
[3] T I Oh et al., Multi-frequency EIT system with radially symmetric architecture: KHU Mark1, Physiol. Meas. 28. (2007) S 183-S196

Abdominal fat thickness measurement using Focused Impedance Method (FIM) - phantom study

Salahuddin Haowlader, Tanveer Noor Baig[1] and K Siddique-e Rabbani
Department of Biomedical Physics & Technology, University of Dhaka, Dhaka 1000, Bangladesh
1: Department of Physics, University of Dhaka, Dhaka

email: rabbani@univdhaka.edu

Abstract. Abdominal fat thickness is a risk indicator of heart diseases, diabetes, etc., and its measurement is therefore important from the point of view of preventive care. Tetrapolar electrical impedance measurements (TPIM) could offer a simple and low cost alternative for such measurement compared to conventional techniques using CT scan and MRI, and has been tried by different groups. Focused Impedance Method (FIM) appears attractive as it can give localised information. An intuitive physical model was developed and experimental work was performed on a phantom designed to simulate abdominal subcutaneous fat layer in a body. TPIM measurements were performed with varying electrode separations. For small separations of current and potential electrodes, the measured impedance changed little, but started to decrease sharply beyond a certain separation, eventually diminishing gradually to negligible values. The finding could be explained using the intuitive physical model and gives an important practical information. TPIM and FIM may be useful for measurement of SFL thickness only if the electrode separations are within a certain specific range, and will fail to give reliable results if beyond this range. Further work, both analytical and experimental, are needed to establish this technique on a sound footing.

1. Introduction

Abdominal fat consisting of both subcutaneous (under the skin) and visceral (around deeper organs) fat have been found to have a bearing on health conditions, and are being used as risk indicators, particularly for Cardiovascular disease and Diabetes [1][2]. Existing reliable and direct methods for measuring fat thickness are MRI and CT scans, which are expensive, and are not available widely. Therefore there is a necessity of a direct measurement method that is reliable and widely usable.

Electrical impedance techniques offer a possible solution and there has been several efforts in this direction [3]. Some of these measure the impedance of the whole thorax while some use localised tetrapolar impedance measurements (TPIM) at the abdomen. The latter is particularly suitable for the subcutaneous fat layer (SFL) and this paper also concentrates on this measurement. However, even a TPIM system has a wide zone of sensitivity and it seemed logical to try the application of a recently developed technique known as Focused Impedance Method (FIM), innovated by our group at Dhaka University[4][5], which may provide an improved performance. One version, known as the 6-electrode FIM, is to be used for this study, which essentially is two orthogonal TPIM systems placed centrally over a common zone, which is the resulting focused zone of interest. FIM therefore shares many inherent features of TPIM. Therefore, the main focus of the present work was in developing some insight and understanding of TPIM towards measurement of abdominal fat, particularly of the

SFL. The distance between various electrodes in TPIM may have a bearing on the measured value of SFL thickness, therefore it seemed justified to study this aspect first, using TPIM in a phantom. The target is to get an optimum electrode separation for TPIM which could then be applied to FIM. As the distance between the current drive electrodes in a TPIM increase, the current density distribution spreads out and it is expected that the depth sensitivity will increase too. Therefore, an objective of this work was also to see if any new technique can be found out for measuring SFL thickness by changing the electrode separations.

2. Methods and Models

Schematic of the phantom designed and developed for this work is shown in figure 1. Electrodes were fixed into a plastic (polystyrene) foam piece floating on saline in a square tank made of transparent acrylic (PERSPEX) measuring 50cm x 50cm in area having a saline depth of 50cm as well. The lower

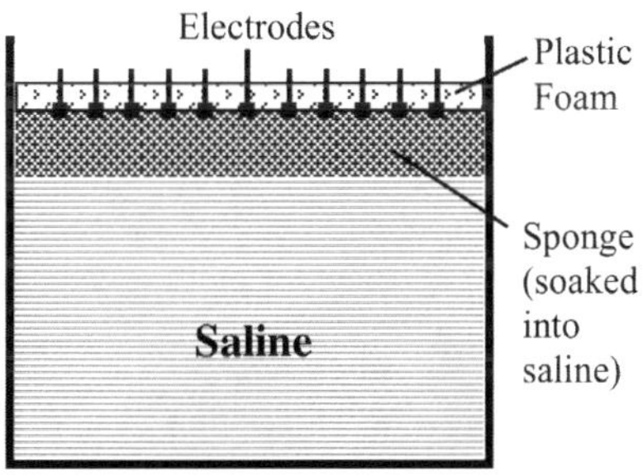

Figure 1. Schematic of phantom

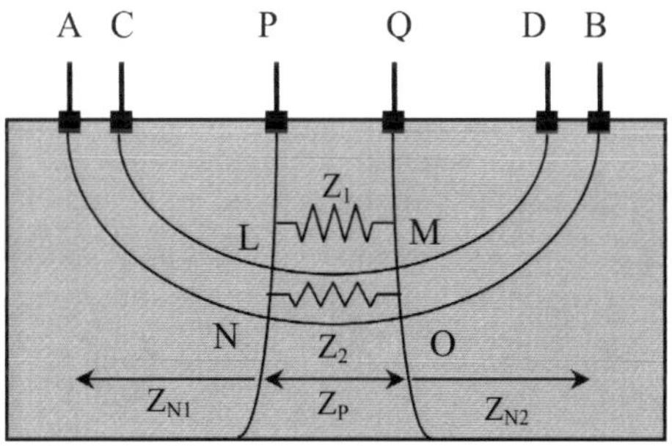

Figure 2. Model to show sensitive zone and variation of impedance with S-S separation

side of the electrodes touched the saline while connections were taken out from the top. To simulate subcutaneous fat layer, one or more layers of plastic sponge were immersed into the saline at the top and then pushed down using the polystyrene foam piece, with appropriate weights placed on the top. The electrode at the centre served as the common terminal for a differential potential measuring arrangement. Symmetrically from the centre, two proximal electrodes were chosen for potential measurement, and two outer ones for current drive (source), to configure a TPIM arrangement. This gave a number of source to source (S-S) electrode and potential to potential (P-P) electrode separations. Current at 10 kHz was used for the measurement, and both current and voltage amplitudes were measured to provide an impedance value. Measurements were carried out for different P-P and S-S separations and for different thickness of the plastic sponge layer. To vary the thickness of the sponge layer, additional sponge pieces were introduced by taking the polystyrene foam sheet out and replacing it again.

The effect of increasing the S-S separation is discussed using a simple intuitive model as shown in figure 2. Suppose the P-P separation (PQ) is kept fixed, while the S-S separation is changed (CD to AB). Because of the spread of current lines the depth sensitivity is expected to be higher for the wider current drive AB than for drive CD, as schematically represented by the approximate semicircular current lines. Therefore the impedance measured with drive CD is that of the region PLMQ with current driven perpendicular to PL and QM, denoted by Z_1. For current drive AB, the impedance Z_P measured is that of the region PNOQ with current driven perpendicular to PN and QO, which can be modelled as a parallel combination of Z_1 and Z_2, where Z_2 is the corresponding impedance of the region LMON, again perpendicular to the current direction. Therefore the value of this combined impedance would be less than Z_1. For a volume conductor the incremental sensitive volume is large and is expected to contribute to a significant reduction in the measured impedance. Thus with increased S-S separation the measured impedance is expected to decrease.

However, in the above description, the negative effect of the impedance ($Z_{N1}+Z_{N2}$) of the zones between the current and potential electrodes has not been considered. In TPIM measurements a significant negative sensitivity has been observed in these zones [4], and needs to be considered in the present case. If an insulator is placed in the central zone within the saline, between electrodes P & Q in figure 2, the bending of equipotentials will give rise to a higher potential difference between P & Q, resulting in increased measured impedance, compared to the background measured impedance due to saline only. On the other hand if an insulator is placed between the source electrode C and the potential electrode P, then the resulting measured impedance will decrease because of reversed bending of equipotential lines. This gives rise to the negative sensitivity. In the present situation, either for the saline alone, or for sponge layers soaked in saline at the top, the impedance is distributed throughout the region of the electrodes. Therefore, the impedance $Z_{N1}+Z_{N2}$ contributes to a reduction in the measured impedance as $Z = Z_P - (Z_{N1}+Z_{N2})$. As can be visualised from figure 2, for an increase in the source to potential electrode separation (S-P separation), the impedance $Z_{N1}+Z_{N2}$ is likely to increase, while due to an increase in the depth sensitivity, this is likely to decrease. So the combined effect would be complex and is likely to vary for different S-P separations. The results of the experiments performed on the phantom are given in the next section.

3. Results and Observations

Figure 3 shows the impedance values as a function of S-S separation for fixed P-P separations of 2 cm, and for two different thickness of the sponge layer (3.5cm and 7.5cm) and one with no sponge (or zero sponge thickness). It can be seen that the impedance increases with increasing thickness of sponge, which is reasonable to expect, and which has been used by other workers [3]. However, the steep slope of fall of the impedance with increasing S-S separation means that small changes in electrode positioning could lead to significant changes in the impedance values. This naturally suggests that TPIM measurement will give different impedance values for the same abdominal SFL thickness for slightly different P-P and S-S separations, and therefore give unreliable results in practice. However, when the phantom experiment was performed for increased values of P-P separation, a horizontal segment began appearing in the beginning of the curves. Figure 4 shows the results for a P-P separation of 4 cm. At this value a horizontal segment was noticeable even without any sponge layer at the top, while larger horizontal segments were visible for increasing sponge layer thicknesses. This new finding seems to provide a usable range within which a small variation in electrode placement may not change the measured impedance value significantly, and may therefore provide a reliable measurement favouring the use of a TPIM system for abdominal SFL thickness.

The reason for the generation of the horizontal segment is not very straightforward and a possible explanation is provided in the next section. However, the above results suggest that the ratio of S-S to P-P separation is to be kept within a certain range if TPIM or FIM are to be useful for measuring SFL

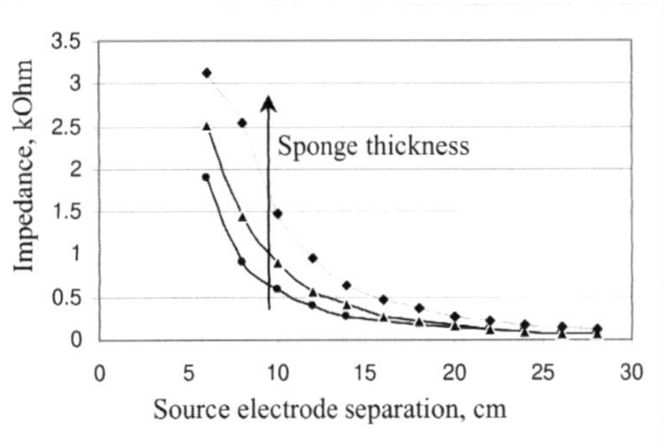

Figure 3. Variation of impedance with S-S separation for a P-P separation of 2 cm. Sponge thickness: 0 (no sponge), 3.5cm, 7.5cm

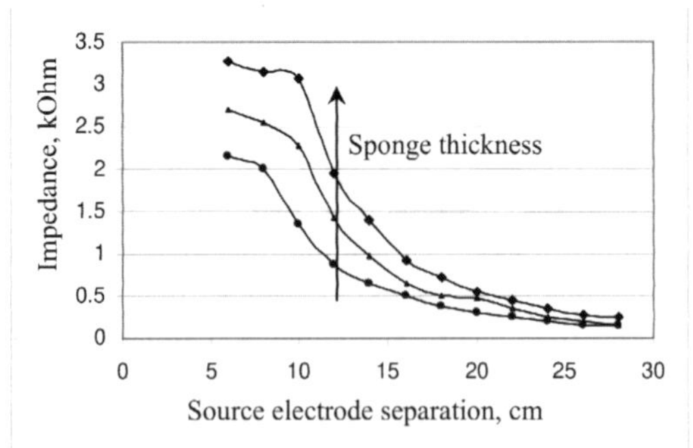

Figure 3. Variation of impedance with S-S separation for a P-P separation of 4 cm. Sponge thickness: 0 (no sponge), 3.5cm, 7.5cm

thickness. From figure 4 it appears that for the phantom used, a maximum S-S separation of 8 cm may be used, corresponding to a maximum S-S/P-P ratio of 2. Of course, for abdominal fat measurement, a proper figure of this ratio has to be obtained through further studies on more realistic phantoms having close values of impedances and for practical ranges of SFL thickness.

Another point worth noting are the values of S-S to P-P ratio at the knees of the curves for the different sponge thicknesses in figure 4, which seem to be increasing with increasing sponge thicknesses. Therefore, such values at knee may be utilised to extract a value of the SFL thickness, however, further studies would be necessary to make it a practicable technique.

4. Discussions

This study gives an important insight into the use of TPIM, and of FIM for studies of abdominal SFL thickness. Since the SFL is spread in three dimensions, it contributes to the measured impedance in a complex way, having both positive sensitivity in the central region between the potential electrodes, and negative sensitivity outside. This work has presented an intuitive physical model based on current density distribution and perturbation of equipotential surfaces, using which a qualitative prediction was made for decreasing impedance with increasing S-S separation. The prediction was borne out by measurements on a phantom having saline and saline soaked plastic sponge layers. However, the measurements produced a new finding, almost unchanging impedance at low values of S-S separation. This was beneficial and important since this gives the possibility of absolute impedance measurement in order to obtain SFL thickness reliably by choosing appropriate P-P and S-S separations. This observation may be explained, again using the negative and positive sensitivity regions mentioned above. At low values of S-P separation the impedance $Z_{N1}+Z_{N2}$ to be subtracted is small in comparison to Z_P, however, as S-P separation increases, $Z_{N1}+Z_{N2}$ is no longer insignificant, and eventually dominates resulting in the very small measured impedance. This phenomenon is not frequently discussed as mostly the target is considered to be localised within the central sensitive volume. However, this negative sensitivity is present in any TPIM measurement, and artefacts with negative sensitivity observed in EIT images are in fact due to this phenomenon. Further work, including an analytical treatment is needed in this direction which will establish both TPIM and FIM techniques on a sound footing.

5. References

[1] Wiklund P, Toss F, Weinehall L et al, 2008 *J of Clin Endocrinology & Metabolism* **93**, 4360-4366
[2] International Association for the Study of Obesity (IASO), 2004 http://www.idf.org/webdata/docs/Diabetes_meta_syndrome.pdf
[3] Scharfetter H, Schlager T, Stollberger R et al, *Int J Obes Relat Metab Disord.* 2001 **25 (4)**:502-11
[4] Rabbani KS, Sarker M, Akond MHR and Akter T 1999 *"Electrical Bioimpedance methods", Annals of the New York Academy of Sciences,* **873** 408 - 20
[5] Rabbani KS, and Karal MAS 2008 *Annals of Biomedical Engg.* **36** 1072-77

A Modified Electrode Configuration for Brain EIT

P. K. Manwaring, R. J. Halter, A. Borsic, A. Hartov

Thayer School of Engineering, Dartmouth College, Hanover, NH 03755, USA

E-mail: preston.k.manwaring@dartmouth.edu

Abstract. Electrical impedance tomography (EIT) of the brain holds great promise for long term non-ionizing detection and imaging of blood flow, ischemia, stroke, and even neuronal activity. One of the most difficult challenges with this modality, however, is overcoming the high impedance of the skull, which severely limits current passage through the intracranial space and "washes out" the tissue property images. There are situations, however, in which invasive electrode configurations are appropriate to overcome this limitation. We propose the use of a central and circumferential-electrode configuration to improve detection and localization of edema, hemorrhage, and ischemia within the cranium. Results from a simulation study and a phantom experiment verifying the simulation are shown.

1. Introduction

Electrical Impedance Tomography is an enticing and safe candidate method to image the human body over an extended period of time. However, the method is diffuse by nature as current prefers to follow the most conductive paths (i.e. not necessarily a straight line). This is especially troublesome if one wants to pass current across the cranium for imaging or detecting blood flow, ischemia, stroke, and even neuronal activity within the head. The skin and muscle surrounding the cranium are very conductive. The cranium, however, is insulative. No matter the electrode configuration on the periphery, most current will pass through the outer tissue rather than take the preferable path through the skull and brain. This leaves a less sensitive center region within the image [1, 2, 3].

To overcome the current passage limitation, one can place electrodes invasively within the cranium as demonstrated by Holder *et al* in his evoked response experiments [4]. Most research shies away from this solution as implanting electrodes in the brain is inherently risky and generally seen as undesirable. There are a few situations, however, in which a centrally-located electrode is an obvious and even desirable choice. One such application includes treatment of traumatic brain injury in which catheters and intracranial pressure (ICP) sensors will be inserted into the brain parenchyma.

It is not difficult to conceive of an ICP sensor or catheter-electrode combination, which could be used to augment present brain-EIT setups. Indeed there are several FDA-approved predicate devices on the market now, two of which include the deep-brain stimulators from Medtronic and impedance-based pressure-volume sensors from Millar. The addition of such an electrode would allow new excitation patterns in both active (current sink/source) and passive sensing modes. In this way one could guarantee more current passage through the cranium to detect worsening injury or even neuronal function by blood flow imaging [4]. This paper will demonstrate our

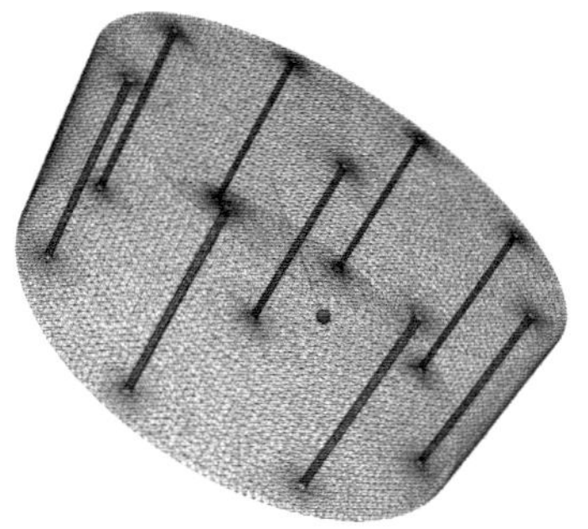

(a)

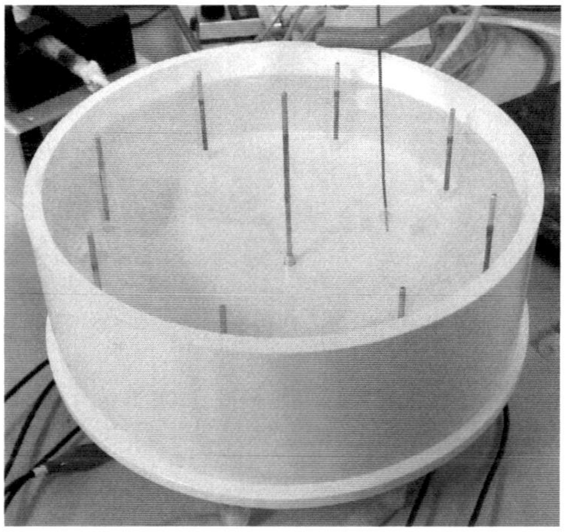

(b)

Figure 1. Image (a) shows the surface of the 159,000 node forward problem mesh. The red sphere in the image is the 100 uL inclusion. Image (b) shows the saline tank setup. Nine electrodes are visible with 8 on the periphery and one in the center. The red catheter has a latex balloon at its tip, which is inflated with 100 uL of fluid.

preliminary look into the use of a centrally-located electrode for brain-based EIT applications through simulation evidence and bench top verification.

2. Methods

2.1. Saline Tank Simulation

The purpose of this simulation is to determine how much a central electrode improves reconstruction of a small 100 uL (roughly one drop of fluid) inclusion within a larger saline $\tilde{2}$ liter tank . A 3-dimensional model was created in SolidWorks (Dassault Systmes SolidWorks Corp., Concord, MA.) with a cylinder of diameter 20.3 cm and a depth of 7.6 cm. Eight 3 mm diameter x 7.6 cm high cylindrical electrodes were placed equidistant from one another 8.9 cm radially from the center. A ninth electrode of the same characteristics is placed in the center of the cylinder. Two meshes are created. The first is a 184,000 node homogeneous mesh for the inverse problem. The second is a 159,000-node mesh with well-defined concentric spherical inclusions of differing volumes centered at 3.2 cm radially outward from the center and 3.2 cm from the bottom (see Fig. 1(a)). These are used for the forward problem. The background conductivity is 0.28 S/m (roughly that of brain), and a 100 uL spherical inclusion is set at 0.80 S/m (roughly that of blood) [5]. The forward solution is solved using tetra-polar excitation patterns for all possible combinations of electrodes. Noise is added to the forward problem solution at 1%, 2%, and 5% to simulate systematic and random errors according to the equation

$$\nu = \nu \frac{\text{std}(d)}{\text{std}(\nu)}$$

where ν is a vector of Gaussian distributed random numbers scaled by the standard deviation of the difference $d = Z_1 - Z_0$. Z_0 and Z_1 are vectors of the electrode impedance measurements for the system with all excitation patterns from the homogeneous and inclusion meshes respectively. Comparisons are made between solutions with and without the central electrode participating. All solutions are computed using NDRM, Dartmouth's custom finite-element software [6].

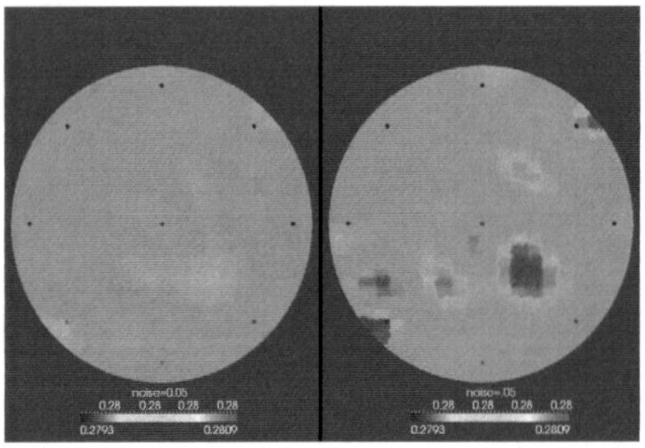

Figure 2. The image on the left shows the results of a simulation using only 8 peripheral electrodes. The 100 uL "blood-like" inclusion at 5 o'clock is washed out and ill-defined. The right-hand image uses the central electrode. Even in the presence of 5% noise, the object is easily identified. Both images are on the same scale. Tikhonov regularization of 1e-4 was used.

2.2. Saline Tank Verification

A saline tank (after which the finite-element simulation is modeled) was constructed to determine the sensitivity a new USB-based EIT system developed at Dartmouth. The system provides the same tetra-polar excitation patterns described above. It is capable of DC to 100kHz excitation frequencies. The drive current is hard-coded to be in line with the IEC 60601-1 guidelines. This experiment uses a frequency of 50 kHz and a current drive of 5 mA. The background conductivity of the saline solution is about 0.1 S/m. A Fogarty catheter (small 1.3 mm diameter catheter with a latex balloon at its tip) is placed into the tank mid-water level at the same location described for the inclusion in the simulation section. It is then inflated with 100 uL of fluid. Figure 1(b) shows the experimental setup. EIT frames are acquired with and without the central electrode at 9 frames per second.

3. Results

3.1. Saline Tank Simulation

The small volume and biological contrast for this simulation were chosen to give us an idea of whether our device is sensitive enough to detect blood flow. The L2-norm was calculated from the difference between the baseline impedance data from the homogeneous background forward problem and that of the forward problem from the inclusion. This yielded 0.15 ohms with the central electrode and 0.13 ohms without. The peak impedance for the homogeneous tank is 100 ohms.

Typical signal-to-noise ratios for this new system ranges between 85 and 95 dB depending upon the current, frequency, and pre-amp scaling. The worst-case scenario would be to use the 85 dB value. This is equivalent to a noise floor at 56 uV. This value is used as a threshold for analysis. Finding all values from a singular-value decomposition of the reconstruction Jacobian greater than this threshold yields 36 values for the central electrode and 28 without. This is an improvement of about 28% over the standard circumferential ring topology. The effects can be seen in Fig. 2 with 5% noise.

3.2. Saline Tank Verification

The difference images are presently reconstructed in 2-dimensions in real-time with minimal averaging. The effect of the Fogarty catheter inflated with 100 uL of fluid is seen in Fig. 3(a) imaged without a central electrode and in Fig. 3(b) with a central electrode. On the same color scale, a peripheral-only electrode topology shows the inclusion with less intense values,

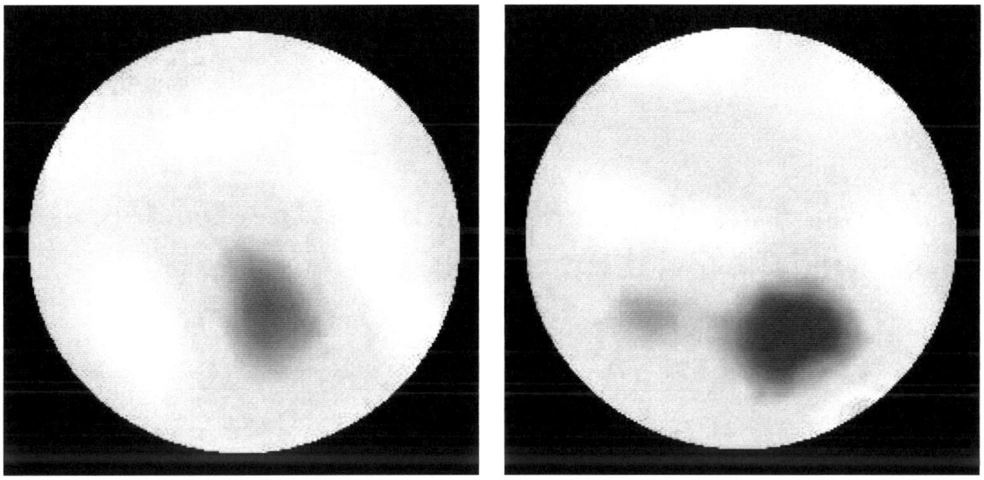

(a) (b)

Figure 3. Image (a) shows the 100 uL balloon reconstructed in real-time without using a central electrode. Image (b) shows the same balloon while using the central electrode.

deformed shape (oblong towards the center), and slightly skewed position. On the other hand, the inclusion is readily seen as a tighter region correctly positioned midway between the center and periphery using the central electrode.

4. Conclusion and Future Work

EIT has the potential to be used as a long-term monitor of status in patients with potentially life-threating brain injuries. We put forth an argument for placement of a centrally located-electrode which is in-line with today's clinical standard of care for such patients. We demonstrated that such an electrode improves detectability of small inclusions in both simulation and a saline tank phantom. *In vivo* studies of 1 month-old piglets have begun using a central electrode in a real-time EIT system. Analysis and comparison of this EIT data to CT/MRI data and other physiologic monitors will continue in an effort determine the efficacy of this modality.

Bibliography

[1] Tidswell A T, Gibson A, Bayford R H and Holder D S 2001 *Physiological Measurement* **22** 167–176
[2] Bayford R H, Boone K G, Hanquan Y and Holder D S 1996 *Physiological Measurement* **17 Suppl 4A** A49–57
[3] Sadleir R J, Tang T, Tucker A, Borum P and Weiss M 2009 *Proceedings of the 31st Annual International Conference of the IEEE EMBS* (Minneapolis, MN, USA: IEEE) pp 3169–3172
[4] Holder D S, Rao A and Hanquan Y 1996 *Physiological Measurement* **17 Suppl 4A** A179–86 journal Article
[5] Latikka J, Kuurne T and Eskola H 2001 *Physics in Medicine and Biology* **46** 1611–1616
[6] Borsic A, Hartov A, Paulsen K D and Manwaring P 2008 *Engineering in Medicine and Biology Society, 2008. EMBS 2008. 30th Annual International Conference of the IEEE* pp 1175–1177

Journal of Physics: Conference Series **224** (2010) 012063 doi:10.1088/1742-6596/224/1/012063

In-Vivo Detection of Bleeding Simulated in a Peritoneal Dialysis Model using a Hemiarray EIT Configuration

Aaron S Tucker[1], Edward Ross[2], Jennifer Paugh-Miller[2] and Rosalind J Sadleir[1]

[1] J. Crayton Pruitt Family Department of Biomedical Engineering, University of Florida, Gainesville FL, 32611, USA
[2] Department of Medicine, Division of Nephrology and Hypertension, University of Florida, Gainesville FL, 32611, USA

E-mail: astuck@ufl.edu

Abstract. A new method to image and quantify intra-abdominal hemorrhage (IAH) using electrical impedance tomography (EIT) was tested. We proposed an electrode hemi-array placed exclusively on the anterior abdomen to monitor a supine patient for IAH without complicating concurrent traumatic injuries. Peritoneal dialysis (PD) was used as a model for IAH to assess the *in vivo* performance of the hemi-array. Using the EPack 2 data acquisition system, EIT measurements were recorded before, during, and after the administration of dialysate. Simultaneously, the amount of dialysate was recorded, synchronous to the EIT measurements, to be used as a control. Tomographic images of impedance change were reconstructed using a weighted, sensitivity-based method and then post-processed to obtain a quantitative estimate of the total dialysate administered. Our preliminary study included two subjects, one male and one female, each participating for two sessions, spaced about six months apart. Data collected from these sessions indicate that with a realized *in vivo* SNR of about 35dB the EPack 2 can detect accumulations larger than about 100mL with a quantification uncertainty of about +/- 50mL. Using these data we have developed new algorithms that can automatically detect the onset of bleeding in less than two minutes. This method shows promise for automated detection of other pathologies, e.g. ascites, and is adaptable for use on other anatomy including the skull, and pelvis.

1. Introduction

Intra-abdominal hemorrhage (IAH) is a life-threatening consequence of bomb blasts, motor vehicle accidents, building collapses and falls. Victims with mild IAH sometimes benefit from nonintervention and heal naturally. However, as total blood loss approaches one liter hypovolemic shock, ischemia and death are inevitable without surgical intervention. Rapid assessment of the amount of bleeding is therefore important.

The noninvasive modalities conventionally used to diagnose IAH are sonography [1] and computed tomography (CT). The former is rapid and somewhat portable and the latter has high sensitivity and specificity. Neither is well suited for continuous monitoring, and the accuracy of both methods depends on a technician well-trained in device operation and interpretation of the results.

Impedance-based methods detect IAH by exploiting the electric conductivity contrast of blood (0.67 S/m) and mean abdominal tissue (0.2 S/m). Tomographic images of dynamic bleeding are reconstructed using electrical impedance tomography (EIT), and post-processed to extract volumetric estimates of the dynamic blood accumulation. This method is autonomous, continuous and does not require subjective operator interpretation.

Abdominal EIT has typically been recorded using 16 electrodes spaced equidistantly around the subject's girth [2]. Sixteen-electrode EIT studies in both porcine [3] and human [4] models of IAH have concluded that bleeding rates exceeding 100mL/min were reliably detectable. But IAH is often accompanied by other traumatic injuries which limit responders' ability to move victims. Accordingly, it is risky to place electrodes around the abdominal circumference without first evaluating concurrent injuries. Instead, in this study electrodes were applied only to the anterior abdomen accessible without moving a supine patient. The performance of the eight electrode hemiarray configuration has been tested *in vitro* [5]; herein we present preliminary *in vivo* results using this configuration.

Peritoneal dialysis (PD) is a useful model for simulating intra-abdominal hemorrhage. PD is an invasive procedure used to clear waste in patients with renal failure [6]. Dialysate solution is percutaneously introduced into the abdomen through a surgically placed catheter. The dialysate absorbs wastes that would normally be extracted by the kidneys, and is then drained.

2. Materials

EIT measurements were recorded using a hemi-circumferential electrode array designed to accommodate immobile patients. Carbon-loaded rubber electrodes (10x5cm) were equidistantly spaced on the anterior half of the abdomen, superior to the umbilicus.

The electrodes are connected to the EPack, a second generation, multiplexed EIT data acquisition system that is similar to the classic Sheffield Mark I [6]. The EPack has a DAC-driven, bipolar Howland current source that injects 1mA peak-to-peak current at 62.5kHz. A 14-bit digital voltmeter measures the magnitude of the complex voltage. The current source and voltmeter are multiplexed to eight electrodes via 1m coaxial cables with driven shields. The EPack collects and preprocesses measurement data, then wirelessly transmits the results to a Bluetooth-capable computer [8]. Data

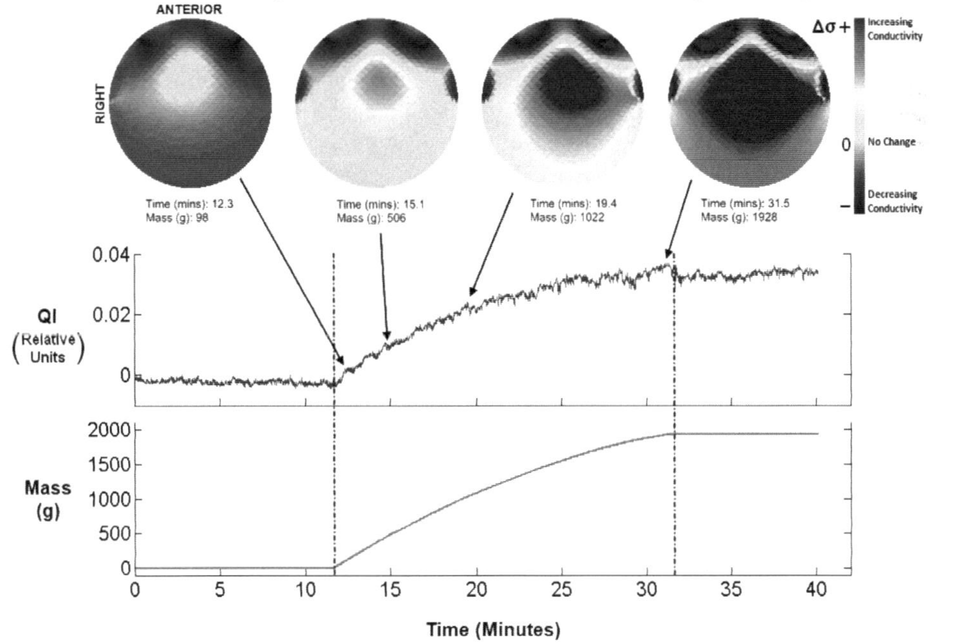

Figure 1. (Top) Reconstructions of dynamic dialysate accumulation at four interesting times. (Middle) The QI is calculated from each reconstruction frame and plotted as a time series. (Bottom) Time series plot of the amount of dialysate in the abdomen, the control variable. Vertical dashes represent the start and stop of fluid flow, respectively.

exchange is controlled by a graphical user interface (GUI) written in Visual C# 2008 (Microsoft, Redmond, VA).

3. Methods

Four subjects undergoing PD at the University of Florida's dialysis clinic were recruited for the study. Of the four, two were excluded based on their medical history. The remaining two subjects, one male and one female, were each monitored using the EPack during a routine dialysis exchange. Six months later, subjects returned to repeat the experiment.

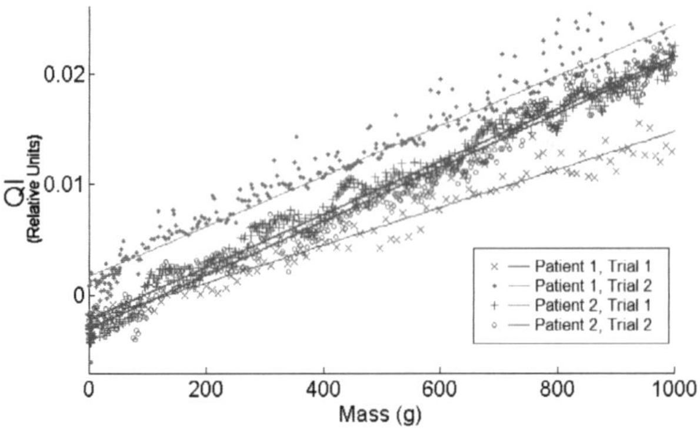

Figure 2. Relationship between QI and mass of dialysate in the abdomen.

EIT recording at a frame rate of 1Hz was started after the electrode array was applied. Before dialysate infusion started, and after it was completed, a baseline no shorter than 10 minutes was recorded. After the baseline period, solution was allowed to flow into the abdomen. As dialysate drained into the abdomen, the mass of liquid remaining in the bag was weighed using a digital hanging scale with +/-0.2g precision. Recordings from the EPack and the scale were synchronized.

Images of the dialysate accumulation were reconstructed using a linearized sensitivity method described by Geselowitz [9] and Lehr [10]. The sensitivity matrix for the system was obtained by forward solution of two-dimensional electrostatic models created using COMSOL Multiphysics (COMSOL, Inc., Burlington, MA) in conjunction with MATLAB (MathWorks, Inc., Natick, MA) using the finite element method (FEM). The abdomen was assumed circular and uniformly conductive [11]. The pseudoinverse of the sensitivity matrix was calculated using truncated singular value decomposition (TSVD). Inspection of the L-curve was used to choose the regularization parameter, k=12 [12]. Preconditioning the sensitivity matrix using the weighted minimum norm method proposed by Clay and Ferree generally improved reconstruction quality [13].

QI is the parameter used to estimate the volume of dialysate within an image. QI is the area-weighted sum of the pixels, $\Delta\sigma_j$, of a 2D EIT reconstruction,

$$QI = \sum_{j=1}^{n_P} A_j \cdot \Delta\sigma_j ,$$

where A_j is the area of the jth pixel, and n_P is the number of pixels in the reconstruction. Here, images are of secondary interest, used to confirm the plausibility of volume change estimates.

Flow information was calculated by first smoothing the QI using a 120-point moving average filter. This reduced noise at the expense of a 120-sample delay. The derivative of the smoothed QI time series was calculated to obtain the flow rate as a function of time. Results of sensitivity analysis were used to convert flow rate from the arbitrary units of QI to SI units.

4. Results and Discussion

Preliminary results in both subjects are similar; representative results from one Patient 2, Trial 1 are illustrated in figure 1. Signal-to-noise ratio (SNR) was calculated during the static baseline over 100 voltage samples. Preliminary results indicate *in vivo* SNR levels on the

Table 1. Quantification sensitivity in two patients.

	Trial	Sensitivity (QI/kg)	Girth (cm)	Weight (kg)
Patient 1	1	0.017	81	46
	2	0.023		
Patient 2	1	0.024	112	78
	2	0.024		

order of 35-40dB, slightly lower than those observed *in vitro* using saline tank phantoms. Baseline drift of QI corresponded to +/- 8g/min.

The dialysate flow rate was measured to decrease as the abdomen filled with dialysate and the dialysate bag emptied. This was also reflected in the QI results. Typically, dialysate accumulations in excess of 100g were distinguishable from background noise in image reconstructions.

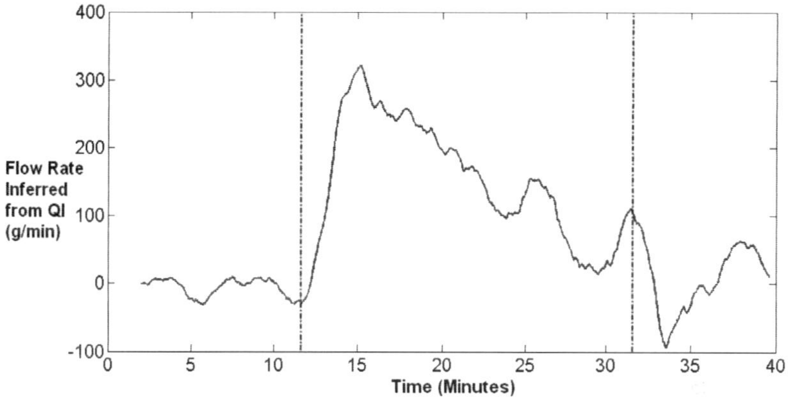

Figure 3. Flow rate measured using EIT.

The reconstruction algorithm is intended for accumulations as large as 1kg; beyond this range nonlinearity (decreasing sensitivity) in QI response was observed, as expected. Figure 2 shows the relationship between QI and the actual fluid mass for this range. QI had excellent linearity with respect to the control ($0.94 < R^2 < 0.98$). Table 1 compares the sensitivities across experiments. There was less dependence on anthropometric variation than expected. A communication glitch during Patient 1, Trial 1 limited the framerate half that used for other trials; this may explain that trial's lower sensitivity.

In contrast with the sensitivity, the offset in QI varied considerably depending on the selection of the reference measurement. Differentiation of QI removed this offset and gave consistent results for each patient. Figure 3 shows the inferred flow data for Patient 2, Trial 1. The flow information is delayed by 120 seconds as a result of the smoothing filter. This estimate is useful for automatic detection of bleeding because it does not rely on careful selection of reference measurement.

5. Conclusion

Results presented herein are considered preliminary; due to the small sample size it would be premature to draw conclusions based on this data. Nevertheless, the results do illustrate the proof of concept and allow us to fine-tune our methods in preparation for larger data collection.

6. Acknowledgements

This work was supported by US Army MRMC under Award No. W81XWH-07-1-0591 to RJS.

References

[1] Rozycki G S, Ochsner M G, Jaffin J H and Champion H R 1993 *J. Trauma* **34**(4)
[2] Sadleir R J, Fox R A and Turner V F 2000 *Rev Sci Instr* **71**(2) 530-5
[3] Wanjun S, Fusheng Y et al. 2008 *Physiol. Meas.* **29**(2) 217-25
[4] Sadleir R J and Fox R A 2001 *IEEE Trans. Biomed. Eng.* **48**(4) 484-91
[5] Sadleir R J, Zhang S U, Tucker A S and Oh S 2008 *Physiol. Meas.* **29**(8) 913-27
[6] Maxwell M H, Rockney R E, Kleeman C R, Twiss M R. 1959 *J. Am. Med. Assoc.* **170**(8):917-24
[7] Brown B H and Seagar A D 1987 *Clin. Phys. Physiol. Meas.* **8**(4A) 91-7
[8] Tang T, Zhang S U, and Sadleir R J 2008 *Proc. 7th Conf. on Biomedical Applications of EIT.* **14** 3980-3
[9] Geselowitz D B, 1971 *IEEE Trans. Biomed. Eng.* **18**(1) 38-41
[10] Lehr J 1972 *IEEE Trans. Biomed. Eng.* **19**(2) 156-7
[11] Meeson S, Killingback A L T, and Blott B H 1995 *Phys. Med. Biol.* **40**(4) 643-57
[12] Hansen P C and O'Leary D P. 1993 *SIAM Journal on Scientific Computing* **14**(6) 1487-1503
[13] Clay M T and Ferree T C, 2002 *IEEE Trans. Med. Imaging* **21**(6) 629-37

Quantification of Intraventricular Hemorrhage is Consistent Using a Spherical Sensitivity Matrix

Te Tang, Rosalind Sadleir

E-mail: tangte@ufl.edu

Abstract. We have developed a robust current pattern for detection of intraventricular hemorrhage (IVH). In this study, the current pattern was applied on two realistic shaped neonatal head models and one head-shaped phantom. We found that a sensitivity matrix calculated from a spherical model gave us satisfactory reconstructions in terms of both image quality and quantification. Incorporating correct geometry information into the forward model improved image quality. However, it did not improve quantification accuracy. The results indicate that using a spherical matrix may be a more practical choice for monitoring IVH volumes in neonates.

1. Introduction

Intraventricular Hemorrhage (IVH) affects 35-50% of very premature human infants to some degree and is a significant cause of death and disability [14]. EIT, as a non-invasive imaging technique, could offer the potential to monitor bleeding continuously. In conventional EIT measurements, currents are injected into a volume through one pair of an electrode array, and the data consists of differential voltages measured between other array electrodes. The conductivity distribution can be reconstructed using a sensitivity-based method [6]. The sensitivity matrix can be formulated using a forward model. Reconstructions incorporating accurate forward models are increasingly employed [1] [2] [4] [5]. However, realistic models need to be used carefully, because with inaccurate prior information, they may yield images worse than reconstructed with a simple forward model [4]. In practice, head geometry varies from patient to patient and during movement, which makes it difficult to accurately specify the forward model. In this study, we focused on investigating the effect of model boundary geometries on reconstructed images. More precisely, we compared image qualities and quantification accuracies in the head using a spherical model and two realistic geometry models.

2. Materials and methods

2.1. Forward models

Two forward models were used to formulate the sensitivity matrices. One was a spherical model with unit radius and homogeneous conductivity distribution (the SP model). The other was a neonatal head model with realistic head geometry and homogeneous conductivity distribution (the RG model). Both models had 16 electrodes placed on them in an EEG-based configuration (figure 1a). Current patterns were applied so all currents were returned via an electrode on the apex of the head. Voltages between adjacent electrodes (in terms of electrode numbering) were measured [12].

2.2. Data generation

We gathered simulated bleeding data using two different head-shaped finite element models and one head-shaped phantom. The first one (the M model) (figure 1b) had a boundary geometry the same as the RG model we used to formulate the RG sensitivity matrix. The background conductivity was assigned to be homogeneous with the conductivity of CSF (1.3 S/m) [8]. To simulate intraventricular bleeding, we created two bleeding sites at positions corresponding to the centers of the left and right ventricles, respectively. Bleeding was simulated by manually setting the corresponding elements of the finite element forward model to the conductivity of blood (0.67 S/m) [3]. The centers of the ventricles were determined from representative neonatal head MRI data. We determined that the ventricles were 18 mm above the anatomical center and were spaced laterally 18 mm apart. We simulated 6 different cases on each side with total blood volume increasing linearly from 0.5 ml to 3 ml.

To test our methods in the presence of realistic noise, we constructed a plastic head-shaped phantom of the UK model, using a 3D printer (Dimension, Inc., MN, USA). The phantom had the same boundary shape as the UK model, but its total volume was 1628 ml, about twice the actual model volume. The phantom had 16 stainless steel electrodes attached as shown in figure 1c. A 0.6% saline solution (the conductivity of which was approximately 1.3 S/m, the conductivity of CSF) was used as the phantom background. Bleeding was simulated using gel anomalies made of TX151, agar and saline [10]. The TX-151 gel conductivity was adjusted to 0.67 S/m. We made 6 cylindrically-shaped anomalies with volumes from 1-6 ml. Following the simulated data generated from model M, we placed gel anomalies at the left and right ventricle centers, which were determined to be 23 mm above the anatomical center and 23 mm apart from each other. Anomalies were placed accurately using a 3D positioner (Velmex, Inc., NY, USA). Data was collected by a KHU Mark1 16-electrode EIT system provided by the IIRC at Kyung Hee University, Korea [7]. We chose to use a single frequency of 10 kHz and a constant current output of 1 mA. Ten data sets were collected and averaged as a reference before anomalies were introduced. In each of the 12 experiments, ten data sets were collected and averaged as the perturbation data for that volume of bleeding.

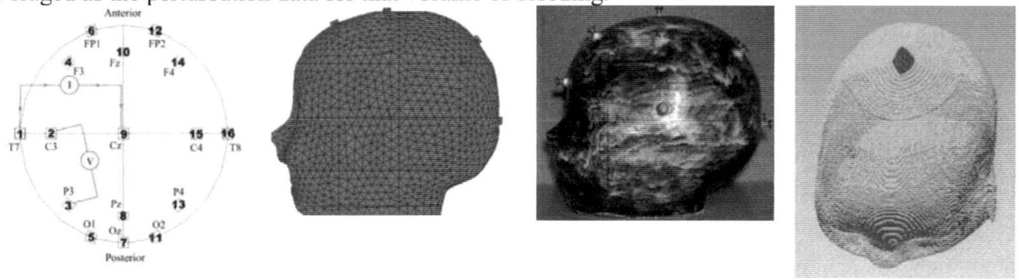

(a) (b) (c) (d)

Figure 1. (a) the EEG current pattern. (b) the UK model. (c) the UK phantom (d) the Belgium model

The other RG model used in this study (model G) (figure 1d) was constructed from the MRI neonatal head data provided by researchers at the University of Gent. The purpose of using this model was to test our two sensitivity matrices on a different 'patient', whose geometry matched neither one of these matrices. In this model, we set the background conductivity to 1.3 S/m and selected two bleeding sites at the ventricle centers. Ventricle center positions were determined using the original MRI data from which this model was constructed. We determined that the two ventricles were 44 mm above the base and 17 mm apart. We simulated six different cases each side with blood volumes increasing linearly from 0.5 ml to 3 ml.

2.3. Reconstruction, Quantification and Image Quality

In EIT difference imaging, the conductivity perturbation can be related to the changes on boundary voltage measurements with a sensitivity matrix S ($\Delta V = S\Delta\sigma$). A truncated singular value decomposition method (TSVD) [13] was used to regularize the pseudo-inversion of S. We chose a truncation value of 50 based on L-curve inspection. This truncation was used for all reconstructions in this study. Quantification was assessed with a characteristic parameter called the quantification index (QI), which has been found to correlate well with volumes of blood-like anomalies using 2D disk–

based sensitivity matrices [9] [18]. QI is defined as $QI = \sum_{i=1}^{ne} \Delta\sigma_i v_i$, where ne is the total number of elements in the mesh, $\Delta\sigma_i$ is the conductivity change reconstructed in the ith element and v is the volume of each element. We evaluated image quality by estimating the resolution R. This was calculated by estimating the portion of volume that contains voxels having reconstructed values more than half of the maximum value, then taking the cube root ($R = \sqrt[3]{\frac{V_{HM}}{V_0}}$).

3. Results

3.1. Image quality

Reconstructions of anomaly 1 are shown for the model M and model G data in figure 2. Only reconstructions of anomaly 6 are shown for model M, because the reconstructed images of anomaly 1 were severely contaminated by noise. Left and right reconstructions were compared in each case to demonstrate asymmetry. As expected, the reconstructions from model M data using the RG matrix had an average $R = 0.1649$ for the left side and $R = 0.1958$ for the right side, which were consistently better than reconstructions using the SP matrix (average $R = 0.2583$ for the left side and $R = 0.2464$ for the right side). The RG matrix also showed an advantage over the SP matrix in reconstructing model M phantom data. However, the RG matrix did not show a consistent advantage in reconstructing data from model G. Instead, the SP reconstructions had consistently better image resolutions ($R = 0.1887$) than the RG reconstructions ($R = 0.2168$) for the right side, although the RG reconstructions showed better image qualities ($R = 0.1836$) than the SP cases ($R = 0.2929$) for the left side.

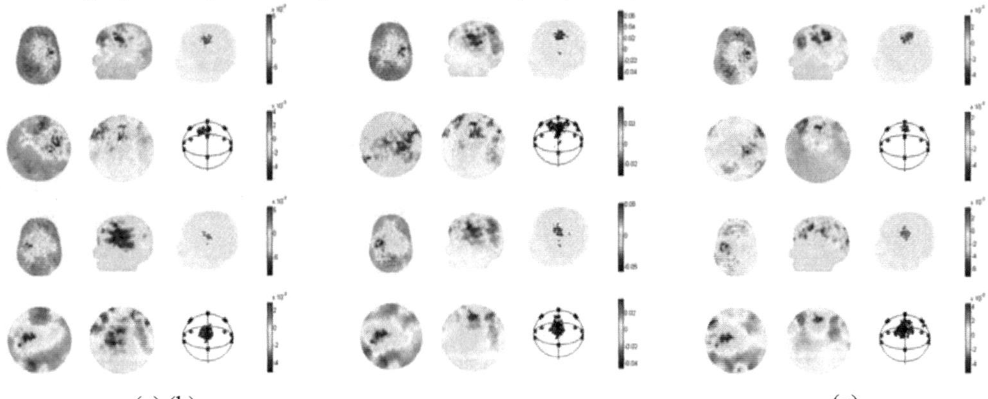

(a) (b) (c)

Figure 2. Reconstructions with the RG matrix and SP matrix using the data from (a) model M, (b) model M phantom, (c) model G. First and second row: anomaly on the right. Third and fourth row: anomaly on the left.

3.2. Quantification accuracy

Quantification indices for the 6 different anomaly sizes on the left side were calculated for each of the three data sources (figure 3). The quantification performances of the two matrices were almost identical for simulated model M data. The RG matrix had a relatively better quantification accuracy for model M phantom data (average error = 0.0025 S.cm^2), compared with the SP matrix (average error = 0.0034 S.cm^2). The RG matrix did not show any significant advantage over the SP matrix for the data from model G.

4. Discussion

Early studies have shown that the EIT image quality is highly dependent on the forward model. Fan et al [17] found that the human lung reconstruction was improved by using a boundary model developed from 3D CT scan data. In our study, we found that better image resolutions were obtained using the

RG matrix when using both simulated and phantom data from model M. This confirmed the findings from early studies, since the RG model had exactly the same boundary shape and electrode locations in model and phantom. When using data from model G, the RG matrix still produced better results for left-located anomalies. This can be explained by the asymmetry of the Cz electrode in the RG model. In the RG model, the Cz electrode was placed to the left of the head apex and therefore produced relatively larger sensitivity and selectivity in the left ventricle region [11]. However, the advantage for the RG matrix was not sustained for anomalies positioned on the right and here the SP matrix showed consistent better image resolutions. Kolehmainen et al [15] and Blott et al [16] concluded that the locations of electrodes and the boundary shape of the forward model brought the most difficulties for the reconstructed images. All the models used in this study had homogeneous backgrounds, which enabled us to investigate the effect from boundary geometries alone. Future work will include comparisons using data generated from models with realistic anatomical structures.

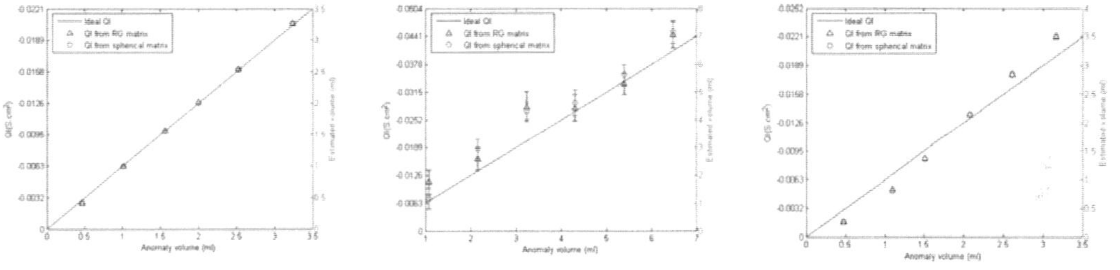

(a) (b) (c)

Figure 3. Quantification indices for the images reconstructed with the RG matrix and SP matrix using the data from (a) model M, (b) model M phantom, (c) model G. Anomalies were on the left.

The RG matrix showed no better performance than the SP matrix in term of quantification. Figure 3a shows that the SP matrix could be calibrated to quantify the anomalies as well as the RG matrix and these QIs matched the ideal QIs. The RG matrix did not show consistent advantage with model M phantom data (figure 3b). An interesting observation was that the QIs calculated from the SP matrix showed smaller error bars in most cases, which indicates that the SP matrix may have an advantage over the RG matrix in a noisy environment. Both matrices quantified data from model G well (figure 3c). In this case, where the object geometry matched neither sensitivity matrix, the RG matrix did not show any advantage over the SP matrix.

5. Conclusion

We compared the performance of two sensitivity matrices in three different situations. It is evident that the matrix developed from a realistic-shaped forward model does not guarantee image quality when the subject has mismatched boundary shape and electrode locations. Moreover, the SP matrix performed better in quantification of phantom experiment data. All our observations indicate that using a simple spherical matrix might be adequate for monitoring IVH in neonates.

Acknowledgments

This work was supported in part by NIH grant RO1EB-002389 to RJS. We also thank Drs Tizzard and Bayford (Middlesex University, London, UK) for providing model M and Ivana Despotovic (University of Gent, Belgium) for providing model G.

References

[1] Abascal Juan-Felipe P J, Arridge S R, Atkinson D, Horesh R, Fabrizi L, De Lucia M, Horesh L, Bayford R H, Holder D S 2008 Use of anisotropic modeling in electrical impedance tomography; Description of method and preliminary assessment of utility in imaging brain function in the adult human head. *NeuroImage*. **43** 258-268

[2] Bagshaw A P, Liston A D, Bayford R H, Tizzard A, Gibson A P, Tidswell A T, Sparkes M K, Dehghani H, Binnie C D and Holder D S 2003 Electrical impedance tomography of human

brain function using reconstruction algorithms based on the finite element method. *NeuroImage.* **20** 752-764

[3] Geddes L A and Baker L E 1967 The specific resistance of biological material – a compendium of data for the biomedical engineer and physiologist. *Med. Biol. Eng.* **5** 271-293

[4] Liston A D, Bayford R H and Holder D S 2004 The effect of layers in imaging brain function using electrical impedance tomography. *Physiol. Meas.***25** 143-158

[5] Meeson S, Killingback A L T and Blott B H 1995 The dependence of EIT images on the assumed initial conductivity distribution: a study of pelvic imaging *Phys. Med. Biol.* **40** 643-657

[6] Murai T and Kagawa Y 1985 Electrical impedance computed tomography based on a finite element model. *IEEE Trans. Biomed. Engr.* **32** 177-184

[7] Oh T I, Woo E J and Holder D 2007 Multi-frequency EIT system with radially symmetric architecture: KHU Mark1. *Physiol. Meas.***28** S183-S196

[8] Oostendorp T, Delbeke J and Stegeman D 2000 The conductivity of the human skull: results of in vivo and in vitro measurements. *IEEE Trans. Biomed. Engr.* **47** 1487-1492

[9] Sadleir R J and Fox R A 1998 Quantification of blood volume by Electrical Impedance Tomography using a tissue equivalent phantom. *Physiol. Meas.***19** 501-516

[10] Sadleir R J, Neralwala F, Tucker A S and Tang T 2009 A controllably anisotropic conductivity or diffusion phantom constructed from isotropic layers. *Ann. Biomed. Eng.* **37** 2522-2531

[11] Sadleir R J and Tang T 2009 Electrode configurations for detection of intraventricular haemorrhage in the premature neonate. *Physiol. Meas.***30** 63-79

[12] Tang T and Sadleir R J 2010 A robust current pattern for the detection of intraventricular hemorrhage in neonates using electrical impedance tomography. in press, *Ann. Biomed. Eng.*

[13] Xu P 1998 Truncated SVD methods for discrete linear ill-posed problems. *Geophys. J. Int.***135** 505-514

[14] York J and DeVoe M 2002 Health issues in survivors of prematurity *Southern Medical Journal* **95** 969-976

[15] Kolehmainen V, Vauhkonen M, Karjalainen P A and Kaipio J P 1997 Assessment of errors in static electric impedance tomography with adjacent and trigonometric current patterns. *Physiol. Meas.* **18** 289-303

[16] Blott B H, Daniell G J and Meeson S 1998 Electrical impedance tomography with compensation for electrode positioning variations. *Phys. Med. Biol.* **43** 1731-9

[17] Wenru Fan, Huaxiang Wang, Xiaoyan Chen and Zhiying Lv 2009 Three dimensional EIT models for human lung reconstruction based on Schur CG algorithm *Complex Medical Engineering. CME. ICME International Conference on*, pp. 1-5

[18] Sadleir R J, Zhang S U, Tucker A S and Oh S 2008 Imaging and quantification of anomaly volume using an eight-electrode 'hemiarray' EIT reconstruction method. *Physiol. Meas.***29** 913-927

Feasibility of electrical impedance tomography in haemorrhagic stroke treatment using adaptive mesh

J Nasehi Tehrani[1], C Anderson[2], C Jin[1], A van Schaik[1], D Holder[3] and A McEwan[1]
[1]School of Electrical and Information Engineering, CARLAB
The University of Sydney, Australia, NSW 2006
[2]The George Institute for International Health,
Royal Prince Alfred Hospital and The University of Sydney, Australia, NSW 2006
[3]Department of Medical Physics and Bioengineering, University College London, Gower St, London, UK.

E-mail: alistair.mcewan@sydney.edu.au

Abstract. EIT has been proposed for acute stroke differentiation, specifically to determine the type of stroke, either ischaemia (clot) or haemorrhage (bleed) to allow the rapid use of clot-busting drugs in the former (Romsauerova et al 2006). This addresses an important medical need, although there is little treatment offered in the case of haemorrhage. Also the demands on EIT are high with usually no availability to take a 'before' measurement, ruling out time difference imaging. Recently a new treatment option for haemorrhage has been proposed and is being studied in international randomised controlled trial: the early reduction of elevated blood pressure to attenuate the haematoma. This has been shown via CT to reduce bleeds by up to 1mL by Anderson et al 2008. The use of EIT as a continuous measure is desirable here to monitor the effect of blood pressure reduction. A 1mL increase of haemorrhagic lesion located near scalp on the right side of head caused a boundary voltage change of less than 0.05% at 50 kHz. This could be visually observed in a time difference 3D reconstruction with no change in electrode positions, mesh, background conductivity or drift when baseline noise was less than 0.005% but not when noise was increased to 0.01%. This useful result informs us that the EIT system must have noise of less than 0.005% at 50 kHz including instrumentation, physiological and other biases.

1. Introduction

Stroke is traditionally defined as the sudden onset of a focal neurological deficit lasting longer than 24 hours caused either by intracerebral haemorrhage or ischaemic infarction. It is the third most common cause of death in Australia. About one-third of victims will die within one month. Intracerebral haemorrhage (ICH) accounts for about 10% of all strokes but for a much higher percentage of deaths due to stroke. The principal cause of intracerebral haemorrhage is elevated blood pressure or 'hypertension'. Bleeding disorders and use of anticoagulants are also increases causes of intracerebral haemorrhage. Control of risk factors is the key approach to management. Most patients with a stroke require urgent imaging to determine the cause and guide treatment.

Investigation is with CT or MRI. Both can distinguish an ischaemic from haemorrhagic stroke, and other structural abnormalities or stroke mimics. Because the development of an acute stroke is a medical emergency, admission to a stroke unit and investigation as soon as possible is advised [1]. EIT has been proposed for acute stroke differentiation, specifically to determine the type of stroke, either ischaemia or haemorrhage to allow the rapid use of clot-busting drugs in the former [2]. This addresses an important medical need, although there is little treatment offered in the case of ICH. Also

the demands on EIT are high with usually no availability to take a 'before' measurement, ruling out time difference imaging [3].
One potential way to manage ICH is with early intensive blood pressure reduction. However, there is much uncertainty about the effects of early lowering of elevated blood pressure after acute ICH. Such an approach appears safe and feasible but a large scale trial is required to determine its effectiveness on substantive clinical outcomes to influence practice [4].Electrical impedance tomography (EIT) may be appropriate for online-monitoring and control of blood pressure reduction without too much inconvenience for patients. A major methodological problem is the poor quality of the conductivity images, which is due to the low spatial resolution and low sensitivity for structures far away from the object's surface as well as large qualitative errors in the reconstruction conductivity values. In this paper we report on a 3D simulation of ICH inside brain to investigate the feasibility of detecting changes in a haemorrhagic lesion located close to scalp.

2. Methods

The electric field was calculated at each node in the mesh using a linear finite element solver written in Matlab (the Math Works Inc., MA). Simulations using the UCL realistic head finite element method (FEM), associated head shaped tank eeg31b protocol, and tissue conductivities from the literature were used to assess the feasibility of EIT in these applications [5], [6], [11]. All meshes were constructed from linear, tetrahedral elements. We used ISO2MESH software and head and brain sample produced by Qianqian Fang for making one coarse and one fine mesh for forward and inverse solution [12]. The mesh for forward solution had 136387 nodes and 770660 elements and the number of nodes and elements for inverse solution was 61496 and 365481 respectively (figure 1).

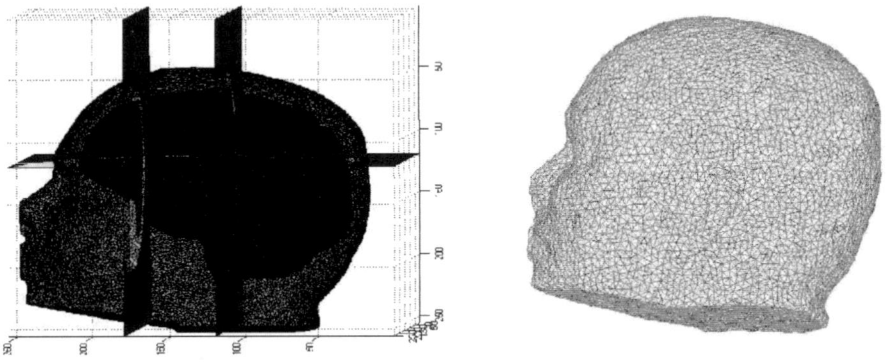

Figure 1. Left side image of mesh for forward solution with 770660 elements, Right side image of mesh for inverse solution with 365481 elements shown.

The conductivity of the skull is that measured by Oostendorp [7], which is significantly higher than the value of Law [8]. The sensitivity at each node was calculated using Geselowitz's theorem [9], [10]. The sensitivity matrix was created by performing the calculation for each node in the mesh and for all combinations of current injection and voltage measurement electrodes. In all cases, point electrodes were assumed, located at the node closest to the actual position of each electrode. The number of electrodes was 31 and the ground index supposed to be on the nose. The exact location of the electrodes has been shown in the figure 2.The diameter of each electrode considered 6.5 mm and the stimulation current source was 400 μA [11].For evaluating EIT in detection of small changes of ICH, we created 1 ml haemorrhagic lesion close to scalp which is shown in figure 2 on the right side. This location was chosen from elements of brain which produced in smaller size in comparison with outside brain elements. The software that we used for image visualizing was MayaVi written by Prabhu Ramachandran. In order to make our simulation result closer to practical experience we added different percentage of noise to measuring signal. Uniformly distributed random noise was generated using the Matlab function 'Rand'.

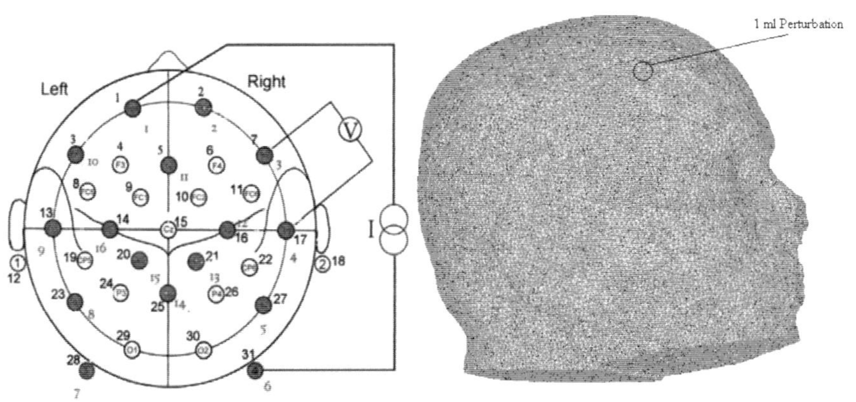

Figure 2. Left side Electrode positions, Black numbers are for EEG31b, red electrodes and numbers are spiral 16 protocol. Electrodes 6 & 7 are positioned 2cm inferior to O1 and O2.Right side location of 1mL Perturbation.

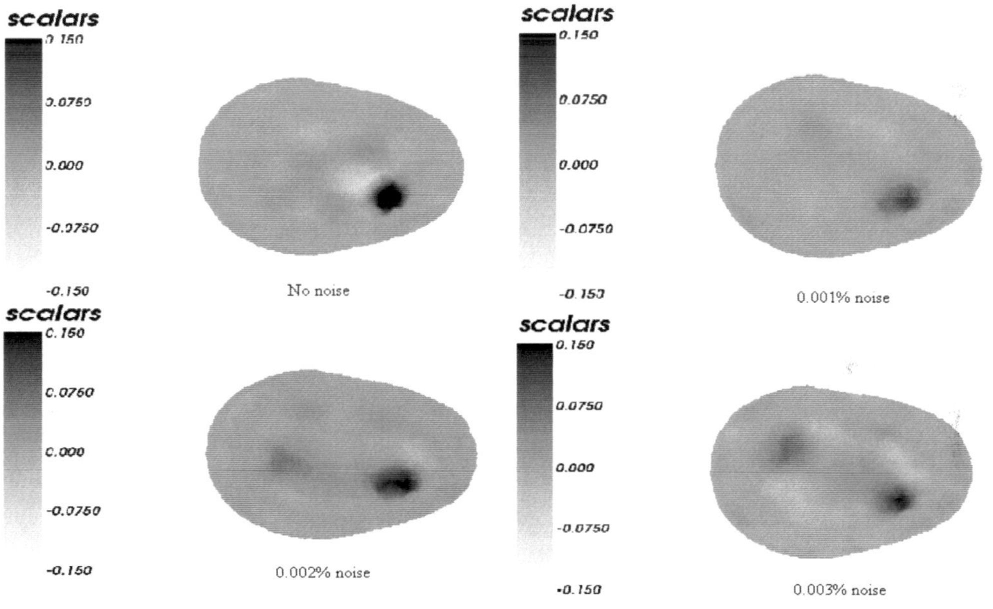

Figure 3. Reconstruction of 1mL perturbation top left no noise, top right 0.001% noise, bottom left 0.002% noise and bottom right 0.003% noise

3. Results

An increase of 1mL ICH caused a boundary voltage change of less than 0.05% at 50 kHz. In figure 3 reconstruction of ICH with zero noise up to 0.003% noise has been shown. The reconstructed images with different noise input were used to compare ability of EIT systems for detection of ICH. It shows in figure 3 that a low noise EIT system can easily detect voltage changes near 0.05%.We changed the percentage of input noise between 0.001% and 0.01%. This observation has been done in a time difference 3D reconstruction without any change in electrode positions, mesh, background conductivity or drift. The experience demonstrated that in images with baseline noise up to 0.005 % small changes in ICH can easily be detected by EIT systems. However, in images with 0.005% or more added noise, detection of small changes near 1ml is more difficult (figure 4). This useful result informs us that the EIT system must have noise of less than 0.005% at 50 kHz including instrumentation, physiological and other biases.

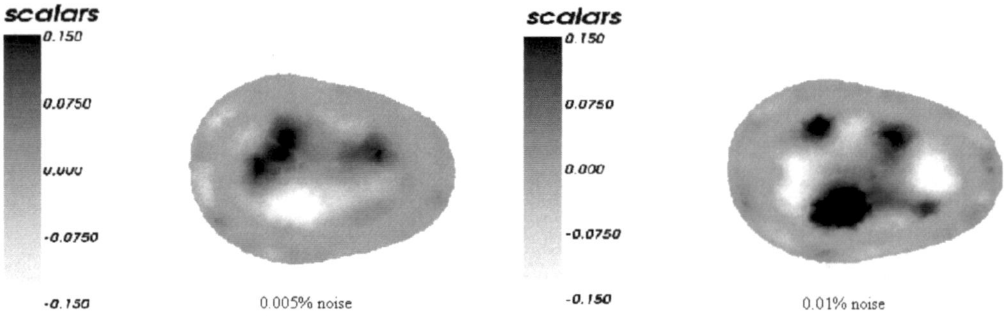

Figure 4. The reconstruction image 0.005% noise (left), 0.01% noise (right).

4. Discussions

This is the first study demonstrating the feasibility of EIT in haemorrhagic stroke treatment.The main attraction of EIT is that it can be easily assessed non-invasively, at low cost and at bedside. EIT techniques can be used to visualize physiological activities in a human body such as respiration, cardiac circulation, brain function, stomach emptying, fracture healing, bladder filling, and others. An optimal treatment requires information on conductivity distribution inside a patient to design effective injection current patterns. EIT systems do exist that have instrumentation noise of less than 0.1% when measured in ideal conditions but not when measuring on a saline tank or real subject [13]. The white noise in this system including 2 more suitable systems for imaging epileptic seizures increased from about 0.03% on the resistor to 0.08% on the human. It increased with load but was independent of use of the multiplexer [14]. Major sources of error in MFEIT are the common mode phenomena, stray capacitances, common mode rejection and the mentioned transient effects due to the multiplexing which lead to a load dependence in the frequency response of systems[15], [16]. A distributed system with digitization at the electrode could be a next step in MFEIT system development. Overall, the results indicate that it is possible to extract small changes in ICH if we are able decrease the noise level.

5. References

[1] Romsauerova, A., et .al. 2006 *Physiol. Meas.* 27 147-161
[2] L. Horesh, et. al. (EMBEC 2005), Prague, Czech Republic, 20-25 November 2005
[3] Murtagh J 2008 *General practice* (Australia: McGraw Hill) pp1362-64
[4] Anderson C et al 2008 *J. Lancet Neural.* **7** 391-9
[5] Polydorides N et al 2002 *J. Meas. Sci. Technol.* **13** 1871-83
[6] Bagshaw A et al 2003 *J. NEUROIMAGE* **20** 752-764
[7] Oostendorp T, Delbeke J and Stegeman D 2000 *IEEE Trans. Biomed. Engr.* **47** 1487–92
[8] Law S 1993 *Brain Topography* **6** 99–109
[9] Geselowitz D B 1971 *IEEE Trans. Biomed. Eng.* **18** 38–41
[10] Dehghani, H Delpy DT and Arridge SR 1999 *Physics in Medicine and Biology* **44** 2897-2906
[11] Horesh L et al 2007 *IFMBE Proceedings* **14** 3874-77
[12] Qianqian F et al 2009 ISBI 2009 1142-45
[13] McEwan A et al 2006 *J. Physiol. Meas.* **27** S199-S210
[14] Fabrizi L, McEwan A, Woo E. and Holder, D. 2007 *J. Physiol. Meas.* **28** S217-236
[15] Hahn G et al 2008 *J. Physiol. Meas.* **29** S163–S172
[16] McEwan A, Cusick, G. and Holder, D.S. 2007 *J. Physiol. Meas.* **28** S197-215

Visualization of the Meridian of Traditional Chinese Medicine with Electrical Impedance Tomography: An Initial Experience

Yanli Cao, Xiaozuo Lu, Xuemin Wang

College of Precision Instrument and Opto-Electronics Engineering, Tianjin University, Tianjin 300072, China

E-mail: xueminw@tju.edu.cn

Abstract. The meridian is a concept central to traditional Chinese medical techniques such as acupuncture. There is no physically verifiable anatomical or histological basis for the existence of meridians. In Chinese medicine, the meridians are channels along which the energy of the psychological system is considered to flow. It has been proven that the resistance along the meridian channels is lower compared to other paths. Based on this knowledge, we proposed using electrical impedance tomography (EIT) to visualize the meridians of human being. A simplified three dimensional (3D) mathematical model of the forearm developed. Current was injected in the direction perpendicular to the cross-section where eight electrodes were equally placed around the surface of the forearm for the voltage measurements. The model was solved using Finite Element Method (FEM) and dynamic image was reconstructed using truncated singular value decomposition (TSVD) regularization method. The conductivity distributions were compared with different current injections, along the meridian channel and channels around respectively. We also conducted experiments on models and the meridians were shown in final reconstructed images.

1. Introduction

There are many points with different physical properties in human body, which is called Meridian [1] in traditional Chinese medicine. A new approach for research on the Meridian characteristics is proposed on the basis of the low flow resistance characteristics: Tracer currents are injected into points and measuring electrodes are used to get voltages. Meridian image is obtained using dynamic imaging which is an image reconstruction of the change of electrical conductivity. It provides a scientific basis for the research on meridian imaging.

2. A simplified physical model

Some prior knowledge is considered to study properties of meridian including structures and electrical conductivity of interior tissue in medical application. In order to study the meridian more easily, combined with shape and position of meridian, both of the simplified forearm and large intestine meridian are modeled as cylindrical uniform dielectric, as shown in Figure 1, according to the low flow resistance characteristics of the meridian.

Figure 1. The physical model of forearm. The large cylinder is forearm tissue, while the small red cylinder is the low resistance channel of large intestine meridian.

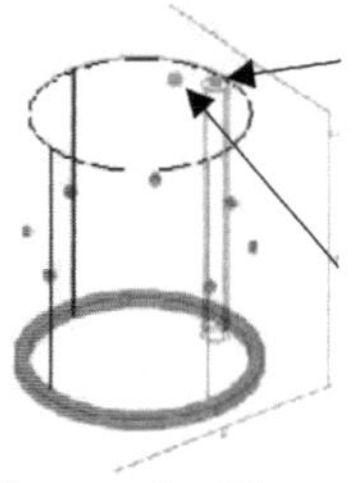

Figure 2. The eight blue electrodes is to measure, the blue circle at the bottom is grounding electrode.

3. Method

3.1. Forward problem

In this model forward problem is that the currents are injected into points vertically and then voltages are measured based on EIT. As illustrated in Figure 2, the red on the cylinder is the exciting electrode where current is injected into. Currents are injected into meridian channel and channels around [2] respectively and difference between them are obtained to study characteristic parameters with the method of dynamic imaging.

As the derivation of inverse problem will use the method of sensitivity analysis, it is necessary to establish sensitivity matrix J which is exactly what forward problem should deal with. Sensitivity matrix is based on the theory that a small change of electrical conductivity could induce a change of the boundary voltage to solve a nonlinear problem in a linear way, and the linear physical model is as follows: $J\delta\sigma = \delta U$

Where J is linear system matrix (sensitivity matrix), $\delta\sigma$ is the difference of electronic conductive when current is injected into meridian channel and around, and δU is the difference of boundary voltage corresponding to $\delta\sigma$.

Electromagnetic finite element method is applied to analyze the distribution of voltage through COMSOL software. In order to reduce ill-conditioned problems for this established model, rectangular mesh generation is used to solve sensitivity matrix. In this way, imaging field is divided into many small ones, and electrical conductivity in every small area is a constant. After some mathematical derivation [3], J is obtained as follows: $J = \frac{\partial V_{dm}}{\partial \sigma_k} = -\frac{1}{I}\int_{\Omega_k} E(\vec{I}^{d}) \cdot E(\vec{I}^{m}) dV$

Where V_{dm} is potential difference between d and m types of current, σ_k is the conductivity in field k, $E(\vec{I}^{d})$ and $E(\vec{I}^{m})$ are distribution of field strength when the currents are type d and type m respectively.

3.2. Converse problem

Solving converse problem is to obtain distribution of the characteristic parameters in the filed. And for this physical model of meridian, it is an ill-posed problem [4] and J is ill-conditioned. So the study theoretically analyzes TSVD [5] (truncated singular value decomposition) regularization method starting with least square method and image reconstruction of EIT on meridian is performed.

Traditional least squares method to solve cannot make a good image. So we try three methods to improve image. (1) normalized singular value method; (2) norm method; (3) L-curve method. The reconstructed images are shown in Figure 3. We can see the image of normalized singular value method cannot reflect the distribution of conductivity; but both norm method and L-curve method give better images, especially when 1% and 5% random noise are added. Hence both TSVD methods which can reduce condition numbers of sensitivity matrix and to a great extent restrain noise manage imaging effect well. They commendable reflect distribution of electrical conductivity in the field even the disturbance exists.

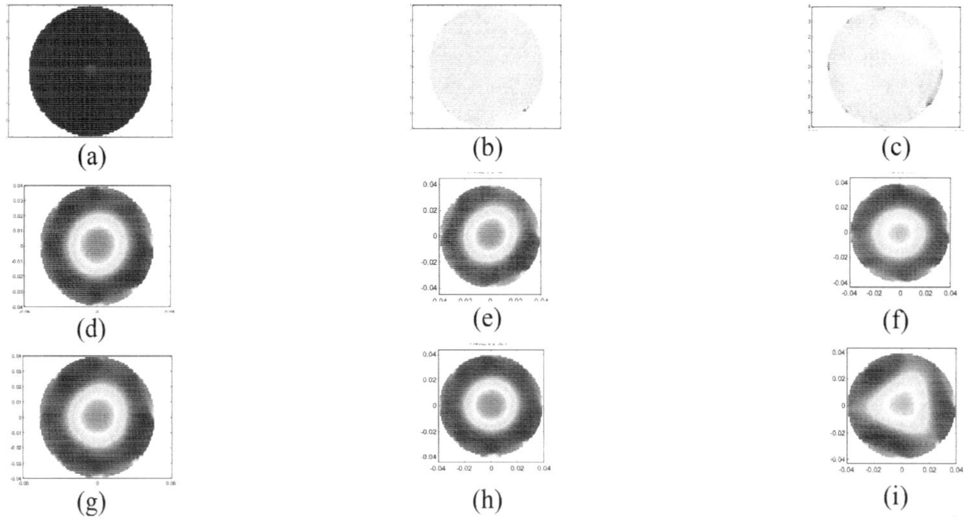

Figure 3. (a) is the distribution of original conductivity. (b) comes from Traditional least squares method. (c) comes from normalized singular value method. (d) comes from norm method; (e) and (f) are reconstructed with norm method while 1% and 5% random noise are added respectively. (g) comes from L-curve method; (h) and (i) are reconstructed with L-curve method while 1% and 5% random noise are added respectively.

4. Experiments

4.1. Experiments design

Considering complexity and instability of human meridian, we didn't experiment on human, but on two simple and steady models: normal saline-resistance wire cylinder model and gel-normal saline model. First, normal saline-resistance wire cylinder model simulate meridian from the perspective of electrical conductivity of tissue. Then gel-normal saline model makes use of model method in physics. This model can simulate not only the low impedance of meridian but also the low flow resistance. What's more, gel is more stable.

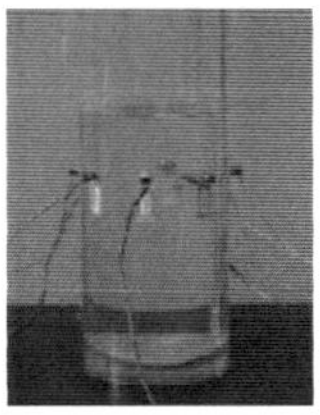

Figure 4. Normal saline-resistance wire cylindrical model.

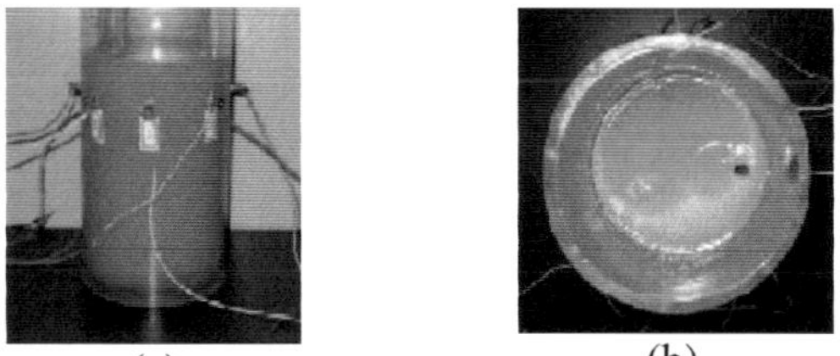

(a) (b)

Figure 5. Gel-normal saline model. (a) is a side view picture and (b) is a top view picture.

4.1.1. 0.9% normal saline is prepared with distilled water and NaCl to simulate human fluid. An organic glass rod with the radius of 3 mm is wound by copper wire with the impedance of 20 Ω and this unit simulates meridian. And there are eight electrodes on the side surface of the cylinder and a circle grounding electrode at the bottom, as shown in Figure 4.

4.1.2. Ager powder and 0.9% normal saline are mixed with a ratio of 3 g: 100ml and heated to be dissolved. Then mixture is put into the prepared cylinder container, at the same time, an organic glass rod with the radius of 3 mm is buried in the mixture. After solidification, the rod is removed. At this time, a narrow channel comes where the rod has been. 4.5% saline is pour into this channel to simulate meridian, as seen in Figure 5.

Signal generator and high precision multimeter Agilent 34401A are applied to generate stable currents and to measure voltages. Eight groups current of 100Hz frequency are generated in the range from 8.5mA to 12mA. Every exciting current is injected into meridian and around of models above respectively and we got arrayed voltages at the same cross section by multimeter.

4.2. Results

We realized dynamic imaging based on the established sensitivity matrix J which is solved by TSVD method with norm method and the results are shown in Figure 6. Through the experiences, we can see that the research method that is injecting current vertically and measuring boundary voltages at a cross section can reflect the meridian characteristic of higher electrical conductivity, so it is feasible for EIT. Especially for gel-normal saline model, imaging consequent is fairly good; Normal saline-resistance wire cylinder model due to ions in solution is not good enough, however it can also reflect the characteristic of meridian to a certain extant.

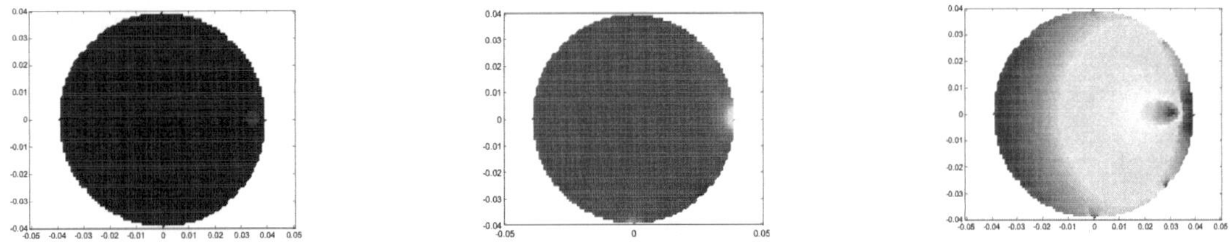

(a) (b) (c)

Figure 6. (a) is original conductivity distribution of meridian model; (b) is reconstructed image in normal saline-resistance wire cylinder model; (c) is the image in gel-normal saline model.

5. Conclusion

This paper proposes EIT of meridian on a basis of low flow resistance and other biophysical properties and it proves feasible through experiences. Because of complexity of human meridian, we did not experimented on human. Still it needs further study, for example, to improve physical model of meridian and reconstruction algorithm.

References

[1] Zhang W B 1997 What is the meridian *Science and Technology of China Press* 116–117,56–62

[2] Lee M S, Jeong S Y and Lee Y H 2005 Differences in electrical conduction properties between meridians and non-meridians *The American Journal of Chinese Medicine* **33(5)** 723-728

[3] Nick Polydorides, William R B and Lionheart 2002 A Matlab toolkit for three-dimensional electrical impedance tomography: a contribution to the Electrical Impedance and Diffuse Optical Reconstruction Software project *MEASUREMENT SCIENCE AND TECHNOLOGY* **13** 1871–1883

[4] Xiao T T, Yu S G and Wang Y F 2003 numerical solution of inverse problems *science press* 4–6

[5] Huang X W, Wu C S and Zhu H P 2005 A new regularization method based on the singular value decomposition for compact operators *Acta Mathematica Scientia* **25A(3)** 331–336

Sensitivity Study of an Ultrasound coupled Transrectal Electrical Impedance Tomography System for Prostate Imaging

Y. Wan[1], R. Halter[1], A. Borsic[1], P. Manwaring[1], A. Hartov[1], K. Paulsen[1]

[1] Thayer School of Engineering, Dartmouth College, NH 03755, USA

E-mail: yuqing.wan@dartmouth.edu

Abstract. In 2009, prostate cancer ranks as the most common cancer and the second most fatal cancer in men in the United States. Unfortunately, the current clinical diagnostic methods (*e.g.* prostate-specific antigen (PSA), digital rectal examination, endorectal MRI, transrectal ultrasound, biopsy) used for detecting and staging prostate cancer are limited. It has been shown that cancerous prostate tissue has significantly different electrical properties when compared to benign tissues. Based on these electrical property findings, a TransRectal Electrical Impedance Tomography (TREIT) system is proposed as a novel prostate imaging modality. The TREIT system is comprised of an array of electrodes interfaced with a clinical TransRectal UltraSound (TRUS) probe. We evaluate this imaging system through series of phantom imaging experiments to assess the system's ability to image high and low contrast objects at various positions. We found that the TREIT system can easily discern high contrast inclusions of 1 cm in diameter at distances centered at 2 times the radius of the TREIT probe away from the probe surface. Furthermore, this technology's ability to detect low contrast inclusions suggests that it has the potential to successfully detect prostate cancer.

Keywords: Electrical Impedance Tomography, Ultrasound, prostate cancer, sensitivity, reconstruction

1. Introduction

In 2009, prostate cancer ranks as the most common cancer and the second most fatal cancer in men in the United States. Approximately 192,280 men in the United States are expected to be diagnosed with prostate cancer in 2009, and 27,360 men are expected to die from it [1]. Currently, clinical diagnostic methods including prostate-specific antigen (PSA) level, digital rectal examination, endorectal MRI, Transrectal Ultrasound (TRUS), biopsy, *etc.*, are limited to detect prostate cancer.

Halter *et al* reported that cancerous tissue has significant different conductivity and permittivity than that of glandular or stroma tissue in the prostate at frequencies ranging from 0.1 kHz to 100 kHz [3]. Based on these findings, we developed a TransRectal Electrical Impedance Tomography (TREIT) modality to distinguish cancerous tissue from surrounding tissue in the prostate by sensing these significant differences in electrical properties.

Borsic *et al* performed a three dimensional simulation-based analysis of a TREIT probe [4] and showed that there was sufficient sensitivity to suggest the feasibility of using this system. In this study, we describe our imaging hardware and analyze the sensitivity of it when combined with the reconstruction algorithm presented by Borsic *et al* [2].

2. Methods

The TREIT system is incorporated into a standard TRUS probe (TargetScan from Envisioneering Medical Technologies, St. Louis, MO, USA) to provide a multi-modal imaging device. The ultrasound (US) provides anatomic information of the prostate and peri-prostatic space, which can be used as *a priori* information to generate finite element method (FEM) meshes of the prostate gland and guide Electrical Impedance Tomography (EIT) image reconstruction. A flexible kapton circuit with thirty electrodes lying along the periphery of the acoustic window without impeding the ultrasound field (shown in Figure 1(a)) bridges the electrodes and the data acquisition (DAQ) system, which is comprised of a custom built front-end switching network interfaced to a NI USB 6259 data acquisition unit (National Instruments, Austin, Texas). The data acquisition scheme involves driving an AC voltage at 3.2 kHz through a pair of electrodes and sensing passive voltages on the remaining 28 electrodes. An average channel SNR is 81 dB.

A suite of phantom experiments were employed to assess the imaging capabilities of this system. The TREIT probe is submerged into a tank filled with a saline solution so that the probe tip points downward. The conductivity of the saline solution was measured with a conductivity meter (CDH222) to be 0.1 S/m. The radius of the tank is 12 cm, and the height of the saline bath is 20 cm. Inclusions are submerged into the saline bath, and positioned in front of the the electrodes for imaging (Figure 1(b)). In order to better describe the position of the inclusion, it is convenient to view the imaging system in a cylindrical coordinate system. Within this system, the inclusion center is defined by its distance from the outer surface of the probe (ρ), the angle spanned between it and the center of the acoustic window (ϕ), and its height relative to the lower edge of the acoustic window (z).

For this study, three specific factors affecting the sensitivity of the TREIT system are analyzed:

(i) Size
Spherical stainless steel balls of four different sizes are used as high contrast inclusions to assess image sensitivity to inclusion size. The radii of the balls are $R_1 = 1.03\,\text{cm}, R_2 = 0.76\,\text{cm}, R_3 = 0.64\,\text{cm}, R_4 = 0.52\,\text{cm}$.

(ii) Position
A spherical stainless steel ball with radius $R_3 = 0.64\,\text{cm}$ is imaged at different positions to assess regional sensitivity. The dimensional sensitivity assessed includes:

International Conference on Electrical Bioimpedance IOP Publishing
Journal of Physics: Conference Series **224** (2010) 012067 doi:10.1088/1742-6596/224/1/012067

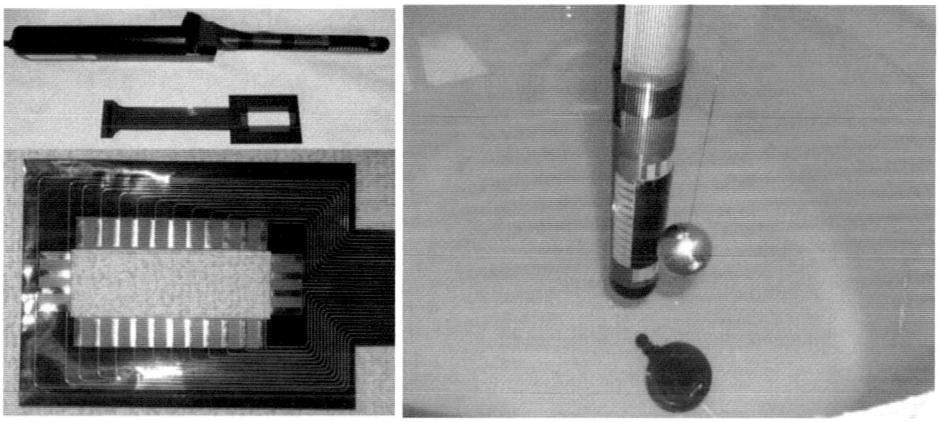

(a) (b)

Figure 1. (a) Upper: flex circuit of electrodes wrapped around the ultrasound probe; lower: magnified view of flex circuit of electrodes laid flat. (b) Imaging configuration: the inclusion is placed immediately in front of the acoustic window..

(a) Radial distance from the probe: the inclusion is placed at $\rho_1 = 0.69\,\text{cm}, \rho_2 = 1.51\,\text{cm}, \rho_3 = 2.30\,\text{cm}, \rho_4 = 3.93\,\text{cm}$, and $\rho_5 = 5.00\,\text{cm}$ with $\phi = 0^\circ$ and $z = 3\,\text{cm}$ fixed.
(b) Angular position: the inclusion is placed at $\phi_1 = 26.85^\circ, \phi_2 = 0^\circ$, and $\phi_3 = -27.58^\circ$ with $\rho = 1.71\,\text{cm}$ and $z = 3\,\text{cm}$ fixed.
(c) Height: the inclusion is positioned at $z_1 = 4.8\,\text{cm}, z_2 = 3\,\text{cm}$, and $z_3 = 1.2\,\text{cm}$ with $\rho = 1.71\,\text{cm}$ and $\phi = 0^\circ$ fixed.

(iii) Conductivity
Inclusions of different conductivity are created by filling an electrically conductive enclosure (natural collagen-based membrane) [5] with saline solutions of various NaCl concentrations to assess sensitivity to inclusion contrast. Eight saline solutions with conductivities ranging from $0.02\,\text{S/m}$ to $1\,\text{S/m}$ are tested and compared in this study.

3. Results

The TREIT reconstructed images are shown in Figure 2. Conductivity is plotted along a horizontal line through the center of the imaged inclusion (Figure 3). The reconstructed inclusion radius decreases linearly with the size of the actual inclusion, indicating that the TREIT system is able to accurately distinguish between different sized high contrast objects. For distant inclusions (ρ_4 and ρ_5), there are no well defined regions of high conductivity observed near where the inclusion is positioned. For nearfield (ρ_1 and ρ_2), the reconstructed inclusions more accurately depict the correct inclusion position. The reconstructions of inclusions at different angular positions and height showed that the TREIT probe has slightly better sensitivity in the middle than at the edge of the electrode array. In addition, the reconstructed inclusions at the edge of the acoustic window appeared closer to the middle than their actual position. In the reconstructed images of different conductivity, it is obvious that the reconstructed conductivity gradually increases with the true inclusion conductivity. For inclusion conductivities of $0.1\,\text{S/m}$ and $0.12\,\text{S/m}$, it is very difficult to discern the inclusion from the background. For higher conductivity contrasts, it is relatively easy to distinguish inclusions from the background in the reconstructed image.

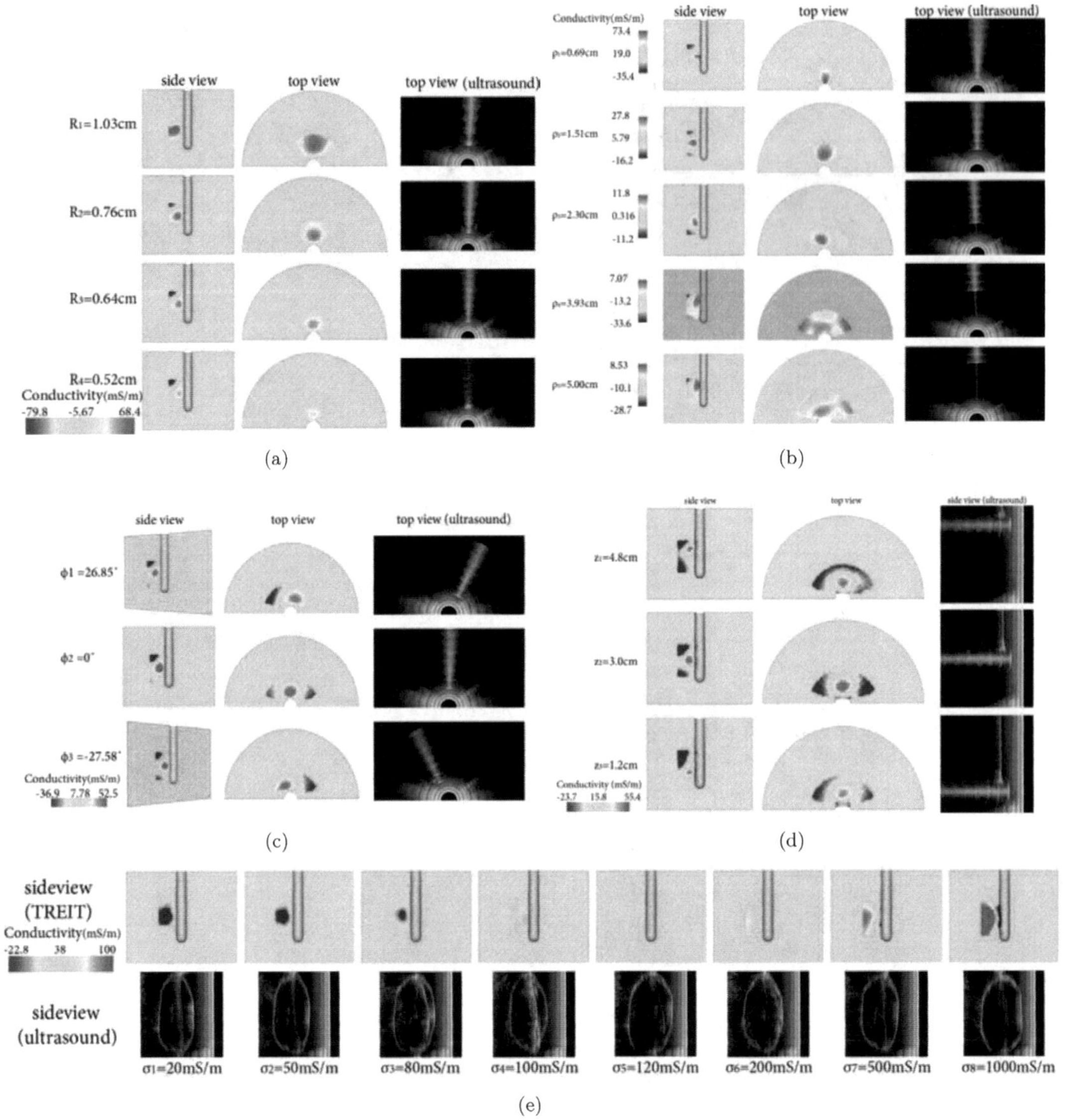

Figure 2. TREIT reconstructed images: (a) with high contrast inclusions of 4 different sizes; (b) with high contrast inclusions at 5 different radial distances from the probe; (c) with high contrast inclusions at 3 different angular positions; (d) with high contrast inclusions at 3 different heights; (e) with inclusions of different contrasts. Transverse ultrasound images are displayed as a reference of inclusion sizes in (a) and inclusion positions in (b) and (c), and side views of ultrasound images are shown as a reflection of the inclusions at different heights in (d) and the shapes and positions of low contrast inclusions in (e). In the ultrasound images in (a), (b), (c) and (d), artifacts (bright tails after inclusions) are induced by mismatch of ultrasound reflection index of stainless steel ball and the saline solution.

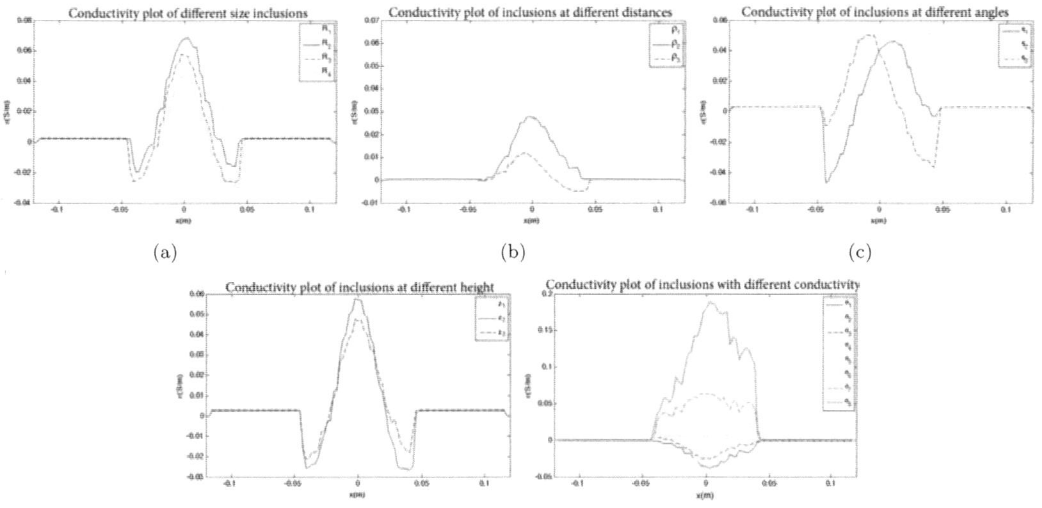

(a) (b) (c)

(d) (e)

Figure 3. Conductivity plot over a line going through the center of the reconstructed inclusion: (a) with high contrast inclusions of 4 different sizes; (b) with high contrast inclusions at 5 different radial distances from the probe; (c) with high contrast inclusions at 3 different angular positions; (d) with high contrast inclusions at 3 different heights; (e) with inclusions of different contrasts.

4. Discussion

For high contrast inclusions of different sizes, the reconstructed object is clearly dependent on the inclusion radius. The ratio between reconstructed inclusion radius and true inclusion radius ranges from 2 to 2.4. High contrast inclusions of 1 cm diameter are easily observed and we expect that smaller high-contrast inclusions will also be detectable. Similarly sized inclusions have been reconstructed with closed-domain EIT systems [6][7].

The TREIT probe loses sensitivity at $\rho_4 = 3.93\,\mathrm{cm}$ and is regarded as an estimate of the limit in radial sensitivity. The conductivity and size of the reconstructed inclusions are somewhat dependent on the distance from the probe surface. Smaller current densities exist far from the probe surface as compared to the probes near field, leading to a decreased far-field sensitivity. Jossinet *et al* [8] found that the radial sensitivity for an open-domain EIT imaging was limited to detecting objects centered at a distance of 2 to 3 radii from the surface of the probe. We found that at a radial distance of 2.30 cm the inclusion was accurately gauged. This distance represents 2 times the radius of the probe (radius $= 1.14\,\mathrm{cm}$). The inclusion appears to be closer to the probe in reconstructed image for ρ_3 than ρ_2, however the conductivity value of the reconstructed inclusion at ρ_3 is much smaller than that at ρ_2. This positional inaccuracy ρ_3 results from a non-unique solution to this ill-conditioned EIT inverse problem. The conductivity distribution in the reconstructed image of ρ_3 produces voltages at the electrodes matching the experimental measurements. Alternatively stated, the positional inaccuracies are compensated for by the decrease in reconstructed inclusion conductivity. Having ultrasound as a guidance with *a priori* geometric information is expected to better condition the problem and improve the accuracy of the reconstructed image.

For the angular positions, the maximum accuracy occurs at 0°, which is in the middle of the electrode window. Similarly, as the z-location of the inclusion was adjusted the maximum accuracy was observed when it was centered within the acoustic window. This occurs because the current density is larger flowing through the central regions. Despite different arrangement of the electrode array, Borsic *et al* [4] showed similar results in their theoretical sensitivity analysis. Specifically, the anterior and lateral aspects of the imaging domain will have the lowest sensitivity, while the central pixels closest to the probe surface have optimal sensitivity.

In the low contrast sensitivity analysis, we are able to assess conductivity variation 2 cm away from probe face as evidenced by the well defined spherical reconstruction. The smallest discernible inclusion conductivity is 0.08 S/m in a saline background solution of 0.1 S/m, and the contrast is $(1.0 - 0.08)/0.1 = 20\%$. This level of contrast is similar to that observed in conductivity between cancerous and stroma tissues (23%) at 10 kHz and is much less than the contrasts observed in permittivity between benign and malignant tissues (43 - 63%), reported by Halter *et al* [3]. As a result, we expect this TREIT system to be able to distinguish cancerous from benign prostate tissue.

5. Conclusion

In conclusion, we demonstrated that our TREIT imaging system is able to image high contrast inclusions of at least 1 cm in diameter, and the effective range is 2 times the radius of the probe. The TREIT probe can easily distinguish high contrast inclusions located at different angular positions as well as different heights. In addition, the imaging system is capable of detecting inclusions of 20% conductivity contrast to the background. Currently, we are collecting clinical data with the system. Future work includes co-registeration of TREIT reconstructions with US images and incorporation of the anatomic information gleaned from US images into the TREIT reconstruction algorithm.

Acknowledgement

This work is supported by the NIH grants 5R01CA124925 and 1RC1EB011000-01.

References

[1] Jemal A, Siegal R, Ward E, Hao Y, Xu J, Thun M 2009: Cancer Statistics, 2009 *CA Cancer J. Clin 2009. 59:225-249.*

[2] A. Borsic, R. Halter, Y. Wan, A. Hartov, K.D. Paulsen 2009: Electrical Impedance Tomography Reconstruction for Three Dimensional Imaging of the Prostate *submitted to Physiological Measurement*

[3] Halter RJ, Schned A, Heaney J, *et* al 2009: Electrical Properties of Prostatic Tissues: I. Single Frequency Admittivity Properties *Journal of Urology 2009-10182:4, 1600-1607.*

[4] A Borsic, R Halter, Y Wan, A Hartov, K D Paulsen 2009: Sensitivity study and optimization of a 3D electric impedance tomography prostate probe *Physiological Measurement 2009 0630:6*

[5] P K Manwearing, Y Wan, R Halter and A Hartov 2009: A natural membrane material enabling multi-layer saline tank phantoms for use in EIT *submitted to Physiological Measurement*

[6] Ryan J. Halter, Alex Hartov, and Keith D. Paulsen, 2008: A Broadband High-Frequency Electrical Impedance Tomography System for Breast Imaging *IEEE Transactions on Biomedical Engineering, VOL. 55, NO. 2, Feburary 2008*

[7] Saulnier, G.J. N. Liu, C. Tamma, H. Xia, T.-J. Kao, J. C. Newell and D. Isaacson 2007: An Electrical Impedance Spectroscopy System for Breast Cancer Detection *Annual International Conference of the IEEE Engineering in Medicine and Biology - Proceedings, p 4154-4157, 2007*

[8] Jacques Jossinet, Emmanuel Marry, Adrien Matias 2002: Electrical Impedance Endotomography *Phys. Med. Biol. 47 2002 21892202*

Using of electrical impedance tomography for diagnostics of the cervix uteri diseases

O V Trokhanova[1], Y A Chijova[1], M B Okhapkin[1], A V Korjenevsky[2] and T S Tuykin[2]

[1]Yaroslavl State Medical Academy, Yaroslavl, Russia
[2]Kotel'nikov Institute of Radioengineering and Electronics of the RAS, Moscow, Russia

E-mail: trokhanova@yandex.ru

Abstract. The report presents results of investigation of the neck of the womb (cervix) of 64 women aged from 19 to 70, that formed 4 clinical groups (1st group: 15 women without any pathology of the cervix, 2nd group: 27 women with the erosion, 3rd group: 11 women with dysplasia, 4th group: 11 women with cancer of the cervix). The aim of this research is to assess the capabilities of electrical impedance tomography in the diagnostics of the cervix diseases. The methods of the research were: visual examination with the speculum, colposcopy, and biopsy of the cervix. Also the new method of visualization of the cervix was used - electrical impedance tomography with the help of gynecological impedance tomograph (GIT). The following results were obtained. The electrical conductivity of the cervix in norm and in different pathology has different indices, which allow differential diagnostics of benign and malignant diseases. Summary: the method of electrical impedance tomography allows not only visually estimate portio vaginalis, but receive indices of electrical conductivity of the cervix on the depth up to 0.8 cm and thereby reveal pathological changes in epithelium without invasive and operative intervention.

1. Introduction

More than 493 000 new cases of cervical carcinoma (8 for 100 000 of population) are being annually registered in the world. Carcinoma of uterine cervix mortality makes up about 280 000 women a year (3.5 for 100 000). Every 2.5 minutes in the world a woman dies from carcinoma of uterine cervix [1, 2]. In Russia more than 125 000 cases of this disease are annually detected (8.7 for 100 000) and about 7000 women a year die from cervical carcinoma [2]. Carcinoma of uterine cervix is the second in frequency women's cancer following cancer of mammary gland It is characterized by bimodal distribution of the patients with the morbidity peak at the age of 35 – 39 and 60 – 64.

Because of any clinical finding absence of precursor cervical cancer the question of the early detection of this pathology is urgent. The method of electrical impedance tomography can be used for imaging and diagnostics of the cervix. For its evaluation the gynecology electrical impedance tomography system has been developed [3]. It allows not only to estimate vaginal part of the neck of womb visually but get electroconductivity characteristics of the cervix at the 8 mm depth and thereby detect epithelium pathological changes without invasive and surgical intervention.

The aim of this research is to assess the capabilities of electrical impedance tomography in the diagnostics of the diseases of the neck of womb.

2. Materials and methods of examination

The report presents main results of a comprehensive examination of the cervix of 64 women from 19 to 70 years.

The following diagnostic methods were applied: visual examination on the uterine speculum, colposcopy, biopsy of the cervix. All examination results were statistically processed by alternative variation method with obtaining a mean value and standard deviation. To check the data reliability we resorted to Student's criterion.

Also the new method of visualization of the cervix was used - electrical impedance tomography with the help of compact gynecological impedance tomograph (GIT). The apparatus used was a 32-electrode 3D electrical impedance tomography system developed by the Institute of Radioengineering and Electronics of the Russian Academy of Sciences. The frequency used was 50 kHz. The work area of the tomography with 32 electrodes is applied to the portio vaginalis. Auxiliary electrode is placed to the woman's wrist. During the scanning process the device, utilizing the 32 contact electrodes of the matrix successively injects into the cervix weak alternative electric current and registers corresponding distributions of electric potentials on its surfaces. The obtained data are used then for reconstruction of the electrical impedance images with the help of the mathematical algorithms utilizing a personal computer to which the device is connected via the USB port. After the reconstruction procedure is over, three images of the cervix, matching the three scanning planes, appear on the screen. The first scanning plane corresponds to the 2 mm depth from surface, the other too planes follow every 3 mm to the depth 8 mm.

In order to define particularities of the cervix electrical conductivity in norm and in case of pathology were subjected to clinical methods of examination, colposcopy and biopsy. As a result the following clinical groups were formed:

- The 1st clinical group composed of 15 women without any pathology of the cervix.
- The 2nd clinical group composed of 27 women with the ectopia of the cylindrical epithelium of the cervix.
- The 3rd clinical group composed of 11 women with dysplasia of the cervix.
- The 4th clinical group composed of 11 women with cancer of the cervix.

3. Examination results

3.1. Electrical impedance characteristics in norm

From the anatomical point of view, the cervix is a smoothly muscular organ covered with the laminated pavement nonkeratinous epithelium. Cervical canal which is situated inside is covered with the cylindrical epithelium. The zone of two kinds of epitheliums crossing corresponds to the external fauces and as a rule, is the place where the pathology of the cervix locates. On the electrical impedance images a zone of the external fauces looks like the hypoimpedance area the size of which is different at different phases of the menstrual cycle. The size of the area is bigger during the first phase when estrogens predominate and it diminishes at the second phase when progesterone influences. The zone of the external fauces of the women of the postmenopausal period is not visualized because of the reduction of the estrogen and progesterone background at this age. The portio vaginalis looks like isoimpedance mass which corresponds to the muscular layer with hypoimpedance inclusions.

The characteristics of the average electroconductivity of the cervix during the different physiological periods of life at the different scanning depth are shown in the table 1. As the table figures show there is no statistically significant difference between the rates of the electrical conductivity of the cervix of the women of different physiological periods in normal state ($p>0.05$). The rates of the electrical conductivity statistically don't have any difference at the various scanning levels ($p>0.05$).

Table 1. The characteristics of the average electroconductivity of the cervix during the different physiological periods of life at the different scanning depth, M±SD.

	1 level (2 mm)	2 level (5 mm)	3 level (8 mm)
Juvenile period	0.96±0.10	0.96±0.10	0.96±0.20
Reproductive period	0.96±0.09	0.95±0.10	0.94±0.10
Postmenopause	1.003±0.005	1.007±0.005	1.007±0.005

3.2. Electrical impedance characteristics in case of pathology of the cervix
The most frequent pathology of the cervix is the ectopia of the cylindrical epithelium and the dysplasia of the cervix which is considered as the precancerous state. So, naturally, the carcinoma of uterine cervix is of the great interest as the state threatening the life of a woman. Having this pathology, we can see visually, in the electrical impedance images, the extension of the hypoimpedance zone of the external fauces or the appearance of the hypoimpedance foci which correspond to the locating of dysplasia or cancer. The characteristics of the average electroconductivity of the cervix with ectopia, dysplasia or cancer in comparison with the normal state are shown in table 2.

Table 2. The characteristics of the average electroconductivity of the cervix during the different physiological periods of life in normal state and during the pathology at the different scanning depth, M±SD.

	1 level(2 mm)	2 level (5 mm)	3 level (8 mm)
Juvenile period	0.96±0.10	0.96±0.10	0.96±0.02
Reproductive period	0.96±0.09	0.95±0.10	0.94±0.10
Postmenopause	1.003±0.005	1.007±0.005	1.007±0.005
Ectopia of the cylindrical epithelium	0.97±0.06	0.96±0.07	0.96±0.08
Dysplasia of the cervix	0.99±0.07	0.98±0.09	0.99±0.10
Carcinoma of uterine cervix	0.99±0.04	0.99±0.03	0.98±0.04

The comparison of the characteristics of the average electroconductivity of the cervix in the normal state and during the pathology shows the slight increase of electroconductivity having dysplasia and the carcinoma of uterine cervix in comparison with the electroconductivity of the cervix in the normal state of the menstruated women. But the difference is not statistically trustworthy ($p>0.05$). There is no statistically trustworthy difference in various scanning levels during both pathology and the normal state ($p>0.05$). But the clear difference appear while estimating the electroconductivity of the hypoimpedance zone of the external fauces, table 3, which corresponds to the two borders between two types of epitheliums and, as a rule, the place where the pathology of the cervix is situated.

Table 3. The characteristics of the electroconductivity of the zone of the external fauces of the cervix in the normal state and the pathology at the different scanning depth, M±SD.

	1 level (2 mm)	2 level (5 mm)	3 level (8 mm)
1 phase MC	1.16±0.03	1.13±0.04	1.04±0.06
2 phase MC	1.050±0.006	1.05±0.10	1.050±0.006
Ectopia of the cylindrical epithelium	1.20±0.06	1.14±0.02	1.07±0.05
Dysplasia of the cervix	1.27±0.03	1.17±0.01	1.06±0.08
Carcinoma of uterine cervix	1.300±0.005	1.280±0.006	1.14±0.02

During the second phase of the menstrual cycle the electroconductivity of the zone of the external fauces is less than in the first one on the first two scanning levels ($p<0.01$) and doesn't differ at different depth ($p>0.05$). In the normal stage at the first phase of the menstrual cycle and at any pathology the characteristics of the electroconductivity of the zone of the external fauces statistically are positively less ($p<0.01$) at the deeper scanning level. There is a statistically trustworthy difference in the characteristics of the electroconductivity in disease in comparison with the normal state at a scanning depth of 2 mm ($p<0.01$, $p<0.05$) and a statistically trustworthy increase of electroconductivity having the cancer of the cervix at a depth of 5 and 8 mm in comparison with the normal state and beginning pathology. The difference in the characteristics of the electroconductivity of the zone of the external fauces depends on the invasion degree and the degree of the epithelium transformation.

4. Conclusions

In spite of the absence of significant difference in the characteristics of the average electrical conductivity of the cervix uteri in the normal state and the pathology, there are clear differences in the characteristics of the electrical conductivity of the zone of the external fauces which can be the criteria of the diagnosis of different pathology of the cervix uteri.

Method of the electrical impedance tomography of the cervix meets all the requirements to the screening diagnostic technique of the precancerous disease: it is simple, accessible, noninvasive, has no any restrictions for repetition of examination tests, doesn't require any special patient's preparation, unlike other methods allows to detect the pathological changes of the cervix at the depth of up to 8 mm without an operative intervention. Because of its safety, it is possible to use the method during pregnancy and for the young nullipara women when biopsy of the cervix uteri is undesirable.

References

[1] Marchetta J, Descamps P 2008 *La Colposcopie* (Paris: Masson)

[2] Gur'ev D, Okhapkin M and Khitrov M 2004 *General problems of oncogynecology. Textbook for physicians and interns* (Yaroslavl: Yaroslavl State Medical Academy)

[3] Korjenevsky A, Cherepenin V, Trokhanova O and Tuykin T 2010 Gynecologic electrical impedance tomograph *Proc. XIVth Int. Conf. Electrical Bioimpedance and 11th Conf. Biomedical Applications of EIT (Gainesville, USA)*

Journal of Physics: Conference Series **224** (2010) 012069 doi:10.1088/1742-6596/224/1/012069

Clinical application of Electrical Impedance Tomography in the Present Health Scenario of India

Dr KL Chakraborti[1] and **Dr W Selvamurthy**[2]

[1]Head, Deptt. Of Radiological Imaging, Institute of Nuclear Medicine & Allied Sciences, Brig. S. K. Mazumdar Road, Timarpur, Delhi-110054, India

[2] Chief Controller (R&D) LS, DRDO HQrs, DRDO Bhawan, N. Delhi-110015, India

E-mail: drklchakraborti@yahoo.co.in

Abstract. Early detection of Breast Cancer is currently emerging as a big clinical entity requiring a non invasive, radiation less, harmless, cost effective diagnostic technique. Survival is improved if detected early. Breast Cancer is the second most common cancer in India. Health corporate system of India is urgently requiring a cost effective, noninvasive novel technique like "Electrical Impedance Tomography (EIT)" for screening large poor rural population of India for early diagnosis of Breast Cancer. EIT is the technique to visualize spatial distribution of Electro-impedance (or conductivity) inside the object, such as human body. A medical device which allows imaging of the distribution of conductivity in 3D in regions below the skin surface has been developed and tested. Its purpose is to enable early detection and preliminary diagnosis of breast tumors. The system uses a planar array consisting of 256 electrodes and enables obtaining images of the three-dimensional conductivity distribution in regions below the skin's surface up to several centimeters deep. The developed measuring system and image reconstruction algorithm can be used for breast tissue imaging and diagnostic, in particular for malignant tumor detection. Initially ten patients as control and ten patients with breast lesions have been studied with this new technique. It was found that electrical impedance mammograms from different groups has clear visual distinctions and statistically significant difference in breast glands conductivity. The results are quiet encouraging. EIT may emerge as the first line noninvasive imaging method of choice for screening large population for early detection of breast cancer.

1. Introduction

Breast cancer is the second most cancer in the women worldwide. Survival is improved if detected early. Therefore, early detection of breast cancer is of great clinical importance both from the angle of treatment and total cure. Various imaging techniques available for diagnosis of breast cancer are x-ray mammography, ductography, xeroradiography, thermography, ultrasound,CT Scan,MR imaging, isotope imaging and PET scanning etc. WHO has introduced different screening procedures for early diagnosis of breast cancer. X-Ray Mammography screening is the only currently known method and also is considered as gold standard. But X-Ray Mammography involves lots of ionizing radiations. Too frequent exposure can also be a cause of breast cancer. It is not advised for young patients below 30 years of age, pregnant women and in screening of large dense breasts. Moreover x-ray mammography/digital mammography machines are very costly and available only in main city areas. Other techniques like xeromammography and thermography are less sensitive. Remaining imaging procedures are too costly. Hence a non-invasive, radiation less cost effective technique like electrical impedance tomography/mammography is welcomed. Electrical impedance tomography (EIT) is an wonderful technique which provides a safe, non invasive way to visualize the electrical properties [1,2] of tissues inside a human body. An ingenious reconstruction algorithm and measuring approach introduced by Barber and Brown has been widely developed since 1980s. Basing on the fact that different tissues of human body have different electrical properties, EIT is able to image an anatomical structures. But till recently medical EIT was able to produce only dynamic images, i.e images of conductivity changes that occurred between two consequent measurements. But for medical diagnosis of the different pathologies requires recapability of EIT system to image static objects inside a human body. The main difficulties in implementing EITs in clinical practice was the uncertainness of the human body shape, electrode spacing and low spatial resolution. Subsequently development of a fast and robust static EIT reconstruction method and efficient measuring equipment has made the implementation of EIT in clinical practice possible.

Application of this method to breast cancer detection requires a special design of the measuring system and image reconstruction algorithm, as imaging of a three-dimensional distribution of conductivity is required. The breast cancer detection device, electric impedance mammography machine designed and developed by the Institute of Radio-Engineering and Electronics of the Russian Academy of Science and is being currently marketed by TD-SIM Technika of Yaroslavl city, Russia. It is a 3D EIT system which consists of a compact of electrodes positioned over the tissues being measured, two additional electrodes placed remotely from the array of electrodes, a source of alternating current (AC), a means to measure potential difference and computing means to reconstruct and visualize the conductivity distribution as stack of tomography images [3,4]. Utilizing this new system, a complete 3-D diagnostic test takes 20 seconds for each breast totaling to 40 seconds for both the breasts. Seven tomographic image slices appear on a computer monitor within 40seconds at depths of 0.4cm to 4.6cm to detect and isolate tumors. These coronal images include tissue near the chest wall. This reconstruction process can be compared to CT or PET scanning but without radiation of any kind and at a significantly lower cost.

2. Materials and Methods

An electrical impedance tomography machine intended for imaging and diagnosis of pathological changes of the breast tissue is used. The system uses a planar array consisting of 256 electrodes and enables obtaining images of the three-dimensional conductivity distribution in regions below the skin's surface up to several centimeters deep. The electrodes are arranged in a round matrix with a diameter of 12 cm. The visual result is several cross sections of the medium, which are parallel to the electrode array and located at different depths. During examination the plane of electrodes is pressed against the breast and this increases the number of electrodes in contact with the breast and decreases the thickness of the tissue layer to be measured. In this study total ten patients as control and ten patients with breast lesions have been studied. The patients were of different age groups ranging from 20 years onwards. To begin with routine clinical history has been taken. Physical examination by palpable method has been done. Following physical examinations , the patients were prepared for electrical impedance mammographic examination. For this, the patients' breasts were evenly moistened with water, avoiding formation of droplets. Along with this the patient's wrists (where the corresponding electrode will be applied later) were moistened. Then electric impedance mammography examination were performed first in supine position followed by standing position. For each breast the study takes 20 seconds i.e., 40 seconds for both the breast study. Though initially the study has been done using 50 KHz electric potential, subsequently it has been seen that the clinical results are better if 10 KHz electric potentials are also used for both the breasts. Therefore, each breast has been studied by using 50 KHz and 10 KHz as well. After the study both visual and quantitative estimation of the data done. The impedance curve indicates the distribution type and quantitative representation of the conductivity of different tissues of the breast during measurement at that particular frequency. Each impedance peak was analyzed for: a) Mean electric conductivity, b) Presence of shift from the central line, c) Whether it is symmetrical or not?, d) Whether it is unimodular or multi-modular?. The patients were studied by ultrasound and at times by ultrasound Doppler. Wherever possible the x-ray mammography correlation has been done.

3. Results and Analysis

It has been found that electrical impedance mammograms from different age groups had clear visual distinctions and statistically significant difference in breast glands' conductivity. In grey scale, changes from dark to light indicates from low electric conductivity to high electric conductivity. Increased conductivity areas has been seen as white spot. The conductivity increased from fat to cancer. For quantitative estimation, the differences of conductivity distribution between the normal group and from the group with the breast tumour/malignancy has been studied. It was found that results were varying along with the different hormonal status during the cycle. The ladies who were taking oral pills/contraceptives the impedance results varied. Out of 10 patients with the breast lesions 3 patients had fibroadenoma which is shown in Figure 1, 3 patients had cystic lesions, 2 patients had ductal carcinoma, 1 had micro calcifications and 1 had breast tuberculosis. In this cases it is found that the patients with the asymmetrical distribution of the conductivity almost always had multi-modular conductivity curve. This type of findings indicated malignancy. Benign lesions had always unimodular curve and symmetrical distribution of the curve, which is depicted in Figure 2. According to conductivity values the inference also could be drawn. The findings were correlated in most of the cases with the ultrasound scanning, an example of which is depicted in Figure 3. In few cases x-ray mammography study also were correlated, as shown in Figure 4. The diagnosis of the breast pathology was confirmed post surgical biopsy specimens/fine needle aspiration cytology. It is been seen the electric impedance findings if correlated with the ultrasound scanning the provisional diagnosis of the breast diseases/pathology could be done.

4. Discussion

It is an well established fact that every organ in the man's body posses its own electrical characteristics, which in turn defines electrical conductivity of the tissues of that organ and surroundings. This characteristics are well defined and varied in electrical properties of both intravascular and extravascular fluid. The electrical impedance mammograph makes it possible to obtain images of electroconductivity distribution of biological tissues along several cross section of a human body, thus defining the areas with abnormal values of electroconductivty [3,4]. Implementation of the popular screening programs also helps to an greater extent. One of the most important Etiology of breast cancer is ionizing radiation. Presently various imaging techniques like x-ray mammography, ductography, xeroradiography, thermography, ultrasound,CT Scan,MR imaging, isotope imaging and PET scanning etc are available. Out of these, x-ray mammography is considered as gold standard for detecting breast cancer. All these techniques are costly and not available at village level. Besides, it becomes unaffordable for public health sector to screen large population at village level involving a high cost.

India is a developing country. The majority of its population are rural based. The costly machine like x-ray mammography (which is considered as gold standard for screening for early detection of breast cancer) is available only in city areas. Moreover the procedure is very expensive and can not be afforded by majority of the village people. It is seen that the village women with the breast complaints whether it is major or minor undergo psychological trauma. Sometimes hormonal fluctuations causing breast problems can force the poor village people to run to city areas to undergo many costly diagnostic techniques and at times surgical biopsy procedures. Ultimately after undergoing all costly diagnostics techniques it is found that this segment of patients did not have any breast tumours/malignancy, it was only a hormonal problem, which did not require at all the help of costly techniques. Hence a screening technique like electrical impedance mammography will have a significant role in removing the fear psychosis of breast cancer from the mind of poor village people. Moreover in countries like India where the electrification in rural areas are not very proper and availability of electricity at all the times are uncertain, there the technique like electrical impedance which can run on UPS system will be of great value. Following screening of large population, only somebody suspected of pathology of breast can be referred for further investigation and management.

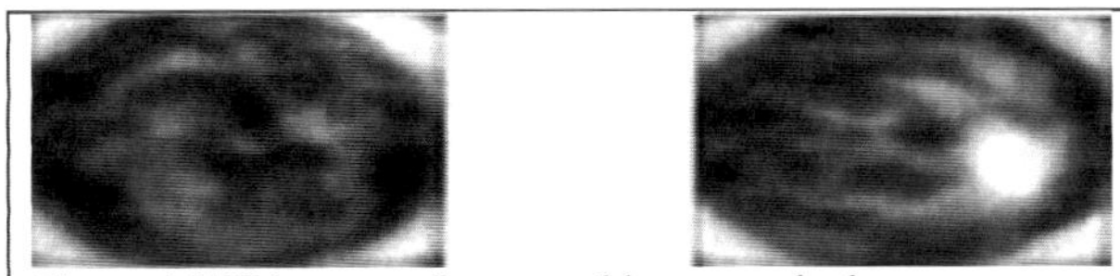

Figure 1 EIT images of a normal breast and a breast mass

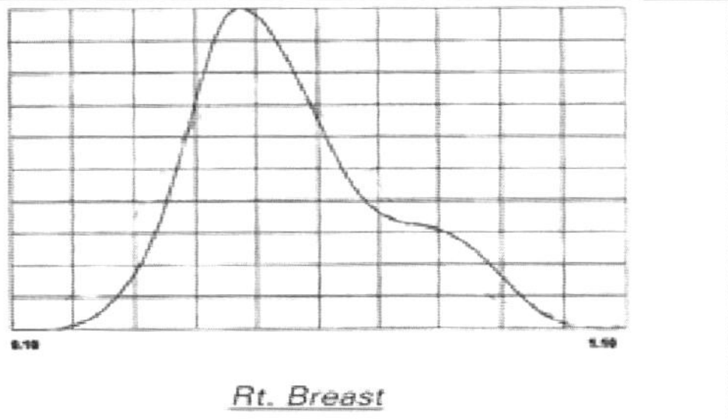

Figure 2 Impedance curve of a breast mass

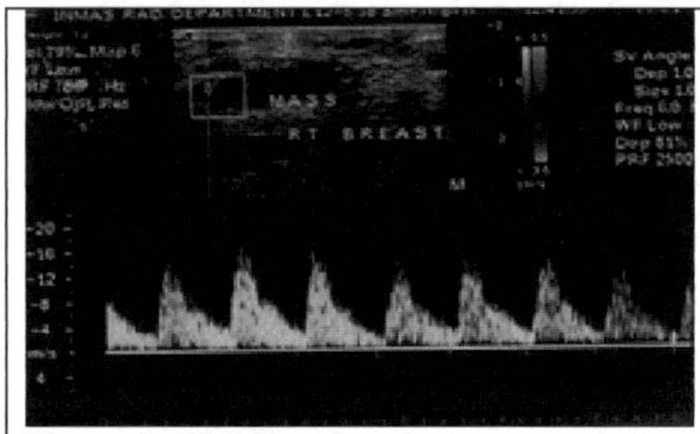

Figure 3. USG Doppler of a breast mass

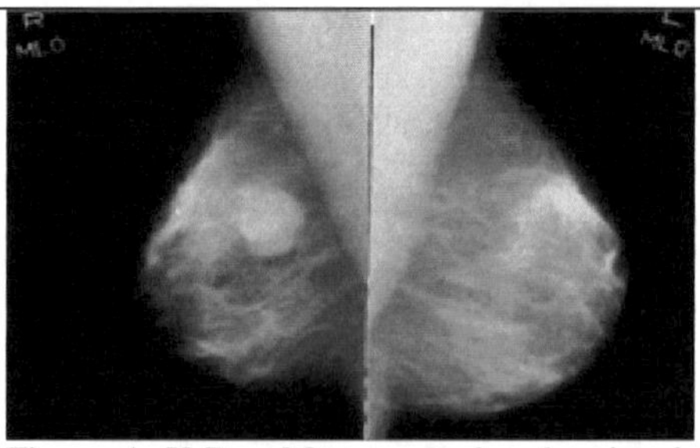

Figure 4. X-Ray Mammography showing breast mass

Acknowledgements

This work was supported by Institute of Radio Engineering & Electronics, Moscow, Russia, under an Indo Russian collaborative ILTP Project No. A-9.20

References

[1] Surowiec J, Stuchly S S, Brar J R, and Swarup A 1988 Dielectric properties of breast carcinoma and the surrounding tissues *IEEE Trans. Biomed. Eng.*, vol. BME-35, pp 257-262

[2] Assenheimer, M, Laver-Moskovitx, O, Malonek, D, Manor, D, Nahaleil, U, Nitzan, R and Saad, A. 2001 The T-SCAN TM technology: electric impedance as a diagnostic tool for breast cancer detection *Physiol. Meas.*, vol 22 pp 1-8

[3] Cherepenin V, Karpov A, Korjenevsky A, Korneinko V, Mazaletskaya A, Mazourov D and Meister D 2001 A 3D electrical impedance tomography (EIT) system for breast cancer detection *Physiol. Meas.*, vol. 22, pp 9-18

[4] Cherepenin V, Karpov A, Korjenevsky A, Korneinko V, Kultiasov Y, Ochapkin M, Trochanova O and Meister D 2002 Three-dimensional EIT imaging of breast tissues: system design and clinical testing *IEEE Trans. Medical Imaging*, vol. 21(6), pp 662-667

Gynecologic electrical impedance tomograph

A Korjenevsky[1], V Cherepenin[1], O Trokhanova[2] and T Tuykin[1]

[1]Kotel'nikov Institute of Radioengineering and Electronics of the RAS, Moscow, Russia
[2]Yaroslavl State Medical Academy, Yaroslavl, Russia

E-mail: korjenevsky@cplire.ru

Abstract. Electrical impedance tomography extends to the new and new areas of the medical diagnostics: lungs, breast, prostate, etc. The feedback from the doctors who use our breast EIT diagnostic system has induced us to develop the 3D electrical impedance imaging device for diagnostics of the cervix of the uterus - gynecologic impedance tomograph (GIT). The device uses the same measuring approach as the breast imaging system: 2D flat array of the electrodes arranged on the probe with handle is placed against the body. Each of the 32 electrodes of the array is connected in turn to the current source while the rest electrodes acquire the potentials on the surface. The current flows through the electrode of the array and returns through the remote electrode placed on the patient's limb. The voltages are measured relative to another remote electrode. The 3D backprojection along equipotential surfaces is used to reconstruct conductivity distribution up to approximately 1 cm in depth. Small number of electrodes enables us to implement real time imaging with a few frames per sec. rate. The device is under initial testing and evaluation of the imaging capabilities and suitability of usage.

1. Introduction

The cervix cancer is the second cancer by frequency at women after the breast cancer. In view of absence of any clinical picture of the disease predecessors, the problem of early detection of this pathology is topical. Carrying out the screening which allows suspecting the disease conditions is necessary because of this. The requirements to the screening procedure are: availability, low cost, informativity (sensitivity should be 60 - 75 % at least), revealing of disease at early stage (preinvasive cancer). Currently there is a set of methods for diagnostics of cervix diseases including biopsy, echography, MRI and so on, but besides the set of advantages, there is a number of the drawbacks limiting usage of these methods for screening examinations.

The feedback from the doctors who use our breast EIT diagnostic system [1 - 3] has induced us to develop the 3D electrical impedance imaging device for diagnostics of the cervix of the uterus - gynecologic impedance tomograph (GIT). The developed device allows not only to estimate visually vaginal part of cervix of the uterus, but also to obtain parameters of its electrical conductivity at depth of almost 1 cm thus revealing pathological changes in epithelium without invasive and operative interventions. In addition the EIT method will probably enable to investigate features of electrical conductivity during background diseases of the cervix.

The electrical impedance tomography of the cervix of the uterus meets all screening method requirements for diagnostics of precancer diseases: it is simple, accessible, noninvasive, it has no restrictions for the frequency of procedure repetition, does not require special preparation of the patient, allows to detect pathological changes in cervix at depth up to 1 cm without surgical intervention, it is informative and painless. Safety of the method allows using it during pregnancy and also at young nonparous women for whom carrying out biopsy is undesirable.

2. Materials and methods

The GIT device uses a method of electrical impedance tomography. It consists of the measuring unit with the built-in microcontroller system, scanning head with 32 electrodes arranged in flat 2D array, see figure 1, and a dual auxiliary electrode connected to the main unit. During investigation the device injects alternating electric current into the body of patient by means of each single electrode of the head sequentially and measures corresponding distributions of electric potentials by means of the rest electrodes.

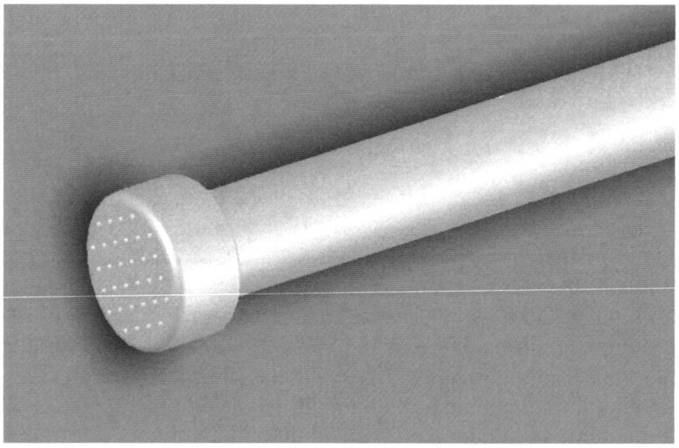

Figure 1. Measuring head of the tomograph with array of 32 electrodes. Diameter of the wide part of the head is 3 cm.

The tetrapolar measurement approach is used: the current is injected through the single electrode of the array and first auxiliary electrode placed on the patient's wrist. The voltage distribution is measured relative to second auxiliary electrode. All possible combinations of injecting and receiving electrodes are used. The redundant data caused by reciprocity are used to improve accuracy of measurements. Received data are used then for solving inverse problem and showing on the PC screen of the electrical impedance tomograms of tissues under investigation by means of the reconstruction algorithm implemented on the personal computer. The measuring unit is connected with PC through USB connection, which provides power supply also. The software provides functions for the analysis, comparison and archiving of the results of measurements and reconstruction. Carrying out of dynamic repeated investigations with the purpose of revealing of the processes dynamics is possible. Use of multifrequency sounding allows GIT device to obtain the additional information on a condition of tissues, considerably raising informative value of the examinations.

The block-diagram of the measuring system is shown on figure 2. The low level hardware functions are controlled by the embedded microcontroller. Both generation of output and demodulation of input signals are carried out digitally. The embedded software and periphery enable measuring and compensation of residual galvanic voltages, detection of bad contacts of the electrodes with the object and self-testing of the whole measuring chain. Important difference between the GIT device and its prototype, the EIT system for breast imaging, is usage of active shields in the cables connecting electrodes of the head with the main measuring unit. Without such shielding, stray penetration of the signals between channels in 1-meter connecting cable may add artifacts in the reconstructed images. The measuring system can be programmatically configured to make measurements at any frequency between 1 and 50 kHz.

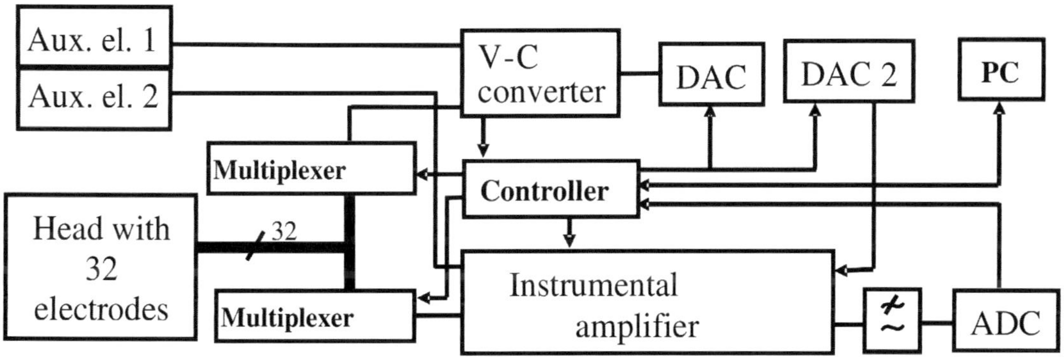

Figure 2. Block-diagram of the measuring system.

The image reconstruction is carried out using 3D weighted back projections along electric field equipotential surfaces. The reconstruction formula and weighting factors are the same as for breast imaging devicc [2]:

$$S(x,y,z) = 1 + W_1(z)\sum_i \frac{1}{\int\limits_{L(x,y,i)} W_2(l)dl} \int\limits_{L(x,y,i)} W_2(l)(E_r(l) - E_m(l))/E_r(l)dl,$$

where $S(x,y,z)$ is reconstructed conductivity, E_m and E_r are the measured and synthesized reference electric intensities at the surface of the object, W_1 is monotone increasing weight function, which equalizes sensitivity in depth *z*, $W_2 = 1/R^4 = 1/\left((x-x_l)^2 + (y-y_l)^2 + z^2\right)^2$, *i* is the number of the injecting electrode, and *L(x,y,i)* is the line of intersection of the equipotential surface with the plane $z = 0$, on which the electrodes are arranged.

The significant difference between the two systems is number of the electrodes in array: 256 in case of breast imaging device and only 32 for the gynecological tomography system. This makes real time imaging possible in the GIT device but limits quality of the images. To make 3D data readable on the 2D PC screen the reconstructed conductivity distributions are represented as set of three 2D slices with increasing depth.

3. Results

The measuring system is under preliminary clinical tests now [4]. Due to relatively small number of electrodes the device can provide near real time visualization with a few frames measured and reconstructed per second. The small number of electrodes provides images with quite poor resolution, see example on the figure 3. Nevertheless, high correlation between electrical properties of the tissues and their physiological state and conditions and relatively simple anatomy of the object under investigation makes such images usable for diagnostics. First tests have shown that cable connecting measuring head with the main unit may cause artifacts on the reconstructed images due to spurious coupling between channels if the wires are not shielded. This effect is strong for the objects with high impedance (skin resistance), when imaging hand, for example. But low impedance mucous membrane of the cervix makes its influence insignificant. Nevertheless using of active shields in the cable seems strongly desirable.

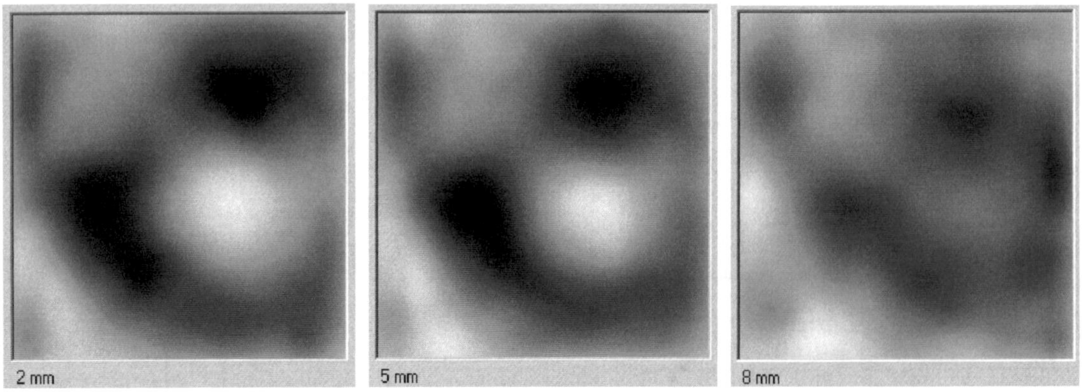

Figure 3. One of the first images obtained with the GIT device. The circular structure with bright area in the centre corresponds to the cervix cross-sections at different depth.

4. Conclusions

The first electrical impedance imaging system for examination of the cervix of the uterus has been developed. The device provides 3D EIT images of the conductivity distribution in a few cross-sections with depth up to 8 mm and with more than 1 frame per second measurement and reconstruction rate. The cervix cross-section anatomy is visible on the images. Further development should include optimization of the measuring head for convenient application and sterilization. The device should be designed as self-contained and independent on external PC, to be approved for most clinical scenarios of usage.

References

[1] Cherepenin V, Karpov A, Korjenevsky A, Kornienko V, Mazaletskaya A, Mazourov D and Meister D 2001 A 3D electrical impedance tomography (EIT) system for breast cancer detection *Physiol. Meas.* **22**(1) 9-18

[2] Cherepenin V, Karpov A, Korjenevsky A, Kornienko V, Kultiasov Y, Ochapkin M, Trochanova O and Meister D 2002 Three-dimensional EIT imaging of breast tissues: system design and clinical testing *IEEE Trans. Medical Imaging* **21**(6) 662-7

[3] Electrical Impedance Mammograph MEM http://medimpedance.com

[4] O V Trokhanova, Y A Chijova, M B Okhapkin, A V Korjenevsky and T S Tuykin 2010 Using of electrical impedance tomography for diagnostics of the cervix uteri diseases *Proc. XIVth Int. Conf. Electrical Bioimpedance and 11th Conf. Biomedical Applications of EIT (Gainesville, USA)*

Memristance in human skin

Ø.G. Martinsen[1,2], S. Grimnes[1,2], C.A. Lütken[1] and G.K. Johnsen[1]

1. Dept. of Physics, University of Oslo, Norway
2. Dept. of Biomed. and Clin. Eng., Oslo University Hospital, Rikshospitalet, Norway

E-mail: ogm@fys.uio.no

Abstract. The memristor is basically a resistor with memory, so that the resistance is dependent on the net amount of charge having passed through the device. It is the regarded the fourth fundamental component, in addition to the resistor, capacitor and inductor, that can be deduced from the four basic circuit variables; current, voltage, charge and magnetic flux. We show that memristors can be used for modelling electrical properties of human skin. In particular is electro-osmosis in human sweat ducts of memristive nature.

1. Introduction to memristance

The four basic circuit variables, current i, voltage v, charge q and magnetic flux φ, give six different, possible combinations. Five of them are the definitions of capacitance, inductance and resistance, the definition of electric current and Faraday's law of induction. The last combination is the ratio of magnetic flux to electric charge, now called *memristance* [1]. However, the memristance is not necessarily linked to magnetic systems and can be regarded as the ratio between the time dependent voltage and current:

$$M \equiv \frac{d\varphi}{dq} = \frac{v \cdot dt}{i \cdot dt} = \frac{v(t)}{i(t)} [\Omega] \tag{1}$$

The memristor can be thought of as a variable resistor where the resistance is dependent on the amount of charge having passed the device in a given direction. Predicted to exist on purely theoretical grounds many years ago [2], memristors did not gain acceptance as a new passive circuit element that complements the conventional ones (RLC) until a team from Hewlett-Packard recently announced that they had invented a nano-memristor that is fully compatible with conventional semi-conductor technology [3].

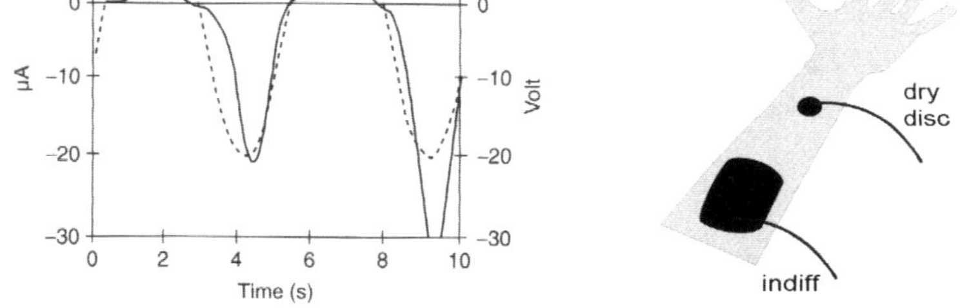

Figure 1: Applied potential (dashed line) and current response on ventral forearm [1,2].

2. Electroosmosis

The measured electrical current response of human skin shown in fig. 1 (reproduced from [4]), which was obtained using dry disc electrodes, was explained in [5] by electro-osmotic transport of fluid through pores in the skin, mostly the sweat ducts. In dry skin most of the ducts are empty or partially filled, but a conductive fluid film normally covers the inner wall of the duct, as shown in fig. 2.

When an electric field is applied the fluid column in the filled part of the duct is dragged along by the mobile part of the double layer near the wall. The amount of liquid dragged along is determined by the electric field strength E and electro-kinetic (zeta-) potential ζ at the interface; the dielectric constant ε, resistivity ρ and viscosity η of the fluid; as well as the geometry of the duct [2]:

$$\frac{dV}{dt} = \alpha A E, \qquad (2)$$

where dV is the volume of liquid transported in time dt under the influence of the electric field E. $A = \pi r^2$ is the cross-sectional area of the duct, and $\alpha = \zeta\varepsilon / 4\pi\eta$.

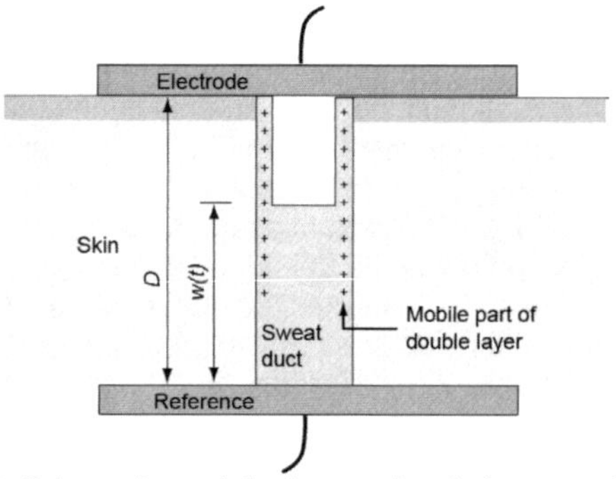

Figure 2: Schematic model of sweat duct in human epidermis.

The average resistance in the filled part of the duct is significantly smaller than in the thin fluid film covering the wall of an otherwise empty duct. Because the location *w(t)* of the fluid/gas interface ("water table") is a function of the electrical history of the duct, this system has some *memory* and is *hysteretic*. The main feature of the mathematical model developed in [5] is that it attempts to take this into account, by coupling the change in conductivity to the motion of the fluid.

3. Memristive model

Our objective is to reexamine and refine this explanation, taking into account some effects that were neglected in previous work [6]. As a first step we include a kind of feedback effect: when the water table moves a large effective resistivity is traded for a smaller one, or *vice versa*. The resistance in the water column is $R_w(t) = \rho w(t)/A$, and the resistance in the film is $R_f(t) = \rho(D - w(t))/a$. D is the total length of the duct and $a \approx 2\pi r d$, with $d << r$ denoting the thickness of the film. The total resistance is then:

$$M = R_w + R_f = R(\tau + (1-\tau)x), \qquad (3)$$

with $R = \rho D / A$, $\tau = r/2d$ and $x = w/D$, whence $0 \le x \le 1$.

This type of immittance is logically and physically distinct from impedance and is called *memristance* [1,7,8]. (The analogue of admittance is called *memductance*.)

We argue that the *memristive paradigm* may be useful also in biophysics. In particular, we show that electro-osmosis in ducts can be usefully thought of as a *memristive system* [7]. This is a generalization of the memristor concept where the memristance is controlled by any number of additional state-variables (in our case just one, $x = x(t)$), which may also be coupled to each other in complicated nonlinear ways. Thus, memristive systems should be rich enough to capture some of the nonlinearities that are ubiquitous in bioelectrics.

Consider first the "linear regime" where the water table remains at a finite distance from either end of the duct, $0 < x < 1$. Nonlinear effects at the ends where the liquid is injected and extracted are neglected here. Since the microscopic width $w(t)$ of the filled region is unobservable, we must relate it to macroscopic state-variables and device geometry that are easier to measure. Because the volume of fluid in our partially filled duct is Aw we have $dV = Adw$, which combined with eq. (2) gives:

$$dx = \alpha \frac{E}{D} dt = \alpha \frac{\rho}{V} dq \tag{4}$$

Here $V = AD$ is the total volume of the duct, and we have assumed that E is constant and that all the current $i(t) = dq / dt$ flows through the fluid column. Integrating eq. (4) gives $x(t) = x_0 + \alpha\rho q(t) / V$, which inserted in eq. (3) and using $\tau >> 1$, gives the charge controlled memristance of this system:

$$M(q) = \tau \; R(c_0 - \alpha\rho q / V), \tag{5}$$

where $0 < c_0 = 1 - x_0 < 1$ is determined by the initial position $w_0 = x_0 / D$ of the water table. Inserting this expression into $M(q)dq = v(t)dt$ and integrating we find the charge

$$q(t) = \frac{c_0 D^2}{\alpha R}\left(1 - \left(1 - \frac{2\alpha\varphi(t)}{\tau \; c_0^2 D^2}\right)^{1/2}\right) \tag{6}$$

as a function of the flux $\varphi(t) = \int_0^t v(s)ds$. This gives the current:

$$i(t) = v(t) \; G_\tau \left(c_0^2 - \frac{2\alpha\varphi(t)}{\tau \; D^2}\right)^{-1/2} \tag{7}$$

where $G_\tau = 1 / \tau \; R$.

With a simple harmonic driving potential $v(t) = v_0 \sin(\omega t)$ this gives the *iv*-characteristics shown in fig. 3 for three different frequencies. The signature of memristance is clearly visible at low frequencies: hysteresis anchored at the origin. At high frequency the memristance degenerates to ordinary resistance, giving the approximately linear characteristic seen in fig. 3.

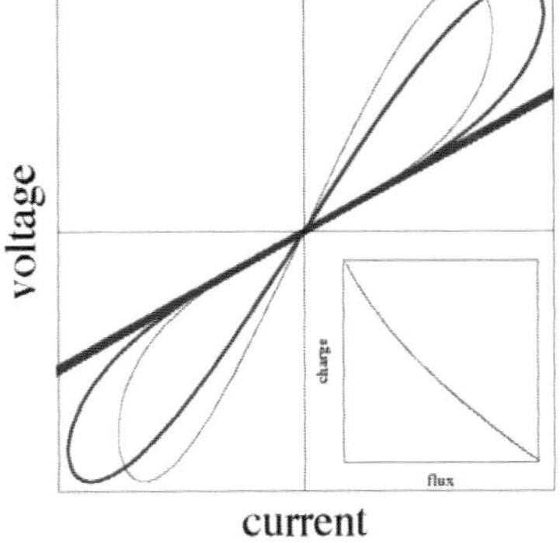

Figure 3: Parametric plot of the voltage and current of the "micro-memristive" model of skin capillaries described in the text, shown for three different frequencies. Inset: flux-charge characteristic.

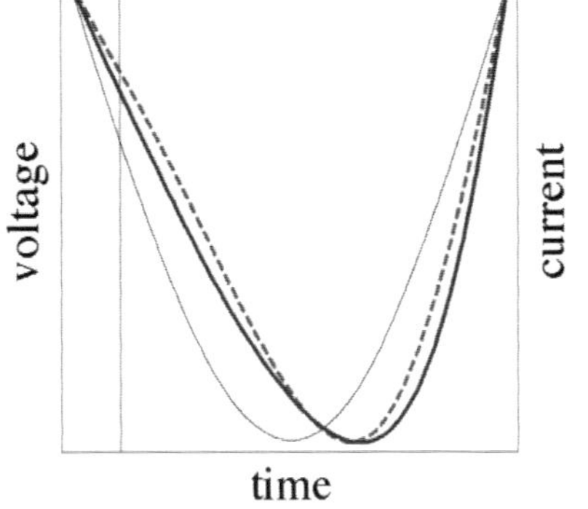

Figure 4: Input (thin line) and output (thick line) signals of the "micro-memristive" model of skin capillaries described in the text. The dashed line shows similar behavior in the model developed in ref.[5].

With this potential the flux is $\varphi(t) = (2v_0 / \omega)\sin^2(\omega t / 2)$, which gives the flux-charge characteristics shown in the inset in fig.3: it is single valued and frequency independent as it must be for a memristive system [1,7,8].

For suitable choices of parameters the current behaves as shown by the solid thick line in fig.4, when the driving potential is negative. For a limited range of parameters this is quite similar to the response found in [5], both experimentally and theoretically – compare the dashed line which shows similar behavior found in the model developed in [5] – thus demonstrating the viability of the memristive paradigm in this regime.

4. Discussion

Comparing the behavior of this memristive model with the data shown in fig. 1, we find that some of the main features agree, including the phase-shift, while others need more work. There are two main properties of the experimental response that remain to be explained: the increasing amplitude of the current with time, and the damping or "clipping" of the positive current.

We will in a future paper show that the increasing amplitude can be explained by "wetting" of the duct wall. As the fluid column oscillates it deposits more fluid on the liquid film coating the wall, thus decreasing the resistance in that part of the duct, at least temporarily. This wetting mechanism alone will only increase the current amplitude from the first to the second period of the applied potential. It will then reach a new, stable value. However, the lack of a reservoir of fluid at the top of the duct will lead to less transport of fluid away from the duct orifice than towards the orifice, and hence a net buildup of fluid in the duct over several cycles of the current. This explains the increasing amplitude with time demonstrated by Grimnes [5].

This other mechanism can be called "clipping": when the potential repels the fluid film it eventually lets go of the surface electrode, and the current is cut. This gives rise to the characteristic spike in the positive current. The liquid then diffuses rapidly back to the electrode which reconnects the circuit, and the process repeats rapidly several times until the potential becomes insufficient to push the fluid away.

Hence, further improvements to the memristive model are possible. We have here also ignored all electrical activity in the *stratum corneum* itself. It is known [4] that this acts like a capacitive load coupled in parallel to the sweat capillaries, albeit a rather unusual one which is traditionally modeled using a device called a "constant phase element" (CPE) [4]. It is known that a CPE cannot be replaced by a finite number of conventional passive circuit elements (RLC). We shall return to this question elsewhere and argue that this activity in the epidermis is also memristive.

5. References

[1] Chua LO 1971 Memristor – the missing circuit element. IEEE Trans. Circuit Theory. CT-18(5):507-509
[2] Glasstone S 1946 Textbook on physical chemistry. Van Nostrand
[3] Strukov DB, Snider GS, Stewart DR, Williams RS (2008) The missing memristor found. Nature 453:80-83
[4] Grimnes S, Martinsen ØG 2008 Bioimpedance and bioelectricity basics. 2. ed. Academic Press
[5] Grimnes S 1983 Skin impedance and electro-osmosis in the human epidermis. Med. & Biol. Eng. & Comput 21:739-749
[6] Grimnes S, Lütken CA, Martinsen ØG 2009 Memristive properties of electro-osmosis in human sweat ducts. WC2009, IFMBE Proceedings. 25/VII: 696-698
[7] Chua LO, Kang SM 1976 Memristive devices and systems. Proc. IEEE 64:209-223
[8] Chua LO 2003 Nonlinear circuit foundations for nano-devices, Part1: the four-element torus. Proc. IEEE 91:1830-1859

From impedance theory to needle electrode guidance in tissue

Håvard Kalvøy[1,2], Per Høyum[2], Sverre Grimnes[1,2], Ørjan G. Martinsen[2,1]

[1]Dept. of Clinical and Biomedical Engineering, Oslo Univ. Hospital, Norway
[2]Dept. of Physics, Univ. of Oslo, Norway

E-mail: havard.kalvoy@fys.uio.no

Abstract. Fast access to blood vessels or other tissues/organs can be crucial in clinical or acute medical treatment. We have developed a method for needle guidance for use in different types of applications. The feasibility of an automatic application for fast access to blood vessels during acute cardiac arrest, based on this method, has been evaluated. Suited electrode setups were found by development of needle electrode models used in simulation and sensitivity analyses. *In vitro* measurements were done both to determine the fundamental properties of the electrodes for use in the models and to confirm the simulation results. Development of algorithms for tissue characterization and differentiation was based on *in vivo* impedance measurement in porcine models and confirmed in human tissue *in vivo*. Feasibility was proven by application prototyping and impedance data presented as invasive Electrical Impedance Tomography (iEIT). Our conclusion is that this method can be utilized in a wide range of clinical applications.

1. Introduction

As participators in an ongoing resuscitation project we wanted to explore the feasibility of using bioimpedance measurements to confirm intravascular needle position. For administration of drugs and liquid for induced hypothermia fast access to blood vessels was required. Earlier we have presented a method for guiding of needle electrodes based on bioimpedance measurements [1]. Further development of the method and gathering of fundamental properties of different needles [2] and electrode systems has now been done. The aim was to determine if our method could be utilized as part of an application for fast vascular access, but also to refine the method in general for use in other types of clinical applications. Different needles and electrode systems was used in measurements *in-vitro* and in *in-vivo* porcine models. Impedance spectra for different types of tissues were measured and compared to earlier results [3,4].

2. Methods

Suited needle electrodes were found after a market survey and testing of available needles. Different special designs of electrodes were also evaluated, but not carried further since commercial available stainless steel electrodes with acceptable properties were found. *In-vitro* testing of electrodes was done to determine electrode properties with focus on spatial sensitivity, stability over time, and influence from Electrode Polarization Impedance (EPI) [5]. Monopolar measurements were obtained by using 3-electrode setups [6] or 2-electrode setups with a very large counter electrode [5]. Pilot studies gave no significant difference between the two as we used them. Stability of electrode properties over time were determined by impedance measurements as function of time and exposure to saline and excitation currents [2]. Repeatability was also tested *in-vivo* by repeated measurement over time in porcine models.

The spatial sensitivity was tested by measurement at different insertion depths in a saline tank [1], and in a piece of bacon from the local grocery store. The bacon was used to obtain a more realistic *in-vitro* model. By use of a servomotor we measured the impedance at 10 kHz as function of time at constant insertion speed. The insertion depth could then be calculated from the time and speed. To illustrate the sensitivity we plotted the measured impedance modulus on a picture of the bacon in the plane of insertion.

A theoretical evaluation of the sensitivity was also done by development of a Finite Element (FE) model of the needle setup. Plotting the equipotential line where the potential had dropped to 3 % of the simulated excitation potential was done to determine the 97 % sensitivity zone.

A laptop/DAQ-card prototype was constructed for *in-vivo* testing of the tissue discrimination method. The results were used to construct an array of needles for insertion towards the femoral artery. Impedance spectra were taken for all needles at different depths (a step length of 3 mm was obtained by insertion of Plexiglas spacers) to obtain a matrix of measurement points covering a sufficient cross section of a volume where the femoral artery was expected to be found. Ultra sound images were taken to aim the array and to be used as a gold standard for verification of the results.

3. Results and discussion

All needles were dominated by EPI at frequencies up to some kHz, where the tissue or saline properties gradually were more dominant as the frequency was increased [1]. The differences in magnitude of the EPI and the stability due to saline exposure were large between the tested needle types [2]. The 97 % sensitivity zone was found to be inside a radius of about 2.6 mm in the saline tank measurements. In the bacon measurements abrupt changes in the impedance modulus were seen when the needle entered from one tissue type to the next (figure 1).

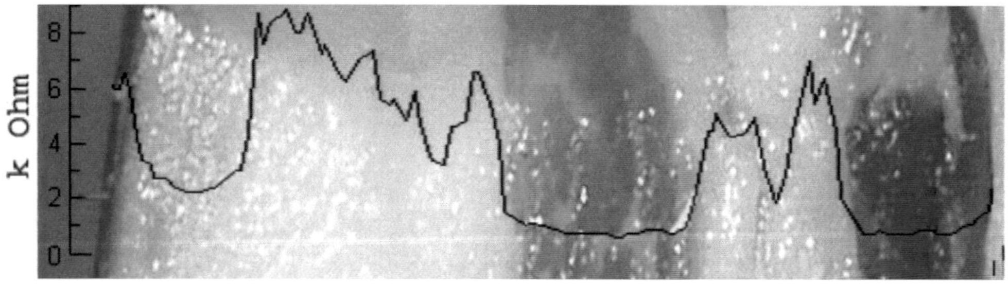

Figure 1. Measured modulus at 10 kHz plotted as function of insertion depth in bacon. The mark at about 2 kΩ indicates the starting point of the horizontal needle insertion path.

The results from the FE-analysis are illustrated in figure 2. The 97 % equipotential line (surface in 3D) was found in a radius of 3.7 mm from the needle tip. This result was not exactly the same as obtained in the *in-vitro* experiment, thus further refinement of the models are planned for a follow-up study. However, the results gave an estimate of the sensitivity zone that was sufficient for use in the design of our needle array. By using 3 mm distance between the needles the conductance of the needles did not interfere significant on the neighbour needle, and still blind volumes between the needles could be avoided. The *in-vivo* measurements on the porcine model gave impedance spectra [1] in accordance with results from earlier investigators [3,4]. Sufficient tissue characteristic data was obtained in the impedance spectra from the array measurement to localize femoral vessels. This result was illustrated by plotting the phase angle at 316 kHz for each pixel in a matrix (described in more detail in a paper under preparation). We look upon this plot as a first approach to construct an invasive Electrical

Impedance Tomograph (iEIT). The feasibility for separating muscle and fat by a method based on needle electrode impedance measurement was showed [1]. Separation of other tissue types may also be possible, but even if we only can discriminate blood, muscle and fat, the method most likely can produce valuable needle guidance for many clinical procedures as surgery or drug administration.

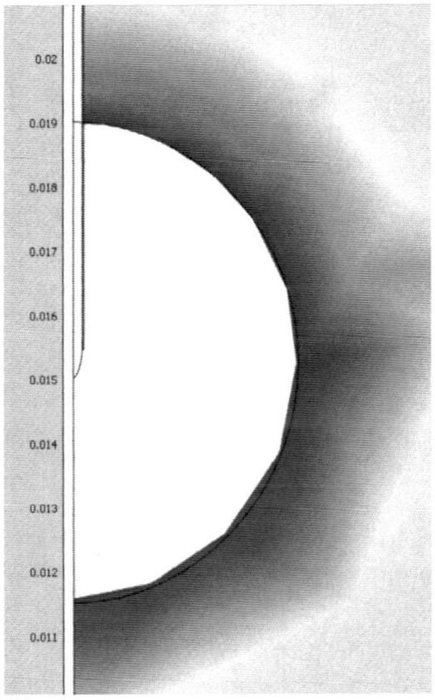

Figure 2: Result form the FE-simulation. The white zone illustrates the volume responsible for 97 % of the measured impedance in a homogeneous medium.

4. Conclusions

The experiences from this project gives reason to believe that further a development of the impedance based method can enable needle guidance in a wide range of clinical applications.

References

[1] Kalvøy H, Frich L, Grimnes S, Martinsen ØG, Hol PK and Stubbhaug A 2009 Impedance based tissue discrimination for needle guidance *Physiol. Meas.* **30** 129-140

[2] Kalvøy H, Nordbotten B, Tronstad C, Martinsen ØG, Grimnes S 2009 Impedance properties of stainless steel needle electrodes *IFMBE Proceedings WC 2009 "World Congress on Medical Physics and Biomedical Engineering"* **25** 380-383 Munich, Germany

[3] Foster K R and Schwan H P 1995 Dielectric properties of tissue *Handbook of Biological Effects of Electromagnetic Radiation 2nd edn*, ed C Polk and E Postow (Boca Raton, FL: CRC Press) 25–102

[4] Gabriel C, Gabriel S and Corthout E 1996 The dielectric properties of biological tissues: I. Literature survey *Phys. Med. Biol.* **41** 2231–49

[5] Grimnes S and Martinsen ØG 2008 *Bioimpedance and Bioelectricity Basics 2nd edn* (San Diego: Academic)

[6] Grimnes S 1983 Impedance measurement of individual skin surface electrodes *Med. Biol. Eng. Comput.* **21** 750–5

Alpha-dispersion in human tissue

Sverre Grimnes, Ørjan G Martinsen
Dept. of Clinical and Biomedical Engineering, Rikshospitalet, Oslo University hospital, Oslo, Norway
Dept. of Physics, Univ. of Oslo, Norway

E-mail: sverre.grimnes@ fys.uio.no

Abstract. Beta dispersion is found in living tissue in the kilohertz – megahertz range and is caused by the cellular structure of biological materials with low frequency properties caused by cell membranes. Alpha dispersion is found in the hertz range and the causes are not so well known. Alpha dispersions are the first to disappear when tissue dies. Tissue data have often been based upon excised specimen from animals and are therefore not necessarily representative for human tissue alpha dispersions. Here we present data obtained with non-invasive skin surface electrodes for different segments of the living human body. We found alpha dispersions in all cases; the ankle-wrist results had the smallest. Large alpha dispersions were found where the distance between the electrodes and muscle masses was small, e.g. on the calf. Further studies on electrode technique and reciprocity, electrode positioning, statistical variations, gender, age and bodily constitutions are necessary in order to reveal more about the alpha dispersion, its appearance and disappearance.

1. Introduction

Schwan [1] introduced three dispersion mechanisms (α, β, γ) to characterize the anomalous electric properties of biomaterials[1]. He linked the β dispersion with the cellular structure of biological materials with low frequency properties caused by cell membranes. He proposed that the α dispersion was due to surface admittance (lateral to the membrane surface). If this is correct the alpha dispersion should be less dependent on the intracellular content but more dependent on the complex membrane surface structures. α dispersion is more sensitive to age and environmental factors than β dispersion [1].

Basically the dispersion concept was defined according to the mechanism causing dispersion. If the generating mechanisms are different it is important to know what dispersion type measured data belongs to. Alpha dispersion can be found down to below 1 Hz and up to 100 kHz [1]. But the mechanisms behind dispersions are often unclear or unknown, and the dispersion grouping has also been based upon simply a defined frequency range. Loosening the link to mechanisms weakens the concept of dispersions and some its usefulness for bioimpedance research may get lost.

It is the purpose of this paper to look for and characterize alpha dispersions in human in-vivo and in-situ data.

2. Method

The following tissue volumes were examined with a 4-electrode technique: The forearm for the short distance to small muscle masses. The thigh for the short distance to large muscle masses and the calf for the shortest distance to large muscle masses; the "whole body" heel-to-wrist segment; the forehead for the small muscles masses and short distance to the low conductivity skull. The 4 electrodes were positioned

[1] Alpha in this paper is the name of a dispersion, but still it is sometimes confused with the alpha angle in Cole and Cole-Cole equations.

on a straight line on the chosen skin area. No skin preparations were used, and the electrodes had a more than 15 min contact period before measurements started.

Two types of AgCl electrodes were used: 1) Solid gel contact disk of radius 1,25 cm and skin contact area 4,9 cm^2 (Tyco, Kendall, KittyCat model 1050NPSM small). The solid gel surface adhered directly to the skin. The solid gel together with a cloth ring formed the rim of the electrode. Used for all results except those shown on figure 1 right and figure 3 left. 2) Wet gel model with skin wetted area $\approx$3 cm^2 and an outer plaster ring (Ambu Medicotest Blue Sensor model Q-00-A).

Transfer complex impedance was measured with a Solartron 1260/1294 system in the frequency range from 1 (or 0,1) Hz to 1 MHz. The current carrying (CC) pair was excited with a constant voltage of 0,5 or 1 V rms. With 1 V rms excitation voltage the excitation current was about 15 μA at the lowest frequencies, increasing to about 1 mA at the highest frequencies. Throughout the frequency spectrum the current was too small to be sensed by the test person. This excitation modus is preferred because the current level is adapted to linearity and current perception thresholds: the lower the frequency the lower the current density threshold of non-linearity [2] and perception [3].

The complex admittance locus curve was shown in the Wessel plane [4] and visually inspected for dispersions in the form of circular arcs. Y-plots were preferred to Z-plots because of the better balance between the size of the alpha and beta arcs.

Quality control

In the Wessel plane plots most admittance loci of figures 1-3 showed negative phase at the highest frequencies. Positive phase impedance (negative admittance) at high frequencies is often found in 4-electrode systems [5]. In the results presented here it is linked with the impedance contribution of the self-inductance of the measured volume and the leads combined with the low transfer impedance values measured. The range of linearity was examined and maximum excitation voltage applied to the CC electrodes was 1 V rms (3V for figure 1 right side). Bode diagrams (not shown here) were controlled with the Kramers-Kronig rule [4] (falling impedance with increasing frequency shall correspond to negative phase angles and visa versa). All results were Kramers-Kronig compatible. Reciprocity was controlled by swapping the PU and CC pairs to check that the same spectra were obtained. Reciprocity failed at the lowest frequencies, the reason for this is not clear and will be pursued further.

3. Results

Forearm

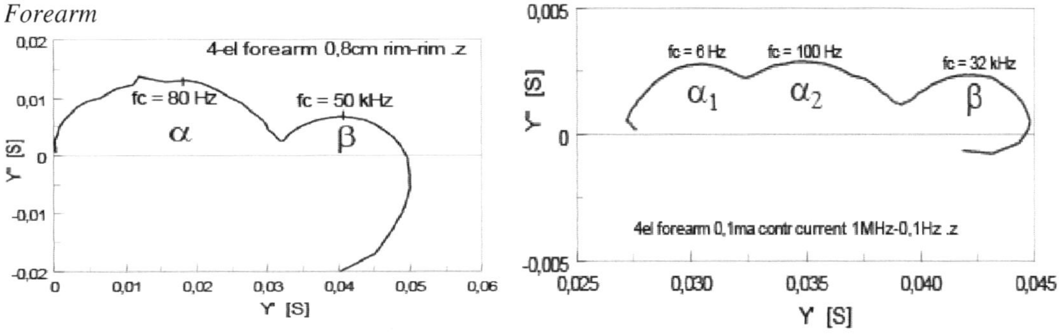

Figure 1 Wessel plane Y-plots. PU and CC pairs in line at the ventral side of the forearm. Left: equal 3,3 cm center to center distance. Right: equal 5 cm center to center distance, wet gel electrodes.

Figure 1 left illustrates the noisier curve form often found at lower frequencies in the alpha region. The low-frequency part of the alpha segment approaches the origin, illustrating small DC conductance as often found when an electrode line is not parallel to e.g. muscle fiber directions (anisotropy).

The plot to the right was obtained driving the instrumentation in controlled current mode at 100 μA. Two clear alpha dispersions at 6 Hz and 100 Hz are seen.

Thigh and ankle-to-wrist segments

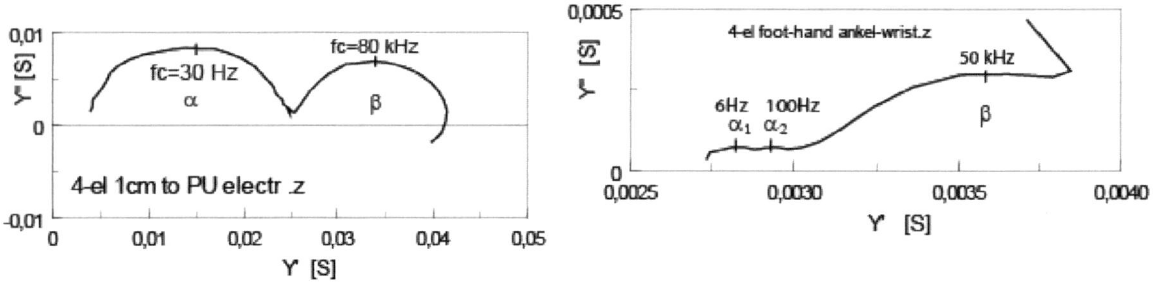

Figure 2 Wessel plane Y-plots. Left: PU and CC pairs placed at top of the thigh, PU electrodes 6 cm apart and CC electrodes at 1 cm rim-to-rim distance from the PU electrodes. Right: PU pair ankle-wrist, CC pair heel-hand.

Figure 2 left side shows a clear dispersion with the alpha dispersion having a sufficient DC conductance to make it probable that the electrode positioning line was parallel to the muscle fibers. Right side is from a large segment approaching a "whole body" segment comprising one arm, one leg and a part of the abdomen. The dispersions are small, the alpha dispersions hardly discernible.

Calf and forehead

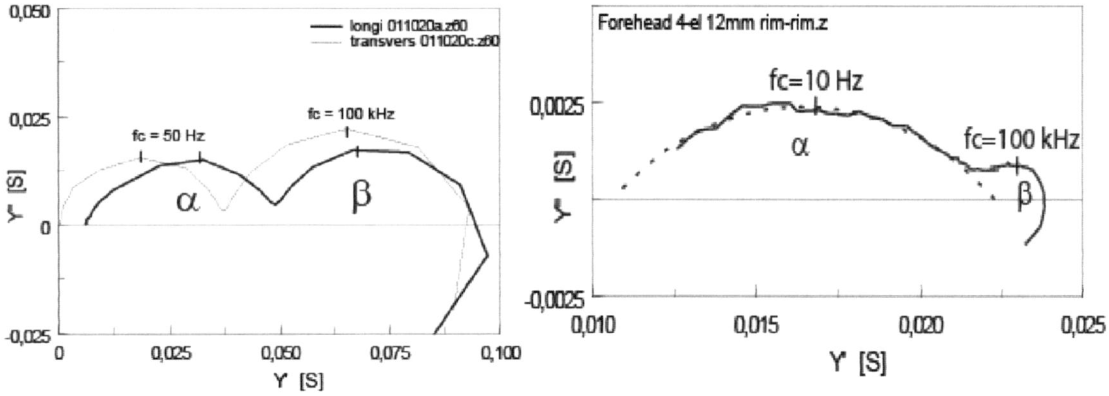

Figure 3 Wessel plane Y-plots. Left: Calf longitudinal: 4 wet electrodes in a line parallel to the muscle fiber direction. Calf perpendicular: 4 electrodes in a line transversal to the muscle fiber direction. Right: Forehead: Solid gel electrodes in line with equal 1,2 cm rim-to-rim distance.

On figure 3 left side the differences between the longitudinal and transversal dispersions are small: they have similar characteristic frequencies, the similar imaginary maximum values and the similar real value increments. The important difference is that the transversal curve has moved to the left approaching the origin at the lowest frequencies, compatible with the often found low DC conductance in transversal measurements.

Figure 3 right side shows a result obtained with 4 electrodes positioned on the forehead with rim-to-rim distances of 1,2 cm. Excitation amplitude was 500 mV rms, and the low frequency part is noisy and a calculated regression circular arc is therefore included in the diagram for easier determination of the characteristic frequency.

4. Discussion and conclusions

Our results are in agreement with the notion that all living tissues have beta dispersions caused by their membranes. Our results with in-vivo non-invasive skin 4-electrode technique did show that all the measured cases also had at least one alpha dispersion even if they could be very small. It is well known and seen as slightly enigmatic that whole blood has beta- and gamma-dispersion, but no alpha dispersion [6]. Martinsen et al [7] measured impedance of haddock muscles for a period of 13 hours after the fish had been sacrificed. Most of the alpha dispersion disappeared after a few hours, while in the same period the low frequency end of the beta-dispersion increased. This is in agreement with our notion that alpha- and beta-dispersions are generated by different mechanisms, but both are related to the cell membranes of living tissue. Beta dispersion used for body composition measurements are based upon the well defined Maxwell-Wagner mechanism. Much of the work with body composition has been based upon beta-dispersion, and it is a question whether alpha dispersion actually gives additional information or on the contrary is just a disturbance for such measurements. Further studies on electrode technique and reciprocity, electrode positioning, statistical variations, gender, age and bodily constitutions are necessary in order to reveal more about the alpha dispersion, its appearance and disappearance.

References

[1] Schwan H P (1957): Electrical properties of tissue and cell suspensions. In: Advances in Biological and Medical physics. Lawrence J H, Tobias C A, Eds. Vol V, 147-209, Acad.Press.
[2] Schwan H P 1992 Linear and non-linear electrode polarisation and biological materials. Ann Biomed Eng, 20, 269-288
[3] Dalziel C F 1954 The threshold of perception currents. AIEE Trans Power App Syst 73, 990-996
[4] Grimnes S, Martinsen ØG 2008 Bioimpedance and bioelectricity basics. Acad.Press
[5] Grimnes S, Martinsen ØG 2007: Sources of error in tetrapolar impedance measurements on biomaterials and other ionic conductors. J Phys D: Appl Phys. 40, 9-14
[6] Foster K R, Schwan H P (1989): Dielectric properties of tissue. CRC Critical Rev Biomed Eng, 17, 25-104.
[7] Martinsen Ø G, Grimnes S and Mirtaheri P (2000): Non-ivasive measurements of post mortem changes in dielectric properties of haddock muscle - a pilot study. J. Food Eng. 43, 189-192.

Journal of Physics: Conference Series **224** (2010) 012074 doi:10.1088/1742-6596/224/1/012074

Mutual localization of electrode pairs in a 4-electrode measuring system

Sverre Grimnes[1,2], Ørjan G Martinsen[1,2] and Gorm K Johnsen[2]
[1]Department of Clinical and Biomedical Engineering, Rikshospitalet, Oslo University hospital, Oslo, Norway
[2]Dept. of Physics, University of Oslo, Norway

E-mail: sverre.grimnes@ fys.uio.no

Abstract. The purpose of this study was to examine the transformation of a 2-electrode system into a 4-electrode system by gradually removing the current carrying (CC) electrode pair from the stationary signal pick-up (PU) pair. Both a Finite Element Method (FEM) analysis on a homogeneous tissue model and in-vivo measurements were used. We found that the general transfer impedance dependence on CC electrode distance was similar for the two methods. With the same excitation current the transfer impedance of a 4-electrode system may be measured with less than 1/10 of the signal amplitude obtained with a 2-electrode system. Also the transfer impedance spectra changed character. The admittance characteristic frequency of the alpha dispersions increased from 32 Hz to 160 Hz and beta dispersions decreased from 100 kHz to 13 kHz with increasing distance.

1. Introduction

A major problem in bioimpedance work has been electrode polarization. This was discussed in detail by Schwan [1]. He listed four different solutions and one of these was the 4-electrode technique. However, these solutions were used for in-vitro work usually with the examined tissue or suspension in a tube where the current flow direction was well defined and parallel to the tube axis. Measuring the impedance of a multilayered tissue volume with current spread from the current carrying (CC) electrodes implies additional problems not treated by Schwan. One problem is the complicated electrode sensitivity distribution (sensitivity field) [2]. Sensitivity S is a factor determining the contribution of a voxel with local resistivity ρ to the total measured transfer resistance.

$$S = \mathbf{J'}_{reci} \cdot \mathbf{J'}_{cc} \qquad [1/m^4]$$

$\mathbf{J'}_{reci}$ is the current density vector with 1 ampere excitation applied to the pick up (PU) electrode pair (reciprocal excitation). $\mathbf{J'}_{cc}$ is the current density vector with 1 ampere excitation applied to the CC electrode pair. S is the vector dot product in a small voxel dv, and ρS is the impedance volume density [Ω/m^3]. The transfer resistance R is:

$$R = \iiint \rho S \, dv \qquad [\Omega]$$

The purpose of this study was to examine the transformation of a 2-electrode system into a 4-electrode system by gradually removing the CC electrode pair from the PU pair by using both a Finite Element Method (FEM) analysis on a homogeneous tissue model as well as in-vivo measurements.

2 FEM analysis

The FEM analysis was carried out by modeling a "human tissue box" with our desired electrode configuration in COMSOL Multiphysics (CM) version 3.5. 4 disc electrodes were located in-line at the centre of the surface of the box, and they were modeled with a conductivity of 6e7 S/m corresponding to copper. The two PU electrodes were at fixed center-to-center distance of 6 cm, whereas the distance between the PU pair and the CC pair was gradually increased from 6 cm (i.e. a two-electrode system) up to 150 cm. The resistance was calculated as a function of distance between the CC electrodes.

The measuring volume, "the human tissue box" with dimensions 0,5x0,5x1,5 m^3 was assumed to be homogeneous and with a conductivity of 1 S/m. All the outer surfaces of the volume were given the conditions $n \bullet J = 0$. At the frequencies of interest (< 1 MHz) the relative permittivity was set to 1 for simplicity. Even if we calculate resistance we also call the parameter impedance Z=R+jX with X ≈0.

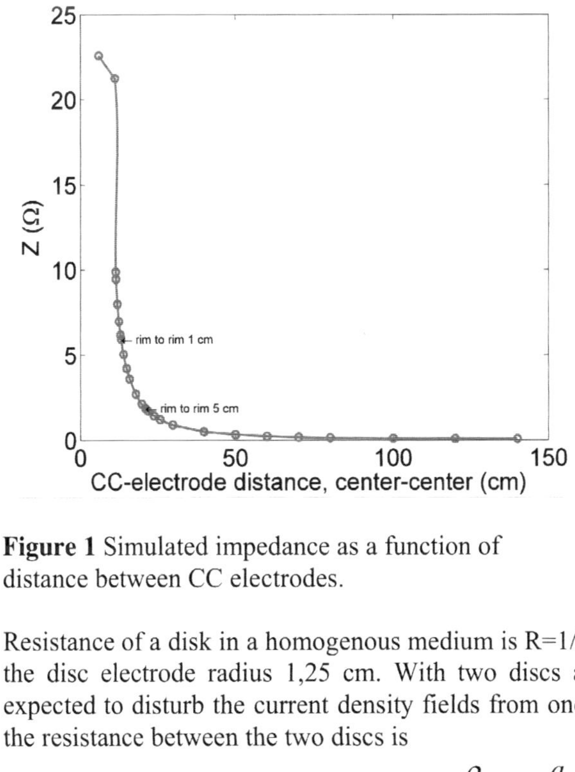

Figure 1 Simulated impedance as a function of distance between CC electrodes.

One result of the FEM simulations is given in figure 1, showing the impedance as a function of the distance between the CC electrodes. The two smallest CC distances correspond to a configuration where the CC and PU electrodes are physically in contact. The first point represents a complete overlap of the two sets of electrodes, which effectively is a two-electrode system, whereas the second point is the limit where the CC and PU are just in physical contact. All remaining values represent a 4-electrode configuration. We note the sudden drop in impedance as the system suddenly transforms from a two –and into a four-electrode setup. Two of the configurations studied in the in vivo measurements are highlighted in figure 1.

Resistance of a disk in a homogenous medium is R=1/(4aσ), which in our case is about 19 ohm, when a is the disc electrode radius 1,25 cm. With two discs at a center-center distance of 6 cm they must be expected to disturb the current density fields from one another. A correction term must be included, and the resistance between the two discs is

$$R = \frac{\rho}{2\pi a}(1 + \frac{a}{L} + (\frac{a}{L})^2 + ...)^{-1}$$

With the PU electrodes 6 cm apart we end up with a resistance of 24,2 ohm. This corresponds quite well with the simulated resistance of 22,6 ohm in the case of a two-electrode setup.

3 In-vivo study

Method

The test person was sitting on a chair with a thigh-leg angle of 90 degrees. All 4 electrodes were solid gel disks with skin-gel contact radius 1,25 cm and contact area A=4,9 cm^2 (Tyco, Kendall, modell KittyCat 1050NPSM Small). The fixed PU electrode pair had a center to center distance of 6 cm along the longitudinal axis of the thigh as shown at figure 2. The two CC electrodes were localized outside the PU pair at four different positions: a) foot-hand (>50 cm), b) knee-hip (≈20 cm), c) 5 cm and d) 1 cm rim-to-rim distance from the respective PU electrodes. Impedance spectra were measured with a Solartron 1260/1284 system. We used a measuring mode with constant excitation voltage creating a measuring current dependent on the impedance of the CC system. Because tissue impedance is frequency dependent also the measuring current was frequency dependent. With 1 V rms excitation voltage the measuring current was typically about 15 uA at the lowest frequencies, increasing to about 1 mA at the highest frequencies.

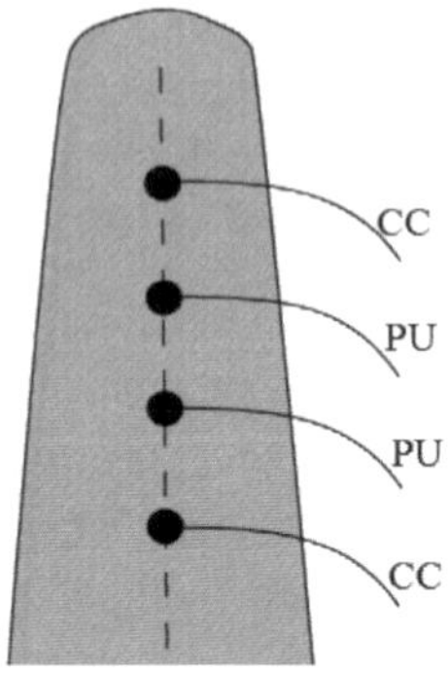

Figure 2 Thigh of a sitting person as seen from above, knee at top of drawing.

Quality control

The Bode plots of figure 3 showed positive phase at the highest frequencies. Phase polarity reversals at high frequencies are often found in 4-electrode systems [3]. We found that the positive phase was not dependent on measuring current level, and it is most likely due to self-induction. The lower the impedance level, the stronger the influence from the inductive impedance Z=ωL where ω=2πf and L the inductance [henry] of the measuring circuit. The relative contribution is larger the lower the measured total impedance, in accordance with figure 3.

Bode diagrams were controlled with the Kramers-Kronig rule (falling impedance with increasing frequency shall correspond to negative phase angles). All results were found compatible with the Kramers-Kronig rule.

Reciprocity was controlled by swapping the PU and CC pairs to check that the same spectra were obtained. Reciprocity criteria were not met at the lowest frequencies when excitation voltages >1 V rms were used. All impedance spectra were therefore performed with 0,5 or 1 volt rms excitation.

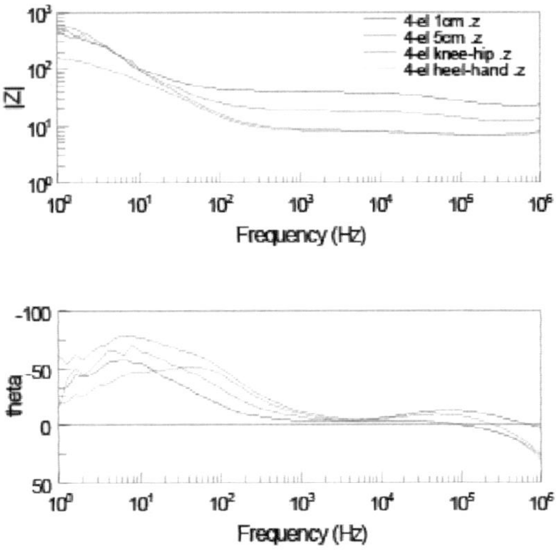

Figure 3 4-electrode transfer impedance spectrum at 1cm, 5cm, 25 cm and >50cm distance between CC and the stationary PU electrode pair positioned at the thigh.

In-vivo results

The PU pair impedance spectrum showed two dispersions with characteristic frequency fc around 1 kHz and 100 kHz. The low frequency part of the spectrum was dominated by the SC impedance of the skin. The lowest impedance result was 340 ohm at 631 kHz, this is the 2-electrode driving point impedance [4] of our in-vivo measuring system.

The 4-electrode transfer impedance obtained with 1 cm distance between the PU and CC electrode pairs is shown on figure 3. The characteristic frequency of the alpha dispersions increased from 32 Hz to 160 Hz and beta dispersions decreased from 100 kHz to 13 kHz with increasing distance.

4 Discussion

In Table 1 the FEM calculated values valid for sigma 1 S/m has been multiplied with a factor of ten and therefore corresponds to a sigma of 0,1 S/m. The FEM values and in-vivo results have similar general trends. A common finding is the abrupt drop in impedance when the two pairs no longer are in direct contact. For the in-vivo case the drop is to less than 1/10 of the 2-electrode value when the CC pair is remote. This means that the recorded signal amplitude also may drop to 1/10, and the signal to noise ratio get poor. This is often experienced in practice, and a usual remedy has been to increase the measuring current. The problem then is that the non-linear zone is easily entered with nobody taking notice. When working with 4-electrode systems it is therefore good practice to perform the reciprocity test to ensure that the measuring current level is sufficiently low.

Table 1 Impedance as a function of distance between PU an CC electrode pairs, FEM model results (sigma = 0,1 S/m) and in-vivo measurement results.

system	ohm in-vivo	ohm FEM results x 10	condition
2-el	340	226	CC and PU pairs concentric in contact
2-el		215	CC and PU pairs in rim-to-rim contact
4-el	38,0	59	CC electrode rims 1 cm from PU rims
4-el	18,1	20	CC electrode rims 5 cm from PU rims
4-el	8,1		CC electrode rims 20 cm from PU rims
4-el	7,8		CC electrode rims >50 cm from PU rims

The differences between the two methods found in Table 1 are due to many factors. The in-vivo results are from the thigh, which has a smaller volume than "the tissue box" where the current is spread out in a larger volume and the current density for a given current will be lower and therefore also the recorded PU signal and the calculated impedance. The in-vivo results are influenced by the long and small volume segments of tissue as well as the tissue complexity.

5 Conclusions

The general transfer impedance dependence on CC electrode distance was similar for the FEM model and the in-vivo results. With the same excitation current the transfer impedance of a 4 electrode system may be measured with less than 1/10 of the signal amplitude obtained with a 2-electrode system. Also the transfer impedance spectra changed character with increasing distance, e.g. the characteristic frequency of the Y-plot alpha dispersions increased from 32 Hz to 160 Hz and beta dispersions decreased from 100 kHz to 13 kHz.

References

[1] Schwan HP 1963 Determination of biological impedances. In: Nastuk WL (Ed), *Physical techniques in biological research,* Vol 6, Academic Press.
[2] Grimnes S, Martinsen ØG 2008 Bioimpedance and bioelectricity basics. *Academic Press*, p.188
[3] Grimnes S, Martinsen ØG 2007 Sources of error in tetrapolar impedance measurements on biomaterials and other ionic conductors. *J. Physics D: Applied Physics,* 40, 9-14
[4] Kuo FF 1966 Network analysis and synthesis. *John Wiley & Sons*

Methods for calculating phase angle from measured whole body bioimpedance modulus

Bernt J. Nordbotten [a], **Ørjan G. Martinsen** [a,b], **Sverre Grimnes** [a,b]

[a] Dept. of Physics, University of Oslo, P.O.Box 1048 Blindern, NO-0316 Oslo, Norway
[b] Dept. of Clin. & Biomed. Eng., Rikshospitalet University Hospital, NO-0027 Oslo, Norway

berntn@fys.uio.no

Abstract. Assuming the Cole equation we have developed a method to calculate the Cole parameters (R_0, R_∞, α, τ_Z) and the phase angle from four frequency measurements of impedance modulus values. The values obtained compare well with impedance measurements obtained using the Solatron 1294/1260 as obtained when making whole body measurements on five persons. We have also performed calculations using an algorithm based on the Kramers-Kronig approach. The results which are presented show that it is possible to obtain complete body impedance data combining relatively simple measurements with advanced calculation using a laptop. This extends the potential of portable equipment, since the measurements will require less instrumentation.

1. Introduction

In bioimpedance measurements we generally need both the impedance modulus and the phase angle values. This is provided using the more advanced, and also more expensive equipment, like the Solatron 1260/1294 system that we have used.

For many measurements somewhat simpler and transportable equipment is used and the measurements can be limited to an impedance modulus measurement only. This however is a strong limitation for the use of the results since the phase angle values are needed for many further calculations.

This paper discusses how impedance modulus measurements together with established principles and formulas can be used to calculate the complete bioimpedance characteristics.

First, both impedance modulus and phase angle are measured using the Solatron 1260/1294 system. The phase angle value is used as a reference for the calculations.

A least squared method is used to calculate the Cole parameters, which then can be used to find the real and imaginary impedance components and from them the phase angle.

In the second method we employ the Kramers-Kronig [1] relation to calculate the phase angle values.

The measurements and calculations are performed for measurements on five persons. The phase angle values thus obtained correspond well with the Solatron 1260/1294 system measurements.

2. Methods for calculation of phase angle

We have used two methods to obtain the phase angle from the impedance modulus, one is based on a Cole equation model while the other is based on an numerical adaption of the Kramers-Kronig equation.

2.1. Cole equation model

From earlier experiments we have seen that all measurements produce circular arcs in the complex impedance plane (Wessel plot) above 1kHz [2]. We can use this to make a model of the Cole equation to calculate the phase angle from the impedance modulus and also the Cole parameters (R_0, R_∞, α, τ_Z) using non-linear least squares method on the Cole equation model. We implemented this using the Python programming language, and using four frequencies; 5kHz, 50kHz, 100kHz and 200kHz. Starting from the Cole equation:

$$Z=R_\infty+\frac{R_0-R_\infty}{1+(j\omega\tau_Z)^\alpha} \qquad \text{Eq. 1}$$

And so introducing the relation:

$$j^\alpha=\cos\left(\alpha\frac{\pi}{2}\right)+j\sin\left(\alpha\frac{\pi}{2}\right) \qquad \text{Eq. 2}$$

The modulus $|Z|$ can then be expressed as:

$$|Z|=\sqrt{\frac{k_1^\alpha R_\infty^2+2R_\infty k_1^{\alpha/2}\cdot k_2 R_0+R_0^2}{1+2k_1^{\alpha/2}k_2+k_1^\alpha}} \qquad \text{Eq. 3}$$

Where k_1 and k_2 are given by:

$$k_1=(2\pi f)^2\cdot\tau_Z^2=\omega^2\tau_z^2 \qquad k_2=\cos\left(\frac{\pi\alpha}{2}\right)$$

The expression for $|Z|$ is determined by the Cole parameters R_0, R_∞, α and τ_z. Thus by making measurements of $|Z|$ at four different frequencies we get four independent equations for determination of the Cole parameters. These can be solved using numerical methods.

2.2. Kramers-Kronig equation

From the Kramers-Kronig theory we have that if the real part of a linear network function of frequency is known over the complete frequency spectrum, it is possible to calculate the imaginary part [1]. We can then use the following relation to calculate the phase angle at a given frequency (ω) from the modulus.

$$\varphi(\omega_1)=\frac{2\omega}{\pi}\int_0^\infty \ln\frac{|Z(\omega)|}{\omega^2-\omega_1^2}d\omega \qquad \text{Eq. 4}$$

This equation can be used for analytical evaluation, but since we are using numerical evaluation we need to make the following sum approximation from Eq. 4:

$$\varphi(\omega_1)=\frac{2\omega}{\pi}\sum_{i=0}^{n}\ln\frac{|Z(\omega)|}{\omega_i^2-\omega_1^2}\Delta\omega=\frac{2\omega}{\pi}\sum_{i=0}^{n}\frac{\ln(|Z(\omega_i)|)-\ln(|Z(\omega_1)|)}{\omega_i^2-\omega_1^2}\Delta\omega \qquad \text{Eq. 5}$$

This sum requires equal and small steps in frequencies, and since we did not do the measurements with these small steps in the frequencies we performed a cubic-spline interpolation using Python. We then obtained a linear distribution of data points, with a step in frequency between each data point of 100Hz.

3. Measurements

Using a Solatron 1260/1294 we used a 4-electrode setup to do whole body measurements on five persons, using measurement frequencies from 100Hz to 500kHz. For the Cole equation model we only used four of the frequencies (5, 50, 100 and 200kHz), while for the Kramers-Kronig equation we have used data from all the frequencies.

Two electrodes, one current carrying (CC) and one pickup (PU), were placed on the right leg with 7 cm between, and similarly on the right arm. The measurement was done using 800mV AC, this amplitude was determined after doing an amplitude sweep from 80mV to 2V using a whole body measurement with a 2-electrode setup to determine the maximum safe amplitude within the linear range. The measured results are shown in figures 1 and 2.

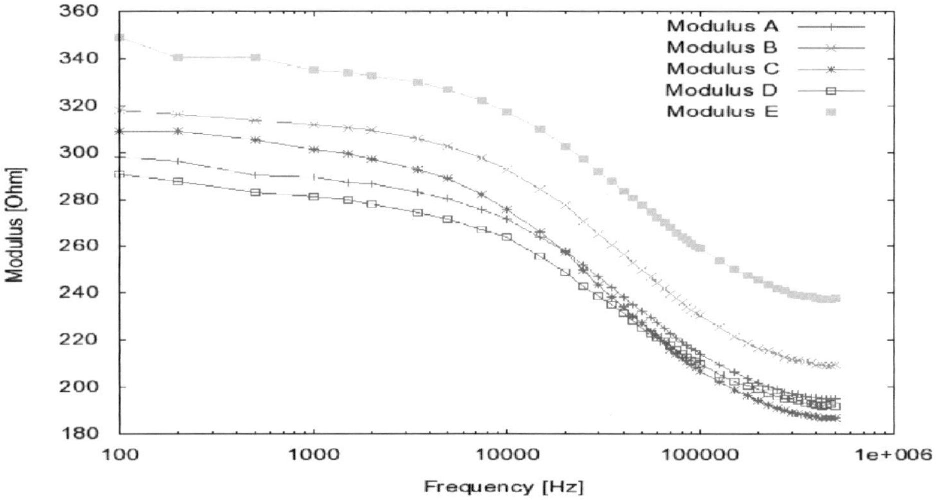

Figure 1. Five whole body impedance modulus measurements

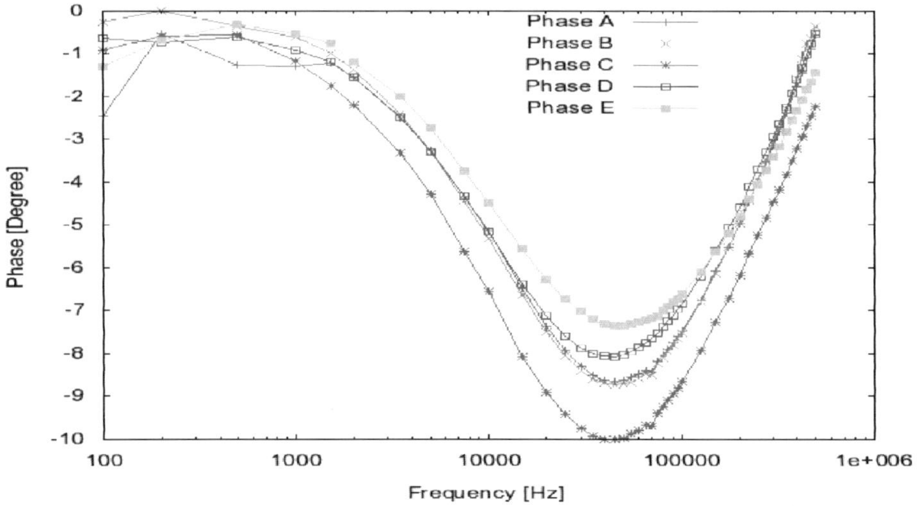

Figure 2. Five whole body phase angle measurements

4 Comparison of measured and calculated results

Table 1 shows the measured and calculated phase angles and also the percentage error results for five different persons A to E at the frequencies 5, 50, 100 and 200kHz. Column M represents the values measured by the Solatron system, here considered as reference values while C and K are the values calculated using the Cole and Kramers-Kronig approaches respectively.

	A			B			C			D			E		
f [kHz]	M [°]	C [°]	K [°]	M [°]	C [°]	K [°]	M [°]	C [°]	K [°]	M [°]	C [°]	K [°]	M [°]	C [°]	K [°]
5	-3.8	-3.52	-3.57	-3.83	-3.89	-3.65	-4.69	-4.58	-4.71	-3.83	-3.97	-3.62	-3.34	-3.47	-3.16
50	-8.46	-8.63	-7.84	-8.53	-8.66	-7.93	-9.61	-10.28	-9.28	-7.92	-7.86	-7.39	-7.92	-7.86	-6.79
100	-7.45	-7,73	-6.90	-7.49	-7.94	-6.83	-8.47	-8.83	-7.86	-6.90	-6.99	-6.27	-6.90	-6.99	-5.80
200	-5.27	-5.83	-4.96	-5.27	-6.25	-4.93	-6.32	-6.43	-5.58	-4.94	-5.41	-4.50	-4.94	-5.41	-4.36

	C [%]	K [%]	C [%]	K [%]	C [%]	K [%]	C [%]	K [%]	C [%]	K [%]
5	-7.37	-6.05	+1.57	-4.70	-2.35	-0.43	+3.66	-5.48	+3.89	-5.39
50	+2.01	-7.33	+1.52	-7.03	+6.97	-3.43	-0.76	-6.69	-0.76	-14.27
100	+10.74	-7.38	+6.01	-9.08	+4.25	-7.91	+1.30	-9.13	-1.30	-11.11
200	+10.63	-5.88	+18.60	-15.94	+1.74	-11.71	+1.82	-8.91	+9.51	-11.34

Table 1. Measured (M), calculated using Cole (C) and calculated phase angles and using Kramers-Kronig (K) for the five whole-body measurements on persons A-E.

The calculated values correspond well with those measured, with the best fit obtained for the Cole equation approach. The results for the Kramers-Kronig approach all have less negative values (up to 1°) for the phase angle than the measured. This is caused by the fact that that the calculation only used values up to 500kHz resulting in a truncation of the summation [3]. Increasing the highest frequency should improve the results.

5 Conclusion

It has been shown that the bioimpedance phase angle can be calculated from simple bioimpedance modulus measurements with good accuracy using both a Cole equation approach giving the Cole parameters and a Kramers-Kronig direct calculation of the phase angle. The best results are obtained using the Cole equation, while the Kramers-Kronig method results in a reduced value due to the truncation of the expansion.

This shows that portable measurement equipment with limited functionality may give complete bioimpedance results when used in combination with a laptop for computation.

References

[1] S. Grimnes, Ø. G. Martinsen, *Bioimpedance and bioelectricity basics,* second edition, Academic Press, London, 2008.

[2] H. Kalvøy, Ø. G. Martinsen, S. Grimnes, *Calculating phase angle from bioimpedance modulus measurements*, 12. Int Conf on Electrical Bioimpedance, Gdansk, 333-336, June 2004.

[3] P. J. Riu, C. Lapaz, *Practical limits of the Kramers-Kronig relationships applied to experimental bioimpedance data*, Ann. NY Acad. Sci., 873, 374-380, 1999.

MAT-MI acoustic source reconstruction using ultrasound B-Scan imaging

L Mariappan, X Li, G Hu, B He

Department of Biomedical Engineering, University of Minnesota

E-mail: binhe@umn.edu

Abstract. We present in this study an ultrasound B-scan based imaging approach for magnetoacoustic tomography with magnetic induction (MAT-MI) to reconstruct electrical conductivity distribution. In MAT-MI acoustic waves are generated in the sample by placing it in a static and a time-varying magnetic field. The acoustic waves from these sources propagate in all directions. In the present approach these acoustic signals are collected with a focused ultrasound transducer which confines the collected signal to that from sources along a line. The focused transducer also gives signal gain in the focus region improving the MAT-MI signal quality. The time-resolved acoustic signals are back projected to form a one-dimensional (1D) image of the source distribution along the line. The complete cross-section of the object is obtained by acquiring 1D images along multiple directions in the cross-sectional plane. A simulation model of the image reconstruction method is developed with ultrasound simulations using the Field II program. The present reconstruction results suggest that acoustic source imaging in MAT-MI can be achieved using the much practical ultrasound B scan imaging technique. The developed method is applied to MAT-MI in experiments. This method should allow combining MAT-MI with clinical ultrasound imaging methods and broadening the potential applicability of the technique.

1. Introduction

Electrical impedance imaging of biological tissue has been of considerable interest in recent years. Several approaches like electrical impedance tomography (EIT) [1],[2] magnetic resonance electrical impedance tomography (MREIT) [3],[4], magnetic induction tomography (MIT) [5],[6] magnetoacoustic tomography [7],[8] and hall effect imaging (HEI)[9],[10] have been explored to image tissue impedance distribution. Magnetoacoustic tomography with magnetic induction (MAT-MI) [11]-[14] is a recently proposed approach to image electrical impedance distribution. In MAT-MI ultrasound is generated in the object to be imaged by placing it in a dynamic and a static magnetic field. Eddy currents are induced in the object due to the dynamic field. The static field leads to generation of acoustic vibrations from Lorentz force on the induced currents. The acoustic vibrations are at the same frequency as the dynamic magnetic field which is chosen to be in the ultrasound frequency range. This allows us to reconstruct the acoustic source distribution in the object using possible ultrasound imaging approaches. The conductivity distribution of the object can then be reconstructed from the obtained acoustic source map [12], [13].

In the current study, a focused transducer is used to reconstruct the MAT-MI ultrasound source distribution. With a specific focusing gain map, the focus transducer receives signal mainly from sources in a relatively narrow beam region along its acoustic axis. These sources are time resolved in the collected signal due to time of flight difference to the transducer. The signal obtained is backprojected to reconstruct acoustic sources in the beam region. However, the signal received from a smoothly distributed acoustic source is the strongest when the transducer axis is perpendicular to the source distribution [15],[16]. Therefore to obtain a complete image of the MAT-MI sources, we need to set the transducer to face the various edges of the source distribution. This is achieved by shifting and rotating the transducer to scan along various lines through the object. The reconstructed images from individual scans are added up to form the final image.

2. Theory

2.1. MAT-MI

In MAT-MI static and dynamic magnetic fields induce ultrasound in the object. The induced Eddy currents (***J***) in the object depends on both the magnetic stimulation and conductivity distribution [13].The eddy currents in the presence of a static magnetic field ($\boldsymbol{B_0}$) are subject to Lorentz force ($\boldsymbol{J}\times \boldsymbol{B_0}$). The generated acoustic waves are governed by the following equation [8]:

$$\nabla^2 \boldsymbol{p} - \frac{\mathbf{1}}{c_s^{\,2}}\frac{\partial^2 \boldsymbol{p}}{\partial t^2} = \boldsymbol{\nabla}\cdot(\boldsymbol{J}\times\boldsymbol{B}_0) \quad (1)$$

where $\boldsymbol{p}$ is the pressure and C_S is the acoustic speed in the medium. Using Green's function the solution to equation (1) can be written as (2) [8], [12]

$$\boldsymbol{p}(\boldsymbol{r},t) = -\frac{1}{4\pi}\int_V d\boldsymbol{r}'\nabla_{\boldsymbol{r}'}\cdot\left[\boldsymbol{J}(\boldsymbol{r}')\times\boldsymbol{B}_0(\boldsymbol{r}')\right]\frac{\delta(t-R/c_s)}{R} \quad (2)$$

where $R=|r-r'|$, and V is the volume containing the acoustic source.

2.2. Image reconstruction with B-Scan

The ultrasound reconstruction with the focussed transducer involves backprojecting the signal along the focus of the transducer. This gives a 1D image of the object also known as B-scan. The complete image is obtained by scanning multiple 1D lines of the object. This is done by performing sector scans at a number of locations around the object to get a complete image [15]. The effect of transducer's directivity is corrected for during backprojection [17].

3. Computer Simulation

Ultrasonic imaging simulations were performed with Field II software. The software allows defining transducers of various geometries. The spatial impulse response of the transducer is calculated in the software by breaking the transducer surface into smaller rectangles. The program also allows simulating the transducer's electromechanical impulse response. In the current simulations this is set to the measured electromechanical impulse response of the transducer obtained with a wide band hydrophone acting as a point receiver.

The MAT-MI acoustic source for simulation was assumed to be from conductivity boundaries in the object. This is consistent with previous MAT-MI simulations and experiment studies [13], [14]. In the current simulation these sources have been assumed to be of equal strength. An object space of 20mm by 20mm was used in the present simulation study. The simulated source distribution is shown in figure 1 (a). A 500 KHz, 4 cm diameter transducer is simulated. The received signal is computed according to equation (2). The image reconstruction is done by backprojection [16], [17]. Sector scans at 11 locations over 180 degree view angle [15] around the object were used in the current setup for reconstruction. The image reconstruction results are shown in figure 1 (b). The resolution of the reconstruction is about 3-4 mm at 500kHz transducer frequency [18].

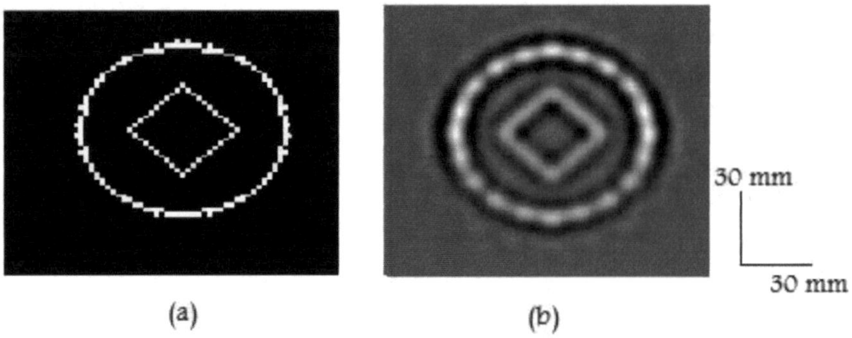

Figure 1 (a) Simulated target MAT-MI acoustic source distribution and (b) reconstructed MAT-MI acoustic source image.

4. Discussion

We have applied to MAT-MI a commonly used ultrasound reconstruction method, i.e. B-scan imaging. This allows using the mature scanning techniques in ultrasonography to reconstruct MAT-MI acoustic sources [19]. The reconstructed images in computer simulation show a good agreement with the simulated target source profiles. Similar results are also obtained in our MAT-MI experiments using a B-scan mode MAT-MI system. In this system, sector scan is achieved by mechanically rotating a focus transducer. In a practical setting, this could also be done by beam forming and beam steering using a transducer array with better efficiency and accuracy. In addition the discrete number of scanning locations needed for reconstruction allows using MA- MI when a limited number of acoustic windows are available to collect ultrasound signal. Thus a method like this promises to broaden the applicability of MAT-MI for human imaging.

Acknowledgement
This work was supported in part by NIH R21EB006070 and NSF BES-0602957.

References

[1] Webster J G 1990 *Electrical Impedance Tomography*(Adam Hilger,Bristol, UK)
[2] Metheral P, Barber D.C, Smallwood R H and Brown B H 1996 *Nature* **380** 509-12
[3] Kwon O, Woo E, Yoon J and Seo J K 2002 *IEEE Trans. Biomed. Eng* **49** 160–7
[4] Gao N, Zhu S and He B 2005 *Phys. Med. Biol* **50** 2675-87
[5] Peyton A J, Yu Z Z, Lyon G, Al-Zeibak S, Ferreira J, Velez J, Linhares F, Borges A R, Xiong H L, Saunders N H and Beck M S 1996 *Meas Sci Technol* **7** 261-71
[6] Griffiths H 2001 *Meas sci tech* **12** 1126-31
[7] Towe B C and Islam M R 1988 *IEEE Trans. Biomed.Eng* **35** 892–94
[8] Roth B J, Basser P J and Wikswo J P Jr 1994 *IEEE Trans.Biomed.Eng* **41** 723–28
[9] Wen H 2000 *Ultrason. Imaging*. **22** 123-36
[10] Montalibet A, Jossinet J, Matias A and Cathignol D 2001 *Med. & Bio. Eng. Comput* **39** 15-20
[11] He B 2005 *Proc. Annu. Int. Conf. IEEE-EMBS* 4178-82
[12] Xu Y and He B 2005 *Phys. Med. Biol* **50** 5175–87
[13] Li X, Xu Y and He B 2007 *IEEE Trans. Biomed. Eng*. **54** 323–30
[14] Xia R, Li X and He B 2007 *Appl Phys Lett* **91** 083903
[15] Xu Y, Wang L H, Ambartsoumian G, and Kuchment P 2004 *Med. Phys.* **31** 724
[16] Xu M H, Ku G, and Wang L H 2001 *Med. Phys.* **28** 1958-63
[17] Yao H and Ebbini ES 2003 *Proc. IEEE Ultrason. Symp.* 1243–46
[18] Li X, Xu Y and He B 2006 *J. Appl.Phys.* **99** 066112
[19] Wells P N T 1999 *Rep. Prog. Physc*. **62** 671-722

Imaging Electric Properties of Human Brain Tissues by B1 Mapping: A Simulation Study

Xiaotong Zhang, and Bin He

Department of Biomedical Engineering, University of Minnesota, 312 Church Street SE, Minneapolis, MN 55455, USA

E-mail: zhan0709@umn.edu and binhe@umn.edu

Abstract. The electric properties (EPs) of biological tissues, i.e., the electric conductivity and permittivity, can provide important diagnostic information in cancer detection and characterization. The conductivity also plays an important role in specific absorption rate (SAR) calculation, which is a major concern in high-field Magnetic Resonance Imaging (MRI). In MRI systems the radio frequency (RF) radiation is dependent on the object's EPs by Maxwell Equations, and we can reconstruct the EPs distribution by measuring the active transverse component of the applied rotating magnetic field in MRI systems (known as B1-mapping). Applying previously proposed Dual-excitation algorithm, we conducted a series of computer simulations on a three-dimensional five-tissue anatomically accurate head model using the Finite Element Method (FEM). The reconstruction results utilizing several RF coils were examined and compared, and the present simulation results indicate that the TEM coil could produce the optimal result. The Specific energy Absorption Rate (SAR) within the head was also estimated using the reconstructed electric properties distribution, and the promising results suggest our approach might provide an efficient way for *in-vivo* SAR determination in high field MRI.

1. Introduction

The electric properties (EPs, conductivity σ and permittivity ε) of biological tissues at radio and microwave frequencies have been the subject of research for over four decades [1]. In the past two decades, many efforts have been made to produce cross-sectional images of EPs *in-vivo* by means of the Electrical Impedance Tomography (EIT) [2] and its variants using magnetic induction (MIT) [3], but these methods are limited by low spatial resolution and need to solve an ill-posed inverse problem. A technique called "Magnetic Resonance Electrical Impedance Tomography (MREIT)" [4], which is based on the Magnetic Resonance Current Density Imaging (MRCDI) [5] technique, has been pursued; but it requires current injection into the body within an MRI scanner. Another recently developed approach, Magnetoacoustic Tomography with Magnetic Induction (MAT-MI) [6], suggests the promises of obtaining high resolution tissue electric conductivity profiles; however, there are no *in vivo* experiments reported so far.

Haacke et al. proposed an iterative method of extracting EPs from MRI images which reflect the disrupted RF profile [7]. Later on, Wen reported a modified Helmholtz equation based non-iterative algorithm and tested in phantom experiments [8]. Recently, using an iterative algorithm and an integration method, Katscher et al. conducted experiments to image the EPs within phantoms and human heads, and they firstly named this approach "Magnetic Resonance Electric Properties Tomography (MREPT)" [9,10].

The B_1-mapping technique was developed to measure the rotating RF field components [11]. MREPT utilizes the measurements to reconstruct the EP values within the biological tissues, and no electrode mounting is required and no external energy is introduced into the body during MRI scanning. In the present study, we have applied our previously proposed Dual-excitation algorithm [13] on a five-tissue 3-D realistic head model to image the tissue EPs. Several RF coils were used to test its performance, and SAR distribution was also estimated within human brain tissues.

2. Method

2.1. Dual-excitation algorithm

Description of Dual-excitation algorithm can be found in [12]. Two sets of $H_x \sim H_y$ data are needed to reconstruct the complex permittivity is $\varepsilon_c = \varepsilon - j\sigma/\omega$, in which ω is the operating angular frequency. We used linear excitation mode to generate different polarizations by switching the current feed point by 90° (excitation II) apart from the previous one (excitation I) to acquire the data sets. All EM fields were calculated by means of finite element method (FEM) using software ANSYS 11.0 (ANSYS, Inc., PA, USA), and the data was post-processed by MATLAB 7.1 (The Mathworks Inc., Natick, MA).

2.2. Finite element model

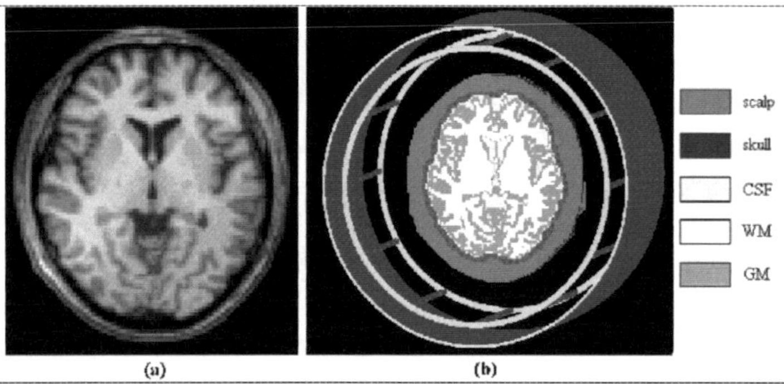

Figure 1. (a) T1-weighted MRI image slice of interest, (b) axial view of the FEM head within the 28-cm long birdcage coil (green: copper shield, cyan: rungs and end-rings; magenta: scalp, purple: skull, yellow: CSF, white: WM, grey: GM).

T1-weighted MRI images (matrix size 128×128×72, voxel size $2\times2\times2mm^3$) of a human head (Fig. 1 (a)) were segmented into five tissues: skull, scalp, cerebrospinal fluid (CSF), white matter (WM) and grey matter (GM). The structure information was imported into ANSYS software, and a five-tissue anatomically accurate head model was constructed. A 28cm-long birdcage coil, a 28cm-long TEM coil and a 12cm-long birdcage coil were modeled with a diameter of 28cm and with 12 rungs. Either coil was enclosed by a cylindrical shield having a diameter of 32cm and a length of 30cm and 14cm, respectively (Fig. 1 (b)).

The equivalent circuit model was utilized to perform a linear excitation. All of the capacitors in the coil were replaced with current sources at required resonant frequency -- 128MHz (3 Tesla). The corresponding EP values for different head tissues were set to be isotropic and derived from the 4-Cole-Cole Model [13].

3. Results and Discussion

In our algorithm, H_z is neglected. We have performed a series of simulations on the three coils to evaluate this assumption (noise free). Simulation results are shown in Fig. 2. It can be observed that, the TEM coil, which has been developed in order to restrain nonproductive end-ring-generated H_z component [14], could produce the optimal result: the relative errors of reconstructed permittivity of CSF, WM and GM are 22.2%, 13.8% and 17.6%, respectively; for the 12cm birdcage coil in which H_z component appears to be more dominant compared with those in the other two coils, the relative errors of reconstructed permittivity are 36.2%, 25.5% and 38.2%, respectively. While the experimental measurement of H_Z component is not feasible in MRI, the simulation results suggest that assuming $H_Z = 0$ within a common volume coil could also provide reasonable structural information of biological tissues.

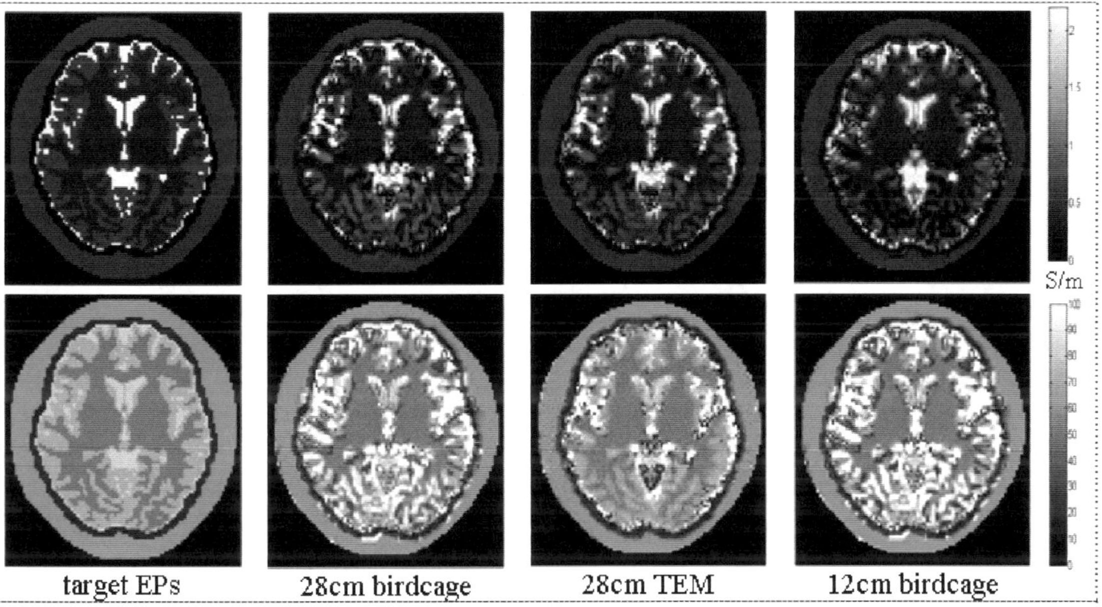

Figure 2. Reconstructed electric property distribution within different coils. Upper row: absolute conductivity, bottom row: relative permittivity. From left to right: target EP distribution, reconstruction results in a 28cm-long birdcage, TEM coil and a 12cm-long birdcage coil, respectively.

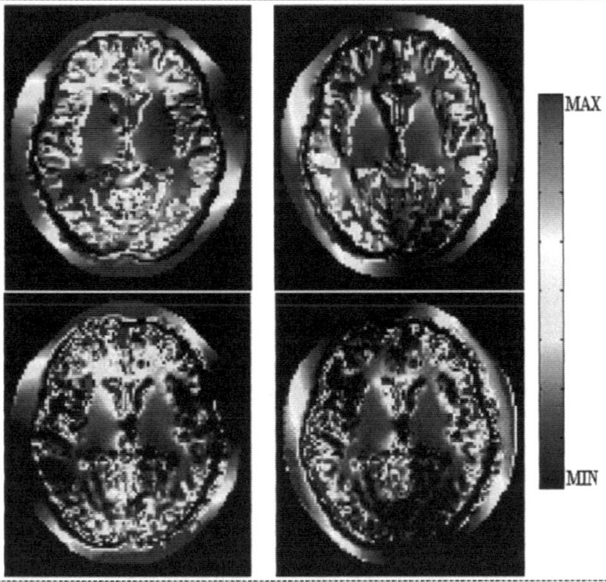

Figure 3. Normalized SAR distribution under each excitation within the 28cm-long birdcage coil. Upper row: theoretical distribution, bottom row: estimated distribution. Left column: under excitation I, right column: under excitation II.

Once we get the reconstructed electric property values, we can estimate the SAR distribution within the corresponding biological tissues via

$$\mathrm{SAR} = \sigma[\,\|\partial H_y / \partial z\|^2 + \|\partial H_x / \partial z\|^2 + \|\partial H_y / \partial x - \partial H_x / \partial y\|^2\,]/(2\rho\omega^2\varepsilon_c^2) \quad (1)$$

in which ρ is the tissue mass density. Normalized SAR distributions under each coil excitation are shown in Fig. 3. Compared with theoretical values, the relative errors of estimated SAR are 19.3% (left) and 27.7% (right), respectively.

4. Acknowledgments

This work was supported in part by NIH R21EB006070, NSF BES-0602957, NIH R01EB007920, and the Supercomputing Institute of the University of Minnesota.

References

[1] E.C. Fear, X. Li, S.C. Hagness, and M.A. Stuchly, "Confocal microwave imaging for breast cancer detection: localization of tumors in three dimensions," *IEEE Trans. Biomed. Eng.*, vol. 49, (no. 8), pp. 812-22, Aug 2002.

[2] P. Metherall, D.C. Barber, R.H. Smallwood, and B.H. Brown, "Three-dimensional electrical impedance tomography," *Nature*, vol. 380, (no. 6574), pp. 509-12, Apr 11 1996.

[3] H. Griffiths, W.R. Stewart, and W. Gough, "Magnetic induction tomography. A measuring system for biological tissues," *Ann. N Y Acad. Sci.*, vol. 873, pp. 335-45, Apr 20 1999.

[4] H.S. Khang, B.I. Lee, S.H. Oh, E.J. Woo, S.Y. Lee, M.Y. Cho, O. Kwon, J.R. Yoon, and J.K. Seo, "J-substitution algorithm in Magnetic Resonance Electrical Impedance Tomography (MREIT): Phantom experiments for static resistivity images," *IEEE Trans. Med. Imaging*, vol. 21, (no. 6), pp. 695-702, Jun 2002.

[5] M. Joy, G. Scott, and M. Henkelman, "In vivo detection of applied electric currents by magnetic resonance imaging," *Magn. Reson. Imaging.*, vol. 7, (no. 1), pp. 89-94, 1989.

[6] Y. Xu and B. He, "Magnetoacoustic tomography with magnetic induction (MAT-MI)," *Phys. Med. Biol.*, vol. 50, pp. 5175-5187, 2005.

[7] E.M. Haacke, L.S. Petropoulos, E.W. Nilges, and D.H. Wu, "Extraction of conductivity and permittivity using magnetic resonance imaging," *Phys. Med. Biol.*, vol. 38, (no. 6), pp. 723-734, 1991.

[8] H. Wen, "Noninvasive quantitative mapping of conductivity and dielectric distributions using RF wave propagation effects in high-field MRI," *Proc. SPIE*, vol. 5030, pp. 471-477, 2003.

[9] U. Katscher, T. Dorniok, C. Findeklee, P. Vernickel, and K. Nehrke, "In vivo determination of electric conductivity and permittivity using a standard MR system," *13th International Conference on Electrical Bioimpedance and the 8th Conference on Electrical Impedance Tomography*, 2007, pp. 508-511.

[10] U. Katscher, T. Voigt, C. Findeklee, P. Vernickel, K. Nehrke, and O. Dossel. "Determination of Electric Conductivity and Local SAR Via B1 Mapping," *IEEE Trans. Med. Imaging*, vol. 28, (no. 9), pp. 1365-1374, Sep 2009.

[11] S. Akoka, F. Franconi, F. Seguin, and A. Lepape, "Radiofrequency Map of an Nmr Coil by Imaging," *Magn Reson Imaging*, vol. 11, (no. 3), pp. 437-441, 1993.

[12] X.T. Zhang, and B. He, "Imaging Electric Properties of Human Brain Tissues by RF Field Mapping in MRI," in *Conf. Proc. Annu. Int. Conf. IEEE Eng. Med. Biol. Soc.*, Minneapolis, 2009, pp. 4432-4435.

[13] S. Gabriel, R.W. Lau, and C. Gabriel, "The dielectric properties of biological tissues: III. Parametric models for the dielectric spectrum of tissues," *Phys. Med. Biol.*, vol. 41, (no. 11), pp. 2271-93, Nov 1996.

[14] J. T. Vaughan, G. Adriany, C. J. Snyder, J. Tian, T. Thiel, L. Bolinger, H. Liu, L. DelaBarre, and K. Ugurbil, Efficient high-frequency body coil for high-field MRI," Magn. Reson. Med., vol. 52, no. 4, pp.851–859, Oct. 2004.

MREIT conductivity imaging of canine head using multi-echo pulse sequence

Y Q Han[1], Z J Meng[1], W C Jeong[1], Y T Kim[1], A S Minhas[1], H J Kim[1], H S Nam[2], O Kwon[2] and E J Woo[1]

[1] Department of Biomedical Engineering, Kyung Hee University, Yongin, Korea
[2] Department of Mathematics, Konkuk University, Seoul, Korea

E-mail: ejwoo@khu.ac.kr

Abstract. In magnetic resonance electrical impedance tomography (MREIT), we measure induced magnetic flux densities subject to multiple injection currents to reconstruct cross-sectional conductivity images. Spin echo pulse sequence has been widely used in MREIT and produce postmortem and *in vivo* conductivity images of animal and human subjects. The image quality depends on the SNR of the measured magnetic flux density image. In order to reduce the scan time and current amplitude while keeping the image quality, we have developed a multi-echo pulse sequence for MREIT. In this study, we show results of canine head MREIT imaging experiments using the multi-echo pulse sequence. Compared to the injection current nonlinear encoding (ICNE) pulse sequence, it provides a higher SNR of MR magnitude images by combining multiple echo signals. Noise in measured magnetic flux density data is significantly reduced due to an increased current injection time over multiple echo signals. These allow us to significantly decrease the total scan time. Reconstructed conductivity images of a canine head show enhanced conductivity contrast between gray and white matter using the multi-echo pulse sequence. In our future work, we will focus on *in vivo* human and disease model animal experiments using the new MREIT pulse sequence.

Keywords: MREIT, conductivity image, multi-echo pulse sequence

1. Introduction

Magnetic resonance electrical impedance tomography (MREIT) utilizes an MRI scanner as a tool to measure a magnetic flux density distribution inside an imaging object induced by an externally injected current [1-4]. The quality of a measured magnetic flux density data is closely related with the adopted MR pulse sequence. The injection current nonlinear encoding (ICNE) pulse sequence, which has been widely used in MREIT, can produce *in vivo* conductivity images of human and animal subjects using imaging currents of 9 mA or less [5-8]. However, in performing *in vivo* human and disease model animal experiments, a long scan time and high injection current are still required to enhance the SNR. To improve the quality of a measured magnetic flux density data so that we can reduce the scan time, we developed a new multi-echo pulse sequence for MREIT.

In this study, we will show the results of canine head MREIT imaging experiments using the new multi-echo pulse sequence for MREIT. Comparing its performance with the conventional spin echo pulse sequence, we analyze advantages and limitations of the proposed sequence.

2. Methods

2.1. Animal preparation

To prevent dribbling during experiments, we injected 0.1 mg/kg of atrophine sulfate. Ten minutes later, we anestheti zed the animal with intra muscular in jection of 0.2 m l/kg T iletamine and Zolazepa m (Zoletil 50, Virbac, France) and then sacrificed it by an intravenous injection of 2 mmol/kg KCl (KCl-40 inj, Daihan Phar macy, Korea). Aft er clipping hair at four locations (dorsal, ventral and bilateral surfaces) on the head, we attached four carbon-h ydrogel electrodes and placed the anim al inside th e bore of our 3 T MRI scanner (Magnum3, Medinus, Korea). The experi mental protocol was approved by the Institutional Animal Care and Use Committee (IACUC) of Konkuk University, Seoul, Korea.

2.2. Imaging experiment

Using our custo m-designed MREIT current source, we injected 20 m A currents as the first current I_1 between one opposing pai r of electrodes. Both spin echo and multi-echo (3 echos) pulse sequences were used for imaging experiments. Imaging parameters were as follows: TR/TE = 900/30 ms, FOV = 180×180 mm^2, slice thickness = 4 mm, NEX = 24, matrix size = 128×128, number of slices = 8 and total imaging time = 100 min. Figure 1 shows the cu rrent injection scheme used in spin e cho and multi-echo sequences. Details of the multi-echo pul se sequence will be de scribed elsewhere. The current injection intervals are denoted by T_{c1} and T_{c2} and TE is the echo tim e. After acquiring the first magnetic flux density (B_z) data set for I_1 in 8 axial sli ces, the second injection current I_2 with the sa me amplitude and p ulse widt h was injected thr ough t he other pair of o pposing electrodes to obtain the second B_z data set.

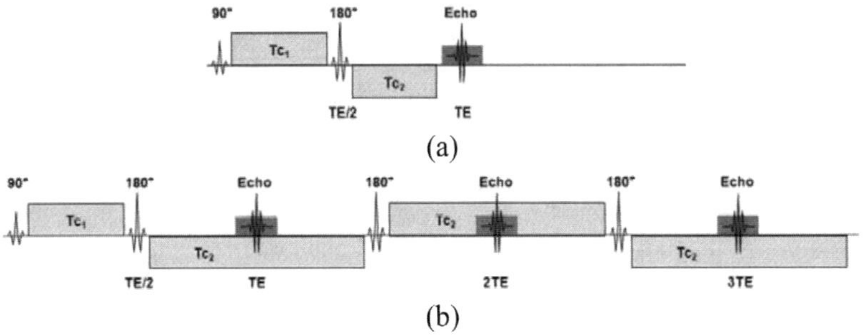

Figure 1. Current injection for MREIT experiment. (a) Spin echo and (b) m ulti-echo pulse sequences.

2.3. Conductivity image reconstruction

We used CoReHA (conductivit y reco nstructor usi ng harmonic algorithm s), which is an integrated software p ackage for MR EIT [9]. It p rovides GUI -based functions for all dat a processing routines needed to produce condu ctivity im ages fro m measured *k*-space data set s. We used the single-step harmonic B_z algorithm implem ented in CoReHA for m ulti-slice conductivit y image reconstructions [10]. All conductivit y im ages presented in this pape r should be interpreted as equivalent isotropi c scaled conductivity im ages provi ding onl y conductivity -based contrast inform ation. Interpretation of these images should be pursued in future work.

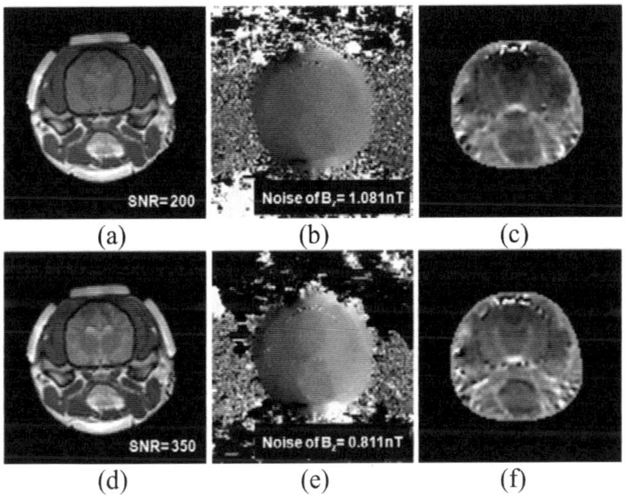

Figure 2. (a), (b) and (c) are MR magnitude, magnetic flux density and conductivity images, respectively, of a canine head using the spin echo pulse sequence. (d), (e) and (f) are images of the same canine head using the multi-echo pulse sequence.

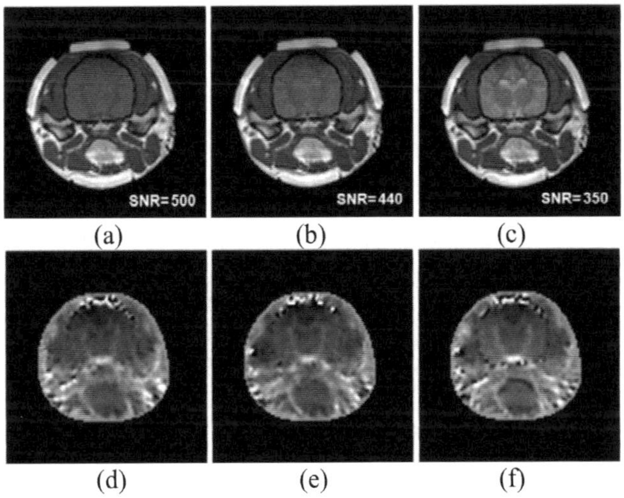

Figure 3. MR magnitude images (a, b, c) and conductivity images (d, e, f) using the multi-echo pulse sequence with 3 different echo spacing of 15, 20 and 30 ms, respectively.

3. Results

Figure 2(a), (b) and (c) are MR magnitude, B_z and conductivity images of a canine head using the spin echo pulse sequence. Figure 2(d), (e) and (f) are images using the multi-echo pulse sequence. Both the SNR of MR magnitude images and noise level in B_z images were significantly enhanced in the multi-echo pulse sequence. Conductivity images using the multi-echo pulse sequence show a better conductivity contrast between gray and white matter. Figure 3 are MR magnitude images (a, b, c) and conductivity images (c, d, e) using multi-echo pulse sequences with different echo spacing of 15, 20 and 30 ms, respectively. Though the magnitude image SNR of 30 ms case was poor, the conductivity image shows an improved contrast between gray and white matter due to its prolonged current injection time. The semi-automatic boundary extraction method implemented in CoReHA often resulted in somewhat distorted shape of the outer most boundary in conductivity images. This minor geometrical distortion has negligible effects in reconstructed conductivity images since the harmonic B_z algorithm is insensitive to geometrical errors.

4. Discussion

We found that the multi-echo pulse sequence outperforms the spin echo pulse sequence. Compared to the spin echo pulse sequence, multi-echo sequence provides a higher magnitude image SNR by combining multiple echo signals. The noise level in measured magnetic flux density data is significantly reduced due to a prolonged current injection time over multiple echo signals, which allows a more phase accumulation. Conductivity images of a canine head show enhanced contrast between gray and white matter using the multi-echo pulse sequence. These will allow us to decrease the total scan time. We expect the multi-echo pulse sequence be useful in experimental MREIT studies where a short scan time is required. In our future work, we will focus on *in vivo* human and disease model animal experiments using the multi-echo pulse sequence.

Acknowledgments

This work was supported by the SRC/ERC program (R11-2002-103) of MEST/NRF.

References

[1] Zhang N 1992 Electrical impedance tomography based on current density imaging *MS Thesis* University of Toronto
[2] Woo E J, Lee S Y and Mun C W 1994 Impedance tomography using internal current density distribution measured by nuclear magnetic resonance *Proc. SPIE* **2299** 377–85
[3] Ider Y Z and Birgul O 1998 Use of the magnetic field generated by the internal distribution of injected currents for electrical impedance tomography (MR-eIT) *Elektrik* **6** 215–25
[4] Woo E J and Seo J K 2008 Magnetic resonance electrical impedance tomography (MREIT) for high-resolution conductivity imaging *Physiol. Meas.* **29** R1–R26
[5] Park C, Lee B I, Kwon O and Woo E J 2006 Measurement of induced magnetic flux density using injection current nonlinear encoding (ICNE) in MREIT *Physiol. Meas.* **28** 117–27
[6] Kim H J, Lee B I, Cho Y, Kim Y T, Kang B T, Park H M, Lee S Y, Seo J K and Woo E J 2007 Conductivity imaging of canine brain using a 3 T MREIT system: postmortem experiments *Physiol. Meas.* **28** 1341–53
[7] Kim H J, Oh T I, Kim Y T, Lee B I, Woo E J, Seo J K, Lee S Y, Kwon O, Park C, Kang B T and Park H M 2008 In vivo electrical conductivity imaging of a canine brain using a 3 T MREIT system *Physiol. Meas.* **29** 1145–55
[8] Kim H J, Kim Y T, Minhas A S, Jeong W C, Woo E J, Seo J K and Kwon O J 2009 In vivo high-resolution conductivity imaging of the human leg using MREIT: the first human experiment *IEEE Trans. Med. Imaging* **28** 1681–7
[9] Jeon K, Kim H J, Lee C O, Woo E J and Seo J K 2009 CoReHA: conductivity reconstructor using harmonic algorithms for magnetic resonance electrical impedance tomography (MREIT) *J. Biomed. Eng. Res.* **30** 279–87
[10] Seo J K, Kim S W, Kim S, Liu J J, Woo E J, Jeon K and Lee C O 2008 Local harmonic B_z algorithm with domain decomposition in MREIT: computer simulation study *IEEE Trans. Med. Imaging* **27** 1754–61

MREIT conductivity imaging based on the local harmonic B_z algorithm: animal experiments

Kiwan Jeon[†], **Chang-Ock Lee**[†], **Eung Je Woo**[‡], **Hyung Joong Kim**[‡], **and Jin Keun Seo**[§]

† Department of Mathematical Sciences, KAIST, Daejeon, Korea
‡ College of Electronics and Information, Kyung Hee University, Korea
§ Department of Computational Science and Engineering, Yonsei University, Korea

Abstract. From numerous numerical and phantom experiments, MREIT conductivity imaging based on harmonic B_z algorithm shows that it could be yet another useful medical imaging modality. However, in animal experiments, the conventional harmonic B_z algorithm gives poor results near boundaries of problematic regions such as bones, lungs, and gas-filled stomach, and the subject boundary where electrodes are not attached. Since the amount of injected current is low enough for the safety for *in vivo* animal, the measured B_z data is defected by severe noise. In order to handle such problems, we use the recently developed local harmonic B_z algorithm to obtain conductivity images in our ROI(region of interest) without concerning the defected regions. Furthermore we adopt a denoising algorithm that preserves the ramp structure of B_z data, which informs of the location and size of anomaly. Incorporating these efficient techniques, we provide the conductivity imaging of post-mortem and *in vivo* animal experiments with high spatial resolution.

1. Introduction

Magnetic resonance electrical impedance tomography (MREIT) attempts to provide static conductivity images of an electrically conducting object with high spatial resolution [1–7]. We sequentially inject multiple currents through chosen pairs of surface electrodes to produce internal current density $\mathbf{J} = (J_x, J_y, J_z)$ and also magnetic flux density $\mathbf{B} = (B_x, B_y, B_z)$ distributions inside the object. In order to reconstruct conductivity images, MREIT uses an MRI scanner to measure a set of B_z data where z is the direction of the main magnetic field of the scanner [8]. The conductivity image reconstruction in MREIT is based on the relationship among the conductivity distribution σ, the applied current I, and the measured B_z data:

$$B_z(\mathbf{r}) = \frac{\mu_0}{4\pi}\int_\Omega \frac{\sigma(\mathbf{r})\,[(x-x')\frac{\partial u}{\partial y}(\mathbf{r}')-(y-y')\frac{\partial u}{\partial x}(\mathbf{r}')]}{|\mathbf{r}-\mathbf{r}'|^3}\,d\mathbf{r}' + \text{lead wire effects}, \quad (1)$$

where u is the electrical potential generated by the injection current I that satisfies the following mixed boundary value problem:

$$\begin{cases} \nabla\cdot(\sigma\nabla u) = 0 & \text{in } \Omega \\ I = \int_{\mathcal{E}^+}\sigma\frac{\partial u}{\partial \mathbf{n}}\,ds = -\int_{\mathcal{E}^-}\sigma\frac{\partial u}{\partial \mathbf{n}}ds & \\ \nabla u\times\mathbf{n} = 0 & \text{on } \mathcal{E}^+\cup\mathcal{E}^- \\ \sigma\frac{\partial u}{\partial \mathbf{n}} = 0 & \text{on } \partial\Omega\setminus\overline{\mathcal{E}^+\cup\mathcal{E}^-} \end{cases} \quad (2)$$

Here, $\mathbf{n}$ is the unit outward normal vector to $\partial\Omega$, $\mathbf{r} = (x, y, z)$, $\frac{\partial u}{\partial \mathbf{n}} = \nabla u \cdot \mathbf{n}$, and μ_0 the magnetic permeability. The first B_z-based MREIT reconstruction algorithm called the harmonic B_z algorithm was proposed in [9]. Numerous numerical and experimental studies have shown that it is quite successful in producing high resolution conductivity images. However, in animal experiments, the conventional harmonic B_z-algorithm shows poor performance in providing accurate conductivity reconstruction near boundaries of problematic regions including bones, lungs, air-filled stomach, and the subject boundary. The recent MREIT algorithm called *local harmonic B_z-algorithm* [10] was designed to deal with image defects caused by the presence of those problematic regions. In the local harmonic B_z-algorithm, the quality of the conductivity image is mainly dependent upon the quality of the B_z data. Recently a new denoising method [11] was proposed to presereve ramp structures in the B_z data, that shows the change of conductivity.

In this paper, we examine the local harmonic B_z-algorithm with the ramp preserving denoising method for the B_z-data from *in vivo* and post-mortem animal experiments.

2. Methods

2.1. The local harmonic B_z−algorithm: sub-regional reconstruction

Let the subject to be imaged occupy a three-dimensional bounded domain $\Omega \subset \mathbb{R}^3$ with a smooth connected boundary $\partial\Omega$. We assume that the subject Ω contains a region with low conductivity. In practice, the region may include bones, lungs, and an air-filled stomach. We denote by $\Omega_{z_0} := \Omega \cap \{z = z_0\} \subset \mathbb{R}^2$, the slice of Ω cut by the plane $\{z = z_0\}$. We assume that the conductivity σ is isotropic. We inject two independent electrical currents I_1 and I_2 through two pairs of surface electrodes $\mathcal{E}_1^{\pm}$ and $\mathcal{E}_2^{\pm}$, respectively. For $j = 1$ and 2, let $u_j[\sigma]$ be a solution of (2) corresponding to I_j. The z-component of the curl of the Ampere law $\nabla \times \mathbf{J} = \frac{1}{\mu_0}\nabla \times \nabla \times \mathbf{B}$ is

$$\langle \nabla\sigma \,,\, \mathbb{L}\nabla u_j[\sigma]\rangle = \frac{1}{\mu_0}\nabla^2 B_{z,j} \tag{3}$$

where $\mathbb{L} = \begin{pmatrix} 0 & 1 & 0 \\ -1 & 0 & 0 \\ 0 & 0 & 0 \end{pmatrix}$. The above identity means that the data $\frac{1}{\mu_0}\nabla^2 B_{z,j}$ can convey any local change of σ along the direction $\mathbb{L}\nabla u_j[\sigma]$. We know that the vector field $\sigma\nabla u_j[\sigma]$ is dictated mainly by injection current I_j and the global geometry of the domain $\partial\Omega$ instead of the local distribution of σ provided the magnitude of $\nabla\sigma$ is small. Indeed, we may approximate $\sigma\nabla u_j[\sigma] \approx \nabla u_j[\sigma_0]$ where $\sigma_0 = 1$. Taking account of this approximation of the vector field $\sigma\nabla u_j[\sigma]$ in our mind, it is desirable that (3) is rewritten in the following from:

$$\langle \nabla \ln\sigma \,,\, \mathbb{L}\sigma\nabla u_j[\sigma]\rangle = \frac{1}{\mu_0}\nabla^2 B_{z,j}\,. \tag{4}$$

Combining the identity (4) for $j = 1, 2$, we have

$$\begin{bmatrix} \frac{\partial \ln\sigma}{\partial x}(\mathbf{r}) \\ \frac{\partial \ln\sigma}{\partial y}(\mathbf{r}) \end{bmatrix} = \frac{1}{\mu_0}\left(\mathbb{A}[\sigma](\mathbf{r})\right)^{-1} \begin{bmatrix} \nabla^2 B_{z,1}(\mathbf{r}) \\ \nabla^2 B_{z,2}(\mathbf{r}) \end{bmatrix}, \qquad \mathbf{r} \in \Omega \tag{5}$$

where $\mathbb{A}[\sigma](\mathbf{r}) = \begin{bmatrix} \sigma(\mathbf{r})\frac{\partial u_1[\sigma]}{\partial y}(\mathbf{r}) & -\sigma(\mathbf{r})\frac{\partial u_1[\sigma]}{\partial x}(\mathbf{r}) \\ \sigma(\mathbf{r})\frac{\partial u_2[\sigma]}{\partial y}(\mathbf{r}) & -\sigma(\mathbf{r})\frac{\partial u_2[\sigma]}{\partial x}(\mathbf{r}) \end{bmatrix}$. Unfortunately, in *in vivo* animal experiments, B_z data is very noisy due to the requirement of low amplitude of injection currents. Hence, before solving the equation (5), we need pre-processing such as harmonic fill-in of defected regions, denoising of B_z, and so on.

2.2. Preprocessing

2.2.1. Harmonic fill-in of defected regions. We first deal with locating defected region R where MR signal is very low. Numerous animal experiments show that the conventional inpainting

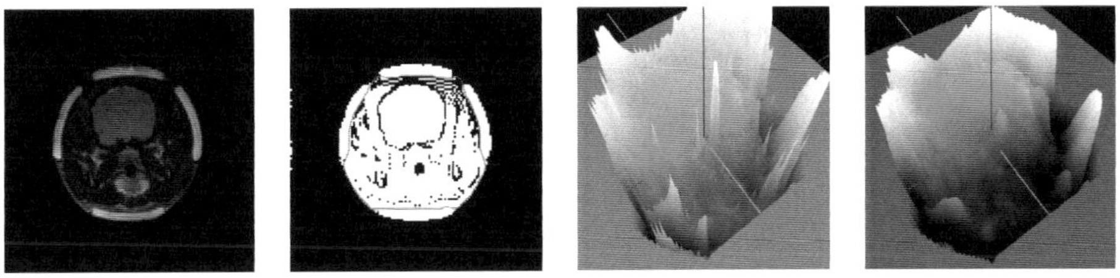

(a) (b) (c) (d)

Figure 1. (a) Conventional MR image, (b) segmentation with the region R, (c) B_z data of which injection current direction is horizontal, (d) its denoised B_z data.

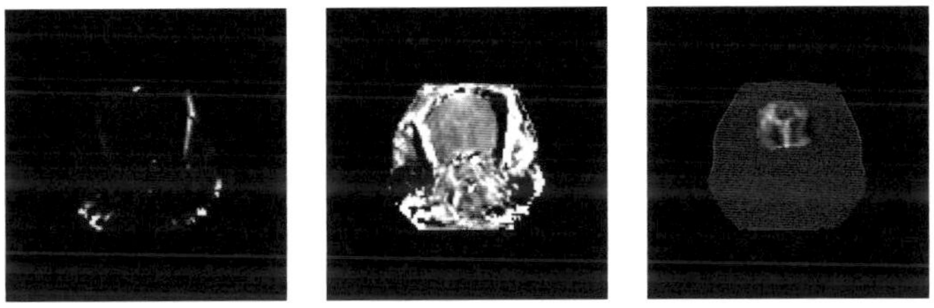

(a) (b) (c)

Figure 2. (a) Reconstructed image using the harmonic B_z-algorithm, (b) Its scaled version for the best view, (c) Reconstructed image using the local harmonic B_z-algorithm.

method [12] of solving Dirichlet problem in the segmented domain R is not appropriate since it causes some artifacts near ∂R and requires additional segmentation process which is highly dependent on user's manual work. One way is to take isotropic diffusion into R from the reliable data in the neighborhood of R. This method does not require to get the boundary ∂R via segmentation for solving the harmonic equation.

2.2.2. Ramp preserving denoising. In [11], authors proposed a new ramp preserving denoising method which utilizes the structure tensor of the image gradient to identify ramp structure of the data image. This method takes advantage of eigenvalue analysis of the structure tensor to eliminate salt-pepper type noise while preserving ramp structure of the noisy data.

3. Results

Based on the reconstruction method mentioned in the previous section, we performed animal experiments with *in vivo* canine head and post-mortem canine pelvis using a 3.0 Tesla MRI scanner. Figure 1 shows the conventional MR image, segmentation with region R, B_z data, and its denoised B_z data of the *in vivo* canine head. Figure 2-(a) shows an autoscale image of the reconstructed conductivity distribution of the *in vivo* canine head by the conventional harmonic B_z-algorithm. Because of serious reconstruction errors near the boundary and in the defected region, the autoscale image in Figure 2-(a) cannot provide any useful information of the conductivity distribution inside the domain. Figure 2-(b) is its scaled image adjusted for the best view. Figure 2-(c) is the result of the local harmonic B_z-algorithm with denoising and the harmonic fill-in of R. In Figure 2-(c), we see the brain of *in vivo* canine through a local window. We see that the local harmonic B_z algorithm combined with denoising and harmonic fill-in provides better image of canine brain than the conventional harmonic B_z algorithm.

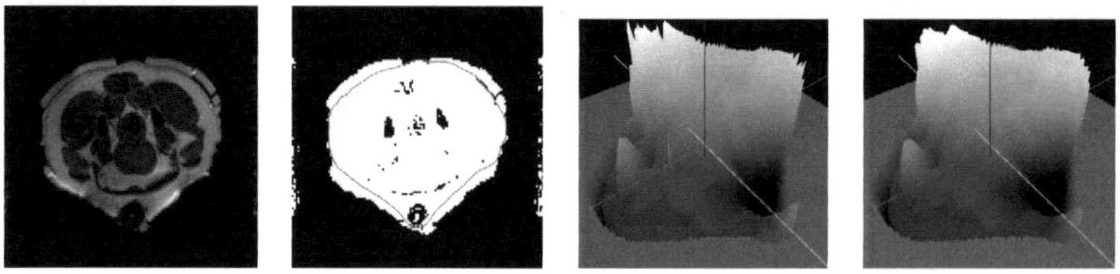

(a) (b) (c) (d)

Figure 3. (a) Conventional MR image, (b) segmentation with the region R, (c) B_z data of which injection current direction is horizontal, (d) its denoised B_z data.

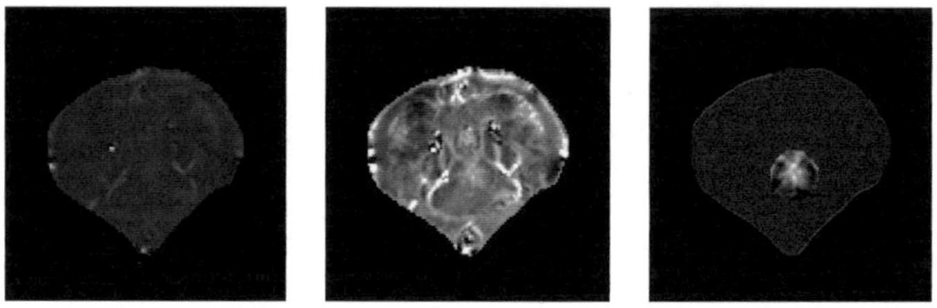

(a) (b) (c)

Figure 4. (a) Reconstructed image using the harmonic B_z-algorithm, (b) Its scaled version for the best view, (c) Reconstructed image using the local harmonic B_z-algorithm.

Figure 3 shows the conventional MR image, segmentation with region R, B_z data, and its denoised B_z data of the post-mortem canine pelvis. Figure 4-(a) shows a reconstructed image by the harmonic B_z-algorithm. Because of the boundary artifact, we can not see the scaled conductivity distribution information in the domain. Figure 4-(b) is its scaled image adusted for the best view. Figure 4-(c) is the result of the local harmonic B_z-algorithm with denoising and the harmonic fill-in of R.

Acknowledgment
This work was supported by the SRC/ERC program (R11-2002-103) of MEST/NRF.

References
[1] Woo E J, Lee S Y and Mun C W 1994 *SPIE* **2299** 377–385
[2] Kwon O, Woo E J, Yoon J R and Seo J K 2002 *IEEE Trans. Biomed. Eng.* **49** 160–167
[3] Oh S H, Lee B I, Woo E J, Lee S Y, Cho M H, Kwon O and Seo J K 2003 *Phys. Med. Biol.* **48** 3101–3116
[4] Ider Y Z, Onart S and Lionheart W R B 2003 *Physiol. Meas.* **24** 591–604
[5] Park C, Kwon O, Woo E J and Seo J K 2004 *IEEE Trans. Med. Imag.* **23** 388–394
[6] Park C, Park E J, Woo E J, Kwon O and Seo J K 2004 *Physiol.Meas.* **25** 257–269
[7] Liu J J, Seo J K, Sini M and Woo E J 2007 *SIAM J. Appl. Maths* **67** 1259–1282
[8] Scott G, R Armstrong M J and Henkelman R 1992 *J. Mag. Res.* **97** 235–254
[9] Seo J K, Yoon J R, Woo E J and Kwon O 2003 *IEEE Trans. Biomed. Eng.* **50** 1121–1124
[10] Seo J K, Kim S W, Kim S, Liu J, Woo E J, Jeon K and Lee C-O 2008 *IEEE Trans. Med. Imag.* **27** 1754–1761
[11] Lee C-O, Ahn S and Jeon K 2009 Denoising of B_z data for conductivity reconstruction in magnetic resonance electrical impedance tomography (MREIT) (*Preprint* `KAIST DMS BK21 Research Report Series/09-16`)
[12] Lee S, Seo J K, Park C, Lee B I, Woo E J, Lee S Y, Kwon O and Hahn J 2006 *IEEE Trans. Med. Imag.* **25** 168–176

Journal of Physics: Conference Series **224** (2010) 012080 doi:10.1088/1742-6596/224/1/012080

Basis for a biomechanical 'field-effect-transistor' mechanism using organic monolayers on silicon(111) revealed by electrical impedance spectroscopy

T C Chilcott, H G L Coster and E L S Wong

School of Chemical and Biomolecular Engineering

terry.chilcott@sydney.edu.au

Abstract. The width of the depletion layer in a silicon wafer and the thickness of an undecanamine organic monolayer covalently attached to the silicon surface was monitored using electrical impedance spectroscopy (EIS) as a function of the bias voltage V applied between the monolayer surface and the base of the wafer. The dependence of the depletion layer width on V was analogous to that dependency in a conventional field-effect-transistor (FET) in the range 800mV > V > -185mV and consistent with the monolayer performing the function of a FET insulating gate. Decreases in the monolayer thickness of approximately 4Å were correlated with approximately 100-fold increases in depletion layer width for fixed V in the range of -185mV > V > -800mV. This illustrates that changes in thickness induced by the immobilization of important biological molecules targeted by reactive groups incorporated into such a monolayer, could also induce field effects with similar amplification potential to the conventional FET. Thus bio-recognition in a FET-biosensor could occur directly via the mechanics of immobilization rather than indirectly via post immobilization electrochemistry that electrically induces the field effect and is highly dependent on biochemical conditions.

1. Introduction

The principal obstacle to the realization of sophisticated devices for sensing, stimulating and mimicking numerous functions performed by biological systems is the long-term robustness of the interface that links biological processes in aqueous environments to electronic circuitry. The interface formed with metals are plagued by notoriously unstable ionic double layers of high impedance that persist even after surfaces are plied with carbon based monolayers to resist corrosion and contamination [1]. Further, the oxide layer on silicon impedes transduction of biological signals.

Present research is directed towards attaching monolayers directly to the hydride-terminated silicon substrate via formation of robust Si-C bonds [2] which facilitates fabrication of monolayer surfaces terminated with reactive groups that can immobilize biologically recognizable molecules. Such biosensors are of special interest because transduction of the immobilization event can occur via an electric field effect similar to that which underlies the amplification mechanism of field effect transistors (FETs) [3] in which the FET gating is evoked electrochemically by a secondary process that is highly dependent on the pH of the biological environment, after immobilization.

Here, mechanical activation of a field effect response is reported for an organic monolayer on silicon that illustrates that direct detection of the immobilization event can occur independently of the pH and chemistry of the biological environment.

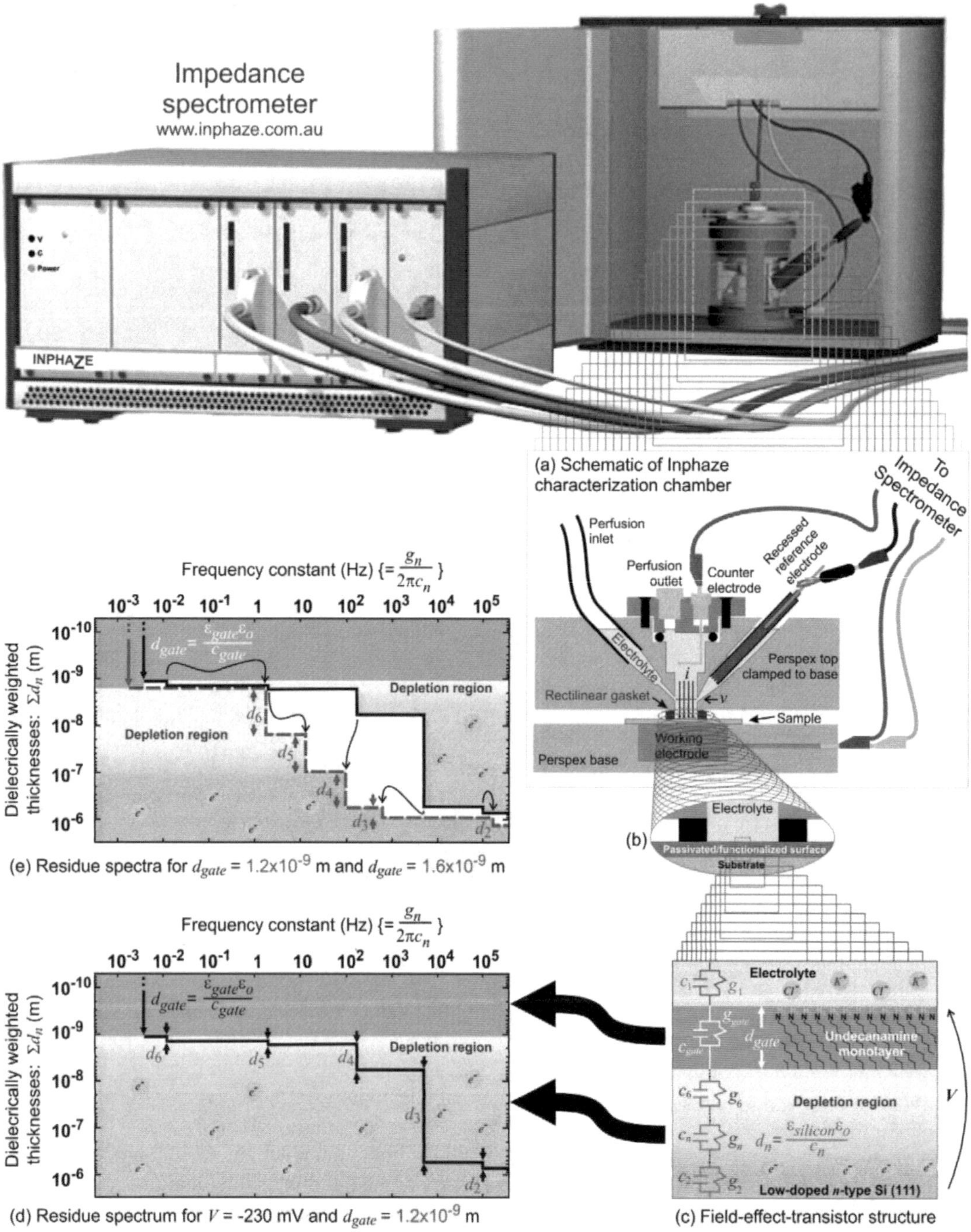

Figure 1. High resolution Inphaze impedance spectrometer and **(a)** schematic of the characterization chamber illustrating the three-terminal configuration for characterizing **(b)** functionalized surfaces on substrates and specifically **(c)** the field-effect-transistor (FET) construct of an undecanamine monolayer on low-doped silicon. Residue lin-spectra reveal a widening of the FET depletion layer arising from a transition in the gate (monolayer) thickness of **(d)** 1.2x10^{-9} m (—) to **(e)** 1.6x10^{-9} m (- - -) at a constant gating voltage V of -230 mV

2. Materials and Methods

Undecanamine monolayers were prepared according to Böcking et al [4]. Si(111) wafers (n-type, 2-8Ωcm) were cleaved into 15 mm square pieces, cleaned in concentrated H_2SO_4 :30% H_2O_2 (3:1, v/v) and rinsed in MilliQ water. Hydrogen terminated Si(111) surfaces were prepared by etching in a deoxygenated 40% solution of NH_4F. Silicon pieces were placed into a Teflon holder inside a custom made reaction chamber with a quartz glass window and backfilled with argon. *Tert*-butoxycarbonyl-protected ω-aminoalkene was injected onto the Si-H terminated surface and irradiation with UV light for 2-3h and rinsed several times with hexane, dichloromethane, tetrahydrofuran and ethanol and blown dry under a stream of nitrogen. The *tert*butoxycarbonyl (*t*-Boc) protecting group was removed by treatment with 25% TFA in dichloromethane for 60 min followed by immersion in 10% NH_4OH for 2-3 min to obtain a surface terminated with unprotonated amino groups.

Gallium-indium eutectic was applied to the underside of silicon pieces to form a low-ohmic contact with the working electrode of the Inphaze characterization chamber shown in figure 1(a). A gearing mechanism regulated by springs (not shown) delicately lowered the gasket of rectilinear cross-section onto the functionalized surface to seal a precisely defined 'active' area of 1.81×10^{-5} m^2 while minimizing distortion of the surface at the gasket perimeter. A 100 mM KCl electrolyte wetted the 'active' surface, the counter electrode and the recessed reference electrode (type LF-2 from Innovative Instruments Inc. 8533 Queen Brooks Ct.Tampa, FL 33637, USA). The counter and working electrodes injected sinusoidal current i of angular frequency ω through the surface while the reference electrode monitored the voltage response v. The amplitude of v divided by that of i yielded the impedance magnitude and the phase of v referenced to that of i yielded the impedance phase ϕ. The conductance and capacitance are given by $\cos(-\phi)/z$ and $\sin(-\phi)/\omega/z$, respectively. The Inphaze chamber design includes special features [5] for ensuring a uniform density of i that enables full utilization of the 24-bit acquisition of v that yields better than $^1/_{1000}$ degrees resolution in ϕ at values of ω as low as $^1/_{1000}$ Hertz [6]. Angstrom resolutions of substructures on silicon [7] and gold [8] have been achieved.

3. Results and Discussion

Figure 2 depicts two quasi steady state capacitance and conductance spectra for the undecanamine monolayer on silicon at a fixed bias. Maxwell-Wagner models fitted to the spectra are shown in figure 2(c). The conductance and capacitance values for the 1st Maxwell-Wagner element, i.e. g_1 and c_1, are consistent with the known properties of the electrolyte. Those values for the 7th element yield thickness values ($d_7 \equiv \varepsilon_r \varepsilon_o / c_7$) consistent with those expected for a monolayer comprised principally of hydrocarbon chains of dielectric constant $\varepsilon_r = 2$ ($\varepsilon_o = 8.85 \times 10^{-12}$ F/m). The 4×10^{-12}m transition in thickness from one quasi state (●) to the other (♦) is consistent with the reported range of likely canting angles for the undecanamine molecules in such a monolayer [4]. Further, the conductivity values (i.e. $\sigma_7 \equiv g_7 d_7$) are consistent with those expected for such a monolayer and for an insulating gate of a FET, e.g. the silicon dioxide layer in a metal-oxide-silicon-field-effect-transistor (MOSFET).

The transition in monolayer thickness correlates with changes in the properties of the 2nd, 3rd, 4th, 5th and 6th elements shown in figure 2(c), which can only be attributed to sub-layers induced in the silicon by the field effect. The depth of penetration of the effect is illustrated in the Residue line-spectra shown in figures 1(d) and 1(e). The line-spectrum in figure 1(d) is that for a monolayer thickness of 12Å (——) and reveals that sub-layers with the lowest frequency constant values, given by $f_i \equiv g_i/(2\pi c_i)$, and hence those corresponding to sub-layers of lowest conductance, are consistent with the known properties of a depletion region. Figure 1(e) illustrates that a 4Å increase in the monolayer thickness ( ) was correlated with increases in the thicknesses of most of the silicon sub-layers resulting in an approximate 100-fold widening of the depletion region to approximately 0.1μm.

This illustrates that the field effect can be induced to a similar extent either physically or electrically. Further, the physical change and the associated field effect are of suitable magnitude for the transduction of the mechanics of immobilization of biologically recognizable molecules to the functionalized surface of an organic gate in a field effect transistor.

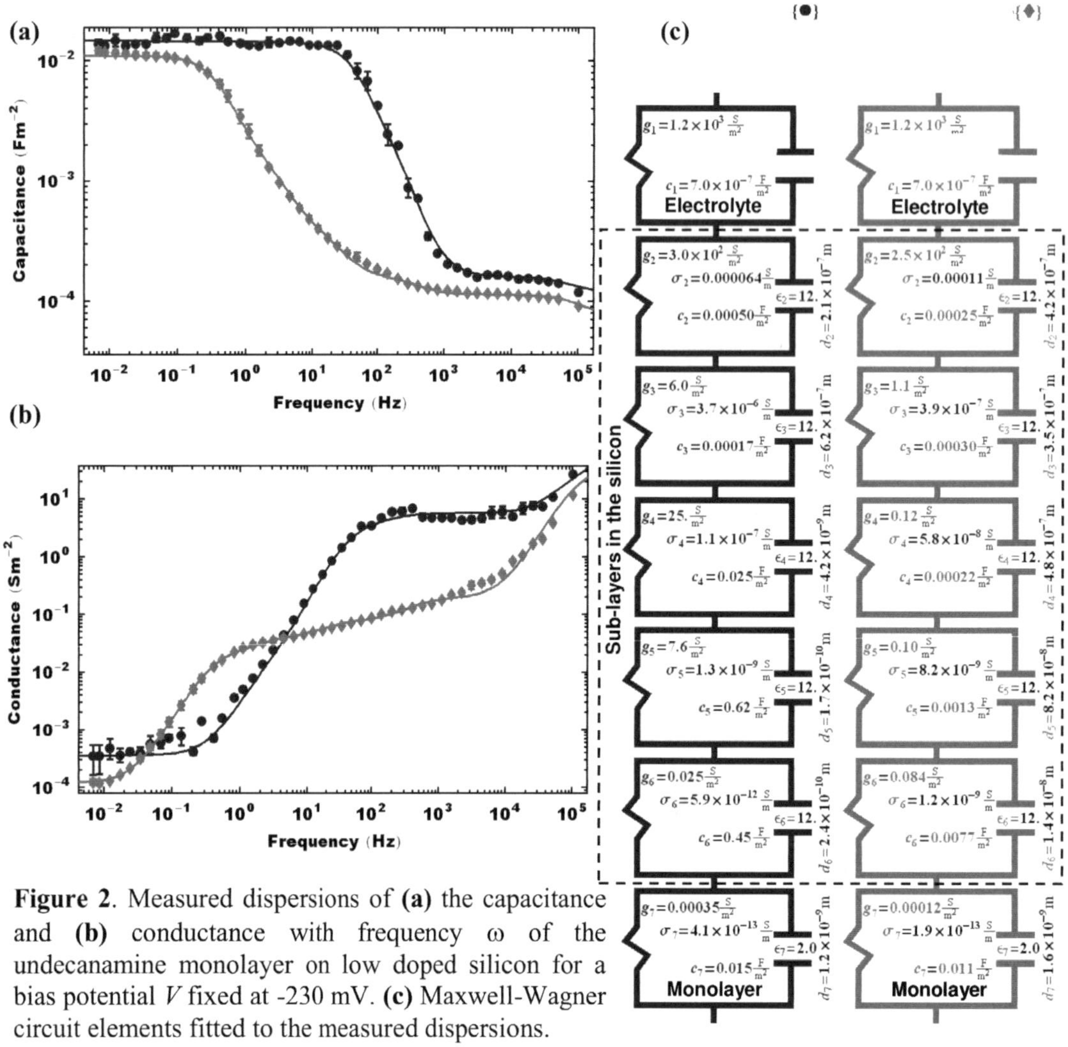

Figure 2. Measured dispersions of **(a)** the capacitance and **(b)** conductance with frequency ω of the undecanamine monolayer on low doped silicon for a bias potential V fixed at -230 mV. **(c)** Maxwell-Wagner circuit elements fitted to the measured dispersions.

References

[1] Cornell, B.A. et al. (1999) US patent PN 5,874,316 (February 23 1999)

[2] Sieval, A.B., Linke, R., Zuilhof, H., Sudholter,. E J.R. (2000) *Advanced Materials* 2000, 12, 1457-1460

[3] Patolsky, P, Lieber, C.M. (2005) Materials Today, April issue 20-27

[4] Böcking T, James M, Coster HGL, Barrow KD and Chilcott TC (2004) *Langmuir* **20(21)** 9227-9235

[5] Coster, H.G.L, and Chilcott, T.C., (2007) "Apparatus for electrochemical characterisations" PCT application filed in Germany(2006), internationally (2007)

[6] Coster, H.G. L. and Chilcott, T. C. (2006) "An improved method of measuring the phase difference between arbitrary periodic waveforms." PCT application filed in Australia.(2006)

[7] Chilcott, T.C., Wong, E.L.S., Böcking, T. and Coster, H.G.L. (2008) *Physiol. Meas.* (29), S307–S319

[8] Chilcott , T.C., Wong, E.L.S., Coster, H.G.L., Coster, A.F.C. and James, M. (2009) *Electrochimica Acta* 54, 3766-3774

Electrical properties of breast cancer cells from impedance measurement of cell suspensions

G Qiao[1], W Duan[2], C Chatwin[1], A Sinclair[2] and W Wang[1]
[1] Biomedical Engineering Group, School of Engineering and Design, University of Sussex, Brighton, BN1 9QT, UK
[2] Department of Biochemistry, School of Life Sciences, University of Sussex, Brighton, BN1 9RH, UK

Email: wei.wang@sussex.ac.uk

Abstract. Impedance spectroscopy of biological cells has been used to monitor cell status, e.g. cell proliferation, viability, etc. It is also a fundamental method for the study of the electrical properties of cells which has been utilised for cell identification in investigations of cell behaviour in the presence of an applied electric field, e.g. electroporation. There are two standard methods for impedance measurement on cells. The use of microelectrodes for single cell impedance measurement is one method to realise the measurement, but the variations between individual cells introduce significant measurement errors. Another method to measure electrical properties is by the measurement of cell suspensions, i.e. a group of cells within a culture medium or buffer. This paper presents an investigation of the impedance of normal and cancerous breast cells in suspension using the Maxwell-Wagner mixture theory to analyse the results and extract the electrical parameters of a single cell. The results show that normal and different stages of cancer breast cells can be distinguished by the conductivity presented by each cell.

1. Introduction

One of the main ways to understand cell function is by the investigation of cell electrical behavior when subjected to an electric field [1] [2]. The cell membrane causes major dispersion in wide frequency bandwidth, while the dispersion is a function of the membrane permittivity. Typically, in the radiofrequency range between 1 kHz and 10 MHz many biological tissue or cells will present significant Beta dispersion [3], such multi-frequency impedance measurement is also known as impedance spectroscopy. The whole Beta dispersion range has been used to characterize cell status, e.g. cell proliferation, viability [4].

The impedance of breast cells has been studied based upon single cell measurements [5]. However, both the large cell size and shapes variation together with single cell measurement handling difficulty would introduce significant errors into the results. These shortcomings present major difficulties when the electric signatures are utilized to distinguish between normal and cancerous cells. In order to overcome these problems cell suspensions are used instead of a single cell. Measurements were taken for cell suspensions for four breast cell lines, namely: MCF-10A, MCF-7, MDA-MB-231 and MDA-MB-435S, which are: normal cell, early stage, invasive and late stage cancer cells, respectively. A special chamber designed with four electrodes connected to an HP impedance analyzer 4194A was used for the impedance measurement of the cell suspensions. After a calibration procedure the impedance results were analyzed according to the mixture theory for two- and three-phase suspension system [6]. The electrical properties of single cell, e.g. whole cell conductivity, membrane capacitance, relaxation frequency, etc. were extracted for each type of breast cell.

2. Methods

In this research the impedance of the cell suspensions was measured using the same system introduced in the reference [6] but with the measuring chamber re-designed. Potential error sources, such as temperature variations, cell viability and volume fraction were strictly controlled during the measurement as they could have major effects on measurement sensitivity and accuracy.

2.1 Cell preparation

MDA-MB-435S, MDA-MB-231 and MDA-MB-7 (CLS-Cell lines service, Germany), and human breast tissue cell line MCF-10A (American Type Culture Collection, USA) were cultured in a DMEM/F12 media supplemented with 10% heat-inactivated fetal bovine serum, 2 mmol/L L-glutamine, 100 IU/mL penicillin, 100 µg/mL streptomycin, 20 ng/mL epidermal growth factor, 500 ng/mL hydrocortisone, 100 ng/mL cholera toxin and 10 µg/mL bovine insulin. For routine growth, the cells were maintained in an incubator at 37°C with 5% CO_2. For routine culture and prior to experiments, the cells were removed from the plastic cell culture dishes by digestion for 5 minutes with trypsin (Invitrogen). Cell viability was determined by counting cells on a haemocytometer using Trypan Blue exclusion (Intvitrogen). The volume of cells in the suspension was measured using a centrifuge tube with 0.5 µL accuracy.

2.2 Measurement chamber

The impedance measurement of cell suspensions was carried out in a specially designed chamber (figure 1) with four electrodes, two disc electrodes for current injection and two needle electrodes for voltage pickup. Some of the chamber specifications are listed in table 1. The volume of the cell suspension required for one test was about 200 µL and the cell density was in the 10^6 scale depending upon the controlled volume fraction of cells in the suspension.

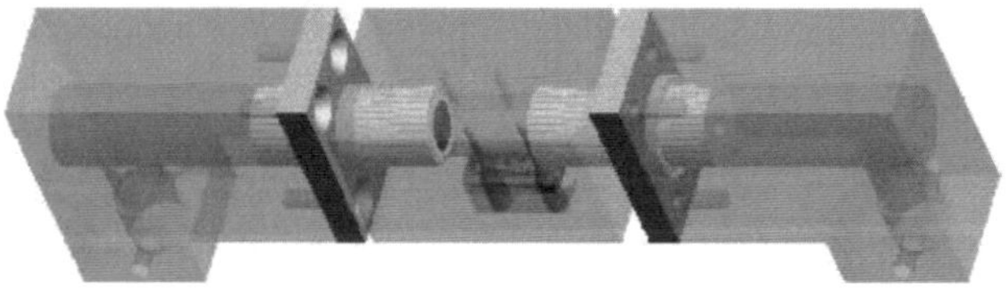

Figure 1. 3D sketch of the impedance measurement chamber for cell suspensions

Table 1. Specifications of measurement chamber

Chamber specifications			
Electrode materials	Stainless Steel	Diameter of chamber:	4mm
Disc electrode distance	11mm	Needle electrode distance	6mm tip-to-tip

2.3 Mixture theory
Theories of the two- and three-phase mixture suspension system have been studied and successfully demonstrated for analyzing cell suspension electrical properties [7] [8]. In this study the volume fraction of cells in the suspension is controlled at around 0.2, Hanai's equation (1), derived from Maxwell-Wagnar equations, which are applicable for diluted suspensions, were adapted for the high volume fraction case. In order to simplify these equations, the conductivity of the cell membrane was assumed to be negligible at low frequency and infinite at high frequency, however its capacitance was treated as an important parameter and the membrane capacitance was calculated from equations 3 and 4.

$$1 - v = \left(\frac{\sigma - \sigma_p}{\sigma_2 - \sigma_p}\right)\left(\frac{\sigma_2}{\sigma}\right)^{1/3} \tag{1}$$

$$\sigma_\infty = \sigma_2\left(1 + 3v\frac{\sigma_i - \sigma_2}{\sigma_i + 2\sigma_2}\right) \tag{2}$$

$$\tau = RC_m\left(\frac{1}{2\sigma_2} + \frac{1}{\sigma_p}\right) \tag{3}$$

$$f_0 = 1/(2\,\pi\tau) \tag{4}$$

In the equations, v is the volume fraction of cells and σ is the conductivity of the cell suspension. Subscripts p, 2 and i are for the suspended cells, buffer and cytoplasm, respectively. σ_∞ is the conductivity of the cell suspension at high frequency and C_m is the membrane capacitance related to the cell radius R and relaxation time τ. f_0 is the relaxation frequency commonly used to describe β dispersion.

3. Results and Discussion

The performance of the measuring chamber was tested using the buffer with no dispersion expected in the frequency range of interest, that is: from 1kHz to 10MHz. The bandwidth of the entire measurement system was monitored and calibrated using impedance data from both buffer and known conductivity saline solutions. The cell viability (figure 2), in the measuring buffer, was tested to make sure breast cells were fully functional during the measurement at a controlled room temperature of 20°C. Rapid data acquisitions, less than one minute, were carried out for cell suspensions from the four breast cell lines.

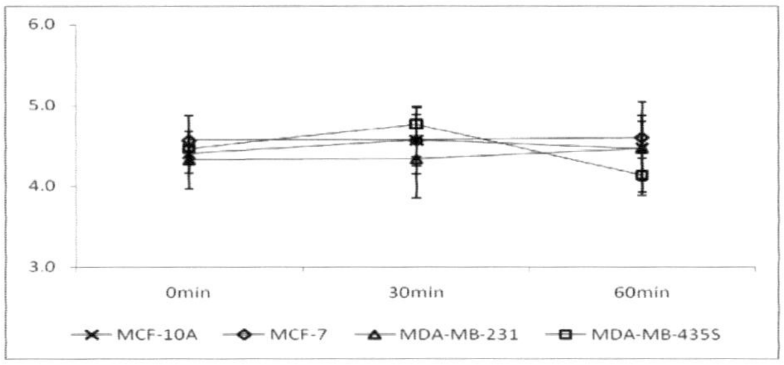

Figure 2 The relative density of four breast cell lines in modified buffer at 20 up to 60 minutes

The results of the viability of the cell lines in modified phosphate-buffered saline (50% (v/v) D-PBS supplemented with 2% serum replacement (S2640 Serum Replacement 3, Sigma) in figure 2 showed that all the four cell lines were kept alive in the modified culture buffer up to at least one hour, at a temperature of 20°C. Electrical properties of four different single cells were then extracted from the impedance of the cell suspension and the results are presented in table 2

Table 2. Electrical properties of single cells

	MCF-10A	MCF-7	MDA-MB-231	MDA-MB-435s
Whole cell conductivity at 50kHz (mS/cm)	5.58	4.44	2.81	3.97
Cytoplasm conductivity (mS/cm)	14.04	12.99	11.68	11.84
Relaxation frequency	310kHz	600kHz	610kHz	1.01MHz
Membrane capacitance ($\mu F/cm^2$)	3.94	1.95	1.81	1.10

The results for the electrical properties of the four different breast cell lines clearly showed that each cell line had a specific electrical signature which could be utilized for identification of cancer cells and differentiation of the pathology stages of malignant cells. From the results it can be seen that the healthy cell has a higher whole cell and cytoplasm conductivity, and higher membrane capacitance than the malignant cells. On the other hand, the relaxation frequency of the four types of cell gradually increased from 300kHz up to 1MHz from normal to the late stage cancer cells, which presents progressing development natures. In the continuing research, equivalent circuits analysis and different mixture theories will be utilized to examine the results. In the future, studies will focus on the developing of this technique for cancer pathology analysis.

4. References

[1] Martinsen O G, Grimnes S and Schwan H P 2002 *Encyclopedia of Surface and Colloid Science* 2643-52
[2] Hoffman R A and Britt W B 1979 *The Journal of Histochemistry and Cytochemistry* **27** 234-240
[3] Sverre G and Orjan M 2000 *Bioimpedance and Bioelectricity: Basic* (New York: Academic)
[4] K'Owino I O and Sadik O A 2005 *Electroanalysis* **17** 2101-13
[5] Han A, Yang L and Frazier A B 2007 *Clinical Cancer Research* **13** 139-143
[6] Qiao G, Wang L, Epstein D and Wang W 2008 *The 9th International Conference on Biomedical Application of Electrical Impedance Tomography, Hanover, USA, June 16 to 18 2008*
[7] Schwan H P *1957 Electrical Properties of Tissue and Cell Suspensions. (Advances in Biological and Medical Physics* vol 5) (New York: Academic) pp147–209
[8] Asami K 2002 *Progress in Polymer Science* **27** 1617-59

Influence of transfection process on single cell impedance

C M Kurz, H Büth and H Thielecke

Biohybrid Systems, Fraunhofer Institute for Biomedical Engineering, Ensheimer Str. 48, 66386 St. Ingbert, Germany

E-mail: hagen.thielecke@ibmt.fraunhofer.de

Abstract. Many new therapeutic approaches are based on gene transfection of cells. Data about dynamic interaction with cell membranes can support the selection of polymers and polyplexes regarding an optimal gene delivery reagent. A microchip/-fluidic system and an impedance-based method for monitoring the effects on individual cells after application of polymeric DNA-delivery systems is developed and evaluated. It was successfully demonstrated that the effects of polymers on the cell membrane can be monitored over time by using impedance measurements.

1. Introduction

Many new therapeutic approaches are based on gene transfection of cells. However, for an optimisation of gene transfection processes methods are needed urgently which can provide information about the interaction of gene delivery systems and cells. A fundamental property of a delivery system is that the gene vectors are able to cross the cell membrane and enter the intracellular space. It was found, that cell membrane damages can be monitored by impedance measurement ([1-3]). Thus it is assumed that the specific interaction with the cell membrane during a transfection process can be monitored by a suitable impedance-based setup.

Different impedance-based methods have been developed on the basis of Micro-Electro-Mechanical Systems (MEMS) in order to manipulate, handle or position single living cells and measure the electrical signal by various microelectrode interfaces. Some well known principles are microfluidic channels with integrated electrodes [4], micro electrode-based chips [5] and cell-filters [6]. These systems are engineered for instance for label-free cell discrimination or for basic studies of electrical and dielectrical properties of biological cells respectively. However, until now there is no existing setup suitable for a sensitive analysis of cell membrane alterations on single cell level in long-term measurements. To address this problem an impedance-based microfluidic unit predicated on a micro hole cell chip [7], [8] is developed. After hydrodynamical positioning of a single cell on a micro hole of a chip the cell adheres onto the biocompatible insulating chip membrane under cell culture conditions.

In this paper the setup for single cell impedance monitoring is evaluated regarding its suitability for measuring the interaction of gene transfer systems and cell membranes according to time. Therefore experiments are carried out with transfecting and non-transfecting polymer/DNA-complexes (polyplexes) as well as pure polymers. To validate the impedance measurements the data are correlated to standard molecular biological methods.

2. Methodology

For the measurement the micro hole cell chip described in detail in [7], [8] was used. The cell chip was integrated into a microfluidic biocompatible cell positioning unit. This micro-scaled unit consisted of one fluidic connection nanoport as well as two gold electrodes. The electrode interface was electrically linked to an impedance analyzer (Solartron 1260 and 1294, Solartron Analytical, Farnborough, UK) for the time monitoring. The two-terminal configuration turned out to be an adequate method for the measurement at 1 kHz according to time, because the electrode polarization was negligible at this frequency. The applied input potential was set to 5 mV. For single cell positioning on the micro hole a fluidic controller was used. This equipment consisted of a negative pressure system and a pressure sensor. To realize the cell culture conditions the impedance-based microchip system was put into a cultivation chamber combined with an integrated inverse microscope. The temperature, humidity and CO_2 were adjusted to 37 °C, 80 % and 5 % respectively. For the experiment, Arpe-19 (retinal pigment epithelial cell line) cells and a culture medium (DMEM/F12 31330, 10 % FCS, 1 % Penicillin/Streptavidin) were prepared. The schematic of the measurement setup is shown in figure 1.

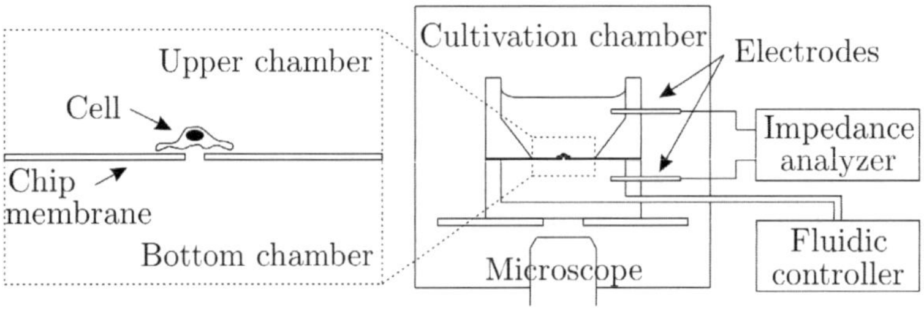

Figure 1. Schematic of the whole impedance measurement setup with the micro hole cell chip. The occupied micro hole is drawn in detail on the left hand side.

After the cell suspension was added to the upper chamber of the system, a gently negative pressure was applied to the bottom chamber to position the single cell on the hole. The experiments showed that only the cell, which occupied the hole, had an influence on the impedance level so other cells on the chip membrane were not rinsed away. The suction pressure was monitored by the integrated pressure sensor. For the cell-positioning a differential negative pressure of around 0.05 mbar was used. For reducing cell movement on the micro hole this value was increased to 0.2 mbar after the cell adhesion process. The differential pressure was measured against atmospheric pressure. These values were adjusted and established, so that during the positioning and the holding of the cell on the micro hole neither membrane- nor cell damages concerning the hydrodynamic stress were observed by the use of fluorescent markers.

3. Results and discussion

3.1. System stability

Figure 2 shows the positioning/adhesion process (figure 2(a)) and the system stability (figure 2(b)).

In figure 2(a) the low impedance magnitude level ($< 125\,\mathrm{k\Omega}$) until 3.3 min described the unoccupied micro hole setup. At 3:3 min the cell was sucked on the hole, which increased the impedance magnitude level to a value around 180 kΩ. After the positioning the cell started to adhere to the biocompatible chip membrane surface, so that the impedance magnitude increased continuously till it reached a stable and maximum value ($\sim 350\,\mathrm{k\Omega}$).

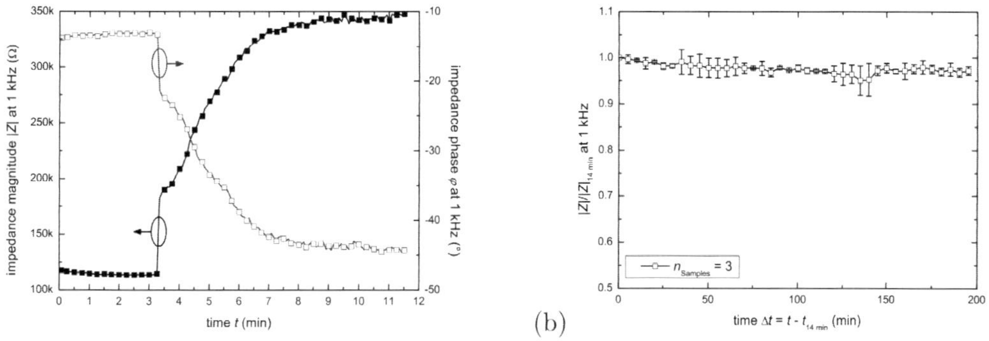

(a) (b)

Figure 2. In (a) the influence of the adhesion process on the impedance level is shown and (b) shows the normalized impedance values of the trapped Arpe-19 cells after the adhesion process.

During the experiments it pointed out, that an equilibrium state and therefore a constant impedance level of this cell line was given 14 min after the cell-positioning. Consequently a reasonable time to start monitoring was 14 min after the cell-positioning ($t_{14\,\mathrm{min}}$). Therefore an impedance normalisation was done as following.

$$\frac{Z}{Z_{14\,\mathrm{min}}} = \frac{|Z|\,\exp(\mathrm{i}\,\varphi)}{|Z_{14\,\mathrm{min}}|\,\exp(\mathrm{i}\,\varphi_{14\,\mathrm{min}})} = \frac{|Z|}{|Z_{14\,\mathrm{min}}|}\,\exp\Big[\mathrm{i}\,(\varphi - \varphi_{14\,\mathrm{min}})\Big] \tag{1}$$

Figure 2(b) shows that a stable impedance level could be guaranteed for more than 3 hours. The mean impedance magnitude values in figure 2(b) varied with 3.12 % in the worst case situation. This variation could not be totally reduced, because of the cell motility on the hole and the morphological changes. The deviation in the stability was an important factor to find out which impedance alteration was caused by the substances or by the movement or the morphological changes of the captured cell.

3.2. Measuring membrane interaction on single cell level

The application of different polymers and polyplexes was done by exchanging the whole cell culture medium in the upper chamber by the polymer or polyplex solution. The time $t_{\mathrm{application}}$ in the figures was the point of medium/(polymer or polyplex) exchanging.

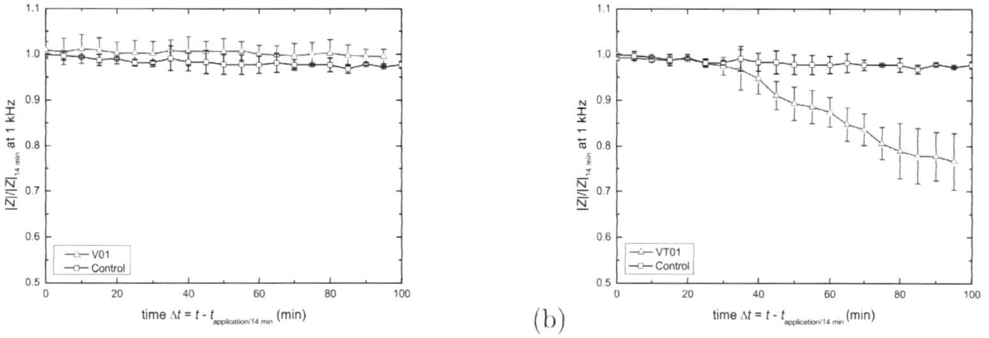

(a) (b)

Figure 3. Comparison of the normalized impedance magnitude between the control (from figure 2(b)) and the influence of the applied V01 (a) and VT01 (b) solution on the trapped cell.

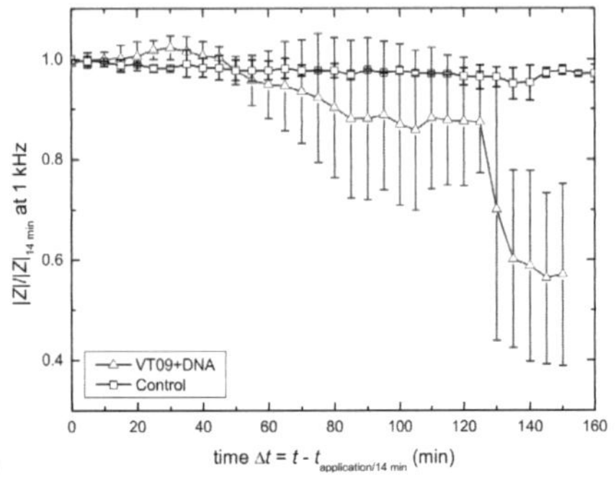

(a)

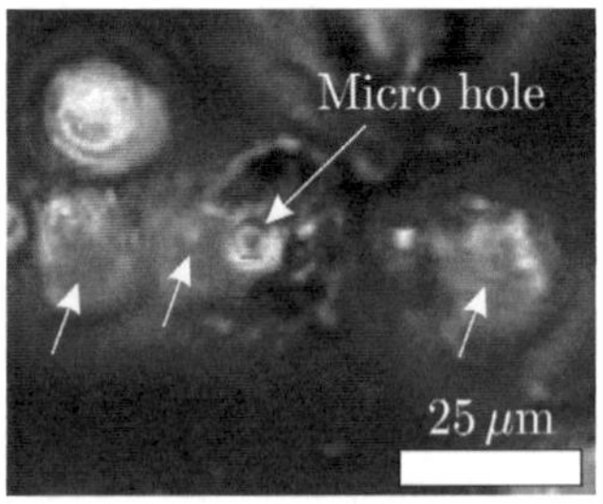

(b)

Figure 4. Comparison of the normalized impedance magnitude between the control (from figure 2(b)) and the influence of the applied VT09+DNA (a) solution on the kept cell. In (b) three Propidium iodide (PI) positive Arpe-19 cells (one covers the micro hole) are shown 145 min after VT09+DNA application (stained nuclei are marked with white arrows).

The measured data in figure 3 and 4 showed at time $t = t_{\text{application}}$ that $|Z| \approx |Z|_{14\,\text{min}}$. Therefore the replacement of medium with the polymer or polyplex solution had no influence on the impedance level. In figure 3(a) no considerable difference in the impedance magnitude between the control and the polymer V01 could be observed. Consequently this reagent solution had no possibility to interact with the cell membrane. Figure 3(b) and 4(a) exhibited a significant decay in the impedance magnitude between the polymer VT01 and the control and the polyplex VT09+DNA and the control after 30 min and 60 min respectively. These data were correlated to the standard biological marker PI to validate the measurement and to figure out the reagent impact on the cell membrane. Figure 4(b) shows three PI positive cells, which means that the polyplex VT09+DNA was able to permeabilize the cell membrane. This membrane permeabilization caused the impedance level decay.

Based on the shown figures it was possible to conclude that the impedance-based micro hole chip setup is a highly stable and suitable method for measuring cell membrane alterations. It was successfully demonstrated that such an arrangement can be used for the selection of a polymer or polyplex concerning an optimal cell membrane crossing.

Acknowledgement
This work was supported by the EU under the Framework VI project PolExGene (contract number: 019114).

References

[1] Höber R 1910 *Arch. Ges. Physiol.* **133** 237-253
[2] Höber R 1912 *Arch. Ges. Physiol.* **148** 189-221
[3] Höber R 1913 *Arch. Ges. Physiol.* **150** 15-45
[4] Schade-Kampmann G, Huwiler A, Hebeisen M, Hessler T and Di Berardino M 2007 *Cell Prolif.* **41** 830-40
[5] Giaever I and Keese C R 1993 *Nat.* **366** 591-2
[6] Bao J Z, Davis C C and Schmukler R E 1993 *IEEE Trans. Biomed. Eng.* **40** 364-78
[7] Thielecke H, Stieglitz T, Beutel H, Matthies T, Ruf H H and Meyer J U 1999 *IEEE Eng. Med. Biol. Mag.* **18** 48-52
[8] Cho S and Thielecke H 2007 *Biosens. Bioelectron.* **22** 1764-68

A hybrid FEM model to simulate the electrical characteristics of biological tissues at the cellular level

H Gagnon[1], R Guardo[1], V Kokta[2] and A E Hartinger[1]

[1] Institut de génie biomédical, École Polytechnique de Montréal, Montréal, H3C 3A7, Canada
[2] Department of Pathology, Sainte-Justine Hospital, Montréal, H3T 1C5, Canada

E-mail: gagnon@igb.polymtl.ca

Abstract. Cancer screening is possible with multi-frequency electrical impedance tomography since specific impedance variations are observable as a function of frequency. Through genetic mutations, malignant cells have different cellular characteristics than benign cells and a different impedance signature as a function of frequency. The objective of this project is to develop a FEM model to simulate the electrical characteristics of benign and malignant cells at the cellular level. Traditional tetrahedral element meshing techniques were first considered but the narrowness of the extracellular space and thinness of cell membranes made it impractical since the number of FEM elements quickly reached values that were computationally unmanageable on a typical workstation. We are therefore proposing a hybrid FEM model which combines a standard tetrahedral element meshing technique with 2D surfaces to model the extracellular fluid and discrete electrical components to represent cell membranes. The proposed hybrid modelling approach was used to develop a 3D FEM model of the skin with cellular-level details that can be used to better understand how impedance measurements performed on the skin are affected by lesions as a function of frequency as well as cellular geometrical and electrical parameters.

1. Introduction

Multi-frequency electrical impedance tomography (MF-EIT) has been proposed as a technique for cancer screening. Biological tissue identification is possible with MF-EIT since specific impedance variations are observable as a function of frequency. At lower frequencies, currents are generally constrained to the extracellular space since cell membranes exhibit high impedances. At higher frequencies, currents are free to flow through the cells since membranes have much lower impedances. Through genetic mutations, malignant cells have different cellular characteristics than benign cells such as increased nuclear to cytoplasmic ratios, larger cell volumes and wider extracellular spaces [1]. Malignant cells have therefore a different impedance signature as a function of frequency which allows MF-EIT to discriminate them [2].

The objective of this project is to develop a 3D FEM model to simulate the electrical characteristics of benign and malignant cells at the cellular level including nuclei, cytoplasm, extracellular fluid as well as nuclear and cytoplasmic membranes. Traditional tetrahedral element meshing techniques were first considered but the narrowness of the extracellular space and thinness of cell membranes made it impractical since the number of 3D FEM elements quickly reached values that were computationally unmanageable on a typical workstation. We

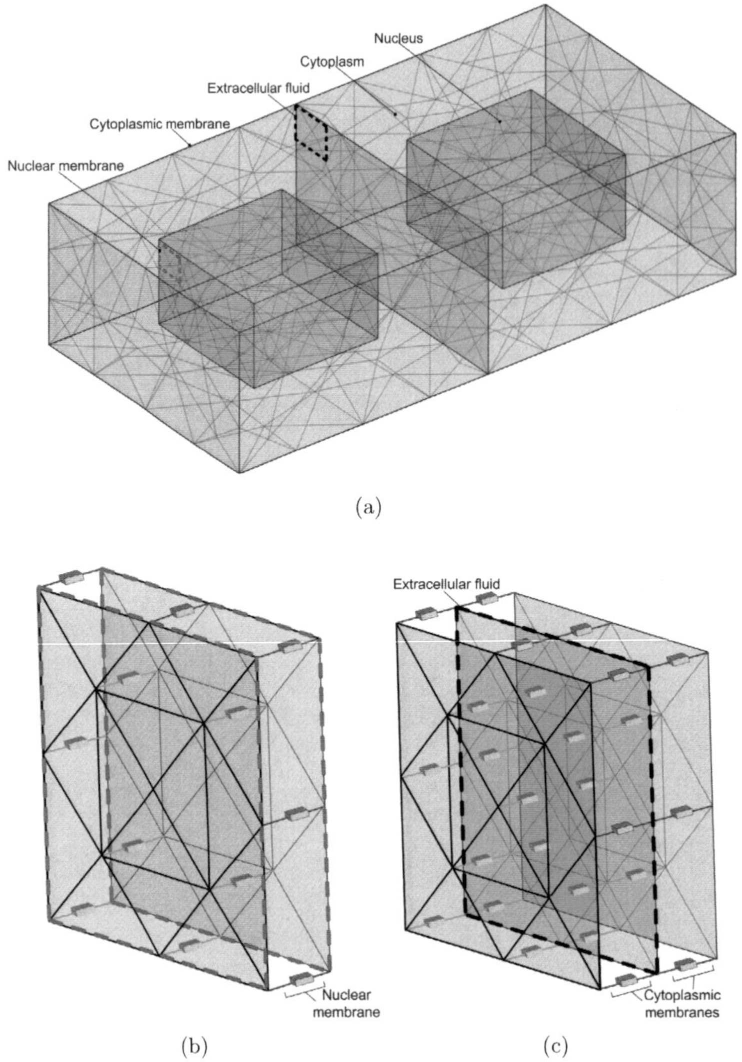

Figure 1. Cellular-level model. (a) Two adjacent cells. (b) Model of a section of the nuclear membrane bordered by a red dotted line in (a). (c) Model of a section of the cytoplasmic membranes and the extracellular fluid bordered by a black dotted line in (a).

are therefore proposing a hybrid 3D FEM model which combines a standard tetrahedral element meshing technique with 2D surfaces to model the extracellular fluid which are interconnected to the tetrahedral mesh using discrete electrical components representing cell membranes. This hybrid modelling approach has been applied to malignant and benign cutaneous tissues to assess the applicability of MF-EIT to skin cancer screening.

2. Methods

As described in previous studies [3, 4], the proposed hybrid modelling approach is realized in two steps: macroscopic and cellular modelling. The complete macroscopic 3D hybrid model is obtained by subdividing the entire medium into subdomains each corresponding to a different biological tissue. For instance, the 3D skin model that will be presented in the next section has been subdivided into the following subdomains: *stratum basalis*, *stratum spinosum*, *stratum granulosum*, *stratum corneum*, the basement membrane and the dermis. The macroscopic model also incorporates different lesion types, such as *in situ* melanoma, malignant melanoma and dysplastic nevus as well as an electrode array. All FEM modelling was performed using the complete electrode model [5]. The anisotropic electrical properties of each subdomain was either obtained from the literature (*stratum corneum* and basement membrane) or from the cellular modelling approach described in the next paragraph.

A cellular-level 3D FEM model was assembled for each biological tissue as a 3D interlaced bricked segment of typical cells, two of which are shown in figure 1(a) and include the following cellular structures: nucleus, cytoplasm, extracellular fluid as well as nuclear and cytoplasmic membranes. The nucleus and cytoplasm were modelled using standard tetrahedral elements as represented in figure 1(a). In previous studies [3, 4], nuclear and cytoplasmic membranes as well as the extracellular fluid were also modelled using standard tetrahedral elements. Since these structures are very thin compared to the typical dimensions of a cell, this approach quickly lead to very large numbers of FEM elements that were unmanageable on a typical workstation. In the hybrid approach we propose, these structures are rather modelled by 2D surfaces and discrete electrical components. As represented in figure 1(b), nuclear membranes are modelled by electrical components which interconnect corresponding nodes from the nucleus and cytoplasm. As shown in figure 1(c), the extracellular fluid is modelled by a 2D surface subdivided in triangular elements. The nodes from these triangular elements are connected to corresponding nodes from the cytoplasm tetrahedral elements through discrete electrical components representing cytoplasmic membranes.

The electrical properties of each cellular structure were obtained from the literature [6] while geometrical properties were extracted from histological images obtained from a dermatopathologist. To obtain the anisotropic electrical properties of each biological tissue, the forward problem of EIT was solved on a 3D bricked segment of geometrically regular cells while applying a current in the longitudinal or transverse direction to simulate anisotropy. The computed voltages were then interpolated onto the vertices of a homogeneous mesh whose dimensions were the same as the 3D bricked segment. The inverse problem of EIT was solved on the homogeneous mesh to obtain the anisotropic electrical properties of each biological tissue that were then incorporated into the macroscopic model described above.

3. Results

The proposed approach was applied to develop a hybrid 3D cutaneous model including all skin layers and epidermis sublayers. The model includes two microinvasive electrodes and lesions such as an *in situ* melanoma, a malignant melanoma and a dysplastic nevus. Figure 2 shows transverse sections that were obtained at 500 Hz and 1 MHz in normal skin and with an *in situ* melanoma whose location is outlined by a black rectangle. Figures (a)-(c) show that at 500 Hz the current density below the basement membrane is quite low compared to 1 MHz (b)-(d). Figures (c)-(d) show that the current density inside the region corresponding to the *in situ* melanoma is higher at each frequency than in normal skin (a)-(b).

4. Discussion

The proposed hybrid modelling approach was used to develop a 3D FEM model of the skin with cellular-level details that can be used to better understand how impedance measurements performed on the skin are affected by lesions as a function of frequency as well as cellular geometrical and electrical parameters. This new approach makes better use of computing resources since it does not require meshing thin cellular structures with standard 3D meshing techniques that would increase the number of FEM elements by a factor of 1 to 10 thousands.

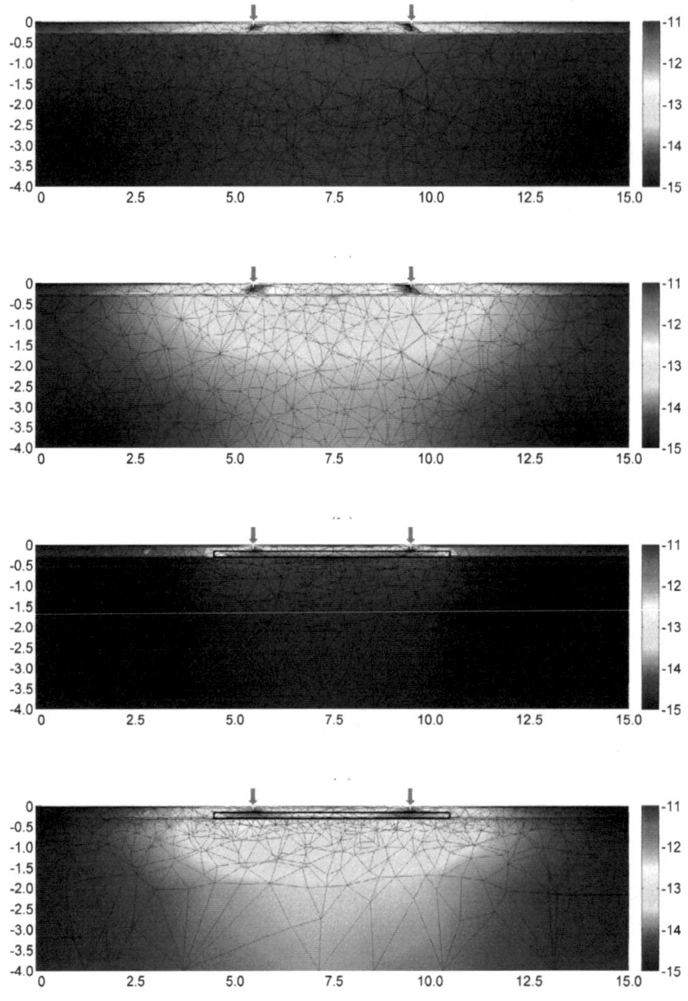

(d)

Figure 2. Transverse section of the skin model showing the modulus of the current density (A/μm^2) distribution flowing between 4-mm-spaced electrodes identified by red arrows at 500 Hz (a)-(c) and 1 MHz (b)-(d). (a)-(b) show the current distribution in normal skin and (c)-(d), an *in situ* melanoma. The colour map is defined on a decimal logarithmic scale. The cutaneous lesion is outlined in black. The x-axis and y-axis are expressed in millimetres.

Acknowledgments
This work was supported in part by the Natural Sciences and Engineering Research Council of Canada and the Fonds québécois de la recherche sur la nature et les technologies.

References
[1] Walker D C, Brown B H, Blackett A D, Tidy J and Smallwood R H 2003 *Physiol. Meas.* **24** 121-35
[2] Aberg P 2004 *Ph.D. thesis* Karolinska Institutet University Stockholm Sweden.
[3] Walker D C, Brown B H, Smallwood R H, Hose D R and Jones D M 2002 *Physiol. Meas.* **23** 159-68
[4] Keshtkar A and Keshtkar A 2008 *IEEE Trans. Biomed. Eng.* **55** 733-8
[5] Soleimani M, Powell C E and Polydorides N 2005 *IEEE Trans. Med., Imaging* **24** 577-83
[6] Kotnik T and Miklavcic D 2000 *IEEE Trans. Biomed. Eng.* **47** 1074-81

Journal of Physics: Conference Series **224** (2010) 012085 doi:10.1088/1742-6596/224/1/012085

Low frequency impedance spectroscopy of cell monolayers using the four-electrode method

H Asfour[1], W Soller[1], NG Posnack[2], AE Pollard[3] and MW Kay[1,2]

[1] Department of Electrical and Computer Engineering
[2] Department of Pharmacology and Physiology
The George Washington University, Washington, DC, USA.
[3] Department of Biomedical Engineering
The University of Alabama in Birmingham, Alabama, USA.
E-mail: asfour@gwmail.gwu.edu

Abstract. We have developed an impedance spectroscopy system for monitoring changes in extracellular conductivity and cell-to-cell coupling in cells cultured on microelectrode arrays (MEAs). Impedance spectra were measured at low frequencies (0.5-10kHz) using the four-electrode technique and standard electrophysiological instruments. A reduction in extracellular conductivity (using isotonic sucrose solution) was observed as an elevation of impedance at all measured frequencies. A reduction in cell coupling (using heptanol) was observed as an elevation of impedance at low frequencies (<100Hz) and small reductions or no change in impedance at higher frequencies. Monolayer confluency and cell adhesion to electrodes were major factors in the measurements. The effect of disrupting cell adhesion (using trypsin) was most apparent at frequencies below 1kHz. Impedance was dependent upon the spacing between stimulating electrodes. Impedance reduced with increasing spacing. Overall, using commercially available MEAs and the four-electrode method provides a fast and reliable way to measure cell monolayer impedances.

1. Background

Electrical impedance spectroscopy (EIS) is emerging as a tool for fast non-invasive monitoring of cell cultures. EIS measurements can be made using two main techniques. One is to use two electrodes, where one electrode is used for sensing and the other is a reference electrode. This is typically used for monitoring cell adhesion, motility, proliferation, and cell toxicity [1]. Electrode surface areas in the two electrodes technique are typically large to reduce electrode impedance, however this hampers spatial sampling and the ability to accurately measure local cell densities and cell migration velocities.

Another technique for EIS is to use four electrodes, where current flows between two outer electrodes (stimulating electrodes) and potential differences are measured between two inner electrodes (measuring electrodes). This reduces effects of electrode impedance and electrode polarization. This is important when using grids of small electrodes (i.e. microelectrode arrays (MEAs) with 10μm dia electrodes are now available) to achieve high spatial resolutions, especially for measurements at low frequencies where electrode-electrolyte impedance effects are significant. Recently the four-electrode technique has been used in multi-electrode systems to provide increased accuracy compared to the two electrode technique [2, 3].

From a theoretical basis, changes in intracellular and extracellular conductivities can be measured using the four-electrode technique with either multi-site stimulation or frequency modulation.

Adequately small spacing between stimulating electrodes is required for this to work using frequency modulation [4-6]. At low frequencies, relatively little current penetrates the cell; it is shunted through the extracellular space. At high frequencies, the capacitive impedance of the membrane is low, so a significant amount of current flows through the cell. Accordingly, frequency dependent impedance changes may indicate relative differences between intra- and extracellular conductivities. Our goal has been to use this concept for assessment of extracellular conductivity and cell-to-cell coupling in cell culture monolayers.

Our objectives were to 1) show the effect of stimulation electrode spacing on four-electrode impedance measurements of fibroblast cell monolayers and 2) measure changes in four-electrode impedance spectra caused by alterations of extracellular conductivity and cell-to-cell coupling. To accomplish these objectives, cell monolayers were grown on commercially available MEAs. Impedance spectra were measured at low frequencies using standard electrophysiological instruments.

2. Materials and Methods

MEAs (Multichannel Systems, Reutlingen, Germany) having either gold (100μm dia) or titanium nitride (10μm dia) contact electrodes with inter-electrode spacings of either 700μm or 30μm (figure 1A) were mounted within an adapter (Multichannel Systems, MEAadapt) that provided pin-outs for each electrode. Current (I_{stim}) was injected between stimulating electrodes (the two outermost pairs, A_i:D_i and A_{iv}:D_{iv}, figure 1B) using an AC/DC current source (Keithley 6220) at frequencies between 0.5Hz and 10kHz (3 points per decade). Extracellular potentials ($\Delta\Phi$) were measured from the two center electrodes (B:C, figure 1B) using a differential amplifier (Dagan EX4-400). Signals were recorded using a laptop PC, DAQ hardware (NI DAQCard-6062E), and custom LabVIEW software. Sampling frequencies ranged from 1000sps to 90Ksps and were set at least 10 times higher than the frequency of current injection. The amplitude of extracellular potentials ($|\Delta\Phi|$) was measured using custom Matlab routines by fitting potentials at each frequency to a sine wave at the frequency of the stimulating current. Impedance magnitude at each frequency was then calculated as $|Z|=|\Delta\Phi| / |I_{stim}|$.

First, impedance was measured for clean MEAs containing 750 μL of either 10% FBS or Tyrode's solution. Current limits were set to prevent solution ionization, usually a max of 100nA. MEAs were then plated with fibroblasts (PT67). Cell concentrations varied between 0.5 and 2 million cells per mL. Cultures were incubated overnight and measurements were acquired the next day at room temperature after replacing the FBS solution with a buffered Tyrode's solution.

Cell-to-cell coupling and extracellular conductivities were then sequentially altered. Heptanol (a gap junction uncoupler) was applied (1mM) and impedance spectra were measured after 10 minutes. Cultures were carefully rinsed three times to remove heptanol and 750 μL of isotonic sucrose solution was added to the culture. Impedance spectra were measured after 10 minutes. Trypsin was then added to cultures, cells were washed from the MEA, and 750 μL of Tyrode's solution was added to the

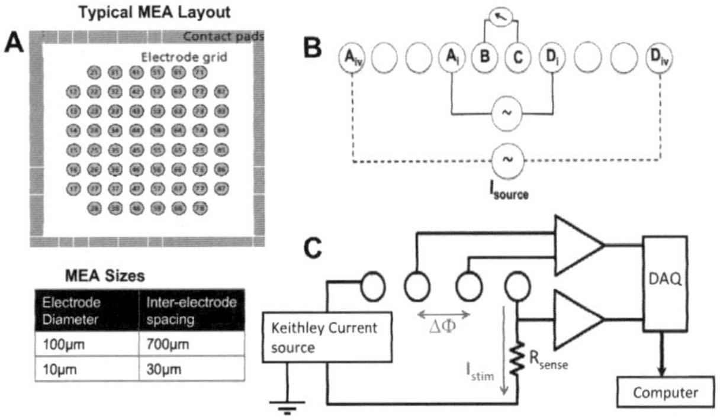

Figure 1. Four electrodes impedance measurement system. **A**: Typical MEA geometry. **B**: Implementation of the four electrodes technique using MEAs. **C**: Equipment for injecting current and measuring $\Delta\Phi$.

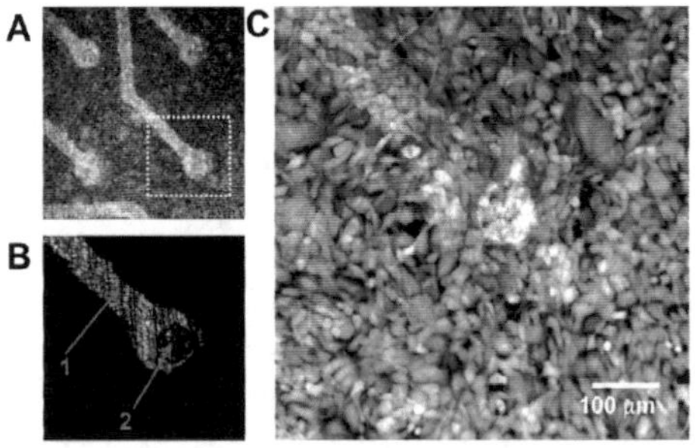

Figure 2. Fibroblasts cultured on top of gold electrodes (100μm dia). **A**: Closeup of the confluent monolayer. **B**: One electrode showing the insulated region (area1) and the exposed region (area2). **C**: Image showing monolayer confluency over the electrode.

MEA. Impedance measurements were repeated. MEAs were then treated with an enzymatic pH neutral detergent (Tergazyme, Alconox Inc.) and impedance measurements were repeated.

3. Results and Discussion

Impedance at 0.5Hz for clean MEAs (100μm electrode dia) having 750 μL of dionized water, isotonic sucrose solution, or FBS solution was approximately 2000Ω, 1000Ω, 200Ω, respectively. Confluent monolayers were obtained a day after cell plating (figure 2). To ensure confluency and electrode coverage, monolayers were inspected using a confocal microscope (Zeiss LSM510) after staining the cells with CellTracker^{TM} (Invitrogen) (figure 2).

Impedance was dependent upon the spacing between stimulating electrodes (figures 3&4). Typically, impedance measured at larger stimulation electrode spacing was lower than impedance measured at smaller stimulation electrode spacing. This effect might be described by simple volume conductor theory *if the supplied current always remained outside of the cells*. However, the validity of our approach relies upon the redistribution of current, in part, to the intracellular volume. In this situation a description that accounts for the separate contributions of the intracellular and extracellular domains becomes necessary. Presently, there is no analytic description of this redistribution, unless an

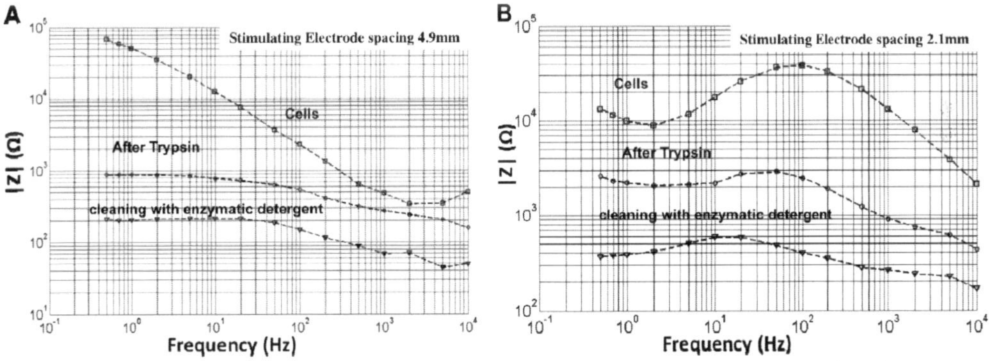

Figure 3. Impedance spectra before and after applying trypsin and after cleaning the MEA (100 μm electrode dia). **A**: Spectra using a stimulation electrode spacing of 4.9mm. **B**: Spectra using a stimulation electrode spacing of 2.1mm.

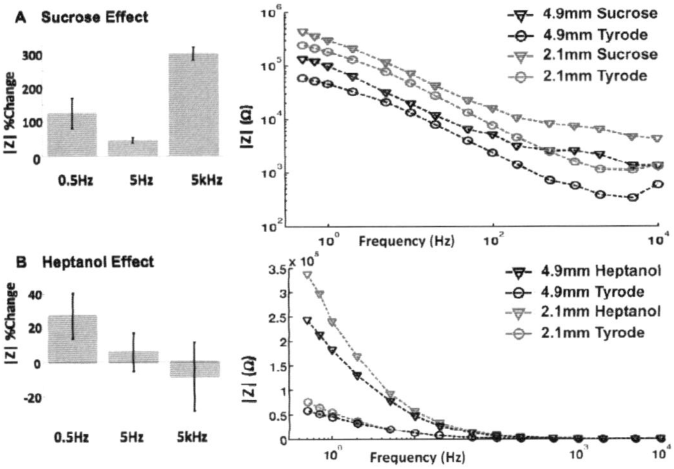

Figure 4. Alterations of extracellular conductivity and cell-to-cell coupling. Average impedance change (left) and typical impedance spectra (right) are shown. Error bars represent MSE. **A**: Replacement of cell culture media with isotonic sucrose solution. **B**: Application of heptanol (1mM).

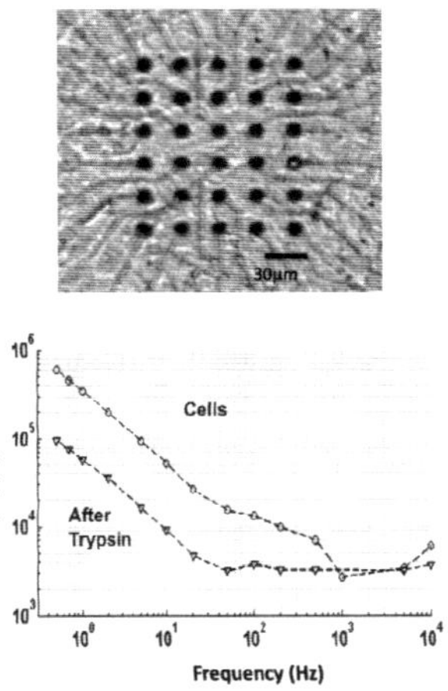

Figure 5. Top: Confluent monolayer on a High-Dense MEA (10 μm dia electrodes). **Bottom**: Impedance spectra (90 μm stim electrode spacing).

assumption of uniform, isotropic conductivities is valid. However, there are compelling reasons to believe that this assumption does not apply to our experimental conditions. A goal of future studies is to develop an analytic framework for describing the redistribution of current in our cell monolayers.

Impedance was drastically reduced after the application of trypsin (figure 3), indicating that cells were adhered to MEA electrodes and that adhesion was disrupted after applying trypsin. The effect of trypsin was most apparent at frequencies below 1kHz. Impedance was further reduced after cleaning the electrodes with an enzymatic detergent, indicating the presence of residual proteins after the detachment of cells (figure 3). Impedance spectra measured from non-confluent monolayers showed little difference before and after the application of trypsin, indicating that measurements are dependent upon adequate coverage of the electrodes with cells.

Reducing extracellular conductivity with isotonic sucrose solution increased impedance at all frequencies below 10kHz (figure 4A). Specific changes were 125%±44% at 0.5Hz, 75%± 17% at 5Hz, and 300%±19% at 5kHz. The magnitude of impedance changes after reducing gap junction conductance were less and appeared to be frequency dependent (figure 4B). Impedance increased at lower frequencies (27%± 13% at 0.5Hz and 16%±12% at 5Hz) while impedance decreased at higher frequencies (-8.4%±20% at 5kHz), but was not significant.

Cells were also plated on MEAs having titanium nitride electrodes (10μm dia, High-Dense MEAs). We found that although the effect of electrode impedance is less using the four-electrode technique, impedance measured using the 10μm dia electrodes was at least an order of magnitude higher than impedance measured using 100μm dia electrodes (figure5). In addition, impedance differences measured before and after applying trypsin to cells were usually smaller for the 10μm dia electrodes.

Impedance measurements required low levels of stimulation currents and measurements of small potential differences, therefore minimization of noise was important. This was done by shielding the equipment and placing the MEA adapter within a Faraday cage. Sine wave fitting to measured potential differences minimized the effect residual noise levels. Raw signal to noise ratios (SNRs) were lowest at low frequencies. SNR at 1Hz was 60dB while SNR at 5kHz was 5dB.

Since changes in phase due to monolayer perturbations might be more sensitive than changes in the impedance magnitude [7], future work will involve the development of techniques for measuring impedance magnitude and phase and for conducting studies within an incubator.

4. Summary and Conclusions

We have shown how MEAs and the four-electrode method can be used for non-invasive measurements of cell monolayer conductivities, with emphasis at low frequencies (0.5Hz to 10kHz). Impedance was dependent upon the spacing between stimulating electrodes. Impedance reduced with increasing spacing. Reducing extracellular conductivity (replacing FBS with isotonic sucrose solution) was observed as an elevation of impedance at all measured frequencies. Reduced cell coupling (using heptanol) was observed as an elevation of impedance at low frequencies (<100Hz) and small reductions or no change in impedance at higher frequencies. Accordingly, monolayer confluency and cell adhesion to electrodes were major factors in the measurements. The use of commercially available MEAs provided a fast and reliable way to measure monolayer impedances.

References

[1] Giaever I and Keese CR 1986 *IEEE Trans. Biomed. Eng.* **33**(2) 242.
[2] Linderholm P, Marescot L, Meng Heng L and Renaud P 2008 *IEEE Trans. Biomed. Eng.* **55**(1):138.
[3] Bragos R, Sarro E, Fontova A, Soley A, Cairo J, Bayes-Genis A and Rosell J 2006 *IEEE EMBS 28th Annual International Conference* 2106-2109.
[4] Pollard AE and Barr RC 2006 *Am. J. Physiol. Heart Circ. Physiol.* **290**(5):H1976.
[5] Le Guyader P, Trelles Francisco and Savard P 2001 *Ann. Biomed. Eng.* **29**(10):862.
[6] Plonsey R and Barr R 1982 *IEEE Trans. Biomed. Eng.* **29**(7):541.
[7] Rahman AR, Lo CM and Bhansali S 2009 *IEEE Trans. Biomed. Eng.* **56**(2):485.

Differentiation of fast and slow muscle fibers by bioimpedance

M-V Moreno[1], N Khider[2], E Ribbe[1] and J-L Damez[3]

1 Research Department, BioparHom, Savoie Technolac, BP238, 73374 Le Bourget-du-Lac Cedex, France
2 Research Department, Bio2M, Parc Technologique Rives de l'Oise, BP50149, 60201 Compiègne Cedex, France
3 UR 370 Qualité des produits Animaux, INRA, F-63122 Saint Genès Champanelle, France

E-mail: marie-valerie.moreno@bioparhom.com

Abstract. The differentiation of fast and slow muscle fibers in vivo still requires constraining equipment (ergometer, biopsy ...) and invasive techniques. These fibers conduct the electrical current differently. Therefore the aim of this study is to see if it is possible to differentiate quickly, by bioimpedance, fast and slow fibers, and firstly muscles which are typical composed by slow or fast fibers. To do this, we used a multifrequency impedancemeter Z-Metrix® (BioparHom© Company, France). We collected the electrical characteristics (Longitudinal and Transversal, from 1 to 1000 kHz) for a population of 20 rats aged 70 days, on *Soleus* muscles (composed principally of slow fibers) and *Extensor Digitroum Longus* (*EDL*) muscles (composed principally of fast fibers). We compared the means of alpha (L/T), R (L/T) and X (L/T) with Wilcoxon tests. We obtained non significant differences between electrical data obtained on *EDL* and *Soleus* muscles, but we could see differences on graphics representation and with the example of one rat. Therefore, we can assume that differentiation, by bioimpedance, of muscles typed slow and fast fibers, could be possible.

1. Introduction

Localized impedance measurements can provide useful approaches for assessing neuromuscular disease [1] and more generally characteristics of muscle. In 2006, Nie showed that it's possible to transfer Electrical Impedance Myography (EIM) from human to animal, especially rats [2]. Recently, Ahad presented a study which indicated that it seems possible to link electrical characteristics of rat muscle and fiber size [3].

Thus, the differentiation of fast and slow muscle fibers *in vivo* still requires constraining equipment (ergometer, biopsy...). Then it's difficult to use this information in clinical or sports routine.

Therefore, the aim of this study was to see if bioimpedance can differentiate quickly, fast and slow fibers, firstly in rats, before transferring to human.

2. Material and method

To do this, we used a multifrequency impedancemeter Z-Métrix® (BioparHom© Company, France) showed figure 1. This device operates on six 1.5 volt batteries with a low injection current with 77 μA.

Figure 1. Multifrequency (1 to 1000 kHz) impedancemeter Z-Métrix®.

Figure 2 indicates the 4 needles-electrodes we used for our experiments. We made sure that the distance (2 mm) between the needles was always constant. Needles are made in stainless steel and they have a length of 5 mm. Their external and inner diameters were respectively about 0.635 mm and 0.318 mm.

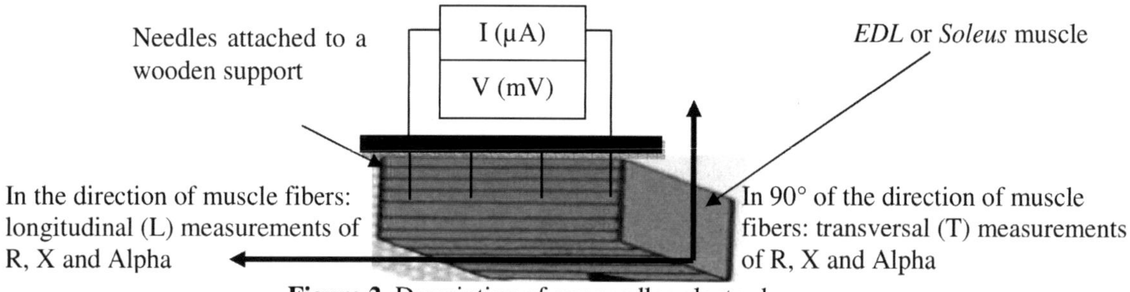

Figure 2. Description of our needles-electrodes.

We collected some electrical characteristics (resistances R and reactances X from 1 to 1000 kHz) and we calculated the phase angle Alpha, in degree, with equation 1:

$$Alpha = ATan(X / R) * \frac{180}{\pi} \quad (1)$$

Where X and R are respectively the reactance and the resistance in ohm.
We did that in longitudinal direction of fibers of muscles (L) and in transversal direction of fibers of muscles (T), as it's explained in figure 2. We calculated the ratio L/T by dividing respectively the measurements of R, X and alpha taken in longitudinal way, by the measurements of R, X and Alpha taken in transversal way.

We explored muscles in a population of 20 rats aged 70 days, on *Soleus* muscles (composed principally of slow fibers) and *Extensor Digitroum Longus* (*EDL*) muscles (composed principally of fast fibers).
Table 1 indicates the composition of each type of muscle obtained with histological sections.

Table 1. Mean and standard deviation (SD) of numbers of slow and fast fibers in each type of muscles explored.

	Number of slow fibers Mean ± SD	%	Number of fast fibers Mean ± SD	%	Total number of fibers Mean ± SD
Soleus	1905.42 ± 409.46	93.81	125.26 ± 26.22	6.19	2030.68 ± 426.47
EDL	100.47 ± 28.67	4.30	2215.16 ± 319.73	95.70	2315.63 ± 339.80

We used non parametric paired test (Wilcoxon) for our statistic tests because of our small sample and non Gaussian distribution.

3. Results and discussion

We can see in table 1 that *Soleus* muscle have about 94% of slow fibers in comparison with *EDL* muscle which have 95.7% of fast fibers. So, for a first step, we can consider that differentiating *Soleus* and *EDL* muscle is about differentiating slow and fast fibers.

Table 2. Mean and standard deviation (SD) of differences of alpha L/T between *Soleus* and *EDL* in percent, for various frequencies

Frequency (KHz)	Mean ± SD	Wilcoxon Tests
10	-33.1 ± 77.4	Ns
50	-46.9 ± 111.8	Ns
100	-56.7 ± 143.2	Ns
200	-42.2 ± 110.8	Ns
300	-32.6 ± 81.6	Ns
500	-2.9 ± 60.5	Ns
1000	12.1 ± 56.7	Ns

As we can see on figure 3, we can observe a difference between the mean of alpha L/T of *Soleus* and *EDL,* especially from 50 to 350 and from 750 to 1000 KHz, even if those differences are not significant as shown in table 2.

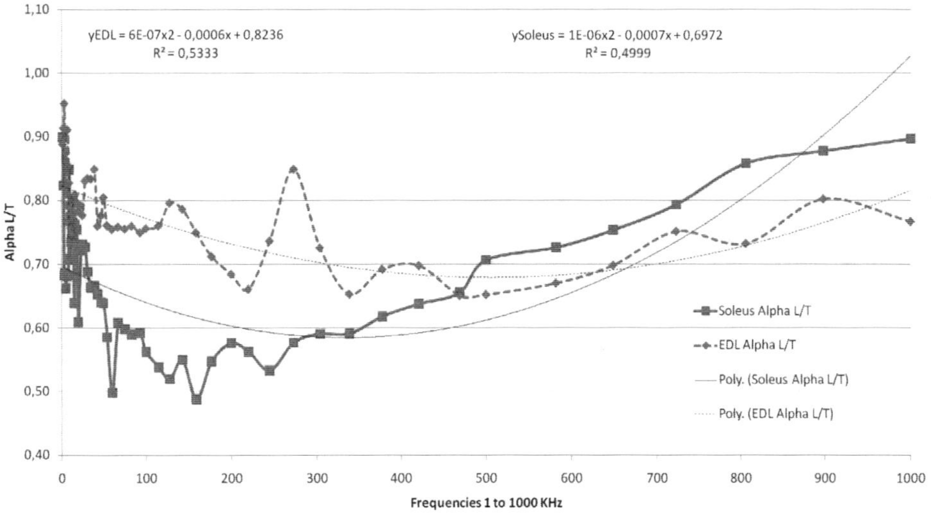

Figure 3. Variation of means of alpha (L/T), for 1 to 1000 kHz, for *Soleus* and *EDL*.

Then we wanted to explore with more details the electricals characteristics of our two types of muscles. Table 3 lists mean and SD of resistances R (L/T) and reactances X (L/T) for various frequencies.

Table 3. Mean and standard deviation (SD) of differences of Resistances R and Reactances X (L/T) between *Soleus* and *EDL* in percent, for various frequencies.

Frequency (KHz)	MeanR ± SD	Wilcoxon Test	MeanX ± SD	Wilcoxon Test
10	58.2 ± 52.5	Ns	121.8 ± 172.5	Ns
50	48.8 ± 112.9	Ns	100.1 ± 221.7	Ns
100	42.6 ± 108.5	Ns	92.7 ± 217.6	Ns
200	31.6 ± 77.8	Ns	98.2 ± 195.8	Ns
304	20.4 ± 77.2	Ns	76.6 ± 196.4	Ns
500	17.9 ± 65.8	Ns	51.6 ± 181.6	Ns
1000	10.4 ± 31.4	Ns	54.5±156.3	Ns

As the first approach with alpha (L/T), we observe, on figure 4, a different dispersion of mean of R (L/T) *vs* mean of X (L/T), between *Soleus* and *EDL*. But as listed in table 3, those differences in resistance and reactance are not significant.

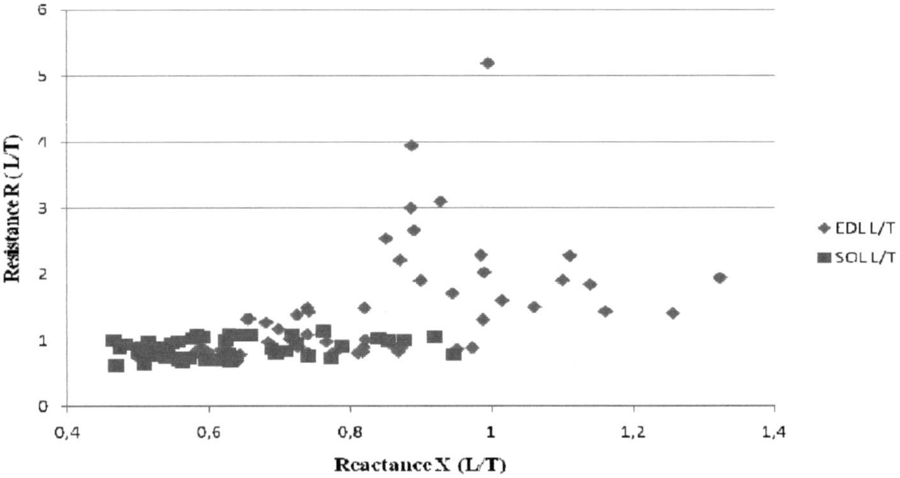

Figure 4. Dispersion of mean of R (L/T) *vs* mean of X (L/T)), for 1 to 1000 kHz, for *Soleus* and *EDL*.

As we saw differences, but not significant, we wanted to try to apply those methods (Alpha (L/T) and R (L/T) vs X (L/T)), for differentiating *EDL* and *Soleus* on one rat, taken randomly. Figure 5 and 6 show differences in the electrical characteristics, alpha (L/T) and R (L/T) *vs* X (L/T), between *EDL* and *Soleus*. This suggests that *EDL* has a larger difference between longitudinal and transversal electrical characteristics as resistance, reactance and phase angle, than *Soleus*, maybe due to the difference in the number of NaK pump.

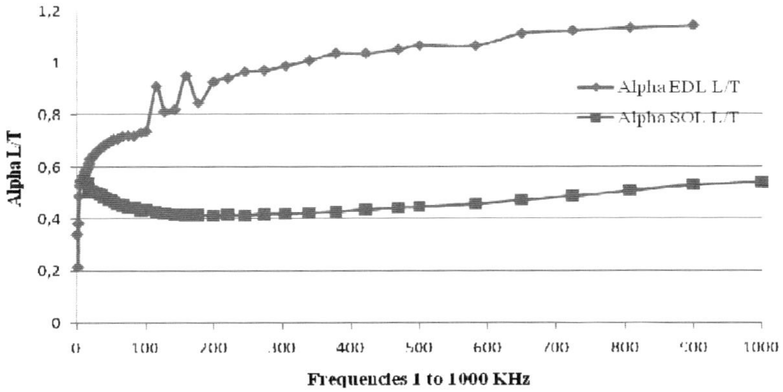

Figure 5. Example of one rat. Variations of alpha (L/T), from 1 to 1000 kHz, for *EDL* and *Soleus*.

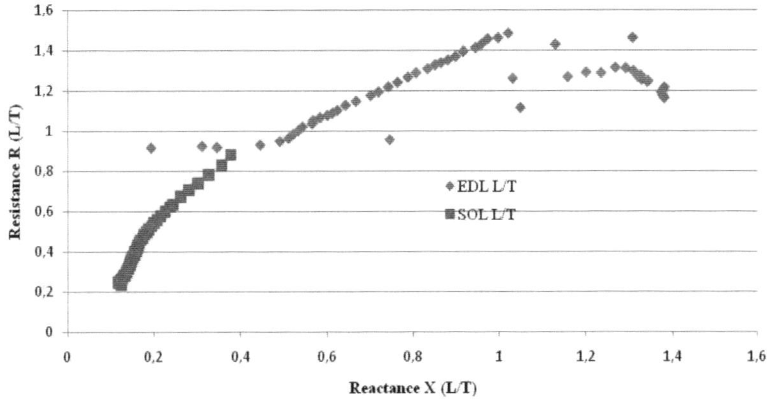

Figure 6. Example of one rat. Variations of R (L/T) *vs* X (L/T), from 1 to 1000 kHz, for *EDL* and *Soleus*.

So, in that rat, we could differentiate *EDL* and *Soleus* muscle by bioimpedance. Therefore, we distinguished fast and slow fibers by bioimpedance.

4. Conclusion

We tried two approaches to characterize electrical data of muscles: alpha (L/T) and R (L/T) *vs* X (L/T). For the two approaches, we could see differences between *Soleus* and *EDL*, but not significant. Maybe we didn't have significant differences because our sample was too small (with important SD), or we used the inappropriate statistic tests. We're going to do a new validation, with a new bigger sample. When we applied those two methods, on one rat, to have an example, we could distinguish *EDL* and *Soleus* electrical characteristics. Then, we can assume that bioimpedance could allow us to differentiate, by extension, fast and slow fibers, quickly, with a non-invasive and low cost tool, ideal for clinical and sports routine.

References

[1] Esper GJ, November 2009 *Muscle Nerve* **34** 595-602.
[2] Nie R, August 2006 Clinical Neurophysiology **117(8)** 1844-9.
[3] Ahad MA, December 2009 Physiology Measurement **30(12)** 1415-27.

Variability of Maximum Systolic Amplitude of ΔZ/Δt Curve in Pregnancy. Perennial Observations.

I Ilyin, A Karpov, M Korotkova
Clinical Hospital 9, Yaroslavl, Russia

E-mail: kb9@mail.ru

Abstract. Maximum systolic amplitude is quite an important component of the impedance cardiogram ΔZ/Δt curve. Its values make it possible to calculate many hemodynamic indices. Therefore it is necessary to keep informed about monthly, annual and perennial maximum systolic amplitude trend. We can produce the measuring data of the maximum systolic amplitude for a fifteen-year period (from 1994 to 2009). The impedance cardiograms were obtained with the help of an electric impedance analyzer "RA-5" (1 mA, 70 kHz) with disk ECG electrodes. The data analyzed were taken from the pregnant women with non-complicated pregnancy (n=5709). We have analyzed the average monthly and annual changes of the maximum systolic amplitude ΔZ/Δt curve. It allowed us to reveal the six-year periodicity of the maximum systolic amplitude changes. There were discovered statistically significant peak values difference of the amplitude ($p<0.001$). The data obtained should be taken into consideration when using impedance cardiography in clinical practice. The article is supplied with tables and diagrams.

1. Introduction

Impedance cardiography is a noninvasive, inexpensive and easy-to-use method of analysis of some cardiac activity and cycle, such as stroke volume and cardiac output. Maximum systolic amplitude (MSA) of ΔZ/Δt curve [1, 2, 3] is used in stroke volume formula. MSA is an important constituent of ΔZ/Δt curve of of the impedance cardiogram. For that reason when using MSA it's necessary to allow for several factors which influence its variability, and to take into consideration the MSA trends for a long-term period. The aim of the research in hand is to evaluate the variability of maximum systolic amplitude ΔZ/Δt curve in pregnancy.

2. Methodology

Maximum systolic amplitude was calculated after the recording of ΔZ/Δt curve. The averages for 4 consecutive cardiac cycles were computed. Maximum systolic amplitude (A1) was measured according to the zero line (figure 1).

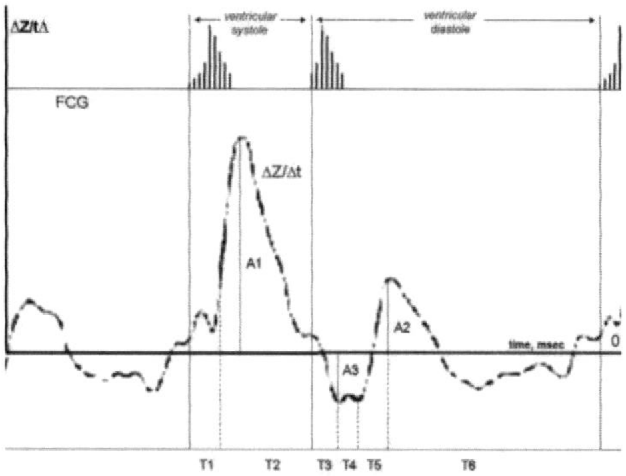

Figure 1. Amplitude-time characteristic of impedance cardiogram.

3. Method

The impedance cardiograms were obtained of with the help of an electric impedance analyzer "PA-5" (1 mA, 70 kHz) with disk ECG electrodes. The electrodes were placed according to the Sramek scheme. Two current electrodes were put on the mastoid bone and on the iliac crest in line with two measuring electrodes which were placed on the lateral lines of the neck and those of the thoracic cage (on the xiphoid level). The parameters were studied and registered in standard conditions: from 9 a.m. to 12 p.m., the pregnant was lying on her left side, after a 15-minute rest before the tests. The results were processed by means of descriptive statistics methods. In order to check the difference significance we used Student's test.

4. Materials

We produce the measuring data of maximum systolic amplitude $\Delta Z/\Delta t$ curve in pregnancy for 15-year period – from 1994 to 2009. We have conducted 5631 studies of pregnant women of 6-44 weeks of gestation, among them 5292 with non-complicated pregnancy (1323 pregnant women at the age of 27±5, 73±14 kg of weight, and 165±6 cm tall), 338 with a serious preeclampsia in accordance with the World Health Organization classification (135 pregnant women at the age of 27±6 77±13 kg of weight, and 163±6 cm tall).

5. Result

We picked out internal and external factors which influence the variability of maximum systolic amplitude $\Delta Z/\Delta t$ curve**.** Anthropometric and hemodynamic characteristics as well as pregnancy and its complications were considered as the internal factors. The influence of the external factors was revealed during the MSA trend analysis for a long-term period.

5.1 Influence of Anthropometric Characteristics (Weight, Height, Body Surface Area)

The variability of MSA is mostly influenced by weight and body surface area of the pregnant (figure 2, 3). Significant difference of MSA was found out between the pregnant in weight range 40-50 kg and 100-110 kg ($p<0.001$). Height of the pregnant as well as the distance between electrodes does not influence significantly the variability of MSA (figure 4, 5).

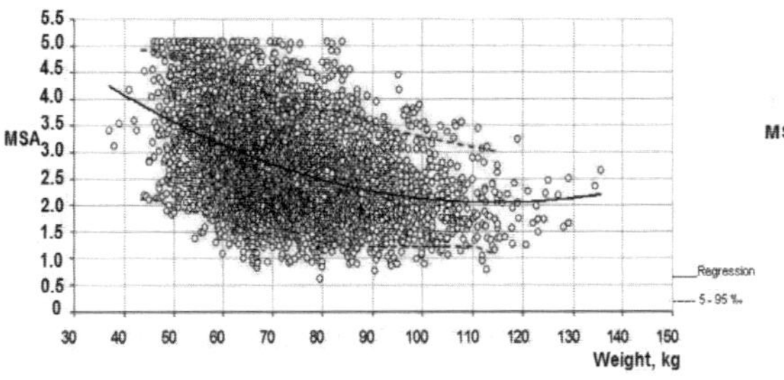

Figure 2. Influence Weight to the Variability of the Maximum Systolic Amplitude (r=-0.42; y=0.0004x-0.0635x+6.8352)

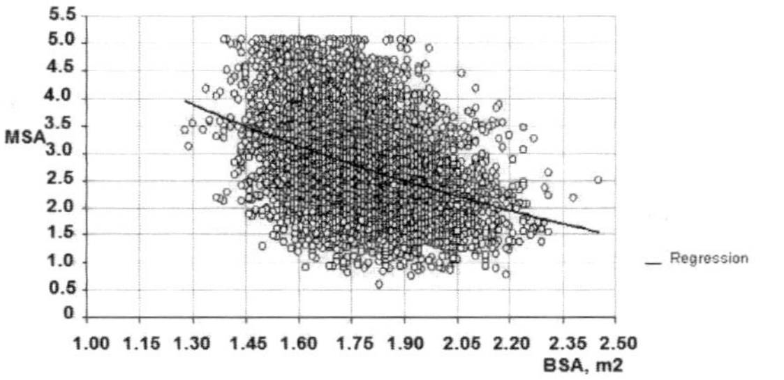

Figure 3. Influence BSA to the Variability of the Maximum Systolic Amplitude (r=0.37; y=-3.6919Ln(x)+4.8645)

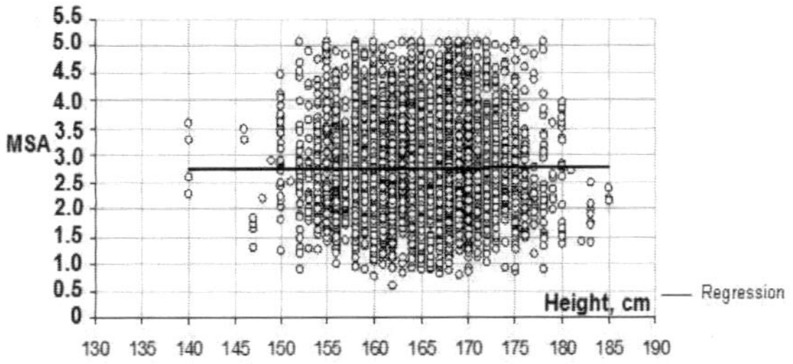

Figure 4. Influence Height to the Variability of the Maximum Systolic Amplitude

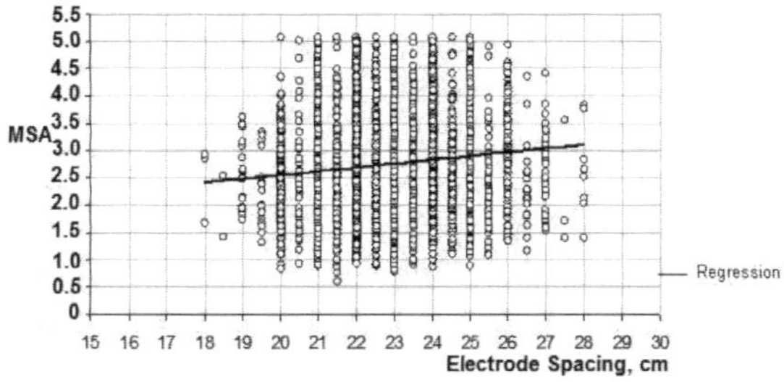

Figure 5. Influence Electrode Spacing to the Variability of the Maximum Systolic Amplitude

5.2 Influence of Hemodynamic Characteristics (blood pressure, AVVST, heart rate)

The variability of MSA is significantly influenced by blood pressure and aorta valve vibration spread time (AVVST) (figure 6, 7). Significant difference of MSA was found out between the pregnant in systolic blood pressure range 70-80 mm Hg and 120-130 mm Hg ($p<0.001$). Heart rate does not influence the variability of MSA significantly (figure 8).

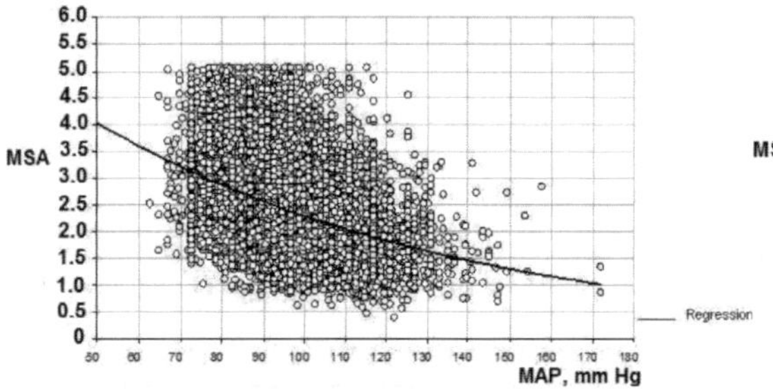

Figure 6. Influence MAP to the Variability of the Maximum Systolic Amplitude ($r=-0.40$; $y=7.1034e^{-0,0113x}$)

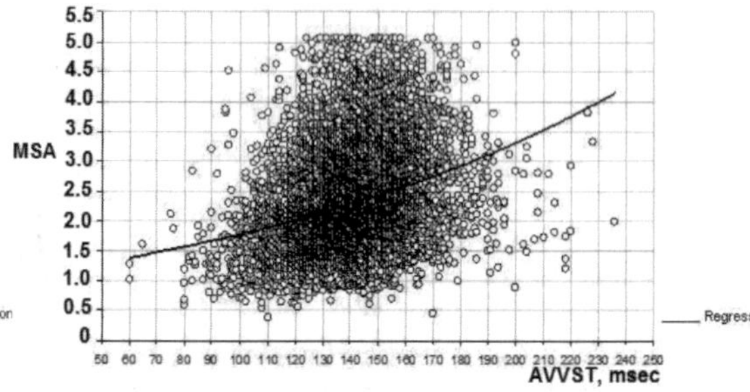

Figure 7. Influence AVVST to the Variability of the Maximum Systolic Amplitude ($r=0.30$; $y=0.954e^{0,0623x}$)

5.3 Influence of Pregnancy

Pregnancy and its complications have telling impact on the variability of MSA [4, 5]. By the end of pregnancy significant decay of MSA is observed (figure 9). But this fact cannot be considered as indicative of left ventricular dysfunction in late pregnancy. As aforesaid the variability of MSA is significantly influenced by weight of the pregnant, which increases progressively during pregnancy.

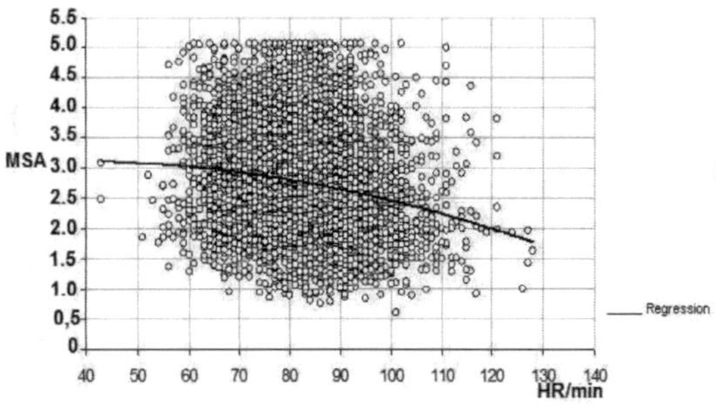

Figure 8. Influence HR to the Variability of the Maximum Systolic Amplitude

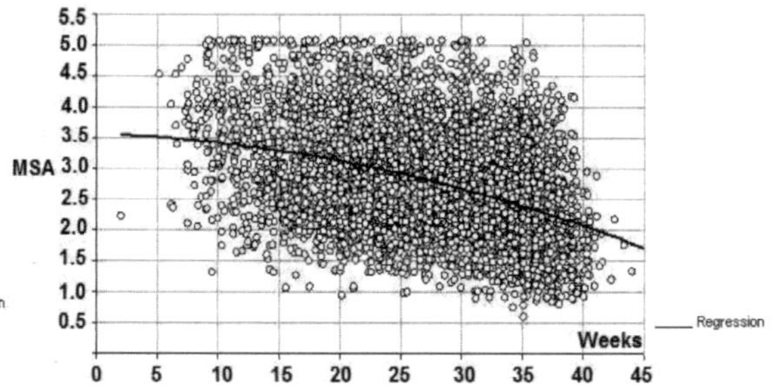

Figure 9. Influence Pregnancy to the Variability of the Maximum Systolic Amplitude ($y=-0.0008x^2-0.0068x+3.562$)

Chest circumference increases simultaneously and has a direct influence on the MSA value. Therefore for each weight group of the pregnant should correspond to certain MSA values within 5 and 95 percentile curves (figure 2). In case of preeclampsia apparent decrease of MSA (A1) is observed. Average amplitude A1 in case of non-complicated pregnancy is 2.555 ± 0.729 and in case of preeclampsia – 1.336 ± 0.481. Significant differences of maximum systolic amplitude (A1) values are found out by means of Student's test ($p<0.001$).

5.4 External Factors Influence

When analyzed the variability of maximum systolic amplitude for a long-term period showed fluctuations with tendencies for repetition in certain time length (figure 10). In figure 11 the average monthly variability of maximum systolic amplitude for 1994-2009 is shown. The average monthly gestational age during the study varied from 26 to 30 weeks, so this condition could not influence the amplitude variability. Maximum amplitude peaks are on years 1997, 2003 и 2009. Significant difference between peak average annual values of maximum systolic amplitude was found out. Average annual

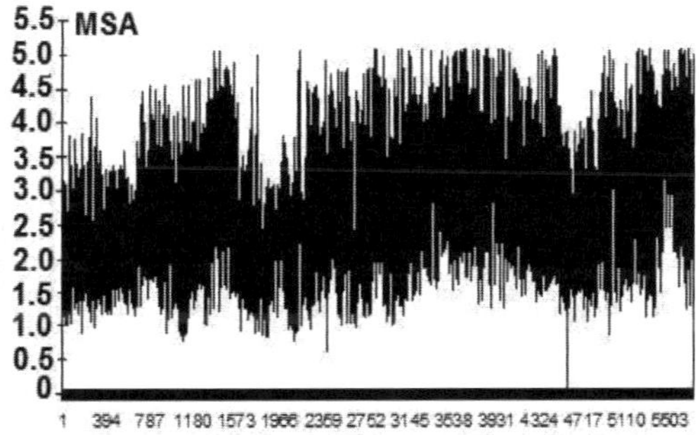

Figure 10. The variability of maximum systolic amplitude for a long-term period showed fluctuations with tendencies for repetition in certain time length

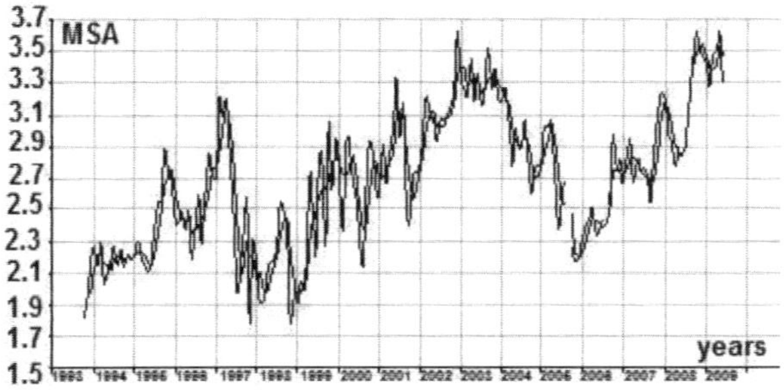

Figure 11. The average monthly variability of maximum systolic amplitude for 1994-2009

maximum systolic amplitude for 2003 is 3.295±0.908 and average annual maximum systolic amplitude for 2006 - 2.565±0.803 ($p<0.001$). The difference significance was checked by means of Student's test ($p<0.001$).

6. Conclusion

Maximum systolic amplitude of $\Delta Z/\Delta t$ curve gives useful diagnostic information. Maximum systolic amplitude of $\Delta Z/\Delta t$ curve proves to be useful for assessment of myocardial contractility of the left ventricle of heart in pregnancy as a self-sufficient criterion. However one shoul take into accont internal and external factors which can influence the variability of maximum systolic amplitude. The internal factors include weight, body surface area and mean arterial blood pressure of a pregnant. For their use some nonparametric criteria can be used such as, for instance, percentile ranges. As for the external factors - further studies are necessary to reveal the reasons for the long-term variability of MSA. Even though the data for the variability of MSA were obtained during the examination of the pregnant, similar regularity can be observed for the nonpregnant. Finding of other factors which influence the variability of maximum systolic amplitude of $\Delta Z/\Delta t$ curve will enable to standardize the usage of this diagnostic criterion.

7. References

1. Kubicek W G, Karnegis J N, Patterson R P, Witsoe D A, Mattson R H 1966 Development and evaluation of an impedance cardiac output system *Aerosp. Med.* **37** 1208-12
2. Sramek B B, Rose D M, Miyamoto A 1983 Stroke volume equation with a linear base impedance model and accuracy, as a compared to thermodilution and magnetic flowmeter techniques in humans and *animals Proc. Of the 6th International Conference on Electrical Bioimpedance Zadar Yugoslavia* p 38-41
3. Bernstein D P 1986 A new stroke volume equation for thoracic electrical bio-impedance: Theory and rationale *Crit. Care Med.* **14** 904-909
4. Karpov A. , Belov N. at al. 1991 The cardiac insufficiency in pregnant women with the initial arterial hypertension and toxemia of pregnancy. *6 meeting european associationof gynaecologists and obstetricians, Moscow,.*
5. Karpov A., Shmelev V., Ilyin I. at al. 1998 Heart's work and vessels contractility during pregnancy. Blood circulation dynamics of pregnant women suffering of some forms of hypertension. *X international conference on electrical bio-impedance. Barcelona, Spain,.*

In vivo cerebral blood flow autoregulation studies using rheoencephalography

M. Bodo[1], F. Pearce, A. Garcia, S. Van Albert, T. Settle, J. Szebeni, L. Baranyi, J. Hartings, R. Armonda[2]
Walter Reed Army Institute of Research, Silver Spring, MD; [2]National Naval Medical Center, Bethesda, MD, USA

E-mail: michael.bodo@med.navy.mil

Abstract. Acute management of patients with traumatic brain/blast injury is a challenge. To minimize secondary injury and improve outcome, it is critical to detect neurological deterioration early, when it is potentially reversible. One potential monitoring method is cerebral electrical impedance (rheoencephalography-REG) because of its non-invasiveness and good time resolution. Reported here are the results of cerebral blood flow (CBF) manipulations comparing electroencephalogram (EEG) with REG (both intra-cerebral) and measuring with surface and skull REG electrodes. Our hypothesis was that REG would reflect spreading depression and CBF autoregulation. Animal experiments were performed using one rat (four trials with intracerebral electrodes), monkeys (n=8, with surface electrodes) and pigs (n = 24 pigs with skull electrodes; 57 trials, 19 types of liposomes). Challenges included intracranial pressure (ICP) elevation, liposome infusion, and hemorrhage. Data were stored on a PC and evaluated off line. CBF autoregulation was evaluated both by visual inspection and by a Matlab script. These studies confirmed that REG reflects CBF autoregulation and that REG is useful for detecting spreading depression (SD), vasospasm and the lower limit of CBF autoregulation. These findings have clinical relevance for use in noninvasive neuro-monitoring in the neurosurgery intensive care and during transportation of patients with brain injury.

1. Introduction

The goal of neurocritical care is the prevention of secondary brain injury. Currently, there is no single modality that can monitor for hypoxia, ischemia, elevated ICP, edema, intracranial hemorrhage, vasospasm, and lack of CBF autoregulation [1]. Monitoring needs to be non-invasive, continuous and convenient. Current brain mapping methods (SPECT, PET scan, MRI) do not address these needs. Measuring brain electrical impedance is a potential technique for neuro-monitoring [2]. In previous studies we reported results of correlative studies indicating that REG has potential for use in neuro-monitoring [3-5]. Here we present results from three additional REG studies (rat, monkey, pig).

2. Methods

Animal experiments were performed using a rat (four trials under anesthesia; Ag/AgCl intracerebral electrodes; ICP elevation was by intra-cerebral probe/balloon inflation); monkeys (n=8, with surface EKG electrodes); and pigs (n = 24 pigs with skull, stainless steel electrodes; 57 trials, 19 types of liposomes). Challenges included elevation of ICP, intravenous liposome infusion, and lethal

[1] Michael Bodo, Naval Medical Research Center, Silver Spring, MD, 20910, USA

hemorrhage. Data were stored on a PC and evaluated off line. Two types of REG devices were used : 1) KR -Ea Rheo Preamp, excitation frequency of 45, kHz time constant 3 s (OTE Galileo, Italy); 2) Cerberus: excitation frequency of 125 kHz, time constant 0.3 s, Quintlab, Hungary).

To calculate the lower limit of CBF autoregulation, waveform data (arterial pressure - SAP, and REG) were averaged in 10 second epochs, and the Pearson correlation coefficient was computed on 30 consecutive points (5 min) and updated every minute with a sliding window. For further details, see [6].

3. Results

3. 1. Rat - Spreading depression

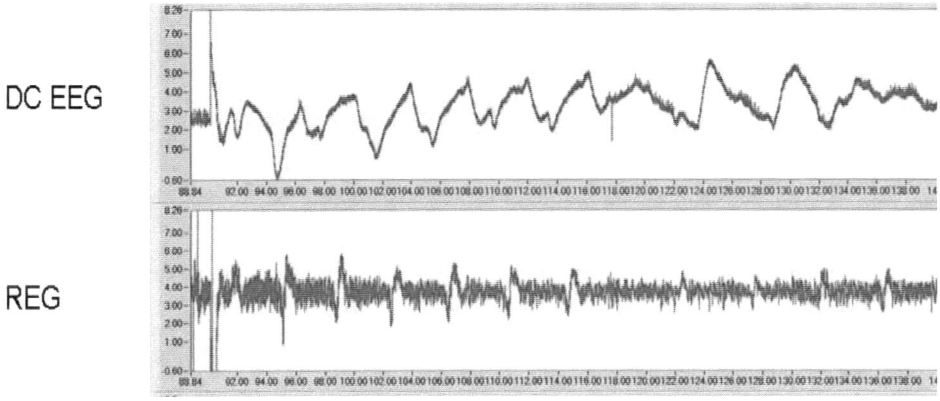

Figure 1. SD was elicited by ICP elevation. ICP elevation caused SD shown as amplitude fluctuation of DC EEG trace. REG showed close identical fluctuation.

3.2. Monkey – Lower limit of CBF autoregulation

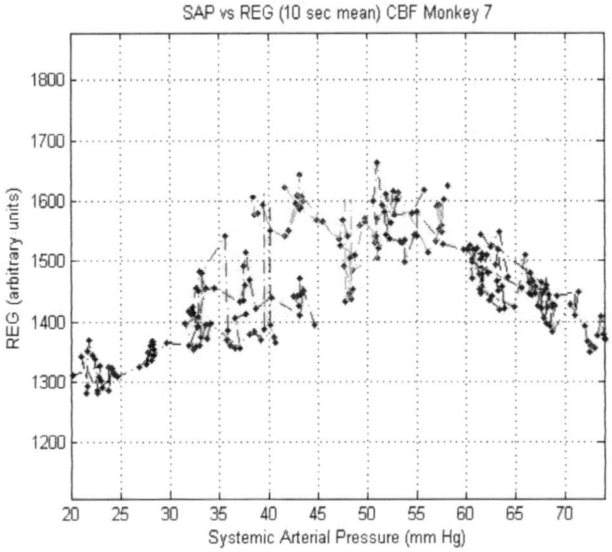

Figure 2. Relationship of REG to systemic arterial pressure during lethal hemorrhage. The lower limit of CBF autoregulation is at about 50 mmHg.

3.3. Pig – Vasospasm

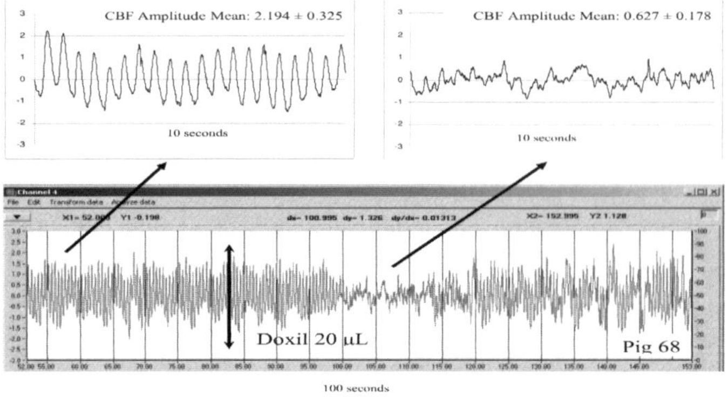

Figure 3. After the liposome injection, a transient decrease in REG pulse amplitude was observed. This result indicates the involvement of a cerebrovascular reaction during liposome/Doxil injection.

4. Discussion

Detection of spreading depression, of the lower limit of CBF autoregulation, and of vasospasm would provide clinical personnel in the ICU unit with information to prevent secondary brain damage in neurosurgery patients. The following in vivo studies of REG monitoring modeled these three clinical conditions.

4. 1. Spreading depression

Measurement of spreading depression, an early warning sign of disturbed brain metabolism, would be ideal for neuro-monitoring if it could be measured non-invasively. In humans, EEG measurements showing SD cannot be accomplished by placing electrodes on the scalp; such measurements require intra - or epicortical electrodes with the concomitant hazard of infection. In the present rat study, REG detected spreading depression with such intra-cerebral electrodes. However, in human measurements, REG can be used to measure non-invasively with surface electrodes, thus eliminating the hazard of infection. The relationship between SD and impedance was described by Olsson et al [7].

4.2. Lower limit of CBF autoregulation

If a patient's CBF autoregulation is impaired, administering of infusion by ICU neurosurgery staff may cause secondary brain damage. Too much infusion may lead to brain edema; too little infusion may cause brain ischemia/hypoxia. The lower limit of CBF autoregulation is typically obtained invasively by measuring such modalities as ICP and SAP [8] to determine the phase relationship of slow oscillations of SAP (transfer analysis). REG conducted on monkeys with surface electrodes on the skin proved that slow wave oscillations were present in the REG signal and were not filtered out by the skull. Therefore, with a specialized algorithm developed for this purpose, REG can be used to detect the lower limit of CBF autoregulation.

4.3. Vasospasm

Vasospasm, the constriction of brain arteries, causes brain ischemia/hypoxia, which may lead to secondary brain injury [9]. Vasospasm after traumatic brain/blast injury can be present without physical impact or visible injury and may develop any time after the blast incident, even weeks later [10]. The present pig study showed that REG measurements can detect vasospasm. Cerebral vascular symptoms (nausea, dizziness, headache) have been described in humans following infusion with the cancer treatment Doxil (doxorubicin), a liposome encapsulated drug. In this study, REG detected transient vasospasm following Doxil infusion. A possible explanation is the release of vasoactive mediators causing a cerebrovascular reaction [11].

5. Conclusion

Brain injured patients are known to experience spreading depression, impaired CBF autoregulation, and vasospasm. Currently there is no non-invasive method in use to assess the risk of implementing a hypotensive resuscitation strategy in brain- injured patients. Measurements with REG can be used to identify the autoregulatory breakpoint for individual patients to determine their limit for permissive hypotension or to adequately administer infusion to increase cerebral perfusion pressure. The results reported here support the application of REG as a neuro-monitoring modality. Additional correlative studies of REG as a non-invasive modality for brain monitoring are required and are in progress.

Acknowledgements

The author thanks former and recent members of Departments of Resuscitative and Veterinary Medicine, and Janice Meer for editorial assistance.

Disclaimer: Material has been reviewed by the Walter Reed Army Institute of Research. There is no objection to its presentation and/or publication. The opinions or assertions contained herein are the private views of the author, and are not to be construed as official, or as reflecting true views of the Department of the Army or the Department of Defense. Research was conducted in compliance with the Animal Welfare Act and other federal statutes and regulations relating to animals and experiments involving animals and adheres to principles stated in the Guide for the Carc and Use of Laboratory Animals, NRC Publication, 1996 edition.

References

[1] Czosnyika M, Hutchinson P, Kirkpatrick P J, Pickard J D 2009 Monitoring of the brain: pressures, flows, and brain tissue probes. *Neurotrauma and Critical Care of the Brain*, ed J Jallo, C M Loftus (New York: Thieme) p 66-86

[2] Rubin M, Yo M, Agostini M A, Madden J, Diaz-Arrastia R R 2009 Noninvasive monitoring. *Neurotrauma and Critical Care of the Brain*. ed J Jallo, CM Loftus (New York: Thieme) p 53-65

[3] Bodo M, Pearce F J, Armonda R A 2004 Cerebral blood flow changes: rat studies in rheoencephalography. *Physiol Meas,* **25** 1371-1384

[4] Bodo M, Pearce F J, Baranyi L, Armonda R A 2005 Changes in the intracranial rheoencephalogram at lower limit of cerebral blood flow autoregulation. *Physiol Meas,* **26** S1-S17

[5] Bodo M, Pearce F, Van Albert S, Armonda R 2007 Rheoencephalogram reflects cerebral blood flow autoregulation in pigs ed H Scharfetter, R Merva *ICEBI 2007*, IFMBE Proceedings **17** (Berlin: Springer) p 695–698

[6] Czosnyka M, Smielewski P, Kirkpatrick P, Laing R J, Menon D, Pickard J D 1997 Continuous assessment of the cerebral vasomotor reactivity in head injury. *Neurosurgery* **41** 11-7

[7] Olsson T, Broberg M, Pope K J, Wallace A, Mackenzie L, Blomstrand F, Nilsson M, Willoughby J O 2006 Cell swelling, seizures and spreading depression: an impedance study. *Neuroscience* **140** 505-15

[8] ICM+ Brain monitoring for neurosurgery and intensive care; web site; http://www.neurosurg.cam.ac.uk/icmplus/index.html

[9] Mendelow A D, Crawford P J. 2005 Primary and secondary brain injury ed P L Reilly, R Bullock *Head injury* (London; Arnold)

[10] Armonda R A, Bell R S, Vo A H, Ling G, DeGraba T J, Crandall B, Ecklund J, Campbell W W 2006 Wartime traumatic cerebral vasospasm: recent review of combat casualties. *Neurosurgery* **59** 1215-25

[11] Bodo M, Szebeni J, Baranyi J, Savay S, Pearce F J, Alving C R, Bünger R 2005 Cerebrovascular involvement in liposome – induced cardiopulmonary distress in pigs. *J. Liposome Res.* **15** 3-14

Monitoring cerebrovascular pressure reactivity with rheoencephalography

KM Brady[1], JO Mytar[1], KK Kibler[1], RB Easley[1], RC Koehler[1], M Czosnyka[2], P Smielewski[2], C Zweifel[3], M Bodo[4], FJ Pearce[4], RA Armonda[5]

[1]Department of Anesthesiology and Critical Care Medicine, Johns Hopkins University School of Medicine, Baltimore, MD.
[2]Department of Academic Neurosurgery, Addenbrooke's Hospital, Cambridge, U.K.
[3]Department of Neurosurgery, University Hospital, Basel, Switzerland.
[4]Department of Resuscitative Medicine, Walter Reed Army Institute of Research, Silver Spring, MD.
[5]Department of Neurosurgery, National Naval Medical Center, Bethesda, MD.

E-mail: kbrady5@jhmi.edu

Abstract. Determining optimal perfusion pressure for patients with traumatic brain injury can be accomplished by monitoring the pressure reactivity index, or PRx, which requires an intracranial pressure monitor. We hypothesized that pressure reactivity could be quantified using a rheoencephalography index, or REGx. We measured the REGx and PRx as repetitive, low-frequency linear correlation between arterial blood pressure and intracranial pressure (PRx) or arterial blood pressure and REG pulse amplitude (REGx) in a piglet model of progressive hypotension. We compared the PRx and REGx against a gold standard determination of the lower limit of autoregulation using laser-Doppler measurements of cortical red cell flux. The PRx produced an accurate metric of vascular reactivity in this cohort, with area under the receiver-operator characteristic curves of 0.91. REGx was moderately correlated to the PRx, (Spearman $r = 0.63$, $p < 0.0001$; Bland-Altman bias-0.13). The area under the receiver-operator curve for the REGx was 0.86. Disagreement occurred at extremes of hypotension.

1. Introduction

Traumatic brain injury (TBI) is a major health concern for children in the developed world, and soldiers exposed to blast injury.[1-2] Cerebral perfusion pressure (CPP) management is critical to the outcome of patients with TBI, but there is debate regarding optimal CPP management strategies.[3-5] The latest guideline for management of TBI from The Brain Trauma Foundation endorses the option to individualize management with ancillary monitoring, specifically citing the use of autoregulation monitoring and the pressure reactivity index (PRx).[3,6] The PRx is a low-frequency linear correlation between mean arterial blood pressure (ABP) and intracranial pressure (ICP)[7] In health, the PRx is negative because intracranial blood volume (and therefore ICP) has slow fluctuations which are phase shifted from similar slow fluctuations of ABP. When vascular reactivity is impaired, the PRx is positive because low-frequency intracranial blood volume changes are in phase with ABP changes.[8] Patients with CPP in a range that optimizes the PRx have better outcomes.[6,9]

We seek to evaluate rheoencephalography (REG) as an alternative to ICP measurements for the purpose of monitoring pressure reactivity. Rheoencephalography is the measurement of impedance across the brain, which is a function of the ratio of insulating (fat) and conducting (blood and cerebrospinal fluid) mass.[10,11] Oscillations in the REG signal occur at pulse and respiratory frequencies, thought to be the result of blood volume shifts during hemodynamic cycles. The rate of

1 Ken Brady, MD, Johns Hopkins University School of Medicine, 600 North Wolfe Street, Baltimore, MD. 21287 USA

change of the REG waveform has been shown to increase with decrements of ABP above the lower limit of autoregulation (LLA), and decrease with decrements of ABP below the LLA.[12] We propose that these changes in the REG waveform are due to vascular resistance changes associated with autoregulatory activity. We therefore hypothesize that slow changes in pulse amplitude of the REG recapitulate similar low-frequency changes in mean ICP. Further, we hypothesized that vascular reactivity can be monitored by correlation of REG pulse amplitude with ABP (REGx). We tested these hypotheses by comparing the REG pulse amplitude with ICP and REGx with PRx.

2. Methods

Hypotension was induced in piglets (5-7 days old, n=9) over 3-4 hours by inflation of a balloon catheter in the inferior vena cava. This method allows for the occurrence of spontaneous slow waves of ABP as hypotension ensues.[13] All animals were monitored with invasive REG (by insertion of stainless steel electrodes 19 mm apart in the right parietal cortex), ICP (by left external ventricular drain), cortical red cell flux (by laser-Doppler), and ABP (by femoral artery catheterization). The standard LLA was determined using plots of laser-Doppler flux as a function of cerebral perfusion pressure according to our previously published method (See figure 1).[13]

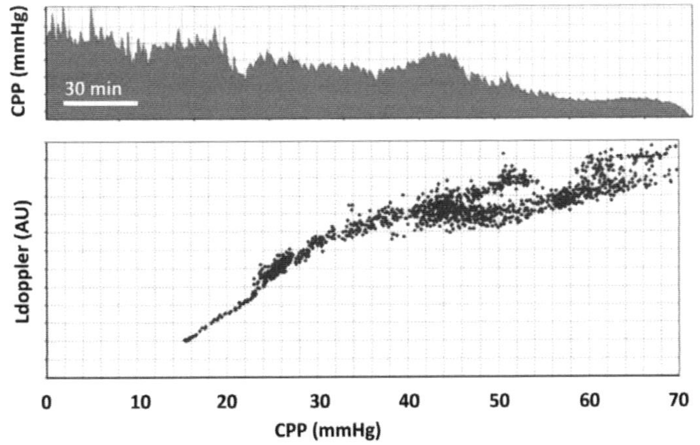

Figure 1 Laser-Doppler flux (Ldoppler) is plotted as a function of cerebral perfusion pressure (CPP) as CPP is lowered to lethality over 4 hours. The lower limit of autoregulation is found at the intersection of two best fit lines describing the data with the lowest residual error squared. In this animal, the lower limit of autoregulation is 30 mmHg.

The fundamental amplitude of the REG waveform at pulse frequency was calculated and updated every 10 seconds. The ICP waveform was low pass filtered by recording 10 second mean values. This operation removes pulse and respiratory frequency oscillations from the ICP, but preserves slow wave activity, which occurs at periods between 20 and 300 seconds (see figure 2).[7]

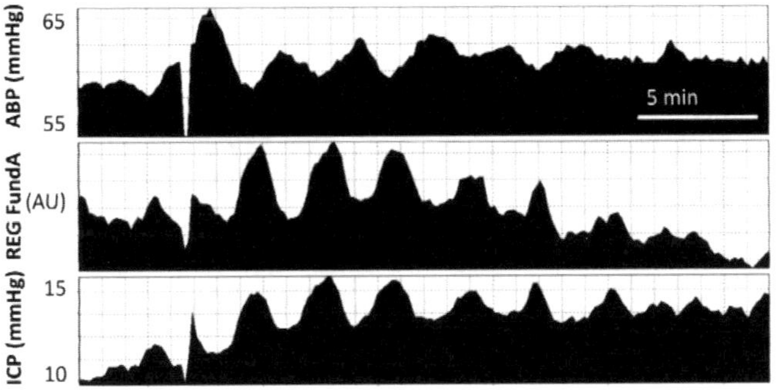

Figure 2 Slow waves of arterial blood pressure (ABP), intracranial pressure (ICP) are compared with a trend of pulse-frequency amplitude changes in rheoencephalography (REG FundA). Slow waves of ICP are phase shifted with respect to slow waves of ABP in this normotensive animal. The ICP slow waves are recapitulated in the REG fundA.

PRx and REGx were recorded as a continuous, moving, linear correlation coefficient between 30 samples of mean ICP and mean ABP (PRx) or between the REG pulse frequency fundamental amplitude and mean ABP (REGx), in overlapping 300 second epochs updated every 10 seconds. All values of PRx and REGx were sorted according to the CPP of their calculation interval, binned and averaged in 5 mmHg increments (See figure 3).

Journal of Physics: Conference Series **224** (2010) 012089 doi:10.1088/1742-6596/224/1/012089

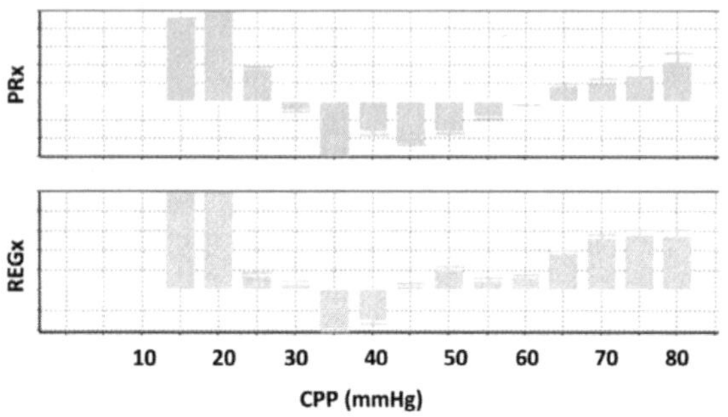

Figure 3 Rheoencephalography index (REGx) and Pressure reactivity index (PRx) recordings from the animal shown in Fig 1 are sorted and averaged according to cerebral perfusion pressure (CPP). The two indices show similar descriptions of autoregulatory activity with pressure passivity above 60 mmHg, and below 30 mmHg.

3. Results

The PRx and REGx both detected the LLA in this cohort of animals. The REGx was less robust at extremes of hypotension. This is shown in plots of REGx and PRx as a function of CPP normalized to the LLA, with receiver operator characteristics (figure 4).

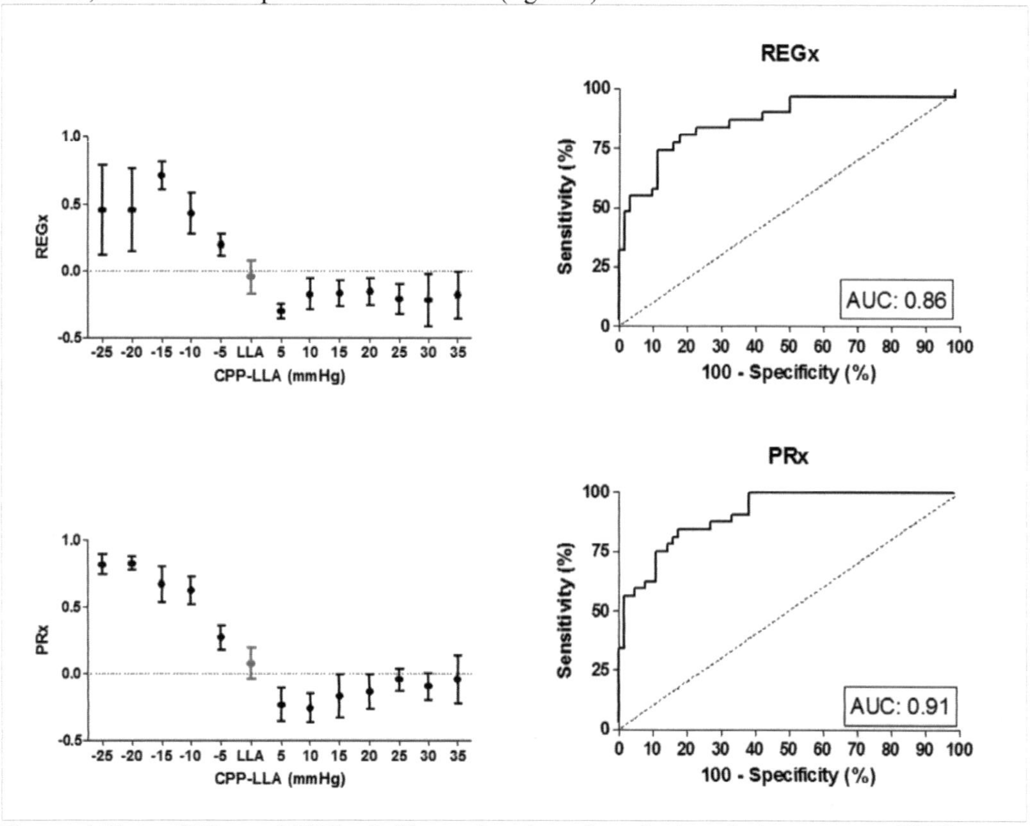

Figure 4 Plots of PRx and REGx (left) as a function of cerebral perfusion pressure (CPP) normalized to the LLA. Values above the LLA are consistently lower than values below the LLA, and this accuracy is quantified with receiver operator characteristics for each index in the right panels (AUC = area under the curve).

Bland-Altman analysis of the PRx and REGx shows that the REGx is slightly biased to a low result (bias = -0.13 when comparing the difference vs. the average), consistent with the low scores seen at extremes of hypotension in Figure 4. Correlation data are shown in figure 5. In this study, the REGx gave the most accurate information at CPP around the LLA.

4. Conclusions

The REGx is a promising modality for measuring cerebrovascular pressure reactivity. The finding that the fundamental amplitude of REG at pulse frequency recapitulates slow waves of ICP is consistent with the hypothesis that pulse amplitude changes in REG are affected by autoregulation-induced changes in cerebral vascular resistance. The REGx has reasonable accuracy for detection of the LLA when compared with a gold standard and good correlation with the PRx. For functional deployment of the modality, non-invasive measurement of the REG should be evaluated.

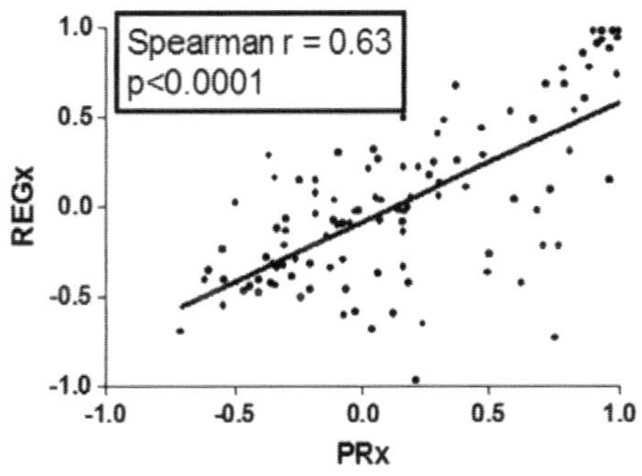

Figure 5 Linear correlation shows good correlation with a slight bias for low REGx at high PRx values.

Disclaimer Material has been reviewed by the Walter Reed Army Institute of Research. There is no objection to its presentation and/or publication. The opinions or assertions contained herein are the private views of the author, and are not to be construed as official, or as reflecting true views of the Department of the Army or the Department of Defense. Research was conducted in compliance with the Animal Welfare Act and other federal statutes and regulations relating to animals and experiments involving animals and adheres to principles stated in the Guide for the Care and Use of Laboratory Animals, NRC Publication, 1996 edition.

References

[1] Okie S 2005 *N. Engl. J. Med.* **352** 2043-7
[2] Langlois J A, Rutland-Brown W and Thomas K E 2004 *Traumatic Brain Injury in the United States: Emergency Department Visits, Hospitalizations, and Deaths.* National Center for Injury Prevention & Control (Atlanta, Georgia: US Dept of Health & Human Services)
[3] Brain Trauma Foundation, American Association of Neurological Surgeons, Congress of Neurological Surgeons, Joint Section on Neurotrauma and Critical Care, AANS/CNS. 2007 *J Neurotrauma* **24** Suppl 1 S59-64
[4] Adelson P D, Bratton S L, Carney N A, et al. 2003 *Pediatr. Crit. Care Med.* **4** S31-3
[5] Naredi S, Koskinen L O, Grande P O, et al. 2003 *Crit. Care Med.* **31** 2713-4
[6] Steiner L A, Czosnyka M, Piechnik S K, et al. 2002 *Crit. Care Med.* **30** 733-8
[7] Czosnyka M, Smielewski P, Kirkpatrick P, Laing R J, Menon D, Pickard J D 1997 *Neurosurgery* **41** 11,7; discussion 17-9
[8] Lee J K, Kibler K K, Benni P B, et al. 2009 *Stroke* **40** 1820-6
[9] Brady K M, Shaffner D H, Lee J K, et al. 2009 *Pediatrics* **124** e1205-12
[10] Moskalenko Y E, Kislyakov Y Y, Vainshtein G B, Zelikson B B 1972 *Am. Heart J.* **83** 401-14
[11] Moskalenko Y E, Cooper R, Crow H J, Walter G 1964 *Nature* **202** 159-61
[12] Bodo M, Pearce F J, Baranyi L, Armonda R A 2005 *Physiol. Meas.* **26** S1-17
[13] Brady K M, Lee J K, Kibler K K, Easley R B, Koehler R C, Shaffner D H 2008 *Stroke* **39** 2531-7

Simultaneous measurement of skin potential and conductance in electrodermal response monitoring

A. Jabbari[1], B. Johnsen[1], S. Grimnes[1,2] and Ø.G. Martinsen[1,2]

[1] Department of Physics, University of Oslo, Oslo, Norway
[2] Dept. of Clin. and Biomed. Eng., Oslo University Hospital, Rikshospitalet, Norway

E-mail: azarj@fys.uio.no

Abstract. Measurement of electrodermal activity (EDA) has been an important tool in psychophysiological research. The emotional sweat activity is very sensitive to psychological stimuli or conditions. The changes are easily detected by means of electrical measurements and since the sweat ducts are predominantly resistive, a low-frequency conductance measurement is appropriate for measurement of skin conductance in electrodermal response. The main purpose of this study was to develop a measuring system where DC current was replaced by a small AC current in a system so the DC potential and AC conductance could be measured simultaneously at the same skin site. A small, battery operated, PDA based instrument has been developed. The preliminary results of this ongoing study show that there is additional information in the DC potential channel and that different stimuli seem to produce slightly different response patterns.

1. Introduction

Electrodermal activity (EDA) is a set of physiological parameters of autonomous activity. It has been used for measurement of psychological and sensory activity. However, the mechanism of its appearance and the phenomenon itself, are not sufficiently clarified yet [1]. Two important parameters in the Electrodermal Activity (EDA) tradition are skin conductance and skin potential, even if the conductance parameter has been much more commonly used than the potential parameter [2-3]. It has already been shown that gender, age and race are important for EDA. Women are more emotionally reactive than men. They demonstrated greater electrodermal activity [4].

In this study we focus on the two parameters: DC potential and AC conductance. We have looked more closely at the possibilities offered by modern electronic and computer technology to see whether it is possible to measure more than one parameter simultaneously at one skin site. Measuring with the same electrode is important because of the often large skin site dependence of levels and response waveforms.

2. New measuring method

The measuring instrument has an analog and a digital part. The analog part consists of two circuit boards to measure the AC conductance, susceptance and potential in the skin. The digital part is a PDA (Personal Digital Assistant – HP iPAQ 214) with a data acquisition card card. A LabVIEW program is used to process the signals in the PDA. The results are shown on the PDA display and saved in the memory card. Three electrodes are connected to the skin when using this instrument. Two of the electrodes are used as measuring electrodes, while the third one is a reference electrode (see figure 1) [modified from ref. 5]. The electrodes used here were Kendall Kittycat 1050NPSM small Ag/AgCl electrodes with a diameter of 25 mm from Tyco Healthcare.

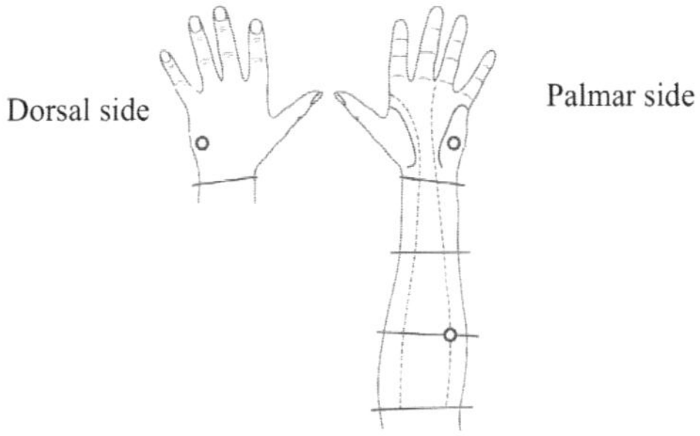

Figure 1: Placement of electrodes

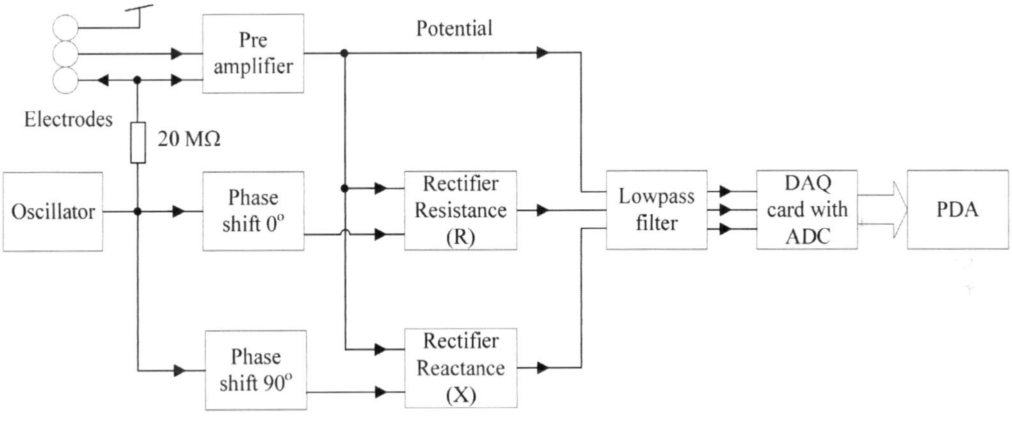

Figure 2: Block Diagram for measurement system and measuring system

Figure 2 shows a block diagram of the instrument [6]. The oscillator produces a voltage of 2.5 V rms and thus sends a 125 nA rms constant current into the skin because of the 20 MΩ series resistor. A preamplifier then picks up the differential voltage between two of the electrodes and these signals are amplified up one hundred times in the preamplifier. Skin impedance is a complex quantity because a biomaterial, in addition to opposing current flow, phase-shifts the voltage with respect to the current in the time-domain [7]. When sweat ducts are filled, conductance is increased. We hence use dc or low frequency ac to measure EDA [8]. The skin conductance is measured by a constant amplitude AC current, 125nA and 24 Hz. The impedance of the skin is measured only underneath the electrode on the hypothenar area, while the measure potential is the voltage difference between the hypothenar area and the upper part of the volar underarm. The latter skin site will in most cases have no EDA. The impedance measured in the skin is decomposed to resistance and reactance by two synchronous rectifiers. AC conductance G and susceptance B can be calculated from these impedance parameters [8]:

$$G = \frac{R}{R^2 + X^2} \quad \text{Conductance [S]} \qquad\qquad B = \frac{-X}{R^2 + X^2} \quad \text{Susceptance [S]}$$

These calculations were performed by LabVIEW program in the PDA. The measured potential goes through a lowpass filter in the analog part to remove the AC signal, and is then saved in the PDA's memory card.

Since both resistance and reactance are needed to calculate the conductance, the susceptance can be calculated without any extra hardware. The low frequency skin susceptance reflects the stratum corneum hydration, and is hence an interesting parameter to include in this instrument [9].

3. Results

The instrument has been tested on nine healthy male volunteers, age 25-29. Measurements were done in a quiet room with about 21 °C for 10 minutes, with six different stimulations; deep breath, fright (strong sound), joy (hearing a joke), cognitive (mathematical calculations), discomfort (touching the person's forehead for three seconds) and pain (pinching the skin). It is important to note that these provocations are just for testing purposes, and are hence not standardized in any way. Each provocation may therefore actually involve several different types of stimuli.

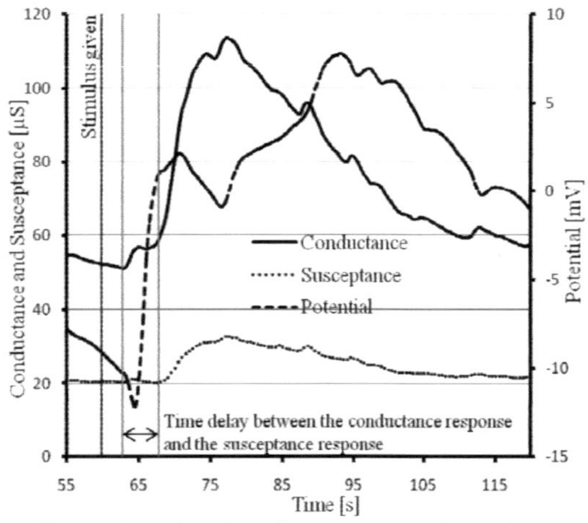

Figure 3: Delay time for responses after a deep breath

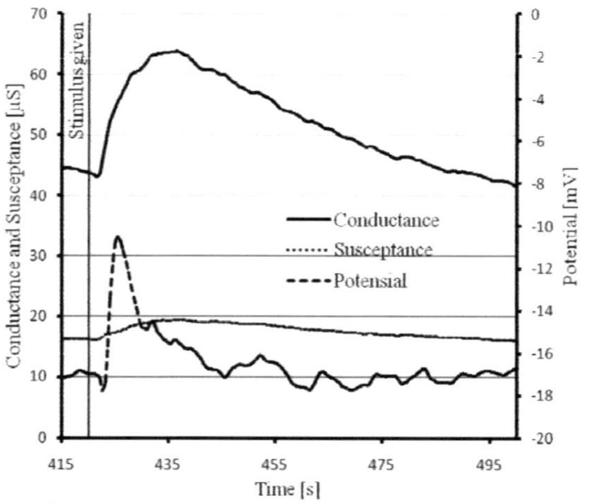

Figure 4: The average value of fright responses

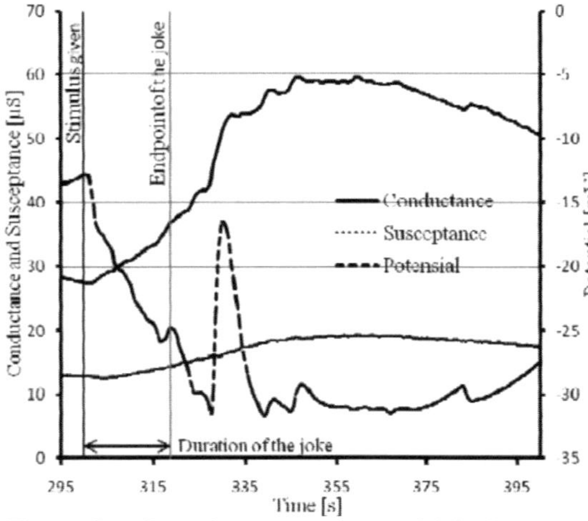

Figure 5: AC conductance and potential for joy

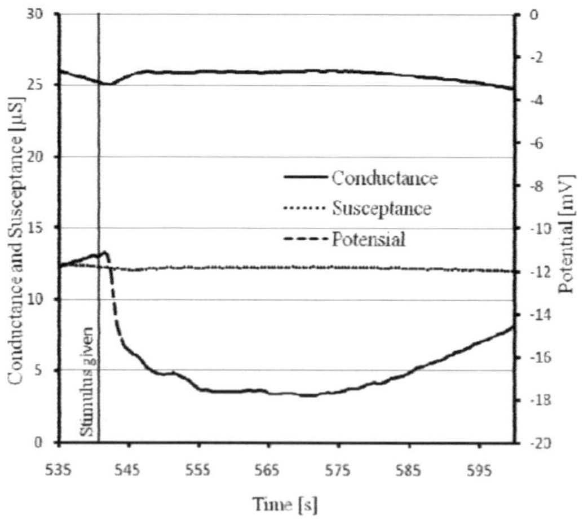

Figure 6: AC conductance and potential for discomfort

Figure 3 shows the results from one test subject taking a deep breath. There is a delay between the AC conductance and potential responses and there is also a delay between the conductance and susceptance responses. Figure 4 shows the average value of fright responses for all nine test subjects. The potential response rose faster than the AC conductance and had a shorter total reaction time.

Figure 5 shows the joy response from one of the test subjects. The potential response decreases and then rises, then quickly decreases again. AC conductance response is rising for about 50 seconds before it decreases. Figure 6 shows the results for discomfort. The operator keeps his hand on the subject's forehead for 3 seconds. The potential shows a significant negative response and the AC conductance has almost no response.

4. Discussion

Our results show that it is possible to measure skin potential and skin AC conductance separately but simultaneously in a low noise system. We can see differences in DC potential and AC conductance responses with different stimulation [10]. Typically, DC potential responses are slightly faster than AC conductance response [11], and there seems to be different response patterns for different types of stimuli. However, more measurements are needed before any conclusions can be drawn.

5. References

1. Shirai K, Yamamoto Y, Nakamura T, Kusuhara T (2006) Formative Mechanism of Skin Potential Activity and Relationships between Skin Potential and Skin Impedance, IFMBE Proc 14:2573-2576
2. Gaviria B, Coyne L, Thetford P E (1969) Correlation of skin potential and skin resistance measures. Psychophysiology, 5, 465-477
3. Boucsein (1992) Electrodermal activity. Plenum Press
4. Chentsova-Dutton YE (2007) Gender differences in emotional response among European Americans and Hmong Americans. Cognition and emotion 21:162-181
5. Fowles DC, Christie MJ, Edelberg R, Grings WW, Lykken DT, Venables PH (1981) Publication Recommendations for Electrodermal Measurements. Psychophysiology 18:232-239
6. Johnsen B (2009) PDA-based instrument for measuring electrodermal activity response (in Norwegian). Master thesis, Department of Physics, University of Oslo, Norway
7. Grimnes S, Martinsen ØG (2006) Bioimpedance. Encyclopedia of Biomedical Engineering (Akay M, ed.) John Wiley, Hoboken, 1:438-447
8. Grimnes S, Martinsen ØG (2008) Bioimpedance and Bioelectricity Basics, 2.ed, Elsevier.
9. Martinsen ØG, Grimnes S, Karlsen J (1995) Electrical methods for skin moisture assessment. Skin Pharmacology 8:237-245
10. Malmivuo J, Plonsey R (1995) Bioelectromagnetism, Principles and applications of bioelectric and biomagnetic fields. Oxford University Press.
11. Jabbari A (2006) Mechanisms behind endogenous and exogenous electrodermal response (in Norwegian). Master thesis, Department of Physics, University of Oslo, Norway

Twelve years evolution of skin as seen by electrical impedance

Ingrid Nicander[1,2], Lennart Emtestam[1], Peter Åberg[2] and Stig Ollmar[2]

[1] Dermatology, Huddinge University Hospital, SE-14186 Huddinge, Sweden
[2] Medical Engineering, Karolinska institutet, SE-14186 Huddinge, Sweden

E-mail: lennart.emtestam@ki.se

Abstract. Twelve years ago we reported an electrical impedance baseline study related to age, sex and body locations. The results showed significant differences between different anatomical locations and ages. In this study, the same participants were recalled to explore how the skin had evolved at the individual level over time. A total of 50 subjects, divided into an older and a younger group, were recalled for measurements of electrical impedance at eight anatomical locations. Readings were taken with an electrical impedance spectrometer. Information was extracted from the impedance spectra using indices based on magnitude and phase at two frequencies as in the earlier study. All included body sites had undergone alterations over time, and the size of the changes varied at different locations. The results also showed that changes in the younger group were different over time compared with the older group. In conclusion: Electrical impedance can be used to monitor skin evolution over time and baseline characteristics differ between various locations.

1. Introduction

There are quite large variations in the skin structure and function depending on anatomical location. The skin on the palmo-plantar surfaces has a rather simple function to protect the underlying living tissue with its very thick stratum corneum (SC) from strong external force and friction. Its barrier function is poor. The SC of the facial skin is thinner, consisting of fewer layers of corneocytes than that of the trunk and limbs. There are regional differences not only in the living skin tissue but also in the thin SC reflecting the function of each anatomical location [1]. These differences in the SC have been mostly disclosed by the use of non-invasive biophysical instruments. The in vivo instrumental measurements have disclosed the presence of remarkable differences in the functional properties of the SC particularly between the face and other portions of the body. They will offer clinical parameters to evaluate skin functions and, more importantly, dysfunctions (e.g. psoriasis, atopic dermatitis, epidermolysis bullosa, burns) that result in excessive water loss and are characterized by defective SC lipid organization. The differences of normal, healthy skin at different anatomical locations indicate that there is no single measurement values, or "normal" values, for the entire skin. As for the important issue "sensitive skin" there is no international consensus on the definition and classification of "sensitive skin," making rational discourse among investigators almost impossible [2].

As an example, a recent study has been published on diaper dermatitis, which affects nearly 50% of infants as well as many bed-ridden adults, particularly elderly (Xhaflaire-Uhoda et al 2009). It was demonstrated that the application of the anhydrous paste resulted in a shielding effect and a trend to normalize electrometric properties of skin, probably reflecting the combination of transepidermal water loss and the water-holding capacity of the stratum corneum [3]. Several methods and instruments are now available for the rapid reliable determination of electrical impedance or related parameters reflecting skin condition, which may become a standard research and clinical tool in dermatology [4-6].

Twelve years ago we reported an electrical impedance baseline study related to age, sex and body locations [7]. The results showed significant differences between different anatomical locations and ages. In this study, the same participants were recalled to explore how the skin had evolved at the individual level over time.

2. Materials and Methods

A total of 50 subjects, divided into an older and a younger group, were recalled for measurements of electrical impedance at various anatomical locations. The test persons were interviewed regarding their tanning habits, in door or outdoor work, et cetera. Readings were taken with an electrical impedance spectrometer with the original non-invasive electrode system. (SciBase II, SciBase AB, Stockholm, Sweden). The technology has been described in [8]. Measurements were made in April and May, which corresponds well to original data collected in May and September due to seasonal variations [9]. Information was extracted from the impedance spectra using indices based on magnitude and phase at two frequencies as in the earlier study. Data were analysed both on the whole cohort, in various subgroups, and on the individual level. In this paper, only a simple magnitude index is used, and p-values calculated. The p-values in the graphs are estimates of the probability that the median MIX in 1995 and 2008 are the same according to paired-sample Wilcoxon signed rank test. The indices are described in e.g. [7] and can also be found in several of our early publications. Here only the so called MIX-index is used, defined as

Magnitude index $$\mathrm{MIX} = \mathrm{abs}\left(Z_{20\mathrm{kHz}}\right)/\mathrm{abs}\left(Z_{500\mathrm{kHz}}\right)$$

Before impedance measurements the skin was soaked with saline solution of physiological concentration for one minute.

3. Results

All included body sites had undergone alterations over time, and the size of the changes varied at different locations. The results also showed that changes in the younger group were different over time compared with the older group, an observation made possible by using paired comparison for the same individual test persons now and 12 years ago. In Figure 1, the evolution of the impedance index MIX is presented from two body locations, i.e. the volar forearm and the cheek, for individuals belonging to older female, older male, younger female and younger male subgroups respectively. The changes of the MIX index over 12 years are significant or highly significant in all cases except for the cheek of older females.

4. Discussion and Conclusion

It is known that the properties of skin changes with age and other factors, such as diseases, toxic reactions, allergic reactions, sun exposure, seasonal variations, mechanical stress, et cetera. In our early studies we investigated the possibility to quantify and classify artificially induced skin reactions using electrical impedance [10-15]. In particular [12] sparked us to look for clinical, even diagnostic, applications of the technology, and it was found that, with more sophisticated data analysis and enhanced electrode systems, it is possible to discriminate between skin cancer and harmless lesions with clinically useful accuracy [16-19].

Future studies aiming at the establishment of the functional mapping in different skin regions will further shed light on more delicate site-dependent differences, as well as explaining the physical causes for the observed phenomena, thereby providing more information to facilitate tailor-made skin care for each location and condition of the body, whether caused by aging, reactions or disease.

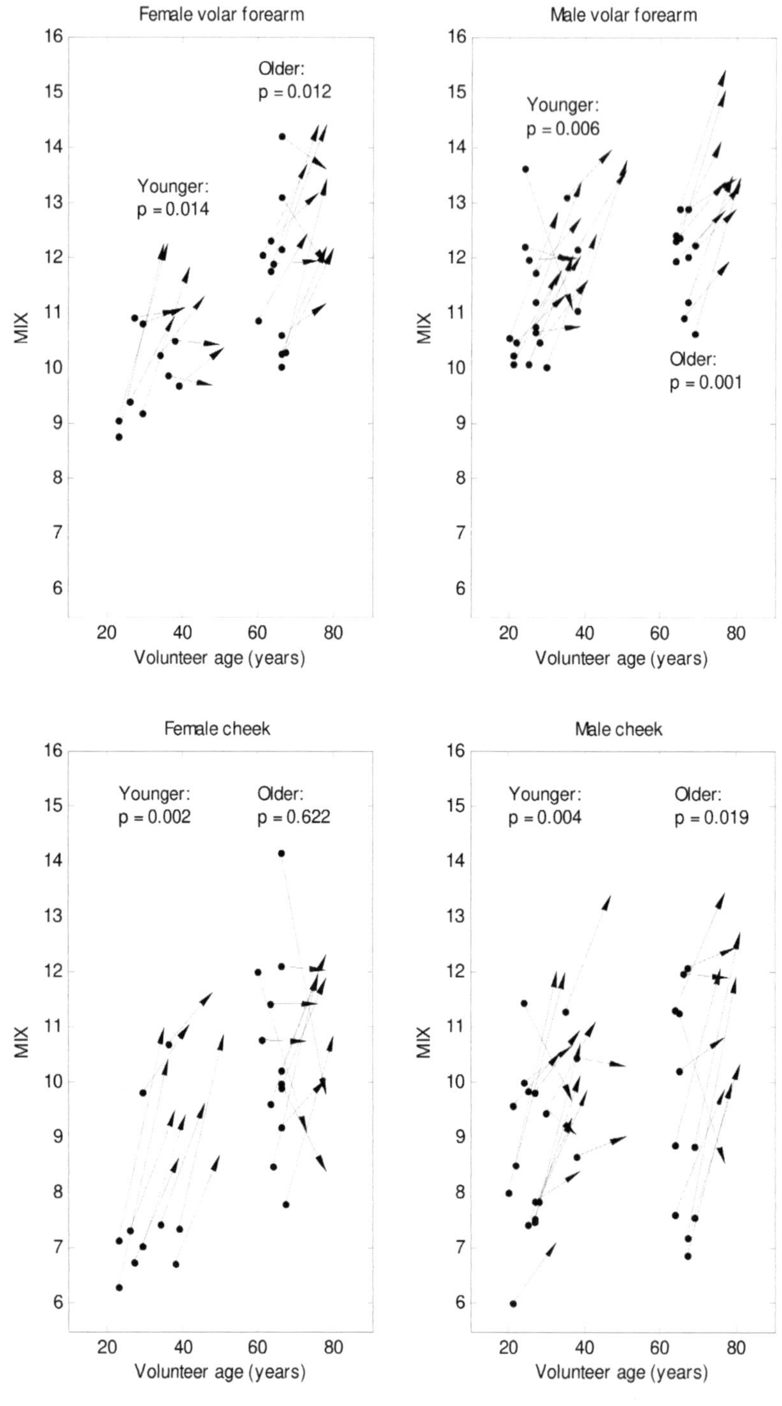

Figure 1. Scatter plots of volunteer age vs. measured MIX on volar forearm and cheek for the younger and older men and women. Each subject is represented with an arrow, where the start and endpoints indicate the volunteer age and MIX values in 1995 and 2008.

5. References

[1] Tagami H 2008. Location-related differences in structure and function of the stratum corneum with special emphasis on those of the facial skin. *Int. J. Cosmet. Sci.* **30** 413-34
[2] Menon GK and Kligman AM 2009. Barrier functions of human skin: a holistic view. *Skin Pharmacol. Physiol.* **22** 178-89
[3] Xhauflaire-Uhoda E, Henry F, Piérard-Franchimont C and Piérard GE 2009. Electrometric assessment of the effect of a zinc oxide paste in diaper dermatitis. *Int. J. Cosmet. Sci.* **31** 369-74
[4] Fluhr JW, Gloor M, Lazzerini S, Kleesz P, Grieshaber R and Berardesca E 1999. Comparative study of five instruments measuring stratum corneum hydration (Corneometer CM 820 and CM 825, Skicon 200, Nova DPM 9003, DermaLab). Part I. In vitro. *Skin Res. Technol.* **5** 161-170
[5] Fluhr JW, Gloor M, Lazzerini S, Kleesz P, Grieshaber R and Berardesca E 1999. Comparative study of five instruments measuring stratum corneum hydration (Corneometer CM 820 and CM 825, Skicon 200, Nova DPM 9003, DermaLab). Part II. In vivo. *Skin Res. Technol.* **5** 171-8
[6] Ollmar S and Nicander I 2005. Within and beyond the skin barrier as seen by electrical impedance (book chapter). *Bioengineering of the Skin: Water and the Stratum Corneum*, ed J Fluhr, H Maibach, E Berardesca and P Elsner (Boca Raton: CRC Press) pp 335-50
[7] Nicander I, Nyrén M, Emtestam L and Ollmar S 1997. Baseline electrical impedance measurements at various skin sites – related to age and sex. *Skin Research Technol.* **3** 252-8
[8] Ollmar S, Eek A, Sundtröm F and Emtestam L 1995. Electrical impedance for estimation of irritation in oral mucosa and skin. *Med. Prog. Technol.* **21** *29-37*
[9] Nicander I and Ollmar S 2000. Electrical impedance measurements at different skin sites related to seasonal variations. *Skin Res. Technol.* **6** 81-6
[10] Ollmar S and Emtestam L 1992. Electrical impedance applied to non-invasive detection of irritation in skin. *Contact Dermatitis* **27** 37-42
[11] Emtestam L and Ollmar S 1993. Electrical impedance index in human skin: Measurement after occlusion, in 5 anatomical regions and in mild irritant dermatitis. *Contact Dermatitis* **28** 104-8
[12] Nicander I, Ollmar S, Eek A, Lundh Rozell B and Emtestam L 1996. Correlation of impedance response patterns to histological findings in irritant skin reactions induced by various surfactants. *Br. J. Dermatol.* **134** 221-8
[13] Nicander I, Ollmar S, Lundh Rozell B and Emtestam L 1997. Allergic contact reactions in the skin assessed by electrical impedance – a pilot study. *Skin Res. Technol.* **3** 121-5
[14] Nicander I and Ollmar S 2000. Electrical impedance measurements at different skin sites related to seasonal variations. *Skin Res. Technol.* **6** 81-6
[15] Nyrén M, Kuzmina N and Emtestam L 2003. Electrical impedance as a potential tool to distinguish between allergic and irritant contact dermatitis. *J. Amer. Acad. Dermatol.* **48** 394-400
[16] Åberg P, Nicander I, Holmgren U, Geladi P and Ollmar S 2003. Assessment of skin lesions and skin cancer using simple electrical impedance indices. *Skin Res. Technol.* **9** 257-61
[17] Åberg P, Nicander I and Ollmar S 2003. Minimally invasive electrical impedance spectroscopy of skin exemplified by skin cancer assessments. *Proceedings Annual Conference of the IEEE Engineering in Medicine and Biology Society (New York: IEEE) pp 3211-4*
[18] Åberg P, Nicander I, Hansson J, Geladi P, Holmgren U and Ollmar S 2004 Skin cancer identification using multi-frequency electrical impedance – a potential screening tool. *IEEE Trans. Biomed. Eng.* **51** 2097-102
[19] Åberg P, Geladi P, Nicander I, Hansson J, Holmgren U and Ollmar S 2005. Non-invasive and microinvasive electrical impedance spectra of skin cancer – a comparison between two techniques. *Skin Res. Technol.* **11** 281-6

Acknowledgement: This study was sponsored by SciBase AB, Stockholm, Sweden.

PC-based instrumentation for electrodermal activity measurement

Christian Tronstad[1], Sverre Grimnes[1,2], Ørjan G. Martinsen[2,1], Vegard Amundsen[2] and Slawomir Wojniusz[3,4]

[1]Department of Clinical and Biomedical Engineering, Oslo University Hospital, Rikshospitalet, Norway.
[2]Department of Physics, University of Oslo, Rikshospitalet, Norway.
[3]Department of Neuropsychiatry and Psychosomatic Medicine, Oslo University Hospital, Rikshospitalet, Norway.
[4]Faculty of Health Sciences, Oslo University College, Norway

E-mail: christian.tronstad@gmail.com

Abstract. A PC-based EDA measuring system is presented. The system is composed of a laptop with a PCMCIA DAQ-card running LabVIEW® software, a small front-end with a dual op-amp IC and a few passive components, and three skin-surface electrodes. The electrode system gives a monopolar measurement below the measuring electrode regardless of the electrode sizes, unless very small. Usage of the system is demonstrated by measurements from a mental stress experiment on 17 volunteers. There was a significant correlation ($R=0.51$, $p<0.001$) between the self-assessed stress-level and the EDA response frequency. The system allows easy on-site customization in software of measuring parameters, signal-quality monitoring and non-linearity detection in real time. We believe that the most suited use for the system is for stationary experimental purposes where this flexibility is desired. The system is easy to reproduce by engineers interested in doing EDA research.

1. Introduction

The activity from the sympathetic nervous system (SNS) regulates the secretory part of the sweat glands, which in turn changes the electrical properties of the skin due to the filling of electrolyte-containing sweat in the ducts. These changes are mostly in the stratum corneum (SC) layer of the skin, and by measuring the SC conductance we can obtain a link to the SNS activity. This phenomenon is usually called electrodermal activity (EDA). The best sites for measuring EDA are known to be the palmar and plantar dermatomes, which probably differ from the innervation of sweat glands on the rest of the body [1]. There is a wide range of uses for EDA recordings, with applications within psychophysiology, ambulatory monitoring, intensive care and biofeedback among others.

EDR instrumentation generally requires very few electronic components and can be built into small devices with low current consumption [2,3]. However, for stationary experimental purposes where a laptop is the monitor of choice, the instrumentation is easily realized by a DAQ card and a few off-the-shelf components. This paper presents one such solution. EDA is known as an indicator of stress [1], and measurements from usage of the system in a mental stress experimental setting are presented.

2. Materials and methods

2.1. Instrumentation

The measuring system is composed of three main parts, illustrated in figure 1:

- Three skin surface gel electrodes.
- Front-end electronics consisting of one dual op-amp IC, 5 resistors and two capacitors.
- A laptop with a National Instruments 6062E DAQ-card and LabVIEW® software for signal processing, display and data storage.

A low amplitude sine voltage generated by the D/A of the DAQ is applied to the skin through the E and R electrodes. Remote sensing at the current-less R electrode causes the right op-amp to drive a current through E which keeps the potential under R equal to the excitation sine, thus eliminating the SC impedance below E [4,5].

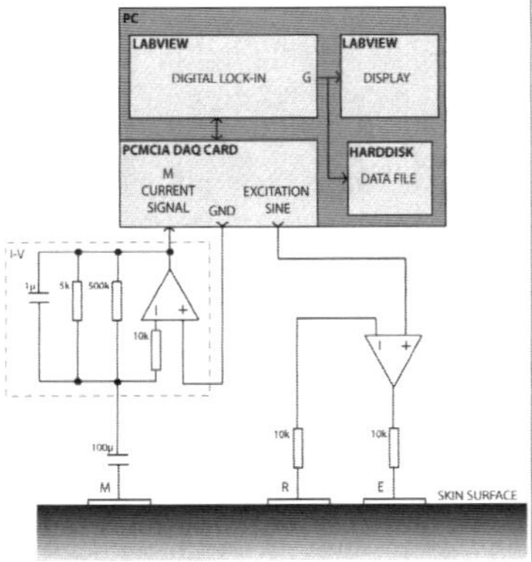

Figure 1. PC-based EDA instrumentation

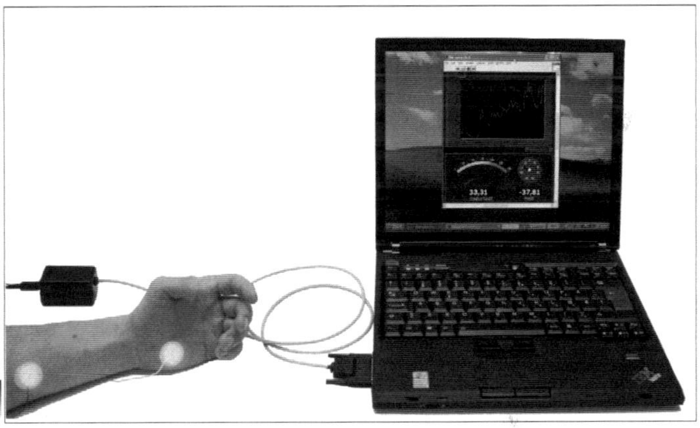

Figure 2. Photo of the system in use

Because the impedance of the wet tissue is negligible compared to that of the SC, the only current-opposing element that is left is the impedance of the SC below the M electrode. Thus, the measurement is monopolar regardless of the effective electrode areas (EEA) (unless very small [4]), and confined to the SC of the EEA below M. The major contributor to changes in this impedance is the filling and reabsorption of sweat in the sweat ducts. Measuring with AC eliminates the endogenous potentials generated by the skin or electrodes. The I-V dashed box on the left constitute a current to voltage converter with a $Fc=200\ Hz$ RC lowpass filter. The 100µ capacitor blocks DC which may saturate the op-amp and its impedance needs to be very small compared to the SC below M, thus the large capacitance. The 10k resistors are for safety in order to reduce the current through the skin should a breakdown in the op-amp occur. The digitized current signal is demodulated by the phase-corrected excitation sine in software, giving the AC conductance G as the real part of the current divided by the excitation voltage. The voltage input and output of the DAQ is synchronized with the same sample clock. The op-amp power is supplied by the DAQ-card. Figure 2 shows a photo of the total system.

2.2. Software

Apart from the front-end with the op-amp IC as the only active component, all the measurement tasks are handled digitally in software. The digital lock-in technique is similar to that presented in [2]. For research purposes, the aim is to preserve as much as possible of the EDR signal information, because the EDR parameters may vary depending on the feature extraction methods when the sampling frequency Fs is decreased below a certain threshold. Very little signal energy is lost when Fs is as low as $1Hz$, but depending on the signal feature extraction method, a higher Fs may be required. The requirement of using a low excitation frequency Fe for the EDA measurement [6] decides the compensation between Fs and the signal quality by means of the number of periods N for one measurement. The software will also need a certain processing time Tp between each measurement for signal processing, storage etc, giving the following limit for the maximum Fs:

$$Fs = \left(\frac{N}{Fe} + Tp\right)^{-1} \tag{1}$$

2.3. Experimental protocol

17 volunteers (11M, 6F) who gave informed consent were subjected to three epochs of stress exposure, each lasting 10 min: Relaxation with classical music, quietly reading a text while provoked by irregular bursts of white noise and anticipation of public speaking while still receiving the white noise. The M electrode was placed on the hypothenar area of the palm, while the R and E electrodes were placed on the underarm, but the placement of these two is not critical. The skin AC conductance was sampled at $Fs=8Hz$.

2.4. Data analysis

As a measure of EDA response frequency, the number of EDA peaks (EDA TC) within a central five minute window within all the epochs was computationally determined by a top counting algorithm (A.1) and compared with a self-assessed stress score (1-10) of the volunteers using the Pearson product moment correlation. Comparison between the epochs was done using one way repeated measures ANOVA and the Holm-Sidak t-test. Normality was assumed based on the Shapiro-Wilk test.

3. Results

The SNR of the conductance measurement was 74 dB with a phase error of 0.02°, tested in a lab with a high degree of electromagnetic interference. The linear dynamic range of the system with the configuration as in fig.1 was 0 to approx 200µS. This also depends on the type of op-amp used.

There was a significant correlation between the EDA TC and the subjective stress-score ($R=0.51$, $p<0.001$). As shown in fig.3, both the EDA TC and the subjective score was significantly higher ($p<0.05$) during epoch 2 and 3 compared to epoch 1. Fig.4 shows an example measurement from epoch 2.

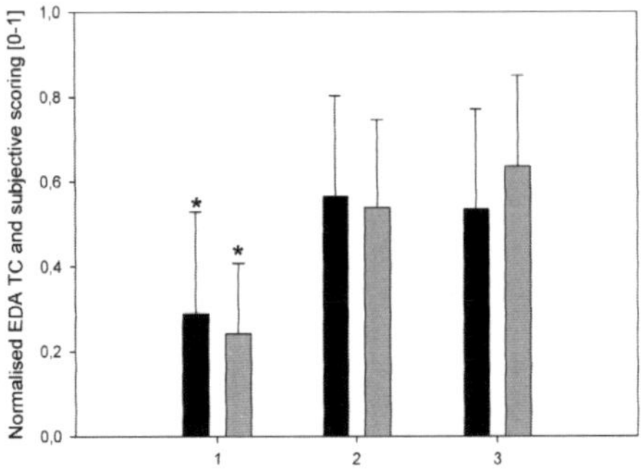

Figure 3. Mean and standard deviation of the EDA TC (black) and the subjective scoring (grey) for the three epochs. *p<0.05.

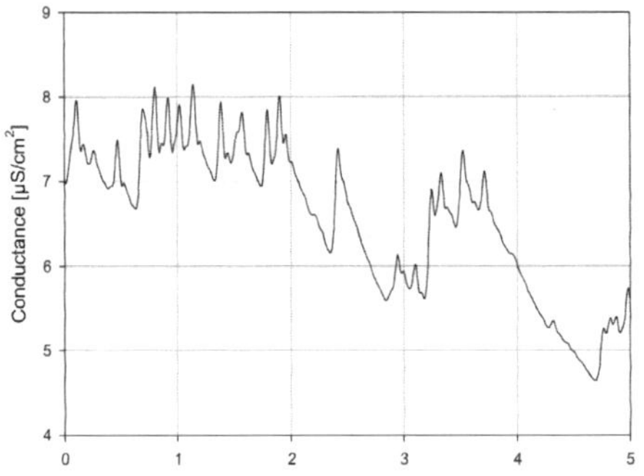

Figure 4. Example measurement from epoch 2.

Discussion

Expectedly, epoch 2 and 3 led to a higher self-assessed stress level than epoch 1. Although the protocol failed to produce a significant difference in mean stress level between the epoch 2 and 3, large intraindividual differences between these epochs were frequently observed. Thus, the correlation between the EDA parameter and the self-assessed score may be a better suited measure of the link between EDA and mental stress than the ANOVA. However, the variation in intersubjective understanding of the scoring will impact the correlation. Although the first 2.5 minutes of the epochs were excluded in the analysis to allow habituation to the noise, part of the EDA responses in epochs two and three may be directly caused by the acoustic stimulus, and whether this represents stress or not can be debated.

With this setup, measuring parameters such as the excitation voltage and frequency, phase correction, filtering and calibration is easily modified by a few mouse-clicks. In addition, the system enables automatic detection of artefacts or non-linearity from the raw input current signal, which can be important for this type of electrode system [4]. The simplicity of a laptop with only a cable, a front-end box which can be made very small and electrodes as the total measuring system makes it convenient for use in an experimental setting where a laptop is the monitor of preference. This system would be easy to reproduce by biomedical engineers interested in doing EDA research and it would work just as well with most other types of DAQ-cards, e.g. the USB-DAQ which has become more popular lately.

For electrical safety, it is important to run the laptop on battery while measuring on live subjects in order to obtain a body floating connection with the equipment.

Appendix

With FFT analysis revealing negligible energy above 1 Hz, the window length was set to $1Hz \cdot Fs = 8$ points and the EDA TC was calculated as:

$$TC = \sum_{i=Fs/2}^{N-Fs/2} (pos\left(G[i] - G\left[i - \frac{Fs}{2}\right] - \Delta G\right)$$

$$\cdot \prod_{j=0}^{\frac{Fs}{2}} (pos(G\,[i-j] - G[i-j-1]) \cdot pos(G[i+j] - G[i+j+1]))) \qquad (A.1)$$

$$Where\; pos(x) = \begin{cases} 1, & x \geq 0 \\ 0, & x < 0 \end{cases}, \Delta G = \frac{0.01\mu S}{cm^2}$$

References

[1] Boucsein W. 1992. *Electrodermal Activity*. NewYork: Plenum Press. 460 pages, ISBN-10: 0306442140, ISBN-13: 978-0306442148.

[2] Tronstad C., Gjein G.E., Grimnes S., Martinsen Ø.G., Krogstad A-L., Fosse E. 2008 Electrical measurement of sweat activity. *Physiol.Meas.* **29** 407-415

[3] Tronstad C., Martinsen Ø.G., Grimnes S. 2008 Embedded instrumentation for skin admittance measurement. *Engineering in Medicine and Biology Society, 2009. EMBS 2008. 30th Annural International Conference of the IEEE, Vancouver.* 2373-2376

[4] Grimnes S., Martinsen Ø.G., Tronstad C. 2009. *Noise properties of the 3-electrode skin admittance measuring circuit. 4th European Conference of the IFMBE, Antwerp. Springer Berling Heidelberd, IFMBE Proceedings* **22** 720-722.

[5] Grimnes S. 1983 Impedance measurement of individual skin surface electrodes. *Med.Biol.Eng.Comput.* **21** 750-755

[6] Martinsen Ø.G., Grimnes S., Haug E. 1999 Measuring depth depends on frequency in electrical skin impedance measurements. *Skin Res. Technol.* **5** 179-81.

Sorption studies of human keratinized tissues

G K Johnsen[1], Ø G Martinsen[1,2] and Sverre Grimnes[2,1]

[1] Department of Physics, University of Oslo, PO 1048 Blindern, 0316 Oslo, Norway.
[2] Department of Clinical and Biomedical Engineering, Rikshospitalet University Hospital, N-0027 Oslo, Norway

E-mail: g.k.johnsen@fys.uio.no

Abstract. Water content is known to be the most important single parameter for keratinized tissue to remain its vital functions. In that sense, a general knowledge of the water binding properties is of great interest, and a reliable measurement setup must be found. Also, revealing the sorption properties of human keratinized tissues is vital towards a calibration of susceptance based skin hydration measurements that already is an important diagnostic tool in clinical dermatology, and we will see that any hysteresis will complicate such a calibration further. In this study we investigated the sorption properties of keratinized tissues such as human epidermal stratum corneum (SC), hair and nail. The study was performed under controlled environmental conditions with a dynamic vapor sorption (DVS) instrument, and the water uptake of the keratinized test samples was measured as the relative humidity in the ambient air was altered step-wisely. In this study, vital and characteristic water sorption properties such as the isotherm, relative water uptake, and hysteresis were investigated and will be discussed.

1. Introduction

Water content and water-holding capacity are known to be very important for the overall function of keratinized tissues such as SC, hair, and nail, and Blank showed that such tissues needed at least 10 per cent water content in order to remain its vital functions [1]. Electrical measurements as a tool for estimating the hydration state of keratinized tissues have been studied and continuously improved over the last 30 years [2, 3, 4, 5, 6, 7], and a low frequency susceptance method has been found to be the most suitable [8]. Water content has been shown to be the most significant parameter governing electrical susceptance of human tissue [9]. The basic knowledge of how absorbed water effects the conductivity of keratinized tissues rests upon insight in the binding mechanisms of water, and therefore it is of general interest to investigate the sorption properties, including any differences, of SC, hair, and nail.

The main target in this study was to investigate the absorption and desorption properties of human keratinized tissues. The findings in this manner will have direct implications on the progress of calibration of susceptance based hydration measurements that, among others, are aiming to better estimate absolute water content of e.g. human skin. Such attempts have been carried out *in vitro* [10] and *in vivo* [11] based on absorption data of SC from the heel. However, any hysteresis, that is differences in hydration level between absorption and desorption at similar environmental conditions, will necessarily influence the final result depending on whether the calibration uses absorption or desorption data, and thus prevents the calibration from being unique.

Although the general structure and properties of SC, nail and hair, all consisting of dead cells tightly cemented together by a group of proteins called keratins, are quite similar, they show different sorption properties, especially at high relative humidities (RH). Also the amount of hysteresis for keratinized tissues may be a measure of structural changes that occur during a full sorption cycle. The differences in water uptake between SC, nail and hair also reveal differences in composition and structure, and our results suggest that hair and nail are similar in structure, however deviating more from epidermal SC.

2. Methods

The keratin samples were taken from healthy and male volunteer test subjects. The two SC pieces were taken from the heel by an Aesculap dermatome (Braun, Tuttlingen Germany), and the skin was gently abraded before the dermatome was used in order to get a smooth surface. The sample pieces were about 1 by 1 cm in size and with a thickness of 0.2 mm which is the smallest adjustment possible by the dermatome. The two samples of toenail were cut in full thickness, and the sample sizes were approximately 3 by 3 mm. Finally, one hair sample was prepared so that both of the two ends, the follicle and the tip, were removed, leaving only the center part as the sample. The water sorption properties of each sample were measured separately by means of a dynamic vapor sorption (DVS) from the Surface Measurement Systems, U.K. The DVS setup provides very delicate and sensitive gravimetric measurements provided by a Cahn microbalance system and ensures a resolution of 0.1 μg. The samples were placed in a closed chamber where both temperature and relative humidity is controlled. Furthermore, the DVS was pre-programmed to increase RH in steps of 10 %, and allowed each of the samples to reach thermodynamical equilibrium with its ambient surroundings in the measurement chamber before a new step in RH was initiated. Starting from completely dry conditions at 0 % the RH was increased in steps up to 90 % and then back to 0 %. The constraint for the sample to be considered in its equilibrium state was that the mass rate was less than 20 ppm per minute over a time period of minimum 10 min. All samples were investigated one by one, and each measurement run lasted from 5 to 7 days.

3. Results

Figure 1 shows the time course for absorption (water gain) and desorption (water loss) for hair. The time courses of the two other sample types showed the same form. The alteration in sample weight corresponds to the water content relative to the dry weight. We note that the change in weight, initiated by an instantaneous step in RH, followed closely an exponential time course, indicating that the mass changed at a rate proportional to the difference between the final, equilibrium mass and the actual mass at the time of interest.

The relative water gains for the various samples as a function of RH in the DVS chamber are given in figure 2. The dotted lines correspond to absorption data, and the continuous lines to desorption data. We note the distinct differences between the different sample types: SC gained nearly no amount of water until 40 % RH was reached, thereafter gaining weight with an increasing rate as RH increased. Nail followed a similar curve as SC until RH was 50 %, but had a much lower water uptake at high RH. Hair, however, gained water quite rapidly at low RH, and had a more or less constant gain rate over the entire span of RH. Nail and hair had similarities in curve form, only with the hair data more elevated. Hysteresis was substantial for hair and nail, but for SC there were only little signs, if any at all, of hysteresis at RH lower than 30 %.

4. Discussion

In this paper we have found the sorption properties of SC, hair, and nail to be quite different, and also that there may be a substantial hysteresis. The hysteresis is large for hair and nail,

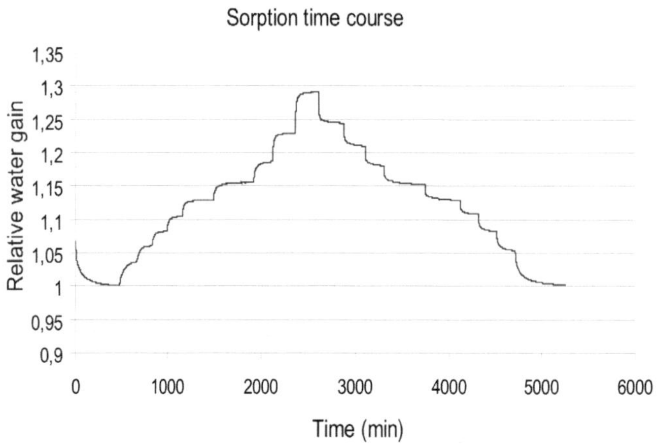

Figure 1. A full sorption time cycle for hair tissue.

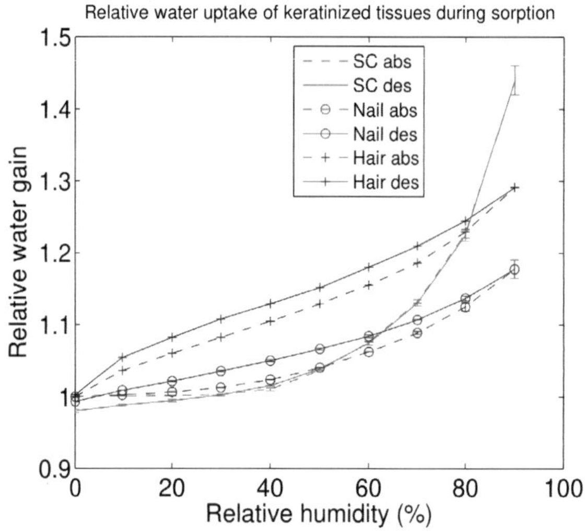

Figure 2. Relative water uptake for keratinized tissue.

but very little for SC where it totally vanishes when RH is larger than about 30 %. This means that calibration of susceptance based hydration measurements, as described by Martinsen et al. [10], will end up with the same result for estimated absolute SC water content independently of whether absorption or desorption data are used as long as the RH is greater than roughly 30 %. For hair and nail, however, the large amount of hysteresis, that is the large gap between absorption and desorption data, will necessarily make such a calibration method ambiguous. The SC hysteresis data are in contradiction to similar results found by Polefka et al. who reported on significant hysteresis in porcine SC [12]. However, the porcine data were gathered using a larger value of the DVS mass rate that is used used to establish the steady state sorption data at a given RH, and therefore the samples have less time to reach their thermodynamical equilibrium state before a step in the RH is initiated, which finally can be expected to result in substantial hysteresis.

The much lower water uptake of hair and nail compared to the SC is best interpreted as differences in the structure and water binding mechanisms of the constituents as well as the amount of lipids. Hair and nail cells are formed by restrictions imposed by the shape of their anatomical site, whereas SC cells do not suffer any restrictions at formation and are therefore free to swell [13, 14, 15].

Further knowledge on the water binding mechanisms and structure may be gathered from the time constants, that is the time needed for water to enter and leave the keratinized samples following a step in the RH. This should be a topic for future studies, and would be of interest for among others the differentiation process which is the overall description of the production process of keratinized tissues in the human integument, as well as for equivalent modeling of electrical properties of SC, hair, and nail.

5. Conclusions

This paper shows that hysteresis may be a complicating factor in calibrations of susceptance based hydration measurements of keratinized tissues. Also, SC, hair, and nail show very different sorption properties, which again may be expected to induce different electrical properties. The differences in sorption properties of the various keratinized sample types may be related to structural differences resulting in differences in water holding capacities.

References

[1] Blank H I 1952 *J. Invest. Dermatol.* **18** 433
[2] Kohli R, Archer W I, Roberts J M C, Cochran A J and Po A L W 1985 *Int. J. Pharmacol.* **26** 275
[3] Blichmann C W and Serup J 1988 *Acta. Derm. Venereol.* **68** 284
[4] Leveque J L and Rigal J D 1983 *J. Cosmet. Chem.* **34** 419
[5] Martinsen Ø G and Grimnes S 1998 *Innov. Technol. Biol. Med.* **19** 395
[6] Clar E J, Her C P and Sturelle C G 1975 *J. Soc. Cosmet.* **26** 337
[7] Martinsen Ø G, Grimnes S and Karlsen J 1995 *Skin Pharmacol.* **8** 237
[8] Martinsen Ø G and Grimnes S 2001 *Dermatology* **202** 87
[9] Martinsen Ø G and Grimnes S 2008 *Bioimpedance and Bioelectricity Basics* (Oxford, UK: Academic Press) pp 102–19
[10] Martinsen Ø G, Grimnes S, Nilsen J K, Tronstad C, Jang W, Kim H, Naderi M and Thielmann F 2008 *IEEE Trans. Biomed. Eng.* **55** 728
[11] Johnsen G K, Martinsen Ø G and Grimnes S 2009 *Open Biomed. Eng.J.* **3** 8
[12] Kilpatrick-Liverman L and Polefka T G 2006 *Skin Res. Technol.* **12** 36
[13] Fluhr J, Elsner P, Berardesca E and Maibach H I 2005 *Bioengineering of the Skin. Water and the Stratum Corneum* (Boca Raton: CRC Press) p 17
[14] Norlen L, Emilson A and Forslind B 1997 *Arch. Dermatol. Res.* **289** 506
[15] Baden H P 1970 *J. Invest. Dermatol.* **55** 115

Journal of Physics: Conference Series **224** (2010) 012095 doi:10.1088/1742-6596/224/1/012095

Intracardiac Impedance as a Method for Ventricular Volume Monitoring - Investigation by a Finite-Element Model and Clinical Data

M Lippert[1], S Berdyshev[2], G Czygan[1], M Bocchiardo[3], B Hensel[2]

[1] Biotronik SE & Co. KG, Erlangen, Germany
[2] Center for Medical Physics and Engineering, University Erlangen-Nürnberg, Germany
[3] Cardiology Division, Ospedale Cardinal Massaia, Asti, Italy.

E-mail: michael.lippert@biotronik.com

Abstract. A method for monitoring left ventricular (LV) volume changes of the human heart by intracardiac impedance measurement was developed. In order to model this method, we simulated the ventricular contraction using a finite-element model (FEM). The myocardium comprised three layers with anatomical fiber orientation. During excitation propagation contraction forces were applied, taking into account the myocardial elastic properties and the blood pressure time course. For a set of 21 contraction stages we calculated the intracardiac impedance Z between the right ventricular (RV) and LV leads for a set of common LV lead positions. The FEM results were compared to clinical data. Impedance and stroke volume were determined during overdrive pacing and end diastolic volume (EDV) at rest in 14 patients. All clinical EDV values were in the range of 147-394ml. Both the clinical data and the FEM in this volume range showed a linear correlation between admittance Y=1/Z and ventricular volume. For a quantitative comparison end diastolic impedance EDZ and the slope dY/dV were calculated. The model results across all LV lead positions were EDZ=0.16-1.2Ω, and dY/dV=3.3-21mS/ml, corresponding to clinical values of EDZ=0.14-1.46Ω and dY/dV=1-64mS/ml. In conclusion, the FEM resembled the clinical measurement data and serves as theoretical basis for ventricular volume monitoring via intracardiac impedance.

1. Introduction

Monitoring of the hemodynamic state of the circulatory system is essential for assessment of cardiovascular diseases. In particular, for heart failure (HF) patients hemodynamic monitoring is of great relevance as the ventricular pumping efficacy is impaired in these patients. A biventricular pacemaker or defibrillator frequently is implanted in HF patients with ventricular asynchrony for cardiac resynchronization therapy (CRT). CRT devices comprise pacing leads implanted in the RV and LV for synchronous stimulation of the ventricles. A method to monitor LV volume changes based on intracardiac impedance measurements between the implanted RV and LV leads was proposed [1]. The LV intracardiac impedance Z or admittance Y, respectively, reflects LV volume and can be used for hemodynamic longterm monitoring with implanted CRT devices.

This paper presents a finite-element model of the human ventricles based on geometry obtained from MRI data. The model served to investigate the impedance method and to evaluate the sensitivity on several influence parameters. LV volume and the related intracardiac impedance were computed for the different contraction states during the heart cycle and compared to clinical data from HF patients. A hybrid model approach was chosen to simulate the heart contraction based on a realistic ventricular geometry and myocardial fiber structure.

2. Methods

2.1. The hybrid model

A tetrahedral mesh obtained from a static 3D-model of the human right and left ventricles in the diastolic state was the basis for modelling the mechanical contraction of the human heart, see figure 1a. The geometry was based on cardiac MRI data. The passive mechanical parameters of the myocardium Young's Modulus E = 600 kPa and a Poisson Ratio $\mu = 0.485$ were used. The myocardial surface was fixed at the rigid valve plane.

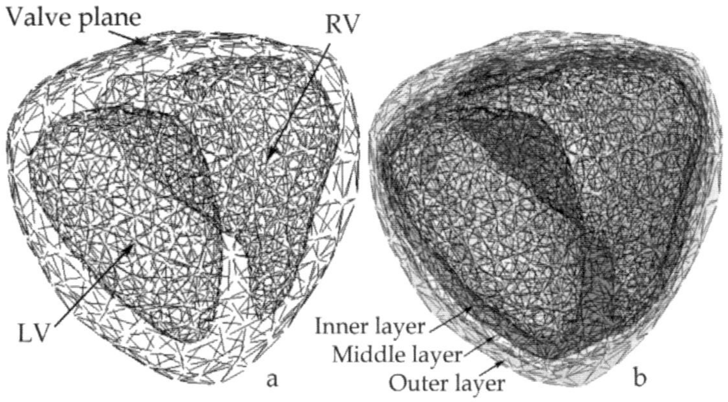

Figure 1. Model of the ventricular myocardium (posterior-inferior view). **a.)** Outer and inner surfaces. **b.)** The myocardium divided into three layers to model the anisotropic muscle fiber structure.

The bulk myocardium was divided into three layers [2], see figure 1b, to simulate the anisotropic spatial arrangement of muscle fibers in the ventricles: an inner layer of longitudinal, a middle layer of circular and an outer layer of diagonal fibers [3,4]. The internal layer boundaries were rigid so that no slip between neighboring layers occured. The fiber structure was contained in a second geometrical model consisting of dense sets of line segments that represent the corresponding local fiber direction, separately for each of the three myocardial layers, see figure 2.

In order to simulate the contraction of the ventricles forces resulting from the isotropic blood pressure in the ventricles were applied on the inner layer boundary, while the outer boundary was free to deform without additional constraints, except for the area of the valve plane, where no displacements could occur. The time courses of the blood pressures in the ventricles were obtained from [5]. The corresponding pressure ranges for the LV were 1.3-18.7 kPa (10-140 mmHg) and for the RV 0.7-3.7 kPa (5-28 mmHg), respectively. The directions of the forces that were applied to the tetrahedral nodes were calculated from the fiber line segments. After deformation of the tetrahedral mesh the fiber structure was morphed in order to adjust the line segments to the new positions of the tetrahedral nodes. This sequence was repeated for all contraction stages. When calculating the direction of the nodal contraction forces only line segments were considered that belonged to the same layer as the node itself. For nodes on the rigid boundaries of the middle layer both adjacent layers contributed to the calculation. The direction was obtained by averaging the individual directions of all fiber segments within a given distance around the node. Having defined the nodal forces in this way the deformation of the tetrahedral mesh was calculated by standard finite element code. In this step the myocardium was considered to be mechanically isotropic, the applied force field, however, was highly anisotropic.

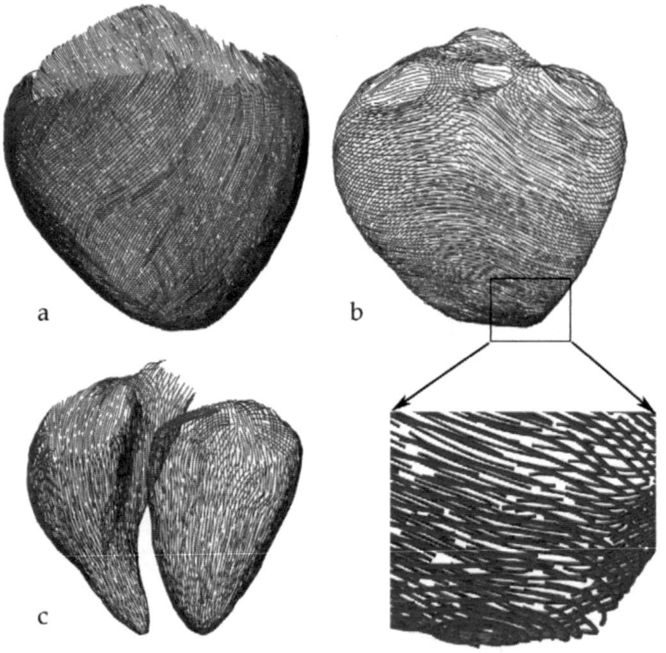

Figure 2. Simulated fiber structures in each myocardium layer. **a.)** Outer longitudinal layer. **b.)** Circular fibers of the middle layer. **c.)** Inner longitudinal layer.

In order to mimic the propagation of the electrical excitation over the heart, the excitation, and thus nodal force generation, started at the nodes nearest to the source of excitation in the RV apex. The excitation propagation was modeled by delaying the excitation according to the geometrical distance to the primary excitation source in the RV apex. The internal ventricular pressure was isotropically applied to the innermost fiber layer during the contraction. Eighteen discrete stages of the active contraction phase of the ventricles and three dilated stages representing heart failure conditions were calculated with a resulting ejection fraction baseline of 55%. Only the active contraction phase of the heart cycle was simulated. Passive filling and isovolumetric relaxation were not included as the main interest was the impedance-volume relation.

For impedance simulation the mechanical hybrid model was embedded in a conducting sphere that simulated the tissues of the thorax. For each step of the mechanical contraction impedance was calculated for the selected current injection configuration and various LV lead positions. As a result dynamic impedance curves were obtained. The current was injected between tip and coil of an intracardiac RV defibrillation lead and the voltage was measured between tip and ring of an epicardial LV lead (config1), see also [1]. In a second configuration current was injected between RV-coil and LV-ring and voltage was determined between RV-tip and LV-tip (config 2). Impedance was simulated for nine LV lead positions: all combinations of anterior/lateral/posterior and basal/medial/apical.

2.2. *Clinical Study*

An acute clinical study was performed to investigate the intracardiac impedance approach in heart failure patients. 14 patients with non-ischemic cardiomyopathy were included. Impedance and stroke volume were measured during graded overdrive pacing. The measurements were repeated with different LV lead positions. EDV was determined via echocardiography. For more details and patient characteristics see [6].

3. Results

The linear correlation coefficient between admittance Y (config 1) and ventricular volume of the model was r = 0.9±0.12, computed for nine different simulated LV lead positions. This result was in excellent agreement with the correlation results found in an earlier experimental animal study r = 0.89±0.15 [1]. The simulation of config2 in mid-lateral LV lead position resulted in a lower correlation coefficient of r = 0.74 compared to r = 0.99 in config1, see figure 3.
The FEM results were compared to a clinical study with 14 heart failure patients [6]. All clinical EDV values were in the range 147-394 ml (mean: 260±70 ml). The model results across all LV electrode positions were EDZ = 0.16-1.2 Ω, and dY/dV = 3.3-21 mS/ml corresponding to clinical values of EDZ = 0.14-1.46 Ω (mean 0.57±0.32 Ω) and dY/dV = 1-64 mS/ml (mean 14±18 mS/ml). Figure 3 depicts the simulation results of the clinically most important lateral LV lead positions in the volume range up to 300 ml.

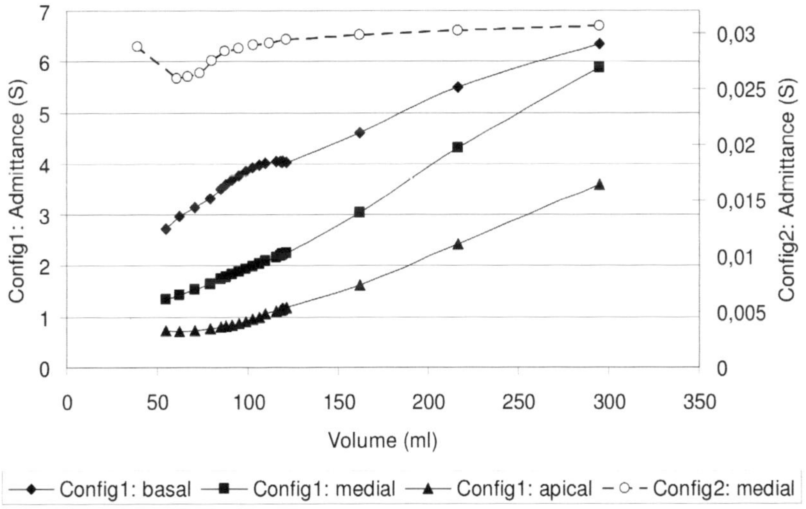

Figure 3. Simulated admittance for config1 and config2 in lateral LV-lead positions.

4. Summary

A hybrid model was constructed to mimic the macroscopic mechanical contraction of the human ventricles to simulate intracardiac impedance signals. In the absolute impedance level and the sensitivity to cardiac volume changes a reasonable agreement between simulation results and clinical impedance measurements was achieved. Thus, the model is suitable for simulating the influence of various physiological and pathophysiological conditions on the contraction and the measured intracardiac impedance.

5. References

[1] Zima E, Lippert M, Czygan G and Merkely B 2006 *Europace* **8** 537-44
[2] Berdyshev S and Hensel B 2005 *Biomed. Technik* **50** 550-1
[3] Bertolini R 1988 *Systematische Anatomie des Menschen* (Berlin: VEB Verlag Volk und Gesundheit) chapter 3 pp 164-70
[4] Kahle W, Leonhardt H and Platzer W 1986 *Color atlas and textbook of human anatomy. Vol. 2: Internal organs* (New York: Thieme) pp 4-14
[5] *Geigy Scientific Tables: Heart and circulation. Vol. 5* 1990 ed Lentner C (West Caldwell: Ciba-Geigy Corporation) p 18
[6] Bocchiardo M, Schimpf T, Meyer zu Vilsendorf D, Militello C, Lippert M, Czygan G, Gaita F, Schauerte P and Stellbrink C 2008 *Europace* **10** i159

Use Of Paravascular Admittance Waveforms To Monitor Relative Change in Arterial Blood Pressure

Todd M. Zielinski, Doug Hettrick, Yong Cho

Medtronic, Inc., Cardiac Rhythm Disease Management, 8200 Coral Sea Street, Mail Stop MVN41, Moundsview, Minnesota 55112, United States.

E-mail for Correspondance: todd.m.zielinski@medtronic.com

Abstract. Non-invasive methods to monitor ambulatory blood pressure often have limitations that can affect measurement accuracy and patient adherence [1]. Minimally invasive measurement of a relative blood pressure surrogate with an implantable device may provide a useful chronic diagnostic and monitoring tool. We assessed a technique that uses electrocardiogram and paravascular admittance waveform morphology analysis to one, measure a time duration (vascular tone index, VTI in milliseconds) change from the electrocardiogram R-wave to admittance waveform peak and two, measure the admittance waveform minimum, maximum and magnitude as indicators of change in arterial compliance/distensibility or pulse pressure secondary to change in afterload. **Methods:** Five anesthetized domestic pigs (32 ± 4.2 kg) were used to study the effects of phenylephrine (1-5 ug/kg/min) on femoral artery pressure and admittance waveform morphology measured with a quadrapolar electrode array catheter placed next to the femoral artery to assess the relative change in arterial compliance due to change in peripheral vascular tone. **Results:** Statistical difference was observed ($p < 0.05$) comparing baseline VTI to phenylephrine VTI (246 ± .05 ms to 320 ± .07 ms) and baseline admittance waveform maximum to phenylephrine admittance waveform maximum (0.0148 ± .002 siemens to 0.0151 ± .002 siemens). **Conclusion:** Chronic minimally invasive admittance measurement techniques that monitor relative change in blood pressure may be suitable for implantable devices to detect progression of cardiovascular disease such as hypertension.

1. Introduction

Noninvasive ambulatory blood pressure monitoring is rapidly expanding as both an instrument in clinical research and a diagnostic tool in clinical practice. However, complications with arm cuff size and arm position relative to the heart often limit the accuracy of ambulatory monitors [1]. Moreover, wearing a blood pressure monitoring device during sleep with cuff inflations programmed to occur at 10-minute or 30-minute intervals causes a reduction of slow-wave (non-REM) sleep and longer nocturnal awakenings [2-4]. External bioimpedance measurements have proved useful for noninvasive monitoring of cardiac hemodynamic parameters, but commercial systems are too cumbersome for ambulatory monitoring [5-7]. Medical device manufacturers have explored the use of implantable pressure transducers to monitor intracardiac pressure but clinical acceptance of these devices have not been promising [8]. Minimally invasive implantable devices may be suitable to monitor pulsatile activity next to an artery that correlates admittance waveform parameters such as maximum, minimum, magnitude and mean values to arterial systolic pressure, diastolic pressure, pulse pressure and mean arterial pressure respectively. The goal of this study was to assess the feasibility of measuring a change in admittance waveform values secondary to change in arterial pressure during an acute vasoconstrictive drug intervention used to elevate arterial blood pressure. We hypothesize that the change in the admittance waveform measured by an acutely implanted paravascular electrode catheter and external commercially available bioimpedance system are measurable and related to change in systemic arterial blood pressure.

2. Methods

Five domestic female pigs (32 ± 4.2 kg) were anesthetized (propofol, 6mg/kg titrated to effect) and mechanically ventilated at a rate to maintain end tidal CO_2 at 30 to 35 mmHg. Surgical bilateral femoral artery cutdowns were then performed to isolate the femoral arteries. The left femoral artery was surgically dissected leaving the arterial facia intact. A 5F quadrapolar catheter (Pices Quad® Model 3487-A, Medtronic, Moundsview, MN, USA) was acutely implanted in the arterial facia parallel to the left femoral artery (Figure 1). The right femoral artery was cannulated and a 8F sheath introduced to allow temporary implant of a 7F pressure transducer micro-tipped catheter (Millar Instruments, Houston, TX, USA) to be positioned in the right femoral artery in approximately the same anatomical location as the quadrapolar electrode catheter positioned on the left femoral artery. The left femoral cut down was sutured closed and a 60 minute stabilization period ensued prior to drug intervention. The quadrapolar electrode catheter was connected to an external BIOPAC®, EBI100C bioimpedance amplifier in line with a BIOPAC®, MP150 data acquisition system (BIOPAC Systems, Goleta, CA USA). Femoral artery pressure and electrocardiogram signals were connected to an analog input channel on the MP150 data acquisition system. All signals, impedance magnitude and phase angle, electrocardiogram and femoral artery pressure were recorded at a sample rate of 200 Hz using the AcqKnowledge® software program for Windows (BIOPAC® Systems) and real time conversion of impedance to admittance was displayed on the data acquisition screen. Femoral artery pressure, impedance and admittance signals were processed with a low pass filter at 15 Hz.

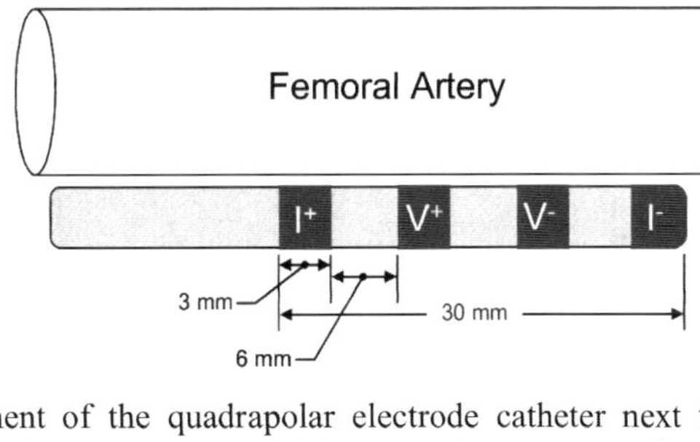

Figure 1. Anatomic placement of the quadrapolar electrode catheter next to the femoral artery. Current injection (12.5 kHz @ 400 uA sinusoid) was applied between the two outer electrodes (I+ and I-). Voltage sense was measured between the two inner electrodes, polarities corresponding with the applicable current injection electrode (V+ and V-).

A constant rate infusion of phenylephrine (1-5 ug/kg/min) was titrated to effect to elevate and maintain systolic arterial blood pressure ≥ 40 mmHg compared to baseline for a minimum duration of 30 minutes.

3. Analysis

Data was analyzed using the AcqKnowledge® software program over a 30 second duration at baseline and a 30 second duration two minutes after arterial blood pressure stabilized at the aforementioned titrated blood pressure. The following parameters were measured and imported into an Microsoft Excel spreadsheet: R to Y_{MAX} : representing the time interval or Vascular Tone Index (VTI) from the ECG R-wave to the maximum admittance waveform value measured from the paravascular electrode array per cardiac cycle, Y_{MAX}: representing the maximum waveform admittance per cardiac cycle, Y_{MIN}: representing the minimum waveform admittance per cardiac cycle. Arterial systolic and diastolic pressures were also measured during the baseline and drug intervention analysis windows.

Statistical analysis comparing baseline to phenylephrine was performed using a student's paired T-test with a p-value < 0.05 considered significant.

4. Results

Mean ± standard deviation is shown for Figures 2 & 3. Each graph element represents data combined for all five animals, 30 seconds of data per animal during baseline and 30 seconds of data per animal during the phenylephrine drug intervention.

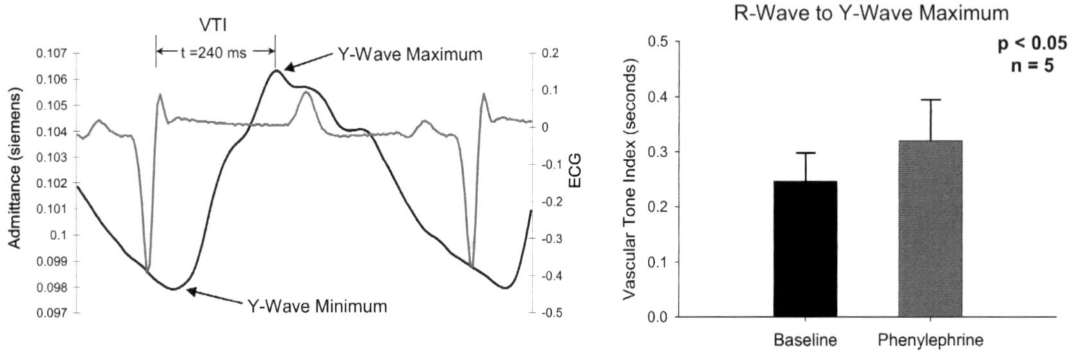

Figure 2. Left: Data depicts and example of an electrocardiogram waveform (red) superimposed on a paravascular admittance waveform (blue). The time duration or vascular tone index (VTI) is shown from the onset of the ECG R-wave to the admittance waveform maximum (Y-Wave Maximum). Right: Data depicts baseline VTI compared to phenylephrine VTI. Statistical difference was observed ($p < 0.05$).

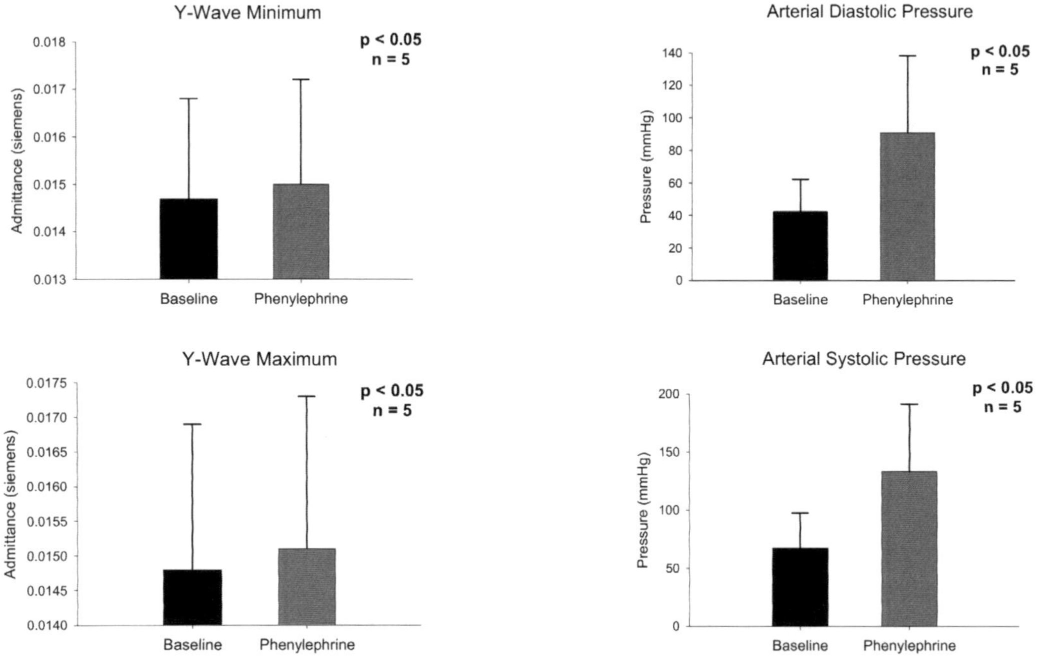

Figure 3. Data depicts admittance waveform minimum (Y-Wave Minimum) and maximum (Y-Wave Maximum) comparing baseline to phenylephrine values. Also shown is the corresponding arterial diastolic and systolic pressure during the same time points. In all comparisions shown in this figure, statistical difference was observed ($p < 0.05$).

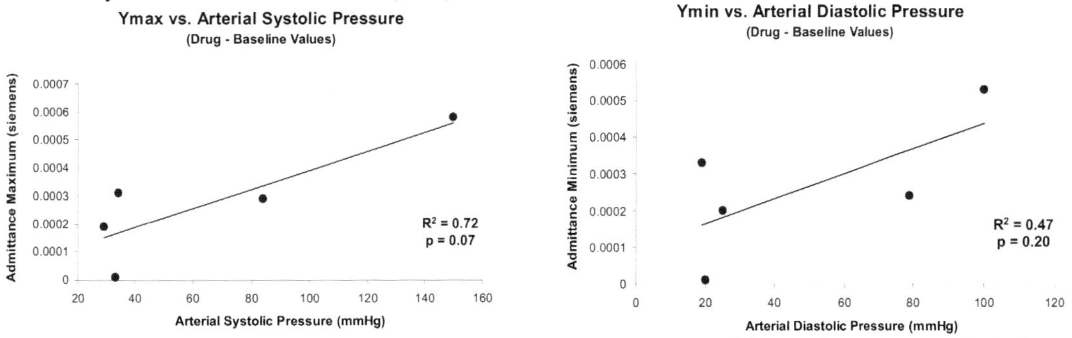

Figure 4. Data depicts admittance waveform maximum (Ymax) vs. arterial systolic pressure and admittance waveform minimum (Ymin) vs. arterial diastolic pressure. Admittance and arterial pressure values plotted represent the difference in values between baseline and phenylephrine drug intervention. A better correlation was observed ($R^2 = 0.72$) comparing Ymax to arterial systolic pressure than Ymin to arterial diastolic pressure ($R^2 = 0.47$) albeit there were no significant relationships between any pair of variables ($p > 0.05$).

5. Discussion

One of the primary observations in this study was the signal fidelity of the admittance waveform acquired from a paravascular electrode array (Figure 1) and the relationship of this waveform to arterial pressure. The observed significant increase ($p < 0.05$) in VTI during the phenylephrine intervention (Figure 2) indicates that the paravascular pulse waveform has a longer transit time from aortic distention to the systemic vasculature thus suggesting that the increased VTI duration is associated with an increased arterial blood pressure. Moreover, there is a good relationship ($p < 0.05$) in the admittance waveform minimum and maximum values and arterial diastolic and systolic values independently comparing baseline to phenylephrine time points respectively (Figure 3). An additional finding in this study was the improved relationship between Ymax and arterial systolic pressure compared to Ymin and arterial diastolic pressure when the difference between baseline and drug intervention values were calculated to determine the magnitude change associated with elevated blood pressure secondary to phenylephrine administration (Figure 4). This observation suggests that Ymax may be more sensitive to change in relative blood pressure due to vessel distention during the arterial pulse with corresponding movement of blood volume, vessel tissue and surrounding tissue in closer proximity to the paravascular electrode array. However, there were no significant relationships ($p > 0.05$) in this latter observation between any pair of variables comparing the change in admittance (min & max) vs. arterial pressure (min & max) when the difference in magnitude values between baseline and phenylephrine time periods are calculated. This may be due to limited sample size (n=5).

6. Conclusion

This acute feasibility study demonstrated that paravascular admittance waveforms may be useful as a pressure surrogate to monitor relative change in blood pressure.

Acknowledgement: The author's would like to acknowledge Mustafa Karamanoglu, PhD for commentary review of this manuscript.

References

1. Jan A. Staessen, Robert Fagard, Lutgarde Thijs, Antoon Amery; A Consensus View on the Technique of Ambulatory Blood Pressure Monitoring. *Hypertension.* 1995;26:912-918.
2. Degaute JP, van de Borne P, Kerkhofs M, Dramaix M, Linkowski P. Does non-invasive ambulatory blood pressure monitoring disturb sleep? *J Hypertens.* 1992;10:879-885.
3. Eoin O'Brien; Neil Atkins; Jan Staessen. State of the Market. A Review of Ambulatory Blood Pressure Monitoring Devices. *Hypertension.* 1995;26:835.
4. Charlotte Richardson, Kaisa Maki-Petaja, Barry McDonnell, Stacey Kickson, Ian Wilkinson, Carmel McEniery. Comparison of estimates of central systolic blood pressure and peripheral augmentation index obtained from the Omron HEM-9000AI and SphygmoCor Systems. *Artery Research* 2009;3:24-31.
5. Patterson RP: Fundamentals of Impedance Cardiography. *Engineering in Medicine and Biology Magazine, IEEE* 1989;8:35-38.
6. Cybulski G, Kozluk E, Michalak E, Niewlamdoski W, Platkowska A. Holter-type Impedance Cardiography Device. A System for Continuous and Non-Invasive Monitoring of Cardiac Hemodynamics. *Kardiol Pol* 2004;61:138-146.
7. Yancy C, Abraham WT. Noninvasive Hemodynamic Monitoring in Heart Failure: Utilization of Impedance Cardiography. *Congest Heart Fail* 2003;9:241-250.
8. Bourge RC, Abraham WT, Adamson PB, Aaron MF, Aranda JM Jr, Magalski A, Zile MR, Smith AL, Smart FW, O'Shaughnessy MA, Jessup ML, Sparks B, Naftel DL, Stevenson LW; COMPASS-HF Study Group. Randomized controlled trial of an implantable continuous hemodynamic monitor in patients with advanced heart failure: the COMPASS-HF study. J Am Coll Cardiol. 2008 Mar 18;51(11):1073-9.

Monitoring of CRT by means of impedance multiple measurements – simulation studies

M Lewandowska[1], J Wtorek, A Bujnowski and L Mierzejewski

Department of Biomedical Engineering, Faculty of Electronics, Telecommunication and Informatics, Gdansk University of Technology, Narutowicza 11/12, 80-233 Gdansk, Poland

E-mail: maglew@biomed.eti.pg.gda.pl

Abstract. Cardiac resynchronization therapy (CRT) is a very promising treatment for patients with congestive heart failure. An Impedance Cardiography (ICG) is used for evaluation of heart mechanical activity. A simulation study with the use of a FEM model was performed. The developed realistic model of the thorax, based on CT data obtained from examination of a 68-year-old man, consisted of 212 000 tetrahedral elements. Different configurations of electrodes allowing extraction of ventricular components from impedance signal were examined. The associated sensitivity distributions were calculated for different phases of heart contraction and blood distribution inside the thorax. These sensitivity distributions were compared to that obtained for a uniform model of the thorax described by average conductivity. The performed simulation studies proved that it is possible to extract ventricular components of impedance signal with a reasonable accuracy.

1. Introduction

The synchronization of the ventricles can be compromised by a condition known as a bundle branch block. The bundle branch block occurs when the conduction of electrical impulse from the right atrium to the ventricles is blocked or slowed down. When this occurs, it may cause the ventricles to become out of synchronization, and when the heart loses its coordination, it cannot pump as well [1]. Cardiac resynchronization therapy (CRT) is a very promising treatment for patients with congestive heart failure. However, patients' classification for CRT as well as evaluation of the therapy effectiveness should be improved.

A magnitude of resynchronization can be achieved by means of measurement of the electromechanical delay. It is defined as the difference between the longest and the shortest activation time in the six basal segments of the left ventricle. Mechanical dyssynchrony can be assessed with the use of imaging modalities, e.g. MRI or ultrasound echo-Doppler [2]. In general, myocardial velocity timing can be evaluated by ultrasound tissue Doppler imaging and myocardial strain timing can be evaluated by MRI tagging [3].

An Impedance Cardiography (ICG) is used for evaluation of heart mechanical activity [4]. However, it reflects conductivity changes which occur in all regions of the thorax. The contribution of particular phenomenon to total impedance change depends on the value of conductivity change, its

[1] To whom any correspondence should be addressed.

Journal of Physics: Conference Series **224** (2010) 012097 doi:10.1088/1742-6596/224/1/012097

volume and associated sensitivity distribution [5, 6]. Thus, a method of extracting only ventricles', or in general selected, components from the total impedance change is needed.

2. Methods

Finite Element Models have been developed to examine a possibility of discriminating between contributions from different parts (ventricles) of the thorax for the assumed electrode configurations.

2.1. 2D model

A simplified 2D model was developed to study basic properties of the considered problem (figure 1). It consisted of four tissues: lung, myocardium, blood and effective one standing for tissue of thoracic wall. Each tissue was described by conductivity, respectively σ_l, σ_m, σ_b, and σ_t. The measured impedance change, due to conductivity changes undergoing in different regions, is described by equations (1-5), where I_ψ – current flowing between current electrodes, hypothetical current I_φ flowing between voltage electrodes (four-electrode method), $\Delta\sigma$ – conductivity change [6, 7].

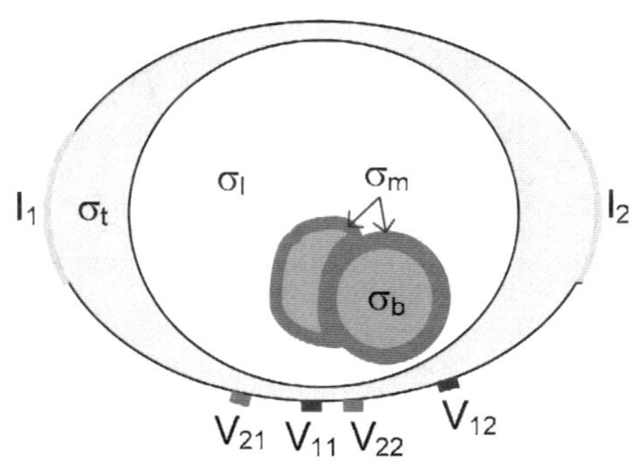

Figure 1. 2D model examined in the study, description in the text.

$$\Delta Z = \int_V -\Delta\sigma_s \frac{\nabla\varphi(\sigma+\Delta\sigma)}{I_\phi} \cdot \frac{\nabla\psi(\sigma)}{I_\psi} dv \quad (1)$$

$$\Delta Z = \sum_i^I K_i \Delta\sigma_i \quad (2)$$

$$K_i = \int_{V_i} -\frac{\nabla\varphi(\sigma+\Delta\sigma)}{I_\phi} \cdot \frac{\nabla\psi(\sigma)}{I_\psi} dv \quad (3)$$

$$\Delta Z_1 = \int_{v1} K_{11}\Delta\sigma dv + \int_{v1} K_{12}\Delta\sigma dv \quad (4)$$

$$\Delta Z_2 = \int_{v1} K_{21}\Delta\sigma dv + \int_{v1} K_{22}\Delta\sigma dv \quad (5)$$

According to the relationship (1) the impedance change for two different electrode's configurations and for object consisting of two regions can be described by the following relationships (4) and (5). Thus, the impedance change measured by a certain electrode probe was a weighted sum of conductivity changes occurring in selected regions. Without loss of generality it could be assumed that the changes of conductivity were sinusoidal. Thus, assuming that both impedance signals are sinusoidal, however, of different amplitude and phase shift, the resulting signal is also sinusoidal i.e. $A_1\sin(\omega t) + A_2\sin(\omega t + \Delta\varphi_{12}) = A\sin(\omega t + \varphi)$. The resulting amplitude and phase can be easily calculated with the use of simple trigonometric relations. Additionally, assuming that relation between amplitudes is a simple ratio, i.e. $A_2 = kA_1$ the resulting amplitude and phase is calculated.

2.2. 3D model

A realistic model of the thorax, based on CT data obtained from examination of 68 years old man was developed [8, 9]. A FEM model included of lungs, heart, thorax wall, liver, skeleton and other anatomical organs was developed to mimic impedance measurements. It consisted of 212 000 tetrahedral elements (figure 2). Three different matrix configurations were examined (figure 3).

The associated sensitivity distributions were calculated for different phases of heart contraction and blood distribution inside the thorax. These sensitivity distributions were compared to that obtained for a uniform model of the thorax described by average conductivity. The value of average conductivity was calculated on the assumption equal impedance for uniform and non-uniform models. It was assumed that in both cases the measured impedance is identical. Similar to equation (4) relations were calculated, however, using a FEM approximation. It allowed estimating sensitivity to conductivity changes for regions occupied by organs included in the model [5, 6, 10]. Three different models were developed: 1. non-uniform with conductivity of heart muscle assumed to be equal to conductivity of blood, 2. non-uniform with conductivity of heart muscle different from that of blood, and 3. uniform.

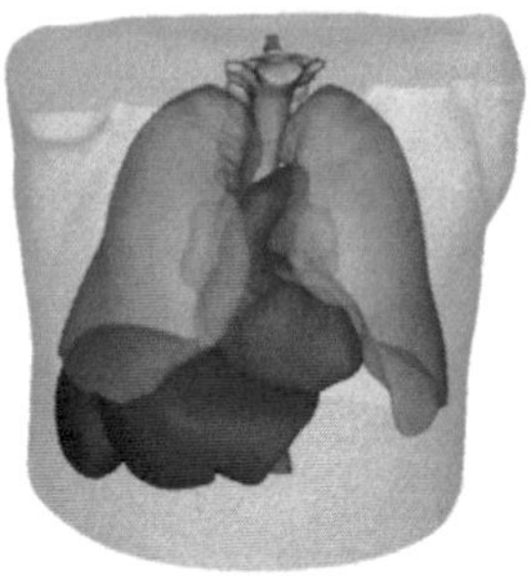

Figure 2. A developed geometric model of the thorax.

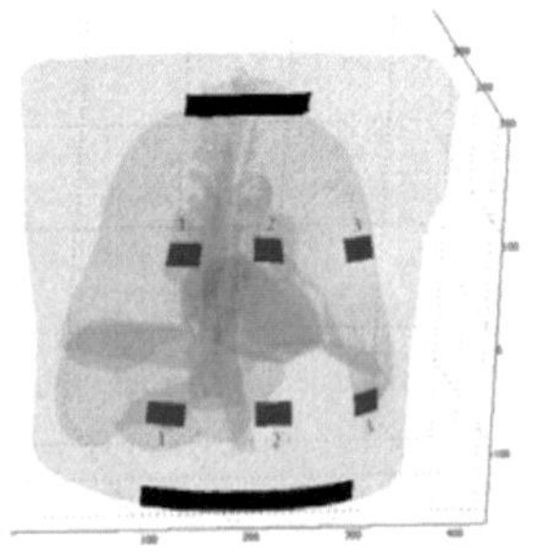

Figure 3. Examined electrode configurations in FEM model.

Conductivity values of different organs are gathered in table 1. Aside conductivity a relative value of electric permittivity used in study is also given [11].

Table 1. Values of conductivity and relative permittivity of tissues considered in the study.

Organ/tissue	σ[S/m]			ε_r
	σ_1	σ_2	$\sigma_{average}$	
Lungs	0.1012	0.1012	0.16	5090.8
Aorta	0.7	0.7	0.16	5211.3
Left Atrium	0.7005	0.7005	0.16	5211.3
Right Atrium	0.7005	0.7005	0.16	5211.3
Left Ventricle	0.7005	0.56	0.16	5211.3
Right Ventricle	0.7005	0.56	0.16	5211.3
Liver	0.0689	0.0689	0.16	12100
Bone	0.0206	0.0206	0.16	281.06
Thoracic wall	0.15	0.15	0.16	5000

3. Results

Estimated values of amplitude and phase of the measured impedance change due to conductivity changes undergoing in regions of ventricles are presented in figure 4. These results have been obtained with the use of a simple 2D model.

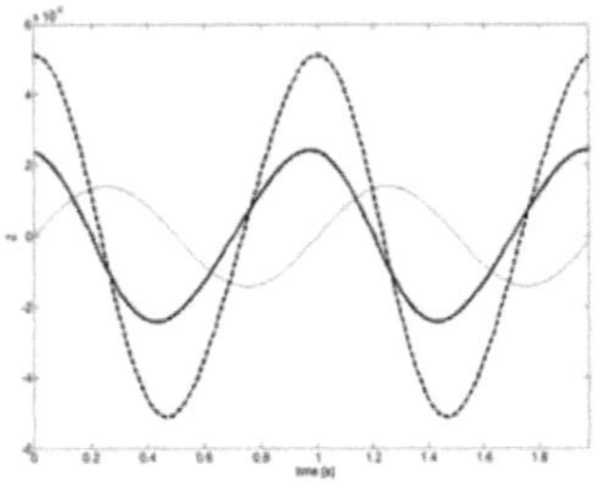

Figure 4. Impedance changes calculated for electrodes V_{11}-V_{12} (black line), V_{21}-V_{22} (dashed line) and conductivity change in region of left ventricle (grey line).

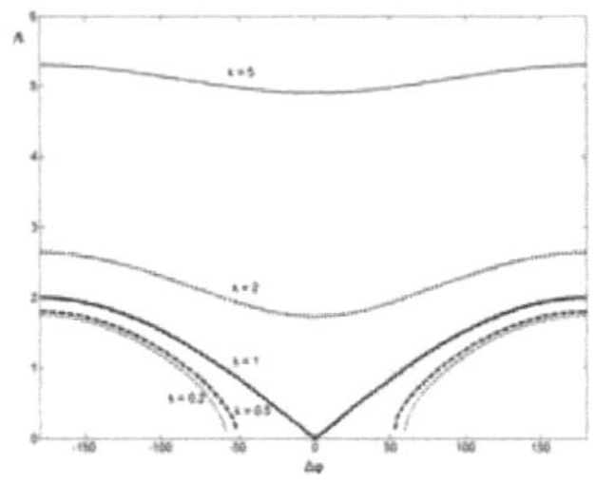

Figure 5. Amplitude, A, of the resulting impedance change due to phase shift between conductivity changes localized in left and right ventricle.

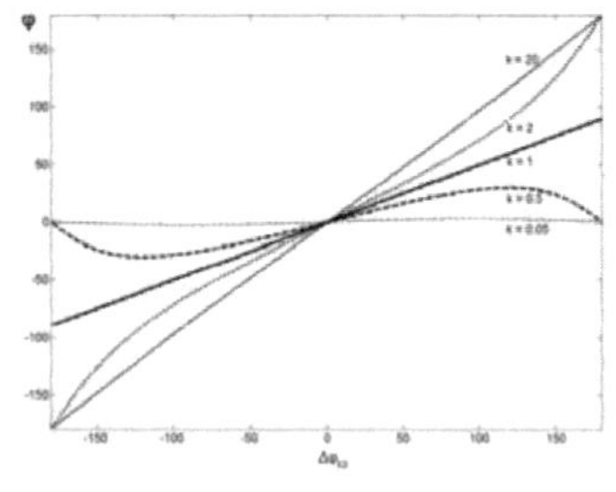

Figure 6. Phase of the resulting impedance change due to phase shift between conductivity changes localized in left and right ventricle.

Conductivity changes in the left and right ventricles had the same frequency, however, they were shifted in phase. Extension of that result, for different phase shift and amplitudes of conductivity changes are presented in figure 5 and figure 6. Sensitivity values of the chosen electrode configuration to conductivity changes located in different organs are gathered in table 2.

Table 2. Sensitivity values for different conductivity distribution, 2nd pair of voltage electrodes.

Organ/tissue	Sensitivity [%]		
	σ_1	σ_2	$\sigma_{average}$
Left Lung	1.601	1.64	7.82
Right Lung	0.91	0.94	6.17
Aorta	3.224	3.229	1.943
Left Atrium	0.059	0.064	0.291
Right Atrium	0.251	0.268	0.917
Left Ventricle	2.655	2.300	2.724
Right Ventricle	5.953	4.892	2.724
Liver	0.765	0.769	9.396
Bone	0.007	0.007	0.662
Thorax	84.566	85.889	68.697

4. Discussion and conclusions

As expected, in case of one dominant component the resulting signal of impedance change almost follows it. Nevertheless, an extraction of the component, which arises from a certain source, is not trivial when amplitudes of all components are comparable. However, the relation between contributions from different organs (regions) strongly depends on conductivity distribution inside the thorax. Multiple simultaneous measurements which use optimally designed electrode configurations allow, in some cases, enhancement of the signal's component from the region of interest.

References

[1] Boriani G, et al. 2006 Cardiac resynchronization therapy in clinical practice: Need for electrical, mechanical, clinical and logistic synchronization *J. Interv. Card. Electrophys.* **17** 215 – 224

[2] Lafitte S, Reant P, Serri K and Roudaut R 2008 Echocardiographic algorithm for cardiac resynchronization *Echocardiography* **25**(9)

[3] Wilinski J, Czarnecka D and Kloch-Badelek M 2008 Resynchronization theraphy in heart inssufficienty treating (in Polish) *Physisian Guide* 23 -34

[4] Braun M U, et al. 2005 Impedance cardiography as a noninvasive technique for atrioventricular interval optimization in cardiac resynchronization therapy *J. Interv. Card. Electrophysiol.* **13** 223-229

[5] Wtorek J 2000 Relations between components of impedance cardiogram analyzed by means of Finite Element Model and sensitivity theorem *Ann. Biomed. Eng.* **28** 1352 – 1361

[6] Wtorek J 2003 *Electroimpedance techniques in medicine* (in Polish) (Gdansk: GUT Publishing Office) Monography Series **43**

[7] Geselowitz D B 1971 An application of electrocardiographic lead theory to impedance pletysmography *IEEE Trans. Biomed. Eng.* **18** 38-41

[8] Mimics® 12.3, 1992-2009 Materialise n.v.

[9] Visible Human Server, Ecole Polytechnique Federale de Lausanne, http://visiblehuman.epfl.ch/index.php

[10] Yang F and Patterson R P 2008 A simulation study on the effect of thoracic conductivity inhomogeneities on sensitivity distributions *Ann. Biomed. Eng.* **36** 762-768

[11] Calculation of the dielectric properties of body tissues, IFAC-CNR, Florence (Italy), 1997-2007 http://niremf.ifac.cnr.it/tissprop/htmlclie/htmlclie.htm

Longitudinal and Transversal Bioimpedance Measurements in Addition to Diagnosis of Heart Failure

N Ribas[1], L Nescolarde[2], M Domingo[1], P Gastelurrutia[1], A Bayés-Genis[1], J Rosell-Ferrer[2]
[1] Service of Cardiology, Hospital de la Santa Creu I Sant Pau, c/ Sant Antoni Maria Claret, 167, 08025 Barcelona, Spain
[2] Department of Electronic, Technical University of Catalonia, Barcelona, Spain

E-mail: lexa@eel.upc.edu

Abstract. Heart Failure (HF) is a clinical syndrome characterised by signs of systemic and pulmonary fluid retention, shortness of breath and/or fatigue. There is a lack of reliable indicators of disease state. Benefits and applicability of non-invasive bioimpedance measurement in the hydration state of soft tissues have been validated, fundamentally, in dialysis patients. Four impedance configurations (2 longitudinal and 2 transversal) were analyzed in 48 HF patients (M=28, F=20) classified according to a clinical disease severity score (CDSS) derived from the Framingham criteria: CDSS≤2 (G1: M= 23, F= 14) and CDSS>2 (G2: M= 5, F= 6). The aim of this study is to analyze longitudinal and transversal bioimpedance measurement at 50 kHz, in addition to clinical diagnosis parameters of heart failure, including: clinical disease severity score (CDSS) and a biomarker concentrations (NT-proBNP). The Kolmogorov-Smirnov test was used for the normality test of all variables. The CDSS, NTproBNP and impedance parameters between groups (G1 and G2) were compared by mean of Mann Withney U-test. The statistical significance was considered with $P < 0.05$. Whole-body impedance measured was analyzed using RXc graph.

1. Introduction

Heart failure (HF) is a clinical syndrome defined by the inability of the heart to provide adequate blood flow to other organs. Consequently, it generates systemic and pulmonary fluid retention. This congestion is the responsible of the symptoms (breathlessness, fatigue, anorexia) and signs (peripheral oedema, hepatomegaly, rales over lungs) that are used in the diagnosis of the exacerbation of HF. However, their lack of sensitivity and the importance to detect congestion in early stages supports the need of more aggressive management of these patients. For this reason, biomarkers concentration, particularly B-type natriuretic peptide (BNP) and its amino-terminal cleavage equivalent (NT-proBNP) [1] helps to the clinical features in the diagnosis of the syndrome. However, they also have a significant heterogeneity, and better markers are needed. The feasibility of segmental bioimpedance measurement [5] in the determination of segmental volumes and regional fluid shifts to overcome the limitations of whole-body impedance methods is an alternative.
The bioelectrical impedance vector analysis BIVA [2] is a non invasive procedure easily applicable in the clinic due to its sensitivity to detect changes in the hydration of soft tissues. This method is unaffected by regression adjustments to estimate body composition, body geometry or hydration state. The aim of this study is to analyze longitudinal and transversal bioimpedance measurement at 50 kHz, in addition to clinical diagnosis parameters of heart failure.

2. Methods

2.1 Patients

Patient clinical status, NT-proBNP levels and bioimpedance measurements were obtained from 48 patients (M= 28; F= 20). The study was approved by the Local Ethics Committee and informed consent was obtained from each patient. Exclusion criteria included the presence of: 1) pacemaker or implantable automatic defibrillator; 2) large metallic prosthesis such as hip or knee prosthesis; and 3) limb amputations. HF clinical disease severity score (CDSS) based on Framingham criteria was used to diagnose destabilized HF. Two groups of patients were identified based on clinical criteria: group G1 of stable HF patients with CDSS ≤ 2, and group G2 of non-stable congestive HF patients with CDSS > 2. NT-proBNP levels were measured by electrochemiluminescence immunoassay using Elecsys 1010 analyzer (Roche Diagnostics GmbH, Mannheim, Germany).

2.2 Bioimpedance Measurements

Impedance measurements were taken with model SFB7 (ImpediMed, Queensland, Australia) at 50 kHz with injected current of 800 μA. The specific measurement error of the system is lower than 1 Ω and 1 ° at 50 kHz using electrical models. Tetrapolar bioimpedance measurements were made in four electrode configurations: whole-body (RS), longitudinal leg (LLEG), transversal bioimpedance measurement in cross-section of the right hemithorax (TTH), and transversal bioimpedance measurement in cross-section of the hepatic abdominal region (TAB). Disposable pre-gelled Ag/AgCl (3M Red Dot, Canada) electrodes were used.

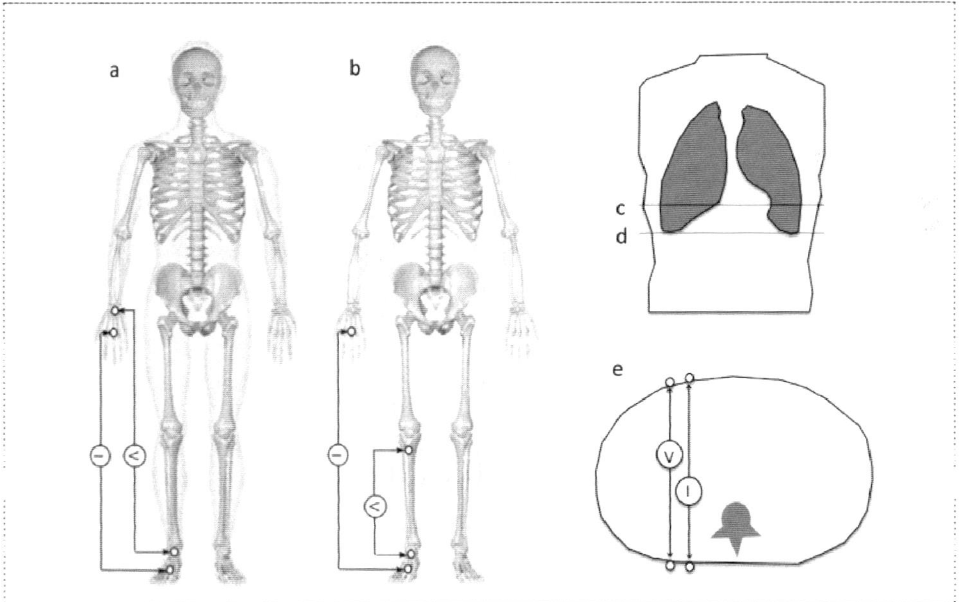

Figure 1. Electrode locations for longitudinal measurements: whole-body (a), longitudinal leg (b). Transversal measurements location: cross-section of the right hemithorax (c), cross-section of the hepatic abdominal region (d). Electrode locations for transversal measurements (e).

2.3 Statistical Methods

Forty-eight HF patients were identified in two groups of patients based on clinical criteria: group G1 of stable HF patients with CDSS ≤ 2, and group G2 of non-stable congestive HF patients with CDSS > 2. All data are expressed as mean values (± SD). Individual bioelectrical impedance vector, in whole-body (RS) was analysed in the RXc-graph [2] with tolerance ellipses (50%, 75% and 95%) of the reference population [3]. Whole-body (RS) impedance measured was normalized by height of the patients (H), longitudinal leg (LLEG) by leg length (LL); transversal bioimpedance measurement in cross-section of the right hemithorax (TTH) by perimeter of the thorax (PT); and finally transversal bioimpedance measurement in cross-section of the hepatic abdominal region (TAB) by abdominal circumference (AC). Kolmogorov-Smirnov test was used for the normality test of clinical, laboratory and bioimpedance vector components. For differences between impedance vector components with clinical and laboratory parameters the Mann-Whitney test was used. Minitab software version 15.1.0 (Minitab, Inc.) was used for statistical analysis. Statistical significance was considered with $p < 0.05$.

3. Results

Table 1 and table 2 show the mean ± standard deviation (SD) of CDSS, NTproBNP, impedance parameters (R and Xc), and Mann Whitney-U test, for male and female sample.

Table 1. Clinical and impedance parameters in male sample

	G1 (CDSS ≤ 2) N=23	G2 (CDSS > 2) N=5	P
CDSS	1.0 ± 0.5	3.3 ± 1.1	0.000**
NTpro BNP	2777 ± 3441	2936 ± 2380	0.589
RS, R/H	304.5 ± 63.0	247.9 ± 50.9	0.081
RS, Xc/H	25.5 ± 5.7	19.6 ± 6.9	0.047*
LLEG, R/LL	398.0 ± 98.0	237.1 ± 48.5	0.004*
LLEG, Xc/LL	30.5 ± 13.8	11.8 ± 3.8	0.005*
TTH, R/PT	42.3 ± 10.4	31.9 ± 14.2	0.093
TTH, Xc/PT	4.7 ± 1.8	4.1 ± 1.6	0.435
TAB, R/AC	44.8 ± 13.2	32.9 ± 13.1	0.150
TAB, Xc/AC	4.6 ± 1.7	3.4 ± 0.5	0.026*

Table 2. Clinical and impedance parameters in female sample

	G1 (CDSS ≤ 2) N=14	G2 (CDSS > 2) N=6	P
CDSS	1.4 ± 0.4	3.1 ± 1.0	0.000**
NTpro BNP	3317 ± 3416	665 ± 167	0.107
RS, R/H	322.2 ± 82.1	320.5 ± 68.8	1.000
RS, Xc/H	25.2 ± 7.0	28.4 ± 7.8	0.713
L-LEG, R/LL	360.9 ± 115.5	396.2 ± 144.5	0.386
L-LEG, Xc/LL	23.0 ± 9.5	30.0 ± 17.6	0.302
T-TH, R/PT	46.6± 12.6	42.2 ± 8.9	0.386
T-TH, Xc/PT	4.8 ± 0.9	5.3 ± 0.9	0.231
T-AB, R/AC	53.0 ± 19.2	50.8 ± 12.9	0.836
T-AB, Xc/AC	4.9 ± 1.0	4.8 ± 2.3	0.483

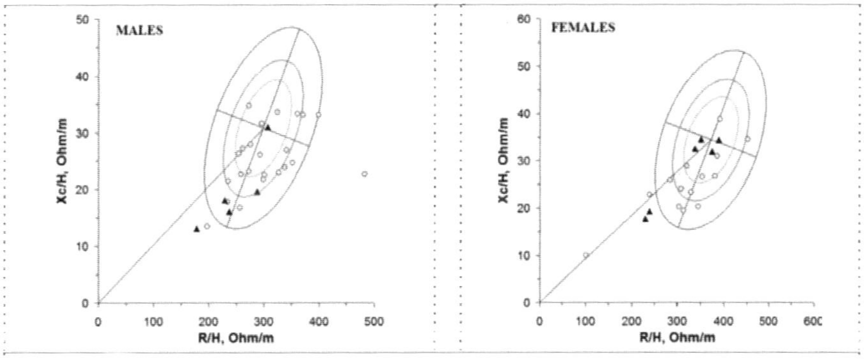

Figure 2. Individual vectors for 48 patients with CDSS≤2 (G1: M= 23, F= 14) and with CDSS>2 (G2: M= 5, F= 6) in the tolerance ellipses for the reference population [3].

4. Discussion and Conclusions

The Mann-Whitney U test results in the male sample (table 1) indicate a statistically significant difference in R and Xc parameters in the L-LEG segment; and Xc parameters in the T-AB and RS segment. In BIVA, R/H provides information of the hydration state and Xc/H the structure of soft tissues. In some previous studies, the inferior pole of the 75% tolerance ellipse was considered the threshold for oedema [4]. Figure 2 shows individual impedance vector in RS configuration. Some patients classified as G1 with CDSS ≤ 2 were located by BIVA in the hyper-hydration region; and some patients classified as G2 with CDSS > 2 were located in the normal-hydration region in males and more prominent in female's sample. This situation and the small sample size might be the reason why the table 2 not shows statistically significant results in any segment. Patterson *et al* 1988 analyzed the feasibility of segmental bioimpedance measurement in the determination of segmental volumes and regional fluid shifts to overcome the limitations of whole-body (RS) impedance methods. This preliminary study suggests that in HF longitudinal bioimpedance measurement in leg segment (L-LEG) provides additional information to clinical examination and could even reflect sub-clinical hyper-hydration.

Acknowledgments

This work has been supported by grants from the Spanish Ministry of Science and Technology and the FEDER project SAF2005-02270 and from the Spanish Ministry of Health, REDINSCOR.

References

[1] Bayes-Genis A, Santalo-Bel M, Zapico-Muńiz E, Lopez L, Cotes C, Beelido J, Leta R, Casan P, Ordońez-Llanos J 2004 N-terminal probrainnatriuretic peptide (NT-proBNP) in the emergency diagnosis and in-hospital monitoring of patients with dyspnoea and ventricular dysfunction. *Eur. J. Heart. Fail.* **6**:301-308.

[2] Piccoli A, Rossi B, Pillon L, Bucciante G 1994 A new method for monitoring body fluid variation by bioimpedance analysis: The RXc graph *Kidney Int.* **46**:534-539.

[3] Piccoli A, Pillon L, Dumler F 2002 Impedance vector distribution by sex, race, body mass index, and age in the United States: standard reference intervals as bivariate Z scores. *Nutrition* **18**:156-170.

[4] Piccoli A, Rossi B, Pillon L 1996 Body fluid overload and bioelectrical impedance analysis in renal patients. *Miner. Electrolyte. Metab.* **22**:76-78.

[5] Patterson R, Ranganathan C, Engel R and Berkseth R 1988 Measurement of body fluid volume change using multi-site impedance measurements. *Med. Biol. Eng. Comput.* **26**: 33-37.

An approach to the diagnosis of metabolic syndrome by the multi-electrode impedance method

N Furuya[1], **K Sakamoto**[2], **H Kanai**[3]
[1] Course of Medical and Welfare Engineering, Tokyo Metropolitan College of Industrial Technology, Tokyo, Japan
[2] Department of Clinical Engineering, Kitasato University Kanagawa, Japan
[3] Institute of Applied Superconductivity, Tokyo Denki University, Chiba, Japan

E-mail: nfuruya@acp.metro-cit.ac.jp

Abstract. It is well known that metabolic syndrome can induce myocardial infarction and cerebral infarction. So, it is very important to measure the visceral fat volume. In the electric impedance method, information in the vicinity of the electrodes is strongly reflected. Therefore, we propose a new multi-electrode arrangement method based on the impedance sensitivity theorem to measure the visceral fat volume. This electrode arrangement is designed to enable high impedance sensitivity in the visceral and subcutaneous fat regions. Currents are simultaneously applied to several current electrodes on the body surface, and one voltage electrode pair is arranged on the body surface near the organ of interest to obtain the visceral fat information and another voltage electrode pair is arranged on the body surface near the current electrodes to obtain the subcutaneous fat information. A simulation study indicates that by weighting the impedance sensitivity distribution, as in our method, a high-sensitivity region in the visceral and the subcutaneous fat regions can be formed. In addition, it was confirmed that the visceral fat volume can be estimated by the measured impedance data.

1. Introduction
To estimate the visceral fat volume, it is necessary to detect conductivity information deep inside the body and it also necessary to remove the influence of subcutaneous fat on the measured impedance. In this research, we propose a multi-electrode impedance method to measure the visceral and subcutaneous fat regions.

2. Simulation studies

2.1 Theoretical basis
The sensitivity theorem in impedance plethysmography was derived by Geselowittz and Lehr [1]. Several research studies have since applied the impedance sensitivity theory.[2],[3] The theorem is stated as follows (see Fig. 1).When there is a difference of conductivity Δg in a specific region, the mutual impedance difference ΔZ for prescribed pairs of current and voltage electrodes can be given as the following volume integration over the region of changed conductivity:

[1] 8-17-1 Minamisenju Arakawa-ku, Tokyo, 116-0003, Japan

$$\Delta Z = \int -\Delta g \mathbf{L}_c(g) \cdot \mathbf{L}_v(g + \Delta g) dv \cong \int -\Delta g \mathbf{L}_c(g) \cdot \mathbf{L}_v(g) dv \quad (1)$$

where $\mathbf{L}_c$ and $\mathbf{L}_v$ denote the lead fields of the current and voltage electrode pairs, respectively: namely, $\mathbf{L}_c$ is the electric field when the unit current is applied to the current electrode pair and $\mathbf{L}_v$ is the electric field when the unit current is applied to the voltage electrode pair. Thus, impedance sensitivity $S(x, y, z)$ can be defined as

$$S(x, y, z) \equiv \mathbf{L}_c \cdot \mathbf{L}_v \quad (2)$$

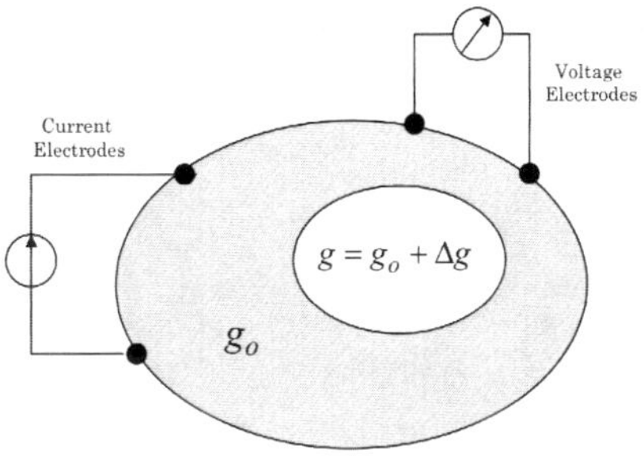

The sensitivity function $S(x, y, z)$ reflects the sensitivity of the measured impedance difference for a given geometry of electrode arrangement to the local conductivity difference at the point of interest (x, y, z). Using the finite element method (FEM) for the lead fields, $S(x, y, z)$ can be numerically calculated for any position in the assumed model.

2.2 Estimation of impedance sensitivity distribution

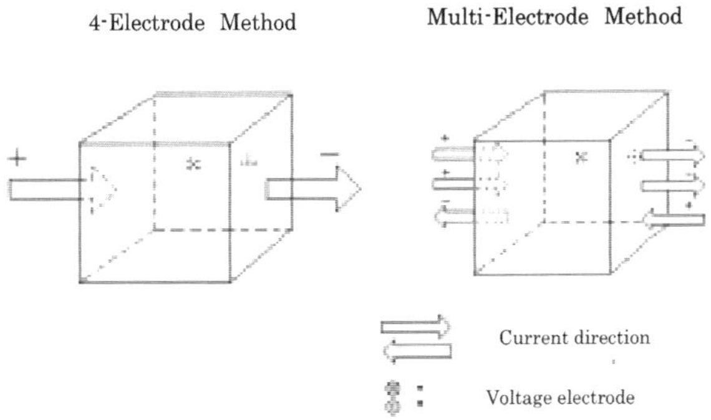

Fig. 2 Electrode arrangement and current injection method

Fig. 2 and Fig. 3 show the electrode arrangements of the 4-electrode method and our multi-electrode method and the current injection method. Simulation studies were carried out with a simple cube model consisting of 4096 finite elements. In our method, each current value is the same and currents are simultaneously applied to plural current electrodes to obtain the appropriate $\mathbf{L}_c$ in (2). A Ch-A voltage electrode pair and a Ch-B voltage electrode pair are located as shown in Fig. 3. Fig. 3 (a) and (b) show a horizontal slice and vertical slice of the simulation model, respectively. The colored areas show the visceral fat (inside cube of

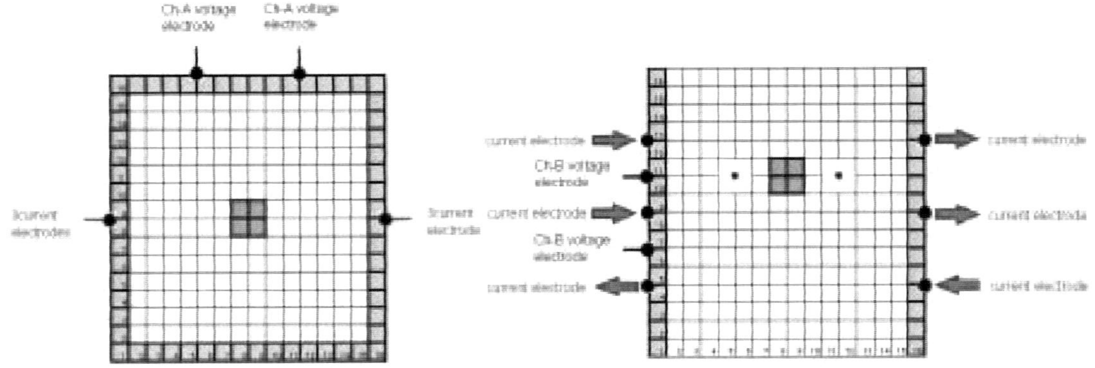

(a) Horizontal slice of the model (b) Vertical slice of the model

Fig. 3 Simulation model

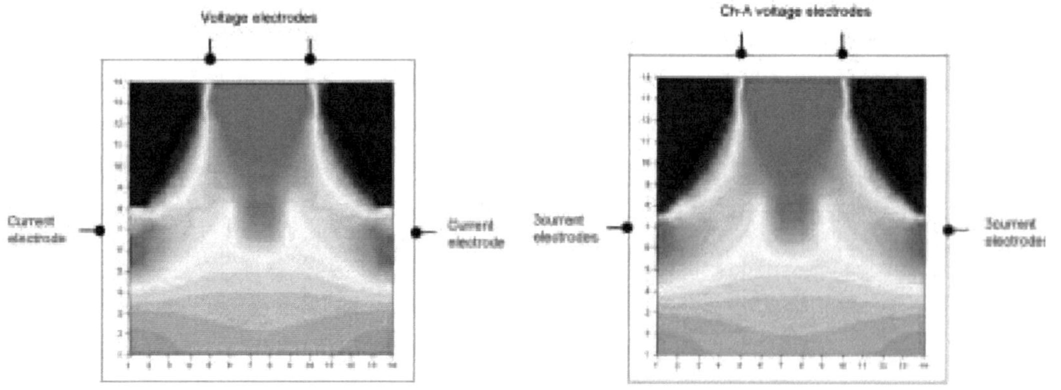

(a) (b)

Fig.4 Impedance sensitivity map
(a) Horizontal slice map with voltage electrodes in the 4-electrode method
(b) Horizontal slice map with the voltage electrodes of Ch-A
(c) Vertical slice map with the voltage electrodes of Ch-B

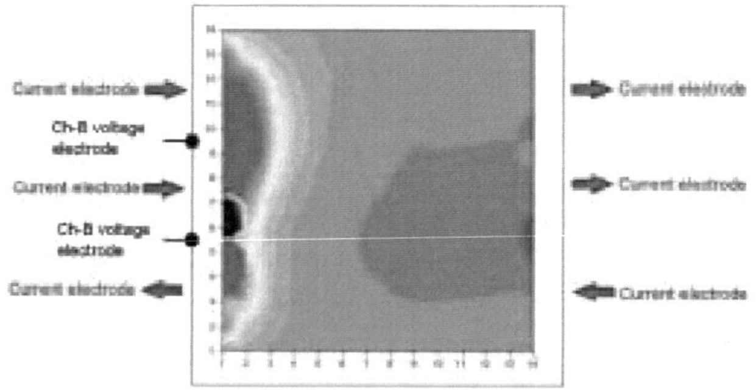

(c)

elements) and subcutaneous fat (border elements) In the calculation, the ratio of fat resistivity and general tissue resistivity is set to 4:1. Fig. 4 shows the calculated impedance sensitivity map in the simulation model.

In Fig. 4, red and blue indicate the plus and minus impedance sensitivity regions, respectively. The color depth indicates the magnitude of the relative value of the impedance sensitivity. As compared with the 4-electrode method, the impedance sensitivity in the deep region is clearly high in Ch-A. Moreover, the impedance sensitivity in Ch-B is high in the subcutaneous fat region.

2.3 Relationship between the visceral fat volume and measurement impedance

Fig. 5 (a), (b) show the results of the 4-electrode method and the results of our method (Ch-A), respectively, as parameters of the subcutaneous fat volume. Here, the form of visceral fat was assumed to be a cube. The percent value indicates the ratio of visceral fat to the whole volume. The gradient of the curves suggests that the impedance value of Ch-A is more sensitive to the visceral fat volume change. However, variation in the impedance value of the subcutaneous fat volume exists in both methods.

2.4 Relationship of the subcutaneous fat volume and measurement impedance (Ch-B)

Fig. 6 shows the result of our method (Ch-B). The result indicates that the visceral fat volume hardly affects the impedance measured by Ch-B. That is, the Ch-B impedance predominantly reflects only the information on the subcutaneous fat.

2.5 Estimation of the visceral fat volume by multi-regression analysis

Thirty patterns of the rectangular parallelepiped form of visceral fat were assumed, and three patterns of the subcutaneous fat form were assumed. Therefore, calculations were performed for a total of 90 patterns. The following regression formula was obtained with correlation coefficient r=0.96 as a result of the multi-regression analysis in the simulation model.

$$\text{Visceral fat volume }(\%) = 0.11 \cdot Z_A - 0.000659 \cdot Z_B - 16.9 \quad (3)$$

where visceral fat volume (%) is the ratio of visceral volume to the whole volume of the model. Z_A and Z_B are the impedance values of Ch-A and Ch-B, respectively

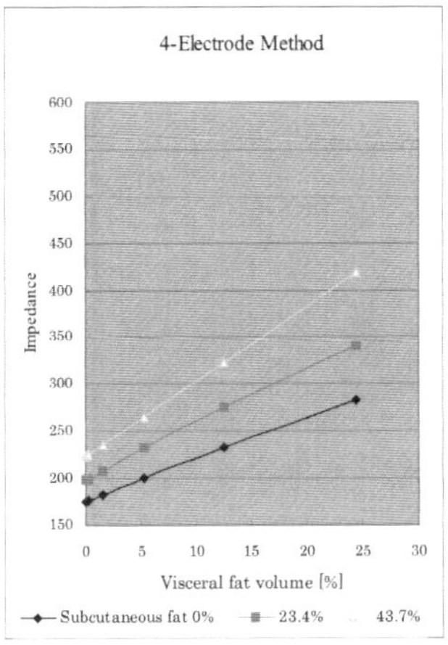

(a)

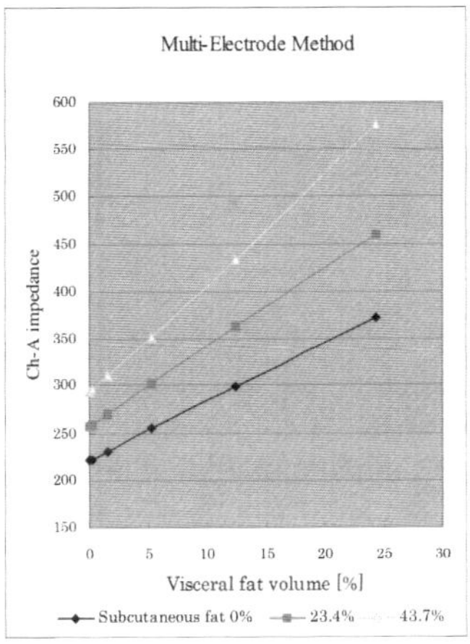

(b)

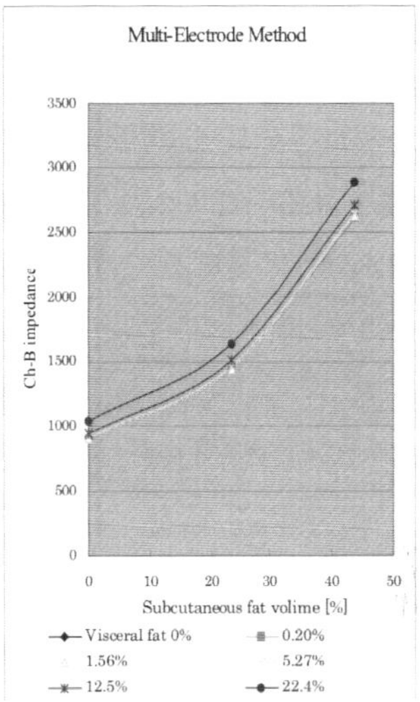

Fig. 5 Relationship between the visceral fat volume and measured impedance
(a) Impedance in 4-electrode method
(b) Impedance of Ch-A

Fig. 6 Relationship between the subcutaneous fat volume and Ch-B impedance

3. Discussions and Conclusion

It can be predicted that high impedance sensitivity areas are located in the regions in the vicinity of the electrodes. We weighted the impedance sensitivity by the proposed multi-electrode method to obtain visceral fat and subcutaneous fat information. Two channels for impedance measurement of the deep region and the shallow region were prepared. Ch-A better reflects the visceral fat volume sensitivity compared with the 4-electrode method, and Ch-B better reflects the subcutaneous fat volume sensitivity because it is not as influenced by the visceral fat volume. To estimate the visceral fat volume by the impedance method, it is necessary to remove the influence of subcutaneous fat. Therefore, multi-regression analysis using two impedance values (Ch-A and Ch-B) was performed. As a result, it was indicated that the visceral fat volume can be described with a simple multi-regression formula. Throughout this study, it was suggested that the real visceral fat volume can be easily presumed by the impedance data in clinical applications, when the relation between the exact visceral fat volume and Ch-A impedance and Ch-B is investigated.

4. References

[1] **Geselowitz, D. B. 1971 An application of electrocardiographic lead theory to impedance pletysmography.** *IEEE Trans.*, **BME-18, 38-41.**

[2] **Nakayama, K., Yagi, W. and Yagi, S. 1981 Fundamental study on electrical impedance CT algorithm using sensitivity theorem on impedance pletysmography.** *Proc. of Vth ICEBI*, **99-102.**

[3] **Furuya, N. and Nakayama, K. 1986 Development of multi electrode impedance pletysmography.** *Medical and Biological Engineering & Computing*, **24, 62-70.**

Electrical properties of rat muscle after sciatic nerve injury: impact on surface impedance measurements assessed via finite element analysis

M A Ahad and S B Rutkove[1]
Department of Neurology, Beth Israel Deaconess Medical Center, Harvard Medical School, 330 Brookline Avenue, TCC-810, Boston, MA, 02215, USA

E-mail: srutkove@bidmc.harvard.edu

Abstract. Tetrapolar surface electrical impedance methods are sensitive to changes in muscle status and can therefore provide a means for studying neuromuscular disease noninvasively. In order to better understand the relationship between surface impedance measurements and the actual muscle electrical properties, we performed measurements on 20 adult Wistar rats, 8 of which underwent sciatic nerve crush. Surface impedance measurements were performed on the left hind limb both before injury and out to 2 weeks after injury. In addition, both normal and sciatic crush animals were sacrificed and the dielectric properties of the extracted gastrocnemius muscle measured. We found that 50 kHz conductivities were greater in the animals that underwent crush than in the animals that did not. The permittivities in both directions, however, showed non-significant differences. In order to analyze the effect of these changes as well as the accompanying reduction in muscle volume, a finite element model of the hind limb was developed based on computerized tomographic imaging. The model successfully predicted the surface impedance values in the animals after crush injury and, by its inverse application, may be used to help determine the underlying electrical properties of muscle in various neuromuscular diseases based on surface impedance data.

1. Introduction

Electrical Impedance Myography (EIM) is a non-invasive electro-physiologic technique to assess muscle status in healthy as well as neuromuscular diseases. In EIM a low-intensity, high-frequency current is passed between two surface electrodes and the voltage pattern is observed by another pair of electrodes residing between the two current electrodes [1], [2]. Although under refinement, our initial work in humans [3], [4] has already shown the potential of this approach for the evaluation of a variety of neuromuscular diseases. In order to improve our understanding of this impedance data and its relationship to the muscle's inherent electrical and pathological properties, over the past several years, we have been pursuing studies in rodents. We have already shown that surface EIM parameters such as resistance, reactance and phase change significantly after nerve injury in the rat [5].

The finite element method is a useful tool to study EIM data since it provides a means for determining how muscle geometry and the inherent electrical properties of the muscle itself impact our surface measurements both in health and in disease states. In this study, we measured the electrical properties such as conductivity and permittivity of rat gastrocnemius muscle for both normal rats and

[1] To whom any correspondence should be addressed.

rats with sciatic nerve injury. A finite element model of rat leg was then developed both for normal and sciatic crush rats by incorporating the measured electrical values of the rat muscle and altering the shape of the leg based on computerized tomographic (CT) imaging.

2. Methods

2.1. EIM measurement

A total of twenty Wister rats weighing 425-475 grams were obtained from Charles River, Wilmington, MA. Of thes e twenty ra ts, 8 underwen t sciatic n erve crush. All s tudies were approved by the Institutional Animal C are an d Use Committee at Beth I srael De aconess Medical Ce nter. Detailed experimental procedures can be found in [5].

Ambu Neuroline 700 surface adhesive Ag-AgCl electrodes (Product # 70010-K/C/12, AMBU Inc., Bethesda, Ma ryland) w ere us ed fo r EIM measurement. Th ese ele ctrodes ar e A g-AgCl w ith a conductive adhesive gel. To facilitate our measurements on rats, these electrodes were resi zed to 2cm by 3.5mm by using a razor press cus tomized for this purpose, which consistently produced electrodes of th ese exact di mensions. The backs of 4 s uch e lectrodes were affixed to a pi ece of a dhesive ta pe, maintaining precisely an inter-electrode distance of 0.75 cm (center to center) between electrodes, thus producing a tetrapolar electrode ar ray. The o uter tw o elec trodes served a s the current ele ctrodes and the inne r tw o electrodes se rved as th e vo ltage e lectrodes. The entire array o f electrodes wa s then placed on the shaved, depilated leg with the elec trode axes perpendicular to the long-axis of the limb, keeping the center of the electrode set aligned with the previously applied tattoo [5].

A multi-frequency lock-in amplifier, Model 7280 of Signal Recovery, Oak Ridge, TN coupled with a very low capacit ance acti ve probe (Model 1103 of Tektronix, Beaverton, OR) was used to measure the impedance of the muscle [5]. The system was calibrated with resistor-capacitor circuits of known character, de signed to mimic to the el ectrical pr operties of muscle; in a ddition, the system was also tested using different concentration of saline solutions to confirm data consistency.

2.1.1. Sciatic crush

A 1 mm length s egment of s ciatic nerve was crushed us ing a jeweler's forceps b y applying 3–4 M Pa pressure for approximately 30 s. The incisions were then sutured closed, and the animals were allowed to recover until they were killed 1–2 weeks later. Baseline EIM measurements were performed on all the rats including the ones that were to undergo sciatic crush. EIM was also measured for all the rats for two successive weeks. Four rats with sciatic crush were killed a week after crush and the remaining four rats with crush were killed two weeks after crush.

2.2. FEM model

Two di stinct cu rrent types, the conductive curr ent and dis placement current, contribute to t he total current that flows through the muscle. The total current in frequency domain is thus given by:

$$J_{total} = \sigma.E + j\omega\varepsilon_0\varepsilon_r E \quad (1)$$

where σ is the conductivity of the tissue and E is the electric field, ε is the permittivity and $\varepsilon = \varepsilon_0\varepsilon_r$ where ε_0 is permittivity of air ($8.854*10^{-12}$) and ε_r is the relative permittivity. Since we are assessing the potential on the surface of the muscle, we must calculate the potential φ, which is directly related to electric field as:

$$E = -\nabla\varphi \quad (2)$$

In this simulation, the magnetic induction effects are negligible and the field is considered quasi static. Conservation of charge requires that:

$$\nabla.J = 0 \quad (3)$$

Thus the equation that needs to be solved for the potential measurement is:

$$\begin{aligned}\nabla.J_{total} &= \nabla.(\sigma.E + j\omega\varepsilon_0\varepsilon_r E) = 0 \\ &= \nabla.[(\sigma + j\omega\varepsilon_0\varepsilon_r)\nabla\varphi] = 0\end{aligned} \tag{4}$$

A 1mA test current was supplied between the two current electrodes. The normal component of the electric current is assumed to be continuous between any two media. Also it is assumed that there is no current flow out of the boundary at the two ends of the model. This assumption is reasonable, as one side of the model ends at ankle and other side ends at knee, both consisting of mostly poorly conductive bone.

The rat finite element model that we have developed, consists of leg from ankle to knee joint which has a skin layer, a sub cutaneous fat layer, the superficial biceps femoris muscle, the gastrocnemius-soleus complex beneath, and two bones (the fibula and the tibia). The shape of the leg was adopted from a computerized tomographic (CT) scan of the rat leg that we had been obtained while maintaining the rat in an identical position as when performing EIM. Angular anisotropy [6] was incorporated for gastrocnemious and bicep-femoris muscle layer. Electrode size and thickness were also modelled according to the actual size of the measuring electrodes used in the experiment.

2.3. Measurement of dielectric properties

A superficial portion of gastrocnemius muscle, approximately 1 cm × 1 cm square was excised with a scalpel, and its height precisely measured. The tissue was kept moist and warm by its being covered with 0.9 % saline-saturated gauze under a heating lamp at 36–37 °C. Dielectric properties were measured with a four electrode measurement technique as previously described [7].

3. Results

Table 1 shows the measured longitudinal (σ_l) and transverse (σ_r) conductivities and longitudinal (ε_l) and transverse (ε_t) permittivities for normal and sciatic crush rats. These values are in good conformity with reported values in [8], [9]. For the purposes of employing the finite element method, isotropic conductivities and permittivities of skin, fat and bone were adopted from [8] and [10]. A complete finite element model of a rat with sciatic crush is shown in Figure 1 with all layers labelled. Adult normal rats showed 50 kHz surface EIM resistance (R) of 73±4 ohms (mean +/- SD) and reactance (X) of 23±2 ohms. Because of the degree of crush, the surface EIM parameters and muscle shape varied considerably between the crushed rats. We incorporated the geometry of one of the sciatic crushed rats. The measured and modelled R and X values for both the normal and sciatic-crushed animals are presented in Table 2.

Table 1. Measured conductivities and permittivities at 50 kHz and level of significance.

	σ_l (S/m)	σ_t (S/m)	ε_l	ε_t
Normal mature rats (N=12)	0.38 ± 0.02	0.15 ± 0.01	47400 ± 3100	47600 ± 1900
Sciatic crushed rats (N=8)	0.55 ± 0.03	0.21 ± 0.01	45800 ± 2700	48500 ± 2000
Significance 0.00	04	0.0002	ns	ns

Table 2. Comparison of experimental EIM data and FEM data at 50 kHz (in ohms).

	Measured R	Modelled R	Measured X	Modelled X
Normal rats	73 ± 4	69	23 ± 2	22
Sciatic rat	64 55	15		12

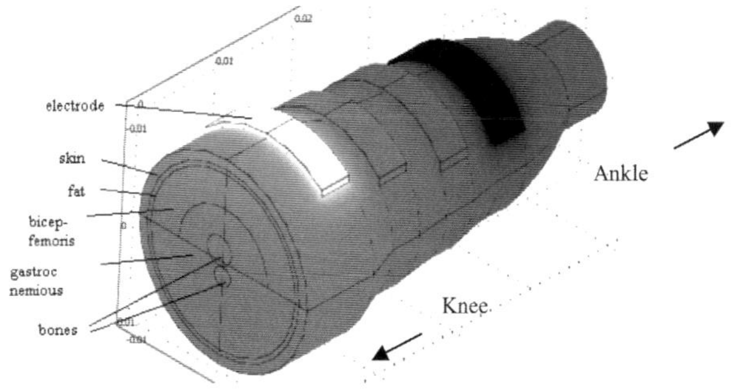

Figure 1. Basic model used for finite element analysis of leg with sciatic crush.

4. Discussion

By employing finite element model based on the actual muscle electrical constants and limb geometry, we have obtained results that are in reasonable agreement with the experimental surface EIM data. There are several factors that impact the outcome of the model results presented here. First, variability in the severity of the experimental crush lesion contributes to some of the variability in the observed measurements of the electrical constants. Second, the conductivities and the permittivities measured depend critically up on accurately measuring the muscle sample size, maintaining a consistent temperature, and performing the measurements quickly after sacrifice, so as to avoid degradation of the tissue. Third, the geometry of the leg incorporated in the model needs to be as accurate as possible. We attempted to model the geometry as by obtaining CT measurements with the leg in the same position as we perform EIM, but inaccuracies are inevitable. Fourth, the actual internal anatomy of the leg, including the orientation of the bones, the skin and fat thickness and the angles of different layers of muscle need to be precise [6]. Finally, the electrical values for fat, skin and bone will impact results.

The dielectric values that we reported here are average values and crush severity plays a role in measured dielectric values. Moreover while developing a finite element model we adopted a particular rat with crush and its own shape derived by CT scan. The shape of crushed leg is more irregular than that of a normal rat leg and thus likely contributed to the observed differences between the experimental and the modelled data. A MRI-based finite element modelling is thus anticipated to be more accurate in predicting the EIM data, with incorporation of each crushed animal's unique geometries. Nevertheless, the results found here provide a basis for this basic analytic approach in neuromuscular diseases, including muscular dystrophy and amyotrophic lateral sclerosis.

5. References

[1] Shiffman C A, Aaron R, Amoss V, Therrien J and Coomler K 1999 *Phys. Med. Biol.*.**44** 2409
[2] Rutkove S B, Aaron R and Shiffman C A 2002 *Muscle Nerve* **25** 390
[3] Rutkove S B, Esper G, Lee K, Aaron R and Shiffman C A 2005 *Muscle Nerve* **32** 335
[4] Rutkove S B, Zhang H, Schoenfeld D A, Raynor E M, Shefner J M, Cudkowicz M E, Chin A B, Aaron R and Shiffman C A 2007 *Clin. Neurophysiol* **118** 2413
[5] Ahad M A and Rutkove S B 2009 *Clin. Neurophysiol.***120** 1534
[6] Ahad M A and Rutkove S B 2009 *Conf. Proc. IEEE Eng. Med. Biol. Soc.* **1** 630
[7] Ahad M A, Fogerson P M, Rosen G D et al B 2009 *Physiol. Meas.* **30** 1415
[8] Gabriel C, Gabriel S and Corthout E 1996 *Phys. Med. Biol.* **41** 2231
[9] Stoy R D, Foster K R and Schwan H P 1982 *Phys. Med. Biol.* **27** 501
[10] Gabriel C 2005 *Bioelectromagnetics* Supplement **7** S12

Acknowledgments

This work was funded by Grant R01055099 from the National Institutes of Health.

Multichannel impedance monitoring for evaluation of alpha-adrenoblocker effect on the ureteral function in patients with stone disease

O I Apolikhin, L A Khodyreva, I S Mudraya[1], V I Kirpatovsky and A A Serdyuk
Research Institute of Urology, Moscow, Russia

E-mail: mis51ser@yahoo.com

Abstract. The study of distal ureter function was carried out on patients with stones in the upper urinary tract, who underwent ureteroscopy and lithotripsy procedures. The parameters of ureteral peristalsis such as peristalsis amplitude, peristalsis rate, ureteral wall tone, contractile wave duration, and its direction obtained by multichannel impedance ureterography were assessed and compared from two groups of patients. The group I patients received tamsulosin in addition to standard regimen, while the group II patients matched according to the stone size and location were managed without tamsulosin medical therapy. In comparison with group II, the group I patients demonstrated smaller average peristalsis amplitude (0.60±0.08 vs 0.81±0.06 Ohm), shorter contractions (7.1±0.3 vs 7.7±0.3 s), greater peristalsis rate (3.3±0.3 vs 2.8±0.2 per minute), and diminished ureteral tone (4.0±0.5 vs 4.7±0.2 Ohm^{-1}). Incidence of the retrograde contractile waves was two-fold greater in the group II, while normal antegrade regular contractions were 30% more frequent in the group I. In addition, our results showed that the effect of tamsulosin on ureteral function was manifested in the patients with different stone size and location in the upper urinary tract, and it depended pronouncedly on individual ureteral tone and contractility parameters.

1. Introduction

Alpha-1 adrenoblocking agents are increasingly used in the treatment of distal ureteral stones. Their beneficial effects include promoting spontaneous stone passage, shortening of the stone expulsion time, and analgesia [1, 2, 3]. These effects are explained by ability of these agents to block specific receptors situated mainly in the distal ureter resulting in moderation of ureteral contractility [2] and by relaxation of smooth muscles in the ureteral wall [4]. However, stone disease management with tamsulosin is characterized with varying rate of success [1-4]. The stone size and location are considered as the most important factors in stone passage [4], while little objective data is available on the ureteral function after alpha-blocker treatment in clinical trials. In our previous studies, the peristalsis of ureter in patients with upper urinary tract stones evaluated by multichannel impedance monitoring significantly differed from the peristalsis of stone-free ureter. We distinguished four types of peristaltic disorders based on quantitative (amplitude and tone) as well as qualitative (shape and direction of the waves) attributes. Those data allowed us to suppose that tamsulosin effect would depend on initially different ureteral conditions in individual patients.

[1] To whom any correspondence should be addressed.

The purpose of this study was to evaluate the peculiarities of ureteral function in patients with stones in the upper urinary tract after alpha-blocking treatment with tamsulosin.

2. Material and methods

The study of distal ureter peristaltic function was carried out on patients with stones in kidney and proximal (n=24) or distal (n=14) ureter, who were subjected to ureteroscopy and lithotripsy procedures. Patients were divided into two groups: the "tamsulosin" group I patients (n=15) treated with tamsulosin (0.4 mg daily) in addition to the standard regimen, and the control group II patients managed without tamsulosin matched according to the stone size and location (n=25).

Ureteral function was assessed by multichannel impedance ureterography with the help of special probe inserted into ureter coupled with RPKA2-01 impedance converter. The following parameters were evaluated: the peristalsis amplitude and rate, ureteral wall tone, and the number of contractile wave characteristics including velocity, duration, shape, and direction of their propagation in the ureter. The measurements of the renal pelvic pressure (RPP) were made electromanometrically in patients with indwelled nephrostomy tube.

3. Results

The average data on peristaltic function after tamsulosin treatment in patients with stone disease compared to the controls are shown in table 1. The average magnitude of peristalsis amplitude was higher by 0.22±0.10 Ohm (27%, $p<0.02$) in patients treated with tamsulosin compared to the control group. In the group I (tamsulosin), the following trends were observed: contractions were shorter (by 9%, $p<0.1$), frequency was greater (by 18%, $p<0.1$), and the wall tone was smaller (by 14%, $p<0.5$) compared to the group II. The retrograde contractile waves occurred twice more often in the control group, while the normal antegrade regular contractions were more frequently observed in the tamsulosin group (by 30%). The contractile wave velocity of antegrade ureteral contractions did not differ.

Table 1. Effect of tamsulosin on the parameters of ureteral function in stone disease patients.

	Tamsulosin group	Control group
Amplitude (Ohm)	0.81±0.06	0.60±0.08
Contractile wave velocity (cm/s)	1.83±0.30	2.12±0.06
Contraction rate (min^{-1})	2.81±0.18	3.32±0.28
Contractile wave duration (s)	7.75±0.26	7.09±0.26
Ureteral wall tone (Ohm^{-1})	4.66±0.25	4.00±0.54

These peristalsis changes in distal ureter were evaluated together with RPP, which is an important urodynamic parameter characterizing the upper urinary tract. Initially, the average mean values of RPP were elevated in patients with stone disease compared to its normal values (10 cm H_2O) reported in literature [5]. In patients treated with tamsulosin, the average RPP values were smaller by 15-28% compared to that in the control group (table 2). These data can be an indicator of improved urine transport along the ureter in patients of the "tamsulosin" group I, because diuresis via the nephrostomy tube was similar in both groups (1.1±0.1 and 0.9±0.1 ml/min).

The obtained average results agree with literature on the mechanisms of alpha-blocker effects including ureter spasm relief and decrement of the contractile amplitude [2, 4]. These factors are proved to be important to promote successful stone passage down the ureter. Another factors postulated as significant for stone elimination are the stone size and location [4]. As the stone size was similar in both groups for proximal (1.14±0.17 vs 0.93±0.15 cm) and distal (1.02±0.16 vs 0.93±0.11

cm) stones, we analyzed whether there was an interrelation between the ureteral function and stone location.

Table 2. Effect of tamsulosin on pressure in renal pelvis in stone disease patients.

	RPP in supine position		RPP in standing position	
	Basal	Peristaltic	Basal	Peristaltic
Tamsulosin group	13.18±0.64	16.57±0.82	19.71±1.06	24.05±1.08
Control group	16.09±0.53	19.54±0.57	27.20±0.63	30.70±0.65

3.1. Tamsulosin effects on ureteral peristalsis in patients with various stone location
In patients with stones located in the proximal region of upper urinary tract, there were no significant differences in parameters of ureteral peristalsis between the examined groups except the lower peristalsis amplitude (by 38% , $p<0.02$) in patients receiving tamsulosin medication. Similar, there was no intergroup difference in renal pelvis pressure.

In the patients with distal stones, no differences were found between the parameters of peristalsis in both groups except RPP, which was significantly lower in the patients of grpou I (tamsulosin) demonstrating smaller basal and peristaltic pressure components, respectively by 47% and 39% ($p<0.01$).

Table 3. Effect of tamsulosin on the functional parameters of ureter and renal pelvis in patients with various stone locations.

	Peristalsis amplitude (Ohm)	Ureteral wall tone (Ohm-[1])	Pressure in renal pelvis (cm H_20)	
			Basal	Peristaltic
Proximal stones				
Tamsulosin group	0.62±0.13	3.86±0.87	14.59±0.87	17.71±1.23
Control group	0.99±0.09	3.92±0.25	14.18±0.58	17.43±0.52
Distal stones				
Tamsulosin group	0.55±0.19	6.16±2.16	10.0±1.55	14.00±2.23
Control group	0.39±0.04	6.1±1.00	18.95±1.71	23.06±2.07

Thus, tamsulosin effect on the upper urinary tract urodynamics depended on the stone location. We can hypothetise that renal and proximal ureteral stones can evoke reflex irritation of the excitatory adrenergic fibers in distal ureter that stimulate its contractions and increase the wall tone; therefore, tamsulosin can moderate ureteral function by blocking alpha-adrenergic receptors. Elevated RPP in patients with proximal stones in the upper urinary tract is urodynamic in nature, i.e. it predominantly results from obstruction with a stone hindering urine outflow from kidney. In the case of distal stones, irritation of distal ureter via adrenergic excitatory mechanism can induce reflex elevation of pressure in the kidney; therefore, tamsulosin can induce reduction in RPP by blocking the specific receptors.

3.2. Individual peculiarities of ureteral peristalsis in tamsulosin-treated patients
In both examined groups, there were patients with aberrant peristalsis in distal ureter characterized by very weak chaotic waves (figure 1, A). The tone of ureteral wall in distal ureteral region was high both in the control (6.40±1.64 Ohm-1) and in the tamsulosin (6.96±1.34 Ohm-1) groups. There were no significant differences between other parameters of ureteral peristalsis in both groups and in the values of pressure in renal pelvis (17.59±2.87 vs 17.90±1.79 cm H_2O). We suppose that such aberrant peristalsis can result from previous general inflammatory and fibrotic changes in ureteral wall that cannot be modulated by autonomic neural influences, and therefore would not be affected by

tamsulosin. These ureters do not actively participate in urine transport, so they should be treated by creating unobstructed low-pressure urine passage.

In patients whose ureters demonstrated evident contractions and moderate ureteral wall tone with preserved antegrade waves (figure 1, B), tamsulosin modified urodynamics in the upper urinary tract. In such patients from the group I compared to group II, the peristalsis amplitude was smaller (0.97±0.25 vs 1.30±0.14 Ohm), the ureteral wall tone was lower (2.30±0.55 vs 3.60±0.56 Ohm-[1]), the contractions were shorter by 17% and faster by 52%, respectively. Measurements of RPP after tamsulosin treatment revealed significantly smaller values in standing position (13.50±0.69 vs 22.90±5.40 cm H_2O) indicating the better urine drainage from the kidney.

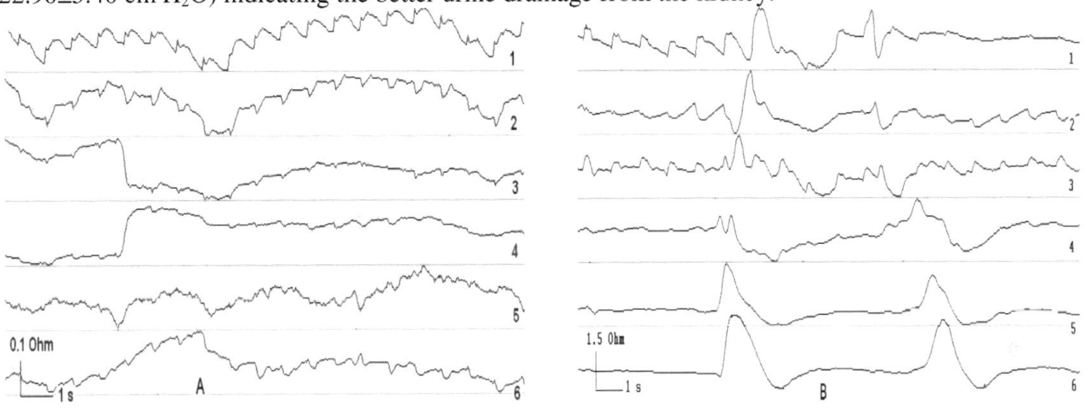

Figure 1. Pattern of aberrant peristalsis (A) and pronounced contractile function of the ureter (B) in two patients with stone disease. The recording channels from 1 to 6 correspond to impedance measurements from the consecutive (proximal to distal) parts of ureter.

4. Summary

Evaluation of ureteral peristalsis by multichannel impedance monitoring is an important supplementary diagnostic tool both for clinical practice and for understanding the mechanisms of urodynamic disorders in patients with stone disease. Multichannel impedance ureterography yields quantitative and qualitative data on ureteral peristalsis and reveals its individual peculiarities in the patients. This method can be useful when choosing a medical treatment for individual patients, and it can predict the efficacy of treatment.

In the present study of tamsulosin effect on function of distal ureter in patients with stone disease, efficacy of alpha-adrenoblocker treatment can be achieved both for proximal and distal stones in the upper urinary tract, and it largely depends on individual ureter status. In patients with functional peristaltic disorders, especially with those provoked by increased adrenergic stimulation, tamsulosin treatment would be helpful in the terms of contractility normalization and ureteral wall relaxation. Pronounced organic changes in ureteral wall caused by previous inflammatory or fibrosis processes characterized by aberrant peristalsis can be responsible for medication inefficacy

5. References

[1] Lipkin M and Shah O 2006 *Rev Urol* **8** (Suppl 4) S35–S42
[2] Troxel S A, Jones A W, Magliola L and Benson J S 2006 *J.Endourol.* **20** 8 565-8
[3] Gravas S, Tzortzis V, Karatzas A, Oeconomou A and Melekos M D 2007 *Urolc Res* **35** 5 231-5
[4] Anderson K, Ogan K and Pearle M S 2006 Ureteral calculi: improving stone passage rates with medical therapy *Contemporary Urology* **8** 2 8-17
[5] Shafik A 1998 Ureteric profilometry. A study of the ureteric pressure profile in the normal and pathologic ureter *Scand. J. Urol. Nephrol.* **32** 1 14-19

Visualizing transplanted muscle flaps using minimally invasive multi-electrode bioimpedance spectroscopy

R Gordon, V Zorkova, M Min, I Rätsep

Department of Electronics, Tallinn University of Technology, Ehitajate tee 5, 19086 Tallinn, Estonia

E-mail: rauno@elin.ttu.ee

Abstract. We describe here an imaging system that uses bioimpedance spectroscopy with multi-electrode array to indicate the state of muscle flap regions under the array. The system is able to differentiate between different health states in the tissue and give early information about the location and size of ischemic sub-regions. The array will be 4*8 electrodes with the spacing of 5mm between the electrodes (the number of electrodes and the spacing may vary). The electrodes are minimally invasive short stainless steel needles, that penetrate 0.3 mm into the tissue with the goal of achieving a wet electric contact. We combine 32 configurations of 4-electrode multi-frequency impedance measurements to derive a health-state map for the transplanted flap. The imaging method is tested on a model consisting of 2 tissues and FEM software (Finite Element Method –COMSOL Multiphysics based) is used to conduct the measurements virtually. Dedicated multichannel bioimpedance measurement equipment has already been developed and tested, that cover the frequency range from 100 Hz to 1 MHz.

1. Introduction

Transplanting tissue for reconstructing damaged areas or filling tissue defects is a common surgical practice. During the transplantation, a tissue graft is harvested from patients own body (special donor-areas [1]) and joined to the required site. Medical staff can externally assess if the operation was successful and occasionally monitor the healing process, but a more reliable system is needed for continuous monitoring of the healing process of the transplanted tissue flap for getting earlier signal if surgical repair is needed. Bioimpedance measurement provides a cheap and handy method for continuous monitoring of such transplants. Measuring bioimpedance has been proposed as a diagnostic method in various medical conditions, but post-operation monitoring of transplanted tissue grafts is easily one of its' most promising medical applications. The visualization method proposed in this work is limited to tissue grafts placed on the body surface.

2. Proposed diagnostic system

We propose a bioimpedance-spectroscopy based system for easy visualization of the health state of a transplanted muscle flap. We use minimally invasive electrode array in 4*8 configuration for acquiring an image of health-state regions under the array. The electrodes are 0.3 mm long needles with thickness up to 0.1 mm. They are spaced 5 mm apart in the regular grid array to form a patch that covers a 15*35 mm area. The electrode array should be flexible and it would be installed on the operation site so that all needles penetrate a little into the tissue to gain a wet electric contact.

The measurement system will perform 4-electrode impedance measurements between all the 32 electrodes with 32 different configurations. The whole set of measurements should be performed under a minute or 10 minutes because the tissue health state is not very rapidly changing. The frequency range is selected to be from 100 Hz to 1 MHz for reliable tissue health characterization. Two measurement devices have been built and tested in our electronics laboratory with the capabilities to perform these measurements with the needed accuracy [2, 3]. The measurement system gathers the frequency characteristics of all 32 configurations and passes the data to the imaging system.

The imaging system determines the state of tissue under each electrode configuration based on the measurement results. It then composes an image, where tissue health areas are clearly visible, for a doctor to further analysis. The image can easily be sent to the doctor's cell-phone by an MMS for example.

3. Locally assessing the tissue health state

Each frequency characteristic of a 4-electrode impedance measurement is compared to a set of existing characteristics (healthy, ischemic, necrosis, other/error) with their lower and upper tolerances. The impedance data for muscle tissue has been acquired from Gabriel et.al. [4]. The data for other essential types has been gathered [5, 6, 7], but not yet composed for this system, because it is inconclusive for our needs. To test our working system, we used 2 distinct tissues - muscle and fat - with their full frequency characteristics. The 4-electrode measurement simulation results of those tissues are shown in Figure 1 with lower and upper tolerances used in a sample system testing.

If the measurement result from one 4-electrode configuration is set within the bounds of healthy muscle, then it will be indicated with value 1 on the final image. If the result accounts as clearly ischemic, then the respective area will be marked with value 2. If other distinguishable states can be identified, other values will be used to mark them.

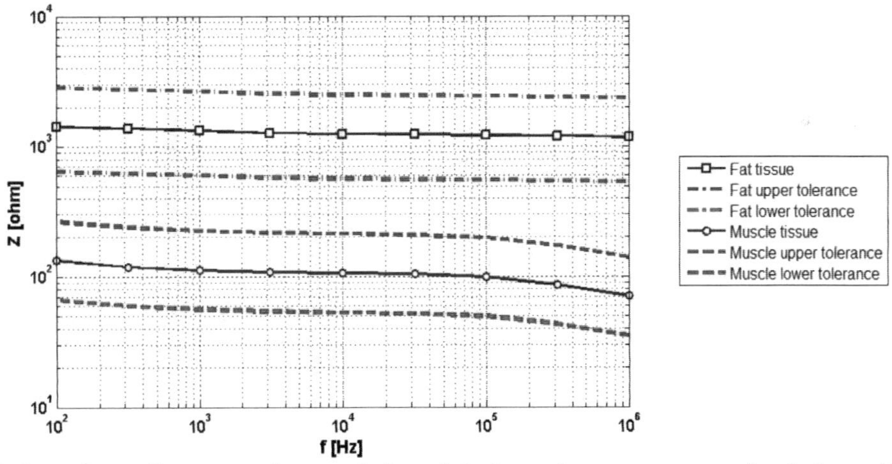

Figure 1. Impedance frequency characteristics of 4-electrode measurement simulation with muscle and fat tissue.

4. Testing the precision of simulations

The bioimpedance measurement was simulated at different FEM-mesh densities to determine, if the mesh resolution is good enough for accurate results. The 4-electrode measurement was simulated with the domain (a muscle flap 100*40*10 mm with 32 electrodes) consisting of 182,000 tetrahedral mesh elements and 360,000 tetrahedral mesh elements.

The results of this test show that there is negligible difference between high-resolution mesh and lower-resolution mesh (less than 0.1 %) in all sections of the frequency characteristic. Therefore the lower resolution mesh was used for all the simulated measurements.

5. Positioning the measurement results

The 4-electrode impedance measurement has positive sensitivity areas (impedance increase locally in that area contributes to signal rise) as well as negative sensitivity areas (impedance increase contributes to signal fall). The sensitivity of a 4-electrode impedance measurement can be calculated as a dot product of 2 electric fields: the lead-field from electrodes 1-4 and the lead-field from electrodes 2-3. The lead-fields of 1-4 and 2-3 were calculated in a 80*80*20 grid and then the dot product of the 2 fields was found in each of the 3D grid points. The sensitivity maps for 4-electrode measurements were calculated with two cases: electrodes in a straight-line (used in 28 measurements, Figure 2) and electrodes in corner formation (used in 4 measurements, Figure 3). The 3D sensitivity maps were used to determine the areas in the image, where the diagnosis of the specific 4-electrode measurement was applied. The diagnosis is one of the following: healthy, ischemic, necrosis or error and in the final image it is being marked with colour codes 1,2,3 or 4 (refer to section 3).

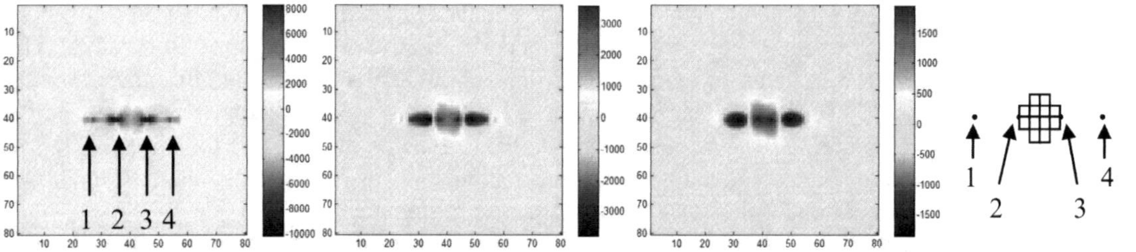

Figure 2. The 4-electrode impedance measurement with linearly positioned electrodes. Sensitivity maps of 3 top layers and the positioning of diagnostic patch (12-pixel patch) on the right.

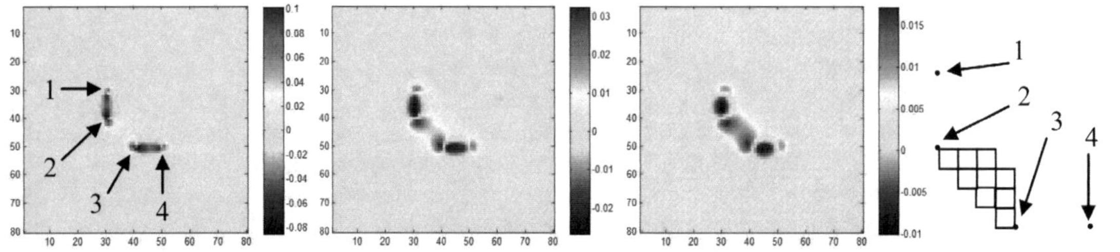

Figure 3. The 4-electrode impedance measurement with corner formation electrodes. Sensitivity maps of 3 top layers (of 20) and the positioning of diagnostic patch (9-pixel patch) on the right.

6. Composing the image

The final image will be composed from all the patches that are acquired from the 32 measurement configurations. The patches can have values 1 (healthy), 2 (ischemic) or other. There will be 28 shape-1 patches (Figure 2, right) to show results of all the linear electrode configurations, and 4 shape-2 patches (Figure 3, right) to fill the corners of the image that would otherwise be left empty. When the patches are positioned on the image, there will be overlaps. The overlapping pixels are taken as average values of patch pixels. When all patches are on the image, the distracting holes between the patches will be filled in with average values of neighbouring pixel values.

7. Results

To test the system, we ran all measurements in a computer as simulations with Comsol Multiphysics software. The object under test was described as 10*4*1 cm block composing of 2 tissues (Figure 4). On that biological object the electrode array was placed. In the modelling domain only the tips of the electrodes are specified, that penetrate into the tissue.

When the measurements are performed on the model, the imaging system effectively composes an image from the patches. The resulting image composition is shown in Figure 5 with 3 steps: 1) the composing of the image with measurements from 28 linear electrode configurations; 2) the corner patches have been added to the image; 3) the holes are filled with average values of neighbours. The two tissue areas are accurately distinguished.

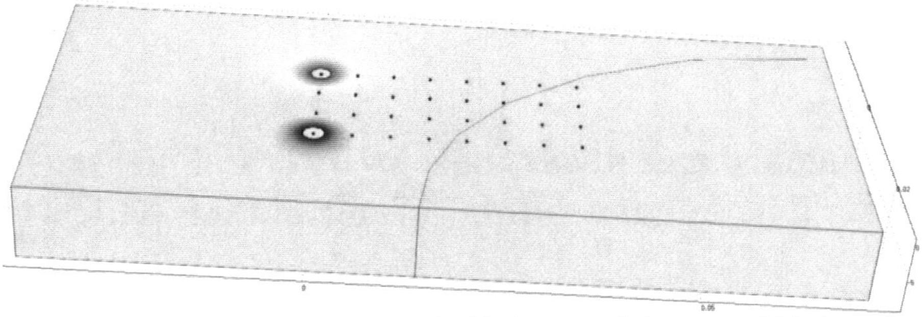

Figure 4. The 3D modelling domain, composed of 2 tissues and the array of 32 electrodes. Electric field from 2 corner-electrodes is shown on the model surface with colour.

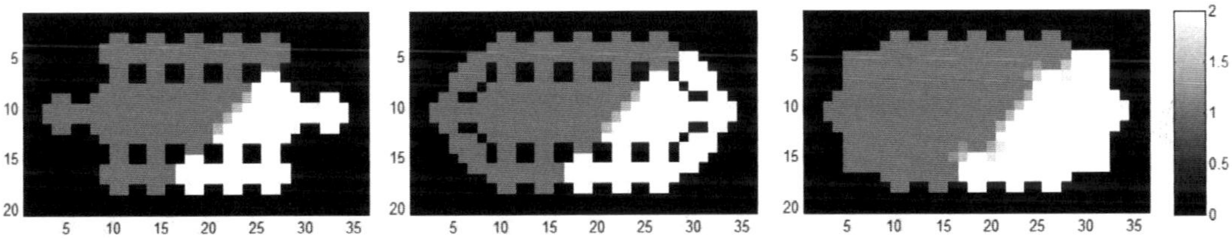

Figure 5. The 3 steps of image composition. The tissues are distinguished with colours 1 and 2.

8. Discussion

The number of electrodes can be increased and the distance between electrodes can be chosen some other value, when physical experiments and clinical trials suggest a need for other configurations. Same applies also for the length and thickness of the needles, which depends on the tissue type that will be targeted with this diagnostic system.

9. Acknowledgements

This research was supported by the European Union through the European Regional Development Fund.

References

[1] Microsurgeon internet site, http://www.microsurgeon.org/
[2] Min M, Parve T, Ronk A, Annus P and Paavle T 2007 Synchronous Sampling and Demodulation in an Instrument for Multifrequency Bioimpedance Measurement *IEEE Transactions on Instrumentation and Measurement* **56** (4) 1365-72
[3] Annus P, Kuusik A, Land R, Haldre E, Min M, Parve T and Poola G 2007 An energy efficient wearable tissue monitor *IFMBE Proc.* **17** *ICEBI 2007 13th Int. Conf. on Electrical Bio-impedance and 8th Conf. on Electrical Impedance Tomography 2007* Graz, Austria, 240-3
[4] Gabriel S, Lau R W and Gabriel C 1996 The dielectric properties of biological tissues: II. Measurements in the frequency range 10 Hz to 20 GHz *Phys. Med. Biol.* **41** 2251-69
[5] Dean D A, Ramanathan T, Machado D and Sundararajan R 2007 Electrical impedance spectroscopy study of biological tissues *Science Direct*
[6] Ristic B, Kun S and Peura R A 1996 Muscle Tissue Ischemia Monitoring Using Impedance Spectroscopy: Preliminary Results *IEEEXplore*
[7] Schaefer M, Gross W, Ackemann J and Gebhard M M 2002 The complex dielectric spectrum of heart tissue during ischemia *Science Direct*

Journal of Physics: Conference Series **224** (2010) 012104 doi:10.1088/1742-6596/224/1/012104

A low bioimpedance phase angle predicts a higher mortality and lower nutritional status in chronic dialysis patients

Francis Dumler MD

Division of Nephrology, William Beaumont Hospital, Royal Oak, MI 48073, USA

fdumler@beaumont.edu

Abstract. Bioelectrical impedance analysis is an established technique for body composition analysis. The phase angle parameter, an index of body cell mass, tissue hydration, and membrane integrity, makes it suitable for assessing nutritional status and survivability. We evaluated the significance of a low phase angle value on nutritional status and mortality in 285 chronic dialysis patients during a longitudinal prospective observational study. Patients in the lower phase angle tertile had decreased body weight, body mass index, fat free mass, body cell mass, and lower serum albumin concentrations than those in the higher tertile (P<001). In addition, mortality rates were significantly lower (P=0.05) in the highest tertile patients. In conclusion, the phase angle is a useful method for identifying dialysis patients at high risk for malnutrition and increased mortality.

1. Introduction

The evaluation of nutritional status is of significance importance in the care of dialysis patients. Even though improvements in management of cardiovascular diseases, infections, and dialysis therapies have been significant, mortality in the chronic kidney disease population continues unabated. Malnutrition, whether alone or in association with a micro inflammatory state, affects 30-70% of patients and is clearly identified as a significant risk factor for survival [1-4]. Anthropometric and biochemical parameters have been traditionally used for nutritional assessment in clinical practice. More recently, bioelectrical impedance analysis (BIA) has established itself as a technique for evaluating body composition and nutritional status [5]. In addition, the phase angle parameter (arctangence of the reactance to resistance ratio) provides an index of tissue hydration, cell membrane integrity, and cell mass [6]. The objective of this study was to determine if a low phase angle is associated with a decreased nutritional status and/or higher mortality in chronic dialysis patients.

2. Patients and methods

This is a prospective longitudinal observational study of 285 chronic maintenance dialysis patients. Bioelectrical impedance analysis was done with a single 50 kHz frequency analyzer using tetrapolar electrodes attached to the ipsilateral wrist and ankle 15 minutes post hemodialysis or before a fill in peritoneal dialysis patients. Even at this low frequency the influence of stray capacitance cannot be neglected. For this reason, a parallel model (Cyprus™ 1.2 RJL Systems, MI, USA) was used for body composition calculations [7]. Briefly, this model assumes a parallel circuit composed of a resistor Rp positioned in parallel with a capacitor Cp, where both electrical processes occur independently, thus allowing evaluation of the bulk electrical properties of tissues.

Parameters: body weight, body mass index and serum chemistries (blood urea nitrogen, creatinine, and albumin) were evaluated monthly. Bioimpedance parameters (phase angle, and derived body cell mass, fat free mass and fat mass) were measured biannually. Fat free mass was calculated as:

Males: $-10.68 + 0.65 \times (\text{Height}^2/\text{Resistance}) + 0.26 \times \text{Weight} + 0.02 \times \text{Resistance}$

Females: $-9.35 + 0.69 \times (\text{Height}^2/\text{Resistance}) + 0.17 \times \text{Weight} + 0.02 \times \text{Resistance}$

Statistical analysis: mean values for the entire observation period were obtained. Patients were divided into tertiles according to phase angle values. Comparisons between tertile groups were made by analysis of variance and post hoc testing.

3. Results

Demographic characteristics were as follow: age: 62 ± 14 years; weight: 75 ± 18; body mass index: 26.3 ± 5.2 kg/m2, 59% male, and 37% black. The prevalence of diabetes was 43%. Dialysis modality was 72% hemodialysis and 28% peritoneal dialysis. The median observation time was 18 months.

Body weight and body mass index values were significantly different in the phase angle tertile groups ($P < 0.001$); with the lesser values corresponding to the lower tertile in both hemodialysis and peritoneal dialysis patients (table 1, table 2). No differences were noted in mean phase angle values between hemodialysis and peritoneal dialysis patients. As expected, peritoneal dialysis patients had a higher body weight than their hemodialysis counterparts. Biochemical profiles were also significantly different in the tertile groups ($P < 0.001$) with the smallest values in the low tertile group in both hemodialysis and peritoneal dialysis patients (table 1, table 2). No differences were noted in mean phase angle values between hemodialysis and peritoneal dialysis patients. As expected, peritoneal dialysis patients had lower serum albumin concentrations than their hemodialysis counterparts.

In hemodialysis and peritoneal dialysis patients significant differences between all tertile groups were noted for body cell and fat free mass ($P < 0.01$). Fat mass was decreased only in the lowest tertile hemodialysis group (table 1). Further analyses were done combining hemodialysis and peritoneal dialysis patients. All nutritional parameters were significantly lower ($P < 0.05$) in the low tertile group when compared to the other groups (table 3). Mortality rates, calculated per 1000 patient months, was significantly decreased ($P< 0.05$) in the upper tertile group (table 3).

4. Discussion

The use of bioelectrical impedance analysis is becoming more widespread in the renal community due to ease of use and a low coefficient of variation for the direct measurements (±1%). In addition, several population reference values have been reported recently [6-10]. This facilitates comparisons between patients and a more specific general (healthy) population. Its reliability and sensitivity also makes it very useful for longitudinal monitoring of chronic kidney disease patients.

This study demonstrates that bioelectrical impedance analysis, in particular the phase angle, is a useful clinical tool for detecting malnutrition in dialysis patients. The phase angle reflects the relative contribution of body fluid (resistance) and cellular membrane integrity (capacitance). Malnutrition reduces cellular membrane mass and integrity and promotes shifts in fluid balance. As a consequence of these changes the phase angle decreases. Conversely, a higher phase angle implies larger body cell mass and preserved membrane integrity.

Patients with smaller phase angle values, defined by < 4.10 (lowest tertile group), had lower body weight and body mass index than their counterparts. In addition, serum creatinine concentration was concomitantly decreased, implying less dietary protein intake and lower muscle mass. Finally, body cell mass and fat free mass were also decreased in patients with a low phase angle.

The finding of a lower mortality rate among patients in the higher phase angle tertile is expected, as this group had a better nutritional status. One large multicenter study has shown a direct relationship between phase angle and survival in hemodialysis patients [1]. Another study in forty eight peritoneal dialysis patients found a close relationship between phase angle and mortality [11]. Our study has the advantage of including a relatively large patient population treated at a single facility by the same health care team.

Uremic malnutrition and chronic inflammation are highly prevalent in maintenance dialysis patients [12]. A low phase angle may be a consequence of both processes. Of interest in this regard is the observation that phase angle is a predictor of all-cause mortality after adjusting for serum C-reactive protein concentrations [13]. The use of ultrapure dialysis fluid and polyamide dialyzers for on-line predilution hemodiafiltration over a six month period has also been linked to a significant decrease in inflammatory markers and improved nutritional parameters [14]. In that particular study a dramatic fall in C-reactive protein was time related to a significant increase in phase angle. Of interest, several studies have shown an inverse relationship between phase angle and survival rates in a variety of disease processes including HIV, post-operative complications, cancer, and cirrhosis [6, 15].

In conclusion, these results suggest that bioelectrical impedance analysis and phase angle measurements are a useful tool for identifying dialysis patients at high risk for malnutrition and/or increased mortality.

References

[1] Chertow GM, Johansen KL, Lew N, Lazarus JM and Lowrie EG 2000 Vintage, nutritional status, and survival in hemodialysis patients *Kidney Int.* **57** 1176-81

[2] Pifer TB, McCullough KP, Port FK, Goodkin DA, Maroni BJ, Held PJ and Young EW 2002 Mortality risk in hemodialysis patients and changes in nutritional indicators: DOPPS Kidney Int. **62** 2238-45

[3] Dwyer JT, Larive B, Leung J, Rocco MV, Greene T, Burrowes J, Chertow GM, Cockram DB, Chumlea WC, Daugirdas J, Frydrych A and Kusek JW for the HEMO Study Group 2005 Are nutritional status indicators associated with mortality in the hemodialysis (HEMO) study? *Kidney Int.* **68** 1766-76

[4] Araújo CI, Kamimura MA, Draibe SA, Canziani MEF, Manfredi SR, Avesani CM, Sesso R and Cuppari L 2006 Nutritional parameters and mortality in incident hemodialysis patients *J. Renal Nutr.* **16** 27-35

[5] Dumler F and Kilates C 2000 Use of bioelectrical impedance techniques for monitoring nutritional status in chronic dialysis patients *J. Renal Nutr.* **10** 116-24

[6] Selberg O and Selberg D 2002 Norms and correlates of bioimpedance phase angle in healthy human subjects, hospitalized patients, and patients with liver cirrhosis *Eur. J. Appl. Physiol.* **86** 509-16

[7] Dumler F and Kilates C 2003 Body composition analysis by bioelectrical impedance in chronic maintenance dialysis patients: comparisons to the National Health and Nutrition Examination Survey III *J. Renal Nutr.* **13** 166-72

[8] Piccoli A, Pillon L and Dumler F 2002 Impedance vector distribution by sex, race, body mass index, and age in the United Status: standard reference intervals as bivariate Z-scores *Nutrition* **18** 153-67

[9] Barbosa-Sila MC, Barros AJ, Wang J, Heymsfield SB and Pierson RN Jr 2005 Bioelectrical impedance analysis: population reference values for phase angle by age and sex *Am. J. Clin. Nutr.* **82** 49-52

[10] Jha V, Jairam A, Sharma MC, Sakhuja V, Piccoli A and Parthasarathy S 2006 Body composition analysis with bioelectric impedance in adult Indians with ESRD: comparison with healthy population *Kidney Int.* **69** 1649-53

[11] Mushnick R, Fein PA, Mittman N, Goel N, Chattopadhyay J and Avram MM 2003 Relationship of bioelectrical impedance parameters to nutrition and survival in peritoneal dialysis patients *Kidney Int.* 64: S53-6

[12] Kalantar-Zadeh K, Ikizler TA, Block G, Avram MM and Kopple JD 2003 Malnutrition-inflammation complex syndrome in dialysis patients: causes and consequences *Am. J. Kidney Dis.* **42** 864-81

[13] Pupim LB, Caglar K, Hakim RM, Shyr Y and Ikizler TA 2004 Uremic malnutrition is a predictor of death independent of inflammatory status *Kidney Int.* **66** 2054-60

[14] Savica V, Ciolino F, Monardo P, Mallamace A, Savica R, Santoro D and Bellinghieri G 2006 Nutritional status in hemodialysis patients: option for on-line convective treatment *J. Renal Nutr.* **16** 237-40
[15] Gupta D, Lammersfeld CA, Vashi P, King J, Dahlk SL, Grutsch JF and Lis CG 2009 Bioelectrical impedance phase angle in clinical practice: implications for prognosis in stage IIIB and IV non-small cell lung cancer *BMC Cancer* **9** 37

Table 1. Nutritional and biochemical parameters in chronic hemodialysis patients.

	Lower Tertile (n=64)	Middle Tertile n=97)	Upper Tertile (n=46)
Phase Angle(°):*	3.64 ± 0.44	5.04 ± 0.53	6.83 ± 0.48
Weight (kg):*	66.1 ± 13.8	76.3 ± 18.5	82.5 ± 16.7
Resistance (ohms):	582 ± 126	540 ± 91	522 ± 111
BMI (kg/m^2):*	23.1± 3.8	26.7 ± 6.0	28.2 ± 6.0
S. Creatinine: (mg/dL)*	7.3 ± 2.2	8.9 ± 2.6	11.9 ± 2.7
S. Albumin (g/L):*	37.1 ± 3.3	39.5 ± 3.2	40.4 ± 2.3
Body Cell Mass (kg)*	19.9 ± 4.8	23.0 ± 5.6	28.5 ± 6.8
Fat Free Mass (kg)*	51.6 ± 11.6	53.2 ± 12.8	59.9 ± 14.4
Fat Mass (kg)	15.1 ± 8.6†	22.8 ± 12.1	22.6 ± 8.3

*P < 0.001 (difference between all groups) †P<0.01(difference between lower and upper tertile groups)

Table 2. Nutritional and biochemical parameters in chronic peritoneal dialysis patients.

	Lower Tertile(n=64)	Middle Tertile (n=97)	Upper Tertile (n=46)
Phase Angle (°):*	3.68 ± 0.38	5.11 ± 0.54	6.60 ± 0.45
Weight (kg):*	76.8 ± 16.4	78.2 ± 14.9	87.9 ± 11.6
Resistance (ohms)	503 ± 126	497 ± 124	465 ± 85
BMI (kg/m^2):	27.2 ± 3.8	26.9 ± 4.7	28.9 ± 4.5
S. Creatinine: (mg/dL)*	7.6 ± 2.7	11.0 ± 4.5	11.7 ± 4.5
S. Albumin (g/L):*	32.5 ± 5.4	35.5 ± 5.4	38.9 ± 3.6
Body Cell Mass (kg)*	21.3 ± 6.4	25.4 ± 6.1	30.2 ± 6.5
Fat Free Mass (kg)*	54.7 ± 16.0	58.0 ± 14.1	63.0 ± 13.0
Fat Mass (kg)	20.9 ± 6.5	25.4 ± 6.1	30.2 ± 6.5

*P < 0.001 (difference between all groups)

Table 3. Bioimpedance, nutritional and mortality parameters in the combined dialysis population.

	Lower Tertile (n=90)†	Middle Tertile (n=132)	Upper Tertile (n=36)
Phase angle (°):	3.65 ± 0.42	5.06 ± 0.53	6.78 ± 0.48
Weight (kg):*	69.2 ± 15.3	76.8 ± 17.6	84.0 ± 15.6
Resistance (ohms):	555 ± 123	528 ± 102	503 ± 105
BMI (kg/m^2):	24.3 ± 4.5	26.9 ± 5.7	28.4 ± 5.6
Fat Free Mass (kg):*	52.3 ± 12.7	55.0 ± 13.4	60.0 ± 14.1
Body Cell Mass (kg):*	20.3 ± 5.2	23.7 ± 5.8	28.6 ± 6.7
Expired:	41	50	13
Patient months:	1,518	2,673	982
Mortality (%):	45	38	18
x1000 pt. months:*	27	19	12

*P<0.05 (difference between all groups) †P<0.01(difference between lower and upper tertile groups)

Bioimpedance Profiling of the Limbs: Update

L C Ward[1], T Essex[2], M Bartlett[3], S Kilbreath[4] & D Brookes[3]

[1]The University of Queensland, Brisbane, Australia, 4072
[2]ImpediMed Limited, 1/50 Parker Ct, Pinkenba, Queensland, Australia, 4008
[3]Royal Brisbane Hospital, Herston, Queensland, Australia, 4006
[4]The University of Sydney, Sydney, Australia, 2006

l.ward@uq.edu.au

Abstract. Bioelectrical impedance spectroscopy (BIS) is now commonly used to assess breast cancer-related lymphoedema. Typically, the ratio of impedances of the two arms, determined at zero frequency (Z0), is used as a quantitative index of the presence of excess lymph. Measurement uses skin electrodes spanning the whole limb. However, lymphoedema may be highly localised and may involve changes other than simple fluid accumulation, e.g. increased fat and fibrosis, that also give rise to changes in impedance-related parameters such as capacitance. We have previously reported (13th ICEBI, Graz, 2007) a prototype mobile electrode probe that replaces the distal sense electrode which, when moved proximally along the arm, provides an impedance profile. We report here the further development of this technology to incorporate real-time measurement of impedance integrated with a digital measuring wheel. This allows exact synchronisation of impedance with position on the arm. A commercial BIS instrument (ImpediMed SFB7) was modified to collect impedance (R and Xc) data every msec and the mean impedance computed for each 10-mm slice. The apparent resistivity values for arm tissue were used to calculate slice volumes. These computed volumes were compared to equivalent slice volumes from perometry and DXA. The system is being further validated by correlating slice impedance parameters with lean tissue volume determined by pQCT (StraTec XCT 3000), for multiple positions along the arm. Ultimately, it is hoped that such measurements will not only allow localised tissue volume measurement but will also provide information of tissue composition in conditions such as lymphoedema.

1. Introduction

Lymphoedema is a chronic debilitating condition that affects millions world-wide. The most prevalent form is secondary lymphoedema as a consequence of damage to the lymphatic system by, for example, parasitic infection, as in the tropical condition filariasis or, and most commonly in the developed world, as a sequela of damage to the lymph system resulting from treatment for cancer. Reported incidence rates for lymphoedema post-breast cancer vary from 25 to 50% [1]. The condition is characterized by accumulation of lymph fluid in the arm on the affected side leading to swelling, immobility, loss of function and reduced quality of life. Lymphoedema is considered incurable but symptomatic relief may be obtained with drug treatment or surgery but physical therapy, massage and the use of compressive bandaging or garments are most commonly used [2].

Treatment is most effective if introduced at the earliest opportunity and, in turn, is reliant upon early diagnosis of the condition. Since lymph is an extracellular fluid, bioelectrical impedance spectroscopy (BIS) has found a place as the method of choice for early detection of lymphoedema [3]. Typically, the impedance of the at-risk arm is compared to that of the unaffected contra-lateral arm, as a ratio, and this is deemed indicative of the presence of lymphoedema if the ratio exceeds a threshold value (mean + 3 standard deviations) determined in a reference population [4]. This approach, while eminently clinically successful [5], is limited to whole arm measurements whereas lymphoedematous swelling may be localized to a limb segment such as the forearm only. Currently localized lymphoedema is best determined as an abnormally large limb circumference in the region of interest. However simple circumferential measurements may not be specific for lymphoedema and there is no universally accepted criterion presumptive of abnormal swelling. We have previously [6] provided a proof of concept for a BIS method for assessing localized lymphoedema. In the present report we describe the further development of this methodology and preliminary data validating this approach using peripheral quantitative computed tomography (pQCT) for comparison.

2. Materials and methods

2.1. Theory

The impedance (*Z*) of a homogeneous conductor varies with the length (*l*) of and inversely with the cross-sectional area (*a*) of the conductor:

$$Z \, \alpha \, \frac{l}{a} \qquad (1)$$

thus if a short segment of the arm is considered to approximate a cylinder the impedance of the segment will reflect the segment length, cross-sectional area and the resistivity (ρ) of the tissue composition of the segment. The whole arm may be considered to be a series of cylinders of changing cross-section and resistivity due to varying tissue composition connected end-to-end. Therefore, the impedance of the arm measured between a fixed point, e.g. the shoulder, and a continually varying point, e.g. an electrode traversing longitudinally along the arm from a distal point, the wrist, will define an impedance profile along the limb. Since the volume of the intervening segments can be determined from the relationship:

$$Vol = \rho \, \frac{l^2}{Z} \qquad (2)$$

if a value for resistivity is known, then the impedance profile may be determined as a series of segment volumes defined by the linear spatial resolution of the impedance measurement.

2.2. Experimental

2.2.1. *Subjects.* For this proof of concept study, subjects were two healthy control males: subject A, aged 58 y, weighing 78.5 kg and body mass index (BMI) of 25.8 kg/m^2 and subject B aged 49 y, weighing 62.3kg with a BMI of 19.7 kg/m^2.

2.2.2. *Impedance measurements.* Impedance was measured using a modified ImpediMed SFB7 tetrapolar impedance monitor. Measurements were obtained along each arm, commencing at the ulnar styloid process, according the principle of equipotentials [7]. The current drive electrodes were placed at the conventional locations of the base of the fingers and the toes on the side to be measured. A sense electrode (Ag-AgCl gel electrode, 23 x 24 mm, Impedimed, Brisbane) was placed on the anterior surface of the contralateral arm at the mid-point between the ulnar styloid processes. The distal sense electrode (that at the wrist) on the ipsilateral arm was replaced by a 1 cm diameter stainless steel disc electrode that could be slid smoothly along the surface of the skin. The electrode probe also housed a purpose-built rotating disc planimeter that interfaced with the SFB7 firmware to record the distance travelled at 1.58 cm resolution. The modified SFB7 hardware and firmware allowed recording of impedance at a single selected frequency in the range 5 kHz to 1 MHz at a sampling rate of 1 reading per msec. Scans along the arm commenced by locating the probe at the ulnar styloid process and after approximately 3 sec, the probe was slid smoothly along the dorsal skin surface towards the acromion where the probe remained stationary, but still recording, for a further 3 sec. Four to eight replicate scans were performed on each arm. A complete scan required approximately 20 seconds. Measurements were performed at 5 kHz, a frequency appropriate for measurement of extracellular volume, including lymph, only [8]. Electrical conductivity for the probe was assured by a very light application of conductivity gel along the arm prior to measurement.

2.2.3. *Tissue volume measurements.* Cross-sectional muscle area and fat area were measured by pQCT (XCT 3000 Stratec, Inc., Pforzheim, Germany) on both arms, in 20 slices at 2 cm intervals commencing at a position distal to the shoulder such that the final slice ended at the wrist. The arm was held horizontal to the body and supported on X-ray transparent bubble wrap in a hemispherical cross-section PVC tube. Each scan took approximately 1 hour. The radiation dose for a single scan was approximately 0.3 μ Sv, which represents less than 5% of the typical natural background radiation dose acquired each day. Scans were performed according to the manufacturer's recommendations. Image processing and calculations of numerical values were performed using the manufacturer's software package (version 5.4, Stratec Inc.). Cross-sectional muscle, bone and fat areas were determined using thresholds of 280 mg/cm^3 for bone, 40 mg/cm^3 for muscle and -80 mg/cm^3 for total tissue area. Adipose tissue area was determined by subtraction. In one subject, total arm volume was also measured by opto-electrical perometry [6].

2.2.4. *Data analysis.* Although transit speed of the probe was not constant, each scan accumulated a total of 10000 data points, equivalent to approximately 400 data points per 1.58 cm slice. The average resistance for each 1.58 cm slice was computed for each scan and the mean and standard deviation computed across all replicate scans. Since pQCT slices were at each 2 cm interval, the resistance for each equivalent slice was computed by interpolation from the resistance slice data and expressed as ohm/cm for each of the 20 pQCT slices along the arm. Volumes of muscle, bone and adipose tissue for each pQCT slice were computed from the slice thickness and measured tissue areas. The ECF volume of each slice was computed from these volume data assuming hydration factors of 0.73 for muscle and 0.15 for adipose tissue and ECF to total fluid ratios of 0.33 for muscle tissue and 0.4 for adipose tissue [9]. Bone was assumed to contain no ECF. The apparent ECF resistivity of each slice was computed from these values.

3. Results and Discussion

Typical, and essentially identical, impedance profiles, along both the left and right arms, for subject A are presented in figure 1. The overall coefficient of variation (CV) between replicate impedance scans

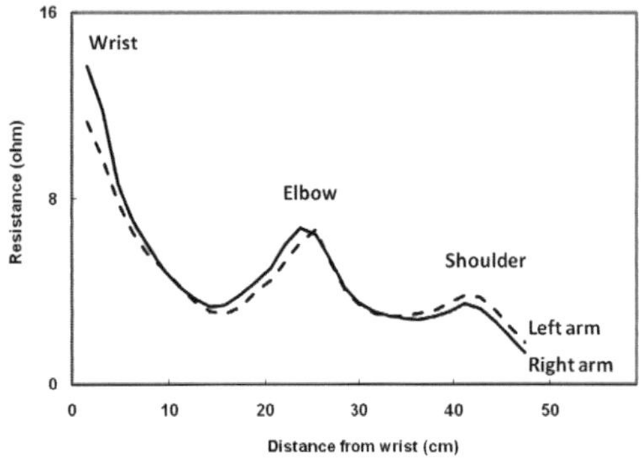

Figure 1. Typical resistance profiles of the right and left arms of subject A. Resistance was measured continuously from the wrist to the shoulder and the mean value for each 1.58 cm slice calculated and plotted.

averaged 7% but was notably higher in the region of the wrist and shoulder. Poorer reproducibility (typical CV of 15%) was obtained for ventral surface scans.

The calculated apparent ECF resistivities for the 20 pQCT slices were: for subject A, mean resistivity was 25.2 ± 3.9 ohm.cm and 25.5 ± 2.5 ohm.cm for subject B. Recalculating the resistivities assuming that the arm was represented as a single cylinder yielded slightly higher values of 27.9 and 27.3 ohm.cm respectively. These computed resistivity values are remarkably similar despite clear differences in tissue composition and overall size of the arm between each subject (figure 2), e.g. percentage muscle masses were 62.7% and 83.7% for subjects A and B respectively. This lends confidence to the view that the method is providing a robust measure of apparent ECF resistivity largely independent of tissue composition and geometry. The mean computed resistivities for each subject were used to predict the ECF volume for each slice of the other subject and these predicted values compared to those determined by pQCT. These data are presented in figure 2. The predicted ECF profile for each subject closely corresponds to that determined by pQCT. The correlations between the measured pQCT ECF volume for each slice with those predicted from impedance were 0.89 and 0.93 for subjects A and B respectively. For subject A, total ECF volume was 338.9 ml determined from impedance and 332.8 ml by pQCT. Corresponding values for subject B were 306.4 ml and 299.8 ml respectively.

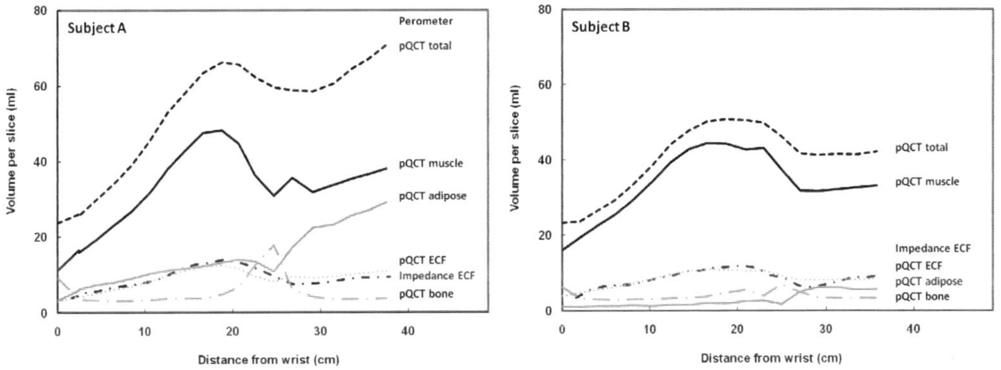

Figure 2. Tissue volume profiles of the left arms for two subjects with differing tissue composition. Predicted ECF volume profile is presented for comparison with that determined by pQCT. For subject A, the correspondence between computed total pQCT tissue volume and volume measured by perometry is also shown.

4. Conclusion

The impedance probe provides a practical method for determining the impedance profile of a limb. Measurement of limb ECF volume by tracer dilution is not feasible. However, pQCT provides an independent method for ECF determination that allows calculation of apparent ECF resistivity, facilitating prediction of ECF volume from impedance measurements. These studies are being extended to the assessment of arms of women with expanded ECF volume due to lymphoedema.

References

[1] Warren AG, Brorson H, Borud LJ and Slavin AS 2007 Lymphedema a comprehensive review *Ann. Plast. Surg.* **59** 464–72

[2] Moseley A, Carati C and Piller N 2007 A systematic review of common conservative therapies for arm lymphoedema secondary to breast cancer treatment *Ann. Oncol.* **18** 639-46

[3] Ward LC 2009 Early detection of lymphoedema: is BIS ready for prime time as the gold standard measure? *J. Lymphoedema* **4** 52-6.

[4] Ward LC 2006 Bioelectrical impedance analysis: proven utility in Lymphedema risk assessment and therapeutic monitoring *Lymphat. Res. Biol.* **4** 51-6

[5] Schonholz SM 2009 Preoperative assessment enables the early detection and successful treatment of lymphedema *Cancer* **115** 909

[6] Ward LC, Kilbreath SL, Lee M-J and York SL 2007 Bioimpedance profiling of limb lymphoedema *Proc. XIII Int Conf Electrical Bioimpedance & VIII Conf on Electrical Impedance Tomography (Graz, Austria, 29 August - 2 September) IFMBE Proc* **17** 624-7

[7] Cornish BH, Jacobs A, Thomas BJ, and Ward LC 1999 Optimizing electrode sites for segmental bioimpedance measurements *Physiol. Meas.* **20** 241-50

[8] Ward LC, Piller N and Cornish BH 2001 Single or multiple frequency bioelectrical impedance analysis for the assessment of lymphoedema? *Proc. XI Int Conf on Electrical Bio-Impedance (Oslo, Norway,17-21June 2001)* Eds. S Grimnes, O Martinsen and H Bruwal. pp373-6

[9] Roche AF, Heymsfield SB and Lohman TG, Eds. 1996 *Human Body Composition* (Champaign: Human Kinetics Press)

Journal of Physics: Conference Series **224** (2010) 012106 doi:10.1088/1742-6596/224/1/012106

Influence of body position, food and beverage consumption on BIS measurements

G Medrano[1], F Eitner[2], A H Ismail[1], R Pikkemaat[1], A Cordes[1], J Floege[2] and S Leonhardt[1]

[1] Philips Chair for Medical Information Technology, RWTH Aachen University, Aachen, Germany

[2] Department of Nephrology (Medical Clinic II), University Hospital, RWTH Aachen University, Aachen, Germany

medrano@hia.rwth-aachen.de

Abstract. Continuous monitoring of fluid changes using bioimpedance spectroscopy (BIS) during hemodialysis could help to predict hypotensive complications and extend the patient's life. Food and beverage consumption during the treatment may influence the measurements and the calculated fluid removal. In the present article the change observed in whole body and segmental (knee-to-knee, abdomen) BIS measurements following a sequence similar to the one of dialysis treatment (lying down, sitting and eating, lying down) on healthy subjects is presented. The measurements have been performed using a commercial bioimpedance device with a frequency range of 5 kHz to 1 MHz. Knee-to-knee measurements seem to be less sensitive to these influences, compared to the standard whole body and the alternative abdomen BIS measurements. The results indicate that the individual influence of both body posture and food and beverage consumption may be superposed when combined.

1. Introduction

Bioimpedance spectroscopy (BIS) measurements enable the determination of body fluid content, constituting a method easier to use and to implement compared to the standard methods [1], [2]. Its implementation in a continuous monitoring application could be combined in future with intelligent textiles [1] for the hemodialysis treatment, where continuous monitoring of body fluids could help to predict hypotensive complications and to extend the life of the patient [3]. Despite these advantages, BIS measurements are mostly limited to laboratory conditions, including a determined body posture, no consumption of food or beverage, etc. These conditions cannot be fulfilled during the dialysis therapy, where the patients remain 3 - 5 hours on a dialysis chair or a bed and change their body posture (sitting or lying down) and sometimes have a small meal and a drink.

The influence of body position on BIS measurements is produced by body fluids redistribution [4], [5], [6]. A change from vertical (or sitting) to horizontal position produces a fluid shift from the extremities into the torso [7], which produces an increase of the extremities' impedance and a decrease of torso impedance [4]. As the torso contributes approx. to only 5% of the whole body impedance, the measured whole body impedance will increase [4], [5]. A change from horizontal to vertical (or sitting) position produces the opposite situation: a fluid shift from the torso into the extremities and a reduction of the whole body impedance [4], [5]. Keeping a specific body position shows no stabilization, even after 4 hours' time [6].

Food and beverage consumption produces an increase of whole body impedance (one hour after consumption) and a decrease two hours later [8]. The continuous increase of body impedance due to ultrafiltration during dialysis seems to be interrupted during food consumption [9].

The purpose of the present article is to compare the magnitude of both influences and of the combined result on whole and segmental BIS measurements on healthy subjects.

2. Basics of bioimpedance spectroscopy (BIS)

Total body fluid (TBF) in humans consists of intracellular (ICF) and extracellular fluids (ECF), separated by the cellular membrane. The ECF, which includes the interstitial fluid and plasma, and the ICF are predominantly electrical resistive entities, whereas the cellular membrane has a capacitive character. Accordingly a low frequency current only flows around the cells through the ECF, whereas a high frequency current will also pass through the cell membrane and the ICF (see figure 1 left).

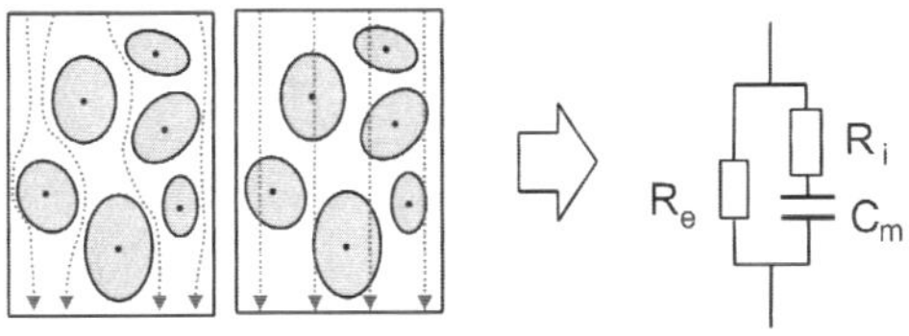

Figure 1. Low and high frequency current flow through body tissue.

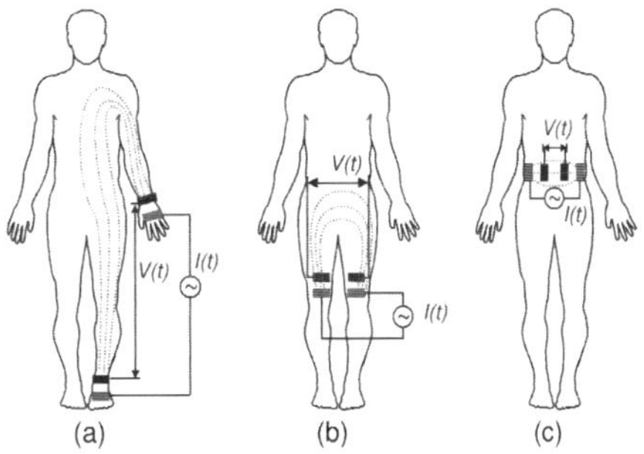

Figure. 2. Location of the electrodes for whole body (a), knee-to-knee (b) and abdomen (c) BIS-measurements.

This phenomenon can be represented by the electrical model given in figure 1 (right) [10]. Taking into consideration a heuristic factor (α) representing different tissues in parallel with specific time constants and a time delay (T_D) produced by the speed of signal transmission through measurement cables, the measured impedance can be written as follows:

$$Z(j\omega) = \left(\frac{R_e}{R_e + R_i}\right)\left(R_i + \frac{R_e}{1 + \left(j\omega C_m (R_e + R_i)\right)^{\alpha}}\right)\left(e^{-j\omega T_D}\right) \tag{1}$$

The values of the electrical model R_e, R_i and C_m can be determined by fitting the modified Cole model [11] of (1) to the impedance measured at different frequencies (ω). Using the Cole parameters (R_e, R_i and C_m), the basics of Hanai Theory [12] and devising the human body as a cylindrical volume, the body fluids volumes (ECF, ICF, TBF) can be calculated [13].

3. Materials and methods

3.1. Whole body and segmental BIS measurements

Whole body and segmental BIS measurements (Xitron Hydra 4200, Xitron Technologies Inc. San Diego, CA, USA) were performed on 4 healthy test subjects, 1 female and 3 males aged between 23 and 27 (see figure 2). Abdomen measurements were performed only on 2 male subjects. Since the expected body fluid changes and redistribution (due to ultrafiltration or changes in body posture respectively) are mainly limited to the extracellular compartment [2], [13], only R_e has been taken into account for the present study.

3.2. Influence of body posture and food and beverage consumption during hemodialysis situation
In order to evaluate the influence of changes in body posture and of food and beverage consumption (as during dialysis) on BIS measured values two sequences were used.
Sequence A: 15 min sitting, 60 min lying down, 60 min sitting and 30 min lying down. No food or beverage consumption was allowed. The bed setup for the lying down and sitting position was registered for every subject at the beginning and was used for the whole experiment.
Sequence B: same as Sequence A, with consumption of food (a pizza of approx. 30 cm of diameter) and beverage (a 200 ml soda drink) being allowed during the second sitting phase (60 min).
The measurements were performed in the dialysis clinic of the university hospital (RWTH Aachen university), room temperature 24 °C, using hospital beds for dialysis treatment. A period of at least 4 hours without eating or drinking and 48 hours without doing sports were prerequisite for the tests.
Assuming a superposition of both influences (body posture and food and beverage consumption) in sequence B, the influence from food and beverage consumption alone is found from the difference between both sequences as follows:

$$R_{e,Food\ Consumption} = R_{e,\ SequenceB} - R_{e,\ SequenceA} \quad (2)$$

4. Results

The relative change of R_e (mean values) for all the measurements is given in figure 3. The maximum relative change in R_e (mean values) for each sequence have been summarized in table 1.

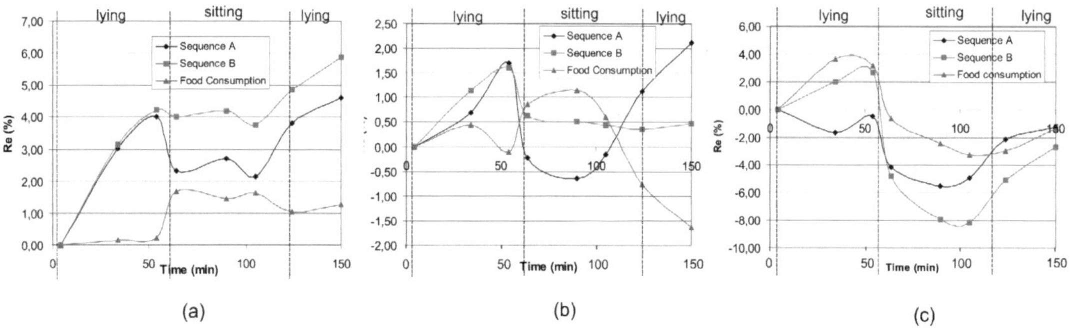

Figure 3. Relative R_e (mean) values for sequences A and B and their difference (as calculated with (2)) on whole body (a), knee-to-knee (b) and abdomen (c) measurements.

Table 1. Maximum relative change (start value as reference) in R_e (in %) during both sequences and their difference.

Measurement	Sequence A	Sequence B	Difference (B-A)
Whole body	+ 4.5	+ 6.0	+ 1.8
Knee-to-knee	+ 2.0	+ 1.6	- 1.6
Abdomen	- 5.0	- 8.5	- 3.0

Sequence A (influence of body posture): R_e values change as expected [5] for whole body BIS measurements; increases during the first lying position, decreases during the sitting position and rises up again during the final lying position. A similar pattern of change is observed with knee-to-knee and abdomen measurements (see figure 3). Magnitudes of change in whole body and abdomen measurements are bigger than on knee to-knee measurements (see table 1).
Sequence B (Influence of body posture combined with food and fluid consumption): the combined influence on R_e relative change can be seen during the sitting and second lying down position. Compared to the values of sequence A, the direction of the change in every body posture remained the same, but the magnitude of the change has been either reduced or increased. For whole body

measurements, the form of the change resembles the one associated to food consumption during a dialysis treatment [9]. Changes in knee-to-knee measurements showed again the smallest change.

The influence of food and beverage consumption alone is calculated according to (2). R_e increases with food consumption in whole body measurements as reported by Fogelholm et al. [8]. The calculated magnitude of change one hour after meal intake (1%) is slightly higher then the reported one (0.6%) [8]. It is not possible to confirm the subsequent reported decrease due to the short time. The other results could not be compared with other investigations, as similar experiments were not found in the literature. The biggest change was observed on the abdomen, where R_e values decrease, most probably due to an increase of local blood perfusion related with digestion. The use of (2) for whole body measurements seems to be validated by the agreement of the calculated change with values of the literature and by the small difference of both sequences in the first period (influence of body posture alone). In the case of abdomen measurements, the validity of the superposition principle seems to be limited: see differences between sequence A and B in the first period (influence of body posture alone). Further validation may consider a higher number of test subjects.

5. Conclusions

The present preliminary study indicates that body posture and food consumption influence BIS measurements in different magnitude, depending on the place measured and probably superposing each other. Knee-to-knee measurements seem to be much less influenced (up to 2 %) in contrast to abdomen and whole body measurements (up to 9 %). Whole body measurements during dialysis may be more influenced by the change of body posture related to the consumption of food than from the consumption of food itself.

Acknowledgment

This work was supported in part by the Mexican Science and Technology Board (CONACYT).

References

[1] Medrano G, Beckmann L, Zimmermann N, Grundmann T, Gries T and Leonhardt S 2007 *4th Int. Workshop on Wearable and Implantable Body Sensor Networks (BSN) (Aachen) IFMBE Proc.*vol **13**, pp. 23-28
[2] Chamney P, Krämer M, Rode Ch, Kleinekofort W, Wizemann V 2002 *Kidney Int.* **61** 2250–58
[3] Zhu F, Leonard E and Levin N 2008 *Physiol. Meas.* **29** 491-501
[4] Zhu F, Schneditz D, Wang E and Levin N 1998 Dynamics of segmental extracellular volumes during changes in body position by bioimpedance analysis. *J. Appl. Physiol.* **85** 497-504
[5] Fenech M and Jaffrin M 2004 Extracellular and intracellular volume variations during postural change measured by segmental and wrist-ankle bioimpedance spectroscopy. *IEEE Trans. Biomed. Eng.* **51** (1) 166-75
[6] Kushner R, Gudivaka R and Schoeller D 1996 *Am. J. Clin. Nutr.* **64 (3)** 423-27
[7] Hinghofer-Szalkay H and Moser M 1986 *Am. J. Physiol.* H68–H75
[8] Fogelholm M, Harri S, Kukkonen-Harjula K, Oja P and Vuori I 1993 Effects of meal and its electrolytes on bioelectrical impedance In: *Human Body Composition: in Vivo Methods, Models and Assessment* (New York: Plenum Press) pp. 331-2
[9] Medrano G, Eitner F, Floege J and Leonhardt S 2010 A novel bioimpedance technique to monitor fluid volume state during hemodialysis treatment *ASAIO J* (on print)
[10] Kanai H, Haeno M and Sakamoto K 1987 *Med. Prog. Technol.* **12** 159-70
[11] Cole, K and Cole, R 1941 Dispersion and absorption in dielectrics. *J. Chem. Phys.* **9**, pp. 341-51
[12] Hanai T 1968 Electrical properties of emulsions in *Emulsions Science*, P. Sherman, Ed. London and New York: Academic Press, pp. 374–475
[13] Matthie J 2005 *J. Appl. Physiol.* **99** 780–81
[14] Maw G, Mackenzie I and Taylor N 1995 *Acta Physiol. Scand.* **155** 157–63

Journal of Physics: Conference Series **224** (2010) 012107 doi:10.1088/1742-6596/224/1/012107

Wavelet-based correlations of impedance cardiography signals and heart rate variability

Sergey Podtaev[1], Andrew Dumler[2], Rodion Stepanov[3], Peter Frick[3] and Kirill Tziberkin[1]
[1]Department of Physics, Perm State University, Bukirev str. 15, Perm 614990, Russia
e-mail: podtaev@termodat.ru
[2]Department of Hospital Surgery, Perm State Medical Academy named after E.A.Vagner, Perm, Kuibishev str. 39, Perm 614990, Russia
[3]Institute of Continuous Media Mechanics, Korolyov str.1, Perm 614013, Russia
e-mail: frick@icmm.ru

Abstract. The wavelet-based correlation analysis is employed to study impedance cardiography signals (variation in the impedance of the thorax $z(t)$ and time derivative of the thoracic impedance $(-dz/dt)$) and heart rate variability (HRV). A method of computer thoracic tetrapolar polyrheocardiography is used for hemodynamic registrations. The modulus of wavelet-correlation function shows the level of correlation, and the phase indicates the mean phase shift of oscillations at the given scale (frequency). Significant correlations essentially exceeding the values obtained for noise signals are defined within two spectral ranges, which correspond to respiratory activity (0.14-0.5 Hz), endothelial related metabolic activity and neuroendocrine rhythms (0.0095-0.02 Hz). Probably, the phase shift of oscillations in all frequency ranges is related to the peculiarities of parasympathetic and neuro-humoral regulation of a cardiovascular system.

1. Introduction

It is well known that the spectral analysis of heart rate variability (HRV) is a simple method for non-invasive quantitative estimation of the autonomic input in the heart. It can also be used as a valuable prognostic characteristic [1]. Just like HRV, impedance cardiography signals (variation in the impedance of the thorax $z(t)$ and time derivative of the thoracic impedance $(-dz/dt)$) have a complex spectral structure, including nonstationary components. Only two of these, viz. cardiac and respiratory, are studied properly. The low frequency (<0.15 Hz) components of impedance cardiography (ICG) signals have not been adequately explored, even though they may contain important physiological information.

The frequency-domain analysis enables decomposition of the HRV signal into high – (HF, 0.15 – 0.4 Hz), low – (LF, 0.04 – 0.15 Hz) and very low – (VLF, < 0.04 Hz) frequency oscillations. The HF oscillations are linked with the respiratory activity and are generally considered as a marker of vagal modulation. The LF range refers to the intrinsic myogenic activity and is modulated by both the sympathetic and parasympathetic nervous systems. The neurogenic (sympathetic) regulation mechanisms manifest themselves within the range 0.02 – 0.06 Hz. The analysis of long-term recordings can yield additional information concerning the ultra low-frequency (ULF < 0.003 Hz) components related to endothelial metabolic activity and neuroendocrine rhythms [2].

Recently, a wavelet-transform-based correlation analysis has been developed. This technique is a method to study not only the spectral composition of unsteady signals, but the degree of correlation of two signals, separately at each timescale, as well [3,4]. The purpose of the present paper is to explore the low frequency components of impedance cardiography oscillations and to perform the wavelet-transform-based correlation analysis of ICG and the HRV signals within a wide frequency range 0.5 – 0.0095 Hz. The correlation function of the measured signals was compared with the correlation of two noisy signals simulated using a Gaussian distribution.

2. Measuring technique and data analysis algorithm

2.1. Measurements

Twenty healthy men at the age of 19 to 22 participated in the study. A method of computer thoracic tetrapolar polyrheocardiography was used for hemodynamic registration [5]. A 20-min ICG and HRV time series were recorded simultaneously and then were analyzed by wavelet transformation. The values of the HRV and impedance cardiography parameters were compared for the supine body positions.

2.2. Wavelet analysis

The continuous wavelet transform decomposes the function of one variable (time) $f(t)$ in the 2D space (time and scale space) as

$$W(a,b) = \frac{1}{a}\int_{-\infty}^{+\infty} f(t)\psi\left(\frac{t-b}{a}\right)dt, \tag{1}$$

where t is the time, b is the time shift (wavelet position), a is the oscillation scale which has the dimension of time and corresponds to the inverse of the frequency $\nu = 1/a$, and $\psi(x)$ is the oscillating function termed the analyzing wavelet. We use here the function, known as the Morlet wavelet,

$$\psi(x) = \exp(2\pi ix)\exp(-x^2/2) \tag{2}$$

which is represented as a product of two functions $\phi(x) = \exp(2\pi ix)$ and $\xi(x) = \exp(-x^2/2)$.

The wavelet image $W(a,b)$ of the analyzed function $f(t)$ will be complex because we use the complex analyzing wavelet. Since the signal has finite length, the gaps on its boundaries produce artifacts. These boundary effects are caused by the violation of the admissibility condition

$$\int_{-\infty}^{+\infty} \psi(x)dx = 0 \tag{3}$$

when part of the wavelet goes outside of the interval under study, or when it overlaps the gap in the signal.

In order to remove artifacts caused by boundary effects, we propose the use of the gapped wavelet technique [5]. When the wavelet is disturbed by a boundary or a gap, the admissibility condition is restored by including a constant shift, *A*, in the oscillatory part of the wavelet as

$$\int_{-\infty}^{+\infty} (\phi(x) + A)\xi(x)dx = 0. \tag{4}$$

The parameter *A* can be determined for each scale, *a*, and position, *b*, with

$$A(a,b) = \int_{-\infty}^{+\infty} \psi\left(\frac{x-b}{a}\right)dx\left(\int_{-\infty}^{+\infty} \xi\left(\frac{x-b}{a}\right)\right)^{-1}. \tag{5}$$

The mathematical properties of the gapped wavelet technique were studied in detail in work [6]. The gapped wavelets not only suppress the noise caused by gaps and boundaries, but also improve the accuracy of frequency determinations for short or strongly gapped signals.

2.3 Wavelet cross-correlation

Let us consider two functions $f_1(t)$ and $f_2(t)$. The cross-correlation function of their wavelet-transforms is defined as a normalized scalar production

$$C(a)=\int_{-\infty}^{+\infty}W_1(a,b)W_2^*(a,b)\,db\left(\int_{-\infty}^{+\infty}W_1^2(a,b)\,db\int_{-\infty}^{+\infty}W_2^2(a,b)\,db\right)^{-1/2}, \tag{6}$$

where $W_1(a,b)$ and $W_2(a,b)$ are the corresponding wavelet transforms and * stands for conjugation. The cross-correlation function $C(a)$ is complex; its modulus shows the level of correlation of oscillations at the given scale (frequency) and takes on a value in the segment [0:1], and the phase of $C(a)$ indicates the mean phase shift of oscillations of the given scale. The wavelet cross-correlation (6) was firstly introduced in [3] for the analysis of correlations of different characteristics of solar activity.

3. Results and discussion

The typical integral wavelet spectra of thoracic impedance oscillations and wavelet-correlations $z(t)$ and HRV signals obtained with the use of gapped wavelets are given in figure 1.

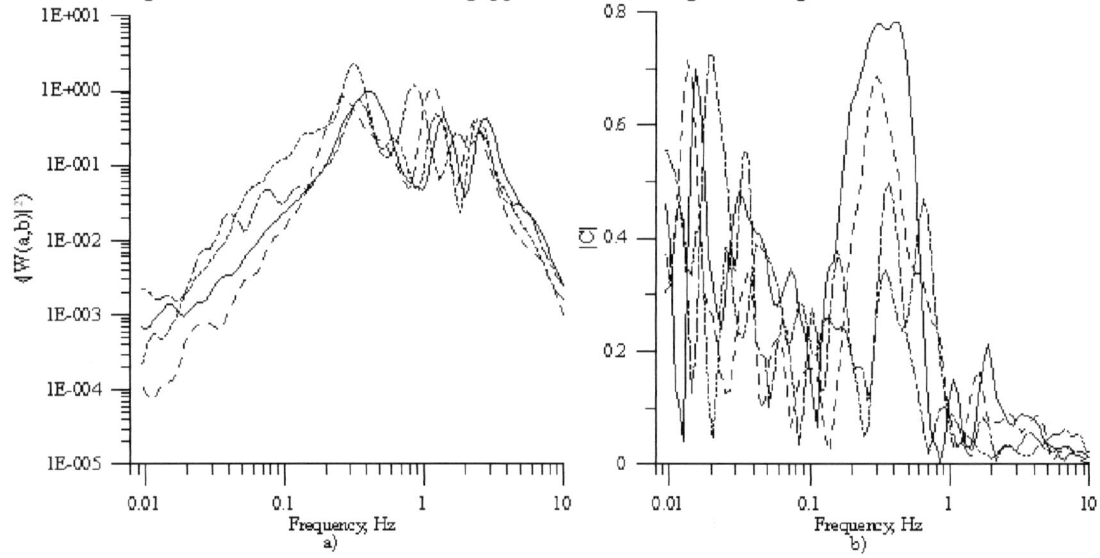

Figure 1. Typical integral wavelet spectra of thoracic impedance oscillations (a) and wavelet-correlations $z(t)$ and HRV signals (b) obtained with the use of gapped wavelets.

For wavelet spectra, five characteristic frequency intervals were defined: I (0.0095-0.02 Hz), II (0.02-0.05 Hz), III (0.05-0.14 Hz), IV (0.14-0.5 Hz), V (0.5-2 Hz). The modulus of wavelet-correlation function shows the level of correlation and the phase indicates the mean phase shift of oscillations at the given scale (frequency). The upper bound in the HRV spectrum is about 1 Hz, and therefore, the wavelet correlations are defined within the range 0.5 – 0.0095 Hz. For comparison, the mean values of correlation were calculated for 15 pairs of noise signals generated according to the Gauss distribution. To verify the hypothesis concerning the difference between the mean values, the non-parametric U-criterion by Mann/Whitney was applied. The statistically valid difference ($p<0.01$) between the values of correlation of the measured signal and those of the noise signals was found for I and IV frequency intervals.

Rhythmic changes in the heart beat within the range 0.14-0.5 Hz are known as RSA (respiratory sinus arrhythmia) show the high values of correlation with ICG signals. It is seen that within this range for ICG signals, the wavelet-spectra have the highly pronounced maximum, which is also caused by the influence of respiratory waves. Of particular interest is a high correlation of signals in the range 0.0095-0.02 Hz. This frequency range has not been extensively studied, and it is supposed to be responsible for rhythmic metabolic activity, in particular, the activity of the endothelium. The mean phase shift over the whole measurement interval of the examined oscillations with a high correlation value ranges from -2.7 to 2.4 radians. The supplementary wavelet analysis methods demonstrate that during measurements the dynamics of the phase shift variation is changed. We can observe different variants of these changes, which may lead to the phase coincidence of the signals and to the motions out of phase. To gain insight into the reason of these differences, it is necessary to conduct additional studies including examination of different functional tests.

4. Conclusions

We have investigated the wavelet-spectra of ICG signals and determined the reliable high values of correlation of these signals with the HRV signals within the frequency range corresponding to the respiratory wave and metabolic activity. Probably, the phase shift of oscillations within these frequency ranges is related to the peculiarities of parasympathetic and neuro-humoral regulation of a cardiovascular system. The proposed approach can be used in clinical practice for early diagnostics of cardiovascular system remodelling in the course of different pathologies. The early detection system for cardiac insufficiency can be offered to practitioners.

Acknowledgements

This work was supported by Russian Foundation for Basic Researches and the Government of Perm Territory under grant RFBR-Ural 09-01-99012.

References

[1] Camm A J, Malik M, Bigger J T, Breithardt G, Cerutti S, Cohen R J et al 1996 (Task Force of the European Society of Cardiology and the North American Society of Pacing and Electrophysiology) Heart rate variability: standards of measurement, physiological interpretation and clinical use *Circulation* **93** 1043–65

[2] Lotrie M B, Stefanovska A, Stajer D and Urbaneie-Rovan V 2000 Spectral components of heart rate variability determined by wavelet analysis *Physiol. Meas.* **21** 441–57

[3] Nesme-Ribes E, Frick P, Sokoloff D, Zakharov V, Ribes J-C, Vigouroux A and Laclare F 1995 Wavelet analysis of Maunder minimum as recorded in Solar diameter data. *Comptes Rendus Acad. Sci. Paris* **321** Serie IIb 525-32

[4] Podtaev S, Morozov M and Frick P 2008 Wavelet-based correlations of skin temperature and blood flow oscillations. *Cardiovascular Engineering* **8** 185–9

[5] Zubarev M, Dumler A, Shutov V and Popov N 1999 Assessment of the left ventricular systolic function and diastolic time intervals by the bioimpedance polyrheocardiographic system. *Ann. N.-Y. Acad. Sci.* **873** 191–6

[6] Frick P, Grossmann A and Tchamichian Ph 1998 Wavelet analysis of signals with gaps. *Journal of Mathematical Physics* **39** 4091–107

Wavelet analysis of bioimpendancometric data

A. Dumler[1], M. Zubarev[1], N. Muraviev[1], A. Mamatova[1], N. Salnikova[1], S. Podtaev[2], R. Stepanov[3], P. Frick[3]

[1] Department of Hospital Surgery, Perm State Medical Academy named after E.A.Vagner, Kuibishev str. 39, Perm 614990, Russia

[2] Department of Physics, Perm State University, Bukirev str. 15, Perm 614990, Russia

[3] Institute of Continuous Media Mechanics, Korolyov str.1, Perm 614013, Russia

E-mail: dumler_a@dom.raid.ru

Abstract. Up-to-date bioimpedancometric methods offer a wide spectrum of data that can be used for complex analysis of cardiovascular system state. Still, the use of appropriate mathematical approaches for data processing and calculation of main parameters is essential for confident diagnosis. The data processing problems are mainly connected with unavoidable noise sources, device noises, necessity to differentiate the registered data, pattern recognition of the structures responsible for specific fragments of the heart cycle and for the integral characteristics. In this work wavelet analysis is offered to resolve the various upcoming problems. Approaches based on decomposition of the analyzed signal on the base of special functions - wavelets - allow filtration of noises, artefacts and trends caused by side processes. They offer a wide spectrum of spectral and correlation analysis of synchronously recorded signals (for polyrheocardiograf those are impedance signals, cardiogram and phonocardiogram). Wavelet decomposition allows to distinguish high-frequency device noise from low-frequency variations caused by breathing, for example. Use of original wavelet differentiation algorithm allows to combine filtration and calculation of the derivatives of rheocardiogram. Time-spectral representation of the data on the surface forms the wavelet-portrait that gives images with relief markers of cardiac cycle phases. Utilization of the offered mathematical method raises the self-descriptiveness of impedancometric examination of cardiovascular system and makes more accurate the definition of traditional hemodynamic parameters.

1. Introduction

Wavelet-based time-frequency analysis is the known approach to select the disturbance free impedance cardiography signal from the noisy input. The possibilities of this method are although much wider and are not limited just by digital filtration of the signals. Objective of this work is an approbation of a new method of processing and interpretation of impedance polirheocardiography data [1] using wavelet analysis. The main goal is to make more accurate the calculation of traditional hemodynamic parameters by removal of negative influence of breathing wave and usage of noise-resistant differentiating algorithms. It is also necessary to develop new diagnostic criteria based on pattern recognition of wavelet images of cardiovascular pathologies. To achieve the goal we have processed signals of differential rheorgram (DR), phonocardiogram (PCG) using program pack of wavelet algorithms.

Wavelet transform presents a kind of "local" Fourier transform, allowing us to isolate a given structure in physical space and the Fourier space. Let us define the wavelet transform of the

Journal of Physics: Conference Series **224** (2010) 012108 doi:10.1088/1742-6596/224/1/012108

analyzing function $F(t)$ as

$$w_F(a,b) = \frac{1}{|a|} \int\limits_{-\infty}^{\infty} F(t)\psi^* \left(\frac{t-b}{a}\right) dt, \qquad (1)$$

where $\psi(t)$ is the analyzing wavelet, a defines the scale and b defines the position of the wavelet. Then the coefficient w_F gives the contribution of corresponding structure into the function F.

The function F can be reconstructed using the inverse transform (see, e.g. [2])

$$F(t) = \frac{1}{C_\psi} \int\limits_{-\infty}^{\infty} \int\limits_{-\infty}^{\infty} \psi\left(\frac{t-b}{a}\right) w_F(a,b) \frac{da\, db}{a^2}. \qquad (2)$$

The reconstruction formula (2) exists under condition that

$$C_\psi = \frac{1}{2} \int\limits_{-\infty}^{\infty} \frac{|\hat{\psi}(k)|^2}{|k|} dk < \infty. \qquad (3)$$

Here $\hat{\psi}(k) = \int \psi(\phi) e^{-\imath k\phi} d\phi$ is the Fourier transform of the analyzing wavelet $\psi(\phi)$.

Let us considered the function $F(t)$ is given by its derivative $G(t)$. Applying differentiation by part to equation (1) we have

$$w_F(a,b) = \frac{1}{|a|} \int\limits_{-\infty}^{\infty} G'(t)\psi^* \left(\frac{t-b}{a}\right) dt = \frac{1}{|a|} \int\limits_{-\infty}^{\infty} G(t)\chi^* \left(\frac{t-b}{a}\right) dt, \qquad (4)$$

where $\chi(t) = -\psi'(t)$ is another wavelet. So the wavelet transform with wavelet family $\chi(t)$ gives the wavelet coefficients of the derivative of the analyzing function. Use of original wavelet differentiation algorithm [3] allows to combine filtration and calculation of the derivatives of rheocardiogram. We use as the analyzing wavelet the so-called Mexican hat $\psi(t) = (1-t^2)\exp(-t^2/2)$ for higher resolution of separated picks in time and the wavelet Morle $\psi(t) = \exp\left(-\frac{t^2}{2} + 2i\pi t\right)$ for better spectral resolution.

2. Some results of the wavelet analysis

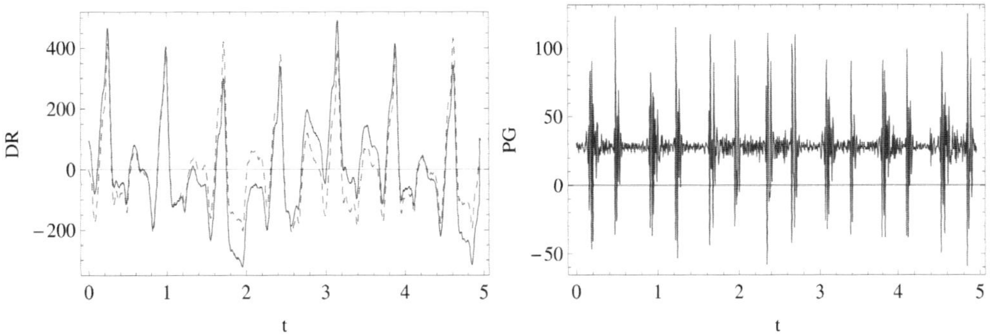

Figure 1. Left panel: original (solid line) and filtered (dashed) DR. Right panel: Original PCG.

We used DR and PCG to calculate the stroke volume. The analyzed signals are shown on figure 1. One can see the results of breathing filtering in DR. PCG demonstrates moments of the valves closing and opening.

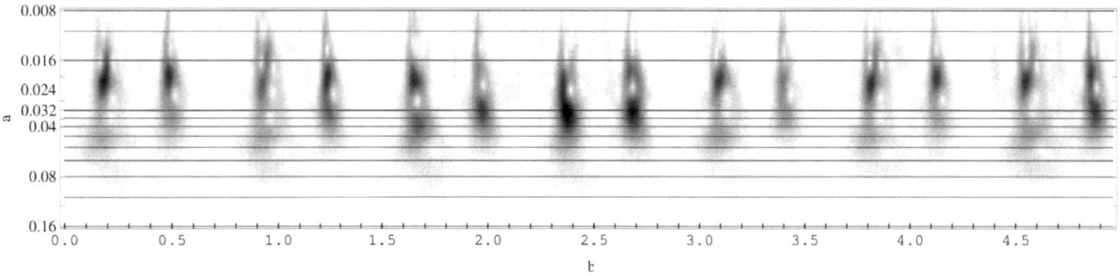

Figure 2. Absolute value of wavelet coefficient distribution $w_{PCG}(a, b)$ for PCG.

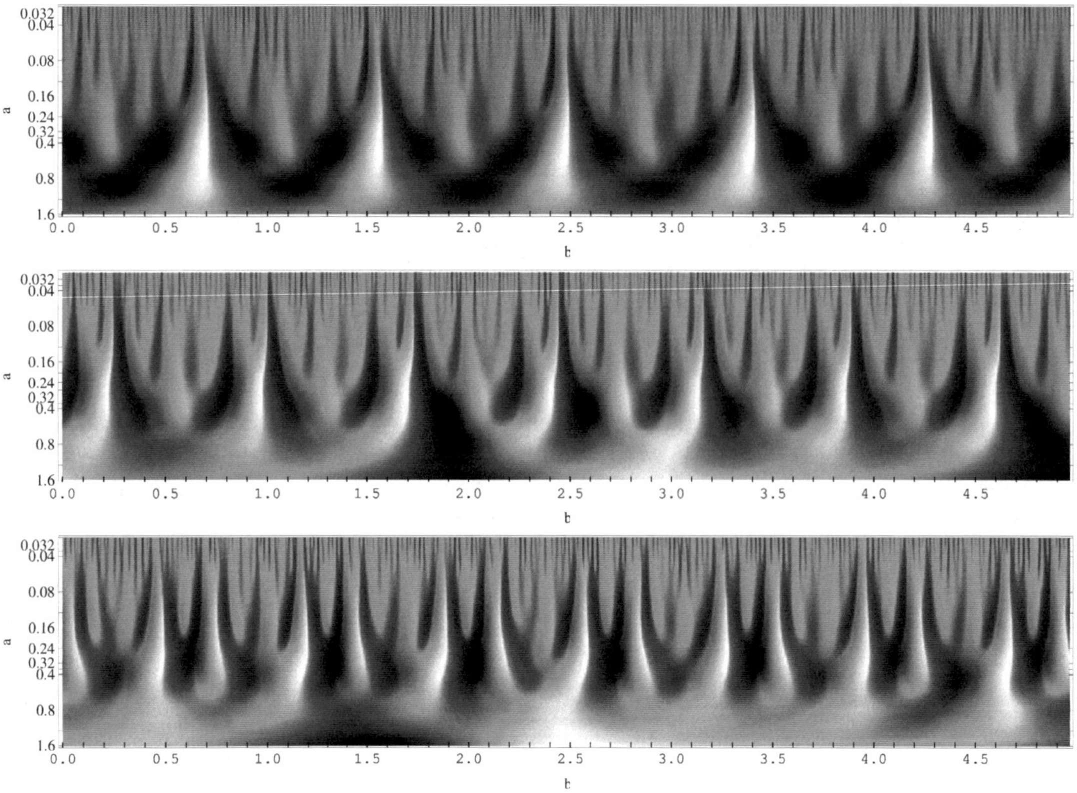

Figure 3. Wavelet coefficients relief of DR for three typical examples: healthy patients (top), hypertensive patient (middle), infarct patient (bottom). White corresponds to the maximum of $|w_{DR}(a, b)|$ and black corresponds to the minimum.

For the better identification of the systolic time interval, we performed the wavelet decomposition of PCG. The wavelet coefficient map of PCG is presented on figure 2. Value of $|w_{PCG}(a, b)|$ gives altitude of the fluctuation with frequency $1/a$ at the time moment b. Maxima in the dependence $|w_{PCG}(a = 0.02, b)|$ can be used for robust definition of the begin and the end of systolic phase.

Figure 3 demonstrates two dimensional patterns of the wavelet images obtained from DR for three typical cases. Healthy patients have distribution consisting of straight structures corresponding to dominant systolic maxima. Same image for the hypertensive patient has fewer pronounced structures. One can see the arising structures in the diastolic phase of the cardiac

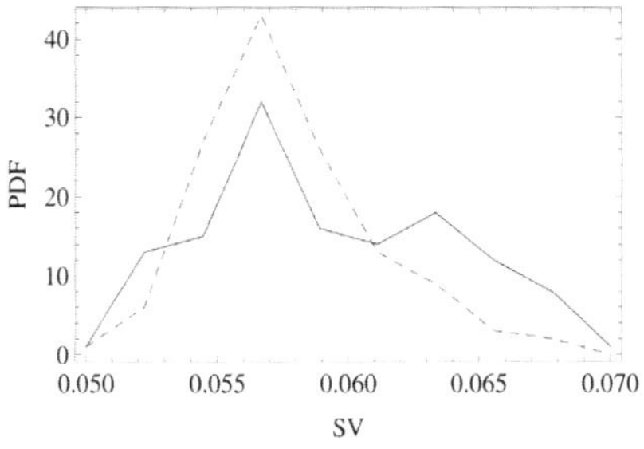

Figure 4. Probability density function of stroke volume using original (solid line) and filtered (dashed) DR.

cycle. Infarct patient has two pronounced structures per one cardiac cycle. The differences are quite enough significant and might be use for pathology recognition.

Maxima of the wavelet coefficient at $a = 0.4$ were used to calculate the systolic amplitude. Probability density function of stroke volume over 100 heart beats is shown on the figure 4. Breathing filtration gives less variation in the stroke volume.

Application of the wavelet analysis seems to be a promising tool for visualization of bioimpendancometric data and for obtaining a robust statistics of the hemodynamic characteristics.

Acknowledgements
This work was supported by Russian Foundation for Basic Researches and the Government of Perm Territory under grant RFBR-Ural 09-01-99012.

References

[1] Zubarev M, Dumler A, Shutov V and Popov N 1999 Assessment of left ventricular systolic function and diastolic time intervals by the bioimpedance polyrheocardiographic system *Electrical Bioimpedance Methods: Applications To Medicine And Biotechnology* (*Annals Of The New York Academy Of Sciences* vol 873) ed Riu, PJ and Rosell, J and Bragos, R and Casas, O (New York Acad Sciences) pp 191–196
[2] Daubechies I 1992 *Applied Mathematics* **61**
[3] Stepanov R, Frick P, Shukurov A and Sokoloff D 2002 *A&A* **391** 361–368 (*Preprint* `arXiv:astro-ph/0112507`)

Broadband spectroscopy of a dynamic impedance

Mart Min[1,3], **Raul Land**[1,3], **Toivo Paavle**[1,2], **Toomas Parve**[1] **and Paul Annus**[1,2]
[1] Department of Electronics, Tallinn University of Technology,
Ehitajate tee 5, 19086 Tallinn, Estonia
[2] ELIKO Competence Centre,
Teaduspargi 6/2, 12618 Tallinn, Estonia
[3] Institute for Bioprocessing and Analytical and Measurement Techniques,
Rosenhof, 37308 Heilbad Heiligenstadt, Germany

E-mail: min@elin.ttu.ee

Abstract. Impedance spectrum of dynamic systems is time dependent. For example, fast impedance changes take place in high throuput microfluidic devices. Also, the impedance of cardiovascular system is dynamic. Measurements must be as short as possible to avoid significant impedance changes during the spectrum analysis and, at the same time, as long as possible for enlarging the excitation energy. The authors propose to use specific short chirp pulses for excitation. Thanks to unique properties of the chirp function, it is possible to meet the needs for spectrum bandwidth, measurement time, and signal-to-noise ratio so that the most accurate impedance spectrogram is obtained. The chirp wave excitation pulse can include thousands of cycles when the impedance changes slowly, but in the case of very high-speed changes it can be even shorter than a single cycle, preserving the same excitation bandwidth. For example, 100 kHz bandwidth can be covered by the chirp pulse with duration from 10 μs to 1 s, only its excitation energy differs also 10^5 times.

1. Introduction

Bioimpedance spectroscopy enables to characterize the biological matter, including label free identification of bioparticles in the microfluidic lab-on-a-chip (LoC) [1, 2], see figure 1. Broadband excitation is preferred when the speed of analysis is important [1-7]. Different signal waveforms are used [3, 4-7], the chirp wave is assumed to be the most suitable because of scalability [3-5].

The result of spectral analysis of dynamic system is not merely a spectrum but spectrogram – discrete time series of spectral snapshots. To avoid dynamic uncertainties, we have to perform short time measurement. On the other hand, we have to use long measurement time for maximizing the signal-to-noise ratio: (a) – through smoothing out the background noise, and (b) – via enlarging the excitation energy. The optimal solution is a balanced trade-off between process dynamics, measurement time, signal-to-noise ratio, and excitation bandwidth. A key factor for finding a successful compromise, is independent scalability in time (duration of excitation) and frequency (excitation bandwidth). The chirp signal provides such the double scalability [2-4]. A simple linear chirp has an instantaneous frequency $\omega(t) = \mathrm{d}\theta(t)/\mathrm{d}t$, which changes linearly during the excitation interval T_{exc}. Expressing $\omega = 2\pi f$ and denoting f_1 as an initial, and f_2 as a final frequency and marking T_{ch} as a duration of the chirp pulse, we obtain the following expression for the linear chirp excitation:

$$v_{ch}(t) = v_{\sin}(t) = A\sin\left(2\pi\left(f_1 t + \left(f_2 - f_1\right)\cdot t^2 / 2T_{ch}\right)\right) \quad (1)$$

The excitation bandwidth $B_{exc}=f_2-f_1$ remains the same [2] when the excitation time $T_{exc}=T_{ch}$ changes. Only the chirping rate $(f_2-f_1)/T_{ch}$ and signal energy $E=(A^2)T_{ch}$ vary together with T_{ch}.

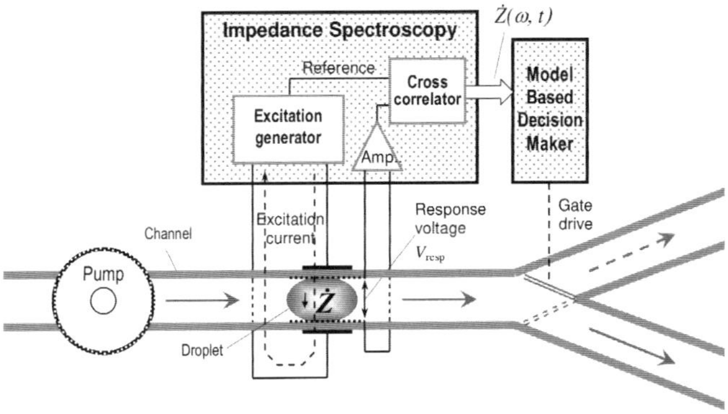

Figure 1. Generalized diagram of the LoC using the impedance spectroscopy method for sorting of bioparticles.

2. Proposed excitation waveforms

Traditionally is assumed that the chirp excitation contains a big number of cycles (rotations by 2π of the signal generating vector). The authors have analyzed shorter excitations and propose to use the chirp pulses having only some of cycles (e.g., 1 to 3) or even a part of a single cycle. They introduce a term "titlet" for denoting such very short chirplet type signals referring to short tweetings of tit (a bird *parus*, in Latin). The titlets allow to match requirements for spectrum bandwidth and excitation time for obtaining accurate impedance impedance spectrograms. As the lowest frequency f_1 is usually much smaller than the highest one, f_2, it is reasonable to introduce a simplification $f_1=0$ into (1):

$$v_{\sin}(t)=\sin\left(2\pi\left(f_2\cdot t^2/2T_{ch}\right)\right) \tag{2}$$

Analysis of (2) shows ($f_1=0$) that the minimal excitation time is limited by the highest frequency f_2 in the excitation bandwidth B_{exc} and by the number of cycles p in the excitation signal:

$$\min T_{exc}=2p/f_2 \tag{3}$$

Moreover, the choice of bandwidth B_{exc}, duration $T_{exc}=T_{ch}$, and number of cycles p of short linear chirps is limited with the discrete values (figure 2). For example, for $p=1$ and $f_2=100$ kHz, the minimal excitation time $T_{exc}=20$ µs. For $p=½$ and $p=¼$, the excitation time reduces to 10 and 5 µs.

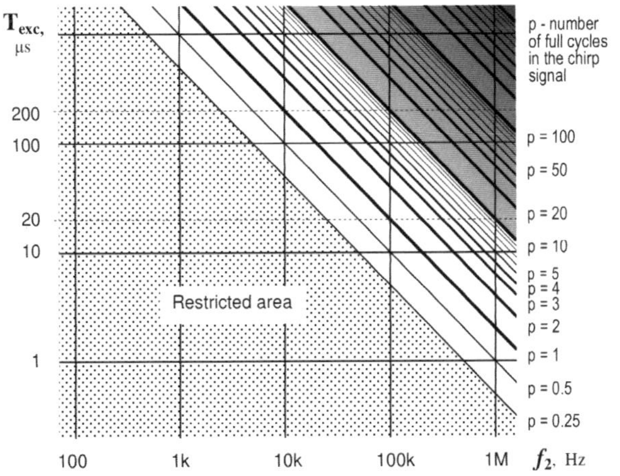

Figure 2. Nomogram of relationships between the excitation time T_{exc} (the same value as duration T_{ch} of the chirp pulse), upper frequency f_2 (practically the same as excitation bandwidth B_{exc}, because the lower frequency $f_1<<f_2$ can be neglected, see (1) and (2)), and the number p of cycles (full-turn rotations by 2π of the generating vector) in the linear chirp pulse excitation signal.
Remark: excitation signals can contain only a fraction of the full cycle, e.g., $p=0.5$ corresponds to a half-turn π, and $p=0.25$ to a quarter of turn $\pi/2$.

3. Waveforms and spectral densities of the short-chirp titlet excitations

The chirp wave excitation pulse with linear frequency sweep can cover a wide frequency span B_{exc} from $f_1 \geq 0$ to f_2, and include lots of cycles if the speed of changes is low. More than 90% of signal energy $E = (A^2/2)T_{exc}$ [2] falls into B_{exc} and, therefore, the power spectral density (PSD) of a voltage signal is near to E/B_{exc}, V^2/Hz. The spectral density of RMS value, RMSSD, is equal to $(PSD)^{1/2}$, $V/Hz^{1/2}$.

Figure 4 describes a chirp with moderate duration (200 μs), which contains $p = 10$ cycles. The shape of its normalised RMSSD (figure 5) is only slightly distorted in comparison with spectral density of long chirps ($p = 1000$ to 100 000) having almost ideally flat spectrum [3, 6] within $B_{exc} \approx f_2 = 100$ kHz.

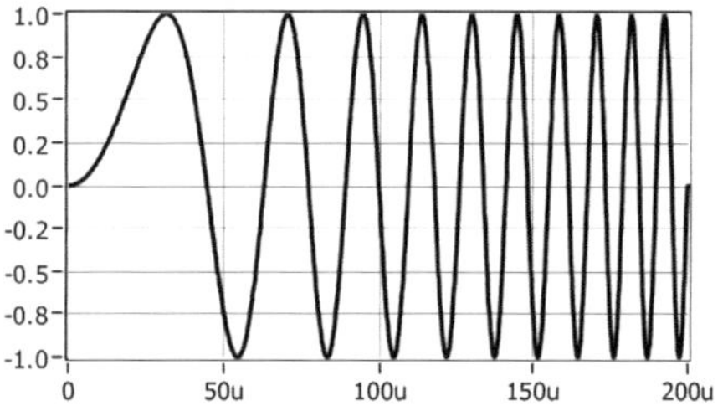

Figure 4. A 10-cycle chirp with duration 200 μs and bandwith $f_2 = 100$ kHz.

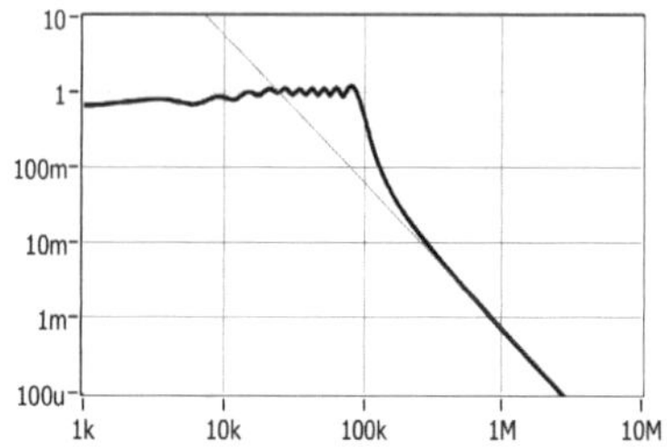

Figure 5. Normalized RMS spectral density of the 10-cycle chirp in figure 4.

In figure 6 we can find a single cycle titlet ($p = 1$, the final value of rotation phase $\theta_{fin} = 2\pi$) with duration 20 μs having the same bandwidth 100 kHz. The flatness of spectrum (figure 7) is somewhat distorted in comparison with the previous case ($p = 10$) and its actual energy is 10 times lower.

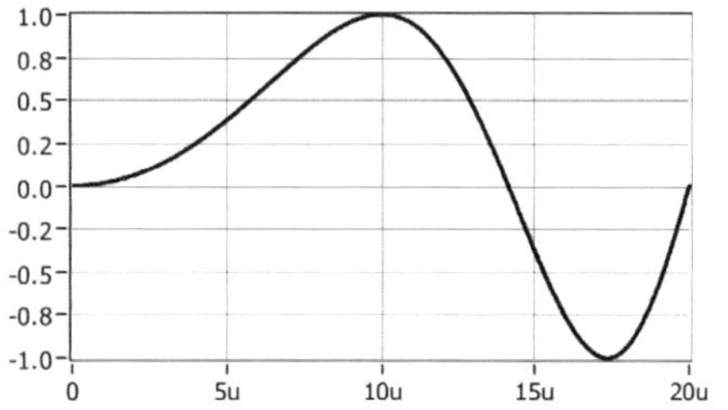

Figure 6. A single cycle ($\theta_{fin} = 2\pi$) chirp (titlet) with duration 20 μs and bandwith $f_2 = 100$ kHz.

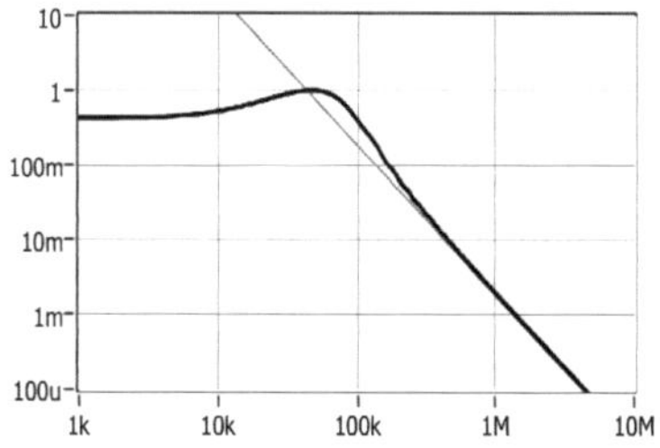

Figure 7. Normalized RMS spectral density of the single cycle titlet in figure 6.

Figure 8 describes a 1/2-cycle titlet ($p = 1/2$, $\theta_{fin} = \pi$) with 10 μs duration, which has almost perfectly flat spectrum within the excitation bandwidth $B_{exc} \approx f_2 = 100$ kHz, only its energy is 2 times lower than in the previous case ($p = 1$). The spectral density drops off steeply (-40 dB/decade) outside the excitation bandwidth.

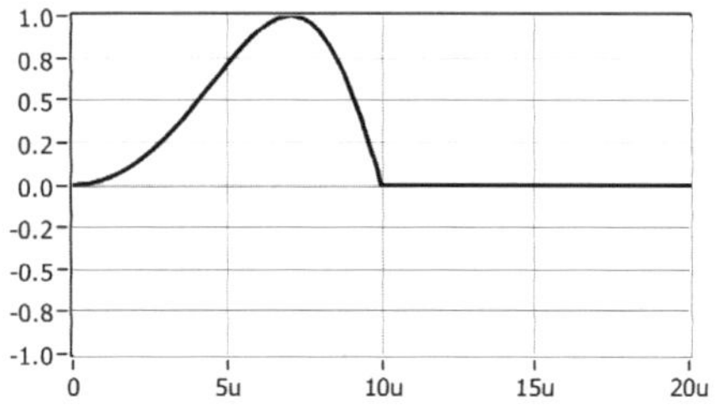

Figure 8. A half-cycle ($\theta_{fin} = \pi$) chirp (titlet) with duration 10 μs and bandwith $f_2 = 100$ kHz.

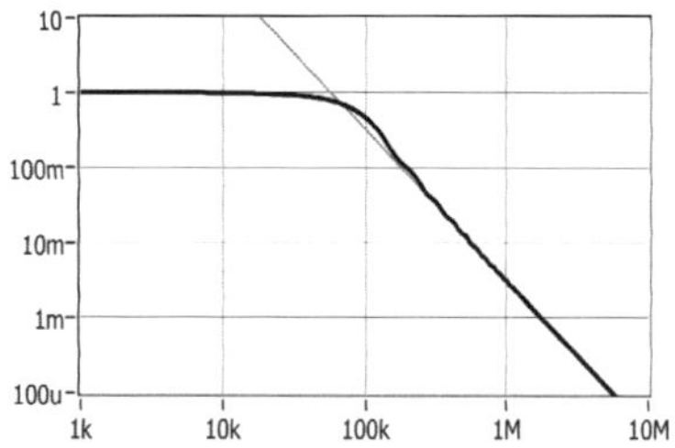

Figure 9. Normalized RMS spectral density of the half-cycle titlet in figure 8.

Finally, figure 10 describes a 1/4-cycle titlet (p=¼, $\theta_{fin}=\pi/2$) with 5 µs duration. Flatness of its spectrum (figure 11) is very near to ideal within the excitation bandwdth $B_{exc}\approx f_2$ = 100 kHz, only the spectral density drops slowly (-20 dB/decade) outside the bandwidth and the signal energy is lower.

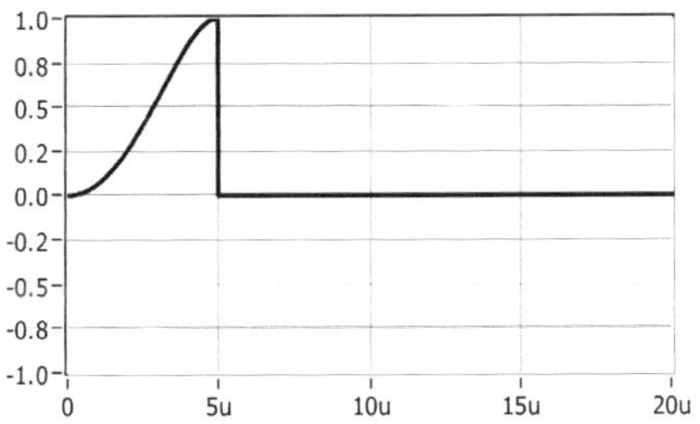

Figure 10. A quarter-cycle ($\theta_{fin}=\pi/2$) titlet with duration 5 µs and bandwith f_2 = 100 kHz.

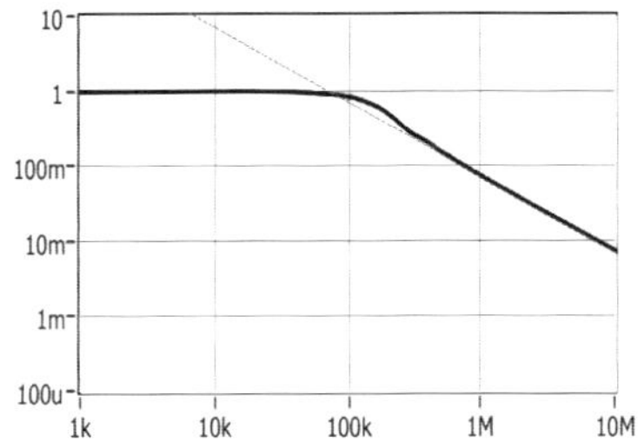

Figure 11. Normalized RMS spectral density of the quarter-cycle titlet in figure 10.

4. Conclusions

The short chirp pulse excitation (known also as a chirplet or titlet) is a perspective candidate for implementation in wideband spectroscopy of dynamic impedances. The spectral density of excitation is flat beginning from very low frequencies (fractions of one Hz or even DC) up to tens of MHz range. More than 90% of generated energy falls into the useful excitation frequency band B_{exc}, which remains invariant when the duration of excitation T_{exc} changes in tens and hundreds of times.

In other words, we can choose and change the excitation bandwidth and the excitation time independently in the best way by using the scalability in time and frequncy domain and taking into account the speed of impedance changing, noise level, and the excitation energy maximally allowed. As a result, we can maximize the amount of information derived from the data of spectral measurements.

5. Acknowledgments

The research was supported by the European Union (EU) through European Regional Development Fund, Enterprise Estonia through the ELIKO Competence Center, and by the EU FP6 Marie Curie Fellowship Transfer of Knowledge Project #29857 "InFluEMP - Integration of Fluidic and Electric Microscaled Principles".

References

[1] Sun T, van Berkel C, Green N G, Morgan H 2009 Digital signal processing methods for impedance microfluidic cytometry *Microfluid Nanofluid* **6** 179-87 DOI 10.1007/s10404-008-0315-3

[2] Min M, Giannitsis A T, Land R, Cahill B P, Pliquett U, Nacke T, Frense D, Gastrock G, Beckmann D 2009 Comparison of rectangular wave excitations in broad band impedance spectroscopy for microfluidic applications *IFMBE Proceedings* **25**/VII *World Congress on Medical Physics and Biomedical Engineering* (Sept.7-12, 2009, Munich) 85-8

[3] Nahvi M and Hoyle B S 2009 Electrical impedance spectroscopy sensing for industrial processes *IEEE Sensors Journal* **9** 1808-16

[4] Nahvi M and Hoyle B S 2008 Wideband electrical impedance tomography *Meas. Sci. Technol.* **19** DOI: 10.1088/0957-0233/19/9/094011 9 pp

[5] Min M, Pliquett U, Nacke T, Barthel A, Annus P, and Land R 2008 Broadband excitation for short-time impedance spectroscopy *Physiological Measurement* **29** S185-S192

[6] Min M, Pliquett U, Nacke T, Barthel A, Annus P, Land R 2007 Signals in bioimpedance measurement: different waveforms for different tasks *IFMBE Proceedings.* **17** *Proc. 13th Int. Conf. on Electrical Bioimpedance* (ICEBI'2007, Graz, Austria, Aug 29–Sept 2, 2007) 181-4

[7] Pliquett U, Gersing E, Pliquett F, 2000 Evaluation of time-domain based impedance measurements on biological tissue *Biomed. Tech.* **45** 6-13

On measurement errors of the impedance spectrum of human body in vivo

A V Smirnov, D V Nikolaev, V A Kolesnikov

Scientific Research Centre 'Medass', Chistoprudny blvd, 12, 101000 Moscow, Russia

Email: avs_ramb@rambler.ru

Abstract. Deviations of the impedance spectrum of human body segments from that predicted by the Cole model are explained by the properties of biological objects and measurement errors. Our analysis of the measurement's equivalent scheme shows an importance of contact impedances and allows for explaining positive reactance of low-resistance body segments. Representation of the measured impedance as a connection of the Cole model impedance and capacity in parallel gives better fitting of measured impedance spectrum. The results of analysis show that said capacity is determined mainly by the properties of body tissues rather than stray capacitance between the body and surrounding objects.

1. Introduction

The frequency dependence of biological objects' complex impedance is usually approximated by the Cole model [1, 2]. This model is used for the description of biological tissues' properties in *in vitro* measurements [3]. However, in *in vivo* measurements, deviations of the impedance spectrum from the circular arc predicted by the Cole model usually occur. Such deviations were explained previously by various factors, such as stray capacity between the body and surrounding objects, contact impedances, cable capacities, and other [4–6]. The aim of our study was to analyze quantitatively various factors influencing the results of impedance measurements and discriminate between biological tissues' effects and that of measurement errors.

2. Device and methods

BIA measurements were conducted by the analyzer ABC-01 'Medass' (Moscow, Russia) that measures complex impedance according to the conventional tetrapolar scheme using multiple (31) frequencies ranged from 5 to 500 kHz [7]. Two kinds of electrodes were used: the disposable bioadhesive electrodes with contact area 2.4×2.4 cm, and reusable AgCl electrodes with contact area 2.5×3.6 cm, denoted as E1 and E2, respectively. E2 electrodes were used with the conductive gel. Patient's cable was 1.8 m in length. The measurement was conducted in standing position distantly enough from the conductive objects with legs and arms moved apart.

The results of measurements were processed using device-supplemented software that allows for best approximation of the impedance spectrum in an operating frequency range as the impedance $\mathbf{Z}_{Cole}$ corresponding to the Cole model with the capacity C_p taken in parallel. This capacity may be influenced by biological object's own capacity, stray capacity, and measurement errors.

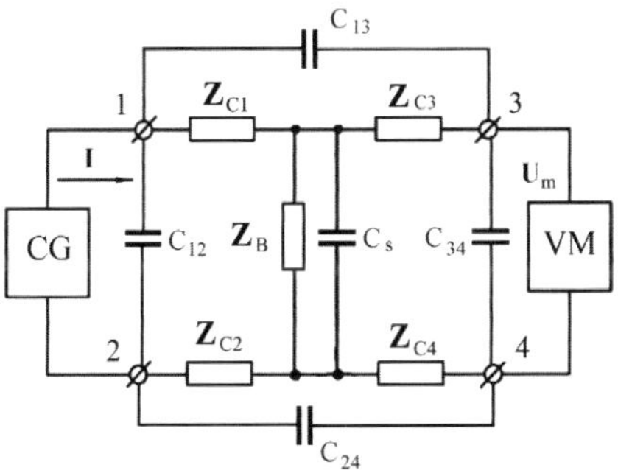

Figure 1. Equivalent scheme of the measurement circuit

BIA measurement errors were analyzed both theoretically, i.e. using equivalent scheme calculations, and experimentally, i.e. by measuring the impedances of reference circuits. Two kinds of errors were considered: calibration errors as well as extra errors due to differences between the parameters of equivalent scheme during measurement compared to those during calibration procedure. Much attention was paid to extra errors as calibration errors are systematic in nature and, hence, their influence can be minimized.

Equivalent measurement circuit is shown on Fig. 1, where 1–4 – electrodes' clamps, CG – current generator, $\mathbf{Z}_B$ – biological object impedance, C_s – stray capacity, VM – voltmeter, $\mathbf{Z}_{C1}$, $\mathbf{Z}_{C2}$, $\mathbf{Z}_{C3}$, $\mathbf{Z}_{C4}$ – contact impedances, C_{12}, C_{13}, C_{24}, C_{34} – interelectrode capacities. Each of impedances $\mathbf{Z}_{C1}$–$\mathbf{Z}_{C4}$ consists of electrode's own impedance, impedance of electrode to body contact, and impedance of body part between current and voltage electrodes. Capacities C_{12}–C_{34} are influenced by the device and cables capacities. It can be shown that the influence of the capacitance C_s is equivalent to that of distributed stray capacitance between the subject and surrounding objects.

We simulated calibration and measurement procedures. For this, $\mathbf{U}_m$ values were calculated using equivalent circuit by the node potentials method. We utilized the same values of C_{12}–C_{34} on calibration and measurements because of using the same cable. For contact impedances $\mathbf{Z}_{C1}$–$\mathbf{Z}_{C4}$, similar assumption is not reasonable because of dependence on the types of electrodes, their positioning, and the properties of skin. The measurements using analyzer ABC-01 "Medass" includes registration of the impedances $\mathbf{Z}_{13}$ and $\mathbf{Z}_{24}$ between 1-3 and 2-4 clamps. These impedances give an approximation of $\mathbf{Z}_{C1}$ and $\mathbf{Z}_{C2}$ impedances used in calculation of measured impedance. Such a correction was also implemented during simulation of measurements. The data on interelectrode capacities and contact impedances $\mathbf{Z}_{C3}$ and $\mathbf{Z}_{C4}$ were obtained from auxiliary experiments.

Verification of the measurement model present in Fig. 1 was carried out by the comparison of calculated and experimentally derived error values. Then, using the obtained error values, possible deviations of impedance spectrum under corresponding measurement conditions were analyzed.

3. Results

The results of segmental BIA measurements of the impedance spectrum for healthy volunteer are presented in Fig. 2. Similar spectrums were obtained for other volunteers. Fig. 3 shows estimates of contact impedances for electrode types E1 and E2.

Fig. 4 shows the results of theoretical and experimental analysis of extra errors. Calculations of expected values of measurement errors of the resistance R and reactance X were performed for a frequency of 500 kHz in the range of R between 50 and 400 Ohms at X/R = 0,1. The following real values of interelectrode capacities were used: C_{12} = 40 pF, C_{34} = 60 pF, C_{13} = C_{24} = 70 pF. The values of impedances that simulated contacts at calibration were $\mathbf{Z}_{C1}$ = $\mathbf{Z}_{C2}$ = 100 Ohms, $\mathbf{Z}_{C3}$ = $\mathbf{Z}_{C4}$ = = 10 Ohms. The values of contact impedances during measurement (based on Fig. 3b) were $\mathbf{Z}_{C1}$ = $\mathbf{Z}_{C2}$ = 170 Ohms, $\mathbf{Z}_{C3}$ = $\mathbf{Z}_{C4}$ = 120 Ohms for E1 electrode type, and $\mathbf{Z}_{C1}$ = $\mathbf{Z}_{C2}$ = 70 Ohms, $\mathbf{Z}_{C3}$ = $\mathbf{Z}_{C4}$ = 40 Ohms for E2 electrode type. Calculations with other parameters show that, in first approximation, the magnitude of errors is proportional to the frequency value and interelectrode capacities. Experimental estimates of measurement errors, denoted as E1exp and E2exp, were obtained using reference circuits that simulated biological object and contacts of E1 and E2 electrodes. Calibration errors were excluded by measurement of reference circuits corresponding to calibration conditions.

Results of approximation of the impedances of measured body segments using the impedance $\mathbf{Z}_{Cole}$ and capacitance C_p taken in parallel are given in Table 1.

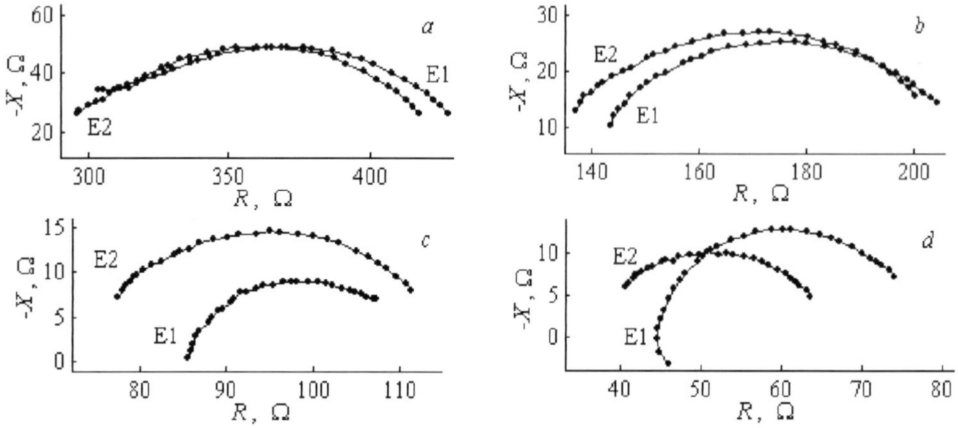

Figure 2. Impedance plots of body segments: wrist-to-ankle (a), elbow-to-knee (b), knee-to-knee (c), proximal forearm (d)

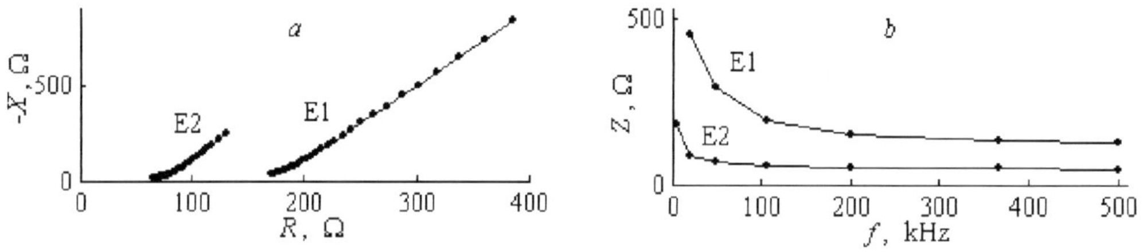

Figure 3. $\mathbf{Z}_{C1}$ and $\mathbf{Z}_{C2}$ impedance plots (a), frequency dependences of modules of $\mathbf{Z}_{C3}$ and $\mathbf{Z}_{C4}$ (b)

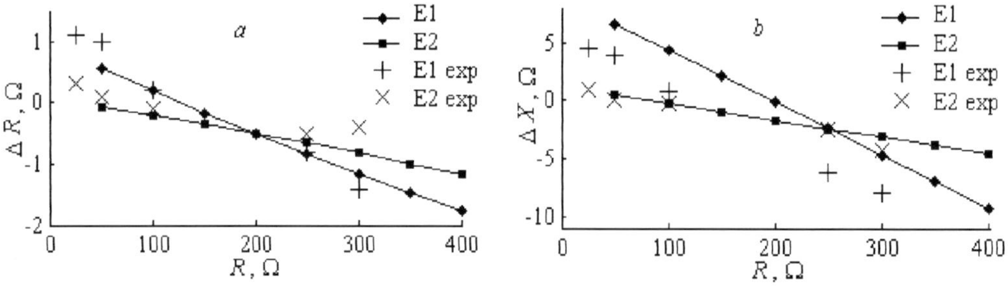

Figure 4. Resistance (a) and reactance(b) measurement errors dependences on resistance value

Table 1. The results of BIA segmental measurements

Body segment	SEE Cole, Ω	Cp, pF	SEE Cp, Ω	R_{500}, Ω	X_{500}, Ω	ΔR_{500}, Ω	ΔX_{500}, Ω	ΔR_{50}, Ω	ΔX_{50}, Ω
Wrist-to-ankle	2,43	60	0,54	297	-27.5	2,2	16.7	0,6	2,2
Wrist-to-wrist	1,03	40	0,51	327	-27.8	1,7	13.5	0,5	1,7
Elbow-to-knee	1,59	180	0,71	135	-12.8	1,1	10.3	0,5	1,5
Knee-to-knee	0,82	320	0,22	76	-7.4	0,7	5.8	0,3	0,8

Results in the Table 1 were obtained using E2 electrode type. Here, SEE Cole and SEE with C_p are standard errors of the impedance approximation as $\mathbf{Z}_{Cole}$ only and $\mathbf{Z}_{Cole}$ with C_p capacitance taken in parallel; R500 and X500 are measured values of resistance and reactance at a frequency of 500 kHz; ΔR500 and ΔX500 are changes in resistance and reactance at a frequency of 500 kHz after elimination of the capacitance C_p influence on measured impedance; ΔR50, ΔX50 are similar changes at a frequency of 50 kHz.

4. Discussion and conclusions

The impedance plots shown in Fig. 2 for body segments of various resistances are similar to those published by other authors [4, 5, 8]. The results of error analysis (Fig. 4) provide an opportunity to explain the differences of impedance spectrum of different body segments. Satisfactory agreement between calculated and experimental error values suggests that such an explanation is correct.

In case of body segments with the resistances over 200 Ohms an effect is prevalent of the low-pass filter of the impedances $\mathbf{Z}_{C3}$, $\mathbf{Z}_{C4}$ and capacitance C_{34} (Fig. 2a). In the resistance range 100–200 Ohms, change of polarity of the reactance measurement error takes place for the E1 electrodes caused by the predominant effect of C_{13} and C_{24} capacitances that transmit voltage from the current generator output directly to the voltage gauge. This leads to the shift of the high-frequency end of an impedance plot in positive direction (Fig. 2b,c). In both ranges effects strongly depend on the contacts impedance. Other values of interelectrode capacities entail changes in the boundaries range.

If absolute value of the impedance is less than 100 Ohms, then the positive error of the reactance measurement with electrodes E1 becomes so large that the measured value of the reactance X reverses its sign (Fig. 2d). The error of resistance R also reverses sign (Fig. 4a). The observed form of the impedance locus is sometimes interpreted as the presence of an inductive component. But this effect is formed by joint action of the interelectrode capacities and contacts impedances. So, when measured low-resistant body segments, the electrodes contact impedance may produce unreliable results.

For sufficiently large object resistances and small contact impedances the measurement provides reliable information on the object impedance. Table 1 show that the values of ΔX500 are significantly higher than the absolute value of the expected error of the reactance when measured using E2 electrodes (Fig. 4b). So, the capacitance C_p is really present in measured impedance, and is not the result of measurement errors. Representation of measured impedance as the parallel connection of $\mathbf{Z}_{Cole}$ and C_p significantly improves an accuracy of approximation. Note also, that excluding of the C_p capacitance exerts little effect on the results of measurements at the frequency 50 kHz.

Since the value of C_p increases with the decrease of resistance of measured segment, one can draw conclusions about the nature of this capacity. Stray capacitance C_s must increase with length and, hence, with the resistance of segment. Yet, the resistance and own capacity of a homogeneous object are linked by obvious relation $RC = \rho\varepsilon$, where ρ is the resistivity, and ε the dielectric constant. Thus, the value of C_p is determined primarily by the capacitive component of the biological object impedance $\mathbf{Z}_B$ rather than by the stray capacitance C_s.

In conclusion, understanding the nature and properties of errors caused by contact impedances and interelectrode capacitances gives a possibility to improve methods of calibration and measurements by the correction of the influence of these elements.

References

[1] Cole K S 1932 *J. Gen. Physiol.* **15** 641–649
[2] Grimnes S, Martinsen O G 2000 *Bioimpedance and Bioelectricity Basics* (London: Academic Press)
[3] Gabriel S, Lau R W, Gabriel C 1996 *Phys. Med. Biol.* **41** 2271–2293
[4] De Lorenzo A, Andreoli A, Matthie J, Withers P 1997 *J. Appl. Physiol.* **82** 1542–1558
[5] Bolton M P, Ward L C, Khan A et al 1998 *Physiol. Meas.* **19** 235–245
[6] Grimnes S, Martinsen O G 2007 *J. Phys. D:Appl. Phys.* **40** 9–14
[7] Nikolaev D V, Smirnov A V, Bobrinskaya I G, Rudnev S G 2009 *Bioelectric impedance analysis of human body composition* (Moscow: Nauka)
[8] Beckmann L, Cordes A, Saygili E et al 2009 IFMBE Proceedings 25/VII, 532–535

Influence of volume and flow change on the electrical impedance signal (in vitro)

M Bodo[1], A Garcia, F Pearce, S Van Albert, R Armonda[2]
Walter Reed Army Institute of Research, Silver Spring, MD; [2]National Naval Medical Center, Bethesda, MD, USA

E-mail: michael.bodo@med.navy.mil

Abstract. On the basis of preliminary results, rheoencephalography (REG) shows promise as a practical, noninvasive and continuous monitoring modality of brain injuries. However, REG literature reflects uncertainty about whether the signal reflects flow or volume. Presented here are results of in vitro studies manipulating flow/volume to model clinical conditions (such as brain ischemia and vasospasm) while recording the electrical impedance signal. A loop was created using tubing filled with 0.9 % NaCl. This loop was comprised of a Doppler in-line flow probe connected to an ultrasound flow meter, a peristaltic pump, a pressure transducer and home-made electrical impedance measuring cell, incorporating a balloon catheter. Bipolar impedance amplifiers were used for measuring impedance pulse waves. Data were stored on a PC and processed off-line. This in vitro study confirmed that 1) Doubling flow rate influenced the pulse amplitude and mean flow of the Doppler signal; 2) Doubling flow rate had no influence on the amplitudes of the pressure or electrical impedance signals; 3) An increase in amplitude was observed in the pressure and electrical impedance signals when the first derivative was taken. 4) Balloon inflation decreased electrical impedance and Doppler flow pulse amplitudes; 5) With balloon inflation, Doppler and electrical impedance signals showed an identical relationship to decreased flow (R^2=0.966).

1. Introduction

Acute management of patients with traumatic brain/blast injury remains a challenge. To minimize secondary injury and improve outcome, it is critical to detect neurological deterioration early, when it is potentially reversible (Rubin et al, 2009). On the basis of preliminary results, rheoencephalography (REG) shows promise as a practical, noninvasive and continuous monitoring modality of traumatic brain and blast injuries. REG literature reveals disagreement about whether the signal reflects flow or volume. The U.S. Food and Drug Administration defines rheoencephalograph as "a device used to estimate a patient's cerebral circulation, without specifying flow or volume. Since electrical impedance is measured in ohms, REG measurements do not quantify flow or volume. The magnitude of REG pulse amplitude is a function of conductive particles between two electrodes. In order to test the clinical applicability of REG, we simulated two pathophysiological changes—flow and volume. Our hypothesis was that electrical impedance (REG) pulse amplitude would decrease as a result of both manipulations (flow and volume decreases). Results of in these vitro studies are reported here.

[1] Michael Bodo, 503 R. Grant Avenue, Silver Spring, MD, 20910, USA

2.Methods

An electrical impedance measuring cell was constructed to simulate changes in flow and volume. A closed loop was created with PVC and C-Flex tubing filled with 0.9 % NaCl. This loop included a Doppler in-line flow probe (4N) connected to an ultrasound flow meter (T201 Ultrasonic Bloodflow Meter, Transonic Systems, Ithaca, NY); a disposable pressure transducer (Argon Medical Devices, Athens, TX) was connected to a Blood Pressure Analyzer (Digi-Med, Micro-Med, Louisville, KY), and two stainless steel metal tubes (14 mm in length and 3 mm internal diameter) were used as electrical impedance electrodes. Signals were stored in a PC using an AD card (PCI-MIO-16XE-10, National Instruments, Austin, TX) with 16 bit resolution. The data sampling rate was 200 Hz. Data were collected and analyzed with DataLyser software.

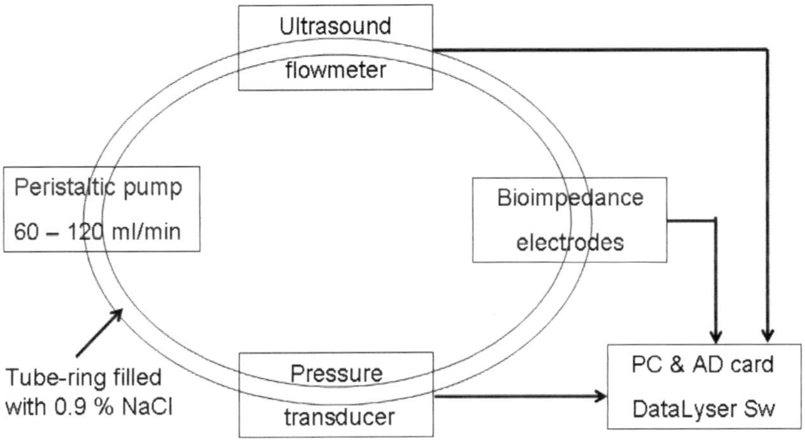

Figure 1. Block schematics of measurements.

2.1. Flow

To simulate flow change, a peristaltic pump (Masterflex, L/S, Cole-Palmer, USA) was connected to the closed loop; flow rate was set first to 60 mL/min, then increased to 120 mL/min. A bipolar impedance amplifier (KR-Ea RHEO Preamp, OTE Galileo, Italy) generated the impedance signal. Measurement frequency was 45 kHz and the resistance between the two electrode tubes was above 10 ohm (maximum balance). Electrical impedance and pressure values were calculated by taking the first derivative.

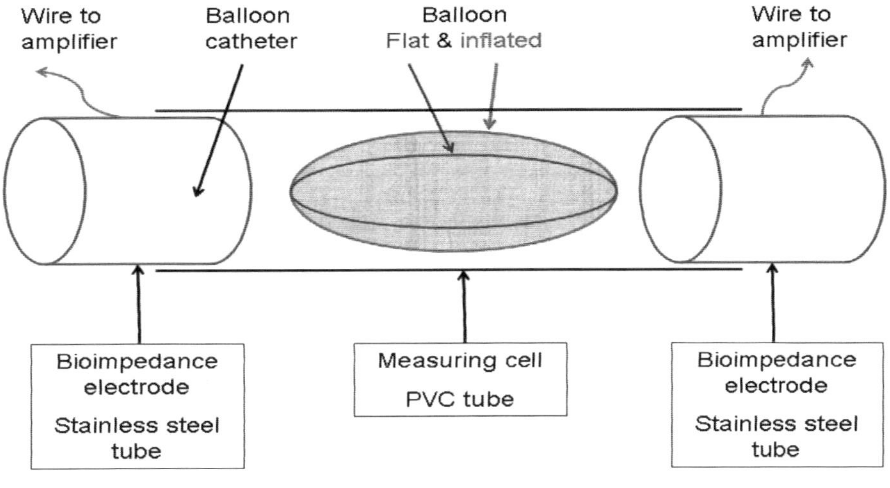

Figure 2. Schematics of electrical impedance measuring cell with balloon inflation.

2.2. Volume

To simulate decreased volume (vasoconstriction/vasospasm), a balloon catheter (PTCA Dilatation catheter, NCR14 9/4.0, Boston Scientific, SCIMED, Maple Grove, MN) was inflated, placed into impedance measuring cell (figure 2) with a syringe in 0.5 cc steps. A pump (model P720, Instech Laboratories, Plymouth Meeting, PA) was connected to the closed loop to generate pulse waves at a constant rate. The bipolar impedance amplifier used to generate the impedance signal was Cerberus (Quintlab, Hungary). Measurement frequency was 125 kHz. Electrical impedance and Doppler flow pulse amplitude measurements (minimum – maximum distance) were made at 0 (no inflation), 0.5, 1.0 and 1.5 cc inflation. Eight pulse amplitudes were measured and averaged.

3. Results

3.1. Flow

Doubling of flow produced corresponding increases in Doppler signals: pulse amplitude and mean flow increased two-fold. First derivative calculations (dZ/dt and dP/dt) made it possible to convert changes in impedance and pressure signals to corresponding amplitude increases.

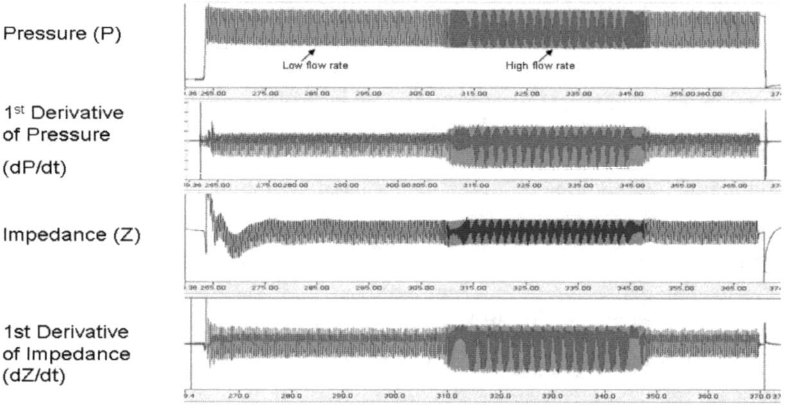

Figure 4. Pressure and impedance traces with low and high flow rate.

Table 1. Comparison of Doppler flow and electrical impedance during low and high flow rates. The effects of increased flow rate were comparable between Doppler flow and impedance when the impedance derivations (first derivative and its integral) were calculated.

		Doppler Flow			Impedance		
	Pump Flow Rate	Display	Amplitude	Mean	Amplitude	1st Derivative (dZ.dt)	Integral of 1st Derivative
	mL/min	mL/min	mL/min	mL/min	a.u	a.u.	a.u.
mean	60	67	68.13	3.58	2.83	0.09	0.44
SD			17.3	0.005	1.51	0.05	0.004
mean	120	124	124.98	4.15	2.83	0.18	0.91
SD			47.9	0.01	1.54	0.09	0.003
ratio	2.00	1.85	1.83	1.16	1.00	2.00	2.07

3.2. Volume

Balloon inflation to simulate vasoconstriction/vasospasm decreased the electrical impedance pulse amplitude, a result identical to that obtained for Doppler pulse change. Pulse amplitude of both the Doppler and electrical impedance signals decreased as a function of balloon inflation (R^2=0.966).

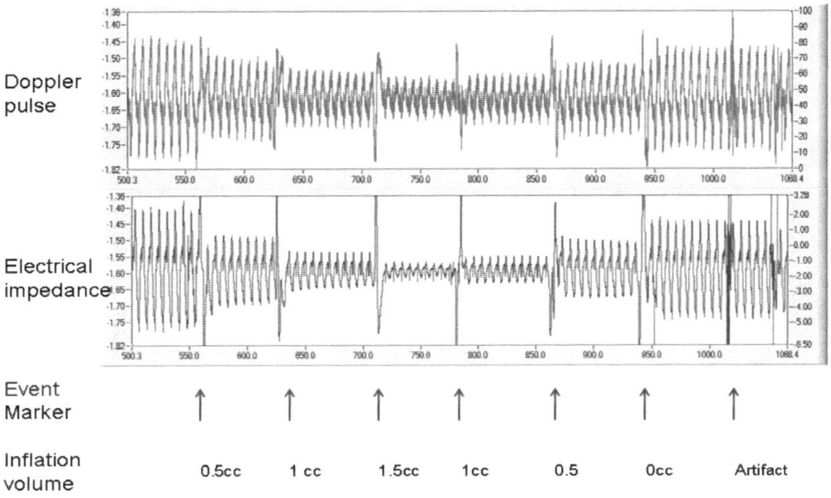

Figure 3. Effect of balloon inflation/deflation on Doppler flow and electrical impedance pulse amplitudes.

4. Conclusion

Although these in vitro electrical impedance studies manipulating flow and volume contribute to an understanding of the basic relationship between electrical impedance and flow/volume changes, biological conditions are much more complex than the simple model discussed here. Other factors influencing the REG signal have been previously summarized [1]. Results of these in vitro electrical impedance studies manipulating flow and volume were as follows:1) Modeling vasospasm/brain ischemia detected decreased pulsatile volume and flow; 2) Doppler and electrical impedance signals showed an identical relationship to decreased flow; 3) Electrical impedance pulse change reflected volume change; however, the increased flow resistance caused by balloon inflation also decreased flow.

5. Reference

[1]. Moskalenko YE ed 1980 *Biophysical aspects of cerebral circulation* (Oxford: Pergamon)

Acknowledgements

Lajos Baranyi for use of DataLyser software and Janice Meer for assistance with the manuscript.

Disclaimer

Material has been reviewed by the Walter Reed Army Institute of Research. There is no objection to its presentation and/or publication. The opinions or assertions contained herein are the private views of the author, and are not to be construed as official, or as reflecting true views of the Department of the Army or the Department of Defense.

Journal of Physics: Conference Series **224** (2010) 012112 doi:10.1088/1742-6596/224/1/012112

Fourier analysis of electrical impedance variations in urinary bladder during changes of intravesical pressure

A V Sivkov, I S Mudraya[1], S V Revenko, A R Ibragimov, A V Nesterov, I Yu Gavrilov, V I Kirpatovsky and M V Stranadko
Research Institute of Urology, National Cardiology Center, Moscow, Russia

E-mail: mis51ser@yahoo.com

Abstract. Original hardware and software system was used to record and analyze the variations of electrical impedance of the urinary bladder in narcotized rats (*n*=7) at rest, during instillation of physiological saline into the bladder resulted in elevation of intravesical pressure, and during subsequent urination accompanied by intravesical pressure release. Fourier analysis of impedance variations revealed three periodic components with the frequencies of heartbeat, respiration, and the Mayer wave (~0.1 Hz). In resting bladder, the amplitude of Mayer and respiratory spectrum peaks were high, and they increased to a different extent during physiological elevation of intravesical pressure, while a small cardiac peak did not changed significantly and tended to decrease at high intravesical pressure. Urination released intravesical pressure and restored all the peaks to the resting level. It is hypothesized that Mayer and respiratory bioimpedance oscillations of urinary bladder are neural in origin, while the cardiac peak is mainly determined by hydrodynamic arterial pulsations. The novel method can assess vesical circulation and its neural control at various phases of bladder activity.

1. Introduction

Urinary bladder function at rest and during micturition is persistently controlled by central and autonomous nervous system, and depends on adequate blood supply [1]. Alterations in these systems are reported for pathological conditions of the bladder and voiding dysfunction examined with various methods in clinical and experimental trials.

The changes in bladder blood flow during its filling were examined in male patients with the help of laser Doppler flowmetry [2]. Vascular response of normal and obstructed rabbit bladders to distension was performed by morphometry and immunohistochemistry [3]. Activation of the afferent signals from urinary bladder during its distension were examined with different methods, including systemic blood pressure and heart rate measurements, neurological study of the reflexes, and direct recording of the neural activity [4-9]. While lots of methods are known to study separately bladder neural activity and circulation, no method is available to evaluate these regulatory factors of bladder function simultaneously.

Our aim was to apply Fourier analysis to variations of urinary bladder impedance for combined assessment of its circulation and autonomic nerve activity under various intravesical pressures.

[1] To whom any correspondence should be addressed.

2. Material and methods

Experiments were performed on rats anesthetized with ether and chloralose. Filling cystometry with electromanometric pressure measurements was performed via cystotomy tube inserted into the bladder dome, while the rats were allowed to urinate spontaneously through intact urethra. Two silver chloride electrodes were attached to the bladder dome and neck in longitudinal direction to measure impedance of entire bladder. Simultaneous recordings of the intravesical (detrusor) pressure (Pdet) and the real (resistive) part of bladder impedance were carried out under the following conditions: (1) resting bladder, (2) during bladder filling (3) and after infusion arrest. Each record was obtained during 4 minutes. Original hardware and software (Biola, Russia) was used to measure the variations of electrical impedance at the measuring frequency of 100 kHz and to process them with Fourier analysis.

3. Results

Infusion of saline into the bladder increased Pdet (the arrow in figure 1, A). Fourier analysis revealed three periodic components of urinary bladder impedance corresponding to the frequencies of the heart beats, respiration, and Mayer wave of about 0.1 Hz (figure 1, B, C). These impedance oscillations changed during different periods of bladder functional activity, as figure 1 shows at rest (B) and at the onset of bladder filling with physiological saline (C). Filling increased Mayer peak (I), drastically decreased respiratory peak (II), and practically unchanged the cardiac peak (III).

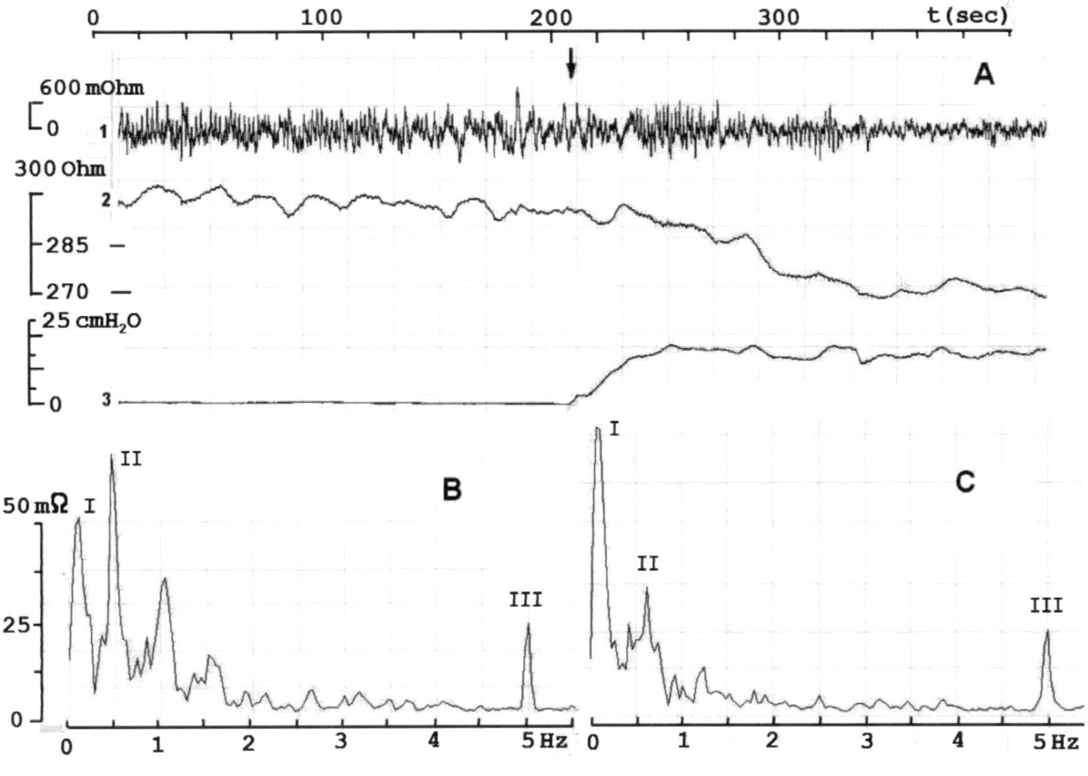

Figure 1. A- simultaneous recordings of rat urinary bladder impedance (variable *1* and basal *2* components) and pressure (*3*). B- spectrum peaks at rest, and C- during bladder filling at 0.01 ml/min.

The basal (total) impedance of bladder (curve *2*, figure 1, A) decreased during bladder filling in parallel with elevation of Pdet (curve *3*), which agrees with clinical urodynamic studies demonstrating negative correlation between electrical impedance and bladder volume [10].

Table 1 shows the average data on bladder physiological characteristics and oscillatory parameters. At rest, the power of Mayer peak was nearly the same as the respiratory one, while the power of cardiac peak was significantly smaller. During bladder filling, Pdet increased 6-fold accompanied by a 32% growth of the power of Mayer peak. The powers of respiratory and cardiac peaks decreased by 73% and 26%, respectively.

Table 1. Interrelation between intravesical detrusor pressure, power of spectrum peaks, and basal impedance of rat bladder.

Bladder status	Mayer (10^{-3} Ohm^2)	Respiratory (10^{-3} Ohm^2)	Cardiac (10^{-3} Ohm^2)	Pdet (cm H_2O)	Z (*Ohm*)
Resting bladder	13.97±0.03	15.81±0.04	0.595±0.04	4.64±1.40	448±19
During bladder filling	18.47±0.09	4.30±0.06	0.443±0.04	28.40±1.76	367±6
After infusion arrest	7.67±0.08	14.50±0.90	0.546±0.09	3.29±0.79	469±38

The Mayer peak attests to neural (sympathetic) control [11] of the empty bladder at rest and increased sympathetic activity during bladder filling (figure 1). Elevation of sympathetic activity during bladder filling is a well-known phenomenon reflected by an increase in systemic blood pressure and heart rate, sympathovagal balance shift in favor of general sympathetic activation, and increased activity in sympathetic neurons, renal, and pelvic nerve [4-6]. In our experiments, the amplitude of Mayer peak increased at the onset of infusion; in the following, it either increased (figure 1) or decreased (figure 2). Supposedly, the rate of infusion and functional state of the bladder can be responsible for these diverse reactions.

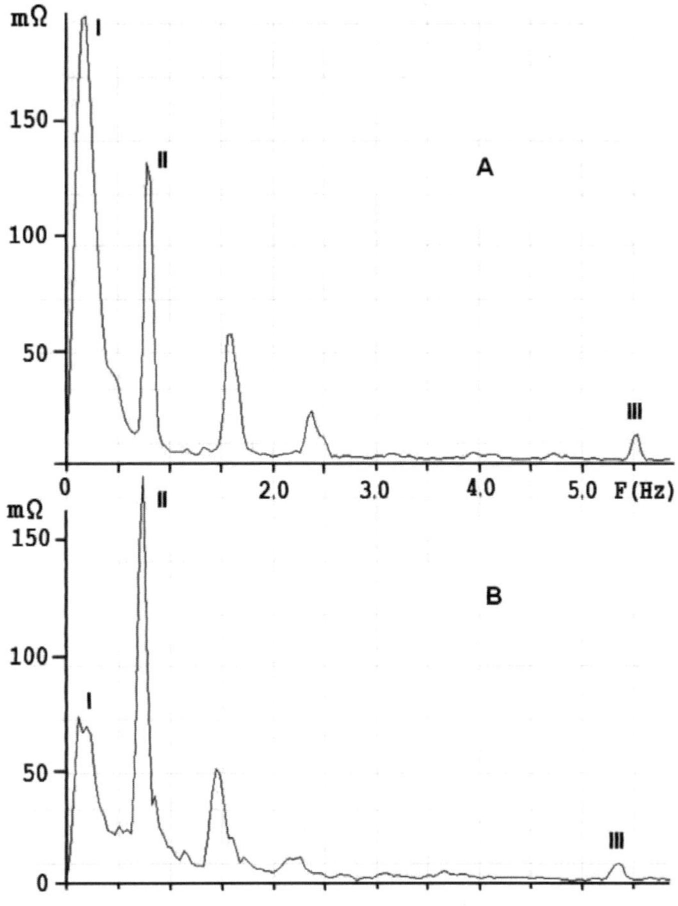

Figure 2. Three periodic components of the bladder impedance variations revealed by Fourier analysis at rest (A) and during high intravesical pressure (52.3 cm H_2O) inducing the contractile responses during bladder filling at 0.1 ml/min (B).

I, II, and III are Mayer, respiratory, and cardiac peak, respectively.

Similar controversial results were reported concerning activity of afferent neurons in the urinary tract evoked by bladder distension, which could trigger both excitatory and inhibitory reflexes [8]. Various changes of bladder afferent activity was revealed by cuff electrode recordings from the pelvic nerve during the changes in bladder pressure [7,9].

We suppose that relation between Mayer and respiratory peaks may reflect the balance of sympathetic and parasympathetic activity in the bladder, because this balance can shift during high intravesical pressure (at maximal bladder capacity) switching the bladder function from urine storage phase (bladder wall relaxation) to active bladder contraction and urination.

The important role of arterial supply to the bladder is evident, since ischemia can provoke detrusor contractile dysfunction. The vascular collapse was established with morphometric method for the bladder at all levels of its distension [3]. Laser Doppler flowmetry demonstrated the changes in vesical blood flow depending on bladder capacity and compliance. The present method can assess arterial bladder supply during different periods of its activity. In resting bladder, the power of cardiac peak was lesser compared to the power of Mayer peak; during bladder filling, the power of the cardiac peak decreased slightly (table 1). These data can be explained by application of physiological pressures in these experiments with urethra opened providing possibility to urinate voluntarily. Under large Pdet, the cardiac peak decreased pronouncedly (figure 2). In contrast, this peak did not significantly vary under low Pdet (figure 1, C).

4. Conclusions

Fourier analysis of bioimpedance variations in urinary bladder revealed three modes of oscillatory processes related to Mayer, cardiac and respiratory activity. These processes are described by the spectrum peaks whose power is determined by intensity of the changes in vascular tone and detrusor functional activity. The opposite changes in the value of Mayer and respiratory peaks attest to their predominantly regional neurogenic origin. In contrast, more stable behaviour of the cardiac peak indicates its predominantly "central" hydrodynamic nature. Therefore, Mayer and respiratory spectrum peaks seem to be important indicators of vascular tone and smooth muscle activity in the bladder. Thus, Fourier analysis of bioimpedance variations is a novel potent tool to examine autonomic neural control simultaneously with blood supply in bladder and other visceral organs.

5. References

[1] Andersson Karl-Erik and Anders Arner 2004 *Physiol. Rev.* **84** 935-86
[2] Kershena R T, Azadzoia K M and Sirokya M B 2002 *J. Urol.* **168** 1 121-5
[3] Matsumoto S, Chichester P, Kogan B A and Levin R M 2003 *Urol.* **62** 6 1129-33
[4] Matsumoto M, Matsukawa K, Murata J and Ninomiya I 2004 *Jap. J. Physiol.* **354** 6 567-74
[5] Mehnert U, Knapp P A, Mueller N, Reitz A and Schurch B 2009 *Neurourol. Urodynam.* **28** 313-9
[6] Fagius J and Karhuvaar S 1989 *Hypertension* **14** 511-7
[7] Rocha I, 2001, Burnstock G and Spyer K M 2001 *Autonomic Neuroscience* **88** 6-15
[8] Habler H J, McLachlan E M, Jamieson J and Davies P J 1999 *J. Urol.* **161** 1666-71
[9] Jezernik S, Wen J G, Rijkhoff N J M, Djurhuus J C and Sinkjær T 2000 *J. Urol.* **163** 1309-14
[10] Kim C T, Linsenmeyer T A, Kim H and Yoon H 1998 *Am. J. Phys. Med. Rehabil.* **77** 6 498-502
[11] Julien 2006 *Cardiovasc. Res.* **70** 1 12-21

Journal of Physics: Conference Series **224** (2010) 012113 doi:10.1088/1742-6596/224/1/012113

Relationship between electrical admittivity and quantitative histopathology in human prostate

Ryan Halter[1,2], Michael Milone[2], Alan Schned[3], John Heaney[3]

[1]Thayer School of Engineering, [2]Dartmouth College and [3]Dartmouth-Hitchcock

ryan.halter@dartmouth.edu

Abstract. Passive bioelectrical properties have been demonstrated to provide sufficient contrast for use in differentiating benign from malignant tissue in a number of different organs including breast, prostate, cervix, bladder, and skin. The underlying microscopic anatomy responsible for these measured differences has been primarily speculative in the past. In this study we recorded electrical conductivity and permittivity spectra (100 Hz – 100 kHz) from 464 three mm diameter circular prostate samples. Each of these tissue specimens were stained with hematoxylin and eosin, processed onto microscopy slides, and digitized using optical microscopy. We used digital imaging processing tools to extract quantitative morphological features including total number of glands, average and total glandular lumen size, shape characteristics of the luminal spaces, and average and total glandular perimeter lengths. Correlative analysis was performed to assess the relationships between the tissue architectural features and the precisely co-registered electrical properties. We report on the findings from this analysis. This statistical assessment aims to provide a valuable piece of new information to help formulate a better understanding of the precise influence morphological architecture has on the flow of current through tissue.

1. Introduction

Several investigators over the last 20 years have verified that the electrical properties of malignant and benign tissues are significantly different and have demonstrated that measurement of these properties has obvious clinical potential. These findings have been confirmed for a number of different organ systems including breast [1], skin [2], cervix [3], bladder [4], and prostate [5], however the translation of this technology to the clinic has yet to be fully realized. The electrical property parameters used to describe tissues, including conductivity, permittivity, resistivity, admittance, impedance, and other model-base parameters are not part of the typical clinician's vernacular. Further, these properties have been shown to differ for various pathologies, and hypothesis have been made regarding the biophysical mechanisms responsible for these differences, but no detailed analysis has been conducted to verify these hypothesis.

To address these issues we have instantiated a program aimed at quantitatively evaluating the relationship between bulk electrical properties and the underlying microscopic tissue morphology. The objective of this analysis is to provide electrical properties in terms of specific biological metrics that are readily appreciable by clinicians. While developing these relationships, a more complete understanding of the precise biophysical mechanisms responsible for observed electric properties will be realized. This early analysis reports specifically on the relationship between microscopic glandular components in prostate tissue and electrical admittance.

2. Methods

2.1. Electrical property sampling

Trans-admittance was recorded from 576 prostate samples using a custom-designed tetrapolar configured probed [6]. Specifically, conductivity and permittivity were sampled at 31 frequencies logarithmically-spaced spanning from 100 Hz to 100 kHz. The tissue samples are placed between two co-aligned circuit boards with coaxial electrode pairs printed on their surfaces. Drops of india ink are deposited into holes on both sides of the electrode pairs to identify the precise region probed. Straight pins are placed through these ink dots when the tissue has been removed from the probe. The tissue-pin combination is submerged in formalin for ~24 hrs to fix the tissue. Following fixation, 5 μm tissue samples are stained with H&E and microscopy slides are processed. The pin holes remain after the fixation process and provide visual landmarks to precisely delineate where electrical properties were recorded. A nonlinear least-squares approach [6] was employed to fit a Cole-Cole model to the 31 conductivity and permittivity values and the Cole parameters $\Delta\sigma$, σ_∞, f_c, and α were extracted.

2.2. Digital image processing (DIP)

Each of microscopy slides were digitized with an Olympus BX51 (Center Valley, PA) microscope fitted with a EvolutionVF digital camera and motor-controlled three-axis stage controlled through ImagePro Plus software (Media Cybernetics, Bethesda, MD). Pin holes were manually identified and an array of high magnification (20x) sub-images encompassing these holes were sampled and automatically merged together to form a large-field composite image for analysis. The pin holes in each of the individual micrographs were identified manually and a mask denoting a circular region of interest representing the probes area of sensitivity was applied to the image. The green channel from the RGB images provided optimal contrast for distinguishing luminal spaces from background tissue and was selected as the processing channel. Image processing of the ROI included contrast enhancement, threshold-based segmentation to separate glandular luminal components from the remaining tissue matrix, and median filtering with a 5x5 kernel to eliminate pixel noise (see figure 1). 5 x 5 pixel kernels of approximately 300 different luminal areas were manually sampled to statistically estimate the threshold levels to use for luminal segmentation. Finally, luminal area, eccentricity, major axis length, minor axis length, and perimeter were computed for each of the individual objects identified. Objects with luminal areas less than 50 μm^2 were identified as artifacts and eliminated from the object list. Matlab's Image Processing Toolbox was used for all digital image processing.

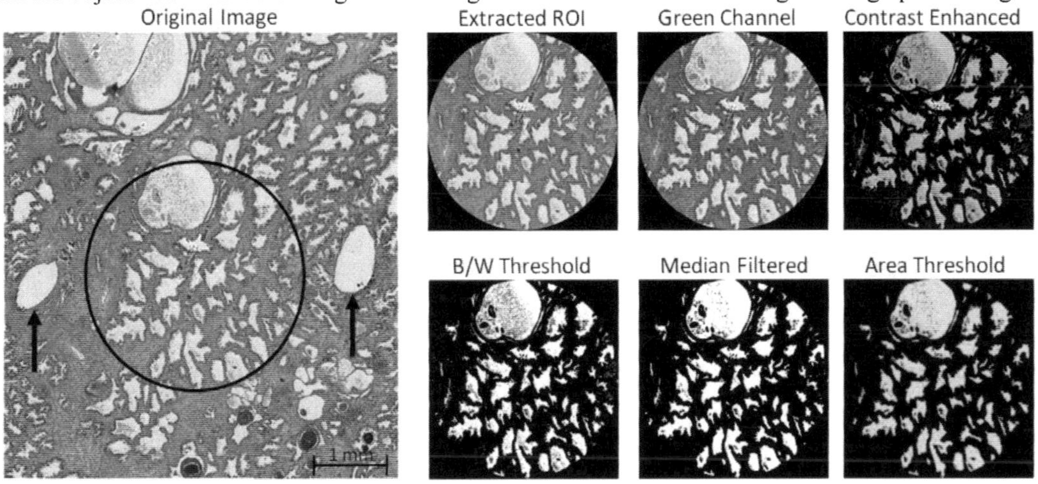

Figure 1. Example digital image processing procedure. Arrows denote visible pinholes. Circular region in original image denotes the selected region of interest (ROI) which corresponds to the precise tissue area gauged with our transimpedance probe. Each of the objects detected denotes a single glandular element.

2.3. Statistical analysis

The sum, mean, and standard deviation of the luminal properties extracted from each ROI were computed. Linear correlation coefficients comparing the electrical properties and the DIP extracted luminal properties were calculated. We report specifically on the electrical property correlations with the total number of glands, total glandular area, mean glandular area, mean luminal eccentricity, mean length of major axis, mean length of minor axis, total sum of luminal perimeters, and mean luminal perimeter of the glandular objects identified within each ROI. Moderate significance was defined as $p<0.05$ and high significance was defined as $p<0.001$. All statistics were computed with Matlab.

3. Results

Of the 576 tissue regions probed and digitized, pin holes were clearly identifiable in 464 microscopy slides. A subset of the electrical parameters including conductivity and permittivity at 10 kHz and 100 kHz, and the four Cole parameters are presented as a scatter plot matrix in Figure 2 with the x-axes denoting the extracted luminal properties. Table 1 lists the correlation coefficients for each of the electrical parameters and luminal properties and notes which of these correlations are significant.

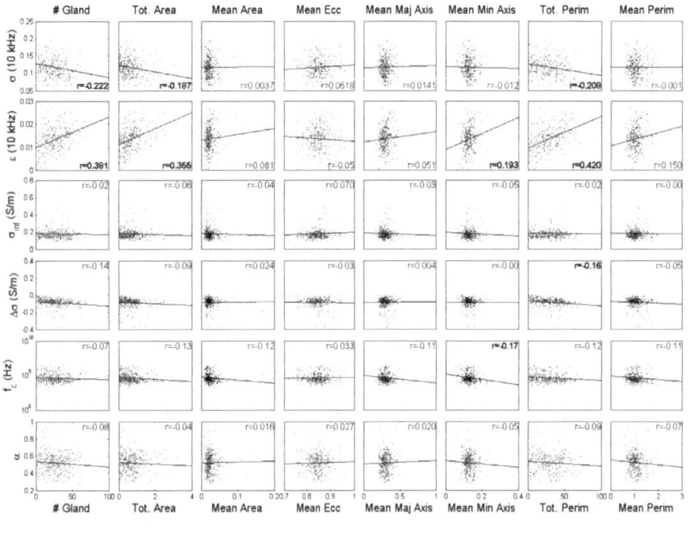

Figure 2. Scatter plots comparing the electrical properties to the extracted glandular properties. Conductivity and permittivity are plotted for 10 kHz only and are presented in units of S/m. The eight glandular properties displayed include the total number of glands, summed area of all luminal spaces, mean area of all glands, mean eccentricity of all glands, mean major axis length, mean minor axis length, summed perimeter of all glandular lumens and mean luminal perimeter from within each ROI. r values denote correlation coefficients with bolded entries representing significant findings ($p<0.001$).

Table 1. Pearson correlation coefficients for digital image processing parameters and electrical properties. Total # denote the total number of glands detected within in an ROI, while all other parameters represent the sum of all or mean of all parameters computed from within a particular ROI. Bolded elements denote correlations reaching a level of significance $p < 0.05$ and those bolded and shaded reached a significance of $p<0.001$.

		Extracted Glandular Properties							
Electrical Properties	Frequency (kHz)	Total #	Total Area	Mean Area	Mean Ecc	Mean Maj Axis	Mean Min Axis	Total Perim	Mean Perim
σ	0.1	**-0.29**	**-0.28**	-0.04	0.04	-0.06	-0.09	**-0.32**	**-0.1**
	1	**-0.28**	**-0.26**	-0.03	0.05	-0.02	-0.07	**-0.3**	-0.06
	10	**-0.22**	**-0.19**	0	0.06	0.01	-0.01	**-0.21**	0
	100	**-0.14**	**-0.1**	0.03	0.08	0.02	0.03	**-0.11**	0.02
ε	0.1	**0.19**	**0.2**	0.04	-0.06	0.05	**0.1**	**0.22**	**0.11**
	1	**0.31**	**0.31**	**0.1**	-0.01	**0.1**	**0.22**	**0.38**	**0.23**
	10	**0.38**	**0.36**	0.08	-0.05	0.05	**0.19**	**0.42**	**0.15**
	100	**0.19**	**0.15**	0.01	0.02	-0.03	0.01	**0.21**	0.01
σ_{∞}		-0.02	-0.06	-0.04	0.07	-0.03	-0.05	-0.02	0
$\Delta\sigma$		**-0.14**	**-0.09**	0.02	-0.03	0	0	**-0.16**	-0.05
f_c		-0.07	**-0.13**	**-0.12**	0.03	**-0.11**	**-0.17**	**-0.12**	**-0.11**
α		-0.08	-0.04	0.02	0.03	0.02	-0.05	-0.09	-0.07

4. Discussion

It is well-established that bioelectrical properties are a function of tissue morphology and the significant correlations observed here support these claims. In prostate, we have previously hypothesized [5-7] that the electrical properties are largely dependent on glandular density with more glandularly-dense tissues exhibiting higher impedances to current flow since charger carriers must traverse around these structures. Glandular lumens in benign prostate consist of a layer of tightly packed columnar-shaped epithelial cells that are lined by a ring of flat basal cells. These cellular structures inhibit low frequency currents from entering the luminal spaces and force current to flow around these spaces. Increases in glandular density provide a smaller area for current and result in a lower overall conductivity.

The significant negative correlations ($p<0.001$) noted between conductivity (0.1 – 10 kHz) and total # of glands, total luminal area, and total luminal perimeter support this hypothesis. These extracted luminal properties specifically tally the total amount of luminal space within a particular ROI. The other extracted parameters describe the characteristics of the average glandular structure within this ROI. For the most part, these average properties do not significantly correlate to the electrical properties; it is the global characteristics of the ROI that are the significant biological drivers dictating electrical properties. Of note, is the decreased conductivity correlation as frequency increases; this likely arises from the lower impedance cell membranes exhibited at higher frequencies.

In terms of permittivity, there is a significant ($p<0.001$) positive linear correlation noted between permittivity and total # of glands, total glandular area, and total luminal perimeters. The cell membranes of the epithelial and basal cells lining the glandular lumens provide a capacitive storage element for charge. The positive relationship noted between glandular content and permittivity suggests that with increased glandular density, cell membranes couple predominantly in parallel since capacitances sum in this configuration and permittivity is directly proportional to total capacitance.

Both $\Delta\sigma$ and f_c were observed to be significantly correlated with the glandular morphology, while σ_∞ and α did not demonstrate significant correlations. $\Delta\sigma$ is a combined measure of both high and low frequency conductivities and tracks the single frequency conductivity findings at 0.1 kHz and 100 kHz. f_c is suspected to be largely cellular membrane dependent, however only moderately significant correlations were noted in this study. In a previous study [7], we have demonstrated that f_c is significantly different in prostatic cancer tissues as opposed to benign tissues. This may suggest that f_c is highly dependent on the type of cell and viability of that cell.

Additional image processing aimed at more precisely characterizing individual cell types, analyzing tissue texture, and quantifying specific tissue concentrations is suggested to more completely describe the relationships between tissue pathology and electrical properties.

5. Conclusion

The approach to assessing the relationship between electrical properties and tissue presented here is an important step toward understanding more completely the biophysical mechanisms response for certain electrical properties. This approach may yield an alternative method for reporting electrical properties so that instead of reporting in terms of electrical properties, we may be able to present doctors with information specifically regarding the pathological status of the tissue.

References

[1] da Silva J, de Sa J and Jossinet J 2000 Med Biol Eng Comput. 38 26
[2] Åberg P, Geladi P, Nicander I, Hansson J, Holmgren U and Ollmar S 2005 *Skin Res Tech* **11** 281
[3] Abdul S, Brown BH, Milnes P and Tidy JA 2006 *Int J Gynecol Cancer* **16**1823
[4] Wilkinson BA, Smallwood RH, Keshtar A, Lee JA and Hamdy FC 2002 *J Urol* **168** 1563
[5] Halter RJ, Schned A, Heaney J, Hartov A and Paulsen KD 2009 *J Urol* **182** 1600
[6] Halter RJ, Hartov A, Paulsen KD, Schned A and Heaney J 2008 Physiol Meas 29 S111
[7] Halter RJ, Schned A, Heaney J, Hartov A and Paulsen KD 2009 *J Urol* **82** 1608

Online bioimpedance feedback for *in vivo* electroporated tissues

J Medrano[1], J I Rey[2], R J Connolly[2], A Anderson[1], M Jaroszeski[2] and R Gitlin[1]

[1]Department of Electrical Engineering, University of South Florida, Tampa, FL 33620 USA
[2]Department of Chemical and Biomedical Engineering, University of South Florida, Tampa, FL 33620 USA

E-mail: jmedrano@mail.usf.edu.

Abstract. Electroporation *in vivo* is a biotechnology method that uses short-duration high intensity electric fields to enhance plasma membrane permeability in living cells in order to facilitate the uptake of drugs, DNA, genes and proteins into the cytoplasm. The degree of permeability is related to the tissue's bioimpedance; hence, accurate impedance evaluation throughout electroporation treatment is essential to 1) avoid over-treating tissues resulting in excessive cell death and 2) under-treating tissues resulting in poor permeability. Cell viability and membrane permeability is based on a number of factors, including: time elapsed after electroporation, electroporation pulse amplitude, tissue type, and so on; thus, efficient feedback protocols must minimize delays between treatment and impedance readings. Current methods of bioimpedance feedback are often cumbersome and impedance analysis devices can be expensive, bulky, and immobile. Emerging technologies facilitate economical methods, fast protocols, and portability to realize bioimpedance measurement and feedback online (i.e. real-time). Consequently, this research uses automation software, logic-biased protocols, an inexpensive commercially available impedance analyzer microchip, and a custom-built hexagonal electrode probe to measure dynamic bioimpedance changes. This work demonstrates how this novel system measures tissue bioimpedance instantly and efficiently before and after electroporation. Additionally this system allows for the comparison of electrode geometries as well as electric field' magnitudes and distributions. Follow up work will pursue the optimization of plasma membrane permeability for several tissue/cell types.

1. Introduction

Bioimpedance frequency spectra of many biological tissues have been widely reported in terms of complex permittivity components (e.g. conductivity). To date, however, there is little data on *in vivo* bioimpedance measurements taken immediately before and after electroporation (EP) to measure cell membrane permeability [1, 2]. This current work presents a new approach to performing online bioimpedance feedback for *in vivo* (and *ex vivo*) electroporated tissues by using inexpensive and portable equipment, faster protocols, and modular components that allow for segregating pulsing and measuring electrodes, as well as the application of unipolar pulsing over two paths to provide extended electric field coverage. Pre and post electroporation resistivity feedback data is immediately analyzed for *ex vivo* bovine kidney tissue and *in vivo* on rodent skin using 150/cm and 200V/cm electric fields in 4 unipolar square pulses of 20ms duration and 1 second intervals form 5kHz to 100 KHz, and two electrode geometries -a needle electrodes and a hexagonal type multielectrode array.

2. Methods and materials

2.1 *Electrical impedance spectroscopy*

Tissue bioimpedance was measured using the instrumentation setup shown in Figure 1. Prior to experimentation, the accuracy of the Analog Devices AD5933 Impedance Analyzer (IA) chip was validated with an Agilent 4294A Precision Impedance Analyzer and calibrated using known RC analog circuits. A validation impedance magnitude of $|Z|=5{,}636\Omega$ resulted in an error of less that 1% as specified by the manufacturer [3].

The equivalent circuit model used for tissue was a series resistor followed by a parallel RC combination known as the Debye single-dispersion electrical model [4]. Given this model, the IA was calibrated for an impedance magnitude of $|Z|=5.5k\Omega$ at a midrange frequency of 40 kHz as shown by Gabriel [5] and in Figure 3b. This skin tissue resistivity data from Gabriel was also used for comparative and validation of the results. The IA was controlled by a PC to set the frequency sweep parameters from 5 kHz to 100 kHz at 1kHz increments with a sample speed of one million samples per second (MSPS), set excitation voltage at $1V_{pp}$, select circuit modeling and calibration gain, measure tissue temperature, and collect data. A BTX EMC 830 Square Pulse Generator was set for 60V or 80V (i.e., 150V/cm or 200V/cm electric fields) and a 20 ms pulse duration with four repetitions every second.

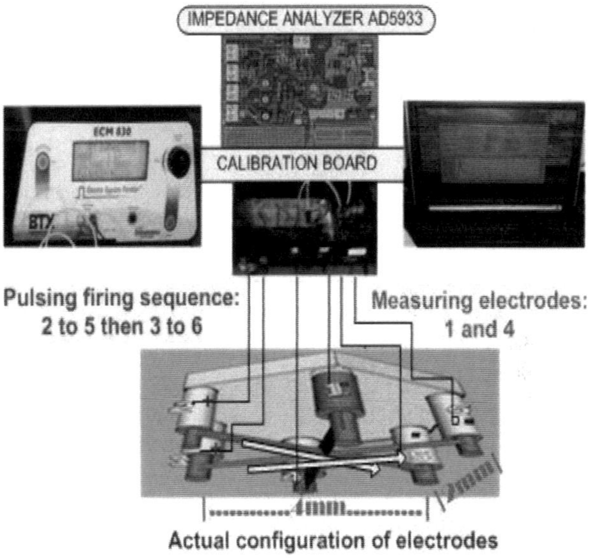

Figure 1: Instrumentation, electrode, and control panel setup for bioimpedance measurements.

2.2 *Electrode geometries*

Two geometries of surface electrodes were used for impedance probing. The first probe was comprised of two needle electrodes of 1mm diameter at a distance of 5mm apart. The second probe contained six needle electrodes of 0.38 mm diameter, spaced 2mm apart and a 4mm diameter as shown in Figure 1. This second electrode configuration allowed for different electrodes to be used for applying pulses and measuring impedance. For skin measurements, conductive gel (Spectra 360 electrode gel) was placed onto each electrode before they were contacted with the stratum corneum. Tissue treatment area was held constant by virtue of the fixed spacing between the electrodes. The K ratio use to convert the impedance measurement to resistivity was calculated as 2.5 cm. However, it should be noted that the focus of this work is on the analysis of the differential resistivity before and after EP and not necessarily as a database for tissue impedances itself.

2.3 *Ex vivo tissue*

Bovine tissue for *ex vivo* electroporation was obtained 'fresh' and allowed to equilibrate to room temperature. Measurements were taken with a 150V/cm field for a total of 12 measurements and 1200 data points were analyzed.

2.4 *In vivo mouse skin tissue*
Mouse skin measurements were obtained on female C57Bl/6 mice (NIH, Bethesda, MD) that were eight weeks of age and weighed 18 grams. Impedance measurements and EP were performed on animals anesthetized with 2% isoflurane in oxygen. Core temperatures were maintained by placing the animal's ventral-side-down on a 32°C heating pad during anesthesia exposure. Measurements were taken on the left and right flank skin, which maintained a constant temperature of 32°C. EP was carried out by exposing the skin to a field strength of 150 V/cm or 200 V/cm applied by one set of electrodes on the applicator. A total of 18 measurements and 1800 data points were acquired and analyzed over nine separate skin areas. Following data acquisition, the animals were humanely euthanized in accordance with an approved Institutional Animal Care and Use Committee protocol.

2.5 *Rat skin tissue*
Measurements were taken from freshly harvested skin tissue from male Sprague-Dawley rats that were 18 weeks of age and weighed between 450 and 550g. Measurements were made at room temperature for EP electric field parameters of 150 and 200 V/cm. 100 data points were sampled during nine measurements over nine rat skin areas of three rats for a total of 8,100 measurements.

3. Results and discussion

The resistivity of bovine liver *ex vivo* before and after EP is shown in Figure 2. Twelve liver samples were measured using the hexagonal type electrode with an electrical field of 150V/cm for a total of 1,200 data points. Bioimpedance after EP was consistently lower than the pre-EP values for the entire spectrum. However, this difference was not statistically significant at any of the points shown in figure 2 (95% level of significance Student's T-test). This result is consistent with previous published data on swine liver electrical resistivity postmortem. It has been noted that *ex-vivo* resistivity decreases when compared to *in vivo* tissue resistivity due to a progressive increase in extracellular fluid due to membrane rupture that begins about 2 hours postmortem [6]. This result suggests that a possible application of bioimpedance probing could be the estimation of the degree of 'freshness' of edible meats.

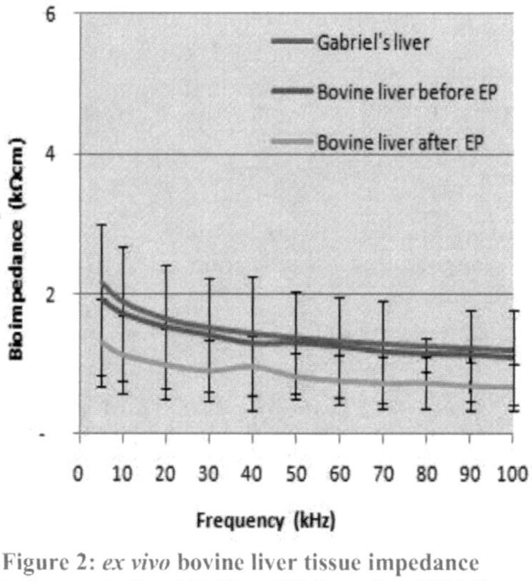

Figure 2: *ex vivo* bovine liver tissue impedance before and after 150 V/cm EP from 5 to 100 kHz.

The resistivity of mouse skin *in vivo* before and after EP is shown in Figure 3a. The same electrodes were used for pulsing and measuring with 150V/cm and 200V/cm electric fields. The pulse duration was 20 ms with four repetitions every second over nine differentiated skin sections on the back of mice. A studentized T-test found a statistically significant reduction in skin impedance following EP for both 150 V/cm and 200 V/cm pulses. For 150 V/cm, the impedance was significantly different for frequencies ranging from 5 to 35 kHz. For 200 V/cm, the impedance drop was only significant from 5 to 100 kHz. These results were from a single set of electrodes used for both pulsing and measuring impedance. Employing this electrode configuration causes a bias in the samples, because measurements obtained with this applicator include those from dead cells caused by electrode polarization.

The resistivity of rat skin *in vivo* before and after EP is shown in Figure 3b. A 150 V/cm electric field and the hexagonal electrode configuration were used over nine separate skin sections on three different mice for a total of 5,400 data points as shown with corresponding standard deviations.

A studentized T-test demonstrated with 95% confidence that EP resulted in a significantly lower bioimpedance in rat skin at frequencies of 5 to 100 kHz.

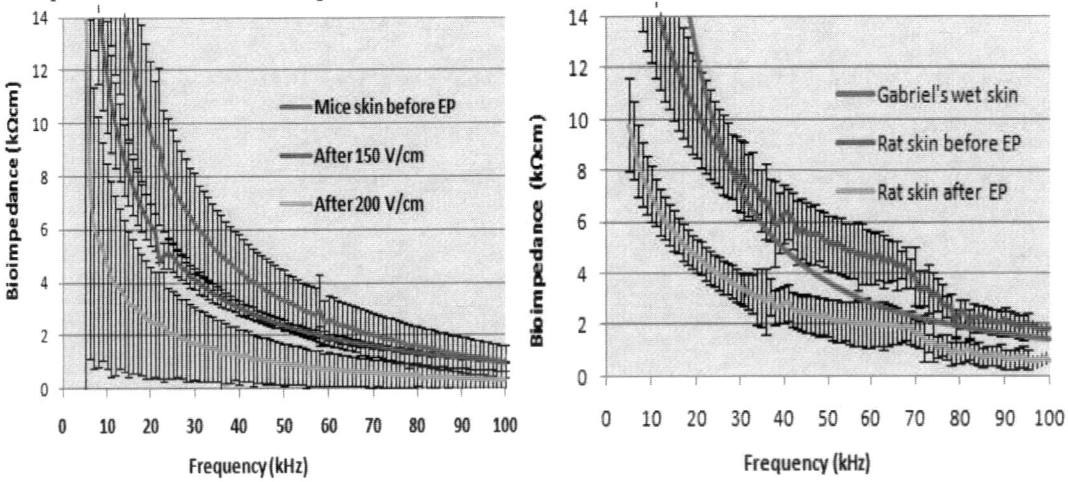

Figure 3: *In Vivo* Tissue resistivity (a) mice skin using the 2-electrode probe (b) rat skin using the 6-electrode probe.

4. Conclusions

This work provides a very cost effective and novel means to accurately measure bioimpedance changes during and after electroporation of tissues *in vivo*. The system proved capable of obtaining accurate measurements at sweep frequencies from 5 to 100 kHz within milliseconds using an inexpensive microchip in parallel with a pulse generator. A custom-built measurement probe consisting of an array of three sets of electrodes was designed and built in order to separate electrode pairs for pulsing and measurement in order to avoid the influence of electrolytic products and tissue damage on the electrodes used for impedance measurements.

As expected, electroporation yielded lower bioimpedance drops for *ex vivo* tissue when compared to *in vivo* tissue. Thus, this experiment suggests that a possible application for bioimpedance measurements could be the estimation of the 'degree of freshness and leanness' of market edible meats. *In vivo* mouse skin impedance measured after EP using two electrical fields (150 and 200 V/cm) showed a larger drop at the lower electric field which may be explained by electrode effects and/or cell death. Similarly, an impedance drop was determined in rat skin as a result of EP at 150 V/cm. .

5. References

[1] Heller L.C., M. Jaroszeski, M., McCray A.N., Hickey J., and Heller R. (2006): 'Optimization of cutaneous electrically mediated plasmid DNA delivery using novel electrode', *Gene Therapy*, pp. 1-6

[2] Rubinsky B., Ivorra B. (2006): 'An Impedance Analyzer for *in vivo* Electroporation Studies'. Proceedings of the IEE EMBS 28th Annual Interantional Confernce, NY 2006

[3] http://www.analog.com/static/imported-files/eval_boards/537700023EVAL_AD5933EB.pdf

[4] Chizmadzhev Y., Indenbom A., Kuzmin P., Galichenko S.,Weaver J., and Potts R. (1998): 'Electrical properties of skin at moderate voltages: contribution of appendageal macropores", *Biophysical Journal,* **74** pp. 843–856

[5] Gabriel C., Gabriel S., and Corthout E. (1996): 'The dielectric properties of biological tissues: literature survey', *Phys. Med. Biol.,* **41** pp. 2231–2249

[6] Haemmerich, D., Ozkan S., Tsai J., Staelin S., and Webster G., (2002): 'Changes in electrical resistivity of swine liver after occlusion and postmortem', *Med. Biol. Eng. Comput.*, **40**, pp. 29–33

Prevalence of stroke/cardiovascular risk factors in Hungary

M Bodo[1], K Sipos, G Thuroczy, G Panczel, L Ilias, P Szonyi, M Bodo,Jr, T Nebella, A Banyasz, Z Nagy
Kalman Santha Foundation, Budapest, Hungary

E-mail: mikebodo@comcast.net

Abstract. A cross-sectional survey was conducted in Hungary using the Cerberus system which includes: 1) a questionnaire addressing the risk factors for stroke/cardiovascular disease; 2) amplifiers to record the pulse waves of cerebral arteries (rheoencephalography) and peripheral arteries, electrocardiogram and electroencephalogram. Additionally, subjects were measured for carotid stenosis by Doppler ultrasound and 12-lead electrocardiogram; subjects were also screened for blood cholesterol, glucose, and triglyceride levels. Prevalence of the following stroke risk factors was identified: overweight, 63.25%; sclerotic brain arteries (by rheoencephalogram), 54.29%; heart disease, 37.92%; pathologic carotid flow, 34.24%; smoking, 30.55%; high blood cholesterol, 28.70%; hypertension, 27.83%; high triglyceride, 24.35%; abnormality in electrocardiogram, 20%; high glucose, 15.95%; symptoms of transient ischemic attack, 16.07%; alcohol abuse, 6.74%; and diabetes, 4.53%. The study demonstrates a possible model for primary cardiovascular disease/stroke prevention. This method offers a standardizable, cost effective, practical technique for mass screenings by identifying the population at high risk for cardiovascular disturbances, especially cerebrovascular disease (primary prevention). In this model, the rheoencephalogram can detect cerebrovascular arteriosclerosis in the susceptibility/presymptomatic phase, earlier than the Doppler ultrasound technique. The method also provides a model for storing analog physiological signals in a computer-based medical record and is a first step in applying an expert system to stroke prevention.

1.Introduction
A multi-faceted survey was conducted in 1992–94 to ascertain the somatic, mental and socio-economic conditions of the residents of a village in eastern Hungary. Here we report data on prevalence of somatic disorders from the survey. Objectives were: a) To collect and compare prevalence of known cardiovascular disease, including stroke risk factors, in a specific population (a Hungarian village); b) to test a computer-based mass screening device ("Cerberus") designed to identify individuals in the test population at high risk for stroke; c) to compare results obtained with Cerberus with known stroke risk/cardiovascular disease factors and traditional medical records.

2.Methods
A sample of 546 non-symptomatic volunteers ranging in age from 14 to 83 years participated in the study. The sample was evaluated according to the full protocol, which included blood chemistry, electrocardiogram (ECG), Doppler control, and Cerberus testing. The analyzed REG sample involved 457 subjects, 309 female and 148 male. For 252 subjects, Doppler ultrasonography (B mode or duplex scan, Ultramark 4 and Apogee, ATL, USA) was used to measure carotid stenosis of internal carotid

[1] Michael Bodo, 2804 Atlanta Drive, Silver Spring, MD, 20906, USA

artery. Doppler B images were obtained for 184 of the 252 subjects. Mean velocity for "out of normal range" (pathological blood flow) was considered as above 40 cm/sec. Psychological tests were performed and reported elsewhere together with other somatic data [2,3].

2.1 Cerberus measurement

The Cerberus system (Quintlab Ltd, Budapest, Hungary) was described in detail elsewhere [1]. Briefly, it is a computer-based inquiry employing Magic, previously Gral software. The examination requires about 20 minutes and can be administered by a technician. A professional trained in vascular neurology interprets the results, contained in a graphically illustrated summary. It is possible to give each subject a report at the conclusion of the examination, indicating 1. healthy, 2. cerebral blood flow disturbance, 3. peripheral blood flow disturbance, 4. neurological or internal medical disturbance, 5. other. First, a questionnaire was administered; blood pressure was measured; and vegetative balance was calculated from blood pressure and heart rate. The questionnaire addressed demographics (12 items), stress (8 items) and anxiety (STAI), stroke risk factors (5 items), and neurological symptoms indicating a possible past transient ischemic attack (21 items). Then, multichannel polygraphic recordings (16 items) were made as follows: 1. ECG; 2. EEG from T5-O1, T6-O2 according to 10/20 system of EEG with ECI System (Electro-Cap Int'l, Eaton, OH); 3. bio-impedance pulses of the head (REG) derived from Fp1-F7 and Fp2-F8 (according to electrode localization in international 10/20 EEG system) and of limbs (both sides) with circumferential electrodes (Rheoscreen, Medis GmbH, Ilmenau, Germany). Following a one minute polygraphic recording, blood pressure was re-measured, and the STAI was repeated. Custom software was developed for Cerberus to perform the following signal processing procedures: analog – digital conversion (275 Hz) and display of analog physiological signals; calculation of fast Fourier transform from EEG; averaging of pulse curves; determination of dominant frequency peak for EEG and of minimum/maximum values for REG and peripheral pulse waves. REG was considered arteriosclerotic if its anacrotic (rising) portion was above the 180 ms threshold, modified after Jenkner [4]. From data collected by measuring blood pressure and heart rate, Cerberus software calculated the ergotrop/trophotrop ratio; modified vegetative (autonomic nervous system) balance by Kerdo and Sipos [1]: 1–(diastolic blood pressure/HR)×100. Body Mass Index (BMI) was calculated as weight (kg) /height (m^2). Overweight was defined as BMI>25 for male and BMI>23.8 for female, respectively.

2.2 Data analysis

Artifact contaminated recordings were excluded from further processing (REG, ECG). For statistical treatment and analysis, Framework, GraphPad Prism, Excel and SAS software were used. In most cases, results for male and female subjects were calculated separately (for REG data, where hemispherical differences were less than 5%, female and male groups were analyzed together). After the automatic measurement of REG, a second REG calculation was made off-line to calculate the first derivative parameter (n=390). $P<0.05$ was considered significant.

3. Results

Dominant risk factors, listed by incidence, were as follows: obesity, smoking, heart disease, and hypertension. There were slight differences between male and female groups. Significant correlations with age were observed for overweight, hypertension, total risk, vegetative balance and sclerotic brain arteries. Blood serum values had no significant correlation to age. Mean vegetative balance was 88.05 (m), 93.54 (f), but these differences were not significant; however, age dependency was significant for females in the youngest and oldest groups, decreasing from 109 (16–25 y) to 83 (over 65 y). Smoking was more frequent among males (39.08%) than females (22.07%). Less than 1% of females and more than 12% of males declared themselves as a "regular drinker". The proportion females and more than 12% of males declared themselves as a "regular drinker". The proportion of "hypertonics" (systolic blood pressure above 135 mmHg) in the self-report questionnaires equaled 22.90% for males and 32.68% for females. Heart disease was mentioned by every third male subject (35.64%) and by

40.22% of females. Less than 5% of both sexes were aware of the presence of diabetes mellitus. For further details see [2,5].

3.1. REG

REG anacrotic time was significantly correlated with age and vegetative balance; there were no gender differences. REG was significantly correlated to systemic arterial pressure in females. The most significant observation in regard to REG anacrotic time was related to age: both REG anacrotic time and the time interval between ECG R peak and peak of REG first derivative increased with age. Lateral differences in REG amplitude and REG anacrotic time were found more frequently in middle aged groups (age 40 to 60) than in younger or older groups. In the whole sample, 70% of the subjects over 60 years of age could be considered as cerebrovascular patients, according to the mentioned 180 ms criteria. For 52.78% (male) and 55.8% (female), a sclerotic REG curve was observed. The slope of regression function for REG vs. age was 3.195±0.45 for females, 2.96±0.62 for males 41. The age dependency of sclerotic brain arteries by REG was highly significant, above all other risk factors. Significant correlations were observed between somatic and psychological variables – for details see [3]. In all subjects who were found arteriosclerotic by REG measurement, the REG pulse wave was altered; however, for most of these patients, the EEG spectral parameters were in the normal range, and the Doppler control showed no pathological alteration. The regression line of the REG anacrotic time vs. age had a slope 10 times sharper than that of Doppler systolic velocity (figure 1).

3.2 Doppler ultrasound

For subjects measured by Doppler ultrasound, pathologic flow (mean systolic velocity) was found in 29.41% of the males and 39.06% of the females; pathological flow values appeared in younger as well as older age groups. Although moderate stenosis (less than 50%) was detected in more than 20% of this group (right side, 35.32%; left side, 25.54%), the differences were not significant. Severe stenosis (higher than 50%) was observed in 3.8% of the measured population, without laterality. Maximum "out of normal range" patients appeared in the 55–65 age group. The slope of regression for IC systolic flow vs. age was –0.45±0.11 (f) and –0.346±0.15 (m).

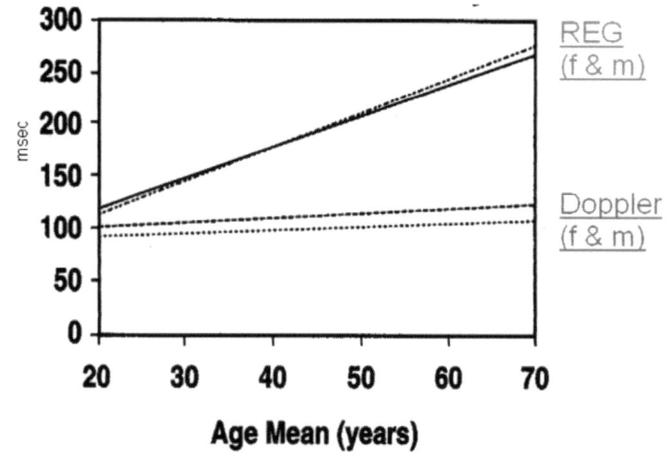

Figure 1. Comparison of slope of age regression lines (REG anacrotic time and Doppler systolic velocity). REG anacrotic time (Y axis in ms: REG anacrotic time) and Doppler systolic flow velocity (Y axis not shown) of internal carotid artery are plotted as a function of age (X axis). The regression lines of male and female groups are similar, but the slope of REG is about ten times steeper than that of the Doppler curve. f: female, m: male. Reprinted courtesy of Health Quest Publications Anti-Aging Medical Therapeutics, Vol. II [5].

4. Discussion

Stroke is unique among neurological diseases, since it has a high prevalence and burden of illness, high economic cost, and is preventable [6]. Epidemiological approaches to stroke prevention include the "high risk" and "mass" approaches. The prevention of stroke can be accomplished by the high-risk or mass approach or a combination of these approaches. The high-risk approach prevents strokes but is also expensive. The mass approach may be more cost-effective, which could lead to substantial savings, but further research is needed to investigate these possible benefits. In this study we introduced a mass approach: REG measurements revealed symptoms of arteriosclerosis in 54.26% of our subjects; within the identical population, the Doppler ultrasound measurements showed 30.43% with arteriosclerosis. The difference of slope of regression lines between REG anacrotic time and Doppler systolic velocity was ten to one, which can be interpreted as increased sensitivity of REG to detect cerebrovascular arteriosclerotic alteration. The present study validated our initial hypothesis – that Cerberus (REG measurement) would predict the presence of cerebral arteriosclerosis in the susceptibility/ pre-symptomatic or "upstream phase", earlier than Doppler ultrasound measurements. The explanation of this difference can be based upon the nature of the development of arteriosclerosis and upon differences between REG and Doppler methods.

5. Conclusion

In summary, our results showed that the earliest manifestation of cerebral arteriosclerosis is the lost elasticity of cerebral arteries, which can be detected by the bioimpedance method (REG). This information is even more practical and more important than the presence and prevalence of a single stroke risk factor. The present study demonstrated the usefulness of the new, multifactorial screening tool, Cerberus, which offers the potential for primary prevention of a general population both for stroke and cardiovascular disorders [7]. People found to be at high risk for stroke would be referred to physicians for further treatment and control of the disease. We propose screening individuals every five years, beginning at age 40, to detect and control the progression of arteriosclerosis.

References

[1] Bodo M, et al 1995 A complex cerebrovascular screening system (Cerberus). *Medical Progress through Technology* **21** 53-66

[2] Bodo M, et al 2008 Prevalence of stroke/cardiovascular risk factors in rural Hungary - a cross-sectional descriptive study. *Ideggyogy. Sz*, **61** 87–96

[3] Sipos K, et al 2008 Risk of mental disorders, their changes and somatic consideration in rural Hungary. *Ideggyogy. Sz*, **61** 97–105

[4] Jenkner F L 1970 Rheoencephalographic differentiation of vascular headaches of various causes, ed S E Markovich International conference on bioelectrical impedance. *Annals of the New York Academy of Sciences* **170/2** p 661-65

[5] Bodo M, Thuroczy G, Brockbank K, Sipos K 1998 Cerebrovascular aging assessment by Cerberus, ed R Klatz, R Goldman *Anti-aging medical therapeutics*, vol. II. (Marina Del Rey: Health Quest) **13** 86-95

[6] Gorelick P B 1994 Stroke prevention. An opportunity for efficient utilization of health care resources during the coming decade. *Stroke* **25** 220-224

[7] de Freitas G, Bogousslavsky J 2001 Primary stroke prevention. *European J. Neurol.* **8** 1-15

Body composition and phase angle in Russian children in remission from acute lymphoblastic leukemia

G Ja Tseytlin[1], I A Khomyakova[2], D V Nikolaev[3], M V Konovalova[1], A Yu Vashura[1], A V Tretyak[2], E Z Godina[2], S G Rudnev[4]

[1] Federal Centre of Pediatric Hematology, Oncology and Immunology
[2] Institute and Museum of Anthropology, Moscow State University
[3] Scientific Research Centre 'Medass'
[4] Institute of Numerical Mathematics, Russian Academy of Sciences
all – Moscow, Russian Federation

Email: rudnev@inm.ras.ru

Abstract. Elevated degree of body fatness and changes in other body composition parameters are known to be common effects of treatment for acute lymphoblastic leukemia (ALL) in children. In order to study peculiarities of somatic growth and development in ALL survivors, we describe the results of BIA body composition analysis of 112 boys and 108 girls aged 5-18 years in remission from ALL (remission time range 1-13 years) compared to data from the same number of age- and sex-matched healthy controls (n=220). Detrimental effect on height in ALL boys was observed, whereas girls experienced additional weight gain compared to healthy subjects. In ALL patients, resistance, body fat, and percent body fat were significantly increased. The reactance, phase angle, absolute and relative values of skeletal muscle and body cell mass were significantly decreased. Principal component analysis revealed an early prevalence of adiposity traits in the somatic growth and development of ALL girls compared to healthy controls.

1. Introduction

Acute lymphoblastic leukemia (ALL) represents the most common form of cancer in children having incidence about 3-4 cases per 100.000/year that peaks around 2-5 years of age. The use of modern treatment protocols involving polychemotherapy and prophylactic central nervous system interventions provides cure rate exceeding 80% [1]. So, estimation of the long-term effects of treatment for childhood ALL has become increasingly important. There exist a substantial number of studies on body composition changes in children in remission from ALL [2-4]. In particular, excessive fatness and significant reduction in bone mineral density are reported as frequent late complications of ALL treatment. To our knowledge, in Russia, such studies have not been previously conducted.

Along with BIA estimates of body composition, there is growing interest for direct use of measured BIA parameters, such as resistance, reactance, and phase angle, in epidemiological and clinical investigations [5-7]. In our recent paper we reported data on phase angle and body composition in healthy Russian children aged 10-16 years [8]. Here, we aim at the evaluation of the long-term effects of ALL and its treatment on the somatic growth and development of children in terms of basic anthropometric and BIA parameters, including height, weight, phase angle and body composition.

2. Subjects and methods

220 children (112 males and 108 females) after treatment for acute lymphoblastic leukemia (ALL) were assessed cross-sectionally in 2008-2009 at an average age 10.6 years (range, 5-18 yr). All patients were in first remission, and none had received any hormone replacement therapy. Remission time ranged from 1 to 13 years. Another 220 age- and sex-matched healthy children were measured who comprised a control group.

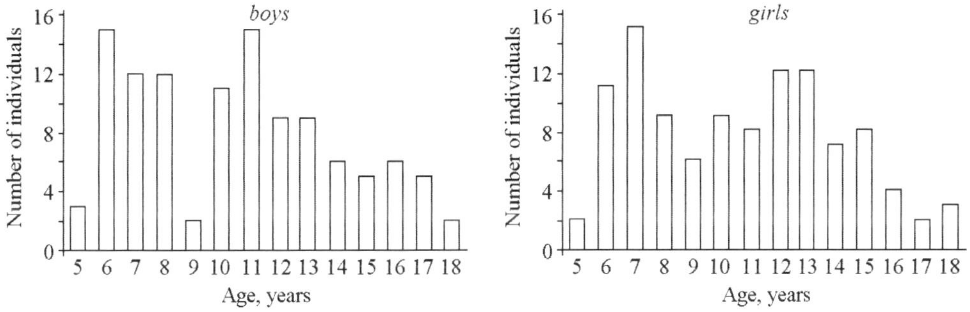

Figure 1. Histograms of ALL patients' and control subjects distribution across age

Table 1. Basic anthropometric characteristics of ALL patients and controls, mean (s.d.)

Parameter	Boys		Girls	
	ALL (n=112)	Controls (n=112)	ALL (n=108)	Controls (n=108)
Age, years	10.5 (3.5)	10.5 (3.5)	10.7 (3.4)	10.7 (3.4)
Body mass, kg	40.1 (15.0)	40.2 (17.7)	40.6 (15.4)	37.8 (14.3)
Body height, cm	143.2 (19.6)	146.8 (22.8)	142.4 (17.1)	144.1 (19.4)
BMI, kg/m^2	18.9 (3.1)*	17.6 (2.8)	19.3 (4.1)*	17.4 (2.6)

* significant difference compared to control group, $p<0.01$

Table 2. BIA estimates of body composition parameters in ALL patients and controls, mean (s.d.)

Parameter	Boys		Girls	
	ALL	Controls	ALL	Controls
R, Ohm	677 (79)*	626 (99)	715 (86)*	678 (84)
X_C, Ohm	62.3 (7.0)*	69.6 (10.0)	65.3 (7.5)*	74.7 (9.4)
PA, grad	5.3 (0.7)*	6.4 (0.7)	5.3 (0.6)*	6.3 (0.7)
TBW, l	22.7 (7.9)	25.2 (10.9)	21.7 (6.4)	22.3 (7.5)
FM, kg	18.9 (3.1)*	17.6 (2.8)	19.3 (4.1)*	17.4 (2.6)
%FM	22.1 (6.5)*	14.1 (4.0)	25.5 (6.8)*	18.0 (5.0)
FFM, kg	30.9 (10.9)	34.4 (14.9)	29.4 (8.9)	30.4 (10.3)
SMM, kg	16.5 (7.3)*	22.5 (7.2)	14.1 (4.7)*	16.8 (4.6)
BCM, kg	15.6 (6.5)*	19.3 (9.1)	14.7 (5.1)*	16.9 (6.1)

* significant difference compared to control group, $p<0.01$

Standing height in ALL patients was measured using a stadiometer, and weight was measured on a digital scale. In a control group, standing height was determined using the GPM anthropometer. Body mass index (BMI) was calculated as body mass (BM) divided by standing height (Ht) squared. Age distributions and basic anthropometric characteristics of patients and controls are shown in Figure 1 and Table 1, respectively. The whole-body impedance was measured on the right side of the body by the bioimpedance analyzer ABC-01 'Medass' (SRC Medass, Russia) according to a conventional tetrapolar scheme at a constant frequency 50 kHz. Phase angle (PA) was calculated as $\arctan(X_C/R)\times180°/\pi$, where X_C is reactance and R the whole-body electric resistance. Fat-free mass (FFM) was assessed using Houtkooper equation [9]: $FFM=0.61\times(Ht^2/R)+0.25\times BM+1.31$, where Ht is measured in cm. Fat mass was calculated as the difference between BM and FFM, and %FM as (FM/BM)×100. Body cell mass (BCM) and skeletal-muscle mass (SMM) were determined using the formulae provided by the manufacturer. Two-sample rank (Mann-Whitney) test of the equality of two population medians was used. A p value of 0.05 was used to define significance.

3. Results

Body mass index was significantly compromised in ALL patients (Table 1). In ALL boys, this effect was observed due to changes in body height only, while in girls both Ht and Wt were affected compared to healthy controls.

Table 3. Correlations between anthropometric and BIA parameters, boys. Above and below the diagonal are the data on ALL patients and controls, respectively

	Age	Ht	Wt	BMI	R	X_C	PA	FM	%FM	FFM
Remission time	0.34	0.39	0.43	0.28	-0.43	n.s.	0.44	0.25	n.s.	0.47
Age	•	0.90	0.82	0.38	-0.44	n.s.	0.54	0.56	n.s.	0.85
Ht	0.89	•	0.94	0.39	-0.47	n.s.	0.46	0.60	n.s.	0.93
Wt	0.85	0.89	•	0.75	-0.64	n.s.	0.50	0.83	0.41	0.96
BMI	0.65	0.69	0.87	•	-0.60	-0.31	0.31	0.87	0.69	0.60
R	-0.68	-0.75	-0.86	-0.89	•	0.40	-0.59	-0.32	n.s.	-0.72
X_C	-0.46	-0.56	-0.63	-0.64	0.72	•	0.48	n.s.	n.s.	n.s.
PA	0.42	0.37	0.46	0.48	-0.52	0.21	•	0.20	n.s.	0.58
FM	0.68	0.76	0.87	0.86	-0.65	-0.49	0.34	•	0.83	0.65
%FM	n.s.	n.s.	0.22	0.38	n.s.	n.s.	n.s.	0.63	•	n.s.
FFM	0.86	0.95	0.99	0.84	-0.88	-0.63	0.47	0.80	n.s.	•

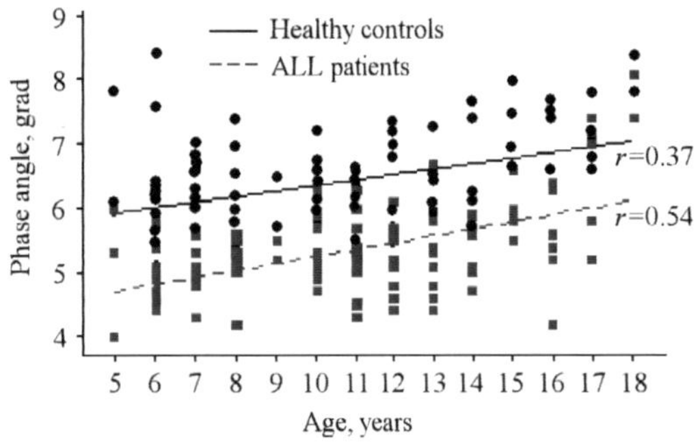

Figure 2. Scatter plots of phase angle across age: remission for acute lymphoblastic leukaemia (■) vs. healthy controls (•), boys

In ALL, the resistance, body fat, and percent body fat were significantly increased (Table 2). The reactance, phase angle, absolute and relative values of skeletal muscle and body cell mass were significantly decreased. Fat-free mass and total body water (TBW) did not differ significantly between groups. Table 3 shows correlations between anthropometric and BIA parameters in boys. BIA estimates of SMM and BCM in children highly correlated with FFM and were excluded from the analysis. One can see stronger associations of BMI with Age, Ht, and FFM in healthy boys compared to ALL patients. Stronger associations are also present for resistance and reactance of healthy controls with the majority of parameters shown in Table 3 compared to ALL subjects. Unlike this, BMI correlates better with %FM in ALL boys. Similar results were obtained for girls (data not shown).

Phase angle (PA) is increasingly used in clinical studies as an indicator of nutritional status, disease severity and life expectancy [6, 10]. In ALL, PA was significantly decreased compared to healthy children and positively correlated with age in both ALL and control groups (see Fig. 2 for boys). After controlling for age, PA was only weakly associated with the remission time (partial r=0.19 for boys).

In boys, principal component analysis with varimax rotation of phase angle with the selected anthropometric and BIA parameters returned two factors responsible for 75% and 71% of the total variance in patients and controls, respectively (Table 4). Factor 1 had maximal loadings on Ht, Wt, and FFM reflecting a principal role of skeletal and muscular traits in the somatic development of boys. In ALL boys, this factor had lesser, but still prevalent, extent. Factor 2 ('adiposity') had maximal loadings on %FM and FM. In girls, the above two factors have changed together with the prevalence of adiposity traits in ALL, and skeletal and muscular traits in healthy controls, respectively. Our previous data showed that the transition to prevalence of adiposity factor is characteristic for healthy girls of more advanced age [8]. Also, unlike healthy boys and girls, in ALL patients, BMI had much greater impact on adiposity rather than on skeletal and muscular development.

Table 4. The results of factor analysis of phase angle with selected anthropometric and body composition parameters

Variable	Boys				Girls			
	ALL		Healthy		ALL		Healthy	
	Factor 1	Factor 2	Factor 1	Factor 2	Factor 1	Factor 2	Factor 1	Factor 2
PA	0.315	-0.056	0.231	0.031	0.191	-0.190	0.113	-0.037
Ht	0.969	0.148	0.982	-0.086	0.242	-0.961	0.933	-0.335
Wt	0.834	0.404	0.905	-0.151	0.605	-0.700	0.830	-0.399
BMI	0.293	0.706	0.603	-0.290	0.776	-0.260	0.553	-0.423
FM	0.496	0.826	0.691	-0.586	0.795	-0.510	0.667	-0.639
%FM	0.055	0.986	0.063	-0.990	0.946	-0.289	0.374	-0.906
FFM	0.899	0.148	0.920	-0.047	0.419	-0.807	0.874	-0.287
Proportion of variance	0.411	0.338	0.504	0.206	0.395	0.359	0.461	0.252

4. Conclusion

Long-term remission represents a common consequence of treatment for acute lymphoblastic leukemia in children. So, the assessment of nutritional status of such children has become a question of significant interest. In this study, we described peculiarities of growth, development and body composition changes in Russian children after treatment for ALL. Taken together, our results show that many anthropometric and BIA parameters of body composition in children in remission of ALL, change significantly compared to healthy ones. The presence of ALL weakens associations of BMI, as well as of resistance and reactance, with Ht, Wt, and other parameters of body composition that reflect the state of an organism's skeletal and muscular development. Phase angle was significantly decreased in ALL, and, after controlling for age, only weakly correlated with the remission time. In ALL girls, we observed early prevalence of adiposity traits in their somatic growth and development compared to healthy controls. Further studies are needed to elucidate other health-related consequences of treatment for ALL in Russian children aimed at optimization of the disease management and rehabilitation strategy.

References

[1] Seibel N L 2008 in: *Hematology* (ASH Education book) 374-380
[2] Delbecque-Boussard L, Gottrand F, Ategbo S et al 1997 *Am. J. Clin. Nutr.* **65** 95-100
[3] Nysom K, Holm K, Michaelsen K F et al 1999 *J. Clin. Endocrinol. Metab.* **84** 4591-4596
[4] Murphy A J, Wells J C K, Williams J E et al 2006 *Am. J. Clin. Nutr.* **83** 70-74
[5] Piccoli A, Rossi B, Pillon L, Bucciante G 1994 *Kidney Int.* **46** 534-539
[6] Wells J C K, Williams J E, Fewtrell M et al 2007 *Int. J. Obes.* **31** 507-514
[7] Nikolaev D V, Smirnov A V, Bobrinskaya I G, Rudnev S G 2009 *Bioelectric impedance analysis of human body composition* (Moscow: Nauka), in Russian
[8] Martirosov E G, Khomyakova I A, Pushkin S V et al 2007 ICEBI 2007, IFMBE Proceedings **17** 807-810
[9] Houtkooper L B, Going S B, Lohman T G et al 1992 *J. Appl. Physiol.* **72** 366-373
[10] Baumgartner R N, Chumlea W C, Roche A F 1988 *Am. J. Clin. Nutr.* **48** 16-23

Relationship between the Initial Systolic Time Interval and RR-interval during an exercise stimulus measured with Impedance Cardiography

Femke Hoekstra[1], Esther Habers[1], Thomas W J Janssen[1,3], Rudolf M Verdaasdonk[2] and Jan H Meijer[2]

[1] VU University Amsterdam, Faculty of Human Movement Sciences,
[2] VU University Medical Center, Dept. of Physics and Medical Technology, Amsterdam, the Netherlands
[3] Rehabilitation Center Amsterdam, Amsterdam, the Netherlands

E-mail: jh.meijer@vumc.nl

Abstract. The Initial Systolic Time Interval (ISTI), obtained from the electrocardiogram and impedance cardiogram, is considered to be a measure for the time delay between the electrical and mechanical activity of the heart and reflects an active period of the heart cycle. The relationship between ISTI and the total heart cycle (RR-interval) was studied in three groups of young, healthy volunteers: low, moderately and highly trained subjects. The three groups were exposed to an exercise stimulus on a cycle ergometer with an increasing work load to increase the heart rate. ISTI was decreased with decreasing RR-interval. However, the relative proportion of ISTI, ISTI/RR, was found to *increase* with decreasing RR-interval. This relationship was found to be inversely proportional. The rate of this increase in ISTI/RR was significantly higher in highly trained subjects. Also, over the whole range of heart rates ISTI was longer in these subjects. It is concluded that ISTI can be used to evaluate cardiac performance during physical exercise non-invasively and in an extramural setting.

1. Introduction

When measuring the electrical impedance of the human thorax, a variation synchronous with the heart activity is observed. The time derivative of this variation is called the Impedance CardioGram (ICG). The ICG-signal represents the mechanical activity of the heart cycle, while the widely used ElectroCardioGram (ECG) represents the electrical activity of the heart cycle. The Initial Systolic Time Interval (ISTI), obtained from simultaneous recordings of the ECG and ICG, can be considered as a measure for the time delay between the electrical and mechanical activity of the heart and reflects an early active period of the heart cycle [1]. Measurement of the ISTI is attractive because it is fast and easy to obtain. Moreover, the ISTI can be measured non-invasively and is not restricted to hospital settings.

The effect of exertion on the ISTI has not been studied explicitly so far. In the present study the relationship between ISTI and RR-interval was investigated by exposing young healthy subjects to an exercise stimulus. The dependence of the relative proportion of ISTI/RR on the RR-interval was studied as well.

Further, it has been reported that long term physical exercise can induce physiological changes of the heart of elite athletes. These adaptations to training can cause an 'Athlete's heart', which is structural and electrical different from a 'normal' heart [2]. Abnormalities in the ECG of these athletes are commonly reported [2] and sometimes associated with a pathological condition [3]. It is possible that these physiological changes also lead to differences in ISTI. Therefore, the ISTI-RR relationships were studied in highly, moderately and low trained subjects to detect possible differences between these groups.

2. Methods

2.1. ICG and ECG recordings

ICG recordings were made using the method described by Meijer et al. [1] (see figure 1). A typical example of a simultaneous recording of the ICG and ECG signal is shown in figure 2. The Initial Systolic Time Interval (ISTI) is the time interval between the R-peak in the ECG and C-wave in the ICG signal, which is depicted in the figure.

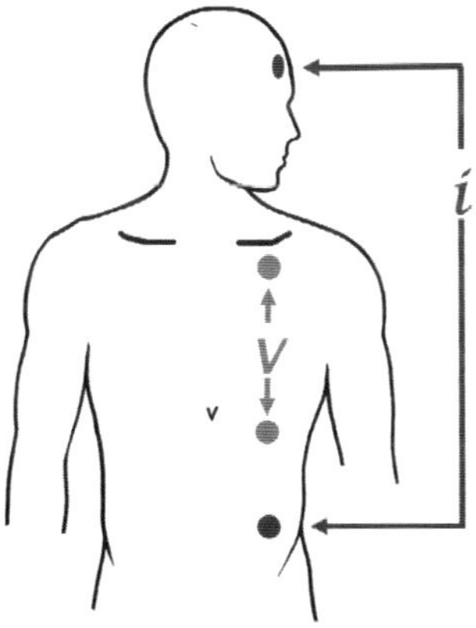

Figure 1. Electrode configuration: A small electrical current (i) of 0.3 mA r.m.s. (64 kHz) is applied between the outer two electrodes. The inner pair of electrodes measure the subsequent electrical voltage difference (V) over the heart. The impedance CardioGram (ICG) is obtained from the time course of this voltage signal.

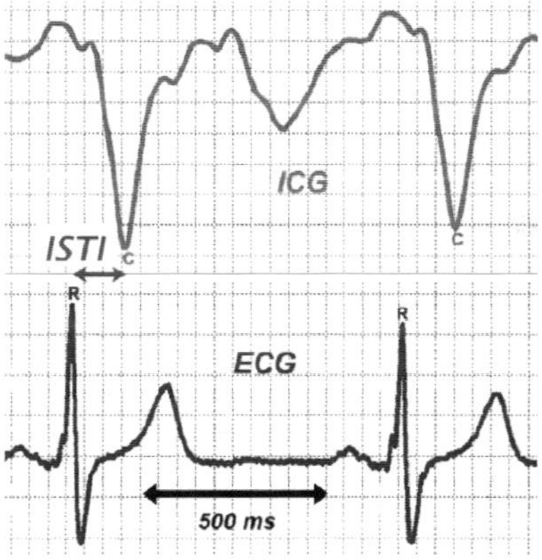

Figure 2. A typical example of simultaneous registration of an Impedance CardioGram (ICG) and an ElectroCardioGram (ECG) (arbitrary units). The Initial Systolic Time Interval (ISTI) is the time interval between the R-peak in the ECG and the C-wave in the ICG signal, which are indicated in the figure.

2.2. Subjects

Twenty young, healthy subjects participated in this study (table 1). The group consisted of 9 males and 11 females with an age (mean ± S.D.) of 21 ± 2 years, a height of 178 ± 10 cm and a body mass of 67 ± 9 kg. Based on the level of exercise, the subject group was divided into three groups: low trained (<4 hrs/wk of exercise, N=7), moderately trained (4 – 10 hrs/wk of exercise, N=7) and highly trained subjects (>10 hrs/wk of exercise, N=6). Most of the highly trained subjects were endurance athletes.

Table 1. Subjects' characteristics (mean ± S.D.). The groups did not differ in gender, age and weight. The highly trained group was significantly ($p<0.05$) taller than the other two groups.

	total (N=20)	**low (N=7)**	**moderate (N=7)**	**high (N=6)**
age (years)	21 ± 2	22 ± 3	21 ± 1	22 ± 3
male(m)/female(f)	9m/11f	3m/4f	2m/5f	4m/2f
height (cm)	178 ± 10	173 ± 12	176 ± 5	186 ± 6
body mass (kg)	67 ± 9	64 ± 10	65 ± 7	72 ± 6

The moderately trained subjects were recreational or at a regional level active in their sport. The highly trained subjects were active at a national or international level. Unpaired t-tests showed that the groups did not differ in gender, age and body mass. However, the highly trained subjects were significantly taller than the other two groups. The study was approved by the ethics committee of the Faculty of Human Movement Sciences of the VU University Amsterdam. Prior to participation subjects signed an informed consent after being informed about the protocol.

2.3. Procedure

A questionnaire assessed the daily physical activity of the subjects and their participation in sports. Before measurement, the subjects rested in supine position during ten minutes. Subsequently, they performed an exercise test on a cycle ergometer with an increasing work load to increase heart rate. The exercise stimulus consisted of 5 or 6 blocks of 3 minutes cycling with an individually adapted, increasing work load. To prevent disturbances caused by movement, ECG and ICG recordings were made during a one-minute break after each block. During the last block of cycling all subjects exercised with an intensity of at least 80% of the maximal heart rate.

2.4. Analysis

The first 15 cardiac cycles after each block of exercise were used to determine the RR-intervals and the corresponding values of ISTI. Periods with too much noise to determine the ISTI were omitted. The relationship between ISTI and RR-interval was established using linear regression analysis according to equation 1: $\mathrm{ISTI} = a + b \cdot \mathrm{RR}$ (1)

The relationship between the ISTI/RR and the RR-interval was observed to be inversely proportional. Equation 2 was used to describe this relationship and to establish the parameters by regression analysis: $\mathrm{ISTI/RR} = a/\mathrm{RR} + b$ (2)

To illustrate differences between the groups at high and low exertion, the individual values of ISTI were calculated from the individual relationships at an RR-interval of 350 ms and 600 ms: ISTI(350) and ISTI(600). Unpaired t-tests were used to detect the differences between the groups in the values of the parameters mentioned above. All statistics was run on SPSS 15.0 for Windows.

3. Results

3.1. ISTI, ISTI/RR and RR-interval

The ISTI decreased significantly in all subjects with a decrease in RR-interval caused by exercise. A typical example of this relationship is shown in figure 3. However, ISTI/RR *increases* with a decrease in RR-interval. Figure 4 presents a typical example of the relationship between ISTI/RR and RR-interval. No differences were found in the mean values of the coefficients a and b when determined according to equation 1 or 2.

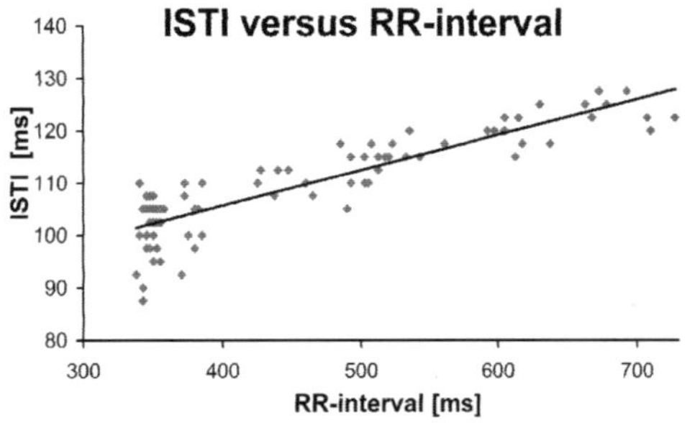

Figure 3. A typical example of the relationship between ISTI and RR-interval. ISTI decreases with decreasing RR-interval caused by exercise.

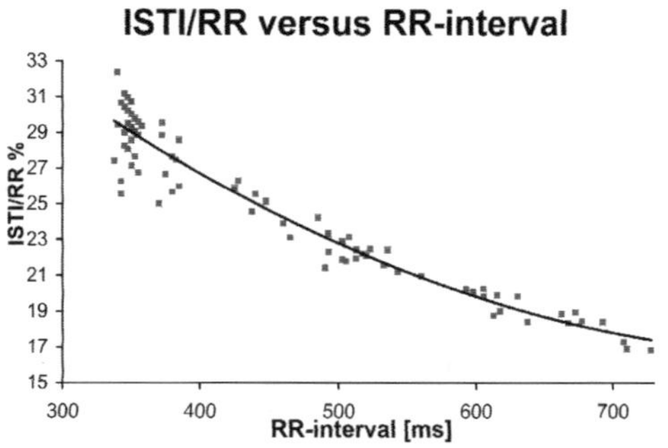

Figure 4. The same subject as in figure 3. The relative proportion of ISTI (ISTI/RR) *increases* with decreasing RR-interval.

3.2. *Differences between groups*

The constant a in the equations 1 and 2 was significantly higher in highly trained subjects than that in the other groups (table 2). No differences between the groups were observed in the coefficients b. Over the whole range of heart rates, ISTI was longer in the highly trained subjects, which is illustrated by significantly higher values for ISTI(350) and ISTI(600) in this group. No differences were found between the low and moderately trained subjects.

Table 2. Mean values (± S.D.) of coefficients a (in s) and b and the mean (± S.D.) ISTI (in ms) at an RR-interval of 350 and 600 ms. The a, ISTI(350) and ISTI(600) were significantly different in highly trained subjects compared with less trained subjects.

Group	a	b	ISTI(350)	ISTI(600)
low	0.06 ± 0.03*	0.09 ± 0.03	90 ± 23*	112 ± 20*
moderate	0.06 ± 0.03*	0.08 ± 0.02	89 ± 23*	108 ± 20*
high	0.10 ± 0.02	0.06 ± 0.04	119 ± 14	133 ± 14

*Significantly different ($p<0.05$) from highly trained subjects

Consistent with the observed differences in length, significant correlations ($p<0.05$) were found between the subjects' height and respectively the coefficients a, ISTI(350) and ISTI(600).

4. Discussion and conclusions

The positive linear relationship between the ISTI and RR-interval demonstrates that with a decrease in the RR-interval, also the ISTI is shortened. The inverse proportional relationship between the ISTI/RR and RR-interval shows that with a decrease in RR-interval, this active phase decreases less than the total cardiac cycle. The observed linear relationship between ISTI and RR-interval is in agreement with Boudoulas et al. [4], who reported linear relationships between the electrical systole versus the heart rate and the electromechanical systole versus the heart rate. Also, it has been reported in early literature that the Preejection Period (PEP), a parameter analogous to the ISTI [1], was shortened with increased heart rate [5].

The coefficients a and the values of ISTI(350) and ISTI(600) are significantly higher in highly trained subjects than in the other groups, indicating that ISTI is longer in this highly trained group at the same length of the cardiac cycle, and that ISTI acquires a longer proportion of this cycle when the cardiac cycle shortens. This finding is consistent with the results of Fardy et al. [5]. They reported that during exercise the PEP was significantly longer in former athletes compared to non-athletes. Also, during exercise the PEP was significantly longer in subjects who were more physical active [5]. However, since the highly trained subjects were taller than the other two groups, it cannot be excluded that the observed difference is influenced by a difference in height.

The longer ISTI and the higher rate of increase of ISTI/RR in highly trained subjects indicate that the time delay between electrical and mechanical activity is longer in highly trained subjects and responds differently to an exercise stimulus as compared to less trained subjects. This may reflect a more efficient use of the heart muscle. Further research is necessary to elucidate these observations. It can be concluded that the ISTI is an attractive measure that can be used to evaluate cardiac performance during physical exercise non-invasively and in an extramural setting. The measurement can be performed fast and easily, not only in sport settings, but also in clinical applications.

References

[1] Meijer J H, Boesveldt S, Elbertse E and Berendse HW 2008 *Physiol. Meas.* **29** S383-S391
[2] Fagard R 2003 *Heart* **89** 1455-1461
[3] Lauschk J, Maisch B 2009 *Clin. Res. Cardiol.* **98** 80-88
[4] Boudoulas H, Geleris P, Lewis R P and Rittgers S E 1981 *Chest* **80** 613-617
[5] Fardy P S, Maresh C M and Abbott R D 1976 *Med Sci Sports* **8** 26-30

Impedance cardiography: What is the source of the signal?

R P Patterson
Professor Emeritus, University of Minnesota, Minneapolis MN 55455, USA

E-mail: patte001@umn.edu

Abstract. Impedance cardiography continues to be investigated for various applications. Instruments for its use are available commercially. Almost all of the recent presentations and articles along with commercial advertisements have assumed that aortic volume pulsation is the source of the signal. A review of the literature will reveal that there is no clear evidence for this assumption. Starting with the first paper on impedance cardiography in 1964, which assumed the lung was the source of the signal, the presentation will review many studies in the 60's, 70's and 80's, which suggest the aorta and other vessels as well as atria and again the lung as possible sources. Current studies based on high resolution thoracic models will be presented that show the aorta as contributing only approximately 1% of the total impedance measurement, making it an unlikely candidate for the major contributor to the signal. Combining the results of past studies along with recent work based on models, suggest other vessels and regions as possible sources.

1. Introduction

Impedance cardiography continues to be used in research and for clinical applications for stroke volume and cardiac output measurements. Almost all of the recent presentations and articles along with commercial advertisements have assumed that aortic volume pulsation is the source of the signal. This paper will review past studies that attempted to determine the physiological origin of the signal and present new modelling data that suggest alternate sources for the signal. The assumption that the aorta is the most significant source of the signal is on very weak grounds.

2. Background and past studies

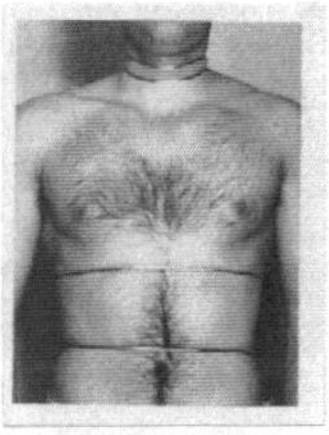

Figure 1. Original electrode band arrangement

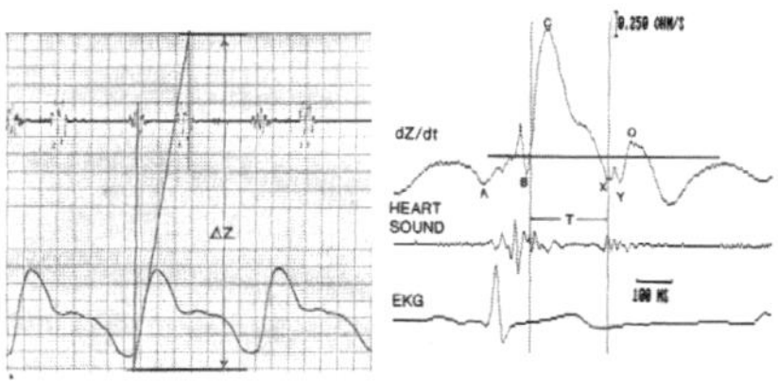

Figure 2. Original ΔZ slope extraction process and the current dZ/dt. Decreasing impedance upward.

In order to develop an electrode configuration that was consistent and repeatable over a wide range of individuals and somewhat independent of the amount of air in the lungs, Patterson et al [1,2] devised an arrangement of band electrodes at the ends of the thorax. More recently this configuration has used spot electrodes in the previous band electrode locations. The waveform from such an arrangement showed decreasing impedance during systole. Based on a volume event this would suggest that the region causing the impedance change was increasing in volume during systole. This would eliminate the ventricles as a source of the signal. At the time of the original development it was assumed that the lungs would be the most likely candidate for the source. The right ventricle delivers the stroke volume to the lungs during systole, but at the same time the pulmonary veins drain blood from the lungs. An empirical forward graphical slope procedure was developed to account for the pulmonary drainage during systole. This resulted in an equation that gives reasonable stroke volume values [1,2]. The graphical slope procedure was later replaced by an equivalent method using the first derivative of the waveform [2].

2.1. Animal studies

In 1965, NASA funded for 5 years, a research group at the University of Minnesota to develop the impedance method to measure cardiac output on the astronauts for the Apollo moon mission. As part of this contract a large series of animal experiments were undertaken to determine the source of the waveform in order to develop a sound formula for cardiac stroke volume. Dogs were studied with pressure catheters placed in all the major vessels and heart chambers. Electromagnetic flow transducers were placed on the pulmonary artery and aorta. Many different manoeuvres were used to change stroke volume. At the start of the studies it was assumed that the lungs were the source of the signals. By accident one day one of the animals developed complete left mechanical alternates. Tracings from the flow probe and impedance only showed a change when the left heart ejected blood. This is shown in Figure 3. In many dogs with complete left mechanical alternans there was no change in delta Z with only the right ventricle ejecting, but in some dogs there was a change of about 50% compared to both ventricles ejecting as shown on the right frame of Fig. 3. This suggested that the signal came from the arterial system. One region considered for the source was the aorta, which was reported [3,4].

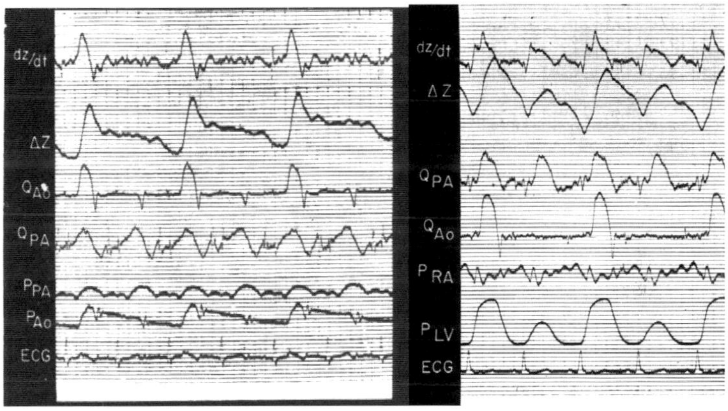

Figure 3. The left graph shows no change in ΔZ without left ventricular ejection whereas in the right graph there is a small change. Q_{Ao} is aortic flow and Q_{PA} is pulmonary artery flow. P_{PA} pulmonary artery pressure and P_{Ao} is aortic pressure.

2.2 Human Studies

Another approach using field mapping was undertaken to understand better the source of the cardiogenic impedance signal. The amplitude of dZ/dt was measured as a function of distance up the thorax. In Figure 4 on the left side is shown the position of strip electrodes on the chest and on the right is shown the increase in amplitude of dZ/dt with distance increasing from the xiphisternal joint up to the neck. The results show the lower part of the thorax contributes very little to the total signal. In the region from the supra-sternal notch down the sternum 10 cm gives a signal amplitude that accounts for about 75% of the total for the full bands.

3. Current model based studies

The new studies involve use our finite difference (FD) electrical model of the human thorax [6]. This model has 3.8 million control volumes or elements. The resolution is 1.5 mm in the x-y direction with 5 mm thick slices. The software used with the model can determine current and voltage distributions from external excitation along with sensitivity analysis, which can determine the contributions of various organs and regions to the total impedance.

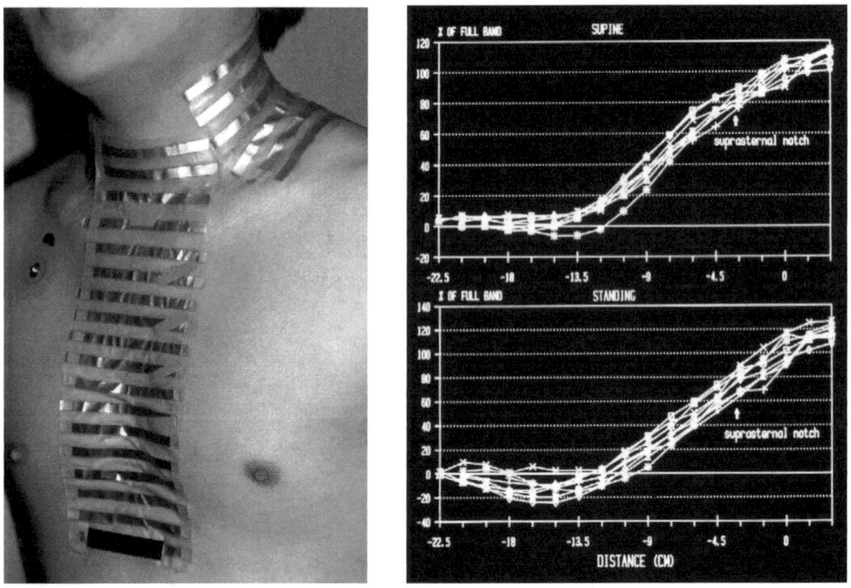

Figure 4. The electrode array for the voltage pick up electrodes is shown on the left. The reference electrode is at the bottom in black. The graph shows the change in the dZ/dt amplitude expressed as a percentage of the full band signal as a function of distance up the thorax from the xiphisternal.

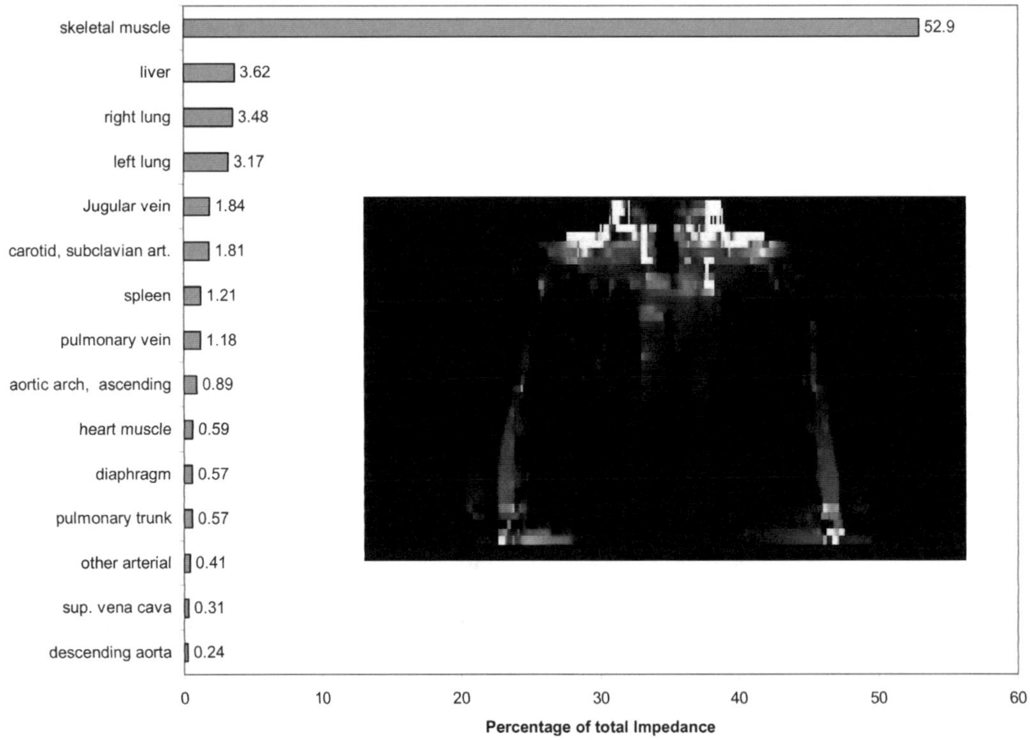

Figure 5. This gives the percentage contribution of various organs and regions to the total impedance as determined with the model along with an insert of showing regions with high sensitivity to local resistivity changes.

4. Discussion and Conclusion

The results of the modelling study agree very closely with Kaupinen et al. [7]. Both show the large contribution of the skeletal muscle to total impedance and a small contribution by the aorta, along with high sensitivity of impedance change in the upper thorax. With the aorta's contribution to total impedance of less than one percent, during systole the aortic volume would have to change more than 60%, which is unreasonable. Also the other arterial and venous vessels in the upper thorax contribute twice as much as the aorta to the total impedance. The fact that the jugular vein contribution is similar to the upper thoracic arteries may explain the distortion in the waveform in patients with conditions like heart failure who frequently have a large pulse in jugular vein.

It is highly unlikely that the aorta is a major contributor to the impedance cardiography waveform and also the venous pulsations from the neck region may cause significant distortions in the waveform leading to errors in the stroke volume determinations.

References

[1] Patterson R, Kubicek W G, Kinnen E, Witsoe D and Noren G 1964 Development of an electrical impedance plethysmographic system to monitor cardiac output *First Annual Rocky Mountain BioEngineering Symposium, May 4-5 U. S. Air Force Academy* p56

[2] Patterson R P 1965 Cardiac output determinations using impedance plethysmography *M. S. Thesis, University of Minnesota*

[3] Kubicek W G, Patterson RP and Witsoe D A 1970 Impedance cardiography as a non-invasive method of monitoring cardiac function and other parameters of the cardiovascular system *Annals of the New York Academy of Sciences* **170 Article 2** 724

[4] Kubicek W G, Kottke F J, Ramos M U, Patterson R P, Witsoe D A, LaBree J W, Remole W, Layman T E, Schoening H, and Garamella J T 1974 The Minnesota impedance cardiograph-theory and applications *Biomedical Engineering* **9** 410.

[5] Patterson R P, Wang L, Raza B, and Wood K 1990 Mapping the cardiogenic impedance signal on the thoracic surface *Physiol. Meas.* **28** 212

[6] Belalcazar A., and Patterson RP 2004 Improved lung edema monitoring with coronary vein pacing leads: a simulation study. *Physiol. Meas.* **25** 475-487

[7] Kauppinen P K, Hyttinen J A and Malmivuo J A 1998 Sensitivity distributions of impedance cardiography using band and spot electrodes analyzed by a three-dimensional computer model *Ann. Biomed. Eng.* **26** 694–702

Dynamic Response of the Initial Systolic Time Interval to a Breathing Stimulus measured with Impedance Cardiography

Jan H Meijer[1], Femke Hoekstra[2], Esther Habers[2], Ruud M Verdaasdonk[1] and Thomas W J Janssen[2,3]

[1]VU University Medical Center, Dept. of Physics and Medical Technology,
[2]VU University Amsterdam, Faculty of Human Movement Sciences,
Amsterdam, the Netherlands
[3]Rehabilitation Center Amsterdam, Amsterdam, the Netherlands

E-mail: jh.meijer@vumc.nl

Abstract. The Initial Systolic Time Interval (ISTI) is a measure for the time delay between the electrical and mechanical activity of the heart. The present study reports about the dynamic response of ISTI to a Valsalva manoeuvre. This response was investigated in 22 young healthy volunteers, having different levels of training in sports. The time course of the ISTI during the Valsalva manoeuvre was found to follow a distinct pattern and to be analogous to the course of the Pre-Ejection Period (PEP), also obtained from ECG and ICG signals, reported earlier. The recordings show a definite influence of the Frank-Starling mechanism and are to some extent consistent with reports on the time course of sympathetic activation. The highly trained subjects showed an ISTI that was systematically longer at all moments of the manoeuvre.

1. Introduction

When measuring the electrical impedance of the human thorax a variation synchronous with the heart activity is observed. The time derivative of this variation is called the Impedance CardioGram (ICG). During the last decades several studies have demonstrated that the *amplitude* of the ICG-signal is too complicated to make a simple physiological model interpretation as cardiac stroke volume [1]. An interesting aspect of the ICG-signal, however, can be found in the *time relationships*, especially when this signal is compared to the Electro CardioGram (ECG). Regardless of the multiple sources of the signal, the ICG reflects the mechanical aspects of the cardiac cycle [2], while the ECG reflects the electrical aspects of the cardiac activity. Therefore, by recording both signals simultaneously, the time difference between electrical and mechanical events in the cardiac cycle can be evaluated in various (patho-) physiological conditions.

The Initial Systolic Time Interval (ISTI), obtained by simultaneous recordings of the ECG and the ICG, has been proposed as a measure for the time delay between the electrical and mechanical activation of the heart [3]. The use of the ISTI in clinical and other diagnostics is attractive because the registration is inexpensive, fast and easy. The measurements are non-invasive, form no burden to the subjects and are not restricted to a hospital environment. However, so far little is known about the behaviour of the ISTI under various circumstances. Theoretically, the ISTI depends on three factors: the preload of the heart by way of the Frank-Starling mechanism, the autonomic nervous control, and the afterload of the heart caused by the peripheral resistance. There is a need for observational research to reveal the behaviour of the ISTI under different circumstances. The present study reports about the dynamic response of ISTI to a Valsalva manoeuvre, which is used in clinical practice to provide measures of cardiovascular autonomic nervous function [4]. In the present study, this response of ISTI to a Valsalva manoeuvre was investigated in a group of 22 young, healthy volunteers, consisting of individuals who were trained at various levels in sports.

2. Methods and subjects

2.1. ICG and ECG signal recordings

ICG recordings were made using a four-electrode system on the left side of the body (figure 1). The outer two electrodes applied a small electrical current through the thorax. The inner two electrodes continuously measured the subsequent electrical voltage difference over the heart, from which the impedance was calculated. The measurement was described in detail by Meijer et al. [3]. The ECG signal was derived simultaneously from the two inner electrodes.

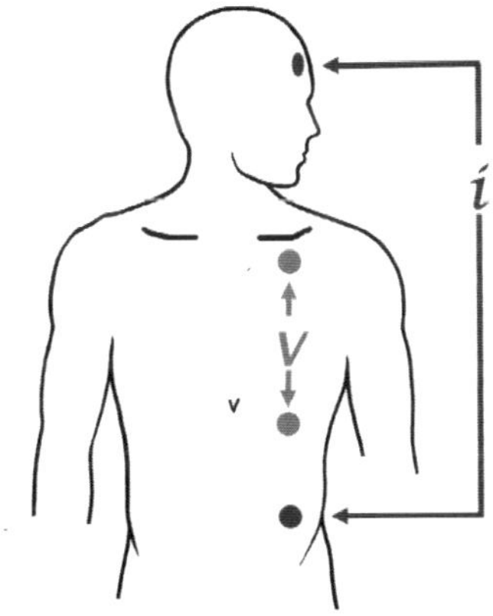

Figure 1. Electrode configuration: A small AC current (0.3 mA, 64 kHz) is applied to the thorax by means of the two outer electrodes. The two inner electrodes measure the subsequent electrical voltage difference over the heart. The impedance recording is obtained from the time course of this voltage signal. The ECG-signal is obtained from the two inner electrodes.

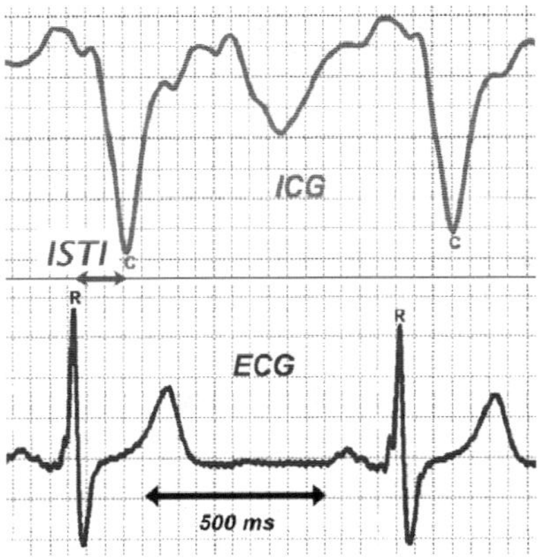

Figure 2. A typical example of simultaneous registration of an Impedance CardioGram (ICG) and an ElectroCardioGram (ECG) (arbitrary units). The marker points R in the ECG and C in the ICG are indicated. From these points the Initial Systolic Time Interval (*ISTI*) is determined, which can be considered as a measure of the time lag between the electrical and mechanical activity of the heart.

Figure 2 shows a typical example of a simultaneous recording of an ICG and an ECG. The ICG-signal is the first time-derivative of the impedance variations across the heart. The R- and C- points, which are used for the determination of the Initial Systolic Time Interval (ISTI), are indicated.

2.2. Subjects

Twenty-two young healthy volunteers (11 males and 11 females), with an age (mean ± s.d.) of 21 ± 2 years (range 19-28 yrs), a height of 180 ± 10 cm (range 158-197 cm) and a weight 68 ± 8 kg (range 50-81 kg) participated in this study. Subjects using medication were excluded from the study, which was approved by the Ethics Committee of the Faculty of Human Movement Sciences of the VU University Amsterdam. All subjects gave informed consent to participate in the test. Based upon their level of training, the group was stratified in three subgroups: highly active sportive (>10 hrs/wk of exercise , N=7) moderately active (4-10 hrs/wk of exercise, N=7) and low active subjects (<4 hrs/wk of exercise , N=8).

2.3. Procedure

After ten minutes of supine rest, the manoeuvre was performed in a sitting position at an expiratory pressure of 40 mmHg during 20s. An analogue manometer provided visual feedback to the subjects. Closure of the glottis was prevented by a small leak in the mouthpiece. ECG and ICG recordings were made from at least 20s before, during and until at least 30s after the manoeuvre. From these recordings the RR-interval and ISTI were calculated off-line and plotted against time.

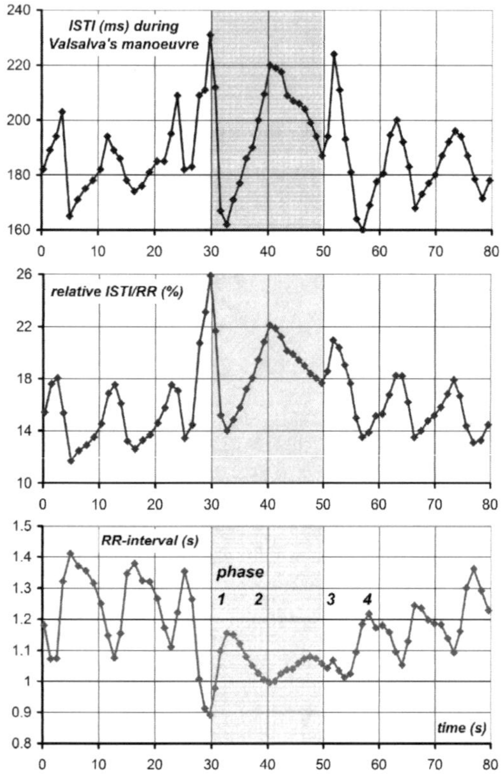

Figure 3. A typical example of the time course of the ISTI (top graph) and the RR-interval (bottom graph) during a Valsalva manoeuvre. The points represent the values of one heart beat and are plotted on the moment of the first R-top in the ECG of a cycle. The time course of ISTI/RR, which represents the proportion of RR occupied by ISTI, is shown in the middle. The successive phases of the Valsalva's manoeuvre are indicated in the lower graph. Before the manoeuvre, the ISTI showed a pattern synchronous with breathing. At the onset of the manoeuvre, there frequently was an increase in ISTI. In phase 1 a sharp decrease in ISTI occurred in all of the subjects. In early phase 2 the ISTI increased in all of the subjects, up to or above the pre-Valsalva level, followed by a decrease in late phase 2. At the relief of pressure, in phase 3, there was a sharp decrease in ISTI, frequently preceded by a sharp increase. In phase 4 no specific dissimilarity in the course of ISTI with the pre-Valsalva period was observed.

3. Results

The time course of the parameters during the Valsalva manoeuvre is described in terms of the standard phases 1 to 4 [5, 6, 7]. A typical example of the time course of the ISTI and the RR-interval is shown in figure 3. In the middle the relative proportion of ISTI, ISTI/RR is expressed as a percentage of the RR-interval. The specific observations are mentioned in the caption of the figure. Based on these observations, five reference moments were selected that were present in the course of the ISTI in each subject: the onset, the sharp minimum in phase 1, the maximum in phase 2a, the minimum in phase 2b and the minimum in phase 3. The mean value (values ± S.D.) of the ISTI was calculated in each group of subjects and is presented in table 1.

Table 1. Mean ISTI (± S.D.) (ms) at five reference moments during the manoeuvre in the three groups of subjects. All maximum values differed from the minimum values in each group (all $p < 0.005$). The mean ISTI of the highly trained subjects was longer than that in the other two groups at all reference moments (all $p < 0.025$).

training level group	onset	minimum phase 1	maximum phase 2a	minimum phase 2b	minimum phase 3	N
low	146 ± 13	108 ± 19	143 ± 16	112 ± 20	116 ± 21	7
moderate	142 ± 13	102 ± 15	139 ± 10	115 ± 17	104 ± 17	7
high	184 ± 25	141 ± 26	187 ± 22	157 ± 28	151 ± 37	8

All minimum values of ISTI differed significantly from the maximum values in each group of subjects (paired t-test, all $p < 0.005$). The mean of ISTI in the highly trained group was significantly longer than that in the other two groups at all reference moments (unpaired t-test, all $p < 0.025$). No significant difference was found between the low and moderately trained groups at any of the reference moments (unpaired t-test, all $p > 0.1$).

4. Discussion and conclusions

The time course of the ISTI before and after the Valsalva manoeuvre showed a cyclic pattern, originating from breathing. It is concluded that also the maxima of ISTI at the onset and just after the Valsalva manoeuvre, present in most subjects, are the consequences of taking a breath. Variations in these maxima are thought to be due to variations in the depth of the breath. Very significant is the sharp decrease in ISTI in phase 1. This occurs within 1-2 heartbeats after the onset of the Valsalva's manoeuvre. This response is too fast to originate from a sympathetic influence. It is the consequence of an extra volume of blood that is pushed out of the lungs towards the heart by the high pressure inside the thorax. This clearly demonstrates the influence of the Frank-Starling mechanism on the ISTI. The cyclic breathing pattern before and after the manoeuvre can be attributed to the same mechanism. The sharp decrease in ISTI at the end of phase 3/early phase 4 can also be attributed to the increased preload of the heart originating from the decrease in pressure inside the thorax, although sympathetic influence cannot be excluded. It coincides with a period of high sympathetic activity [5, 6, 7]. The decrease of ISTI in phase 2b cannot be attributed to preload, because of the low venous return in this phase. It corresponds with a high level of sympathetic activity in this phase [5, 6, 7]. Therefore, ISTI is influenced by at least two factors: preload and sympathetic control.

The time course of the ISTI during the Valsalva manoeuvre was similar to the course of the Pre-Ejection Period (PEP), also obtained from the ICG and ECG, reported by Ermishkin et al. [8]. The course of the PEP showed a comparable pattern of maxima and minima. This means that PEP and ISTI are expected to measure similar (patho-)physiological factors. The only difference observed in this study, is that ISTI was not systematically shorter in phase 4 than during the pre-Valsalva period as was reported for the PEP [8]. Measurement of the PEP has a disadvantage that it is based on points in the recordings (Q- and B-points) that are difficult to trace and that are strongly susceptible to noise, especially in a hospital environment. This makes automated processing of the PEP almost impossible. Recording of the ISTI is considered to be a good alternative.

The long ISTI observed in highly trained subjects during the whole manoeuvre demonstrates a difference in functioning of the heart in these subjects. Further research is needed to elucidate the significance of this observation.

References

[1] Bernstein P D 2010 *J. Electr. Bioimp.* **1** 2-27
[2] Lababidi Z, Ehmke D A, Durnin R E, Leaverton P E and Lauer R M 1970 *Circulation* **41** 651-658
[3] Meijer J H, Boesveldt S, Elbertse E and Berendse H W 2008 *Physiol. Meas.* **29** S383-S391
[4] Gerritsen J, Ten Voorde B J, Dekker J M, Kingma R, Kostense P J, Bouter L M and Heethaar R M 2003 *Diabetologia* **46** 330-338
[5] Smith M L, Beightol L A, Fritsch-Yelle J M, Ellenbogen K A, Porter T R and Eckberg D L 1996, *Am J Physiol* **271** (*Heart Circ. Physiol.* **40**) pp H1240-H1249
[6] Cooke W H, Reynolds B V, Yandl M G, Carter J R, Tahvanainen K U O and Kuusela T A 2002 *Med. Sci. Sports Exerc.* **34** pp 928-935
[7] Cooke W H, Carter J R and Kuusela T A 2003 *Aviat. Space Environ. Med.* **74** pp 731-737
[8] Ermishkin V V, Lukoshkova E V, Bersenev E Yu, Saidova M A, Shitov V N, Vinogradova O L and Khayutin V M 2007 *Proc. XIIIth International Conference on Electrical Bio Impedance (Graz), IFMBE Proc.* **17**, eds H Scharfetter and R Merwa, (Springer Verlag Berlin Heidelberg), pp 655-658

Journal of Physics: Conference Series **224** (2010) 012120 doi:10.1088/1742-6596/224/1/012120

Transoesophageal detection of heart graft rejection by electrical impedance: using Finite Element Method simulations

G.Giovinazzo[1], **N.Ribas**[2], **J.Cinca**[2] **and J.Rosell-Ferrer**[1]

[1] Center for Research in Biomedical Eng. (CREB). Dept. of Electronic Eng., UPC, Barcelona, Spain

[2] Hospital de la Santa Creu i Sant Pau, UAB, Barcelona, Spain

E-mail: ggiovinazzo@eel.upc.edu

Abstract. Previous studies have shown that it is possible to evaluate heart graft rejection level using a bioimpedance technique by means of an intracavitary catheter. However, this technique does not present relevant advantages compared to the gold standard for the detection of a heart rejection, which is the biopsy of the endomyocardial tissue. We propose to use a less invasive technique that consists in the use of a transoesophageal catheter and two standard ECG electrodes on the thorax. The aim of this work is to evaluate different parameters affecting the impedance measurement, including: sensitivity to electrical conductivity and permittivity of different organs in the thorax, lung edema and pleural water. From these results, we deduce the best estimator for cardiac rejection detection, and we obtain the tools to identify possible cases of false positive of heart rejection due to other factors. To achieve these objectives we have created a thoracic model and we have simulated, with a FEM program, different situations at the frequencies of 13, 30, 100, 300 and 1000 kHz. Our simulation demonstrates that the phase, at 100 and 300 kHz, has the higher sensitivity to changes in the electrical parameters of the heart muscle.

1. Introduction

Organ rejection may cause edema, leucocytary infiltration, inflammation, ischemia and, in some cases, fibrosis or cell necrosis. That is why there are changes in the electrical properties of the myocardial tissues after transplanted heart rejection and for this reason electrical impedance measurement can be used to detect cases of heart rejection. In fact, previous studies [1][2][3][4] have shown that it is possible to evaluate heart graft rejection level using a bioimpedance technique by means of intramural electrodes or using an intracavitary catheter.

However, this last technique medical approach does not have important advantages compared to biopsy, because in both cases the patient has to be measured in a catheterization room under fluoroscopic guidance. Moreover, the cost of the percutaneous catheter is comparable to the catheter used for biopsies. For this reason, we propose to use a less invasive technique that consists in the use of a transoesophageal catheter and two standard ECG electrodes on the thorax.

The aim of this work is to evaluate, using Finite Element Method, several parameters affecting the impedance measurement, including: sensitivity to electrical conductivity and permittivity of different organs in the thorax, changes in module and phase due to a lesion producing a scar, a global ischemia of the heart and pleural effusion in the lungs. From these results we deduce the best estimator for cardiac rejection detection and we obtain the tools to identify possible cases of false positive of heart rejection due to other factors.

2. Methods

In order to create the model (fig. 1), we used computer assisted tomography (CT) and magnetic resonance (MR) images of the trunk from [6]. Conductivity and permittivity values are taken from [5]. We developed an assembly of a thorax model, consisting on a combination of parts that are: thoracic muscles, subcutaneous fat, ribs and sternum, heart, lungs, blood inside the heart, spinal column, aorta, superior cava vein and, finally, surface electrodes. We choosed not to detail further the created model for two reasons: 1) the large differences in electrical conductivity and permittivity data for different tissues published in various studies, that will be a major source of error and 2) the design with SolidWorks must be simple in order to allow Comsol Multiphysics to import it and produce a Finite Element (FE) mesh with a limited number of elements.

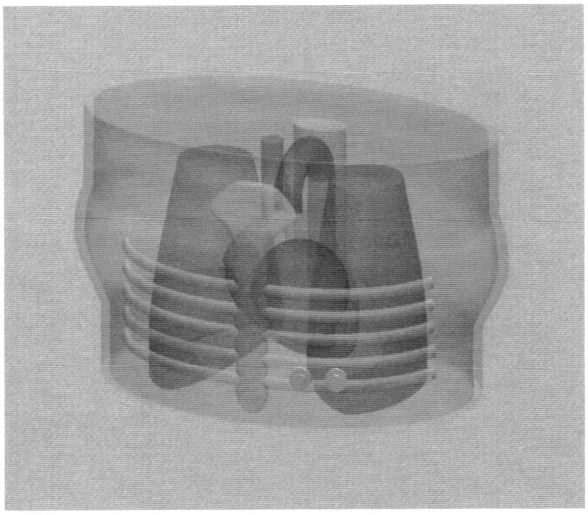

Figure 1. Design of a human thorax developed with SolidWorks.

3. Simulation results

3.1. Impedance changes in presence of ischemia or necrosis (scar) of heart tissues

The aim of this study is to verify the maximum expected changes in transoesophageal bioimpedance measurement considering the extreme cases of a global acute ischemia and a global scar of the cardiac tissue.

Our simulation demonstrates (fig. 2) that, comparing a transoesophageal measurement with normal values of electrical conductivity and permittivity with those in presence of a global cardiac ischemia, the maximum change in the module will be at a frequency of 13 kHz (0.82 Ω), while the phase will change more at a frequency of 100 kHz (-0.81^o).

Whereas, in presence of a global scar in the cardiac muscle, the maximum change of the module (-2.15 Ω) will be at a frequency of 13 kHz, while the phase will vary more at a frequency of 300 kHz (1.34^o).

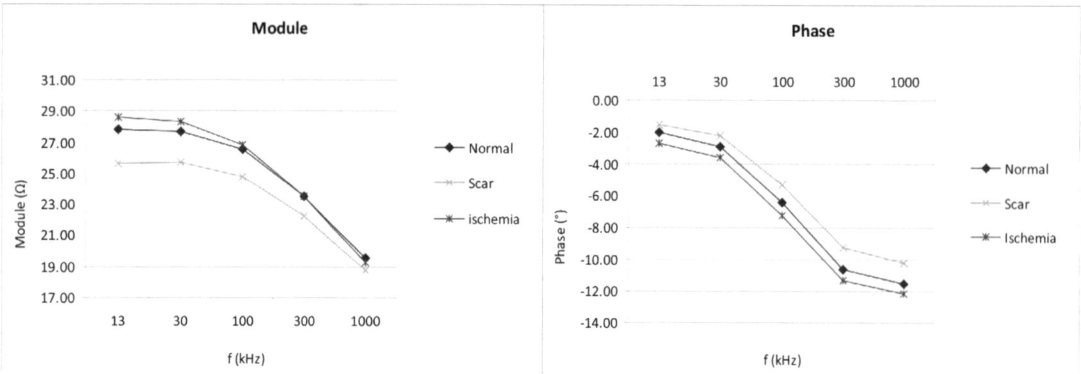

Figure 2. Simulation of the electrical impedance variation due to a change in the electrical parameters of the heart.

3.2. Sensitivity to conductivity and permittivity changes of thoracic tissues

In figure 3 there are summarized the values of the relative sensitivity of the module ($Sr = \frac{\delta|Z|/|Z_{\sigma 0}|}{\delta\sigma/\sigma_0} \times 100$) and the variation of the phase, to an increase in the electrical conductivity and to a decrease in the electrical permittivity of 10% respectively, considering all the tissues in our model.

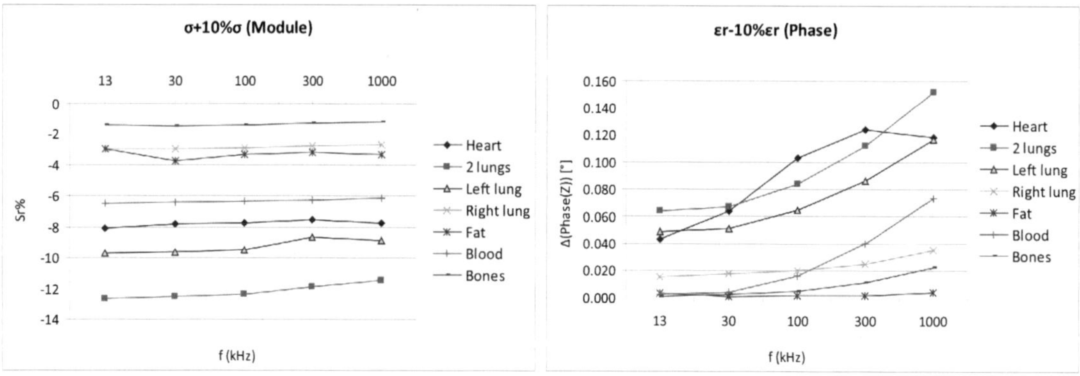

Figure 3. Summary of the simulation of the percentage relative sensitivity of the module to an increase in the electrical conductivity of 10%, in the different tissues.

The left graph in figure 3 shows that the tissues causing the biggest change in the module, are, in order, lungs (mainly the left one), heart and blood; while the module and the phase are less sensitive ($S_r < 4\%$ at all frequencies) to a change in the conductivity of subcutaneous fat, right lung and bones.

The right graph shows that at 100 kHz and 300 kHz we are more sensitive to a change of the relative permittivity of the heart, in the phase, if compared to the same percentage change in other tissues.

In figure 4, we can see the graphs of electrical impedance (module and phase), with inflated and deflated lungs, as well as in presence of a left and right pleural effusion.

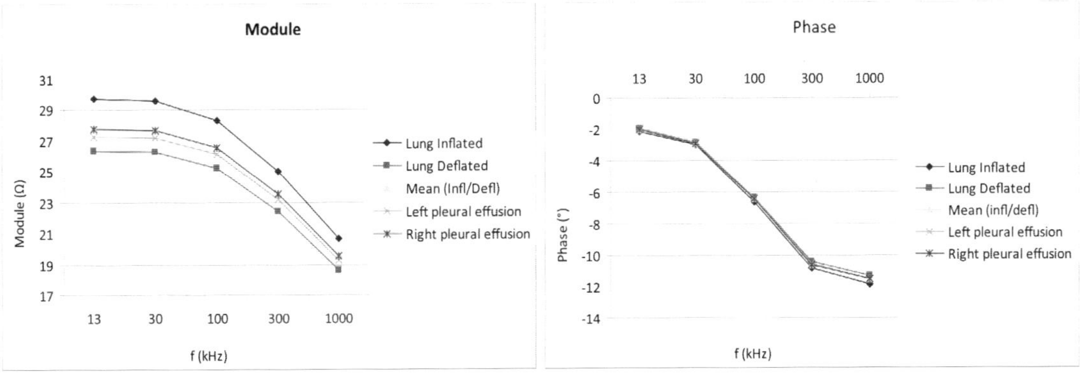

Figure 4. Simulation of the electrical impedance variation due to a change in the electrical parameters of the lungs.

4. Conclusions

Electrical impedance is more sensitive to a change in the electrical conductivity of lungs compared to other tissues, as happens, for example, during normal respiration. For this reason we propose to carry out a transoesophageal bioimpedance measurement during several respiratory cicles, with the purpose to reduce the effect due to respiration.

Moreover our simulation demonstrates that the phase, at the frequency of 100 and 300 kHz, varies more due to a percentage change in the electrical parameters of the heart muscle, if compared with the same percentage changes in other tissues.

Besides, from results not shown in this work, the phase is less influenced than the module to a displacement of the catheter position inside the oesophagus, as highlighted both from human measurement and FEM simulations (about 0.8 Ω/cm for the module while the phase change is almost comparable to the noise of the measurement device, less than 0.1 degrees).

Acknowledgments

This work has been supported by grants from the Spanish Ministry of Science and Technology and the FEDER project SAF2005-02270 and from the Spanish Ministry of Health, Redes de Investigacin del Instituto de Salud Carlos III (REDINSCOR).

References

[1] Cinca J, Ramos J,Rosell-Ferrer J et al 2008 *Changes in myocardial electrical impedance in human heart graft rejection* vol. 10 no. 6 (*Eur J Heart Fail*) pp. 594-600
[2] Grauhan O, Muller J, Knosalla C, Hetzer R et al 1996 *Electric myocardial impedance registration in humoral rejection after heart transplantation* vol. 15 no. 2 (*J Heart Lung Transplant*) pp. 136-143
[3] Grauhan O, Muller J, Pfitzmann R, Hetzer R et al 1997 *Humoral rejection after heart transplantation: reliability of intramyocardial electrogram recordings (imeg) and myocardial biopsy* vol. 10 no. 6 (*Transpl Int*) pp. 439-445
[4] Pfitzmann R, Muller J, Grauhan O and Hetzer R 2000 *Intramyocardial impedance measurements for diagnosis of acute cardiac allograft rejection* vol. 70 no. 2 (*Ann Thorac Surg*) pp. 527-532
[5] `http://niremf.ifac.cnr.it/docs/DIELECTRIC/AppendixB1.html#B01`
[6] Moeller T B, Reif E 2007 *Pocket Atlas of Sectional Anatomy Computed Tomography and Magnetic Resonance Imaging* vol 2 (Stuttgart - New York, Thieme) pp. 2–28

Journal of Physics: Conference Series **224** (2010) 012121 doi:10.1088/1742-6596/224/1/012121

Bioimpedance spectroscopy in haemodynamic analysis

W A M^c^Cullagh and L C Ward

The University of Queensland, Brisbane, Australia, 4072

l.ward@uq.edu.au

Abstract. Venous insufficiency is estimated to affect between 10% and 35% of the population in the US with symptoms ranging from mild discomfort to chronic ulceration which may reduce quality of life. Early diagnosis is the key to pre-emptive treatment. We have previously reported [5] an impedance technique for measurement of calf muscle pump function although it was noted that the results could be confounded by change in limb geometry during the exercise protocol. We report here a modified protocol to account for change in limb geometry. Impedance of a 20 cm segment of the calf was continuously recorded using an SFB7 bioimpedance spectrometer whilst subjects performed a sequence of manoeuvres: a) supine with leg raised (for 10 min); b) standing (4 s); c) plantar flexion (tiptoe, 4 s); d) standing elevated on one leg removing tension on the measured leg with the foot horizontal (4 s); e) with the leg relaxed (4 s) and then the sequence repeated. The impedance ratio, k = (e-d)/(c-b), was the proportion of the impedance change, occurring during calf muscle pumping, due primarily to change in limb shape only, i.e. independent of the impedance change due to ejection of blood as a function of calf muscle pump action. Thus (1-k) can be used to correct the ejection fraction (%) calculated as (c-b)/(a-b)*100 for the confounding effect of change in limb geometry. Ejection fractions calculated by this method in 10 control subjects were 51.5 ± 30.1% with no values greater than 100% as found previously.

1. Introduction

Disorders of the peripheral vascular system are commonplace. Venous insufficiency is estimated to affect between 10% and 35% of the population in the US. Poor venous drainage results in venous hypertension, increased transmural pressure with consequent oedema that frequently leads to inflammation, infection and ulceration. Symptoms range from varicose veins, which are present in approximately 30% of adult women and 20% of adult men, to chronic ulceration, with a prevalence of ~1% [1]. These may result from deep vein thrombosis (DVT) which affects approximately 84 in every 100,000 adults [2]. The direct costs of venous disorders have been estimated at over $1 billion for the Unites States alone [3].

Since the venous system contains approximately 70% of the total blood volume, any alteration to the capacity or efficacy of the venous system can affect venous return and cardiac output which should be the same under steady state conditions. The compliance of deep veins of the legs is controlled by the surrounding skeletal muscles since the deep veins have little sympathetic innervation [4]. Venous valves, present in leg veins, assisted by the massaging effect of the calf muscle pump on the veins, ensure that there is a flow of blood towards the heart and reflux is prevented when changing posture from the supine position to standing. The action of the calf muscle pump (CMP) has been implicated as a critical “anti-oedema” mechanism by enhancing fluid flow to the lymphatics thereby decreasing tissue fluid pressures [4]. A deficient calf muscle pump has also been identified as probably the key factor in venous insufficiency in the elderly [1].

Various methods for haemodynamic analysis exist including, ambulatory venous pressure measurements (AVP), photoplethysmography, air plethysmography (APG), ambulatory strain gauge plethysmography (SPG), foot volumetry, continuous-wave Doppler or duplex scanning ultrasound and impedance plethysmography (IPG) with the latter having been identified as a potential criterion diagnostic method for DVT [2]. In the present study, we measured CMP activity by an IPG technique during a calf muscle pumping protocol modified from that used in APG and SPG.

2. Materials and methods

2.1. Principle of the method

In the normal upright posture, despite the venous return of the circulatory system, blood tends to pool in feet and lower portions of the leg due to gravity. When a supine position is assumed with the foot and lower leg elevated above the level of the heart (> 20 cm height difference), the blood is drained from the limb and attains a minimum volume (mv) after about 10 min. When the upright posture, without weight-bearing, is reassumed, blood volume rapidly achieves a maximum (MV) due to venous filling. The difference (MV-mv) represents the functional venous volume (VV). If the subject now undertakes a tip-toe manoeuvre, calf muscle contraction activates the muscle pump, ejecting blood from the lower leg, i.e. the ejected volume (EV). These volume changes can be measured as changes in impedance of the lower leg. However, the measured impedance is a not solely a function of volume change but also the shape change of the calf associated with muscle contraction, confounding the measurements and potentially contributing to anomalous results noted previously [5] where the ejected volume exceeded the functional volume, i.e. EV > VV, in some subjects. We hypothesised that this shape change could be simulated under non-contracting muscle conditions (no CMP activity) by the relaxed leg with an unsupported downward hanging foot. Measurement of impedance under these conditions would approximate that due to geometric change alone and could be used to correct the impedance measured during the tip-toe manoeuvre.

2.2. Experimental

2.2.1. *Subjects* Ten self-reported healthy volunteers (7 males and 3 females) with body mass indices (BMI) ranging from 21 to 28 kg/m^2 participated.

2.2.2. *Impedance measurements* Impedance, deconvoluted into its components of resistance and reactance, was measured using an ImpediMed SFB7 impedance analyser modified to record impedance at a sampling rate of one reading per ms at a frequency of 5 kHz.

2.2.3. *Protocol* Subjects were weighed and their height measured. While standing upright, the point of maximal calf circumference was determined and measured. Points 10 cm either side of this mid-point along the mid-line of the calf were located, the circumferences measured and sense electrodes (Ag-AgCl gel) applied. Current drive electrodes were located at the base of the fingers and toes. All measurements were performed on the right side. The subject reclined supine for 10 min with the foot raised 40 cm above heart level, supported on a box. Impedance was continuously recorded for 10 s after 1 min and 5 min recumbency and 10 seconds immediately before standing. After 10 min, the subject stood, initially taking their weight on the left foot and stabilising themselves with a walking frame until weight was evenly distributed on both feet. Measurement continued during attainment of the upright position and for a further 20 s after the stable standing position was achieved. The subject remained standing for 4 min and with impedance measurements again being made while the subject remained standing for 10 s and during tip-toe movements at approximately 2 s intervals. Measurements were performed in triplicate. The subject then stood on a box, again taking their weight on the left foot and supported on by the walking frame. The right leg was held horizontal for 5 s and then allowed to hang fully relaxed for 4 s over the side of the box whilst measurements were made. Measurements were made in triplicate at 5 s intervals.

2.2.4. *Data analysis* Resistance data were processed through a Butterworth filter and plotted against time of measurement. The mean resistance was determined during: a) lying supine with leg raised; b) standing; c) plantar flexion (tiptoe); d) standing elevated on the opposite leg, thereby removing tension on the measured leg, with the foot horizontal and e) with the leg relaxed and foot hanging. VV is represented by (a-b) whilst (c-b) is the resistance change due to change in blood volume and segment shape. This was corrected for the proportion of the impedance change due primarily to change in limb shape only by the impedance ratio, k = (e-d)/(c-b). The volume of ejected blood (EV) may be calculated according to the well known Nyboer equation:

$$EV = \frac{-\rho_b L^2 \Delta R}{R_0^{\,2}} \qquad (1)$$

where ρ_b is the resistivity of blood (150 ohm.cm [6]); L is the length of the segment; ΔR is the change in resistance related to the change in blood volume; R_0 is the resistance of the calf segment before the tip-toe manoeuvre. The volume (V) of the leg segment being measured was calculated from the measured circumferences using the equation for a truncated cone assuming that the calf was best represented by two opposed cones [5].

3. Results

A typical time course profile for the change in resistance of the calf is presented in Figure 1. Resistance decreased (i.e., volume increased due to venous filling) as the subject transfered from recumbency with the leg elevated to the fully upright standing position with the weight evenly distributed on both feet. Resistance increased with each tiptoe movement as CMP activity pumped blood from the calf. The resistance change during this manoeuvre was presumed to include a contribution due to geometric change estimated from the shape change occurring in the relaxed unsupported leg (figure 1 d).

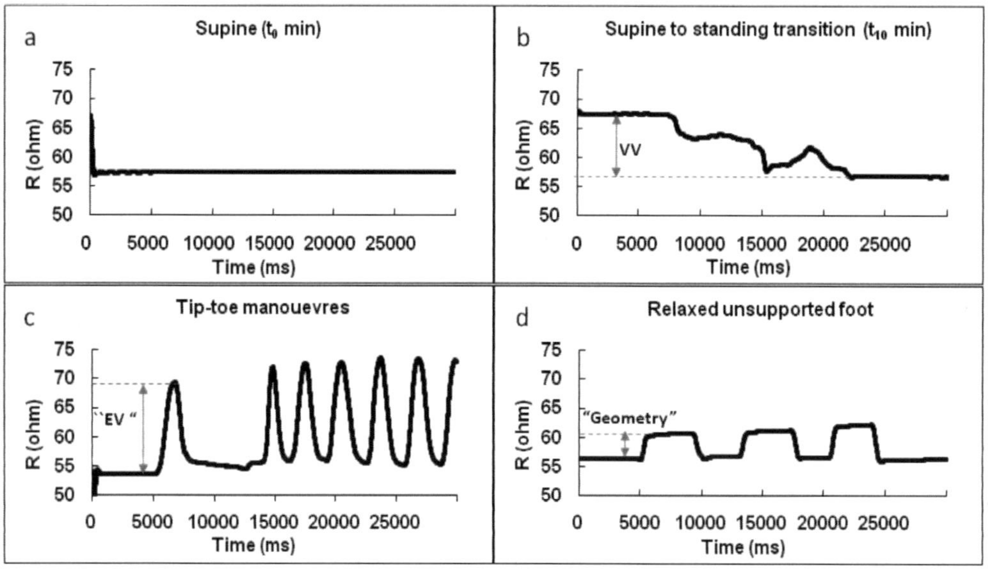

Figure 1. Time course change in resistance (R, ohm) from supine (a) to standing (b), during tiptoe movements (c) and when the foot is relaxed and unsupported while standing (d). VV is the difference between the maximum and the minimum volumes. "EV" is the difference between the maximum volume and the mean volume of the tiptoe movements with the contribution due to shape change of the calf (d). Corrected EV = "EV" – "Geometry".

Impedance profiles were reproducible within subjects, coefficients of variation for replicate measurements within subjects ranged from 1 to 5 %. Profiles were similar between subjects but varied in magnitude, e.g., supine impedance varied from 49 to 70 ohm, and were correlated with calf volume (r = 0.7). Calculated CMP haemodynamic parameters, corrected for geometric change, are presented in Table 1. The maximum EF observed for any subject in any trial was 87%.

Table 1. Haemodynamic parameters (mean ± SD, n = 10) calculated from impedance measurements of the calf. Ejected volume data are corrected for changes in limb geometry.

Venous volume (ml)	Ejected volume (ml)	Ejected fraction (%)	Ejected volume per 100 ml tissue (ml)
191.2 ± 56.1	93.7 ± 52.3	51.5 ± 30.1	4.1 ± 1.7

4. Discussion

The quantitative measurement of CMP activity is notoriously difficult [2] and there are relatively few published data. Nevertheless, the ejection volumes and fractions are similar to those obtained by others using methods such as APG [7] and ejection volume per unit volume of tissue fall within the range expected for healthy subjects.

Change in limb shape and geometry are recognized to affect haemodynamic measurements. In an attempt to overcome these problems, measurement at the ankle rather than at the calf has been proposed [5]. In the present study, the alternative approach of correcting for shape change was adopted. A number of observations attest to the support of this approach. Firstly, data comparable to published normal data was obtained. Secondly, no ejection fractions greater than 100% were observed unlike previously where values of greater than 150% were found [5]. Thirdly, ejection volume was calculated from the total change in impedance during a tip-toe manoeuvre rather than proportion of this value [4] as used in standard APG and SPG protocols.

Calculation of volume from resistance was based on the Nyboer equation (equation 1). This equation has been criticized as not taking into account the non-homogeneous nature of limb tissue and should be corrected using the muscle volume to total limb volume ratio, ~0.6 [7]. Application of this correction decreases the mean ejected volume from 93.7 ml to 62.1 ml. However, application of both this and a geometric correction may lead to underestimation since it is likely that the volume correction partially corrects for geometric changes [7]. We used a value of 150 ohm.cm for blood resistivity, although widely used in the Nyboer equation, resistivity varies with haematocrit and blood flow. More recent estimates [8], suggest that a value of 135 ohm.cm may reflect better the value found *in vivo*. This would decrease the volume estimates proportionately.

In conclusion, with further development and validation, quantitative impedance plethysmography could become the preferred method for haemodynamic analysis due to its convenience and low cost.

References

[1] Nicolaides AN 2000 Investigation of chronic venous insufficiency. A consensus statement *Circulation* **102** e126-63

[2] Kearon C, Julian JA, Newman TE and Ginsberg MD 1998 Noninvasive diagnosis of deep vein thrombosis *Ann. Intern. Med.* **128** 663-77

[3] Goddard AA, Pierce CS and McLeod KJ 2008 Reversal of lower limb edema by calf muscle pump stimulation *J. Cardiopulmon. Rehab. Prevent.* **28** 174-179

[4] Buckley JC, Peshock RM and Blomqvist CG 1988 Deep venous contributions to hydrostatic blood volume change in the human leg Am. J. Cardiol. **62** 449-53

[5] McCullagh WA, Ward LC, Shirer, W and Chetham S 2007 Bioelectrical impedance analysis measures the ejection fraction of the calf muscle pump *Proc. XIII Int Conf Electrical Bioimpedance & VIII Conf on Electrical Impedance Tomography (Graz, Austria, 29 August - 2 September) IFMBE Proc* **17** 616-619

[6] Anderson FA 1984 Impedance plethysmography in the diagnosis of arterial and venous disease *Ann. Biomed. Eng.* **12** 79-102

[7] Bermudez K, Knudson M, Marabito D and Kessel O 1998 Fasciotomy, chronic venous insuffuciency, and the calf muscle pump *Arch. Surg.* **133** 1356-61

[8] Traugott FM, Quail AWE and White SW 1981 Evaluation of blood resistivity in vivo for impedance cardiography in man, dog and rabbit *Med.Biol. Eng. Comput.* **19** 547-52

Prediction and monitoring of fluid responsiveness after coronary bypass surgery using the Initial Systolic Time Interval: Preliminary Results

A Smorenberg[1], E J Lust[1], R M Verdaasdonk[2], A B J Groeneveld[1], and J H Meijer[2]

[1]VU University Medical Center, Department of Intensive Care for Adults,
[2]VU University Medical Center, Department of Physics and Medical Technology,
Amsterdam, the Netherlands

E-mail: jh.meijer@vumc.nl

Abstract. The objective of the study is to develop a non-invasive method to optimize the assessment of cardiac preload and therapeutic fluid administration after coronary artery bypass surgery. Previous studies have reported that the pre-ejection period (PEP), obtained from the electro-cardiogram (ECG) and from the invasively measured arterial pressure P_a, can be used for this assessment as it is dependent on the cardiac preload. The Initial Systolic Time Interval (ISTI), obtained non-invasively by simultaneous measurement of the Electro-CardioGram (ECG) and Impedance CardioGram (ICG), is expected to depend on the cardiac preload as well. 16 patients, admitted to the Intensive Care Unit after coronary artery bypass surgery and presumably hypovolaemic, were measured during administration of 2x250 ml of an isosmotic colloidal fluid solution. The parameters PEP and ISTI were determined before and after the administrations and compared with the change in cardiac output (CO), obtained by a thermodilution technique. Preliminary results show significant relationships between ISTI and CO and between changes in both of these variables before and after fluid administration.

1. Introduction

The objective of this pilot-study is to investigate whether the Initial Systolic Time Interval (ISTI) has the potential to be used as a non-invasive measure to optimize the assessment of cardiac preload and therapeutic volume expansion after coronary artery bypass surgery. Intravenous fluid administration is accepted universally as a treatment for hypotension after cardiac surgery. The accuracy of indices, used to assess preload and predict fluid responsiveness has been questioned [1]. Previous studies have reported that the pre-ejection period (PEP), obtained from the ECG and the invasively measured arterial pressure P_a, can be used for this assessment as it is dependent on the cardiac preload [2, 3]. The PEP is also regarded as a measure for the time delay between the electrical and mechanical activation of the heart. The Initial Systolic Time Interval (ISTI), obtained non-invasively by simultaneous measurement of the ECG and the Impedance CardioGram (ICG) has been reported to be related to the PEP and proposed as an alternative [4]. Therefore, it is expected to depend on the cardiac preload as well. The registration of ISTI is inexpensive, fast, and easy which makes its use in diagnostics attractive, in clinic as well as in an extramural setting. The measurements are non-invasive and form no burden to the patients. The present study investigates whether the ISTI can be used to predict responsiveness to fluid administration in patients admitted to the intensive care unit after coronary artery bypass surgery, by comparing ISTI and PEP with the cardiac output (CO) measured by thermodilution.

2. Methods and patients

2.1. ICG and ECG signal registration

ICG recordings were made using a four-electrode system on the left side of the body (figure 1). The outer two electrodes applied a small electrical current (0,3 mA r.m.s., 64 kHz) through the thorax. The inner two electrodes continuously measured the subsequent electrical voltage difference over the heart, from which the impedance was calculated. The measurement has been described in detail by Meijer et al. [4]. The ECG signal, used for the determination of ISTI, was derived simultaneously from the two inner electrodes.

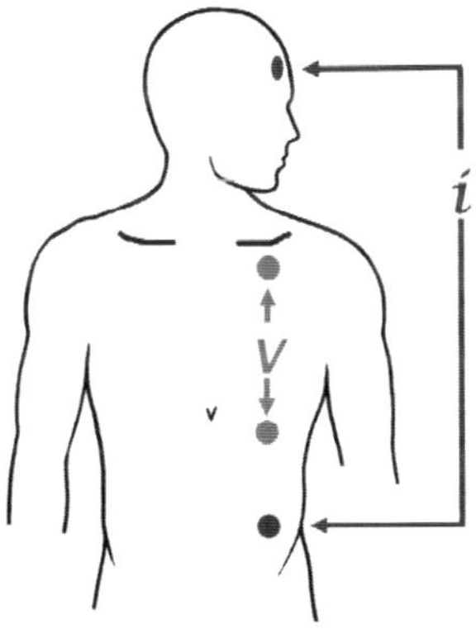

Figure 1. Electrode configuration: A small AC current (0.3 mA, 64 kHz) is applied to the thorax by means of the two outer electrodes. The two inner electrodes measure the subsequent electrical voltage difference over the heart. From the time course of this voltage signal the impedance signal is obtained.

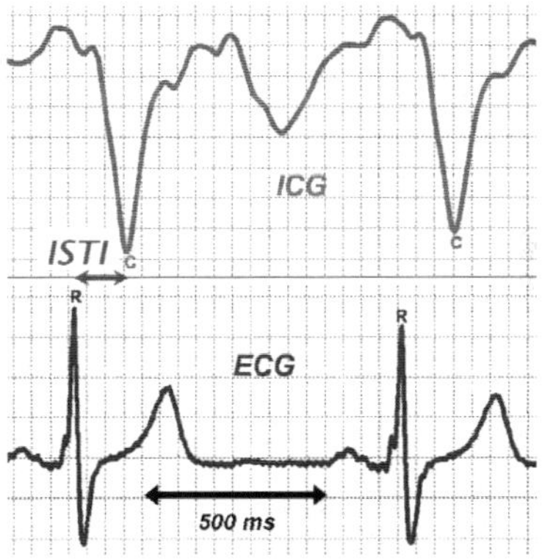

Figure 2. A typical example of simultaneous registration of an Impedance CardioGram (ICG) and an ElectroCardioGram (ECG) (arbitrary units). The marker points R in the ECG and C in the ICG are indicated. From these points the Initial Systolic Time Interval (*ISTI*) is determined, which can be considered as a measure of the time lag between the electrical and mechanical activity of the heart.

Figure 2 shows a typical example of a simultaneous recording of an ICG and an ECG. The ICG-signal is the first time-derivative of the impedance variations across the heart. The R- and C-points, which are used for the determination of the Initial Systolic Time Interval (ISTI), are indicated.

The pre-ejection period (PEP), the time from the onset of ventricular depolarization to the beginning of left ventricular ejection, was obtained by simultaneous ECG recording and arterial pressure wave tracing (P_a), obtained from the radial artery, which are commonly monitored in critically ill patients. Cardiac output (CO) was measured in order to establish the effect of volume expansion. The mean of three CO-measurements was used, obtained by a thermodilution technique which is part of the standard clinical practice, using a catheter in the *arteria pulmonalis*.

2.2. Patients and procedures

16 patients who were administered to the Intensive Care Unit after coronary artery bypass surgery and who were presumed to be hypovolaemic, were measured during intravenous administration of 2x250 ml of a Gelofusine® solution. The patients were considered to be hypovolaemic if either: a) systolic bloodpressure < 110 mmHg, b) CVP < 10 mmHg, c) PCWD < 12 mmHg, d) central or mixed venous SO_2 < 70 %, or e) if isotropic or vasopressor drugs were being used. The measurements of CO,

ISTI and PEP were performed at three moments: before infusion, after the first and after the second infusion of 250 ml each. The group of patients consisted of 14 males and 2 females, having an age (mean ± s.d.) of 68 ± 11 years (range 44-84 yrs), height 174 ± 10 cm (range 151-188 cm) and weight 84 ± 15 kg (range 54-125 kg). Reasons for exclusion were an abnormal cardiac rhythm or an artificial pacemaker, relevant alterations in medication, and any medical, ethical or practical drawbacks or objections to perform the measurements. The study was approved by the Ethics Committee of the VU University medical center Amsterdam. All subjects gave informed consent to participate in the test.

Table 1. Mean values (± S.D.) of the Cardiac Output (CO), Pre-Ejection Period (PEP) and Initial Systolic Time Interval (ISTI) at three moments, before and after intravenous administration of 2x250 ml of a Gelofusine® solution in two steps of 250 ml each.

	before infusion	after 250 ml	after 500 ml
CO (L/min)	4.7 ± 1.2	4.9 ± 1.3	4.9 ± 1.3
PEP (ms)	190 ± 9	184 ± 10	179 ± 9
ISTI (ms)	148 ± 30	151 ± 28	151 ± 26

3. Results

The mean values (± S.D.) of the three parameters at the three moments are listed in table 1. Cardiac output increased in 7 patients, remained unchanged (change in CO less than 5%) in 6 patients and decreased in 3 patients after the full administration. The mean value of change in cardiac output between moments 1 and 3 was: $\Delta CO_{1-3} = -0.2 \pm 0.5$ l/min. No significant relationship between CO and PEP and between the changes in CO and PEP were detected. However, a significant relationship between ISTI and CO was found at each of the three moments. These relationships are presented in table 2. Because of the consistency between the observed relationships at the three different moments, the relationship for the pooled data was established. This relationship is presented in the last row of table 2 and in figure 3.

Table 2. Relationships between the Cardiac Output CO (L/min), and Initial Systolic Time Interval ISTI (ms) at the three moments before and after intravenous fluid administration. The coefficients of correlation r and the levels of significance p are also given. The last row gives the relationship for the pooled data (N = 48).

moment	relationship ISTI-CO	r	p
before	ISTI = -13.5*CO + 211	0.5453	< 0.025
after 250 ml	ISTI = -10.1*CO + 201	0.4920	< 0.05
after 500 ml	ISTI = -9.5*CO + 197	0.4609	< 0.05
pooled data	ISTI = -10.7*CO + 202	0.4917	< 0.005

The relationship between the absolute and relative change in ISTI and the relative change in CO (%) before and after 500 ml of fluid administration (moment 1 and 3) was also found to be significant: $\Delta ISTI = -0.70 * \Delta CO + 6.8$ % ($r = 0.5304$; $N = 16$; $p < 0.025$). The relationship between both relative changes between the moments 1 and 2 was found to be similar, however not significant. Between the moments 2 and 3 the effect of fluid administration on all parameters was minimal. The cardiac output did hardly change: $\Delta CO_{2-3} = -0.0 \pm 0.3$ l/min.

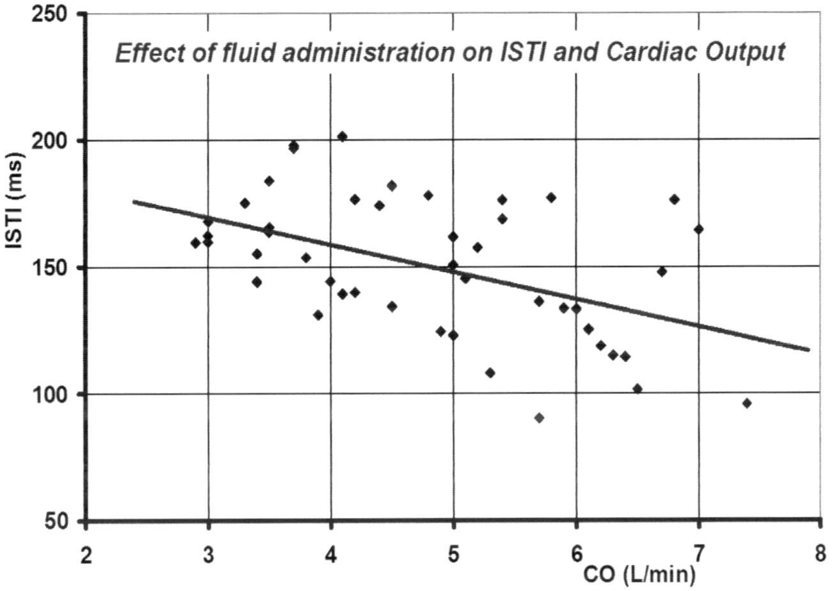

Figure 3. The administration of fluid resulted in a significant correlation between the Initial Systolic Time Interval (ISTI) and the Cardiac Output (CO).

4. Discussion and conclusions

There is a need for parameters that assess preload and predict fluid responsiveness in patients in an intensive care unit. Common methods are invasive and the accuracy has been questioned [1]. In the present study, the use of PEP and ISTI was investigated by comparing these parameters with CO responsiveness. No significant or systematic relationship between PEP and CO and ΔPEP and ΔCO was found. Such relationships have been reported earlier [2, 3]. Probably, this absence originates from the limited nature of this study. The number of patients who responded to the fluid administration was small. The mean response in cardiac output to fluid administration was low and the assessment of CO by thermodilution is of limited accuracy. Moreover, the accuracy of measurement of PEP by means of a catheter in the *arteria radialis* is limited. The time needed for the systolic pressure wave to reach the catheter tip is substantial and will vary between individuals. This suggests that measurement of the PEP is of limited value for the clinical assessment of preload and of the effect of volume expansion.

However, the preliminary results of this pilot study show significant relationships between the ISTI and the cardiac output at any moment during the fluid administration procedure. Further, a significant relationship between ΔISTI and ΔCO before and after 500 ml of fluid administration was found. This indicates that the ISTI is dependent upon preload via the Frank-Starling mechanism, and that ISTI has the potential to be used as a clinical parameter assessing preload. Future research will clarify the significance and practical usefulness of this non-invasive parameter in monitoring preload and predicting fluid responsiveness.

References

[1] Reuter D A, Felbinger T W, Schmidt C, Kilger E, Goedje O, Lamm P, and Goetz A E 2002 *Intensive Care Med.* **28** 392-398
[2] Bendjelid K, Suter P M, and Romand J A 2004 *J. Appl. Physiol.* **96** 337–342
[3] Feissel M, Badie J, Merlani PG, Faller JP, and Benjelid K 2005 *Crit Care Med.* **33**(11): 2534-9
[4] Meijer J H, Boesveldt S, Elbertse E, and Berendse H W 2008 *Physiol. Meas.* **29** S383-S391

Journal of Physics: Conference Series **224** (2010) 012123 doi:10.1088/1742-6596/224/1/012123

Over-Hydration Detection in Brain by Magnetic Induction Spectroscopy

César A González[1,2], María Pérez[3], Nidiyare Hevia[4], Fernándo Arámbula[4], Omar Flores[2], Eliot Aguilar[2], Ivonne Hinojosa[2], Leo Joskowicz[5] and Boris Rubinsky[5]

[1] Universidad del Ejército y FAM/EMGS. DF, México.
[2] Instituto Politécnico Nacional/Escuela Superior de Medicina. DF, México.
[3] Hospital Central Militar-Sección de Imagen Seccional, D.F., México.
[4] Centro de Ciencias Aplicadas y Desarrollo Tecnológico-UNAM. DF, México
[5] University of California-Berkeley/Mechanical Engineering and Bioengineering Department, Berkeley CA, USA

E-mail: c.cesar.gonzalez@gmail.com

Abstract. Detection and continuous monitoring of edema in the brain in early stages is useful for assessment of medical condition and treatment. We have proposed a solution in which the bulk measurements of the tissue electrical properties to detect edema or in general accumulation of fluids are made through measurement of the magnetic induction phase shift between applied and measured currents at different frequencies (Magnetic Induction Spectroscopy; MIS). Magnetic Resonant Imaging (MRI) has been characterized because its capability to detect different levels of brain tissue hydration by differences in diffusion-weighted (DW) sequences and it´s involve apparent diffusion coefficient (ADC). The objective of this study was to explore the viability to use measurements of the bulk tissue electrical properties to detect edema or in general accumulation of fluids by MIS. We have induced a transitory and generalized tissue over-hydration condition in ten volunteers ingesting 1.5 to 2 liters of water in ten minutes. Basal and over-hydration conditions were monitored by MIS and MRI. Changes in the inductive phase shift at certain frequencies were consistent with changes in the brain tissue hydration level observed by DW-ADC. The results suggest that MIS has the potential to detect pathologies associated to changes in the content of fluids in brain tissue such as edema and hematomas.

1. Introduction

Edema is a medical condition in which the relative amount of liquid in tissue increases. It is of substantial concern when it occurs in the brain. The characteristic of brain edema is that it develops in a delayed fashion, over a period of hours or days, after a brain trauma or event has occurred and is a cause of substantial mortality [1]. Detection and continuous monitoring of edema in the brain is essential for assessment of the medical condition and treatment.

We have suggested the use of MIS as a potential tool for non-invasive detection of tissue edema using multy-frequency induced magnetic fields. In earlier studies we have shown that measuring induction phase shift throughout the bulk of the tissue in time and in a broad range of frequencies could be used as an alternative technique for detection of the tissue edema formation [2] and [3]. Detecting the change in phase shift in time after the occurrence of a suspect episode could serve as a first order clinical warning to detect the presence and progression of edema.

MRI-DW has been used for the evaluation of acute cerebral stroke, intracranial tumours, demyelinating diseases and edema [4]. A water loading in healthy young adults have shown a measurable increase in brain tissue T2 relaxation times moderately correlated to the peripheral osmolality [5]. The diffusion of water molecules in the human brain is restricted by the tissue microstructure and water content. An increase in water ADC in the lentiform nucleus region of the brain has been reported as an early warning of brain edema detectable by conventional MRI [6].

The goal of this study was to test experimentally the feasibility to use MIS to detect the water content in the brain of young volunteers influenced by an over-hydration condition. The experimental system employs a two-coil induction system configuration to detect volumetric fluid changes by inductive phase shift measurements. MIS and DW-ADC data were estimated and compared.

2. Material and Methods

2.1. Biophysical Considerations for Inductive Phase Shift Measurements.

We consider the human head/coil configuration shown in figure 1. An alternating current Ie^{jwt} flowing through the inductor coil generates a magnetic field **B** surrounding the head volume. The induced eddy currents inside the brain promote a perturbation of the primary magnetic field **ΔB** as a function of the tissue electrical properties (σ, ε). The composite magnetic field given by **B**+**ΔB** is detected in the sensor coil as an inductive phase shift. We define a basal induced current argument in the sensor coil as ($\omega t + \Theta$) and the argument influenced by an over-hydration tissue condition as ($\omega t + \Theta_1$). To estimate the inductive phase shift ($\Delta\theta$) we can use the argument differences at specific frequency and time according to the following expression:

$$\Delta\theta=(\omega t+\theta_1)-(\omega t+\theta)=\theta_1-\theta \tag{1}$$

2.2. Experimental Inductive Spectrometer

An experimental Inductive Spectrometer was designed and constructed. Figure 1 shows a block diagram of the experimental prototype. The system consists of five modules: function generator, transceiver, phase detector, data acquisition and a PC as data processing unit. The function generator was implemented by a commercial radiofrequency signal generator (Rohde & Schwarz SMP02), which supplies an alternating current Ie^{jwt} of approximately 10mA rms in the range of 10 to 200 MHz. The transceiver consists of two different diameter coils coaxially centered at a distance d=10cm with radii of R1=2 cm and R2=8.5 cm. Both coils were built from ten turns of magnet wire AWG22 rolled on a plastic harness specifically designed for an adult human head. The coil inductances, as calculated on the basis of Faraday's law, are approximately 44.4μH and 7.73mH for the inductor and sensor coils respectively. The inductor coil induces a current in the sensor coil by magnetic induction, it involves an inductive phase shift ($\Delta\theta$) in the sensor coil as a function of the brain electrical properties. The phase detector was designed on the basis of the AD8302 (Analog Devices Inc. Norwood, MA USA). The AD8302 is a fully integrated RF IC for measuring differences in phase between two signals with a resolution of 10mV/degree. The signals of the inductor and sensor coils were connected directly to the phase detector module as shown in figure 1. The data acquisition module uses the A/D converter USB-1408FS (Computing Measurements Inc. Norton, MA USA) with a sample rate of 48kS/s. For each frequency, the data is an average of 1024 measurements. The data processing system is controlled by a personal computer Dell Precision 490 (Dell Inc. Round Rock, Texas USA). The fundament hypothesis in this study is that the changes in phase measured with the detection coil will be representative of the changes in electrical conductivity and permittivity inside the brain.

2.3. Diffusion-weighted MRI and ADC estimations
DW images with slice thickness of 7 mm and interslice gap of 1 mm were obtained in the axial plane by using a single-shot spin-echo echo planar imaging (SE EPI) sequence (repetition time (TR) 6000 ms, echo time (TE) 80 ms), 20 slices of 256x256 matrix, with b values of 0 and 1000 s/mm^2. DW images were spatially registered and the ADC maps were produced at the base-line and in hydration state to each subject. The ADC value was calculated with an algorithm implemented according to the following equation:

$$ADC\left[\frac{mm^2}{s}\right] = \frac{1}{b1}\ln\left(\frac{IS(bo)}{IS(b1)}\right) \qquad (2)$$

where IS(*b0*) and IS(*b1*) are the signal intensities voxel by voxel in the DW images obtained with two different gradient factors (*b0* and *b1*). All regions-of-interest (ROIs) were 1 cm^3 volume approximately localized in the lentiform nucleus region for the measurement of ADC values. For each ROI the mean ADC value was calculated.

2.4. Experimental Design
The experimental concept is shown in Figure 2. It consists of the induction of a generalized and transitory over-hydration condition in ten healthy adult volunteers by ingesting 1.5 to 2 litters of water (25 ml/kg) in ten minutes. Spectra of $\Delta\theta$ in the range of 10 to 200 MHz at 20 pre-programmed frequencies (lineally spaced) were measured by the spectrometer prototype described in subsection 2.2. DW images and water ADC values were estimated in the ROIs brain of the volunteers as described in previous section by the use of a 1.5-T General Electric system (Fairfield CT, USA). Spectra of $\Delta\theta$ and DW-ADC measurements were taken in normal brain tissue and thirty minutes after over-hydration as basal and over-hydration conditions respectively. Changes in $\Delta\theta$ and ADC at specific frequencies and ROIs respectively were compared in basal *vs* over-hydration conditions by a t-student paired ($P<0.5$). All statics analysis was done by a STATISTICA V7.0 (Stat Soft. Inc). The experimental protocol was developed in the Sectional Image Department of the Military Hospital of the Mexican Army and was previously approved by the Committees of Research and Bio-ethical of the Institution.

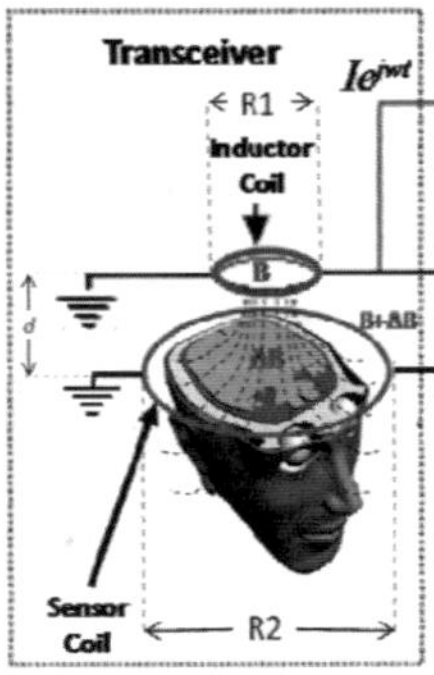

Figure 1. Head/coil configuration and block diagram of the experimental inductive spectrometer prototype.

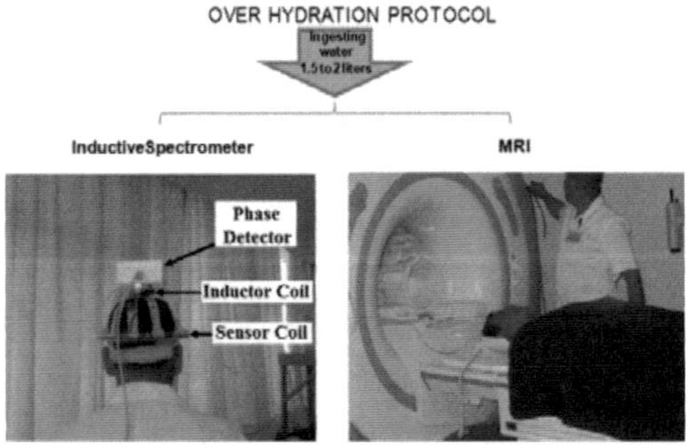

Figure 2. Experimental concept. Over-hydration induction in healthy volunteers by water ingests. MRI *vs* MIS data are compared.

3. Results
Figure 3 shows axial slice diffusion MRI images in basal and over-hydration conditions in the brain of one volunteer as representative behavior of the physiological event. A widespread change in the intensity of the brightness is evident. Figure 4 shows the average spectra of inductive phase shift values estimated in all volunteers in basal and over-hydration conditions. Significant inductive phase shift changes after 100MHz are evident and an important increment of the sensitivity around 150 MHz highlights. Figure 5 shows a box plot of the average assessment water ADC in basal and over-

hydration conditions estimated in all volunteers. A moderated increment of the average water ADC value estimated in the ROIs after the over-hydration induction with respect to basal conditions is observed, nevertheless; a t-student paired test not shows significant changes of water ADC values in the evaluated ROIs.

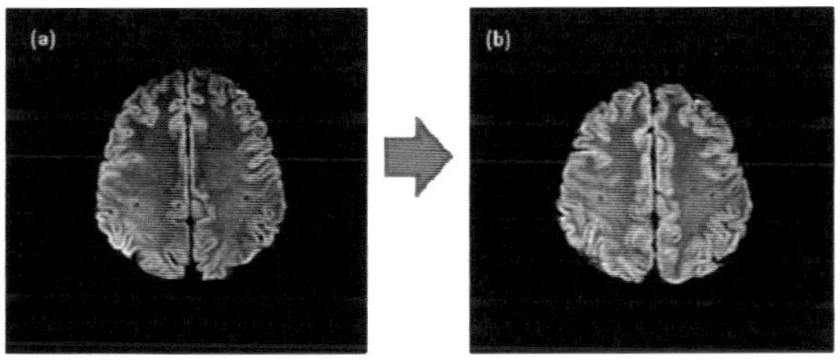

Figure 3. Box plot of the average assessment water ADC in basal and over-hydration conditions estimated in all volunteers.

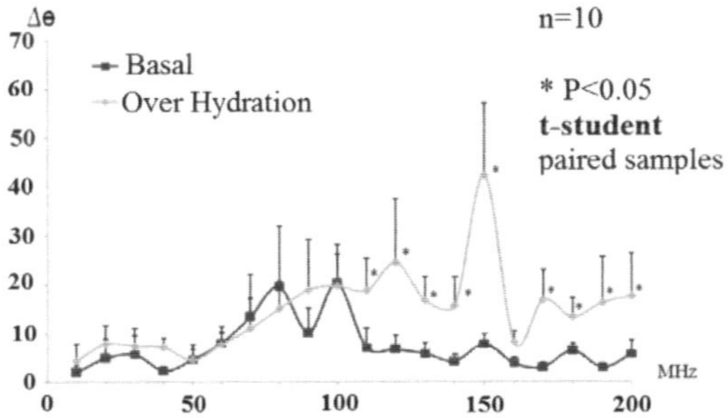

Figure 4. Average spectra of inductive phase shift values estimated in all volunteers in basal and over-hydration conditions. Bars represent one standard deviation.

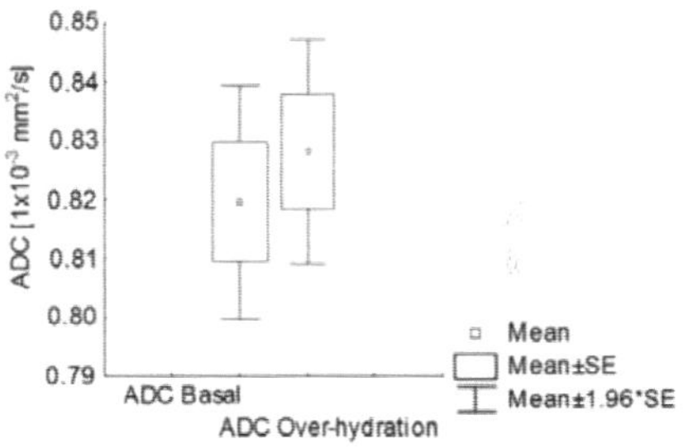

Figure 5. Box plot of the average assessment water ADC in basal and over-hydration conditions estimated in all volunteers.

4. Discussion and Conclusions

The comparisons of basal *vs* overhydration volumetric inductive phase shift values were significant at high frequencies consistently with previous studies [2] and [3]; such condition is explained by the MIS sensitivity to the water dipole effect. The comparison of ADC values not show significant changes in the evaluated ROIs, it´s explained in part by a moderate change in the water content of the brain as the experimental model is intended to induce and in agreement with Rossmiller et al [5]. The results suggest that MIS has the potential to detect in early stages pathologies associated to changes in the content of fluids in brain tissue such as edema and hematomas.

References

[1] Ayata C and Ropper A H. "Ischaemic brain oedema" Journal of Clinical Neuroscience 2002. 9(2): 113-124.

[2] Gonzalez C A and Rubinsky B. "A Theoretical Study on Magnetic Induction Frequency Dependence of Phase Shift in Oedema and Haematoma". Physiol. Meas. 2006. 27 829-838.

[3] Gonzalez C A and Rubinsky B. "Detection of Brain Oedema with Frequency Dependent Phase Shift Electromagnetic Induction". Physiol. Meas. 2006. 27 539-552.

[4] Castillo M and Murkherji SK. Diffusion-weighted imaging in the evaluation of intracranial lesions. Semin Ultrasound CT MR 2000;21:405–16.

[5] Rossmiller S R, Farace E, Sirnnad V I, Siragy H, Manning C A, Barrett E J and Knight-Scott J. "H-MRS T2 Measurements of Water in the Brain Tissue of Hyperhydrated Healthy Young Adults" Proc. Intl. Soc. Mag. Reson. Med. 10 (2002).

[6] Lodi R, Stracciari A, Iottil S, Donati G, Guarino M, Clementi Z V, Bolondi L and Barbiroli B. "Diffusion MRI shows increased water ADC in the brains of cirrhotic patients" Proc. Intl. Soc. Mag. Reson. Med. 10 (2002).

Assessment of degree of hydration in dialysis patients using whole body and calf bioimpedance analysis

F Zhu[1, 2], P Kotanko[1, 2], G J Handelman[1,3], J Raimann[1, 2], L Liu[1, 2], M Carter[1,2], M K Kuhlmann[1] , E Siebert[1, 2], E F Leonard[4], N W Levin[1, 2]

[1]Renal Research Institute
[2]Division of Nephrology and Hypertension, Beth Israel Medical Center New York, NY
[3]University of Massachusetts, Lowell, MA
[4] Department of Chemical Engineering, Columbia University, New York, NY

Email: fzhu@rriny.com

Abstract. Prescription of an appropriate post hemodialysis (HD) dialysis target weight requires accurate evaluation of the degree of hydration. The aim of this study was to investigate whether a state of normal hydration as defined by calf bioimpedance spectroscopy (cBIS) could be characterized in HD and normal subjects (NS). cBIS was performed in 62 NS (33 m/29 f) and 30 HD patients (16 m / 14 f) pre- and post-dialysis to measure extracellular resistance. Normalized calf resistivity at 5 kHz ($\rho_{N,5}$) was defined as resistivity divided by body mass index. Measurements were made at baseline (BL) and at a state of normal hydration (NH) established following the progressive reduction of post-HD weight over successive dialysis treatments until the $\rho_{N,5}$ was in the range of NS. Blood pressures were measured pre- and post-HD treatment. $\rho_{N,5}$ in males and females differed significantly in NS (20.5 ± 1.99 vs 21.7 ± 2.6 10^{-2} $\Omega m^3/kg$, $p<0.05$). In patients, $\rho_{N,5}$ notably increased and reached NH range due to progressive decrease in body weight, and systolic blood pressure (SBP) significantly decreased pre- and post-HD between BL and NBH respectively. This establishes the use of $\rho_{N,5}$ as a new comparator allowing the clinician to incrementally monitor the effect of removal of extracellular fluid from patients over a course of dialysis treatments.

1. Introduction

In conventional hemodialysis (HD), excess body fluid is removed during 3-4 hours of ultrafiltration per treatment. An accurate measure of hydration status is essential in determining the prescriptions of target weight necessary over a period of time to achieve normal body hydration (NH) [1].Current available methods to assess hydration status are not clinically useful because of limited accuracy and practicality [2].

Bioimpedance techniques measuring body fluid content have the advantages of being noninvasive, low cost and convenient. Bioimpedance using the 50 kHz current frequency vector methods have been proposed to indicate both hydration and nutrition [3]. Multi-frequency bioimpedance spectroscopy (BIS) measures extracellular and intracellular fluid volumes [4]. Both of these measurements may be useful in population studies but are less useful when applied to the individual HD patient because of problems inherent in whole body measurements [5]. During dialysis in ambulatory patients, the percentage decrease in ECV in the lower limbs is larger than in other body segments, presumably due to the prior effect of gravity on distribution of extracellular fluid [6, 7, 8]. We propose as a practical approach for evaluating degrees of hydration the measurement of change in electrical resistivity in the calf, as calculated from specific resistance per unit area.

The main aim of this study was to evaluate whether this method can be used to indicate degrees of body hydration in dialysis patients whose ECV is gradually reduced over a course of dialysis treatments by comparing calf resistivity, pre and post dialysis to that of normal subjects (NS) and also accompanying blood pressure changes.

2. Materials and Methods

2.1 Study protocol

This study was approved by the Institutional Review Board of Beth Israel Medical Center. Informed consent was obtained from all subjects. Calf BIS (cBIS) was measured in 62 apparently normal subjects (NS) (33 male; 29 female). In 30 stable HD patients (16 male; 14 female) three baseline (BL) pre-HD and post-HD cBIS measurements were performed in sequential dialyses. If normalized calf resistivity at BL was lower than the minimal level (Mean minus SD) of NS, indicating a state of overhydration, post-HD dialysis target weights were reduced by 0.2 - 0.3 kg per treatment, until normalized calf resistivity was above the minimal level of NS range, the definition of a state of normal hydration (NH).

2.2 Measurements

cBIS using the Xitron 4200 device was measured continuously during HD to obtain resistance and resistivity with patients in a sitting position with legs horizontal. Four electrodes were placed on the lateral side of calf to inject current (0.8 mA) and to measure voltage [26]. One sensing electrode (E_{S1}) was placed on the area of maximal calf circumference (C_{Max}); while the other sensing electrode (E_{S2}) was placed 10 cm below the E_{S1} in the circumference denoted as C_{Min} (Fig.1). The two current-injecting electrodes (E_{I1} and E_{I2}) were placed on the same side of the calf 5 cm above and below E_{S1} and E_{S2}, respectively. C_{Max} and C_{Min} were measured using a soft measurement tape with 0.1 cm accuracy. Body mass (BM) was measured using a calibrated electric scale. Blood pressures were measured in NS and in patients pre- and post-HD using a calibrated electronic blood pressure device (A&D Engineering, Inc, Milpitas CA).

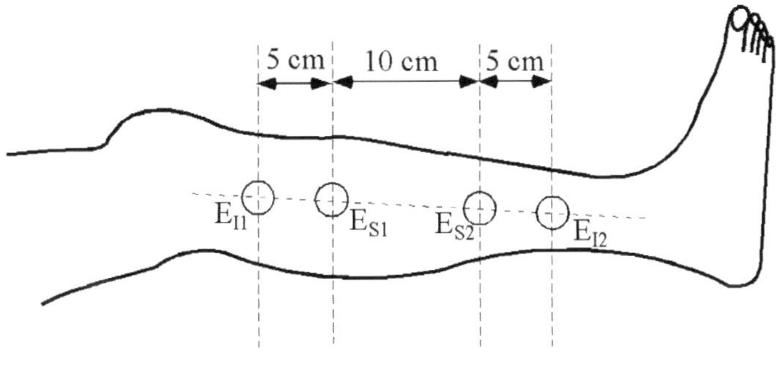

Figure.1 Calf bioimpedance measurement with four electrodes

2.3 Calculations

Extracellular resistance (R_E) and intracellular resistance (R_I) were calculated based on the Cole-Cole model using raw data of resistance and reactance with frequencies from 5 kHz to 1000 kHz. Calf extracellular (cECV) and intracellular fluid volumes (cICV) were calculated using Xitron software. Total calf fluid volume (TCV) was defined as cECV + cICV. The ratios of cECV/ TCV were calculated as markers of degree of hydration. Resistivity (ρ_5) was calculated from resistance (R_5) at 5 kHz by Eq.1 as follows.

$$\rho_5 = \frac{R_5 A}{L} \qquad (1)$$

where A and L are the cross sectional areas of the calf and the distance between the sensing electrodes, respectively. The cross sectional area was calculated by Eq 2, where C_{ave} was the mean of the two measured calf circumferences ($C_{ave} = (C_{Max} + C_{Min})/2$).

$$A = \frac{C_{Ave}^2}{4\pi} \qquad (2)$$

To standardize for the effects of differences in body composition, resistivity ρ_5 was normalized by the body mass index (BMI) calculated as BM [kg] divided by height^2 [m]), and reported as normalized resistivity ($\rho_{N,5}$) in units of $10^{-2}\Omega m^3/kg$ (Eq 3).

$$\rho_{N,5} = \frac{\rho_5}{BMI} \quad (3)$$

2.4 Data analysis

$\rho_{N,5}$ and blood pressure levels were compared between patients and NS. Data are presented as mean ± SD. The values between groups were compared by *t*-test using GraphPad Prism 4 (Graphpad Inc., San Diego, CA). A difference of $p<0.05$ was considered significant. The paired *t*-test was used to compare differences over time in the same patients.

3. Results

Table 1 summarizes results obtained in NS and patients post-HD. Body height, body mass, systolic blood pressure (SBP), BMI, C_{ave} and calf cross-sectional area (A) were higher in males. Calf resistance (R_5), resistivity (ρ_5) and normalized resistivity ($\rho_{N,5}$) were significantly lower in males.

Table 1 Summary of results from normal subjects (NS) and patients (P) in men (M) and females (F)

		M-NS	F-NS	M-P-BL	M-P-NH	F-P-BL	F-P-NH
BM	kg	80.3±17[a]	59.1±8.8	74.4±14[b]	73.2±14	72.6±18.9[b]	70.8±18
BMI	kg/m^2	25.4±4.9[a]	22.3±3	24.8±5[b]	24.4±4.9	27.6±7.3[b]	26.9±7.1
C_{ave}	cm	35.4±3.6[a]	32.0±2.1	31.8±3.9	31.6±4.1	31.3±3.7	31.1±4.1
cR_5	Ω	51.9±6.8[a]	59.1±8.6	54.8±10[b]	62.8±11	62.9±11[b]	72.6±12.4
ρ_5	Ω cm	518.6±100	482.7±83	437±82[b]	494.1±95	494.6±131[b]	557±129
$\rho_{N,5}$	$10^{-2}\Omega$ m^3/kg	20.5±1.99[a]	21.7±2.6	17.4±2.8[b]	20.±1.5	18.4±4.1[b]	20.9±1.8
cECV/(TCV)		0.26±0.03[a]	0.3±0.03	0.29±0.08[b]	0.27±0.08	0.33±0.1	0.32±0.11

[a] Difference between males and females in NS
[b] Difference between BL and NH state in patients post-HD

The standard deviations and coefficient of variation (CV) for normalized resistivity ($\rho_{N,5}$) in NS were less than the non-normalized values (ρ_5) (Fig.2). Mean of $\rho_{N,5}$ increased significantly in HD patients from BL to NH state pre- and post-HD and no difference in $\rho_{N,5}$ between patients at post-HD NH and NS (Fig.3).

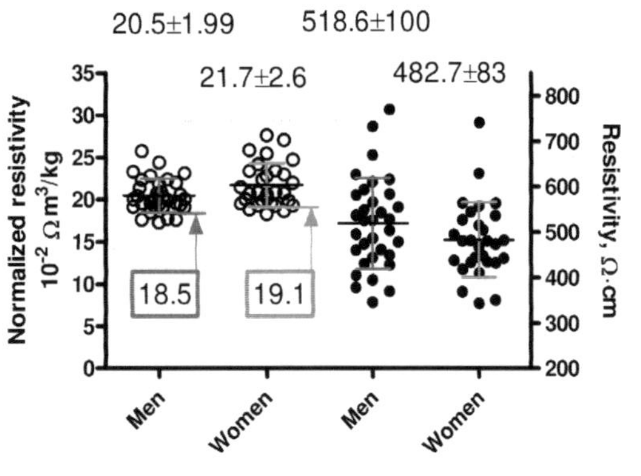

Figure 2 Normalized resistivity and resistivity in NS

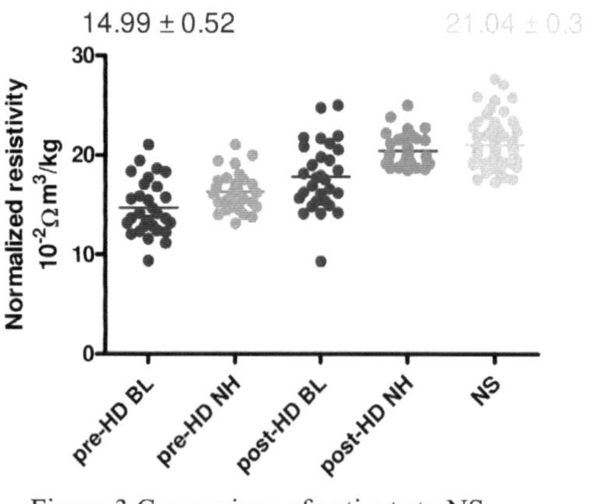

Figure 3 Comparison of patients to NS

Values of the mean minus the standard deviation are defined as minimal levels to establish the range of hydration in normal males (18.5 10^{-2} $\Omega m^3/kg$) and in females (19.1 10^{-2} $\Omega m^3/kg$) respectively

(Fig.2). Normalized resistivity ($\rho_{N,5}$) in NS differed significantly between males and females but there were no significant differences between age groups.

With decrease in BM pre- and post-HD, calf resistance at 5 kHz (R_5), resistivity (ρ_5) and normalized resistivity ($\rho_{N,5}$) of patients increased significantly post-HD in males and females between BL and NBH (Table 1). Mean of ($\rho_{N,5}$) was no difference between NS and patients post-HD (Fig.3). cECV decreased significantly in both genders but cECV/TCV decreased significantly only in male patients between BL and NH. Pre-HD systolic blood pressure decreased (131.1± 13 vs 123. 8±14 mmHg, $p<0.05$) and post-HD (124.4±20 vs 114.0±17 mmHg, $p<0.05$) between BL and NH.

3. Discussion

This study illustrates that a gender-specific range for normalized calf resistivity ($\rho_{N,5}$) can be established in healthy control subjects (Fig. 2), and used as a reference to estimate the hydration status of dialysis patients. Fluid removal from BL to NH by ultrafiltration during dialysis was accompanied by increases in calf normalized resistivity (Fig.3) and significant decreases in mean SBP in all patients which suggest that this measurement can be used to provide guidance for prescription for fluid removal during the course of dialysis treatments.

Because the lower limbs are more hydrated than other body segments, calf bioimpedance provides more accurate information about body hydration than other body segments as excess fluid is finally removed. In addition, segmental measurement of the calf is preferable to that of the whole body because the geometric structure of the calf is uniform, resembling the segment of a cylinder. Resistivity at 5 kHz (ρ_5) varies widely in normal subjects (Fig. 2, right). Variability may be explained in part by the resistance measured at the skin surface of the calf reflecting the different electrical properties of components of body composition, such as fat, muscle and fluid. To minimize this variability, resistivity was normalized to body mass index (BMI), giving a new variable designated as normalized resistivity ($\rho_{N,5}$) (Fig.2, left). The unit of $\rho_{N,5}$ $\Omega m^3/kg$, can be interpreted also for its physical meaning. Since m^3/kg is equal to $1/kg/m^3$, which is the reciprocal of density, the parameter $\rho_{N,5}$ can be interpreted as ohm per density of tissue, reflecting the relationship of hydration to body density.

Calf $\rho_{N,5}$ significantly increases in HD patients following progressive decrease in body water by ultrafiltration, with measurement variance being reduced by normalizing to BMI. Comparison of $\rho_{N,5}$ with healthy subjects was useful to evaluate degrees of hydration in HD patients. The parameter $\rho_{N,5}$ could also be applied to chronic kidney disease (CKD) patients to detect degrees of overhydration.

References

[1] Charra B 1998 'Dry weight' in dialysis: the history of a concept. *Nephrol Dial Transplant* **13** 1882-5.
[2] Jaeger JQ 1999 Mehta RL, Assessment of dry weight in hemodialysis: an overview. *J Am Soc Nephrol* **10** 392-403.
[3] Piccoll A 1998 Identification of operational clues to dry weight prescription in hemodialysis using bioimpedance vector analysis. *Kidney Int* **53** 1036-1043.
[4] Chamney PW, Kramer M, Rode C, et al 2002 A new technique for establishing dry weight in hemodialysis patients via whole body bioimpedance. *Kidney Int* **61** 2250-2258.
[5] Zhu F, Schneditz D, Wang E, Levin NW 1998 Dynamics of segmental extracellular volumes during changes in body position by bioimpedance analysis. *J Appl Physiol.* **85** 497-504
[6] Zhu F, Leonard EF, Levin NW 2008 Extracellular fluid redistribution during hemodialysis: bioimpedance measurement and model. *Physiol Meas* **29** S491-501.
[7] Kuhlmann MK., Zhu F, Seibert E et al 2005 Bioimpedance, dry weight and blood pressure control: new methods and consequences. *Current Opinion Nephrology and Hypertension* **14** 543-549.
[8] Zhu F, Kuhlmann MK, Kotanko P, Seibert E, Leonard EF and Levin NW 2008 A method for the estimation of hydration state during hemodialysis using a calf bioimpedance technique. *Physiol Meas* **29** S503-516

Assessment of vasomotor oscillations with Fourier analysis of biological tissue impedance

A Nesterov, I Gavrilov, L Selector, I Mudraya and S Revenko[1]
National Cardiology Center, Research Institute of Urology, Moscow, Russia

E-mail: s_revenko@mail.ru

Abstract. Fourier analysis revealed a number of periodicities in small variations of bioimpedance of human finger including the major spectrum peaks at the frequencies of heart beats, respiration, and Mayer wave (0.1 Hz). These periodic variations of bioimpedance were detected under the normal conditions and during blood flow arrest in the hand by a pneumatic cuff placed on the arm. They are explained by periodic variations in systemic blood pressure and by oscillations of regional vascular tone resulted from neural vasomotor control. During normal blood flow, the greatest variations in bioimpedance were observed at the heart rate, and their amplitude surpassed by an order of magnitude the amplitudes of respiratory oscillations and Mayer wave. In contrast, during blood arrest, the largest amplitude of rhythmical changes of the impedance characterized the oscillations at respiration rate, while the amplitude of oscillations at the heart rate was the smallest. During normal respiration and circulation, two side cardiac peaks were revealed in bioimpedance amplitude spectrum which disappeared during respiration arrest and thought to reflect the amplitude respiratory modulation of the cardiac output via sympathetic influences. During normal breathing, the second and the third harmonics of the cardiac spectrum peak were split reflecting frequency respiratory modulation of the heart rate by parasympathetic influences. The results favour applicability of Fourier analysis of bioimpedance variations in assessment of regional neural influences and neurogenic modulation of cardiac activity.

1. Introduction

A spectacular progress in microelectronics resulted in appearance of low-noise chips and powerful computers which made it possible to measure and analyze small bioimpedance variations with a laboratory-made high-resolution impedance converter and original software. We analyzed bioimpedance variations in human finger with Fourier transform in the frequency band of 0.08-15.0 Hz under the normal conditions and during circulation arrest in the arm optionally combined with expiratory delay for 40 sec. The aim was to assess the neurogenic contribution to bioimpedance variations, which probably result from vasomotor activity of sympathetic nerve system.

2. Methods

Bioimpedance was measured with original impedance converter based on synchronous detection principle resulting in total ("basic") real part of impedance in the frequency band of 0–15 Hz ($R_{=}$) and the variable component of this real part in the frequency band of 0.08–15 Hz ($R_{\sim}$). The measuring

[1] To whom any correspondence should be addressed.

frequency was 100 kHz. The resolving power of $R_\sim$ channel was 250 microOhm in the range of ±4 Ohm. To obtain the amplitude Fourier spectrum, $R_\sim$ record was cut into 25.6-sec fragments containing 4096 points. FFT was calculated with Hanning window. The amplitude spectra of 2 to 4 such fragments were averaged. The impedance converter was connected to the proximal phalanx of the middle human finger (n=14) with two output current wires and two input voltage wires coupled to two Ag/AgCl wire electrodes made of silver wire (0.4 mm in diameter) covered with narrow gauze bandages soaked in physiological saline. The distance between these electrodes was about 2 cm. To damp the mechanical vibrations caused by heart and lungs, the arm was fixed to a massive support. Circulation in the arm was temporarily arrested with a pneumatic cuff inflated to at least 2-fold systolic pressure to prevent the spread of hydraulic wave in the blood vessels of the arm.

3. Results

Figure 1 show basic impedance $R_=$ (*b*) and variable impedance component $R_\sim$ measured under normal conditions (*a*) and during circulation arrest in the left arm (*c*). While the large oscillations of bioimpedance in (*a*, *b*) resulted mostly from pumping action of the heart, the smaller ones (*c*) observed during circulation arrest were not cardiac reflecting hypothetically the neural vasomotor activity.

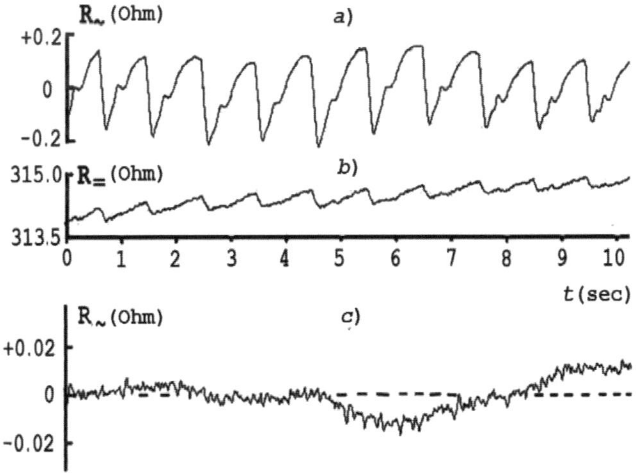

Figure 1. Variable component (*a*) and basic (total) real part impedance (*b*) in human finger. (*c*) Variable component during circulation arrest in the arm.

The nature of these oscillations can be elucidated by frequency-domain analysis. Figure 2 shows amplitude Fourier spectrum during intact circulation. Peak *1* was observed at the frequencies of 0.1–0.2 Hz. Oscillations of arterial pressure at this rate are known as "Mayer wave" [1] so the corresponding peak is termed here as Mayer's. The frequencies of peaks *2* and *3* coincided with respiration and heart rate, correspondingly. The large cardiac peak 3 and its harmonics 3'', 3''', 3'''' spaced equidistantly along the frequency axis represented the large-scale oscillations clearly seen in the raw graphic data (figures 1, *a*, *b*). Other bioimpedance oscillations revealed by Fourier analysis cannot be directly observed. The cardiac peak and its harmonics were accompanied by two smaller left (*L*) and right (*R*) peaks spaced symmetrically from the central peak at the distance equal to the respiration rate. These side peaks and the respiratory peak *2* simultaneously disappeared during respiration arrest for 40 sec (figure 3). Thus, the cardiac side peaks *3L* and *3R* depend on respiration.

The trident spectrum structure with a central and two symmetrical side peaks is well known in physics as reflecting the amplitude modulation of one periodic process (here the heart beats) with another (and slower) periodic process (respiration). The amplitude of the side peaks is proportional to the "depth" of such modulation. Thus, the amplitude spectrum of bioimpedance variations makes it possible to assess the intensity of respiratory control over the force of the heart beats [2, 3].

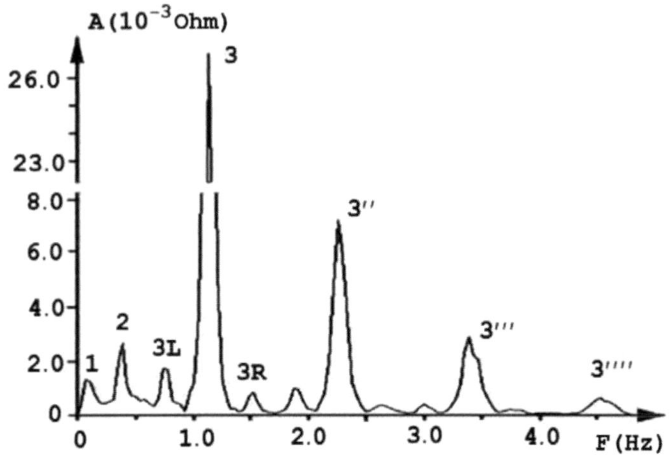

Figure 2. Amplitude spectrum of bioimpedance variations in human finger over broad frequency range with Mayer's peak (*1*), respiratory peak (*2*), four cardiac harmonics (*3* to *3''''*), and the side peaks (*3L, 3R, etc*). Note the break in ordinate.

Another important observation is splitting of higher harmonics of the cardiac peak observed under normal conditions in some cases (figure 3, *a*). Such splitting was far less pronounced under respiration delay (figure 3, *b*). Splitting of a peak can be explained by frequency modulation of basic oscillatory process with another process going on at a smaller rate; in this case, splitting can result from vagal modulation of the heart rate at the respiration frequency. In this experiment, the oscillations were cut off below 0.3 Hz, so no Mayer peak is seen in figure 3.

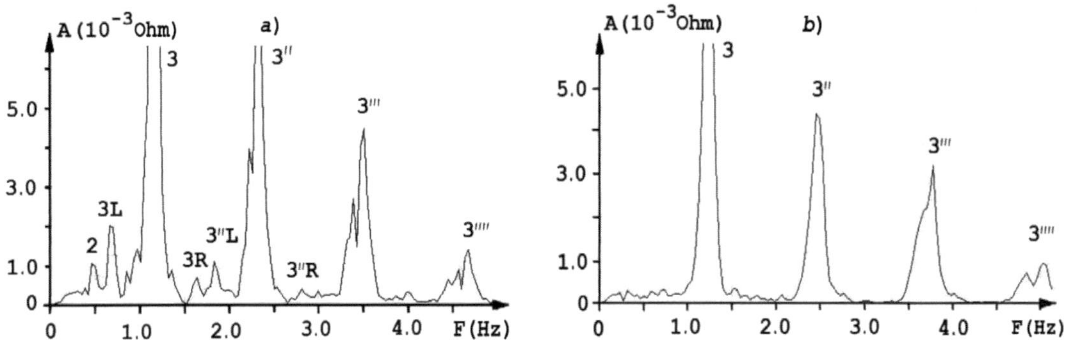

Figure 3. The respiratory (*2*) and side cardiac peaks (*L* and *R*) are observed under normal conditions (*a*) but disappeared during expiratory delay (*b*).

In order to assess the changes in bioimpedance of non-pulsatile (non-cardiac) origin, we compared the bioimpedance spectra during normal circulation (figure 4, *a*) and during blood arrest in the arms (figure 4, *b*). Circulation arrest did not eliminate oscillations of bioimpedance although diminished their amplitude (figure 4, *b*). Under these conditions, the periodic bioimpedance oscillations could be produced either by neurogenic vasomotor influences or by spontaneous vasomotions. The latter are characterized with a broad range at the low frequencies of 0.008 – 0.11 Hz [4]. The narrow shape of all peaks in Fig. 4 does not favour the hypothesis of spontaneous nature of the corresponding bioimpedance oscillations. Thus, all the peaks in figure 4, *b* are probably neurogenic and vasomotor in nature.

Arrest of circulation dramatically reduced the cardiac peak, which is a trivial consequence of the fact that during normal circulation the major variations in bioimpedance are produced by pumping action of the heart. Surprisingly, the respiratory peak *2* and Mayer's peak *1* decreased by no more than 2-fold. It means that during normal circulation, the heart and systemic blood pressure are not the major players who control these rhythmic variations of bioimpedance. During circulation arrest, the respiratory peak *2* surpassed the cardiac peak in amplitude; moreover, in contrast to the cardiac peak, it was accompanied by the second and third harmonics indicating that neurogenic respiratory oscillations are far more powerful than the cardiac ones.

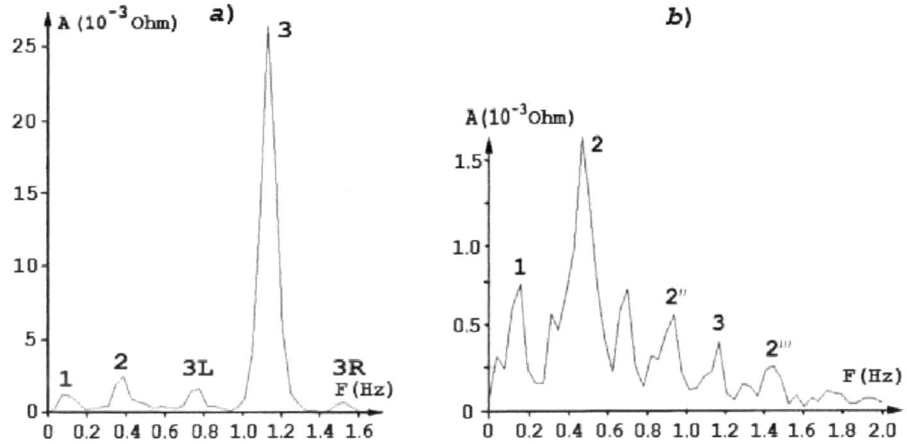

Figure 4. Amplitude spectra of bioimpedance variations under normal conditions (*a*) and during circulation arrest (*b*). Note different vertical scales.

Combined respiration and circulation arrest resulted in elimination of all peaks except the Mayer's one, which however decreased 3-fold. This is another surprising observation indicating the major role of respiratory process in sympathetic vasomotor supply, since respiration delay can eliminate neurogenic influences normally observed at the cardiac rate.

4. Discussion

Fourier analysis of bioimpedance variations reflecting blood filling and fluid redistribution in biological tissue can assess activity of periodic neurogenic vasomotor influences both during normal and arrested circulation. During normal circulation, the neurogenic vasomotor supply to human finger is predominantly modulated at Mayer's and respiratory rate, while the overwhelming part of the heart-rated bioimpedance variations reflects the pumping action of the heart. During arrested circulation, the bioimpedance variations seem to be exclusively neurogenic in nature with the largest contribution of oscillations at the respiration rate. The maximum frequency of bioimpedance variations under these conditions (1.4 Hz, the third respiratory harmonic) can assess the upper limit of frequency range of neurogenic vasomotor control.

The trident structure of cardiac peak and the amplitude of the side peaks can assess the depth of respiratory modulation of the stroke volume. The width of splitting of the higher cardiac harmonics reflects the depth of frequency modulation of the heartbeats by respiratory system. Thus, Fourier analysis of bioimpedance variations seems to be a new powerful tool to examine the regional neurogenic vasomotor influences and central interaction between cardiovascular and respiratory systems.

5. References

[1] Julien C 2006 *Cardiovasc. Res.* **70** 1 12–21
[2] Bernardi L, Porta C, Gabutti A, Spicuzza L and Sleight P 2001 *Auton. Neurosci.* **90** 1-2 47–56
[3] Taylor E W, Jordan D and Coote J H 1999 *Physiol. Rev.* **79** 3 855–916
[4] Christ F, Raithel P, Gartside I B, Gamble J, Peter K and Messmer K 1995 *J. Physiol.* **487** 1 259–272

Journal of Physics: Conference Series **224** (2010) 012126 doi:10.1088/1742-6596/224/1/012126

A Novel Approach for Removing the Hook Effect Artefact from Electrical Bioimpedance Spectroscopy Measurements

R Buendia[1,2], F Seoane[1,3] and R Gil-Pita[2]

[1]School of Engineering, University of Borås, SE-501 90 Borås, SWEDEN

[2]Department of Signal Theory and Communications, University of Alcala, ES-28871, Madrid, SPAIN

[3]Department of Signal & Systems, Chalmers University of Technology, SE-41296, Gothenburg, SWEDEN

E-mail: fernando.seoane@hb.se

Abstract. Very often in Electrical Bioimpedance (EBI) spectroscopy measurements the presence of stray capacitances creates a measurement artefact commonly known as *Hook Effect*. Such an artefact creates a hook-alike deviation of the EBI data noticeable when representing the measurement on the impedance plane. Such Hook Effect is noticeable at high frequencies but it also causes a data deviation at lower measurement frequencies. In order to perform any accurate analysis of the EBI spectroscopy data, the influence of the Hook Effect must be removed. An established method to compensate the hook effect is the well known *Td compensation*, which consists on multiplying the obtained spectrum, $Z_{meas}(\omega)$ by a complex exponential in the form of exp[jωTd]. Such a method cannot correct entirely the *Hook Effect* since the hook-alike deviation occurs a broad frequency range in both magnitude and phase of the measured impedance, and by using a scalar value for *Td*. First a scalar only modifies the phase of the measured impedance and second, a single value can truly corrects the *Hook Effect* only at a single frequency. In addition, the process to select a value for the scalar *Td* by an iterative process with the aim to obtain the best Cole fitting lacks solid scientific grounds. In this work the *Td compensation* method is revisited and a modified approach for correcting the *Hook Effect* including a novel method for selecting the correcting values is proposed. The initial validation results confirm that the proposed method entirely corrects the *Hook Effect* at all frequencies.

1. Introduction

Electrical Bioimpedance Spectroscopy is a typical approach currently and potentially in use in several applications of Electrical Bioimpedance (EBI) analysis like total body composition [1], electronic biopsies [2], as well as pulmonary edema [3].

In order to perform any useful data analysis, in addition to use an appropriate analysis method the data should be free from interferences and from artefacts. It is rather common to obtain EBI measurements affected by a characteristic deviation especially noticeable at high frequencies [4] and [5]. Such a deviation is commonly known as *Hook Effect* because it reassembles to a hook when the impedance spectrum is represented on an impedance plot. Its origin is related to parasitic capacitances influencing on the EBI measurements and its compensation has been studied previously [6].

The *Hook Effect* artefact distinctly modifies the impedance spectra at high frequencies but it affects the complex EBI data at any AC frequency and consequently interfering any posterior EBI data analysis; especially Cole-based analysis that requires to fit the corrupted data with two dominant dispersions into a single-dispersion system: The Cole function.

Currently, the method for correcting the influence of the *Hook Effect* on EBI data is the so-called Td compensation [6], which is worldwide spread and has several limitations.

2. Methods and Model

The error on the estimation of the impedance in a simple measurement model, including a stray capacitance in parallel with the Tissue Under Study (TUS), has been analytically studied using the software packages of Mathematica and Matlab. The *Hook Effect* introduced in the impedance spectrum by the parasitic capacitance has been observed and the compensation of such estimation error by the *Td* compensation method has been analyzed.

Combining the models proposed by Scharfetter in 1997 for artefacts [6] and Martinssen in 2004 [7] in admittance and neglecting the electrode polarization impedance of the later, it is possible to obtain a simple model equivalent to a current divider. The model used for the study is depicted in Figure 1.

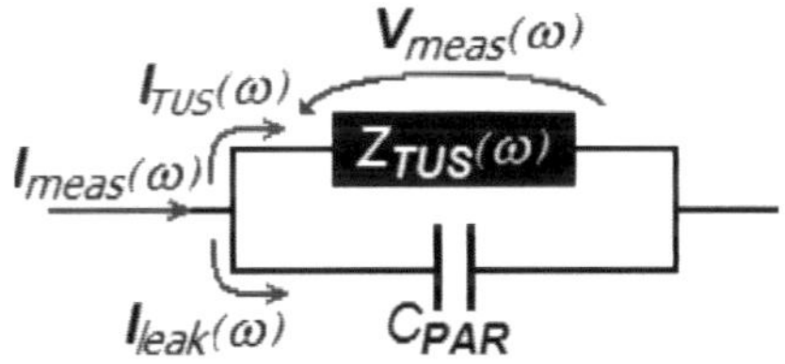

Figure 1. Model of study

The TUS impedance models an experimental measurement obtained with 4-Electrode wrist to ankle EBI measurement fitted to a Cole function (1) with the following Cole parameters R_0=449.6 Ω, R_∞=296.7 Ω, α=0.7186 and τ = 5.2727x10^{-6}, equivalent to a characteristic frequency of 30.2 kHz.

$$Z_{TUS} = R_\infty + \frac{R_0 - R_\infty}{1+(j\omega\tau)^\alpha} \quad (1)$$

$$Z_{meas}(\omega) = \frac{V_{meas}(\omega)}{I_{meas}(\omega)} = \frac{I_{TUS}(\omega) * Z_{TUS}(\omega)}{I_{TUS}(\omega) + I_{leak}(\omega)} \quad (2)$$

The measured impedance is obtained by expression in (2). Note that the current causing the voltage sensed by the impedance meter, $V_{meas}(\omega)$ is not the same current generated by the impedance meter.

3. Hook Effect

1.1. Origin and deviation of EBI spectrum

According to the models proposed by [6] and [7], and the model depicted in Figure 1, electrical current intended for stimulating the TUS leaks away from the measurement load through parallel electrical pathways enabled by parasitic capacitances. Such current leakage introduces an impedance estimation error that is frequency dependant [8], this impedance estimation error produces a deviation in the impedance spectrum, especially noticeable at high frequencies since it is at high frequencies when the parasitic leakage pathways become more conductive dragging more current away from the TUS. The produced deviation affects the EBI spectrum at all AC frequencies and it deviates both real and imaginary parts of the EBI spectrum. This means that both module and phase of the EBI spectra are deviated.

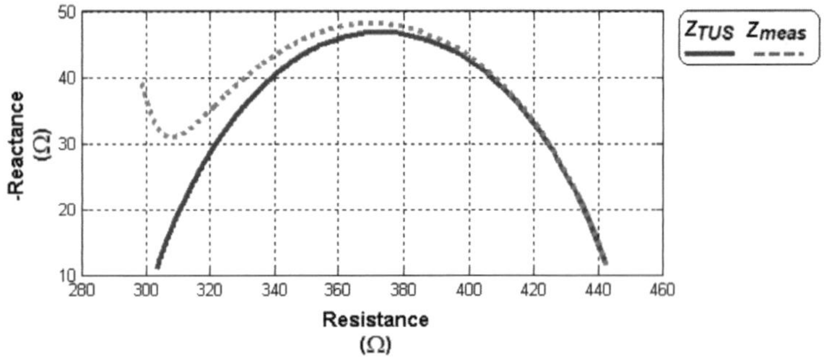

Figure 2. Plot of the TUS impedance (1) against the measured impedance (2) with a C_{PAR} of 50 pF

The deviation is especially noticeable in the spectra of both reactance and phase at high frequencies, such a deviation is what creates the so-called *Hook Effect* easily identified in the impedance plot of Figure 2.

4. Td Compensation

Td Compensation is a well known and spread approach used nowadays for correcting the *Hook Effect*, it consists on multiplying the obtained measured EBI spectra by a complex exponential in the form of $exp[j\omega Td]$.

The main limitation of this approach is that *Td* is a scalar and consequently multiplying a complex magnitude like the impedance by $exp[j\omega Td]$ only modifies the phase of the measured impedance, while the *Hook Effect* affects both the phase and the module of the impedance spectrum. This implies that for a proper compensation of the produced error on the impedance estimation the value of *Td* should be complex.

From the analysis of the *Td compensation* a second limitation arises that for a complete correction of the impedance estimation error, the value of *Td* cannot be just a number but a function of frequency. This way *Td compensation* with *Td* scalar only corrects the deviation produced on the phase by the *Hook Effect* at a single frequency. In addition a very important limitation of the *Td compensation method* is that there is no solid scientific method published for the selection of the value for *Td*.

5. The Hook Correction Method

In order to completely correct the impedance estimation error causing the *Hook Effect*, the value of *Td* in $exp[j\omega Td]$ should be complex and function of the frequency instead than just an scalar. A new method developed on the same basis than the *Td compensation* method is proposed together with a methodology to obtain the values of the correction function.

1.2. Correction Function

From the mathematical analysis performed on the effect caused on the impedance estimation error by the complex exponential $exp[j\omega Td]$, it was obtained that a mathematical expression could indeed eliminate the impedance estimation error caused by the parasitic capacitance of the model C_{PAR}, correcting completely the *Hook Effect*. The Correction function $F_{Corr}(\omega)$ in (3) is a logarithmic complex function dependent on the natural frequency ω, C_{PAR} and $Z_{meas}(\omega)$, that when substitutes *Td* on the complex exponential $exp[j\omega Td]$ and multiplies the obtained complex EBI spectra $Z_{meas}(\omega)$, produces the following final expression (4) for correcting the *Hook Effect*:

$$F_{Corr}(\omega) = -j\frac{Log[1 - j\omega C_{PAR} Z_{meas}(\omega)]}{\omega} \quad (3)$$

$$Z_{TUS}(\omega) = Z_{meas}(\omega) * e^{-Log[1 - Z_{meas}(\omega) * j\omega C_{PAR}]} \quad (4)$$

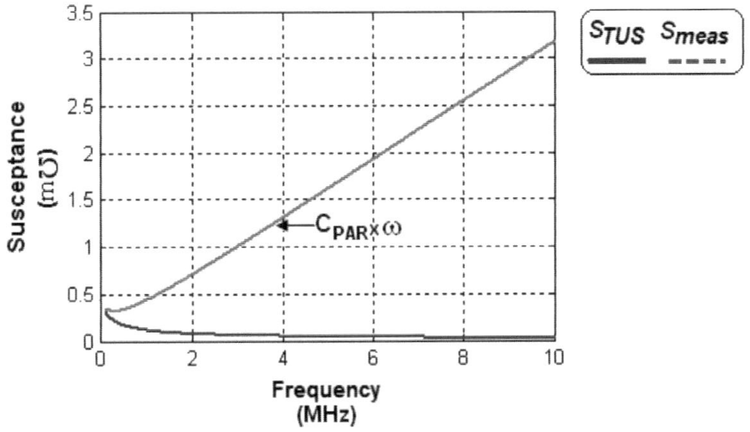

Figure 3. Susceptance plot of the TUS impedance (1) against the measured impedance (2) with a C_{PAR} of 50 pF.

1.3. Parasitic Capacitance Estimation

The admittance of the model depicted in Figure 1, can be written as in (5). Analyzing the frequency dependence of the imaginary part of equation 5 i.e. the susceptance of measurement, it is known that $SC_{PAR}(\omega)$ increases linearly with frequency in the form $SC_{PAR}(\omega)=\omega C_{PAR}$ while $S_{TUS}(\omega)$ decreases at high frequencies. See Figure 3. Therefore at high frequencies the value of susceptance of the TUS is practically negligible and it is possible to estimate the value of the parasitic capacitance C_{PAR} using (6) from the measurement.

$$Y_{\text{meas}}(\omega) = G_{TUS}(\omega) + j\left[S_{TUS}(\omega) + Sc_{PAR}(\omega)\right] \quad (5)$$

$$C_{par} \approx \frac{Lim[Y_{\text{meas}}(\omega)]\Big|_{\omega \to \infty}}{\omega} \quad (6)$$

6. Validation of the approach

Figure 4 shows the correction effect caused by the correction function for a $Z_{meas}(\omega)$ obtained with the previously introduced model and parasitic capacitance C_{PAR} = 50 pF. In the figure it can be observed that the *Hook Effect* is completely removed, obtaining a $Z_{CORR}(\omega)$ identical to $Z_{TUS}(\omega)$.

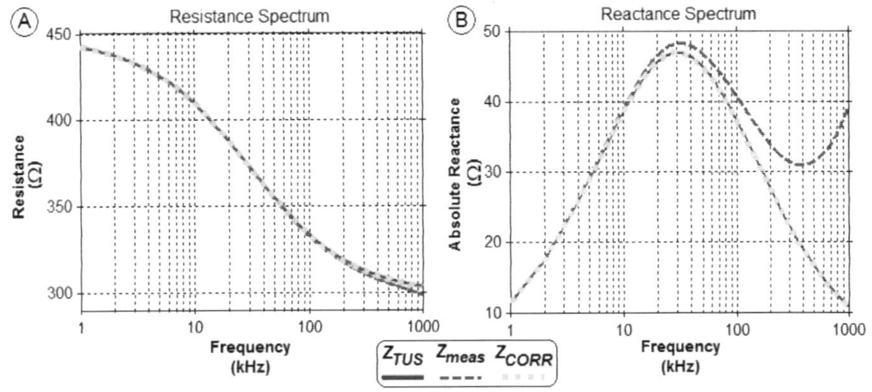

Figure 4. Correction effect of *FCorr(ω)* over *Zmeas(ω)* with a C_{PAR} of 50 pF.

7. Discussion & Conclusion

The proposed method corrects the *hook effect* caused by the leaking of measurement current away from the measurement load through parasitic pathways at all frequencies in the complex impedance, i.e. both in magnitude and phase. In this way, the mathematical intrinsic limitation of *Td compensation*, the currently in use correction approach, is overcome. In addition, the proposed method is based on a well-accepted model of artefacts, so its implementation is not arbitrary, which it is another advantage over the *Td compensation* method.

The main limitation of the proposed method resides in the estimation of the parasitic capacitance from the susceptance of the measurement, which requires performing EBI measurements up to very high frequency, theoretically the best estimation is done with measurements up to ∞. Ongoing work with experimental measurements suggest that the Correction Function is valid and that works fine with EBI measurements obtained up to 600 kHz.

8. References

[1] U. M. Moissl, P. Wabel, P. W. Chamney *et al.*, "Body fluid volume determination via body composition spectroscopy in health and disease," *Physiol Meas,* vol. 27, no. 9, pp. 921-33, Sep, 2006

[2] Aberg P, Nicander I; Hansson J; Geladi P; Holmgren U and Ollmar S 2004 Skin cancer identification using multifrequency electrical impedance–a potential screening tool *IEEE Transactions on Biomedical Engineering* **51** 2097- 2102

[3] Beckman DL, Mehta P; Hanks V, Rowan W, H Liu L 2000 Effects of Peroxynitrite on Pulmonary Edema and the oxidative state *Experimental Lung Research*, **26** 349-359

[4] Scharfetter H, Hartinger P, Hinghofer S H and Hutten H 1997 A model of artefacts produced by stray capacitance during whole body or segmental bioimpedance spectroscopy *Physiol. Meas.* **19** 247-261.

[5] Van Marken Lichtenbelt W D, Westerterp K R, Wouters L and Luijendijk S C M 1994 Validation of bioelectrical-impedance measurements as a method to estimate body-water compartments *Am. J. Clin. Nutr.* **60** 159–66

[6] Scharfetter H, Monif M, László Z, Lambauer T, Hutten H, Hinghofer-Szalkay H. Effect of postural changes on the reliability of volume estimations from bioimpedance spectroscopy data. *Kidney Int.* 1997 **51** 1078-87

[7] Mirtaheri P, Grimnes S, Martinsen Ø and Tønnessen T I A new biomedical sensor for measuring PCO2 *Physiol. Meas.* **25** 421-436

[8] Buendia R 2009 Hook Effect on Electrical Bioimpedance Spectroscopy Measurements. Analysis, Compensation and Correction. *Master Thesis Borås* Borås University

Influence of ambient temperature on whole body and segmental bioimpedance spectroscopy measurements

G Medrano, R Bausch, A H Ismail, A Cordes, R Pikkemaat and S Leonhardt

Philips Chair for Medical Information Technology, RWTH Aachen University, Aachen, Germany

medrano@hia.rwth-aachen.de

Abstract. Bioimpedance spectroscopy (BIS) measurements are easy to implement and could be used for continuous monitoring. However, several factors (e.g. environment temperature) influence the measurements limiting the accuracy of the technology. Changes in skin temperature produced by changes in ambient temperature are related with changes in skin blood flow and skin impedance. It is assumed that skin impedance change is responsible for the error observed in whole body and segmental measurements. Measurements including body parts more distant from the torso seem to be more affected. In the present article skin and segment impedance have been performed on healthy subjects under extreme changes in environment temperature (13-39 °C). A commercial BIS device with a range between 5 kHz and 1 MHz has been used for the measurements. The results indicate that not only skin impedance, but also impedance of deeper tissue (e.g. muscle) may be responsible for the influence of environment temperature on BIS measurements. Segmental (knee-to-knee) BIS measurements show a relative change of only 2 %, while forearm and whole body impedance changed 14 % and 8 % respectively.

1. Introduction

Bioimpedance spectroscopy (BIS) constitutes an easy to use and easy to implement technology, which could be used together with intelligent textile electrodes [1] to monitor fluid changes during hemodialysis treatment or during sports. Despite of these advantages, several factors (e.g. environment temperature) affect the measurements [2], therefore requiring trained personnel and laboratory conditions, which can not be fulfilled in continuous monitoring applications all the time.

Changes in ambient temperature activate several mechanisms in order to preserve the body 'core' temperature [3]. Skin temperature of body parts distant from the torso (e.g. hands, forearms, feet and underlegs) can experience, on the contrary, extreme changes [4]. Changes of skin temperature produced by ambient temperature are related to changes in skin blood flow [5] and impedance. It is supposed that changes in skin impedance are responsible for the influence of ambient temperature on BIS measurements. BIS measurements including peripheral limbs are more influenced by temperature changes than measurements excluding them [6].

The first purpose of the present study was to investigate the magnitude of impedance change at the skin-electrode interface compared to the whole segment. The second purpose was to compare the magnitude of change produced by ambient temperature on measurements excluding peripheral limbs (e.g. knee-to-knee) and other including them (e.g. whole body and forearm).

2. Theoretical background

2.1. The skin-electrode interface

Electrodes constitute an interface between electron current (measurement circuit) and ionic current (tissue). In a routine BIS procedure, a four point measurement will be used (see figure 1). Two electrodes are used for the current injection I(t) and the other two for the voltage measurement V(t). In doing so, the influence of the skin-electrode impedance ($Z_{Skin\text{-}Electrode}$) on the measurement is reduced ($I_{Bias} * Z_{Skin\text{-}Electrode} << I_{Meas} * Z_{Body}$) and the following equation and approximation is valid:

$$Z_{4\,point} = \frac{V(t)}{I(t)} = \frac{I_{Meas} \cdot Z_{Body} + 2 \cdot Z_{Skin-Electrode} \cdot I_{Bias}}{I(t)} \approx \frac{I_{Meas} \cdot Z_{Body}}{I(t)} \approx Z_{Body} \qquad (1)$$

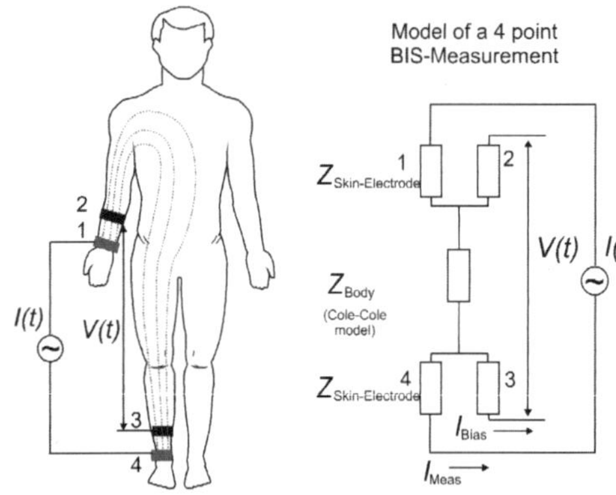

Figure 1. Tetrapolar configuration for a typical BIS measurement (left) and equivalent circuit (right).

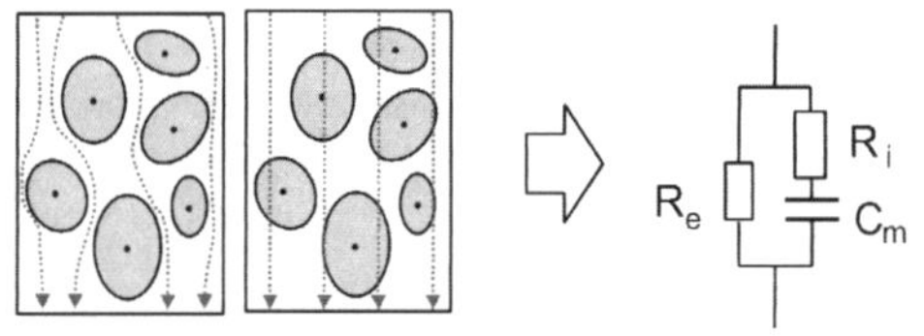

Figure 2. Low and high frequency current flow through body tissue

2.2. Intracellular (R_i) and extracellular (R_e) resistances.

A biompedance spectroscopy (BIS) measurement on tissue can be represented by an electrical model composed of two resistances (R_e and R_i) representing the extracellular and intracellular fluids respectively and a capacitor representing the cellular membrane [7], [8]. Accordingly a low frequency current only flows around the cells through the extracellular fluid and resistance, whereas a high frequency current will also pass through the cell membrane and the intracellular fluid and resistance (see figure 2). Once R_e and R_i have been calculated from a bioimpedance measurement, the amount of extracellular and intracellular fluids can be calculated (see [9] for further details).

3. Materials and methods

3.1. Whole body and segmental BIS measurements

Four points whole body and segmental BIS measurements (see figure 3) were performed on 5 healthy test male subjects aged between 25 and 30, wearing only short pants and t-shirt. The measurements were performed with a commercial bio-impedance device (Xitron Hydra 4200, Xitron Technologies Inc. San Diego, CA, USA) in a temperature controlled room. After a first 30 minute phase with room temperature (24°C), a 30 minute heating phase (up to 39 °C) and a 30 minute cooling phase (up to 13°C) followed. The subjects remained in the room during the whole experiment. BIS measurements were performed 6 min after recumbency and 2-4 min after the desired temperature ±1°C was achieved. Five and 3 hours without food and beverage consumption, respectively, as well as 24 hours without performing any sports were a prerequisite for the measurements. Commercial disposable (hydrogel-aluminum) bioimpedance electrodes approx. 9x2 cm^2 in size (company Fresenius Medical Care, Bad Homburg, Germany) were used.

 doi:10.1088/1742-6596/224/1/012128

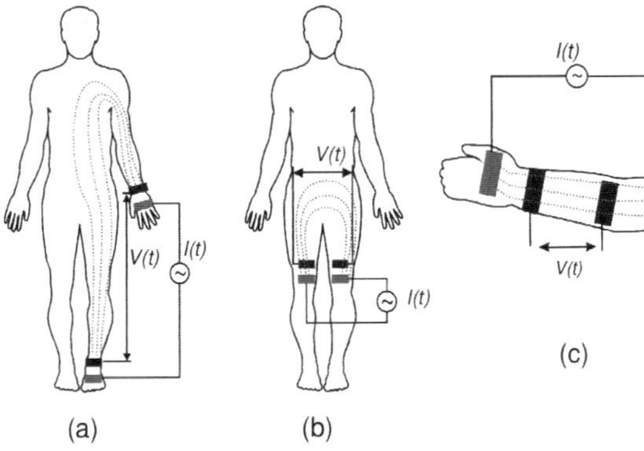

Figure 3. Whole body (a), knee-to-knee (b) and forearm (c) BIS measurements.

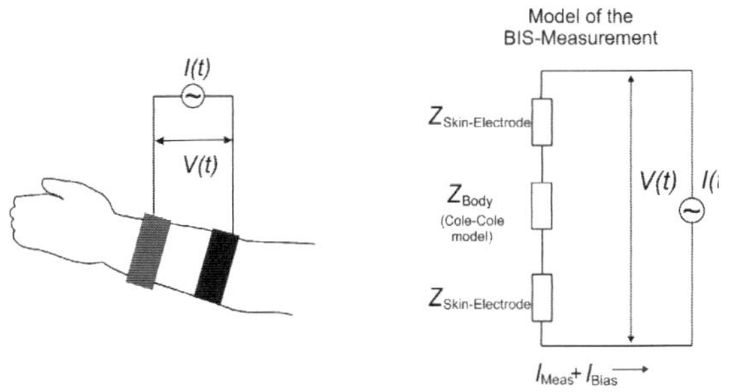

Figure 4. Bipolar electrode configuration (left) and equivalent circuit (right) for the measurement of the skin-electrode impedance.

3.2. Electrode-skin impedance measurement

Additionally to the 4 point measurements, two point measurements were performed on the forearm (see figure 4). The electrode-skin impedance was calculated as:

$$Z_{Skin-electrode} = \frac{Z_{2\,point} - Z_{4\,point}}{2} \qquad (3)$$

4. Results

4.1. Influence of ambient temperature on calculated intra (R_i) and extracellular (R_e) resistances

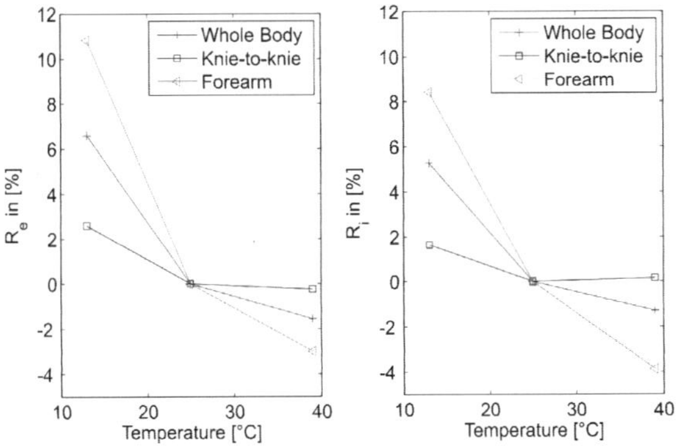

Figure 5. Relative changes in R_e and R_i (mean values) for high (39°C) and low (13°C) ambient temperatures (reference values at 24°C) in whole body and segmental BIS measurements

The relative values of R_e and R_i for whole and segmental measurements decrease during the heating phase and increase during the cooling phase (see figure 5). Changes in R_e at the arm and whole body measurements are much larger compared to changes in knee-to-knee (14 and 8% compared to 3%). Relative changes in R_i are slightly smaller than R_e changes, but with the same order of magnitudes (11, 6 and 2 % for forearm, whole body and knee-to-knee BIS measurements, respectively). The relative low sensibility of knee-to-knee measurements to external changes of ambient temperature, may be caused by the tendency of the body to keep central parts at constant temperature [3], [4]. Wearing of clothes by the test subjects could also have a slight influence on the results.

4.2. Influence of ambient temperature on skin-electrode ($Z_{skin-electrode}$) and segment ($Z_{4\,point}$) impedances

For the real part, the change observed in $Z_{skin-electrode}$ is only slightly higher than for the segment. For the imaginary part, changes at the segment impedance are approx. three times higher than the changes for segment impedance. As local changes in skin-electrode impedance have a low influence on the segmental measurement when a 4 point measurement setup is used [10], the observed increase in $Z_{skin\text{-}electrode}$ (18.1 and 15.1 %) should do not account for the observed change in the segment (12.1 and 15.1 %). The validity of (3) could be affected by measuring two different volumes in $Z_{2\ point}$ and $Z_{4\ point}$ (e.g. due to current distribution between the current electrodes). However, an impedance change should indicate a change of the segment itself. Therefore, the results suggest that the ambient temperature has an effective influence on the segment impedance and not only on the skin-electrode impedance. As ambient temperatures should affect mainly skin blood flow [3], changes on segment impedance could be due to effect of temperature on electrolyte conductivity of the fluids contained in the segment.

Table 1. Relative change (mean values) in segment and skin-electrode impedance (50 kHz) by changing from high (39°C) to low (13°C) ambient temperatures compared to measurements at 24°C.

Temperature	Forearm (segment) Real part (%)	Forearm (segment) Imaginary part (%)	Skin-electrode Real part (%)	Skin-electrode Imaginary part (%)
39-13°C	↑ 12.1	↑ 15.1	↑ 18.1	↑ 5.1

5. Conclusions

A first result of this experiment is that knee-to-knee measurements are in comparison to whole and forearm impedance measurements almost insensitive to changes in environment temperature. This could be important for measurements were the change of ambient temperature can not be avoided.
The second result of the experiment indicates that both the calculated skin-electrode impedance and the segment impedance itself are affected by changes in ambient temperature in almost same order of magnitude. This may indicate that ambient temperature has also an influence on the impedance of tissue under the skin.

Acknowledgment

This work was supported in part by the Mexican Science and Technology Board (CONACYT).

References

[1] Medrano G, Beckmann L, Zimmermann N, Grundmann T, Gries T and Leonhardt S 2007 *4th Int. Workshop on Wearable and Implantable Body Sensor Networks (BSN) (Aachen) IFMBE Proc.*vol **13**, pp. 23-28
[2] Kushner R, Gudivaka R and Schoeller D, et al. 1996 *Am. J. Clin. Nutr.* **64 (3)**, pp. 423-27
[3] Guyton A and Hall J, 2000 *Textbook of medical physiology* (Philadelphia: WB. Saunders Company)
[4] Schmidt R and Thews G, (1997) *Physiologie des Menschen* (German Version) (Berlin: Springer)
[5] Buono M, Burke S, Endemann S, Graham H, Gressard C, Griswold L and Michalewicz B 2004 The effect of ambient air temperature on whole-body bioelectrical impedance. *Physiol. Meas.* **25** 119-23
[6] Gudivaka R, Schoeller D and Kushner R 1996 Effect of skin temperature on multifrequency bioelectrical impedance analysis. *J. Appl. Physiol.* **81 (2)** 838-45
[7] Kanai H, Haeno M and Sakamoto K 1987 *Med. Prog. Technol.* **12**, pp. 159-70
[8] Cole, K and Cole, R 1941 Dispersion and absorption in dielectrics. *J. Chem. Phys.* **9**, pp. 341-51
[9] Medrano G, Eitner F, Ismail AH, Pikkemaat R, Cordes A, Floege F and Leonhardt S 2010 Influence of body position, food and beverages consumption on BIS measurements*Int. Conf. on Electrical Bioimpedance (Florida) J. Phys.: Conf. Ser. (on print)*
[10] Cornish B, Thomas B and Ward L 1998 Effect of temperature and sweating on bioimpedance measurements *Appl. Radiat. Isot.* **20** 489–91

Study of body composition in small animals by a multifrequency impedancemeter

E RIBBE[1], N KHIDER[2] and M V MORENO[1]
[1] Research Department, BioparHom, Savoie Technolac, BP238, 73374 Le bourget du Lac Cedex, France
[2] Research Department, Bio2M, Parc Technologique des Rives de l'Oise, BP50149, 60201Compiègne Cedex, France

E-mail: eva.ribbe@gmail.com

Abstract. Bioimpedance is essentially used today to study the body composition in the human body but not really in small animals. The aim of this paper is to develop a model for body composition in rats to help pharmaceutical labs assessing effects of medicine on rats. We propose a non invasive, rapid and scientific method. With a multifrequency impedancemeter, Z-Métrix® (BioparHom© Company France), resistances and reactances are measured at 55 frequencies for a population of 40 rats (males and females). With our model, derived from Cole-Cole model, resistances of extracellular (Re) and total body (Rinf) compartment are extrapolated. Three methods were applied: posterior to posterior leg, anterior to posterior leg on the left and on the right side. Measurements by CT imaging were performed on the anesthetized population to determine Fat Mass (FM), Lean Body Mass (LBM) and Bone Mineral Content (BMC), as our reference measurements. With electrical data, age, sex and weight, equations are created to calculate FM, LBM and BMC with the three methods. Graphs of correlation, between tissue masses calculated by bioimpedance and obtained with scanner, indicate that measurements with posterior to posterior leg are better. Moreover, there is no significantly difference between tissue masses measured by bioimpedance and with the scanner.

1. Introduction

Bioimpedance is a common technique for accessing body composition in human. However, it isn't a case in small animals, which are usually sacrificed. Moreover, many laboratories need to know the effects of drugs, hormones, vaccines on body composition, or to model the physiology of human or other animals. [1] To do so, they use small animals included rats. Body composition analysis by bioimpedance is a convenient diagnostic tool as it is non invasive, rapid and can be repeated frequently. Using this method can avoid sacrificing the animal and permit monitoring data during a study and not only at the end and at the beginning. The aim of this paper is to develop a model for body composition in rats.

2. Materials and methods

This study has been carried on a 40 rats 'population: 10 males aged 63 days, 10 males aged 70 days, 10 females aged 63 days and 10 females aged 70 days. Males weigh 363.52 ± 17.11 grams and females weigh 224.62 ± 7.13 grams. Rats have been anesthetized with isofluran and placed in dorsal decubitus.

In order to measure fat mass (FM), lean body mass (LBM) and bone mineral content (BMC) as reference measurements, we used a scanner GE HiSpeed NX/I Pro. After we have taken the images, we also have to quantify the volume of the tissues. We have converted volumes in masses with density of BMC, LBM and FM which are respectively 1.6667 g.cm^3, 1.1364 g.cm^3 and 0.9375 g.cm^3. These densities have been calculated by samplings.
We used a multifrequency impedancemeter Z-Métrix® (BioparHom©, France) measuring electrical characteristics of the whole body with two electrodes placed on the right hand and two placedon the right foot. This device operates on six 1.5 volt batteries. Injection current is low with 77 μA. Resistances and reactances have been measured at 55 frequencies between 1 and 1000 kHz. The device uses patented equations to calculate the resistance of extracellular compartment (Re) and the resistance of total body compartment (Rinf) using a new model derived from Cole-Cole model. [2]
We chose three methods to collect electrical data and to calculate tissue masse: posterior to posterior leg ("posterior" method), anterior to posterior legs on the right side ("right side" method) and on the left side ("left side" method). Table 1 gives number of animals in the three methods.

Table 1: Number of animals in the three methods: posterior, right side and left side.

	Right side	Left side	Posterior
Male	16	15	14
Female	13	14	16

Using electrical data and rats' age and weight, we developed multivariate equations in order to calculate tissue masse with bioimpedance. We based our model on the whole population.
Then FM, LBM and BMC have been calculated, using theses equations, in order to compare with reference measurements in a sample of 10 rats from the population.
We used paired Student T-test in order to compare tissue masses obtained by bioimpedance and our reference.

3. Results and discussion

Mean values and standard deviations (SD) for FM, LBM and BMC calculated by bioimpedance in population (male and female) with the three methods and measured with the scanner, are summarized in tables 2 to 4.

Table 2: FM, LBM, BMC measured by scanner and by bioimpedance in male and female population with method "right side" (in grams).

Male	Scanner	Bioimpedance	P. T test
	Mean ± SD	Mean ± SD	
FM	41.26 ± 8.54	41.27 ± 9.27	0.997
LBM	222.01 ± 10.25	222.05 ± 11.59	0.966
BMC	36.09 ± 1.35	36.11 ± 1.56	0.929
Female	Scanner	Bioimpedance	P. T-test
	Mean ± SD	Mean ± SD	
FM	29.23 ± 3.48	29.17 ± 3.92	0.910
LBM	121.90 ± 7.27	121.87 ± 7.80	0.971
BMC	27.52 ± 0.69	27.57 ± 0.69	0.597

Table 3: FM, LBM, BMC measured by scanner and by bioimpedance in male and female population with method "left side"(in grams).

Male	Scanner	Bioimpedance	P. T-Test
	Mean ± SD	Mean ± SD	
FM	40.63 ± 7.32	40.63 ± 7.49	0.998
LBM	212.31 ± 9.14	212.30 ± 8.75	0.992
BMC	35.32 ± 0.98	35.30 ± 1.11	0.909
Female	Scanner	Bioimpedance	P. T-Test
	Mean ± SD	Mean ± SD	
FM	28.09 ± 3.41	27.84 ± 3.56	0.829
LBM	119.88 ± 2.89	120.14 ± 1.77	0.857
BMC	26.77 ± 0.41	26.77 ± 0.48	0.788

Table 4: FM, LBM, BMC measured by scanner and by bioimpedance in male and female population with method "posterior" (in grams).

Male	Scanner	Bioimpedance	P. T-Test
	Mean ± SD	Mean ± SD	
FM	42.46 ± 8.67	42.35 ± 9.02	0.930
LBM	212.55 ± 11.32	212.69 ± 11.30	0.900
BMC	35.33 ± 1.41	35.41 ± 1.60	0.616
Female	Scanner	Bioimpedance	P. T-Test
	Mean ± SD	Mean ± SD	
FM	28.94 ± 3.83	28.81 ± 4.15	0.809
LBM	120.85 ± 7.05	120.85 ± 7.33	0.989
BMC	27.38 ± 0.62	27.43 ± 0.67	0.577

Means and SD of FM, LBM and BMC obtained by bioimpedance and with the scanners are substantially identical whichever population and method. Tissue Mass are not significantly different with P T-test from 0.597 to 0.998.
Then, we are able to measure FM, LBM and BMC in rats by bioimpedance equally as with a scanner. But, we are aware of having validated our calculation model on the population which is used to develop equations.
However, theses results aren't sufficient because we can't see which method is better in order to estimate tissue mass in small animals.

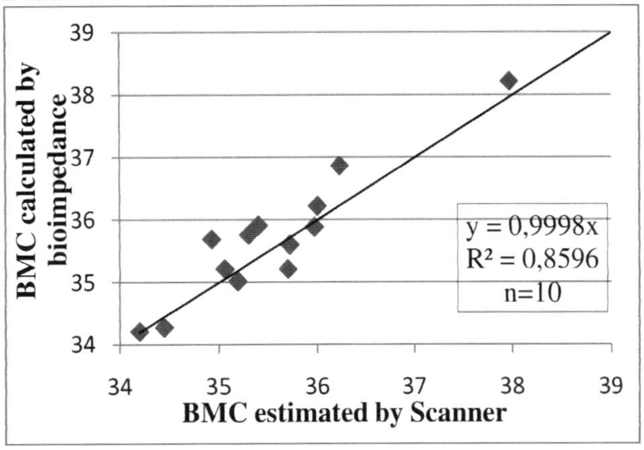

Figure 1: Example of obtaining error rate for the BMC male, right size with the relation between BMC measured by bioimpedance and BMC estimated by the scanner.

Figure 1 shows that bioimpedance underestimate BMC about 0.02% by scanner with a coefficient of linear regression (R^2) of 0.8596. Error rate has been calculated from the difference of director coefficients between the bisector and the linear regression of the population.

Table 5: Error rates (%) and R^2 for the two populations calculated with graphics representations, with the three methods.

		Error rate Male	Error rate Female	R^2 Male	R^2 Female
Right side	FM	3.29	0.29	0.601	0.739
	LBM	0.03	0.03	0.894	0.879
	BMC	0.02	0.01	0.860	0.802
Left side	FM	0.5	0.6	0.814	0.445
	LBM	0.01	0.03	0.959	0.729
	BMC	0.02	0.01	0.882	0.869
Post-erior	FM	0.09	0.45	0.699	0.620
	LBM	0.03	0.02	0.903	0.704
	BMC	0.02	0.1	0.812	0.728

Table 5 shows error rates and R^2, for FM, LBM and BMC, using the same approach of figure 1, for the three methods of measurements. Method "right side" shows error rates between 3.29% for the male FM and 0.01% for the female BMC. Method "left side" lists error rates between 0.6% for the female FM and 0.01% for the male LBM and female BMC. Method "posterior measure" gives error rates between 0.45% for female FM and 0.02% for male BMC and female LBM.
For the three methods, coefficients of linear regression are between 0.445 and 0.903.

With theses results, we can see that error rates are lower with the "posterior" method. We can say that this method of measurement is better than right and left side.
Moreover, calculation method shows no significantly difference between tissue masses obtained by the reference and assessed by bioimpedance.

Moreover, we are aware that a second population would be interesting to validate clearly our model for assessing BMC, FM and LBM in rats.

4. Conclusions

This study shows that bioimpedance with theses equations can be relevant to assess tissue masses in small animals. Using this model could avoid the animal being sacrificed. Moreover, it's a quick, non invasive and low cost method in front of the scanner and it can be used in clinical routine by laboratories for example. This study focus on tissue masses but the next step is validation of metabolic and fluids characteristics with the same device.

5. References

[1] Eisen E J, October 1989 *Livestock Production Science* **23** 17-32
[2] Moreno M V; May 2008 *tel-00282690* 52

Bioimpedance spectroscopy as technique of hematological and biochemical analysis of blood

M V Malahov[1], A V Smirnov[2], D V Nikolaev[2], A A Melnikov[1], A D Vikulov[1]

[1] State Pedagogical University, Respublikanskaya str. 108, 150000, Yaroslavl, Russia
[2] Scientific Research Centre 'Medass', Moscow, Russia

Email: a.melnikov@yspu.yar.ru

Abstract. Bioimpedance spectroscopy may become a useful method for the express analysis and monitoring of blood parameters. The aim of this study was to identify biochemical and hematological parameters of blood that can be accurately predicted by means of bioimpedance technique. Hematological (red blood cell and white blood cell parameters) and biochemical (total proteins, albumins, fibrinogen, sodium, potassium, chloride ion concentrations in plasma) parameters were measured with a hematological analyzer and routine methods. Bioimpedance spectroscopy of the whole blood (1.5 ml) in frequency range 5–500 kHz (31 frequencies) was performed using BIA analyzer ABC-01 "Medass". Frequency relationships of resistance and reactance of the whole blood and the parameters of the Cole model were investigated. Close simple and multiple correlations of bioimpedance indices were observed only with erythrocyte parameters (Ht, Hb, RBC). Thus bioimpedance analysis of the whole blood can accurately predict red cell parameters but it is less effective for estimation of plasma biochemical and white cell parameters.

1. Introduction
Investigation of biochemical and hematological blood parameters is important for physiology and medicine. It aims at searching for more accessible, inexpensive and exact methods of estimation of composition of blood. For this, manual direct microscopy of blood cells, automated blood cell counters [1] and routine biochemical techniques [2] are used. Many of them are labour-consuming and expensive. Bioimpedance spectroscopy of blood may be useful and convenient methods for estimation of blood parameters. It was shown that BIA parameters of blood are related to various morphological and biochemical indices of blood [3, 4], as well as to biochemical modification of blood during storage [5]. The purpose of our work was to identify biochemical and hematological parameters of blood that can be accurately predicted by means of bioimpedance technique.

2. Materials and methods

2.1. Blood samples
Blood samples were collected from 46 donors and placed into 4 plastic tubes (Improvacuter). For bioimpedance analysis 3 ml blood sample was drawn into tubes containing heparin. For biochemical serum analysis 4 ml blood sample was drawn into empty tubes. For fibrinogen measurement 2 ml blood sample was drawn into tubes containing sodium citrate, and for hematological analysis 2.5 ml of blood was drawn into tubes with EDTA.

2.2. BIA measurements of blood

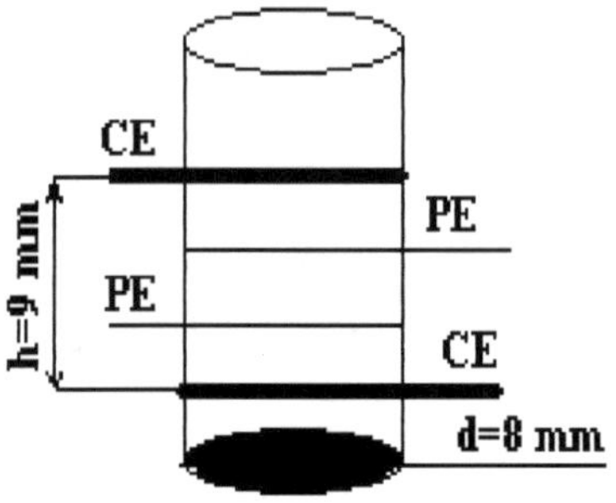

Figure 1. The conductivity cell

The electrical measurements of blood (1.5 ml) were done in a conductivity cell (Fig. 1). The conductivity cell is a uniform cylindrical plastic tube 8 mm in diameter (d) with four gilded copper wire electrodes and inner interelectrode height (h) 9 mm. Current (CE) and potential electrodes (PE) of 3 mm and 0.5 mm in diameter were applied by piercing the plastic tube. The interelectrode distance between CE and PE was 2.5 mm, and 3 mm between PE and PE. The electrodes were connected to BIA analyzer ABC-01 "Medass" (SRC Medass, Russia). Measurements of impedance (Z), resistance (R), reactance (Xc), and phase angle (Phi) of blood samples were carried out at room temperature (21±1°C) in the frequency range from 5 kHz to 500 kHz (31 frequencies). For statistical analysis, the values of phase angle at 10, 50 and 500 kHz (Phi10, Phi50 and Phi500 respectively) were used. The Cole model parameters of extracellular (Re) and intracellular resistance (Ri), as well as parameter Alpha and characteristic frequency (Fch) were calculated by the analyzer software. Example of the impedance plot (the frequency dependence of reactance from resistance) for the whole blood sample with a hematocrit of 47% is shown in Fig. 2. Coefficient of variation for all parameters obtained by this method was less than 1.0%.

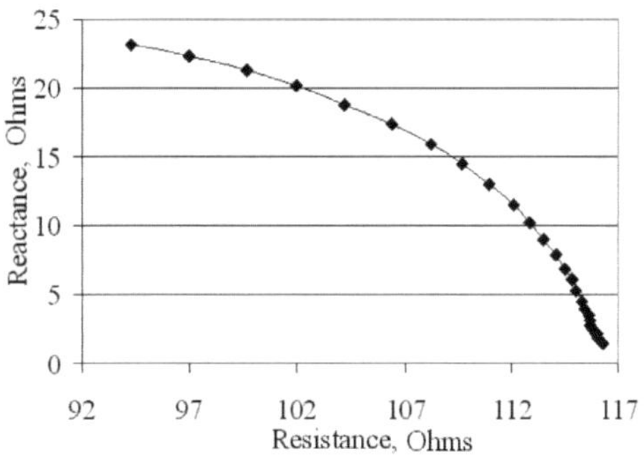

Figure 2. The impedance plot of whole blood

As BIA measurement was done in the vertical conductivity cell, the parameters Re, Phi, Alpha have changed in time due to sedimentation of red blood cells between potential electrodes. We estimated the rate of change of Re (ΔRe) and Alpha (ΔAlpha) for 1 minute in percents as follows, ΔRe (%) = (Re4-Re64)*100/Re64, where Re4 and Re64 are Re at 4 and 64 seconds of BIA monitoring, respectively. The similar formula was used for calculating ΔAlpha. Descriptive statistics for ΔRe and ΔAlpha are presented in Table 1.

2.3. Biochemical and hematological measurements of blood

Biochemical parameters, total protein concentration (TP), albumin (Alb) and globulin (Glb) concentrations, sodium (Nai), potassium (Ki), and chloride (Cli) ion concentrations were measured according to conventional methods using the Sapphire 400 chemical analyzer. Plasma fibrinogen level (Fib) was determined by the method of A. Clauss.

Table 1. Descriptive statistics for the rate of change of blood bioimpedance parameters Re (ΔRe) and Alpha (ΔAlpha) for 1 minute (n=17)

	Mean	Std. Dev.	p		Mean	Std. Dev.	p
Re4, Ohm	93,9	±10,5		Alpha4	0,937	±0,015	
Re64, Ohm	95,8	±11,4		Alpha64	0,922	±0,014	
ΔRe, Ohm	1,9	±1,6	0.0001	ΔAlpha	-0,015	±0,016	0,0008

Hematological parameters, hematocrit (Ht), mean corpuscular volume (MCV), red blood cells (RBC), mean corpuscular hemoglobin concentration (MCHC), mean content of hemoglobin (MCH), leucocytes (WBC), neutrophils (Neu), lymphocytes (Lym), monocytes (Mon), platelets (Plt) in blood were analyzed using the MEK-8222J/K hematological analyzer. Reticulocytes (Ret) and erythrocytes sedimentation rates (ESR) for 1 hour at room temperature were determined by routine methods.

Correlations were tested by linear and multiple regression analysis. As the distribution of ESR was very skew, correlation analysis was performed after logarithmic transformation of ESR.

3. Results

The results of correlation analysis between hematological and bioimpedance blood parameters are represented in Table 2. As is shown, hematological parameters RBC, Hb, Ht, MCV, MCH, MCHC were significantly related to Re, Ri, Phi10, Phi50, Phi500, Fch and Alpha. All correlations of the indices with Re, Phi10, Phi50, Phi500 were positive, but with Ri, Fch and Alpha were negative. Concentration of Ret was weakly correlated to Re only and WBC was weakly related to Re and to Phi500 (all p<0.05). Neu, Lym, Mon and Plt were not related to bioimpedance blood parameters.

Table 2. Correlations (r) between hematological and bioimpedance blood parameters (n=46)

	RBC	Hb	Ht	MCV	MCH	MCHC	WBC	Ret
Re	0,64***	0,78***	0,80***	0,51***	0,50***	0,32*	0,34*	0,46**
Ri	-0,54***	-0,75***	-0,75***	-0,59***	-0,59***	-0,39*	-0,11	-0,11
Phi10	0,33*	0,46**	0,43**	0,29*	0,35*	0,32*	-0,09	0,08
Phi50	0,56***	0,73***	0,72***	0,50***	0,52***	0,40**	0,26	0,26
Phi500	0,68***	0,82***	0,83***	0,51***	0,53***	0,39**	0,32*	0,33
Fch	-0,44**	-0,56***	-0,54***	-0,36*	-0,40**	-0,37*	-0,28	-0,19
Alpha	-0,16	-0,47***	-0,47***	-0,59***	-0,55***	-0,28	-0,04	-0,12

(*, **, *** - p <0.05, <0.01, <0.001 respectively)

Multiple regression analysis showed that a sum of bioimpedance parameters (Ri, Fch, Phi500, Re in relation to RBC and to MCHC; Ri, Fch, Phi500, Re, Alpha in relation to Ht, Hb, MCV, MCH) explained 82% of Ht dispersion (R^2=0.82, p<0.001), 78% of Hb dispersion (R^2=0.78, p<0.001), 61% of RBC dispersion (R^2=0.61, p<0.001), 47% of MCV dispersion (R^2=0.47, p<0.001), 45% of MCH dispersion (R^2=0.45, p<0.001), and 25% of MCHC dispersion (R^2=0.25, p=0.018). Thus the best multiple correlation was obtained between a sum of bioimpedance blood parameters (Ri, Fch, Phi500, Re, Alpha) and hematocrit. Correlation between calculated Ht and Ht measured by hematological analyzer was r=0.91 (p<0.001, Fig. 3).

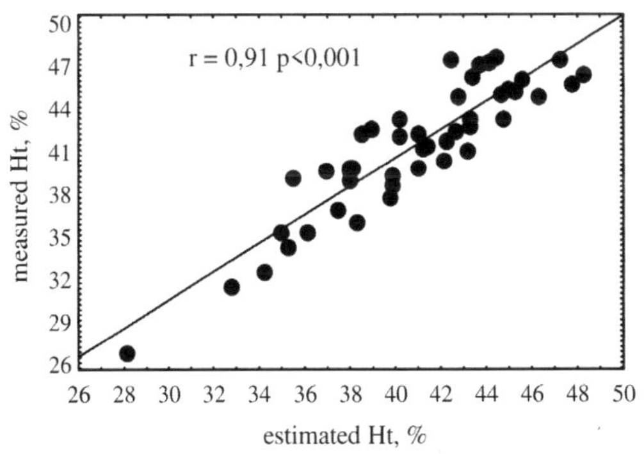

Figure 3. Correlation between estimated and really measured Ht

The results of bivariate correlation analysis between biochemical and bioimpedance blood parameters are shown in Table 3. Of biochemical parameters, only Glb, Fib, TP, Ki and ESR weakly correlated with the bioimpedance parameters. Higher correlations of Glb were with Phi50 (r=0.42, p=0.004) and Fch (r=-0.44, p=0.002). Similar result was obtained for Fbr (r=0.44, p=0.003 and r=-0.54, p<0.001 with Phi50 and Fch respectively) and TP (r=0.40, p=0.006 and r=-0.40, p<0.006 with Phi50 and Fch respectively).

Table 3. Correlations (r) between biochemical and bioimpedance blood parameters (n=46)

	Alb	Glb	Fib	TP	Ki	Nai	Cli	ESR	ESR/Ht
Re	0,13	0,16	0,27^	0,20	0,01	0,18	0,09	-0,20	-0,37**
Ri	-0,03	-0,23	0,06	-0,23	-0,35*	-0,10	-0,24	0,32*	0,49***
Phi10	-0,16	0,40**	0,21	0,32*	0,07	0,03	0,04	0,08	-0,05
Phi50	0,03	0,42**	0,44**	0,40**	0,00	0,09	0,05	-0,10	-0,28
Phi 500	0,06	0,30*	0,31	0,31*	0,04	0,17	0,10	-0,23	-0,40**
Fch	0,02	-0,44**	-0,54*	-0,40**	0,14	-0,06	0,07	0,00	0,15
Alpha	-0,14	-0,08	-0,07	-0,12	-0,25	0,03	-0,12	0,01	0,17

(^, *, **, *** - p <0.1, <0.05, <0.01, <0.001 respectively)

Multiple regression analysis showed that both of the parameters Phi50 and Fch explained 19% Glb dispersion (R^2=0.19, p=0.010), 29% of Fbr dispersion (R^2=0.29, p=0.001) and 17% TP dispersion

(R^2=0.17, p=0.018). Ions Nai, Cli were not associated with the bioimpedance parameters. Only serum Ki was significantly associated with Ri (r=-0.35, p=0.017).

Correlations of the parameters ΔRe and ΔAlpha with biochemical and hematological parameters are presented in Table 4. As is shown, TP (p=0.009), Glb (p=0.013), Fbr (p=0.012) and ESR (p=0.046) were all positively related to the rate of change of blood Re for 1 minute (ΔRe). On the contrary, TP (p<0.1) and Glb (p=0.046) were negatively related to the rate of change of Alpha for 1 minute (ΔAlpha).

Table 4. Correlations (r) of the change of Re (ΔRe) and Alpha (ΔAlpha) with biochemical and hematological parameters (n=17)

	RBC	Ht	MCV	ESR	Alb	TP	Ki	Nai	Fbr	Glb
ΔRe	0,35	0,50*	0,19	0,49*	0,24	0,61**	-0,10	-0,19	0,59*	0,59*
ΔAlpha	-0,33	-0,40	-0,16	-0,22	0,05	-0,46^	-0,40	-0,19	-0,09	-0,49*

(^, *, ** - p <0.1, <0.05, <0.01 respectively)

4. Discussion and conclusions

Our data show that the bioimpedance parameters of the whole blood correlate closely with the red blood indices: Ht, Hb, RBC. Similar correlations with Ht were obtained elsewhere [3-5]. All BIA parameters of blood were moderately associated with characteristics of single erythrocytes: MCV, MCH and especially MCHC. These low correlations were obtained because the impedance parameters of our BIA method reflect integral characteristics of whole blood: the total red cell count and plasma volume in a zone of measurement, instead of characteristics of single blood cells. Possibly, reduction of the sizes of conductivity cell till the sizes of one erythrocyte allows estimating characteristics of a single blood cell. So the BIA technique needs further development for estimating these blood indices. Hence, the bioimpedance analysis of whole blood can be used for fast and accurate determination of total red blood cell parameters (Ht, RBC, Hb) only. Electrical parameters (Re, Ri, Alpha, Phi) of the whole blood poorly reflected the concentration of white blood cells. This is possibly due to a comparatively low concentration of leukocytes in blood. Our results show that the extracellular fluid resistance (Re) of the whole blood is not related to the main extracellular plasma ions: sodium and chloride ion concentrations. This could be ascribed to a stronger influence of plasma proteins on the impedance parameters. Phase angle of blood at low frequencies (Phi50), characteristic frequency (Fch), as well as the rate of changes of Re and Alpha (ΔRe, ΔAlpha) in vertical measuring cells for 1-2 minute, can roughly estimate the concentration of macromolecular plasma globulins (gamma-globulins, alpha2-globulins and fibrinogen) and ESR. The observed changes in extracellular resistance and parameter Alpha possibly reflect erythrocyte aggregation process in blood caused by large protein molecules. This assumption requires additional tests of the blood bioimpedance characteristics along with measurements of aggregation parameters of red blood cells

References

[1] Pohland D 1989 *J. Clin. Chem. Clin. Biochem.* **27.** 41-47
[2] Faulkner W.R., Meites S.1982 *Selected methods of clinical chemistry* (Washington DC)
[3] Zhao T X 1996 *J. Med. Eng. Technol.* **20** 115-120
[4] Zhao T X 1993 *Physiol. Meas.* **14** 299-307
[5] Ulgen Y, Sezdi M 2007 *Med. Biol. Eng. Comput.* **45** 653-660

Characterization of the mechanical behavior of human skin by means of impedance spectroscopy

N Pavšelj[1], M Mitar[1], FX Hart[2], D Miklavčič[1]

[1]University of Ljubljana, Faculty of Electrical Engineering, Tržaška 25, 1000 Ljubljana, Slovenia

[2]The University of the South, Department of Physics, 735 University Avenue, Sewanee, TN 37383, USA

E-mail: natasa.pavselj@fe.uni-lj.si

Abstract. There is increased interest for the use of impedance spectroscopy to measure skin dielectric properties in vivo. The aim of such measurements can be either to evaluate the hydration state of the skin, to detect diseased states such as skin cancer, to follow the progress of transdermal drug delivery, or simply to gather data on skin tissue impedance to be used in theoretical studies. However, obtaining reliable data can be difficult. Namely, skin is a highly nonhomogeneous multi-layered structure whose composition and dimensions differ depending on the location on the body and interindividual variations. Also, impedance measurements on skin are accompanied by a number of artefacts. We performed a series of impedance measurements using an Agilent/HP 4284A precision LCR meter with parallel plate electrodes pressed on the skin, at different locations on the body. We observed substantial impedance changes over the course of the measurement. These changes can be mainly attributed to skin deformation caused by the electrodes pressing against skin. The analysis showed that skin mechanical properties and layer thicknesses can be inferred from these temporal changes. Such data on mechanical properties of skin tissue give valuable extra information, crucial for successful estimation of the impedance of different skin layers.

1. Introduction

There is increased interest for the study of different physico-chemical properties of human skin in vivo. These properties, as well as their changes, reflect the skin structure and functions, which can be used in dermatology, cosmetology, pharmaceutical sciences and oncology. We can evaluate the hydration state of the skin, its response to a specific treatment, detect diseased states such as skin cancer, follow the progress of transdermal drug delivery, and assess modifications caused by aging or external assault [1], [2]. The bulk properties of biological materials can also be used in theoretical studies, studying the effects of electromagnetic fields on cells and tissues by means of numerical modeling [3], [4]. These properties dictate the current densities and pathways that result from an applied stimulus and are thus very important in the analysis of a wide range of biomedical applications. To analyze the response of a tissue to electric current, we need data on the conductivities (when studying steady-state response) and relative permittivities (when alternating current is used or when studying transients of the modeled system) of the tissues or organs.

2. Theory

Obtaining reliable data on the electrical properties of skin can be difficult. Namely, skin is a highly nonhomogeneous multi-layered structure whose composition and dimensions differ depending on the location on the body and interindividual variations. Generally, skin consists of three main layers: the epidermis, dermis, and subcutaneous tissue. The epidermis is composed of more sub-layers, but the one that defines its electrical properties the most is the outermost layer, the stratum corneum, composed of dead, flat skin cells that shed about every two weeks. Although very thin (typically around 20 μm), it contributes significantly to the electrical properties of the skin. Its high resistivity makes skin one of the most resistive tissues in the human body. Deeper layers, including the rest of the epidermis (important in the immune response), the dermis (which gives firmness and elasticity) and the subcutaneous tissue (fat, connective tissue, larger blood vessels and nerves), all have much lower resistivities. The impedance of skin is thus dominated by the stratum corneum even though this layer is very thin [5].

For any theoretical study on skin, this layered structure has to be taken into account. However, little data exists on the electrical properties of the separate skin layers. The contributions of different skin layers towards skin impedance can only be estimated. Especially in the low frequency range (under 10 kHz), the impedance of skin is dominated by the stratum corneum. Studies show that for frequencies under 10 kHz the share of stratum corneum in the total impedance of skin is around 50%, but at 100 kHz drops to around 10% [6].

On the other hand, the mechanical properties of skin show a similar layer-dependence. The outermost stratum corneum is very firm (its hardness is comparable to glass), underneath is the less rigid viable epidermis, while the major mechanical component is the dermis - the elastic part of the skin. When skin is submitted to sustained stress, its mechanical response can be divided into three phases: i) the immediate, purely elastic phase, ii) the visco-elastic phase of variable creep, and iii) the entirely viscous phase of constant creep (see Figure 1) [7].

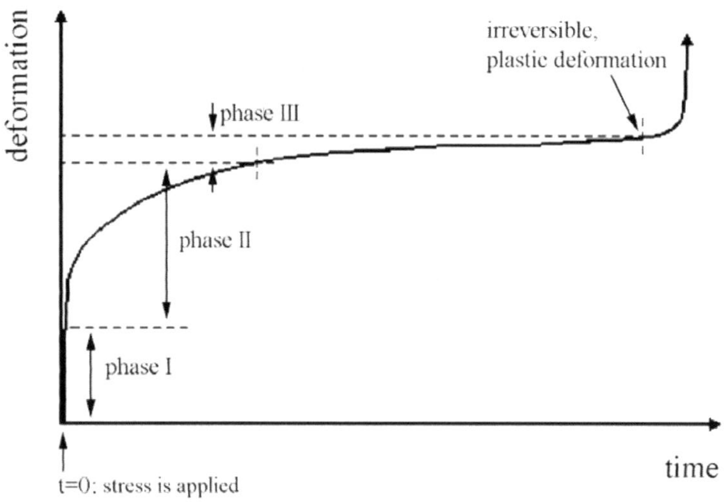

Figure 1. Skin deformation versus time for constant applied stress. Phase I: immediate, elastic deformation, phase II: visco-elastic deformation, phase III: viscous deformation

3. Materials and methods

Impedance measurements on skin were performed by means of precision LCR meter HP4284A in the 20 Hz to 1 MHz measurement frequency range. Special parallel plate electrodes were designed that allowed for easy changing of the distance between the electrodes. The plates were 1 mm thick and 1 cm wide and were made of stainless steel. They were pressed perpendicularly against the skin for the duration of the measurement. Also, as the pressure on electrodes influences the measurement, a special housing with spring-loaded electrodes was designed to keep constant pressure during the measurements. We analyzed the acquired impedance data in terms of an equivalent parallel resistance-capacitance impedance model. The measurements were done on 10 volunteers, both male and female, aged 23-67, on five different locations on the body.

4. Results

The focus of our measurements was the connection between the mechanical properties of tissue and its impedance. Namely, in all instances, parallel capacitance was rising with time for the course of the measurement, faster at the beginning, slowing down over time, showing a transient phenomenon. Figure 2 shows capacitance over time, measured at 1 kHz on volar forearm for 450 seconds, for two subjects, males, aged 67 and 26.

If we compare the theoretical mechanical response (Figure 1) with the changes in capacitance over time due to mechanical deformation under plate electrodes (Figure 2), we observe a similar time dependent curve. Unfortunately, with our measuring system we didn't record the impedance in the first few seconds after the fixation of the electrodes, where the fast elastic phase of the skin mechanical deformation takes place. Biologically speaking, this elastic phase is dictated by the straightening of the naturally furrowed stratum corneum, and by the dermal elastic network. In the visco-elastic phase, the elastic fibers are stretched and collagen network elongated, however, this is dampened by the viscosity of interstitial ground substance (proteoglycans), difficult flow of liquids in the tissue, friction between large molecules. Skin viscosity is decreased after the age of 60, which means that the time constant of the visco-elastic creep will be lower as well.

Looking at capacitance over time during the visco-elastic and viscous phase (Figure 2), we observe distinct differences between subjects. For skin of younger subjects, the ratio between the starting and the final value of the conductance seems to be consistently lower than in older subjects (due to limited space, not all data are shown, but the results are consistent for all body locations), and the time constant seems to be higher. Also, the skin of young smokers behaves closer to aged skin. These results indicate that this measured capacitance over time may reflect mechanical properties of skin. Namely, in a mechanical creep test, higher time constant also means younger skin (time constant of the visco-elastic phase is a direct measure of viscosity), and lower ratio between the starting and the final value of the visco-elastic deformation means higher electrode penetration depth, which is also a feature of younger skin. The absolute value of the capacitance is higher in older subjects, which is probably due to the lower hydration and thinner skin [7].

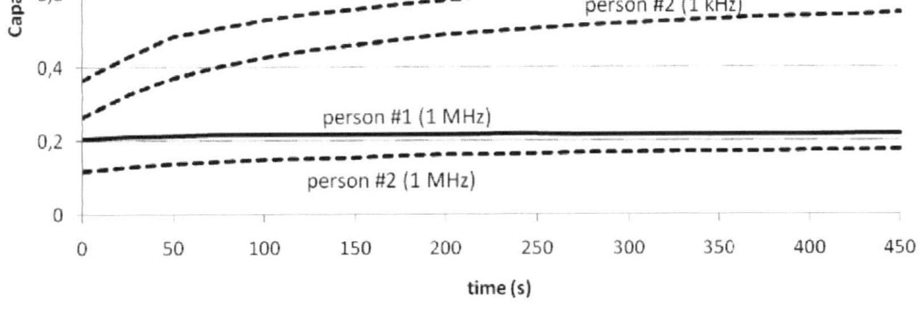

Figure 2. The change in capacitance with time, at 250 Hz, 1 kHz and 1 MHz on volar forearm. Data for two different subjects are shown here: Person 1: male, age 67, Person 2: male, age 26.

In the paper by Yamamoto and Yamamoto [5], the tape stripping of the stratum corneum had almost no effect on the parallel capacitance of the skin. The stratum corneum plays only minor role in the total skin capacitance, which is dominated by deeper tissues, at any frequency. Thus the impedance measurements may have potential in measuring the skin's mechanical properties. Changes of capacitance over time may not be due only to the moistening of the stratum corneum because of sweat accumulation, but due to mechanical deformation of deeper, softer layers.

5. Conclusion

We measured skin impedance in the frequency range from 20 Hz to 1 MHz, with parallel plate electrodes, pressed against the skin. The acquired impedance data were analyzed in terms of an equivalent parallel resistance-capacitance impedance model. We noticed similarity between the theoretical mechanical temporal response of skin subjected to constant pressure and the measured capacitance over time caused by mechanical deformation of skin during the measurement. Differences in mechanical behavior of skin of different subjects, inferred from impedance measurements were in good agreement with their biological background (in terms of age and lifestyle).

These were only preliminary measurements, and several limitations of the measurement system were discovered. First, the measurement should start at the onset of force, i.e. immediately after the fixation of the electrodes, in order to record the elastic phase of the mechanical deformation. Also, temporal spacing between consecutive measurements should be decreased, especially for the duration of the elastic phase. Also, it would be beneficial to be able to change and measure the pressure under the electrodes, as well as the indentation.

Our next step will be combining our impedance measurement with purely mechanical, suction device testing of the skin of different subjects. Our aim is to check for consistency in the agreement between the impedance and the mechanical method for measuring the mechanical response of skin. Such measurements can be useful for the estimation of skin mechanical properties, to evaluate the state of the skin (aging, chemical assault, diseased state) or its response to a specific treatment. Also, combined with numerical modeling of mechanical changes and impedance response, the thicknesses of separate skin layers can be assessed. Such data, together with the variation of electrode separation, frequency and different locations on the body give valuable extra information, crucial for successful estimation of the impedance of different skin layers. In the future, this will be achieved by constructing a series of numerical models representing the measurement circumstances and employing inverse numerical modeling.

References

[1] Keshtkar A and Keshtkar A 2008 Modeled current distribution inside the normal and malignant human urothelium using finite element analysis *IEEE Trans. Biomed. Eng.* **55(2)** 733-8

[2] Woolfson AD, Moss GP, McCafferty DF, Lackermeier A and McAdams ET 1999 Changes in skin AC impedance parameters in vivo during the percutaneous absorption of local anesthetics *Pharm. Res.* **16(3)** 459-62

[3] Pavšelj N, Bregar Z, Cukjati D, Batiuskaite D, Mir LM and Miklavčič D 2005 The course of tissue permeabilization studied on a mathematical model of a subcutaneous tumor in small animals *IEEE Trans. Biomed. Eng.* **52** 1373-81

[4] Pavšelj N, Préat V and Miklavčič D 2007 A numerical model of skin electropermeabilization based on in vivo experiments *Annals Biomed. Eng.* **35** 2138-44

[5] Yamamoto T and Yamamoto Y 1976 Electrical properties of the epidermal stratum corneum *Medical and Biological Engineering* **14(2)** 151-8

[6] Martinsen OG, Grimnes S and Haug E 1999 Measuring depth depends on frequency in electrical skin impedance measurements *Skin Res. Technol.* **5** 179-81

[7] Agache P and Varchon D 2004 Skin Mechanical Function *Measuring the skin* ed Agache P and Humbert P (Berlin: Springer-Verlag) p 429

Journal of Physics: Conference Series **224** (2010) 012132 doi:10.1088/1742-6596/224/1/012132

Effect of metal fragments in brain on electrical monitoring: In vitro and in vivo rat studies

A Ahmed[1], M Bodo[1], R A Armonda[2]

Walter Reed Army Institute of Research, Silver Spring, MD; [2]National Naval Medical Center, Bethesda, MD, USA

Email:Ambereen.Ahmed@us.army.mil

Abstract. Preliminary results showed, measurements by rheoencephalography (REG) very promising as a practical, noninvasive continuous monitoring modality of traumatic brain/blast injuries. As the impact of metal fragments on the REG signal is unknown, we report here results of our study .The in vitro study confirmed that impedance pulse amplitude waves do not change in the presence of metal (needles) placed between electrodes. In vivo studies: rats under anesthesia (10 rats, 101 trials) were measured after implantation of EEG and REG electrodes in the brain. Metal fragments were represented by 18 g needles inserted and removed between EEG and REG electrodes. Data were stored in a PC. EEG recording typically showed amplitude decrease; REG showed transitory amplitude increase after placement of a needle into either hemisphere. Removal of needles caused a decrease in REG amplitude after a transitory increase. The change in REG amplitude statistically was non-significant. Cerebral blood flow (CBF) autoregulation(AR) persisted following placement of metal fragments in rat brain.

1. Introduction

1.1 Clinical relevance

A major diagnostic limitation for the blast induced head injured patient is the inability to image the cranium using MRI due to the possibility of embedded metal fragments from improvised explosive devices. Additionally CT angiography sometimes fails to detect vasospasm due to the associated metal artifact. On the basis of preliminary results, REG seems to be a practical noninvasive and continuous monitoring modality of traumatic brain and blast injuries. This situation makes bioimpedance method (Rheoencephalogram – REG) particularly relevant due to the lack of ability to obtain transcranial Doppler assessments, angiography or cerebral blood flow measurements in the combat zone. Typically, CBF is monitored in neurosurgery intensive care units once a day by Doppler ultrasound, which in this case measures CBF on the middle cerebral artery; however, this measurement is not a continuous procedure. The probability of detecting vasospasm in recent clinical practice is low, although such a capability would be lifesaving [1]. To accomplish our objectives we conducted this in vitro and an in vivo study, modeling the metal fragments in brain while recording REG and EEG signals.

1.2 REG (rheoencephalogram)

REG is being tested in this study as a potential gl obally applicable CBF reactivi ty monitoring modality. REG is a device used to estimate a patient's cerebral circulation (blood flow in the brain) by electrical impedance methods with direct electrical connections to the scalp or neck area [2]. REG pulse amplitude changes refl ect brain vasoconstriction and vaso dilatation as well as cerebral volum e changes. The physical basis of the el ectrical impedance method is based on the fact that blood and cerebro spinal fluid are better conductors than the brain and other 'dry' tissue [3].

2. Methods

2.1 In vitro study

A loop was created using rubber and plastic tubes filled with 0.9 % NaCl (Figure 1). This loop involved a Doppler in-li ne flow probe (4N) connected to an ultrasound flo w meter (T2 01 Ultrasonic Blood flow Meter, Transonic Sy stems, Ithaca, NY); a dispos able pressure transducer (Argon Medical Devices, Athens, TX) connected to a Blood Pressure Analy zer (Digi-Med, Micro-Med, Louisville, KY); Cod man Micro sensor (Codm an IC P Express, Codm an & Shurtle ff, Raynha m, MA), an infusion pump (Powe r infuser, Infusion D ynamics, Plymouth Meeting, PA), a tem perature probe: MT-29/1 Needle microprobe (Time constant: 0.125 sec; dia meter: 29 ga BAT-12 Mi croprobe Thermometer, Physitemp (Clifton, NJ); and two home- made bioi mpedance measuring cells: 1) the firs t was a polyethy lene tube , im pedance electrodes (2 stainless steel needles with plastic tube fittings with luer connections). 2) The second was the si milar p olyethylene t ube withim pedance electrod es (2 copper metal tubes) connected to a bipolar impedance amplifier (Cerberus, Quintl ab, Hungary).Metal fragments were rep resented by 18 g stainless steel ne edles inserted bet ween bioi mpedance electrodes half w ay. Signals w ere stored in a PC using DataLyser software (B aranyi). The pump, flow wa s set to 4 L/h. For num erical co mparison, 50-second periods were measured with DataLy ser; pulse wave peak am plitudes and FFT peak am plitudes were compared. Further data analysis was performed in Excel (Microsoft, Redmond, WA).

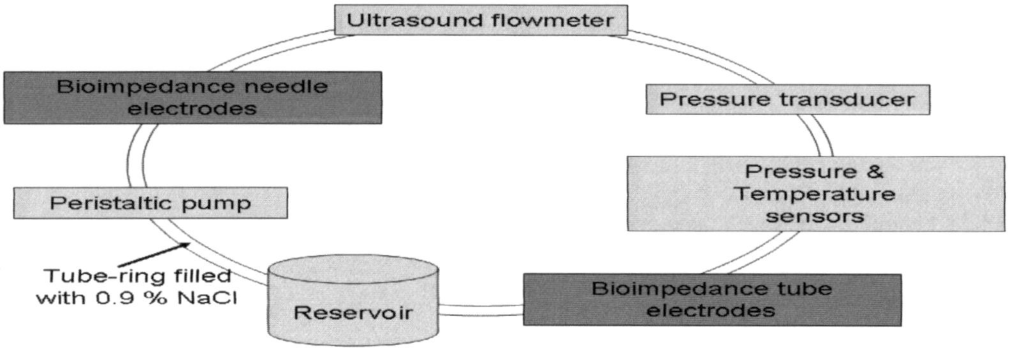

Figure 1. Block schem atics of an in vitro m easurement.The reservoir (600 m l Pyrex glass container) was uncovered, having free inflow.

2.2. In vivo - Rat measurement
Sprague-Dawley rats (n=12; weighted 463 ± 29 g; n=101 trials) were anesthetized. Cath eters were inserted surgically into femoral artery and trachea. Animal experiments were performed using rats placed into a stereotaxic frame. REG and EEG electrodes were implanted. Tests were to insert and remove 18 G needles (about 1 % of brain volume) between REG and EEG electrodes. The electrodes were connected to

the cables and to EEG and REG amplifiers. Systemic arterial pressure (SAP), exhaled CO_2 concentration, electrocardiogram were also recorded. CBF autoregulation was tested by CO_2 inhalation and SAP changes. Data acquisition was realized with 200 Hz with the same PC and software as in vitro. REG measurement was based on its first derivative; additional measurements and calculations were made by DataLyser and Excel software. REG pulse amplitude values were measured by cursor operation. Data were copied into an Excel spreadsheet. Values of baseline (before any CBF manipulation) and during CBF manipulation were compared. Bio-impedance (REG) data were stored and calculated in Volt units.

3. Results

3.1. In vitro

There was a difference between peak to peak amplitudes: the REG needle electrode voltage was ± 0.2 V and the REG tube voltage was ± 0.4 V; additionally, the REG needle signal was noisy. The two peaks with different amplitudes (Y1, Y2) and frequencies (X1, X2) were result of an unequal power of infusion pump, also every second pulse was higher in both measuring cells. The temperature sensitivity of this in vitro measurement was characterized by the linear regression lines of measured modalities, demonstrating that the pressure probes (SAP and ICP) show identical slopes to the needle cell, while the tube cell correlation was modest.

Table 1. Results of calculation of pulse peak amplitudes in presence of needle and in absence of inter-electrode needles (left 3 shaded columns) and FFT peaks (right 4 columns). Samples were characterized by the time (in seconds) of recordings (first column).ICP: Codman sensor; SAP: Digi-Med Pressure analyzer and Flow: Doppler flow meter. Numerical values beyond 2 digits were not shown. First column (N) showed the number of peaks involved in the calculation. Time window of analysis was 50 seconds.

Modality	Pulse Peak			FFT Peaks			
Group	N	Mean V	SD V	X1 Hz	Y1	X2 Hz	Y2
REG tube							
present	94	0.37	0.06	0.94	0.00	1.90	0.01
absent	94	0.39	0.06	0.94	0.00	1.88	0.02
REG needle							
present	98	0.14	0.02	0.94	0.00	1.88	0.00
absent	94	0.15	0.02	0.94	0.00	1.90	0.00
ICP							
present	93	0.42	0.21	0.94	0.01	1.88	0.04
absent	93	0.43	0.21	0.94	0.01	1.88	0.02
absent	93	0.44	0.21	0.94	0.01	1.88	0.05
SAP							
present	93	0.48	0.19	0.94	0.01	1.88	0.03
absent	93	0.49	0.19	0.94	0.01	1.88	0.02
absent	93	0.48	0.19	0.94	0.01	1.88	0.04
Flow							
present	103	-0.17	0.00	0.94	0.00	1.88	0.00
absent	113	-0.17	0.00	0.94	0.00	1.88	0.00
absent	113	-0.17	0.00	0.94	0.00	1.90	0.00

3.2 In vivo
A qualitative summary of the m easurement showed that 10 rats had 101 i nter-electrode i mmersion and removal. In few ca ses the re was bleedi ng following the immersi on and rem oval of needle which was stopped. In a subgroup of inter EEG electrode needle insertion the EEG amplitude decreased in 69% of the trials while at the sam e time REG amplitude i ncreased transiently in 92 % of trials. During the needle insertion between REG electrodes (n=12 trials), EEG amplitude decreased and REG am plitude increased, both in 75% of trials. The difference between signifi cant and non significant groups were non-significant (P=0.68). CBF autoregulation persisted following placement of needle in rat brain.

4. Discussion

The advantage of REG monitoring is that it is continuous and non-invasive. The drawback is that no clear pathophysiological background is know n to have multiple influencing factors. An extensive overview of such influencing factors was bey ond the scope of the present study but was detailed previousl y by Moskalenko and Jenkner [4, 5]. It was shown that REG reflects lower limit of CBF autoregulation [6, 7]. However, the presence of metal f ragments may i mpact RE G sig nal. On the basis of our studies REG signal seems to be insensitiv e to the presence of metal frag ment. Po ssible explanation is that the used measuring frequency (125 KHz) was too low to cause any thermal eff ect; a microwave o ven works by usually at a fr equency of 2.45 GHz. Further studies are needed to answer the siz e dependency of metal in the brain.

Acknowledgement.
The author thanks A.Garcia, S. Van Albert for th eir technical assistanc e a nd L.Barany i for use of DataLyser in this study.

Disclaimer: Material has been reviewed by the Walter R eed Ar my Institute of Resear ch. There is no objection to its presentation and/or publication. The opinions or assertions contained herein are th e private views of the author, and are not to be constr ued as offici al, or a s refl ecting true views of th e Department of the Arm y or the Department of Defe nse. Research was conducted in com pliance with the Animal Welf are Act and other federal statutes an d regulations relating to animals and experiments involving ani mals and ad heres to prin ciples stated in the Guide for the Care and Use of Laborator y Animals, NRC Publication, 1996 edition.

References.
[1] Armonda RA et al. 2006 Wartim e traumatic cerebral vasospasm: recent rev iew of combat casualties. *Neurosurgery* **59** 1215-1225.
[2] Anony mous (1997 Rh eoencephalograph (a) Iden tification *Code of Federal Regulations* Title 21, **8**, Sec 882.1825, Revised as of April 1, 1997 (Washington DC: US Government Printing Office)
[3] Ny boer J 1970 Non-invasive approaches to card iac, vascular, and pulm onary dynamics. *Mich. Med.* **69** 991-1000.
[4] Moskalenko Y E, (ed). 1980 Biophysical aspects of cerebral circulation (Pergamon: Oxford)
[5] Jenkner F L 1986 Clinical Rheoencephalography (Ertldruck: Vienna).
[6] Bodo M, Pearce F, Van Albert S, Arm onda R. Rheoencephalogram reflects cerebral blood flow autoregulation in pigs. In: Scharfetter H, Robert Merva R (E ds.): ICEBI 2007, *IFMBE Proceedings* 17, pp. 695–698, 2007 Springer-Verlag Berlin Heidelberg 2007.
[7] Brady KM, et al. Mon itoring cerebrovascular pressure reactivity with rheoencephalography. *Journal of Physics: Conference Series*, Proc eedings of the International Conference on Electrical Bioimpedance, 2010.

Journal of Physics: Conference Series **224** (2010) 012133 doi:10.1088/1742-6596/224/1/012133

Bioelectrical impedance analysis for bovine milk: preliminary results

P Bertemes-Filho, R Valicheski, R M Pereira and A S Paterno

Department of Electrical Engineering, Santa Catarina State University, Campus Universitário Prof. Avelino Marcante s/n, 89223-100 Joinville, Brazil

E-mail: dee2pbf@joinville.udesc.br

Abstract. This work reports the investigation and analysis of bovine milk quality by using biological impedance measurements using electrical impedance spectroscopy (EIS). The samples were distinguished by a first chemical analysis using Fourier transform midinfrared spectroscopy (FTIR) and flow citometry. A set of milk samples (100 ml each) obtained from 17 different cows in lactation with and without mastitis were analyzed with the proposed technique using EIS. The samples were adulterated by adding distilled water and hydrogen peroxide in a controlled manner. FTIR spectroscopy and flow cytometry were performed, and impedance measurements were made in a frequency range from 500 Hz up to 1 MHz with an implemented EIS system. The system's phase shift was compensated by measuring saline solutions. It was possible to show that the results obtained with the Bioelectrical Impedance Analysis (BIA) technique may detect changes in the milk caused by mastitis and the presence of water and hydrogen peroxide in the bovine milk.

1. Introduction

Over a decade the Bioelectrical Impedance Analysis (BIA) is being investigated as a potential technique for tissue characterization [1, 2]. BIA technique has also been widely used to evaluate body composition and the quality of food, water and milk [3, 4, 5]. BIA is a non-invasive technique in which an impedance spectrum of the material under study is calculated. Most BIA systems drive a constant sinusoidal current over a wide frequency range through two electrodes and then the resulting differential voltage is measured by other two electrodes. The impedance is then related to an equivalent electrical model of the material in order to fit it to the measured data [6].

The dairy industry uses many techniques to increase food quality, as, for example, the somatic cell counting (SCC) and conductivity measurements. However, improvements in nutrition and animal health studies require a more detailed evaluation. For many reasons, consumers ask for information about the food they are purchasing. This is specially the case for bovine milk. It is natural to consider that milk and its products require a high degree of quality, ensuring consumer health and increasing the competitiveness of dairy products. A high concentration of somatic cells and organisms in the milk may be caused by an inflammation of the teat, which would transmit such infectious organisms to the milk during or after the milking process. The current methods to determine the SCC may be time consuming and quite expensive to be provided as a permanent monitoring tool during the milking process. Most of the cases of mastitis are caused by bacterial infection (infective mastitis), which causes a decrease in the lactose concentration encountered in milk. Furthermore, there would be a decrease in the concentration of potassium while an increase in sodium and chloride ions would cause cellular damage [7, 8]. Ions concentration can also be significantly changed by the addition of distilled water and hydrogen peroxide. The goal of this work would be to investigate the adulteration in raw bovine milk by using electrical impedance measurements and also to identify the presence of somatic cells in milk samples by comparing the results obtained with chemical analysis.

2. Methodology

In a local farm, 17 sample sets of fresh raw milk from different Holstein cows in lactation were collected and three of the cows had infective mastitis. Three milk samples of each cow with approximately 100 ml each were collected and then storaged in a refrigerator at 4°C. After four hours the impedance measurements were made. Two samples of each cow had 10 ml distilled water and 1 ml hydrogen peroxide added to the samples. The samples were sent to an accredited laboratory in order to determine the amount of fat, proteins, lactose and solids by FTIR spectroscopy according to the International IDF Standard 141C:2000 and to count the somatic cells and bacteria by using flow cytometry, according to the International IDF Standards 148:2006 and 196:2004, respectively.

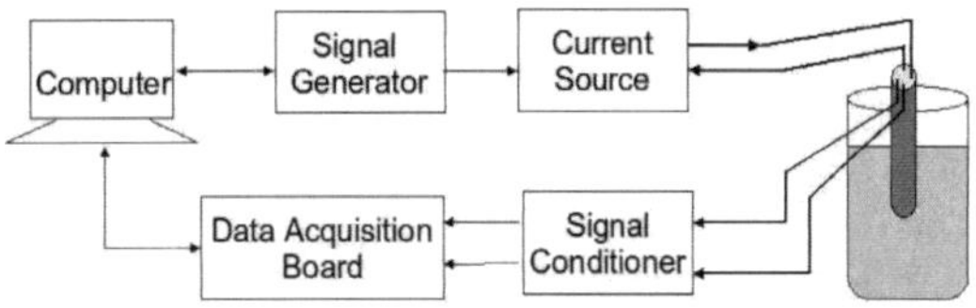

Figure 1. Block diagram of the implemented EIS system connected to a tetrapolar probe.

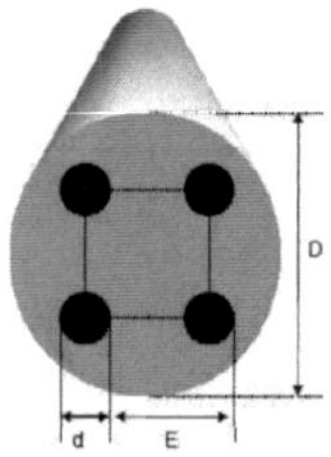

Figure 2. Tetrapolar probe used in the detection.

Bioimpedance measurements were performed in the frequency range from 500 Hz up to 1 MHz by using an Electrical Impedance Spectroscopy (EIS) system developed for this purpose. The diagram of the implemented system is shown in fig. 1, and the tetrapolar probe used in the measurements is depicted in fig. 2 where d = 1.0 mm, E = 2.5 mm and D = 4.5 mm. The system is basically composed by a voltage-to-current converter, a data acquisition board and a computer control system with a graphical interface. More details of the measurements are given in the following.

A voltage sinusoidal signal is supplied by a signal generator board PCI-5401 (National Instruments). In practice, 33 discrete frequencies are generated within the frequency range. The voltage signal is converted into current by a Howland bipolar current source [9] which generates an output current of 1 mA over the operation range.

The resulting voltages were buffered in order to prevent load effects by the acquisition system. The measurements were made by a Data Acquisition (DAQ) (model PCI-6259) manufactured by National Instruments. The board was configured to work as a single differential channel, performing a maximum sampling rate of 1.25 MS/s. The measurements were repeated for 5 times for each sample at the 33 discrete frequencies and then the data was averaged. The system was configured to provide at least 10 complete cycles of a sinusoidal waveform during signal excitation and acquisition for each frequency. The amplitude of the measured signal was detected after each cycle and then all values were averaged and the impedance spectra were obtained. Before performing milk impedance measurements, the phase shift of the EIS system was measured in saline solutions with known conductivities and used in the system's calibration.

3. Results and discussion

The results from Fourier transform infrared spectroscopy and flow cytometry analysis are shown in table 1. It is important to mention that a milk sample with a SCC concentration higher than $600/\mu l$ is considered a milk with a considerable concentration of organisms resulting from mastitis, according to the international food standards. It is not possible to infer that there is a correlation between the SCC and the concentration of fat, proteins, lactose and solids.

Table 1. Milk sample set data provided by an accredited laboratory: percentage of fat, proteins, lactose, solids, single cell count (SCC) in units per microliter and its corresponding milk type.

Sample	*Fat* (%)	*Proteins* (%)	*Lactose* (%)	*Solids* (%)	*SCC* (*units*/μl)	*Milk Type*
1	2.84	3.32	4.44	11.5	1274	Mastitis
2	3.57	4.42	3.68	12.29	1203	Milk
3	3.62	3.57	4.14	12.07	621	
4	2.95	3.12	4.81	11.99	188	
5	3.72	4.53	4.1	13.27	186	
6	2.12	2.86	4.23	9.87	176	
7	2.91	3.62	4.6	12.19	104	
8	3.13	3.64	4.55	12.35	84	
9	2.51	3.78	4.38	11.63	80	Full
10	2.28	3.07	4.51	10.75	53	Milk
11	3.07	3.1	4.55	11.65	33	
12	3.49	3.09	4.68	12.27	27	
13	2.7	2.99	4.73	11.45	23	
14	4.27	2.83	4.47	12.41	21	
15	4.05	3.44	4.57	13.08	12	
16	1.91	2.84	4.42	9.97	5	
17	3.13	3.13	4.39	11.49	3	

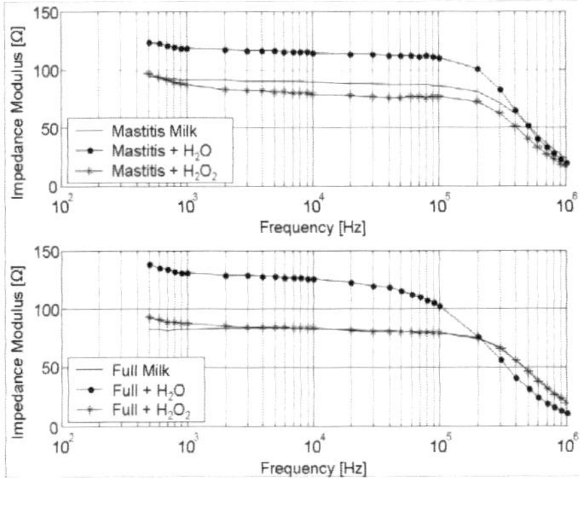

Figure 3. Measured impedance spectrum of the bovine full and mastitis milk.

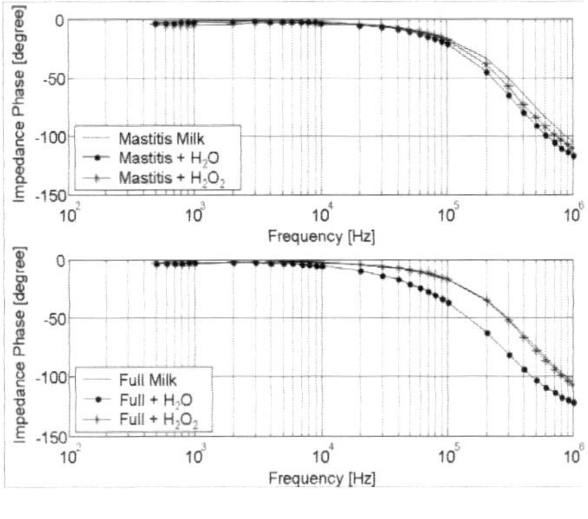

Figure 4. Measured impedance phase spectrum of the bovine full and mastitis milk.

The adulterated milk was investigated by measuring the impedance spectrum. The impedance modulus of both full and mastitis milk with and without adulteration were measured and are shown in fig. 3 where a significant difference in the impedance modulus is observed between adulterated and unchanged milk samples. The full milk impedance Z_{fm} is approximately 82Ω in the frequency range 1 kHz to 100 kHz and 89Ω for the mastitis milk (Z_{mm}). It can be observed that the milk with mastitis had its modulus impedance decreased to approximately 82Ω after the addition of hydrogen peroxide while the Z_{fm} remained approximately the same. The decrease in impedance may be explained by the fact that the dissolution of the hydrogen peroxide added to the milk increases the concentration of free ions in the sample. This causes the conductivity to increase and the impedance to decrease, specially at low frequencies where the conducting current is dominant.

However, when distilled water is added to both types of milk, the impedance modulus increases, specially at lower frequencies. It can be seen in fig. 3 that the $|Z_{mm}|$ increased to approximately 114Ω in the frequency range 1 kHz to 100 kHz while $|Z_{fm}|$ increased to 131Ω at 1 kHz. This might be explained by the fact that there is a decrease in the concentration of ions when water is diluted in the sample then the conductivity of the solution is also decreased and, consequently, the impedance increases. The impedance modulus curves were fitted to a polynomial function and then the -3 dB frequency (cut-off frequency) were calculated. The results show that the full milk has a cut-off frequency of approximately 373 kHz without distilled water and 136 kHz with the added water. It was observed that the cut-off frequency of the mastitis milk without distilled water is approximately equal to the full milk, namely 377 kHz if compared to the 307 kHz with distilled water in the milk sample.

In order to investigate the complete impedance frequency response of the milk obtainable with the EIS system, the phase of the impedance was also measured for the six milk samples. The results are shown in fig. 4. It can be seen that the milk frequency response of the impedance is approximately equal with and without adding hydrogen peroxide. However, there is a significant difference in the phase spectrum when distilled water is added to the milk, if compared to the phase obtained for the unchanged milk sample.

4. Conclusions

Impedance spectrum measurements of both full and mastitis bovine milk were analyzed and also somatic cell counting was performed for comparison purposes. Differences between milk types were found in the impedance spectra, extending the idea of using conductivity measurements for milk quality analysis. It is shown that the technique is a potential tool for detecting adulteration in milk samples with water or hydrogen peroxide. It is shown that a high sensitivity is needed to identify changes due to hydrogen peroxide using only the modulus and phase of milk impedance.

Acknowledgments

The authors wish to acknowledge financial assistance from the Santa Catarina State University.

References

[1] Abdul S, Brown B H, Milnes P and Tidy J A 2006 *Int J Gynecol Cancer* **16** 1823–1832
[2] Keshtkar A 2007 *Meas* **40** 585–590
[3] Piton C, Dasen A and Bardoux I 1988 *Lait* **68** 467–484
[4] McDougalla S, Murdough P, Pankey W, Delaney C, Barlow J and Scruton D 2001 *Small Ruminant Res* **40** 245–254
[5] Yang L and Bashir R 2008 *Biotech Adv* **26** 135–150
[6] Kyle U G *et al* 2004 *J Clin Nutr* **23** 1226–1243
[7] Kitchen B G 1981 *J Dairy Res* **48** 167–188
[8] Harmon R J 1994 *J Dairy Sci* **77** 2103–2112
[9] Bertemes-Filho P, Paterno A S and Pereira R M 2009 *IFMBE Proc* **25** 657–660

Optimization of Interdigitated Electrode (IDE) Arrays for Impedance Based Evaluation of Hs 578T Cancer Cells

Frank Alexander Jr., Dorielle T. Price, Shekhar Bhansali

BioMEMS and Microsystem Lab, Dept. Of Electrical Engineering, University of South Florida

Email: faalexan@mail.usf.edu

Abstract. This paper examines the effect of electrode width and spacing of interdigitated electrodes (IDEs) for impedance-based cancer detection and characterization. IDEs are desired for bioimpedance measurements because their fabrication process is simple and inexpensive, and the geometry presents a potential for improved sensitivity over other microelectrode designs. Optimizing the geometry will eliminate this problem and increase the sensitivity of these devices for bioimpedance measurement applications. This paper evaluates the effect of IDE geometry on the sensitivity of breast cancer cell impedance measurements. Equivalent circuit data analysis was conducted to quantify and characterize the cells.

1. Introduction

Increasing demand for faster and more accurate diagnostic procedures has sparked trends in the development of MEMS devices for biosensing and medical diagnostics. MEMS are desirable for use within medical diagnostics for their ability to manipulate and analyze small volumes of biological materials. Biosensors can be incorporated into portable lab-on-a-chip devices which will be able to quickly run diagnostics on samples. These techniques improve on current cancer detection techniques because they do not require surgery or chemical tags and can shorten the time to diagnosis.

Bioimpedance-based sensors offer an inexpensive, label-free solution for MEMS-based medical diagnostics and cell and tissue research which will enable non-invasive, real-time monitoring. Bioimpedance monitoring has already been employed to study cellular kinetics [1], cancer drug screening [2], and cellular adhesion [3]. One of the most important factors to be considered when designing cellular impedance sensors is electrode geometry.

Currently, interdigitated electrodes (IDEs) are implemented in various sensing devices including surface acoustic wave (SAW) sensors, chemical sensors as well as current MEMS biosensors [4]. IDEs have been optimized for a variety of sensing applications including biosensors sensors, acoustic sensors, and chemical sensors; however, optimization for cancer cell detection has yet to be reported. The output signal strength of IDEs is controlled through careful design of the active area, width, and spacing of the electrode fingers. Min et al., for example, examined the relationship between signal to noise ratio and different aspects of IDE geometry and composition for optimizing oxidation and reduction reactions of potassium ferro/ferrihexacyanide. It was found that increasing the number of electrode fingers and their finger widths yielded a proportional increase in overall signal. Also, it was found that increasing electrode thickness yielded an increase in current and overall signal with small increases in signal noise. This increase in current is thought to be due to increased surface area [5]. Another example of IDE optimization was performed by Radke et al. in [6]. Through simulations, it was found that the optimal IDE electrode finger width and spacing were 3 and 4um, respectively, for enhanced detection of bacteria. Radke et al. demonstrated that careful design of the electrode spacing influenced the penetration depth of the electric field thus concentrating it around the captured bacteria instead of the bulk solution.

Research indicates that currently, a design model for IDEs with an application in cancer cell detection and characterization does not exist. Because design criteria for IDEs vary by application a design rule for cell monitoring and characterization must be provided in order to maximize the sensitivity of bioimpedance measurements. In order to address this need, microelectrode devices were fabricated with constant electrode finger widths and varying electrode finger spacings of 15, 30 and 60 microns. Bioimpedance measurements were then made on carcinoma cells deposited on these devices. This paper examines the impact of varying electrode spacing on the measured bioimpedance of the cancerous cellular monolayer.

2. Materials and Methods

2.1 IDE Fabrication

Gold IDEs were patterned on 4-inch glass substrates using photolithography and metal deposition techniques. Fig. 1 illustrates the fabricated interdigital electrodes. First the wafers were cleaned using acetone and methanol. Next, the interdigital electrodes and contact pads were patterned using 3000PY photoresist. A thin layer of chromium (15nm) then gold (20nm) was electron beam evaporated onto the substrate. The photoresist layer was then removed using acetone and ultrasonic agitation. Finally, a passivation layer was patterned using SU-8 photoresist. This step leaves only the electrodes and contact pads exposed. This layer of photoresist is then hard baked to create an inert polymer resin. The devices were then diced and 6mm cultivation chambers were attached and sealed using slowly heated photoresist around the outer edges of the chamber.

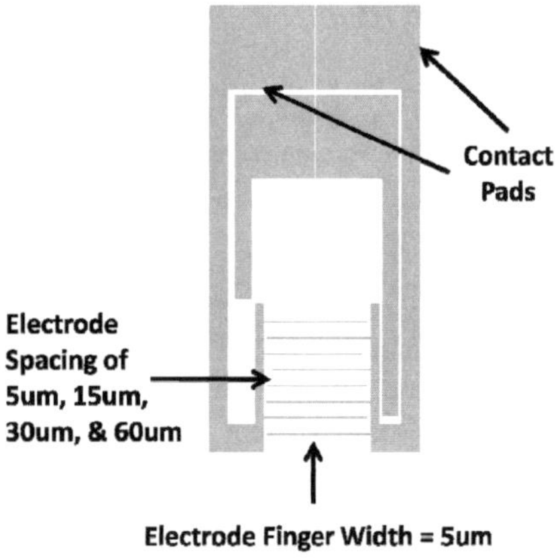

device

2.2 Cell Culturing Protocol and Sample Preparation

HS 578T breast tumor cells (ATCC) were cultured at 37°C and 5% CO_2 in Dulbecco's Modified Eagle Medium (DMEM) supplemented with 10 ng/ml epidermal growth factor (EGF), gentamicin, and 10% fetal bovine serum at final concentration. The cells were subcultured at 95% confluence and medium was changed every 2-3 days. Devices were cleaned with detergent, sterilized using 70% IPA and rinsed with Dulbecco's Phosphate Buffer Saline (DPBS). Next, the devices were precoated with .02% gelatine solution. The solution was allowed to coat the electrodes for 20 minutes and removed prior to cellular inoculation. Cells were tripsinized using a 25% trypsin solution and harvested. Devices were then inoculated with approximately 134ul of cell suspension with a concentration of 14.67×10^4 cells/ml. 200ul of medium was then added to the culture chambers. A minimum incubation period of 24 hours was allowed to ensure proper adherence of the cells to the electrodes prior to impedance measurements.

2.3 Impedance Testing Procedure

Impedance measurements were made using an Agilent 4294A impedance analyzer. A voltage of 10mV was applied to the contact pads using gold probe tips. No DC bias was applied and impedance magnitude and phase were collected over a frequency range of 100 Hz to 10 MHz. Standard calibration of the system was done using protocol from the manufacturer. Baseline measurements were performed first using only medium, followed by cellular impedance measurements. The increase of impedance was observed as well as the phase shift.

3. Results and Discussion

Measurements were first performed with medium only in the devices. Figure 2 is a bode magnitude plot of the devices with spacings of 15, 30, and 60 μm. The slope, covering the majority of the frequency spectrum, is characteristic of the double layer capacitance at the electrode interface, caused by the absorption of ions/molecules from the medium onto the surface of the electrodes. At the high frequency end, a plateau begins to form, which is representative of the solution resistance. It can be seen from the plot that the devices with larger spacings have the greatest impedance. This is expected because the total electrode area exposed to medium is less than that of the devices with smaller spacings. Equation (1) is a simplified equation for theoretical calculation of the double layer capacitance. When the surface area increases, the capacitance increases, and the impedance decreases, since it is inversely proportional to the capacitance: $C = \frac{\varepsilon_o \varepsilon_r A}{d}$. The solution resistance also increases with increased device spacing, in accordance with the simplified equation: $R_s = \frac{\rho l}{A}$.

As the spacing between the IDE fingers increase, the solution resistance should also increase; as demonstrated in the bode magnitude plot in Figure 2 and in Table 1. Parameters were extracted using a constant phase element (CPE) and resistor in series, representing the double layer capacitance and the solution resistance.

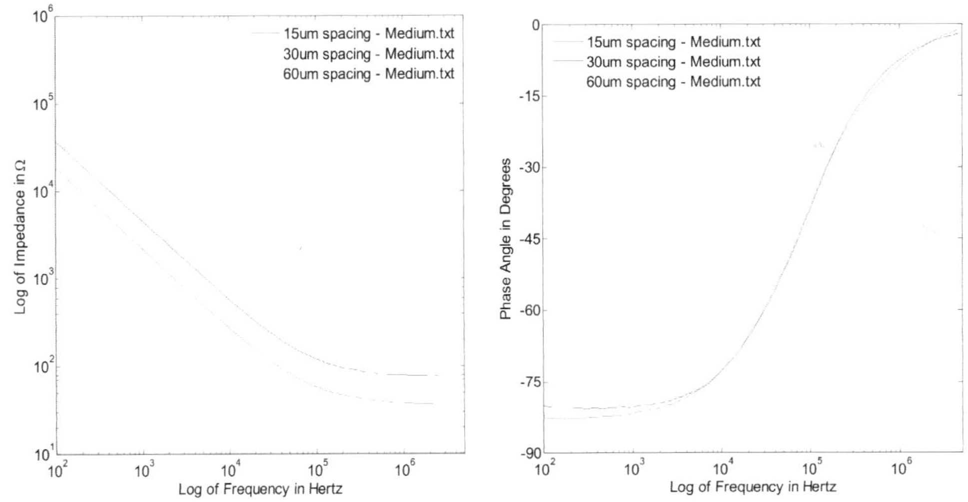

Figure 2: Bode magnitude plot of medium across IDEs with different spacings

Table 1: Extracted parameters for medium and cells, and the difference between the measured data

	Medium			Cells			Diff
	Rs (Ω)	**Cdl (F)**	**ndl**	**Rs (Ω)**	**Cdl (F)**	**ndl**	**Rs (Ω)**
15um spacing	37.1	1.61E-07	0.91	223	1.63E-07	0.91	186
30um spacing	77.4	8.29E-08	0.90	198	9.56E-08	0.88	121
60um spacing	155	7.83E-08	0.90	154	1.42E-07	0.83	-1.0

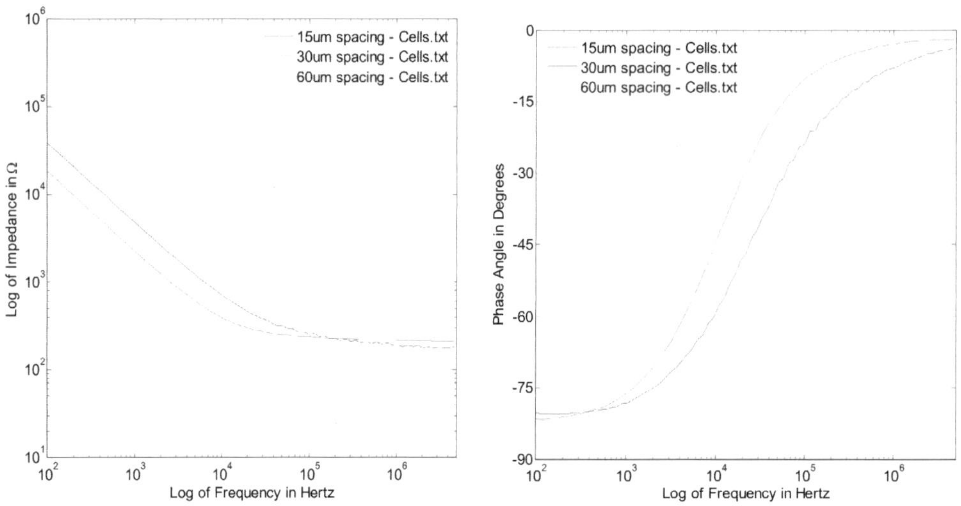

Figure 3: Bode magnitude plot of cells across IDEs with different spacings

With the presence of cells, the solution resistance increases, as seen in Figure 3 and Table 1. The 'diff' column in Table 1 shows the calculated difference between measurements with and without cells. The largest change in resistance is seen in the 15µm-spacing devices. This experimentally validates Radke's theory [6] that smaller IDE electrode spacing (on the order of cell size) is more sensitive to the presence of cells due to the concentrated penetration depth.

4. Conclusion

It can be seen from the results that the measured solution resistance increases along with the effective area of the interdigitated electrodes. By decreasing the electrode spacing the effective electrode area is increased causing an increase in the sensitivity of the device. This increase in sensitivity can be utilized to more accurately quantify modifications within cellular culture environment due to contamination, the introduction of experimental compounds and population changes.

References

[1] J. Wegener, C. R. Keese, and I. Giaever, "Electric Cell-Substrate Impedance Sensing (ECIS) as a Noninvasive Means to Monitor the Kinetics of Cell Spreading to Artificial Surfaces," *Experimental Cell Research,* vol. 259, pp. 158-166, 2000.

[2] D. Kloß, R. Kurz, H.-G. Jahnke, M. Fischer, A. Rothermel, U. Anderegg, J. C. Simon, and A. A. Robitzki, "Microcavity array (MCA)-based biosensor chip for functional drug screening of 3D tissue models," *Biosensors and Bioelectronics,* vol. 23, pp. 1473-1480, 2008.

[3] C. R. Keese and I. Giaever, "A biosensor that monitors cell morphology with electrical fields," *Engineering in Medicine and Biology Magazine, IEEE,* vol. 13, pp. 402-408, 1994.

[4] A. V. Mamishev, "Interdigital Sensors and Transducers," *Proceedings of the IEEE,* vol. 92, pp. 808-845, August 26, 2009 2004.

[5] J. Min and Antje J. Baeumner, "Characterization and Optimization of Interdigitated Ultramicroelectrode Arrays as Electrochemical Biosensor Transducers," *Electroanalysis,* vol. 16, pp. 724-729, 2004.

[6] S. M. Radke and E. C. Alocilja, "Design and fabrication of a microimpedance biosensor for bacterial detection," *Sensors Journal, IEEE,* vol. 4, pp. 434-440, 2004.

Bioimpedance in medicine: measuring hydration influence

J Hlubik[1], P Hlubik[2], L Lhotska[1]

[1] Gerstner Laboratory, Faculty of Electrical Engineering, Czech Technical University in Prague, Czech Republic

[2] Faculty of Military Health Sciences Hradec Králové, University of Defense, Czech Republic

hlubikjan@post.cz

Abstract. The aim of the paper is to present results of our ongoing research focused on the influence of body hydration on the body impedance measurements and also on the influence of the frequency used for the measurement. The question is why to measure human body composition and if these values have beneficial results. First goal of the work deals with a question of measuring human body composition. The performed measurements showed certain influence which must be verified by repeated experiments.

1. Introduction

The most popular and publicly well-known are the studies using the Body Mass Index (BMI) as a single parameter indicating the degree of obesity (or slimness). However it cannot tell anything about the body composition. This parameter is very specific because it does not take into account human body composition. This fact means that 180cm, 90kg sports man and 180cm, 90kg lazy man can have the same BMI. The BMI is the same but just at glance we can tell who is healthier. Because of that body composition is very important, with values of body composition it is easier to predict and treat diseases connected with obesity.

It is known that density of bones and quality of muscles influences the weight to a certain degree. That means that two persons having the same height and waist size might have relatively high difference in weight due to the differences in body composition. Thus, a method allowing measurement of more informative quantities is necessary in order to evaluate the amount of body fat more objectively and precisely. A method that would be able to measure important body composition more precisely and cheaper is needed.

There are several methods to determine body composition such as DXA that is considered being the reference method. Despite all benefits, this method has also disadvantages. A practical limitation of DXA is exposure of the subject to ionizing radiation (<5 μSv). Another disadvantage is price of this method. Other methods are e.g.: W/H circumference, measurements with calipers and so on.

The Body Impedance Analysis (BIA) represents one of the methods for classification of body composition. This method uses electrical current that passes through the body. This current is then measured and from these values there are calculated parts of body composition such as FFM (fat free mass), TBW (total body water), ICW (intracellular water), ECW (extracellular water), fat% and so on. At present the most frequently used frequency for elementary devices is 50 kHz. The question is whether this frequency is correct. Or would be another one better? BIA can be performed on various devices. Starting with basic single frequency bipedal devices and ending with high end devices that are using multi frequency tetra polar electrodes system. It is crucial to know which value is needed and due to this choose right device. For example, for basic examination basic device is enough, but these devices could be often biased. So it is crucial to use proper device.

This is very true for obesity as obesity does not refer to excessive body weight but it refers to the condition in which the individual has an excessive amount of body fat. Many laboratory and field assessment techniques exist for estimating a person's body composition.

Obesity is characterized by Body Mass Index (BMI) > $30 kg/m^2$ and by body fat increase. It is possible to use a number of methods including the measurement of body impedance for observing the body fat mass. Body impedance depends on body composition, i.e. on the amount of water in individual tissues.

Many of the studies [1,2] performing the BIA for various purposes (e.g. relevance to diagnosing metabolic disorders, cardiovascular risk, assessment of degree of obesity, etc.) usually do not mention level of hydration of examined subjects and frequency used during the measurement. This led us to formulation of the following three goals of our study.

2. Goals

The first goal is to find out what effect liquid consumption has to the measurement and to which degree the measured values are influenced by level of hydration. Impedance has the highest value in fat tissue that contains only 10 – 20 % of water. In fat-free mass that contains 70 – 75 % of water the values are lower. For the measurement under standard conditions it is assumed that the body is within normal hydration ranges. If a person is dehydrated then the amount of fat tissue can be overestimated.

The second goal is to identify importance of frequency on measuring. The reason is that the recently developed devices for clinical purposes allow multi-frequency measurement. It is assumed that it may lead to more precise measurements of specific parts of the body, and that each frequency can have specific importance.

The third goal of this work was to compare two devices InBody 720 and Tanita MC-180 MA in extreme body weight. We consider that every device uses specific software (equations) for calculating body parameters.

3. Methods

For the first measurement there was taken a group of 15 individuals. They received 1,3 l of water to find out the importance of body hydration. Measurements were taken before drinking, 40 minutes after drinking, then 30 minutes later and the last measurement was taken before going to toilet.

In the second experiment the group of 15 individuals was measured. Two devices InBody 720 and Tanita MC-180 MA were used. Tanita MC-180 MA device is able to measure on frequencies 5 kHz, 50 kHz, 250 kHz and 500 kHz. InBody 720 device is able to measure on frequencies 1 kHz, 5 kHz, 50 kHz, 250 kHz, 500 kHz and 1 MHz. We used for comparison only four frequencies available on both devices. We studied changes of impedance values in dependence on changing frequency.

An individual with BMI > 35 kg/m^2 was measured to compare values of these two devices in extreme body composition conditions.

4. Results

Quantitative results are shown in Table 1, Table 2, and Figure 1.

Table 1, 2: Values of impedance (Ω) during hydration measurements on both devices

Tanita	Meas 1	Meas 2	Meas 3
1	208,9	209,5	210
2	227,3	224,2	226,6
3	277,1	277,5	281
4	183,8	184,5	185,2
5	219,9	217,3	218,5
6	251,7	241,3	240,2
7	256,6	252	254,3
8	239,6	245,4	242,9
9	229,3	235,9	235,5
10	226,9	229,2	231,7
11	264,6	261	256,4
12	204,2	198,9	196,5
13	219,2	221,7	224
14	228,1	228,6	226,1
15	243,7	242,3	241,7
Average	232,06	231,28667	231,37333
sd	23,516173	23,023897	23,129994

InBody	Meas 1	Meas 2	Meas 3
1	273,81	285,47	266,21
2	325,22	317,87	363,05
3	363,03	382,17	429,36
4	188,49	181,58	220,11
5	280,29	273,2	290,91
6	310,78	311,19	357,77
7	308,69	311,59	339,89
8	291,24	308,53	348,49
9	277,97	291,55	332,26
10	268	296,66	343,73
11	283,69	284,62	324,93
12	275,22	279,76	325,86
13	282,85	300,78	329,78
14	314,26	334,63	376,33
15	328,39	328,48	374,73
Average	291,462	299,20533	334,894
sd	37,403705	40,895356	47,52057

Figure 1: History of impedance (Ω).

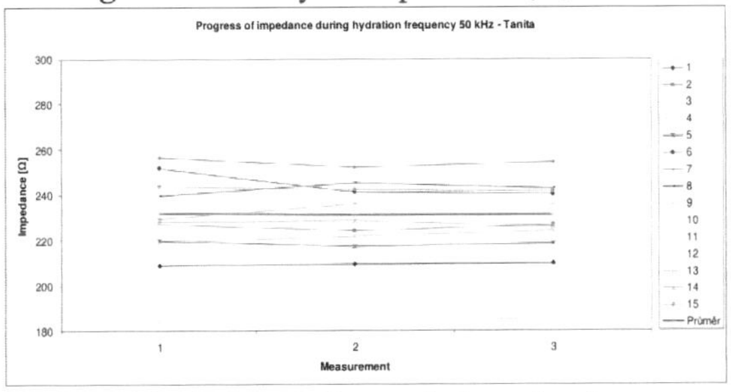

Figure 2: History of TBW (kg).

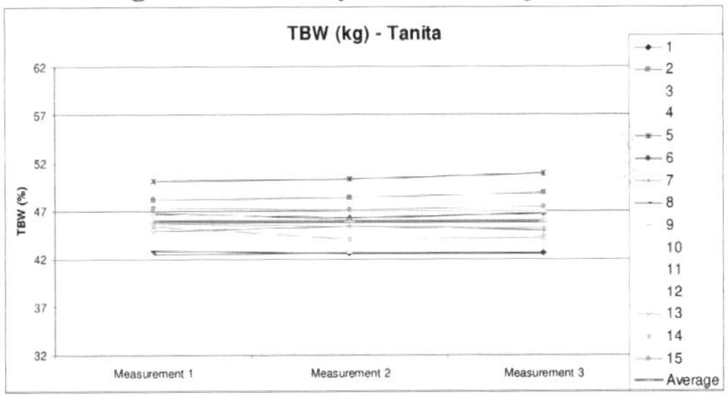

On table 1, 2 and figure 1 there is shown progress of impedance during hydration. From these outcomes it is obviously seen that consumption of water is noticeable but it depends on subject. And it also depends on each used device. From our research it is shown that in general consuming 1.3 l of mineral water does not have so high importance, but only when we use comparable devices. In figure 2 there is shown progress of TBW during hydration.

Next goal was following progress of impedance during change of frequency. These trends are shown in figure 3.

Figure 3: Progress of impedance Tanita and Inbody.

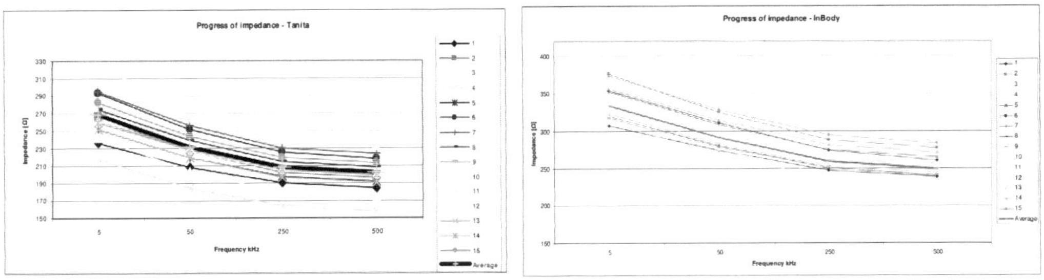

In this figure there is shown progress of impedance dependent on frequency on both devices. The dependence of impedance on frequency was very significant. However, further studies are needed to confirm which frequency is best for measurements or if all four frequencies are necessary.

The last goal was to compare two devices to find out which device is more suitable for measuring and whether there is a significant difference in these devices. We used extreme conditions. The result of this comparison is shown in figure 4.

Figure 4: Comparison of devices – segmental impedances on 50 kHz.

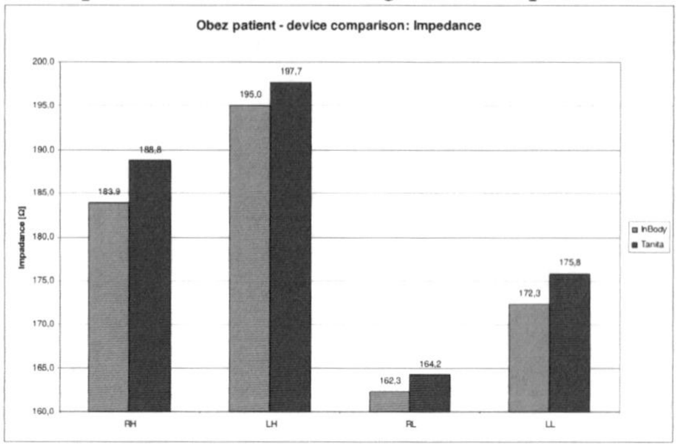

From figure 4 it is obvious that there is a difference between these two devices but the difference is not so significant. So it is possible to replace these devices and outcome values should not differ so much.

5. Conclusions

The outcome of this study is that these devices are comparable for measurements of extreme conditions. Correlation of impedances and other body composition values were very high. The dependence of impedance on frequency was very significant. Therefore a deep analysis is necessary which frequencies to use. Third outcome was that hydration has not significant influence on changes of body impedance. Work still continues to prove whether hydration of an individual has certain influence on body composition, such as FFM, TBW.

Acknowledgement
Research described in the paper has been supported by the research program No. MSM 6840770012 "Transdisciplinary Research in Biomedical Engineering II" of the CTU in Prague.

References

[1] Alcibiades Lara Lafargue, Luis Bergues Cabrales, and Ricardo Morales Larramendi 2002 *Bioelectrical Parameters of the Whole Human Body Obtained Through Bioelectrical Impedance Analysis*. Bioelectromagnetics 23:**450-454**

[2] Ursula G. Kylea, Ingvar Bosaeusb, Antonio D. De Lorenzoc, Paul Deurenbergd, Marinos Eliae, Jose Manuel Gomezf, Berit Lilienthal Heitmanng, Luisa Kent-Smithh, Jean-Claude Melchiori, Matthias Pirlichj, Hermann Scharfetterk, Annemie M.W.J. Scholsl, Claude Pichardm, Composition of the ESPEN Working Group (2004) *Bioelectrical impedance analysis part I: review of principles and methods.* Clinical Nutrition 23, **1226–1243**

S. Grimnes and Ø. G. Martinsen 2000 *Bioimpedance and Bioelectricity Basics. San Diego*, CA: Academic Press

H.C. Lukaski 2009 *Evaluation of body composition: why and how?* Mediterr J Nutr Metab 2:**1–10**, DOI 10.1007/s12349-009-0042-x

S. Grimnes and Ø. G. Martinsen, *Wiley Encyclopedia of Biomedical Engineering*, Copyright & 006 John Wiley & Sons, Inc.,

J. Hlubik 2009, *Imedane in medicine: measurin specific body impedance,* Masterss degree thesis

J. Rosell-Ferrer, L. Nescolarde, and J. Elvira, *Total Body Water Changes Using Segmental Bioimpedance in Healthy Population with Similar Anthropometry,* IFMBE Proceedings 25/VII, p. 60 ff.

L Nescolarde, M A García-González, J Rosell-Ferrer, T Doñate and U Querfeld, *Thoracic versus whole body bioimpedance measurements: the relation to hydration status and hypertension in peritoneal dialysis patients,* L Nescolarde *et al* 2006 *Physiol. Meas.* **27** 961-971

Monitoring of breast tissue thermo-ablation by means of impedance measurements

K Lukaszewicz[a,1], **J Wtorek**[a], **A Bujnowski**[a], **and J Skokowski**[b]
[a]Department of Biomedical Engineering,
Gdansk University of Technology, Narutowicza 11/12, 80-233 Gdansk, Poland
[b]Department of Surgical Oncology, Medical University of Gdansk, Dębinki 7, 80-233 Gdansk, Poland,
E-mail: lukaszewiczk@gmail.com

Abstract. A usefulness of a developed probe combining power delivery to ablated breast tumour and simultaneous impedance measurements is discussed in the paper. It consists of a cylindrical electrode used for applying ablation current and four rings electrodes. They are located, in pairs, at both ends of the ablation electrode. The probe properties were examined using a simulation studies - Laplace and bio-heat equations were solved. An associated distribution of power absorbed by tissue was calculated. Temperature dependence of tissues' conductivity was assumed. This dependence involved an iteration procedure for solving the considered problem. The four-electrode probe enabled monitoring the insertion and the ablation process.

1. Introduction

A breast cancer takes the second place on a list of the most common types of cancer. Moreover, it is a fifth reason of carcinomas-induced death worldwide. In 2004 this disease caused 519,000 deaths all around the world which is 7% of cases involved by cancer [1]. There are many treatment procedures applied e.g. chemotherapy, hormone therapy, radiation and surgery. It is mainly due to a diversity of breast cancer types. They are dependent on carcinoma stage, its aggressiveness or genetic makeup. There is a trend observed to use minimally invasive procedures and to avoid, if it is possible, surgical operations like mastectomy.

Radiofrequency thermo-ablation (RFA) is minimally invasive surgical technique which applies frequency alternating electrical current to destroy the pathogenic tissue. An electrical current of 500 kHz frequency while flowing through the biological tissue causes its heating. If it is strong enough it may cause a coagulative necrosis of the tissue, assuming that it is being heated above 50 Celsius degree. Nowadays, the method is commonly utilized in treatment of non-resectable hepatic tumours, tumours of bone, kidney, lung and also breast. It is also used to destroy abnormal electrical pathways in heart tissue.

The surgical intervention relies on accurate insertion of working electrode (needle) into the treated tissue and application of enough RFA power to destroy it. The procedure is performed under guidance using imaging techniques, e.g. X-Ray, CT or ultrasound. Recently, commonly discussed problems are how to reliably estimate the necrosis area caused by radiofrequency ablation current and how to find out if the changes in carcinoma tissue are irreversible one [4].

[1] To whom any correspondence should be addressed.

The objective of the presented study was to investigate usefulness of a developed probe combining power delivery to ablated breast tumor and simultaneous impedance measurements. The trial to use impedance measurement to evaluate the size of tumor before and after RFA was made in wide frequency range and in different temperatures of sample tissues. Electrical impedance measurements were also used for monitoring the position during an insertion procedure of the ablation catheter.

2. Methods

2.1. Problem statement and a probe construction

The developed probe is composed from one cylindrical electrode used as RFA current inlet and four rings electrodes. Ablation current is flowing between the cylindrical and the dispersive electrodes. In addition, these two electrodes enable electrical impedance measurement. The rings electrodes, which form a four-electrode probe, are used for measuring impedance of heated tissue. They are located, in pairs, symmetrically at both ends of ablation electrode (figure 1).

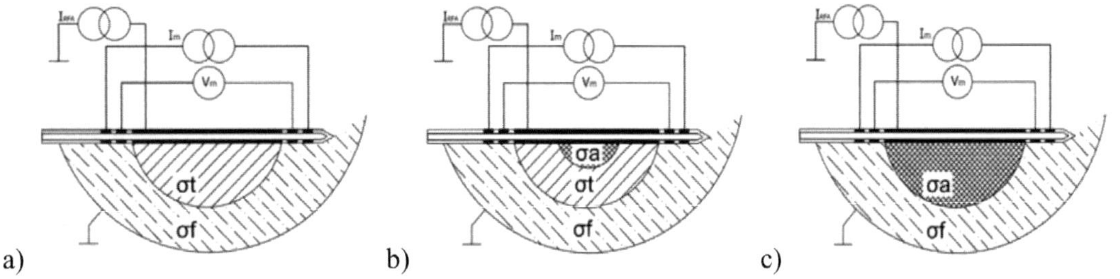

Figure1. Cases considered in the study a) beginning of ablation, b) intermediate state of ablation, c) final state after ablation, σ_t – conductivity of tumour tissue, σ_a – conductivity of ablated tissue, σ_f – conductivity of surrounding tissue, I_m – measurement current I_{RFA} – ablation current

2.2. Theoretical model

Proposed probe was modelled using finite element method (FEM) in Matlab environment. It was assumed that RF ablation process would be described by bio-heat equation in thermal domain [4, 5]:

$$\rho c \frac{dT}{dt} = \nabla(k(T)\nabla T) + q + \rho_b C_b \omega (T - T_{bl}) + Q_m , \qquad (1)$$

where ρ - the tissue density (kg/m3), c - the heat capacity (J/kg°C), k - the thermal conductivity (W/°C m), q - the electrical energy source, ρ_b - the density of blood, C_b - the heat capacity of blood, ω - the blood perfusion, T_{bl} - the basal temperature of the blood, and Q_m - the metabolic heat source.
In electric field domain Laplace equation is used [4, 5]:

$$\nabla(\sigma(T)\nabla\varphi) = 0 , \qquad (2)$$

where: $\sigma(T)$ - electrical conductivity, φ potential.
Above equations have been solved using appropriate boundary and initial conditions. Electrical and thermal conductivity of investigated tissue were assumed to be temperature dependent. Power distribution absorbed by tissue volume was calculated according the following formula:

$$\int q dV = \int \mathbf{E} \cdot \mathbf{J} dV = \int \frac{\mathbf{J}^2}{\sigma(T)} dV = \int \sigma(T)\mathbf{E}^2 dV \qquad (3)$$

where: **E** – electric filed, **J** – current density.

Difference in electrical conductivity between normal, carcinoma and ablated tissue was found as potentially useful when electrode positioning and estimating of tumour and ablated tissue region [3, 7].

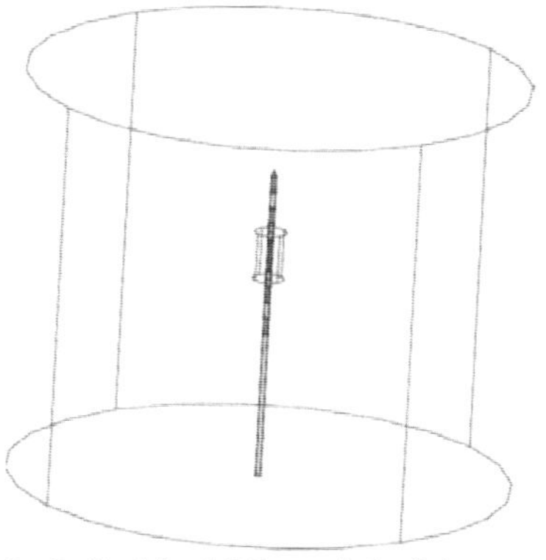

Figure 2. Cylindrical 3D model of the considered problem

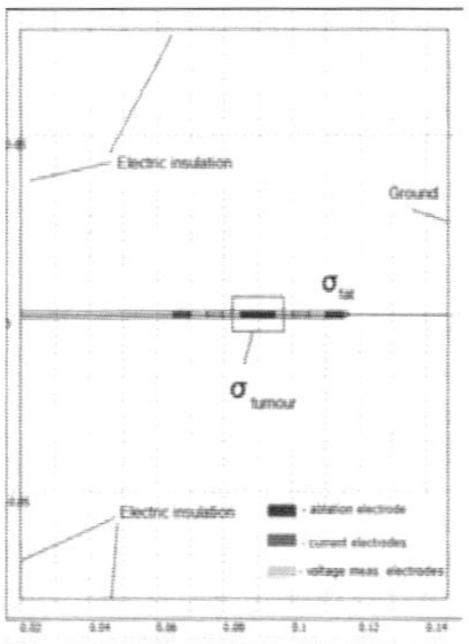

Figure 3. Axial cross-section of the model, conductivity of regions, boundary conditions and electrodes are also presented

A relatively precise model including all electrodes was developed using Finite Element Method (figure 2) [4, 5, 6]. Then, it was implemented and solved in Matlab. The measurement current was of frequency not exciding 1 MHz. The electrode placement was so chosen that the created spatial sensitivity was optimal to monitor conductivity change undergoing in the vicinity of ablation electrode (figure 3).

2.3. Experimental measurements
Conductivity of biological materials as a function of temperature was measured. A measurement system and a custom probe consisting of a glass pipe filled with saline and covered by accurately mounted four silver rings were developed. The probe was inserted into examined material filling a thermally isolated box. The selected temperature of examined material was set by passing a current through the saline. Electrical impedance module |Z| and phase φ were registered in wide frequency range of the current, from 10 Hz to 1 MHz, and in different temperatures of the examined sample.

To confirm the simulation results two measurements procedures were developed and applied. The measurements were performed using two- and four-electrode systems. Two-electrode system was composed of ablation and dispersive electrodes while the latter one of ring electrodes. Ablation and measurement currents were applied interchangeably. In case of ablation period the Rita system [2] was connected to the electrodes while during a measurement period Solartron SI1260 was used.

3. Results
The results of the current and temperature distributions depend on electrical properties of modelled tissues. However, spatial character of distributions was similar (figure 4).

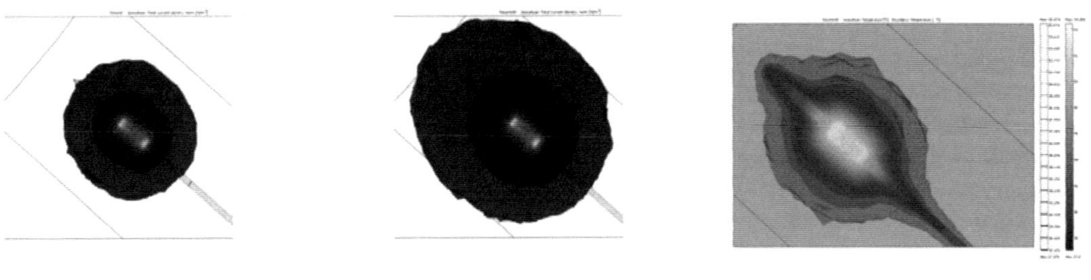

a) b) c)

Figure 4. Current density distribution a) before ablation, b) after ablation and c) temperature distribution at the end of thermo-ablation procedure

Also potential distribution when measuring impedance by means of the four – electrode probe was as expected one (figure 5). A shape of the destroyed tissue is given in figure 6.

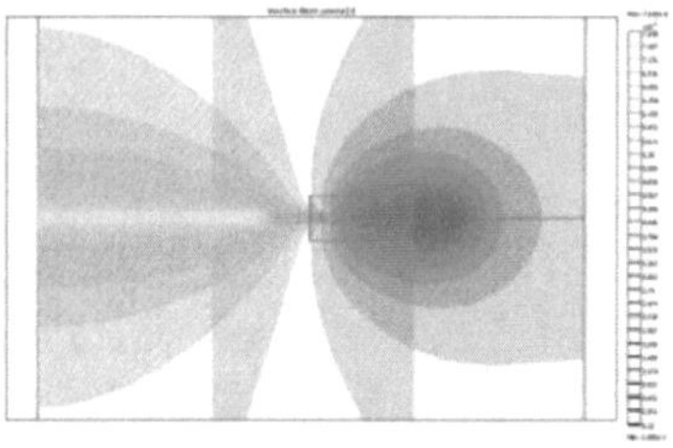

Figure 5. Potential distribution involved by measurement current when the four – electrode probe is utilized

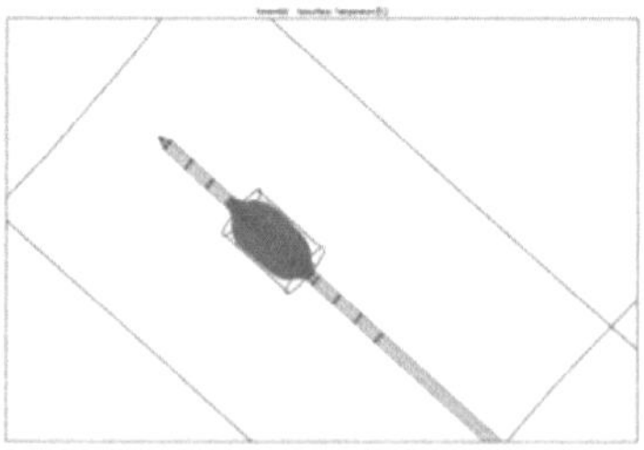

Figure 6. A shape of the ablated tissue assuming that a certain value of temperature inside of tumour tissue is reached

The values of measured impedance module and phase between life and coagulated tissues (potato, liver, muscle) were found to be visibly different (figure 7). The four-electrode technique was used.

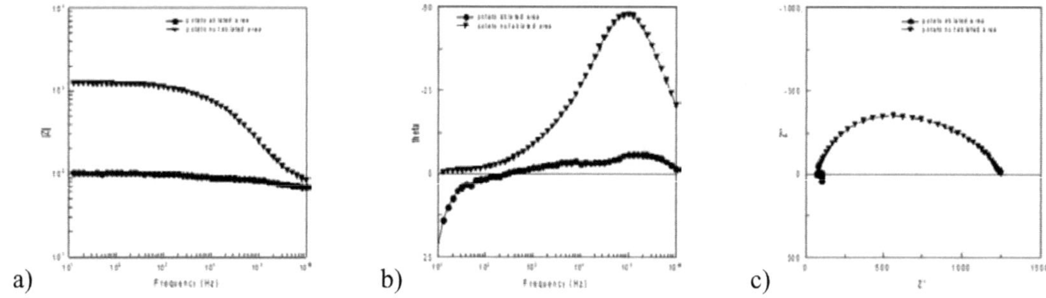

Figure 7. Electrical properties of biological tissue examined a) impedance modulus, b) phase function, and c) presentation of data on a complex plane. The tissue properties before applying ablation procedures are marked by triangles

4. Discussion and conclusion

Both theoretical and experimental studies were performed. The examined models have shown that maximum RFA current density occurs in area close to active electrode. The ablated tissue was also examined experimentally. The results obtained from experiments prove that tissue was destroyed in estimated theoretically region. Thus, it was possible to estimate the size of tissue characterized by an assumed value of temperature. It is expected that a further improvement of accuracy will be achieved due to introduction into estimation method the knowledge of ablation electrode temperature and more reliable thermal characteristics of tissues' conductivity.

References

[1] Breast Cancer *Merck Manual Online* www.merck.com/mmpe/print/sec18/ch253/ch253e.html
[2] Angiodynamics, Rita Model 1500X Generator and Uniblade RFA electrode user guide
[3] C Gabriel et al 2009, Electrical conductivity of tissue at frequencies below 1 MHz *Phys. Med. Biol.* 54 4863-4878
[4] Ekstrand V et al 2005 Influence of electrical and thermal properties on RF ablation of breast cancer: is the tumour preferentially heated? *BioMedical Engineering OnLine* 2005, 4:41
[5] Chang I 2003 Finite Element Analysis of Hepatic Radiofrequency Ablation Probes using Temperature-Dependent Electrical Conductivity. *Biomed Eng Online*, 2:12.
[6] Haemmerich D et al. 2002 Finite-element analysis of hepatic multiple probe radio-frequency ablation *IEEE TBME* 49:836-42.
[7] Hammerich D, Ozkan O R, Tsai J Z, Staelin S T, Tungjitkusolmun S, Mahvi D M and Webster J G 2002 Changes in electrical resistivity of swine liver after occlusion and postmortem *Med. Biol. Eng. Comput.* 40: 29–33

Evaluation of bioelectrical impedance analysis (BIA) to measure condition and energy allocated to reproduction in marine fishes

G R Fitzhugh[1], M J Wuenschel[2] and R S McBride[2]

[1]National Marine Fisheries Service, Southeast Fisheries Science Center, Panama City, FL 32408, USA
[2]National Marine Fisheries Service, Northeast Fisheries Science Center, Woods Hole, MA 02543, USA

Email: Gary.Fitzhugh@noaa.gov, Mark.Wuenschel@noaa.gov, Rich.McBride@noaa.gov

Abstract: Reliable estimates of fish energy density at specific times prior to spawning may provide suitable proxies for egg production, and thereby help to explain some of the observed annual variation in recruits per spawner. Our goal is to develop and test modifications of BIA technology to measure energy allocation to reproduction for a variety of marine fishes. To date, a newly developed measuring board and probe system stabilized readings, which was demonstrated by a significant reduction in the coefficients of variation for impedance measures. Total body water, wet and dry weights could be predicted with very good precision (r^2 =0.92–0.99) using BIA measures of reactance or resistance for a number of finfish species. While constituent relationships (e.g. body water- body mass functions) did not differ seasonally, we did find that BIA measures are sensitive to body composition changes related to the seasonal spawning cycle. In an examination of monthly samples of tilefish, phase angle decreased below 15° in post-spawning (regressed) females. Such a monthly trend, which suggests available energy had decreased following the spawning season, was not evident from other, more traditional measures of condition including body-muscle water content, Fulton's *K* or ordinal measures of fat deposition (such as mesenteric fat). These preliminary results show that BIA technology is a promising application for tracking and efficiently predicting energetic condition of marine fishes.

1. Introduction

In many fishes, the energy allocated to growth and reproduction (surplus energy) is accumulated from pulses of environmental productivity. Estimates of fish condition are typically derived from length-weight data, and assume that heavier fish of a given length are in better condition [1]. However, methods based on whole body weight may obscure more complex (asynchronous and/or inverse) seasonal patterns of energy allocation to different organs and tissues of the body (e.g. gonads, liver, muscle) that occurs in many fishes [2,3]. Recent studies have proposed bioelectrical impedance analysis (BIA) as a rapid, non-lethal method to determine the proximate composition and energy density of fishes [4,5,6]. Our objective is to apply this technology to predict energy density in marine groundfishes and to explore BIA as an ecosystem indicator of environmental productivity and reproductive cycles. An ultimate goal would be to establish linkages between BIA measures of condition and the energy available for reproduction to provide alternative (non-fecundity) measures of reproductive potential for stock assessments. Our initial efforts to determine BIA accuracy, precision, and cost-effectiveness have required equipment modifications to improve repeatability of measured values. Herein, we outline these modifications and present preliminary data demonstrating good precision of BIA estimates in relation to water content, dry weight, and predicted energy density.

2. Methods

Measures of resistance and reactance were made following Cox and Hartman [4]. In addition, a simple device was constructed to facilitate use of a commercially-available BIA system (RJL Sytems Inc.) for fishes. The system consists of a nonconductive fish measuring board with moveable articulating electrode holders. The holders can accommodate different types of needle/electrodes, easy replacement of electrodes, and adjustment of electrode depth. These properties make this system easily adaptable to different species and sizes of fishes. The performance of this system was evaluated by repeated measures on a number of individuals of various species. A metabolic condition index, based upon phase angle (degrees) was determined from the BIA measures of resistance and reactance using the following equation:

$$PHASE\ ANGLE\ (°) = \left(\arctan\left(\frac{REACTANCE}{RESISTANCE}\right)\right) \times \frac{180°}{\Pi}$$

Phase angle is sensitive to body condition and may be sensitive to medium-term metabolic states (e.g., integrated over weeks) as well [4,7]. The advantage of phase angle measurements in either human or animal applications is that BIA technology may be used effectively without the need for expensive calibrations and regression to estimate proximate composition.

3. Results and Discussion

Using previously published methods for applying BIA technology to relatively small-sized freshwater fishes [4], our initial attempts to apply BIA to medium to large-sized marine groundfishes identified several shortcomings which might lead to sources of error. Therefore, we designed and built a prototype BIA measuring board (figure 1) to circumvent these problems and reduce the level of 'noise' in BIA readings.

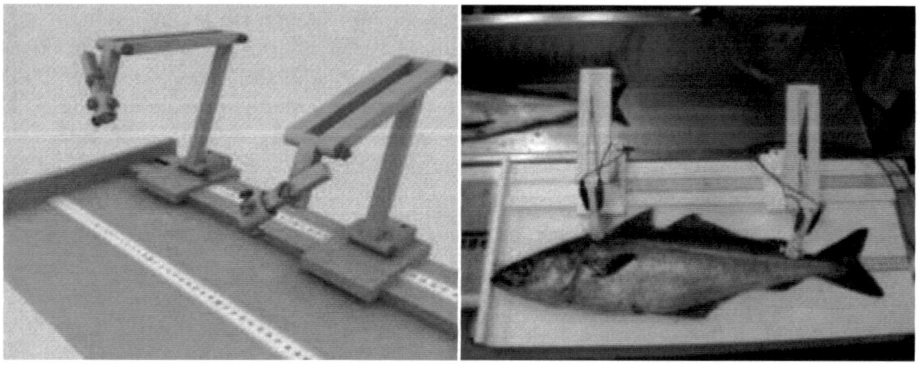

Figure 1. Technical rendering of the newly designed BIA measuring board and probe assembly (left) and the prototype in use to measure BIA on pollock at sea.

Side-by-side comparisons of measures – between the previous probe configurations and our new probe and measuring board assembly on the same individuals – indicated a significant reduction in the coefficient of variation in BIA readings with the new board and probes. We have also completed repeated measures analysis on groups of fish to determine how the length of time after capture and period on ice affects BIA measures. This information further enhanced our ability to establish protocols for applying this technology to specimens sampled under various field conditions (i.e. at fishing ports, in fish houses, onboard ships etc.).

The newly developed measuring board and probe system produced stable and reliable readings from bioelectrical impedance analyzers, and it is relatively inexpensive (and easy) to construct. This new apparatus was tested on a number of fish species, and it has several advantages over earlier devices. To date, we have collected BIA data on the following species from a variety of sampling activities (surveys and cooperative study fleets); red hake (*Urophycis chuss*), Acadian redfish (*Sebastes fasciatus*), haddock (*Melanogrammus aeglefinus*), Atlantic cod (*Gadus morhua*), Pollock (*Pollachius virens*), winter flounder (*Pseudopleuronectes americanus*), American plaice (*Hippoglossoides platessoides*), yellowtail flounder (*Limanda ferruginea*), Atlantic herring (*Clupea harengus*), tilefish (*Lopholatilus chameleonticeps*), and black sea bass (*Centropristis striata*). Preliminary analyses demonstrated high r^2 values (0.92–0.99) for relationships of total body water, wet and dry weights predicted from BIA measures to observed weights for most species, with the exception of flatfishes which had generally lower r^2 values. Although the r^2 values were high for many species, the slopes differed from 1. Therefore, we conclude that previously published regressions, primarily for freshwater species, cannot be routinely applied to marine fishes. Moreover, in the case of flatfishes, a new model formulation may be needed. Nonetheless, new and useful calibration models appear attainable because constituent relationships (e.g. body water- body mass functions) determined in different seasons did not vary significantly for the groundfishes examined so far.

While constituent relationships did not differ seasonally, we did find that BIA measures are sensitive to seasonal cycles. In an examination of monthly samples of tilefish, BIA indicated that composition changes had occurred over the spawning cycle and were detectable among females with differing reproductive states. Phase angle decreased below 15° following the peak spawning period (gonadosomatic index, GSI) and was distinctly lowered by August in post-spawning (regressed) females (figure 2). Phase angle below 15° may be a common indicator of poor condition in fish [7]. This monthly trend, which suggests available energy had decreased, was not evident from other, more traditional measures of condition including body-muscle water content, Fulton's *K* and hepatosomatic index (HSI). In addition, tilefish exhibited no obvious visual cycles of fat deposition (such as mesenteric fat), which makes a tool like BIA worth full investigation for its utility in tracking energetic condition.

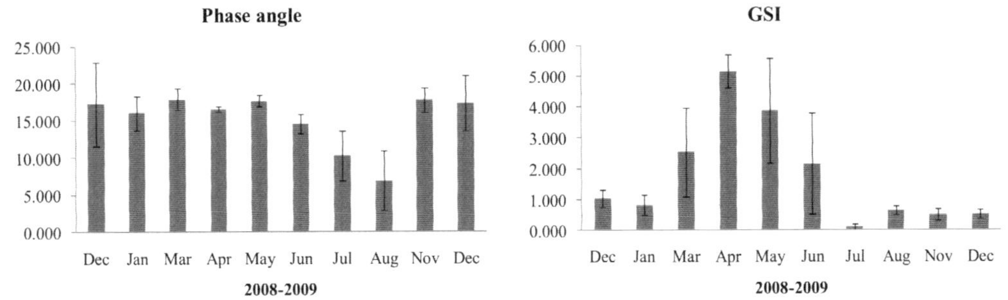

Figure 2. BIA estimated phase angle and gonadosomatic (GSI) indices from monthly collections of female tilefish.

Our progress in modifying and testing this devise on economically-important marine species is encouraging. Using black seabass, we have begun a laboratory experiment to test the effect of food consumption rate as the energetic mechanism affecting energy density as measured by BIA. The experiment correlates BIA measures to reproductive condition of individuals at various points in time. Additional activities underway include a broader field application of BIA across species and seasons, and calibration of BIA indices to laboratory analyses of proximate composition.

References

[1] Lambert Y and Dutil J D 1997 *Canadian Journal of Fisheries and Aquatic Science* **54** (Suppl. 1) 104–112
[2] Koops M A, Hutchings J A and McIntyre T M 2004 *Fish and Fisheries* **5** 120-130
[3] Kennedy J, Gundersen A C and Boje J 2009 *Fisheries Research* **100** 260-265
[4] Cox M K and Hartman K J 2005 *Canadian Journal of Fisheries and Aquatic Science* **62** 269-275
[5] Duncan M, Craig S R, Lunger A N, Kuhn D D, Salze G and Mclean E 2007 *Aquaculture* **271** 432-438
[6] Willis J and Hobday A J 2008 *Fisheries Research* **93** 64-71
[7] Cox M K and Heintz R 2009 *Fishery Bulletin U.S.* **107** 477-487

Application of Longitudinal and Transversal Bioimpedance Measurements in Peritoneal Dialysis at 50 kHz

L Nescolarde[1], T Doñate[2], R Casañas[3], J Rosell-Ferrer[1]
[1] Department of Electronic, Technical University of Catalonia, Barcelona, Jordi Girona 1-3, Edifici C4, 08034, Barcelona, Spain
[2] Service of Nephrology, Puigvert Foundations, Barcelona, Spain
[3] Faculty of Medicine, Central University of Venezuela, Caracas, Venezuela

E-mail: lexa@eel.upc.edu

Abstract. More relevant information of the fluid changes in peritoneal dialysis (PD) might be obtained with segmental bioimpedance measurements rather than whole-body measurement, who hidden information of body composition. Whole-body and segmental bioimpedance measurements were obtained using 5 configurations (whole-body or right-side (RS), longitudinal-leg (L-LEG), longitudinal-abdomen (L-AB), transversal-abdomen (T-AB), and transversal-leg (T-LEG)) in 20 patients: 15 males (56.5 ± 9.4 yr, 24.2 ± 4.2 kg/m2) and 5 females (58.4 ± 7.1 yr, 28.2 ± 5.9 kg/m2) in peritoneal dialysis (PD). The aim of this study is to analyze the relationship between whole-body, longitudinal-segmental (L-LEG and L-AB) and transversal-segmental (TAB and TLEG) bioimpedance measurement at 50 kHz, with clinical parameters of cardiovascular risk, dyslipidemia, nutrition and hydration. The Kolmogorov-Smirnov test was used for the normality test of all variables. Longitudinal bioimpedance parameters were normalized by the height of the patients. The Spearman correlation was used to analyze the correlation between bioimpedance and clinical parameters. The statistical significance was considered with $P < 0.05$. Transversal bioimpedance measurements have higher correlation with clinical parameters than longitudinal measurements.

1. Introduction

Body composition analysis is based nowadays mainly based on body weight and an examination of soft tissues. One of the alternative techniques for body composition assessment in clinical practice is bioelectrical impedance, as a simple, non-invasive, innocuous and repeatable method [1]. Moreover, bioelectrical impedance vector analysis [2] does not require any assumption about constant soft tissue hydration and body composition and is independent of the body weight.

The patient in peritoneal dialysis session represents an ideal model for studying localized changes in body fluid. Each body segment such as: leg, arm, thorax, abdomen, has its own characteristic pattern of fluid shift in response to PD fluid exchange [3, 4]. The value of the whole-body impedance is the addition of the right arm, the thorax, the abdomen and the right leg impedances. In PD patients, the approximate percentage contributions to whole-body resistance (longitudinal measurement) were: thorax 1%, abdomen 5%, leg 43% and arm 51% [5, 6].

[1] Lexa Nescolarde: lexa@eel.upc.edu

The aim of this study is to analyze the relationship between whole-body, longitudinal-segmental (L-LEG and L-AB) and transversal-segmental (T-AB and T-LEG) bioimpedance measurement at 50 kHz, with clinical parameters of cardiovascular risk, dyslipidemia, nutrition and hydration.

2. Methods

2.1 Patients

Twenty patients in continuous ambulatory peritoneal dialysis (CAPD): 15 males (56.5 ± 9.4 yr, 24.2 ± 4.2 kg/m2) and 5 females (58.4 ± 7.1 yr, 28.2 ± 5.9 kg/m2) at the Service of Nephrology of the Fundació Puigvert (Barcelona, Spain) were analyzed by bioimpedance. Clinical evaluation, laboratory determinations, and bioimpedance measurement were carried out in the morning, before the second daily fluid drainage of the abdominal cavity.

The following clinical parameters were measured: systolic and diastolic blood pressure (SBP and DBP), mean blood pressure (BPmean), haematocrit, cholesterol, triglycerides, low-density lipoprotein (LDL, col), high-density lipoprotein (HDL, col), total protein (Tprot), albumin and C-reactive protein (CRP).

2.2 Bioimpedance Measurements

Impedance measurements were taken with model STA-BIA (Akern, Florence-Italy) at 50 kHz with injected current of 800 μA. The specific measurement error of the system is lower than 1 Ω and 1 ° at 50 kHz using electrical models. Disposable pre-gelled Ag/AgCl (3M Red Dot, Canada) electrodes were used.

The tetrapolar measurements (two electrodes for injecting current (I) and two for sensing voltage (V)) were located on the patients in supine decubitus. Electrode positions are shown in figure 1. Longitudinal current injection electrodes were located in 1 and 4 for RS, L-AB and L-LEG, and transversal current injection electrodes were located in 11 and 14 for T-AB and 7-10 for T-LEG. Electrodes for sensing the voltage are:

1. RS: whole-body; the standard whole-body impedance measurement (2-3).
2. L-AB: longitudinal-abdomen; the voltage was measured by using the electrode on the left-foot articulation and on the xiphoid (5-6).
3. L-LEG: longitudinal-leg; this represents the addition of both previous segments from the inguinal region to the malleolar articulation (6-3).
4. T-AB: transversal-abdomen; at the level of the umbilicus (12-13).
5. T-LEG: transversal-right-leg; located in the thigh femoral region (8-9).

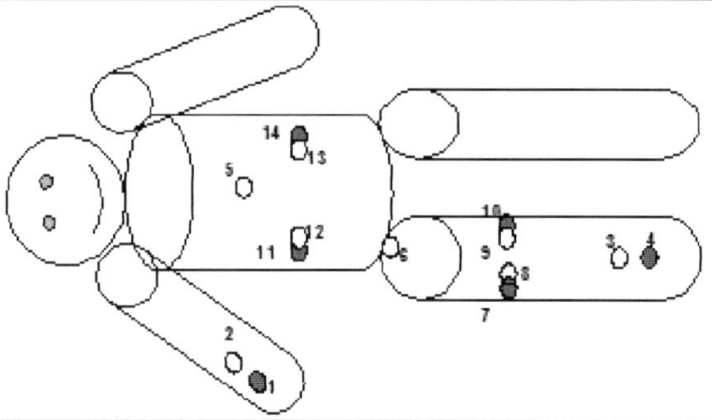

Figure 1. Electrode locations for longitudinal and transversal measurements.

2.3 Statistical Methods

The Kolmogorov-Smirnov test was used for the normality test of all variables. All data are expressed as mean values (± SD). The Spearman correlation was used to analyze the correlation between bioimpedance and clinical parameters. In longitudinal bioimpedance segments, the impedance parameters were normalized by the height (H) of the patients. Minitab software version 15.1.0 (Minitab, Inc.) was used for statistical analysis. Statistical significance was considered with $p < 0.05$. Female sample were not analyze statistically due to the small sample size.

3. Results

Table 1 shows the mean ± standard deviation (SD) for clinical laboratory variables in male and female samples.

Table 1. Clinical laboratory results.

	Males *N*=15 *mean ± SD*	Females *N*=5 *mean ± SD*
BMI (kg/m^2)	24.2 ± 4.2	28.2 ± 5.9
SBP (mmHg)	133.0 ± 15.6	135.0 ± 14.1
DBP (mmHg)	78.7 ± 10.4	80.1 ± 7.1
BPmean (mmHg)	96.8 ± 11.2	98.3 ± 7.1
Hematocrit (g/L)	37.3 ± 4.4	35.2 ± 5.3
Cholesterol (mmol/L)	4.2 ± 0.8	4.8 ± 1.5
Triglycerides (mmol/L)	1.4 ± 0.6	2.1 ± 1.9
LDL, col (mmol/L)	2.5 ± 0.7	2.2 ± 1.2
HDL, col (mmol/L)	1.1 ± 0.3	1.2 ± 0.4
T.Protein (g/L)	62.4 ± 6.3	63.0 ± 3.1
Albumin (g/L)	35.8 ± 5.3	37.5 ± 3.2
CRP (mg/L)	2.7 ± 1.7	3.7 ± 3.9

Table 2 shows the mean ± standard deviation (SD) of impedance parameter; resistance (R) and reactance (Xc) divided by height of the patients in the longitudinal measurements (RS, L-LEG and L-AB) and in transversal impedance segment (T-LEG, T-AB).

Table 2. Mean ± SD of impedance for each segment.

	Males *N*=15		Females *N*=5	
	R/H	Xc/H	R/H	Xc/H
RS	308.9 ± 91.1	31.8 ± 10.2	331.4 ± 57.7	31.1 ± 7.2
L-LAB	24.8 ± 5.8	-	24.2 ± 3.3	-
L-LEG	218.5 ± 74.5	22.2 ± 14.4	221.8 ± 31.3	18.8 ± 5.4
	R	Xc	R	Xc
T-AB	30.5 ± 10.4	-	41.6 ± 17.0	-
T-LEG	26.8 ± 7.4	-	30.4 ± 6.2	-

Table 3 shows the Spearman correlation results between the bioimpedance and clinical parameters in the male sample. Only segments and clinical variables with significant correlations are shown.

Table 3. The Spearman correlation between impedance and clinical parameters for the male sample.

	SBP (mmHg) Rho P	DBP (mmHg) Rho P	BPmean (mmHg) Rho P	T.Protein (g/L) Rho P	Albumin (g/L) Rho P	CRP (mg/L) Rho P	Triglyc. (mmol/L) Rho P	LDL, col (mmol/L) Rho P
R,T-LEG	-0.643 0.015	-0.575 0.033	-0.617 0.019	0.570 0.033	0.642 0.013	-0.642 0.013	-	-
R, T-AB	-	-	-	0.610 0.021	-	-	0.660 0.010	-
Xc/H-L-LEG	-	-	-	-	-	-	-	0.632 0.015

4. Discussion and Conclusions

The real part of the complex impedance (resistance, R) provides information of the hydration state, and imaginary part (reactance, Xc) the structure of soft tissue [2]. The resistance (R), in transversal impedance measurements in right leg (T-LEG), shows high correlation with parameters of cardiovascular risk such as systolic and diastolic blood pressure, and as consequence with mean blood pressure; also with hydration and nutrition state such as, total protein, albumin and C-reactive protein. The resistance parameters, in transversal impedance measurements in the abdomen segment (T-AB), show high correlation with nutrition and dyslipidemia parameters such as total protein and triglycerides respectively. The imaginary part of the complex impedance, reactance (Xc) normalized by the height of the patients, in longitudinal impedance measurement in right leg (L-LEG), show high correlation with parameter of dyslipidemia, such as low-density lipoprotein (LDL, col). These results are in agreement with the complex changes and alterations in body composition due to renal insufficiency. The patients with terminal chronic renal failure present dysfunctions in the distribution of the body water (fluid-overloaded), malnourishment and muscle wasting. Sub-clinical volume overload is a frequent complication of peritoneal dialysis patients associated with hypertension and increased cardiovascular risk. A careful choice of electrode placement will enable the user to focus the measurements on the desired structure of the human body.

Acknowledgments

This work has been possible thanks to collaboration of the patients and the unmeasured support of the nurse Rosa Pallás from the Service of Nephrology, Puigvert Fundation.

References

[1] Ellis KJ 2000 Human body composition: in vivo methods *Physiol Rev.* **80**:649-680

[2] Piccoli A, Rossi B, Pillon L, Bucciante G 1994 A new method for monitoring body fluid variation by bioimpedance analysis: The RXc graph *Kidney Int.* **46**:534-539

[3] Song JH, Lee SW, Kim GA and Kim MJ 1999 Measurement of fluid shift in CAPD patients using segmental bioelectrical impedance analysis *Perit. Dialysis Int.* **19:**386-390

[4] Woodrow G, Oldroyd B, Turney JH, Smith MA 1996 Segmental bioelectrical chronic renal failure impedance in patients with chronic renal failure *Clinical Nutrition* **15:**275-279

[5] Heitman BL 1994 Impedance: a valid method in assessment of body composition? *Eur J Clin Nutr.* **48:**228-240

[6] Nescolarde L, Doñate T, Piccoli A, Rosell-Ferrer J 2008 Comparison of segmental with whole-body impedance measurements in peritoneal dialysis patients. *Med. Eng. and Physics* **30:** 817-824

Finite element model of needle electrode sensitivity

P. Høyum[1], H. Kalvøy[2], Ø. G. Martinsen[1,2], S. Grimnes[1,2]

[1] University of Oslo, Department of Physics, Oslo, Norway .
[2] Rikshospitalet, Dept. of Biomedical & Clinical Engineering, Oslo, Norway

E-mail: per@hoyum.net

Abstract. We used the Finite Element (FE) Method to estimate the sensitivity of a needle electrode for bioimpedance measurement. This current conducting needle with insulated shaft was inserted in a saline solution and current was measured at the neutral electrode. FE model resistance and reactance were calculated and successfully compared with measurements on a laboratory model. The sensitivity field was described graphically based on these FE simulations.

1. Introduction

The main objective for this work was to describe the sensitivity field of a needle electrode with an insulated shaft. The description of sensitivity field for this type of electrode is important in the ongoing project [1] describing a new method for anatomical needle positioning in the clinic.

As shown below, the sensitivity field S can be expressed by the following [2]:

$$S =$$

(J'_{reci} is the reciprocal current density vector and J'_{cc} the needle current density vector.)
Since this is a two electrode system, S becomes:

$$S = |J'|^2 \qquad [1/m^4]$$

To obtain the necessary information in order to express the sensitivity field of a needle electrode system, a Finite Element (FE) model was used. This FE-model was physically based on an actual lab-model with a needle in saline solution used by Kalvøy *et al.* [1].

The needle was modeled (in the FE application) with an insulated shaft and had electrode polarization impedance (EPI) incorporated based upon measurements on the lab-model. A literature study showed that a description of the EPI as a layer with a certain thickness, permittivity and conductivity was not easily obtained. Martinsen & Grimnes [2] stated that EPI thickness could be expected to be from 1 to 10 nm depending on the concentration of the electrolyte solution.

2. Methods

To establish a FE-model able to simulate the properties of the laboratory needle electrode setup, we first had to determine the main properties of the real setup used in the laboratory. The FE-application Comsol Multiphysics (CM) was used to establish and solve this FE-model.

The real setup consisted of three main parts: active electrode surface (Double layer/polarization impedance and of course the metallic part of the needle), needle insulation and the saline solution.

Because of the low system frequency, we chose to treat the bulk saline in the tank as a pure resistance [3].

When the non-insulated active electrode surface is brought into contact with biological tissue or an electrolytic solution, electrochemical phenomena will occur and can be observed as electrode polarisation impedance (EPI).

The main challenge in modelling EPI was the small dimensions of geometry. Grimnes & Martinsen describe [2] that layer thickness approximately equals the Debye-length, meaning that small dimensions which will give an unnecessary complicated FE-model. To get a functional FE-model we had to set the FE EPI-layer as a thicker layer, but with electrical properties corresponding to the actual EPI-layer. This resulted in a FE-model with a layer thickness of 0.5 μm with an adapted conductivity/permittivity giving the correct values for the conductance/susceptance. The electrical properties of the EPI-layer were given by the measurements on the LAB-model.

The saline solution was modelled as a cylinder with height 35 mm and radius of 52.5 mm. Anything between this cylindrical boundary and the needle was defined as saline (conductivity σ = 1.3 S/m [2]).

We defined the needle insulation as a cylindrical capacitor with the Teflon insulation as the dielectric. The capacitance will then depend on the insertion depth according to the area of insulation exposed to the saline solution.

The needle "Disposable Monopolar Needle Electrode, 37 x 0.33 mm, Medtronic Inc, Minneapolis, US" had an active electrode area of 0.3 mm^2. The thickness of this insulation was estimated to 26 μm based on information from the manufacturer.

Figure 1 The needle (37 X 0,33 mm) [1]

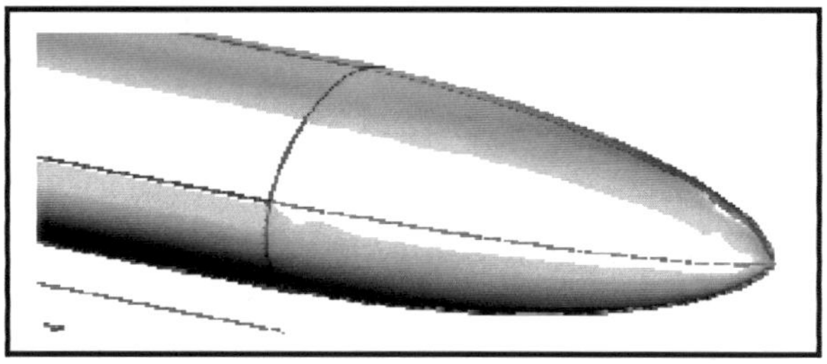

Figure 2 The needle electrode geometrical model

Figure 3 the complete model. The needle tip is marked with a square. The contents of this square can be seen in the next figure.

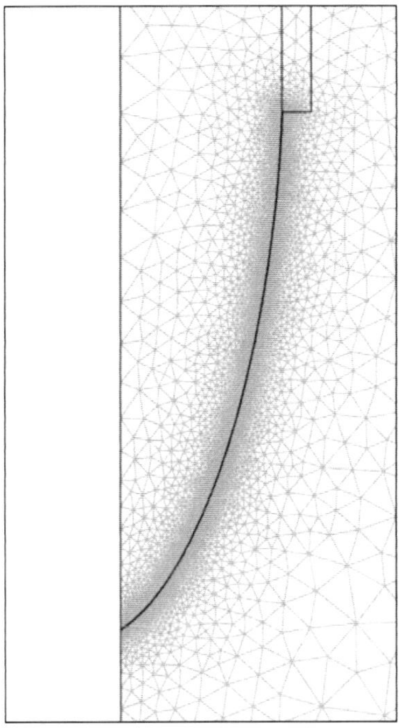

Figure 4 A magnified section of the mesh from the proximity of the needle. Due to the small dimensions the 0,5 μm EPI layer cannot be seen on this figure.

The two previous figures show the axis-symmetric FE-model. This model is a 2D-model until the set of valid equations are to be solved. CM then revolves the model 360 degrees and the solution becomes a 3D-solution.

A series of FE-simulations were made by reducing the distance between the needle tip and the bottom of the saline tank. Model impedances were recorded and compared with the measurements on the LAB-model from Kalvøy & al [1].

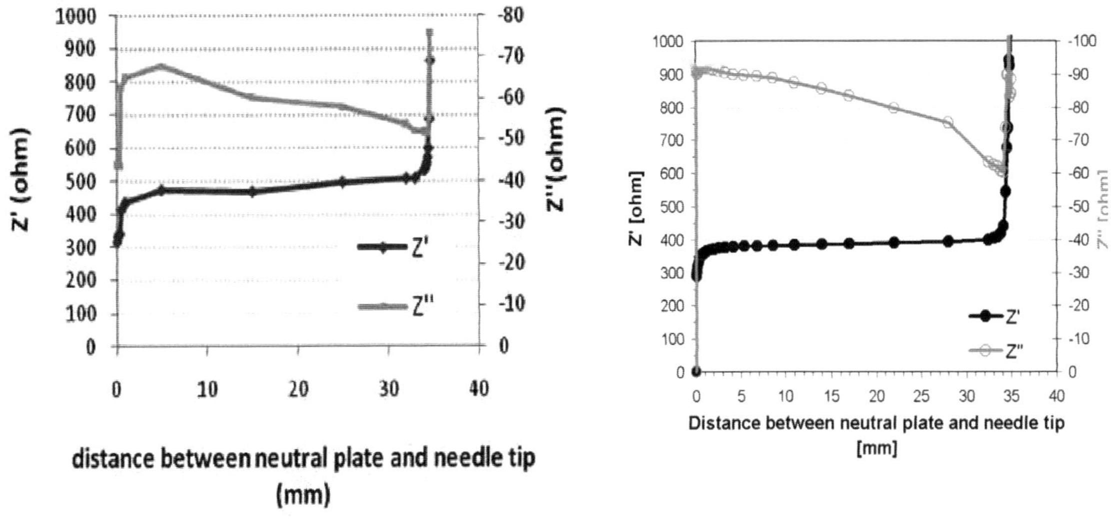

Figure 5 Comparation of FE-model and Lab model. Z' and Z". FE-model to the left.

3. Results and Discussion

The recorded current density vectors and potentials from the FE-simulations compared well with the measurements on the LAB-model. Calculation of the sensitivity zone was done by plotting the potential distribution. This showed that the sensitivity zone for this needle electrode in saline solution had a spherical shaped sensitivity zone with a radius of approx 3.75 mm. This figure is in accordance with the observations from the LAB-model measurements when the needle either approaches the bottom of the saline tank or the surface of the saline solution.

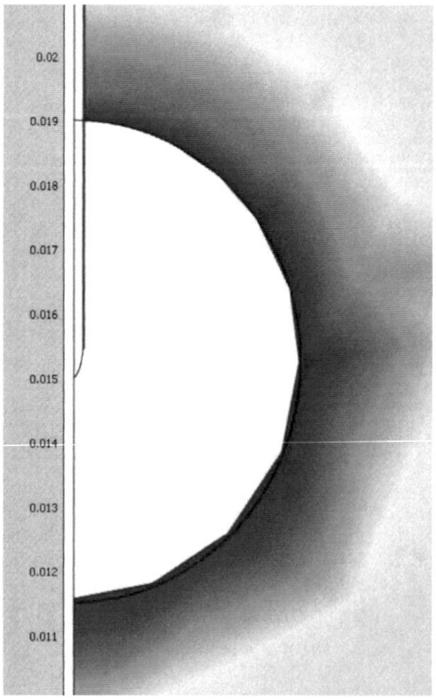

Figure 6 "97 %-zone" i.e. potentials and sensitivity field. This figure shows that the sensitivity field is a sphere with radius r=3.75 millimeters.

Some deviances between the FE-model and the LAB-model are observed and closely related to the necessary simplifications of needle electrode geometry in order to establish a well performing FE-model. The main reason is the simplifications of the EPI-thickness and adaption. This will be subject to further work.

4. Conclusion

It has been shown that sensitivity field in a measuring system involving a needle electrode can be modelled, solved and described with a Finite Element model. In this system sensitivity field has been estimated to be a spherically shaped zone with a radius of approximately 3.75 mm from needle metal surface. This result complied with the measurements on the needle-lab-model.

References

[1] Kalvøy H, Frich L, Grimnes S, Martinsen ØG, Hol PK and Stubbhaug A 2009 Impedance based tissue discrimination for needle guidance *Physiol. Meas.* **30**, 129-140

[2] Grimnes S, Martinsen ØG (2008) Bioimpedance & Bioelectricity Basics, Academic Press, San Diego

[3] Cooper R 1946, The electrical properties of salt-water solutions over the frequency range 1-400 Mc/s, J inst Elect Eng, 93, 69-75

Comparison of Dry-Textile Electrodes for Electrical Bioimpedance Spectroscopy Measurements.

J C Márquez [1,2], F Seoane[1,2], E Välimäki[3,4] and K Lindecrantz[5]

[1]School of Engineering, University of Borås, SE-501 90 Borås, SWEDEN

[2]Department of Signal & Systems, Chalmers University of Technology, SE-41296, Gothenburg, SWEDEN

[3]Swedish School of Textiles, University of Borås, SE-501 90 Borås, SWEDEN

[4]Institute of Fibre Material Science, Tampere University of Technology, Tampere, 33720, FINLAND

[5]School of Technology and Health, Royal Institute of Technology, SE-141 52 Huddinge, SWEDEN

E-mail: juan_carlos.marquez@hb.se

Abstract. Textile Electrodes have been widely studied for biopotentials recordings, specially for monitoring the cardiac activity. Commercially available applications, such as Adistar T-shirt and Textronics Cardioshirt, have proved a good performance for heart rate monitoring and are available worldwide.

Textile technology can also be used for Electrical Bioimpedance Spectroscopy measurements enabling home and personalized health monitoring applications however solid ground research about the measurement performance of the electrodes must be done prior to the development of any textile-enabled EBI application.

In this work a comparison of the measurement performance of two different types of dry-textile electrodes and manufacturers has been performed against standardized RedDot 3M Ag/AgCl electrolytic electrodes. 4-Electrode, whole body, Ankle-to-Wrist EBI measurements have been taken with the Impedimed spectrometer SFB7 from healthy subjects in the frequency range of 3kHz to 500kHz. Measurements have been taken with dry electrodes at different times to study the influence of the interaction skin-electrode interface on the EBI measurements.

The analysis of the obtained complex EBI spectra shows that the measurements performed with textile electrodes produce constant and reliable EBI spectra. Certain deviation can be observed at higher frequencies and the measurements obtained with Textronics and Ag/AgCl electrodes present a better resemblance.

Textile technology, if successfully integrated it, may enable the performance of EBI measurements in new scenarios allowing the rising of novel wearable monitoring applications for home and personal care as well as car safety.

1. Introduction

Textile technology might play an essential role enabling home and personal healthcare systems and applications. Measurements of Electrical Bioimpedance (EBI) are being uses for cardiovascular monitoring[1], body composition assessment [2] and other monitoring applications that could benefit significantly from smart textile technology. The reliability of physiological measurements obtained with textile-enabled measurements systems must be ensured prior obtaining any satisfactory optimal integration of textile technology in a health-related measurement system.

Recent developments in textile technology have made available textile electrodes [3] that can function as sensors in non-invasive physiological measurements[4]. In an EBI measurement the electrode has a determinant role in the system due to its double function in the system: as potential sensing elements and as electrical charge interface between the measurement system and the body. The absence of an electrolyte compound in the composition of textile electrodes may impede the charge transfer from the current injecting terminals into the body, thus affecting the measurement.

In this work the performance of two different kinds of dry-textile electrodes from two manufacturers are compared against the performance of typical Ag/AgCl electrolytic electrode, observing the possible influence of time in the obtained EBI complex spectra.

2. Material & Methods

In this study ankle-to-wrist EBI spectroscopy measurements have been done on three healthy subjects lying supine in a resting state at different times, using two types of textile electrodes and conventional electrolytic electrodes. The EBI spectra obtained with textile and electrolytic electrodes have been studied in order to assess their resemblance.

2.1. Electrodes

For this comparative, three different types of electrodes have been used: textile bracelets, wrist cuffs both shown in Figure 1 A) and B) respectively and sticky electrolytic Ag/AgCl pads.

2.1.1. Textile bracelet.- The bracelet has been custom made with a width of 2.5 cm and adjustable length. The material of the inner surface is the electrode sensor manufactured by Clothing+ with synthetic wrap knitted textile material with silver fibre as a conductive element. The outer material of the bracelet is knitted cotton with spandex and Velcro for fastening it around the wrist and ankle.

2.1.2. *Wrist/Ankle Cuffs Electrode.-* The cuffs are manufactured by Textronics Inc. and made of polyamide (nylon) 15%, conductive fibres 30%, Spandex 20% and polypropylene 35%. The conductive textile material is knitted in the inner surface of the cuff electrode.

2.1.3. Electrolytic electrode.- The electrode is a repositionable electrode of the RedDot series manufactured by 3M with conductive and adhesive hydro gel. The electrode patch is rectangular with an area of 10.1cm^2 a snap-button connector.

2.2. Measurements & Analysis

EBI measurements with textile electrodes have been taken at four different times in a time window of 12 minutes with the Impedimed SFB7 spectrometer in the frequency range 3-500 kHz. The obtained EBI measurements have been compared with a final measurement taken with electrolytic electrodes. The comparison, based in a relative error analysis, has been made studying the deviation observed from the complex spectral values obtained with the electrolytic electrodes.

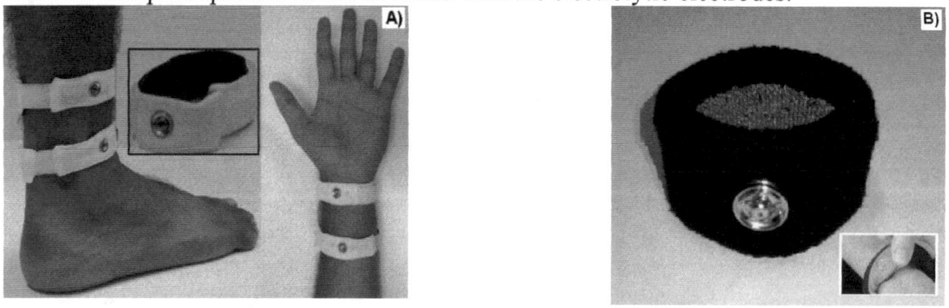

Figure 1. Textile electrodes. Textile bracelet in A) and Wrist cuffs in B).

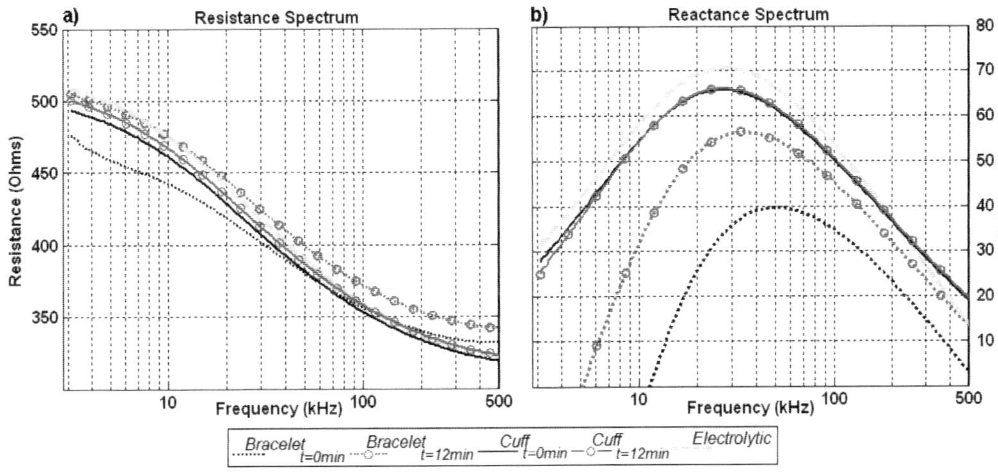

Figure 2. Spectra obtained from the EBI measurements with both textile electrodes at different times. The spectrum obtained with the electrolytic electrode is shown as a reference value.

3. Results

3.1. Complex EBI Spectrum

Figure 1 contains the EBI spectra obtained with both textiles, dotted trace for the bracelet and continuous trace for the cuff, at two different times indicated by a circular marker. The EBI spectra obtained with electrolytic electrodes is included for comparison purposes with dashed trace. In the case of the resistance spectrum, depicted in Figure 2.a), the only spectrum that is remarkably different from the rest is the one obtained with the bracelet electrodes for t=0. The same EBI measurement taking 12 minutes later exhibit a better resemblance. In the case of the reactance spectrum, depicted in Figure 2.b), only the measurements obtained with the elastic cuff electrodes produce an EBI spectra that closely resemblances the spectrum obtained with the electrolytic electrodes.

While Figure 1 contains the obtained spectra from the measurements on a single subject, Figure 2 presents the averaged relative error of complex impedance spectra. The error has been calculated considering the spectra obtained with the electrolytic electrode as reference.

It is clearly observed in Figure 2, that the deviation from the reference value obtained with the cuff electrodes is much smaller than the deviation obtained with the bracelet electrodes. The deviation obtained with the cuff electrodes exhibit also a smaller frequency dependency.

In Figure 3, it is also possible to observe certain time dependency. The relative error obtained with the bracelets exhibit a denoted time dependency that suggests that the error decreases with time. In the

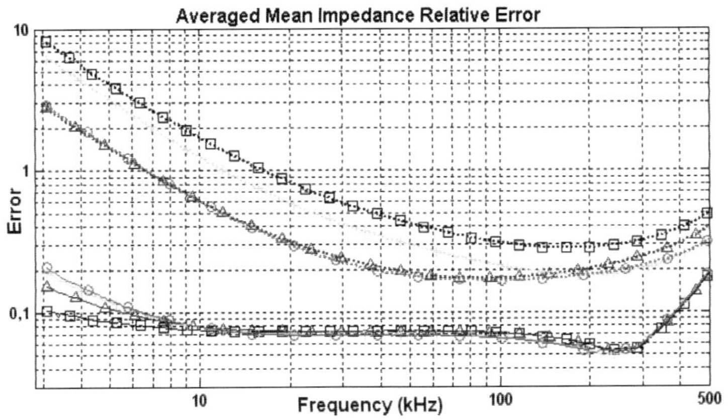

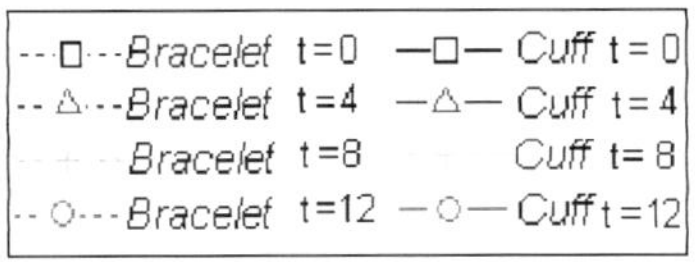

Figure 3. Mean relative error of the resistance and reactance averaged among three subjects for both textile electrodes at different times

Table 1. Averaged Sum of the Relative Error of the Resistance and Reactance Spectrum for both Textile Electrodes at Several Times.

RESISTANCE SPECTRUM							REACTANCE SPECTRUM					
Subject 1		Subject 2		Subject 3		Time	Subject 1		Subject 2		Subject 3	
Bracelet	*Cuffs*	*Bracelet*	*Cuffs*	*Bracelet*	*Cuffs*	(min)	*Bracelet*	*Cuffs*	*Bracelet*	*Cuffs*	*Bracelet*	*Cuffs*
0,022	0,056	0,128	0,090	0,039	0,022	0	0,259	0,095	6,767	0,125	0,969	0,077
0,019	0,058	0,123	0,080	0,065	0,016	4	0,239	0,102	2,567	0,147	0,151	0,093
0,015	0,056	0,112	0,072	0,045	0,015	8	0,218	0,111	5,585	0,154	0,135	0,092
0,022	0,054	0,115	0,076	0,024	0,010	12	0,218	0,092	2,397	0,187	0,377	0,085
0,020	0,056	0,120	0,080	0,043	0,016	**Mean**	0,234	0,100	4,329	0,153	0,408	0,087

case of the cuff electrodes such trend cannot be confirmed, since in most of the frequency range the deviations obtained are very similar.

The mean impedance relative error depicted in Figure 3 is obtained as the mean of the relative errors for the reactance and the resistance, averaged for the three subjects. Therefore, the differences between resistance and reactance for each of the subjects cannot be evaluated from Figure 3 alone. In order to study the error obtained in the resistance and the reactance as well as the time dependency for each of the subjects, the averaged sum of the relative errors have been calculated for both resistance and reactance spectra specifically at four different times for both textile electrodes. The obtained values are presented in Table 1.

In Table 1, the averaged sum values of the resistance and reactance relative errors are presented. From this table is possible to observe that in terms of the averaged sum of relative error the deviation obtained from the resistance spectrum is smaller than the produced in the reactance. The deviation obtained from de resistance measurements with both type of textile electrodes follow the same general trend seen in Figure 1, time reduces the obtained deviation. Therefore the averaged sum of the relative error for the measurements taken after minute 12 are smaller than the ones obtained at t=0. In the case of the reactance spectrum, the bracelet electrodes follow the general trend previously indicated while the deviation obtained with the cuff electrodes does not exhibit any specific trend.

When comparing the deviation obtained with both type of electrodes, it is easy to notice that for all cases with the exception of one, the deviation obtained with the cuff electrodes is smaller.

4. Discussion

The results obtained showed that the EBI measurements done with both textiles electrodes present reliable and repetitive measurements for the resistance spectrum. The error obtained with the bracelet electrodes for the reactance spectrum is very high while the reactance spectra obtained with the cuffs present exhibit a relatively low deviation.

Certain improvements can be observed on the performed spectroscopy measurements. In the case of the measurements taken with the bracelets an improvement can be observed especially in the reactance spectrum, where the initial error is significantly high. In the case of the measurements obtained with the electrode cuff such improvement cannot be notice, but this can be due to the fact that the error is much smaller. Such a smaller error can be due to the fact that the textile sensor used in the cuff exhibits a higher conductivity and the surface of the conductive sensor is also much larger than in the case of the bracelets. Such higher conductivity combined with a larger surface may improve significantly the electrode-skin interface.

5. Conclusion

The results obtained suggest a general good performance of textile electrodes when measuring resistance. The reactance spectrum obtained with the bracelets is not reliable enough while the reactance spectrum obtained with the cuff electrodes is more reliable. Moreover to be able to assess on their reliability for EBI applications for body composition further analysis must be done, especially regarding the estimation of the Cole parameters like in [5]. With Home monitoring applications in mind, further improvements in the electrode ergonomics and textile design should be done aiming to increase their usability.

The cuff electrodes apparently exhibit a better performance and stability in time and frequency than the bracelet electrodes. As this study gives no clue to whether the difference is due to material or design of the electrode further studies may point out ways to improve the performance of textile electrodes for recording of Electrical Bioimpedance.

In any case the resistance spectra from the EBI measurements obtained with the textile and electrolytic electrodes exhibit a close resemblance. Recent work done on Cole parameters estimation from EBI measurements suggest that EBI applications using Cole model-based analysis can be implemented by only measuring the spectrum of the reactance [6]. Therefore if the latter it is confirmed the implementation Textile-enabled EBI application is not so far ahead.

References

[1] Habetha J 2006 The MyHeart project--fighting cardiovascular diseases by prevention and early diagnosis *Int. Conf Proc IEEE Eng Med Biol Soc* **Suppl**: p. 6746-9

[2] Kyle UG, et al. 2004 Bioelectrical impedance analysis-part II: utilization in clinical practice *Clin Nutr* **23**(6): p. 1430-53

[3] De Rossi D and Lymberis A 2005 New generation of smart wearable health systems and applications *IEEE Trans Inf Technol Biomed*, **9**(3): p. 293-4

[4] Paradiso R, Loriga G, and Taccini N 2005 A wearable health care system based on knitted integrated sensors *IEEE Trans Inf Technol Biomed*, **9**(3): p. 337-44

[5] Marquez JC, et al. 2009 Textile electrodes in electrical bioimpedance measurements - a comparison with conventional Ag/AgCl electrodes *Int Conf Proc IEEE Eng Med Biol Soc*, **1**: p. 4816-9

[6] Ayllon D, Seoane F, and Gil-Pita R 2009 Cole equation and parameter estimation from electrical bioimpedance spectroscopy measurements - A comparative study *Int Conf Proc IEEE Eng Med Biol Soc*, **1**: p. 3779-82

Bioimpedance microelectronics in a 24-microwell plate with metabolic-sensors for testing chemosensitivity of tumor cells and tissues

T. Schwarzenberger[1,2], F. Demmel[1,2], B. Becker[1,2], M. Zottmann[1,2], P. Wolf[1,2], R. Kleinhans[1], M. Brischwein[1] A. Otto[1] and B. Wolf[1]
[1]Heinz Nixdorf-Lehrstuhl für Medizinische Elektronik, TU München, Germany
[2]HP Medizintechnik GmbH, Oberschleißheim, Germany

E-mail: schwarzenberger@tum.de

Abstract. Living cells react to external influences such as pharmacological agents in an intricate manner due to their complex internal signal processing. Cell reactions are an impact on vitality, cell-cell or cell-matrix interaction and morphological changes. A number of published techniques on impedance spectroscopy (IS) of adherent cells with planar electrodes address these changes. However, IS can merely serve as an indicator of cellular events rather than provide detailed information on a specific cell process. Thus our approach is a 24-microwell sensor-plate with impedance-electrodes in parallel to pH- and O_2-sensors, capable of being integrated into a fully automated screening system. For the purpose of IS, high precision impedance-electronics have been developed based on integrated circuits and validated against a Solartron 1260 impedance analyzer. IS data is correlated to the metabolic-sensors and additionally compared with cell images shot by an inverse optical microscope which is also part of the screening system. Proof of principle is demonstrated by experimental growth monitoring of a MCF-7 culture and cellular response to chemotherapeutics. Furthermore, the potential to monitor living tissue probes is presented for the first time.

1. Introduction to IS for label-free cellular monitoring in cell culture plates

In standard assays, the cellular response is determined by end-point methods usually involving labeling and optical methods. However, cell labeling is expensive, it is a possible source of artifacts, and long-term monitoring is hard to realize. Thus, the kinetics of an effect remains unclear in many cases and recovery effects may completely escape detection. For these reasons impedance spectroscopy (IS) appears to be a suitable method for monitoring cultured cells label-free and in real-time. A significant scientific interest arose when a first report on IS in cell culture appeared in 1984 [1]. This led to commercialization of products which are now in some extent automated screening systems.

A brief review of three systems is provided by [2]: ECIS [3], RT-CES [4] and CDS [5] are purely impedance based technologies utilizing different electrode arrangements, materials and measurement strategies. Each of these systems shares basic principles in IS [6]: In a parallel approach, a plausible physical model has been constructed and correlated with the measured impedance data. When based on a profound theory, mathematical models or electrical equivalent circuits can be parameterized. When modeling cultured cells on electrodes, one has to account for electrode impedance in series with the impedance contribution of the cellular layer. A lot is known on both domains, but one or the other is often neglected. Although modeling is thought to be descriptive, combining these impedance fractions increases complexity enormously. Since these systems suffer from data indexation, limited bandwidth or even predefined frequencies, they remain unique and studies are not easily comparable.

The concept of an outright novel screening system called "intelligent microwell-plate reader" (IMR), in which an IS module is embedded, has previously been published elsewhere [7,8]. It has been clearly shown that micrometabolic sensor technology and image documentation provide compounded insight into distinct aspects of cellular life and cellular responses, which can be correlated with data about cell morphological changes obtained by impedance measurements. The development of miniaturized impedance electronics to address planar electrodes in microwell-plates is essential, since there is no measurement apparatus available, which is sufficiently packaged to integrate all system components in a small unit, meets the needs for non-invasive cellular measurements and provides complete impedance information over a large frequency range.

2. Materials and Methods

Core of the IMR (figure 1) is a sensor-equipped microwell-plate (MWP). Together with the IMR system components, this plate is placed into a standard cell culture incubator. Integrated microelectronics measure the complex cellular impedance with planar interdigitated electrodes (IDES, figure 2). This information is correlated with data on cell metabolic rates retrieved with optochemical pH- and pO_2-sensor spots. These are read out by a moveable sled-mounted process-microscope in parallel to image capturing. Cell culture media (and drug solutions) are regularly replaced by an automated pipetting-robot. A central control unit ensures long-term experiments up to days and weeks.

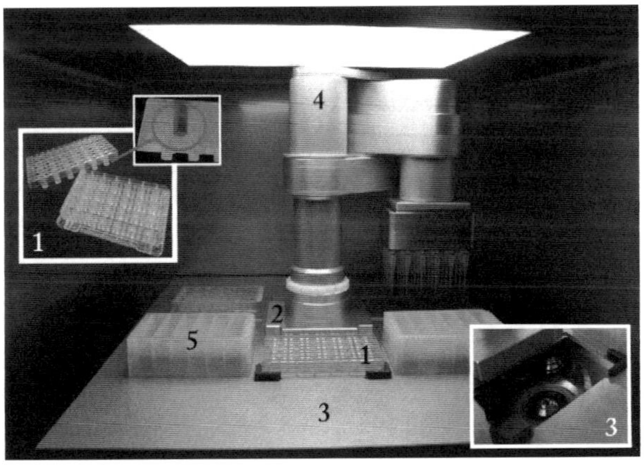

Figure 1. Intelligent Microwell-plate Reader (IMR): Microwell-plate with cover plate (1), impedance electronics (2), inverse microscope (3), pipetting robot (4) and cell-culture medium reservoirs (5).

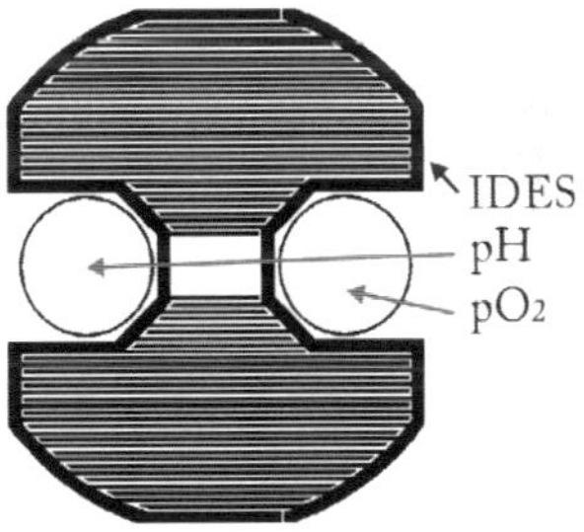

Figure 2. Interdigitated electrode structure with optochemical sensor spots and window for microscopy.

2.1. *Microwell-plate*

The MWP is a composed of a 24-well polymer corpus and a glass bottom plate with 24 IDES and optochemical sensor spots. Each cell culture well has a diameter of 7 mm. IDES are fabricated by thin film structuring of Pt, both electrode width and distance are 50 µm. Two side chambers connected to the culture well at the bottom serve as access points to the pipetting robot. Equilibration of hydrostatic pressure differences thereby result by medium flowing through the culture well. A lid with inserts to each well confines the assay volume to 23 µl. For the first time, Au-plated inserts are utilized for vertical impedance measurements against bottom IDES in cell culture wells (figure 1).

2.2. *Microelectronics*

Novel high precision IS microelectronics have been developed on the basis of the AD5933 impedance converter IC from Analog Devices. This impedance analyzer in a chip derives complex impedance by a DFT-algorithm from a sinusoidal stimulus signal. Measurement accuracy of 0.5% could be attested from 5 kHz to 100 kHz for high impedances greater than 1 kΩ. To overcome bandwidth limitations, to be able to measure small impedances and to avoid interference with other electrical sensors, several modifications have been implemented in our design (figure 3). These include flexible clocking, galvanic decoupling and the development of an analog frontend. Because a true two-electrode measurement is performed on IDES, stimulus voltage and resulting current are regulated well below

26 mV or 4 mA/cm^2, respectively, in order to monitor a quasi-linear system [9]. As a result, impedance spectra of 24 wells from 500 Hz to 100 kHz can be measured within 40 s.

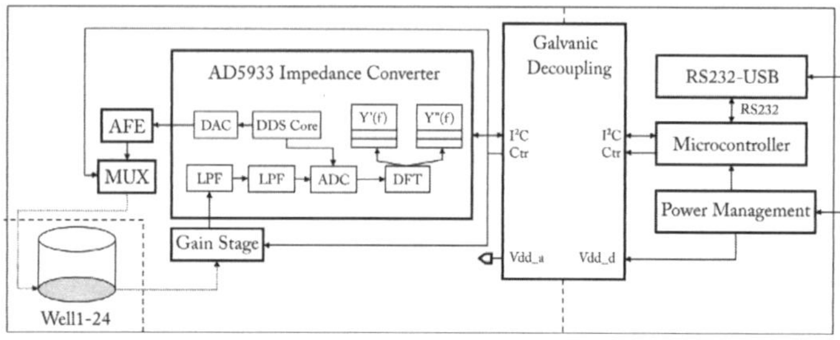

Figure 3. AD5933 bioimpedance electronics peripheral block diagram

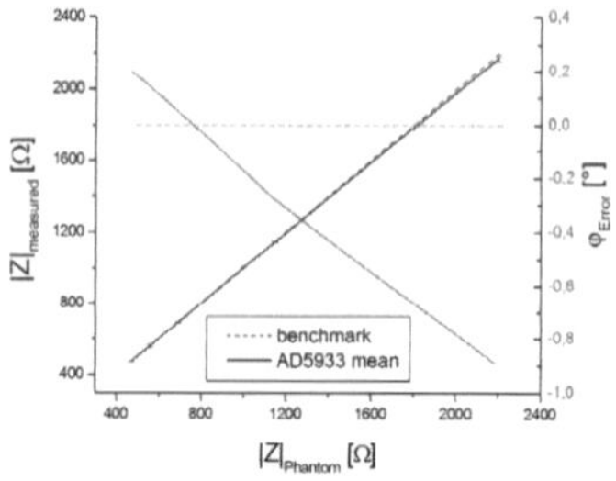

Figure 4. System accuracy (10kHz)

As a gold-standard the frequency-response analyzer Solartron 1260, which yields 0.1% accuracy within a frequency range from 10 μHz up to 32 MHz, was chosen. System accuracy was determined to be less than 2% for either magnitude or phase with the use of R-C impedance phantoms as compared with the reference (figure 4). Dummy circuits were therefore assembled based on experimental results to match true cell impedance in the frequency bandwidth of interest. Working bandwidth of 50 μm Pt-IDES was found to be from 100 Hz to 10 kHz which overlaps with the operation range of the analyzers. Moreover, β-dispersion of biological tissue is expected in the same region [10].

2.3. Cell measurements

MCF-7 cells (human breast cancer cells) were inoculated into the central chamber of the MWP with 500 μl Dulbecco's Modified Eagle Medium containing 5% fetal calf serum. Since a medium with reduced pH-buffer capacity was used for enhanced pH-sensitivity, the cell culture medium was exchanged in 10-20min cycles. Impedance data was interpreted in magnitude $|Z(\omega)|$ and phase $\varphi(\omega)$. For simple detection of events, change in impedance $\Delta|Z(\omega)|=[(|Z(\omega)|-|Z_0(\omega)|)]/|Z_0(\omega)|$ was the normalized cell impedance, with Z_0 being the impedance of pure medium. As experimental compounds, cytochalasin B and cisplatin were added with the pipetting robot from a separate medium reservoir. Cytochalasin B is an alkaloid drug known to inhibit glucose import into the cell (thereby altering the cell metabolic profile) and to act on actin filaments in the cellular cytoskeleton. Cisplatin is a cytostatic drug used for cancer chemotherapies. Addition of the non-ionic detergent Triton X-100 yields baseline values of sensor date validates reversibility and correctness of the signals.

3. Results and Discussion

Experiments were performed to monitor cellular growth on electrodes (figure 5) whilst verifying the proliferation process with microscopy (figure 6). After attaining a confluent cell layer, cytochalasin B was applied for three hours and withdrawn thereafter to investigate the impact on cell morphology and potential for recovery. In a second assay, cells were treated with cisplatin to compare the effect on cell morphology with the effect on cell metabolism (figure 7). To examine the responses of cells within their native 3D tissue architecture, pieces of explanted sheep pancreas were analysed within the system (figure 8).

The results show the successful integration of IS microelectronics for non-invasive cell culture monitoring and its benefit compared to established laboratory methods. Parallel live-cell imaging is a valuable tool for the validation of sensor data interpretation. As a next step, it is envisaged to use sensor data for image acquisition. Furthermore the attempt of applying vertical IS in culture-wells appears promising for monitoring cultured tissue probes. In conjunction with metabolic monitoring, a more complete picture of cellular physiology and its responses to experimental treatment is obtained. A particular focus of application is the use of the system for improving and individualising cancer chemotherapies by on-chip cell culturing systems.

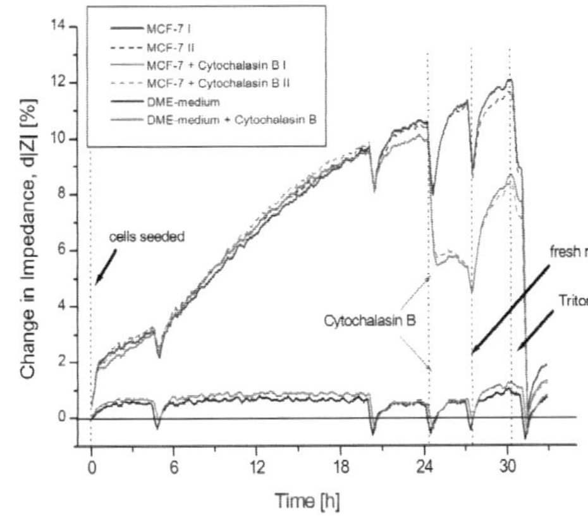

Figure 5. Parallel assays of MCF-7 cells (8x10^4 per well) monitoring proliferation and morphological changes after addition and removal of 6μM cytochalasin B.

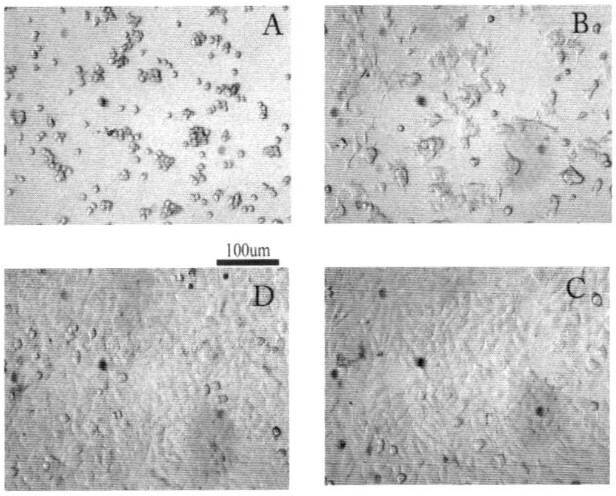

Figure 6. Automated picture documentation: Cells after seeding (0h, A) and spreading (5h, B), confluent cell layer (20h, C) and cells after agent addition (27.5h, D).

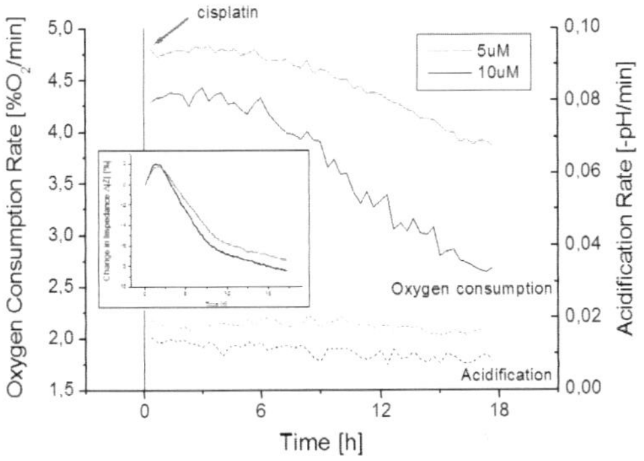

Figure 7. Metabolic reaction in comparison to the change in impedance during long-term treatments of MCF-7 assays with 5μM and 10μM cisplatin.

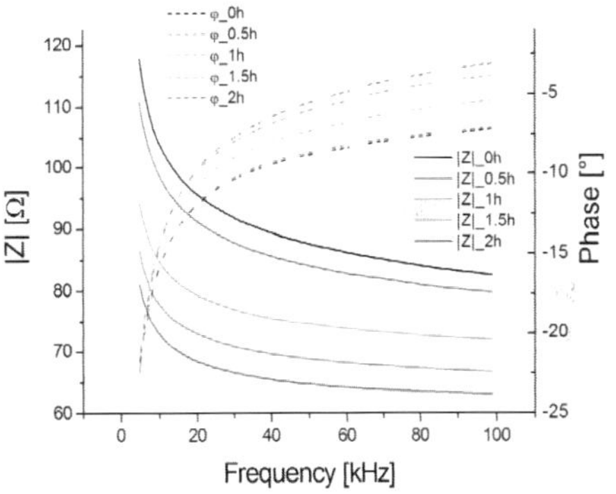

Figure 8. Triton X-100-induced cell destruction of sheep pancreas tissue in a microwell-plate: Impedance spectrum vertically recorded against gold-plated lid.

References

[1] Giaever I and Keese C 1984 *Proc. Nat. Acad. Sci.* **81** 3761
[2] Cooper MA 2006 *Drug Discovery Today* **11** 1068
[3] Giaever I and Keese C 1995 *Biophysiological Journal* **69** 2800
[4] Abassi Y 2008 *Journal of Biomolecular Screening* **10(8)** 796
[5] McGuiness R 2007 *Current Opinion in Pharmacology* **7** 535
[6] Barsoukov E and Macdonald R 2005 Impedance Spectroscopy *Wiley*
[7] Becker B 2008 *IFMBE Proceedings* **20** 1680
[8] Wolf B 2008 *Cell Based Screening* **1** 21
[9] Schwan H P 1968 *Ann. N. Y. Acad. Sci.* 191
[10] Schwan H P 1994 *Proc. 16th Annu. Int. Conf. IEEE Eng. i. Med. a. Biol. Soc.* **1** A70–A71

Non-destructive determination of impedance spectrum of fruit flesh under the skin

E Vozáry[1,3] and P Benkő[2]

[1] Department of Physics, Faculty of Food Science, Corvinus University of Budapest, Somlói st. 14-16. H-1118 Budapest, Hungary

[2] Faculty of Electrical Engineering and Informatics, Budapest University of Technology and Economics, Műegyetem rkp. 3-9. H-1111 Budapest, Hungary

eszter.vozary@uni-corvinus.hu

Abstract. Impedance spectrum of fresh (intact) apples and of artificially bruised (pressed) apples was determined on the surface of skin with ECG electrodes (Fiab Spa). The magnitude and the phase angle of impedance were measured with a HP 4284A precision LCR meter. The open-short corrected spectra were approached a model consisting of serial resultant of an ohmic resistance and three distributed elements. Approach was performed with complex non-linear least squares method by MathLab program. Variance analysis was performed ($P<0.05$) on impedance parameters (SPSS 12.0 for Windows). Parameters of the first distributed element can describe the impedance of apple skin, and parameters of the second and the third element can characterize the impedance of apple flesh. Parameters of the second and the third element are in good agreement with impedance parameters obtained from spectra measured directly on apple flesh without skin. The value of parameters resulted from measurement on apples with skin are sensitive to the degree of artificial bruises.

1. Introduction

The electrical impedance spectrum – in low frequency range – of biological tissues depends on the state of cellular structure [1], therefore the parameters evaluated from measured spectrum can be used for quality assessment of fruits and vegetables [2]. In the majority of works the impedance is measured with pin electrodes [2], [4], [5] and there are few measurements with non-destructive electrodes [3].

In this work an attempt is made at determination of impedance spectrum of apple flash from the spectrum measured on the surface of the whole apple covered with skin. The impedance parameters obtained from this mathematical method are compared with impedance parameters resulted from direct measurement on the flesh tissue without skin. The sensitivity of apple flesh impedance parameters to the level of mechanical injury of tissue under the skin is also investigated.

2. Materials and methods

Idared apples were purchased on the local market. The impedance spectrum on the surface of intact whole objects with skin was measured. After this the apples were peeled and the impedance of the flesh was direct measured at the same place, where the impedance on skin was measured.

[3] To whom any correspondence should be addressed.

The bruises of apples under the skin were artificially established. Apple flesh named as "pressed" was caused by press of apple skin with force less than force belonging to the bioyield point. Under such low force only the shape of cells changes and the cell membrane remains unbroken [6]. "Pulpy" apple flesh was caused by pressing force much higher than force at bioyield and lower than force at rupture point. In this case the cell membrane is broken and the intercellular and intracellular parts are mixed.

Impedance of ten apples from each group – intact, pressed and pulpy with and without skin– was determined. The magnitude and the phase angle of impedance were measured with an HP 4284A precision LCR meter in a frequency range from 30 Hz up to 1 MHz. The measuring voltage was 1 V. The measured spectra were open and short corrected, to eliminate the stray capacities and inductances. Fiab Spa Ag/AgCl ECG electrodes with 10 mm diameter were used. The electrodes were touched to apple skin or to apple flesh, and the distance between the centres of electrodes was 2 cm. A special Signal gel was put between the electrodes and the skin of investigated objects for the good electrical contact.

The corrected impedance spectra of whole objects with skin were approached with an electrical model consisting of serial resultant of three distributed elements and one ohmic resistance

$$R+\frac{R_1}{1+(i\tau_1\omega)^{\psi_1}}+\frac{R_2}{1+(i\tau_2\omega)^{\psi_2}}+\frac{R_3}{1+(i\tau_3\omega)^{\psi_3}}, \quad (1)$$

where R, R_1, R_2 and R_3 are resistances, τ_1, τ_2 and τ_3 are relaxation times, ψ_1, ψ_2 and ψ_3 exponents characterize the distribution of relaxation times, $\omega = 2\pi f$ and f is the measuring frequency and $i=\sqrt{-1}$ is the imaginary unit. A model consisting of only two distributed elements and one resistance was fit to the spectra of apple flesh. The approaching procedure was performed with complex non-linear least squares (CNLS) method by MathLab program.

Variance analysis, a Dunnett T3 test was performed (P<0.05) on impedance parameters (SPSS 12.0 for Windows, SPSS Inc., Chicago, IL, USA).

3. Results and discussion

The measured and the corrected spectra of apple with skin (Figure 1.) were approached with expression (1). Impedance spectra measured on peeled apples were approached with a model consisting of only two distributed elements in a serial connection with each other and with one ohmic resistance. The first distributed element can characterize the impedance of apple skin. The second and the third elements can give the impedance parameters of the apple flesh tissue. The ohmic resistance, R, gives the resistance of measured object at infinite high frequency.

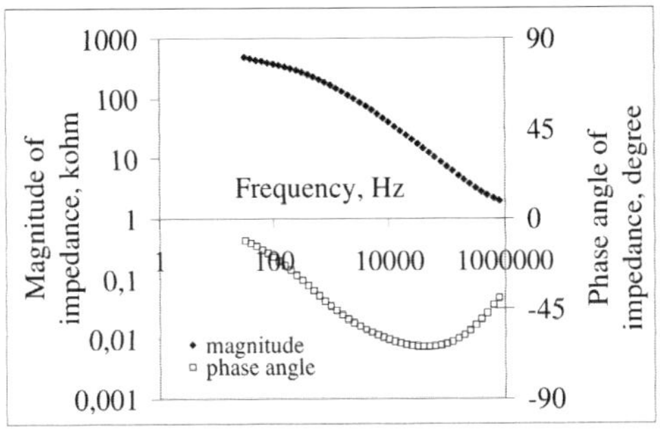

Figure 1. Typical measured impedance magnitude and phase angle on apple with skin

A typical approach of spectrum measured on apple with skin can be seen on the Figure 2. Similar good curve fitting (not shown) was resulted from approaching the impedance of apple flesh without skin, but in this case the calculated impedance contained only two components.

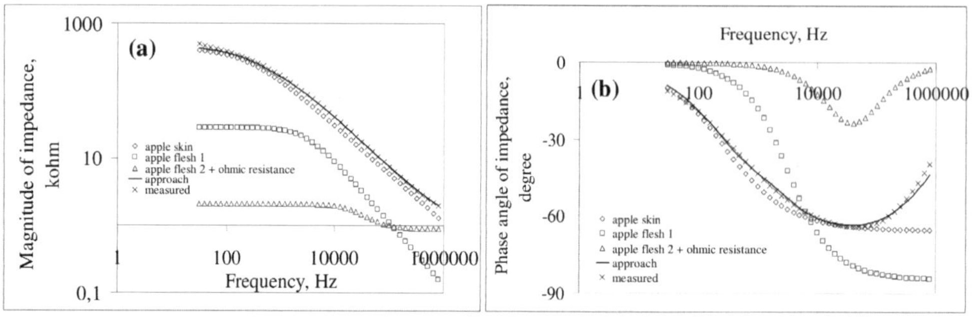

Figure 2. Typical approach of magnitude **(a)** and phase angle **(b)** of measured apple impedance

In intact apples R parameter was higher for apple flesh (Table 1.) than for apple with skin. This difference can be explained by a slightly mixing if intercellular and intracellular parts after peeling the apple. The strong decrease of R value with increasing bruise of apple flesh can be caused by more extensively mixing of inter- and intracellular parts. Practically for the pulpy apple flesh there is not difference between the two R values measured by the two methods.

The resistance and the relaxation time of apple skin decreased as the degree of bruise increased, but the exponent does not depend on it (Table 1.). It seems that the press of apple skin with relative low force can cause structure changes in it.

Table 1. R parameter and impedance parameters of apple skin (average and standard deviation). The same letter in a column means there is no significant difference (P<0,05) in the values of parameters.

	R (kohm)		R_1 (kohm)	τ_1 (ms)	ψ_1
State of apple flesh	Measured on apple flesh	Measured on apple with skin	Measured on apple with skin		
Intact	2,26±0,54 a	1,05±0,26 b	3500±2204 a	6,15±5,93 a	0,778±0,033 a
Pressed	1,79±0,15 a	0,79±0,90 b	700±184 b	0,50±0,23 b	0,747±0,017 a
Pulpy	0,70±0,24 b	0,61±0,07 b	36,5±35,1 c	0,03±0,03 c	0,751±0,060 a

The resistance, R_2 is a little higher in apple with skin than in apple flesh for the all three groups of apple tissue (Figure 3. (a)). The relaxation time parameter also is higher for apple flesh with skin, than for peeled apple tissue (Figure 3. (b)). The standard deviation is higher for both resistances and relaxation times obtained from approach on apple with skin in comparison with standard deviation of these parameters from spectrum of peeled apples (Figure 3. (a) and (b)).

The resistance, R_3 is higher for peeled apple flesh, than for apple flesh with skin (Figure 4. (a)), but the relaxation time and exponent are higher for unpeeled apple tissue (Figure 4. (b) and (c)). These differences also can be understood by the mixing of intra and intercellular tissues after peeling or after artificial bruise.

4. Conclusion

In conclusion it can be stated that this mathematical procedure resulted similar parameters for apple flesh from measured spectra on apple with skin to parameters from directly apple flesh measurements.

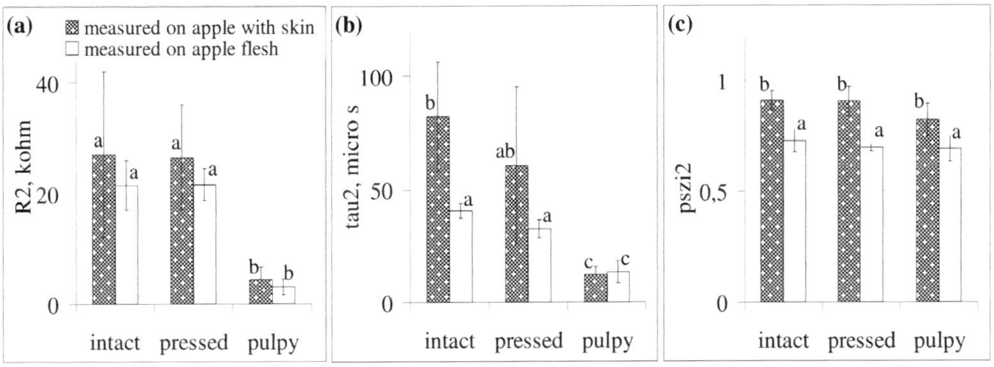

Figure 3. Impedance parameters (average with standard deviation) of second distributed element characterizing the apple flesh impedance measured on apple with and without skin. The state of apple flash was intact, pressed with force lower bioyield and pulpy – pressed with force higher than bioyield. (**a**): Resistance, R_2, (**b**): relaxation time, τ_2, (**c**): exponent, ψ_2 characterizing the distribution of relaxation times. The same letter in each picture means there is no significant difference (P<0.05) in the values of parameter.

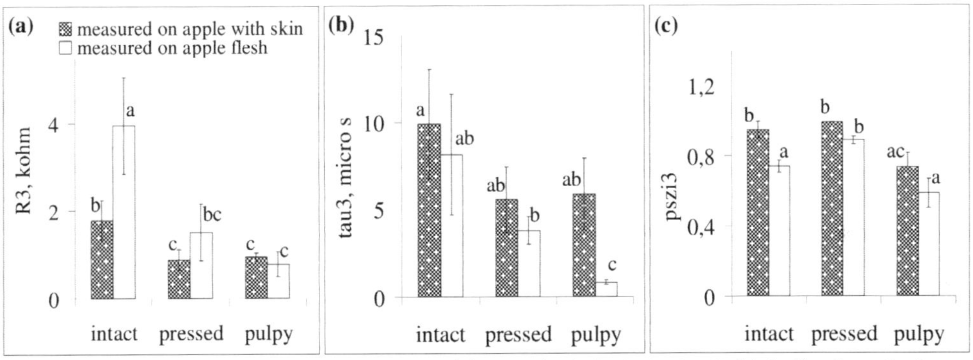

Figure 4. Impedance parameters (average with standard deviation) of third distributed element characterizing the apple flesh impedance measured on apple with and without skin. The state of apple flash was intact, pressed with force lower bioyield and pulpy – pressed with force higher than bioyield. (**a**): Resistance, R_3, (**b**): relaxation time, τ_3, (**c**): exponent, ψ_3 characterizing the distribution of relaxation times. The same letter in each picture means there is no significant difference (P<0.05) in the values of parameter.

The resistance and relaxation time parameters of the second distributed element can be used for characterizing the state of apple flesh.

References

[1] Grimnes S and Martinsen O G 2000 *Bioimpedance and Bioelectricity Basics* (New York, Academic Press)
[2] Harker F R and Maindonald 1994 *Plant Physiol.* **106** 165
[3] Repo T, Paine D H and Taylor A G 2002 *Seed Sci. Res.* **12** 17
[4] Zhang M I N and Willison J H M 1991 *J. Exp. Bot.* **42** 1465
[5] Zhang M I N and Willison J H M 1993 *J. Exp. Bot.* **44** 1369
[6] Mohsenin N N 1986 Physical properties of plant and agricultural material (New York,London, Gordon and Breach)

Spread Spectrum EIT by Code Division Multiplexing

A McEwan[1] J Tapson[2], A van Schaik[1] and D Holder[3]
[1]School of Electrical and Information Engineering, CARLAB
The University of Sydney, Australia, NSW 2006
[2]Department of Electrical Engineering, University of Cape Town,
Cape Town, South Africa.
3 Department of Medical Physics and Bioengineering,University College London,
Gower St, London, UK.
alistair.mcewan@sydney.edu.au

Abstract Code division multiplexing is proposed as a new EIT method that can provide simultaneous impedance measurements of the multiple channels. Code division provides clear advantages of a wide frequency range at reduced cost and reduced complexity of sources. A potential drawback is the lack of perfectly orthogonal code sets. The method described provides images and spectra which are equivalent to the conventional time multiplexed method, with increases in frequency resolution and measurement speed which may be of benefit in some applications of EITS.

1. Introduction

The purpose of this manuscript is to describe a new method of EITS called Code Division Multiplexing (CDM) EITS, and to demonstrate that CDM-EITS is an inexpensive method with higher frequency resolution than current systems. Furthermore, CDM-EITS may be used measure the impedance spectrum from multiple channels simultaneously to form EITS images at a faster rate. The principle of CDM is that a signal through a particular channel is modulated using a unique binary digital code. A receiver which is receiving several channels simultaneously can separate the data for each channel by demodulating the measured signal using a copy of each of the modulation codes. There are a number of different codes which may be used. The basic requirements are that there should be at least one code per channel; that the codes should be orthogonal, or nearly so; and that the autocorrelation functions of the codes should be flat with a single sharp peak (in the ideal case, approximating a delta function) [1]. The codes which we have used in this work are called Gold codes, which are also used in the Global Positioning System (GPS) to encode the signals from the GPS satellite constellation. The codes are bit sequences which appear to be random, but in fact are deterministically generated, usually by means of a modulo-addition of bits in a shift register. Codes of this type are generally referred to as pseudorandom noise (PRN) sequences, as they appear to be random and have the characteristics of noise, including a broad spectral content. For EIT the use of CDM allows us to take measurements over all the channels simultaneously. In addition, for EITS the spectral characteristics of the CDM input signal effectively interrogate the sample over a wide range of frequencies and the output signals can be transformed to produce a spectrum, giving us simultaneous wide-band spectroscopy on all channels. A similar technique has been used previously in a single channel where a maximum length sequence (MLS) was used to detect changes in impedance spectrum in single cell analysis [2]. The use of MLS, but not PRN, in EIT has been proposed in [3], but no working system has been reported. Only a single non-zero MLS exists for any given code length, so for simultaneous measurement we are forced to use larger sets of PRN sequences, such as the Gold codes. These have similar properties to a MLS, but their autocorrelation deviates more from the ideal Kronecker delta function, i.e., they are sets of nearly orthogonal codes. The size of the PRN code set and the orthogonality increases with code length [1], so longer codes are preferred. Longer codes also conveniently increase the resolution and bandwidth of the demodulated spectrum at the cost of a longer acquisition time.

2. Methods

2.1 Serial Imaging of a 16 electrode saline tank

A serial EITS system, the UCL Mk2.5 [4], was reprogrammed with the 1st 1023-chip-length code of a set of eight Gold codes. Only one code was used as the UCL Mk2.5 is based on a single channel of the Sheffield Mk3.5 system multiplexed to 32 electrodes. In normal operation the UCL Mk2.5 system measures the spectrum using a 2048 samples long composite waveform of 10 frequencies of equal amplitude, logarithmically spaced between 2kHz and 1MHz [4,5]. The waveform is normally repeated in 3 epochs with a shift of 1 octave to give 30 frequencies; however to compare with our new method we only use one 10 frequency epoch. The Gold code was sampled at 2 samples/chip to meet the Nyquist criterion. As the lengths of both test waveforms are the same, and the timing of the system clock (2MHz) was unchanged, they span the same frequency range of 2kHz-1MHz. For simplicity, and so that the instrumentation did not saturate, the amplitude of the Gold code waveform was set to the maximum of the composite waveform. Firstly a reference data set is collected with only saline in the tank, by sequentially injecting a current modulated by the Gold code from each electrode while measuring the voltages at the other electrodes. The observed signals were correlated with the Gold code to obtain the transfer functions from each electrode to all other electrodes and the magnitude of the Fourier transform was taken as a measure of the impedance between electrodes as a function of frequency. In a second step, the object is introduced and another data set taken. These two data sets are subtracted which removes small yet significant errors due to geometry and electrode positions. The difference data, which ideally should only represent the impedance changes, is then used in the image reconstruction algorithm to create a 'time difference' image. In keeping with previous EITS measurements we chose to use a piece of banana 2 cm long and 1 cm in diameter located at 8 o'clock and 1cm from the edge of the tank as our object. A piece of banana changes the local impedance in the tank by over 100% between 2kHz and 1MHz. The saline solution's impedance will remain relatively unchanged over the same frequency range and hence it is possible to determine the spectrum of the object inside the tank from a sequence of images at the measured frequencies.The tank was cylindrical with a diameter of 10cm and filled with 0.1% saline solution. It had 16 stainless steel electrodes on its perimeter. We used polar (opposite) current injection and time-difference image reconstruction was done using a linear solver (EIDORS, [6]) and a 15,000 element FEM mesh of the tank.

2.2 Simultaneous Imaging of an 8 electrode saline tank

To investigate if the technique could be used to successfully acquire simultaneous images, a multichannel, time multiplexed, Sheffield Mk3.5 EIT system [5] was reprogrammed to activate two sources simultaneously, either with orthogonal frequencies, or the 1023 chip Gold codes. Three voltmeter channels measured the impedance change from introducing a cylindrical piece of banana, 2cm diameter at 2 o'clock, and 4cm from the edge of a cylindrical tank (of 10cm diameter) which was filled with 0.2% NaCl. The resulting signal was demodulated with each orthogonal frequency, or Gold code, and an impedance image formed at 10kHz and 50kHz.

3. Results

Boundary voltages (figure 1), Serial images (figure 2) and extracted spectra (figure 3a) were obtained with the composite waveform and Gold codes, and simultaneous images were obtained with FDM and CDM (figure 3b).

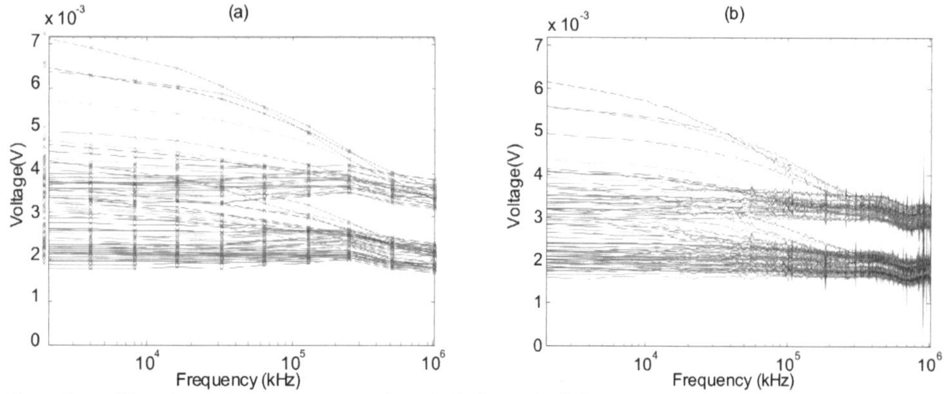

Fig. 1. Comparison of boundary voltage measurements in a single frame. Each line is a different electrode combination measurement. Combinations more sensitive to the perturbation show a greater change over frequency. Figure (a) shows a composite representation of the output of the standard 10 frequency system. Figure (b) shows the output of the new system, using PRN codes for excitation and demodulation.

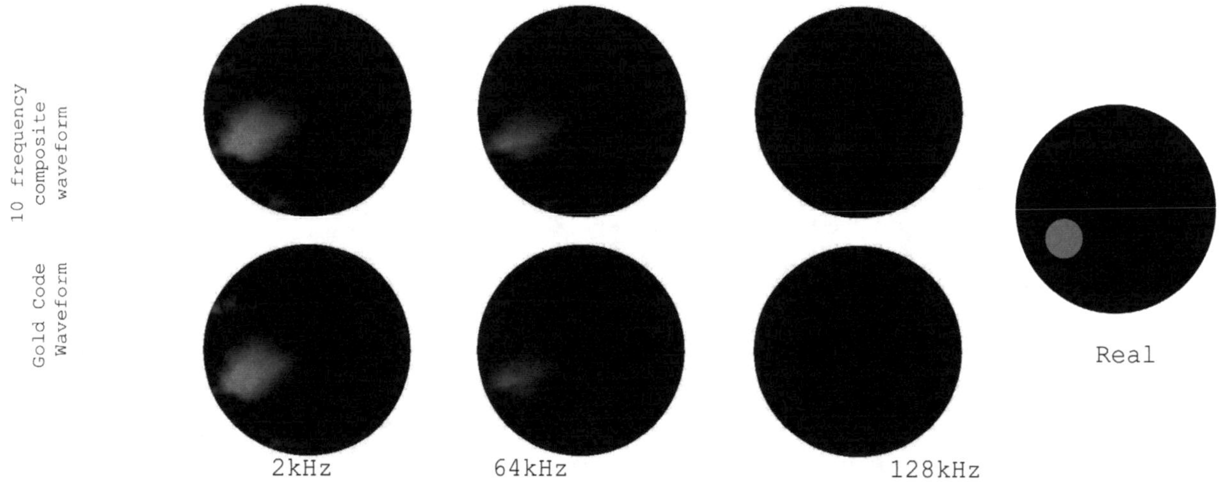

Fig. 2. Serial imaging on a 16 electrode tank. Time difference images collected at three frequencies using the two waveforms. As the frequency increases, the contrast (impedance difference) between perturbation and background decreases.

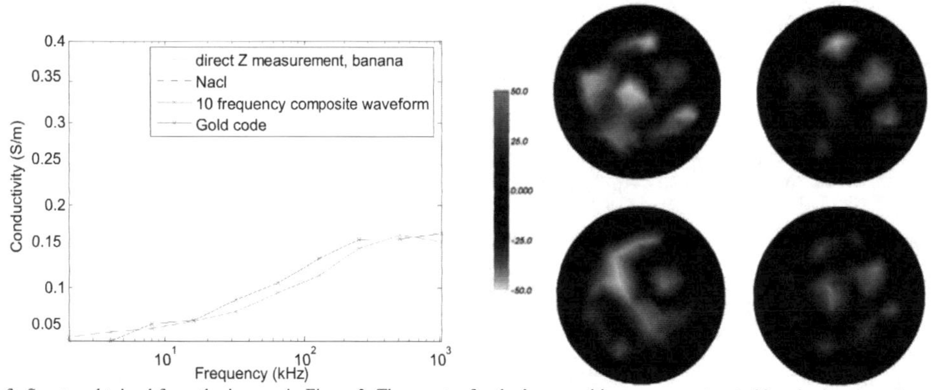

Fig. 3 Left: Spectra obtained from the images in Figure 2. The spectra for the banana object were constructed by plotting the value of a pixel at the known position of the object at 10 different frequencies. These spectra are similar to those that have been previously obtained from the Mk2.5 system [4]. The new method produces a spectrum which is very similar to the standard method. Both EITS spectra deviate from a direct measurement at frequencies above 128kHz, for reasons explained below. Right: imaging on an 8 electrode tank. Left column: 10kHz, right column: 50kHz. Row 1: FDM, Row 2: CDM.

4. Discussion

Using CDM the frequency resolution was increased to 1023 points from the 10 points available in the system before modification. Parameters of interest such as corner frequencies can be correctly derived. Imaging could be performed twice as fast as TDM or FDM as two sources could be used simultaneously. Future systems might be able to use many more codes simultaneously. FDM does offer truly orthogonal signals by using exact harmonics of the data acquisition period [15]. However there are a limited number of exact harmonics for any given bandwidth, so unlike CDM, FDM could not be extended to high electrode count systems. In addition FDM is not able to measure the spectrum at each electrode simultaneously. The boundary voltages measurements obtained using the Gold codes are similar to those obtained with a 10 frequency composite waveform, as shown in Figure 1. The Gold code voltage measurements are 1mV lower, due to the lower current per frequency component. As this is constant between the perturbation and reference frames, it is cancelled out in the subtraction process and is not apparent in the images. The Gold code boundary voltages appear to be noisier, particularly at higher frequencies. In practice these spectra would be smoothed, thus reducing the noise. The 1023 point CDM waveform and 10 point TDM waveform were formed from the same number of points and driven at the same clock frequency leading to an equal acquisition time. Therefore the CDM waveform was able to measure the same 96 combinations in the 16 channel tank with over 10x the frequency resolution in the same acquisition time. The Mk2.5 EIT system has limited performance at frequencies above 128kHz (figure 3a) due to the effect of 0.5m long unscreened cables [4]. These are likely to be the cause of the increased noise seen in the Gold code spectra. This is not seen with the 10 frequency composite waveform, due to the lower spectral resolution of this method. High frequencies are difficult to apply in EITS systems due to the effects of stray capacitance, cables and contact impedances. In this design we lessened the effect by measuring the injected current and using this signal to demodulate the recorded voltages. However this is not only an issue for CDM-EITS; the upper bandwidth requirements are exactly the same for the maximum frequency of TDM and FDM systems. The simultaneous images (figure 3b) were qualitatively similar, but the spatial resolution was constrained in our experiments with the currently available system, since only five effective channels were available, i.e., two sources and three differential measurement channels. In this experiment we chose to use separate electrodes for current injection and voltage measurement in a similar way to that used to demonstrate the FDM approach [7-8]. However a recent conference paper on electrode circuits for FDM and CDM EITS show that the same electrodes may be used [9]. Additionally when moving to multiple current sources, the sources must be well matched adding circuit complexity which may impact the advantages of CDM depending on the application.

5. Conclusion

Code division multiplexing using pseudo-random noise sequences can be used to measure high resolution impedance spectra on multiple channels simultaneously, thereby increasing the acquisition speed of electrical impedance spectroscopy measurements. However, the quality of the spectra produced is slightly degraded due to the lack of completely orthogonal code sets. This will increase as the number of simultaneous channels goes up. From a commercial perspective, the codes are cheaper and easier to generate than sinusoidal waveforms as they are binary sequences that may be generated with digital hardware, usually by modulo-addition of bits in a shift register. Therefore, the CDM method should be pursued for biomedical and industrial EITS in some applications where spectral information is required in a short time frame. This potentially powerful technique might also be applicable to rapid measurement of the transfer functions of any multichannel network.

6. References

[1] D.V. Sarwate, and M.B. Pursley, *Proc. IEEE*, 68, 593-619,1980.
[2] S. Gawad et al, *Rev. Sci. Instrum.*, 78, 2007.
[3] I. Schneider, *IEEE EMBS Conference*,1996, Vol. 5, pp.1934–1935.
[4] A. McEwan et al 2006 *J. Physiol. Meas.* **27** S199-S210
[5] A.J. Wilson et al, Physiol. Meas., 22, 49-54, 2001.
[6] N Polydorides, WRB Lionheart, *Measurement science and technology*, 13,12, 1871-1883, 2002.
[7] G. Teague, PhD thesis, University of Cape Town, 2002.
[8] Y. Granot, et al *Int J Biomed Imaging*. 2007.
[9] A. McEwan. et al. BIOCAS, 2007. IEEE pp 130 – 133.

Journal of Physics: Conference Series **224** (2010) 012144 doi:10.1088/1742-6596/224/1/012144

Comparison of different coil positions for ventilation monitoring with contact-less magnetic impedance measurements

A Cordes, D Pollig and S Leonhardt

Chair for Medical Information Technology, RWTH Aachen University, Aachen, Germany

E-mail: cordes@hia.rwth-aachen.de

Abstract. For monitoring the health status of individuals, proper monitoring of ventilation is desirable. Therefore, a continuous measurement technique is an advantage for many patients since it allows personal home care scenarios. As an example, monitoring of elderly people at home could enable them to live in their familiar environment on their own with the safety of a continuous monitoring. Therefore, a measurement technique without the restriction of mobility is required. Since it is possible to monitor ventilation with magnetic impedance measurements without conductive contact, this technique is well suited for the mentioned scenario. Integrated in a chair, a person's health state could be monitored in many situations, e.g. during meals, while watching TV or reading a book. In this paper, we compare different positions of coil arrays for a magnetic impedance measurement system integrated in a chair in order to monitor ventilation continuously. For limiting the costs and technical complexity of the magnetic impedance measurement system, we have a focus on coil configurations with one RF channel. To limit the needed space and thickness of the array in the backrest, planar gradiometer coil setups are investigated. All measurements will be performed with a new developed portable magnetic impedance measurement system and a standard office chair.

1. Introduction

One application for magnetic impedance measurements is monitoring vital parameters [1]. Both ventilation and heart activity influences the local impedance distribution over the time. This is a reaction of well conductive blood transport or poor conductive air shift in the human chest. Therefore, it is possible to monitor ventilation or perfusion by measuring the changing eddy currents. In this article we focus on ventilation monitoring. The benefit of this technology is the complete contact-free measurement setup. Necessary coils can be easily integrated in a bed or in chairs. With this setup monitoring in a day-to-day basis is applicable without the need of cables that could limit the users mobility. For example over night monitoring during sleep phases or monitoring bus drivers and pilots could help to reduce accidents or alarm paramedics just in time. For the mentioned scenario, a magnetic impedance measurement device is needed as well as a coil setup designed for this special application. In our work, we focus on a coil array with only one RF channel and two coils (one for the excitation signal and two as a planar gradiometer). To reduce the needed size we test three different planar coil arrays at different positions in the backrest of a chair and determine which position or array is best suited for continuous ventilation monitoring. In the literature, a similar monitoring setup with only one RF coil was presented in [2]. One alternative measurement solution based on preasure measurements in a chair was presented in [3].

2. Measurements

2.1. Portable Impedance Measurement System

For a complete chair-integrated measurement setup a small and lightweight portable magnetic impedance measurement device is needed. First investigations regarding this scenario were done with the Multi-Channel-Simultaneous-Magnetic-Impedance-Tomography-System (MUSIMITOS), presented in [5]. However, due to size and weight of the device, we decided that it is not adequate for a chair-integrated measurement solution. Thus, we reduced the number of excitation channels (now one) and the number of measurement channels (now two) and implemented the software IQ-demodulation and the control functions for the DDS on a Blackfin DSP from Analog Devices Inc. The new developed device called "Portable-Impedance-Measurement-System (PIMS)" (see [6], system overview in figure 1 and foto in figure 2) is now battery driven and fits into a 10 inch housing. Another group uses the same concept togehter with NI PXI-5105 ADCs [8]. To expand the scenario, data visualisation was done on a notebook

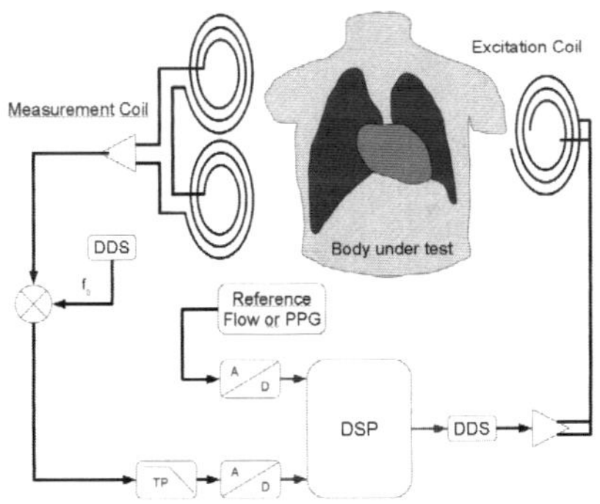

Figure 1. Overview of the portable measurement system

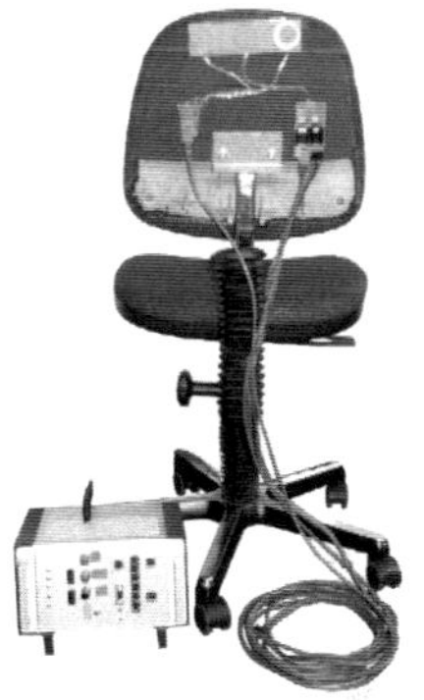

Figure 2. Foto of the measurement system from backside (here: planar gradiometer)

computer via a common PC Network to show the data from more than one measurement device on one screen. This is ideal for situations in which all passengers in a bus or plane should be monitored. To validate the measured data, a flow or pulse reference sensor can be attached to the device. Some first measurements for estimation ventilation with this device are presented in [7].

2.2. Coil configurations

As measurement coils we used three planar gradiometer configurations (see figure 3). The diameter of the measurement and excitation coils are Ds = Dm = 35mm. The distance between them varies for each gradiometer. For the coil configuration 1 we used 25 mm, for 2 40 mm and for 3 10 mm. This leads to a tradeoff between coil diameter and size of the complete coil arrays. All coils are maid of FR4 PCB on two layers. This makes the gradiometers very thin. They are driven with symmetrical amplifiers, placed on the backrest very close to the coils, and are connected via short cables.

2.3. Setup

Together with the portable magnetic impedance measurement system we used the three coil configurations described above and a common office chair as a test scenario. For each of the

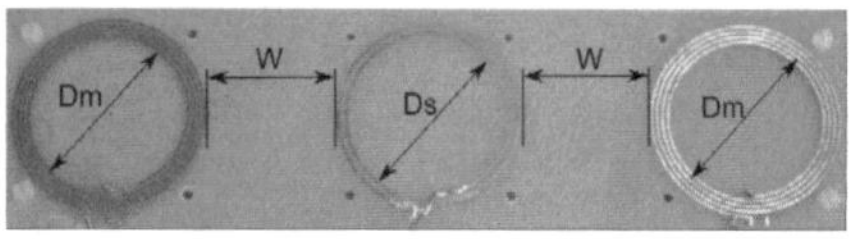

Figure 3. Planar coil array: Ds=Dm = 35 mm; Coil 1 W=25 mm; Coil 2 W=40 mm; Coil 3 W=10 mm

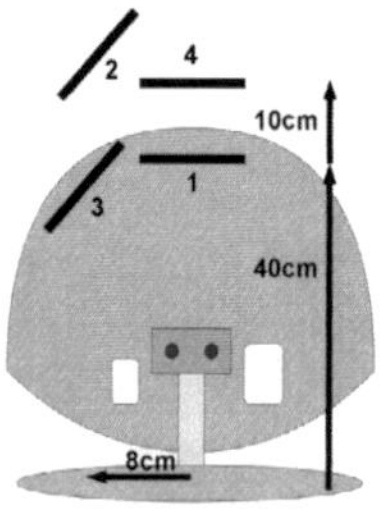

Figure 4. Different positions of the coil array in the backrest of the chair

coil configurations we measured changes in the magnetic impedance caused by respiration at four different positions on the backrest of the office chair and the flow signal from a differential pressure sensor together with a breathing mask as a reference. The positions of the coils are shown in figure 4. Assuming that it is a symmetrical problem, only one side of the chair was investigated. Due to ethical reasons we performed the measurement on one male volunteer of our project group. The candidate was in the age of 28 and healthy. We advised him to breath as normal as he could and in the same way for all measurements.

2.4. Results

Compared to the results in [7] the actual investigated measurement scenario is more ill posed. In this approach the measurement coils are not in front of the person. By measuring from the backside the spine covers the lung and decreases the reinduced fields. This fact also reduces the SNR. Hence it is necessary to verify the HF respiration signal measured with the magnetic impedance method and the flow signal again. A sample for this test is presented in figure 5

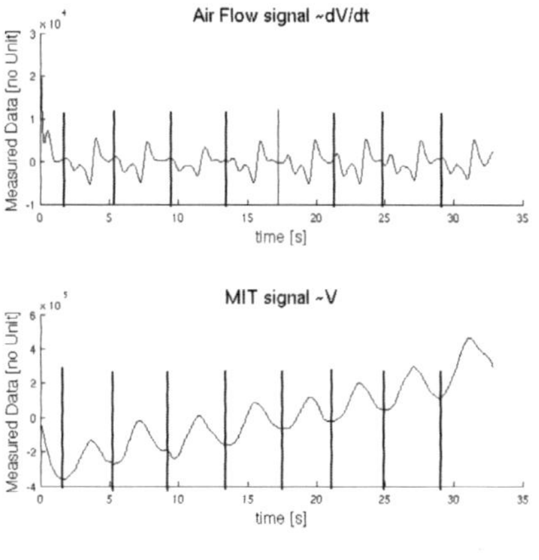

Figure 5. Measured HF and Flow Signal

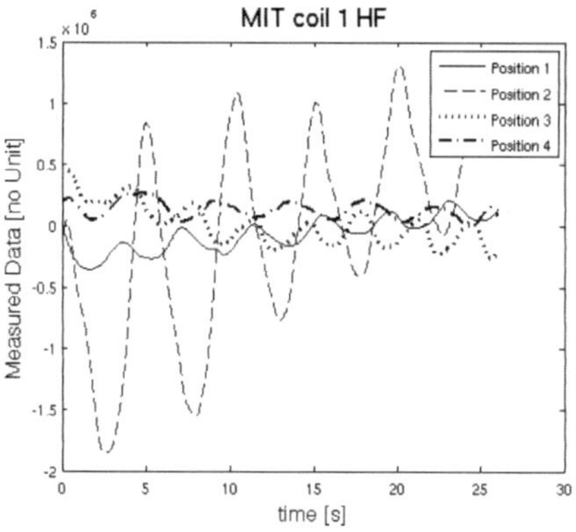

Figure 6. Measurement results for all 4 positions with coil configuration 1

for coil 1 on position 1. As shown in this example, breath strokes correspond well. However, this was not the case for all measurements. Only for coil configuration number one, the MIT respiration signal correlates with the flow reference signal. The results for these measurements are presented in figure 6. For coil configuration 2, the measurement position two and three are the only one for which the measured data represents respiration. For the other two positions the data contains noise and artefacts (see figure 7 for details). With coil setup number three it was not possible to detect any respiration signal with the magnetic impedance measurements. The

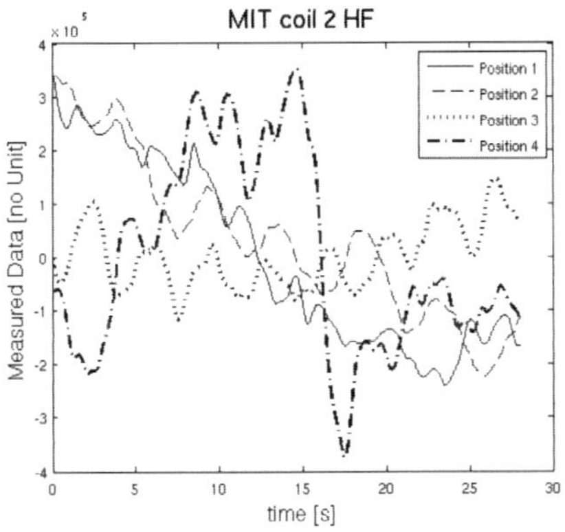

Figure 7. Measurement results for all 4 positions with coil configuration 2

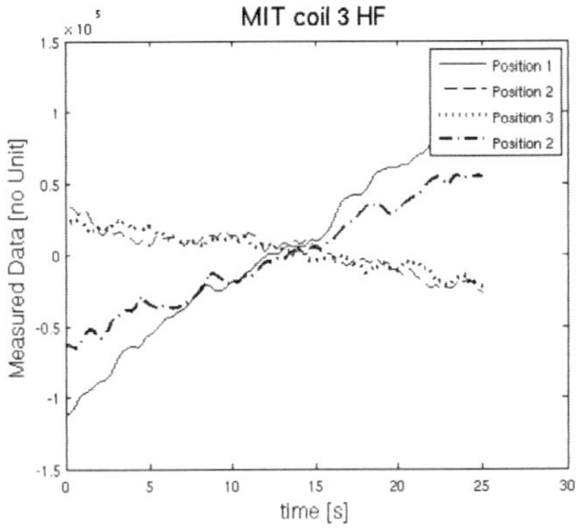

Figure 8. Measurement results for all 4 positions with coil configuration 3

data plotted in figure 8 is only noise. In total only 50% of the respiration could be monitored. As shown in the figures, the amplitude of the signal depends on position and coil configuration and contain drift movements.

3. Conclusion

As presented, the dimensions of a planar coil configuration and the position at the backrest are both important for proper respiration detection using magnetic impedance measurements. However, it is possible to detect respiration with a thin planar gradiometer from behind the patient. Since these measurements are the result of a data acquisition with only one volunteer, it is necessary to investigate the scenario with more volunteers within an clinical trial. For the future, measurements with other types of coil configurations are also planned in order to further improve the monitoring results. With saline solution it would be possible to evaluate drifts and noise and give a perform index of the system.

References

[1] Griffiths H 2001 Magnetic induction tomography *Measurement Science and Technology* **12** 1126-31

[2] Richer A and Adler A 2005 Eddy Current Based Flexible Sensor for Contactless Measurements of Breathing *IMTC 2005*

[3] Tohru Takarada et al. 2006 An Alternative Approach to the Monitoring of Respiration by Dynamic Air-Pressure Sensor *Anesth. Prog.* **54** 2-6

[4] Ko Keun K, Yong Kyu L, and Kwang Suk P 2005 Common Mode Noise Cancellation for Electrically Non-Contact ECG Measurement System on a Chair *Proceedings of the 2005 IEEE Engineering in Medicine and Biology 27th Annual Conference Shanghai* 5881-3

[5] Steffen M, Heimann K, Bernstein N, Leonhardt S 2008 Multichannel simultaneous magnetic induction measurement system (MUSIMITOS) *Physiological Measurement* **29** 291-306

[6] Foussier J, Cordes A and Leonhardt S 2009 Development of a portable impedance tomography system (PITS) *Medical Physics Biomedical Engineering World Congress*

[7] Cordes A, Foussier J and Leonhardt S 2009 Breathing Detection with a Portable Impedance Measurement System: First Measurements *31st Annual International Conference of the IEEE Engineering in Medicine and Biology Society*

[8] McEwan A, Hamsch M, Igney C and Kahlhert 2009 A comparison of two phase measurement techniques for Magnetic Impedance Tomography *11th World Congress on Medical Physics and Biomedical Engineering* 4-6

Some numerical results on the influence of measurement strategies and load patterns in the EIT inverse problem

Franciane C. Peters and Luis Paulo S. Barra

Federal University of Juiz de Fora, Graduate Program on Computational Modeling, CEP 36036-330, Minas Gerais, Brazil

E-mail: franciane.peters@engenharia.ufjf.br, luis.barra@ufjf.edu.br

Abstract. It is well known that the Electrical Impedance Tomography reconstructed images are very sensitive to noise in the electrical potential measures at the external boundary of the domain. Furthermore, there are a variety of strategies to inject current and measure these values. Thus, the aim of this work is to verify how much some of these strategies affect the quality of the reconstructed image and the sensitivity of each strategy to the existence of noise in the measured data. For a known position and shape of a single homogeneous inclusion inside a conductor domain, measured data is numerically simulated by the forward problem solution via Boundary Element Method (BEM). Sixteen electrodes placed at the outer boundary of the domain are used with different load patterns and measurement strategies to obtain the exact data for the numerical experiments. In order to simulate noisy data, Additive White Gaussian Noise proportional to the exact measured value is used on each potential measure. The inverse problem is treated as a data fitting problem solved via Levenberg-Marquardt method, that minimizes the difference between measures data and computed potential values. The coordinates of the control points of a spline function that defines the boundary of the inclusion are used as minimization parameters. In each step of this procedure, the computed potential values are obtained via BEM. The identification of inclusions of different sizes, shapes and positions is presented allowing some conclusions about the influence of the studied parameters.

1. Introduction

The aim of the Electrical Impedance Tomography (EIT) is to reconstruct an image of the conductivity or resistivity distribution of a body using the electrical potential measured on the external boundary of this body. The features of this technique have motivated the study of its use in a large variety of applications [1, 2], and the improvement of the reconstruction algorithms, the systems of data collection and other aspects [3, 4]. It is known that this inverse problem is ill-conditioned and therefore, the reconstructed images are very sensitive to noise. So, the aim of this work is to present a study about the influence of the strategies to inject current and measure potential values on the quality of the obtained images. In order to do this, results of inclusion identification obtained with two strategies, the adjacent method and the opposite method [5], are compared.

2. The inverse problem

The inverse problem consists on the identification of inclusions, regions with different electrical property, in a two dimensional conductive domain. In this work, the external boundary of the inclusions are approximated by curves defined by a mathematical model called Extended X-Splines [6]. Then, with this parametrization, the variables of the reconstruction problem are the coordinates and the shape parameter of the control points of the spline. In this paper, the reconstruction problem is solved as a data fitting problem, in which one wants to find the values of the vector of parameters that produces the boundary potentials that best fit the measured data. This optimization problem is solved by the Levenberg-Marquardt method [7] via the MINPACK-1 implementation [8]. The forward problem is governed by Laplace's equation and it is solved by the Boundary Element Method. Using 16 electrodes placed as presented in Figure 1, two strategies are used to generate the potential values to be fitted. In the adjacent strategy (AS), used in the Sheffield system [9], the current is injected between adjacent pairs of electrodes and the potential difference between the other adjacent pairs are measured. In the opposite strategy (OS), the current is injected between "diametrically" opposite electrodes (i and $i + 8$) and the potential measurements are taken between the reference electrode ($i + 4$), and the others. In both patterns, measurements are not used on driven electrodes.

3. Numerical Experiments

Three single inclusion identification problems in two different domains were proposed. The inclusions have conductivity 10 times greater then the rest of the domain. Figure 1 presents the domains and the inclusions to be identified, called targets.

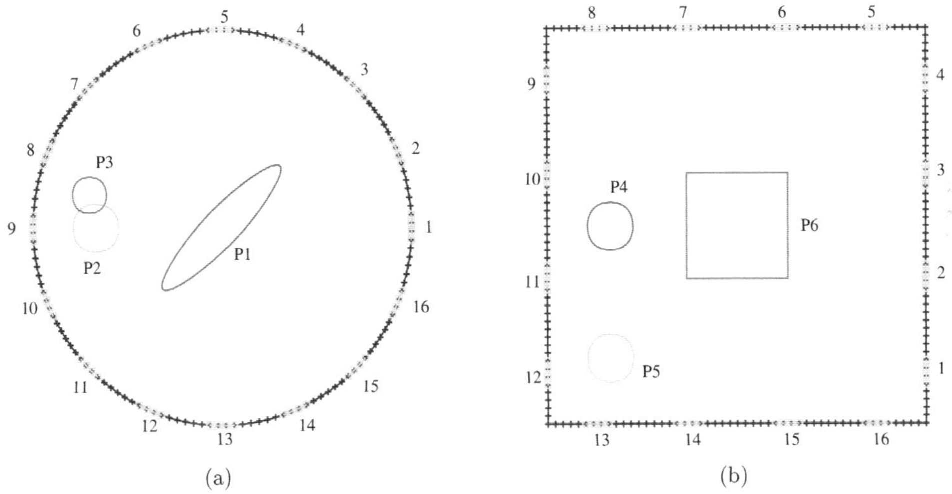

Figure 1. The target inclusions, the circular and the square domain.

For each problem, potential data were simulated via the forward problem solution using the adjacent and the opposite strategy. These noise-free data were used to solve the inverse problem and the results are presented in Table 1, where the "objective function" is the Euclidean norm of the difference between the measured potential values and the computed ones for the solution and the "geometrical error" is the average of the distance between the nodes of the discretized boundary of the identified inclusion and the target, divided by the perimeter of the target. After that, the Additive White Gaussian Noise was use to simulate 60 sets of noisy data for 3 different levels of magnitude that were used to solve the identification problem. Each data set is characterized by its signal-to-noise ratio (SNR). A total of 2160 (6 problems $\times$ 2 strategies $\times$

3 levels × 60 sets) runs of the reconstruction algorithm were done and the corresponding results are shown in Figure 2. For the smallest noise level, near to the noise magnitude typical of the EIT systems [10], the best and the worst results for problems 2 and 5 with both measurement strategies are presented in Figure 3.

Table 1. Identification results with noise-free data.

Problem	Adjacent Strategy Objective Function	Adjacent Strategy Geometrical Error	Opposite Strategy Objective Function	Opposite Strategy Geometrical Error
P1	$1.80 . 10^{-8}$	$5.65 . 10^{-4}$	$5.80 . 10^{-8}$	$4.75 . 10^{-4}$
P2	$4.70 . 10^{-3}$	$5.60 . 10^{-1}$	$1.20 . 10^{-7}$	$1.57 . 10^{-3}$
P3	$2.60 . 10^{-3}$	$1.11 . 10^{0}$	$8.80 . 10^{-7}$	$7.32 . 10^{-3}$
P4	$4.20 . 10^{-6}$	$6.44 . 10^{-2}$	$3.40 . 10^{-7}$	$3.06 . 10^{-3}$
P5	$2.20 . 10^{-2}$	$1.12 . 10^{0}$	$3.90 . 10^{-7}$	$1.78 . 10^{-3}$
P6	$3.60 . 10^{-8}$	$4.19 . 10^{-4}$	$3.90 . 10^{-7}$	$8.57 . 10^{-4}$

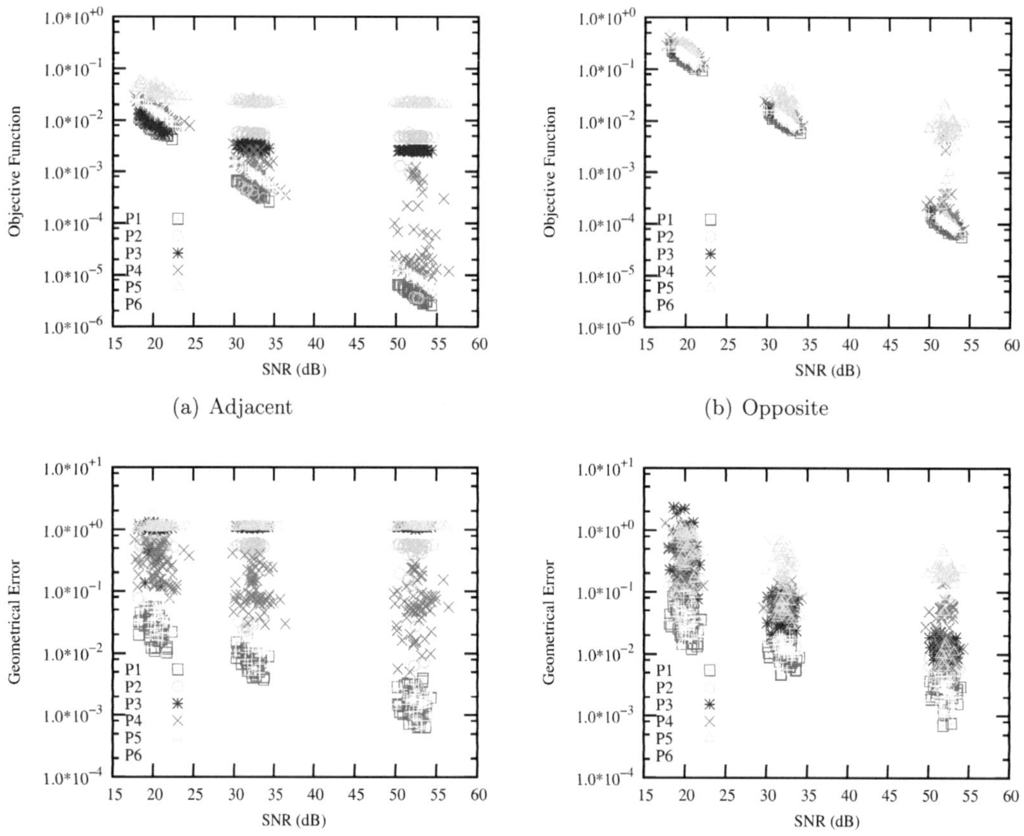

(a) Adjacent (b) Opposite

(c) Adjacent (d) Opposite

Figure 2. Identification results with noisy data.

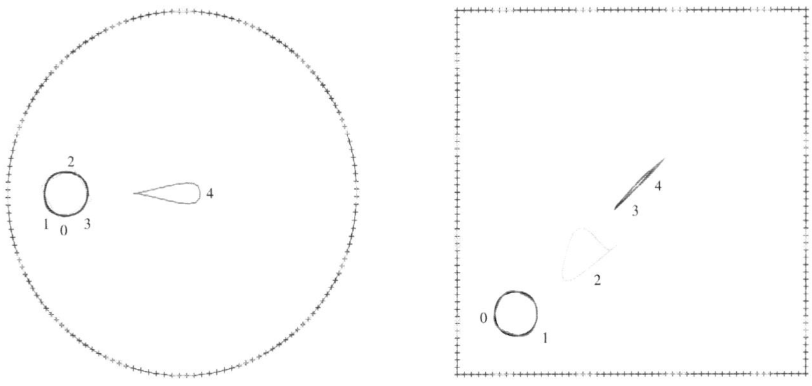

(a) Problem 2 (b) Problem 5

Figure 3. Results: **0** target; **1** best of OS; **2** worst of OS; **3** best of AS; **4** worst of AS.

4. Conclusions

With noise-free data, Table 1, the values of the objective function indicate that all problems were correctly solved with OS, while AS have failed in problems 2, 3 and 5. The geometrical errors confirm these findings and also show that last strategy for the problem 4 lead to a poor solution. Figures 2(c) and (d) show that, with OS, great part of the results have an geometrical error smaller than the ones obtained with AS. Figure 3(a) and (b) show that even with noisy data OS was able to achieve solutions with small geometrical errors even for problems that were not well solved for noiseless data with AS. So, the small set of problems here analysed indicate that the opposite strategy is more suitable than the adjacent strategy for the solution of the inverse problem as proposed. Although it is expected that this conclusion is independent of the optimization algorithm, this independence should be verified in future work.

Acknowledgments

The authors would like to thank FAPEMIG for the financial support.

References

[1] D. C. Barber and B. H. Brown. Applied potential tomography. *Journal of Physics E-Scientific Instruments*, 17(9):723–733, 1984.
[2] R. H. Bayford. Bioimpedance tomography (electrical impedance tomography). *Annual Review of Biomedical Engineering*, 8:63–91, 2006. Bayford, R. H. 85 ANNUAL REVIEWS PALO ALTO 081CI.
[3] Margaret Cheney, David Isaacson, and Jonathan C. Newell. Electrical impedance tomography. *SIAM Review*, 41(1):85–101, 1999.
[4] D. S. Holder. Electrical Impedance Tomography: Methods, History and Applications. *Medical Physics*, 32:2731–+, 2005.
[5] Jaakko Malmivuo Plonsey and Robert. *Bioelectromagnetism: Principles and Applications of Bioelectric and Biomagnetic Fields.* Oxford University Press, 1995.
[6] C. Blanc and C. Schlick. X-splines: A spline model designed for the end-user. In *SIGGRAPH*, pages 377–386, Los Angeles, EUA, 1995. ACM - Association for Computing Machinery.
[7] R. Fletcher. *Practical methods of optimization.* Wiley-Interscience, New York, NY, USA, 1980.
[8] Jorge J. Moré, Burton S. Garbow, and Kenneth E. Hillstrom. User guide for MINPACK-1. Technical Report ANL-80-74, Argonne National Laboratory, Argonne, IL, USA, August 1980.
[9] B H Brown and A D Seagar. The sheffield data collection system. *Clinical Physics and Physiological Measurement*, (4A):91, 1987.
[10] D R Stephenson, R Mann, and T A York. The sensitivity of reconstructed images and process engineering metrics to key choices in practical electrical impedance tomography. *Measurement Science and Technology*, (9):094013, 2008.

Journal of Physics: Conference Series **224** (2010) 012146 doi:10.1088/1742-6596/224/1/012146

Experimental Reconstruction Images of Tissue Phantom by Diffuse Optical Tomography

Young Sik Jun and Woon Sik Baek

Department of Electronics and Radio Engineering, Kyung Hee University, South Korea.

E-mail: wsbaek@khu.ac.kr

Abstract. As one of the new modality to imaging in vivo, Diffuse Optical Tomography(DOT) is using near-infrared(NIR) light sources and detectors by which the photon propagation is dominated by scattering rather than absorption in human tissue. In this paper, we reconstructed images of tissue phantom from measured data that is transmitted through highly scattering turbid media, using DOT experimental set up. In imaging procedure, we used FEM based forward and inverse solver and it is very difficult to obtain a good inverse matrix, since the Jacobian matrix which is composed of differential components of optical parameters(scattering coefficient and diffusion coefficient) is ill-conditioned by limited number of detectors. Therefore, we have to update these parameters repeatedly to obtain a reasonable optical parameters. The optimization of this updating process can be achieved by regularization by which the weighted diagonal matrix are added to the Jacobian matrix. And we present the preliminary experimental reconstruction images of optical parameters for the liquid phantom of human breast tissue.

1. Introduction

In Diffuse Optical Tomography light is considered as particles with energy and velocity, that is photons. These photons are scattered or absorbed in turbid media such as biological tissues and are reflected at boundaries according to the Fresnel's law[1]. The operation between photons and turbid media can be described by the probability per unit length of a photon being scattered or absorbed. And these parameters denoted as scattering coefficient μ_s and absorption coefficient μ_a respectively.

In this paper, we used the diffusion equation that is approximated Boltzmann transport equation which is well describing the light transport in biological tissues. And the FEM forward solver was developed using finite-element method(FEM) to solve the diffusion equation numerically. And calculated images of tissue phantom was reconstructed. Preliminary experimental results with tissue phantom would be presented.

2. FEM forward solver for frequency-domain DOT

2.1. Frequency-domain diffusion equation

Photon transport in highly scattering media such as biological tissue is well modeled by the Boltzmann transport equation. However, it has many variables for consideration and it is very difficult and time consuming to obtain the solution of Boltzmann transport equation. For the diffuse optical applications in which the light transport in medium is dominated by scattering rather than absorption, it can be simplified with diffusion approximation. Using P_1-approximation for the radiance, diffusion equation can be obtained[1],

$$\nabla\cdot(D(r)\nabla\Phi(r,t))-\mu_a(r)\Phi(r,t)-\frac{1}{v}\frac{\partial}{\partial t}\Phi(r,t)=-S(r,t) \tag{1}$$

where v is the speed of light in the medium, r is the position vector and t is time. $\Phi(r,t)$ is the photon fluence rate [Watt/cm^2] which is defined as integral of the radiance over all solid angle, $D(r)$ is the diffusion coefficient and $S(r,t)$ is the isotropic light source [Watt/cm^3].

In the frequency-domain DOT, fluence rate can be assumed $\Phi(r,t)=\Phi(r)e^{-j\omega t}$ and the $e^{-j\omega t}$ terms have factored out[2]. Therefore frequency-domain diffusion equation can be obtained,

$$\nabla\cdot(D(r)\nabla\Phi(r))-\left(\mu_a(r)-\frac{i\omega}{v}\right)\Phi(r)=-S(r) \tag{2}$$

where ω is the light source modulation angular frequency.

2.2. Finite element method

To solve the diffusion equation we adopted the finite element method. If the medium is homogeneous, the optical properties μ_a and D has uniform distribution over all region and it only needs the discretized photon fluence rate Φ and photon flux $F=-D\nabla\Phi$. If the medium is inhomogeneous, the optical properties has local heterogeneities and also needs to be discretized. Therefore we can obtain the discretized diffusion equation,

$$\sum_{j=1}^{N}\Phi_j\left[\left\langle-\sum_{k=1}^{K}D_k\psi_k\nabla\phi_j\cdot\nabla\phi_i\right\rangle-\left\langle\left(\sum_{l=1}^{L}\mu_l\varphi_l-\frac{i\omega}{c}\right)\phi_i\phi_j\right\rangle\right]=-\langle S\phi_i\rangle+\sum_{j=1}^{M}F_j\oint\phi_j\phi_i ds \tag{3}$$

where ϕ, ψ, and φ are the Lagrangian basis functions, $\langle\cdot\rangle$ indicates integration over all element and $\oint$ represents integration over the boundary surface. This complicated equation by the FE approximation can be written in the matrix form,

$$\begin{bmatrix}A_{bb} & A_{bi}\\ A_{ib} & A_{ii}\end{bmatrix}\begin{Bmatrix}\Phi_b\\ \Phi_i\end{Bmatrix}=\begin{bmatrix}B_{bb} & 0\\ 0 & 0\end{bmatrix}\begin{Bmatrix}F_b\\ 0\end{Bmatrix}+\begin{Bmatrix}C_b\\ C_i\end{Bmatrix} \tag{4}$$

where the subscripts i and b denote interior and boundary nodes of the medium respectively.

3. Inverse solver

Generally, human tissues are inhomogeneous and if the rate of change of the D and μ_a can be found, we can obtain the image in vivo using these parameters[3]. The numerical solutions(i.e., photon fluence rate Φ and photon flux F) of the diffusion equation which is approximated by FE can be Taylor expanded in terms of absorption and diffusion coefficient as follows,

$$\Phi(\overline{D},\overline{\mu}_a)=\Phi(D,\mu_a)+\frac{\partial\Phi}{\partial D}\Delta D+\frac{\partial\Phi}{\partial\mu_a}\Delta\mu_a+\cdots\ , \tag{5}$$

$$F(\overline{D},\overline{\mu}_a) = F(D,\mu_a) + \frac{\partial F}{\partial D}\Delta D + \frac{\partial F}{\partial \mu_a}\Delta\mu_a + \cdots \quad . \tag{6}$$

If we define the vector $\Delta\chi$ which including ΔD and $\Delta\mu_a$, then we can derive eq. (8) with the Jacobian matrix of eq. (7),

$$J = \begin{bmatrix} \frac{\partial\Phi_1}{\partial D_1} & \frac{\partial\Phi_1}{\partial D_2} & \cdots & \frac{\partial\Phi_1}{\partial D_K} & \frac{\partial\Phi_1}{\partial\mu_1} & \frac{\partial\Phi_1}{\partial\mu_2} & \cdots & \frac{\partial\Phi_1}{\partial\mu_L} \\ \frac{\partial\Phi_2}{\partial D_1} & \frac{\partial\Phi_2}{\partial D_2} & \cdots & \frac{\partial\Phi_2}{\partial D_3} & \frac{\partial\Phi_2}{\partial\mu_1} & \frac{\partial\Phi_2}{\partial\mu_2} & \cdots & \frac{\partial\Phi_2}{\partial\mu_L} \\ \vdots & \vdots & \ddots & \vdots & \vdots & \vdots & \ddots & \vdots \\ \frac{\partial\Phi_M}{\partial D_1} & \frac{\partial\Phi_M}{\partial D_2} & \cdots & \frac{\partial\Phi_M}{\partial D_K} & \frac{\partial\Phi_M}{\partial\mu_1} & \frac{\partial\Phi_M}{\partial\mu_2} & \cdots & \frac{\partial\Phi_M}{\partial\mu_L} \end{bmatrix}, \tag{7}$$

$$J\Delta\chi = \Phi^o - \Phi^c, \tag{8}$$

where Φ^o and Φ^c denote observed and calculated photon fluence rates respectively. Since the detectors are located only at boundaries, the number of measurements are limited. Consequently, the Jacobian matrix is ill-conditioned. Therefore, we have to update these parameters repeatedly to acquire reasonable $\Delta\chi$. The optimization of this updating process can be achieved by regularization in which the weighted diagonal matrix are added to the Jacobian as below,

$$(J^T J + \lambda I)\Delta\chi = J^T(\Phi^o - \Phi^c). \tag{9}$$

And we utilized the Levenberg-Marquardt algorithm that minimize least-square errors between observed and calculated values by updating λ.

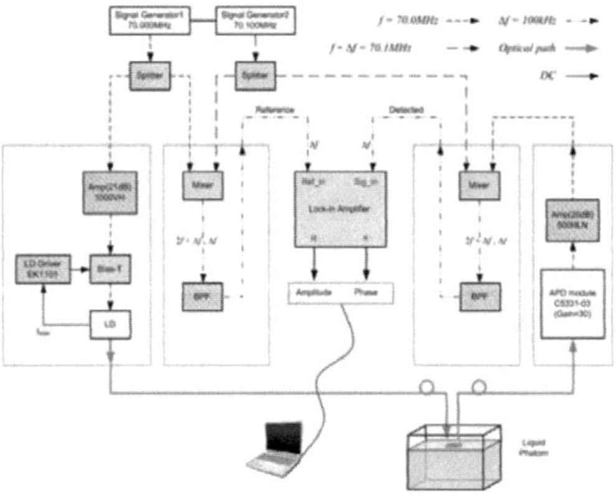

Figure 1. Diagram of the single channel frequency-domain DOT setup.

4. Experimental setup

Fig. 1 shows the diagram of the single channel frequency-domain DOT setup we developed. We used 830nm, 30mW diode laser(Sanyo, DL-5032-001) as a light source which was directly modulated with 70MHz sinusoidal radio-frequency by Bias-Tee(Mini-Circuits, ZFBT-4 R2G) and the modulation depth was greater than 90 percent. The modulated light was delivered to an optical fiber by LD-to-Fiber coupler(OzOptics, LDPC-01-830) and the coupling efficiency was greater than 75 percent. We adopt maximally-flat RF band-pass filter to ensure long-term stability.

We could obtain amplitude and phase of propagated light by detecting the photons that scattered out around the boundaries of medium with APD(Hamamatsu, C5331-03) and Lock-in amplifier(SRS,

SR830 DSP Lo ck-in Amplifier). Co nsequently, we c an obtain un known op tical p roperties b y substitute these data into inverse solver.

5. Results

Fig. 2 sho ws the photon fluenc e rate dist ribution i n the m edium and it h as the homogeneous background properties of $\mu_a = 0.5\,[\text{cm}^{-1}]$, $\mu_s = 140\,[\text{cm}^{-1}]$, $g = 0.9$ and the index of refraction $n = 1.33$.

First, w e ass umed that the anomaly has $\mu_s = 14\,[\text{cm}^{-1}]$ (smaller than backg round) an d t he o ther parameters a re equal to th e background. Second, we a ssumed th at the anomaly has $\mu_s = 1400\,[\text{cm}^{-1}]$ (larger than background). Fig. 2 (a) an d (b) shows the fluence rate distribution by above assumptions respectively. It was ob served that the smaller sc attering coe fficient in the an omaly m ade the light propagate further.

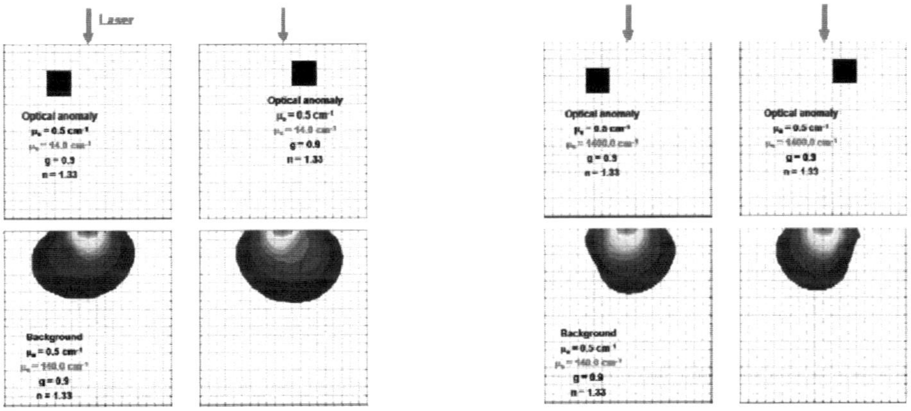

(a) (b)

Figure 2. Fluence rate distribution

6. Conclusions

We developed the FEM forward so lver an d sh owed the distribution of the ph oton fluence rate in the medium according to the anomaly's scattering coefficient changes. And we also developed the inverse solver for image reconstruction using Jacobian matrix. By updating the calculated data from the FEM forward solv er, we sh owed t hat the image re construction c ould b e ach ieved with th e experimental setup o f frequency-domain DO T we developed. A nd we could obtain an elec trical drift characteristic under 0 .3 [de gree/hour] b y ado pting the maximally-flat ban d-pass filter an d tha t was reaso nable for stable phase detection.

Acknowledgements

This work was partly supported by a grant from the Kyung Hee University in 2004 (KHU-20040079) and was partly supported by the Korea Science and Engineering Foundation(KOSEF) grant funded by the Korea government(MEST) (R11-2002-103-07002-0).

References

[1] A.J. Welch, M. J. C. van Gemert, *Optical-Thermal Response of Laser-Irradiated Tissue*, Plenum Press, 1995.

[2] R. Choe, *Diffuse Optical tomography and spectroscopy of breast cancer and fetal brain*, Ph.D thesis, University of Pennsylvania, 2005.

[3] K. D. Paulsen, H. Jiang, *Spatially varying optical property reconstruction using a finite element diffusion equation approximation*, Med. Phys., vol. 22, pp. 691-701, 1995.

Design and Implementation of a DSP-based Digital Phase Sensitive Demodulation for an EIT System

S A Hamidi[1], R Jafari[1], A Moosavi Nia[1] and M Soleimani[2]

[1]K. N. Toosi University of Technology, Tehran, Iran
[2]Invert Centre, Dept. of Electronics and Electrical Engineering, University of Bath, Bath, UK

jafari@eetd.kntu.ac.ir

Abstract. Electrical Impedance Tomography (EIT) is a non-invasive imaging method that can generate image of impedance distribution inside an object. In this imaging method, measured voltages are complex quantities that contain real and imaginary parts. To increase the accuracy of image reconstruction in EIT systems, it is necessary to utilize both real and imaginary parts of measured voltages. Phase sensitive demodulator can accurately specify the two parts of measured voltages. In this paper, a digital phase sensitive demodulator (DPSD) has been designed and implemented using a DSP board. Floating point 32-bit arithmetic with high speed of 225 MHz of the DSP core allows this digital demodulator to accurately measure real and imaginary parts of measured voltages with a desired SNR in an EIT system. This paper describes the theory and implementation of DPSD on a TMS320C6713 DSP board. Next, simulation data generated by the Code Composer Studio software and the data of an EIT phantom are applied to the DPSD. The simulation data results show a 0.12 degree phase error and a 0.37 % amplitude error with high SNR of 130.6 dB. The EIT phantom results present the 0.76 degree phase error and the 0.91 % amplitude error.

1. Introduction

Electrical impedance tomography (EIT) is a non-invasive imaging technique that provides image of impedance distribution insides an object, based on the knowledge of the applied current patterns and the measured surface voltages [1]. In this imaging method, measured voltages are complex quantities that contain real and imaginary parts. In EIT systems, a small perturbation on the boundary measured data will cause large changes in the solution of update impedance distribution resulting the image reconstruction problem is ill-posed. Therefore, boundary voltages must be measured with high accuracy. On the other hand, the number of measurements is limited; it is then necessary to utilize both real and imaginary parts of measured voltages. In other bio-impedance measurements, it is needed to accurately measure amplitudes and phases of voltages when applying current sources. In order to obtain accurate bio-impedance measurements, high precision phase sensitive demodulator is required.

Phase sensitive demodulation is a very powerful technique for accurately measuring in-phase and quadrature components of the voltage signal. The amplitude and phase of the signal can be computed from the two components. Most EIT systems use this demodulating technique for the boundary voltage measurements [2-7]. The demodulator affects the output signal to noise ratio (SNR) so it is necessary to design the demodulator with a desired SNR in an EIT system.

This paper describes the theory and implementation of DPSD on a TMS320C6713 DSP board. Simulation data generated by the Code Composer Studio software and the data of an EIT phantom are applied to the DPSD. After the description of the design and the data, the experimental results are given in this paper.

2. Digital Phase Sensitive Demodulation

2.1. DSP-based DPSD Implementation

For the implementation of the Digital Phase Sensitive Demodulator, we have used a DSP board from Texas Instrument TMS320C6713 DSK. It has a 32-bit floating point DSP running at 225 MHz (TMS320C6713 Digital Signal Processor) with 16 MB of on-chip memory (static RAM). The DSK uses a Texas Instruments AIC23 (TLV320AIC23) stereo codec for input and output of signals with the following on-chip features: a. Two 16-bit ADCs and two 16-bit DACs, with a maximum sample rate of 96 kilo samples per second (ksps), b. Programmable Gain Amplifiers (PGA) and antialiasing filters for the inputs, c. Smoothing filters and programmable attenuators for the outputs. The codec communicates using two serial channels: McBSP0 is used to send commands to the codec control interface while McBSP1 is used for digital audio data. These characteristics make this codec suitable for our range of frequencies and required resolution. A schematic block diagram of the DSK board is shown in Figure 1.

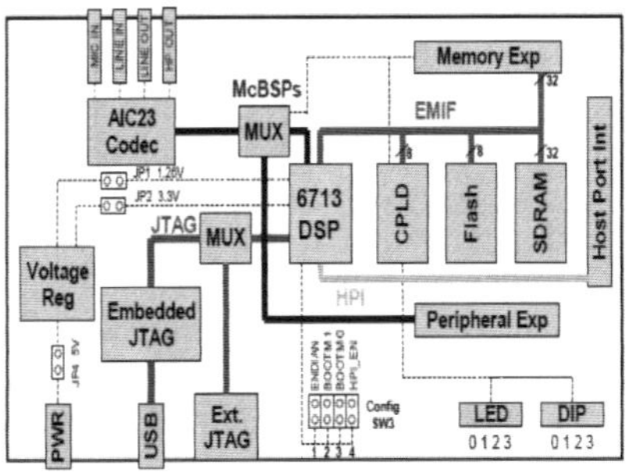

Figure 1. Block Diagram of C6713 DSK board [8].

The codec has two 16-bit input channel Line in Left and Right that can operate separately. The measured voltage signal and sinusoidal reference signal convert to digital samples through two codec input channels, line in left and right, simultaneously. At first, input signals is amplified by an Programmable Gain Amplifier (PGA), to reach an acceptable level. The signals, have also passed through an antialiasing filter. Resultant signals, then convert to digital samples by means of a ADC that utilize sigma-delta technique. These samples are sent to DSP through McBSP1 channel. Sampled measured voltage signal, $V_{in}(i)$ are multiplied by reference signals. Cosine reference signal, also can be provided by a delay from sinusoidal reference signal as follows [9-11]:

$$\cos\frac{2\pi}{N}i = \sin\frac{2\pi}{N}[i + \frac{N}{4}] \quad (1)$$

After digitizing input analog signal and sinosuidal reference signal with a high speed and high resolution analog to digital converter, samples of measured signal are multiplied by corresponding cosine and sinusoidal reference signal samples. Then the outputs can be achieved by filtering the DC frequency. The resulting DC components at the output of the lowpass filtering stage contain the phase and amplitude information of the measured voltage. These data are saved to on-board DSP memory. Figure 2 shows a block diagram of a DSP-based DPSD.

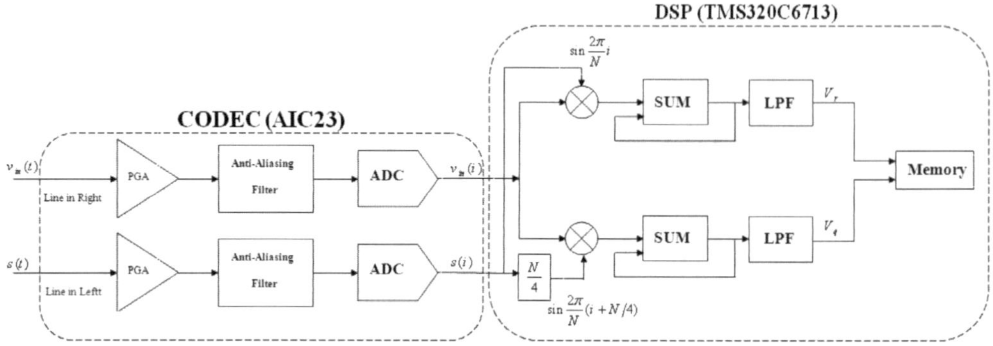

Figure 2. Schematic block diagram for the DSP-based DPSD. The dotted-line shows the operations performed by the DSP.

3. Exprimental Results

3.1. Emulation

To test the designed DPS D system, a reference si nusoidal signal with unit amplitude and zero initial phase and an input sinusoidal si gnal with an amplitude of 20 00 mV and a phase delay of 30 degrees both at fre quency of 24 kHz are produced in Code Composer Studio software. These si gnals are used as inputs of the DPSD system. The sampling frequency of the codec set to maximum value of 96 kHz. To increase the SNR, 1000 samples are used fro m 250 cycles (4 samples per cycles). Simulation data results, with 6 times repetition of the experiment, show a 0.12 degree average phase error and a 0.37 % amplitude error with high SNR of 130.6 dB. Table 1 shows the si mulation data results of the DPSD system together with the simulation data results from a FPGA-based DPSD [12].

Table 1. The s imulation data results of the demodulation system

	DPSD Emulation	[12]
N (Samples)	1000	200
Phase- Error(deg)	0.12	1.03
Amplitude- Error(%)	0.37	3.1
SNR(dB)	130.6	>100

3.2. Phantom

The second experiment is carried on a phantom (P.V.C. cylinderical tank) with prototype circuit board and the results are sent to a PC. The phantom has 15 cm radius and 35 cm height, and 32 Ag/AgCl electrodes. It is filled with 3 % standard saline water. The four electrodes, located in the middle of side wall of the phantom, are used in t he measurement. A hardware system si mulator to [13] is used to generate th e current s ource. A di gital oscill oscope (GW Instek GDS-2204) is also used to compare with DPSD results. The average of measured results of DPSD, with 10 times repetition experiments, show a phase delay of 23.43 degrees and magnitude of 575.17 mV with 130.6 dB SNR. These results present a phase d elay error of 0.76 degree and a magnitude error of 0. 91 %. Table 2 depicts th e measured results with DPSD and oscilloscope.

Table 2. DPSD results compared with the oscilloscope.

	Oscilloscope	DPSD	Error
Mag(mV)	570	575.17	5.17(0.91 %)
Phi(deg)	24.19	23.43	0.76(3.1 %)

4. Conclusion

This paper has presented a design of a digital phase sensitive demodulator and implemented using digital signal processing technology for an EIT system. The simulation data results show a 0.12 degree phase error and a 0.37 % amplitude error with high SNR of 130.6 dB. The EIT phantom results present the 0.76 degree phase error and the 0.91 % amplitude error.

Acknowledgment
This work was supported by Iran Central Oil Fields Company.

References

[1] Saulnier G J, Blue R S, Newell J C, Isaacson D and Edic P M 2001 Electrical impedance tomography *IEEE Signal Processing Mag.* **18** 6 31-43
[2] He C, Zhang L, Liu B, Xu Z and Zhang Z 2007 A digital phase-sensitive detector for electrical impedance tomography *Int. Conf. IEEE EMBS*
[3] Zarifi M H, Frounchi J, Asgarifar S and Baradaran Nia M 2008 (Canada) FPGA implementation of a fully digital demodulation technique for biomedical application *IEEE CCECE/CCGEI*
[4] Murphy D and Rolfe P 1989 Aspects of instrumentation design for impedance imaging *Clin. Phys. Physiol. Meas. Rev. A* **9** 5-14
[5] Smith R W, Freeston I and Brown B 1989 (USA) Digital demodulator for electrical impedance imaging *IEEE Proc. vol 6 Enginering in Medicine & Biology 11th Conf.* pp 1744-1745
[6] Osypka M J and Schafer E E 1998 (Spain) Impedance cardiography: advancement in demodulation techniques *10th Int.Conf. on Electrical Bioimpedance* pp 437–440
[7] Goharian M, Soleimani M, Jegatheesan A, Chin K and Moran G R 2008 A DSP Based Multi-Frequency 3D Electrical Impedance Tomography System *Annals of Biomedical Engineering* **36** 1594-1603
[8] Texas Instruments 2003 *TMS320C6713 DSK, Technical Reference* (Stafford, TX: Spectrum Digital)
[9] Cook R D, Saulnier G J, Gisser D G, Goble J C, Newell J C and Isaacson D 1994 ACT3: A high-speed, high-precision electrical impedance tomography *IEEE Trans. on Biomedical Engineering*, **41** 8 713-22
[10] Masciotti J M, Lasker J M and Hielscher A H 2008 Digital lock-in detection for iscriminating multiple modulation frequencies with high accuracy and computational efficiency *IEEE Trans. on Instrumentation and Measurement* **57** 1
[11] Oh T I, Lee J W, Kim K S, Lee J S and Wu E J 2003 (Mexico) Digital phase-sensitive demodulator for electrical impedance tomography *Proc. 25th Int. Conf. IEEE EMBS*
[12] Ge K and Lifeng R 2007 FPGA-based digital phase-sensitive demodulator for EIT *IEEE Int. Conf. on Electronic Measurement and Instruments,* pp 845-848
[13] Soleimani M 2006 Electrical impedance tomography system: an open access circuit design *BioMed. Central, BioMedical Engineering OnLine,* **5** *28*

In vivo conductivity imaging of human knee using 3 mA injection current in MREIT

W C Jeong[1], Y T Kim[1], A S Minhas[1], T H Lee[1], H J Kim[1], H S Nam[2], O Kwon[2] and E J Woo[1]

[1] Department of Biomedical Engineering, Kyung Hee University, Yongin, Korea
[2] Department of Mathematics, Konkuk University, Seoul, Korea

E-mail: ejwoo@khu.ac.kr

Abstract. Recent *in vivo* human leg MREIT experiments showed successful conductivity image reconstructions using carbon-hydrogel electrodes and optimized RF coils. However, it is still difficult to perform *in vivo* human and disease model animal experiments primarily due to a long scan time and high injection current of about 9 mA. Compared to previous MREIT pulse sequences, a newly developed multi-echo pulse sequence provides a higher SNR of MR magnitude image and better quality of magnetic flux density data. Unlike the human calf, the knee has sensitive nerve bundles and mainly consists of the bone. In this study, we tried to obtain high-resolution conductivity images of *in vivo* human knees using the multi-echo pulse sequence. We injected as much as 3 mA current in the form of an 81 ms pulse into the knee without producing a painful sensation and motion artifacts. Reconstructed conductivity images well distinguish different parts of the subcutaneous adipose tissue, muscle, synovial capsule, cartilage and bone inside the knee. Considering clinical applications, future work should be focused on *in vivo* human and disease model animal experiments.

Keywords: MREIT, conductivity image, human knee

1. Introduction

Magnetic resonance electrical impedance tomography (MREIT) has been recently proposed for cross-sectional conductivity image reconstructions with a spatial resolution of a few millimetres [1-4]. Incorporating a constant current source to an existing MRI scanner, the MREIT technique is expected to provide new contrast information. Postmortem and *in vivo* animal imaging studies demonstrated that we can produce high-resolution conductivity images of intact animals [5-7]. Though the first trial of human imaging experiments has been lately reported [8], this new imaging method requires further human studies focusing on organs of interest in terms of their conductivity values.

In MREIT, it is still difficult to perform *in vivo* human and disease model animal experiments primarily due to a long scan time and high injection current of about 9 mA. Compared to previous MREIT pulse sequences, a newly developed multi-echo pulse sequence provides a higher SNR of MR magnitude image and better quality of magnetic flux density data [9]. Since the knee has sensitive nerve bundles, we tried to reduce the current amplitude by adopting this new pulse sequence. The purpose of this study is to show the potential of the MREIT technique as a new clinically useful imaging modality through human imaging experiments. Describing the imaging method, we will show cross-sectional conductivity images of the human knee.

2. Methods

2.1. Normal volunteers

Four healthy volunteers (2 males, 2 females, 25 to 32 years old) participated in the human experiments. All human experiments were performed following the protocol approved by the Institutional Review Board (IRB). After attaching four electrodes around the knee, we chose one pair of electrodes to inject current. Starting from zero mA, we gradually increased the amplitude of the injection current pulse. The subject's oral response was recorded to mark the current amplitudes for the thresholds of sensation and pain. The same procedure was repeated for the other pair of electrodes. The amplitude of imaging current was determined as the 90% of the smaller pain threshold.

2.2. Imaging experiment

We attached four carbon-hydrogel electrodes around the knee to inject currents (figure 1). We injected 3 mA currents in two mutually perpendicular directions using two pairs of electrodes. Subjects were examined on a 3.0 Tesla MRI scanner (Magnum 3, Medinus, Korea) with a 22 cm STR coil. Imaging parameters using multi-echo (3 echos) spin echo pulse sequence were as follows: TR/TE = 900/30 ms, flip angle = 70°, FOV = 220×220 mm^2, slice thickness = 6 mm, NEX = 12, matrix size = 128×128, number of slices = 8 and total imaging time = 100 min.

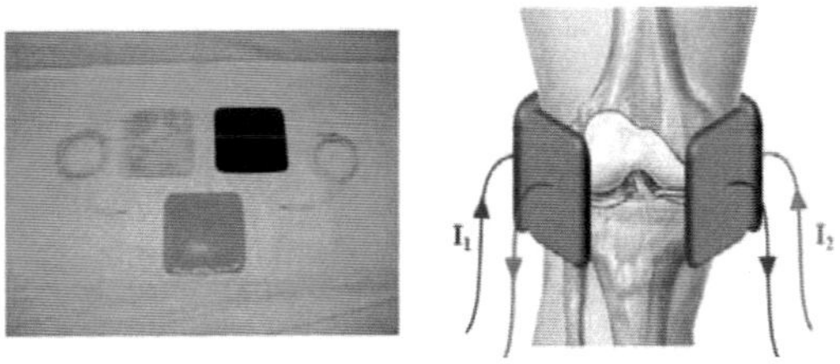

(a) (b)

Figure 1. (a) Carbon-hydrogel electrodes and (b) electrode attachment around the human knee.

2.3. Conductivity image reconstruction

We used CoReHA (conductivity reconstructor using harmonic algorithms), which is an integrated software package for MREIT [10,11]. It provides GUI-based functions for all data processing routines needed to produce conductivity images from measured *k*-space data sets. We used the single-step harmonic B_z algorithm implemented in CoReHA for multi-slice conductivity image reconstructions [12]. All conductivity images presented in this paper should be interpreted as equivalent isotropic scaled conductivity images providing only conductivity-based contrast information. Interpretation of these images should be pursued in future work.

3. Results

The mean threshold current by the subject's oral response was 3.5 ± 0.2 mA. We injected as much as 3 mA current in the form of an 81 ms pulse into the knee. Figure 2(a) is a MR magnitude image of the knee from the normal male subject. Figure 2(b) and (c) show B_z images in the same imaging slice subject to the horizontal and vertical current injections, respectively. Figure 3 compares the anatomy of the knee in (a) with an MR magnitude image in (b) and a reconstructed conductivity image in (c). Conductivity images of the knee well distinguished different parts of the subcutaneous adipose tissue, muscle, synovial capsule, cartilage and bone inside the knee.

Figure 4 shows the anatomy of the knee in (a) with an MR magnitude image in (b) and a reconstructed conductivity image in (c) from the normal female subject. In (b), we can see that MR signal void occurred at the outside of the bones. The conductivity image in (c) exhibits spurious noise spikes there. We observed no significant conductivity difference between male and female subjects.

The semi-automatic boundary extraction method implemented in CoReHA often resulted in somewhat distorted shape of the outermost boundary in conductivity images. This minor geometrical distortion has negligible effects in reconstructed conductivity images since the harmonic B_z algorithm is insensitive to geometrical errors.

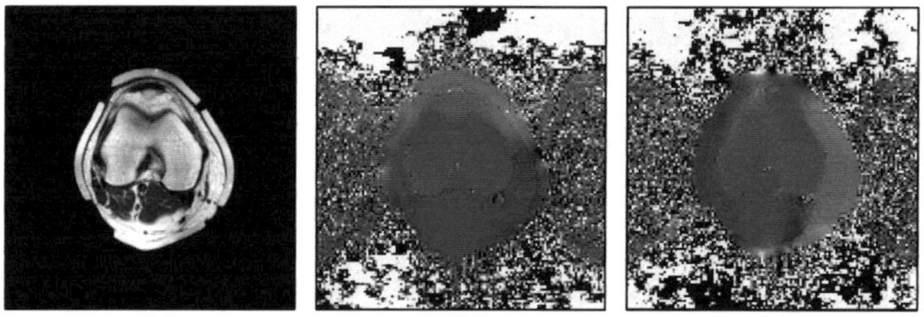

(a) (b) (c)

Figure 2. (a) MR magnitude image of the knee from the male subject. (b) and (c) are magnetic flux density images subject to the horizontal and vertical injection currents, respectively, using 3 mA imaging currents.

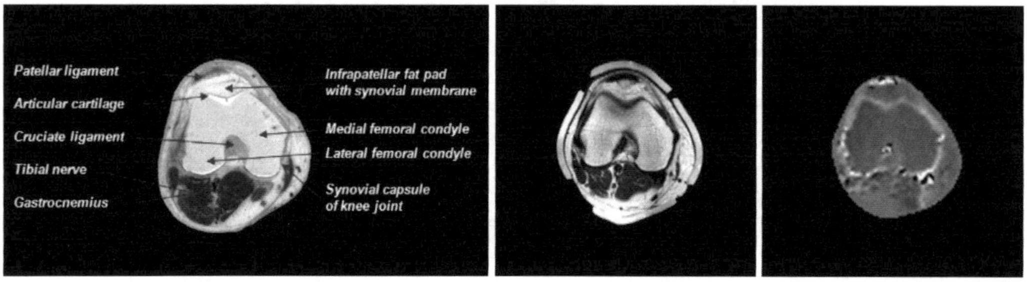

(a) (b) (c)

Figure 3. (a) Anatomy of the male knee. (b) and (c) are MR magnitude and reconstructed conductivity images, respectively.

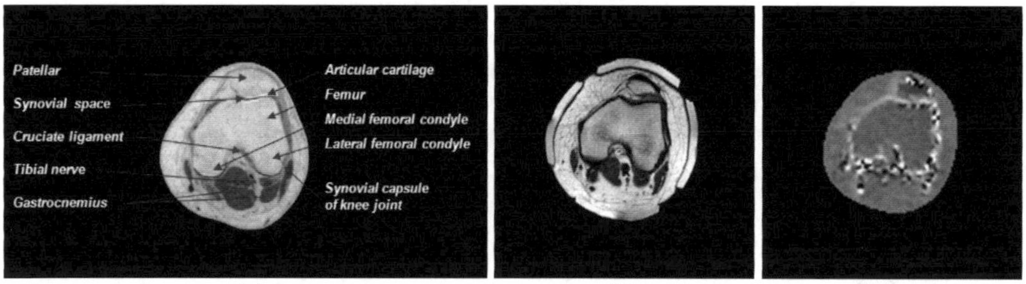

(a) (b) (c)

Figure 4. (a) Anatomy of the female knee. (b) and (c) are MR magnitude and reconstructed conductivity images, respectively.

4. Discussion

Numerous previous MREIT experiments of postmortem and *in vivo* animals enabled us to design the human experiment and successfully produce *in vivo* conductivity images of the human knee with a pixel size of about 1.7 mm. Though the first trial of human imaging experiments has been lately reported, it is still difficult to perform *in vivo* human and disease model animal experiments primarily due to a long scan time and high injection current. In this paper, we could reduce the current amplitude to 3 mA by using a multi-echo pulse sequence. We expect that the multi-echo pulse sequence is advantageous for further *in vivo* conductivity imaging of human subjects using 3 mA or less imaging currents. We plan to perform numerous *in vivo* animal and human experiments to advance the MREIT technique to the stage of clinical uses.

Acknowledgments

This work was supported by the SRC/ERC program (R11-2002-103) of MEST/NRF.

References

[1] Zhang N 1992 Electrical impedance tomography based on current density imaging *MS Thesis* University of Toronto
[2] Woo E J, Lee S Y and Mun C W 1994 Impedance tomography using internal current density distribution measured by nuclear magnetic resonance *Proc. SPIE* **2299** 377–85
[3] Ider Y Z and Birgul O 1998 Use of the magnetic field generated by the internal distribution of injected currents for electrical impedance tomography (MR-eIT) *Elektrik* **6** 215–25
[4] Woo E J and Seo J K 2008 Magnetic resonance electrical impedance tomography (MREIT) for high-resolution conductivity imaging *Physiol. Meas.* **29** R1–R26
[5] Muftuler L T, Hamamura M J, Birgul O and Nalcioglu O 2006 In vivo MRI electrical impedance tomography (MREIT) of tumors *Technol. Cancer. Res. Treat.* **5** 381–7
[6] Kim H J, Lee B I, Cho Y, Kim Y T, Kang B T, Park H M, Lee S Y, Seo J K and Woo E J 2007 Conductivity imaging of canine brain using a 3 T MREIT system: postmortem experiments *Physiol. Meas.* **28** 1341–53
[7] Kim H J, Oh T I, Kim Y T, Lee B I, Woo E J, Seo J K, Lee S Y, Kwon O, Park C, Kang B T and Park H M 2008 In vivo electrical conductivity imaging of a canine brain using a 3 T MREIT system *Physiol. Meas.* **29** 1145–55
[8] Kim H J, Kim Y T, Minhas A S, Jeong W C, Woo E J, Seo J K and Kwon O J 2009 In vivo high-resolution conductivity imaging of the human leg using MREIT: the first human experiment *IEEE Trans. Med. Imaging* **28** 1681–7
[9] Feinberg D A, Mills C M, Posin J P, Ortendahl D A, Hylton N M, Crooks L E, Watts J C, Kaufman L, Arakawa M, Hoenninger J C and Brant-Zwadwaski M 1985 Multiple spin-echo magnetic resonance imaging *Radiology* **155** 437–42
[10] Jeon K, Minhas A S, Kim Y T, Jeong W C, Kim H J, Kang B T, Park H M, Lee C O, Seo J K and Woo E J 2009 MREIT conductivity imaging of postmortem canine abdomen using CoReHA *Physiol. Meas.* **30** 957–66
[11] Jeon K, Kim H J, Lee C O, Woo E J and Seo J K 2009 CoReHA: conductivity reconstructor using harmonic algorithms for magnetic resonance electrical impedance tomography (MREIT) *J. Biomed. Eng. Res.* **30** 279–87
[12] Seo J K, Kim S W, Kim S, Liu J J, Woo E J, Jeon K and Lee C O 2008 Local harmonic B_z algorithm with domain decomposition in MREIT: computer simulation study *IEEE Trans. Med. Imaging* **27** 1754–61

Journal of Physics: Conference Series **224** (2010) 012149 doi:10.1088/1742-6596/224/1/012149

Using Distmesh as a Mesh Generating Tool for EIT

Solomon Fung, Andy Adler, Adrian D.C. Chan

Department of Systems and Computer Engineering, Carleton University, Ottawa, Canada

sfung@connect.carleton.ca

Abstract. We present a collection of automatic mesh generation software tools for electrical impedance tomography (EIT) finite element models (FEMs). Creating an image in EIT involves retrieving conductivity distribution from surface electrode measurements. The initial forward problem of this process requires estimates of electric potential according to the current sources. With non-trivial geometries, this task is difficult analytically and demands a numerical method such as FEM. The areas nearest to electrodes have the highest current density and are the most significant measurement points. For our FEM model, these areas should have the greatest mesh density for the purposes of usefulness and accuracy. The research focuses on Distmesh, an open source MATLAB mesh generator, for creating mesh shapes with non-uniform mesh densities for EIT software. The inherent simplicity of the Distmesh code was the rationale for its evaluation. A mesh shape is defined from specifying its signed distance function and its mesh size distribution is defined with an element length function. We created circular meshes with mesh refinement towards endpoints of distributed electrodes. In three dimensions, cylinders were produced with mesh refinement towards rings of points for each electrode. Finally, to accommodate organic shapes, meshes for closed contours were created using elliptic Fourier descriptors. Because Distmesh was not designed for efficiency, the generation of well-defined meshes is fairly slow. For the contours, Distmesh forms reasonable meshes but is unable to improve node locations upon successive iterations. Distmesh is not a suitable mesh generator for EIT software unless its code is optimized for speed.

1. Introduction

Electrical impedance tomography (EIT) is an imaging technology where images are constructed from impedance distributions. In contrast with conventional medical imaging technologies, EIT is a low-cost, portable and risk-free solution [1]. It does not involve any ionizing radiation, nor depend on limited materials or novel parts. Periphery electrodes are used to inject current across a medium and measure the resulting electric potential. A reconstruction algorithm relates the boundary values to the internal distribution of conductance or impedance. Several limitations in modern EIT technology such as low spatial resolution and noise sensitivity [2] have prevented its adoption in ordinary clinical use.

The EIT image reconstruction process consists of two parts. First, the current flow and voltage measurements are estimated for a current injection pattern. This is known as the forward model. The second part is the inverse solution, where the forward model is used to calculate the impedance distribution for constructing the image. The geometries used for the forward model should be as accurate as possible by considering irregularities and anisotropy [3]. An analytic solution is appropriate for simple geometries, as for a circle or sphere where current and voltage can be simply calculated. However for complex, realistic shapes, a unique and stable solution may not exist. In this case a numerical approximation may provide a solution for this ill-posed problem.

Finite element method (FEM) is a suitable numerical technique for solving impedance for EIT. A field variable normally assumes an infinite number values over a domain. FEM divides the domain into a manageable number of elements, forming a mesh. For each element an interpolation function approximates the value of the field variable. By assembling the element properties and imposing the boundary conditions, a system equation is constructed whose solution describes the behavior of the field variable over the entire domain. FEM is able to handle complex and irregular domains with ease.

Our research involves the use of Distmesh as a mesh generator for EIT models. Distmesh provides a set of MATLAB functions for creating high quality tetrahedral meshes in two or three dimensions. It is capable of accepting different shape boundaries and custom defined mesh densities. The tool was deliberately designed with short, simple code to encourage open source development [4]. Therefore, it is a worthy candidate as a mesh generator for free EIT software. Our research mainly focuses on experimenting with distance and edge length functions to define different shapes and mesh densities appropriate for EIT models.

2. Methods

The primary meshing function of Distmesh is composed of only a few dozen lines of code. At the beginning of the mesh process, the boundary outline is defined by a signed distance function [4]. The signed distanced function calculates the shortest distance between the shape boundary and any point. Its sign is positive for points outside the boundaries and negative for those inside. After the initial randomized distribution of nodes, Distmesh will eliminate nodes outside of the shape according to its sign. The inner space is then triangulated with a Delaunay algorithm. The resulting mesh can be analogously treated as a mechanical truss structure. Its topology is optimized by iterative node displacement and retriangulation to achieve force equilibrium [4]. The iterations halt once node displacements reach a minimum tolerance. A thorough explanation is given in [4].

The Distmesh package includes functions for simple shapes and is capable of producing mesh shapes given a signed distance function. Distmesh accepts an input value that specifies the element edge length in the initial mesh iteration. It also accepts an edge length function that dictates how node distances are modified upon further iterations. The default function is a uniform element size, producing a consistent density of nodes with the same edge length. In our case, we desire more elements at important regions. This increases accuracy since more variable estimates are calculated per unit area. For EIT, the electrode regions are the most significant areas of concern as they are the origins of the applied currents and locations of voltage measurements. With Distmesh, we passed on distance functions to produce mesh shapes and then passed on edge length functions dependant on point distances to refine mesh densities near electrode locations.

2.1. Circular meshes

The simplest distance function with Distmesh creates a circular mesh. A circle with its centre at the origin has a distance function of $d - r$, with d being the distance of any point to the origin and r being the radius of the circle. We produced a circular mesh function including inputs for circular radius, edge length, number of electrodes and electrode length. These function handles were passed on as arguments in Distmesh's mesh generator. The endpoints of each electrode arc length were used as refinement points and each pair was distributed evenly around the circumference according to the total number of electrodes. Element size changed proportionally to the distance from the refinement points using an edge length function. Optional inputs for this include a maximum and minimum value to control the size extremes for the edge lengths.

2.2. Cylindrical meshes

Distmesh supports n-dimensional geometries and shapes resulting from set operations of distance functions, allowing us to proceed from a circular mesh to a cylindrical one. A cylinder can be created by the intersection of a circular distance function with two planes. The circular distance function is identical to the 2D one, since it is independent from the orthogonal *z*-axis. Two upper and lower planes are simply *x*-*y* planes of constant *z* values.

The ideal electrode arrangement for a cylindrical body is rows of vertically aligned rings [5]. We gave the option of inputting three values giving the number of rows and columns and the number of refinement points per electrode. The points for each electrode were distributed about the boundary of a disc curved along the cylindrical surface. The disc points were rotationally duplicated to create a row of electrode regions and then duplicated with vertical shifts for additional columns.

2.3. Meshes using elliptic Fourier descriptors (EFDs)

Circular boundaries are appropriate for experimental modeling but do not accurately depict the irregularities of organic shapes. Thus, we need a mathematical method of describing simple, closed curves in 2D. A series of elliptic Fourier descriptors can efficiently describe these types of contours [6]. EFD has been used for describing and classifying organic shapes in morphological analyses [7]. Joeng and Radke experimented with Fourier descriptors to model sampled slices for medical imaging [8].

The process of elliptic Fourier analysis is described as follows. The varying x and y coordinates about the contour are expressed as a periodic function for one complete revolution. These functions are based on a parameter t which begins at 0 and reaches 2π after one revolution. This parameter does not imply that the contour tracing arm progresses angularly. A Fourier transform of the periodic functions provides a set of harmonic values that can be used to faithfully reconstruct the contour through Fourier series expansion. Elliptic Fourier descriptors are rotationally invariant, making shape description independent of contour orientation.

We used David Thomas' MATLAB functions of forward and inverse elliptic Fourier transforms for handling shape boundaries. The forward transform accepts a set of boundary outline points and calculates the values for a desired number of harmonics. The inverse transform reconstructs the boundary outline based on provided Fourier harmonic values for the parametric x and y expressions. These transforms enable us to convert between elliptic contours and Fourier harmonics, but cannot produce a mesh in Distmesh without first deriving a distance function.

For this distance function, we began with a general distance function for a reference point to any point on the contour given the value for parameter t, which can be thought of as time from 0 to 2π. The derivative of this function with respect to t gives the velocity of the contour point tracing the outline. A change from negative to positive velocity indicates a local minimum distance. Velocities about the contour were sampled at distributed t intervals and the adjacent negative-to-positive value pairs were selected as brackets for the false position method to find the roots. Each root is a value of t giving the point of zero velocity, indicating a local minimum distance. Using our general distance function, the distances at these roots were calculated and the lowest value is selected as the minimum distance. However, the sign of the distance values still needs to be resolved for node elimination. This is essentially a point-in-polygon problem. We decided to use a winding number calculation that measures the number of revolutions traversed around a point while tracing a contour line. Any point with a winding number less than one with respect to a shape must be outside that shape. This calculation for our contours is faster and more accurate than MATLAB's built-in point-in-polygon function. Our mesh densities were uniform as we did not yet pursue electrode refinement points.

Elliptic Fourier descriptors can accurately construct boundaries that adhere to anatomical sections. This capability is valuable for our mesh generation since boundary shape errors adversely affect reconstructed EIT images [9]. Boundaries that are more irregular may require more harmonic descriptors for shape description.

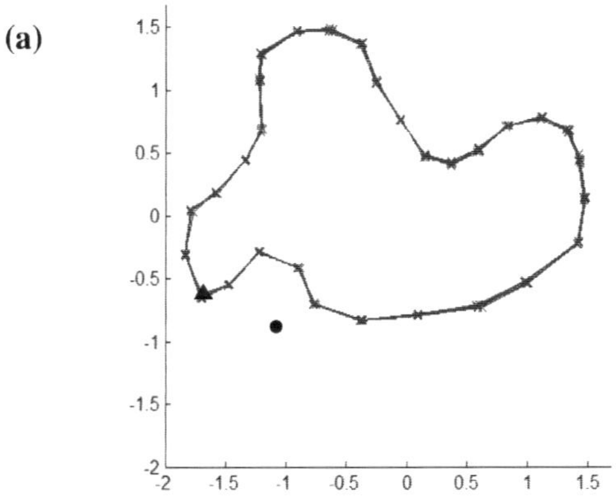

Figure 1. (a) shows an outline formed by connecting points from mouse input. The point is the selected reference point for distance and the contour begins at $t = 0$ at the triangle and proceeds clockwise. (b) is the plot of contour distance from the point as t reaches 2π. (c) is the plot of contour velocity. Zero crossings during increasing velocity indicate t values where a local minimum distance is found. Calculating distances at these t values with (b) gives us a set to compare for smallest distance.

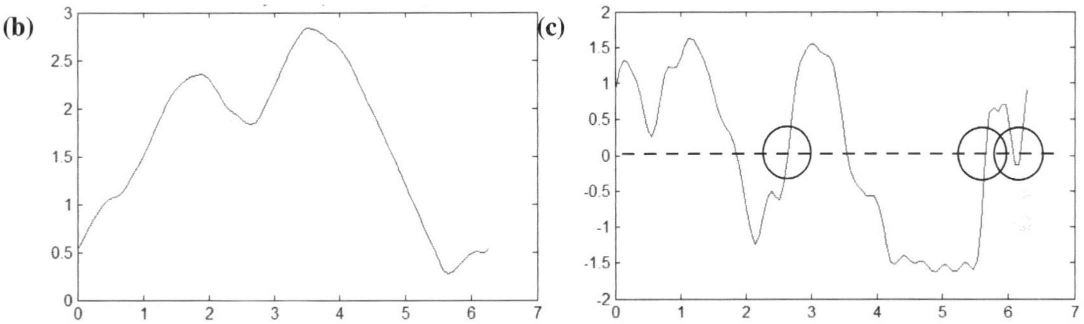

3. Results

The following figures show some of our Distmesh FEM models. Each one was produced by the separate functions previously described. The function producing EFD meshes uses a contour outline formed by user mouse input. Our EFD distance function is then passed onto Distmesh for mesh construction. Using twenty or more harmonics will produce shapes that accurately depict the original outline. Unfortunately, we were not able to reach the default iterative stop condition due to unremitting node displacement. Relaxing the displacement tolerance will comprise mesh quality for dense meshes.

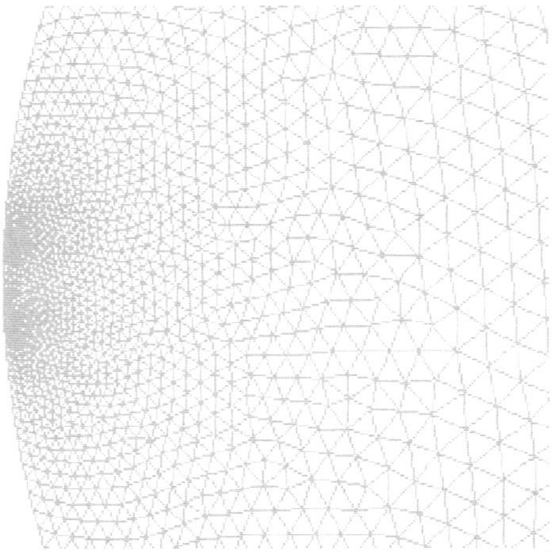

Figure 2. A portion of a circular mesh with refinement towards a pair of electrode endpoints. The entire mesh contains over 10,000 elements with 8 electrode regions.

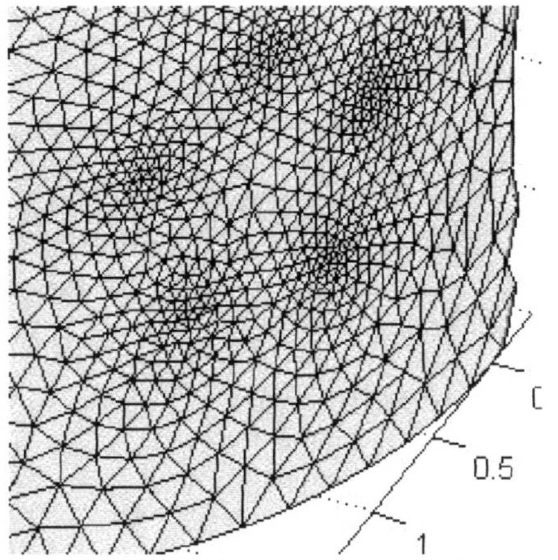

Figure 3. This close-up image shows the mesh surface of an electrode region with 5 endpoints on a 3D cylinder. This mesh has two rows and six columns of electrode regions (not shown).

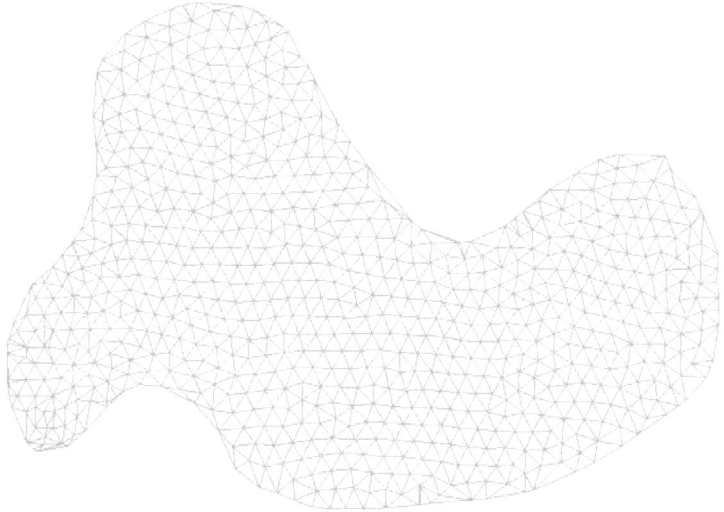

Figure 3. This mesh is formed by passing our EFD distance function to Distmesh. With twenty Fourier harmonics, the mesh boundaries correspond well with the inputted outline from figure 1. Note the inconsistency of the element sizes in the mesh's lower left region. Distmesh was not able to resolve this with additional iterations.

4. Conclusion

Using Distmesh, we were able to create a variety of FEM meshes for EIT models. Our shape boundaries were based on providing signed distance functions. The ability of Distmesh to adjust element sizes based on edge length functions permitted meshes with very fine and dense elements near electrode points. We also used elliptic Fourier descriptors to express curved boundary contours. The signed distance function for EFD was based on finding roots of the contour velocity function then calculating the winding number to solve the point-in-polygon problem. Combining our EFD mesh function with a boundary extraction tool for medical images should produce useful FEM meshes for EIT forward models.

Since the design of Distmesh was intended for simplicity and ease of understanding, the code does not perform efficiently or quickly. Useful 2D EIT meshes contain tens of thousands of mesh elements and perhaps millions for 3D meshes. This can take days for Distmesh to produce quality mesh density refinement in circles and cylinders of many elements. Although our EFD mesh function creates reasonable shapes based on a custom outline, Distmesh is unable to make improvements after successive iterations. This may be caused by numerical approximations in our EFD signed distance function.

Without enhancements to the main code, Distmesh is prohibitively slow for producing complex meshes with many point refinements. Therefore it is not an ideal solution for an EIT meshing tool. Relaxing displacement tolerances will decrease the number of processed iterations at the expense of lower quality mesh triangulation. Porting a C++ version may perhaps bring substantial improvements in speed. However, this is a much more difficult undertaking than what was done so concisely and comprehensibly in MATLAB. We hope to implement our Distmesh functions in the open source EIT and diffuse optical tomography reconstruction software (EIDORS) for further development and experimentation.

5. Acknowledgements

I would like to thank the Natural Sciences and Engineering Research Council (NSERC) of Canada for their support through an undergraduate research award.

References

[1] Soleimani M 2006 Electrical impedance tomography imaging using a priori ultrasound data *Biomed. Eng. Online* **5**

[2] Oh S, Tang T and Sadleir R J 2007 Quantitative analysis of shape change in electrical impedance tomography (EIT) *IFMBE Proc.* **17** 424-27

[3] Bagshaw B, Liston A, Bayford R H, Tizzard A, Gibson A, Tidswell A, Sparkes M, Dehghani H, Binnie C and Holder D S 2006 Electrical impedance tomography of human brian function using reconstruction algorithms based on the finite element method *Physiol. Meas.* **27** S1-11
[4] Persson P and Strang G 2004 A simple mesh generator in MATLAB *SIAM Rev.* **46** 329-45
[5] Graham B M, Adler A 2007 Electrode placement configurations for 3D EIT *Physiol. Meas.* **28** S29-44
[6] Kuhl F and Giardina C 1982 Elliptic Fourier features of a closed contour *Computer Graptics and Image Processing* **18** 236-58
[7] Rohlf F J and Archie J W 1984 A comparison of Fourier methods for the description of wing shape in mosquitos (diptera: culicidae). *Syst. Zool.* **33** 302-317
[8] Jeong Y and Radke R J 2007 Reslicing axially sampled 3D shapes using elliptic Fourier descriptors *Med. Image Anal.* **11** 197-206
[9] Lionheart W R B 1998 Boundary shape and electrical impedance tomography *Inverse Problems* **14** 139-47

Journal of Physics: Conference Series **224** (2010) 012150 doi:10.1088/1742-6596/224/1/012150

Development of a Low Noise MREIT Current Source

Young Tae Kim, Pil Joong Yoo, Tong In Oh and Eung Je Woo

Department of Biomedical Engineering, Kyung Hee University, Yongin, Korea

E-mail: ejwoo@khu.ac.kr

Abstract. In MREIT conductivity imaging experiments of animal and human subjects, we should minimize the noise level in measured magnetic flux density data induced by injection currents with low amplitude. Since noise and artifact from an MREIT current source directly affect the quality of the data, a low-noise current source is desirable. In order to be compatible with various MREIT pulse sequences, it should be also programmable. We have developed a new MREIT current source, which is controlled by a PC program for flexibility. We designed it in such a way that it is located inside the shield room of an MRI system. To minimize noise and artifact, we adopted an optical link for the connection to the PC outside the shield room. The enclosure of the new current source provides a magnetic as well as electric shielding to prevent high frequency switching noise of the current source from interfering with the scanner. It is powered by a rechargeable battery so that the entire current source is isolated from the ground. Equipped with automatic lead switching capability, it simplifies and automates MREIT imaging experiments. Our experimental results show that its performance is superior to the previous version, which is located outside the shield room.

Keyword: MREIT, constant current source, low noise, optical link

1. Introduction

Cross-sectional imaging of electrical conductivity has been pursued in magnetic resonance electrical impedance tomography (MREIT). The MREIT data acquisition system consists of an MRI scanner, surface electrodes and a constant current source. The current source comprises analog and digital circuits to produce injection current, a dc power supply and connectors for lead wires. It is usually located outside the shield room and there are numerous sources noise and artifact including long cables from the current source to electrodes. In most experimental MREIT studies, currents have been injected as rectangular pulses synchronized with a chosen MR pulse sequence. From numerous experimental works in MREIT, we found that a proper design and implementation of the current source is crucial to minimize the amounts of artifact and noise in measured magnetic flux density (B_z) data.

Recently new pulse sequences have been introduced in MREIT including the spin echo with multiple echo data acquisitions, balanced steady sate free precession (b-SSFP) pulse sequence and also echo planar imaging (EPI) type of data acquisition with switching of phase gradient in a single repetition time (TR) [1]. In order to accommodate various MR pulse sequences, we need to develop a new low-noise MREIT current source, which is programmable.

In this paper, we improved the existing MREIT current source [2] by removing numerous sources of noise and artifact. The new current source is located inside the shield room and connected to a PC and MR spectrometer through optical links. Adopting electromagnetic shielding and battery power, we designed the current source to have a low noise level. We describe the new MREIT current source, which is programmable by a remote PC for its flexibility.

2. Method

Injection current must be synchronized with the RF pulse of the MR system. Since the MR spectrometer provides trigger signals, we must connect it to the current source. Figure 1 shows the structure of the new MREIT current source equipped with an optical link to a PC and a separate optical link to the MR spectrometer for triggering. We developed an optical-USB converter for the PC and an optical-Tx converter for the spectrometer. This provides a complete isolation of the current source from other equipments and allows us to put the current source inside the shield room. We used two 10-meter optical cables to connect the current source to the PC and spectrometer outside the shield room. The current source consists of a microcontroller, a USB controller, a waveform generator using a 16-bit DAC, a voltage-to-current converter, a lead switching module, an ADC for voltage measurement, a dc power supply and a magnetic shielding case.

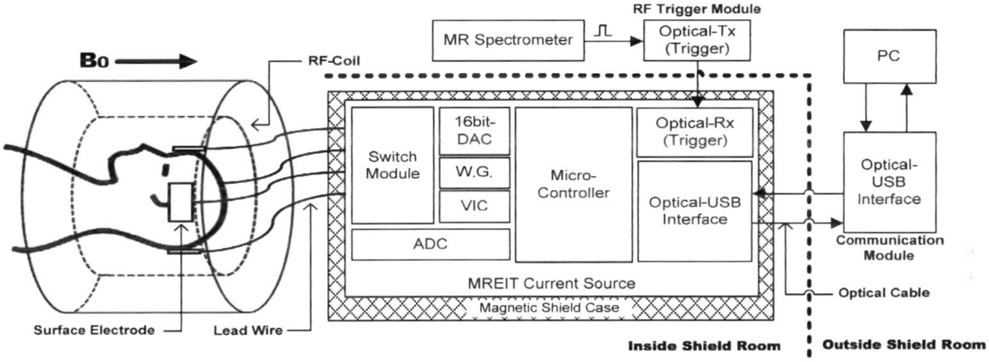

Figure 1. Structure of the new MREIT current source (W.G. is a waveform generator and VIC is a voltage-to-current converter).

2.1 PC communication and RF trigger modules

We used fiber-optic devices (HFBR T-1521 and R-2521, Agilent technologies, USA) for optical links with a PC program and also for trigger signals from the spectrometer. Using the optical links, we reject incoming noise from the PC and other electromagnetic interferences from outside the shield room. On the PC side, we powered the optical link using a separate rechargeable battery since the USB power was not enough. The current source must be triggered by the MR spectrometer. The RF trigger module at the spectrometer side converts TTL trigger pulses to optical signals and this rejects noise from the spectrometer outside the shield room.

2.2 MREIT current source

Serial communication module inside the MREIT current source exchange commands and data with the PC. It receives control signals from the PC and sends measured voltage data to the PC. It measures voltage using a 12-bit ADC operating at 25 MHz. This allows us to monitor injection currents and induced voltages in real time.

The main controller is a DSP (TMS320F2812, Texas instruments, USA) operating at 150 MHz. It controls lead switching though its digital output ports. It has a sequence of instruction in a form of a table including trigger source selection, delay time, pulse width, amplitude and lead switching configuration. When it starts to inject current, the main controller controls all operations using the instruction table. We may inject current in an arbitrary waveform since we can store waveform data in the main controller. One can program current source in a flexible way using a PC by writing a script file for the instruction table.

We used a 16-bit DAC (AD768, Analog Devices, USA) to generate an arbitrary waveform with fine resolution. Injection current amplitude is controlled by a digital potentiometer (DS1267, Dallas Semiconductor, USA) and switches. High-power operational amplifier (OPA452, Burr-Brown, USA) is adopted to produce output current with its maximal amplitude of 50 mA.

2.3 PC control program

At the PC side, we control the current source by using a script file including information about current waveform, current amplitude, offset, repetition time, polarity and triggering conditions. All of these parameters must be determined based on a pulse sequence to be used for an intended MREIT imaging experiment. Table 1 shows an example of a script file. It has a similar structure with a pulse sequence script file in an MRI system. Pulse sequence programmers have the option of changing the script file according to their pulse sequence.

Table 1. Example of script file.

MODE (*R, F, M*)	Delay Time (*ms*)	Pulse Width (*ms*)	Amplitude (*mA*)	Switch Index
R	10	12.4	20	1
R	10	14.7	-20	2
F	0	12.4	30.5	1
F	0	14.7	30.5	2
M	30	10	50	3
M	30	15	42	4
R	0	20	15	5
F	0	20	-15	5

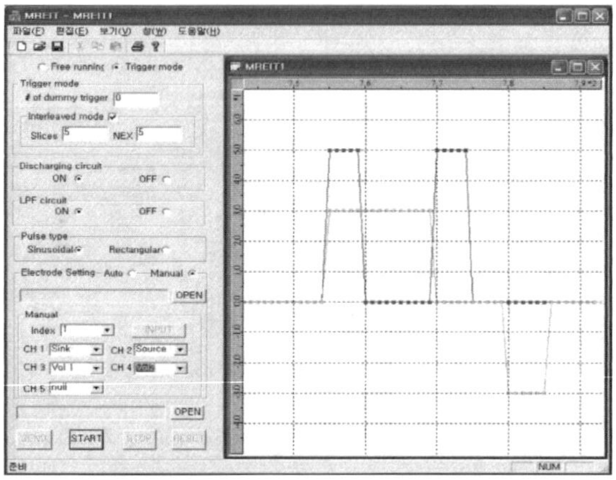

Figure 2. PC control program

3. Results

3.1 Developed MREIT current source

Figure 3. Implemented MREIT current source

Figure 2 and 3 present the PC control program and the developed MREIT current source. Current outputs from the previous and new MREIT current sources are shown in figure 4. Using the new MREIT current source, we could obtain an MR magnitude image with its SNR of 456.5. When we used the previous current source under the same condition, the magnitude image SNR was 330. This indicates that the new current source improves the MR image SNR by 38%. Figure 5 illustrates a simple MREIT imaging experiment using the new current source. We observed SNR improvements in all three images of the MR magnitude in (a), magnetic flux density in (b) and also reconstructed conductivity image using CoReHA in (c) [3].

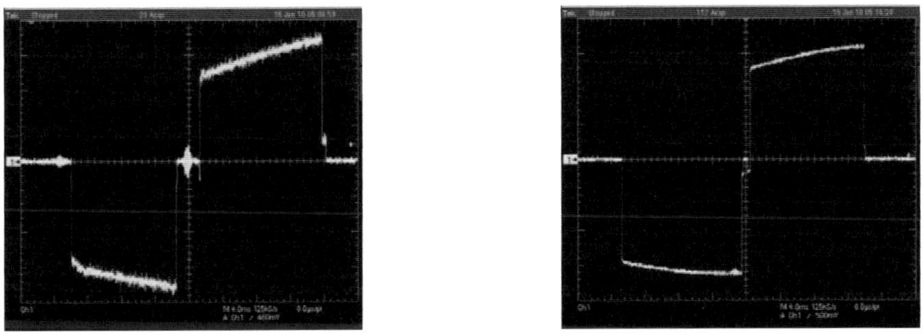

(a) (b)

Figure 4. Injection current waveforms from the (a) previous and (b) new current sources.

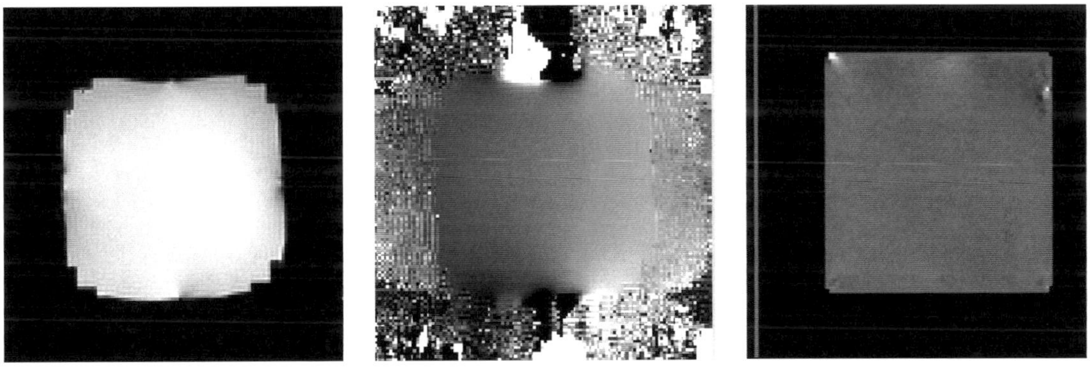

(a) (b) (c)

Figure 5. (a) MR magnitude image of saline phantom. (b) Magnetic flux density image for vertical current injection. (c) Reconstructed conductivity image using CoReHA.

4. Conclusion and discussion

We designed and implemented a new MREIT current source. Using optic links, electromagnetic shielding and battery power, we could place it inside the shield room and minimize noise and interference. We found that SNR improvements in MREIT images are significant when we compare the new current source with the previous one. Since it is programmable by a remote PC, we may use it for any MR pulse sequence. In our future studies, we plan to use the new MREIT current source to develop new MREIT pulse sequences. Human imaging experiments will be also conducted by using the new current source.

Acknowledgments

This work was supported by the SRC/ERC program (R11-2002-103) of MOST/NRF.

Reference

[1] Sersa I 2009 Current density imaging sequences with separation of mobile-ion current from immobile-ion current *Journal of Magnetic Resonance* **196** 33-38

[2] Kim H J, Kim Y T, Minhas A S, Woo C J, Woo E J, Seo J K and Kwon O J 2009 In vivo high-resolution conductivity imaging of the human leg using MREIT : the first human experiment *IEEE Trans. Med. Imag.* **28** 1681-7

[3] Jeon K, Minhas A S, Kim Y T, Jeong W C, Kim H J, Kang B T, Park H M, Lee C O, Seo J K and Woo E J 2009 MREIT Conductivity imaging of the post-mortem canine abdomen using CoReHA *Physiol. Meas.* **30** 957-966

Journal of Physics: Conference Series **224** (2010) 012151 doi:10.1088/1742-6596/224/1/012151

The influence of frequency separation on imaging properties in DFEIT

A Bujnowski, J Wtorek and M Lewandowska

Biomedical Engineering Department, Gdansk University of Technology, Narutowicza 11/12 80-952 Gdansk, Poland

E-mail: bujnows@biomed.eti.pg.gda.pl

Abstract. A Dual Frequency EIT is an extension of a traditional EIT that uses two sinusoidal signals for imaging. Appropriate selection of signals' frequency allows to achieve reasonable contrast of imaged structure. It has already been shown that frequency of the signals should cover a selected dispersion range (usually a beta dispersion) for living biological objects. Thus, in respect to different application the upper frequency may achieve (or at least it should achieve) a relatively large values. The upper limit of frequency used in DFEIT is considered in paper. An application of Laplace equation for solving of the forward model as well for a sensitivity calculation is verified and Maxwell equations are suggested for high frequency component. However, there is a limitation in increasing frequency resulting from current leakages and hardware properties. Increasing frequency above a certain one, determined both by object and measurement system properties, is worthless. A current-to-potential and a potential-to-current mappings were compared from a perspective of achievable properties of existing hardware solutions. The comparison was made using SPICE simulations. Additionally, the coupling between object and environment was considered for the high frequency component.

1. Introduction

Electrical Impedance Tomography (EIT) is an imaging technique that produces images of electrical properties distribution within an object on the basis of the knowledge of currents and potentials at its boundary. From a technical point of view it uses a set of cross-impedances measured between electrodes attached to boundary of the object.

Frequency-Difference Electrical Impedance Tomography (DFEIT) is a version of the EIT that utilizes data collected at two frequencies of applied current or voltage in reconstruction procedure [1, 2]. Obtained image presents the difference of the objects conductivity measured at these two frequencies. This imaging method is particularly useful for dispersive objects visualization e.g. biological tissues. Measurements, depending on the hardware complexity, may be made sequentially for each frequency of applied current or simultaneously. In the later case the applied signal is a sum of two sinusoidal signals $U(t) = A_1 \cos(\omega_1 t) + A_2 \cos(\omega_2 t)$. In general, there are two approaches to use this signal. A first method utilizes sum of two signals of equal amplitudes (i.e. $A_1 = A_2$) and images are reconstructed using backprojection. However, weighted logarithms of measured values at two frequencies are used as measurement data [3]. Second approach is based on a weighted sum of two sinusoids. The frequency and amplitude of each component are selected appropriately to the objects spectral properties. The aim of a selection procedure is to achieve possibly the highest value of the contrast [1, 4]. It has been proved that selection of the frequencies when using DFEIT is essential for the whole reconstruction process. It has an influence on the procedures, accuracy of the measurements and calculations of the forward and inverse problems. An example of the measurement procedure fulfilling above mentioned requirements is shown in [1, 4]. It follows, both, from the literature and our studies that frequency should be chosen for the object presenting the highest contrast while another for the object behaving as uniform (or almost uniform). Typically, for the biological materials, the former frequency is lying in the range of kilohertz while the latter one should be as high as possible. However, there are several factors, both theoretical and experimental, arising when using this approach. It is forbidden increasing the measurement current frequency without changing a mathematical description of the problem. Furthermore, an increased level of errors in measurement data should be also expected. In the paper we consider three essential factors: the properties of the measurement strategies, the forward problem approximation and the coupling of the object with surroundings.

2. Materials and methods

2.1. Measurement strategy

There are two main measurement strategies used in EIT: a potential distribution is measured as a result of applied current pattern or vice-versa: the current division between electrodes is measured in response to applied voltage pattern (fig. 1).

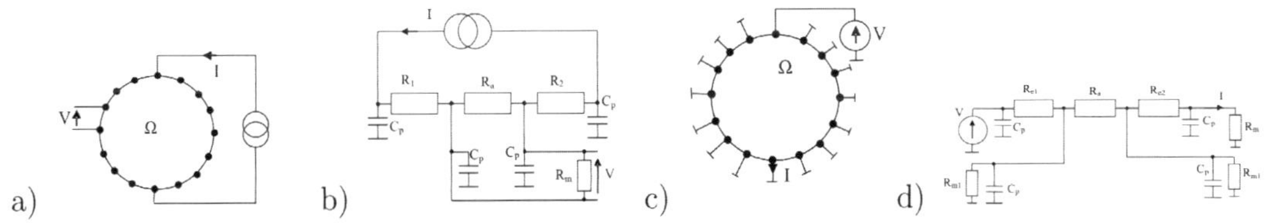

Figure 1. The measurement strategies: a) the current application and voltage measurement, b) an equivalent circuit model for single measurement, c) the voltage application and resulting current distribution between electrodes measurement, and d) equivalent circuit of voltage-current strategy.

Similar models have been used in both cases (along with identical resistance of electrodes). The voltmeter of input resistance equals 10 MΩ and parallel capacitance of 30 pF has been used in current-voltage strategy. An ideal current to voltage converter has been assumed for voltage current strategy. Ideal current and voltage sources have been assumed in both cases. It allowed calculation of frequency characteristics for both strategies.

2.2. The forward problem

In general, the electrical current distribution within the object is described by Maxwell equations [5]. Maxwell equations reduce to Laplace relationship where a quasi-static condition is fulfilled:

$$\omega l \left(\frac{\mu_0 \varepsilon'}{2} \left(\sqrt{1 + \left(\frac{\sigma_{DC} + \omega \varepsilon''}{\omega \varepsilon'} \right)} + 1 \right) \right)^{1/2} << 1 \tag{1}$$

where σ_{DC} - constant-current conductivity, $\varepsilon^* = \varepsilon' + j\varepsilon''$ - the complex permittivity, l - the largest dimension of the object. If the condition (1) is not fulfilled a set of Maxwell equations should be used to solve the forward problem [5].

Journal of Physics: Conference Series **224** (2010) 012151 doi:10.1088/1742-6596/224/1/012151

A model containing a circular object of radius equal 0.16 m and surrounding medium (air or other objects) has been developed (figure 2). Sixteen equally spaced electrodes have been attached to the boundary of the object. Conductivity of the object has been assumed to be independent on frequency and equal 10 S/m while the surrounding medium has been described by permittivity equal to that of free space. A potential distribution inside the object and in surroundings has been calculated.

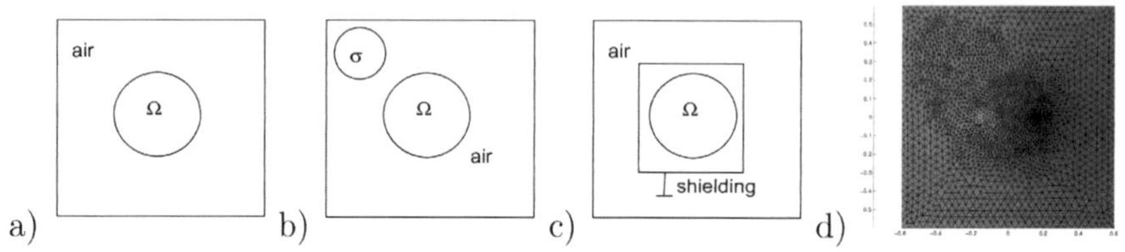

Figure 2. The model considered in the study: a) the object surrounded by air, b) the object and another one located in the vicinity both surrounded by air, c) the object surrounded by shielding surface and d) a developed FEM mesh.

In presence of two conductive objects conductivity of both has been assumed to be similar. Conductivity of the shielding screen attached to ground potential has been equal 10^4 S/m for the last modeled case.

3. Results

Assuming that the examined object is resistive, the potential distribution or the measured current flowing into electrodes should be independent of applied signal frequency. However, the current leakage to the surrounding can evolve and as a result the final signals are subject to change (figure 3).

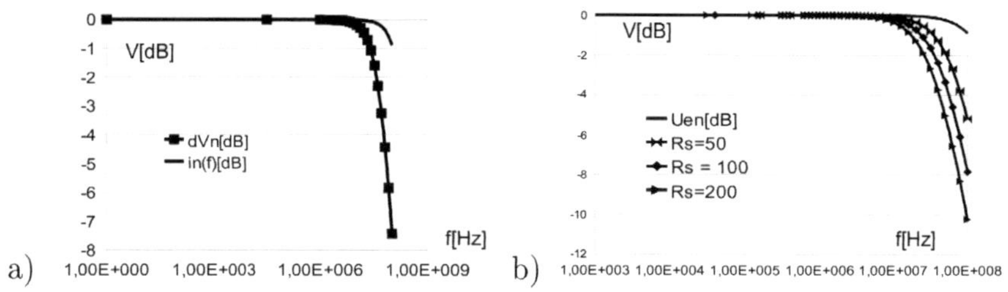

Figure 3. . Influence of measurement circuit properties on achieved frequency band: a) comparison of current and voltage application strategies, b) dependence of frequency band on a quality of the voltage source

The potential distribution was calculated using the Laplace and Maxwell equations (fig 4). Thus, it was possible to compare obtained solutions. A significant influence of measurement strategy was also noted. It was due to different arrangement of boundary conditions which were specific for each considered strategy. For a low frequency the obtained solutions were the same. However, increase in frequency led to big discrepancies (figure 5). A norm of voltage (current) difference $||v(f) - v(1kHz)||$ has been calculated as well for both strategies (figure 6).

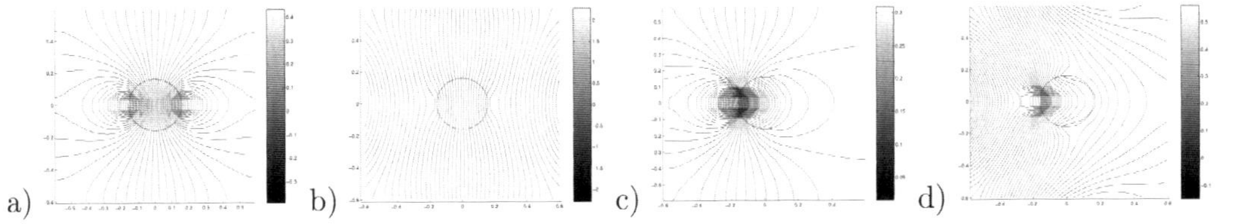

Figure 4. Voltage distribution for model presented in figure 2a): a) current strategy 1 kHz, b) current strategy 10 MHz, c) voltage strategy 1 kHz and d) voltage strategy 10 MHz

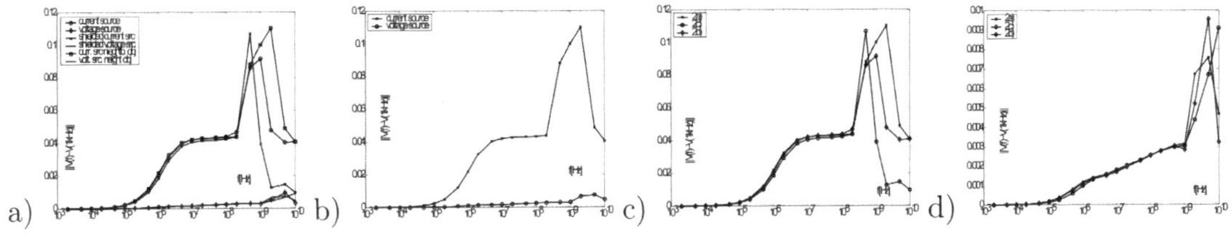

Figure 5. Difference between results obtained using Maxwell and Laplace solutions. Norm of difference between the potential distribution inside the object is presented: a) all cases considered, b) voltage and current strategies for model presented in figure 2a), c) the current strategy for all cases considered presented in figure 2a-2c, d) the voltage strategy for the cases as in c)

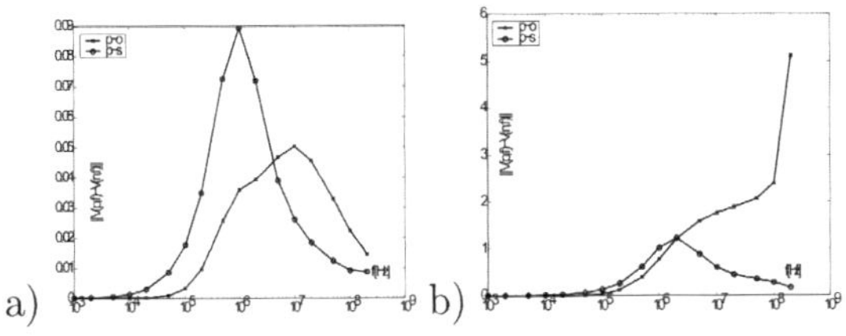

Figure 6. Influence of the object's surrounding on potential distribution inside it. Curves express the norm of potential difference between model 2a, 2b (p-o) and 2a, 2c (p-s) for: a) voltage application strategy, b) current application strategy

4. Conclusions

The obtained results suggest that increase frequency of applied signal involve a necessity of changing type of forward problem. There is also a relatively significant difference between potential distributions when using different signal application strategies. It follows from the fact that the considered strategies inherently create different boundary conditions.

References

[1] Bujnowski A 2007 *Roznicowa tomografia impedancyjna (differential impedance tomography)* Ph.D. thesis Gdansk University of Technology
[2] SCh Jun and J Kuen and J Lee and EJ Woo and D Holder and JK Seo 2009 *Physiol. Meas.* **30** 1087–1099
[3] Griffiths H and Ahmed A 1987 *Clin. Phys. Physiol. Meas.* **8** A103–A107
[4] Bujnowski A and Wtorek J 2007 *Proc of XIII ICEBI & VIII EIT Conference, Graz* 4
[5] Soni N, Paulsen K, Deghani H and Hartov A 2006 *IEEE Trans. Med. Imaging* **25** 55—61

Frequency-difference electrical impedance tomography: phantom imaging experiments

Sujin Ahn[1], Sung Chan Jun[1,5], Jin Keun Seo[2], Jeehyun Lee[2], Eung Je Woo[3], and David Holder[4]

[1]Department of Information & Communications, Gwangju Institute of Science and Technology, Gwangju, Korea

[2]Department of Mathematics, Yonsei University, Seoul, Korea

[3]Department of Biomedical Engineering, Kyung Hee University, Yongin, Korea

[4]Department of Medical Physics, University College London, London, UK

E-mail: scjun@gist.ac.kr

Abstract. Frequency-difference electrical impedance tomography (fdEIT) using a weighted voltage difference has been proposed as a means to provide images of admittivity changes at different frequencies. This weighted difference method is an effective way to extract anomaly information while eliminating background effects by unknown boundary geometry, uncertainty in electrode positions and other systematic measurement artefacts. It also properly handles the interplay between conductivity and permittivity in measured boundary voltage data. Though the proposed fdEIT algorithm is promising for applications such as detection of hemorrhagic stroke and breast cancer, more validation studies are needed. In this paper, we performed two- and three-dimensional numerical simulations and phantom experiments. Backgrounds of imaging objects were either saline or carrot pieces suspended in saline. We used carrot pieces to simulate a more realistic frequency-dependent admittivity distribution. Test objects were banana, potato or conductive gel with known admittivity spectra. When the background was saline, both simple and weighted difference approaches produced reasonably accurate images. The weighted difference method yielded better images from two-dimensional imaging objects with background of carrot pieces. For the three-dimensional head-shaped phantom, the advantage of the weighted frequency difference method over the simple difference method is not as obvious as in the case of the two-dimensional phantom. It is unclear if this is due to measurement errors or limitations in the linear algorithm. Further refinement and validation of the frequency difference image reconstructions are currently in progress.

1. Introduction

Frequency-difference EIT (fdEIT) reconstructs images of admittivity differences between two frequencies using simultaneously acquired boundary current-voltage data. We use it when time-referenced data are not available, as in acute ischemic stroke in the brain or breast cancer. As a simple linear fdEIT method has been shown to be ineffective in adult human stroke, we have developed a weighted voltage difference method, in which changes in the background over frequency are corrected [1, 2]. The goal of this study was to evaluate its performance. Images produced with it, time difference

[5] To whom any correspondence should be addressed.

and simple frequency difference have been assessed with simulated data and that acquired in two- or three-dimensional tanks filled with saline or a background of complex conductivity.

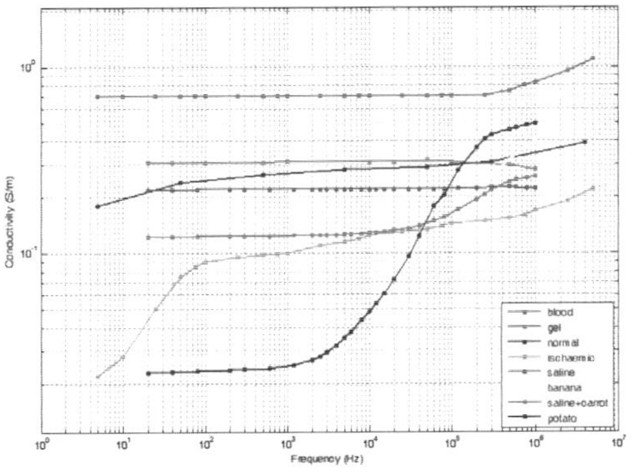

Figure 1. Conductivity spectra of blood [3], normal and ischemic brain tissues [3], and banana, potato, carrot, and gel.

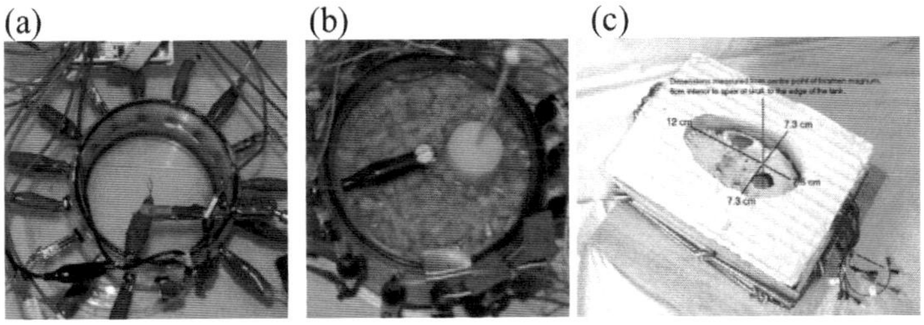

Figure 2.
(a) Two-dimensional cylindrical tank filled with saline, (b) the same tank filled with carrot pieces in saline and (c) three-dimensional head shaped tank.

2. Experimental design and methods

Images were reconstructed using time-difference (TD), simple frequency-difference (FD) and weighted frequency difference (WFD) methods from data produced either by simulation at 1, 5, 10, 50 and 100 kHz or from a two-dimensional cylindrical phantom and a three-dimensional head-shaped phantom filled with either saline or carrot pieces suspended in saline and with anomalies of a cylindrical piece of banana, potato or gel, with the diameter of each anomaly was 20% of the diameter of the tank. Biological materials are adopted since they have frequency-dependent conductivity spectra. The spectral properties of the mixture of saline and carrot pieces, potato and gel resembled those of normal, ischemic and haemorrhagic brain tissues, respectively (Figure 2). For the weighted frequency difference method, the reference frequency of $\omega_0/2\pi$ = 1 kHz was used to estimate an equivalent homogeneous admittivity. Tank data were acquired using a UCH Mk2.5 EIT system operating from 20 Hz to 1 MHz.

2.1 Direct impedance measurement

We measured conductivity spectra of saline, banana, potato and gel using an impedance analyzer (4284A, Agilent, USA) from 20 Hz to 1 MHz (Figure 2). Conductivity values of banana and carrot pieces in saline increased as frequency increased and those values of banana crossed that of saline at around 250 kHz. Conductivity of potato started increasing early and crossed that of saline at around 50 kHz.

2.2 Tank preparation

The diameter and height of the two-dimensional cylindrical tank were 10 and 8 cm, respectively. There were 16 electrodes equally spaced around its circumference. We filled it with either 0.1% saline or a mixture of carrot pieces (60%) and 0.1% saline (40%). We placed the banana anomaly as in figure 2(b) at the position of (2.5, 0) cm when the center of the phantom is denoted as (0, 0) cm. We placed the potato or gel anomaly at (-2.5, 0) cm. A head-shaped tank was made of silicon rubber and had 31 electrodes located at positions of the 10-20 EEG electrode placement system. It did not include a real human skull. The same anomaly settings including the banana anomaly in saline and the potato anomaly in the mixture of carrot pieces with saline were used.

3. Results

3.1. Numerical simulations

For images reconstructed from simulated data, acceptable images were produced for all modelled tanks by TD and WFD; with FD, images with the complex background were blurred for the two-dimensional case and inaccurate for the three-dimensional case.

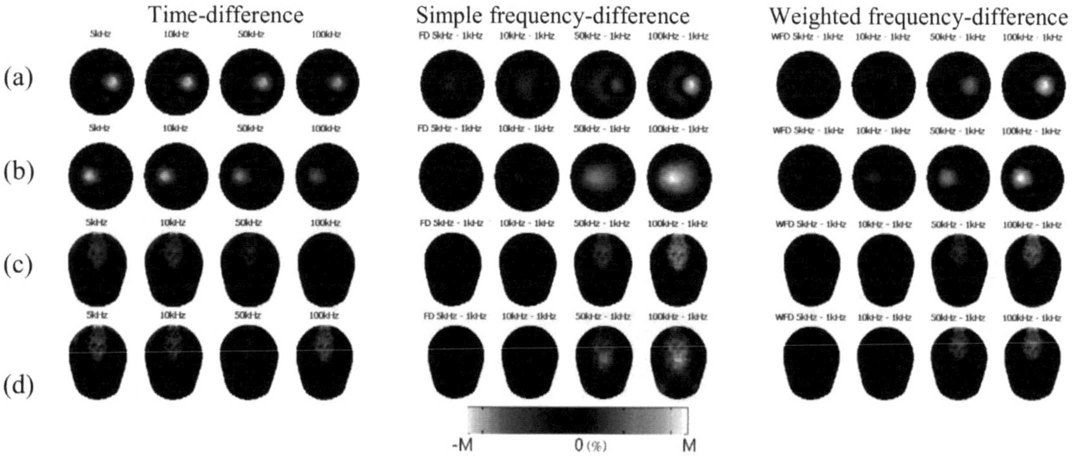

Figure 3. Reconstructed images from numerical simulation. The twelve images in each row are from the same anomaly setting. (a) The banana anomaly in the two-dimensional cylindrical phantom with the saline background. (b) The potato anomaly in the same two-dimensional phantom with the mixture of carrot pieces and saline as the background. (c) The banana anomaly in the saline background for the case of the three-dimensional head phantom. (d) The potato anomaly in the background of carrot pieces suspended in saline for the same head phantom. The pixel values represent percentile changes of conductivity.

3.2. Phantom experiments

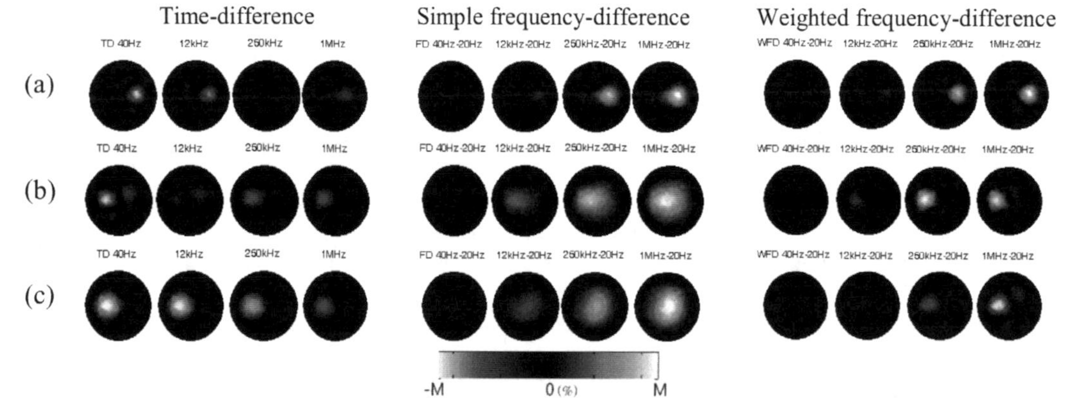

Figure 4. Reconstructed images of the two-dimensional cylindrical phantom. The twelve images in each row are from the same anomaly setting. (a) The banana anomaly in the saline background. (b) The potato anomaly in the background of carrot pieces suspended in saline. (c) The gel anomaly in the background of carrot pieces suspended in saline.

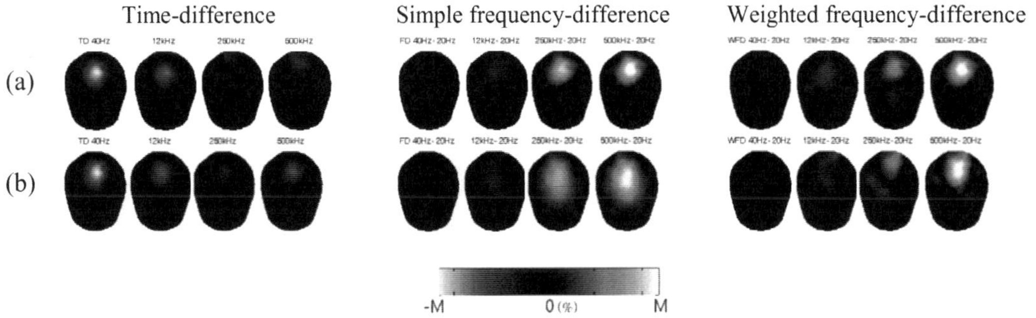

Figure 5. Reconstructed images of the three-dimensional head-shaped phantom. Twelve images in each row are from the same anomaly setting. (a) The banana anomaly in the saline background. (b) The potato anomaly in the background of carrot pieces suspended in saline.

For images collected in tanks, the findings are similar (Figure 4 and 5). With a saline background, the simple and weighted difference methods produced similar images. For the complex background, the weighted frequency-difference method performs better than the simple frequency-difference method. Blurred artifacts occurred around the perimeter of the anomaly since the conductivity values of the background as well as thc anomaly changcd nonlinearly with respect to frequency (Figure 4(b)). The weighted frequency-difference method minimized the artifacts and produced better localization of the anomaly. For the head-shaped phantom with one posterior anomaly, in the posterior the weighted frequency difference method again produced better images but the advantage is not as obvious as the two-dimensional case.

4. Conclusion

When the background conductivity does not change with frequency, the performance of the simple frequency difference method is comparable to that of the weighted frequency difference method. For the case of frequency-dependent background conductivity, the weighted frequency difference method produces better images in terms of its ability to localize an anomaly. However, its performance is degraded for the case of the three-dimensional head-shaped phantom. We plan to investigate effects of boundary geometry and electrode configuration. As shown in Jun et al [2], we could confirm that the weighted frequency-difference method is superior to the simple difference method in two-dimensional cases, but we need to further validate its applicability in three-dimensional case. As there were similar results with simulated and tank data, this is probably an intrinsic effect rather than due to limitations of the instrumentation.

Acknowledgement
This work was supported by NRF grant (2009-0071225) and the BioImaging Research Center at GIST. E J Woo was supported by the SRC/ERC program of MOST/KOSEF (R11-2002-103) and J K Seo and J Lee were supported by the WCU program (R31-2008-000-10049-0).

References
[1] J K Seo et al., Frequency-difference electrical impedance tomography (fdEIT): algorithm development and feasibility study, Physiol. Meas. 29 (2008) 929–944.
[2] S C Jun et al., Frequency-difference EIT (fdEIT) using weighted difference and equivalent homogeneous admittivity: validation by simulation and tank experiment. Physiol. Meas. 30 (2009) 1087–1099.
[3] C Gabriel et al., The dielectric properties of biological tissues: I, II, III., Phys.Med.Biol. 41 (1996) 2231-2293.
[4] L Fabrizi et al., An electrode addressing protocol for imaging brain function with electrical impedance tomography using a 16-channel semi-parallel system, Physiol. Meas. 30 (2009) S85-S101.

A fast BiCG solver for the isotropic Poisson equation in the forward EIT problem in cylinder phantoms

V M Volkov[1], A A Zherdetskij[1], S I Turovets[2,3] and A D Malony[2]

[1]Department of Mathematics, Belarusian State University, Minsk 220050, Belarus
[2]Neuroniformatics Center, University of Oregon, Eugene, OR 97403, USA
[3]Electrical Geodesics, Inc, Eugene, OR 97403, USA

E-mail: sergei@cs.uoregon.edu

Abstract. We describe an efficient numerical method for solving the isotropic inhomogeneous 3D Poisson equation in cylindrical coordinates as applied to analysis of EIT phantom experimental systems. The proposed approach is based on the second order accuracy finite-difference scheme and the BiCG iterative method with the FFT preconditioner. Extensive validation, numerical examples and comparison with experimental results are given. The performance and accuracy of the proposed numerical method are investigated and compared with the similar BiCG method with the Jacoby preconditioner.

1. Introduction

Inverse problems in neuroscience such as electroencephalography (EEG) and magnetoencephalograpy (MEG) source localization, electrical impedance tomography (EIT) or diffuse optical tomography (DOT) are inherently non-linear, underdetermined and ill-posed and require high accuracy in measurements and PDE modeling [1,2]. The emerging new reconstruction algorithms usually need a laboratory experiment on physical phantoms for testing and validation [3,4]. When a physical model is three-dimensional and geometrically complex, like the human brain, the high-resolution forward solution can be difficult to construct and compute. While the high-resolution forward solvers capable to run in parallel in multicore cluster environments are available these days [5,6], in many cases the geometry of phantoms is chosen to be simple, for example , of cylindrical shape, and the symmetry properties of electrode placement can be utilized for building a fast specific solver of the forward problem. In this paper we consider an EIT cylinder phantom experimental system and describe a fast finite difference solver based on the family of the preconditioned conjugate gradient (PCG) algorithms [7]. The major benefit of the proposed solver as a research tool is its simplicity, flexibility and ability to reach good accuracy at a relatively sparse grid. It does not require separate mesh generating software tools, which makes inverse modeling affordable even within the Matlab platform and allows quick validation and prototyping in tomographic experiments.

2. Statement of the Problem and the Numerical Scheme

For simulations of the electric potential distribution in an isolated cylindrical domain, it is natural to use the Poisson equation with the Neumann boundary conditions in the cylindrical coordinates:

$$\frac{1}{r}\frac{\partial}{\partial r}\left(r\sigma\frac{\partial u}{\partial r}\right)+\frac{1}{r^2}\frac{\partial}{\partial\varphi}\left(\sigma\frac{\partial u}{\partial\varphi}\right)+\frac{\partial}{\partial z}\left(\sigma\frac{\partial u}{\partial z}\right)=f(r,\varphi,z),\ r<R,\ 0<z<Z,\ \frac{\partial u_{r=R}}{\partial r}=\frac{\partial u_{z=0,Z}}{\partial z}=0. \quad (1)$$

Here $\sigma=\sigma(r,\varphi,z)$ is conductivity, $f(r,\varphi,z)$ is a function describing injection current density. Because of conditional correctness of the Neumann problem, we replace the boundary conditions at the top and the bottom of the cylinder by the Dirichlet conditions $u_{z=0-\delta,Z+\delta}=0$, with sufficiently low conductivity in the part of the extended domain: $\sigma(r,\varphi,z)\le 10^{-6}$, $z\in[-\delta,0]\cup[Z,Z+\delta]$, $0<\delta<<Z$. The standard second order finite-difference discretization of Equation (1) on the uniform grid in φ and z directions and the half mesh shifted grid in r direction, $r_i=(i-0.5)h_r$, $i=1,2,\ldots N_r$ (see, for example [7]), results in a system of linear algebraic equations

$$Au=f \quad (2)$$

with sparse 7th-diagonal matrix $A\in R^{M\times M}$, $M=N_r\times N_\varphi\times N_z$, where N_r, N_φ, and N_z are numbers of the grid points in the corresponding directions.

In the case of the homogeneous conductivity, $\sigma(r,\varphi,z)=const$, the discrete problem can be solved effectively applying the two-dimension fast discrete Fourier transforms in φ and z directions. As a result, the original problem (2) is reduced to a systems of equations with the bloc-diagonal matrix A_C: $A_CU=F$, where U and F are vectors of the Fourier coefficients of the solution and the right hand side of system (2) correspondingly. Diagonal blocs of matrix $A_C=\{C_{km}\}$ are three-diagonal matrices $C_{km}\in R^{N_r\times N_r}$, $k=\overline{0,N_\varphi-1}$, $m=\overline{0,N_z-1}$, so the reduced problem is solved easily by direct methods with computational complexity $O(M)$. Elements of the matrices C_{km} are calculated as follows

$$c_{k,m}^{n,n-1}=\frac{r_n-h_r/2}{r_nh_n^2},\ c_{k,m}^{n,n+1}=\frac{r_n+h_r/2}{r_nh_n^2},\ c_{k,m}^{n,n}=-c_{k,m}^{n,n-1}-c_{k,m}^{n,n+1}-\frac{4}{h_\varphi^2}\sin^2\frac{h_\varphi k}{2}-\frac{4}{h_z^2}\sin^2\frac{\pi h_z m}{\overline{Z}},$$

where $h_\varphi=2\pi/N_\varphi$, $h_z=\overline{Z}/N_z$, $\overline{Z}=Z+2\delta$.

Note that the zero Neumann boundary conditions leads to singularity of matrix A_C. The problem can be easily regularized by replacement of the zero conditions by sufficiently small flux conditions.

In the case of inhomogeneous conductivity the Fourier based approach described above is not applicable directly to Equation (2), but it can be used as an efficient preconditioning technique for iterative methods. In particular, for numerical treatments of the problem under consideration we use the bi-conjugate gradient iterative method (BiCG) [8] with a preconditioner in the form of the matrix *A* corresponding to discretization of Equation (1) with the constant conductivity coefficients. Due to use of the fast Fourier transform technique for inversion of such kind of preconditioners they are refereed as the FFT preconditioners.

3. Simulation results and comparison with experiments

The experimental EIT setup is based on a cylindrical phantom which is approximately of the typical human head size (20 cm height and 18 cm diameter) and having 128 electrodes uniformly distributed on the lateral surface in 8 rows and 16 columns. The cylinder is filled by saline with conductivity $\sigma = 1.3$ S/m. Two agar gel insertions of the smaller size with conductivity 0.48 S/m (cylinder and rectangular rods) are placed inside the cylinder in arbitrary positions. At a time only one pair of the lateral electrodes is used for the current injection while others are used for measurements of the impressed electrical potentials. At the first stage of the experiments our goal is to reproduce by means of the numerical simulations the electric potential distribution when injection current and conductivities are known (the forward problem).

Results of the numerical simulations and comparison with experimental data are presented in Figure 1. One can see that the numerical model is quite adequate to the experimental setup and simulation results fit well to the measured data for conductivity values determined independently prior to the experiment. Inverse search in a simple case of homogeneous conductivity has converged to the error function minimum exactly at 1.3 S/m at the relatively sparse grid size 32x32x72 nodes.

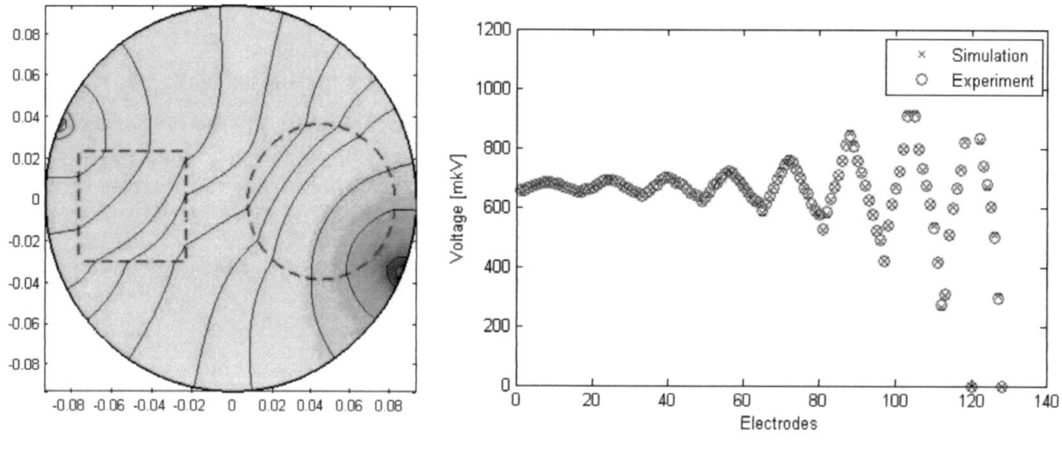

Figure 1. Distribution of electric potential on a cross section of the phantom in the plane of the injection electrodes (left). Dashed lines show contours of the interior bodies with lower conductivity. The measured and calculated potentials on the lateral electrodes (right).

Performance characteristics of the numerical technique used in simulations are illustrated in Figure 2. We have compared the convergence rate and total computational complexity of the BiCG iterative method with different types of preconditioners. The presented results show that BiCG FFT method provides convergence to the approximate solution with the relative residual $2 \cdot 10^{-5}$ in 20-30 iterations and the convergence rate is practically independent on the grid size. In the case of the Jacoby preconditioner the number of iterations to convergence is linearly increasing with the grid size. The computational complexity of the iterative methods is estimated as computation time divided by the total number of the grid points. It can be seen in Figure 2 that efficiency of the Jacoby preconditioner is degraded rapidly with use of denser grids. The Seidel and incomplete factorization preconditioners have no advantages over the Jacoby method. Our BiCG FFT method demonstrates stable performance and is more than seven times faster at the grid size 128x128x78. The typical time for one forward solution in Matlab 7.9 for the problem size of 128x128x78 nodes is about one minute at a desktop PC and scales as N^3 for the lower resolution. In the future work we plan to compare the performance and accuracy of the proposed numerical method with the finite element methods on the adaptive grid like EIDORS [9] and the ADI finite difference method in the Cartesian coordinates [6].

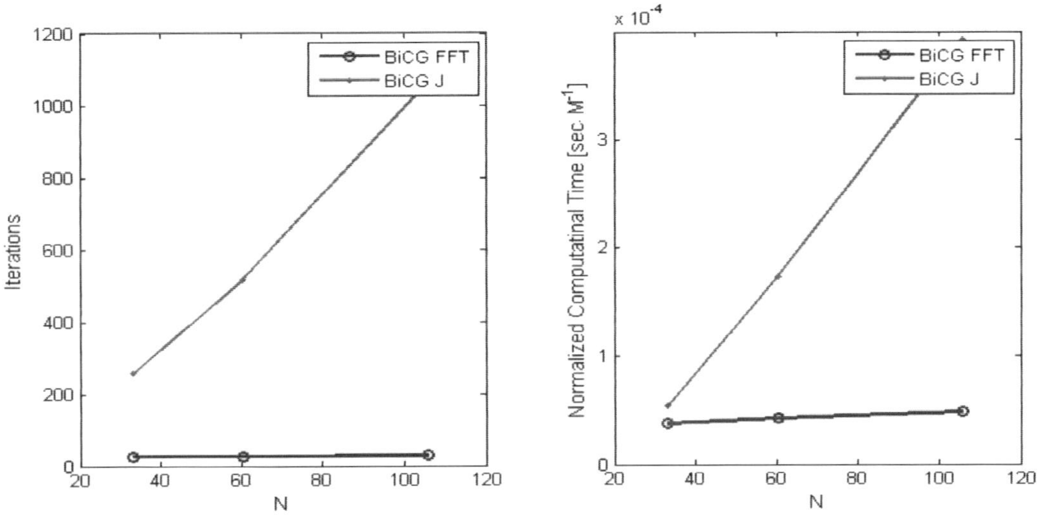

Figure 2. Number of iterations to convergence for BiCG iterative method with FFT and Jacoby preconditioners (left) and normalized computational time of the compared methods (right), $N = \sqrt[3]{N_r N_\varphi N_z}$.

Acknowledgments
This study was supported in part by National Institute of Health grant number 5R44NS056758-04.

References

[1] Arridge S R 1999 Optical tomography in medical imaging Inverse Problems **15** R41-R93
[2] Holder D S 2005 *Eletrical Impedance Tomography* (Bristol: Institute of Physics Publishing)
[3] Oh T I, Koo H, Lee K H, Kim S M, Lee J, Kim S W, Seo J K and Woo E J 2008 *Physiol. Meas.* **29** 295-307
[4] Sadleir R J, Neralwala F, Te T and Tucker A (2009) *Ann. of Biomed. Engineering.* **37** 2522-31.
[5] Hallez H et al 2007 Review on solving the forward problem in EEG source analysis. J. Neuroeng. Rehab **4** 46
[6] Salman A, Turovets S, Malony A and Volkov V 2008 Multi-cluster, mix-mode computational modeling of human head conductivity: Proc. IWOMP 2005/2006. LNCS **4315** pp 119-130
[7] Samarski A A, Andreev V B 1976 Finite-difference methods for elliptical equations (Moscow: Nauka)
[8] Barrett R, Berry M, Chan T F et al 1994 Templates for the Solution of Linear Systems: Building Blocks for Iterative Methods (Philadelphia,SIAM)
[9] Adler A and Lionheart R B 2006 Psysiol Meas **27** S25-S42

Journal of Physics: Conference Series **224** (2010) 012154 doi:10.1088/1742-6596/224/1/012154

Evaluation of the effects of the screen based on an analytical solution of a simplified MIT system

W Yin[2], B Dekdouk[1], C Ktistis[1], A J Peyton[1]

[1]School of Electrical and Electronic Engineering, University of Manchester, M60 1QD,
[2]School of Electrical Engineering and Automation, Tianjin University, China, 300072

wuliang.yin@manchester.ac.uk

Abstract. Magnetic induction tomography (MIT) is a technology that reconstructs cross sectional conductivity distribution of an object from mutual impedance measurements of coils distributed around the object. In high frequency and low conductivity applications, an outer screen is generally used to confine the magnetic fields and to prevent electromagnetic interference from outside. However, the screen will alter the sensing and excitation field, hence the sensitivity distribution of the coil array. Therefore, the design parameters of the screen (thickness, distance to coil, materials) are important to the performance of the sensor system. This paper presents a simple method based on an analytical solution for the evaluation of the effects of the screen. The advantage of the approach includes efficient modelling of thin screens and physical insights into the effects of the screen.

1. Introduction

Magnetic Inductance Tomography (MIT) has been investigated during the last decade for industrial process applications (e.g. visualising industrial processes such as those in metal production [1]) and for biomedical applications (e.g. determining body composition [2], imaging human thorax and head [3] and imaging brain oedema [4] etc). A set of coils are employed and distributed around the object being imaged and measurements of mutual inductive coupling between the coils are taken and used for image reconstruction. The inductive coupling between the coils changes as the objects are subject to eddy current effects and generate secondary fields. In some applications, an outer screen may be used to confine the magnetic fields and to prevent electromagnetic interference from outside. In addition, it provides a ground to reduce the capacitive coupling between the coils. The introduction of the screen also alters the sensing and excitation field, hence the sensitivity distribution of the coil array. In order to evaluate the effects of the screen and its design parameters (thickness, distance to coil, materials), finite element models (FEM) have been used [5]. In this paper, we present an analytical model for a simplified MIT setup for the evaluation of the effects of the screen. The advantages of the approach include efficient modeling of thin screens and physical insights into the effects of the screen.

2. The simplified MIT and model description

A realistic FEM model could be built to evaluate any EM effects in MIT including the effects of the screen given sufficient computational power. However, such a model may be computationally intensive. Since FEM involves dissecting the domain into small elements and for structures with large geometrical ratio in one direction compared to the other, the general FEM is not efficient. Furthermore, for eddy current problems such as those encountered in MIT, elements sizes have to been smaller or at least comparable to the skin depth to gain acceptable results. This poses difficulties especially for high frequency MIT where frequency reaches 10 MHz and the skin depth for aluminum or copper is less than 100 μm.

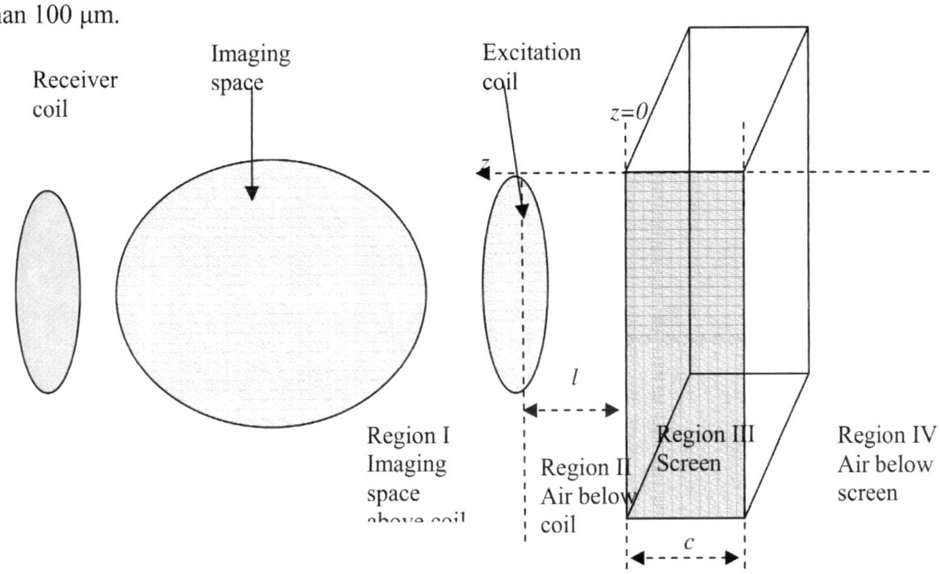

Figure 1. The simplified MIT and model description

Hence, an analytical model with certain simplified features but are capable of modeling the thin structures accurately would be useful to gain physical insight into the effects of such structures. A simplified model is illustrated in Figure 1. The following assumptions are made.

1. The coils are circular and filamentary.
2. Screens are made of homogenous metal.
3. Screens are treated as flat surfaces. This is valid for screens comprising of multi-flat pieces or for circular screens with much larger diameters than the size of the coils.

Based on above assumptions, the model in Figure 1 was used to represent the simplified MIT.

We start from the magnetic vector potentials in regions I, II and III shown in Figure 1. Ignoring the so called 'wave propagation' effects, the magnetic vector potentials in each region are as follows [6].

$$A^{(1)}(r,z)=\frac{\mu_0 I r_0}{2}\int_0^\infty J_1(\alpha r_0)J_1(\alpha r)e^{-\alpha_0|l-z|}\frac{\alpha d\alpha}{\alpha_0}$$

$$+\frac{\mu_0 I r_0}{2}\int_0^\infty J_1(\alpha r_0)J_1(\alpha r)e^{-(\alpha_0 l+\alpha_0 z)}\frac{(\alpha_0+\alpha_1)(\alpha_1-\alpha_2)+(\alpha_0-\alpha_1)(\alpha_1+\alpha_2)e^{2\alpha_1 c}}{(\alpha_0-\alpha_1)(\alpha_1-\alpha_2)+(\alpha_0+\alpha_1)(\alpha_1+\alpha_2)e^{2\alpha_1 c}}\frac{\alpha d\alpha}{\alpha_0} \tag{1}$$

$$A^{(2)}(r,z)=\mu_0 I r_0\int_0^\infty J_1(\alpha r_0)J_1(\alpha r)e^{-\alpha_0 l}\alpha\frac{(\alpha_1+\alpha_2)e^{2\alpha_1 c}e^{\alpha_1 z}+(\alpha_1-\alpha_2)e^{-\alpha_1 z}}{(\alpha_0-\alpha_1)(\alpha_1-\alpha_2)+(\alpha_0+\alpha_1)(\alpha_1+\alpha_2)e^{2\alpha_1 c}}\frac{\alpha d\alpha}{\alpha_0} \tag{2}$$

$$A^{(3)}(r,z)=\mu_0 I r_0\int_0^\infty J_1(\alpha r_0)J_1(\alpha r)e^{-\alpha_0 l}\alpha\frac{2\alpha_1 e^{(\alpha_1+\alpha_2)c}e^{\alpha_2 z}}{(\alpha_0-\alpha_1)(\alpha_1-\alpha_2)+(\alpha_0+\alpha_1)(\alpha_1+\alpha_2)e^{2\alpha_1 c}}\frac{\alpha d\alpha}{\alpha_0} \tag{3}$$

Where μ_0 denotes the permeability of free space; while l denotes the height of the bottom of the excitation coil; and c denotes the thickness of the media. $\alpha_0=\alpha$; $\alpha_1=\sqrt{\alpha^2+j\sigma\mu_1\omega}$; $\alpha_2=\alpha$. $J_1(x)$ is a first order Bessel function of the first kind. I is the current flowing through the excitation coil. r_0 is the radius of the coils.

The kth sensitivity map (k[th] column of the Jacobian matrix) is expressed as the dot product of electrical fields produced by the excitation coil and that by the receiving coil as if it is excited by a unit current:

$$\mathbf{S}_k = -\omega^2 \mathbf{A}_i \cdot \mathbf{A}_j \tag{4}$$

where $\mathbf{A}_i$ and $\mathbf{A}_j$ denote the vector potentials when sensing coils i and j are excited with unit current respectively. They can be obtained using equation (1).

3. The effects of the screen for different setups

Figure 2 lists three types of sensitivity maps (self coil, 90 degree coil and 180 degree coil) for different setups.

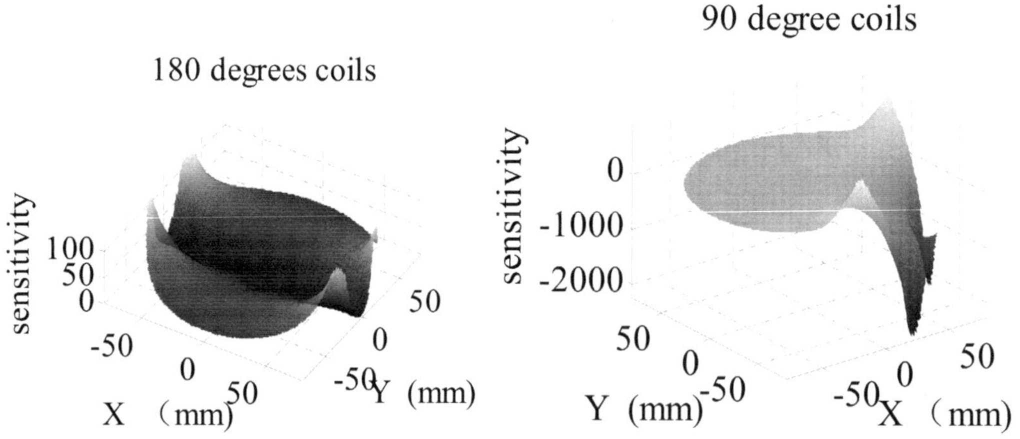

Figure 2. Sensitivity maps

Figure 3 indicates:

1. That the screen has the effect of reducing the sensitivity;

2. That the screen needs to be at least three times the diameter away from the coil to avoid significant reduction of the sensitivity.

3. That as expected, as the distance between the coil and the screen reduces, and as the thickness of the coil increases, and as the conductivity of the screen increases, the effect of screen becomes more pronounced.

4. That even the thickness of the screen is much less than the skin depth, the screen still has a significant effect on the sensitivity maps.

4. Conclusions and Discussion

This paper presents a simplified model for the evaluation of the screen effects in MIT. The advantages of the model include its capabilities to model very thin screens and have less computational costs. Although in this case, the coil was treated as filamentary, analytical model of coils of finite sizes is available; therefore, the model can be extended to account for a realistic circular coil. The limitations of the model include its simplified constraints on the geometry of the screen and the coil, e.g. the screen has to be flat and the coil has to be circular to make the solution valid. Nevertheless, with this model, it has been possible to gain useful physical insights into the screen effect in MIT without resorting to a full FEM model. In addition, the model might provide information for specifying more accurate impedance boundary conditions required in FEM models.

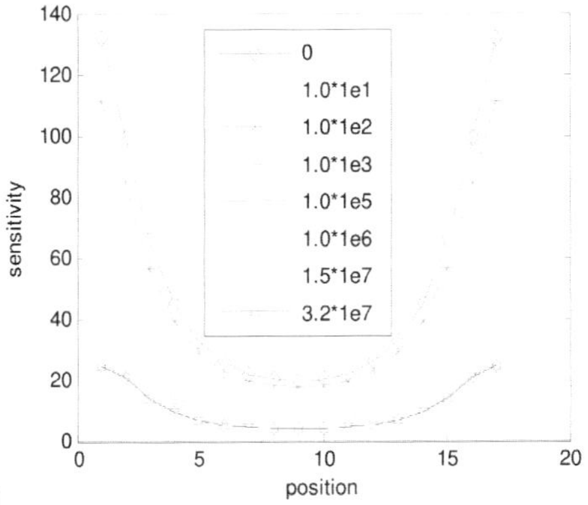

(a)

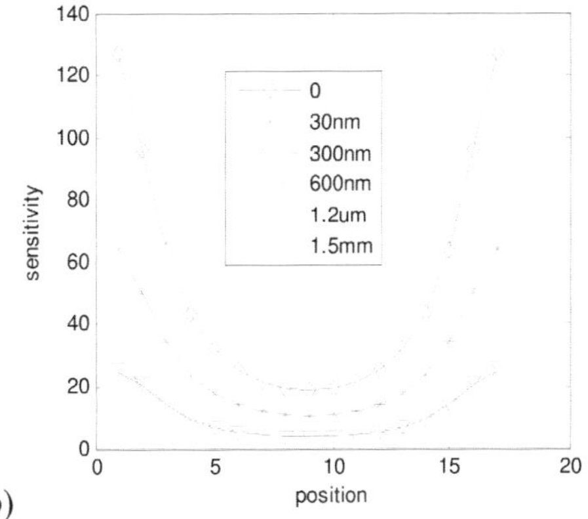

(b)

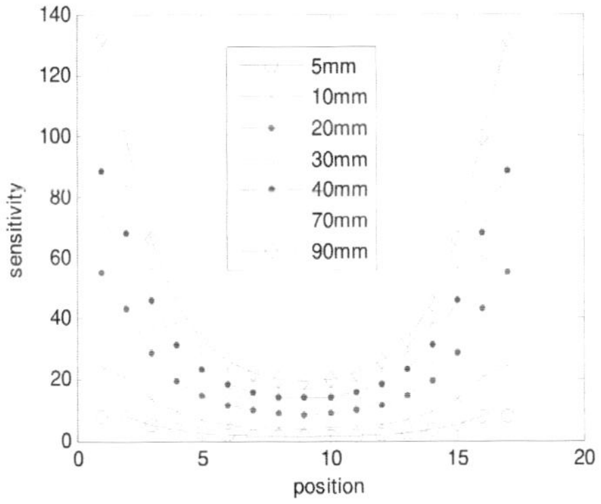

(c)

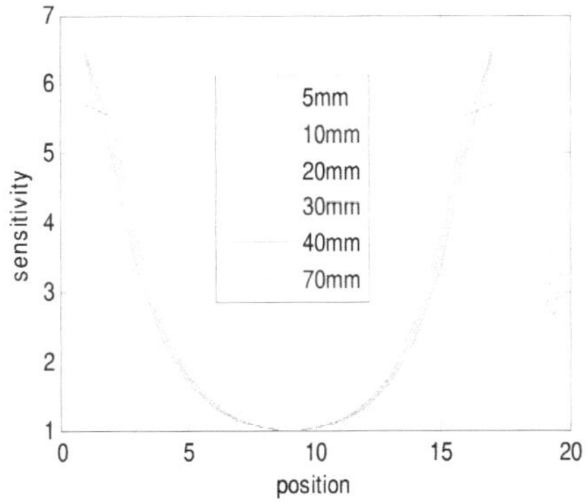

(d)

Figure 3 plots the sensitivities along the line connecting the coils in cases of 180 degree coil. (a) different conductivities, (b) different thickness (c) different distance (d) normalized of (c) with minimum as 1.

Acknowledgements

EPSRC support is acknowledged. W Yin thanks CNSF 60910001.

References

[1] Binns R, Lyons A R A, Peyton A J and Pritchard W D N, Imaging molton steel flow profiles, Meas. Sci. Technol. 12 1132 -1138

[2] Peyton A, Mackin R, Goss D, Wan-Daud W, Crescenzo E, Saunders N, Tapp H, 2003, Addressing the difficulties in using inductive methods to evaluating human body composition, Biométrie Humaine et Anthropologie 21 63-71

[3] Merwa R, Hollaus K, Oszkar B, Scharfetter H, 2004, Detection of brain oedema using MIT: a feasibility study of the likely sensitivity and detectability. Physiol. Meas. 25 347-54

[4] Griffiths H 2001 Magnetic induction tomography Meas. Sci. Technol. 12 1126-1131

[5] Peytion A J， Watson S， Williams R J, Griffiths H, Gough, W, Chrastering the effets of the extrenal electroagentic sgield on a magnnetic induction tomogrpahy sensor, WCIPT 3, 352-7

[6] Yin W, Peyton A J, Zysko G and Ktistis C, Evaluation of the radiative wave propagation effect in EMT, International Journal of Information System Science, 2/4, 575-584, 2006.

[7] Dodd C. V. and Deeds W. E., Analytical solutions to eddy-current probe-coil problem, J. Appl. Phys., vol. 39, pp. 2829–2839, 1968.

Correlation of cerebrovascular disorder and anxiety: The Kecskemet study

Kornel Sipos[1], **Michael Bodo**[2], **Piroska Szalay**[1], **Attila Szucs**[3]

[1]Semmelweis University, Budapest, Hungary, [2]JMMB,LLC, Silver Spring, MD, USA, [3]Kecskemet Municipal Hospital, Hungary

E-mail: sipos@mail.hupe.hu

Abstract. In order to test the hypothesis that anxiety is a risk factor for cardiovascular disease, specifically stroke, we simultaneously measured anxiety and cerebral vascular alternation, using a computer-based system, "Cerberus." Sixty nine psychiatric patients (including an alcoholic subgroup) were selected as subjects for measurements conducted in Kecskemet, Hungary. The five-item short form of anxiety test (STAI) was administered twice during the same session. Between each test, brain pulse waves were recorded by rheoencephalogram (REG). A REG peak time above 180 milliseconds was considered a cerebrovascular alteration (modified after Jenkner). Data were sorted into two groups: low anxiety (N=10) and high anxiety (N=10). Significant differences were found between cardiovascular risk factors ($p< 0.001$), REG peak time ($p<0.043$), and heart rate ($p< 0.045$). Six subjects showed cerebrovascular alteration in the high anxiety group, and two in the low anxiety group. For the two anxiety groups, there were no significant differences in body mass index, cardiovascular sympathetic-parasympathetic balance, age and symptoms of transient ischemic attack. The correlation of REG and age was significantly different only for the alcoholic subgroup (Szalay et al, 2007). These data support the hypothesis that a correlation exists between cerebrovascular disorder and anxiety in the studied population.

1. Introduction

Since arteriosclerosis is not a mono-causal disease, the detection of the presence of a single risk factor cannot accurately predict the development of the disease. A complex, multi-factorial approach can produce a more exact diagnosis. We utilized a device, "Cerberus," a computer-based system developed for primary stroke prevention [1]. The Cerberus system incorporates all known risk factors for arteriosclerosis/stroke and includes a short form of an anxiety test (STAI) [2]. The STAI Form Y is the definitive instrument for measuring anxiety in adults. It clearly differentiates between the temporary condition of "state anxiety" and the more general and long-standing condition "trait anxiety." Vegetative balance was also determined [1,3,4]. Vegetative balance is the measurable status of both the sympathetic nervous system (ergotroph) and the parasympathetic nervous system (trophotroph). To estimate a patient's cerebral circulation, the Cerberus system uses a non-invasive, continuous technology known as REG [5,6]. In this study we tested the hypothesis that anxiety is a

[1] Kornel Sipos, MD, PhD. Semmelweis University, Faculty of Physical Education and Sport Sciences
1123 Budapest, Alkotas st. 44. Hungary

risk factor for cardiovascular disease, specifically stroke, by measuring psychiatric patients for anxiety and cerebral vascular alteration.

2. Methods

The Cerberus system involves the measurement of pulsatile electrical impedance in the brain (REG), detailed elsewhere [1]. A/D sampling rate was 275 Hz. A one minute REG recording was used for signal averaging, triggered by the ECG R wave. The left and right side REG data were averaged for each person. REG peak (anacrotic or rising) time was measured and evaluated as a symptom of cerebrovascular alteration [7]. A short form of STAI was used containing five "state anxiety" items. Cerberus software calculated the ergotroph/trophotroph ratio (modified vegetative balance by Kerdo and Sipos), in which: 1- (diastolic blood pressure /heart rate) x 100 [1]. Values above 100 indicate ergotrop; values below 100 are considered a trophotroph vegetative state. Body mass index (BMI: kg/m^2) was also calculated by Cerberus. Student t-test in Excel (Microsoft, Redmond, WA) was used for statistical evaluation; $P< 0.05$ was considered significant.

2.1. Subjects

Three patient groups were combined (drug abuse, alcohol abuse, depression; N: 42 male, 27 female). The alcohol abuse patients, middle-aged inpatients identified as having alcohol problems for more than ten years (N=44, mean age: 41), were participating either in detoxification (Dg [8]: F10.30, F10.40) or abstinence treatment programs (Dg: F10.10, F10.20, F10.80). Age varied from 27 to 54 years for males and from 27 to 56 years for females. The drug abuse patient group included two subjects with drug dependency (F12.20, F13.20) and four subjects with a diagnosis of abusive disorders (F11.10, F18.10, F19.10) (N=6; mean age: 31). The depressive subjects (N=6, mean age: 51) (F32.00, F32.20, F333.20) were non-psychotic. In the drug patient group, the youngest and oldest subjects were males (19 and 49 years of age). The male/female ratio for alcoholic patients was 32:12; for drug patients, 4:2; and for depressive patients, 1:5; mirroring the general tendency of gender distribution for these mental disorders.

2.2. REG

REG electrodes were placed on the fronto-temporal area, within the electrode cap; localization: Fp1-F7; Fp2-F8, according to 10-20 system of EEG electrode placement. Each REG electrode was a 15-mm diameter tin disc attached to traditional EEG electrodes (ECI Electro-Cap Electrode System, Electro-Cap International, Eaton, OH). Electrode gel was used to improve skin-electrode contact.
The REG sclerosis standard used was based on Jenker's work [7]. A REG peak time longer than 180 msec was considered sclerotic in our evaluation, see figure 1.

3. Results

Significant differences were found between cardiovascular risk factors ($p< 0.001$) and REG peak time ($p< 0.043$). Although the second anxiety level (following a one minute polygraphic recording) was typically lower than the first, this difference was not significant. In the high anxiety group, six subjects (60 %) showed cerebral vascular alteration; in the low anxiety group, two subjects (20 %) showed cerebrovascular alteration. For the two anxiety groups, there were no significant differences in body mass index, cardiovascular sympathetic-parasympathetic balance, age and symptoms of transient ischemic attack The correlation of REG and age was significantly different only for the alcoholic subgroup, detailed elsewhere [9].

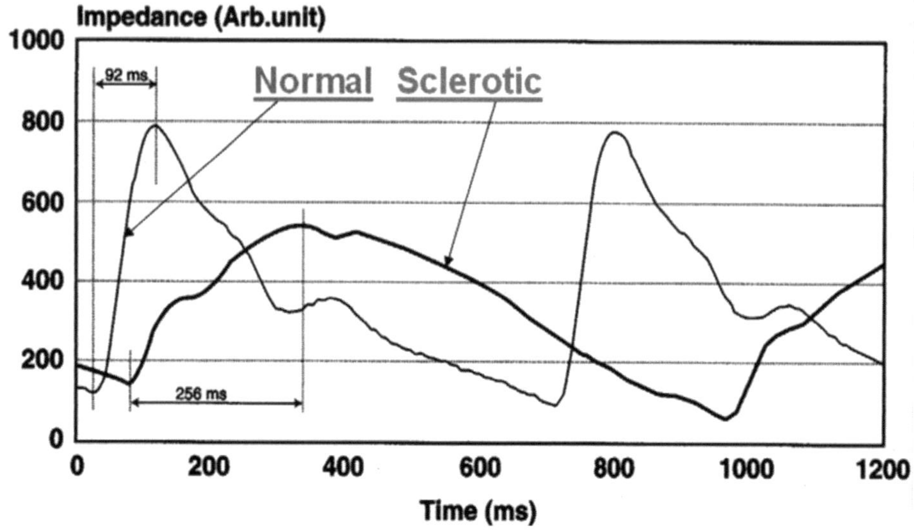

Figure 1. Typical REG curve of a healthy (normal) and a sclerotic person. Note the difference in peak time of REG curves for normal patient (92 milliseconds) and sclerotic patient (256 milliseconds). This difference has pathological meaning, independent of heart rate [10].

Table 1. Comparison of low and high anxiety groups. We found significant differences between low anxiety groups (upper part of table) and high anxiety groups (lower part of table) in cardiovascular risk factors (red column) and REG peak time (yellow column). Both groups consisted of 10 subjects.

Low anxiety (N=10)												
	Male	Female	Kerdo ind	TIA sympt	Risk Fact	Anxiety	Anxiety	Age	REG pk tm	HR	BMI	
	n	n		n	n	1	2	y	ms	1/min	kg/m2	
	8	2	98,90	0,90	0,50	9,00	8,30	39,40	98,53	81,00	23,30	**mean**
			9,67	1,37	0,53	1,94	2,16	7,49	59,38	14,53	2,63	**sd**
									2			**>180 REG**
High anxiety (N=10)												
	6	4	100,00	3,31	**1,92**	19,69	18,92	43,23	**134,81**	88,38	22,77	**mean**
			10,37	3,47	**0,86**	0,48	1,89	11,76	**66,25**	12,02	4,26	**sd**
			0,66	0,08	**0,001**	0,00000	0,00000	0,18	**0,043**	0,45	0,88	t-test P
									6			**>180 REG**
total: 69	**42**	**27**										

4. Discussion

In the literature, there are many supportive findings of a correlation between psychological and somatic factors[11]. The wider spectrum of somatic and psychological factors correlates with our recent findings. In another study we conducted involving 546 subjects, we found significant correlations among somatic and psychologic variables. Cerebrovascular alteration (REG-Doppler-and symptoms of transient ischemic attack) was in a bridge position between measured somatic variables and psychological variables [12]. These results are consistent with our recent findings that anxiety should be considered as a cardiovascular and stroke risk factor. The potential anatomical substrate of this relationship is the vascular endothelium, which plays an essential role in regulation of cerebral blood flow--a fine tuned mechanism. Although the measurement of cerebrovascular reactivity by REG has been known for decades, it was recently documented that REG reflects cerebral blood flow autoregulation [13]. This fact explains findings where cerebral blood flow changes were detected by REG during transcendental meditation [14] and eye opening and closing [15].

5. Conclusion

On the basis of previous and recent findings, we suggest that anxiety should be considered as a stroke risk factor. The potential anatomical substrate of this relationship is the vascular endothelium, which plays an essential role in regulation of cerebral blood flow. These results confirm a correlation between anxiety and sclerotic alteration of coronary and carotid arteries. Spielberger and others observed relationships among anxiety, anger and cardiovascular risk many years ago. Here, we present new evidence supporting this observation.

Acknowledgements: The authors thank volunteers measured and Janice Meer for editorial assistance.

References

[1] Bodo M, et al 1995 A complex cerebrovascular screening system (Cerberus). *Medical Progress through Technology* **21** 53-66
[2] Hanin Y L, Spielberger C D 1983 The development and validation of the Russian Form of the State-Trait Anxiety Inventory, ed C D Spielberger & R Diaz-Guerrero *Cross-cultural anxiety* 2 (Washington: Hemisphere Publishing Corporation) p 15-26
[3] Kerdo I 1966 Study on effect of meteorological and climatic circumstances on human body: some application's possibility; Dissertation in Hungarian (Hungarian Academy of Sciences: Budapest)
[4] Kerdo I 1966 An index for the evaluation of vegetative tonus calculated from the data of blood circulation. *Acta Neuroveg.* (Wien) **29** 250-68
[5] Anonymous 2008 Rheoencephalography. In: *Bioimpedance and bioelectricity basics* ed S Grimnes and O G Martinsen (Amsterdam: Elsevier) p 351-353
[6] Anonymous 1997 Rheoencephalograph. 21CFR882.1825. *Code of Federal Regulations*, Title 21, Volume 8, Parts 800 to 1299. Page 377 (U.S. Government Printing Office via GPO Access: Washington, DC)
[7] Jenkner F L 1970 Rheoencephalographic differentiation of vascular headaches of various causes, ed S E Markovich International conference on bioelectrical impedance. *Annals of the New York Academy of Sciences* **170/2** 661-65
[8] Anonymous 1992 The ICD-10 International Statistical Classification of Diseases and Related Health Problems (10th Revision) WHO Geneva, Switzerland. http://www.who.int/msa/mnh/ems/icd10/icd10.htm
[9] Szalay P, Sipos K, et al. 2006 REG can be used to detect cerebrovascular alteration caused by alcoholism. *Acta Physiol. Hung.* **93** 117-30
[10] Bodo M, Thuroczy G, Brockbank K, Sipos K 1998 Cerebrovascular aging assessment by Cerberus, ed R Klatz, R Goldman *Anti-aging medical therapeutics*, vol. II. (Marina Del Rey: Health Quest) **13** 86-95
[11] Player M S, King D E, Mainous A G 3rd, Geesey M E 2007 Psychosocial factors and progression from prehypertension to hypertension or coronary heart disease. *Ann. Fam. Med.* **5** 403-11
[12] Sipos K, Bodo M, May Z, Lendvai B, Piros A, Spitzer N, Pataky I, Nagy Z, Banyasz A 2008 Risk of mental disorders, their changes and somatic consideration in rural Hungary. *Ideggyogy. Sz.* **61** 97-105
[13] Bodo M, Pearce F, Van Albert S, Armonda R. 2007 Rheoencephalogram reflects cerebral blood flow autoregulation in pigs. In: H Scharfetter, R Merva R Ed ICEBI 2007, *IFMBE Proceedings* 17 (Berlin: Springer) p 695–698
[14] Jevning R, Anand R, Biedebach M, Fernando G. 1996 Effects on regional cerebral blood flow of transcendental meditation. *Physiol. Behav.* **59** 399-402
[15] Jacquy J, Charles P, Piraux A, Noël G 1980 Relationship between the electroencephalogram and the rheoencephalogram in the normal young adult. *Neuropsychobiology* **6** 341-348

3D sensitivity of 6-electrode Focused Impedance Method (FIM)

A H Masum Iquebal[1] and K Siddique-e Rabbani
Department of Biomedical Physics & Technology, University of Dhaka, Dhaka, Bangladesh
1: Presently at Department of Physics, Dhaka College, Dhaka.

email: rabbani@univdhaka.edu

Abstract: The present work was taken up to have an understanding of the depth sensitivity of the 6 electrode FIM developed by our laboratory earlier, so that it may be applied judiciously for the measurement of organs in 3D, with electrodes on the skin surface. For a fixed electrode geometry sensitivity is expected to depend on the depth, size and conductivity of the target object. With current electrodes 18 cm apart and potential electrodes 5 cm apart, depth sensitivity of spherical conductors, insulators and of pieces of potato of different diameters were measured. The sensitivity dropped sharply with depth gradually leveling off to background, and objects could be sensed down to a depth of about twice their diameters. The sensitivity at a certain depth increases almost linearly with volume for objects with the same conductivity. Thus these results increase confidence in the use of FIM for studying organs at depths of the body.

1. Introduction

The 6 electrode Focused Impedance Method (FIM) developed by Dhaka University [1][2] is mainly targeted at probing single objects in 3D inside the human body by placing electrodes on the skin surface. FIM consists of essentially the sum of two orthogonal Tetrapolar Impedance Measurement (TPIM) systems placed symmetrically around a central zone. A conventional TPIM system has four electrodes in a row; the outer pair (to be called current electrodes) drives a current through the test medium while the central pair (to be called potential electrodes) is used to make a potential measurement. FIM gives an enhanced sensitivity within the zone surrounded by the potential electrodes. Two conventional TPIM configurations for FIM would require 8 electrodes, however, by placing 2 potential electrodes diagonally in the central region the requirement of the number of electrodes have been reduced to 6, since these two potential electrodes can work with both the orthogonal measurements. Sensitivity mapping in 2D saline phantoms with cylindrical objects of different conductivity placed at different points has shown success in the focusing effect.

As mentioned before, the main applications envisaged with FIM will try to measure the impedance of single organs at a depth using FIM electrodes on the skin surface, which essentially is a measurement in the 3rd dimension. Besides, FIM would be used to look for changed impedances in an organ, thus eliminating the effect of surrounding tissues and organs whose impedances do not change. This change in impedance may be either due to physiological changes with time, or due to change in tissue impedance with measuring frequency of the instrument. FIM has been applied successfully for the measurement of stomach emptying, the results agreeing very well to that obtained using region of interest analysis of EIT images [2]. It also shows promise in the measurement of ventilation of lungs

[3]. Bladder emptying, tissue temperature monitoring during therapy procedures in cancer, and diagnosis of epithelial cancers, like cervical and oral cancer, may be some other areas where FIM may find useful application.

Numerical analysis performed on FIM at Warwick University,UK, using Geselowitz's lead field method has confirmed the focusing effect of 6-electrode FIM [4] in a medium with uniform conductiviy. A maximum in the average sensitivity of a plane was observed at $^1/_3$ of the drive-receive electrode spacing. Negative sensitivity, usual with all tetrapolar impedance systems, appeared down to ½ of the drive-receive electrode spacing. No attempts were made to study the effect of a differential measurement with and without an object with a changed conductivity within the focused zone. However, it may be anticipated that all other sensitivity values, including the negative ones, will cancel out except that at the object if it is placed within the focused zone.

Since most applications would involve placing the FIM electrode on the skin surface targeting at objects at a depth, a detailed understanding of the 3D sensitivity is necessary to utilize the method with confidence. As can be envisaged, the measured impedance would depend on the conductivity, depth and volume of the target object provided the geometry of the FIM system and all other parameters are kept unchanged. Therefore effects of changing these parameters need to be understood independently. The present work was taken up to estimate the 3D sensitivity of an object in such a differential measurement placed centrally in the focused zone, through measurements in a phantom with fixed FIM geometry.

2. Methods and measurements

FIM measurements were carried out at 10kHz using a saline filled 3D phantom , made of transparent acrylic, of dimension 50cm x 50cm x 50cm (related to actual depth of water in the tank) as schematically depicted in figure 1. Electrodes (shown by black dots) were fixed on one side while a spherical object (shown as a dark circle) under measurement was hung from the top using a thin thread, guided by calibrated scales. The spacing of the current drive electrodes in each pair was 18cm, while the spacing between the diagonal potential measuring electrodes was 5 cm. The latter corresponds to a focused zone with the dimension of about 3.5cm square at the surface. In 3D the focused zone is expected to increase slightly because of the curvature of equipotential surfaces. Care was taken so that the object is placed centrally within the focused zone while hanging it at different depths from the electrode plane. The position of the spherical object was referred to by the position of its centre. Since the current amplitude is constant, measured potential is proportional to impedance, and this is presented as a value of impedance, in arbitrary units. Insulating and conducting spheres of various diameters were used for the measurements. Potatoes cut to make spheres of various diameters were also used to get objects with intermediate conductivity.

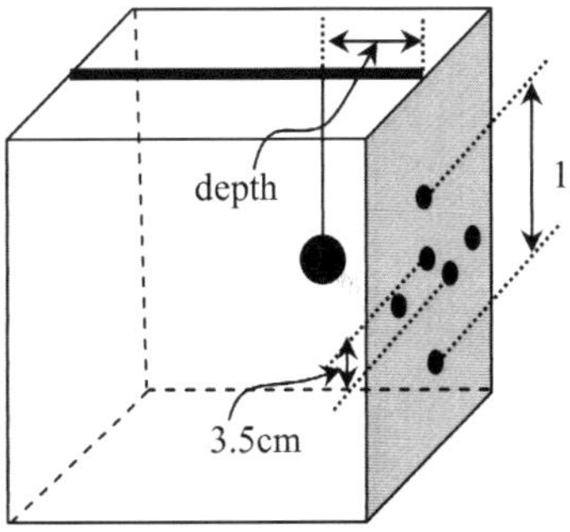

Figure 1. Design of the 3D phantom, and the positioning of the electrodes and the test object

3. Results

The potentials obtained in a constant current measurement system are proportional to the impedance, and hence the measured potential values are directly shown as impedance in arbitrary units in the figures presented in this section. These are shown in the same scale for all the objects, and therefore the values may be compared from the magnitudes shown. Figure 2 shows the measured impedance for insulating spheres while the impedance variation with volume at a certain depth (5 cm) is shown in figure 3.

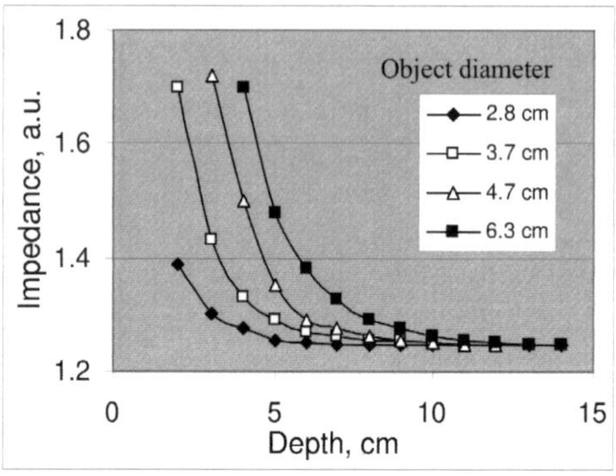

Figure 2. Depth sensitivity of focused impedance (a.u: arbitrary unit) for insulating spherical objects of different diameters (shown inset)

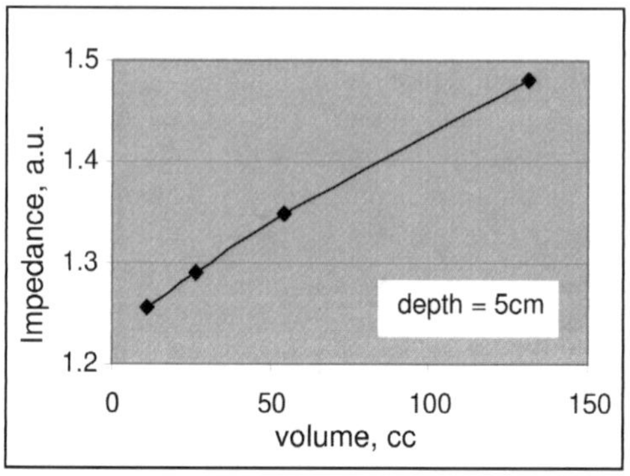

Figure 3. Variation of focused impedance (a.u: arbitrary unit) with volume of insulating objects, at a depth of 5 cm.

It can be seen from figure 2 that the measured impedance falls sharply with depth, till it levels off with the background value. Besides, the impedance values are greater for larger objects, and correspondingly, the depth of sensitivity also increases with object size. Figure 3 shows that at a certain depth, the impedance increases almost linearly with volume. Variation of impedance with depth for different spheres of a conducting object and potato are shown in figures 4 and 5 respectively. For the conductor the curves bend in the opposite direction, showing reduced impedance as expected. Potato has greater impedance than the saline used in the phantom, therefore it has impedance change in the positive direction. Distinction from the background (at about 1 to 2 % level) could be made at depths of more than twice the diameter of the objects. As mentioned before, the geometry of the FIM electrodes were the same for all the above measurements.

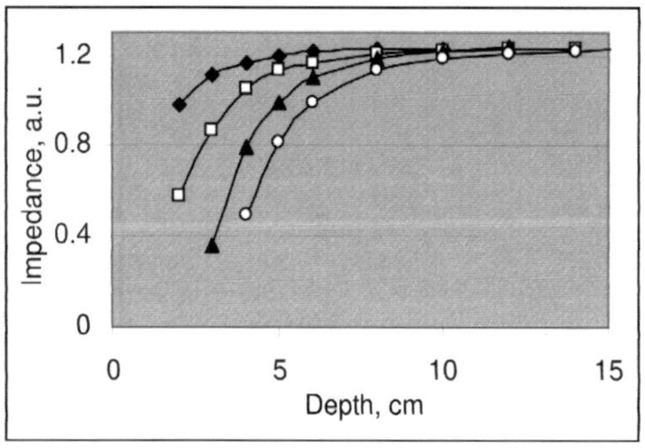

Figure 4. Depth sensitivity of focused impedance (a.u: arbitrary unit) for conducting spherical objects of diameters, in cm: 2.8, 3.7, 4.7, 6.3

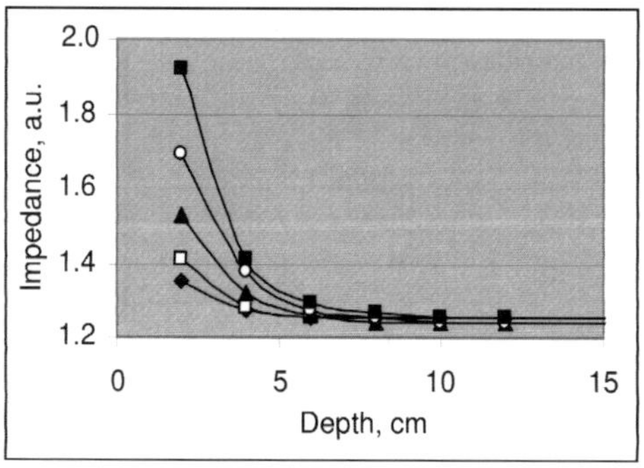

Figure 5. Depth sensitivity of focused impedance (a.u: arbitrary unit) for pieces of spherically cut potato of diameters, in cm: 2.4, 2.7, 3.6, 4.4, 4.8

4. Discussions

A visualization of the depth sensitivity may be obtained from figure 6 showing a cross section in the 3rd dimension with electrodes fixed in the topmost surface as indicated by the small rectangles. The middle two are the potential electrodes and the equipotential surfaces going through these two electrodes define the limits of the sensitive zone, shown gradually shaded, darker shade representing higher sensitivity, schematically. Two factors contribute to the sensitivity of measurement at the top electrodes. Firstly the current density at the position of the object, and secondly, the perturbation in equipotentials created by the object at the potential measuring electrodes. Because of this, the object O1 will have higher sensitivity than O2. This may also be viewed from another angle. Because of curvature of equipotential surfaces in 3D, as indicated in the figure, the size of the focused zone is expected to increase with depth. Therefore, an object of a specific size takes up less and less fraction of the focused zone as it is taken to greater depths, with consequent decrease in sensitivity. Of course sensitivity would also depend on the spacing of the current and potential electrodes and the conductivity of the object, which need to be studied as well.

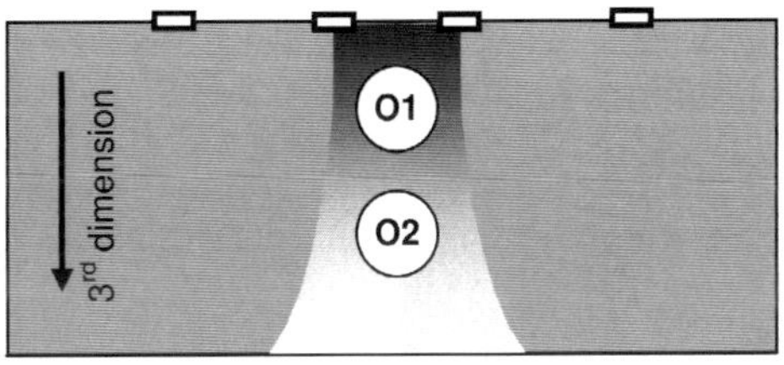

Figure 6. Equipotentials and sensitivity in the 3rd dimension

The present work gives a confidence in the use of the essentially 2D FIM system for measurements of objects in 3D and justifies the use of the system for studying organs at depths from the skin surface. Having an assessment of the sensitive depths for organs of different sizes, FIM may be applied judiciously for different practical implementations. As mentioned before, FIM is to be applied for measuring changes in impedance of a particular organ, either under different physiological conditions, or with different frequencies of measurement. If neighbouring organs do not have impedance changes under these situations then it is only the target organ that would be 'visible' to FIM and its relative changes in impedance can be reliably measured. Of course one has to take note that measuring absolute values of impedance would be difficult, if not impossible. However, relative measurement of impedance change can give a host of diagnostic information which would be easy to measure. Thus the present work paves the way to a better understanding of the FIM for clinical applications.

5. References

[1] Rabbani KS, Sarker M, Akond MHR and Akter T 1998 *Proceedings, X Int. Conf. Elect. Bioimpedance, Barcelona, Spain,* 31-34

[2] Rabbani KS, Sarker M, Akond MHR and Akter T 1999 *"Electrical Bioimpedance methods", Annals of the New York Academy of Sciences,* **873** 408 – 200

[3] Kadir MA, Baig TN and Rabbani KS 2009 *Proceedings (on-line), 10th International Conference on Biomedical Applications of Electrical Impedance Tomography (EIT 2009), Manchester, UK,* http://www.maths.manchester.ac.uk/eit2009/abstracts/kadir.pdf

[4] Islam N, Rabbani KS and Wilson A 2009 *Proceedings, 10th Int Con, Biomed Appl of El Imp Tomography (EIT2009) and Workshop on Electromagnetic Inverse Problems,* http://www.maths.manchester.ac.uk/eit2009/abstracts/islam.pdf

Journal of Physics: Conference Series **224** (2010) 012157 doi:10.1088/1742-6596/224/1/012157

Electric Impedance Imaging of the Mammary Gland in the Case of Mastitis

M Korotkova[1,2], **A Karpov**[1,2]
[1]Clinical Hospital 9, Yaroslavl, Russia
[2]"SIM Technika" PKF, Yaroslavl, Russia

E-mail: kb9@mail.ru
sim-tech@net76.ru

Abstract. The electric impedance mammography technique has been applied for several years. The aim of the research in hand is to reveal the peculiarities of the electric impedance imaging in various stages of the inflammatory process in the mammary gland. We have conducted an examination of twenty six patients: five of them in the stage of arterial hyperemia, eight in the stage of infiltration, three of them in the stage of abscess and ten in the stage of cicatrization. The examination was carried out on the "MEIK" (version 5.6) potencial electric impedance computer mammograph. The weighted reciprocal projection method was used to reconstruct the 3-D electric conductivity distribution of the examined organ. Any inflammatory process is phasic and always attended by the complex vascular alterations with exudation of liquid components of plasma, blood cells outwandering and stromal cells proliferation. Pathophysiological and histopathological peculiarities of each stage of the inflammatory process are well reflected in the electric impedance images. This fact enabled the authors of the research to define the electric impedance imaging as the histofunctional scanning.

1. Introduction

Electric impedance potential mammography method has been applied in the obstetric division of Clinical Hospital 9 (Yaroslavl) for several years already. The most commonly encountered mammary gland pathology in the perinatal period is inflammation – mastitis. Inflammation (Lat. inflammatio – ignite, set on fire) - pathological process characterized by the development of local alterative, vascular and proliferative reactions of an organism to ill effects. In the inflammatory process the following stages are distinguished: alteration, disturbed circulation and microcirculation, proliferation [1, 2, 3].

The research in hand aims to reveal the peculiar features of the mammary gland electric impedance image in different stages of the inflammatory process.

2. Methods and Materials

We have conducted an examination of 26 patients, 5 of them - in the stage of active (arterial) hyperemia, 8 - in the stage of infiltration, 3 – in the stage of abscess formation, 10 - in the stage of cicatrization (scarring). The examination was carried out on the electric impedance computer mammograph "MEIK" v.5.6, in which modificated backprojection method is used to reconstruct the image (50 k Hz, 0.5 m A). The computational result of surface intensity for the data acquired on a homogenious object was used as the referential measurement. Throughout the instrumentality of these measurements the electi cal conductivity index was calculated (expressed in conditi onal units). When the reconstruction is finished, seven consecutive scan planes are created, which correspond to sections of the mammary gland in axial (base) view with the depth increasing fro m 0.4 to 4.6, i.e. from the areolar area to the retromammary space.

3. Result

Inflammatory processes in the mammary gland develop in loose connective tissue and are accompanied by the complex of vascular changes with exudation of liquid constituents of the blood, blood cells migration and proliferation of stroma cells.

3.1 Alteration Stage

This short-term stage quickly develops into the second one - the stage of disturbed circulation and microcirculation. That is why usually alteration stage cannot be registered.

3.2 Disturbed Circulation and Microcirculation Stage

3.2.1 Active (Arterial) Hyperemia Stage. During this stage linear and volume blood velocity increases at the expense of a short-term vascular spasm. Arteriotony in the capillaries and veins increases too.

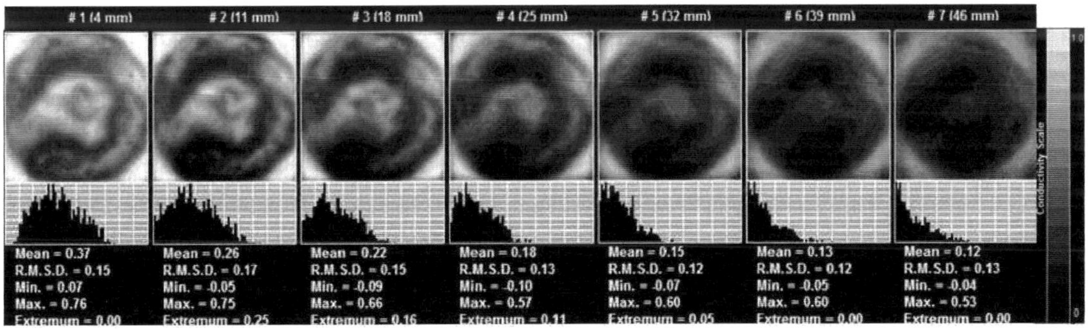

Figure 1. Disturbed circulation and microcirculation stage. Active (arterial) hyperemia stage. A significant hypoimpedance mass (1-4 scan planes) with the hyperimpedance contour is situated in the center of the mammogram (the first row - scanning planes, the second row - electroimpedance tomograms, the third row – histogram of electroconductivity distribution, the fourth row – qualitative assessment, on the right – conductivity scale).

High blood pressure leads to the structural alteration of endothelial capillary walls. It activates the process of water molecules, ions, plasma proteins (i.e. exudation). Extracellular oedema arises in the nidus of inflammation [1, 2]. These processes result in decrease of active component of impedance and in increase of conductivity in the inflammation area. During the stage of vascular changes hyperaemia area and oedema can be observed clinically. On the electric impedance mammograms hypoimpedance areas can be visualized, they possess high conductivity index and their location corresponds with the oedema area contoured by the hyperimpedance on the periphery (figure 1). The histogram of electrical conductivity distribution which is characteristic of this stage of inflammation process has unimodal symmetric form and refers to the class of distributions which are subjects to the normal distribution law. The difference between the mean electrical conductivity index and the mode is insignificant (figure 1).

3.2.2 Passive (Venous) Hyperemia Stage. As the inflammation oedema grows, it impedes blood outflow through venous system and arterial hyperemia comes to venous hyperemia. Blood pressure increases in the venous part of capillars of the inflamed tissue. Blood flow slows down all the way to its arrest. These processes contribute to leukoc yte exit from the bloodstream and to l eukocytic infiltrate formation. Leukocytic infiltrate emerges in just 6 hours after the inflammation starts [2]. Thereby the area of cell membranes grows significantly and this fact results in increase of reactive component of impedance and in decrease of conductivity in the inflammation area. During the migration of leukocytes to the inflammation area the formation of infiltrate can be observed clinically.

In the electric impedance mammograms infiltrate can be visualized as a homogeneous well-defined hyperimpedance area with low conductivity index (Figure 2). The histogram of electrical conductivity distribution during the stage of inflammation process usually has unimodal asymmetric shape (Figure 2). The difference between the mean electrical conductivity index and the mode as well as the shape of

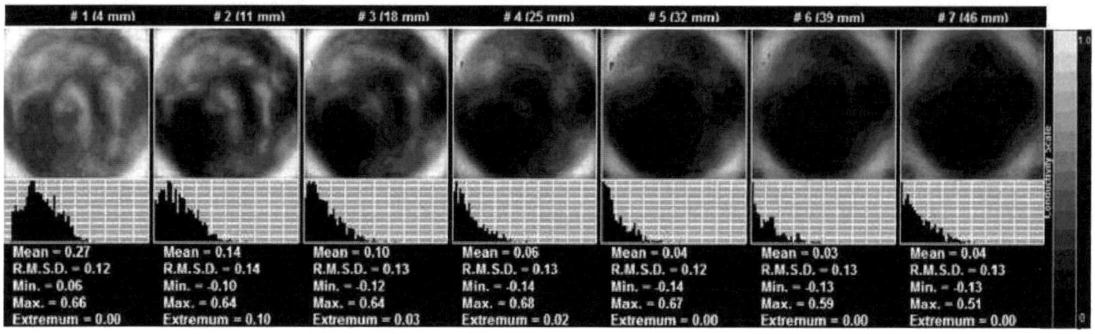

Figure 2. Disturbed circulation and microcirculation stage. Infiltration. The homogeneous hyperechoic mass can be visualized (1-4 scan planes) over an area from 7 to 9 on the clock dial (the first row - scanning planes, the second row - electroimpedance tomograms, the third row – histogram of electroconductivity distribution, the fourth row – qualitative assessment, on the right – conductivity scale).

the histogram depends on the distribution of the infiltrate. When inflammation makes progress, leukocytoclasis and pus formation take place [2]. These processes result in decrease of reactive component of impedance and in increase of conductivity in the abscess formation area. Clinically, abscess formation manifests itself in softening of the inflamed area. Leukocytic infiltrate loses its hyperimpedance homogeneousness.

Clusters of pus manifest themselves by high conductivity. In the electric impedance mammograms the process of pyogenic dissolution of infiltrate is accompanied by the loss of homogeneity and by the emergence of hypoimpedance areas with high conductivity index corresponding with the location of pus clusters (Figure 3). The difference between the mean electrical conductivity index and the mode is significant. The histogram of electrical conductivity distribution during the abscess formation has a multimodal shape.

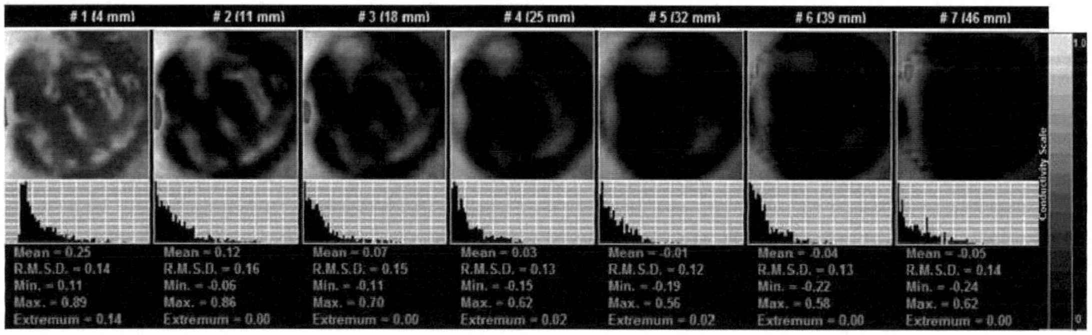

Figure 3. Disturbed circulation and microcirculation stage. Abscess formation. A heterogeneous impedancee structure with a hypoimpedance part in its center can be visualized (1-3 scan planes) over a vast area from 12 to 4 on the clock dial (the first row - scanning planes, the second row - electroimpedance tomograms, the third row – histogram of electroconductivity distribution, the fourth row – qualitative assessment, on the right – conductivity scale).

3.3 Proliferation Stage

Proliferation is the last stage of the inflammation process. During inflammation the tissue always dissolves to a greater or lesser extent. The dissolution which is caused by the inflammation process and finished during the infiltration stage, ends with the complete recovery of damaged tissue structures and thus does not lead to dysfunction [1, 2]. Infiltrate regression is accompanied by its substitution by connective tissues and by the recovery of electrical properties of damaged tissues.

The dissolution of tissues reaches its maximum in the case of suppurative inflammation. A cavity is formed in the place of the former inflammation. In what follows, this tissue defect gradually fills in at the expense of growth of local cells of the connective tissue – fibroblasts, but tissue structures and functions recovery never takes place. When defect healing by primary intention, the cicatrix is formed which fully consists of collagen fibers [1, 2]. In the electric impedance mammograms a cicatrix can be visualized as a hyperimpedance often irregular-shaped strip, it possesses low conductivity index (Figure 4).

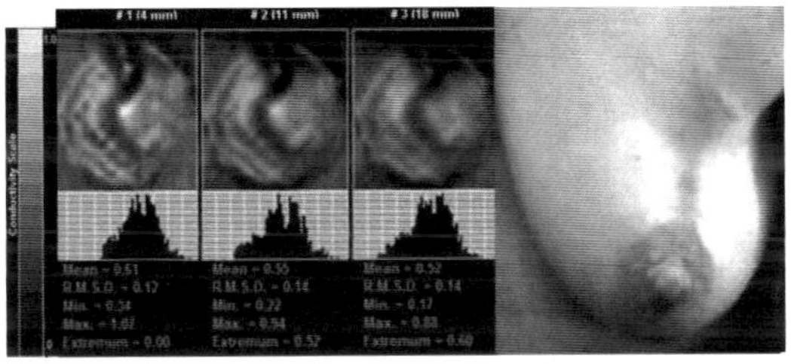

Figure 4. Electric impedance images of the mammary gland (3 scan planes).In the central upper area one observes a hyperimpedance linear zone which correponds with the real scar in shape and size.

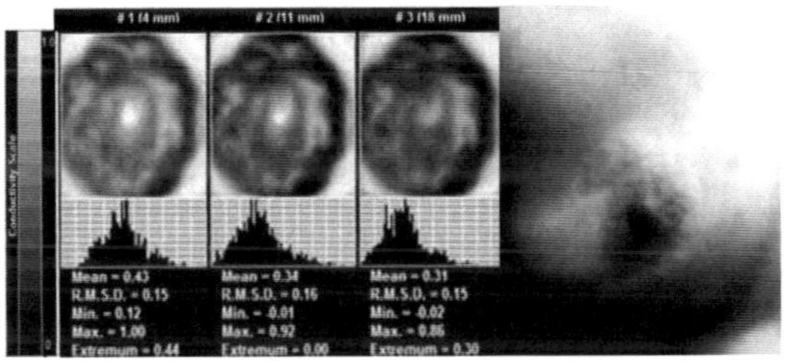

Figure 5. Electric impedance images of the mammary gland (3 scan planes). In the central upper area one observes a hyperimpedance linear zone which corresponds with the real scar in shape and size.

When healing by the second intention, the defect is articulated from the rest of the tissue by collagen fibers, and the nidus itself consists of amorphous substance of the connective tissue [1, 2]. In this case the cicatrix is visualized as hyperimpedance strips with low conductivity index and an isompedance area with medium conductivity index which is located between the stripes (Figure 5). The difference between the mean electrical conductivity index and the mode is minimal. The histogram of electrical conductivity distribution during the proliferation stage has a unimodal symmetrical shape.

4. Conclusion

Pathophysiological stages of the inflammation process are accompanied by the alteration of the electrical properties of damaged tissues of the mammary gland. Stage-dependent alterations of the electrical properties of mammary gland tissues define the peculiar features of the mammary gland electric impedance image. That is why electric impedance mammography method can be defined as a method of histofunctional scanning. The imaging of different inflammation process stages by means of electrical impedance tomograms allows using this method to diagnose mastitis, choice of treatment and its monitoring.

5. References

1. Afanasiev I 2002 *Histology, Citology and Embriology* Moscow pp 199-223 719-725
2. Ado A 1973 *Physiopathology* Moscow pp 160-175
3. Trufanov G 2006 *Radiodiagnostics of diseases the mammary gland* St Petersburg pp 105-117

Measuring the multi-frequency electrical impedance of the mouse gastrocnemius muscle using a tetrapolar technique

J Li, P M Fogerson and S B Rutkove[1]
Department of Neurology, Beth Israel Deaconess Medical Center, Harvard Medical School, 330 Brookline Avenue, TCC-810, Boston, MA, 02215, USA

E-mail: srutkove@bidmc.harvard.edu

Abstract. Electrical impedance methods can be used to evaluate and monitor neuromuscular disease states. Recently, we have applied tetrapolar surface electrical impedance methods to the gastrocnemius muscle of the rat for this purpose and substantial changes in the impedance parameters after sciatic nerve crush can be identified. In order to be able to study additional animal models of nerve and muscle disease, however, it would highly desirable to be able to perform such impedance measurements in the mouse. Yet the small size of the mouse presents a substantial technical challenge. In this study, we evaluate a basic approach for performing such measurements. A series of thin, stainless steel strip electrodes affixed to the gastrocnemius and interfaced via a separate connector to the Imp SFB7® (Impedimed, Inc), provided an effective means for obtaining impedance data in the 20-500 kHz range. After two weeks, test-retest reproducibility was good, with intra-class correlation coefficients as high 0.84 and variability as low as 12.86 ± 6.18% in the 15 mice studied. Using this approach, it may now be possible to study impedance changes in a variety of mouse models of neuromuscular disease, including amyotrophic lateral sclerosis, spinal muscular atrophy, muscular dystrophy and Charcot-Marie-Tooth disease.

1. Introduction

Electrical impedance myography (EIM) is an impedance-based technique for the non-invasive assessment of muscle [1]. Earlier EIM work on human [2] and rats [3] has revealed that EIM holds the promise of serving as an indicator of neuromuscular disease progression while also potentially providing diagnostic information. However, to further broaden the application of EIM to this area, it would be ideal to be able to extend impedance measurements to the mouse, as many additional mouse models of neuromuscular diseases are available, including muscular dystrophy, spinal muscular atrophy, various forms of amyotrophic lateral sclerosis, and hereditary polyneuropathies (e. g., Charcot-Marie-Tooth disease). Thus, in this study, we assess a basic technique for performing localized impedance measurements on the gastrocnemius muscle in a group of healthy, adult mice.

2. Methods

2.1. Mice

All animal work was approved by the Institutional Animal Care and Use Committee of the Beth Israel Deaconess Medical Center, Boston, MA. Fifteen healthy male Swiss Webster mice at 12-weeks of age

[1] To whom any correspondence should be addressed.

were obtained from Charles River Laboratories, Wilmington, MA and acclimated for one week after arrival at our animal facility before any testing was initiated.

2.2. Mouse setup

For consistency, the left hind limb of all the mice was used for all studies. After being placed under isoflurane anesthesia delivered via vaporizer, the mouse was placed in a prone position (see Figure 1 of [4]) with its left leg outstretched at approximately a 45° angle away from its body. The mouse was kept in this position by a retractor attached to a piece of cloth with a hole through which the left hind limb was slipped. The foot of this leg was taped down. This overall approach is similar to that being performed in rats [4]. A depilatory cream was applied to the distal left hind leg for 90 s to remove fur. A heating lamp above the body and a heating pad under the body were applied to maintain the body temperature at 37° C (surface temperature being monitored continually).

A pinpoint tattoo was placed close to the center of the gastrocnemius muscle at a point approximately 2/3 of the distance from the midpoint of the lumbar spine and 1/3 of the distance from the base of the heel pad of each mouse. The purpose of the tattoo was to assist in repeat placement of electrodes.

2.3. Tetrapolar electrode array

The tetrapolar electrode array consisted of 20 mm long and 2 mm wide stainless steel strips, with wires soldered onto the ends that extended a short distance to a set of connectors that were interfaced to the impedance-measuring device. The four strips were placed in a line onto the adhesive surface of medical adhesive tape (3 M Micropore, 3 M Health Care, St. Paul, Minnesota) with an inter-electrode separation distance of 2 mm. As is standard in tetrapolar measurements, the two middle electrodes served as voltage electrodes and the outer two electrodes served as the current-injecting electrodes. Distilled water was used to clean the stainless steel strips between EIM measurements.

2.4. Electrical impedance measurement

Gauze, damp with 0.9% saline, was lightly dabbed on the leg skin to prepare the area prior to measurement. After waiting 30 s for excess saline to evaporate, the four-electrode array was placed along the length of the leg, centered about the tattoo. In this way, the proximal current electrode was close to the knee-joint.

Data were collected using the Imp SFB7® bioimpedance spectroscopy device (ImpediMed, San Diego, CA), in a range from 4 kHz to 1 MHz, although for the analyses that follow, only values from 20 kHz to 500 kHz were utilized.

2.5. Weighting

In addition to testing with adhesive tape alone (0 g), external weights of 6 g, 12 g, and 24 g, each with a foot print of 39 mm x 18 mm, were also placed on top of the electrode array. In addition to potentially improving electrode contact, the weighting also flattened out the measurement surface to some extent thus reducing interfering geometric inconsistencies. The weights were thoroughly insulated with non-conductive tape.

2.6. Reproducibility measurements

Reproducibility was conducted and examined at three different time points. For immediate reproducibility, the electrode array was detached and the animal removed from the set-up entirely, picked up, and weighed briefly on a nearby scale. The animal was then immediately brought back to the testing area and set up again with cleaned electrodes and the measurements repeated. Reproducibility was also measured at 1 day and at 2 weeks. During each series of reproducibility experiments, the first measurement was always the one without weight, followed in order of increasing weight. Based on our immediate reproducibility data, however, we excluded the 6 and 24 g weights from the 1-day and 2-week reproducibility studies.

2.7. Data analysis

Several different impedance parameters were utilized. First, we assessed the 100 kHz resistance (R_{100}), 100 kHz reactance (X_{100}), and 100 kHz phase (P_{100}). In addition, we were interested in evaluating the potential benefit of using "collapsed" parameters that attempt to represent the character of the multifrequency spectrum in a single value. One was the log-resistance slope (i.e., LogRslope), calculated by the taking the resistance data from 20-500 kHz, performing a log transformation of both the measured values and the applied current frequencies, plotting the points, and taking the slope of a linear regression fitting of those points. The other was the phase slope (i.e., Pslope), obtained by performing a linear regression of the phase values in the 20-100 kHz range and calculating its slope generally showing a near-linear rise in values over that range. (The corresponding reactance slope was more variable and is thus not reported here.)

Reproducibility was assessed by calculating intra-class correlation coefficients (ICCs) and mean variability in per cent (as calculated by taking the absolute value of the difference of the first and second values and dividing by their mean x 100).

3. Results

3.1. Immediate reproducibility

Tables 1 and 2 provide the ICCs and mean variability (± standard deviation for the 15 animals studied), respectively, for the immediate reproducibility measurements. As can be seen, weighting substantially improve the reproducibility of the measurements, with only minor differences between the different weights being observed. Except for the Pslope, all the parameters have comparable variability.

Table 1. Summary of immediate reproducibility as measured by ICCs.

Weight	R_{100}	X_{100}	P_{100}	Log Rslope	Pslope
0 g	0.66	0.37	-0.15	-0.23	0.51
6 g	0.81	0.78	0.77	0.80	0.79
12 g	0.77	0.79	0.80	0.83	0.85
24 g	0.73	0.79	0.83	0.81	0.82

Table 2. Summary of immediate reproducibility as measured by per cent variability.

Weight	R_{100}	X_{100}	P_{100}	Log Rslope	Pslope
0 g	7.51 ± 9.67	18.14 ± 20.83	16.12 ± 23.39	16.12 ± 23.66	34.41 ± 31.49
6 g	4.67 ± 7.21	13.92 ± 14.01	10.37 ± 9.48	9.49 ± 8.80	24.39 ± 24.71
12 g	5.20 ± 6.57	12.41 ± 16.13	8.93 ± 11.02	8.12 ± 10.60	23.24 ± 26.07
24 g	5.26 ± 6.53	12.40 ± 18.06	8.96 ± 12.41	8.94 ± 13.59	25.20 ± 31.53

3.2. One-day reproducibility

Table 3 provides the 1-day reproducibility data for 15 mice, now using only the 0 and 12 g weights. As anticipated, the variability is greater than in the immediate reproducibility studies.

Table 3. Summary of ICCs and mean variability for 1-day reproducibility.

	R_{100}	X_{100}	P_{100}	Log Rslope	Pslope
ICC (0 g)	0.40	0.42	0.54	0.60	0.32
Variation (0 g)	17.97 ± 11.16	32.15 ± 24.36	20.19 ± 16.99	18.70 ± 16.69	82.13 ± 102.81
ICC (12 g)	0.45	0.65 0.73		0.77	0.79
Variation (12 g)	17.15 ± 9.46	28.73 ± 22.42	17.32 ± 15.16	16.49 ± 14.42	50.35 ± 74.33

3.3. Two-week reproducibility

Table 4 provides the 2-week reproducibility data for the 15 mice. Interestingly the values are somewhat better than those observed at 1-day, with slightly higher ICCs and lower variability.

Table 4. Summary of ICCs and mean variability for 2-week reproducibility.

	R_{100}	X_{100}	P_{100} Log	Rslope	Pslope
ICC (0 g)	0.56	0.48	0.54	0.49	-0.11
Variation (0 g)	13.79 ± 10.26	23.19 ± 19.07	15.04 ± 11.86	16.17 ± 12.82	77.25 ± 96.40
ICC (12 g)	0.55 0.73 0.84 0.			84 0.43	
Variation (12 g)	13.07 ± 9.67	21.20 ± 13.38	12.86 ± 6.18	13.12 ± 6.14	64.21 ± 71.67

4. Discussion

These results demonstrate that by using a basic tetrapolar set-up of four electrode strips placed over the mouse gastrocnemius muscle, we are able to attain relatively good reproducibility of measurements for a period as long as two weeks. Although the reproducibility is not as good as that achieved in humans [5] or in rats [4], it is likely sufficient and may possibly be further improved by additional modifications. The collapsed LogRslope parameter faired well, although Pslope was less consistent. Other work (unpublished results) in humans, appears to suggest good consistency to these other collapsed parameters, so it is not clear why the results here were not as reproducible.

In most of our work to date, we have relied upon disposable, adhesive Ag-AgCl electrodes. However, such electrodes cannot be utilized in the relatively confined space of the mouse hind limb. One obvious advantage to using stainless steel strips is that they are reusable. A disadvantage is the edges are difficult to smooth and irritate the animal's skin more readily. It is likely this skin irritation that caused the 1-day reproducibility to be somewhat worse than the 2-week reproducibility. At two-weeks, the animals' skin had entirely recovered from the earlier measurements. A variety of other reusable materials may be less injurious to the skin and these are being studied.

The weighting also appeared to help. It is most likely that this improved the contact between the relatively inflexible stainless steel strips and the mouse skin. However, it is possible that the weighting also helped to smooth out variations in the shape in the posterior leg muscles due to slight inconsistencies in positioning of the animal.

In conclusion, it is possible to obtain good reproducibility of localized impedance measurements from the mouse gastrocnemius muscle over periods as long as two-weeks. The technique may be useful for the assessment of mouse models of neuromuscular disease and warrants further refinement and application.

5. References

[1] Rutkove S B 2009 *Muscle Nerve* **40** 936
[2] Esper G J, Shiffman C A, Aaron R, Lee K S and Rutkove S B 2006 *Muscle Nerve* **34** 595
[3] Ahad M A, Fogerson P M, Rosen G D, Narayanaswami P and Rutkove S B 2009 *Physiol. Meas.* **30** 1415
[4] Ahad M A and Rutkove S B 2009 *Clin. Neurophysiol.* **120** 1534
[5] Rutkove S B, Lee K S, Shiffman C A and Aaron R 2006 *Clin. Neurophysiol.* **117** 1244

Acknowledgments
This work was funded by Grant R01055099 from the National Institutes of Health.

Journal of Physics: Conference Series **224** (2010) 012159 doi:10.1088/1742-6596/224/1/012159

High Precision Phase Measurement Technique for Cell Impedance Spectroscopy

Dennis Trebbels, Daniel Woelki, Roland Zengerle

HSG-IMIT, Institut für Mikro- und Informationstechnik,
Wilhelm-Schickard-Str. 10, 78052 Villingen-Schwenningen, Germany

E-Mail: dennis.trebbels@hsg-imit.de

Abstract. This paper presents a new approach for high precision phase measurement. The new system is developed for biomedical applications such as complex cell impedance measurement and dielectric tissue analysis. In many of the named applications it is necessary to measure the complex dielectric constant of a sample as a function of frequency. Therefore the developed system is capable of measuring amplitude and especially high precision phase of the measurement signal over a wide frequency range from 10 Hz to 10 MHz. The experimental result of the new method shows a phase resolution of up to 0.01 degree at 1 MHz and 0.1 degree at 10 MHz. The excellent phase measurement resolution is achieved by a time transformation of the signal using a digital sampling circuitry. The functional principle of the digital sampling circuitry is based on Delta – Modulation and is implemented inside a cheap standard FPGA. The new system is successfully used in blood analysis applications and allows for a practical implementation of cost efficient capacitive hematocrit sensors for the first time.

1. Introduction

This article presents a new measurement system for measuring amplitude and especially high precision phase shifting caused by biological cells and tissue which serve as complex dielectric inside a measurement capacitor. Figure 1 shows a typical measurement setup for such a biological test cell, as it is used in many biological and biomedical applications [1, 2]. The measurement setup must be capable of generating a sinusoidal test signal over a wide range of frequencies. In addition the measurement setup must have a data acquisition unit for sampling the signal which is distorted in amplitude and phasing by the device under test (DUT). The sampled raw data is then filtered inside a signal processing unit and output to the user for further analysis and interpretation.

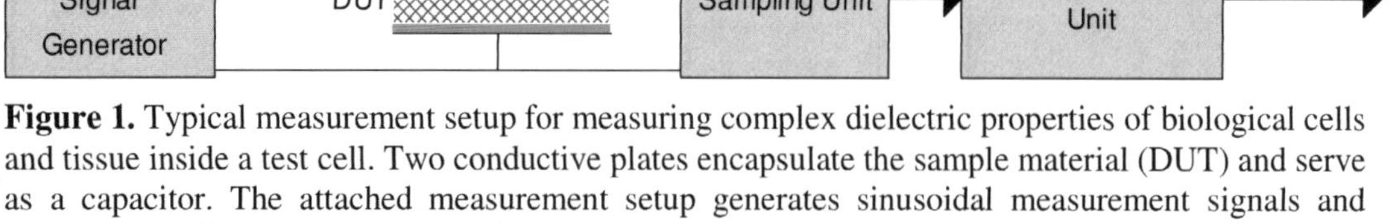

Figure 1. Typical measurement setup for measuring complex dielectric properties of biological cells and tissue inside a test cell. Two conductive plates encapsulate the sample material (DUT) and serve as a capacitor. The attached measurement setup generates sinusoidal measurement signals and measures amplitude distortion and phase shift caused by the sample material inside the test cell.

2. System Concept

This section describes the developed new system architecture which allows for high precision phase measurement. The corresponding block schematic is shown in figure 2.

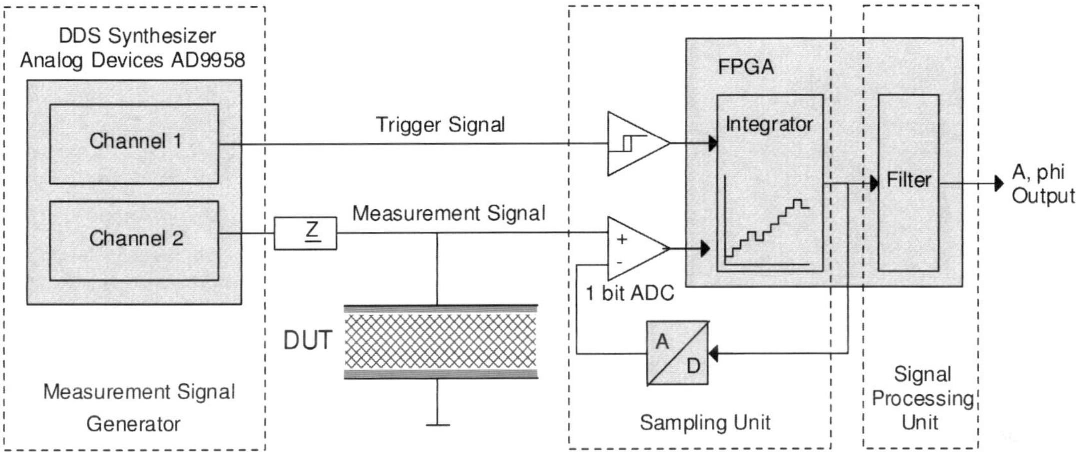

Figure 2. Block schematic of the new phase measurement system based on Delta-Modulation. The measurement signal is generated by a DDS-Synthesizer. The sampling unit consists of a comparator which acts as a 1 bit A/D-Converter followed by a digital integrator / counter which is triggered by an independent channel of the DDS-Synthesizer. The integrators curve is filtered by a spline interpolation algorithm inside the FPGA and output to the user for further analysis.

2.1. Measurement Signal Generator

The measurement signal generator module is implemented by using an off-the-shelf Direct Digital Synthesizer (DDS) Chip. The Synthesizer provides two independent output channels. Channel 2 generates the actual sinusoidal measurement signal while channel 1 generates an adjustable trigger signal for the integrator circuit inside the FPGA (Xilinx Spartan3 XC3S400). The DDS-Synthesizer is internally driven by a 500 MHz clock signal. The phase accumulator width is N = 32 bit for both channels which leads to a fine frequency resolution of each output channel of 0.12 Hz [3]. The achieved precision frequency resolution is a key point for the sampling process as discussed later.

2.2. Sampling Unit

The developed digital sampling circuitry is the main component of the new measurement system. The sampling circuitry is based on a pure digital Delta-Modulator (DM) structure and replaces a conventional A/D-Converter. It digitizes the sinusoidal measurement signal which has been affected by the DUT inside the test cell. The DM-module mainly consists of a digital 12-bit counter inside the FPGA which serves as an integrator. Depending on the output state of the external comparator the counter counts either up or down on each rising edge of the trigger signal provided by channel 1 of the DDS-Synthesizer. The actual state of the integrator is converted into a corresponding analogue voltage signal and fed back to the inverting input of the comparator. The high impedance positive input of the comparator is directly connected to the voltage divider of the reference impedance Z and the test cell. As a matter of principle the integrator is forced to track the sinusoidal signal on the comparators positive input. An example of the typical pattern of the integrator curve is shown in figure 3. By setting the trigger frequency, the number of samples taken in each period of the measurement signal can be adjusted. Using appropriate values, an oversampling rate of up to 10^6 can be achieved which leads to a high theoretical signal to noise ratio (SNR) of up to 120 dB for the DM-structure [4].

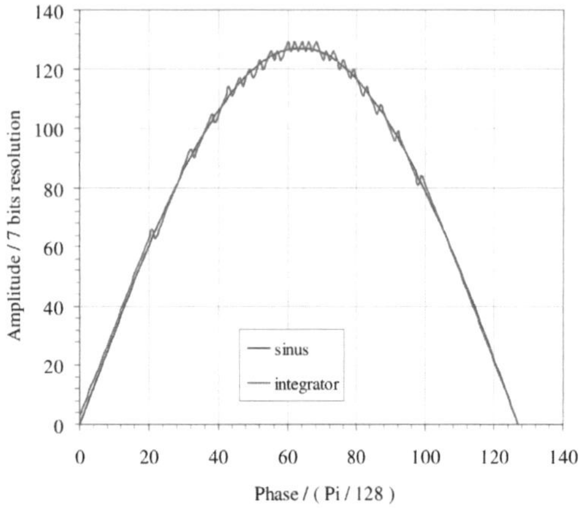

Figure 3. Simulation of the delta modulator illustrates the basic behavior of the integrator curve which tracks the ideal sinusoidal input signal. The simulation is performed for a 7 bit amplitude resolution system for better visualization.

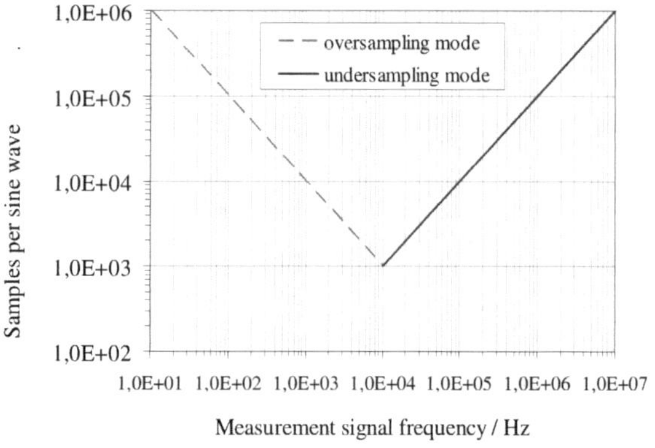

Figure 4. The graph shows the number of samples taken per sine wave as a function of the measurement frequency. The trigger frequency is assumed to be constant at 10 MHz for the oversampling mode. For the undersampling mode the delta frequency is assumed to be constant at 10 Hz between the trigger- and the measurement frequency.

2.3. Signal Processing Unit

The output data of the Delta – Modulator is the integrators curve which represents the measured signal. Since the comparator inside the sampling system acts as a 1-bit A/D-converter the sampled raw data on the output of the integrator contains high frequency components. In figure 3 these high frequency components appear as ripple on the sampled sine wave. Since the high frequency components are not of interest a method known as noise-shaping is applied to the sampled raw data. The integrator works as a high pass filter for high frequency components in the sampled signal. A digital filter implemented inside the FPGA removes these high frequency components and therefore increases the overall SNR [5]. The used low pass filter is based on an on-line Bezier-Spline interpolation algorithm which can be implemented in hardware inside the FPGA. [6].

3. Operation Theory

The phase measurement system is used to measure amplitude and phasing of sinusoidal signals in a frequency range from 10 Hz to 10 MHz. Due to this wide range of six decades, the operation mode of the sampling unit has to be switched between a conventional oversampling mode for digitizing low measurement frequencies and an undersampling mode for high measurement frequencies.

3.1. Oversampling Mode

The oversampling mode is used to digitize signals at low frequencies in the range of 10 Hz to 10 kHz. In this operation mode the output frequency of channel 1 is set to the absolute maximum which can be processed by the FPGA. In this operation mode *all measurement samples are taken during a single period of the sinusoidal measurement signal* generated by channel 2. Since the trigger signal frequency is always set to a constant maximum value which is limited by the FPGA speed, this sampling method causes a varying number of samples taken during a single sinusoidal measurement signal period dependent on the measurement signal frequency. The graph in figure 4 illustrates this effect for an assumed maximum frequency of 10 MHz which can be handled by the used FPGA logic.

3.2. Undersampling Mode

The undersampling mode is used if the measurement frequency is in the range of 10 kHz to 10 MHz. In this operation mode channel 1 also generates the measurement signal which is directed to the DUT. In contrast to the oversampling mode channel 1 generates a trigger signal frequency just slightly above the measurement frequency. Due to the fine frequency resolution of the DDS-Synthesizer the delta frequency between both channels can easily be adjusted to values in the range of a few Hz or even below 1 Hz. For the undersampling mode the graph in figure 4 assumes the trigger frequency to be exact 10 Hz above the measurement signal frequency. As a matter of principle *in the undersampling mode the shape of the measurement signal is captured by sampling many repetitive measurement signals*. On each incoming sine wave of the measurement signal exactly one sample is taken at a slightly delayed point. The trigger process is illustrated in figure 5. The undersampling mode allows for digitizing the sinusoidal measurement signal with a virtual high temporal resolution depending on the settings of the trigger frequency and the measurement frequency. Amplitude and phasing are measured with high accuracy by transforming the real measurement frequency to a lower frequency. Figure 6 shows precision measurement results derived by the new measurement technique in comparison to state of the art equipment (Agilent E4980A) which has been used for reference.

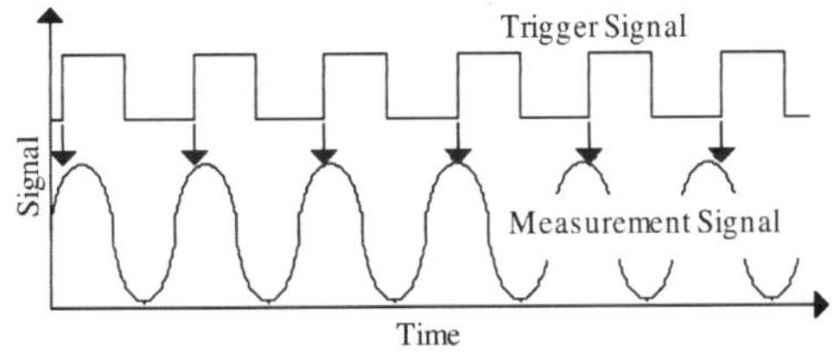

Figure 5. Sampling technique in the under-sampling mode for capturing the signal out of multiple repetitive sine waves.

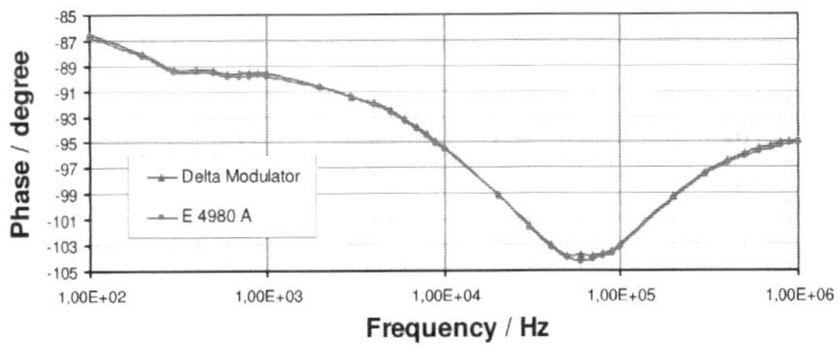

Figure 6. Experimental measurement data derived with blood in the test cell. As reference method an LCR Meter (Agilent E 4980 A) is used.

4. Conclusion

The developed precision phase measurement technique is based on a time transformation performed in a Delta-Modulator sampling circuit. Especially high frequency measurement signals in the range of 100 kHz to 10 MHz can be measured with an outstanding phase resolution of better than 0.1 degree (for 10 MHz) which is in the range of conventional expensive laboratory equipment such as the Agilent E 4980A LCR-Meter. The employed Delta-Modulator circuitry allows for reconstructing high frequency measurement signals out of many repetitive sine waves which is equal to a time transformation of the measurement signal. The implementation of the system can be done by using cheap digital standard components which allows for a wide-spread use of an impedance spectroscopy measurement system in many cost sensitive biomedical applications for the first time.

References

[1] Capacitive on-line hematocrit sensor design based on impedance spectroscopy for use in hemodialysis machines, D. Trebbels, D. Hradetzky, R. Zengerle, IEEE EMBC 09, Minneapolis
[2] Bioimpedance and Bioelectricity Basics, O.G. Martinsen, S. Grimnes, 2008, ISBN: 0123740045
[3] Datasheet AD9958 DDS-Synthesizer, Analog Devices, Norwood, 2008
[4] Spatial Time Domain Reflectometry for Monitoring Transient Soil Moisture Profiles, R. Becker, Dissertation, University of Karlsruhe, 2004
[5] Demystifying Sigma-Delta ADCs, Maxim, Application Note 1870, 2003
[6] Bezier Spline Algorithms in VHDL, B. Szuecs, Masterthesis, University of Furtwangen, 2009

Wideband Bio-impedance Spectroscopy using Voltage Source and Tetra-polar Electrode Configuration

Pil Joong Yoo, Dae Hyun Lee, Tong In Oh and Eung Je Woo

Department of Biomedical Engineering, Kyung Hee University, Korea

E-mail: ejwoo@khu.ac.kr

Abstract. Most bio-impedance spectroscopy (BIS) systems inject sinusoidal current with a variable frequency into a sample with a known geometry through a pair of electrodes. Adopting the so-called tetra-polar configuration, it measures induced voltage data on a separate pair of electrodes. Impedance spectra are plotted in a certain range of frequency. We found that its accuracy decreases at high frequencies primarily due to the deteriorated performance of the constant current source at high frequencies. Using a previous BIS system we developed, we found that the overall performance can be kept high up to several hundred kHz. In this study, we propose a design of a wideband BIS system using a constant voltage source. It is based on the simple voltage division between an internal resistor and an external sample or load. We switch the value of the internal resistor (R_s) so that the source voltage is divided more or less equally. Two pairs of electrodes are attached to the sample. Two independent voltmeters are used to separately measure two voltages across the chosen internal resistor and the sample. The voltage measurement across the sample is done between the second electrode pair only. This enables us to adopt the tetra-polar configuration to avoid the problem related with contact impedances. We describe the design, construction and performance of the new BIS system with 468 Hz to 2.2 MHz bandwidth. We will compare the results with those using the impedance analyzer.

Keyword: bio-impedance spectroscopy, voltage source

1. Introduction

A bio-impedance spectroscopy (BIS) system measures admittivity spectra of biological tissues. Using an RC circuit model, we obtain conductivity and permittivity of the sample. Repeating the measurement at multiple frequencies, we may produce admittivity spectra of the sample. Since 1960s, numerous studies have shown that there exist large variations in admittivity spectra of different biological tissues [1-3]. Results of BIS studies have significantly contributed to our understandings of electromagnetic phenomena inside the human body.

These may be classified into two types according to the source. In the first type, active current source injects current into a tissue sample with a known geometry and measure induced voltage to find its admittance or impedance [4,5]. The second type uses voltage source instead of using current source. Some calibration methods are applied to the result for minimizing the impact of stray capacitance, drift and/or non-ideality [6]. In order to improve the measurement accuracy, the amount of injecting current is measured with sense resistor at the output stage and then, admittance is calibrated with this value.

We have developed a BIS system using a balanced current source and a voltmeter in the frequency range of 10 Hz to 500 kHz [7]. One pair of electrodes was used to inject current to a sample and the other pair measured induced voltage to remove the contact impedance terms in the measurement. For frequencies above 5 kHz, we adopted multiple generalized impedance converter (GIC) circuits to minimize the output capacitance of the current source. Although we calibrated the balanced current source for its maximal output impedance, its performance was degraded at frequencies above 500 kHz primarily due to effects of stray capacitances. In this study, we suggest improving the performance at high frequencies using a method based on a voltage source and two voltmeters. We adjusted the sense resistor at the output stage to get better accuracy.

2. Method

2.1 KHU Mark2 BIS system design

We may consider our previous BIS system comprising a balanced constant current source and a voltmeter as a one-channel multi-frequency EIT system [7]. Because it is based on the current source, the highest operating frequency was limited by 500 kHz. In designing a new BIS system, we adopted a voltage source and tried to minimize the size so that it is portable.

Figure 1 shows the structure of the new BIS system including a voltage source and two voltmeters. We apply a sinusoidal voltage with a variable frequency to the series connection of the internal current-sensing resistor R_s and the sample. Measuring the voltage across R_s by using one voltmeter, we know how much current is being injected. We use the second voltmeter to measure the induced voltage across the sample. Using two separate pairs of electrodes attached on the sample, we can adopt the tetra-polar electrode configuration to reject effects of unknown contact impedances. We expect that the constant voltage source is advantageous at frequencies above 500 kHz.

The FPGA (Cyclon III EP3C10F256C8N, Altera, USA) controls two ADCs (AD9235BRUZ-65, Analog Devices, USA) and one DAC (AD9783, Analog Devices, USA) for digital phase-sensitive demodulations and voltage waveform generation. The DSP (TMS320F2812, Texas Instruments, USA) exchanges commands and data with each modules implemented within the FPGA through 2 SCI ways to control all functions of the impedance measurement. A PC with BIS software is connected to the BIS system through an isolated USB port for graphical user interface and output display.

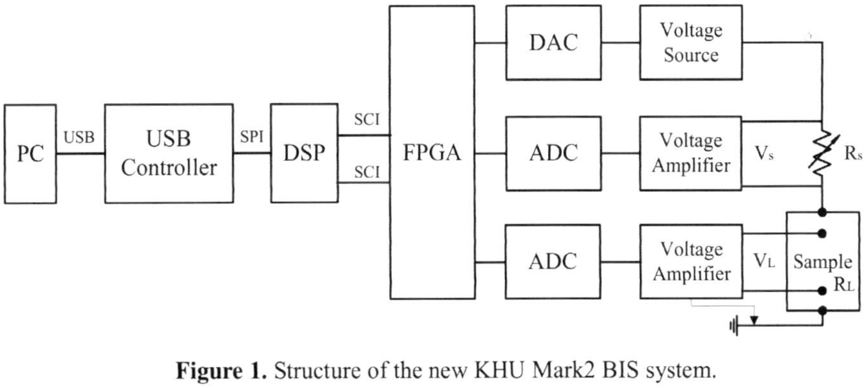

Figure 1. Structure of the new KHU Mark2 BIS system.

2.2 Bio-impedance measurement

The voltage from the constant voltage source is divided by the internal current-sensing resistor R_s and the sample. We measure induced voltages across R_s and the sample simultaneously. Since we adjust the value of R_s so that magnitudes of two voltages are comparable, every impedance measurement includes two measurements. We used a digital potentiometer (DS1267, Dallas semiconductor, USA) to control the value of R_s. If internal current-sensing resistor (R_s) is too smaller than that of sample (R_L), V_s has worse SNR. It can also have bad effects to the result when we calibrate it using V_s. On the contrary to this, total resistance may be too large because internal current sensing resistance is bigger than sample. Output characteristic of voltage source will be worse. So that, we have to adjust R_s until it will have similar value as the sample resistance. In the new design, we needed to increase the bandwidth of voltage amplifiers for impedance measurements at frequencies beyond 500 kHz. We also increased the sampling frequency of the ADC.

We use two independent voltmeters to separately measure two voltages across R_s and the sample. Figure 2 shows two tetra-polar electrode configurations. Since there is no contact impedance across the internal current-sensing resistor, we directly measure voltage across it. For the voltage measurement across the sample, we always use a separate pair of electrodes where no probing current flows. This enables us to adopt the tetra-polar configuration to avoid the problem related with contact impedances.

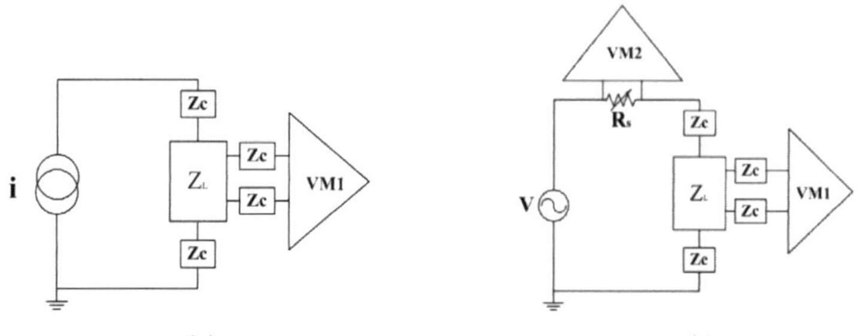

Figure 2. Tetra-polar electrode configuration for (a) current source and (b) voltage source. VM is a voltmeter and Z_C is a contact impedance.

2.3 Digital control and communication
The DSP operates at 150 MHz clock speed to increase the overall speed of the system. We adopted USB technology for communicating with PC. The USB port is electrically isolated from the BIS system. The single FPGA performed multiple functions including data communication, demodulations, waveform generation and control.

3. Result

Figure 3 shows voltage spectra of a parallel RC components of 100 Ω and 10, 100, 2000 nF. We measured load resistance for comparing results in BIS system and impedance analyzer (SI1260, Solartron). We can see that measured data using the voltage source closely follows the numerical simulation results and measurement data using the impedance analyzer.

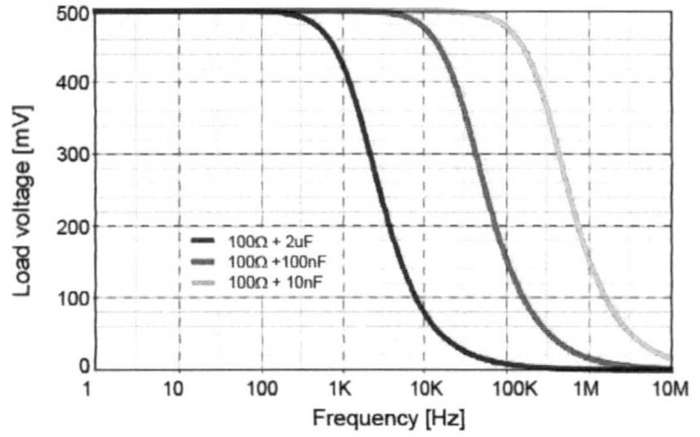

Figure 3. PSPICE simulation of load voltage across a parallel RC circuit (blue line: 100Ω+2uF, red line: 100Ω+100nF, green line: 100Ω+10nF).

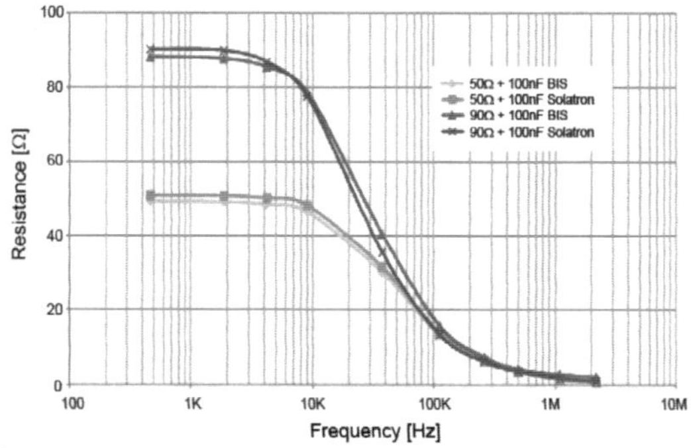

Figure 4. Comparison of the resistance in BIS system and impedance analyzer.

Figure 5 shows the developed new BIS system implemented in a single PCB. We tried to measure the resistivity of carrot with BIS and impedance analyzer as shown in figure 6. Both results have similar trend of spectra.

Figure 5. BIS Mark2 system.

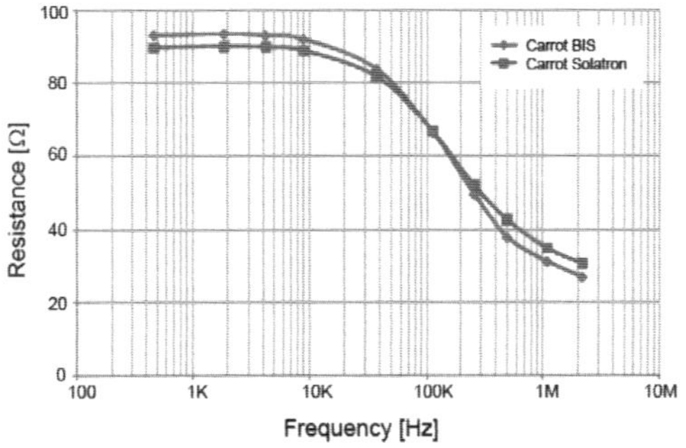

Figure 6. Comparison of the resistance of carrot in BIS system and impedance analyzer.

4. Conclusion and discussion

We developed a wideband BIS system using a voltage source and two voltmeters. We used voltage measurements across an internal current-sensing resistor to calibrate the impedance of the sample. Simultaneously measuring two voltages from the internal resistor and the sample, we adjusted the resistance value of the internal resistor for better measurement accuracy. The tetra-polar electrode configuration was still possible using the voltage source. We need to look at the performance of the developed BIS system by comparison with BIS system which uses current source. Since we have minimized the size of the system, we plan to construct a portable BIS system for application in electrical bio-impedance studies.

Acknowledgement

This work was supported by the SRC/ERC program (R11-2002-103) of MOST/NRF

Reference

[1] Geddes, Baker Geddes L A and Baker L E 1967 The specific resistance of biological material: a compendium of data for the biomedical engineer and physiologist *Med. Biol. Eng.* **5** 271–293
[2] Gabriel S, Lau R W and Gabriel C 1996 The dielectric properties of biological tissues: II. Measurements in the frequency range 10 Hz to 20 GHz *Phys. Med. Biol.* **41** 2251–2269
[3] Grimnes S and Martinsen O G 2000 Bioimpedance and Bioelectricity Basics *London, UK: Academic*
[4] Hong H, Rahal M, Demosthenous A and Bayford R H, Comparison of a new integrated current source with the modified Howland circuit for EIT applications, *Physiol. Meas.* **30**, 999-1007.
[5] Hartov A, Mazzarese R A, Reiss F R, Kerner T E, Osterman K S, Williams D B and Paulsen K D 2000 A multichannel continuously selectable multifrequency electrical impedance spectroscopy measurement system *IEEE Trans Biomed Eng.* **47**, 49-58.
[6] Saulnier G J, Ross A S and Liu N 2006 A high-precision voltage source for EIT, *Physiol. Meas.* **27**, S221-S236.
[7] T. I. Oh, H. Koo, K. H. Lee, S. M. Kim, J. Lee, S. W. Kim, J. K. Seo and E. J. Woo 2008 Validation of a multi-frequency electrical impedance tomography (mfEIT) system KHU Mark1: impedance spectroscopy and time-difference imaging *Physiol. Meas.* **29**, 295-307.

Development of a prototype micro-EIT system using three sets of 15×8 array electrodes

Qin Liu[1], Hun Wi[1], Tong In Oh[1], Eung Je Woo[1] and Jin Keun Seo[2]

[1] Department of Biomedical Engineering, Kyung Hee University, Korea
[2] Department of Mathematics, Yonsei University, Korea

E-mail: ejwoo@khu.ac.kr

Abstract. We developed a new microscopic electrical impedance tomography (micro-EIT) system to visualize admittivity distributions within a miniature hexahedral container, where we place small biological samples with a background solution or gel. Each of two facing sides (left and right) of the container is fully covered by a solid metal electrode. We inject current between them, thereby producing a uniform parallel current flow along the longitudinal direction inside the container. Each of three sides at the bottom, front and back is equipped with a 15×8 array of voltage-sensing electrodes. Switching modules are located underneath the container so that we can measure voltage between any neighbouring pair of electrodes. Three switching modules are connected to a 16-channel multi-frequency EIT system to collect induced voltage data from the three sets of 15×8 array electrodes subject to the single fixed current injection. Voltage data set from 360 voltage-sensing electrodes on three sides are utilized to produce cross-sectional images of the admittivity distribution. We describe the design and construction of the new micro-EIT system. Our future work should include development of a customized image reconstruction algorithm for the micro-EIT system and experimental validation.

Keywords: Micro-EIT, admittivity image, biological tissue sample

1. Introduction

To monitor the behavior or functionality of a biological tissue and cells, many researchers have developed numerous techniques using miniature semiconductor sensors, optical sensors and so on [1,2]. Microelectrode systems have been successfully used to investigate growth and migration of monolayer cell cultures. By applying a voltage across a sample and then measuring the current flow using different kinds of measurement method, various cell functions have been monitored [3-5]. Methods such as scanning impedance imaging have also been applied to cell-culture studies [6,7]. Impedance monitoring is potentially useful in monitoring cell proliferation, morphology and motility since impedance changes occur in cells in response to physical and also chemical changes. Impedance changes could be monitored during cell growth and used to provide information about cell functions [8]. Responses to the stress or chemical changes can be measured in terms of impedance changes. Impedance imaging could be used in developing an automated cell based system for detection of infection or effects of drugs against diseases [9].

We have developed a microscopic impedance imaging system called the micro-EIT system with a goal to noninvasively visualize admittivity distributions of miniature biology tissue samples and cell cultures. As a prototype of a true micro-scale admittivity imaging system, we constructed a micro-EIT system with three sets of 15×8 electrodes. The system is based on the lately developed parallel multi-frequency EIT system KHU Mark2. We describe its development for detecting impedance changes related with cell growth and other factors.

2. Methods

2.1. System design

The prototype micro-EIT system is based on the KHU Mark2 EIT system as illustrated in figure 1. It includes a main controller, a digital backplane, impedance measurement modules (IMM), a switch controller, an analog backplane, switching circuits and a sample container with current-injection and voltage-sensing electrodes.

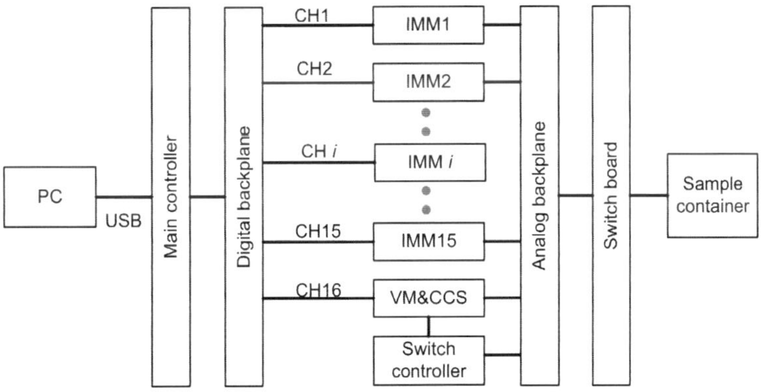

Figure 1. Structure of micro-EIT system.

2.2. Sample container

The sample container has voltage-sensing electrodes on three sides. On each side at the front, bottom and back, there are 15×8 array electrodes. The total number of voltage-sensing electrodes is 15×8×3=360. Each electrode is coated with gold and has a diameter of 2 mm. The electrodes are placed on a 28×50 mm^2 rectangular substrate and are separated by a gap of 2.54 mm as shown in figure 2. Though this sample container is not suitable for microscopic imaging, we decided to use it for preliminary testing of our micro-EIT system.

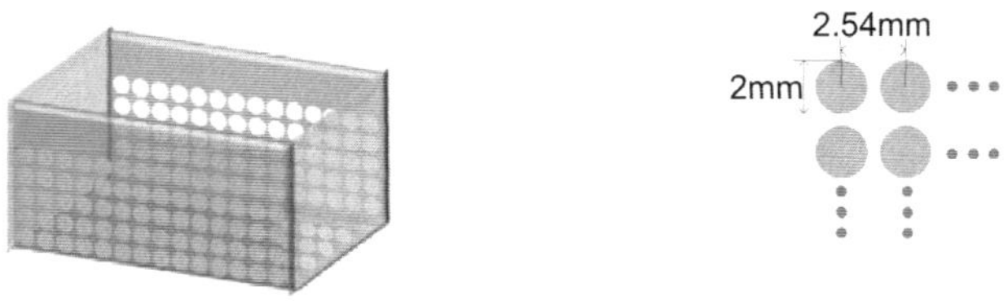

Figure 2. Sample container with three sets of 15×8 array electrodes.

2.3. Switching method for selecting a pair of voltage-sensing electrodes

We are aiming to measure induced voltage at any pair of neighbouring electrodes. Uniform parallel current flows in the sample container since we inject current between large current injection electrodes at the right and left sides. The voltage distribution inside the container with a fixed shape and size is determined by its internal admittivity distribution. We need to obtain as much information on the admittivity distribution as possible from three sets of 15×8 voltage-sensing electrodes.

We divide voltage-sensing electrodes into two groups. The first group includes all electrodes from the 1st to 8th columns. The second group includes electrodes from the 9th to 15th columns. In each group, we measure voltages between pairs of electrodes from the even and odd columns. As the 16th column in the second group, we used the common ground electrode for current injection. We implement a switch controller using a CPLD (EPM3064A, Altera, USA) which controls multiplexers (ADG708, Analog Device, USA). It chooses 15 electrode pairs among the 360 electrodes at the same time for simultaneous acquisition of 15 complex voltage data.

We select one electrode per column along transversal direction in the electrode array. It means 24 electrodes can be chosen in one column. Every 8 electrodes are selected by one multiplexer which needs 4 control signals. Every column (8×3 electrodes, one column of container) needs 24 control signals. This allows us to select any neighbouring electrode pair as shown in figure 3(a). To minimize the number of switches and control signals, we used the same control signals for odd and even numbered columns in both groups. The total number of control signals is 48, which are supplied by the switch controller. Though we can choose only a neighbouring measurement pair along the longitudinal direction, we can choose any pair in the transversal direction. Figure 3(b) shows an example of such switch selection.

Among 16 IMMs in the 16-channel KHU Mark2, the 16th IMM is used to inject current between two facing sides (left and right) of the container. We also measure voltage between current-injection electrodes using the voltmeter in the 16th IMM as shown in figure3(c).

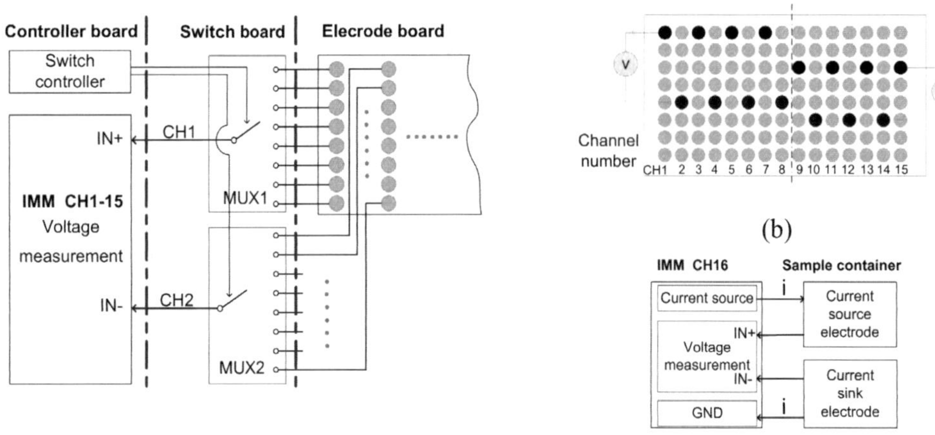

(a) (b) (c)

Figure 3. (a) Selection of an electrode pair in the 1st to 15th columns. (b) Example of an electrode selection. (c) Current injection using the 16th IMM.

3. Results

Figure 4 shows the constructed sample container for the micro-EIT system based on the KHU Mark2 16-channel EIT system. From preliminary testing, we could observe voltage changes from the container filled with saline as we move an anomaly in it. We plan to perform numerous imaging experiments for static, time-difference and also frequency-difference imaging.

4. Conclusion and discussion

We described a design of the micro-EIT system based on the parallel multi-frequency EIT system KHU Mark2. The sample container with electrodes and switching circuits is the key component of the micro-EIT system. The present implementation incorporates 360 voltage-sensing electrodes within the

$50\times44\times28$ mm^3 container. We could control all switching operations using 48 control signals. We plan to apply a customized image reconstruction algorithm to acquired data from the micro-EIT system to validate its performance. We will fabricate a miniaturized sample container for true microscopic admittivity imaging in our future study.

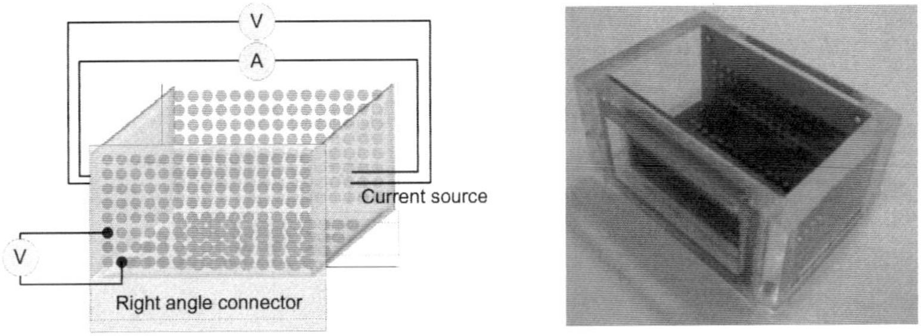

Figure 4. Developed sample container to be used with the 16-channel parallel multi-frequency EIT system KHU Mark2.

Acknowledgment

This work was supported by the SRC/ERC program (R11-2002-103) of MOST/NRF.

References

[1] T. York, L. Sun, C. Gregory and J. Hatfield (2004) Silicon-based miniature sensor for electrical tomography *Sensors and Actuators A.* **110**, 213-218.

[2] T. Ludwig, S. P. Ttmann, H. Bertram, L. Tatenhorst, W. Paulus, H. Oberleithner and V. Senser (2005) Functional measurement of local proteolytic activity in living cells of invasive and non-invasive tumors *Journal of Cellular Physiology* .**202**, 690-697.

[3] B. G. Buss, D. N. Evans, H. Z. Liu, T. Shang, T. E. Oliphant, S. M. Schultz, A. R. Hawkins (2007) Quantifying resistivity using scanning impedance imaging *Sensors and Actuators A.* **137**, 338-344.

[4] A. R. A. Rahman, J. Register, G. Vuppala and S. Bhansali (2008) Cell culture monitoring by impedance mapping using a multielectrode scanning impedance spectroscopy system (CellMap) *Physiol. Meas.* **29**, 227-239.

[5] P. Linderholm, L. Marescot, M. H. Loke, and P. Renaud (2008) Cell culture imaging using microimpedance tomography *IEEE Transactions on Biomedical Engineering.* **55**, 138-146 .

[6] P. Linderholm, A. Bertsch and P. Renaud (2004) Resistivity probing of multi-layered tissue phantomsusing microelectrodes *Physiol. Meas.* **25**, 645-658.

[7] T. E. Oliphant, H. Z. Liu, A. R. Hawkins and S. M. Schultz (2006) Simple linear models of scanning impedance imaging for fast reconstruction of relative conductivity of biological samples *IEEE Transactions on Biomedical Engineering.* **53**, 2323-2332.

[8] X. Q. Huang, D. Nguyen, D. W. Greve, and M. M. Domach (2004) Simulation of microelectrode impedance changes due to cell growth *IEEE Sensors Journal* .**4**, 576-583.

[9] H. Wi and E. J. Woo (2009) Design of KHU mark2 multi-frequency EIT System *Proc. Int. Conf. on Biomedical Applications of Electrical Impedance Tomography (Mancheter)*

Analysis of thoracic regional impedance changes using PCA approach

M Lewandowska[1], J Wtorek, A Bujnowski and L Mierzejewski

Department of Biomedical Engineering, Faculty of Electronics, Telecommunication and Informatics, Gdansk University of Technology, Narutowicza 11/12, 80-233 Gdansk, Poland

E-mail: maglew@biomed.eti.pg.gda.pl

Abstract. A multichannel impedance and one lead electrocardiographic recording system has been developed. The impedance measurement unit is built using a four-electrode technique. It consists of a sinusoidal, 40 kHz, current source and five measurement channels. Circumferential current electrodes have been located at a neck and an abdomen. Voltage measurement disk electrodes formed five two-electrode measurement ports. The impedance change waveforms (ICG) have been recorded simultaneously with ECG. The ICG waveforms have been analyzed using Principal Component Analysis (PCA). The examinations have been done for different ports configurations in relation to the heart. A dependence of obtained components on ports configurations has been found. Similar results have been obtained when examining healthy (test) persons.

1. Introduction

Impedance signals taken from the thorax are widely used for different purposes [1]. However, the most widespread application is the measurement of cardiac output [1, 5]. This application was developed under an assumption that the main source of the measured impedance changes was associated with the changing volume of aorta.

The evaluation of impedance changes associated with blood flow in the thorax requires a special measurement technique allowing measurement of very small signal in the presence of a very big one. Typically, the evaluated changes are very often not bigger than 1% of the measured signal.

PCA is a multivariate statistical procedure where the vector containing measured signal is presented as a weighted sum of orthogonal basis vectors. The central idea in PCA is to reduce the dimensionality of the data set, while retaining as much as possible of the information in the original data [4]. Thus, it allows reducing the dimentionality of data set containing a large number of interrelated variables, while preserving as much as possible of variation present in the data set. It is achieved by transforming the input data set into a new coordination system.

Generally, PCA is motivated by the two following problems: 1. given random vectors described by finite second order moments and zero mean find the linear subspace of reduced dimension that minimizes the expected distance of data from the subspace, and 2. given random vectors find the linear subspace that captures most of data variance. This problem is related to feature extraction, where the

[1] To whom any correspondence should be addressed.

objective is to reduce the dimension of the data while retaining the most of its information content. Both problems have the same optimal solution, which is obtained due to knowledge of data covariance matrix.

2. Methods

2.1. A measurement system

Data for analysis have been collected using a specially developed system (figure 1). It consisted of a sinusoidal, 40 kHz, current source I(ω) and five identical measurement channels. The measurement system, known as a four-electrode one, consisted of two current electrodes, marked I_1 and I_2, and two voltage electrodes, marked V_{i1} and V_{i2}, where subscript *i* stood for current number of the measurement channel. The difference of potential measured using voltage electrodes after amplification was demodulated using a synchronous technique and the result was filtered (using Low Pass Filter, LPF) and dc component was removed (DCR). Only alternating part of the signal was amplified using a dc amplifier (Amp_{dc}) and sampled. A sampling operation was synchronous for all measurement channels. Then, the sampled signals were converted (AD) into digital ones and sent successively to the host computer (H.C.) via USB.

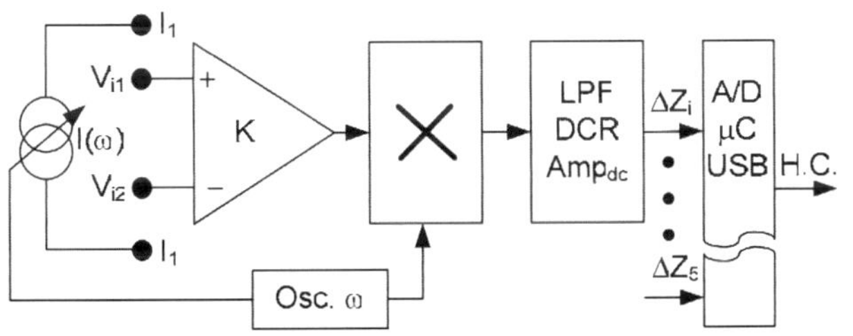

Figure1. Schematic diagram of the measurement system, a detailed description in the text.

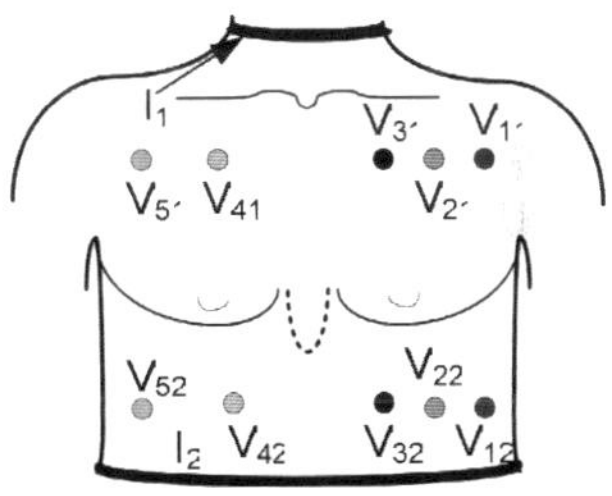

Figure 2. Localization of the electrodes on the thorax.

2.2. Collection of data

A specially developed system (figure 1) was used to collect the data by means of electrodes placed on thorax (figure 2). The current was passing between electrodes I_1 and I_2. Each pair of electrodes with identical first subscript formed a measurement lead, e.g. the electrodes V_{21} and V_{22}. According to sensitivity theorem developed by Geselowitz [2, 5] different localization of voltage measurement electrodes in spite of utilizing common, circumferential, current electrodes located on the neck and abdomen results in different distribution of spatial sensitivity. Thus, even assuming a uniform spatial distribution of current density involved by current flowing between current's electrodes the resulting sensitivity is non-uniform.

2.3. Data analysis

Data were analyzed using a Principal Component Analysis (PCA). The following algorithm adopted from [6, 8] was used:

1. Data preprocessing: The recorded signal is segmented into vectors having length M. The N similar (associated with the activity of the heart described by the same rate) are chosen. These data are formed into a single matrix **T** of dimensions N × M, so that N is the observations and M is the dimensions.
2. The mean along each dimension m = 1...M is calculated. Afterwards, all computed mean values are placed into an empirical mean row vector **u** of dimensions M.
3. The empirical mean row vector **u** is subtracted from each row of the data matrix **T**. Then a new mean-subtracted data matrix is derived.
4. The M × M empirical covariance matrix is calculated using the zero-centered data matrix **B**.

5. Calculation of the eigenvectors and eigenvalues of the covariance matrix. This step requires the use of a computer-based algorithm for computing the eigenvalue matrix **D** and the eigenvector matrix **V** of the covariance matrix. Diagonal matrix **D** having dimensions M × M and matrix **V** also of dimension M × M are obtained. The matrixes **D** and **V** contain, respectively, eigenvalues and eigenvectors of the covariance matrix **C**. The eigenvalues and eigenvectors are ordered and paired. The mth eigenvalue corresponds to the mth eigenvector.
6. The columns of the eigenvector matrix **V** and eigenvalue matrix **D** are sorted out in order to maintain the correct pairings between the columns in each matrix.
7. Data are converted to a new basis. The new basis is denoted as PCA-scores or the reconstruction parameter vectors (RPV). The projected vectors are the columns of the matrix **Z** (N×M). The matrix **Z** is calculated by multiplying the eigenvector matrix with the zero-mean data matrix. The rows of **Z** matrix correspond to the observations, whereas the columns refer to the components or dimensions.

3. Results

Example of waveforms recorded for a healthy person show a regular character. Moreover, they contain specific information (figure 3). A characteristic time shift between waveforms recorded at different localization of left and right sides of the thorax is easily seen. The signals recorded in the vicinity of heart (i.e. ΔZ_2 and ΔZ_3) are morphologically different from others.

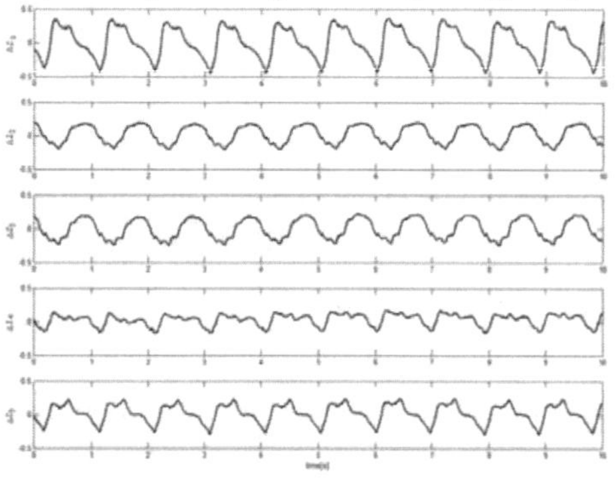

Figure 3. The measured ΔZ waveforms.

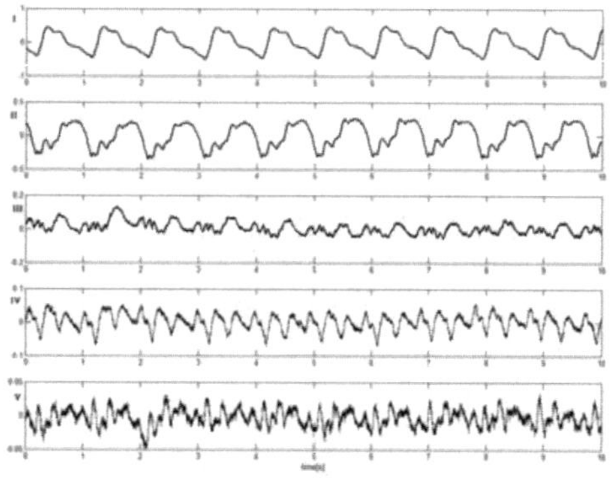

Figure 4. PCA analysis of the signals presented in the figure 3.

PCA of the set of signals presented in figure 3 shows a distinct three components exhibiting a periodic and regular character (figure 4).

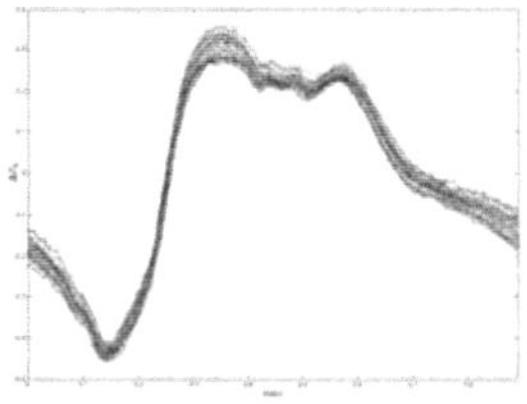

Figure 5. Selected part of the ΔZ_1 waveform recorded for channel 1 (shown in figure 3).

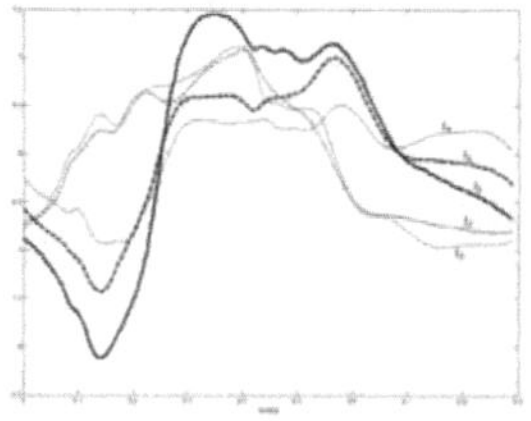

Figure 6. The first components of the signals presented in figure 3.

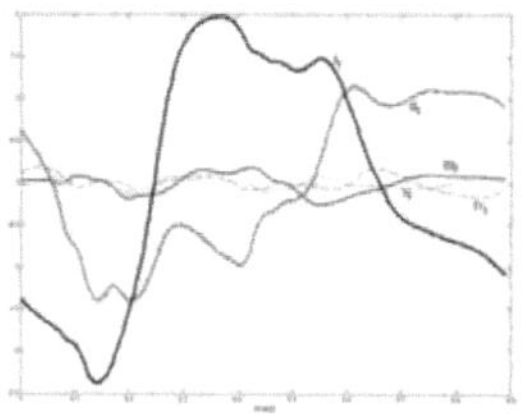

Figure 7. Result of PCA for the set of data composed by the first PC calculated for each channel.

A set of one-cycle waveforms selected from the recording has been used as data for applying PCA. Example of such data is presented in figure 5. It shows a relatively small variability. However, it should be underlined that only samples of equal length (time duration) have been chosen and the signal has been recorded during rest.

4. Discussion and conclusions

PCA is a widely used technique in data processing. It has been shown that this approach is useful when estimating influence of respiration on ECG or ICG [7, 9], removing noise from data [8], analyzing both single-lead and multi-lead signals, compressing the data, separating signals or extracting features [8].

The recorded data show dependence of theirs morphology on localization of recording electrodes. A simple analysis of results obtained and presented in figure 4 reveals that the first and the second components present different phenomena. An initial part of the second PC resembles the corresponding part of the first one. Then, both signals changes in opposite direction. It suggests that they are dominated by different phenomena.

The PCA enables to evaluate local phenomena. In fact, according to Geselowitz [2] it is an obvious remark. The recorded data and PCA analysis allows differentiate between atrial and ventricular activity.

More exhaustive studies led on different groups including diagnosed heart diseases or disorders have to be performed.

References

[1] Wtorek J 2000 Relations between components of impedance cardiogram analyzed by means of Finite Element Model and sensitivity theorem *Ann. Biomed. Eng.*, **28** 1352 – 1361

[2] Geselowitz D B 1971 An application of electrocardiographic lead theory to impedance plethysmography *IEEE Trans. Biomed. Eng.* **18** 38 – 41

[3] Shlens J 2005 A tutorial on principal component analysis http://www.snl.salk.edu/~shlens/pub/notes/pca.pdf

[4] Tarvainen M P, Koiistinen A S, Valkonen-Korhonen M., Partanen J and Karjalainen P A 2008 Analysis of galvanic skin responses with principal components and clustering techniques *IEEE Trans. Biomed. Eng.* **48** 1071 – 9

[5] Valentinuzzi M E 1996 Bioimpedance measurement: general concepts *Crit. Rev. Biomed. Eng.* **24** 223 – 255

[6] Khawaja A, Butrous G, Doessel O 2006 Detecting predisposition to torsade de points usinf a PCA-based method *Comput. Cardiol.* **33** 161 – 164

[7] Langley P, Bowers E J, Murray A 2008 Principal component analysis as a tool for analysing beat-to-beat changes in electrocardiogram features: application to electrocardiogram derived respiration *IEEE Trans. Biomed. Eng.* **PP**(99) 1 – 1

[8] Castells F, Laguna P, Sornmo L, Bollman A and Roig J M 2007 Principal component analysis in ECG signal processing *EURASIP Journal on Advances in Signal Processing* **2007** 98 – 98

[9] Stamkopoulos T, Diamantaras K, Maglaveras N, and Strintzis M 1998 ECG analysis using nonlinear PCA neural networks for ischemia detection *IEEE Trans. Sign. Process.* **46**(11) 3058 – 3067

Performance of an implantable impedance spectroscopy monitor using ZigBee

P Bogónez-Franco[1], A Bayés-Genís[2], J Rosell[1], R Bragós[1]

[1] Electronic Engineering Department. Universitat Politècnica de Catalunya. Campus Nord. Edifici C4. C. Jordi Girona 1-3. 08034 Barcelona. Spain.
[2] Cardiology Service. Hospital de la Santa Creu i Sant Pau. C. Sant Antoni Maria Claret 167. 08025 Barcelona. Spain.

E-mail: bogonez@eel.upc.edu

Abstract. This paper presents the characterization measurements of an implantable bioimpedance monitor with ZigBee. Such measurements are done over RC networks, performing short and long-term measurements, with and without mismatch in electrodes and varying the temperature and the RF range. The bioimpedance monitor will be used in organ monitoring through electrical impedance spectroscopy in the 100 Hz - 200 kHz range. The specific application is the study of the viability and evolution of engineered tissue in cardiac regeneration in an experimental protocol with pig models. The bioimpedance monitor includes a ZigBee transceiver to transmit the measured data outside the animal chest. The bioimpedance monitor is based in the 12 Bit Impedance Converter and Network Analyzer AD5933, improved with an analog front-end that implements a 4-electrode measurement structure and allows to measure small impedances. In the debugging prototype, the system autonomy exceeds 1 month when a 14 frequencies impedance spectrum is acquired every 5 minutes. The receiver side consists of a ZigBee transceiver connected to a PC to process the received data. In the current implementation, the effective range of the RF link was of a few centimeters, then needing a range extender placed close to the animal. We have increased it by using an antenna with higher gain. Basic errors in the phantom circuit parameters estimation after model fitting are below 1%.

1. Introduction

Electrical impedance monitoring is a valuable technique used to monitor the tissue and organ state and electrical impedance spectroscopy (EIS) has been used for many years for cardiac tissue characterization to detect acute ischemia [1], [2], old infarction scar [3] and rejection in transplanted heart [4], [5]. Measurements are performed in open thorax set-ups [2], through catheter [2], [3] or using a pacemaker-type device [4].

We are now working with the Cardiology Department of Santa Creu i Sant Pau Hospital in non-invasive graft rejection detection and regenerative tissue engineering. In the last case a non-functional myocardium area is intended to regenerate by using myocardial cell precursors derived from stem cells. In order to monitor the viability and evolution of the engineered tissue after its implantation in an animal model heart (pig or sheep), an implantable device should be developed. In this paper we present the preliminary results of a device which is capable to acquire an impedance spectrum every 5 minutes in the range of 100 Hz – 200 kHz during 1 month [6]. Some characterization measurements are presented.

2. Materials and methods

2.1. Bioimpedance monitor

The bioimpedance monitor developed [6] measures the electrical impedance at 14 frequencies, stores the data collected and transmits them to an external computer using a ZigBee protocol [7]. The system has a size of 35 mm x 37 mm x 8 mm. The bioimpedance monitor is powered from a NiMH battery. The system makes a measurement every 5 minutes and remains the most part of time in sleep mode, increasing the autonomy. When the system is in sleep mode the current consumption is 0.2 mA and when it is doing a measurement the current consumption is 35 mA. The bioimpedance monitor is designed to measure in the range 10 Ω to 1 kΩ, at a programmable set of frequencies.
The connection of the bioimpedance monitor with the needle electrodes is done using short and thin coaxial cables. Figure 1 shows a system diagram.

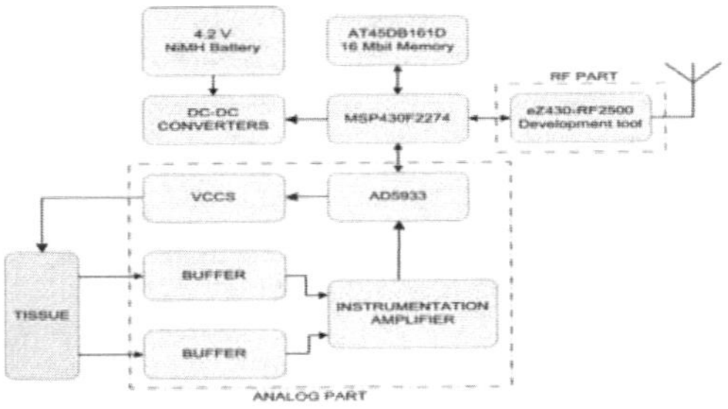

Figure 1: System diagram.

2.2. Measurements

All characterization measurements were done using a series RC network with the following values, Rs = 68 Ω, Rp = 130 Ω and C = 100 nF, and over a 330 Ω resistor and with the electrode impedance network attached. The electrode impedance network was modeled as a parallel RC network with the following values, Rp = 10 kΩ, Rs = 240 Ω and Cs = 18 nF. The measurements performed were short-term, medium-term, with and without electrode mismatch and temperature changes. The temperature change was between 25 °C and 40 °C results. For long-term measurements are shown in [6].

3. Results

3.1. Short-term and mid-term measurements

The short-term measurements where done over a period of time of 5 minutes. With this measurement we would like to see the effect of drifts during the warm-up time. The medium-term measurements where done over a period of time of 5 hours. Figure 2 shows the error at each frequency when the RC network is measured.

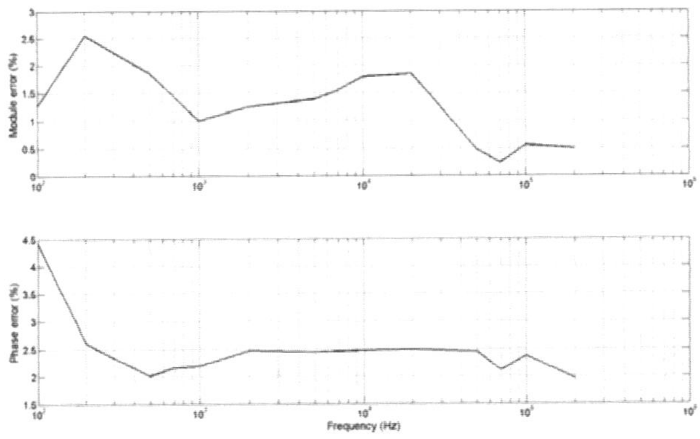

Figure 2: Error at each frequency when a RC network is measured.

3.2. Electrode mismatch

To see the effect of electrode mismatch in the measurement we produce an increment of 100% of the electrode impedance in each lead of the impedance monitor. Electrode mismatch is a very common problem in all impedance measurements and is caused by a poor contact between the electrode and the tissue under measurement. Normal needle electrode impedance is in the range of 10 kΩ or more at low frequencies, whereas at high frequencies its value is around 100 Ω. When the electrode mismatch is produced in the injecting leads, the error is in the order of 3.5% and has periodicities at each decade. When the electrode mismatch is produced in the sensing leads the effect is negligible, less than 1%.

Temperature profile

We measure the temperature effect in module and phase measuring a 330 Ω resistor. The profile consists in three steps of an increment of 5 °C each, see figure 3.

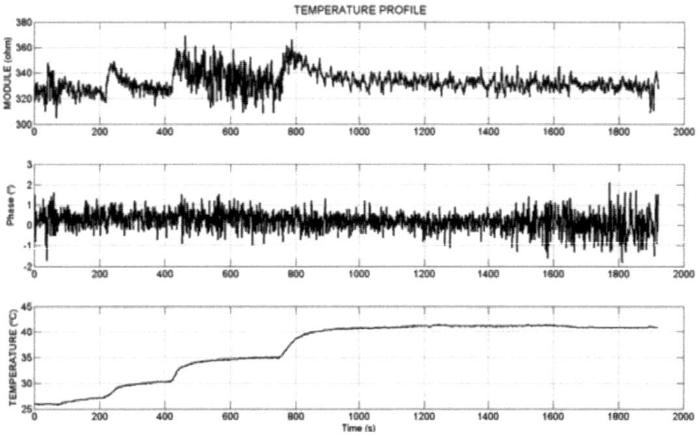

Figure 3: Variations in module and phase over a 330 Ω ohms resistor to a temperature change.

3.3. RF Range

The range measurements were done with the impedance monitor enclosed in a plastic container and immersed into a fishbowl (30 cm x 15 cm x 15 cm) filled with saline solution with a conductivity of 16 mS/cm. As we exposed in a previous work [6] the plastic container was moved at different locations inside the fishbowl to measure the distance when the RF link was broken. In order to increase the range we changed the SMD antenna of the Access Point (AP) for a 5-element Yagi antenna. The figure 4 shows the RF range using the Yagi antenna.

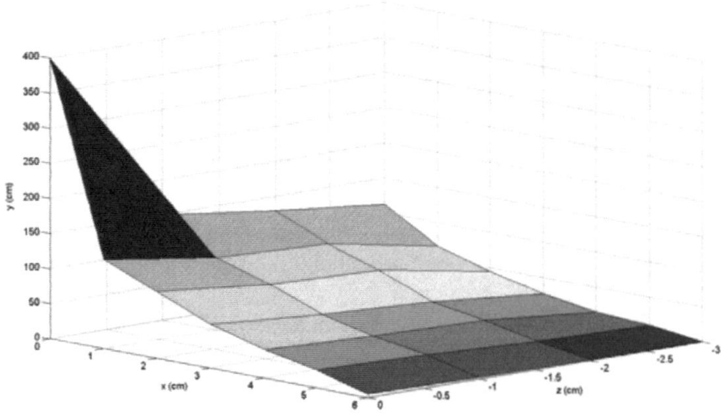

Figure 4: Graph of the impedance monitor RF range. The x-axis corresponds to the longitudinal distance of the plastic container from the border of the fishbowl. The y-axis corresponds to the distance at which we measure the range. The z-axis corresponds to the depth of the plastic container inside the fishbowl.

4. Discussion

The change in the common mode voltage induced by an electrode mismatch in the injecting leads, combined with the output impedance of the current source, has a higher effect in the measurement than the same mismatch produced in the sensing leads, due to the high impedance of the input stage.

The most important effect of temperature changes is due to the slope of the profile than the increment itself. As the temperature remains stable the value of the module returns to its original value. This may be produced by the heterogeneous warming of the internal integrated circuits causing the in-balance in current sources. The phase is less sensitive to temperature variations.

Adding an antenna optimized for the ZigBee band the range can be increased 50% the worst case. of the RF link. This could allow us to remove the range extender (RE) that we have planned to use in the outer surface of the animal.

Acknowledgements
This work has been supported by grants from Spanish Ministry of Science and Innovation SAF 2008-05144, with FEDER funds, and from the Spanish Ministry of Health project REDINSCOR.

References

[1] Gebhard M M, Gersing E, Brockhoff C J, Schnabel P A, Bretschneider H J 1987 Impedance spectroscopy: a method for surveillance of ischemia tolerance of the heart *Thorac. Cardiovasc. Surg.* **35** 1 26-32

[2] Salazar Y, Bragos R, Casas O, Cinca J, Rosell J 2004 Transmural versus nontransmural in situ electrical impedance spectrum for healthy, ischemic, and healed myocardium *IEEE Trans. Biomed. Eng.* **51** 8 1421-7

[3] Warren W, Bragós R, Casas O, Rodriguez-Sinovas A, Rosell J, Anivarro I, Cinca J 2000 Percutaneous electrocatheter technique for on-line detection of healed transmural myocardial infarction *Pacing Clin. Electrophysiol.* **23** 8 1283-7

[4] Pfitzmann R, Müller J, Granhan O, Hetzer R 2000 Intramyocardial impedance measurements for diagnosis of acute cardiac allograft rejection *Ann. Thorac. Surg.* **70** 2 527-32

[5] Cinca J, Ramos J, Garcia M A, Bragós R, Bayés-Genís A, Salazar Y, Bordes R, Mirabet S, Padró J M, Picart J G, Viñolas X, Rosell-Ferrer 2008 Changes in myocardial electrical impedance in human heart graft rejection *Eur. J. Heart Fail.* **10** 6 594-600.

[6] Bogónez-Franco P, Bragós R, Bayés-Genís A, Rosell J 2009 Implantable bioimpedance monitor using ZigBee *Engineering in Medicine and Biology Society EMBC 2009 Annual International Conference of the IEEE* 4868-71

[7] SimpliciTI 2009 *www.ti.com*

The flexible and configurable Sussex EIM MK4 using PCI eXtensions for Instrumentation (PXI)

Nevis Béqo, Gerald Sze, Nick Huber and Wei Wang
Biomedical Engineering Group, School of Engineering and Design, University of Sussex, Brighton, BN1 9QT, UK

E-mail: wei.wang@sussex.ac.uk

Abstract. The Sussex Electrical Impedance Mammography (EIM device is a novel imaging system developed at the University of Sussex for the detection of breast lesions *in-vivo* using quadrature detection of impedance. The device uses a fully programmable planar array of electrodes. The control method employed is such that any two electrodes apply current to the breast whilst any other two, similarly non-invasive, electrodes detect the developed potential difference under the direction of proprietary software. The currents injected are provided by a signal source with available output frequencies that are continuous over a wide bandwidth. The system is a prototype device for research and clinical evaluation, therefore maximum flexibility is required for hardware and software settings. This paper describes how we have achieved a totally configurable system with a quick turn-around in development time using off-the-shelf hardware and a 3U PXI Chassis. The system is based around a National Instruments PXI (PCI eXtensions for Instrumentation) chassis which is a modular electronic instrumentation platform. The modules are register-based products, which use software drivers hosted on a PC to configure them as useful instruments, taking advantage of the increasing power of PCs to improve hardware access and simplify embedded software in the modules.

1. Introduction

Electrical Impedance Tomography (EIT) entered as a medical imaging technique in the early 80s with the work of Barber and Brown[1] and has subsequently been applied to a variety of monitoring and diagnostic situations [4]. At the University of Sussex we are investigating the use of EIT as a breast cancer imaging and diagnostic tool. The current Sussex Mk4 Electrical Impedance Mammography (EIM) device is facing the issue of providing a solid apparatus with minimal measurement error and satisfactory calibration [2][3] values.

This can be achieved initially with phantom measurements [3] but its use *in-vivo* is inevitable. By introducing a human factor to the equation a series of unseen obstacles start to appear which can be categorised in two parts: 1) Technical and 2) Clinical. In the technical part; the vast variety of shapes, sizes and biological tissue distribution (e.g. fat) within the human breast requires hardware that is flexible and adaptive to a considerable dynamic range [9]. In the clinical part; before any *in-vivo* tests are performed on humans, it is required that the device be built to comply with health and safety requirements (e.g. IEC 60601-1-x) [8]. It is easier to achieve compliance if the individual parts of the device already are compliant with some recognised standards and/or certifications (e.g. CE marking or FDA approval).

With the Sussex Mk4 system we achieved a high specification, flexible, software-configurable system to be used for clinical trials in the United Kingdom by using off-the-shelf hardware from National Instruments (NI). This paper explains in more detail the set-up and configuration of the system.

2. Sussex Mk4 System

The Sussex Mk4 system uses a quadrature method for impedance detection [9]. The device uses a fully programmable planar array of electrodes. The control method employed is such that any two electrodes apply current to the breast whilst any other two, similarly non-invasive, electrodes detect the developed potential difference under the control of proprietary software. The currents injected are provided by a signal source with available output frequencies that are continuous over a wide bandwidth. Table 1 shows the specifications for the Mk4 acquisition specifications.

Table 1. Sussex EIM acquisition system specifications

Requirements:	
Bandwidth	100Hz – 10MHz
Peak-to-peak current	~0.01 - 1mA
Signal-to-Noise Ratio	>60dB (1kHz – 1MHz) >46dB (1MHz – 5MHz)
CMRR	> 46dB SNR (@1-5MHz)
System accuracy	>1%
Imaging Frame Rate	< 1 Frame / sec / Freq
Detection Method	Quadrature
Minimum electrodes	Up to 265
Calibration	Automatic, SW-based
Channel selection mode	Single End/Differential (Selectable)
Frequency selection	SW Programmable
Electrode selection	SW Programmable
ADC	Minimum 12-bit
Digital addressing protocol	32-bit
Data Storage	PC hard drive

Due to the fully programmable nature of the electrode selection, if the system is equipped with **n** – number of electrodes it is theoretically possible to select **n*(n-1)*(n-2)*(n-3)** electrode combinations. The selection of the combinations is performed using a 32-bit digital protocol. For each of the electrode combinations selected a number of user-programmable frequencies, generated by a dedicated signal source, are injected sequentially. The developed potentials are measured via a multiplexer (part of the custom PCB board design) by a dedicated high specification Analog-to-Digital Converter with the digitized signal streamed onto the hard drive. The created file is then available for data processing, image reconstruction and analysis.

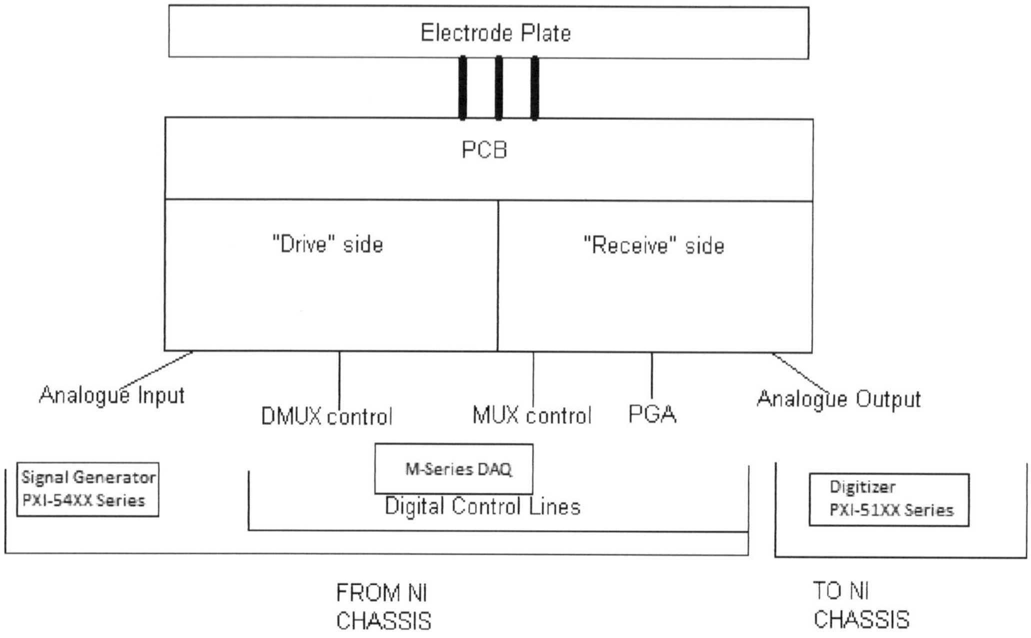

Figure 1. Sussex Mk4 EIM Data Acquisition block diagram

3. PXI Set-up

The Data Acquisition System (DAS) is based on a PXI (PCI eXtensions for Instrumentation) Chassis with a number of dedicated PXI modules controlled by customer-written software. PXI is a modular instrumentation platform originally introduced in 1997 by National Instruments [5][6]. PXI is promoted by the 54-member PXI Systems Alliance (PXISA) and has become an industry standard in computer-based instrumentation with over 1150 different modules, varying from Data Acquisition, Signal Generators to Motor Controllers and Image Grabbers[5]. The modules are categorized in Board Series depending on their purpose (e.g. Data Acquisition, Motion etc). Each Board Series has a Product Family categorized on the hardware capabilities of the modules (e.g. DAQ 12-bit, 14-bit) etc. Naturally the higher the hardware specification the more expensive the module is. For the purpose of this paper we will demonstrate the system using the Product Family (e.g. PXI-62XX) so that it complies with the purpose of maximum flexibility for an EIT system. Being based on CompactPCI, PXI offers all of the benefits of the PCI architecture including performance, industry adoption, COTS technology. Most PXI instrument modules are register-based products, which use software drivers hosted on a PC to configure them as useful instruments, taking advantage of the increasing power of PCs to improve hardware access and simplify embedded software in the modules. The use of this open architecture allows hardware to be reconfigured to provide new facilities and features that are difficult to emulate in comparable bench instruments [5]. The wide variety of the off-the-shelf high specification modules allowed us to concentrate our efforts on the unique features of the EIM system.

For the Sussex Mk4 EIM system a PXI54XX series function generator is used to provide signals to a voltage-to-current convertor for application to the body. The exact electrode combination is selected via a PCI-62XX module. Each electrode can be either configured, using a 32-bit parallel port, for current injection or voltage measurements; but each electrode can *only* be configured for either differential current injection (I+ or I-) or voltage measurement (V+ or V-) at one time. Due to the frequency range used the developed potentials are digitized via a PXI-51XX Digitizer Module for streaming to the computer.

The controlling software is written in LabVIEW™ (Laboratory Virtual Instrumentation Engineering Workbench), a fast prototyping, graphical programming language based on data flow concepts which provides excellent user interface development features [5]. The PXI modules come equipped with driver installation software and an Application Programming Interface (API) which is called and controlled by LabVIEW™. This provides for a reduction in development time and cross-module integration compatibility with the software installed on a dedicated, high specification Microsoft Windows personal computer architecture.

4. Conclusion

The Sussex Mk4 EIM system achieved the required specifications. The use of a PXI-based set-up proved successful in building a working system in record time. The module swapping capability has proved vital in the calibration process. Different module set-ups, to increase performance, are being investigated at present. Due to the modular nature of the PXI system such changes will not require major modifications. Based upon our experience we advocate that a PXI system would be very beneficial to the further study of EIT.

5. References

[1] Barber D C and Brown B H 1984 Applied Potential Tomography (Review Article). *J. Phys. E. : Sci. Instrum.* **17** 723-733

[2] Oh T I *et al* 2007 Calibration methods for a multi-channel multi-frequency EIT system *Physiol. Meas.* **28** 1175-1188

[3] Qiao G, Wang W, Wang L, He Y, Bramer B and Al-Akaidi M 2007 Investigation of biological phantom for 2D and 3D breast EIT images *IFMBE Proceedings, ICEBI 2007* **17** 328-331

[4] Holder D S 2004 *Electrical Impedance Tomography: Methods, History and Applications* (London: Institute of Physics) ISBN 0-7503-0952-0

[5] http://www.ni.com (National Instruments)

[6] http://www.pxisa.org (industry consortium)

[7]http://www.fda.gov/MedicalDevices/DeviceRegulationandGuidance/GuidanceDocuments/ucm073778.htm

[8] http://www.iec.ch (International Electrotechnical Commission)

[9] Wang W, Huber N, Tunstall B, Sze G, Qiao G, Béqo N and Chatwin C 2010 The system performance of Sussex Mk4 Electrical Impedance Mammography (EIM) system for the detection of breast lesions (ICEBI 2010 conference proceedings)

Journal of Physics: Conference Series **224** (2010) 012165 doi:10.1088/1742-6596/224/1/012165

A flexible and configurable hardware for the Combined EIM and Ultrasound device

Nevis Béqo, Gerald Sze, Benjamin Tunstall, Guofeng Qiao, Ali Zarafshani and Wei Wang
Biomedical Engineering Group, School of Engineering and Design, University of Sussex, Brighton, BN1 9QT, UK

E-mail: wei.wang@sussex.ac.uk

Abstract. One of the current methods in breast cancer detection is Sonography, also known as Ultrasound imaging. Sonography is effective for imaging soft tissues of the body and returns a high resolution image. Its operation is highly subjective depending on the operator and the images do vary with positioning. At the University of Sussex we have successfully built an automated and repeatable Ultrasound scanner and combined and reconstructed the data in 3D. Ultrasound imaging does have its limitations when it comes to cancer detection and diagnosis. The Sussex EIM (Electrical Impedance Mammography) device is a novel imaging system developed at the University of Sussex for the detection of breast lesions *in-vivo* using quadrature detection of impedance. Combining the high resolution images of Ultrasound with the parametric data of EIT would give a more precise diagnosis. This paper describes how we have achieved a totally configurable system where we have combined our current EIT technology for breast cancer detection to the 3D Ultrasound scanner we have built. This was done with a quick turn-around in development time using off-the-shelf hardware and a 3U PXI Chassis. The system is based around a National Instruments PXI (PCI eXtensions for Instrumentation) chassis which is a modular electronic instrumentation platform.

1. Introduction

Ultrasound is a high frequency sound, above the human audible threshold, which travels freely through fluid and soft tissues but is reflected when it reaches a change in 'acoustic density'. Sonography, also referred to as Ultrasound imaging, is a widely used medical imaging and diagnostic technique. [1] Ultrasound imaging is non-invasive [1][2] and is effective in imaging soft body tissue. Subcutaneous body structures such as muscles, tendons, breasts etc. are normally scanned at high frequencies (7-18MHz) [1][4], which provide high axial and lateral resolution. Internal organs are scanned at lower frequencies (1-6MHz) [1][4], which results in a lower axial and lateral resolution but greater penetration. Typically the same Ultrasound machine can do both high and low frequency scans by using interchangeable transducers which are placed directly on, and moved over, the patient by a trained professional. This introduces a subjective factor during a scan, which may have negative implications in the diagnoses. By sweeping the transducer on the patient a 2D image representation of a slice into the body is created. 3D images can be generated by acquiring and collating a series of 2D images. To eliminate operator error and introduce consistency, the 3D scanning process can be mechanized.

At the University of Sussex we have been working on the development of an automatic 3D scanner for breast cancer detection using commercially available control hardware. The system combines electrical impedance measurements, [3][5][8] which is able to provide parametric imaging [6][7][8] at a modest resolution, with 3-dimensional ultrasound measurements. This 3-dimensional [9][10] multi-modality will produce superimposed high resolution images with tissue property information therefore complementing the two technologies and achieve a more accurate diagnostic result.

With the Sussex Mk4 EIM system we managed to achieve a high specification, flexible, software configurable system to be used for clinical trials in the United Kingdom by using off-the-shelf hardware from National Instruments (NI). Taking advantage on the modular architecture of the NI hardware we have merged the control and acquisition of the 3D Ultrasound scanner to the existing Sussex Mk4 device. This paper will try to explain in more detail the set-up and configuration of the extended dual modality system.

2. PXI Set-up

The system set-up, is based on a PXI (PCI eXtensions for Instrumentation) Chassis with a number of dedicated PXI modules controlled by proprietary software. PXI is a modular instrumentation platform originally introduced in 1997 by National Instruments [11]. PXI is promoted by the 54-member PXI Systems Alliance (PXISA) and has become an industry standard in computer based instrumentation with over 1150 different modules, varying from Data Acquisition, Signal Generators to Motor Controllers and Image Grabbers [12]. PXI is based on CompactPCI, and offers all of the benefits of the PCI architecture including performance, industry adoption, COTS technology. Most PXI instrument modules are register-based products, which use software drivers hosted on a PC to configure them as useful instruments, taking advantage of the increasing power of PCs to improve hardware access and simplify embedded software in the modules. The modules are categorized in Board Series depending on their purpose (e.g. Data Acquisition, Motion etc). Each Board Series has a Product Family categorized on the hardware capabilities of the modules (e.g. DAQ 12-bit, 14-bit) etc. Naturally the higher the hardware specification the more expensive the module is. The open architecture further allows hardware to be reconfigured to provide new facilities and features that are difficult to emulate in comparable bench instruments [11]. The wide variety of the off-the-shelf high specification modules allowed us to concentrate our efforts on the requirements of the integrated system. For the purpose of this paper we will demonstrate the system using the Product Family (e.g. PXI-62XX) so that it complies with the purpose of maximum flexibility.

The software is written in LabVIEW™ (Laboratory Virtual Instrumentation Engineering Workbench), a fast prototyping, graphical programming language based on data flow concepts and good user interface development features [11]. The PXI modules come equipped with driver installation software and API which is easily called and controlled by LabVIEW™. This speeded up development time and ensured full system integration compatibility in our system. The software was installed on a dedicated, high specification Microsoft Windows based Personal Computer

3. EIM System

The Sussex Mk4 system uses a quadrature method for impedance detection [6]. The device uses a fully programmable planar array of electrodes. The control method employed is such that any two electrodes apply current to the breast whilst any other two, similarly non-invasive, electrodes detect the developed potential difference under the direction of proprietary software. The currents injected are provided by a signal source with available output frequencies that are continuous over a wide bandwidth.

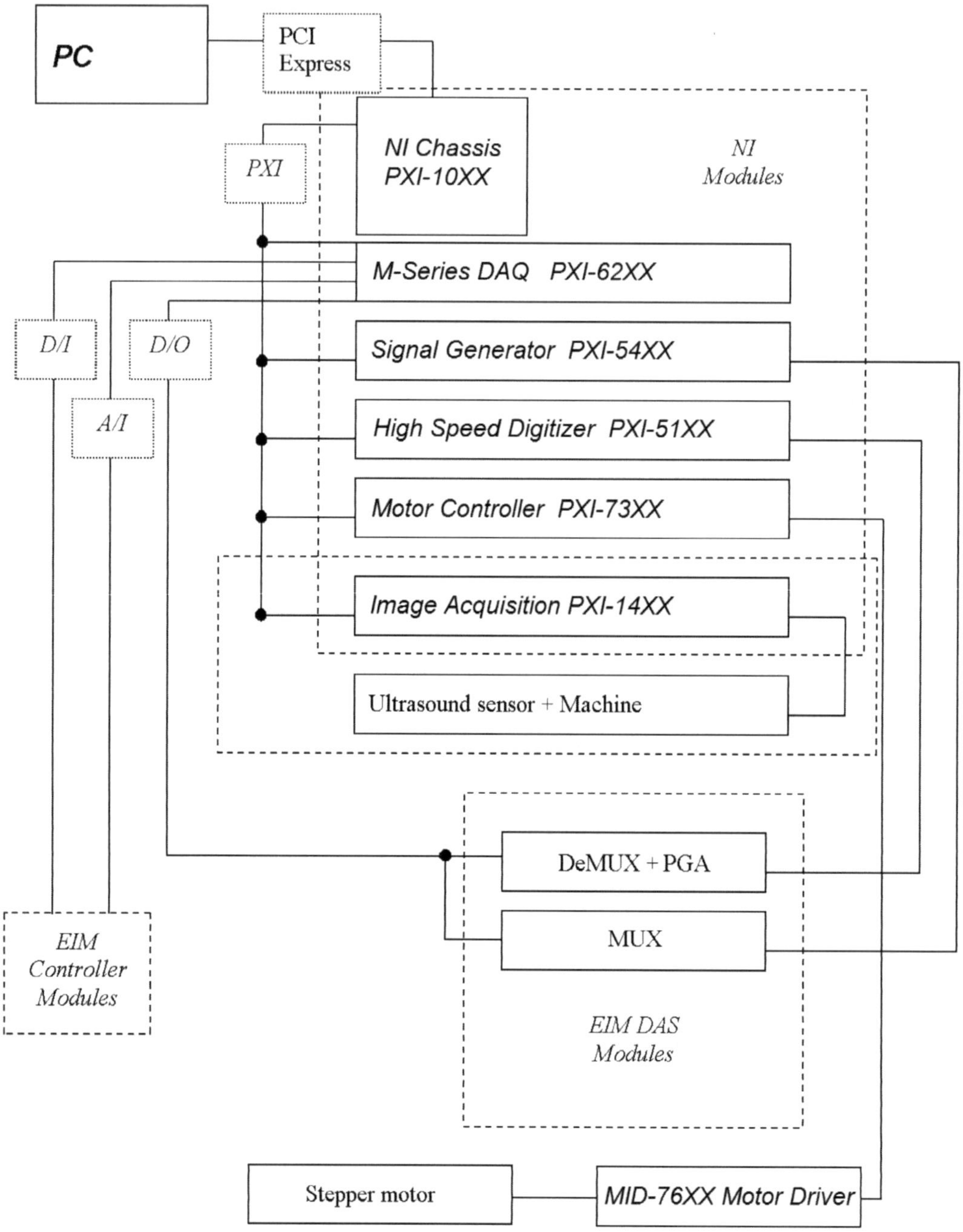

Figure 1. System Control and Acquisition diagram

Within the Sussex Mk4 EIM system a PXI54XX series function generator is used to provide signals to a voltage-to-current convertor for application to the body. The exact electrode combination is selected via a PCI-62XX module. Each electrode can be either configured, using a 32-bit parallel port, for current injection or voltage measurements; but each electrode can *only* be configured for either differential current injection (I+ or I-) or voltage measurement (V+ or V-) at one time. Due to the frequency range used the developed potentials are digitized via a PXI-51XX Digitizer Module for streaming to the computer.

4. Ultrasound

2D slices are captured as images via the video-out port available on standard medical ultrasound system in CCIR TV format (25Hz frame rate).Video capturing is done using a PXI-14XX Analogue Frame Grabber.

The movement of a 2D ultrasound transducer, permitting a 3D scan, is achieved by using a Motion Controller PXI-73XX connected to a MID-76XX drive unit to power the stepper motors.

5. Conclusion

The use of a PXI based setup proved successful in building a working system in record time. The module swapping and adding capability has proved very crucial in the integration of a dual modality scanner. Different module setups to increase the performance are being considered at the moment without major modifications. Based on our experience we advocate that a PXI system would be very beneficial in the further study of EIT and Ultrasound, or other, multi-modalities. By reducing time in the development of the scanner, we were able to concentrate more on the merger of Ultrasound high resolution images with EIM tissue property information therefore complementing the two technologies and achieving a more accurate diagnostic result.

6. Acknowledgements

We would like to cordially thank the Department of Trade and Industry of the United Kingdom for supporting this project (DTI Project: TP/6/ICT/6/S/K1517L) in the 2005-2010 period. We would also like to thank our Industrial Project Partners for their dedication.

7. References

[1] Udupa J K and Herman G T 2000 *3D Imaging in Medicine* 2nd edn, (Florence:CRC Press)

[2] Herman G T 2009 *Fundamentals of computerized tomography: Image reconstruction from projection* 2nd edn (Berlin: Springer)

[3] Henderson R P and Webster J G 1978 An Impedance Camera for Spatially Specific Measurements of the Thorax *IEEE Trans. Biomed. Eng.* **25**: 250-254

[4] Lines D 2003 The application of real-time medical ultrasonic techniques to NDT *Proceedings of NDT2003, Worcester, 2003* ISBN No: 0 903132 31 1 pp189-192

[5] Barber D C and Brown B H 1984 Applied Potential Tomography (Review Article). *J. Phys. E.: Sci. Instrum.* **17** 723-733

[6] Wang W, Tang M, McCormick M, Dong X 2001 Preliminary results from an EIT breast imaging simulation system *Physiological Measurement* **22** (1) 39-48

[7] Yang Jieqiu 2006 *The Investigation and Implementation of Electrical Impedance Mammography Hardware System* PhD Thesis, De Montfort University, UK

[8] Holder D S 2004 *Electrical Impedance Tomography: Methods, History and Applications,* (London: Institute of Physics) ISBN 0-7503-0952-0

[9] William Robert 1990 *Design of a Real-Time Impedance Imaging System for Medical Applications* PhD Thesis, University of Sheffield, UK

[10] Cherepenin V *et al* 2001 A 3D electrical impedance tomography (EIT) system for breast cancer detection, *Physiol. Meas.* **22** 9-18

[11] http://www.ni.com (National Instruments)

[12] http://www.pxisa.org (industry consortium)

Further investigation of a contactless patient-electrode interface of an Electrical Impedance Mammography system

N Huber, N Béqo, C Adams, G Sze, B Tunstall, G Qiao and W Wang

Biomedical Engineering, School of Engineering and Design, University of Sussex, Brighton, UK, BN1 9QT

E-mail: wei.wang@sussex.ac.uk

Abstract. The Sussex Mk4 Electrical Impedance Mammography (EIM) system is a novel instrument, designed for the detection of early breast cancer, based upon Electrical Impedance Tomography (EIT). Many innovations in the field have been incorporated in the design improving both signal distribution and response. This paper investigates the behaviour of the contactless patient-electrode interface. The interface was studied in detail using phantom and healthy volunteer, *in-vivo*, data. Our findings show the necessity for the careful design of electrode enclosure so that the response of the system is not affected by the unpredictable positioning of the breast; it closely mimics those conditions seen when using the phantom. The paper includes a number of possible designs and their individual characteristics. In addition an explanation on the unanticipated effects and solutions for such are described.

1. Introduction

Studies have shown that a steady decrease of 2.0% – 3.2% in breast cancer related death rates per year since 1990 can be attributed to improvements in breast cancer treatments and early detection [1]. Electrical Impedance Mammography (EIM) has been recognized as a possible diagnostic tool for early breast cancer detection. By providing a spatial map of the electrical impedance distribution of the breast tissue, EIT can provide a non-invasive and non-ionizing screening method [2]. EIM has already been suggested as an adjunct to better established screening techniques [3].

One of the main challenges of EIT technology is that of the interface between patients and electrodes. The contact impedance and the current density at the contact point are both high, and any variability at this point will cause abnormal measurements [4]. Any such 'drift', especially if it occurs between measurement frames, is known to be responsible for artifacts appearing in the reconstructed conductivity images [5]. This effect is even more pronounced in absolute imaging, as the forward solution can be heavily dependent on the electrode-skin contact impedances, and other geometric factors [6][7].

Wang *et al.* [8] showed that a system whereby direct contact of the electrodes with the skin is avoided can result in significantly reduced artifacts, and also assist in the repeatability of EIT testing. The current paper presents further investigation into the use of this contactless interface, especially when applied to *in-vivo* studies of the human breast utilizing a planar array of electrodes. We will show how patient movement can negatively affect measurements, if proper care is not taken for the design of the electrode plate.

2. The Mk4 'Wet' Electrode Interface

The term 'wet' electrode, first coined in [8] denotes the constant existence of an intermediary contact medium, e.g. saline, between the electrode and the patient. Conduction of the current is then only ionic, not galvanic. Assuming the measurements are carried out in a tank, this can be achieved by embedding, or in-setting, the electrodes by a very shallow ($\sim 2mm$) depth into the electrode support structure, effectively forming an electrode cavity. Saline is used to fill up the tank before the test is carried out, and hence offer a uniform buffer medium between skin and electrode. In figure 1 is shown the implementation of this concept in the Sussex Mk4 EIM system [9], which utilizes a planar electrode array.

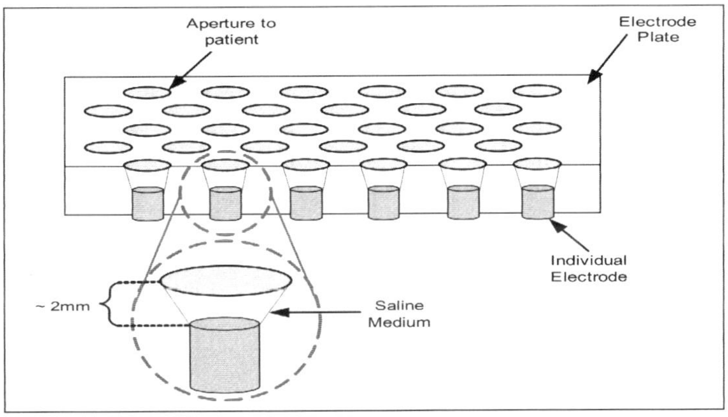

Figure 1. The Sussex Mk4 EIM planar electrode array, showing detail of the 'wet' electrode interface

The electrode plate lies at the bottom of a saline-filled tank, and the patients position themselves in the prone position. The breast comes in contact with the electrode plate and is partially compressed. This spreads the breast tissue over a larger percentage of the electrode array, while at the same time decreases the distance between the chest wall and the electrodes, allowing for higher coverage of the breast volume by the measurement system.

3. Unforseen *In-Vivo* Effects

In-vivo tests were carried out on volunteer patients in the course of the calibration of the Mk4 system against real-life loading conditions. During these tests, artifacts were detected in the form of saturated measurements coupled with measurements at or below the noise floor (effectively non-responsive channels). These artifacts, it was concluded, were not caused by systematic errors, as occurrences of artifacts were registered at different channels between scans of breasts, and the amount of artifacts greatly varied between subjects. In figure 2 is shown the left breast of a healthy subject. The graph shows clear signs of saturation, mixed with non-measurements (measurements at 0V). This corrupt data dominates the back-projection image, obscuring the outline of the breast entirely. The image was reconstructed using a back-projection algorithm [10] for demonstration, as no reconstruction algorithm can handle such saturations and non-measurements.

4. Loss of Contact

The Mk4 system was tested using saline, non-biological, and agar phantoms as part of the calibration procedure. During these tests, no unexpected artifacts were noted. Our initial assumption was that the Mk4 current source suffered from oscillations due to heavy loading. We, hence, devised a phantom based on the design found in [11], representing a very high load compared to the expected range for adipose and stroma. The results of this particular phantom test did not exhibit any saturated measurements either (see figure 4), and hence the phantom (in blue) can clearly be differentiated from the surrounding saline

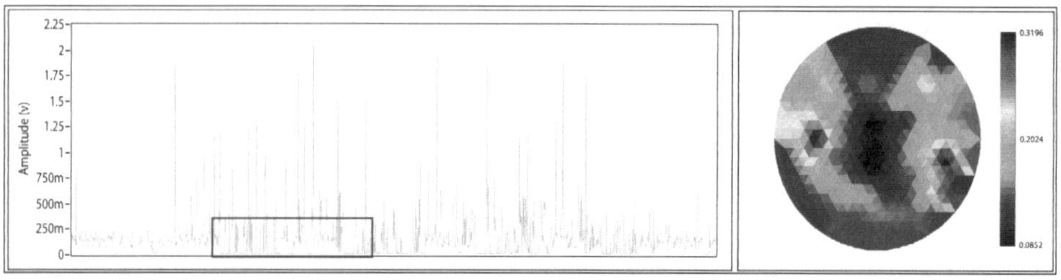

Figure 2. The breast scan shows clear signs of saturation that dominate the back-projection image. In the highlighted area, significant numbers of measurements were at 0V, effectively non-measurements.

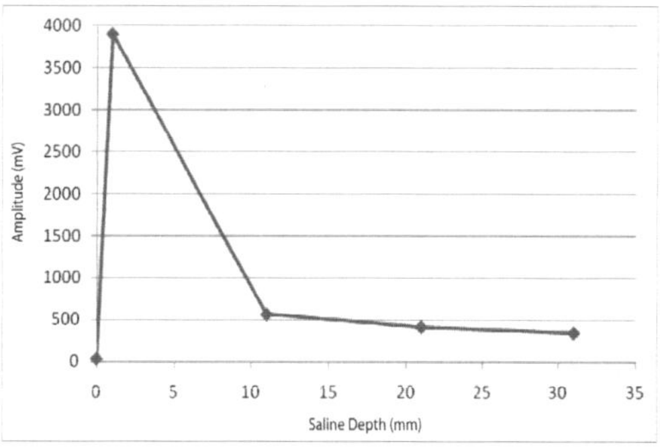

Figure 3. Measured values of single electrode combination against depth of saline.

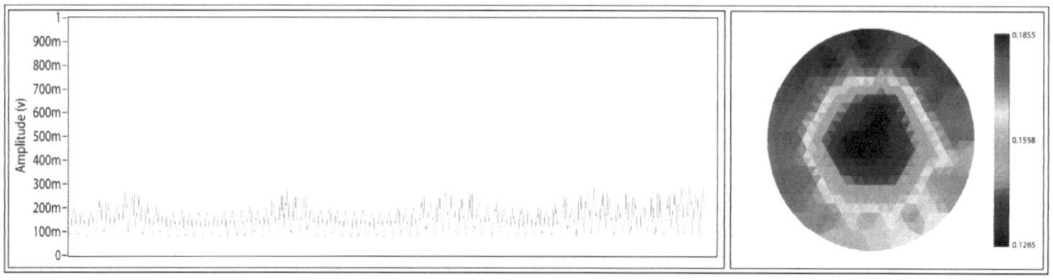

Figure 4. Agar phantom with conductivity structure similar to that of breast tissue, displaying no artifacts.

(in red). Further to this, when the system was tested with a very heavy load (that of a pure insulator), saturations were noted, as expected, but there were no non-measurements.

We next attempted to physically simulate a saturated or non-measurement channel. The depth of saline will normally affect the received output, due to variations in the loading. This test was now extended to verify the effect of lack of saline on the electrodes. Figure 3 shows that a marginal amount of saline (2*mm*) will exhibit saturation, while no saline will null the measurement by lack of loading of the current source.

The above tests led us to the conclusion that the positioning of the patient breast is responsible for the undesired effects. We have surmised that the skin of the patient is entering the electrode aperture and displacing the saline solution present, leaving minimal saline left, or none at all. A simple test of sealing a small number of electrodes was carried out to see if we could reproduce the desired effect, and the results were very similar to those in figure 2. We are concluding that, as agar phantoms are more solid than human skin, the agar would not enter the electrode cavity. Moreover, the 'wet' electrode testing in [8] was performed using the arm of a healthy male subject, again a more solid material.

5. Issue Resolution

Multiple ways of resolving the above mentioned problem can be suggested. One solution would be the use of a layer of material that is absorbent and conductive. The use of absorbent material, however, will introduce non-determinism in the measurements, as the structure of the material cannot be fully controlled, and hence its response cannot be fully modelled, and thus it is not under consideration. Alternatively, we can reduce the diameter of the electrodes significantly, so as to avoid any possibility of such unwanted effects. On the other hand, this would increase the difficulty and complexity (and cost) of producing both the electrodes and the interface.

The solution that was adopted at this stage was for the increase in the aperture size, and the further recessing of the electrodes (to $\sim 5mm$). Results of this test are shown in figure 5.

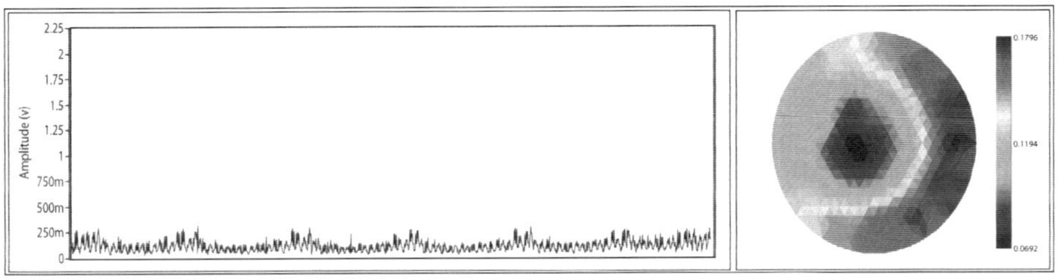

Figure 5. Scan of patient using different design of electrode plate, with further recessing. The breast can clearly be seen (blue in center and 9 o'clock position) surrounded by saline (red in 3 o'clock area).

6. Conclusion

Artifacts due to the electrode-patient interface have been one of the main challenges in EIT. Our research has shown that this is also the case with planar array of electrodes, even with recessed electrodes. The natural elasticity of breast tissue and patient movement can affect the contact impedance between patient and electrodes, leading to saturated channels and non-measurements. A simple solution has been found, but work is in progress for a more advanced solution that will not affect the detection depth.

7. References

[1] American Cancer Society 2009 *Breast cancer facts & figures* 2009-2010

[2] Holder D S 2005 *Electrical Impedance Tomography: methods, history and applications* (London: Institute of Physics Publishing)

[3] Tunstall B, Wang W, McCormick M, Walker R and Rew D A 1997 Preliminary in vitro studies of electrical impedance mammography (EIM): a future technique for non-invasive breast tissue imaging? *The Breast* **6** (4) 253

[4] Hua P, Woo E J, Webster J G, Tompkins W J 1993 Finite element modeling of electrode-skin contact impedance in electrical impedance tomography *IEEE Trans. Biomedical Engineering* **40** (4) 335-343

[5] Boone K G and Holder D S 1996 Effect of skin impedance on image quality and variability in electrical impedance tomography: a model study *Med. and Biol. Eng. and Comput.* **34** (5) 351-354

[6] Kolehmainen V, Vaukhonen M, Karjalainen P A and Kaipio J P 1997 Assessment of errors in static electrical impedance tomography with adjacent and trigonometric current patterns *Physiol. Meas.* **18** 289-303

[7] Nissinen A, Heikkinen L M, Kolehmainen V and Kaipio J P 2009 Compensation of errors due to discretization, domain truncation and unknown contact impedances in electrical impedance tomography *Meas. Sci. Technol.* **20** (10) 13 pages

[8] Wang W *et al* 2007 Study into the repeatability of the electrode-skin interface utilizing electrodes commonly used in Electrical Impedance Tomography *IFMBE Proceedings, ICEBI 2007* **17** (10) 336-339

[9] Wang W *et al* 2010 The system performance of Sussex Mk4 EIM Electrical Impedance Mammography (EIM) system for the detection of breast lesions *IFMBE Proceedings, ICEBI 2010* (To be published)

[10] Barber D C and Brown B H 1983 Imaging spatial distributions of resistivity using applied potential tomography *IEEE Electron. Lett.* **19** (22) 933-935

[11] Qiao G, Wang W, Wang L, He Y, Bramer B and Al-Akaidi M 2007 Investigation of biological phantom for 2D and 3D breast EIT images *IFMBE Proceedings, ICEBI 2007* **17** (10) 328-331

Journal of Physics: Conference Series **224** (2010) 012167 doi:10.1088/1742-6596/224/1/012167

Comparison of Howland and General Impedance Converter (GIC) circuit based current sources for bio-impedance measurements

T R Qureshi, C R Chatwin, N Huber, A Zarafshani, B Tunstall and Wei Wang

Biomedical Engineering, School of Engineering and Design, University of Sussex, Brighton BN1 9QT, UK

Email: wei.wang@sussex.ac.uk

Abstract: The current source is a key component in bio-impedance measurement systems. The accuracy of the current source can be measured in terms of its output impedance together with other parameters, with certain applications demanding extremely high output impedance. This paper presents an investigation and comparison of different current source designs based on the Enhanced Howland circuit combined with a General Impedance Converter (GIC) circuit using both ideal and non-ideal operational amplifiers. Under differing load conditions two different settings of the GIC are evaluated and the results are compared to show its performance settings. Whilst the study has shown that over a wide bandwidth (i.e. 100Hz – 100MHz) the output impedance is limited, operation over a more limited range offers output impedance in the Giga-ohm range, which can be considered as being infinite.

1. Introduction

In any bio-impedance measurement system, unknown impedance can be measured by two possible methods: either by the application of a known voltage across the subject, and then measuring the current through it, or by injecting a known current into the subject and measuring the voltage across it. Unknown contact impedance exists between the electrode and the subject. It will degrade the operation of voltage source having low output impedance. Therefore we must preferably use the second method – to inject current and measure voltage across the object [1].

An important parameter of a current source is its output impedance which is frequency dependent. Ideally the output impedance should be infinitely large; so primarily this paper will investigate ways to achieve the highest possible value of output impedance of current source. Bandwidth of the current source is also an important factor. EIT systems require a wide range of frequencies. Apart from output impedance this paper will also investigate the bandwidth of current source under different loads.

The current source which is mostly used in bio-impedance measuring systems is the Enhanced Howland circuit. According to the available literature [2] [3], the Enhanced Howland circuit has limited output impedance which is consistent across a wide bandwidth. At higher frequencies, the output impedance degrades due to the presence of output capacitance and by cancelling this output capacitance higher output impedance can be achieved. The placement of a GIC in parallel with an Enhanced Howland circuit is known to improve the output impedance of the circuit [4].

In this paper, the Enhanced Howland-GIC circuit is simulated with ideal and non-ideal operational amplifiers. This paper is a comparison and simulation of the Enhanced Howland-GIC circuit with different devices currently available in the market to find the theoretically maximum achievable output impedance. To reduce the output capacitance of the system, different settings of the GIC are simulated, and its output impedance and bandwidth performance are established.

2. Current Source Architecture

2.1. Enhanced Howland Circuit

To remove the common mode voltage or Loading effects in the Basic Howland circuit a resistor is added between the non-inverting and the output terminals to improve the performance of the circuit (see fig. 1). This Enhanced Howland circuit is used as a current source in most EIT systems due to its simple architecture and better performance [1][2][3][5][6]. In terms of output impedance, the Enhanced Howland circuit does not give our desired high output impedance due to the presence of output capacitance. The Output current of the Enhanced Howland circuit for an ideal op amp is given as,

$$I = -V_i \cdot \frac{R_2}{R_5 \cdot R_1} + \frac{R_3}{R_5 \cdot (R_3 + R_4)} \left(\frac{R_2}{R_1} \frac{R_4 + R_5}{R_3} \right) \cdot V \qquad (1)$$

2.2. Generalized Impedance converter (GIC)

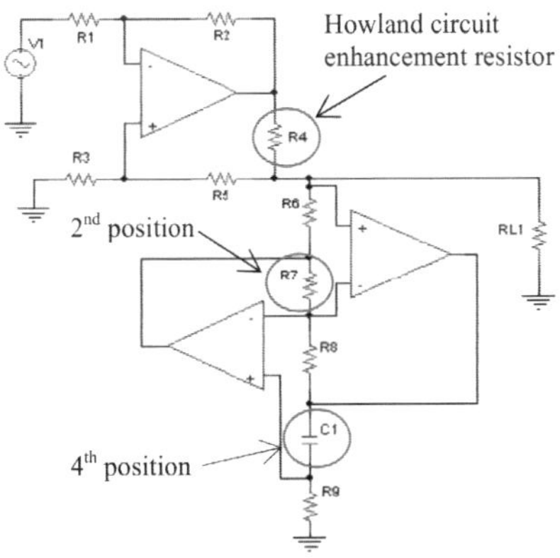

Figure 1: Enhanced Howland-GIC circuit, highlighting possible capacitor positions

GIC circuit is from Northrop [7]. It consists of two operational amplifiers and five discrete components which can be used to synthesize various impedances. These components are either resistors or capacitors. There are several topologies to implement this circuit. In our case, we want to cancel the output capacitance of the current source circuit by configuring the GIC to effectively produce inductance. Therefore, for inductive output the GIC should have two operational amplifiers, four resistors and one capacitor [7]. The inductance produced will have the value:

$$L = \frac{R_1 R_3 R_5 C_2}{R_4} \qquad (2)$$

$$L = \frac{R_1 R_3 R_5 C_4}{R_2} \qquad (3)$$

3. Methodologies

A GIC circuit is placed in parallel with Enhanced Howland circuit for output capacitance cancellation as shown in figure 1. The circuit is simulated using PSpice over a wide range of frequencies i.e. up to 1GHz. Output impedance of the circuit is calculated up to 10MHz over different loads i.e. 1k, 2k, 4k, 10k, 20k, 40k. Initially, the Enhanced Howland-GIC circuit is tuned to a single resistor and capacitor setting to achieve maximum possible performance with a single configuration. It is noticed that this configuration will have high output impedance at lower frequencies but the output capacitance degrades the performance at higher frequencies. On the other hand, it covers a wide range of frequencies, as defined by the amplifier bandwidth specifications. This design is called the *single network configuration* (see figure 2).

The Enhanced Howland-GIC circuit can also be tuned to achieve higher output impedance, but over a limited frequency range. This kind of design we have named *multiple network configuration* (see figure 3 for a 1MHz spot frequency performance).

4. Results and Discussion

The circuit was simulated using an ideal operational amplifier present in PSpice libraries. This paper gives simulation results to show maximum output impedance achieved from group of selectable devices. Either 2nd or 4th component in GIC can be a capacitor. In this paper we had placed capacitor at both location and have noticed its results.

Non-ideal operational amplifiers (OPA656, THS4304 and THS3201 by Texas Instruments and AD8055 by Analog Devices) are simulated using single & multiple circuit configurations as well as different capacitor configurations. These operational amplifiers are selected due to wide bandwidth, higher slew rate, low distortion and simple pin configuration. Table 1 shows the maximum output impedance achieved from all selected non-ideal devices at a particular frequency. e.g. Table 1 shows that maximum output impedance of 4GΩ at 100Hz is achieved using OPA656 device for 2nd position of capacitor, while at same frequency 26GΩ output impedance is achieved using AD8055 for 4th position of capacitor. This difference in output impedance is due to characteristics of operational amplifier used and tuning of GIC until we achieve maximum current injection. Using single network configuration gives us better bandwidth but the output impedance decreases at 1MHz and above. Table 1 shows the output impedance achieved using different non-ideal amplifiers. Multiple circuit configurations can be used to achieve high output impedance at higher frequencies. Both possible capacitor settings highlighted in figure 1 can be individually used because the system is tuned for very narrow frequency bands at which we achieve maximum current injection. Table 2 shows the output impedance possible using non-ideal amplifiers.

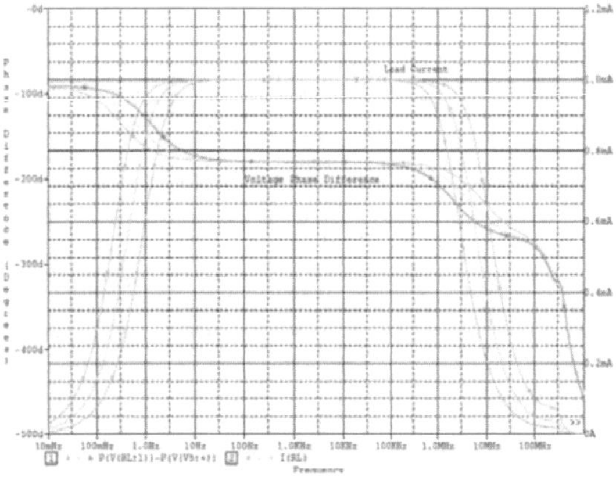

Figure 2: Current and phase difference curves for Enhanced Howland-GIC circuit at 10k, 20k and 40k loads

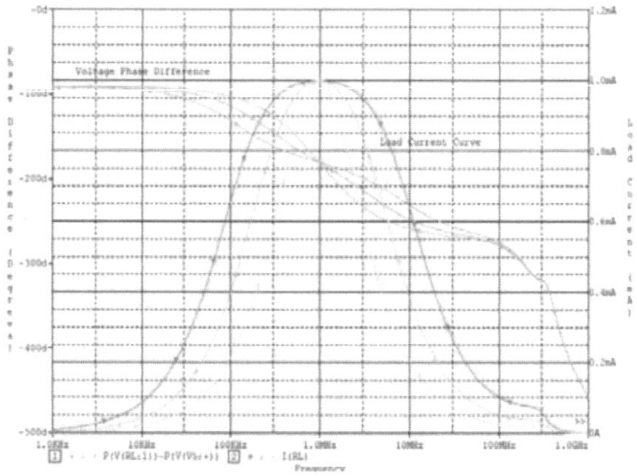

Figure 3: Current and phase difference curves for Enhanced Howland-GIC circuit at 1MHz using 10k, 20k, and 40k loads

For single network configuration, there is also some phase difference between the load voltage and input voltage. From 100Hz to 100kHz phase difference is 180°. Above 1MHz, phase difference increases and the maximum phase difference at 40k load is 259°. For multiple network configurations, phase difference is 180° at 1MHz.When this design is implemented in practice, the phase differences introduced need to be taken into consideration during reconstruction, or calibrated out of the measurements.

5. Conclusion

In this paper, Enhanced Howland-GIC circuit has been simulated using ideal and non-ideal operational amplifiers. Both possible capacitor positions in GIC circuit are simulated. Two different configurations are discussed to cover maximum number of frequencies. Multiple network configurations can be helpful to achieve higher output impedance at higher frequencies, tuned to a narrow band. Results show that, most non-ideal devices can give an acceptable level of output impedance, but THS3201 only achieves low output impedance. However, this same device gives better bandwidth as compared its counterparts. If we want to achieve higher output impedance (i.e. in the GΩ range) with a single configuration at higher frequencies (i.e. > 1MHz), then we have to sacrifice circuit bandwidth.

As mentioned earlier, this paper only gives the simulation results to find maximum achievable output impedance from a specific group of devices while ignoring the parasitic effect of cables, capacitors, resistor, inductor etc. Once the device which gives acceptable output impedance is decided, this circuit will be simulated including the parasitic effects and will be tested practically. These approaches can further be improved using better operational amplifiers when available.

Table 1: Enhanced Howland-GIC output using the single network configuration at different loads

Frequency (Hz)	Maximum Output Impedance (Z_O) (Ω)							
	Ideal		Non Ideal (OPA656, THS4304, THS3201, AD8055)					
	Using capacitor on2nd Position	Using capacitor on 4th Position	Using capacitor on 2nd Position			Using capacitor on 4th Position		
			Max Z_O & Corresponding Load		10k Load	Max Z_O & Corresponding Load		10k Load
100	6.25G	16.66G	4G	40k	24M	26G	40k	23M
10k	33G	33G	13G	40k	23M	4G	40k	22M
100k	33G	33G	400M	4k	12M	6.15G	4k	13M
1M	33G	33G	20M	4k	2.06M	18.32 M	4k	2.1M
5M	33G	33G	672k	4k	286k	734k	4k	300k
10M	33G	33G	204k	2k	81.26k	188k	2k	85k

Table 2: Enhanced Howland-GIC output using multiple network configurations at different loads

Frequency (Hz)	Maximum Output Impedance (Z_O) (Ω)				
	Ideal		Non Ideal (OPA656, THS4304, THS3201, AD8055)		
	Using capacitor on 2nd Position	Using capacitor on 4th Position	Using Capacitor on 4th Position		
			Max Z_O & Corresponding Load		10k Load
100	6.25G	16.66G	4G	40k	25M
10k	33G	33G	4G	40k	25M
100k	33G	33G	4G	40k	25M
1M	33G	33G	4G	40k	20M
5M	33G	33G	1.2G	10k	1.2G
10M	33G	33G	500M	10k	500M

References

[1] Yang J 2006 *The Investigation and Implementation of Electrical Impedance Mammography Hardware System* PhD Thesis, De Montfort University, UK
[2] Wang W 2007 *ICEBI, IFMBE Proceedings* **17** 213-216
[3] Alexander S R 2003 *Physiol. Meas.* **24** 509-516
[4] Franco S 2002 *Design with operational analog integrated circuits* (New York: McGraw-Hill)
[5] Bertemes-Filho P, Brown B H and Wilson A J 2000 *Physiol. Meas.* **21** 1-6
[6] Oh T I, Lee K H, Kim S M, Koo H, Woo E J and Holder D S 2007 *Physiol. Meas.***28** 1175-88
[7] Northrop R B 1997 *Introduction to Instrumentation and measurements* (Florence: CRC)

Signal calibration for an Electrical Impedance Mammography system

G Sze, N Béqo, N Huber, B Tunstall, R C D Young, C R Chatwin and W Wang
Biomedical Engineering Group, School of Engineering and Design, University of Sussex, Brighton, BN1 9QT, UK

Email: wei.wang@sussex.ac.uk

Abstract. Electrical Impedance Tomography (EIT) technology has been applied clinically since the 1980s. Numerous papers have addressed a variety of systematic error sources and indicated different calibration methods. The Sussex Mk4 Electrical Impedance Mammography (EIM) system has been developed for the investigation of early stage breast lesions. Investigations have shown that the system performance is subjected to a number of systematic errors: frequencies-dependant noise level due to both internal and external sources; stray capacitance within both PCB tracks and cable connections; and artefacts generated by patient movement during scanning etc. This paper reports upon several traditional and novel calibration methods utilized to reduce some of these errors in the acquired signals before image reconstruction. Techniques used include frequency spectrum analysis, filtering, phase calibration and other means of noise reduction. Results of both before and after calibration are presented and analyzed. The conclusion is reached that the signal quality of the Sussex Mk4 EIM system is such that the system is, post-calibrated, capable of producing images for the diagnosis of breast cancer.

1. Introduction

Signals acquired from an electrical impedance measurement system usually contain noise, systematic errors and interference. These errors, which are caused by internal sources such as ICs thermal noise, stray capacitance and unreliable connections, or external source such as movement artefacts generated from the measurement subject, and EM field interference generated by electronic devices have to be corrected or reduced to a minimum. Most noise sources are constant and their repeatable nature can be easily filtered out with traditional signal processing techniques. However some noise, or interference, is random and requires the use of threshold values applied to filters so that it can be removed. In the worst case the measurement frame may have to be replaced or removed manually after data collection.

1.1 The Sussex Mk4 EIM system

In this paper, signals collected from the Sussex Mk4a EIM system are presented. Noise appearing in these signals can be grouped according to two categories: a) internal noise and b) external interference. In addition to traditional filtering methods we propose using a method that differential the signal with simulated saline signal. Images reconstructed both with and without this method were compared showing that the Sussex Mk4 EIM system is able to produce stable signal of sufficient quality for image reconstruction for diagnosis of breast disease.

The Sussex Mk4 EIM system has been developed for detection of human breast cancer. The system was designed with a planar electrode array used to both inject current and measure the resulting voltage. The data acquisition system (DAS) can generate both the sinusoidal injected signal and provides digital signals to control the multiplexor for channel switching. Both reference and measurement signals are collected during data acquisition (as shown in figure 1) and are used in a digital demodulation, one is for reference with the other being the transfer impedance related voltage measurement. Magnitude and phase errors are calibrated using the reference signal before processing image. In the ideal case the magnitude and phase changes, between the reference and measured signals, should represent the conductivity changes within the measured volume, but noise from many sources could corrupt the collected signal thus affecting the conductivity calculation.

1.2 Noise
The sources of noise found in the Sussex Mk4 EIM system can be grouped into two types:

1.2.1 Internal noise. Caused by the internal electronic systems, these can be generated by the ICs and any stray capacitance within both PCB tracks and inappropriate cable connections. Most of these sources are repeatable. They could cause frequency dependent DC offset and phase shift within the measured signal and, acting as a low-pass filter, change the magnitude of the received signal at different frequencies. An example of the affected single frequency signal that forced to become a composite signal was shown in figure 1.

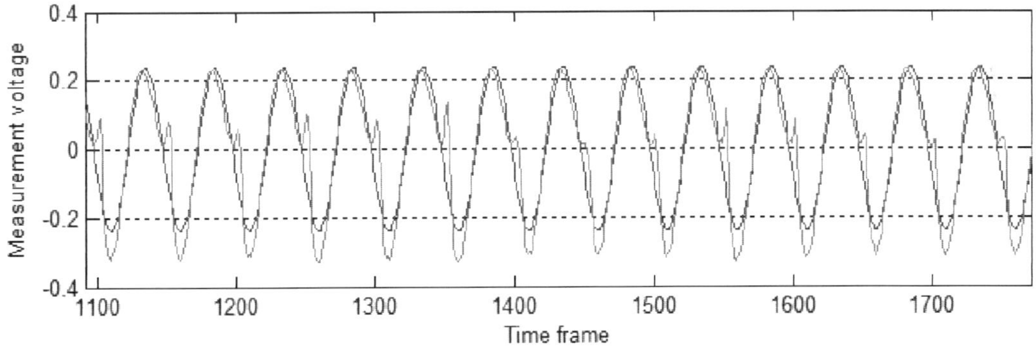

Figure 1. Reference signal in blue and measurement signal in red.

1.2.2 External interference. Caused by all other sources, these are unpredictable and unrepeatable. Typical examples are artefacts caused by patient movement, the effects of nearby X-CT scanning, and mobile telephone signals etc., resulting in sudden changes to the detected impedance magnitude. Most of these external sources are random and therefore more difficult to filter out due to their wide frequency bandwidth.

2. Method
This section lists a number of methods used to calibrate data collected from the Sussex Mk4 EIM system. A calibration method, that differential the signal with simulated saline signal, filters internal noise and is highlighted in section 2.1.2. Images reconstructed both with and without this method were compared using an agar phantom to model the human breast.

2.1 Calibration methods
Firstly collected reference and measurement signal were calibrated using frequency spectrum analysis. We applied Discrete Fourier Transform (DFT) to the measured signal. In the frequency domain, we obtain the magnitude "*Mm*" and phase "*Pm*" from the complex value of the applied frequency. We apply the same method to reference signal and get "*Mr*" and "*Pr*". The phase values between "*Pm*" and "*Pr*" are subtracted to obtain "*Pc*", this aligns phase for every signal generated.

2.1.2 Internal noise. We have "*Mm*" and "*Pc*" from a measurement with object (e.g. human breast), and we need one more set of "*Mm_s*" and "*Pc_s*" from pure saline measurement to calibrate error internally by the system.

"*Pc*" is subtracted by "*Pc_s*" and obtain "*Pc_sc*" to calibrate phase, and magnitude is calibrated by a method that differential the signal with simulated saline signal, so except "*Mm*" and "*Mm_s*", we need to generate a simulated voltage measurement "*Mm_ss*". This method filters the magnitude error, that caused by internal noise, by comparing saline data to the ideal simulated data. It requires the saline data to have a high signal-to-noise ratio [1], and the noise has to be repeatable. The method is implemented by formula $Mm \div Mm\_s \times Mm\_ss$, where "*Mm_ss*" is the simulated saline measurement created by solving the forward problem using a uniform conductivity model. Results from an agar phantom are presented in section 3 to investigate the performance of this algorithm.

2.1.3 Unpredictable random noise. Measured frames with spikes and non-responsive measurement channels are removed from the signal following the application of the DFT as shown in figure 2, the measurement frame 500 can be identified as faulty by the application of a DFT extracted voltage threshold that higher than 120. Any faulty frames are replaced or, *in extremis*, removed. The voltage threshold is calculated by the formula mean(V_{dft})±std(V_{dft})×Factor where V_{dft} is the DFT extracted voltage array and Factor depends upon the acquired signal.

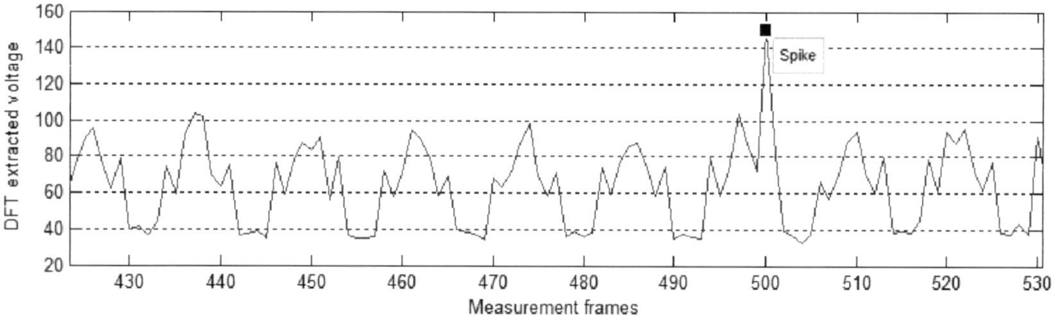

Figure 2. Plotting of DFT extracted voltage signal by measurement frames with spikes.

2.2 Agar model

Data has been collected from an agar breast model set up as in figure 3. It is composed of adipose-simulating material with lowest conductivity 0.43mS/cm shown in black, a stroma analogue with a mid-range conductivity' and two cancer-mimicking objects (highest conductivity) shown in white, located at 10 o'clock (bigger) and 2 o'clock (smaller) position. The agar phantom was 3cm high so a 3-layers 3D mesh was used to reconstruct image slices with each layer representing a distance of 1cm perpendicular to the electrode array. Note that right layer is touching the electrode plane and is therefore the bottom layer. Saline, with the same conductivity as the adipose-simulating material, was used as a contact medium between the agar phantom and the electrodes.

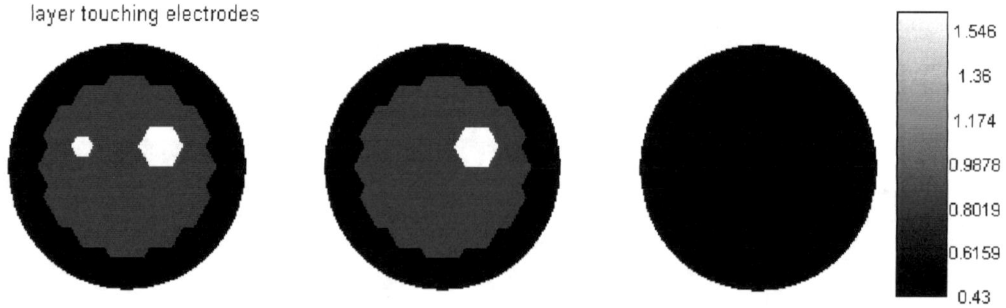

Figure 3. Agar model showing the phantom's conductivity distribution in mS/cm.

3. Results and analysis

3.1 Results from signal without differentiated with simulated saline signal

- Left layer is dominated by noise, almost obscuring the objects, difficult to distinguish between noise and cancer object.
- Object in the middle layer is not clear.
- Conductivity value is shifted to negative, it doesn't directly related to the true values.

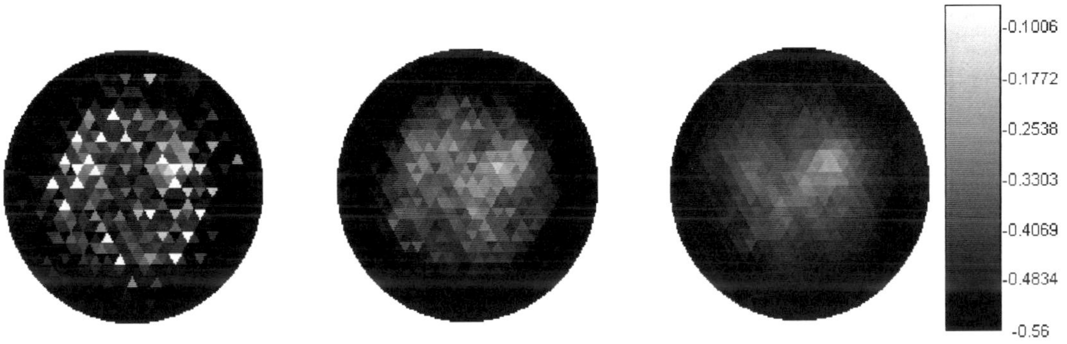

Figure 4. Image reconstructed from signal without differentiated with simulated saline signal.

3.2 Calibration with differential the signal with simulated saline signal

- Noise obviously was reduced.
- Object conductivity is higher than noise, which could be easily extracted from the image.
- Conductivity value is close to the true values.

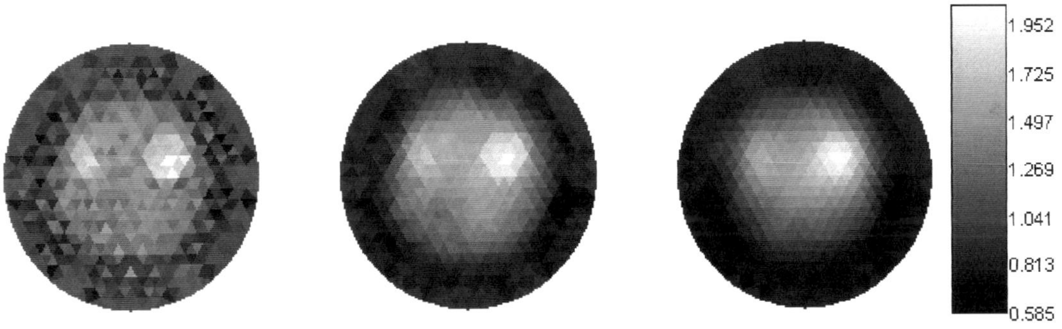

Figure 5. Image reconstructed from signal differentiated with simulated saline signal.

4. Conclusion

This paper presents the calibration of the effects of internal noise and external interference. As can be seen in figures 4 and figure 5, signal differentiated with simulated saline signal is a necessary process to calibrate the data collected. This method for signal calibration is straightforward and essential. Other calibration methods are currently under investigation for further improvement of reconstructed images.

5. Reference

[1] Wang W, Brown B H, Leathard A D and Lu L 1994 *Physiol. Meas.* **15** A211-A216.

Journal of Physics: Conference Series **224** (2010) 012169 doi:10.1088/1742-6596/224/1/012169

A Flexible Low-Cost, High-Precision, Single Interface Electrical Impedance Tomography System for Breast Cancer Detection Using FPGA

A. Zarafshani, N. Huber, N. Béqo, B. Tunstall, G. Sze, C Chatwin and *Wei Wang
Biomedical Engineering, School of Engineering and Design, University of Sussex, Brighton BN1 9QT, UK

E-mail: wei.wang@sussex.ac.uk

Abstract. Typically, in multi-frequency Electrical Impedance Tomography (EIT) systems, a current is applied and the voltages developed across the subject are detected. However, due to the complexity of designing stable current sources, there has been mention in the literature of applying a voltage to the subject whilst measuring the consequent current flow. This paper presents a comparative study between the two techniques in a novel design suitable for the detection of breast cancers. The suggested instrument borrows the best features of both the injection of current and the application of voltage, circumventing their limitations. Furthermore, the system has a common patient-electrode interface for both methodologies, whilst the control of the system and the necessary signal processing is carried out in a field programmable gate array (FPGA). Through this novel system, wide-bandwidth, low-noise, as well as high-speed (frame rate) can be achieved.

1. Introduction

Electrical Impedance Tomography (EIT) is a technique that determines the electrical conductivity distribution within the volume of a subject under test, by collecting measurements of current or measurements taken at its surface, based on induced stimuli. The solution of this inverse problem is further used to reconstruct an image of this distribution [1]. In general, there are two methodologies for the design of the signal source, and the relevant measurement systems, associated with the development of EIT devices [2]. First, using a voltage source can reduce the difficulty and cost of the design, it is relatively easy and quick to calibrate, but suffers from increased sensitivity to high frequency noise [3]. The second possibility is using a current source, which, for stability, requires complex circuitry for trimming to achieve high output impedance, [3], as a precision current source for EIT requires output impedance much higher than that of the load, over a significant frequency bandwidth. This is by far the most common method of choice, due to its high noise immunity [4]. Ideally, one would like the simplicity of voltage sources with the noise advantages of a current source. As discussed, in this paper we will be presenting a system that is under development that uses the approach of both kinds of signal sources, controlled by an FPGA, so that it can meet very stringent specifications.

2. System overview

We are developing an EIT system based on both voltage and current source methods in a single patient interface. Both systems share a common main platform for signal generation and control, based on a field programmable gate array (FPGA) with embedded digital signal processing (DSP). The analogue subsystem comprises (in order) a DAC, the signal generators (voltage and current), a single patient-electrode interface for both types of signal, a PGA and finally an ADC. The system can accurately measure a load impedance of 100Ω to 50 KΩ, with a 1% resolution as the minimum difference that can be differentiated. The frequency range the system is expected to cover is 10Hz to 20MHz. In order to measure impedances in several frequencies, we implemented a multi frequency waveform generator and a phase-sensitive demodulator inside a Xilinx Virtex-5 FPGA for design flexibility and scalability, as well as for control of the multiplexers to select electrodes to achieve a system with low noise and high speed.

3. Analogue sub-system

The developed analogue sub-system in the EIT system shown in Figure 1 consists of a DAC, a current injection and a separate voltage applying sub-system that are linked to separate measurement systems. Suitable converters from measured current to voltage are also part of the system, and the results are taken to programmable gain amplifiers before being passed to the ADC, and finally to the FPGA for demodulation. Work is in progress on the design a multi-frequency constant current and voltage source, V-I converter suitable for the current source and an I-V converter for current measurement. Figure.2 illustrates the system architecture. Please note the separate voltage and current sources [6].

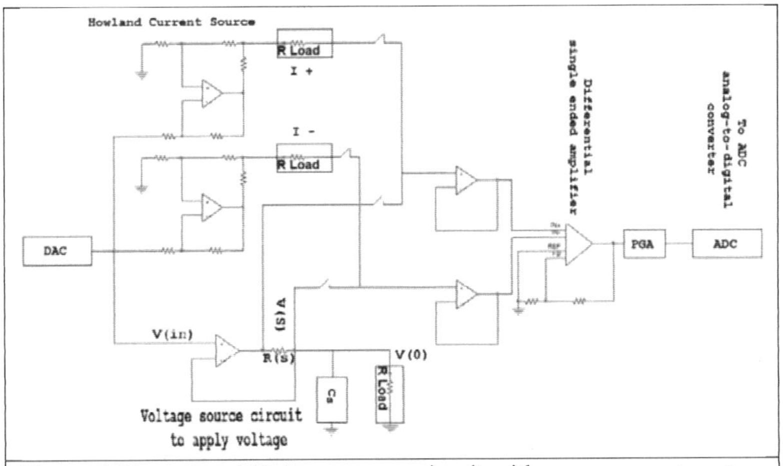

Figure 1 Current and Voltage source circuit with measurement system

The output of the DAC is fed to either the voltage or current sources. When current or voltage is applied to the electrode, voltages (or current I-V converters from R_s) are produced. In the acquisition system, the voltmeters measure single-ended differential voltages, which are developed between electrodes. Then, the voltage difference through a PGA feeds the result to a 16-bit ADC [7] [8]. Then the output is transferred to the FPGA for digital PSD and then by the digital matched filter to extract the real and imaginary components.

4. FPGA control and generator module

4.1. FPGA design

As illustrated in figure 2, in the digital sub-system, the design consists of a signal generator based on a direct digital synthesizer (DDS) module for multi-frequency signal generation by a look-up table (LUT). The DDS is also used to produce the reference signal for the phase-sensitive demodulation (PSD) block. To improve the speed of the system, the digital (PSD) is realized in the FPGA with digital embedded signal processor components (DSP48). The variable precision that can be achieved by the use of an FPGA ensures that our system can reach the accuracy required by EIT.

4.2. Signal generator

The control module of the direct digital synthesizer (DDS) is used to generate a suitable phase argument that is mapped by the look-up table to the desired output waveform, a sine-wave in this case. After that the DDS presents these samples to a digital-to-analogue converter (DAC), external to the FPGA, and a low-pass filter to obtain an analogue waveform with the specific frequency structure. Output frequency, f_{out}, of the DDS waveform is a function of the system clock frequency f_{clk}, and the number of bits $B_{\Theta(n)}$ in the phase increment value $\Delta\theta$; that is,[5]

$$f_{out} = f(f_{clk}, B_{\Theta(n)}, \Delta\theta) \quad \text{Thus,} \quad f_{out} = f_{clk}\, \Delta\theta / 2^{[B_{\Theta(n)}]}\ \text{Hz} \qquad (1)$$

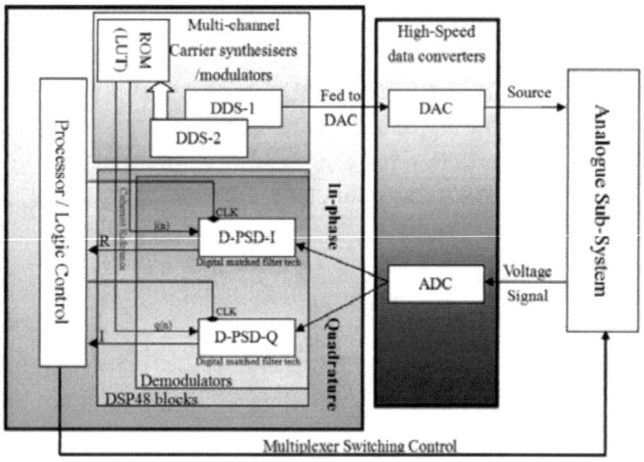

Figure 2 Digital Sub-system and FPGA block diagram

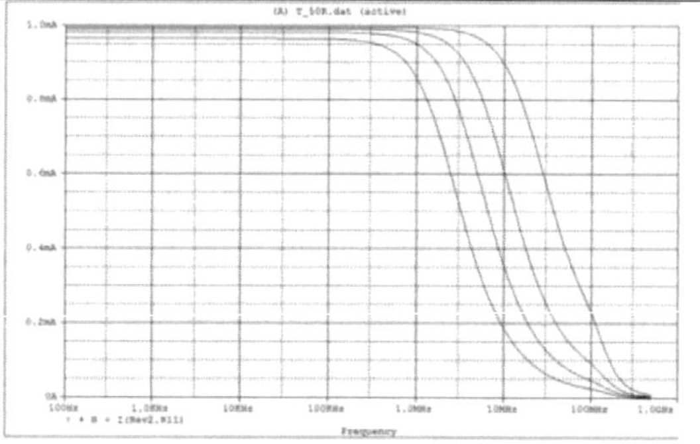

Cutoff_Lowpass_3dB(i(Rload))
R(L) 20k: 20.92856meg
R(L) 10k: 7.77795meg
R(L) 5k: 3.85653meg
R(L) 2k: 1.94947meg

Figure 3 Bandwidth of Current source circuit

The off-the-shelf Xilinx Core associated with signal generation runs at a speed of up to 445MHz. As this can be clocked inside the FPGA by a dedicated clock, this speed is more than adequate to provide a signal of 20MHz to meet EIT requirements.

The latency that may be expected by a digital system like the one defined above can affect the acquisition speed of the overall system. The most significant contribution to this latency internally to the FPGA comes from the DDS Core. This latency is the delay from the assertion to the first valid output and the delays from changes to programmable phase increment and offset to a change in output values and especially for multiple channels.

Table 1 frequency latency

$f_{out\ (Hz)}$	**10 KHz**	**100 KHz**	**1 MHz**	**5 MHz**	**10 MHz**	**20 MHz**
Cycles per frequency	10	10	10	10	10	10
Period per cycle	100μsec	10μsec	1μsec	200nsec	100nsec	50nsec
Total latency	1113.5μsec					

For the total latency per frame, the calculation is further complicated by the fact that the circuit sequentially selects electrodes by controlling the channel-selecting multiplexers. Assuming a 4-electrode measurement method is used, and N number of electrode combinations per frame, the latency is equal to the sum of all the internal delays and external delays per electrode combination, times the number of combinations per frame. As the FPGA is highly parallel, though, it is possible to minimize the external delays caused by the control of the switching mechanism, by pipelining the sending of the command words to the multiplexers with an actual, single electrode combination capture time. As an approximation, the total latency per frame should not exceed 25msec.

4.3. Further FPGA Tasks

The demodulation can again be carried out in the FPGA, and in this case the method to be used is digital quadrature demodulation, returning the In-phase, $i(n)$, and the Quadrature, $q(n)$, components of the complex measurements. Further processing, using the FPGA, returns the amplitude and phase of the measurement.

Within the FPGA the control logic required for the setting of the multiplexers responsible for the channel switching is also implemented.

A single digital sub-system can be developed using a FPGA for both current injection and applying voltage. The advantages of this are a fast system and also the ability to control switching. Ideally, a reconstruction method that can combine both types of measurement should be used, as in trans-admittance systems. However, all the necessary post-processing before reconstruction can and should be run in the FPGA for overall system speed increase. At the same time, the FPGA, ideally suited for a control device, can offer switching of high-channel count systems, in an integrated package.

5. Conclusion

As discussed above, a novel EIT system can be achieved with wide-bandwidth, low-noise, and high-speed by the use of digital techniques. Moreover, this EIT system is not restricting the FPGA only to signal generation and demodulation, but also the full control of the system, especially the electrode switching between channels. In effect, an FPGA can be used to act as the heart of a fast, versatile and scalable EIT system.

6. References

[1] Hartov A. Demidenko E 2002 Using Voltage source as current drivers for electrical impedance tomography, *Measurement Science and Technology*, **13** 1425-1430

[2] Ross A S 2003 *An adaptive Current Tomography for Breast Cancer Detection*, PhD Thesis Rensselaer Polytechnic Institute Toro, New York

[3] Holder David S, 2005 Electrical Impedance Tomography: Methods, History and Applications *Med. Phys.* **32** (8) 70-90

[4] Isaacson D 1986 Distinguishability of conductivities by electric current computed tomography, *IEEE Trans on Medical Imaging* MI-5(2)-92-95

[5] Xilinx Logic Core 2009 *DS Compiler V3.0-DS558,* product specification. *http://www.datasheetarchive.com/datasheet-pdf/055/DSA0038499.html*

[6] Ross A S 2006 A high-precision voltage source for EIT *Physiol. Meas.* **27** S 221-236

[7] Wang W 2007 *A Comprehensive Study on Current Source Circuits* ICEBI 17, pp 213-216

[8] Wang W 1998 The Design of De Montfort Mk2 Electrical Impedance Mammography System, *IEEE Eng in Med and Bio Soc* **20** (2)

AUTHOR INDEX

AUTHOR INDEX

AUTHOR INDEX

AUTHOR INDEX